MACROSCOPIC ELECTRODYNAMICS

An Introductory Graduate Treatment

Second Edition

MACROSCOPIC ELECTRODYNAMICS

An Introductory Graduate Treatment

Second Edition

Walter Wilcox
Baylor University, USA

Chris Thron
Texas A&M University, USA

Special contribution,
"Radiating cavities, scattering resonances, and quasinormal modes"
by
Lamis Al Sheikh
Institut de Mathématiques de Bourgogne, France

World Scientific

NEW JERSEY • LONDON • SINGAPORE • BEIJING • SHANGHAI • HONG KONG • TAIPEI • CHENNAI • TOKYO

Published by

World Scientific Publishing Co. Pte. Ltd.
5 Toh Tuck Link, Singapore 596224
USA office: 27 Warren Street, Suite 401-402, Hackensack, NJ 07601
UK office: 57 Shelton Street, Covent Garden, London WC2H 9HE

Library of Congress Control Number: 2023944496

British Library Cataloguing-in-Publication Data
A catalogue record for this book is available from the British Library.

MACROSCOPIC ELECTRODYNAMICS
An Introductory Graduate Treatment
Second Edition

ISBN 978-981-127-502-9 (hardcover)
ISBN 978-981-127-631-6 (paperback)
ISBN 978-981-127-503-6 (ebook for institutions)
ISBN 978-981-127-504-3 (ebook for individuals)

For any available supplementary material, please visit
https://www.worldscientific.com/worldscibooks/10.1142/13373#t=suppl

Printed in Singapore

To the courageous and freedom-loving people of Ukraine. (WW)

To the peace-loving people of Sudan. (CT)

To the free minds who try to see behind labels and facades. And to my beloved Damascus. (LAS)

To the memory of my mother, Alice Kathryn Wilcox, and of my father, Dr. Marion Walter Wilcox. (WW)

To my parents, Ann and Dennis Thron. (CT)

Preface to the Second Edition

Our aims and purposes in the second edition of *Macroscopic Electrodynamics* (*ME*) are the same as before: to provide a comprehensive graduate level textbook in electrodynamics for beginning graduate students, emphasizing principles and practical methods of analysis. We continue to be inspired by the beauty of the theory and the usefulness of its many applications.

We have made a number of improvements in this second edition of *ME*, as listed below.

(1) Instructors are always in search of new and interesting problems! We have added 129 new and 11 modified end of chapter exercises. This represents an approximate 33% increase; there are now a total of 520 exercises in *ME*. These are again integrated into the textual material by being associated with chapter and section number for convenience. They are added at the end of existing exercise section lists so that the previous numbering is unchanged. These exercises, as before, mainly originate in WW's classroom exercises at Baylor University.

(2) A new *Fundamentals Concepts in Electrodynamics* appendix (Appendix B), has been added, along with the associated *Index*. This is in recognition that the undergraduate preparation of many students may not be ideal as we emerge from pandemic-caused educational disruptions. We offer this as an additional educational tool for the beginning graduate student in physics. Here the student will find a précis of the complete list of undergraduate topics in E&M, from electrostatics to special relativity Lorentz transformations.

(3) We have retired the wireless communication section in Chapter 14 (14.3) and added a new section on quasinormal modes, generously provided by Dr. Lamis Al Sheikh, who is an expert on resonance problems in electromagnetics and gravitational physics. This material is foundational for recent technological innovations in the field of nanoparticles, as well as the study of gravitational waves. Homework problems are included for the student's benefit.

(4) Typos and misstatements have been corrected. We have also updated the *ME Instructor's Solutions Guide.*

We are thankful to Lydia Thron for her work on correcting typos. We are also thankful to World Scientific for their flexibility and understanding of compressed academic schedules.

Walter Wilcox and Chris Thron; June, 2023.

Preface to the First Edition

Macroscopic Electrodynamics (*ME*) is a comprehensive two-semester introductory graduate level textbook for use in physics and engineering programs. It is designed to be accessible to students with a mathematical background covering advanced calculus, linear algebra and variational methods on the one hand, and a physics background including at least an introductory electrodynamics course and its prerequisites, on the other.

ME emphasizes principles and practical methods of analysis. It covers all the fundamental topics at an advanced level and is written directly to the student in a clear-minded but informal and friendly way. Illustrative examples are carefully chosen to promote the students' physical intuition, and are worked out in detail to give students a thorough grounding in solution techniques. In combination with the resources listed in the *Going Deeper* sections, to be discussed later, it can also be used to initiate more in-depth studies of particular topics.

The title of our book is intended to bring to mind large-scale manifestations of the theory as well as to allude to the macroscopic form of the Maxwell equations as applied to idealized media, which forms the core of our presentation. There are in addition a number of other ways this concept is brought forward in the book. We make use of the macroscopic form of the Maxwell equations in the description of magnetic fields from current loops as well as in the description of classical magnetic monopoles. There is a strong emphasis placed in Chapter 7 on macroscopic Maxwell justification formalisms and on macroscopic force law evaluations in Chapters 5, 6 and 7. In a different sense, the term "macroscopic" also brings to mind the amazingly universal manifestations of the theory in our everyday world, even though it actually originates at the smallest scales.

A unifying theme throughout is the Green function technique. This topic is notable for its elegance and adaptability to a wide variety of situations. However, rather than taking a formal mathematical approach, we emphasize the basic mathematics involved and the role of physical intuition. This is a technique that was emphasized by the Nobel Prize winning physicist Julian Schwinger, whose many ideas and methods have also influenced the present work.

In a topic as broad and deep as macroscopic electrodynamics, it is impossible to give a comprehensive guide to the literature. In keeping with the introductory graduate nature of this work, we have chosen a number of relevant topics to supplement each chapter in our *Going Deeper* sections. These entry points will allow students a place to linger on an interesting subject, and will provide a beginning resource to the modern literature for both physicists and engineers.

The exercises lie close to the heart of the purpose for our book. They incorporate our philosophy of a creative and idea-centered approach to teaching physics. We believe that one need not and should not confront the beginning graduate student with complicated harmonic variations until the melody is appreciated. The exercises are well-integrated into the text, and are carefully designed to reinforce concepts and to challenge students at the appropriate level of difficulty. They are designed to promote conceptual understanding and physical intuition rather than calculational virtuosity. Many of the final answers in the problems are given or helpful hints are included. Most of the exercises were written to test student's understandings on exams given in class. This enforced a discipline which required the problems to manifest a core principle in a minimum amount of time. The exercises are numbered and arranged by chapter and section number for convenience of access and ease of evaluation. The section chosen for each problem is not necessarily the primary one for that particular idea, but the last section necessary to completely answer all parts of the question.

We consider all the chapters in *ME* to be essential, although the instructor can design a course to suit their needs by choosing appropriate sections. Each chapter begins with a short description of the content, similar to a journal-type abstract; the Table of Contents of course gives a more complete description of topics. Rather than repeat these descriptions, let us simply hit some of the highlights here. Chapter 1 provides an orientation to the subject. It ends with useful sections giving notational symbols, mathematical identities and operators, as well as figures illustrating coordinate system aspects. The topics covered in Chapters 2 through 6 cover basic aspects of electrostatic and magnetostatic fields and interactions. Chapters 7 and 8 provide a transition to time-dependent phenomena and formalisms. Chapter 9 is an intensive physics chapter covering the basics of the interactions of electromagnetic waves in matter. Chapter 10 is essential for engineering aspects of guided electromagnetic waves and resonators, and requires all the mathematical machinery developed in earlier chapters. Chapters 11 and 12 are closely related and present advanced treatments of radiation by material particles and the scattering of electromagnetic waves. Chapter 13 is the capstone chapter covering relativity and electromagnetic formalisms. Chapter 14 is the "fun" chapter, which is the payoff for two semesters well-spent. It can truly be used in a creative manner. One may introduce the topic of radiation reaction, connect to more advanced theoretical formalisms, or investigate the fascinating world of electronic communications. We think there is something for everyone here!

We have not given short shrift to the essential topic of electromagnetic units in *ME*. An intensive Appendix on units is included, given that we are using Gaussian units in a world and academic environment dominated by the Système Internationale d'Unités (SI). In addition, we have developed an online application to convert equations from one system to the other, which is more completely described in the Appendix.

Our text is accompanied by a *Solutions Guide*. For a book and topic as large as ours, we intentionally chose not to give complete solutions, but simply provide outlines as guides for the experienced. These outlines, however, are sufficiently detailed to help those with compressed teaching responsibilities to transition to our text as easily as possible.

The lectures on which the text is based are online and are part of Walter Wilcox's *Open Text Project®* series, and are freely available in the associated Baylor University website. This Baylor website evolved from the desire to develop high-quality lecture notes into a convenient, accessible form for students, especially in the developing world. It is open to all who wish to contribute.

We thank the many students, faculty and other workers who have contributed to the completion of this work. First of all we thank the many students who have taken our courses and given valuable feedback on the form, ideas and content of our text. We also thank the many who have helped develop and maintain the texts on the Baylor Open Text Project website; acknowledgements are listed there. We are grateful to Melissa

Hammond who created Powerpoint versions of the figures, and to Johnny Watts who worked on figures and assisted (along with Esther and Lydia Thron) in the tedious process of converting the text to LaTeX format. We are also grateful to the initial readers of the text who gave us valuable responses and suggestions. In this regard we thank Tykon Bykov, Rudolf Fiebig, and Randy Lewis. We thank the administrators and faculty of home institutions for their support and encouragement. Finally, we thank the editors and staff at World Scientific, who have generously allowed our schedule to slip and accommodated our many changes.

Walter Wilcox (Baylor University, Waco, TX)

Chris Thron (Texas A&M University-Central Texas, Killeen, TX)

Contents

Preface to the Second Edition vii

Preface to the First Edition ix

1. Introduction and Perspectives 1

1.1 Maxwell's equations 1
1.2 Relativistic and quantum considerations 4
1.3 Macroscopic Maxwell's equations 6
1.4 Boundary conditions on fields 8
1.5 Two-dimensional electrodynamics and boundary conditions 14
1.6 Going Deeper 19
1.6.1 Mathematical methods 19
1.6.2 Electromagnetic units 20
1.7 Exercises 21
1.8 Useful identities 22

2. Introduction to Electrostatics 29

2.1 Electric field: definition 29
2.2 The Dirac delta function and singular charge distributions 31

2.3 Line and surface delta functions 37
2.4 Gauss' law and solid angles 38
2.5 Verification of the inverse square law: the Cavendish experiment 42
2.6 Surface charge and dipole layers 45
2.7 Boundary conditions and uniqueness of solutions 49
2.8 Dirichlet and Neumann Green functions 51
2.9 One-dimensional Green function examples 58
2.10 Electrostatic energy 60
2.11 Normal force on a charged surface 63
2.12 Capacitance 65
2.13 Going Deeper 68
2.13.1 Photon mass limits 68
2.13.2 Capacitance 68
2.14 Exercises 69

3. Boundary Value Problems in Electrostatics 87

3.1 Conducting plane: Green functions and method of images 87
3.2 Reduced Green function technique applied to flat conductor 90
3.3 Method of images: conducting sphere 94
3.4 Charged conducting sphere force example 98
3.5 Separation of variables: conducting box 100
3.6 Eigenfunction expansion of Green function for a conducting box 101
3.7 Separation of variables in polar coordinates 107
3.8 Corner problems in polar coordinates 108
3.9 Cylindrical halves at different potentials 109
3.10 Variational methods 111
3.11 Conformal mapping techniques 113

3.12 Going Deeper 118
3.12.1 Variational methods 118
3.12.2 Conformal mapping 118
3.13 Exercises 120

4. Electrostatics in Cylindrical and Spherical Coordinates 135

4.1 Cylindrical coordinates and Bessel functions 135
4.2 Completeness of Bessel functions 141
4.3 Zeros and orthogonality properties of Bessel functions . . 142
4.4 Reduced Green function for the conducting cylinder . . . 146
4.5 Potential inside a cylinder as a boundary value problem . 148
4.6 Bessel functions of imaginary argument; asymptotic forms of Bessel functions 149
4.7 Cylindrical free-space Green function using the Wronskian technique 151
4.8 Reduced Green function method for the conducting wedge 155
4.9 Schwinger's construction of spherical harmonics 157
4.10 Orthogonality properties of spherical harmonics 165
4.11 The Coulomb expansion, completeness of spherical harmonics and the "Addition Theorem" 166
4.12 Green function for concentric spheres 170
4.13 Potential of conducting sphere in a uniform field via separation of variables 173
4.14 Method of last resort: eigenfunction expansions 175
4.15 Going Deeper 179
4.15.1 Special functions 179
4.15.2 Eigenfunction expansions 179
4.16 Exercises 180

5. Multipoles, Electrostatics of Macroscopic Media, Dielectrics 203

5.1 Cartesian and spherical multipole expansions 203
5.2 Multipole energy expansions 207
5.3 External fields and forces on multipole distributions . . . 211
5.4 Electric polarization and the displacement field 214
5.5 Green functions in the presence of linear dielectrics 217
5.6 Green function for the dielectric slab 218
5.7 Green function for the dielectric sphere 223
5.8 Field energy and dielectrics 226
5.9 Bulk forces on dielectrics: theory 230
5.10 Nonlinear dielectric example: leading logarithm model . . 232
5.11 Bulk forces on dielectrics: examples 235
5.12 Going Deeper 241
5.12.1 Dielectric materials 241
5.13 Exercises 242

6. Magnetostatics 267

6.1 Analogy to electrostatics 267
6.2 General equations for magnetostatics 273
6.3 Ampère's law; vector potentials 274
6.4 Surface current considerations 276
6.5 Solid angle formula for $\vec{B}$ 278
6.6 Circular current loop: solution using solid angle result for $\vec{B}$ 281
6.7 Circular current loop: direct solution 286
6.8 Current distributions and magnetic moments 290
6.9 External fields and forces on magnetic multipole distributions 296
6.10 Introduction of "magnetization" and the $\vec{H}$ field 298
6.11 Boundary conditions at material interfaces 303

6.12 Image method for magnetostatics 304

6.13 Intrinsic and induced magnetization: theory and example 311

6.14 Going Deeper 315

6.14.1 Magnetic materials 315

6.15 Exercises 316

7. Time Varying Fields I 337

7.1 Plausibility argument leading to Maxwell's equations . . . 337

7.2 Faraday's law 343

7.3 Derivations of macroscopic Maxwell equations 344

7.3.1 Schwinger model of macroscopic medium 345

7.3.2 Jackson model of macroscopic medium 352

7.3.3 Summary of macroscopic equations 355

7.4 Second-order formulation of the vacuum Maxwell equations 356

7.5 Magnetostatic field energy 359

7.6 Magnetostatic field energy and forces on current loops . . 363

7.7 Bulk forces on magnetic materials: theory and examples . 368

7.7.1 Energy method 368

7.7.2 Force method 371

7.8 Magnetic charge and the macroscopic Maxwell equations . 374

7.9 Going Deeper 385

7.9.1 Macroscopic Maxwell equations 385

7.9.2 Magnetic charge 386

7.10 Exercises 387

8. Time Varying Fields II 411

8.1 Conservation of energy; energy flux 411
8.2 Conservation of momentum; Maxwell stress tensor 417
8.3 Conservation of angular momentum; shear tensor 422
8.4 Effective conservation laws for macroscopic media 423
8.5 Maxwell stress tensor example: dielectric slab 425
8.6 Retarded potentials 427
8.7 Green function for half-infinite geometry 434
8.8 Transformation properties of electromagnetic quantities 439
8.9 Going Deeper 447
8.9.1 Green functions 447
8.10 Exercises 448

9. Plane Electromagnetic Waves and Propagation in Matter 461

9.1 Plane waves in dielectric media 461
9.2 Reflection and refraction of plane waves from dielectric interfaces I: $E_\perp$ polarization 465
9.3 Reflection and refraction of plane waves from dielectric interfaces II: $B_\perp$ polarization 473
9.4 Brewster's angle and total internal reflection 475
9.5 Simple model for constitutive relations 477
9.6 Modeling of plasmas, metals and dielectrics 482
9.7 Kramers-Kronig relations 485
9.8 Dispersion in one-dimension: theory and example 491
9.9 Charged particle energy loss in materials 496
9.10 Going Deeper 501
9.10.1 Plasma physics 501
9.10.2 Dissipative charged particle energy loss 501
9.11 Exercises 503

10. Waveguides and Resonant Cavities 527

10.1 Boundary conditions near an imperfect conductor 527

10.2 General considerations for waveguides of arbitrary cross-section 535

10.3 TE and TM modes for rectangular waveguides with perfectly-conducting walls 540

10.4 Orthogonality properties for modes in ideal waveguides . . 541

10.5 Reflection and impedance properties of thin obstructions in waveguides 547

10.6 Variational examples: parallel rectangular strips in rectangular waveguides 552

10.7 Power loss in non-ideal waveguides: power method 556

10.8 Power loss in waveguides: perturbation of boundary conditions method 560

10.9 TE and TM modes in an ideal rectangular resonant cavity 565

10.10 Eigenmode expansion for ideal spherical resonant cavity (I): spherical Bessel functions 567

10.11 Eigenmode expansion for ideal spherical resonant cavity (II): vector spherical harmonics 573

10.12 Energy loss and frequency shift in non-ideal resonant cavities 579

10.13 Appendix: waveguide modes for rectangular and cylindrical waveguides 584

10.13.1 Rectangular waveguide modes 584

10.13.2 Cylindrical waveguide modes 586

10.14 Going Deeper 587

10.14.1 Electromagnetic waveguides and cavities 587

10.15 Exercises 589

11. Radiation of Systems and Point Particles 615

11.1 Electromagnetic radiation of systems: harmonic formalism 615

11.2 Electromagnetic radiation of systems: real source formalism . . . 622

11.3 Frequency distribution of radiated energy; impulsive scattering . . . 626

11.4 Multipole expansion: physical interpretation . . . 629

11.5 Dipole and quadrupole contributions to radiated power . 631

11.6 Point particle radiation: Larmor and Liénard results . . . 639

11.7 Nonrelativistic treatment of bremsstrahlung . . . 648

11.8 Radiation from periodic trajectories: general considerations . . . 655

11.9 Synchrotron radiation from circular periodic motion . . . 662

11.10 Going Deeper . . . 669

11.10.1 Transition radiation . . . 669

11.10.2 Bremsstrahlung . . . 669

11.10.3 Synchrotron radiation . . . 670

11.11 Exercises . . . 671

12. Scattering and Diffraction 697

12.1 Definition of polarized scattering cross-section . . . 697

12.2 Kirchhoff identity for scattering . . . 702

12.3 Short wavelength scattering for a conducting sphere . . . 705

12.4 The optical theorem . . . 713

12.5 Partial wave expansion for a conducting sphere . . . 714

12.6 Scalar diffraction theory . . . 720

12.7 Scalar diffraction examples . . . 724

12.8 Vector diffraction theory . . . 728

12.9 Going Deeper . . . 733

12.9.1 Scattering and diffraction theory . . . 733

12.10 Exercises . . . 734

13. Relativistic Formulations of Electrodynamics 747

13.1 The exact relativistic transformation between comoving frames . . . 747
13.2 Lorentz transformations: matrix and index notation . . . 750
13.3 Relativistic form of Maxwell's equations . . . 756
13.4 Interval invariance, restricted Lorentz transformations and causality . . . 760
13.5 Discrete transformations and the full Lorentz group . . . 763
13.6 Energy/momentum aspects of Lorentz transformations . . 767
13.7 Relativistic kinematics in the context of linear and circular particle accelerators . . . 769
13.8 Relativistic Lagrangian formalism . . . 772
13.9 Construction of Lagrangian for the electromagnetic field . 776
13.10 Covariant form of the energy-momentum tensor . . . 779
13.11 Going Deeper . . . 782
13.11.1 Relativistic formulations . . . 782
13.11.2 Particle accelerators . . . 783
13.12 Exercises . . . 784

14. Special Topics 801

14.1 Radiation reaction force . . . 801
14.1.1 Classical renormalization and relativistic generalization . . . 806
14.1.2 Wheeler-Feynman electrodynamics . . . 810
14.2 Connections between classical electrodynamics and quantum field theory . . . 814
14.2.1 Massless scalar field and the Feynman propagator 815
14.2.2 Composition rule for Feynman propagator 818
14.2.3 The photon field . . . 819
14.3 Radiating cavities, scattering resonances, and quasinormal modes (*by Lamis Al Sheikh*) . . . 820

14.3.1 Practical applications 821
14.3.2 Definition of scattering field and quasinormal modes 823
14.3.3 Physical applications of quasinormal modes 824
14.3.4 Mathematical properties of quasinormal modes . . 826
14.3.5 Solution methods for quasinormal modes 829
14.3.6 Hyperboloidal slices approach to quasinormal modes 831
14.3.7 Numerical calculation of QNMs via the spectral method 835
14.3.8 Pseudospectrum and the stability of eigenvalues . 841
14.4 Going Deeper 844
14.4.1 Radiation reaction 844
14.4.2 Quantum field theory 845
14.4.3 Radiating cavities and quasinormal modes 846
14.5 Exercises 848

Appendix A Appendix on Electromagnetic Units 859

A.1 Exercises 862

Appendix B Fundamental Concepts in Electrodynamics 867

B.1 Preface 867
B.2 Electrostatics 867
B.3 Magnetostatics 881
B.4 Electromagnetic synthesis 891
B.5 Electromagnetic waves 898
B.6 Potentials and fields 913
B.7 Radiation 918
B.8 Special relativity 923

Fundamental Concepts Index 931

Index 933

Chapter 1

Introduction and Perspectives

Classical electrodynamics is certainly one of the most elegant and beautiful physical theories ever devised, and has arguably had a greater impact on the development of physics than any other theory. Starting from a ridiculously simple set of fundamental equations, it explains with incredible accuracy a bewildering variety of physical phenomena, from compass needles to electric circuits to colors to radio. Though classical in form, electromagnetic theory paved the way for all of the great non-classical innovations of the 20th century including relativity, quantum mechanics, and quantum field theory.

In this chapter we will present a top-level overview of electrodynamics, including some of its capabilities and its relation to other physical theories.

1.1 Maxwell's equations

Classical electromagnetism describes the *electric field* (denoted by $\vec{E}$) and *magnetic field* (denoted by $\vec{B}$) in the presence of charges and currents that act as *sources* of the vector fields $\vec{E}$ and $\vec{B}$. The relation between fields and their sources is expressed by *Maxwell's equations*, which relates various derivatives of $\vec{E}$ and $\vec{B}$ to the existing distribution of sources. Table 1.1 presents the Maxwell equations in vector form, while Table 1.2 presents the quantities involved and their units. Section 1.8 gives a glossary of symbols used in the text, along with a summary of various vector, differential, integral and trigonometric identities. Note that all fields and all source terms are functions of space and time $(\vec{x}, t)$.

The equations in Table 1.1 utilize the vector differential operators divergence $(\vec{\nabla} \cdot \vec{F})$ and curl $(\vec{\nabla} \times \vec{F})$ on a general vector field, $\vec{F}$. Section 1.8 gives

Table 1.1 Maxwell's equations. ($c \equiv 29,979,245,800$ cm/s).

Equation name	Gaussian form	SI form	Components
Coulomb	$\vec{\nabla} \cdot \vec{E} = 4\pi\rho$	$\vec{\nabla} \cdot \vec{E} = \rho/\epsilon_0$	1
Ampère-Maxwell	$\vec{\nabla} \times \vec{B} - \frac{1}{c}\frac{\partial \vec{E}}{\partial t} = \frac{4\pi}{c}\vec{J}$	$\vec{\nabla} \times \vec{B} - \epsilon_0\mu_0 \frac{\partial \vec{E}}{\partial t} = \mu_0 \vec{J}$	3
Faraday	$\vec{\nabla} \times \vec{E} + \frac{1}{c}\frac{\partial \vec{B}}{\partial t} = 0$	$\vec{\nabla} \times \vec{E} + \frac{\partial \vec{B}}{\partial t} = 0$	3
No-monopole	$\vec{\nabla} \cdot \vec{B} = 0$	$\vec{\nabla} \cdot \vec{B} = 0$	1

Table 1.2 Electromagnetic variables and constants ($c \equiv 299,792,458$). See Tables A.1–A.4 of the Appendix for more detailed explanations and considerations.

Quantity name	Symbol	Gaussian	SI
Electric field	$\vec{E}(\vec{x},t)$	statvolt cm^{-1}	$c \times 10^{-4}$ volt m^{-1}
Magnetic field	$\vec{B}(\vec{x},t)$	gauss	10^{-4} tesla
Charge density	$\rho(\vec{x},t)$	statcoul cm^{-3}	$c^{-1} \times 10^5$ coulomb m^{-3}
Current density	$\vec{J}(\vec{x},t)$	statamp cm^{-2}	$c^{-1} \times 10^3$ ampere m^{-2}
Vacuum permittivity	ϵ_0	—	$(4\pi c^2)^{-1} \times 10^7$ farad m^{-1}
Vacuum permeability	μ_0	—	$4\pi \times 10^{-7}$ henry m^{-1}

these operators in the three most common coordinate systems (rectangular, cylindrical, and spherical).

The unit systems in Table 1.2 are called *Gaussian*, which uses centimeter-gram-second or CGS units, and *International System* (in French: *Système International*) (SI), which uses meter-kilogram-second-ampere or MKSA units. From the theoretician's point of view, Gaussian units are greatly to be preferred; unlike SI, they require no conversion factors that complicate the equations while adding nothing to the physics. However, for practical calculations the SI equations are the more widespread. Indeed, our present international measurement system is actually defined in the SI unit system, and so in that sense it is the more fundamental. We hope the beginning student will read further in the Appendix to consider the question of electrodynamic units and conversions to learn more about these issues.

We note parenthetically that the vector equations in Table 1.1 are not the original form of Maxwell's 1861–1862 equations,[1] nor are they the most concise expression; they can be re-expressed as two *tensor* equations, where $\vec{E}$ and $\vec{B}$ are taken as components of a second-rank tensor.[2] Nonetheless, the vector equations in Table 1.1 (due to Heaviside) are the most widely used.

At this point we will not attempt to justify Maxwell's equations. Later on we will show that the tensor form of the equations can be derived from a least-action principle. But for now, we merely wish to point out some important general features of the equations.

Table 1.1 represents a first order formulation (meaning first order in derivatives) of the Maxwell equations. Considering the left-hand side the "field" side and the right-hand side the "source" side, we notice that we have four equations with sources and four without. The four without sources can be viewed as constraints. The remaining four equations are the correct number to define a four dimensional potential or *4-potential* (Φ, a scalar potential, and $\vec{A}$, a vector potential) in a second-order (in derivatives) formulation. The $\vec{E}$ and $\vec{B}$ fields can be expressed as derivatives of this 4-potential (Exercise 1.1.2).

Table 1.1 shows how the $\vec{E}$ and $\vec{B}$ fields arise from charges and currents: but we also need an equation that expresses how charges and currents respond to the presence of $\vec{E}$ and $\vec{B}$ fields. The force that electromagnetic fields exert on charges is expressed by the *Lorentz force law*:

$$\vec{F} = q\left(\vec{E} + \frac{\vec{v}}{c} \times \vec{B}\right). \tag{1.1}$$

Equation (1.1) can be derived from energy-momentum conservation, when the theory is recast in tensor form (a conserved "energy-momentum tensor" may be defined).

The familiar property of charge conservation (that is, charge is neither created nor destroyed) actually follows from Maxwell's equations. Taking

[1] James Clerk Maxwell, "On Physical Lines of Force", Philosophical Magazine **21** series 4 (1861); **23** series 4 (1862). The proposal that light was an electromagnetic phenomenon was given in: James Clerk Maxwell, "A Dynamical Theory of the Electromagnetic Field", Philosophical Transactions of the Royal Society of London **155**, 459–512 (1865).

[2] A *second-rank tensor* is a 4×4 matrix with certain transformation properties. Tensors are described in Chapter 7. The two tensor equations which comprise Maxwell's equations are (13.59) and (13.72).

the divergence of the Ampére-Maxwell relation gives

$$\underbrace{\vec{\nabla}\cdot\vec{\nabla}\times\vec{B}}_{=0}-\frac{1}{c}\vec{\nabla}\cdot\left(\frac{\partial\vec{E}}{\partial t}\right)=\frac{4\pi}{c}\vec{\nabla}\cdot\vec{J},$$

while taking the time derivative of the Coulomb relation gives

$$\frac{\partial}{\partial t}\left(\vec{\nabla}\cdot\vec{E}\right)=4\pi\frac{\partial\rho}{\partial t}.$$

Substitution gives the *continuity equation*,

$$\frac{\partial\rho}{\partial t}+\vec{\nabla}\cdot\vec{J}=0. \tag{1.2}$$

Equation (1.2) can be seen as a conservation law as follows. Integrating (1.2) over a given volume and using the divergence theorem gives

$$-\frac{\partial}{\partial t}\int_V d^3x\ \rho(\vec{x},t)=\int_V d^3x\ \vec{\nabla}\cdot\vec{J}(\vec{x},t)=\oint_S d\vec{s}\cdot\vec{J}(\vec{x},t). \tag{1.3}$$

The leftmost side of (1.3) is the rate of change in the total amount of charge enclosed within surface S; while the rightmost side is the net flow of charge through the surface S. It follows that any charge decrease in one region of space must be exactly balanced by a charge increase in some other region. No exceptions to this law have ever been detected experimentally.

1.2 Relativistic and quantum considerations

Electrodynamics is a "classical" theory in the sense that it relates to observable phenomena that are neither relativistic nor quantum in nature. However, it is a very different sort of classical theory than Newton's theory of gravity. Newton's theory flagrantly violates causality by assuming instantaneous action at a distance. Electrodynamics on the other hand is actually a fully relativistic theory, although this was not really appreciated until Einstein in 1905. Indeed it was Einstein's intuition that Maxwell's equations are *invariant* (that is, having the same form) in all inertial reference frames that brought him to the theory of relativity in the first place. Apparently electrodynamics is a very fundamental theory.

While Maxwell's equations themselves are not quantized, they can easily accommodate the phenomenon of *charge quantization* — as far as freely-moving objects are concerned (including elementary particles), all such objects carry a charge that is a (positive or negative) integer multiple of the charge on an electron, denoted by e:

$$|e| = 4.80320450(11) \times 10^{-10} \text{ statcoulombs.}^3$$

There are of course non-free particles (quarks) that have fractional charges. Quark charges are either $\frac{2}{3}|e|$ (for up, charmed, and top) or $-\frac{1}{3}|e|$ (for down, strange, and bottom.)

So far, no fundamental magnetic charges (also known as *magnetic monopoles*) have ever been observed. This is reflected in the fact that half the equations in Table 1.1 have no source terms. As we will see in a later chapter, if all the particles in the universe had the same ratio of magnetic to electric charge, the fields and sources could then be redefined such that magnetic charge would be zero. However, there is no known principle of nature which prevents other ratios from existing.

A completely quantum description of electromagnetic phenomena requires the *quantization* of Maxwell's equations (analogous to the quantization of Hamilton's equations of motion to obtain the Schrödinger equation for single particles). The resulting theory is known as *quantum electrodynamics* (or QED), which is indisputably the most accurate scientific theory known to mankind. The theory requires the existence of a massless particle of light, called the "photon", which is responsible for the force between charged particles. A striking example of the accuracy of QED is the theoretical prediction of the gyromagnetic ratios of the electron and muon. Without going too deeply into the details, the magnetic moment $\vec{\mu}$ of a particle of charge q and mass m is related to its angular momentum $\vec{S}$ by:

$$\vec{\mu} = g \cdot \frac{q}{2m} \vec{S}. \tag{1.4}$$

A classical computation yields $g = 1$, while a semiclassical treatment of the electron or muon (using the Dirac equation) yields $g = 2$. Table 1.3 shows the results of a full QED computation of the quantity $g/2 - 1$ for the electron[4] and muon, compared to experimental measurements.[5] (Actually, there are inputs other than QED in the muon calculations.) The electron result is accurate to about one part in 10^{12}, whereas the muon result is

[3]K. Olive *et al.* (Particle Data Group), Chin. Phys. C **38**, 090001 (2014); `http://pdg.lbl.gov`; Review, "Physical Constants". The reliability of the physical value is indicated by the number in parentheses, which denotes the uncertainty in the last two digits of the value.

[4]T. Aoyama, "Tenth-Order QED Contribution to the Electron g − 2 and an Improved Value of the Fine Structure Constant", Phys. Rev. Lett. **109**, 111807 (2012).

[5]K. Olive *et al.* (Particle Data Group), Chin. Phys. C **38**, 090001 (2014); `http://pdg.lbl.gov`; Review, "Electroweak Model and Constraints on New Physics".

Table 1.3 Gyromagnetic ratios: theory versus experiment.

Particle	$\frac{g}{2} - 1$ **(experiment)**	$\frac{g}{2} - 1$ **(theory)**
electron	$(1159652180.76 \pm 0.27) \times 10^{-12}$	$(1159652181.78 \pm 0.77) \times 10^{-12}$
muon	$(1165920.89 \pm 0.63) \times 10^{-9}$	$(1165918.03 \pm 0.49) \times 10^{-9}$

accurate to about one part in 10^9. (Note that your authors are combining some error bars in quadrature in Table 1.3.) Although the muon result is less accurate, it is more sensitive to new physics effects. In this respect, note the marginal discrepancy between experiment and theory for this case.

Because we will not be working with the quantum theory, we will frequently employ non-quantum idealizations in our characterizations of macroscopic situations. Some examples include:

(a) Conductors connected by a wire are assumed to be at the same potential.
(b) Surface charges and currents on conductors or media interfaces are modeled as strictly 2-dimensional.
(c) Fields inside materials which contain atoms are modeled as smoothly varying.

1.3 Macroscopic Maxwell's equations

Macroscopic electromagnetic phenomena usually involve fields within material media. In order to apply Maxwell's equations to these phenomena, we have to adopt idealization (c) above, namely that the "clumpy" electric and magnetic fields produced by atoms within the medium can be accurately approximated by smooth, averaged fields. The effects of these averaged fields are included by defining two new fields known as the *electric displacement* or *D-field* (denoted by $\vec{D}$) and *H-field* (denoted by $\vec{H}$), which are discussed in Chapter 7. Table 1.4 summarizes the macroscopic version of Maxwell's equations. Notice that the unsourced macroscopic equations are identical to the corresponding fundamental equations in Table 1.1; only the sourced equations are changed.

To solve the macroscopic Maxwell equations, additional information about relations between the $\vec{E}, \vec{B}, \vec{D}$, and $\vec{H}$ fields is needed. In practice,

Table 1.4 Macroscopic Maxwell's equations.

Equation name	**CGS**	**MKS**
Coulomb	$\vec{\nabla}\cdot\vec{D}=4\pi\rho$	$\vec{\nabla}\cdot\vec{D}=\rho$
Ampère-Maxwell	$\vec{\nabla}\times\vec{H}-\frac{1}{c}\frac{\partial\vec{D}}{\partial t}=\frac{4\pi}{c}\vec{J}$	$\vec{\nabla}\times\vec{H}-\frac{\partial\vec{D}}{\partial t}=\vec{J}$
Faraday	$\vec{\nabla}\times\vec{E}+\frac{1}{c}\frac{\partial\vec{B}}{\partial t}=0$	$\vec{\nabla}\times\vec{E}+\frac{\partial\vec{B}}{\partial t}=0$
No-monopole	$\vec{\nabla}\cdot\vec{B}=0$	$\vec{\nabla}\cdot\vec{B}=0$

the fields $\vec{D}$ and $\vec{H}$ are related to the applied electric and magnetic field by the so-called *constitutive relations*, that are characteristic of the medium in question. Many media are *linear* and *local*, so that the constitutive relations take the form[6]

$$D_j(\vec{x},t)=\sum_{k=1}^{3}\epsilon_{jk}(\vec{x})E_k(\vec{x},t);\qquad B_j(\vec{x},t)=\sum_{k=1}^{3}\mu_{jk}(\vec{x})H_k(\vec{x},t).\tag{1.5}$$

The 3×3 matrices ϵ_{jk} and μ_{jk} are called the *permittivity tensor* (or *dielectric tensor*) and *permeability tensor*, respectively. We generally assume that the inverses of the permittivity and permeability tensors exist, so that Equations (1.5) can be inverted to solve for $\vec{H}$ and/or $\vec{E}$ in terms of $\vec{B}$ and/or $\vec{D}$.

Linear local media that are *isotropic* satisfy $\vec{D}(\vec{x},t)=\epsilon(\vec{x})\vec{E}(\vec{x},t)$ and $\vec{B}(\vec{x},t)=\mu(\vec{x})\vec{H}(\vec{x},t)$, where ϵ and μ are scalars (as opposed to 3×3 tensors). *Homogeneous* linear local media have permittivity and permeability tensors that are constants independent of $\vec{x}$: in this case ϵ and μ are called the *dielectric constant* and *permeability* respectively.

The situation is more complicated in the case of *nonlocal* linear media, which have constitutive relations as follows:

$$D_j(\vec{x},t)=\sum_{k=1}^{3}\int d^3x'\int dt'\epsilon_{jk}(\vec{x}',t')E_k(\vec{x}-\vec{x}',t-t'),\tag{1.6}$$

$$B_j(\vec{x},t)=\sum_{k=1}^{3}\int d^3x'\int dt'\mu_{jk}(\vec{x}',t')H_k(\vec{x}-\vec{x}',t-t').\tag{1.7}$$

Equations (1.6) and (1.7) model a medium in which the $\vec{D}$ and $\vec{B}$ fields at $(\vec{x},t)$ depend on $\vec{E}$ and $\vec{H}$ fields for neighboring points or earlier times. We

[6]Note that by convention, (1.5) expresses $\vec{B}$ as a function of $\vec{H}$ rather than vice versa.

will explain this more fully for space and time nonlocality in Chapters 7 and 9, respectively.

Besides the constitutive relations, conducting media contain currents that are due to applied electric and magnetic fields. The current density within a conductor typically depends locally on the applied fields. The well-known *Ohm's Law*, which is a good approximation for many conducting materials at rest:

$$\vec{J}(\vec{x},t) = \sigma \vec{E}(\vec{x},t), \tag{1.8}$$

where σ is the *conductivity*. In phenomenological applications using harmonic fields σ may be a complex quantity, manifesting the phase difference between field and current.

Macroscopic constitutive relations are designed to model certain aspects of material behaviors, and predict others. The types of such behaviors span the space of mathematical possibilities and can involve medium anisotropy, field nonlinearity, irreversible behaviors, polarization dependence and many other phenomena. The reader may explore some of these possible behaviors in the upcoming *Going Deeper* Sections 5.12.1 and 6.14.1.

1.4 Boundary conditions on fields

Neither Maxwell's equations in Table 1.1 nor the macroscopic Maxwell's equations in Table 1.4 plus constitutive relations are sufficient to completely specify the fields in a physical system. In addition to these equations, *boundary conditions* are needed.[7] We shall consider the boundary conditions for the macroscopic equations in Table 1.4: the boundary conditions for the vacuum equations in Table 1.1 are easily obtained by noting that in the vacuum $\vec{D} = \vec{E}$ and $\vec{H} = \vec{B}$.

We will require two key mathematical relations, which are used throughout the course:[8]

[7]This is always true with differential equations: some kind of extra information is needed to determine the constants of integration.

[8]These two relations are actually special cases of a single equation known as the *generalized Stokes' theorem*. See for example: Michael Spivak, *Calculus on Manifolds: A Modern Approach to Classical Theorems of Advanced Calculus*, 5th ed., Westview Press, 1971.

The *divergence theorem* (a.k.a. *Gauss' theorem*):

$$\int_V d^3x\ \vec{\nabla}\cdot\vec{A} = \oint_S da\ \vec{A}\cdot\hat{n}. \tag{1.9}$$

In (1.9), $\vec{A}(\vec{x})$ is a continuously-differentiable vector field defined within a region $V \subset \mathbb{R}^3$ with closed surface S and outward-pointing normal $\hat{n}$ at each surface point. Here by "closed surface" we mean a surface with no bounding edges (see Figure 1.1(a)).

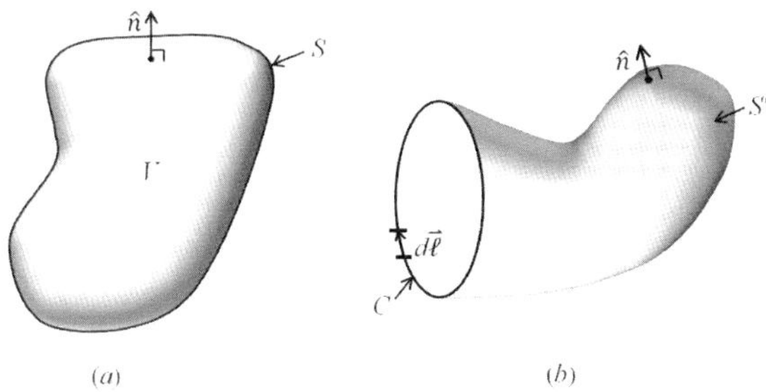

Fig. 1.1 (a) Closed surface S; (b) Open surface S', with boundary contour C.

Stokes' theorem:

$$\int_{S'} da\ (\vec{\nabla}\times\vec{A})\cdot\hat{n} = \oint_C d\vec{\ell}\cdot\vec{A}. \tag{1.10}$$

In (1.10), S' is an "open" surface, with boundary contour C. The directions of surface normal $\hat{n}$ and vector line element $\vec{\ell}$ along C are related by the *right-hand rule*: if you stick your right thumb up into the "hole" bounded by C, then your fingers will curl along the direction of $\vec{\ell}$ (see Figure 1.1(b)).

Using Gauss' and Stokes' theorems, the vector operations of divergence and curl can be interpreted geometrically as follows. Given a point $\vec{x}$ in $\mathbb{R}^3$ let V_r be the ball of radius r centered at $\vec{x}$, and let S_r be its surface (see Figure 1.2). For a continuously differentiable vector field $\vec{A}$, Gauss' theorem (together with the mean value theorem) implies that

$$\vec{\nabla}\cdot\vec{A}(\vec{x}) = \lim_{r\to 0}\frac{1}{V_r}\oint_{S_r} da\ \vec{A}\cdot\hat{n}. \tag{1.11}$$

Equation (1.11) shows that divergence may be interpreted as a measure of the "sourcedness" of a vector field. On the other hand, letting S'_r be a disc centered at $\vec{x}$ with unit normal $\hat{n}$ and boundary C_r (see Figure 1.3), we have

$$(\vec{\nabla}\times\vec{A})\cdot\hat{n} = \lim_{r\to 0}\frac{1}{S'_r}\oint_{C_r} d\vec{\ell}\cdot\vec{A}, \tag{1.12}$$

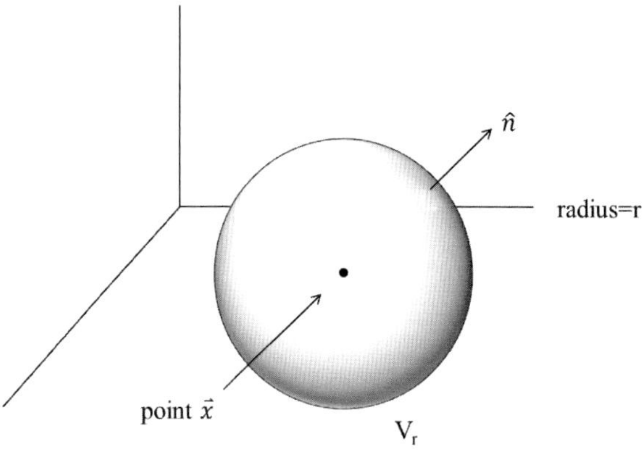

Fig. 1.2 "Sourcedness" interpretation of divergence.

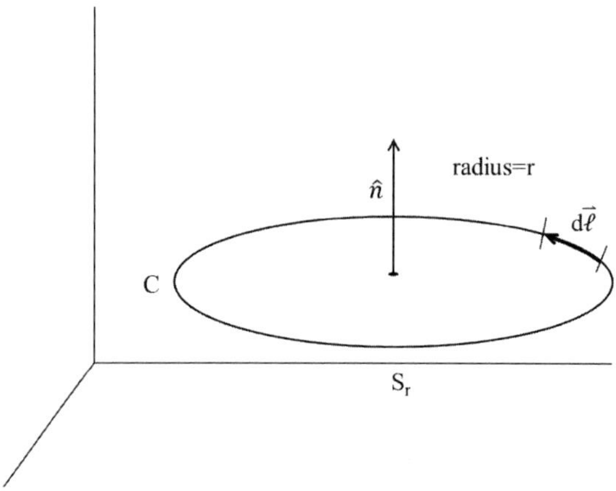

Fig. 1.3 Circulation interpretation of curl.

which shows that curl may be interpreted as a measure of the "circulation" of the vector field $\vec{A}(\vec{x})$ at each point $\vec{x}$.

This technique of integrating over infinitesimal volumes and surfaces can be used to derive boundary conditions for Maxwell's equations as follows. Applying either Gauss' or Stokes' theorem to each of the macroscopic

Maxwell equations in Table 1.4 we obtain:

$$\oint_S da\, \vec{D} \cdot \hat{n} = 4\pi \int_V d^3x\, \rho, \tag{1.13}$$

$$\oint_C d\vec{\ell} \cdot \vec{H} = \int_{S'} da \left(\frac{4\pi}{c} \vec{J} + \frac{1}{c} \frac{\partial \vec{D}}{\partial t} \right) \cdot \hat{n}, \tag{1.14}$$

$$\oint_C d\vec{\ell} \cdot \vec{E} = \frac{1}{c} \int_{S'} da\, \frac{\partial \vec{B}}{\partial t} \cdot \hat{n}, \tag{1.15}$$

$$\oint_S da\, \vec{B} \cdot \hat{n} = 0. \tag{1.16}$$

To obtain boundary conditions, we carefully choose infinitesimal volumes and surfaces that straddle the physical boundary. For instance, consider the infinitesimal cylinder (a.k.a. "pillbox") shown in Figure 1.4. Notice that the fields and their derivatives are continuous above and below the physical boundary, but may be discontinuous across the physical boundary. When we integrate over the surface of the pillbox in Figure 1.4, the contribution of the lateral surface vanishes because the fields are virtually identical on the opposite sides due to continuity, and the directions of integration are opposite. The only nonzero contribution is due to the two flat surfaces.

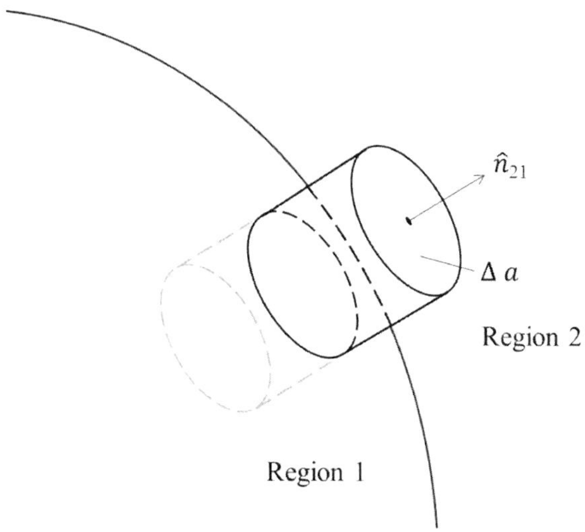

Fig. 1.4 Infinitesimal "pillbox" for computing boundary conditions using the divergence theorem.

Applying this observation to the $\vec{D}$ and $\vec{B}$ fields gives:

$$\int_S da\ \vec{D}\cdot\hat{n} = (\vec{D}_2 - \vec{D}_1)\cdot\hat{n}_{21}\Delta a, \tag{1.17}$$

$$\int_S da\ \vec{B}\cdot\hat{n} = (\vec{B}_2 - \vec{B}_1)\cdot\hat{n}_{21}\Delta a. \tag{1.18}$$

If there is a surface charge density, σ, on the tiny piece of the boundary within the pillbox, then we have:

$$4\pi\int_V d^3x\,\rho = 4\pi\sigma\Delta a. \tag{1.19}$$

Combining (1.13) and (1.16) with (1.17)–(1.19) then yields boundary conditions on $\vec{D}$ and $\vec{B}$:

$$(\vec{D}_2 - \vec{D}_1)\cdot\hat{n}_{21} = 4\pi\sigma, \tag{1.20}$$

$$(\vec{B}_2 - \vec{B}_1)\cdot\hat{n}_{21} = 0. \tag{1.21}$$

To obtain boundary conditions on $\vec{E}$ and $\vec{H}$, we consider the infinitesimal loop shown in Figure 1.5. Note that the direction of $\hat{t}$ is related to the direction of $d\vec{\ell}$ according to the right-hand rule: a right-hand thumb along $\hat{t}$ means that the fingers curl in the direction of $d\vec{\ell}$. Note also the direction of $\hat{\ell}$ is defined in terms of $\hat{t}$ and $\hat{n}_{21}$ by $\hat{\ell} = \hat{t}\times\hat{n}_{21}$. When we perform a line integral of the $\vec{E}$ or $\vec{H}$ field around the infinitesimal loop in Figure 1.5, the

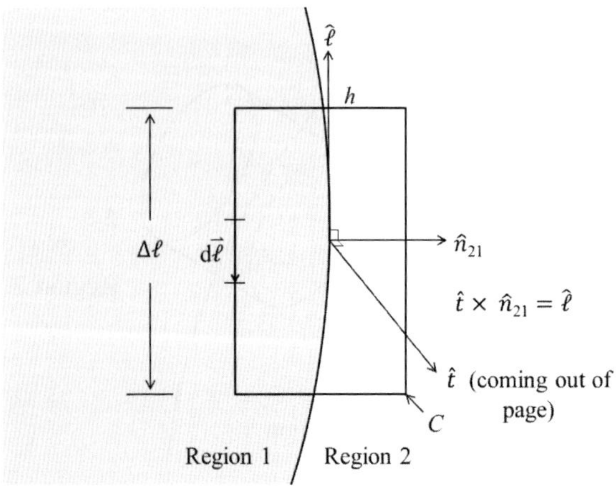

Fig. 1.5 Infinitesimal loop for computing boundary conditions using Stokes' theorem. The lateral dimension, h, is considered vanishingly small compared to $\Delta\ell$.

contribution of the upper and lower edges of the loop cancel, again because of continuity. So Stokes' theorem (1.10) applied to this situation gives:

$$\oint_C d\vec{\ell} \cdot \vec{H} = (\vec{H}_2 - \vec{H}_1) \cdot \hat{\ell}\Delta\ell, \tag{1.22}$$

$$\oint_C d\vec{\ell} \cdot \vec{E} = (\vec{E}_2 - \vec{E}_1) \cdot \hat{\ell}\Delta\ell. \tag{1.23}$$

Consider now the right-hand side of (1.14) which has two terms: a current term and a displacement term. A surface current density $\vec{K}$ on the tiny section of the boundary passing through the loop gives rise to a current term

$$\frac{4\pi}{c} \int_{S'} da\ \vec{J} \cdot \hat{t} = \frac{4\pi}{c} \vec{K} \cdot \hat{t}\ \Delta\ell. \tag{1.24}$$

On the other hand, if we make the physically-reasonable assumption that $\vec{D}$ and its derivatives are *bounded*, then the displacement term will make a contribution proportional to the area of the loop in Figure 1.5 which is vanishingly small as the lateral dimension, h, shrinks to zero. A similar argument shows that the right-hand side of (1.15) is also vanishingly small. Combining (1.14) and (1.15) with (1.22)–(1.24) gives boundary conditions on $\vec{H}$ and $\vec{E}$:

$$\left(\vec{H}_2 - \vec{H}_1\right) \cdot \hat{\ell} = \frac{4\pi}{c} \vec{K} \cdot \hat{t}, \tag{1.25}$$

$$\left(\vec{E}_2 - \vec{E}_1\right) \cdot \hat{\ell} = 0. \tag{1.26}$$

Equations (1.25) and (1.26) may be recast into an alternate form by making use of the "triple product" identity:

$$\vec{v}_1 \cdot (\vec{v}_2 \times \vec{v}_3) = \vec{v}_2 \cdot (\vec{v}_3 \times \vec{v}_1). \tag{1.27}$$

Recalling that $\hat{\ell} = \hat{t} \times \hat{n}_{21}$, we therefore have (using (1.25)–(1.27)):

$$\hat{t} \cdot \left(\hat{n}_{21} \times \left(\vec{H}_2 - \vec{H}_1\right)\right) = \frac{4\pi}{c} \vec{K} \cdot \hat{t}, \tag{1.28}$$

$$\hat{t} \cdot \left(\hat{n}_{21} \times \left(\vec{E}_2 - \vec{E}_1\right)\right) = 0. \tag{1.29}$$

Relations (1.28) and (1.29) must hold for *every* tangential vector $\hat{t}$, which implies:

$$\hat{n}_{21} \times \left(\vec{H}_2 - \vec{H}_1\right) = \frac{4\pi}{c} \vec{K}, \tag{1.30}$$

$$\hat{n}_{21} \times \left(\vec{E}_2 - \vec{E}_1\right) = 0. \tag{1.31}$$

To gain some physical insight into these boundary conditions, we consider some special cases. Suppose we have two adjacent media with $\sigma = 0$ and $\vec{K} = 0$ on the boundary. Then the boundary conditions reduce to continuity conditions on the fields:

$$E_t, D_n, B_n, H_t \text{ are continuous across the boundary,} \tag{1.32}$$

where the subscripts t and n denote components tangential and normal to the boundary, respectively. We further specialize to sources external to a neutral medium of very large dielectric constant ϵ ($\epsilon \to \infty$), which we suppose exists in a vacuum ($\epsilon = 1$). Since $\vec{D} = \epsilon\vec{E}$, in order for $\vec{D}$ to remain finite we must have $\vec{E} \to 0$ within the material. It follows from (1.32) that $E_t \to 0$ immediately adjacent to the boundary in the vacuum. This is the well-known result that the electric field is perpendicular to the surface of a conductor, which is therefore rendered an equipotential. These are called "homogeneous Dirichlet boundary conditions" if the equipotential value is zero.[9] Although it is not physically realistic, we could also consider a dielectric with $\epsilon \to 0$ in a vacuum, again with $\sigma = 0$ and external sources. In this case the finiteness of $\vec{E}$ implies $\vec{D} \to 0$ within the dielectric, so the continuity of D_n implies that $E_n = D_n = 0$ in the vacuum immediately outside the conductor. Thus, the field lines flow around the object like a fluid. These are called "homogeneous Neumann boundary conditions". We will re-visit both of these boundary condition types in Section 2.7.

1.5 Two-dimensional electrodynamics and boundary conditions

As an exercise, let us consider macroscopic electrodynamics in situations where the fields, charges, and currents depend only on x, y and not on z. We shall also assume the media are linear, isotropic, and local, so that $\vec{D}$ and $\vec{B}$ are proportional to $\vec{E}$ and $\vec{H}$ respectively. We shall develop two-dimensional versions of the Maxwell equations under these conditions, as well as boundary conditions between media. Note that this situation is *different* from restricting three-dimensional electrodynamics to a two-dimensional slice (à la Flatland). It is helpful to think of two-dimensional electrodynamics as describing *line charges*, where each line charge has a constant linear

[9] A dielectric with $\epsilon \to \infty$ (see Section 9.6) corresponds to a neutral rather than grounded conductor solution, at least for finite surface areas which do not extend to infinity. We will encounter an explicit example in Section 5.7 where this occurs.

charge density which runs parallel to the z-axis, and can only move perpendicular to the z-axis. Thus we are restricting the type of sources allowed rather than the space of interest, and as a result all fields are independent of z.

Consider first the requirement that the charges in our two-dimensional universe are not allowed to move in the z direction. From the Lorentz force law (1.1) it then follows for all (x, y) that

$$D_z(x, y) = E_z(x, y) = 0, \tag{1.33}$$

$$H_x(x, y) = H_y(x, y) = B_x(x, y) = B_y(x, y) = 0. \tag{1.34}$$

We also have

$$J_z(x, y) = 0. \tag{1.35}$$

Table 1.5 gives the resulting two-dimensional macroscopic Maxwell equations. Note that:

- The vector derivative operators $\vec{\nabla}\cdot$ and $\vec{\nabla}\times$ signify two-dimensional analogues in which $\dfrac{\partial}{\partial z} \to 0$.
- σ and $\vec{K}$ are the renamed z-independent charge and current densities, respectively.
- The no-monopole equation $\vec{\nabla} \cdot \vec{B} = 0$ has disappeared, since it follows automatically from (1.34).

Altogether, Table 1.5 gives three sourced equations and one sourceless equation. The sourceless equation can again be viewed as a constraint. It can be rewritten as

$$\frac{\partial E_y}{\partial x} + \frac{\partial(-E_x)}{\partial y} + \frac{1}{c}\frac{\partial B_z}{\partial t} = 0, \tag{1.36}$$

Table 1.5 Two-dimensional macroscopic Maxwell's equations.

Equation name	**Vector equation**	**Components**
Coulomb	$\vec{\nabla} \cdot \vec{D} = 4\pi\sigma$	1
Ampère-Maxwell	$\vec{\nabla} \times \vec{H} - \dfrac{1}{c}\dfrac{\partial \vec{D}}{\partial t} = \dfrac{4\pi}{c}\vec{K}$	2
Faraday	$\vec{\nabla} \times \vec{E} + \dfrac{1}{c}\dfrac{\partial \vec{B}}{\partial t} = 0$	1

which implies that $(E_y, -E_x, B_z)$ can be written as the curl of a 3-component vector potential (analogous to the 4-potential that gives rise to the three-dimensional Maxwell equations) if ct is interpreted as the third "direction". The remaining three equations match this counting of components.

In order to obtain boundary conditions, the two-dimensional versions of Gauss' and Stokes' laws are required:

2-D Gauss' theorem:

$$\int_S dx\, dy\, (\vec{\nabla} \cdot \vec{A}) = \oint_C d\ell\, \hat{n} \cdot \vec{A}. \tag{1.37}$$

2-D Stokes' theorem:

$$\int_S dx\, dy\, (\vec{\nabla} \times \vec{A}) \cdot \hat{z} = \oint_C d\vec{\ell} \cdot \vec{A}. \tag{1.38}$$

Figure 1.6 is a two-dimensional analogue to Figure 1.4. The boundary between regions is now a curve, which we suppose to have line charge and currents ξ and L respectively. Applying (1.37) to the small square loop gives:

$$\begin{aligned} \int_S da\, \vec{\nabla} \cdot \vec{D} &= 4\pi \int_S da\, \sigma \\ \Longrightarrow\ (\vec{D}_2 - \vec{D}_1) \cdot \hat{n}_{21} \Delta\ell &= 4\pi\xi\Delta\ell, \\ \Longrightarrow\ (\vec{D}_2 - \vec{D}_1) \cdot \hat{n}_{21} &= 4\pi\xi. \end{aligned} \tag{1.39}$$

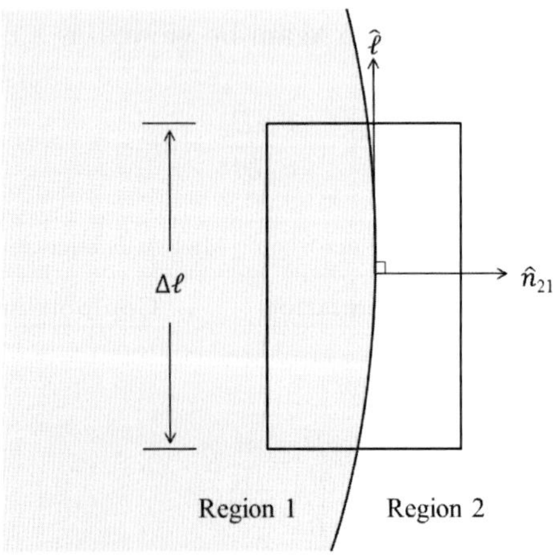

Fig. 1.6 Determination of 2-dimensional boundary conditions.

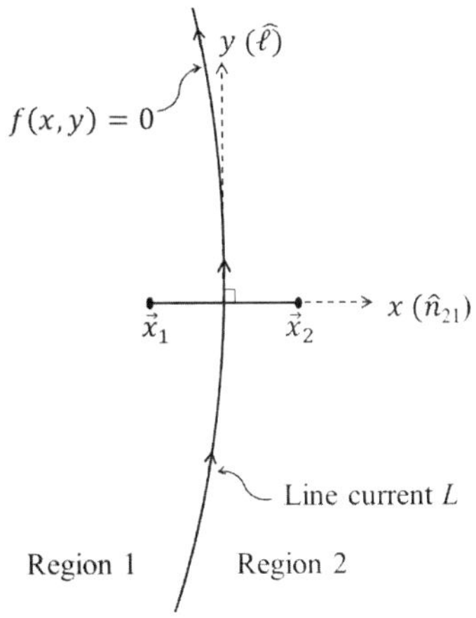

Fig. 1.7 Integration path for 2-dimensional boundary conditions.

Likewise, applying (1.38) to the 2-D Ampère-Maxwell equation in Table 1.5 gives (with the aid of (1.33)):

$$\int_S da\, \left[(\vec{\nabla}\times\vec{H})_z - \frac{1}{c}\frac{\partial D_z}{\partial t}\right] = \int_S da\, \frac{4\pi}{c}K_z, \tag{1.40}$$
$$\Longrightarrow \oint_S d\vec{\ell}\cdot\vec{H} = \int da\, \frac{4\pi}{c}K_z,$$
$$\Longrightarrow 0 = 0.$$

In other words Stokes' law applied to the loop in Figure 1.6 provides no new information. However, this should not be surprising because (1.40) relates the z-components of fields that have only x and y components.

This motivates us to seek another integration path that relates the nonzero components of the fields. Consider for instance the path characterized by $f(x,y) = 0$ in the x, y plane in Figure 1.7. To simplify the notation, we use the symbol ϕ to denote H_z. We set up a local coordinate system at a point on the boundary with x aligned perpendicular to the boundary (in the $\hat{n}_{21}$ direction) and y parallel to it (in the $\hat{\ell}$ direction). Integrating the y-component of the Ampère-Maxwell equation across the boundary ($\vec{x}_1$ to $\vec{x}_2$) gives:

$$\int dx \left(-\frac{\partial\phi}{\partial x} - \frac{1}{c}\frac{\partial D_y}{\partial t}\right) = \int dx\, \frac{4\pi}{c}K_y. \tag{1.41}$$

A line current flowing along the boundary between the two regions will give a contribution

$$\int dx\ K_y = L, \tag{1.42}$$

where L is the boundary line current. On the other hand, the left-hand side of (1.41) can be integrated directly:

$$\begin{aligned} \int dx \left(-\frac{\partial \phi}{\partial x} - \frac{1}{c}\frac{\partial D_y}{\partial t} \right) &= \phi|_2^1 , \\ \Longrightarrow \phi_1 - \phi_2 &= \frac{4\pi}{c} L. \end{aligned} \tag{1.43}$$

The D_y contribution vanishes because of continuity and the assumption that all fields are bounded.

Finally, we find boundary conditions associated with the Faraday equation in Table 1.5, which has only one nontrivial component (the z-component). Using the 2-D Stokes law on the contour in Figure 1.6 gives

$$\begin{aligned} &\int da\ (\vec{\nabla} \times \vec{E})_z + \int da\ \frac{1}{c}\frac{\partial \phi}{\partial t} = 0 \\ &\Longrightarrow \oint_C d\vec{\ell} \cdot \vec{E} = 0 \\ &\Longrightarrow (\vec{E}_2 - \vec{E}_1) \cdot \hat{\ell} = 0. \end{aligned} \tag{1.44}$$

Notice the dimensional differences between the boundary condition derivations for two and three dimensions. In three dimensions, we used pillboxes with Gauss' law and loops with Stokes' law: while in two dimensions, we used loops and straight-line integrals.

Table 1.6 Two and three-dimensional boundary conditions.

Three-dimensional b.c.	**Two-dimensional b.c.**
$(\vec{D}_2 - \vec{D}_1) \cdot \hat{n}_{21} = 4\pi\sigma$	$(\vec{D}_2 - \vec{D}_1) \cdot \hat{n}_{21} = 4\pi\xi$
$(\vec{B}_2 - \vec{B}_1) \cdot \hat{n}_{21} = 0$	–
$\hat{n}_{21} \times (\vec{H}_2 - \vec{H}_1) = \frac{4\pi}{c}\vec{K}$	$(H_{z,1} - H_{z,2}) = \frac{4\pi}{c} L$
$\hat{n}_{21} \times (\vec{E}_2 - \vec{E}_1) = 0$	$(\vec{E}_2 - \vec{E}_1) \cdot \hat{\ell} = 0$

Table 1.6 summarizes the differences between two and three-dimensional boundary conditions. Actually, these two-dimensional boundary equations

may be obtained from the three-dimensional equations using (1.33) and (1.34) and a little dimensional analysis;[10] nevertheless, this has been a useful exercise. Two-dimensional electrodynamics will appear repeatedly as special cases and as examples in the following chapters.

1.6 Going Deeper

1.6.1 *Mathematical methods*

(1) Abramowitz and I. A. Stegun (eds.), *Handbook of Mathematical Functions with Formulas, Graphs and Mathematical Tables*, National Bureau of Standards, 10th printing (Washington D. C.) 1972; Courier Dover reprint; online copy at `http://www.nr.com/aands`.

(2) F. W. J. Olver, D. W. Lozier, R. F. Boisvert and C. W. Clark (eds.), *Handbook of Mathematical Functions*, Cambridge Univ. Press (New York) 2010.

(3) A. B. O. Daalhuis, D. W. Lozier, B. I. Schneider, R. F. Boisvert, C. W. Clark, B. R. Miller and B. V. Saunders (eds.), *NIST Digital Library of Mathematical Functions*, `http://dlmf.nist.gov`.

(4) G. B. Arfken, H. J. Weber and F. E. Harris, *Mathematical Methods for Physicists*, 7th edn., Academic Press (Waltham, MA) 2013.

(5) R. Courant and D. Hilbert, *Methods of Mathematical Physics, Vols. I and II*, Wiley-VCH (Weinheim) 1989.

(6) I. S. Gradshteyn and I. M. Ryzhik, *Table of Integrals, Series and Products*, A. Jeffrey and D. Zwillinger (eds.), 7th edn., Elsevier Academic Press (Burlington, MA) 2007.

(7) P. M. Morse and H. Feshbach, *Methods of Theoretical Physics, Parts I and II*, McGraw-Hill (New York) 1953; also available at Feshbach Publishing, `http://www.feshbachpublishing.com`.

---

[10] It is more consistent to use $2\pi\xi$ in the D_n equation in a geometrical 2-D context and $4\pi\xi$ for an infinite line charge in 3-D. See Exercises 1.5.1 and 2.4.1.

(8) A. D. Polyanin (ed.), "EqWorld: The World of Mathematical Equations", `http://eqworld.ipmnet.ru/index.htm`.

(9) E. T. Whittaker and G. N. Watson, *A Course in Modern Analysis*, 4th edn., Cambridge University Press, reprinted in the Cambridge Mathematical Library series (Cambridge) 1996.

1.6.2 *Electromagnetic units*

(1) G. J. Aubrecht II, "Changes Coming to the International System of Units", *Phys. Teach.* **50**, 338 (2012).

(2) R. T. Birge, "On the Establishment of Fundamental and Derived Units, with Special Reference to Electric Units. Part II", *Am. J. Phys.* **2**, 41 (1934).

(3) Y.-T. Lau, "An Easy Method for Converting Equations Between SI and Gaussian Units", *Am. J. Phys.* **56**, 135 (1988).

(4) J. D. Jackson, *Classical Electrodynamics*, 3rd edn., John Wiley & Sons (New York) 1999; Appendix on Units and Dimensions.

(5) E. S. Weibel, "Dimensionally Correct Transformations Between Different Systems of Units", *Am. J. Phys.* **36**, 1130 (1968).

1.7 Exercises

Exercise 1.1.1. Verify explicitly:

(a) $$\vec{A} \times (\vec{B} \times \vec{C}) + \vec{B} \times (\vec{C} \times \vec{A}) + \vec{C} \times (\vec{A} \times \vec{B}) = 0.$$

(b) $$\nabla^2(\phi\psi) = \phi\nabla^2\psi + \psi\nabla^2\phi + 2\vec{\nabla}\phi \cdot \vec{\nabla}\psi.$$

(c) $$\sum_i A_i \vec{\nabla} B_i = \vec{\nabla}(\vec{A} \cdot \vec{B}) - (\vec{B} \cdot \vec{\nabla})\vec{A} - \vec{B} \times (\vec{\nabla} \times \vec{A}).$$

Exercise 1.1.2. Given the definitions

$$\vec{E} \equiv -\vec{\nabla}\Phi - \frac{1}{c}\frac{\partial \vec{A}}{\partial t}, \qquad \vec{B} \equiv \vec{\nabla} \times \vec{A},$$

where Φ and $\vec{A}$ are called the *scalar* and *vector potentials*, respectively, show that two of the Maxwell equations in Table 1.1 become satisfied identities, and that the other two give coupled differential equations satisfied by Φ and $\vec{A}$. Compute and simplify these differential equations.

Exercise 1.1.3. For charge and current densities of the form

$$\rho(\vec{r}, t) = ef(\vec{r} - \vec{R}(t)),$$
$$\vec{J}(\vec{r}, t) = e\frac{d\vec{R}}{dt}f(\vec{r} - \vec{R}(t)),$$

where $f(\cdots)$ is an arbitrary function, verify the conservation of charge statement:

$$\frac{\partial \rho(\vec{r}, t)}{\partial t} + \vec{\nabla} \cdot \vec{J}(\vec{r}, t) = 0.$$

Exercise 1.4.1.

(a) A planar surface S with area A and boundary C lies in the xy-plane. Consider a coplanar vector

$$\vec{v} = -y\hat{\imath} + x\hat{\jmath},$$

and prove that

$$\oint_C \vec{v} \cdot d\vec{\ell} = \oint_C (\hat{\imath}\, dx + \hat{\jmath}\, dy) \cdot \vec{v} = 2A.$$

(b) Use the equation of an ellipse

$$\frac{x^2}{a^2} + \frac{y^2}{b^2} = 1,$$

and the result of part(a) to find the area of the ellipse. [Adapted from B. DiBartolo, *Classical Theory of Electromagnetism*, 2nd edn., World Scientific (Singapore) 2004, Exercise 1.14.]

Exercise 1.4.2. Show that the following are equivalent expressions of Gauss' theorem:

(a)
$$\int_V d^3x\, \vec{\nabla}\phi = \oint_S da\, \hat{n}\, \phi.$$

(b)
$$\int_V d^3x\, \vec{\nabla} \times \vec{A} = \oint_S da\, \hat{n} \times \vec{A}.$$

Exercise 1.4.3. Show that the following is an equivalent expression of Stokes' theorem:

$$\int_S da\, \hat{n} \times \vec{\nabla}\phi = \oint_C d\vec{\ell}\, \phi.$$

Exercise 1.5.1.

(a) Use Gauss' theorem to derive the radial dependence of the electric field in two dimensions for a point charge q.
(b) Show that in two dimensions Gauss' theorem and Stokes' theorem are equivalent.

1.8 Useful identities

Glossary of symbols

In the text and the identities below we use the following symbols and notation:

- $\vec{A}$, $\vec{B}$, $\vec{C}$: vectors fields (functions of position) that are differentiable

- ϕ, Φ, Ψ: scalar fields (functions of position) that are differentiable
- $\hat{i}$ ($\hat{x}$, $\hat{e}_1$), $\hat{j}$ ($\hat{y}, \hat{e}_2$) and $\hat{k}$ ($\hat{z}$, $\hat{e}_3$): unit vectors along coordinates x (x_1), y (x_2) and z (x_3).
- $\times$, $\cdot$: cross product, dot product, respectively
- $\epsilon \to 0^+$: taking ϵ to zero from positive values
- Kronecker delta:

$$\delta_{ij} = \begin{cases} 0 & \text{for } i \neq j \\ 1 & \text{for } i = j \end{cases}$$

- Levi-Civita (permutation) symbol:

$$\epsilon_{ijk} = \begin{cases} 1 & \text{if } (i,j,k) \text{ is } (1,2,3), (2,3,1) \text{ or } (3,1,2) \\ -1 & \text{if } (i,j,k) \text{ is } (3,2,1), (2,1,3) \text{ or } (1,3,2) \\ 0 & \text{if } i = j \text{ or } j = k \text{ or } i = k \end{cases}$$

- $\vec{\nabla}$: gradient operator; $\nabla_x = \nabla_1 = \partial_x = \dfrac{\partial}{\partial x}$ are partial derivatives with respect to coordinate x
- d^3x, da: infinitesimal volume and area elements, respectively
- $\hat{n}$: given a closed surface S, $\hat{n}$ is the unit outward normal vector ($da\,\hat{n}$ is sometimes written as $d\vec{s}$)
- $\dfrac{\partial}{\partial n}$: alternate notation for the outward normal derivative $\hat{n} \cdot \vec{\nabla}$.
- $\oint_S d\vec{s}$: closed surface integral with $d\vec{s}$ in the outward direction
- $\oint_C d\vec{\ell}$: given an open surface S with boundary C, the direction of $d\vec{\ell}$ in the contour integral is related to $\hat{n}$ for S according to the right-hand rule.

Vector identities

$$\vec{A} \times (\vec{B} \times \vec{C}) = \vec{B}(\vec{A} \cdot \vec{C}) - \vec{C}(\vec{A} \cdot \vec{B}) \qquad (BAC\text{-}CAB \text{ rule})$$

$$\vec{A} \cdot (\vec{B} \times \vec{C}) = \vec{B} \cdot (\vec{C} \times \vec{A}) = \vec{C} \cdot (\vec{A} \times \vec{B}) \qquad \text{(triple product)}$$

$$\vec{A} \times (\vec{B} \times \vec{C}) + \vec{B} \times (\vec{C} \times \vec{A}) + \vec{C} \times (\vec{A} \times \vec{B}) = 0 \qquad \text{(Jacobi identity)}$$

Summation identities

$$\sum_j \delta_{ij}\delta_{jk} = \delta_{ik}$$

$$\sum_i \epsilon_{ijk}\epsilon_{imn} = \delta_{jm}\delta_{kn} - \delta_{jn}\delta_{km}$$

$$\sum_{m,n} \epsilon_{imn}\epsilon_{jmn} = 2\delta_{ij}$$

First-order differential identities

$$\vec{\nabla}\cdot(\phi\vec{A}) = \vec{A}\cdot\vec{\nabla}\phi + \phi\vec{\nabla}\cdot\vec{A}$$

$$\vec{\nabla}\times(\phi\vec{A}) = \vec{\nabla}\phi\times\vec{A} + \phi\vec{\nabla}\times\vec{A}$$

$$\vec{\nabla}(\vec{A}\cdot\vec{B}) = (\vec{A}\cdot\vec{\nabla})\vec{B} + (\vec{B}\cdot\vec{\nabla})\vec{A} + \vec{A}\times(\vec{\nabla}\times\vec{B}) + \vec{B}\times(\vec{\nabla}\times\vec{A})$$

$$\vec{\nabla}\cdot(\vec{A}\times\vec{B}) = \vec{B}\cdot(\vec{\nabla}\times\vec{A}) - \vec{A}\cdot(\vec{\nabla}\times\vec{B})$$

$$\vec{\nabla}\times(\vec{A}\times\vec{B}) = \vec{A}(\vec{\nabla}\cdot\vec{B}) - \vec{B}(\vec{\nabla}\cdot\vec{A}) + (\vec{B}\cdot\vec{\nabla})\vec{A} - (\vec{A}\cdot\vec{\nabla})\vec{B}$$

Second-order differential identities

$$\vec{\nabla}\times\vec{\nabla}\phi = 0$$

$$\vec{\nabla}\cdot(\vec{\nabla}\times\vec{A}) = 0$$

$$\vec{\nabla}\times(\vec{\nabla}\times\vec{A}) = \vec{\nabla}(\vec{\nabla}\cdot\vec{A}) - \nabla^2\vec{A}$$

$$\nabla^2(\phi\psi) = \phi\nabla^2\psi + \psi\nabla^2\phi + 2\vec{\nabla}\phi\cdot\vec{\nabla}\psi$$

Integral identities

$$\int_V d^3x\, \vec{\nabla} \cdot \vec{A} = \oint_S da\, \vec{A} \cdot \hat{n} \qquad \text{(Gauss Theorem)}$$

$$\int_S da\, \hat{n} \cdot (\vec{\nabla} \times \vec{A}) = \oint_C d\vec{\ell} \cdot \vec{A} \qquad \text{(Stokes Theorem)}$$

$$\int_V d^3x (\phi \nabla^2 \psi + \vec{\nabla}\phi \cdot \vec{\nabla}\psi) = \oint_S da\, \phi(\hat{n} \cdot \vec{\nabla}\psi) \qquad \text{(Green's first identity)}$$

$$\int_V d^3x \left(\phi \nabla^2 \psi - \psi \nabla^2 \phi\right) = \oint_S da\, \Big(\phi(\hat{n} \cdot \vec{\nabla}\psi) - \psi(\hat{n} \cdot \vec{\nabla}\phi)\Big) \qquad \text{(Green's second identity)}$$

Trigonometric identities

$$\begin{aligned}
\sin(x+y) &= \sin x \cos y + \cos x \sin y \\
\cos(x+y) &= \cos x \cos y - \sin x \sin y \\
\sin x \sin y &= \frac{1}{2}\left(\cos(x-y) - \cos(x+y)\right) \\
\cos x \cos y &= \frac{1}{2}\left(\cos(x-y) + \cos(x+y)\right) \\
\sin x \cos y &= \frac{1}{2}\left(\sin(x-y) + \sin(x+y)\right)
\end{aligned}$$

Vector operators in various coordinate systems

Gradient ($\vec{\boldsymbol{\nabla}}\boldsymbol{\Phi}$):

$$\begin{aligned}
\text{rectangular}: &\quad \frac{\partial \Phi}{\partial x}\hat{i} + \frac{\partial \Phi}{\partial y}\hat{j} + \frac{\partial \Phi}{\partial z}\hat{k} \\
\text{cylindrical}: &\quad \frac{\partial \Phi}{\partial \rho}\hat{\rho} + \frac{1}{\rho}\frac{\partial \Phi}{\partial \phi}\hat{\phi} + \frac{\partial \Phi}{\partial z}\hat{k} \\
\text{spherical}: &\quad \frac{\partial \Phi}{\partial r}\hat{r} + \frac{1}{r}\frac{\partial \Phi}{\partial \theta}\hat{\theta} + \frac{1}{r\sin\theta}\frac{\partial \Phi}{\partial \phi}\hat{\phi}
\end{aligned}$$

Divergence **($\vec{\nabla} \cdot \vec{A}$):**

$$\text{rectangular}: \quad \frac{\partial A_x}{\partial x} + \frac{\partial A_y}{\partial y} + \frac{\partial A_z}{\partial z}$$

$$\text{cylindrical}: \quad \frac{1}{\rho}\frac{\partial(\rho A_\rho)}{\partial \rho} + \frac{1}{\rho}\frac{\partial A_\phi}{\partial \phi} + \frac{\partial A_z}{\partial z}$$

$$\text{spherical}: \quad \frac{1}{r^2}\frac{\partial(r^2 A_r)}{\partial r} + \frac{1}{r\sin\theta}\frac{\partial(\sin\theta A_\theta)}{\partial\theta} + \frac{1}{r\sin\theta}\frac{\partial A_\phi}{\partial\phi}$$

Curl **($\vec{\nabla} \times \vec{A}$):**

$$\text{rectangular}: \quad \left(\frac{\partial A_z}{\partial y} - \frac{\partial A_y}{\partial z}\right)\hat{i} + \left(\frac{\partial A_x}{\partial z} - \frac{\partial A_z}{\partial x}\right)\hat{j} + \left(\frac{\partial A_y}{\partial x} - \frac{\partial A_x}{\partial y}\right)\hat{k}$$

$$\text{cylindrical}: \quad \left(\frac{1}{\rho}\frac{\partial A_z}{\partial \phi} - \frac{\partial A_\phi}{\partial z}\right)\hat{\rho} + \left(\frac{\partial A_\rho}{\partial z} - \frac{\partial A_z}{\partial \rho}\right)\hat{\phi} + \frac{1}{\rho}\left(\frac{\partial(\rho A_\phi)}{\partial \rho} - \frac{\partial A_\rho}{\partial \phi}\right)\hat{z}$$

$$\text{spherical}: \quad \frac{1}{r\sin\theta}\left(\frac{\partial(A_\phi \sin\theta)}{\partial\theta} - \frac{\partial A_\theta}{\partial\phi}\right)\hat{r} + \frac{1}{r}\left(\frac{1}{\sin\theta}\frac{\partial A_r}{\partial\phi} - \frac{\partial(rA_\phi)}{\partial r}\right)\hat{\theta} + \frac{1}{r}\left(\frac{\partial(rA_\theta)}{\partial r} - \frac{\partial A_r}{\partial\theta}\right)\hat{\phi}$$

Laplacian **($\nabla^2\Phi$):**

$$\text{rectangular}: \quad \frac{\partial^2\Phi}{\partial x^2} + \frac{\partial^2\Phi}{\partial y^2} + \frac{\partial^2\Phi}{\partial z^2}$$

$$\text{cylindrical}: \quad \frac{1}{\rho}\frac{\partial}{\partial\rho}\left(\rho\frac{\partial\Phi}{\partial\rho}\right) + \frac{1}{\rho^2}\frac{\partial^2\Phi}{\partial\phi^2} + \frac{\partial^2\Phi}{\partial z^2}$$

$$\text{spherical}: \quad \frac{1}{r^2}\frac{\partial}{\partial r}\left(r^2\frac{\partial\Phi}{\partial r}\right) + \frac{1}{r^2\sin\theta}\frac{\partial}{\partial\theta}\left(\sin\theta\frac{\partial\Phi}{\partial\theta}\right) + \frac{1}{r^2\sin^2\theta}\frac{\partial^2\Phi}{\partial\phi^2}$$

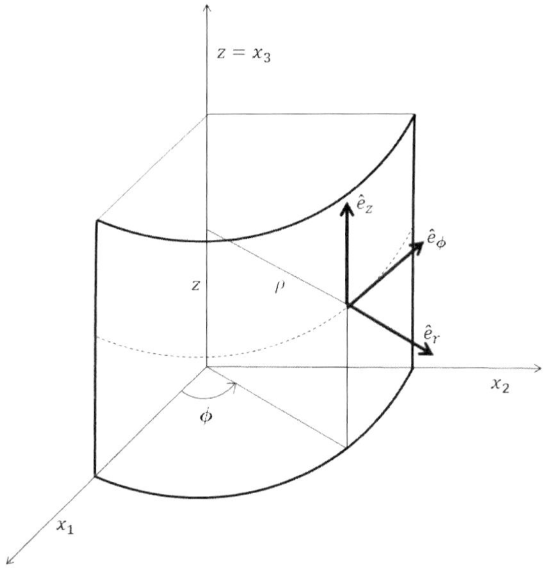

Fig. 1.8 Cylindrical coordinates: variables.

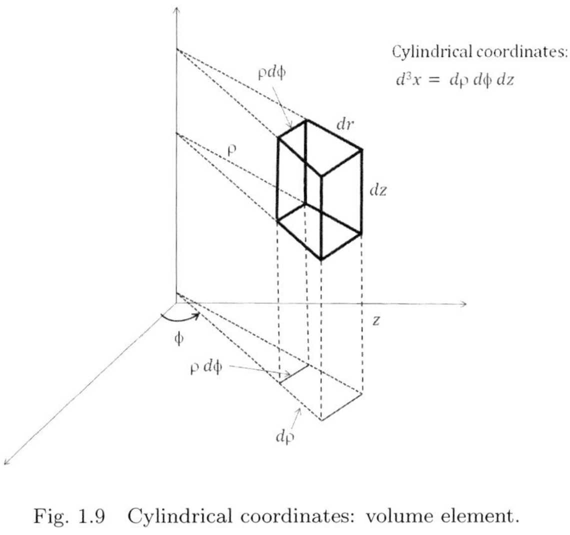

Fig. 1.9 Cylindrical coordinates: volume element.

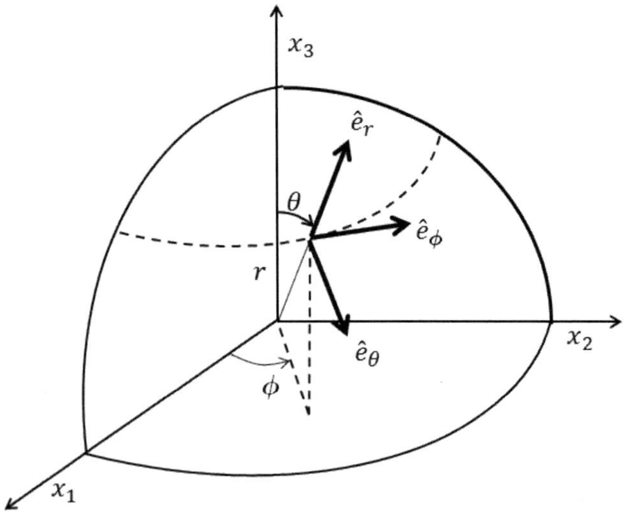

Fig. 1.10 Spherical coordinates: variables.

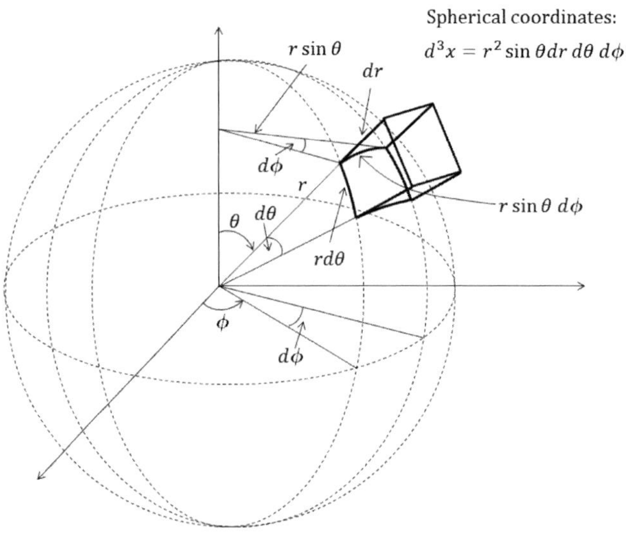

Fig. 1.11 Spherical coordinates: volume element.

Chapter 2

Introduction to Electrostatics

Charges produce electric fields; and conversely, electric fields exert forces on charges. Historically, the nature of these fields and forces were first studied using stationary (or nearly stationary) charges. Accordingly, in this section we will study electric fields produced by "stationary" charge configurations. Of course, relativity tells us that nothing is "stationary": so to be precise, the formulas in this section are valid for charges that are stationary relative to each other, and to the observer.

2.1 Electric field: definition

Consider a system consisting of two highly localized[1] charges (referred to as "point charges") of magnitudes q and q', located at positions $\vec{x}$ and $\vec{x}\,'$ respectively as shown in Figure 2.1. According to *Coulomb's Law*, the electric force that charge q' exerts on charge q is:

$$\vec{F}_{q' \text{ on } q} = qq' \frac{\vec{x} - \vec{x}\,'}{|\vec{x} - \vec{x}\,'|^3}. \tag{2.1}$$

We may describe this situation by saying that charge q' produces an *electric field*

$$\vec{E}_{q'}(\vec{x}) = q' \frac{\vec{x} - \vec{x}\,'}{|\vec{x} - \vec{x}\,'|^3}, \tag{2.2}$$

so that the electric force of q' on q is given by

$$\vec{F}_{q' \text{ on } q} = q\vec{E}_{q'}(\vec{x}). \tag{2.3}$$

[1] "Highly localized" means that the dimensions of each charge are tiny compared to the distance between charges.

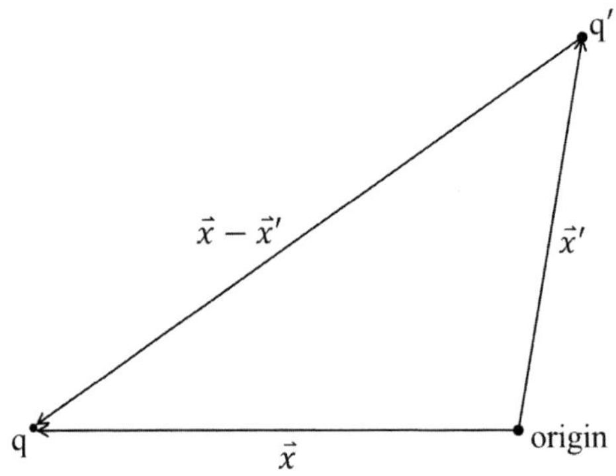

Fig. 2.1 Point charges in space.

According to this description, q' is taken as the field's *source*, and $\vec{x}\,'$ is a source point; while q is taken as a *test charge*, and $\vec{x}$ is a field point.

If we let $q_1 = q'$, $\vec{x}_1 = \vec{x}\,'$, and additional point charges $q_2 \dots q_N$ are present at locations $\vec{x}_2, \dots, \vec{x}_N$, then the electric field at point $\vec{x}$ is the sum of the fields associated with each charge $q_1, \dots, q_N$:

$$\vec{E}_{tot} = \sum_{n=1}^{N} \vec{E}_{q_n} = \sum_{n=1}^{N} q_n \frac{\vec{x} - \vec{x}_n}{|\vec{x} - \vec{x}_n|^3}. \tag{2.4}$$

We also want to consider electric fields produced by charges that are not highly localized. In this case, we describe the charge configuration in terms of a *charge density* which gives the charge per volume (in statC/cm^3) at each location in space. In other words, if ρ is the charge density function, then the charge within volume element d^3x' located at $\vec{x}\,'$ is equal to $\rho(\vec{x}\,')d^3x'$. Finding the total electric field requires summing over all volume elements, which is tantamount to an integration over all space:

$$\vec{E}_{tot}(\vec{x}) = \int d^3x' \rho(\vec{x}\,') \frac{\vec{x} - \vec{x}\,'}{|\vec{x} - \vec{x}\,'|^3}. \tag{2.5}$$

Naturally regions of space where no charges exist have $\rho = 0$, and do not contribute to the integral. On the other hand charge densities which fall off too slowly can lead to singular results. One needs $\rho(\vec{x}) \sim 1/(r)^{1+\epsilon}, \epsilon > 0$ at large distances for the integral in (2.5) to converge.

2.2 The Dirac delta function and singular charge distributions

Although (2.4) and (2.5) both characterize electric fields, the forms of the two equations are considerably different. It's better to have a single mathematical formalism that accommodates both situations, especially in cases in which both discrete and continuous charges are present simultaneously. In this section we introduce the concept of the *Dirac delta function*, which enables a unified treatment of discrete and continuous charges, as well as surface and line charges. We shall find it incredibly important to have a correct understanding of these matters, and so shall spend more time than usual in appropriately characterizing these important constructs.

The definition of the Dirac delta function requires a limiting process. Consider the unit step function on the interval $[-\frac{1}{2}, \frac{1}{2}]$, which we denote by $\Delta(x)$:

$$\Delta(x) = \begin{cases} 0 & \text{if } |x| > \frac{1}{2}, \\ 1 & \text{if } |x| \leq \frac{1}{2}. \end{cases} \tag{2.6}$$

Note that the integral of $\Delta(x)$ is equal to 1. One way to define the delta function $\delta(x)$ is as the "limit" of rescaled versions of $\Delta(x)$:

$$\delta(x) \equiv \lim_{\epsilon \to 0^+} \frac{1}{\epsilon} \Delta\left(\frac{x}{\epsilon}\right). \tag{2.7}$$

The original $\Delta(x)$ and rescaled versions $\epsilon = \frac{1}{2}$ and $\epsilon = \frac{1}{4}$ are shown in Figure 2.2.

Two key properties of the delta function are:

(1) $\int_{-\infty}^{\infty} dx\, \delta(x) = 1$;
(2) $\delta(x) = 0$ for $x \neq 0$.

Property (1) is true since $\int_{-\infty}^{\infty} dx\, \epsilon^{-1} \Delta(x\epsilon^{-1}) = 1$ for any value of ϵ. Note that this implies that the one-dimensional Dirac delta function has engineering units of $[\text{Length}]^{-1}$; n-dimensional generalizations likewise have dimensions of $[\text{Length}]^{-n}$. This also implies the constant ϵ in (2.7) is a length. Property (2) is true because $\Delta(x\epsilon^{-1})$ is only nonzero for $x \in [-\epsilon/2, \epsilon/2]$, which shrinks towards 0 as $\epsilon \to 0^+$. Here we are being somewhat vague about the sense of "limit": practically, you may consider $\delta(x)$ as equal to $\epsilon^{-1}\Delta(x\epsilon^{-1})$ for a value of ϵ that is much smaller than any physical length scale in the system being studied. To be mathematically precise,

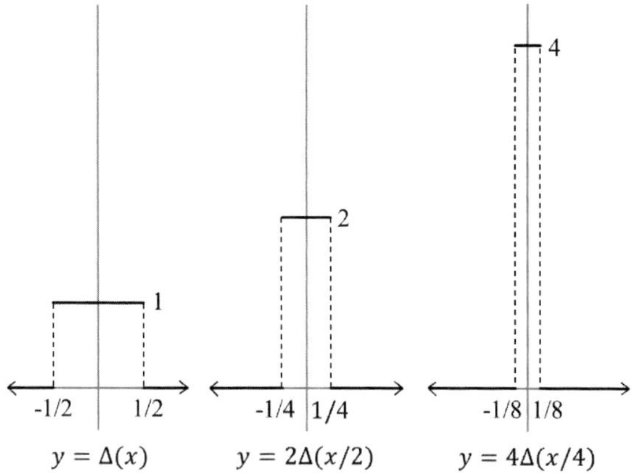

Fig. 2.2 The delta function as a "limit" of step functions.

the limit is a "weak limit": the functions $\phi_\epsilon(x) \equiv \epsilon^{-1}\Delta(x\epsilon^{-1})$ are considered as linear functionals acting on the set of continuous functions $f(x)$ via $\langle\phi_\epsilon, f\rangle \equiv \int_{-\infty}^{\infty} \phi_\epsilon(x) f(x) dx$. It can be shown that $\lim_{\epsilon\to 0^+}\langle\phi_\epsilon, f\rangle = f(0)$.[2] In the present context, this means

$$\int_{-\infty}^{\infty} dx\ f(x)\delta(x) = f(0). \tag{2.8}$$

Other functions $f(x)$ can replace $\Delta(x)$ in definition (2.7):

$$\delta(x) \equiv \lim_{\epsilon\to 0^+} \frac{1}{\epsilon} f\left(\frac{x}{\epsilon}\right), \tag{2.9}$$

as long as $f(x)$ is nonnegative and

$$\int_{-\infty}^{\infty} dx\, f(x) = 1. \tag{2.10}$$

A common choice for $f(x)$ is the *Gaussian distribution*

$$f\left(\frac{x}{\epsilon}\right) = \frac{1}{\sqrt{2\pi}} e^{-x^2/2\epsilon^2}. \tag{2.11}$$

Another example is given in Exercise 2.2.2.

Two additional properties of the Dirac delta function can be quickly established from this representation. First, from (2.6) or (2.11)

$$\delta(-x) = \delta(x), \tag{2.12}$$

[2] I. M. Gel'fand and G. E. Shilov, *Generalized Functions, Vol. I*, Academic Press, 1964.

which characterizes it as an even function of its argument. Using this we also have

$$\begin{aligned}\delta(ax) &= \lim_{\epsilon\to 0^+} \frac{1}{\epsilon}\Delta\left(\frac{|a|x}{\epsilon}\right) \\ &= \lim_{\epsilon'\to 0^+} \frac{1}{|a|\epsilon'}\Delta\left(\frac{x}{\epsilon'}\right) \\ &= \frac{1}{|a|}\delta(x), \end{aligned} \tag{2.13}$$

where $\epsilon' \equiv \epsilon/|a|$.

We can also consider delta functions of functions. We will only get contributions from $\delta(f(x))$ each time $f(x) = 0$. Assuming there are n (non-repeating or simple) roots to this equation $x_1 \ldots, x_n$, we expect that

$$\delta(f(x)) = \delta\left(\sum_{i=1}^{n} \left.\frac{df}{dx}\right|_{x_i} (x - x_i)\right), \tag{2.14}$$

from a Taylor series expansion of $f(x)$ near $x - x_i$. This gives

$$\delta(f(x)) = \sum_{i=1}^{n} \delta\left(\left.\frac{df}{dx}\right|_{x_i} (x - x_i)\right), \tag{2.15}$$

since the sum of separate zeros of $f(x)$ give separate delta function contributions. Then using (2.13), we obtain

$$\delta(f(x)) = \sum_{i=1}^{n} \frac{1}{\left|\left.\frac{df}{dx}\right|_{x_i}\right|}\delta(x - x_i). \tag{2.16}$$

Switching to a three-dimensional context, suppose we treat a "point" charge of magnitude 1 as a uniform continuous charge localized in a very small region. If the charge is centered at the origin and is confined to a tiny box with sides dx', dy', and dz', then the charge density must be $(dx'dy'dz')^{-1}$ in order to obtain total charge 1. The charge density $\rho(\vec{x})$ therefore corresponds to a product of rescaled characteristic functions, one in each dimension:

$$\begin{aligned}\rho(\vec{x}) &= \frac{1}{dx'dy'dz'}\Delta\left(\frac{x}{dx'}\right)\Delta\left(\frac{y}{dy'}\right)\Delta\left(\frac{z}{dz'}\right) \\ &= \frac{1}{dx'}\Delta\left(\frac{x}{dx'}\right)\frac{1}{dy'}\Delta\left(\frac{y}{dy'}\right)\frac{1}{dz'}\Delta\left(\frac{z}{dz'}\right) \\ &\approx \delta(x)\delta(y)\delta(z).\end{aligned}$$

To save time we typically write $\delta(\vec{x})$ to denote $\delta(x)\delta(y)\delta(z)$; and $\delta(\vec{x})$ is referred to as a "three-dimensional delta function".

We may translate the delta function by shifting its argument. Thus, the charge density for a point charge q located at $\vec{x}\,'$ is $q\,\delta(\vec{x}-\vec{x}\,')$. Assigning a collection of N point charges $\{q_n\}$ located at $\{\vec{x}_n\}$ gives the charge density

$$\rho(\vec{x}\,') = \sum_{n=1}^{N} q_n \delta(\vec{x}\,' - \vec{x}_n). \tag{2.17}$$

The integration over the volume element

$$\int d^3x'\,\delta(\vec{x}\,' - \vec{x}_n) = 1, \tag{2.18}$$

then provides the connection between Equations (2.4) and (2.5).

We shall often find it convenient to work in non-cartesian coordinate systems such as cylindrical or spherical coordinates. For this purpose, we will need an expression for the delta function in alternative coordinate systems. For a general rescaling of distance units by a factor m, the three-dimensional delta function version of (2.13) is:

$$\delta\left(m(\vec{x} - \vec{x}\,')\right) = \frac{1}{|m|^3} \cdot \delta(\vec{x} - \vec{x}\,'). \tag{2.19}$$

Note that the factor on the right is just the ratio of the original volume to the mapped volume element. Now suppose that we perform a linear mapping $x_i \to \sum_j M_{ij}x_j$ in three dimensions, symbolized by $\vec{x} \to M\vec{x}$, where M_{ij} is a nonsingular 3×3 matrix. From linear algebra we know that such a linear mapping maps a unit cube to a parallelepiped with volume equal to $|\det M|$. The ratio of mapped volumes then implies:

$$\delta(M\vec{x} - M\vec{x}\,') = |\det M|^{-1}\delta(\vec{x} - \vec{x}\,'). \tag{2.20}$$

We now consider a three-dimensional *nonlinear* mapping from $\vec{u} = (u, v, w)$ to $\vec{x} = (x, y, z)$, given by three coordinate functions:

$$x = F_1(u, v, w),$$
$$y = F_2(u, v, w),$$
$$z = F_3(u, v, w),$$

so that each of x, y, z is considered to be a function of the variables u, v, w. We will symbolize this mapping as

$$\vec{x} = \vec{F}(\vec{u}). \tag{2.21}$$

Vector calculus tells us that in a small neighborhood of $\vec{u}\,' = (u', v', w')$ the mapping can be well-approximated by the linear function:

$$\vec{x} - \vec{x}\,' \approx \left.\frac{\partial(x, y, z)}{\partial(u, v, w)}\right|_{\vec{u}\,'} (\vec{u} - \vec{u}\,'). \tag{2.22}$$

The *Jacobian matrix* is given by

$$\left.\frac{\partial(x, y, z)}{\partial(u, v, w)}\right|_{\vec{u}\,'} \equiv \begin{pmatrix} \frac{\partial x}{\partial u} & \frac{\partial x}{\partial v} & \frac{\partial x}{\partial w} \\ \frac{\partial y}{\partial u} & \frac{\partial y}{\partial v} & \frac{\partial y}{\partial w} \\ \frac{\partial z}{\partial u} & \frac{\partial z}{\partial v} & \frac{\partial z}{\partial w} \end{pmatrix}, \tag{2.23}$$

where all derivatives are evaluated at $\vec{u}\,' = (u', v', w')$. It follows that the integration measures in the two coordinate systems are related by

$$d^3x = \left| \det \left.\frac{\partial(x, y, z)}{\partial(u, v, w)}\right|_{\vec{u}\,'} \right| d^3u, \tag{2.24}$$

and the generalization of (2.20) to nonlinear mappings is:

$$\delta(\vec{x} - \vec{x}\,') = \left| \det \left.\frac{\partial(x, y, z)}{\partial(u, v, w)}\right|_{\vec{u}\,'} \right|^{-1} \delta(\vec{u} - \vec{u}\,'). \tag{2.25}$$

Note that the product of the measure and the delta function preserves the three-dimensional integration result (2.18).

Equation (2.25) can be used to find the delta function in alternative coordinates. For instance, cylindrical coordinates correspond to the following mapping from (ρ, ϕ, z) to (x, y, z):

$$x = \rho \cos\phi, y = \rho \sin\phi, z = z. \tag{2.26}$$

Equation (2.25) then gives

$$\begin{aligned} \delta(\vec{x} - \vec{x}\,') &= \left| \det \left.\frac{\partial(x, y, z)}{\partial(\rho, \phi, z)}\right|_{\rho_0, \phi_0, z_0} \right|^{-1} \delta(\rho - \rho_0)\delta(\phi - \phi_0)\delta(z - z_0) \\ &= \rho_0^{-1}\delta(\rho - \rho_0)\delta(\phi - \phi_0)\delta(z - z_0). \end{aligned} \tag{2.27}$$

We can check this relation by integrating over all space using cylindrical coordinates:

$$\begin{aligned} \int d^3x\, \delta(\vec{x} - \vec{x}\,') &= \int_{-\infty}^{\infty} \rho d\rho \int_0^{2\pi} d\theta \int_0^{\infty} dz\, \rho_0^{-1}\delta(\rho - \rho_0)\delta(\theta - \theta_0)\delta(z - z_0) \\ &= \int_0^{\infty} d\rho \frac{\rho}{\rho_0}\delta(\rho - \rho_0) \int_0^{2\pi} d\theta\, \delta(\theta - \theta_0) \int_{-\infty}^{\infty} dz\, \delta(z - z_0) \\ &= 1. \end{aligned}$$

The delta function integrates to 1, as expected.

An easy way to calculate the Jacobian for *orthogonal coordinate systems* is the use of *scale factors*. Here the Jacobian can be evaluated as

$$\left|\det \frac{\partial(x,y,z)}{\partial(u,v,w)}\bigg|_{\vec{u}'}\right| = UVW, \tag{2.28}$$

where the scale factors U, V and W are given as

$$U \equiv \left|\frac{\partial \vec{x}}{\partial u}\right|, V \equiv \left|\frac{\partial \vec{x}}{\partial v}\right|, W \equiv \left|\frac{\partial \vec{x}}{\partial w}\right|. \tag{2.29}$$

In this form $\vec{x}$ is considered a function of u, v, w. These factors scale the length contributions to the infinitesimal length element $ds^2 \equiv dx^2 + dy^2 + dz^2$ as

$$ds^2 = U^2\, du^2 + V^2\, dv^2 + W^2\, dw^2, \tag{2.30}$$

which follows from the assumption of coordinate orthogonality and the definition of the scale factors in (2.29). In the case of the transformation from cartesian to cylindrical coordinates, $(u, v, w) = (\rho, \phi, z)$ according to (2.26), we have $U = 1, V = \rho_0, W = 1$, giving the same result for the delta function as in (2.27).

Finally, the delta function also has an integral representation in terms of *Fourier transforms*. Here we will use the following conventions for the Fourier transform of $f(x)$,[3]

$$f(k) \equiv \int_{-\infty}^{\infty} dx\, f(x) e^{-ikx}, \tag{2.31}$$

and Fourier inverse transform of $f(k)$,

$$f(x) = \frac{1}{2\pi} \int_{-\infty}^{\infty} dk\, f(k) e^{ikx}. \tag{2.32}$$

Of particular interest is the case where $f(x) = \delta(x)$, in which case the Fourier transform is unity. The corresponding inverse transform gives the formal representation

$$\delta(x) = \frac{1}{2\pi} \int_{-\infty}^{\infty} dk\, e^{ikx}, \tag{2.33}$$

which is extremely useful in physics applications. The n-dimensional integral form of the Dirac delta function is:

$$\delta^{(n)}(\vec{x}) = \frac{1}{(2\pi)^n} \int_{-\infty}^{\infty} d^n k\, e^{i\vec{k}\cdot\vec{x}}. \tag{2.34}$$

We will use the notation $\delta(\vec{x})$ for the three-dimensional form.

[3] Warning! Another common Fourier transform normalization, which we occasionally use, has constant factors of $1/\sqrt{2\pi}$ on both (2.31) and (2.32).

2.3 Line and surface delta functions

Line and surface charges can also be expressed in terms of delta functions in a three-dimensional environment. For example, a line charge with linear charge density $\lambda(z)$ confined to the z-axis can be expressed as $\lambda(z)\delta(x)\delta(y)$. Similarly, a surface charge with surface charge density $\sigma(x, y)$ confined to the xy-plane can be expressed as $\sigma(x, y)\delta(z)$. The quantities $\delta(x)\delta(y)$ and $\delta(z)$ are examples of line or surface delta functions, respectively, in a three-dimensional context. These line delta functions may be recovered by integration from the three dimensional delta function $\delta(\vec{x} - \vec{x}\,')$ defined in the previous section. For example, the line delta function is recovered by

$$\delta(x)\delta(y) = \int_{-\infty}^{\infty} dz'\delta(\vec{x} - \vec{x}\,'), \tag{2.35}$$

where the straight line $\vec{x}\,' = (0, 0, z')$ is specified. Similarly, the surface delta function can be recovered from

$$\delta(z) = \int_{-\infty}^{\infty} dx' \int_{-\infty}^{\infty} dy'\delta(\vec{x} - \vec{x}\,'), \tag{2.36}$$

along the planar surface $\vec{x}\,' = (x', y', 0)$. These constructions may be generalized to "directed" vector quantities for arbitrary lines and surfaces, as in:

$$\text{Line delta function along } C' : \int_{C'} d\vec{x}\,'\delta(\vec{x} - \vec{x}\,'); \tag{2.37}$$

$$\text{Surface delta function on } S' : \int_{S'} d\vec{s}\,'\delta(\vec{x} - \vec{x}\,'). \tag{2.38}$$

The vector line segment $d\vec{x}\,'$ in (2.37) is taken parallel to the line path. The surface integration measure used in (2.38) is $d\vec{s}\,' \equiv \hat{n}'\,da'$ where $\hat{n}'$ is a surface normal and da an infinitesimal surface area. To recover (2.35) (multiplied by the unit vector $\hat{z}$) we would take $d\vec{x}\,' = \hat{z}\,dz'$ in (2.37), and to recover (2.36) (also multiplied by $\hat{z}$) we would take $d\vec{s}\,' = \hat{z}\,dx'\,dy'$ in (2.38). Effectively, we are "smearing" the delta function over a directed line or a surface.

An important property of the directed surface delta function concerns

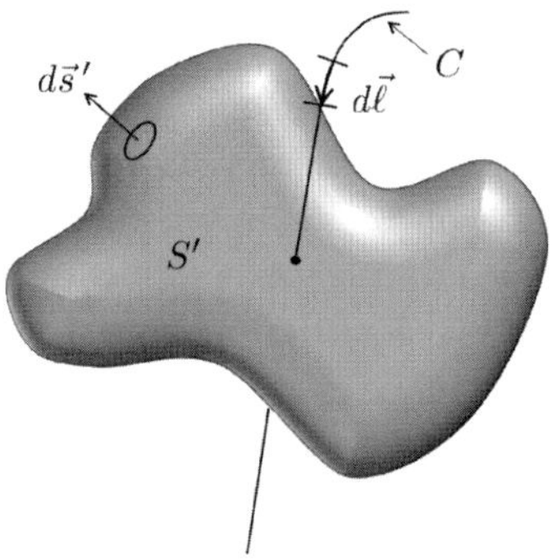

Fig. 2.3 Membrane δ-function; integration from "outside" to "inside" case.

the integration over a line or contour C:

$$\int_C d\vec{\ell}\cdot\int_{S'} d\vec{s}\,'\delta(\vec{x}-\vec{x}\,') = \int_C\int_{S'} \underbrace{d\vec{\ell}\cdot d\vec{s}\,'}_{\text{effectively}=\pm d^3x'} \delta(\vec{x}-\vec{x}\,')$$

$$= \begin{cases} 1 & \text{integrate "inside" to "outside"} \\ -1 & \text{integrate "outside" to "inside"} \\ 0 & \text{misses surface} \end{cases} \tag{2.39}$$

The integration in (2.39) is illustrated in Figure 2.3. Any local integration path C that impinges the membrane will have $d\vec{\ell} = \pm\hat{n}'dz + \hat{x}\,dx$ near the surface, where coordinate z is perpendicular to the surface and x is any coordinate parallel to the surface, and so nicely gives only the three possible results above. We will encounter these types of delta functions in Sections 6.5 and 7.8 where they will be useful in magnetostatics and magnetic monopole discussions. The directed surface delta function will also be referred to as a *membrane delta function* in these sections.

2.4 Gauss' law and solid angles

Given a set of point charges $\{q_j, j = 1\ldots n\}$ located in space as in Figure 2.4, the integral of the normal component of the resulting electric field over any closed, finite surface may be computed as follows:

$$\oint_S da\;\hat{n}\cdot\vec{E} = \sum_j q_j \oint da\;\hat{n}\cdot\frac{\vec{x}-\vec{x}_j}{|\vec{x}-\vec{x}_j|}\frac{1}{|\vec{x}-\vec{x}_j|^2}. \tag{2.40}$$

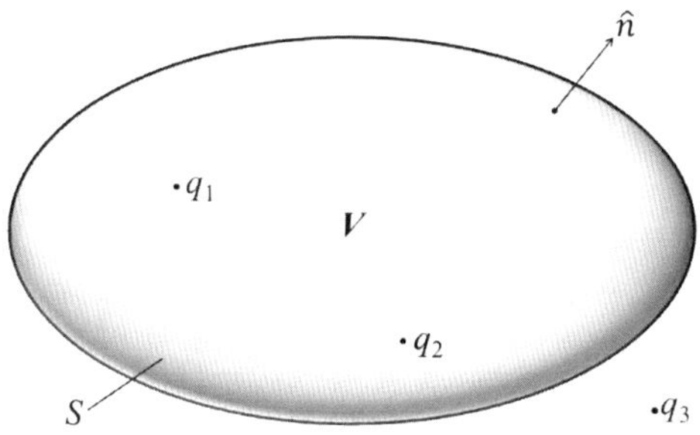

Fig. 2.4 Surface integral in the presence of a charge distribution.

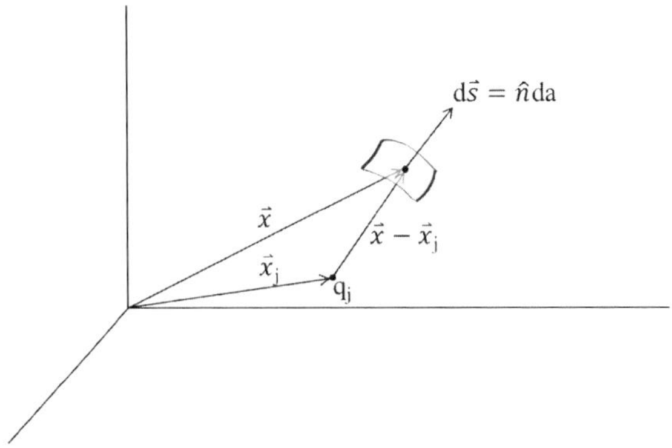

Fig. 2.5 Notation for Equation (2.40).

In (2.40) $da\ \hat{n}$ can be identified as the outward-directed element of surface area $d\vec{s}$, while $(\vec{x}-\vec{x}_j)/|\vec{x}-\vec{x}_j|$ is a unit vector pointing from the jth charge to the point $\vec{x}$ on the surface (see Figure 2.5).

It follows that

$$da\ \hat{n} \cdot \frac{\vec{x}-\vec{x}_j}{|\vec{x}-\vec{x}_j|} = d\vec{s} \cdot \frac{\vec{x}-\vec{x}_j}{|\vec{x}-\vec{x}_j|} = ds_{||}, \tag{2.41}$$

where $ds_{||}$ is the component of directed area that is parallel to the "line of sight" from the jth charge.

The solid angle subtended by $d\vec{s}$ as seen from $\vec{x}_j$ is $ds_{||}/r^2$ (see Figure 2.6), so that

$$d\Omega_j = da\ \hat{n} \cdot \frac{\vec{x}-\vec{x}_j}{|\vec{x}-\vec{x}_j|^3}. \tag{2.42}$$

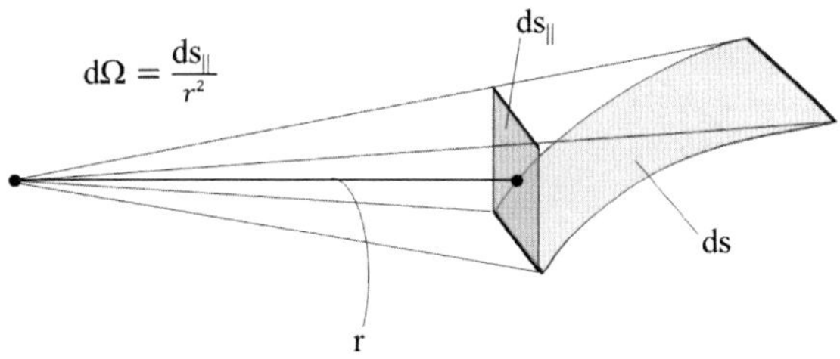

Fig. 2.6 Solid angle subtended by surface element.

Note $d\Omega_j$ is positive or negative depending on whether $\vec{x}_j$ is in "back" or in "front" of the plane determined by $d\vec{s}$ and the point $\vec{x}$.

Combining Equations (2.40) and (2.42) gives

$$\oint_S da\ \hat{n} \cdot \vec{E} = \sum_j q_j \oint_S d\Omega_j, \tag{2.43}$$

where $\oint_S d\Omega_j$ is the net solid angle subtended by S as seen from $\vec{x}_j$. Now if $\vec{x}_j$ is *outside* the surface S, the net solid angle is zero because of cancellation of positive and negative contributions, as shown in Figure 2.7. On the other hand, if $\vec{x}_j$ is *inside* the surface S, the net solid angle is 4π because the entire horizon is subtended; see Figure 2.8. In summary, we have the integral form of Gauss' Law:

$$\oint_S da\ \hat{n} \cdot \vec{E} = 4\pi \sum_{q_j \text{ inside } S} q_j. \tag{2.44}$$

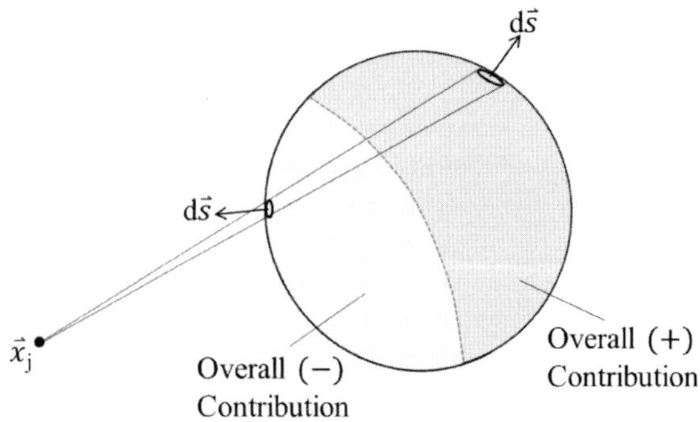

Fig. 2.7 Electric field surface integral for charges outside of surface S.

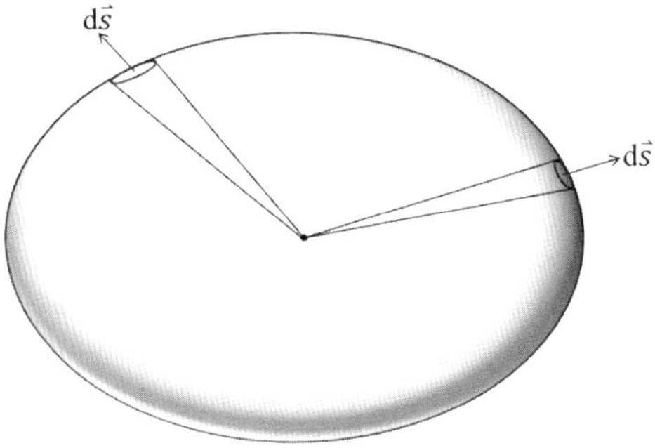

Fig. 2.8 Electric field surface integral for charge inside of surface.

Equation (2.44) characterizes the *global* behavior of $\vec{E}$; but to solve problems, we need to characterize the *local* behavior of $\vec{E}$. Specifically, we need differential equations (to be supplemented with boundary conditions).

We can easily extend (2.44) to the case of a continuous charge density. Replacing q_j with $\rho(x)d^3x$ and the sum over j with an integral gives

$$\oint_S da\ \hat{n} \cdot \vec{E} = 4\pi \int_V d^3x\, \rho(x), \tag{2.45}$$

where V is the volume bounded by S. On the other hand, the divergence theorem (1.9) gives

$$\int_V d^3x\, \vec{\nabla} \cdot \vec{E} = \oint_S da\ \hat{n} \cdot \vec{E}. \tag{2.46}$$

Since both Equations (2.45) and (2.46) are true for arbitrary volumes, we may equate the integrands:[4]

$$\vec{\nabla} \cdot \vec{E} = 4\pi\rho(\vec{x}). \tag{2.47}$$

It is useful to express the vector function $\vec{E}(\vec{x})$ in terms of a scalar function. Notice that

$$-\vec{\nabla}\frac{1}{|\vec{x} - \vec{x}\,'|} = \frac{\vec{x} - \vec{x}\,'}{|\vec{x} - \vec{x}\,'|^3}. \tag{2.48}$$

Note $\vec{\nabla}$ is with respect to $\vec{x}$, while $\vec{x}\,'$ is treated as a constant. We may

[4]This equating of integrands is intuitive, but technically requires mathematical proof (which we will leave to the mathematicians).

plug (2.48) into (2.5) to obtain:[5]

$$\vec{E}(\vec{x}) = -\int d^3x' \rho(\vec{x}\,')\vec{\nabla}\frac{1}{|\vec{x}-\vec{x}\,'|} \tag{2.49}$$

$$= -\vec{\nabla}\int d^3x' \rho(\vec{x}\,')\frac{1}{|\vec{x}-\vec{x}\,'|}. \tag{2.50}$$

This motivates the definition of *scalar potential* $\Phi(\vec{x})$ as

$$\Phi(\vec{x}) \equiv \int d^3x' \frac{\rho(\vec{x}\,')}{|\vec{x}-\vec{x}\,'|}, \tag{2.51}$$

so that

$$\vec{E}(\vec{x}) = -\vec{\nabla}\Phi(\vec{x}). \tag{2.52}$$

Plugging (2.52) into (2.47) gives

$$\nabla^2\Phi(\vec{x}) = -4\pi\rho(\vec{x}), \tag{2.53}$$

which is known as *Poisson's equation.* Another key result of (2.52) follows from $\vec{\nabla}\times\vec{\nabla}\Phi = 0$:

$$\vec{\nabla}\times\vec{E}(\vec{x}) = 0. \tag{2.54}$$

Equation (2.53) provides a useful result on delta functions. When applied to the charge density $\rho(\vec{x}) = \delta(\vec{x}-\vec{x}\,')$ it gives

$$\nabla^2\frac{1}{|\vec{x}-\vec{x}\,'|} = -4\pi\delta(\vec{x}-\vec{x}\,'), \tag{2.55}$$

which reflects the fact that $1/|\vec{x}-\vec{x}\,'|$ is the potential of a unit point charge located at $\vec{x}\,'$. Equation (2.55) is a special case of a more general relation involving the double derivative of this potential we will encounter in Section 6.8.

2.5 Verification of the inverse square law: the Cavendish experiment

We began this chapter with a statement of Coulomb's law, which asserts that the magnitude of an electric field varies as the inverse square of the distance to the source charge. In this section we analyze mathematically

[5]The interchange of differentiation and integration in (2.49) requires mathematical justification, but here (as elsewhere) we will assume its validity without proof.

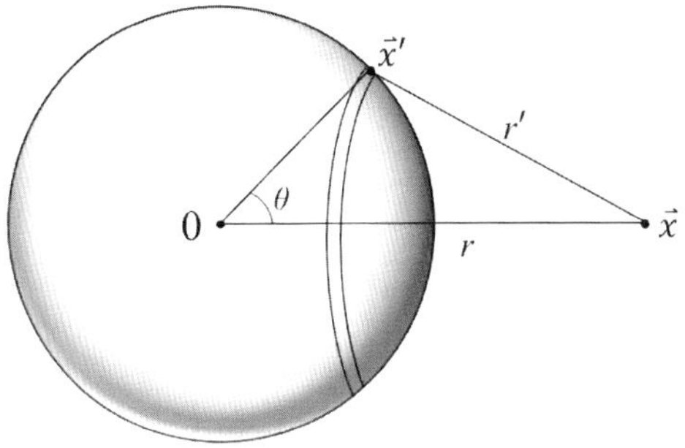

Fig. 2.9 Charged shell for the integration in Equation (2.60).

one early experiment (performed by Cavendish[6]) that verified the inverse-square relationship.

Suppose that contrary to Coulomb's law the actual force between charges is given instead by:

$$\vec{F}_{q' \text{ on } q} = qq' \frac{\vec{x} - \vec{x}\,'}{|\vec{x} - \vec{x}\,'|^{3+\epsilon}}, \tag{2.56}$$

where $|\epsilon| \ll 1$ reflects a deviation from inverse-square dependence. Much of the discussion in Sections 2.1 and 2.4 is still valid, with minor changes. In particular, the electric field due to charge density ρ is (in analogy to (2.5))

$$\vec{E}_{tot}(\vec{x}) = \int d^3x' \rho(\vec{x}\,') \frac{\vec{x} - \vec{x}\,'}{|\vec{x} - \vec{x}\,'|^{3+\epsilon}}. \tag{2.57}$$

Furthermore, the electric field is still given by a scalar potential, since (2.48) modifies to

$$-\vec{\nabla} \frac{(1+\epsilon)^{-1}}{|\vec{x} - \vec{x}\,'|^{1+\epsilon}} = \frac{\vec{x} - \vec{x}\,'}{|\vec{x} - \vec{x}\,'|^{3+\epsilon}}, \tag{2.58}$$

which implies $\vec{E}(\vec{x}) = -\vec{\nabla}\Phi$ where

$$\Phi(\vec{x}) = \frac{1}{1+\epsilon} \int d^3x' \frac{\rho(\vec{x}\,')}{|\vec{x} - \vec{x}\,'|^{1+\epsilon}}. \tag{2.59}$$

We consider the potential due to a uniformly charged spherical shell of radius a and total charge q_a, as in Figure 2.9.

[6]For more details, see for example, C.Jungnickel and R. McCormmach, *Cavendish, The Experimental Life*, Bucknell Univ Press, 1999.

By symmetry, it is clear that the potential depends only on $|\vec{x}| = r$. Since the charge is concentrated on the surface of the sphere, the volume integral in (2.59) reduces to an area integral:

$$\Phi(r) = \frac{1}{1+\epsilon}\int ds' \frac{\sigma}{r'^{1+\epsilon}}, \qquad (2.60)$$

where $\sigma = q_a/(4\pi a^2)$ and $r' = |\vec{x} - \vec{x}\,'|$.

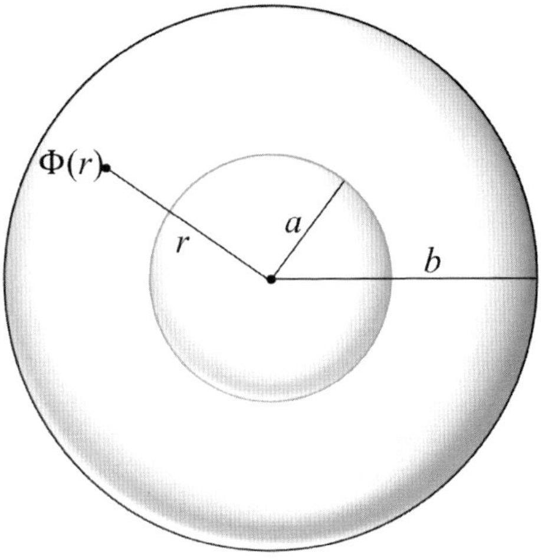

Fig. 2.10 Geometry for Cavendish experiment.

Finally, if we add a second concentric spherical shell of radius $b > a$ (see Figure 2.10), it is possible to show (see Exercise 2.5.1) that

$$\Phi(r) = \Phi_{out}(r) + \Phi_{in}(r), \qquad (a \le r \le b) \qquad (2.61)$$

where

$$\Phi_{out} = \frac{q_a}{(1-\epsilon^2)}\frac{1}{2ar}\left[(a+r)^{1-\epsilon} - (r-a)^{1-\epsilon}\right], \qquad (2.62)$$

$$\Phi_{in} = \frac{q_b}{(1-\epsilon^2)}\frac{1}{2br}\left[(b+r)^{1-\epsilon} - (b-r)^{1-\epsilon}\right]. \qquad (2.63)$$

In practice, Cavendish used conducting spheres, so that an arbitrary charge placed on a sphere is free to distribute itself throughout the sphere without resistance. In this case, a uniform charge distribution is guaranteed on each sphere. Furthermore, if the two spheres are joined by a conducting wire then the electric field between the two spheres must be zero, since otherwise the charge distribution would change due to the electric force on charges in the wire. Since the electric field is the gradient of the potential, it follows that the potential must be constant between the spheres and thus

$$\Phi(a) = \Phi(b). \tag{2.64}$$

In Exercise 2.5.1 you are asked to show that (2.61), (2.62), (2.63) and (2.64) lead to the approximate relation

$$q_a \approx \frac{q_b \epsilon}{2(a-b)} \left[b \ln\left(\frac{b-a}{a+b}\right) + a \ln\left(\frac{4b^2}{b^2-a^2}\right) \right], \tag{2.65}$$

so that the charge on the inner sphere vanishes when $\epsilon = 0$ (inverse square law). Cavendish's 1772 experiment was consistent with ϵ being zero. In the manuscript compiled by Maxwell much later, Cavendish gave the measurement of the inverse square coefficient as $2 \pm .02$.[7]

There have been many modern investigations of the inverse square dependence of Coulomb's law. One was the high-frequency alternating voltage Cavendish type experiment of Williams, Fowler and Hill.[8] Actually, this and other investigations can most naturally be interpreted as giving a measurement of the mass of the photon, the quantized electromagnetic force carrier, which should strictly be zero if the inverse square law holds exactly. However, using the dimensions of the experimental apparatus used, the Williams result can be expressed as the measurement $\epsilon = (2.7 \pm 3.1) \times 10^{-16}$, consistent with ϵ being zero.

2.6 Surface charge and dipole layers

We saw in Section 1.4 that

$$(\vec{D}_2 - \vec{D}_1) \cdot \hat{n}_{21} = 4\pi\sigma, \tag{2.66}$$

$$(\vec{E}_2 - \vec{E}_1) \times \hat{n}_{21} = 0. \tag{2.67}$$

If both volumes are vacuum, then

$$(\vec{E}_2 - \vec{E}_1) \cdot \hat{n}_{21} = 4\pi\sigma \tag{2.68}$$

$$(\text{alternatively}: \quad (E_2 - E_1)_n = 4\pi\sigma)$$

[7] J. Clerk Maxwell, F. R. S., Ed., "The Electrical Researches of the Honourable Cavendish, F.R.S.", Cambridge University Press, p. 112 (1879).

[8] E. R. Williams, J. E. Fowler and H. A. Hill, "New Experimental Test of Coulomb's Law: A Laboratory Upper Limit on the Photon Rest Mass", *Phys. Rev. Lett.* **26**, 721 (1971).

and the potential is

$$\Phi(\vec{x}) = \int_S da' \frac{\sigma(\vec{x}\,')}{|\vec{x} - \vec{x}\,'|}. \tag{2.69}$$

At first sight it may seem that (2.68) and (2.69) are inconsistent, since one describes a discontinuity in $\vec{E}$ and the other provides an apparently continuous expression for Φ. However, they are actually consistent: if we take for instance a flat sheet with surface charge σ, then the normal vector $\hat{n}_{21}$ is constant and the normal component of the electric field is

$$\begin{aligned} E_n &= -\hat{n}_{21} \cdot \vec{\nabla}\Phi \\ &= -\int_S da' \sigma(\vec{x}\,')\hat{n}_{21} \cdot \vec{\nabla} \frac{1}{|\vec{x} - \vec{x}\,'|}. \end{aligned} \tag{2.70}$$

But we have already seen (cf. (2.42))

$$da'\hat{n}_{21} \cdot \vec{\nabla} \frac{1}{|\vec{x} - \vec{x}\,'|} = d\vec{s}\,' \cdot \frac{(\vec{x}\,' - \vec{x})}{|\vec{x} - \vec{x}\,'|^3} = d\Omega'. \tag{2.71}$$

As shown in Figure 2.11, the solid angle is measured from $\vec{x}$ (the field position) rather than $\vec{x}\,'$ as previously. Note that $d\Omega'$ is positive or negative depending on whether $\vec{x}$ is "inside" or "outside", respectively. Using (2.42) we may therefore re-express E_n for flat surfaces as

$$E_n(\vec{x}) = -\int_S d\Omega' \sigma(\vec{x}\,'), \tag{2.72}$$

where it is understood that $\vec{x}$ is fixed and $\vec{x}\,'$ is considered the variable in the integration.

Now consider that $\vec{x}$ is very close to an arbitrary surface. We can imagine partitioning the surface into an infinitesimal disk and a remainder, as shown in Figure 2.12. Linearity of the fields implies that

$$\vec{E}(\vec{x}) = \vec{E}_{\text{remainder}}(\vec{x}) + \vec{E}_{\text{disk}}(\vec{x}). \tag{2.73}$$

But since $\vec{x}$ is close to the disk's surface it appears nearly "flat", so that $\hat{n}_{21}$ is nearly constant over the integration over the disk. Furthermore σ can be considered as constant over the infinitesimal disk, so that

$$(E_{2,\text{disk}})_n = 2\pi\sigma, \tag{2.74}$$

$$(E_{1,\text{disk}})_n = -2\pi\sigma. \tag{2.75}$$

Since the electric field due to the remainder is continuous, (2.74) and (2.75) imply (2.68): hence (2.68) and (2.69) are consistent.

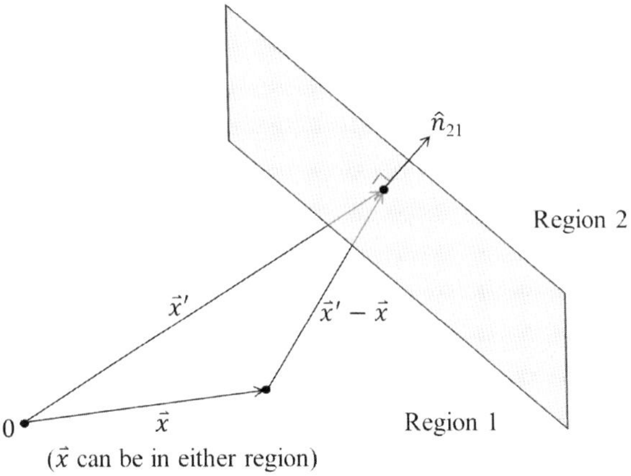

Fig. 2.11 Variables in Equation (2.71).

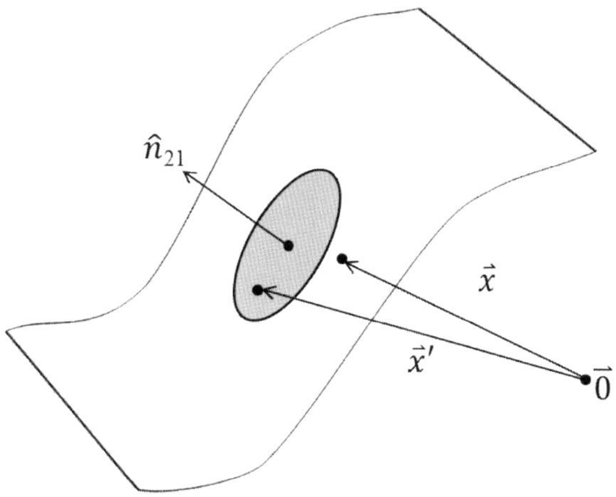

Fig. 2.12 Geometry when $\vec{x}$ is close to a given surface.

Another type of surface distribution is the *dipole layer*. In this case, we model the surface as two "faces" separated by a nonzero thickness $d(\vec{x}\,')$, as shown in Figure 2.13. The dipole layer idealizes the case where the surface is very thin and the surface charge densities are equal and opposite at corresponding points of the two faces. We furthermore assume that the surface charge density is very large, so its product with the thickness is

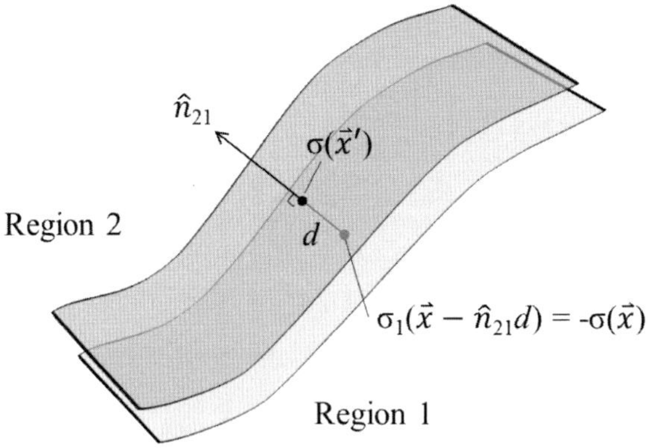

Fig. 2.13 Dipole layer.

finite:

$$\underbrace{D(\vec{x}\,')}_{\text{finite}} = \underbrace{d(\vec{x}\,')}_{\text{infinitesimal}} \cdot \underbrace{\sigma(\vec{x}\,')}_{\text{huge}}. \tag{2.76}$$

In the case where $d(\vec{x}\,') = d$ is constant and a line connecting corresponding charge density points are located along a common perpendicular, one can show (the derivation is described in Exercise 2.6.2) that the limit gives

$$\Phi(\vec{x}) = \int_S da' D(\vec{x}\,')\,\hat{n}_{21}(\vec{x}\,') \cdot \vec{\nabla}' \frac{1}{|\vec{x} - \vec{x}\,'|}, \tag{2.77}$$

which together with (2.71) implies

$$\Phi(\vec{x}) = -\int_S d\Omega' D(\vec{x}\,'). \tag{2.78}$$

If D is a constant and if S is open, the potential of a dipole layer depends only on the boundary of S. (For a closed surface see Exercise 2.6.4.) If $\vec{x}$ is near S, then we may break S up into an infinitesimal disk close to $\vec{x}$ plus the remainder of S (just as in Figure 2.12). The "remainder" portion makes a continuous contribution to Φ, while the infinitesimal disk makes potential contributions:

$$\Phi_{2,\text{disk}} = 2\pi D; \qquad \Phi_{1,\text{disk}} = -2\pi D, \tag{2.79}$$

which gives

$$(\Phi_2 - \Phi_1)|_S = 4\pi D. \tag{2.80}$$

This is true locally at each point on the surface; here D is not a constant.

Another way of obtaining this last result is by treating each face of the dipole layer as a surface in its own right and evaluating the electric field in the small gap between the two surfaces. From (2.74) and (2.75) applied appropriately to the interior, we have that the $E_n^{total} = -4\pi\sigma$ between the layers, so that

$$\Delta\Phi|_1^2 = -\int_1^2 \vec{E}\cdot d\vec{\ell} = 4\pi\sigma d \underset{d\to 0}{\longrightarrow} 4\pi D(\vec{x}). \tag{2.81}$$

2.7 Boundary conditions and uniqueness of solutions

Using Coulomb's law we already have derived two differential equations satisfied by the electric field, which we rewrite here for convenience:

$$\vec{\nabla}\cdot\vec{E} = 4\pi\rho \qquad \text{from (2.47)},$$
$$\vec{\nabla}\times\vec{E} = 0 \qquad \text{from (2.54)}.$$

These equations may be re-expressed in terms of the potential Φ as

$$\nabla^2\Phi = -4\pi\rho, \qquad \vec{\nabla}\Phi = -\vec{E}. \tag{2.82}$$

We saw in Section 1.4 that when multiple media are present, boundary conditions are required in addition to the differential equations. Mathematically, for a set of equations to specify a well-defined solution, the solution must both exist and be unique. Leaving aside the question of existence, we now discuss the uniqueness of solutions to (2.82) together with various types of boundary conditions.

Using Gauss' theorem we may derive *Green's first identity*:

$$\int_V d^3x\left(\phi\nabla^2\psi + \vec{\nabla}\psi\cdot\vec{\nabla}\phi\right) = \oint_S da\,\phi\frac{\partial\psi}{\partial n}. \tag{2.83}$$

Suppose we have two solutions Φ_1, Φ_2 to the potential equation (2.82) with the *same* boundary conditions. To show uniqueness, we only need to show that Φ_1 and Φ_2 must be equal. Also suppose for instance that both potentials satisfy the same *Dirichlet* boundary conditions:

$$\Phi(\vec{x})|_S = f(\vec{x}), \tag{2.84}$$

where f is a pre-specified function. Letting $U = \Phi_1 - \Phi_2$, we have $\nabla^2 U = 0$ in V and $U|_S = 0$. Setting $\phi = \psi = U$ in (2.83) gives

$$\int_V d^3x\left(U\nabla^2 U + \vec{\nabla}U\cdot\vec{\nabla}U\right) = \oint_S da\,U\frac{\partial U}{\partial n}. \tag{2.85}$$

From (2.84) we know $U = 0$ on S, so the right-hand side of (2.85) is zero. We are left with:

$$\int_V d^3x |\vec{\nabla} U|^2 = 0 \implies \vec{\nabla} U \equiv 0 \text{ in } V \tag{2.86}$$

$$\implies \vec{\nabla}\Phi_1 = \vec{\nabla}\Phi_2 \implies \vec{E}_1 = \vec{E}_2. \tag{2.87}$$

A similar argument can be used to show $\vec{E}_1 = \vec{E}_2$ for *Neumann* boundary conditions,

$$\left.\frac{\partial \Phi}{\partial n}\right|_S = g(\vec{x}), \tag{2.88}$$

(where g is pre-specified) and for *mixed* boundary conditions,

$$\left.\begin{array}{l} \Phi|_{S'} = f(\vec{x}) \\ \left.\frac{\partial\Phi}{\partial n}\right|_{S''} = g(\vec{x}) \end{array}\right\} \text{ where } S' \cup S'' = S \text{ and } S' \cap S'' = \emptyset. \tag{2.89}$$

In the case of Dirichlet or mixed boundary conditions, we have $U = 0$ on the boundary (or part of the boundary) as well as $\vec{\nabla} U = 0$ in the interior, which implies that $U = 0$ or $\Phi_1 = \Phi_2$. For Neumann boundary conditions, we have only that $\vec{\nabla} U = 0$ which implies that Φ_1 and Φ_2 can differ by a constant. Note that when $f(\vec{x})$ in (2.84) or $g(\vec{x})$ in (2.88) are zero, these boundary conditions are termed "homogeneous".

So far in this section we have treated the different boundary conditions in a purely mathematical way: let us now consider some physical examples. In Section 1.4 we showed that the electric field at the surface of a conductor is normal to the surface S. This implies $\vec{\nabla}\Phi \perp S$, so that Φ is constant on the surface, corresponding to Dirichlet boundary conditions (2.84) with $f(\vec{x}) = C$. From (2.68) we have

$$E_n^{(\text{outside})} - E_n^{(\text{inside})} = 4\pi\sigma, \tag{2.90}$$

where σ is the surface charge density on S. As described in Section 1.4, the electric field inside the conductor is taken as zero, so that

$$E_n^{(\text{outside})} = 4\pi\sigma, \tag{2.91}$$

or solving for the charge density and relating the outside electric field to its potential,

$$\sigma = \left.\frac{1}{4\pi}\frac{\partial\Phi}{\partial n}\right|_S, \tag{2.92}$$

where n denotes the outer normal to the volume, which, for exterior problems, is directed inward to the conductor. Neumann boundary conditions

(2.88) on the other hand do not lend themselves easily to a physical interpretation within electrostatics. In the case of homogeneous Neumann conditions we have that $E_n = 0$ at the surface. If S is the boundary of a "Neumann conductor" in free space within which $\vec{E} = 0$, then it must be the case that $\Phi = C$ within S. But if there are charges in the exterior region, it is not possible to have both $E_n = 0$ and $\Phi = C$ at the exterior surface. This implies that Φ must be discontinuous as in (2.80). Since $\Phi_1 \equiv C$, the dipole density is determined up to a constant. Thus, since E_n gives the surface charge density, homogeneous Neumann boundary conditions can be interpreted as those for objects with a zero charge density but a non-vanishing dipole density. Electric field lines appear to be "repelled" by the object. We will see an application of this field configuration in Section 5.10 and again in Section 6.11, but for magnetic, not electric, fields.

2.8 Dirichlet and Neumann Green functions

The so-called *Green function* method is a powerful technique for solving *all* potential problems with a given set of boundary conditions for a fixed geometry, rather than trying to solve for each charge distribution individually.

Green functions can be motivated as follows. Suppose that we can solve the equation

$$\nabla^2 G(\vec{x}, \vec{x}\,') = -4\pi\delta(\vec{x} - \vec{x}\,'). \tag{2.93}$$

In this form one should always keep in mind that when the Green function is considered as the potential due to a point charge, the *first* variable (in this case $\vec{x}$) is considered the field point and the *second* variable (in this case $\vec{x}\,'$) is considered the source point. Multiplying both sides by $\rho(\vec{x}\,')$ and integrating over all space gives

$$\nabla^2 \int d^3x'\, G(\vec{x}, \vec{x}\,')\rho(\vec{x}\,') = -4\pi\rho(\vec{x}). \tag{2.94}$$

Equation (2.94) provides a solution of Poisson's equation for *any* $\rho(\vec{x})$ in free space. (We also obtained (2.51) and (2.53) as a result of generalizing point charge contributions to a continuous distribution.) The question is whether we can find a similar solution with boundary conditions present: in other words, can we find an integral expression which solves Poisson's equation with boundary conditions in terms of a function $G(\vec{x}, \vec{x}\,')$ that satisfies (2.93)? For the boundary conditions introduced in the previous section we will show the answer is yes, but with a twist: it turns out that the integration variable is the *field* variable $\vec{x}$ rather than the *source* variable $\vec{x}\,'$.

Our starting point is *Green's second identity*,

$$\int_V d^3x \left(\psi \nabla^2 \phi - \phi \nabla^2 \psi\right) = \oint_S da \left(\psi \frac{\partial \phi}{\partial n} - \phi \frac{\partial \psi}{\partial n}\right), \tag{2.95}$$

which follows from (2.83) by interchanging ϕ and ψ and subtracting the two equations. Setting $\psi = G(\vec{x}, \vec{x}\,')$ and $\phi = \Phi(\vec{x})$ gives

$$\begin{aligned} &\int_V d^3x \left(G(\vec{x}, \vec{x}\,')\nabla^2 \Phi(\vec{x}) - \Phi(\vec{x})\nabla^2 G(\vec{x}, \vec{x}\,')\right) \\ &\quad = \oint_S da \left(G(\vec{x}, \vec{x}\,')\frac{\partial \Phi(\vec{x})}{\partial n} - \Phi(\vec{x})\frac{\partial G(\vec{x}, \vec{x}\,')}{\partial n}\right). \end{aligned} \tag{2.96}$$

Note that $\hat{n}$ denotes the *outward* normal from the volume V. Using Equations (2.53) and (2.93) gives (after integrating and rearranging)

$$\Phi(\vec{x}\,') = \int d^3x\, G(\vec{x}, \vec{x}\,')\rho(\vec{x}) + \frac{1}{4\pi} \oint_S da \left(G(\vec{x}, \vec{x}\,')\frac{\partial \Phi(\vec{x})}{\partial n} - \Phi(\vec{x})\frac{\partial G(\vec{x}, \vec{x}\,')}{\partial n}\right). \tag{2.97}$$

At this point the particular boundary conditions come into play. If Φ satisfies Dirichlet conditions, then $\Phi(\vec{x})$ is known on the boundary but $\partial\Phi(\vec{x})/\partial n$ is not. In order to make the normal derivative term vanish on the right-hand side of (2.97) we require $G_D(\vec{x}, \vec{x}\,') = 0$ for $\vec{x}$ on S, which gives (relabel $\vec{x} \leftrightarrow \vec{x}\,'$ for convenience and future reference)

$$\Phi(\vec{x}) = \int d^3x'\, G_D(\vec{x}\,', \vec{x})\rho(\vec{x}\,') - \frac{1}{4\pi} \oint_S da'\, \Phi(\vec{x}\,')\frac{\partial G_D(\vec{x}\,', \vec{x})}{\partial n'}. \tag{2.98}$$

We have thus expressed the solution $\Phi(\vec{x})$ in terms of integrals involving the Dirichlet Green function $G_D(\vec{x}\,', \vec{x})$ and known functions.

Let us summarize the equations that define $G_D(\vec{x}, \vec{x}\,')$:

$$\nabla^2 G_D(\vec{x}, \vec{x}\,') = -4\pi\delta(\vec{x} - \vec{x}\,'), \tag{2.99}$$

$$G_D(\vec{x}, \vec{x}\,')|_{S(\vec{x})} = 0. \tag{2.100}$$

Equation (2.100) corresponds to *homogeneous* Dirichlet boundary conditions. These equations imply that $G_D(\vec{x}, \vec{x}\,')$ is the unique solution to a potential problem in the variable $\vec{x}$ for a unit point charge at $\vec{x}\,'$ in a fixed geometry bounded by grounded conductors. A portion or the whole of the surface at infinity may be one of these.

Note the surprising fact that the integration variable in both terms of (2.98) is the *field* variable rather than the *source* variable, in contrast to (2.94). This could easily lead to confusion over which of the variables in $G(\vec{x}, \vec{x}\,')$ corresponds to the source variable or field. Fortunately, as we shall soon see Green functions are, or can be chosen, to be symmetric (see (2.107)), so the ordering of the arguments makes no difference.

For Neumann boundary conditions there are additional complications. We would like to make the third term on the right-hand side of (2.97) vanish, but unfortunately this turns out to be impossible as we shall now show. Since the Neumann Green function $G_N(\vec{x}, \vec{x}\,')$ satisfies (2.93), we have

$$\int_V d^3x\, \vec{\nabla} \cdot \vec{\nabla} G_N(\vec{x}, \vec{x}\,') = -4\pi, \tag{2.101}$$

and Gauss' theorem implies

$$\oint_S da \frac{\partial G_N(\vec{x}, \vec{x}\,')}{\partial n} = -4\pi. \tag{2.102}$$

Hence it is not possible to choose $\partial G_N(\vec{x}, \vec{x}\,')/\partial n|_{S(\vec{x})} = 0$ on the surface. However, we can require that this quantity be a constant. Note that the normal derivative specifies the electric field. If the area of S is finite, (2.102) implies that this constant must be $-4\pi/S$. Then (2.97) becomes (again relabel $\vec{x} \leftrightarrow \vec{x}\,'$)

$$\Phi(\vec{x}) = \int d^3x'\, G_N(\vec{x}\,', \vec{x})\rho(\vec{x}\,') + \frac{1}{4\pi} \oint_S da'\, G_N(\vec{x}\,', \vec{x}) \frac{\partial \Phi(\vec{x}\,')}{\partial n'} + \langle \Phi \rangle_S, \tag{2.103}$$

where $\langle \Phi \rangle_S$ is the average of Φ over S,

$$\langle \Phi \rangle_S = \frac{1}{S} \int_S da\, \Phi(\vec{x}). \tag{2.104}$$

Equation (2.103) reflects the fact that the solution Φ for a Neumann problem is arbitrary up to an overall constant.

In summary, the defining equations for $G_N(\vec{x}, \vec{x}\,')$ are:

$$\nabla^2 G_N(\vec{x}, \vec{x}\,') = -4\pi\delta(\vec{x} - \vec{x}\,'), \tag{2.105}$$

$$\left.\frac{\partial G_N(\vec{x}, \vec{x}\,')}{\partial n}\right|_{S(\vec{x})} = -\frac{4\pi}{S}. \tag{2.106}$$

Equation (2.106) corresponds to *non-homogeneous* Neumann boundary conditions. These equations give $G_N(\vec{x}, \vec{x}\,')$ as the unique solution of a

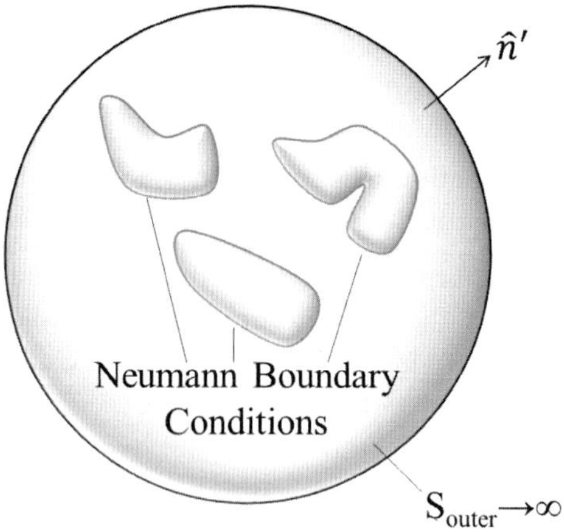

Fig. 2.14 The Neumann exterior problem.

potential problem in the variable $\vec{x}$ for a unit point charge at $\vec{x}\,'$ in a fixed geometry bounded by walls endowed with a constant normal electric field.

As an example of a Neumann problem, consider solving for the field outside of a given set of finite surfaces (the *exterior problem*), as depicted in Figure 2.14. If we add the imaginary surface S_{outer} at $|\vec{x}| = R$ with $R \to \infty$ as an outer boundary, then the boundary of our region (comprised of finite surfaces plus S_{outer}) has infinite total area. Accordingly, the right-hand side of (2.106) gives the normal field derivative of $G_N(\vec{x}, \vec{x}\,')$ as zero on all boundary surfaces even though (2.102) still holds. This means the boundary conditions become homogeneous. If we require the potential to vanish at infinity so that $\Phi \to 0$ on S_{outer}, then the $\langle \Phi \rangle_S$ term vanishes from (2.103) and the solution is unique.

An important aspect of Green functions is the *symmetry property*, which we alluded to earlier:

$$G(\vec{x}, \vec{x}\,') = G(\vec{x}\,', \vec{x}), \tag{2.107}$$

which holds for Dirichlet and for *some* Neumann Green functions. To show (2.107), first recall that $G(\vec{x}\,', \vec{x}\,'')$ is interpreted as the potential at $\vec{x}\,'$ due to a point charge at $\vec{x}\,''$. It follows that we may apply the Green function integral equation (2.97), with the potential $\Phi(\vec{x}\,') \equiv G(\vec{x}\,', \vec{x}\,'')$, obtaining

$$G(\vec{x}\,', \vec{x}\,'') = \int d^3x\, G(\vec{x}, \vec{x}\,')\delta(\vec{x} - \vec{x}\,'') + \frac{1}{4\pi}\oint_S da \left(G(\vec{x}, \vec{x}\,')\frac{\partial G(\vec{x}, \vec{x}\,'')}{\partial n} - G(\vec{x}, \vec{x}\,'')\frac{\partial G(\vec{x}, \vec{x}\,')}{\partial n} \right). \tag{2.108}$$

If $G = G_D$, then when $\vec{x}$ is on S we have $G_D(\vec{x}, \vec{x}\,') = 0$, so the area integrals vanish and (2.108) simplifies to (2.107) (after replacing $\vec{x}\,''$ with $\vec{x}$). If we attempt the same thing for the Neumann case we obtain after rearrangement

$$G_N(\vec{x}\,', \vec{x}\,'') - \langle G_N(\vec{x}, \vec{x}\,'')\rangle_{S(\vec{x})} = G_N(\vec{x}\,'', \vec{x}\,') - \langle G_N(\vec{x}, \vec{x}\,')\rangle_{S(\vec{x})}, \tag{2.109}$$

after substituting the boundary condition (2.106) into (2.108). For the exterior Neumann problem shown in Figure 2.14 the averages also vanish since ($R \equiv |\vec{x}|$)

$$\langle G_N(\vec{x}, \vec{x}\,'')\rangle_{S(\vec{x})} = \frac{1}{S}\oint da\, G_N(\vec{x}, \vec{x}\,'') \overset{R\to\infty}{\longrightarrow} 0, \tag{2.110}$$

$$G_N(\vec{x}, \vec{x}\,'') \sim \frac{1}{R} \text{ for } R \gg R'',\ da = R^2 d\Omega, S = 4\pi R^2,$$

and (2.107) is recovered for G_N. (Adding an overall constant to $G_N(\vec{x}, \vec{x}\,'')$ will not affect (2.107) of course.)

The interior Neumann problem however is a different case. Note that the left-hand side of (2.109) is equal to the right-hand side with the arguments $\vec{x}\,'$ and $\vec{x}\,''$ interchanged. Now by definition, $G_N(\vec{x}\,', \vec{x}\,'')$ is a solution to the Neumann problem in field variable $\vec{x}\,'$ with unit point charge at $\vec{x}\,''$. We saw in Section 2.7 that this solution is unique up to a constant: but since we have a different Neumann problem for each $\vec{x}\,''$, we are free to choose a different "constant" for each $\vec{x}\,''$. Thus given any $G_N(\vec{x}\,', \vec{x}\,'')$ we may add an arbitrary function of $\vec{x}\,''$ and obtain another equally valid Green function! In particular, we may define

$$G_N^{\text{symm}}(\vec{x}\,', \vec{x}\,'') \equiv G_N(\vec{x}\,', \vec{x}\,'') - \langle G_N(\vec{x}, \vec{x}\,'')\rangle_{S(\vec{x})}, \tag{2.111}$$

which as we have noted is symmetric in $\vec{x}\,'$ and $\vec{x}\,''$.

In summary, every Dirichlet Green function is symmetric, and it is always possible to *choose* a symmetric Green function for any Neumann problem. The symmetry of the various Green functions has a natural physical interpretation in terms of energy, as we shall show in Section 2.10.

We close this section by resolving an apparent paradox that gives further insight into the nature of Green functions. Returning to the Neumann problem, we rearrange and average both sides of (2.109) over the surface $S(\vec{x}\,')$ to give

$$\begin{aligned}\langle G_N(\vec{x}\,'',\vec{x}\,')\rangle_{S(\vec{x}\,')} &- \langle G_N(\vec{x}\,',\vec{x}\,'')\rangle_{S(\vec{x}\,')} \\ &= \langle G_N(\vec{x},\vec{x}\,')\rangle_{S(\vec{x}),S(\vec{x}\,')} - \langle G_N(\vec{x},\vec{x}\,'')\rangle_{S(\vec{x})}, \qquad (2.112)\end{aligned}$$

which after cancellation implies that $\langle G_N(\vec{x}\,'',\vec{x}\,')\rangle_{S(\vec{x}\,')}$ is a constant independent of the field point $\vec{x}\,''$. Taking derivatives with respect to the field variable at the boundary (and changing variables for convenience), this implies

$$\frac{\partial}{\partial n}\langle G_N(\vec{x},\vec{x}\,')\rangle_{S(\vec{x}\,')}\bigg|_{S(\vec{x})} = 0. \qquad (2.113)$$

On the other hand, from the boundary conditions (2.106) we apparently have for any surface S'

$$\begin{aligned}\frac{\partial G_N(\vec{x},\vec{x}\,')}{\partial n}\bigg|_{S(\vec{x})} = -\frac{4\pi}{S} &\Longrightarrow \left\langle \frac{\partial G_N(\vec{x},\vec{x}\,')}{\partial n}\bigg|_{S(\vec{x})}\right\rangle_{S'(\vec{x}\,')} = -\frac{4\pi}{S} \\ &\Longrightarrow \frac{\partial}{\partial n}\langle G_N(\vec{x},\vec{x}\,')\rangle_{S'(\vec{x}\,')}\bigg|_{S(\vec{x})} = -\frac{4\pi}{S}, \qquad (2.114)\end{aligned}$$

which appears to contradict (2.113) if we take $S' = S$. The apparent contradiction is due to the fact that the derivation of (2.106) presumes that the source point $\vec{x}\,'$ is in the *interior* of V, so it is not valid to take $S' = S$ in (2.114). We may however amend (2.114) by adding a surface delta function that shows up when the source point is first moved to the boundary: in terms of limits, this means

$$\lim_{\vec{x}\to S}\frac{\partial G_N(\vec{x},\vec{x}\,')}{\partial n}\bigg|_{S(\vec{x}\,')} = -\frac{4\pi}{S} + 4\pi\delta^{(S)}(\vec{x}-\vec{x}\,'). \qquad (2.115)$$

Since the vector $\hat{n}$ is pointing out of the volume, the left-hand side of (2.115) represents the normal electric field on the surface. The left-hand side may be visualized as first moving the source point $\vec{x}\,'$ to the boundary, then evaluating the normal electric field there. Equation (2.115) will be verified in Exercise 3.2.2 for the case when S is an infinite plane and Exercise 4.12.3 when S is a sphere. For now we note that since (2.115) holds for all points $S(\vec{x}\,')$ on the boundary, the additional delta function term on the right-hand side changes (2.114) to

$$\lim_{\vec{x}\to S}\left\langle\frac{\partial G_N(\vec{x},\vec{x}\,')}{\partial n}\right\rangle_{S(\vec{x}\,')} = -\frac{4\pi}{S} + \frac{4\pi}{S} = 0, \qquad (2.116)$$

which is now consistent with (2.113).

We can also show that the surface delta function in (2.115) is consistent with our earlier expression (2.103) for Φ. We use the symmetric form of the Green function so that we may switch the order of its arguments (see Exercise 2.8.2):

$$\Phi(\vec{x}) = \int d^3x'\, G_N^{\text{symm}}(\vec{x},\vec{x}\,')\rho(\vec{x}\,') + \frac{1}{4\pi}\oint_S da' G_N^{\text{symm}}(\vec{x},\vec{x}\,')\frac{\partial\Phi(\vec{x}\,')}{\partial n'} + \langle\Phi\rangle_S, \tag{2.117}$$

which for $\vec{x}\to S$ implies that

$$\begin{aligned}\lim_{\vec{x}\to S}\frac{\partial\Phi(\vec{x})}{\partial n} &= \lim_{\vec{x}\to S}\int d^3x'\,\frac{\partial}{\partial n}G_N^{\text{symm}}(\vec{x},\vec{x}\,')\rho(\vec{x}\,') \\ &\quad + \lim_{\vec{x}\to S}\oint_S da'\,\frac{\partial}{\partial n}G_N^{\text{symm}}(\vec{x},\vec{x}\,')\frac{\partial\Phi(\vec{x}\,')}{\partial n'}. \end{aligned}\tag{2.118}$$

We may further simplify this equation using our conjecture (2.115) which from (2.111) we can see still holds for G_N^{symm}:

$$\begin{aligned}\left.\frac{\partial\Phi(\vec{x})}{\partial n}\right|_{S(\vec{x})} &= \int d^3x'\left(-\frac{4\pi}{S}+4\pi\delta^{(S)}(\vec{x}-\vec{x}\,')\right)\rho(\vec{x}\,') \\ &\quad + \frac{1}{4\pi}\oint_S da'\left(-\frac{4\pi}{S}+4\pi\delta^{(S)}(\vec{x}-\vec{x}\,')\right)\frac{\partial\Phi(\vec{x}\,')}{\partial n'} \\ \implies \left.\frac{\partial\Phi(\vec{x})}{\partial n}\right|_{S(\vec{x})} &= -\frac{4\pi Q}{S}+4\pi\int d^3x'\,\delta^{(S)}(\vec{x}-\vec{x}\,')\rho(\vec{x}\,') \\ &\quad - \frac{1}{S}\oint_S da'\frac{\partial\Phi(\vec{x}\,')}{\partial n'} + \left.\frac{\partial\Phi(\vec{x})}{\partial n}\right|_{S(\vec{x})}. \end{aligned}\tag{2.119}$$

But from Green's first identity (2.83) we have

$$\begin{aligned} -\frac{1}{S}\oint da'\frac{\partial\Phi(\vec{x}\,')}{\partial n'} &= -\frac{1}{S}\int d^3x'\,\nabla'^2\Phi(\vec{x}\,') \\ &= \frac{4\pi}{S}\int d^3x'\,\rho(\vec{x}\,') \\ &= \frac{4\pi Q}{S}. \end{aligned}\tag{2.120}$$

Therefore two terms on the right hand side cancel, and we have

$$\left.\frac{\partial\Phi(\vec{x})}{\partial n}\right|_{S(\vec{x})} = 4\pi\int d^3x'\,\delta^{(S)}(\vec{x}-\vec{x}\,')\rho(\vec{x}) + \left.\frac{\partial\Phi(\vec{x})}{\partial n}\right|_{S(\vec{x})}. \tag{2.121}$$

Since the volume integral in (2.121) is over the *interior* of V, the surface delta function integrates to zero and consistency is maintained.

Actually, surface delta functions appear with Dirichlet boundary conditions as well. Recall the Dirichlet integral equation (2.98):

$$\Phi(\vec{x}) = \int d^3x'\, G_D(\vec{x}\,', \vec{x})\rho(\vec{x}\,') - \frac{1}{4\pi}\oint da\,'\Phi(\vec{x}\,')\frac{\partial G_D(\vec{x}\,', \vec{x})}{\partial n'}.$$

In the case where $\vec{x} \to S$, it follows that $G_D(\vec{x}\,', \vec{x}) = 0$ and

$$\lim_{\vec{x}\to S} \Phi(\vec{x}) = \lim_{\vec{x}\to S} -\frac{1}{4\pi}\oint da\,'\Phi(\vec{x}\,')\frac{\partial G_D(\vec{x}\,', \vec{x})}{\partial n'}. \tag{2.122}$$

If $\Phi(\vec{x})$ is an arbitrary potential on S, this implies (relabelling $\vec{x} \leftrightarrow \vec{x}\,'$)

$$\lim_{\vec{x}\,'\to S} \left.\frac{\partial G_D(\vec{x}, \vec{x}\,')}{\partial n}\right|_{S(\vec{x})} = -4\pi\delta^{(S)}(\vec{x} - \vec{x}\,'). \tag{2.123}$$

This is similar to (2.115) but with three important differences: the absence of the $-4\pi/S$ term, the sign in front of the surface delta function and the order of the limits. In Section 3.2 we will show that the surface delta function has a physical interpretation.

2.9 One-dimensional Green function examples

Enough of the theory, we need examples of these quantities! The simplest place to go to illustrate Green function properties is in a one-dimensional context. Let us start with the Dirichlet case in the simplest possible geometry: a finite line segment of length L. The one-dimensional Green function obeys

$$\frac{\partial^2 G(x, x')}{\partial x^2} = -\delta(x - x'), \tag{2.124}$$

for a unit charged particle at x' with Dirichlet boundary conditions

$$G_D(0, x') = 0, \quad G_D(L, x') = 0, \tag{2.125}$$

at $x = 0$ and $x = L$. The one-dimensional potential function $\Phi(x)$ obeys the differential equation

$$\frac{d^2\Phi(x)}{dx^2} = -\lambda(x), \tag{2.126}$$

where $\lambda(x)$ is considered a one-dimensional density of charge. This gives rise to the analog of Equation (2.98):

$$\Phi(x) = \int dx' G_D(x, x')\lambda(x') - \left.\left(\Phi(x')\frac{\partial G_D(x, x')}{\partial x'}\right)\right|_{x'=0}^{x'=L}, \tag{2.127}$$

where we have taken the liberty of using the symmetry $G_D(x', x) = G_D(x, x')$. The general solution for $G_D(x, x')$ is simply

$$G_D(x, x') = \begin{cases} C_1 x + C_2 & \text{if } x > x', \\ C_3 x + C_4 & \text{if } x < x', \end{cases} \tag{2.128}$$

where $C_{1,2,3,4}$ are constants. The continuity condition between the two branches,

$$G_D(x'^{+}, x') = G_D(x'^{-}, x'), \tag{2.129}$$

yields

$$C_1 x' + C_2 = C_3 x' + C_4. \tag{2.130}$$

We also have the discontinuity condition,

$$\left.\frac{\partial G_D(x, x')}{\partial x}\right|_{x'^{-}}^{x'^{+}} = -1, \tag{2.131}$$

which gives

$$C_1 - C_3 = -1. \tag{2.132}$$

We also have the two boundary conditions,

$$G_D(0, x') = 0 \implies C_4 = 0, \tag{2.133}$$

$$G_D(L, x') = 0 \implies C_1 L + C_2 = 0. \tag{2.134}$$

This results in the construction

$$G_D(x, x') = \begin{cases} x'\left(1 - \dfrac{x}{L}\right) & \text{if } x > x', \\ x\left(1 - \dfrac{x'}{L}\right) & \text{if } x < x', \end{cases} \tag{2.135}$$

or more succinctly

$$G_D(x, x') = x_<\left(1 - \frac{x_>}{L}\right), \tag{2.136}$$

where $x_>$ is the greater of x, x' and $x_<$ is the lesser. This completes the construction of $G_D(x, x')$ for the Dirichlet case.

With the insertion of appropriate factors of 4π, one may alternately interpret the one-dimensional Dirichlet Green function as the potential of a uniformly charged *sheet* of charge between conducting walls embedded in a three-dimensional context. In the Neumann case, however, one must strictly maintain the one-dimensional geometry restriction since there is no such alternate three-dimensional interpretation. One may replace reference to the surface S in (2.106) by the modified Green function boundary conditions:

$$\left.\frac{\partial G_N(x,x')}{\partial x}\right|_{x=0} = \frac{1}{2}, \quad \left.\frac{\partial G_N(x,x')}{\partial x}\right|_{x=L} = -\frac{1}{2}. \tag{2.137}$$

The analog to Equation (2.103) in this case is

$$\Phi(x) = \int dx' G_N^{\text{symm}}(x,x')\lambda(x') + \left.\left(G_N^{\text{symm}}(x,x')\frac{\partial \Phi(x')}{\partial x'}\right)\right|_{x'=0}^{x'=L} + <\Phi>, \tag{2.138}$$

using the symmetrical Green function, where $<\Phi>$ is an arbitrary constant representing the average value of the potential on the two end points. Starting with the general solution, Equation (2.128), and using the boundary conditions (2.137) one now obtains an unsymmetrical Green function. Forming the symmetrical version gives

$$G_N^{\text{symm}}(x,x') = -\frac{1}{2}|x-x'| + C, \tag{2.139}$$

where the constant C is arbitrary. (Using the analog to Equation (2.111) gives a specific value of this constant.) This construction and other investigations in this geometry are left to the Exercises.

2.10 Electrostatic energy

We now turn to a discussion of electrostatic energy, which plays a key role in the physics of electrostatic systems.

The work required to move a test charge q_i from ∞ to a location $\vec{x}_i$ within a potential field Φ (with $\Phi(\infty) = 0$) is given by:

$$W_i = q_i \Phi(\vec{x}_i). \tag{2.140}$$

If Φ is due to $i-1$ point charges $q_1, \ldots q_{i-1}$ at locations $\vec{x}_1 \ldots \vec{x}_{i-1}$, we have

$$\Phi(\vec{x}_i) = \sum_{j=1}^{i-1} \frac{q_j}{|\vec{x}_i - \vec{x}_j|}. \tag{2.141}$$

Thus assembling a configuration of n charges one by one will lead to a total energy of

$$W = \sum_{i=1}^{n}\sum_{j<i} \frac{q_i q_j}{|\vec{x}_i - \vec{x}_j|}, \tag{2.142}$$

which may be rewritten as

$$W = \frac{1}{2} \sum_{i \neq j} \frac{q_i q_j}{|\vec{x}_i - \vec{x}_j|}. \tag{2.143}$$

Passing to the continuous limit, this becomes

$$W = \frac{1}{2} \int \int d^3x \, d^3x' \frac{\rho(\vec{x})\rho(\vec{x}\,')}{|\vec{x} - \vec{x}\,'|}. \tag{2.144}$$

An alternative form of (2.144) using (2.51) is

$$W = \frac{1}{2} \int d^3x \, \rho(\vec{x})\Phi(\vec{x}). \tag{2.145}$$

But we may make use of Poisson's equation (2.53) and various differential identities to obtain

$$\begin{aligned} W &= \frac{1}{2} \int d^3x \left[-\frac{1}{4\pi} \nabla^2 \Phi(\vec{x}) \right] \Phi(\vec{x}). \\ &= -\frac{1}{8\pi} \int d^3x \left[\vec{\nabla} \cdot (\Phi \vec{\nabla} \Phi) - \vec{\nabla}\Phi \cdot \vec{\nabla}\Phi \right] \Phi(\vec{x}) \\ &= -\frac{1}{8\pi} \oint da \, \Phi \frac{\partial \Phi}{\partial n} + \frac{1}{8\pi} \int d^3x \, \vec{E}^2. \end{aligned} \tag{2.146}$$

However as $R \to \infty$ we have $da = R^2 d\Omega, \Phi = \mathcal{O}\left(R^{-1}\right)$, and $\partial\Phi/\partial n = \mathcal{O}\left(R^{-2}\right)$, so the first term vanishes and we have

$$W = \int d^3x \; w, \quad \text{where } w \equiv \frac{1}{8\pi} \vec{E}^2. \tag{2.147}$$

Here w is interpreted as the *energy density* of the electric field. Although these expressions for W are developed as work done to establish a given charge configuration, it is manifested in the electric fields thus generated. We will therefore equivalently regard W as the *field energy* of the configuration. In establishing an equivalence between these two concepts, we are relying on the energy conservation concept at a very basic level.

The field energy associated with an isolated charge distribution is referred to as its *self-energy*. As a consequence of passing to the continuum limit, note that in applying (2.147) to a system of point charges, this equation now includes self-energies, while (2.143) does not. Let's try to calculate the simplest possible self-energy: that of a stationary point charge, q, located at the origin. We have

$$w = \frac{1}{8\pi} \frac{q^2}{|\vec{x}|^4}, \tag{2.148}$$

for the energy density. This implies ($u \equiv 1/|\vec{x}| = 1/r$)

$$\begin{aligned} W_{\text{self}} &= \lim_{b\to\infty} \frac{q^2}{8\pi} \int_{1/b}^{\infty} \frac{d^3x}{r^4} \\ &= \frac{q^2}{2} \lim_{b\to\infty} \int_0^b du. \end{aligned} \tag{2.149}$$

This is an example of a *linear divergence*.[9] As far as practical computations are concerned we need not worry about the infinite self-energy of point charges, since their self-energies remain constant and have no effect on measurable quantities.

The equation $W_i = q_i\Phi(\vec{x}_i)$ gives physical insight into the fact that we may always choose $G(\vec{x}, \vec{x}\,') = G(\vec{x}\,', \vec{x})$ for Green functions. We have seen that $G(\vec{x}, \vec{x}\,')$ represents the potential at $\vec{x}$ of a unit charge at $\vec{x}\,'$ with given boundary conditions. Because of this connection between interaction energy and potential, we see that the Dirichlet or Neumann Green function can also be considered the *interaction energy* of two unit charges at $\vec{x}$ and $\vec{x}\,'$ with the given boundary conditions. Since the two charges are equivalent, the Green function must be symmetric in its arguments. This argument fails in the case of interior Neumann boundary conditions, because one can imagine redefining the zero of potential for the different source positions. This is not possible in the Dirichlet case since the potentials are specified by the boundary conditions.

An important energy theorem for charged surfaces is *Thompson's theorem*. Here we state and prove this theorem for a special case, namely a single closed charged conducting surface.

Theorem. *If a closed surface is fixed in position and a given total charge is placed on it, then the electrostatic energy in the region bounded by the surface and infinity is an absolute minimum when the charges are placed so that the surface is an equipotential.*

The proof of this theorem is as follows. Assume Φ satisfies the conditions of the theorem and that $\Phi' \equiv \Phi + \Delta\Phi$ represents any other potential subject

[9] The divergence is at small distances, which means high energies in quantum mechanics. In quantum field theories the self-energy of a point charge is also divergent, but because of the Heisenberg uncertainty principle the divergence is less extreme: logarithmic instead of linear. This milder divergence permits quantum renormalization, a topic beyond the scope of this course. (See also Section 14.1 for thoughts along these lines.)

to the condition that the total charge on the surface S (see (2.92))

$$Q = \frac{1}{4\pi}\oint_S da\,\frac{\partial\Phi}{\partial n} = \frac{1}{4\pi}\oint_S da\,\frac{\partial\Phi'}{\partial n}, \tag{2.150}$$

be fixed. The equipotential condition is

$$\Phi\big|_S = V, \tag{2.151}$$

where V is a constant. Both Φ and Φ' are assumed to satisfy Laplace's equation

$$\nabla^2\Phi = \nabla^2\Phi' = 0 \implies \nabla^2(\Delta\Phi) = 0, \tag{2.152}$$

in charge-free space. If W is the energy associated with the field configuration Φ and W' is the energy associated with Φ', then from (2.147),

$$\begin{aligned} W' &= \frac{1}{8\pi}\int d^3x\left(\vec{\nabla}\Phi'\right)^2 \\ &= W + \frac{1}{8\pi}\int d^3x\left(\vec{\nabla}(\Delta\Phi)\right)^2 + W'', \end{aligned} \tag{2.153}$$

where the second term on the right is positive definite, and where the cross term is

$$W'' \equiv \frac{1}{4\pi}\int d^3x\,\vec{\nabla}\Phi\cdot\vec{\nabla}(\Delta\Phi). \tag{2.154}$$

Green's first identity now shows that

$$W'' = -\frac{1}{4\pi}\int d^3x\,\Phi\nabla^2(\Delta\Phi) + \frac{V}{4\pi}\oint_S da\,\frac{\partial\Delta\Phi}{\partial n} = 0, \tag{2.155}$$

from Laplace's equation (2.152) and the fixed charge condition. Thus $W' > W$ and the theorem is proved.

The above argument may straightforwardly be generalized to include additional closed surfaces held at various constant potentials. In Section 3.10 we will prove a more general version, which can be used to numerically estimate capacitances as well as the fields produced by charged conductors.

2.11 Normal force on a charged surface

Let us turn now to compute the force per area on the surface of an arbitrary conductor. We shall denote the force per area by the symbol $\mathcal{F}$. We can imagine partitioning the surface into an infinitesimal disk and a remainder,

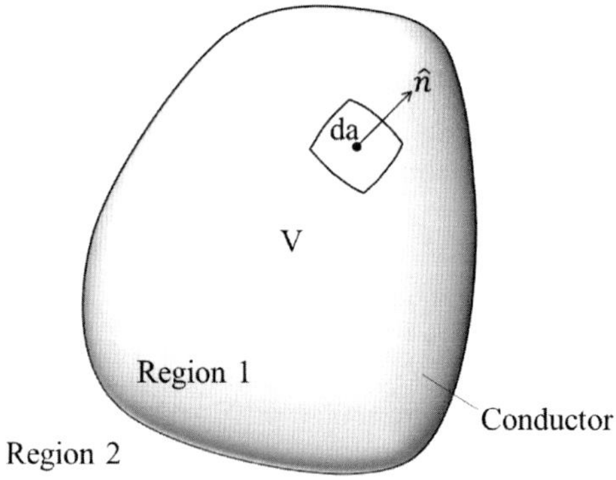

Fig. 2.15 Conducting surface.

as in the Section 2.6 discussion. With reference to Figure 2.15, letting σ be the surface charge density we have

$$\vec{\mathcal{F}}^{\text{disk}} = \sigma\vec{E}_{\text{remainder}} = \sigma\left(\vec{E}_{2,\text{total}} - \vec{E}_{\text{disk}}\right). \tag{2.156}$$

In addition,

$$\vec{E}_{1,\text{total}} = \vec{E}_{\text{remainder}} - \vec{E}_{\text{disk}}. \tag{2.157}$$

Thus in general,

$$\vec{\mathcal{F}}^{\text{disk}} = \sigma\frac{1}{2}\left(\vec{E}_{1,\text{total}} + \vec{E}_{2,\text{total}}\right). \tag{2.158}$$

Note that (2.158) is equivalent to subjecting the charged element σda to the averaged field at the surface. The average removes the self-field of the disk itself. In particular for the normal force per area,

$$\hat{n}\cdot\vec{\mathcal{F}} = \sigma\frac{1}{2}\left(\vec{E}_{1,\text{total}} + \vec{E}_{2,\text{total}}\right)\cdot\hat{n}. \tag{2.159}$$

For a conductor in Region 1,

$$\vec{E}_{1,\text{total}} = 0, \quad \hat{n}\cdot\vec{E}_{2,\text{total}} = 4\pi\sigma, \tag{2.160}$$

from (2.68), so that the normal force per area is given by

$$\hat{n}\cdot\vec{\mathcal{F}} = 2\pi\sigma^2. \tag{2.161}$$

Notice that $\hat{n}\cdot\vec{\mathcal{F}} > 0$, so the force is always directed outward (as expected, since like charges repel). We shall see in Section 5.9 that (2.159) holds also for non-conducting surfaces when $\hat{n}\cdot\vec{E}_{1,\text{total}} \neq 0$.

2.12 Capacitance

In elementary electrostatics you were introduced to the notion of capacitance as the ratio of charge to potential. We have stated that the Green function contains all the information to solve an electrostatic problem in a given geometry — and we will now reinforce our statement by expressing capacitance in terms of the Green function.[10]

We consider the case of a charge-free region bounded by N conducting surfaces $S_1 \ldots S_N$, where the jth surface is held at potential V_j. Thus $\rho(\vec{x}\,') = 0$ and Dirichlet boundary conditions apply, so from (2.98) we have

$$\begin{aligned}\Phi(\vec{x}) &= -\frac{1}{4\pi}\oint_{S_1 \cup \ldots \cup S_N} da'\,\Phi(\vec{x}\,')\frac{\partial G_D(\vec{x}\,',\vec{x})}{\partial n'} \\ &= -\frac{1}{4\pi}\sum_{j=1}^{N} V_j \oint_{S_j} da'\,\frac{\partial G_D(\vec{x}\,',\vec{x})}{\partial n'_j}.\end{aligned} \tag{2.162}$$

Let us now evaluate the charges on the different surfaces by integrating the surface charge densities. We may use the second equation in (2.160) to express the surface charge densities in terms of the potential Φ (an outer volume normal points into the conductor):

$$\begin{aligned}\sigma_i(\vec{x}) &= \frac{1}{4\pi}\frac{\partial \Phi}{\partial n_i}, \\ &= -\frac{1}{16\pi^2}\sum_{j=1}^{N} V_j \oint_{S_j} da'\,\frac{\partial}{\partial n_i}\frac{\partial}{\partial n'_j}G_D(\vec{x}\,',\vec{x}).\end{aligned} \tag{2.163}$$

The total charge on the jth surface is found by integration:

$$Q_i(\vec{x}) = -\frac{1}{16\pi^2}\sum_{j=1}^{N} V_j \oint_{S_i,S_j} da\,da'\,\frac{\partial}{\partial n_i}\frac{\partial}{\partial n'_j}G_D(\vec{x}\,',\vec{x}). \tag{2.164}$$

We can write this more simply as

$$Q_i = \sum_{j}^{N} C_{ij} V_j, \tag{2.165}$$

where the *capacitances* $\{C_{ii}\}$ and *capacitive induction coefficients* $\{C_{ij}, i \neq j\}$ are given by

$$C_{ij} = -\frac{1}{16\pi^2}\oint_{S_i,S_j} da\,da'\,\frac{\partial}{\partial n_i}\frac{\partial}{\partial n'_j}G_D(\vec{x}\,',\vec{x}). \tag{2.166}$$

[10]This section largely based on J. Schwinger *et al.*, *Classical Electrodynamics*, Perseus Books (Reading, MA) 1998, Sec. 24.6.

The capacitance of a conductor is therefore the charge on the conductor when it is maintained at unit potential, all other conductors being grounded. A similar statement holds for the induction coefficients, except the total charge is measured on a conductor other than the one being held at unit potential. The coefficients are symmetric in their indices,

$$C_{ij} = C_{ji}. \tag{2.167}$$

because of the symmetry of the Dirichlet Green function.

Given the energy expression for surface charges,

$$\begin{aligned} W &= \frac{1}{2}\int d^3x\ d^3x' \frac{\rho(\vec{x})\rho(\vec{x}\,')}{|\vec{x}-\vec{x}\,'|} \\ &\to \frac{1}{2}\int da\ da' \frac{\sigma(\vec{x})\sigma(\vec{x}\,')}{|\vec{x}-\vec{x}\,'|} \\ &= \frac{1}{2}\int da\ \Phi(\vec{x})\sigma(\vec{x}), \end{aligned} \tag{2.168}$$

we have, in the case of multiple conductors

$$W = \frac{1}{2}\sum_i V_i \oint da_i \sigma_i(\vec{x}) = \frac{1}{2}\sum_i V_i Q_i \tag{2.169}$$

or

$$W = \frac{1}{2}\sum_{i,j} C_{ij} V_i V_j, \tag{2.170}$$

which gives a different perspective on the C_{ij}'s.

We have to be careful when the surface at infinity is included in our analysis, since its capacitance and coefficients of induction can only be defined in a limiting sense in a given problem. With this understanding then, we can now show that

$$\sum_i C_{ij} = 0, \tag{2.171}$$

for both interior and "exterior" problems as follows:

$$\sum_i C_{ij} = -\frac{1}{16\pi^2}\oint_{S_j} da_j' \frac{\partial}{\partial n_j'}\left(\sum_i \oint_{S_i} da_i \frac{\partial}{\partial n_i} G_D(\vec{x},\vec{x}\,')\right). \tag{2.172}$$

However from Gauss' theorem we have

$$\sum_i \oint_{S_i} da_i \frac{\partial}{\partial n_i} G_D(\vec{x},\vec{x}\,') = \int d^3x\, \nabla^2 G_D(\vec{x},\vec{x}\,') = -4\pi. \tag{2.173}$$

(2.173) in (2.172) implies that $\sum_i C_{ij} = 0$. This means that

$$Q \equiv \sum_i Q_i = \sum_{i,j} C_{ij} V_j = 0, \tag{2.174}$$

as it must since we are imagining a closed system with no volume charge. Equation (2.171) also shows that only the relative potential values determine the total charge, for

$$\sum_j C_{ij}(V_j + \text{constant}) = \sum_j C_{ij} V_j. \tag{2.175}$$

The result (2.171) may be extended to the surface at infinity as long as one requires $V_\infty = 0$. Then for a single isolated conductor, labelled "1", with charge Q and potential, V, one has a single coefficient

$$C \equiv C_{11} = -C_{\infty 1} = -C_{1\infty} = C_{\infty\infty}, \tag{2.176}$$

where

$$Q_\infty = -Q = C_{\infty 1} V \implies Q = CV. \tag{2.177}$$

Applying (2.171) for $j = 1$ in the case where there are additional conductors inside the "1" surface yields a sum rule,

$$C_{\infty\infty} = \sum_{i=1,2,\ldots} C_{i1}, \tag{2.178}$$

relating external and internal capacitances. This can be useful in checking the results of the interior calculations, as in Exercise 2.12.3.

As a simple application of capacitance, we consider two parallel conducting plates at potentials V_1 and V_2 respectively, separated by a distance a. Since the C_{ij} are symmetric in i and j and must sum to zero on either index, this means

$$C_{11} = -C_{21} = -C_{12} = C_{22} \equiv C. \tag{2.179}$$

Also

$$Q_1 = -Q_2 = C(V_1 - V_2) = C\Delta V. \tag{2.180}$$

The field energy is

$$W = \frac{1}{2}\sum_{i,j} C_{ij} V_i V_j = \frac{1}{2} C\Delta V^2. \tag{2.181}$$

These are the usual expressions for a parallel plate capacitor. C can be shown from Gauss' law to be (for a finite surface A)

$$C = \frac{A}{4\pi a}. \tag{2.182}$$

This can also be directly evaluated from the Green function; see Exercise 3.2.3.

Note that it is usually difficult to get exact solutions for the Green function for a given geometry. Typically, approximation methods must be used in computing the C_{ij}. In the simplest case of isolated conducting surfaces see Section 3.10 for a variational method.

2.13 Going Deeper

2.13.1 *Photon mass limits*

(1) A. F. Goldhaber and M. M. Nieto, "Photon and Graviton Mass Limits", Rev. Mod. Phys. **82**, 939 (2010).

(2) K. Olive *et al.* (Particle Data Group), Chin. Phys. C **38**, 090001 (2014); `http://pdg.lbl.gov`; "γ mass".

2.13.2 *Capacitance*

(1) G. T. Carlson and B. L. Illman, "The Circular Disk Parallel Plate Capacitor", Am. J. Phys. **62**, 1099 (1994).

(2) C. O. Hwang and M. Mascagni, "Electrical Capacitance of the Unit Cube", J. Appl. Phys. **95**, 3798 (2004).

(3) W. R. Smythe, *Static and Dynamic Electricity*, 3rd edn., Hemisphere Publishing (New York) 1989; Chs. 2 and 5.

(4) J. Van Bladel, *Electromagnetic Fields*, 2nd edn., IEEE Press (Piscataway) 2007; Sections 4.2 and 4.3.

2.14 Exercises

Exercise 2.1.1.

(a) A vector field is defined by $\vec{E} = -\vec{\nabla}\Phi$, with

$$\Phi = \vec{d} \cdot \frac{\vec{x}}{r^3},$$

where $r \equiv |\vec{x}|$ and $\vec{d}$ is a constant vector. Derive an expression for $\vec{E}$.

(b) Prove that $\vec{E}$ in part (a) can be written as $\vec{E} = \vec{\nabla} \times \vec{A}$, where

$$\vec{A} = \vec{d} \times \frac{\vec{x}}{r^3}.$$

[Note: This problem neglects certain subtleties that occur at $\vec{x} = 0$; see Section 6.8. Adapted from B. DiBartolo, *Classical Theory of Electromagnetism*, 2nd ed., World Scientific (Singapore) 2004, Exercise 1.18.]

Exercise 2.2.1.

(a) Prove that the image of a unit cube under the linear transformation $x'_i = \sum_j M_{ij} x_j$ has volume $|\det M|$, where M is a nonsingular 3×3 matrix. [*Hint*: Column elements of M constitute the new edge vectors, $\vec{v}_i$, of the image parallelepiped whose volume is $|\vec{v}_3 \cdot (\vec{v}_1 \times \vec{v}_2)|$.]

(b) Using part (a) or any other means argue for orthogonal coordinate systems that

$$|\det M| = UVW,$$

where U, V, W are the scale factors in Equation (2.29).

Exercise 2.2.2.

(a) Show that

$$\lim_{\epsilon \to 0^+} \frac{1}{\pi} \frac{\epsilon}{x^2 + \epsilon^2} = \delta(x).$$

(b) Use the same reasoning in a two-dimensional context to show that $(\vec{x} = x\,\hat{\imath} + y\,\hat{\jmath})$

$$\lim_{\epsilon \to 0^+} \frac{1}{2\pi} \frac{\epsilon}{(\vec{x}^2 + \epsilon^2)^{3/2}} = \delta^{(2)}(\vec{x}).$$

Exercise 2.2.3. The $\vec{x}$ position vector in spherical coordinates is

$$x = r\sin\theta\cos\phi,$$
$$y = r\sin\theta\sin\phi,$$
$$z = r\cos\theta.$$

Taking $u = r, v = \cos\theta, w = \phi$ as the new coordinates, find the form of the delta function using:

(a) the determinant of the transformation, (2.25)
(b) scale factors.

[*Ans.*: $\delta(\vec{x} - \vec{x}\,') = \frac{1}{r^2}\delta(r - r')\delta(\cos\theta - \cos\theta')\delta(\phi - \phi')$.]

Exercise 2.2.4. Find the form of the Dirac delta function in *oblate spheroidal coordinates*. One form is ($\xi \geq 0$, $0 \leq \theta \leq \pi$, $0 \leq \phi \leq 2\pi$)

$$x = R\sqrt{\xi^2 + 1}\sin\theta\cos\phi,$$
$$y = R\sqrt{\xi^2 + 1}\sin\theta\sin\phi,$$
$$z = R\xi\cos\theta,$$

where R is a positive constant and $u = \xi, v = \cos\theta, w = \phi$ are the new coordinates. (The ϕ coordinate corresponds to the usual spherical azimuthal angle. Note however that θ corresponds to the usual spherical polar angle and $R\xi$ corresponds to spherical coordinate distance r *only* for $\xi \gg 1$. See P. M. Morse and H. Feshbach, *Methods of Theoretical Physics, Part I*, McGraw-Hill (New York) 1953.)

Exercise 2.4.1.

(a) Given in *two* spatial dimensions the form

$$\vec{E}(\vec{x}) = \int da'\sigma(\vec{x}\,')\frac{\vec{x} - \vec{x}\,'}{|\vec{x} - \vec{x}\,'|^2},$$

for the electric field, argue as in the text that one has

$$\oint d\ell\ \hat{n}\cdot\vec{E} = \sum_{j\ \text{in}\ S} 2\pi q_j,$$

for enclosed point charges within a surface area S bounded by C, and

$$\vec{\nabla}\cdot\vec{E} = 2\pi\sigma(\vec{x}),$$

for the electric field due to a *continuous* charge density $\sigma(\vec{x})$.

(b) For a point charge ($q = 1$) at the origin in two dimensions, show that (a) gives

$$\nabla^2 \ln\left(\frac{|\vec{x} - \vec{x}\,'|}{K}\right) = 2\pi\delta^{(2)}(\vec{x} - \vec{x}\,'),$$

where $\delta^{(2)}(\cdots)$ is a two-dimensional delta function, and K is an arbitrary positive constant.

(c) Confirm this delta function representation by putting $\vec{x}\,' = 0$ and integrating both sides of (b) about an area surrounding the origin.

(d) Part (c) identifies the potential of a point charge in 2 dimensions as $-\ln(|\vec{x} - \vec{x}\,'|/K)$, up to an overall constant. However, the potential of a *line* charge in 3 dimensions with unit line density of charge is $-2\ln(|\vec{x} - \vec{x}\,'|/K)$. Explain where the extra 2 comes from.

Exercise 2.4.2.

(a) Find the potential everywhere from two perpendicular line charges of length $2L$ with uniform linear charge density λ that cross at the origin, making a '+':

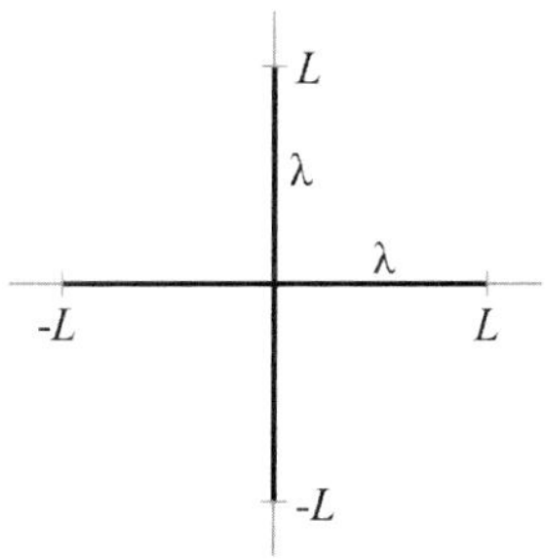

Fig. 2.16 Line charges in Exercise 2.4.2.

Get the exact potential $\Phi(x, y, z)$, then show for $L \gg |x|, |y|, |z|$ that

$$\Phi(x, y, z) \approx -\lambda \ \ln\left(\frac{(x^2 + z^2)(y^2 + z^2)}{16L^4}\right).$$

In light of the previous exercise, why does this answer make sense?

(b) For the opposite extreme, $L \ll |x|, |y|, |z|$, show that you get the expected answer for an effective point charge.

(c) Now modify the line charge so that they extend only along the positive y- and x-axes, and the x-axis line charge density is $-\lambda$ instead of λ. Again get the exact expression for the potential, then show for $L \gg |x|, |y|, |z|$ that

$$\Phi(x, y, z) \approx -\lambda \ln\left(\frac{r-y}{r-x}\right),$$

where $r = \sqrt{x^2 + y^2 + z^2}$.

Exercise 2.4.3.

(a) Find the electric field E_z along the axis of symmetry (z) of a uniform cylinder of radius b and thickness a with constant charge density ρ (see Figure 2.17).

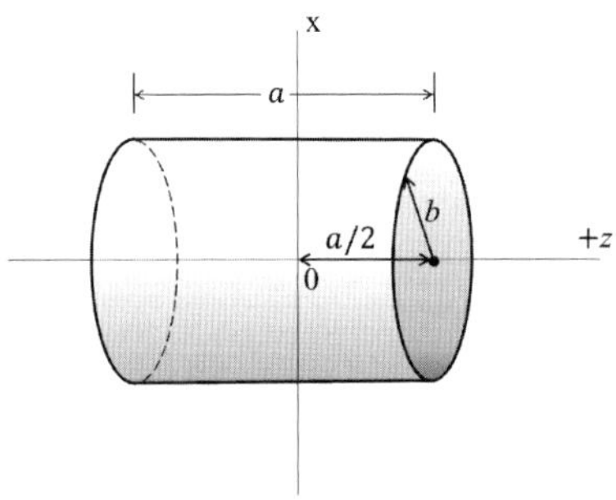

Fig. 2.17 Charged cylinder in Exercise 2.4.3.

(b) For arbitrary a, b, show that near the geometric center (origin O) $E_z =$ Constant $\cdot z$. Find the constant. Can it easily be predicted in the limit $b \gg a$?

Exercise 2.4.4.

(a) A circular loop of radius R with a uniform linear charge density λ is oriented in the x-y plane. Show that the electric field along the z-axis is given by

$$E_z = 2\pi\lambda \frac{Rz}{[R^2 + z^2]^{3/2}}.$$

(b) A truncated circular cone with opening angle α has its symmetry axis oriented along the z-axis, as shown in Figure 2.18. The origin is taken at the theoretical tip of the cone. However, the end of the cone is shaved off at a distance L_1 along the z-axis from the origin, and the distance from the origin to the top of the cone along z is L_2. The cone has a uniform charge density σ smeared on its surface. Using part (a) evaluate E_z at the origin (theoretical tip).

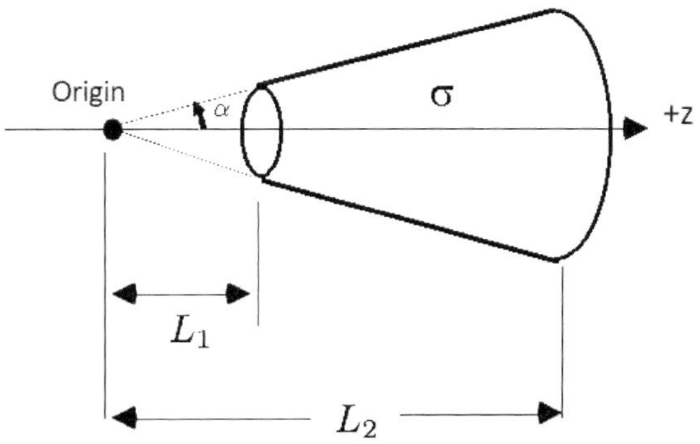

Fig. 2.18 Geometry for Exercise 2.4.4(b).

Exercise 2.5.1.

(a) For a potential of the form ($|\epsilon| \ll 1$)

$$\Phi(\vec{x}) = \frac{1}{1+\epsilon} \int d^3x' \frac{\rho(\vec{x}\,')}{|\vec{x} - \vec{x}\,'|^{1+\epsilon}},$$

show that the potential between two concentric conducting spheres with charges q_a, q_b and radii a, b respectively ($a < b$) is given for $a \le r \le b$ by $\Phi(r) = \Phi_{out}(r) + \Phi_{in}(r)$, using the expressions in Equations (2.62) and (2.63).

(b) After the spheres are set at the same potential (by connecting them with a wire), show that the charge on the inner sphere is given by

$$q_a \approx \frac{q_b \epsilon}{2(a-b)} \left[b \ln\left(\frac{b-a}{a+b}\right) + a \ln\left(\frac{4b^2}{b^2 - a^2}\right)\right].$$

[*Hint*: The approximation $x^{1-\epsilon} = xe^{-\epsilon \ln x} \approx x(1 - \epsilon \ln x)$ is useful here.]

(c) Is there a quick way of establishing the sign of the induced charge for given signs of ϵ and q_b?

Exercise 2.5.2. Let's say Cavendish considered a point potential of the form

$$\Phi(\vec{x}) = \frac{1}{|\vec{x} - \vec{x}\,'|} \exp\left(-\frac{|\vec{x} - \vec{x}\,'|}{R}\right),$$

where R is a very large distance. Consider the experimental setup as in Figure 2.10.

(a) Given a charge q_b on the outer sphere, calculate the charge q_a on the inner sphere, if the two spheres are connected by a thin wire. [*Hint*: When integrating over the spheres, think perfect differential. *Ans.*:

$$q_a = q_b \frac{\frac{a}{b}\left(1 - \exp\left(-\frac{2b}{R}\right)\right) - \exp\left(\frac{a-b}{R}\right) + \exp\left(-\frac{a+b}{R}\right)}{\frac{b}{a}\left(1 - \exp\left(-\frac{2a}{R}\right)\right) - \exp\left(\frac{a-b}{R}\right) + \exp\left(-\frac{a+b}{R}\right)}.]$$

(b) For $R \gg a, b$ show that the charge on the inner sphere in this case is approximately

$$q_a \approx q_b \frac{ab}{6R^2}\left(1 + \frac{a}{b}\right).$$

Note that q_a and q_b have the same sign in this case. Setting $R = h/(m_p c)$ gives us the connection to the photon mass, m_p.

Exercise 2.6.1. A thin isolated metallic disk of radius R is charged until it has acquired a potential V. The solution for the electrostatic potential is found most easily in oblate spheroidal coordinates ξ, θ and ϕ, given in Exercise 2.2.4, and is given by

$$\Phi = A \tan^{-1}\left(\frac{1}{\xi}\right),$$

where A is an unknown constant. (Take your origin of coordinates O at the center of the disk and the z-axis perpendicular to the plane of the disk.)

(a) Change variables to show that the potential may be written in spherical coordinates ($r^2 = x^2 + y^2 + z^2$) as

$$\Phi = A \tan^{-1}\left(\frac{\sqrt{2}R}{\sqrt{(r^2 - R^2) + \sqrt{(r^2 - R^2)^2 + 4R^2z^2}}}\right).$$

(b) By considering the limit $z \to 0$ when $r < R$, argue from consistency that

$$A = \frac{2V}{\pi}.$$

(c) Show that the total surface charge density on the disk $\sigma(\rho)$ is given by

$$\sigma(\rho) = \frac{V}{\pi^2}\frac{1}{\sqrt{R^2 - \rho^2}}.$$

[Note: $d\tan^{-1}(y)/dx = (1 + y^2)^{-1}dy/dx$.]

(d) What is the capacitance, C, of the disk?

Exercise 2.6.2. In deriving (2.77) in the text, consider two close surfaces, 1 and 2, with associated surface charge densities $\sigma_1(\vec{x}\,')$ and $\sigma(\vec{x}\,'')$, respectively; see Figure 2.13. The separation between the surfaces is such that the points at which the charge densities are equal and opposite are separated a uniform distance d along the common perpendicular; that is, $\sigma_1(\vec{x}\,' - \hat{n}_{21}d) = -\sigma(\vec{x}\,')$. We may characterize the potential in the limit of close separation of the two surfaces as,

$$\Phi(\vec{x}) = \int_{S_2} \frac{\sigma(\vec{x}\,')}{|\vec{x} - \vec{x}\,'|}da' - \int_{S_1} \frac{\sigma_1(\vec{x}\,'')}{|\vec{x} - \vec{x}\,''|}da''.$$

Using the relation

$$\vec{x}\,'' = \vec{x}\,' - \hat{n}_{21}d,$$

and the expansion, to lowest order in $\vec{a}$,

$$\frac{1}{|\vec{x} - \vec{a}|} \approx \frac{1}{|\vec{x}|} + \vec{a}\cdot\vec{\nabla}\left(\frac{1}{|\vec{x}|}\right),$$

derive (2.77) in the text.

Exercise 2.6.3. *(Previous exercise, continued)* Now consider the slightly different situation shown in Figure 2.19. Two regions in free space (3 dimensions) are separated by a type of dipole surface. The dipole surface is constructed by placing equal and opposite surface charge densities on complementary open surfaces, which are then brought together such that a small constant separation $\hat{x}d$ remains.

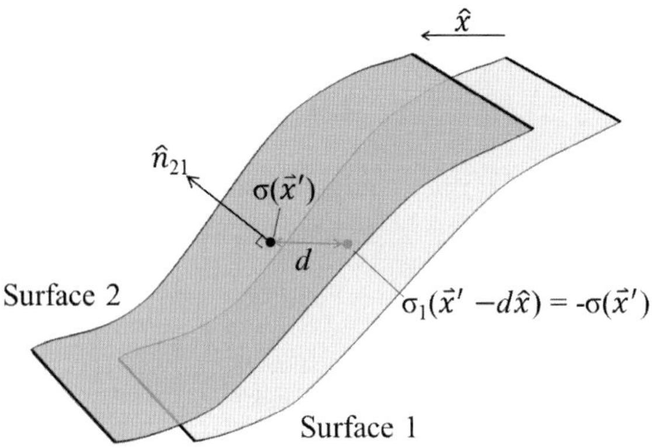

Fig. 2.19 Dipole surface for Exercise 2.6.3.

(a) Show that the potential in both regions can be written as

$$\Phi^d(\vec{x}) = \hat{x} \cdot \vec{E}_{(2)}(\vec{x})d,$$

where $\hat{x}$ is a unit vector in the separation direction and $\vec{E}_{(2)}$ is the electric field due to surface 2.

(b) Show that the discontinuity of the potential in the normal direction (to surface 2) across the surfaces is given by

$$\Phi_2^d - \Phi_1^d = 4\pi D(\vec{x})\hat{x} \cdot \hat{n}_{21},$$

where $D(\vec{x}) \equiv \sigma(\vec{x})d$.

Exercise 2.6.4.

(a) A hemisphere of radius a has a constant dipole surface density D. D is defined relative to the unit vector, $\hat{n}$, as shown in Figure 2.20(a). Using the coordinates shown with the origin O at the hemisphere's center, find

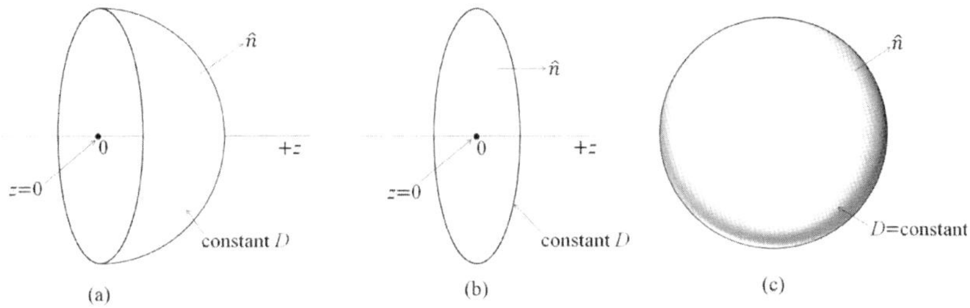

Fig. 2.20 Configurations for parts (a), (b), and (c) of Exercise 2.6.4.

the electric field everywhere along the z-axis ($-\infty \le z \le \infty$). Then using spherical coordinates, find the electric field components E_r and E_θ far from the region of the sphere, $r \gg a$:

$$E_r = \frac{2D\pi a^2}{r^3}\cos\theta,$$

$$E_\theta = \frac{D\pi a^2}{r^3}\sin\theta.$$

(We will see this field again in the multipole discussion of Section 5.1, Equation (5.15).)

(b) Same questions as in part (a), but for flat disk of radius a (see Figure 2.20(b)).

(c) A sphere of radius a is given a constant dipole surface charge density D. D is defined relative to the sphere's outward-pointing normal, $\hat{n}$, as shown in Figure 2.20(c). Find the electric field both inside and outside the sphere.

Exercise 2.7.1. Prove Green's first identity (2.83) using the Gauss theorem.

Exercise 2.7.2.

(a) Prove *Green's reciprocation theorem*: For two different potential problems with exactly the same geometry where the bounding surfaces are conducting, one has

$$\oint da\ \sigma\,\Phi' + \int d^3x\,\rho\,\Phi' = \oint da\ \sigma'\,\Phi + \int d^3x\,\rho'\,\Phi,$$

where Φ, ρ and σ are the potential, volume charge density and surface charge density of the first problem, respectively, while Φ', ρ' and σ' are similar quantities for the second problem.

(b) A point charge q is located somewhere between two grounded concentric spherical conducting shells of radii a, b ($a < r < b$) as shown in Figure 2.21.

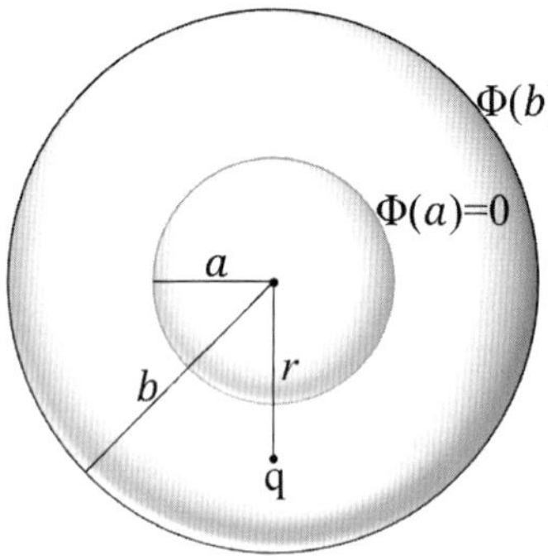

Fig. 2.21 Spherical shells in Exercise 2.7.2 part (b).

Find the charge induced on the inner and outer conducting shells by use of the theorem of part (a).

Exercise 2.7.3.

(a) Prove the electrostatic mean value theorem (MVT):
For charge-free space the value of the electrostatic potential at any point is equal to the average of the potential over the surface of any sphere centered on that point.

(b) Use the MVT to provide an alternate proof of the uniqueness theorem of Section 2.7 for the electrostatic potential $\Phi(\vec{x})$ for Dirichlet boundary conditions.

Exercise 2.8.1. Given a region with boundary surface S, show that the Neumann and Dirichlet Green functions are related by

$$G_N(\vec{x}, \vec{x}\,'') = G_D(\vec{x}\,'', \vec{x}) - \frac{1}{4\pi} \oint_S da' G_N(\vec{x}\,', \vec{x}\,'') \frac{\partial G_D(\vec{x}\,', \vec{x})}{\partial n'}.$$

Exercise 2.8.2. Considering the interior Neumann problem, show that using Equation (2.111) in Equation (2.103) leaves the potential $\Phi(\vec{x})$ unchanged up to a constant:

$$\Phi(\vec{x}) = \Phi^{\text{symm}}(\vec{x}) + \langle\Phi\rangle - \langle\Phi^{\text{symm}}\rangle,$$

where $\Phi(\vec{x})$ uses the unsymmetrical Green function and $\Phi^{\text{symm}}(\vec{x})$ the symmetrical form.

Exercise 2.8.3. (Compare Jackson problem 2.28.) A closed volume is bounded by conducting surfaces that are the n sides of a regular polyhedron ($n = 4, 6, 8, 12, 20$). The n surfaces are at different potentials V_i, $i = 1, \ldots, n$. Using the Dirichlet Green function expression for the interior potential, show that the potential at the center of the polyhedron is the average of the potentials on the n sides.

Exercise 2.9.1.

(a) Using the one-dimensional Dirichlet Green function (2.136), find the force $\left(F = E_{avg}q, E = -\dfrac{dG_D(x,x')}{dx}\right)$ on the unit charged particle at position x'.
(b) Find the induced charges on the endpoints at $x = 0$ and L for this same situation.
(c) Using this Green function and Equation (2.127), find the potential everywhere inside the one-dimensional box when there is no volume charge $\lambda(x)$ and the potential at the endpoints are $\Phi(0) = V$ and $\Phi(L) = 0$.

Exercise 2.9.2.

(a) Complete the derivation of the symmetric Neumann Green function (2.139).
(b) Find the force $\left(F = E_{avg}q, E = -\dfrac{dG_N^{\text{symm}}(x,x')}{dx}\right)$ on the Neumann Green function unit charged particle at position x'.

(c) Using this Green function and Equation (2.138), find the potential everywhere inside the one-dimensional box when there is no volume charge $\lambda(x)$ and the electric fields at the endpoints are $E(0) = E(L) = V/L$. (Note that the requirement of a neutral isolated system gives $E(x)|_0^L = Q$, where Q is the total volume charge.)

Exercise 2.9.3. Consider an electrodynamic problem in one-dimensional space with conducting walls at $x = 0$ and $x = L$. The potential satisfies

$$\frac{d^2\Phi}{dx^2} = -\lambda(x),$$

and

$$\Phi(0) = V, \Phi(L) = -V,$$

on the boundaries, and the charge density, $\lambda(x) = \lambda$ is a constant in space. Use the Dirichlet Green function

$$G_D(x, x') = x_< \left(1 - \frac{x_>}{L}\right),$$

for this geometry to find $\Phi(x)$ for $0 < x < L$. (You may verify your Green function solution with a direct solution if you like, but you must first use the Green function approach for full credit.)

Exercise 2.9.4.

(a) Solve the 1D Helmholtz equation

$$\left[\frac{d^2}{dx^2} + k_0^2\right] \Phi(x) = -\lambda(x),$$

in the variable range $x \in [0, L]$ using the Dirichlet Green function $G(x, x')$ which satisfies

$$\left[\frac{d^2}{dx^2} + k_0^2\right] G(x, x') = -\delta(x - x').$$

Show that

$$\Phi(x) = \int_0^L dx' G(x, x')\lambda(x') - \left(\Phi(x')\frac{dG(x, x')}{dx'}\right)\Bigg|_{x'=0}^{x'=L},$$

using the symmetry $G(x, x') = G(x', x)$, the same as in Equation (2.127).

(b) Using one-dimensional Dirichlet boundary conditions, $G(0, x') = G(L, x') = 0$, show that this Green function is given by

$$G(x, x') = \frac{\sin k_0 x_< \sin k_0(L - x_>)}{k_0 \sin k_0 L},$$

where $x_>$ ($x_<$) is the greater (lesser) of x and x'.

Exercise 2.10.1. Prove the following theorem: If a number of conducting surfaces are fixed in position with a given total charge on each, the introduction of an uncharged, insulated conductor into the region bounded by the surfaces lowers the electrostatic energy.

Exercise 2.10.2. Investigate the self-energy of a one-dimensional straight string of charge with length a and linear charge density $\lambda = Q/a$. Is this quantity divergent, and if so, is it quadratically, linearly or logarithmically divergent?

Exercise 2.10.3. Consider an infinitely thin square sheet of side R with constant surface charge density σ and total charge Q. Show that its self-energy is

$$W_{square} = 1.48660\ldots \frac{Q^2}{R}.$$

[*Hint*: We found that a "center-of-mass" change in variables was useful. Our result was determined numerically, but see O. Ciftja, *Adv. Math. Phys.* **2016**: `https://doi.org/10.1155/2016/7207536` for a more comprehensive approach.]

Exercise 2.10.4.

(a) Show that the self-energy W_{sphere} of a spherically symmetric ball of charge with charge density $\rho(r) = Cr^\alpha$, radius a, and total charge Q is given by

$$W_{sphere} = \frac{3 + \alpha}{5 + 2\alpha} \frac{Q^2}{a}.$$

Show that only for $\alpha > -\frac{5}{2}$ does one obtain a finite result.

(b) Use this to calculate the radius of a particle with uniform charge distribution whose self-energy equals $E = mc^2$, where m is the electron mass (cf. discussion in Section 14.1).

Exercise 2.11.1.

(a) A sphere of radius R has a constant volume charge density ρ and a uniform surface charge density, σ. Find the electric field $\vec{E}$ both inside and outside the sphere.
(b) Find the outward electrostatic force per unit area F/A on the surface of the sphere.

Exercise 2.12.1. In a system of N conductors only the voltages V_A and V_B are not equal to zero. Show, based upon positivity of the field energy, that the capacitance induction coefficient, C_{AB}, and the capacitances, C_{AA} and C_{BB}, obey the inequality

$$C_{AB}^2 < C_{AA}C_{BB}.$$

Exercise 2.12.2.

(a) Given a system with N capacitors, use Green's reciprocation theorem (see Exercise 2.7.2) to prove

$$C_{ij} = C_{ji},$$

for any two capacitors i and j $(i \neq j)$.
(b) Show the same result from (2.166) of the text.
(c) Show $C_{ii} > 0$ from first principles.

Exercise 2.12.3. Using $Q_i = \sum_j C_{ij}V_j$ and the definition of the C_{ij}, find the six independent coefficients of capacitance for three co-centered spherical conducting shells having radii a, b and c $(a < b < c)$. [Call these $C_{aa}, C_{bb}, C_{cc}, C_{ab}, C_{ac}$ and C_{bc}.] Verify the result of Exercise 2.12.1 applied to the three inductances C_{ab}, C_{ac}, C_{bc}. Also verify that the two sums in Equations (2.171) (for $j = a, b$) and the sum in (2.178) are all correct for this system of capacitors.

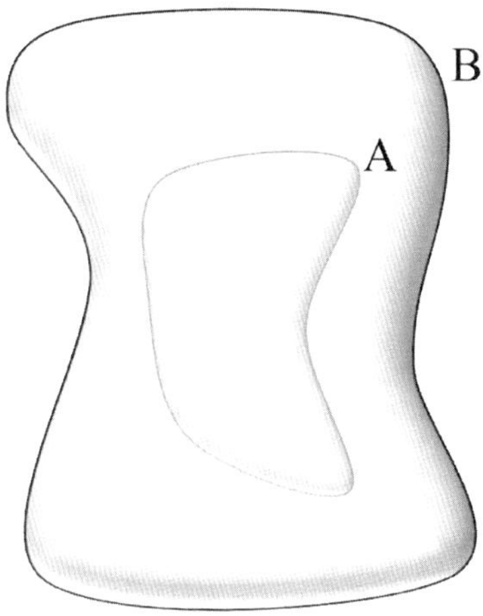

Fig. 2.22 Conductors in Exercise 2.12.4.

Exercise 2.12.4. Given a system of two conductors of arbitrary shape, where conductor B is hollow and contains conductor A (see Figure 2.22), show that the capacitances satisfy $C_{BB} > C_{AA}$.

Exercise 2.12.5.

(a) Consider two arbitrary closed conductors A and B in free space with charges Q and $-Q$, respectively. Define

$$C \equiv \frac{Q}{V_A - V_B},$$

where V_A and V_B are the potentials of conductors A and B. Show that

$$C = \frac{\det \begin{pmatrix} C_{AA} & C_{AB} \\ C_{AB} & C_{BB} \end{pmatrix}}{\sum_{i,j} C_{ij}}.$$

(b) Show that the field energy, W, in this case is given by

$$W = \frac{Q^2}{2C}.$$

Exercise 2.12.6. Consider an isolated conductor A, with capacitance C_A. Let C'_{AA} be the capacitance of A when a second conductor B is introduced in the vicinity of A. Show that the introduction of the second conductor B raises the capacitance of the first conductor, so that $C'_{AA} > C_A$. [*Hint*: The results from Exercises 2.10.1 and 2.12.1 are helpful.]

Exercise 2.12.7. Consider two identical thin co-axial circular conducting parallel plates of radius R situated a distance d apart. Label the plates as A and B. Evaluate the *leading order* expression for the system capacitance, $C(d)$, defined in Exercise 2.12.5, for

(a) the case $d \ll R$;
(b) the case $d \gg R$. Note that $C_{cir} = 2R/\pi$ for the isolated capacitance of a thin circular disk.

Exercise 2.12.8. Consider again the two identical thin co-axial circular conducting parallel plates of radius R situated a distance d apart. Given that the isolated capacitance, C_{cir}, of one of the plates is given by $2R/\pi$, show that the system capacitance, $C(d)$, defined in Exercise 2.12.5, is approximately given by

$$C(d) \approx \frac{R}{\pi}\left(1 + \frac{2R}{\pi d}\right),$$

for $d \gg R$. [*Hint*: When the capacitors are very far apart, they appear to be point charges to each other.]

Exercise 2.12.9. Two *arbitrarily shaped* conductors are located a great distance $d \gg$ (*any internal distance*) from one another. The conductors have *isolated* (non-interacting) capacitances C^0_{11} and C^0_{22} ($C^0_{11} \neq C^0_{22}$ in general).

(a) Present an argument that the approximate induction coefficient of the combined system, $C_{12}(d) = C_{21}(d)$, is given by

$$C_{12}(d) \approx -\frac{C^0_{11}C^0_{22}}{d}.$$

[We will see this result is consistent with the conducting spheres in Exercise 3.3.4(a).]

(b) Using part (a) present an argument that the approximate interacting capacitance $C_{22}(d)$ is given approximately by

$$C_{22}(d) \approx \frac{C_{22}^0}{1 + \dfrac{C_{12}(d)}{d}}.$$

(c) Apply these results to two identical thin co-axial circular conducting parallel plates separated by the distance d, and re-derive the system capacitance result of Exercise 2.12.8.

Exercise 2.12.10.

(a) A volume V in vacuum is bounded by a surface S consisting of several separate conducting surfaces S_i. One conductor is held at *unit* potential and all the other conductors are at zero potential. From Equations (2.147) and (2.181) we have that the capacitance of the one conductor is given by

$$C = \frac{1}{4\pi} \int_V d^3x \, |\vec{\nabla}\Phi|^2,$$

where Φ is the solution for the potential. Now show that the actual capacitance is always less than or equal to the quantity

$$C[\Psi] \equiv \frac{1}{4\pi} \int_V d^3x \, |\vec{\nabla}\Psi|^2,$$

where Ψ is any trial function satisfying the potential boundary conditions.

(b) Use the trial function for a thin cylindrical capacitor of length L, with inner radius a and outer radius b,

$$\Psi = \frac{b - \rho}{b - a},$$

where ρ is the radial distance in cylindrical coordinates, to compute the approximate capacitance. Compare with the exact answer

$$C = \frac{L}{2\ln(b/a)} \approx 0.721L,$$

when $b = 2a$. (*Comment*: Variational methods which fix the potential boundary conditions give *upper limits* to the capacitance, whereas methods which fix the charge (see upcoming Section 3.10) give *lower limits*.)

Exercise 2.12.11. A system of three long parallel cylindrical conductors of common radius a have charge per length $-\lambda$, $-\lambda$, and 2λ on surfaces 1, 2 and 3, respectively (see Figure 2.23). The long cylinders form an equilateral triangle with distance d between their centers. Assume $d \gg a$.

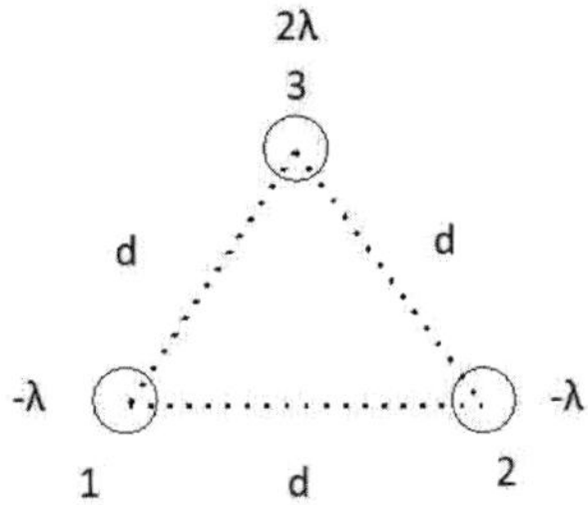

Fig. 2.23 Conductors in Exercise 2.12.11.

(a) Defining the capacitance per unit length C_L as

$$C_L \equiv \frac{\lambda}{2V_3 - V_1 - V_2},$$

show that the system energy per unit length w can be written in standard form as

$$w = \frac{\lambda^2}{2\,C_L}.$$

(b) Show that the capacitance per unit length C_L may be written as

$$C_L = \frac{1}{N \ln(d/a)}.$$

Determine N.

Chapter 3

Boundary Value Problems in Electrostatics

In this chapter we will introduce several useful techniques for solving electrostatic boundary-value problems, including method of images, reduced Green functions, expansion in orthogonal functions, variational methods, and conformal mapping. In some cases we will solve the same problem by multiple methods, in order to highlight the applicability and relative advantages of each method.

3.1 Conducting plane: Green functions and method of images

In three dimensions the free Green function is

$$G_f(\vec{x}, \vec{x}\,') = \frac{1}{|\vec{x} - \vec{x}\,'|}. \tag{3.1}$$

Now consider the situation of a unit point charge near an uncharged conducting plane located at $z = 0$, as illustrated in Figure 3.1. The potential associated with the electric field to the right of the conducting plane in Figure 3.1 is the Dirichlet Green function for the conducting plane, $G_D(\vec{x}, \vec{x}\,')$. It will be a combination of the free potential or Green function and a potential coming from the induced charges on the surface of the conductor.

We showed in Section 1.4 that the electric field at the surface of the conducting plane is perpendicular to the plane (so that the plane is at constant potential). However, the same is true for the electric field produced by the two unit charges shown in Figure 3.2: The plane is at zero potential because no work is done by moving a charge in from ∞ along the plane.

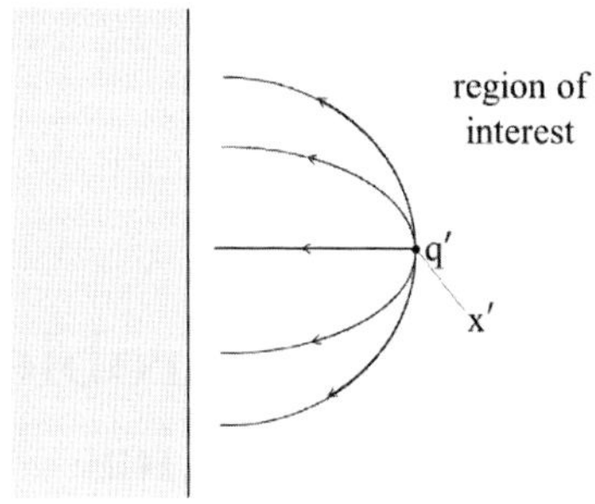

Fig. 3.1 Unit point charge and conducting plane.

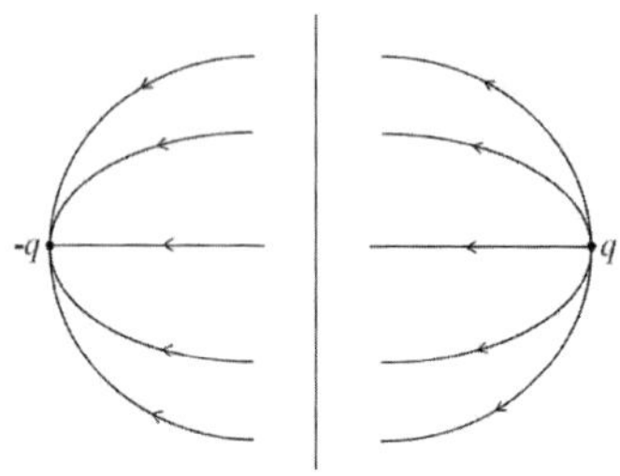

Fig. 3.2 Point charge and image charge.

By uniqueness (see Section 2.7)

$$G_D(\vec{x},\vec{x}\,') = \frac{1}{|\vec{x}-\vec{x}\,'|} - \frac{1}{|\vec{x}-\vec{x}\,''|}, \tag{3.2}$$

where $\vec{x}$ is the field point and $(z' > 0)$

$$\vec{x}\,' = x'\hat{\imath} + y'\hat{\jmath} + z'\hat{k} \text{ is the source point} \tag{3.3}$$

$$\vec{x''} = x''\hat{\imath} + y''\hat{\jmath} - z'\hat{k} \text{ is the image point.} \tag{3.4}$$

The reader may verify from (3.2) that $G_D(\vec{x},\vec{x}\,')|_{z=0} = 0$ and that $G_D(\vec{x},\vec{x}\,') = G_D(\vec{x}\,',\vec{x})$. The image method can also be used to establish the Neumann Green function for the same geometry as

$$G_N(\vec{x},\vec{x}\,') = \frac{1}{|\vec{x}-\vec{x}\,'|} + \frac{1}{|\vec{x}-\vec{x}\,''|}. \tag{3.5}$$

We may verify Neumann boundary condition:

$$\left.\frac{\partial G_N(\vec{x},\vec{x}\,')}{\partial n}\right|_{z=0} = 0.$$

In Section 1.5 we presented two-dimensional electrodynamics as describing three-dimensional systems in which sources are required to be line charges parallel to the z-axis. In this interpretation, the z-integrals in Equation (2.97) get absorbed into a new two-dimensional Green function. For example, in free space with a charge density of length L along the z-axis given by

$$\rho(\vec{x}) = \begin{cases} \sigma(\vec{x}_\perp), & |z| < \frac{L}{2}, \\ 0, & |z| > \frac{L}{2}, \end{cases} \tag{3.6}$$

one may write exactly

$$\int d^3x' \, \frac{\sigma(\vec{x}\,'_\perp)}{|\vec{x} - \vec{x}\,'|} \equiv \int d^2x' \, \sigma(\vec{x}\,'_\perp) I(\vec{x}_\perp, \vec{x}\,'_\perp), \tag{3.7}$$

where

$$\begin{aligned} I(\vec{x}_\perp, \vec{x}\,'_\perp) &= \int_{-L/2}^{L/2} \frac{dz'}{\sqrt{(\vec{x}_\perp - \vec{x}\,'_\perp)^2 + (z - z')^2}} \\ &= \ln\left(\frac{\frac{L}{2} - z + \sqrt{(\frac{L}{2} - z)^2 + (\vec{x}_\perp - \vec{x}\,'_\perp)^2}}{-\frac{L}{2} - z + \sqrt{(\frac{L}{2} + z)^2 + (\vec{x}_\perp - \vec{x}\,'_\perp)^2}} \right). \end{aligned} \tag{3.8}$$

However, in the region where $|z|, |\vec{x}_\perp - \vec{x}\,'_\perp| \ll L$ we have

$$I(\vec{x}_\perp, \vec{x}\,'_\perp) \longrightarrow -2\ln\left(\frac{|\vec{x}_\perp - \vec{x}\,'_\perp|}{L} \right). \tag{3.9}$$

The result is in agreement with the 2-D point-source potential found in Exercise 2.4.1, and gives a physical interpretation to the length K there. We therefore use

$$\Phi(\vec{x}_\perp) = \int d^2x' \, \sigma(\vec{x}\,'_\perp) G_f(\vec{x}_\perp, \vec{x}\,'_\perp), \tag{3.10}$$

$$G_f(\vec{x}_\perp, \vec{x}\,'_\perp) \equiv -2\ln\left(\frac{|\vec{x}_\perp - \vec{x}\,'_\perp|}{L} \right). \tag{3.11}$$

These are approximately valid in the limited spatial range, even though $G_f(\vec{x}_\perp, \vec{x}\,'_\perp)$ does not vanish for $|\vec{x}_\perp| \to \infty$. Indeed, 2-D analogs of Equations (2.98) and (2.103) may be defined in the same manner for analogous geometries. For example, with a conductor interface in (x, y) space defined by $x = 0$ and a source at $\vec{x}\,'_\perp = (x', y')$, we again have from the image method

$$G_D(\vec{x}_\perp, \vec{x}\,'_\perp) = -2\ln\left(\frac{|\vec{x}_\perp - \vec{x}\,'_\perp|}{|\vec{x}_\perp - \vec{x}\,''_\perp|} \right), \tag{3.12}$$

$$G_N(\vec{x}_\perp, \vec{x}\,'_\perp) = -2\ln\left(\frac{|\vec{x}_\perp - \vec{x}\,'_\perp||\vec{x}_\perp - \vec{x}\,''_\perp|}{L^2} \right), \tag{3.13}$$

where $\vec{x}\,''_\perp = (-x', y')$. These are the 2-D analogs of Equations (3.2) and (3.5).

3.2 Reduced Green function technique applied to flat conductor

We may solve this same three-dimensional problem from an alternative point of view by treating directly the Green function equation

$$\nabla^2 G_D(\vec{x},\vec{x}\,') = -4\pi\delta(\vec{x}-\vec{x}\,'), \tag{3.14}$$

subject to the boundary condition

$$G_D(\vec{x},\vec{x}\,')\big|_{z=0} = 0. \tag{3.15}$$

In order to gain some experience, let us first solve (3.14) in the easier case where the point charge is in free space. Of course we already know the solution to (3.14) is Equation (3.1). This can also be derived directly using the method of Fourier transforms, already encountered in Section 2.2. The three-dimensional Fourier transform of (3.14) with respect to $\vec{x}-\vec{x}\,'$ is:

$$\begin{aligned}\int d^3x\,\nabla^2 G_f(\vec{x},\vec{x}\,')e^{-i\vec{k}\cdot(\vec{x}-\vec{x}\,')} &= -4\pi\int d^3x\,\delta(\vec{x}-\vec{x}\,')e^{-i\vec{k}\cdot(\vec{x}-\vec{x}\,')}\\ &= -4\pi. \end{aligned}\tag{3.16}$$

Green's second identity can be used on the left-hand side to obtain

$$-k^2G(k) = -4\pi, \tag{3.17}$$

where $k\equiv|\vec{k}|$ and $G(k)$ is the Fourier transform of $G_f(\vec{x},\vec{x}\,')$:

$$G(k) \equiv \int d^3x\,G_f(\vec{x},\vec{x}\,')e^{-i\vec{k}\cdot(\vec{x}-\vec{x}\,')}. \tag{3.18}$$

Note that $G(k)$ is isotropic: this is a consequence of the fact that $G_f(\vec{x},\vec{x}\,')$ depends only on $\vec{x}-\vec{x}\,'$, which in turn reflects the fact that free space is translationally invariant. We then solve for $G_f(\vec{x},\vec{x}\,')$ by taking the inverse Fourier transform:

$$\begin{aligned}G_f(\vec{x},\vec{x}\,') &= \int\frac{d^3k}{(2\pi)^3}G(k)e^{i\vec{k}\cdot(\vec{x}-\vec{x}\,')}\\ &= \frac{4\pi}{(2\pi)^3}\int d\phi\,d\cos\theta\;k^2dk\frac{e^{ik|\vec{x}-\vec{x}\,'|\cos\theta}}{k^2}\\ &= \frac{1}{\pi}\int dk\int_{-1}^{1}d\cos\,\theta e^{ik|\vec{x}-\vec{x}\,'|\cos\theta} &(3.19)\\ &= \frac{2}{\pi|\vec{x}-\vec{x}\,'|}\int_0^\infty dk\,\frac{\sin k|\vec{x}-\vec{x}\,'|}{k}. &(3.20)\end{aligned}$$

Using the result

$$\int_0^\infty dx \, \frac{\sin x}{x} = \frac{\pi}{2}, \tag{3.21}$$

which can be obtained by contour integration, we obtain the free Green function of Equation (3.1). An alternative derivation is given in Exercise 3.2.1.

Let us return to our half-plane geometry. The geometry here is translationally invariant also, but only in the x and y directions; this means that $G_D(\vec{x}, \vec{x}\,')$ is a function of z, z' and $(\vec{x} - \vec{x}\,')_\perp \equiv (x - x', y - y', 0)$. Taking Fourier transforms of the Green function equation (3.14) with respect to $(\vec{x} - \vec{x}\,')_\perp$, we obtain (using Green's second identity)

$$\begin{aligned}
&\int d^2(\vec{x} - \vec{x}\,')_\perp \, \nabla^2 G_D(\vec{x}, \vec{x}\,') e^{-i\vec{k}\cdot(\vec{x}-\vec{x}\,')_\perp} \\
&= -4\pi \int d^2(\vec{x} - \vec{x}\,')_\perp \, \delta\left((\vec{x} - \vec{x}\,')_\perp\right) \delta(z - z') e^{-i\vec{k}\cdot(\vec{x}-\vec{x}\,')_\perp} \\
&\Longrightarrow \left(k^2 - \frac{\partial^2}{\partial z^2}\right) g(z, z', k) = \delta(z - z'),
\end{aligned} \tag{3.22}$$

where $\vec{k} \equiv (k_x, k_y, 0), k \equiv |\vec{k}|$, and

$$g(z, z', k) \equiv \frac{1}{4\pi} \int d^2(\vec{x} - \vec{x}\,')_\perp \, G_D(\vec{x}, \vec{x}\,') e^{-i\vec{k}\cdot(\vec{x}-\vec{x}\,')_\perp}. \tag{3.23}$$

The function $g(z, z', k)$ is called the *reduced Green function*. The boundary conditions on $G_D(\vec{x}, \vec{x'})$ lead to the condition

$$g(z, z', k)\big|_{z=0} = 0. \tag{3.24}$$

Equation (3.22) is an ordinary differential equation in the variable z. For $z \neq z'$ we have

$$\left[k^2 - \frac{\partial^2}{\partial z^2}\right] g(z, z', k) = 0, \tag{3.25}$$

which has solutions of the form

$$g(z, z', k) = C(z') e^{\pm kz}.$$

A physical solution must be bounded as $z \to \infty$, so

$$g(z, z', k) = \begin{cases} C_1 e^{-kz} & z > z', \\ C_2 e^{kz} + C_3 e^{-kz} & z < z'. \end{cases} \tag{3.26}$$

The boundary condition (3.24) gives $C_2 = -C_3$. Another condition on the constants is obtained from the continuity of g at $z = z'$:

$$C_1 e^{-kz'} = C_2 e^{kz'} + C_3 e^{-kz'} = C_2 \left(e^{kz'} - e^{-kz'} \right). \tag{3.27}$$

Furthermore, integrating (3.22) across $z = z'$ gives

$$\begin{aligned} \int_{z'-}^{z'+} dz \left[k^2 - \frac{\partial^2}{\partial z^2} \right] g(z, z', k) &= \int_{z'-}^{z'+} dz\, \delta(z - z') \\ \Longrightarrow \; -\frac{\partial}{\partial z} g(z, z', k) \Big|_{z'-}^{z'+} &= 1 \\ \Longrightarrow \; C_1 k e^{-kz'} + C_2 k (e^{kz'} + e^{-kz'}) &= 1. \end{aligned} \tag{3.28}$$

Solving (3.27) and (3.28) gives

$$C_1 = \frac{1}{2k}(e^{kz'} - e^{-kz'}); \qquad C_2 = \frac{1}{2k} e^{-kz'}. \tag{3.29}$$

Plugging the constants into (3.26) and simplifying gives the expression (valid for $z, z' > 0$)

$$g(z, z', k) = \frac{1}{2k}(e^{-k|z-z'|} - e^{-k|z+z'|}). \tag{3.30}$$

It follows from the definition of $g(z, z', k)$ in (3.23) and the Fourier inversion formula that

$$\begin{aligned} G_D(\vec{x}, \vec{x}\,') &= \int \frac{d^2k}{(2\pi)^2} e^{i\vec{k}\cdot(\vec{x}-\vec{x}\,')_\perp} 4\pi g(z, z', k) \\ &= \int \frac{d^2k}{(2\pi)^2} e^{i\vec{k}\cdot(\vec{x}-\vec{x}\,')_\perp} \frac{4\pi}{2k}(e^{-k|z-z'|} - e^{-k|z+z'|}). \end{aligned} \tag{3.31}$$

We may break the integral into the sum of two terms, the first of which is ($\rho \equiv |\vec{x} - \vec{x}\,'|_\perp$)

$$I_1 \equiv \frac{4\pi}{2(2\pi)^2} \int_0^{2\pi} d\phi \int_0^\infty dk\, e^{ik\rho\cos\phi} e^{-k|z-z'|}, \tag{3.32}$$

which simplifies to

$$I_1 = \frac{2|z - z'|}{\pi} \int_0^{\pi/2} d\phi \frac{1}{\rho^2 \cos^2\phi + (z - z')^2}. \tag{3.33}$$

The indefinite integral may be found via integral tables or computer algebra:

$$\begin{aligned} I_1 &= \frac{2|z - z'|}{\pi} \left[\frac{1}{|z - z'|\sqrt{\rho^2 + (z - z')^2}} \arctan\left(\frac{|z - z'| \tan\phi}{\sqrt{\rho^2 + (z - z')^2}} \right) \right] \Bigg|_0^{\pi/2} \\ &= \frac{1}{\sqrt{\rho^2 + (z - z')^2}}. \end{aligned} \tag{3.34}$$

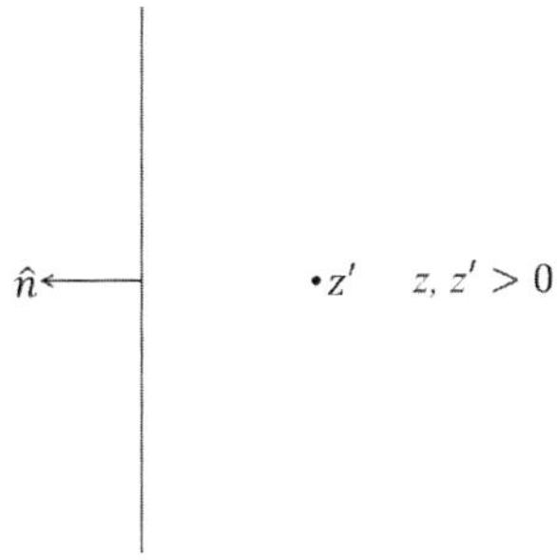

Fig. 3.3 Notation for Equation (3.37).

The second term in (3.31) is identical to I_1 with $z' \to -z'$. Recalling that $\rho = |(\vec{x} - \vec{x}\,')_\perp|$, we obtain

$$G_D(\vec{x}, \vec{x}\,') = \frac{1}{|\vec{x} - \vec{x}\,'|} - \frac{1}{|\vec{x} - \vec{x}\,''|}, \tag{3.35}$$

where $\vec{x}\,'' \equiv (x', y', -z')$. Explicitly

$$\begin{aligned} G_D(\vec{x}, \vec{x}\,') = &\frac{1}{[(x - x')^2 + (y - y')^2 + (z - z')^2]^{1/2}} \\ &- \frac{1}{[(x - x')^2 + (y - y')^2 + (z + z')^2]^{1/2}}. \end{aligned} \tag{3.36}$$

From (3.36) we may confirm the presence of the surface delta function mentioned in Section 2.8. With reference to Figure 3.3, we have

$$\begin{aligned} \left.\frac{\partial G_D}{\partial n}\right|_{S(\vec{x})} &= -\left.\frac{\partial G_D}{\partial z}\right|_{z=0} \\ &= \left\{ \frac{(z - z')}{[(x - x')^2 + (y - y')^2 + (z - z')^2]^{3/2}} \right. \\ &\qquad \left. - \frac{(z + z')}{[(x - x')^2 + (y - y')^2 + (z + z')^2]^{3/2)}} \right\}_{z=0} \\ &= -\frac{2z'}{[\rho^2 + z'^2]^{3/2}}. \end{aligned} \tag{3.37}$$

From Exercise 2.2.2(b) we have

$$\lim_{z' \to 0^+} \frac{1}{2\pi} \frac{z'}{((\vec{x}_\perp)^2 + (z')^2)^{3/2}} = \delta^{(2)}(\vec{x}_\perp), \tag{3.38}$$

which leads to

$$\lim_{z' \to 0^+} \frac{1}{4\pi} \left.\frac{\partial G_D}{\partial n}\right|_{z=0} = -\delta^{(S)}(\vec{x}_\perp - \vec{x}\,'_\perp). \tag{3.39}$$

This derivation suggests a physical interpretation for the surface delta function. Recall that $G_D(\vec{x},\vec{x}\,')$ is the potential at $\vec{x}$ due to a point charge at $\vec{x}\,'$, so $-\vec{\nabla}G_D(\vec{x},\vec{x}\,')$ is the corresponding electric field. We know from Equation (2.92) that the normal electric field at $z=0$ is proportional to the surface charge:

$$\left.\frac{\partial G_D}{\partial n}\right|_{z=0} = 4\pi\sigma_{\text{surface}}. \tag{3.40}$$

Comparing (3.39) and (3.40) and taking the magnitude and sign of the charge into account indicates that in the limit as $z' \to 0^+$, the surface delta function can be thought of as arising from the image charge coming to the surface. Of course the image charge is not real, and (3.37) actually indicates that half of the surface charge contribution comes from each term in (3.35).

Alternatively, we may derive the surface delta function from the Fourier integral representation (3.31):

$$G_D(\vec{x},\vec{x}\,') = 4\pi \int \frac{d^2k}{(2\pi)^2} e^{i\vec{k}\cdot(\vec{x}-\vec{x}\,')_\perp} \frac{1}{2k}\left(e^{-k|z-z'|} - e^{-k|z+z'|}\right).$$

From this expression we may compute (when $z'>0$)

$$-\left.\frac{\partial G_D(\vec{x},\vec{x}\,')}{\partial z}\right|_{z=0} = 4\pi \int \frac{d^2k}{(2\pi)^2} e^{i\vec{k}\cdot(\vec{x}-\vec{x}\,')_\perp} \frac{1}{2}\left(-e^{-kz'} - e^{-kz'}\right), \tag{3.41}$$

and taking limits as $z' \to 0^+$ gives

$$\begin{aligned}\lim_{z'\to 0^+} \frac{1}{4\pi} \left.\frac{\partial G_D(\vec{x},\vec{x}\,')}{\partial n}\right|_{z=0} &= -\int \frac{d^2k}{(2\pi)^2} e^{i\vec{k}\cdot(\vec{x}-\vec{x}\,')_\perp} \\ &= -\delta^{(S)}(\vec{x}_\perp - \vec{x}\,'_\perp),\end{aligned} \tag{3.42}$$

where we have used (2.33) generalized to two dimensions. The result agrees with (3.39). The Neumann Green function for this geometry (3.5) can similarly be verified using the reduced Green function method.

3.3 Method of images: conducting sphere[1]

The Dirichlet Green function outside a conducting sphere may also be found straightforwardly via the method of images. With reference to Figure 3.4,

[1]When treating situations with spherical symmetry, we shall typically use $\vec{r}$ instead of $\vec{x}$ to denote position vectors.

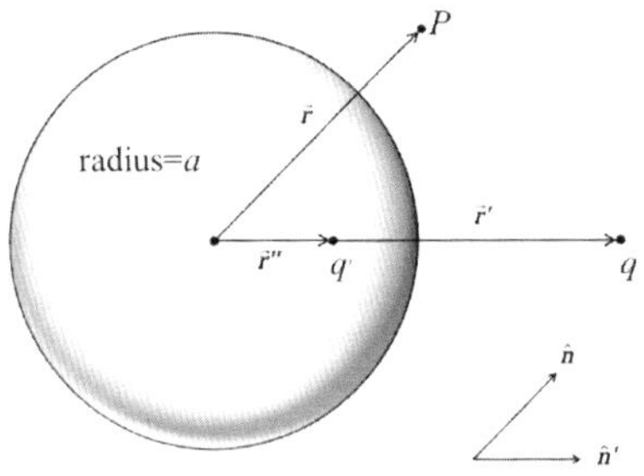

Fig. 3.4 Notation for Green function computation.

we seek an image charge q' located at $\vec{r}''$ such that the potential $\Phi^{\text{sphere}}(\vec{r})$ of a point charge q at $\vec{r}$ may be written as

$$\Phi^{\text{sphere}}(\vec{r}) = \frac{q}{|\vec{r}-\vec{r}'|} + \frac{q'}{|\vec{r}-\vec{r}''|}, \tag{3.43}$$

which can be rewritten using the notation of Figure 3.4 as

$$\Phi^{\text{sphere}}(\vec{r}) = \frac{q}{|r\hat{n}-r'\hat{n}'|} + \frac{q'}{|r\hat{n}-r''\hat{n}'|}. \tag{3.44}$$

For a grounded sphere we want

$$\Phi^{\text{sphere}}(\vec{r})\Big|_{r=a} = \frac{q}{a|\hat{n}-\frac{r'}{a}\hat{n}'|} + \frac{q'}{r''|\hat{n}'-\frac{a}{r''}\hat{n}|} = 0, \tag{3.45}$$

which can be achieved by setting

$$\frac{q}{a} = -\frac{q'}{r''}, \ \frac{r'}{a} = \frac{a}{r''} \implies q' = -\frac{a}{r'}q, \ r'' = \frac{a^2}{r'} \tag{3.46}$$

$$\implies \Phi^{\text{sphere}}(\vec{r}) = \frac{q}{|\vec{r}-\vec{r}'|} - \frac{qa}{r'|\vec{r}-\frac{a^2}{r'^2}\vec{r}'|}. \tag{3.47}$$

The Green function for this situation now requires $q = 1$ in $\Phi^{\text{sphere}}(\vec{r})$. Thus explicitly,

$$G_D(\vec{r},\vec{r}') = \frac{1}{(r^2+r'^2-2rr'\cos\gamma)^{1/2}} - \frac{1}{(\frac{r^2r'^2}{a^2}+a^2-2rr'\cos\gamma)^{1/2}}, \tag{3.48}$$

where, using the usual spherical polar θ and azimuthal ϕ angles,

$$\cos\gamma \equiv \hat{n}\cdot\hat{n'} = \cos\theta\cos\theta' + \sin\theta\sin\theta'\cos(\phi-\phi'). \tag{3.49}$$

From (3.48) we can see explicitly that $G_D(\vec{r},\vec{r}') = G_D(\vec{r}',\vec{r})$. We can also show that the gradient on the surface (*inward* normal to the sphere) is given by

$$\frac{\partial G_D}{\partial n}\Bigg|_{r=a} = -\frac{\partial G_D}{\partial r}\Bigg|_{r=a} = -\frac{(r'^2-a^2)}{a(r'^2+a^2-2ar'\cos\gamma)^{3/2}}. \tag{3.50}$$

As in the flat conductor case discussed in Section 3.2, it is possible to show that

$$\lim_{r'\to a^+} -\frac{1}{4\pi}\frac{\partial G_D}{\partial n}\bigg|_{S(\vec{r})} = \delta^{(S)}(\vec{r}\,' - \vec{r}), \tag{3.51}$$

as follows. It is not difficult to see from (3.50) that

$$\frac{\partial G_D}{\partial n}\bigg|_{r=a} \to 0 \qquad \text{when } r' \to a \text{ and } \gamma \neq 0,$$

and taking the z-axis along $\vec{r}\,'$ and using $d\Omega = \sin\gamma\, d\gamma\, d\phi$, we have

$$\begin{aligned}\oint_{r=a} \frac{\partial G_D}{\partial n} da &= \frac{-a(r'^2 - a^2)2\pi}{2ar'} \int_0^\pi \frac{2ar' \sin\gamma d\gamma}{(r'^2 + a^2 - 2ar' \cos\gamma)^{3/2}}, \\ &= \frac{-\pi(r'^2 - a^2)}{r'} \frac{-2}{(r'^2 + a^2 - 2ar'\cos\gamma)^{1/2}}\bigg|_0^\pi, \\ &= 2\pi\frac{(r'^2 - a^2)}{r'}\left\{\frac{1}{r'+a} - \frac{1}{r'-a}\right\}, \\ &= \frac{-4\pi a}{r'}, \end{aligned} \tag{3.52}$$

which goes to -4π as $r' \to a^+$, thus completing the verification of (3.51). We will show this more simply again in Section 4.12, using the reduced Green function approach.

The image method also works for obtaining the Green function in cylindrical geometry; see Exercise 3.3.5.

Now that we have the Green function, we can solve other problems for this geometry. In particular, consider the case of a sphere held at fixed potential V in the presence of a charge at $\vec{r}\,'$, shown in Figure 3.5. The

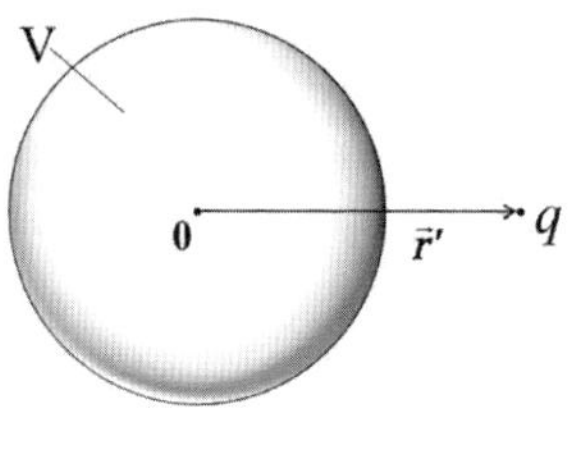

Fig. 3.5 Dirichlet problem outside sphere.

usual Green function expression (2.98) gives

$$\begin{aligned}\Phi(\vec{r}) &= \int d^3x''\, G_D(\vec{r},\vec{r}\,'')\rho(\vec{r}\,'') - \frac{1}{4\pi}\oint_S da''\Phi(\vec{r}\,'')\frac{\partial G_D(\vec{r},\vec{r}\,'')}{\partial n''} \\ &= qG_D(\vec{r},\vec{r}\,') - \frac{V}{4\pi}\oint_S da''\frac{\partial G_D(\vec{r},\vec{r}\,'')}{\partial n''} \\ &= qG_D(\vec{r},\vec{r}\,') + \frac{Va}{r},\end{aligned} \tag{3.53}$$

where (3.53) makes use of (3.52).

The last term in (3.53) has a simple interpretation as the potential due to uniformly-distributed surface charge on the sphere. The total charge on the surface due to both terms in (3.53) is given by integrating the surface charge (2.92) over the surface:

$$Q = -\frac{qa}{r'} + Va. \tag{3.54}$$

The first term on the right-hand side comes from (3.52). The charge Q in (3.54) is determined because V is given. If on the other hand the total charge Q is given, we may solve (3.54) for Va to obtain

$$\Phi(\vec{r}) = qG_D(\vec{r},\vec{r}\,') + \frac{(Q + qa/r')}{r}. \tag{3.55}$$

Another situation involving a conducting sphere is shown in Figure 3.6. Specifying zero potential on the sphere and using

$$\rho(\vec{r}) = Q\left[\delta\left((\vec{r} - (0,0,-R)\right) - \delta\left(\vec{r} - (0,0,R)\right)\right], \tag{3.56}$$

gives

$$\begin{aligned}\Phi(\vec{r}) = &\frac{Q}{(r^2+R^2+2rR\cos\theta)^{1/2}} - \frac{aQ}{R(r^2+a^4/R^2+(2a^2r/R)\cos\theta)^{1/2}} \\ &- \frac{Q}{(r^2+R^2-2rR\cos\theta)^{1/2}} + \frac{aQ}{R(r^2+a^4/R^2-(2a^2r/R)\cos\theta)^{1/2}}.\end{aligned} \tag{3.57}$$

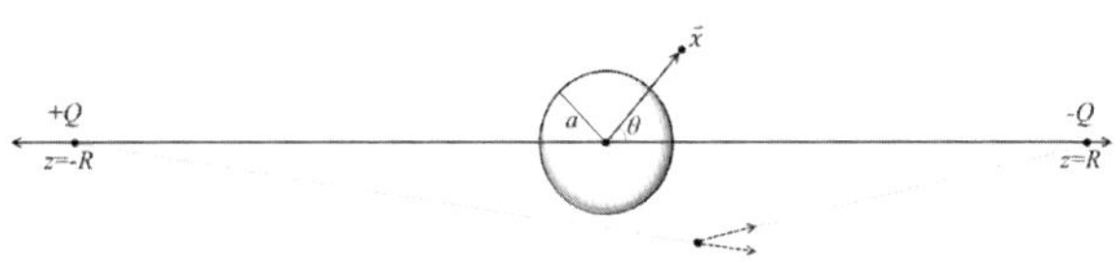

Fig. 3.6 Two distant point charges outside a conducting sphere.

In the case where $a \ll R$, we may make the following approximations valid for $R \gg r > a$,

$$\frac{1}{(R^2 + r^2 \pm 2rR\cos\theta)^{1/2}} \approx \frac{1}{R}\left(1 \mp \frac{r}{R}\cos\theta\right), \tag{3.58}$$

$$\frac{1}{(r^2 + a^4/R^2 \pm (2a^2 r/R)\cos\theta)^{1/2}} \approx \frac{1}{r}\left(1 \mp \frac{a^2}{Rr}\cos\theta\right). \tag{3.59}$$

These imply

$$\Phi(\vec{r}) \approx -2Q/R^2\left(r - \frac{a^3}{r^2}\right)\cos\theta. \tag{3.60}$$

We denote $2Q/R^2$ by E_0: it is the electric field at the origin due to the two charges. This leads to

$$\Phi(\vec{r}) = -E_0\left(r - \frac{a^3}{r^2}\right)\cos\theta = -E_0 z\left(1 - \frac{a^3}{r^3}\right). \tag{3.61}$$

In the limit as $R \to \infty$, the field due to the point charges approaches $E_0\hat{z}$ at finite distances from the sphere. Therefore (3.61) gives the potential due to a grounded conducting sphere in a uniform electric field. Of course, if we were to specify a nonzero potential on the surface, we would get an additional term Va/r as before.

3.4 Charged conducting sphere force example

We can use our result (3.61) to compute the forces on the upper and lower hemispheres (which are separated by a small gap) of an uncharged conducting sphere in a uniform electric field, as shown in Figure 3.7. The induced surface charge density is

$$\sigma = -\frac{1}{4\pi}\frac{\partial\Phi}{\partial r}\bigg|_a = \frac{3}{4\pi}E_0\cos\theta. \tag{3.62}$$

By symmetry, the force is in the z-direction. The total force on the upper hemisphere is obtained by integrating the z-projection of (2.161) over the hemisphere:

$$\begin{aligned} F_z = 2\pi\int_{\text{upper}}\sigma^2\cos\theta da &= 2\pi\int_0^{\pi/2}\int_0^{2\pi}\frac{9}{16\pi^2}E_0^2\cos^3\theta\sin\theta a^2 d\theta d\phi, \\ &= \frac{9(2\pi)^2}{16\pi^2}E_0^2 a^2\int_0^{\pi/2}\cos^3\theta\sin\theta d\theta, \\ &= \frac{9E_0^2 a^2}{16}. \end{aligned} \tag{3.63}$$

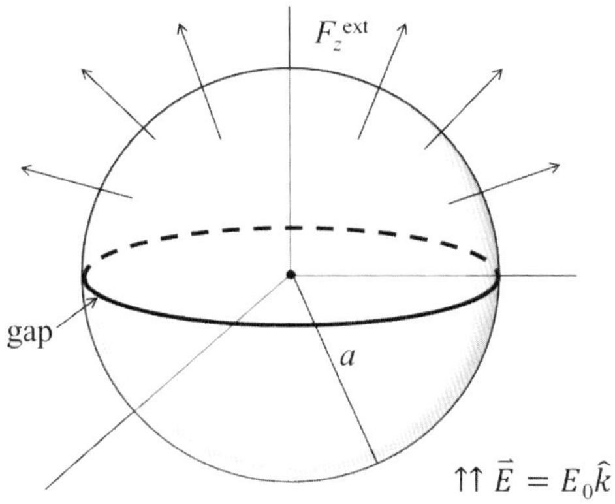

Fig. 3.7 Hemispheres (separated by a small gap) in an electric field.

Clearly the force on the lower hemisphere will be equal and opposite.

If we modify the situation so that the sphere has total charge Q, then the total forces on the two hemispheres will no longer be equal and opposite. The potential in this case is

$$\Phi = \Phi_{\text{uncharged}} + \frac{Q}{r}, \tag{3.64}$$

since the charge Q will distribute itself uniformly over a spherical equipotential surface. This gives

$$\sigma = \frac{3}{4\pi}E_0\cos\theta + \frac{1}{4\pi}\frac{Q}{a^2}, \tag{3.65}$$

so that (the forces on the upper and lower hemispheres, are denoted by F_z^u and F_z^ℓ, respectively)

$$\begin{aligned} F_z^u &= (2\pi)^2a^2\int_0^{\pi/2} d\theta\sin\theta\cos\theta\left(\frac{3}{4\pi}E_0\cos\theta + \frac{1}{4\pi}\frac{Q}{a^2}\right)^2 \\ &= \frac{9E_0^2a^2}{16} + \frac{1}{2}E_0Q + \frac{Q^2}{8a^2}, \end{aligned} \tag{3.66}$$

$$\begin{aligned} F_z^\ell &= (2\pi)^2a^2\int_{\pi/2}^{\pi} d\theta\sin\theta\cos\theta\left(\frac{3}{4\pi}E_0\cos\theta + \frac{1}{4\pi}\frac{Q}{a^2}\right)^2 \\ &= -\frac{9E_0^2a^2}{16} + \frac{1}{2}E_0Q - \frac{Q^2}{8a^2}. \end{aligned} \tag{3.67}$$

Notice that the total force (sum of forces on upper and lower hemispheres) is equal to E_0Q as expected. On the other hand, to keep the hemispheres together requires a force on *each* hemisphere equal to

$$\frac{F_z^u - F_z^\ell}{2} = \frac{9E_0^2a^2}{16} + \frac{Q^2}{8a^2}. \tag{3.68}$$

3.5 Separation of variables: conducting box

Next we shall solve the potential inside an empty rectangular box, with boundary conditions as shown in Figure 3.8. In this case, we need to solve Poisson's equation with $\rho = 0$, which is also known as *Laplace's equation*:

$$\frac{\partial^2 \Phi}{\partial x^2} + \frac{\partial^2 \Phi}{\partial y^2} + \frac{\partial^2 \Phi}{\partial z^2} = 0. \tag{3.69}$$

We apply the technique of *separation of variables*, and assume that Φ has the form

$$\Phi(x, y, z) = X(x)Y(y)Z(z). \tag{3.70}$$

Substitution and algebra yields

$$\frac{1}{X}\frac{d^2X}{dx^2} + \frac{1}{Y}\frac{d^2Y}{dy^2} + \frac{1}{Z}\frac{d^2Z}{dz^2} = 0. \tag{3.71}$$

Note that (3.71) has total derivatives instead of partials. Since each term in (3.71) involves a single variable, all terms must be constant:

$$\frac{1}{X}\frac{d^2X}{dx^2} = -\alpha^2, \frac{1}{Y}\frac{d^2Y}{dy^2} = -\beta^2, \frac{1}{Z}\frac{d^2Z}{dz^2} = \alpha^2 + \beta^2, \tag{3.72}$$

where α and β can be real or imaginary. The general solution to (3.72) is

$$X(x) = C_1 \sin \alpha x + C_2 \cos \alpha x, \tag{3.73}$$

$$Y(y) = C_3 \sin \beta y + C_4 \cos \beta y, \tag{3.74}$$

$$Z(z) = C_5 \sinh \gamma z + C_6 \cosh \gamma z, \tag{3.75}$$

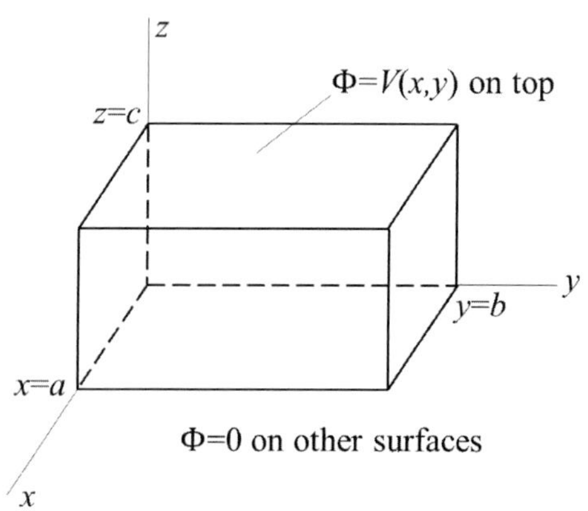

Fig. 3.8 Rectangular box with fixed potential on sides.

where

$$\gamma \equiv \sqrt{\alpha^2 + \beta^2}. \tag{3.76}$$

We use the homogeneous boundary conditions to obtain conditions on the variable-separated solutions:

$$\Phi\big|_{x=a} = 0 \implies \alpha = \frac{n\pi}{a}, n = 1, 2, 3, \ldots \tag{3.77}$$

$$\Phi\big|_{y=b} = 0 \implies \beta = \frac{m\pi}{b}, m = 1, 2, 3, \ldots \tag{3.78}$$

$$\Phi\big|_{x=0} = \Phi\big|_{y=0} = \Phi\big|_{z=0} = 0 \implies C_2 = C_4 = C_6 = 0. \tag{3.79}$$

We still have the inhomogeneous boundary condition at $z = c$ to contend with. We seek an overall solution that is a sum of the variable-separated solutions:

$$\Phi(x, y, z) = \sum_{n,m=1}^{\infty} A_{mn} \sin(\alpha_n x) \sin(\beta_m y) \sinh(\gamma_{mn} z), \tag{3.80}$$

where

$$\alpha_n \equiv \frac{n\pi}{a};\ \beta_m \equiv \frac{m\pi}{b};\ \gamma_{mn} = \pi\sqrt{\frac{n^2}{a^2} + \frac{m^2}{b^2}}. \tag{3.81}$$

Plugging $z = c$ into (3.80) gives the condition

$$V(x, y) = \sum_{n,m} A_{mn} \sin(\alpha_n x) \sin(\beta_m y) \sinh(\gamma_{mn} c). \tag{3.82}$$

Making use of the orthogonality relation

$$\int_0^a dx \, \sin\left(\frac{n\pi x}{a}\right) \sin\left(\frac{n'\pi x}{a}\right) = \frac{a}{2}\delta_{nn'}, \tag{3.83}$$

we may solve (3.82) for the constants A_{mn}:

$$A_{mn} = \frac{4}{ab\sinh(\gamma_{mn}c)} \int_0^a dx \int_0^b dy \, V(x, y) \sin(\alpha_n x) \sin(\beta_m y). \tag{3.84}$$

3.6 Eigenfunction expansion of Green function for a conducting box

The conducting box problem that we solved in the previous section, and all other electrostatic problems with this geometry, may also be solved by finding the Green function. In this section, we will introduce a strategy for finding Green functions that has wide applicability, and will be used repeatedly throughout the book. The idea is to write both the Green function and the delta function in the Green function equation (2.99) as expansions in *eigenfunctions* of the Laplacian. This expansion makes it possible to *reduce* the problem to a one-dimensional form — and the solution to the one-dimensional problem is the reduced Green function that we have seen already in Section 3.2. The eigenfunctions used in a particular situation are chosen according to the geometry and the boundary conditions.

The functions $\cos kx$ and $\sin kx$ are eigenfunctions for all k, since they satisfy

$$\nabla^2 \cos kx = -k^2 \cos kx; \qquad \nabla^2 \sin kx = -k^2 \sin kx, \tag{3.85}$$

and of course similar equations hold when x is replaced by y or z. Now when expanding the Green function as a functional series, it is expedient to use functions that satisfy the same boundary conditions. Since the Green function $G_D(\vec{x}, \vec{x}\,')$ is zero on the boundaries $x = 0$ and $x = a$, we may use the *Fourier sine series* that uses sine terms that also vanish at $x = 0$ and $x = a$. In general, the Fourier sine series of a function $f(x)$ $(0 \le x \le a)$ is given by[2]

$$f(x) = \sum_{n=1}^{\infty} b_n \sin\left(\frac{n\pi x}{a}\right) \qquad (0 \le x \le a), \tag{3.86}$$

where (using (3.83) and integrating)

$$b_n = \frac{2}{a} \int_0^a f(x) \sin\left(\frac{n\pi x}{a}\right) dx\,. \tag{3.87}$$

According to Dirichlet's convergence criterion, if $f(0) = f(a) = 0$ and $f(x)$ is continuous in $0 \le x \le a$ with piecewise continuous first and second derivatives, then the expansion (3.86) converges for all $0 \le x \le a$.

The Dirichlet Green function equation can be written as

$$\nabla^2 G_D(\vec{x}, \vec{x}\,') = -4\pi\delta(x - x')\delta(y - y')\delta(z - z'), \tag{3.88}$$

with homogeneous boundary conditions at the boundary of the box. For the delta function $\delta(x - x')$ with source point $0 < x' < a$, we may easily find

[2]Note that the Fourier sine series is *not* the same as selecting only the sine terms from a conventional Fourier sine-and-cosine series on the same interval; this selected series would not converge to the original function. However, the Fourier sine series of $f(x)$ on $[0, a]$ is identical to the conventional Fourier series of the odd extension of $f(x)$ to the interval $[-a, a]$.

the Fourier sine series coefficients using (3.87), leading to the expansion

$$\delta(x-x') = \frac{2}{a}\sum_{n=1}^{\infty} \sin\left(\frac{n\pi x}{a}\right) \sin\left(\frac{n\pi x'}{a}\right) \qquad (0 \le x \le a,\ 0 < x' < a). \tag{3.89}$$

(Note that this expansion is not necessarily valid outside the interval $0 \le x \le a$ — we will return to this point later.) We may then rewrite (3.88) as

$$\nabla^2 G_D(\vec{x}, \vec{x}\,') = -\frac{16\pi}{ab} \sum_{n,m=1}^{\infty} \sin\left(\frac{n\pi x}{a}\right) \sin\left(\frac{n\pi x'}{a}\right) \times \sin\left(\frac{m\pi y}{b}\right) \sin\left(\frac{m\pi y'}{b}\right) \delta(z-z'). \tag{3.90}$$

In accordance with our strategy, we write the Green function also as an expansion in the same eigenfunctions in the x and y dimensions:

$$G_D(\vec{x}, \vec{x}\,') = \frac{16\pi}{ab} \sum_{m,n} \sin\left(\frac{n\pi x}{a}\right) \sin\left(\frac{n\pi x'}{a}\right) \times \sin\left(\frac{m\pi y}{b}\right) \sin\left(\frac{m\pi y'}{b}\right) g_{nm}(z,z'). \tag{3.91}$$

Plugging this form into (3.90) yields

$$\left[\frac{n^2\pi^2}{a^2} + \frac{m^2\pi^2}{b^2} - \frac{\partial^2}{\partial z^2}\right] g_{nm}(z,z') = \delta(z-z'). \tag{3.92}$$

This is the reduced Green function for this problem. For each fixed z', (3.92) can be considered an ordinary differential equation for the function $g_{nm}(z,z')$ in the variable z that can be solved by conventional methods. The solution can be written in the form

$$g_{nm}(z,z') = \begin{cases} C_1(z')\sinh\gamma_{nm}z + C_2(z')\cosh\gamma_{nm}z & \text{if } z \le z' \\ C_3(z')\sinh\gamma_{nm}(z-c) + C_4(z')\cosh\gamma_{nm}(z-c) & \text{if } z \ge z' \end{cases} \tag{3.93}$$

where

$$\gamma_{nm} \equiv \pi\sqrt{(n/a)^2 + (m/b)^2}. \tag{3.94}$$

The homogeneous boundary conditions give $C_2 = C_4 = 0$, while the continuity and delta function conditions give

$$g_{nm}(z,z')\Big|_{z'-}^{z'+} = 0, \tag{3.95}$$

$$-\frac{\partial}{\partial z} g_{nm}(z,z')\Big|_{z'-}^{z'+} = 1. \tag{3.96}$$

The solution is

$$g_{nm}(z,z') = \frac{\sinh(\gamma_{nm} z_<)\sinh[\gamma_{nm}(c-z_>)]}{\gamma_{nm}\sinh(\gamma_{nm}c)}, \tag{3.97}$$

where $z_< \equiv \min(z,z')$ and $z_> \equiv \max(z,z')$. The complete Dirichlet Green function is

$$G_D(\vec{x},\vec{x}\,') = \frac{16\pi}{ab}\sum_{m,n}\sin\left(\frac{n\pi x}{a}\right)\sin\left(\frac{n\pi x'}{a}\right)\sin\left(\frac{m\pi y}{b}\right)\sin\left(\frac{m\pi y'}{b}\right)$$
$$\times\,\frac{\sinh(\gamma_{nm} z_<)\sinh[\gamma_{nm}(c-z_>)]}{\gamma_{nm}\sinh(\gamma_{nm}c)}. \tag{3.98}$$

Since there is no charge inside the box, the solution for Φ is given by the surface integral term in (2.98), applied to the boundary conditions shown in Figure 3.8:

$$\Phi(\vec{x}) = -\frac{1}{4\pi}\int dx'dy'V(x',y')\frac{\partial G_D(\vec{x},\vec{x}\,')}{\partial n'}\bigg|_{z'=c} \tag{3.99}$$

$$= -\frac{4}{ab}\sum_{n,m}\int dx'dy'V(x',y')$$
$$\times\sin\left(\frac{n\pi x'}{a}\right)\sin\left(\frac{n\pi x}{a}\right)\sin\left(\frac{m\pi y'}{b}\right)\sin\left(\frac{m\pi y}{b}\right)\frac{\partial g_{nm}(z',z)}{\partial z'}\bigg|_{z'=c}. \tag{3.100}$$

For $z' > z$ we have $z = z_<$ and $z' = z_>$ so that from (3.97)

$$\frac{\partial g_{nm}(z,z')}{\partial z'}\bigg|_{z'=c} = -\frac{\sinh(\gamma_{nm}z)}{\sinh(\gamma_{nm}c)}. \tag{3.101}$$

Plugging (3.101) into (3.100) gives the same result that was obtained in Section 3.5 (see (3.80) and (3.84)).

This solution via reduced Green function is closely related to the method of images. Consider the function

$$Q_a(x,x') \equiv \frac{2}{a}\sum_{n=1}^{\infty}\sin\left(\frac{\pi n x}{a}\right)\sin\left(\frac{\pi n x'}{a}\right), \qquad (-\infty < x < \infty) \tag{3.102}$$

which is the extension of the delta function sine series expansion (3.89) to the entire x-axis. To evaluate $Q_a(x,x')$ outside of $0 \le x \le a$, note first that expansion (3.102) implies that $Q_a(x,x')$ is an odd function of x, so that we can immediately extend Q_a to the interval $-a \le x \le a$:

$$Q_a(x,x') = \delta(x-x') - \delta(x+x') \qquad (-a \le x \le a). \tag{3.103}$$

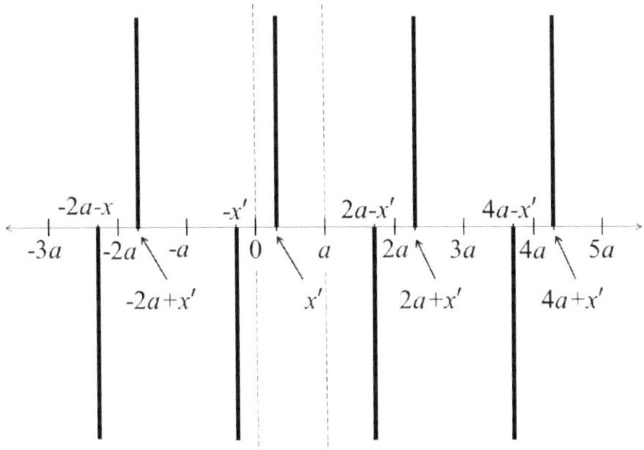

Fig. 3.9 Image charge distribution for a one-dimensional "box" with zero potential at $x = 0$ and $x = a$; the vertical dimension represents charge density.

Furthermore, $Q_a(x, x')$ is periodic in x with period $2a$, so that

$$Q_a(x + 2na, x') = Q_a(x, x'),\ n = 0, \pm 1, \pm 2, \ldots. \tag{3.104}$$

Equations (3.103) and (3.104) are sufficient to specify Q_a for all values of x:

$$Q_a(x, x') = \sum_{n=-\infty}^{\infty} \left(\delta\left(x - x' + 2na\right) - \delta\left(x + x' + 2na\right)\right) \quad (-\infty < x < \infty). \tag{3.105}$$

The function $Q_a(x, x')$ is shown in Figure 3.9 (the positive and negative delta functions are represented as vertical lines). Essentially, we have an infinite sum of point charges, where only one of these charges is in the physical region $0 \leq x \leq a$. It follows that the solution $G_D(\vec{x}, \vec{x}\,')$ to

$$\nabla^2 G_D(\vec{x}, \vec{x}\,') = -4\pi Q_a(x, x')Q_b(y, y')\delta(z - z') \tag{3.106}$$

can be interpreted as a potential defined in *all* space, with unit image charges of sign$(-1)^{k+\ell}$ at the lattice points

$$\left(2na + (-1)^k x', 2mb + (-1)^\ell y'\right) \qquad (n, m = 0, \pm 1, \pm 2, \ldots; k, \ell = 0, 1), \tag{3.107}$$

as shown in Figure 3.10.

Since (3.106) is identical to (3.90), the Green function solution (3.98) corresponds to this potential. The fact that G_D satisfies zero boundary conditions on the lateral surfaces of the box follows from the arrangement of image charges: for each lateral face and image charge there is a corresponding image charge of opposite sign that cancels the potential on the

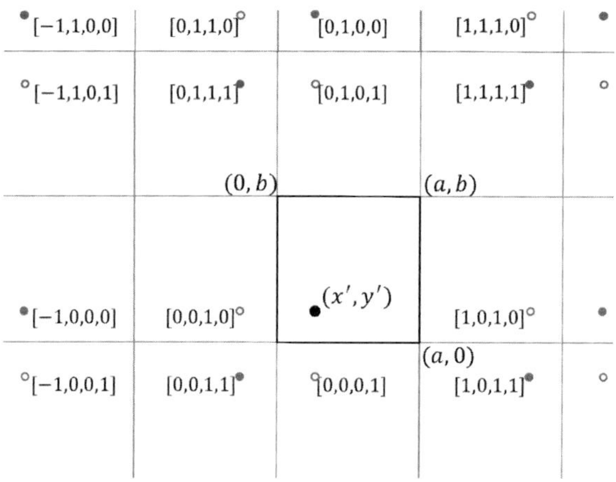

Fig. 3.10 Image charges for a two-dimensional box at zero potential due to point charge at (x', y'). The image charges are labeled with $[n, m, k, \ell]$ values corresponding to Equation (3.107). Solid/hollow dots denote positive/negative image charges, respectively.

given face. In other words, each image charge has an image in each face — just like a house of mirrors.

When $x' = y' = z' = a/2$ the arrangement of image charges is reminiscent of a cubic "crystal" with positive and negative charges at alternating lattice sites. The so-called *Madelung constant* gives the electrostatic interaction energy for an atom in a crystal lattice such as this. It takes on an especially simple form for the NaCl crystal, which has cubic symmetry with the Na and Cl atoms at alternating vertices of the cube. These atoms have alternating signs of electric charge: Na^+ and Cl^-. If one takes Na^+ as the reference atom, this Madelung constant is given by

$$M_{NaCl} \equiv \sum_{i,j,k}{}' \frac{(-1)^{i+j+k}}{\sqrt{i^2+j^2+k^2}} = -1.747564594\ldots, \tag{3.108}$$

for integers $i, j, k \in 0, \pm1, \pm2, \pm3, \ldots$. The interaction energy for unit charges is $W_{int} = M_{NaCl}/a$ if the distance between nearest neighbors is a. The prime on the sum means the origin term ($i = j = k = 0$) is left out. Interestingly, the sums in this expression are only *conditionally convergent*, meaning that convergence is dependent upon the order in which the sums are done. You can try your hand at approximating this constant in Exercise 3.6.1.

3.7 Separation of variables in polar coordinates

Typically, the best strategy for solving two-dimensional electrostatics problems is to adapt the solution method to the geometry of the situation. For instance, if cylindrical symmetry is present then the solution is independent of z (the coordinate along the axis of symmetry) and the problem can be framed as a two-dimensional problem in polar coordinates ρ, ϕ. The Laplace equation reduces to (see Section 1.8)

$$\frac{1}{\rho}\frac{\partial}{\partial\rho}\left(\rho\frac{\partial\Phi}{\partial\rho}\right)+\frac{1}{\rho^2}\frac{\partial^2\Phi}{\partial\phi^2}=0. \tag{3.109}$$

We apply the separation of variables technique that was introduced in Section 3.5, and assume potential solutions of the form

$$\Phi(\rho,\phi)=R(\rho)F(\phi). \tag{3.110}$$

Substituting into (3.109) and multiplying by ρ^2/Φ gives

$$\frac{\rho}{R}\frac{d}{d\rho}\left(\rho\frac{dR}{d\rho}\right)=-\frac{1}{F}\frac{d^2F}{d\phi^2}. \tag{3.111}$$

It follows that both sides of (3.111) must be constants independent of ρ, ϕ:

$$\frac{\rho}{R}\frac{d}{d\rho}\left(\rho\frac{dR}{d\rho}\right)=\nu^2, \tag{3.112}$$

$$\frac{1}{F}\frac{d^2F}{d\phi^2}=-\nu^2. \tag{3.113}$$

These equations have the following solutions:

$$\nu\neq 0:\qquad R(\rho)=a_\nu\rho^\nu+b_\nu\rho^{-\nu},\; F(\phi)=A_\nu\cos\nu\phi+B_\nu\sin\nu\phi, \tag{3.114}$$

$$\nu=0:\qquad R(\rho)=a_0+b_0\ln\frac{\rho}{\rho_0},\; F(\phi)=A_0+B_0\phi. \tag{3.115}$$

When solving for a problem where the full azimuthal range of ϕ is allowed, then continuity mandates that $F(0)=F(2\pi)$:

$$\nu\neq 0:\qquad \sin(0)=\sin(2\pi\nu)\implies \nu=\pm1,\pm2,\ldots, \tag{3.116}$$

$$\nu=0:\qquad B_0\cdot 0=B_0\cdot 2\pi\implies B_0=0. \tag{3.117}$$

In this case, the full solution is

$$\Phi(\rho,\phi)=a_0+b_0\ln\frac{\rho}{\rho_0}+\sum_{n=1}^{\infty}a_n\rho^n\sin(n\phi+\alpha_n)$$

$$+\sum_{n=1}^{\infty}b_n\rho^{-n}\sin(n\phi+\beta_n). \tag{3.118}$$

The solution is therefore reduced to a Fourier series problem.

3.8 Corner problems in polar coordinates

In some problems in polar coordinates, the full range of ϕ is not required in the solution. One example is the corner problem shown in Figure 3.11 with opening angle β. Since $\rho = 0$ is included in the region, we must have $b_\nu = 0$ in (3.114), (3.115) for all ν. Since there is no radial dependence when $\phi = 0$ and $\phi = \beta$, all of the $\nu \neq 0$ terms must vanish at these two angles. These conditions imply $A_\nu = 0$ for all $\nu \neq 0$, and $B_\nu = 0$ except when $\sin \nu\beta = 0$. The boundary conditions applied to the $\nu = 0$ term give $a_0 A_0 = V$ and $B_0 = 0$. Altogether, the general solution at point p in the figure is

$$\Phi(\rho, \phi) = V + \sum_{m=1}^{\infty} a_m \rho^{m\pi/\beta} \sin\left(\frac{m\pi\phi}{\beta}\right). \tag{3.119}$$

The requirement that the potential remain bounded implies that this representation is only valid in the charge-free region near the corner, $\rho = 0$. Near this corner, the term with the lowest power will dominate. Assuming that $a_1 \neq 0$ we have

$$\Phi(\rho, \phi) \approx V + a_1 \rho^{\pi/\beta} \sin\left(\frac{\pi\phi}{\beta}\right). \tag{3.120}$$

The electric field near the corner has components

$$E_\rho = -\frac{\partial \Phi}{\partial \rho} \approx -\frac{\pi a_1}{\beta} \rho^{\pi/\beta - 1} \sin\left(\frac{\pi\phi}{\beta}\right), \tag{3.121}$$

$$E_\phi = -\frac{1}{\rho}\frac{\partial \Phi}{\partial \phi} \simeq -\frac{\pi a_1}{\beta} \rho^{\pi/\beta - 1} \cos\left(\frac{\pi\phi}{\beta}\right). \tag{3.122}$$

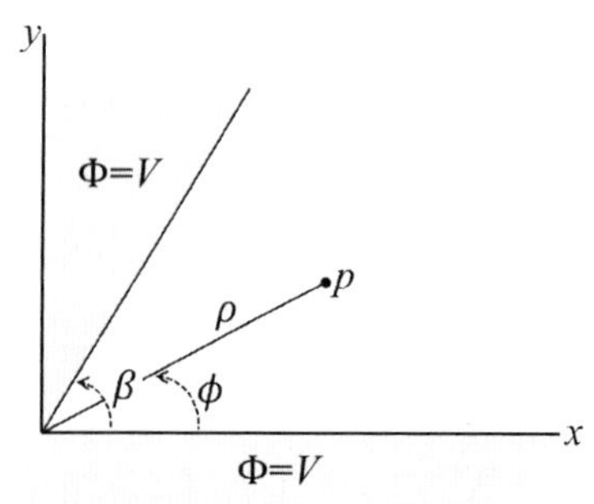

Fig. 3.11 Geometry of corner potential problem.

Notice that

$$\sqrt{E_\rho^2 + E_\phi^2} = \frac{\pi|a_1|}{\beta}\rho^{\pi/\beta - 1}. \tag{3.123}$$

This implies that for an obtuse opening angle $\beta > \pi$, the electric field magnitude is concentrated near the tip for $\rho \to 0$, whereas for an acute angle $\beta < \pi$, the region is partly shielded. In Section 4.8 we will solve for the Green function for this geometry.

3.9 Cylindrical halves at different potentials

Another instructive example with cylindrical symmetry is the interior cylinder problem shown in Figure 3.12. We are solving for the potential inside a cylinder of radius b with the cylinder split into two halves held at potentials V_1 (for $-\pi/2 < \phi < \pi/2$) and V_2 (for $\pi/2 < \phi < 3\pi/2$). In this case, the resulting Fourier series can be summed exactly.

Rewriting (3.118) we have

$$\Phi(\rho, \phi) = a_0 + b_0 \ln \rho/\rho_0 + \sum_{n=1}^{\infty} a_n \rho^n \sin(n\phi + \alpha_n) + \sum_{n=1}^{\infty} b_n \rho^{-n} \sin(n\phi + \beta_n). \tag{3.124}$$

Since $\rho = 0$ is included in the region, it follows that $b_n = 0$ for all n. By integrating around the perimeter, we can immediately establish that a_0 represents the average value of the potential on the surface, so

$$a_0 = \frac{V_1 + V_2}{2}. \tag{3.125}$$

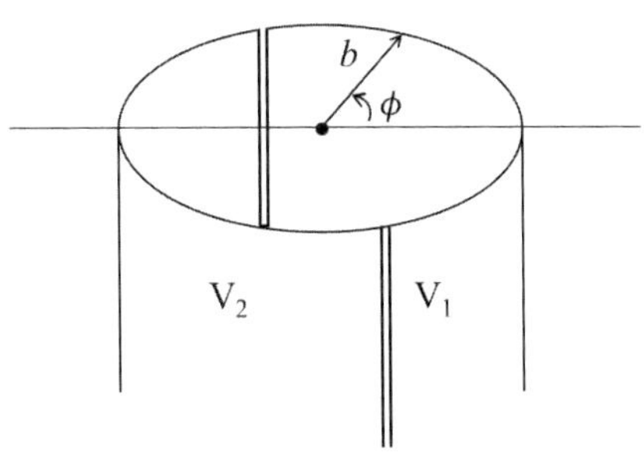

Fig. 3.12 Half-cylinders at different potentials.

We may write the boundary conditions as:

$$\Phi(b,\phi) = \frac{V_1+V_2}{2} + \begin{cases} (V_1-V_2)/2, & -\pi/2 < \phi < \pi/2, \\ -(V_1-V_2)/2, & \pi/2 < \phi < 3\pi/2. \end{cases} \tag{3.126}$$

The form of (3.126) shows that Φ is a constant plus a function that is odd about $\phi = \pi/2$. In order to exploit this symmetry we may change variable: $\phi' \equiv \phi - \pi/2$. We may then rewrite (3.124) as

$$\begin{aligned} \Phi(b,\phi') - \frac{V_1+V_2}{2} &= \sum_{n=1}^{\infty} \left(c_n b^n \sin(n\phi' + \alpha_n')\right) \\ &= \sum_{n=1}^{\infty} \left(c_n' b^n \sin n\phi' + d_n' b^n \cos n\phi'\right), \end{aligned} \tag{3.127}$$

where $\Phi(b,\phi') - (V_1+V_2)/2$ is now an odd function of ϕ'. It follows that all of the cosine terms in the Fourier expansion (3.127) vanish, so $d_n' = 0$ for all n. We may now evaluate the c_n' by plugging in $\Phi(b,\phi') = V_2$ on the interval $0 < \phi' < \pi$ and integrating:

$$\int_0^\pi d\phi' \, \frac{V_2 - V_1}{2} \sin m\phi' = \sum_{n=1}^{\infty} c_n' b^n \int_0^\pi d\phi' \sin n\phi' \sin m\phi', \tag{3.128}$$

which leads to the solution (using the orthogonality relation (3.83) with $a = \pi$)

$$c_n' = \begin{cases} 0, & n \text{ even}, \\ \dfrac{2}{\pi}\dfrac{(V_2-V_1)}{nb^n}, & n \text{ odd}. \end{cases} \tag{3.129}$$

Solving for c_n' and plugging into (3.127) gives a series solution for Φ:

$$\Phi = \frac{V_1+V_2}{2} - \frac{2(V_1-V_2)}{\pi} \sum_{n=0}^{\infty} \left(\frac{\rho}{b}\right)^{2n+1} \frac{\sin[(2n+1)(\phi - \pi/2)]}{2n+1}. \tag{3.130}$$

The series in (3.130) may be evaluated in closed form (see Exercise 3.9.3), leading to the final result

$$\Phi(\rho,\phi) = \frac{V_1+V_2}{2} + \frac{V_1-V_2}{\pi} \arctan\left(\frac{2(\rho/b)\cos\phi}{1-(\rho/b)^2}\right). \tag{3.131}$$

The corresponding surface charge is

$$\sigma = \frac{1}{4\pi} \frac{\partial \Phi}{\partial \rho}\bigg|_{\rho=b} = \frac{V_1-V_2}{4\pi^2} \frac{1}{b\cos\phi}. \tag{3.132}$$

3.10 Variational methods

Unless the geometry of a conductor is one which can be simply characterized in a separable coordinate system (e.g., $\rho = b$ for cylindrical coordinates as we just saw), we can not find the field configuration as a boundary value problem. However, at the end of Section 2.10 we covered Thompson's theorem for closed surfaces, which gave upper limits for the field energy for certain geometries. For general surfaces, one can use a closely related variational method to minimize the energy and consequently find good estimates for the surface charge distribution, and thus also for the field configuration. This leads to variational estimates for capacitances, as was suggested at the end of Section 2.12. To obtain these results, we will need the following general version of Thompson's theorem:

Theorem. *Given a fixed surface in free space with fixed total charge, then the electrostatic field energy produced by the charged surface is minimized when the charges are placed so that the surface is an equipotential.*

Note two important differences between this version of Thompson's theorem and the earlier version. First, the surface is no longer assumed to be closed; and second, the field energy is minimized over *all* space, and not just over the region exterior to the surface.

The proof is as follows. Denote the surface in question by S, and let σ_1 be the correct physical charge density for a conducting surface, which makes S an equipotential. Since the surface is located in open space, we may use the free Green's function

$$G_f(\vec{x}, \vec{x}\,') = \frac{1}{|\vec{x} - \vec{x}\,'|} \tag{3.133}$$

to obtain the potential and charge on S as

$$\Phi_1|_S \equiv \int da' \frac{\sigma_1(\vec{x}\,')}{|\vec{x} - \vec{x}\,'|} = V, \tag{3.134}$$

$$Q = \frac{1}{4\pi} \int_S da \, \frac{\partial \Phi_1(\vec{x})}{\partial n}. \tag{3.135}$$

The field energy is

$$W_1 \equiv \frac{1}{2} \int_S da \int_{S'} da' \frac{\sigma_1(\vec{x})\sigma_1(\vec{x}\,')}{|\vec{x} - \vec{x}\,'|}. \tag{3.136}$$

One may now consider the field energy expression for any other charge density σ:

$$W[\sigma] \equiv \frac{1}{2}\int da \int_{S'} da'\frac{\sigma(\vec{x})\sigma(\vec{x}')}{|\vec{x}-\vec{x}'|} = \frac{1}{2}\int da\,\sigma(\vec{x})\Phi(\vec{x}). \tag{3.137}$$

Given $\vec{E}(\vec{x}) = -\vec{\nabla}\Phi(\vec{x})$, we may write

$$\frac{1}{8\pi}\int d^3x\,\vec{E}^2 = \frac{1}{8\pi}\int d^3x\left[\vec{\nabla}\cdot(\Phi\vec{\nabla}\Phi) - \Phi\nabla^2\Phi\right], \tag{3.138}$$

where the integration is over all space. The first term on the right can be shown to be zero with the use of Gauss' theorem on the surface at infinity. For the other term we may use

$$\nabla^2\Phi(\vec{x}) = -4\pi\sigma(\vec{x})\int_{S'} da'\delta(\vec{x}-\vec{x}'). \tag{3.139}$$

Here we are using the type of smeared delta function discussed in Section 2.3. $\delta(\vec{x}-\vec{x}')$ is a three-dimensional delta function integrated over the surface, S', yielding an open surface delta function. Equation (3.138) now yields

$$\frac{1}{8\pi}\int d^3x\,\vec{E}^2 = \frac{1}{2}\int da \int da'\frac{\sigma(\vec{x})\sigma(\vec{x}')}{|\vec{x}-\vec{x}'|} > 0, \tag{3.140}$$

where Equation (3.137) has been used. The reasoning now is along the same lines as the earlier proof in Section 2.10. For fixed charge Q and potential V, we then have

$$W[\sigma_1+\delta\sigma] = W_1 + W', \tag{3.141}$$

where

$$W' \equiv \frac{1}{2}\int da \int da'\frac{\delta\sigma(\vec{x})\delta\sigma(\vec{x}')}{|\vec{x}-\vec{x}'|} > 0, \tag{3.142}$$

by Equation (3.140). This means that

$$W[\sigma_1+\delta\sigma] > W_1, \tag{3.143}$$

and the proof is complete.

We can use this result to provide bounds and estimates in the following manner. Using a parametrized form for the charge density, one obtains a best guess for the actual charge density in the given parameter space by minimizing the field energy as a function of the parameters. Also, recall that for a single isolated conductor in open space the capacitance and field energy are related via Equation (2.170). This leads to the variational expression

$$C^{-1}[\sigma] \equiv \frac{2W[\sigma]}{Q^2}, \tag{3.144}$$

which gives a lower bound on the true capacitance from the upper bound produced by $W[\sigma]$. However, we will have to wait until Sections 4.7 and 4.11 before we have the mathematical machinery to try examples; see Exercises 3.7.1 and 4.11.4. We will also encounter a variational approach to waveguide energy transmission problems in Sections 10.5 and 10.6.

3.11 Conformal mapping techniques

An important technique for solving two-dimensional potential problems is conformal mapping, which is based on the theory of complex variables. Define $z \equiv x + iy$, and let $f(z) = u(x, y) + iv(x, y)$ be an analytic function of z. Then the *Cauchy-Riemann equations* give the result:

$$\frac{\partial u}{\partial x} = \frac{\partial v}{\partial y}, \qquad \frac{\partial u}{\partial y} = -\frac{\partial v}{\partial x}. \tag{3.145}$$

By taking additional derivatives and adding we may obtain

$$\nabla^2 u = \frac{\partial^2 u}{\partial x^2} + \frac{\partial^2 u}{\partial y^2} = \frac{\partial}{\partial x}\left(\frac{\partial v}{\partial y}\right) - \frac{\partial}{\partial y}\left(\frac{\partial v}{\partial x}\right) = 0, \tag{3.146}$$

$$\nabla^2 v = \frac{\partial^2 v}{\partial x^2} + \frac{\partial^2 v}{\partial y^2} = \frac{\partial}{\partial y}\left(\frac{\partial u}{\partial x}\right) - \frac{\partial}{\partial x}\left(\frac{\partial u}{\partial y}\right) = 0, \tag{3.147}$$

so both $u(x, y)$ and $v(x, y)$ can be identified as charge-free potentials in the region where $f(z)$ is analytic. Notice also that (3.145) implies that

$$\vec{\nabla} u \cdot \vec{\nabla} v = \frac{\partial u}{\partial x}\frac{\partial v}{\partial x} + \frac{\partial u}{\partial y}\frac{\partial v}{\partial y} = -\frac{\partial u}{\partial x}\frac{\partial u}{\partial y} + \frac{\partial u}{\partial y}\frac{\partial u}{\partial x} = 0. \tag{3.148}$$

It follows that the electric fields $\vec{\nabla} u$, $\vec{\nabla} v$ corresponding to potentials u, v respectively are perpendicular at every field point. This in turn implies that the curves $v =$ constant coincide with electric field lines of u, and vice versa.

The basic theorem of conformal mapping may be stated as follows (see Figure 3.13 and Ref. 1 of *Going Deeper* Section 3.12.2):

Theorem. *Let $f(z) = u(x, y) + iv(x, y)$ be an analytic function that maps an arc C in the $z = x + iy$ plane onto an arc Γ in the $w = u + iv$ plane, such that $f'(z) \neq 0$ for all z on C. Let $h(u, v)$ be a differentiable real-valued function on a domain that includes Γ, and let $H(x, y) = h[u(x, y), v(x, y)]$. Then if $h = c$ on Γ, it follows that $H = c$ on C. Furthermore, if $dh/dn = 0$ along Γ, then $dH/dn = 0$ along C.*

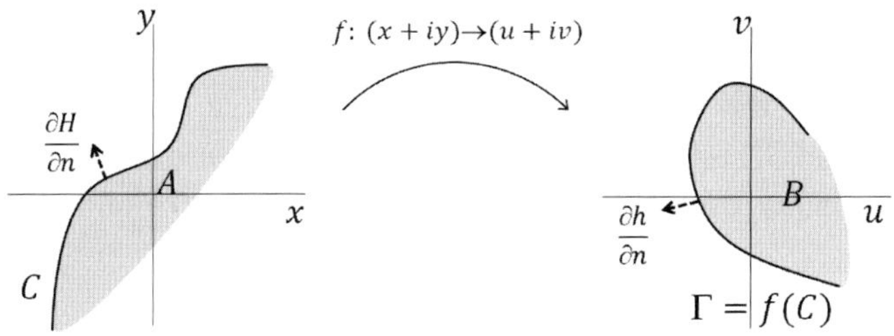

Fig. 3.13 Conformal mapping of boundary curves. If $h(u, v) = c$ on Γ then $H(x, y) = c$ on C; and if $\partial h/\partial n = 0$ on Γ, then $\partial H/\partial n = 0$ on C.

In other words, if region A maps conformally to region B, then a differentiable function in B that satisfies certain boundary conditions "pulls back" via the conformal map to a differentiable function on A that satisfies the same boundary conditions (see Figure 3.13). Note the applicable boundary conditions include constant Dirichlet and homogeneous Neumann conditions; constant Neumann conditions are not included.

As an example, consider the mapping

$$z = e^w, \tag{3.149}$$

which maps the horizontal strip $0 < \operatorname{Im} w < \pi$ to the half-plane $\operatorname{Im} z > 0$ as shown in Figure 3.14. The inverse mapping is $w = \log z = \ln|z| + i \arg z$, where the branch $0 < \arg z < \pi$ is taken. Now consider the function $h(w) = iE_0 w + V$, where $V > 0$ and $E_0 \equiv 2V/\pi$. We define the potential $\Phi(u, v) \equiv \operatorname{Re} h(u, v)$, so that $\Phi(u, v) = -E_0 v + V$ and $-\vec{\nabla}\,\Phi(u, v) = E_0 \hat{v}$.

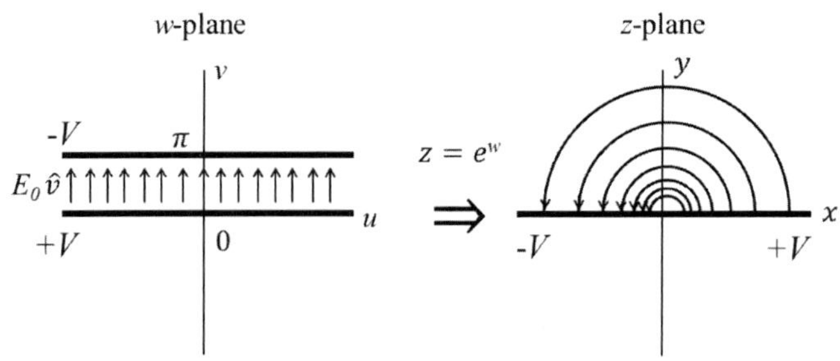

Fig. 3.14 Conformal mapping of strip to half-plane.

It follows that

$$\begin{aligned} H(z) \equiv h[w(z)] &= iE_0 \ln z + V \\ &= iE_0 \ln(x + iy) + V \\ &= iE_0[\ln \sqrt{x^2 + y^2} + i \arctan(y/x)] + V. \end{aligned} \tag{3.150}$$

$H(z)$ is an analytic function, so $\Phi(x, y) \equiv \operatorname{Re} H(x, y)$ is a potential. Replacing $E_0 = 2V/\pi$ gives

$$\Phi(x, y) \equiv \operatorname{Re} H = V\left(1 - \frac{2}{\pi} \arctan\left(\frac{y}{x}\right)\right) = V\left(1 - \frac{2}{\pi}\phi\right), \tag{3.151}$$

where ϕ is the azimuthal angle in cylindrical coordinates. We may easily verify that $\Phi(x, y) = \pm V$ on the positive/negative x-axes. Equation (3.151) is actually a special case of

$$\Phi(x, y) = V\left(1 - \frac{2}{\beta}\phi\right), \tag{3.152}$$

for an opening angle β between the conducting plates; this can be established with the methods of Section 3.7. The electric field lines implied by (3.151) are semicircles, with

$$E_x = -\frac{\partial \Phi}{\partial x} = -\frac{E_0 y}{x^2 + y^2}; \quad E_y = -\frac{\partial \Phi}{\partial y} = \frac{E_0 x}{x^2 + y^2}. \tag{3.153}$$

Notice the interesting inverse radial dependence for the magnitude of the electric field: $|\vec{E}| = (E_x^2 + E_y^2)^{1/2} = |E_0|/\rho$ where $\rho \equiv \sqrt{x^2 + y^2}$. We stated the transformation (3.149) in this case in dimensionless form for simplicity. To map the region $0 < v < a$ to $y > 0$, one only has to make the substitutions $w \to w\pi/a$ and $z \to z/L$ in the mapping. After this the results (3.151) and (3.153) still hold for the new E_0.

There are many other ways to derive the same result. For instance, using the Green function for the half-plane geometry we have:

$$\begin{aligned} G_D(\vec{x}, \vec{x}\,') = &\frac{1}{\sqrt{(x - x')^2 + (y - y')^2 + (z - z')^2}} \\ &- \frac{1}{\sqrt{(x - x')^2 + (y + y')^2 + (z - z')^2}}. \end{aligned} \tag{3.154}$$

To make use of the Green function integral equation for Φ (see Equation (2.98)), we must first evaluate

$$\left.\frac{\partial G}{\partial n'}\right|_s = -\left.\frac{\partial G}{\partial y'}\right|_{y'=0} = \frac{-2y}{((x - x')^2 + y^2 + (z - z')^2)^{3/2}}. \tag{3.155}$$

We then have

$$\begin{aligned}\Phi(\vec{x}) &= -\frac{1}{4\pi}\oint_S da'\Phi(\vec{x}\,')\frac{\partial G_D(\vec{x},\vec{x}\,')}{\partial n'}\\ &= -\frac{yV}{2\pi}\left\{\int_{-\infty}^{0} dx' \int_{-\infty}^{\infty} dz' \frac{1}{((x-x')^2+y^2+(z-z')^2)^{3/2}}\right.\\ &\quad \left. -\int_{0}^{\infty} dx' \int_{-\infty}^{\infty} dz' \frac{1}{((x-x')^2+y^2+(z-z')^2)^{3/2}}\right\}. \end{aligned} \tag{3.156}$$

The z' integration gives

$$\begin{aligned}&\int_{-\infty}^{\infty} dz' \frac{1}{((x-x')^2+y^2+(z-z')^2)^{3/2}}\\ &\quad = \int_{-\infty}^{\infty} dz'' \frac{1}{((x-x')^2+y^2+z''^2)^{3/2}}\\ &\quad = \left(\frac{1}{(x-x')^2+y^2}\right)\frac{z''}{\sqrt{(x-x')^2+y^2+z''^2}}\Bigg|_{-\infty}^{\infty}\\ &\quad = \left(\frac{2}{(x-x')^2+y^2}\right). \end{aligned} \tag{3.157}$$

The two x' integrations are

$$\begin{aligned}\int_{-\infty}^{0} dx' \left(\frac{1}{(x-x')^2+y^2}\right) &= -\int_{\infty}^{x} dx'' \frac{1}{x''^2+y^2}\\ &= -\frac{1}{y}\left(\arctan\left(\frac{x}{y}\right) - \pi/2\right), \end{aligned} \tag{3.158}$$

and similarly

$$\int_{0}^{\infty} dx' \left(\frac{1}{(x-x')^2+y^2}\right) = -\frac{1}{y}\left(-\pi/2 - \arctan\left(\frac{x}{y}\right)\right). \tag{3.159}$$

Plugging (3.157)–(3.159) into (3.156) gives finally

$$\begin{aligned}\Phi(\vec{x}) = \frac{2V}{\pi}\arctan\left(\frac{x}{y}\right) &= \frac{2V}{\pi}\left(\pi/2 - \arctan\left(\frac{y}{x}\right)\right)\\ &= V - \frac{2V}{\pi}\arctan\left(\frac{y}{x}\right), \end{aligned} \tag{3.160}$$

which agrees with (3.151).

Conformal methods can also give us asymptotic results in some nontrivial geometries. Consider the region near the edge of two parallel plates, as shown in Figure 3.15. We will use the mapping

$$z = e^w + w, \tag{3.161}$$

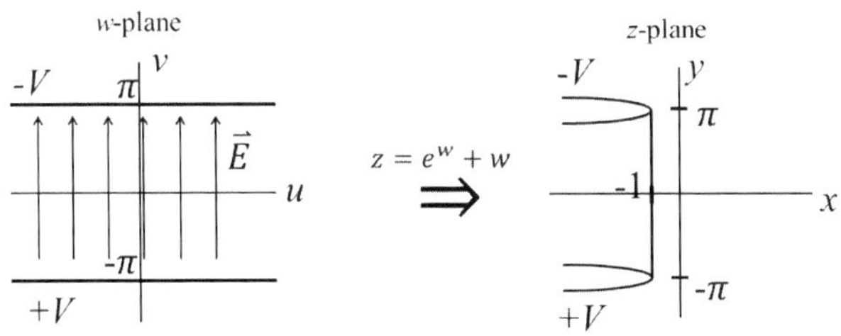

Fig. 3.15 Conformal mapping for finding potential near the edge of two parallel plates.

and the function $h(w) \equiv iVw/\pi$ so that we obtain $\Phi(u,v) \equiv \operatorname{Re} h(u+iv) = -Vv/\pi$. In order to find the potential via conformal mapping, we would need to invert:

$$x = e^u \cos v + u, \tag{3.162}$$

$$y = e^u \sin v + v. \tag{3.163}$$

This is impossible to do analytically. However, we can still obtain the electric field along the x-axis ($y = 0$) without inverting the function. We have that

$$E_y\big|_{y=0} = -\frac{\partial}{\partial y}\left(-\frac{V}{\pi}v(x,y)\right)\bigg|_{y=0} = \frac{V}{\pi}\frac{\partial v}{\partial y}\bigg|_{y=0}. \tag{3.164}$$

Differentiation of the mapping Equation (3.163) by y gives

$$1 = e^u \sin v \frac{\partial u}{\partial y} + e^u \cos v \frac{\partial v}{\partial y} + \frac{\partial v}{\partial y}. \tag{3.165}$$

The line $y = 0$ is the image of $v = 0$, so (3.165) gives

$$1 = e^u \frac{\partial v}{\partial y}\bigg|_{y=0} + \frac{\partial v}{\partial y}\bigg|_{y=0} \implies \frac{\partial v}{\partial y}\bigg|_{y=0} = \frac{1}{e^u + 1}. \tag{3.166}$$

Equation (3.162) with $v = 0$ gives

$$x = e^u + u. \tag{3.167}$$

Comparing (3.167) and (3.166), we find

$$\frac{1}{e^u + 1} \to \begin{cases} 1, & x \to -\infty, \\ \dfrac{1}{x}, & x \to \infty. \end{cases} \tag{3.168}$$

This gives the asymptotic behavior of the electric field as

$$E_y \to \begin{cases} \dfrac{V}{\pi}, & x \to -\infty, \\ \dfrac{V}{\pi}\dfrac{1}{x}, & x \to \infty. \end{cases} \tag{3.169}$$

The field within the plates is roughly constant as expected, while away from the plates the field shows a falloff of $O(1/x)$ in the fringing field along the symmetry axis. We again stated the transformation (3.161) in dimensionless form. To instead map the plates at $v = -a, a$ to $y = -L, L$ one may make the substitutions $w \to w\pi/a$ and $z \to z\pi/L$ in the mapping.

3.12 Going Deeper

3.12.1 *Variational methods*

(1) C. A. Brau, *Modern Problems in Classical Electrodynamics*, Oxford University Press (Oxford) 2004; Section 3.2.7.

(2) L. Cairo and T. Kahan, *Variational Techniques in Electromagnetism*, Gordon and Breach (New York) 1965.

(3) J. D. Jackson, *Classical Electrodynamics*, 3rd ed., John Wiley & Sons (New York) 1999; Section 1.12.

(4) M. N. O. Sadiku, *Numerical Techniques in Electromagnetics*, 2nd ed., CRC Press (Boca Raton) 2001; Chapter 4.

(5) J. Van Bladel, *Electromagnetic Fields*, 2nd ed., IEEE Press (Piscataway) 2007; Chapter 2.

3.12.2 *Conformal mapping*

(1) J. W. Brown and R. V. Churchill, *Complex Variables and Applications*, 9th ed., McGraw-Hill (Boston) 2014; Chapter 10 and Appendix 2.

(2) P. M. Morse and H. Feshbach, *Methods of Theoretical Physics, Parts I and II*, McGraw-Hill (New York) 1953; Section 4.7.

(3) M. R. Spiegel *et al.*, *Complex Variables*, in *Schaum's Outline Series*, 2nd ed., McGraw-Hill (New York) 2009; Chapters 8 and 9.

(4) W. R. Smythe, *Static and Dynamic Electricity*, 3rd ed., Hemisphere Publishing (New York) 1989; Chapter 4.

(5) E. W. Weisstein, *Conformal Mapping*, MathWorld–A Wolfram Web Resource:
`http://mathworld.wolfram.com/ConformalMapping.html`.

3.13 Exercises

Exercise 3.1.1. Consider the three-dimensional infinite parallel conducting plate interior problem shown in Figure 3.16 for a unit point charge between the plates, as shown. Find the image charge locations and values of all the image charges needed to satisfy the boundary conditions for the Dirichlet Green function. Write the Green function formally as a sum over these charges. Show that your expression converges absolutely.

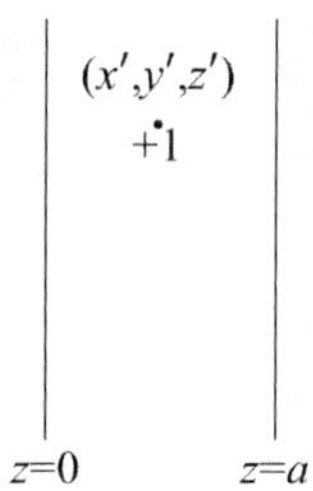

Fig. 3.16 Dirichlet problem between parallel plates.

Exercise 3.1.2.

(a) Use the image method in Figure 3.17 to find the Dirichlet Green function $G_D(\vec{x}, \vec{x}\,')$ in Cartesian coordinates for the three-dimensional corner geometry. The geometry is translationally invariant along the z-axis and the source is outside of the conductors in the $x > 0, y > 0$ quadrant.
(b) Find the work W required to remove the $+1$ charge to spatial infinity.

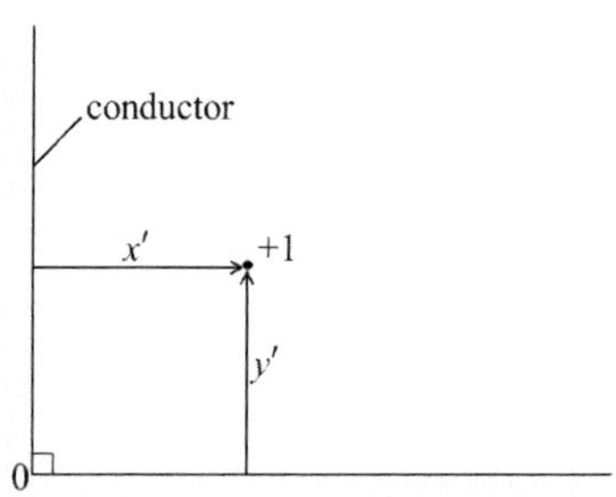

Fig. 3.17 Dirichlet corner problem between two plane conductors.

(c) Using the Green's function from part (a), find the potential for the situation where the vertical wall is at potential $-V$ and the horizontal wall is at V.

Exercise 3.1.3. An infinitely long unit line charge is located with cylindrical coordinates (ρ', ϕ') parallel to the line at which perpendicular conducting planes intersect (see Figure 3.18). Using the image method, find the Dirichlet Green function. Show that the result can be expressed in cylindrical coordinates as

$$G_D(\vec{x}, \vec{x}\,') = \ln\left[\frac{\rho^4 + \rho'^4 - 2\rho'^2\rho^2\cos(2(\phi + \phi'))}{\rho^4 + \rho'^4 - 2\rho'^2\rho^2\cos(2(\phi - \phi'))}\right].$$

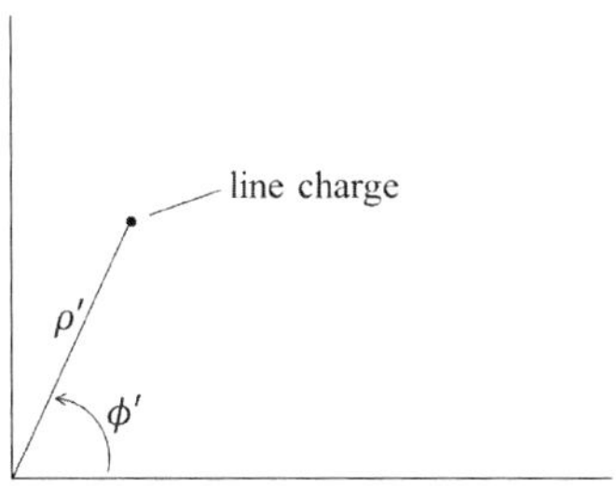

Fig. 3.18 Perpendicular conducting plates and line charge for Exercise 3.1.3.

Exercise 3.1.4. A charged particle of charge e in a weightless environment is attached to a massless string of length L and oscillates above a semi-infinite conducting plane held at zero potential that fills the entire region below the origin O (see Figure 3.19). The distance from the point of

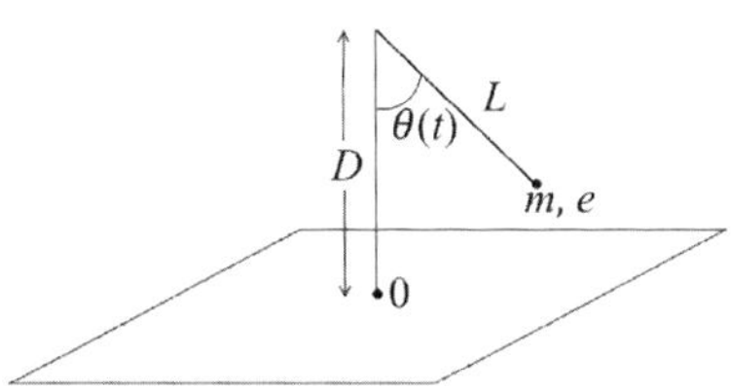

Fig. 3.19 Oscillating charge above conducting plane.

attachment to the plane is D ($D > L$). By computing the potential energy of the system to lowest order in the angle $\theta(t)$ (or other means), find the frequency of small oscillations of the charge.

Exercise 3.1.5. An arbitrary patch of area on the $z = 0$ plane is raised to a constant potential V, as shown in Figure 3.20. The rest of the plane (extending to infinity) is specified to have $V = 0$. Using the Dirichlet Green function for this geometry, show that the electric potential Φ at an arbitrary position above or below the plane can be written

$$\Phi(\vec{x}) = \frac{V}{2\pi}|\Omega(\vec{x})|,$$

where $\Omega(\vec{x})$ is the solid angle subtended by the patch at the observation position, $\vec{x}$.

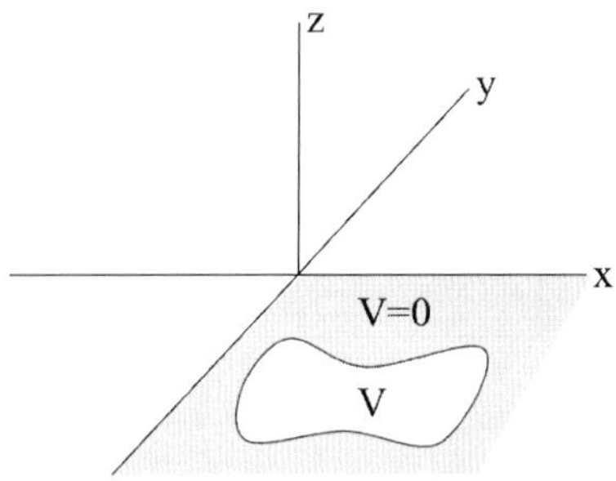

Fig. 3.20 Geometry for Exercise 3.1.5.

Exercise 3.2.1. In Section 3.2, an integral expression for the Green function $G_f(\vec{x}, \vec{x}\,')$ was given in (3.19) as

$$G_f(\vec{x}, \vec{x}\,') = \frac{1}{\pi}\int dk \int_{-1}^{1} d\cos\theta e^{ik|\vec{x}-\vec{x}\,'|\cos\theta}.$$

Evaluate this integral by (i) including a convergence factor $e^{-k\epsilon}$ in the integrand, (ii) interchanging the order of integration (i.e. perform the k integral first) and (iii) taking the limit $\epsilon \to 0^+$.

Exercise 3.2.2. For the half-space ($z > 0$) in three dimensions shown in Figure 3.21:

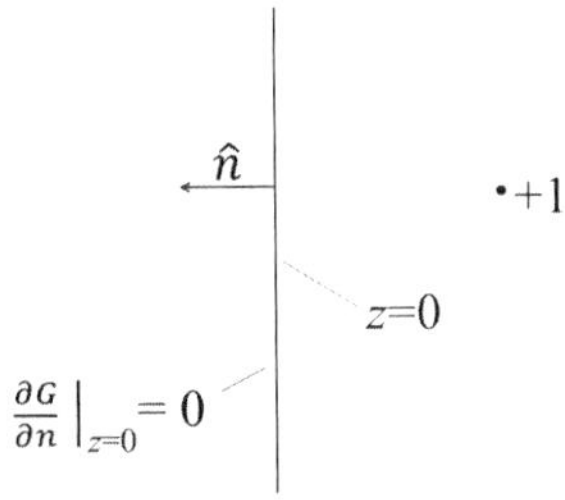

Fig. 3.21 Neumann Green function for half-space $z > 0$ in Exercise 3.2.2.

(a) Use the reduced Green function method to find the Neumann Green function. [You can leave the result in integral form.]
(b) Confirm the presence of a surface delta function in $\partial G_N(\vec{x}, \vec{x}\,')/\partial n|_{z'=0}$ as in Equation (2.115). [This is most easily done using the part (a) representation.]

Exercise 3.2.3. With reference to the Dirichlet Green function for infinite parallel conducting plates (see Exercise 3.1.1 above).

(a) Show that

$$G_D(\vec{x}, \vec{x}\,') = 4\pi \int \frac{d^2k}{(2\pi)^2} e^{i\vec{k}\cdot(\vec{x}-\vec{x}\,')_\perp} g(k; z, z'),$$

where

$$g(k; z, z') \equiv \frac{\sinh(kz_<)\sinh[k(a - z_>)]}{k\sinh(ka)},$$

and where $z_<$ ($z_>$) is the lesser (greater) of z and z'.
(b) Using the result of part (a), verify the result (which can also be shown using the reciprocation theorem of Exercise 2.7.2),

$$Q\big|_{z=0} = -\left(1 - \frac{z'}{a}\right),$$

$$Q\big|_{z=a} = -\frac{z'}{a},$$

where z' is the source position.
(c) Use the explicit expression for the capacitance derived in Chapter 2 and the result of (b) to show that the capacitance is $C = A/(4\pi a)$. [*Hint*: Take one of the surface integrals to include only a finite area A.]

Exercise 3.2.4. A type of boundary condition that combines Dirichlet and Neumann boundary conditions is

$$\left(G_R + h\frac{\partial G_R}{\partial n}\right)\Big|_S = 0,$$

(n is the outward normal) where h is a nonzero constant, called the *Robin boundary conditions.* Starting with the form,

$$G_R(\vec{x}, \vec{x}\,') = 4\pi \int \frac{d^2k}{(2\pi)^2}\, e^{i\vec{k}\cdot(\vec{x}-\vec{x}\,')_\perp} g(k; z, z'),$$

use the reduced Green function method to solve for the function $g(k; z, z')$ in the region $z > 0$ for a source in the region $z' > 0$ for Robin boundary conditions on the $z = 0$ plane.

Exercise 3.3.1.

(a) Find the Dirichlet Green function solution for a unit point charge located inside a hollow, grounded, perfectly conducting spherical shell of radius a. Use the image method and find the potential both inside and outside the conducting shell.
(b) Consider a unit point charge q located inside a thin spherical conducting sphere of radius a. In this case the conductor is neutral (zero net charge). Find the potential due to the charge q everywhere inside and outside the sphere.
(c) In part (b) can one tell where the charge is inside the sphere from the outside field? Generalize your conclusion to a point charge inside a neutral arbitrarily shaped conducting hollow object.

Exercise 3.3.2.

(a) A hollow hemisphere of radius a is placed atop an infinite two dimensional plane, as shown in Figure 3.22. All surfaces are conducting. Using a set of image charges, find the Dirichlet Green function at all points above the conducting surfaces.
(b) Find the Dirchlet Green function for the vacuum region *inside* the hollow hemisphere. (The solution to Exercise 3.3.1(a) is needed.)

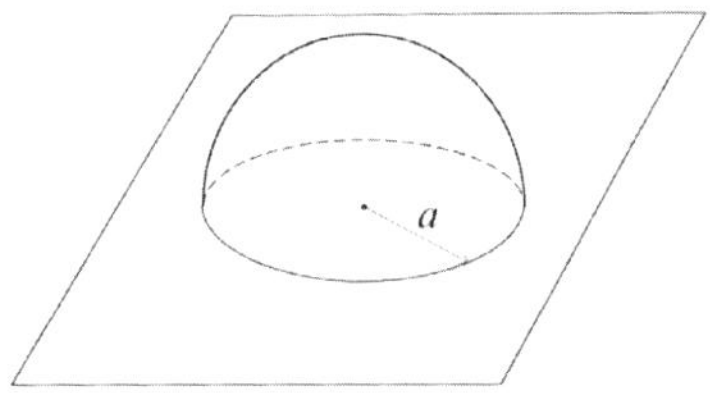

Fig. 3.22 Geometry for Exercise 3.3.2.

Exercise 3.3.3.

(a) An arbitrary charge density $\rho(\vec{x})$ exists outside of a spherical conductor of radius a. Show that the total field outside the conductor can be written as the direct field due to $\rho(\vec{x})$ and an image charge density $\rho^*(\vec{x})$ located inside the sphere, where the image charge density is related to the direct charge density in spherical coordinates centered on the sphere's center by

$$\rho^*(r, \theta, \phi) = -\left(\frac{a}{r}\right)^5 \rho\left(\frac{a^2}{r}, \theta, \phi\right),$$

or

$$\rho^*\left(\frac{a^2}{r}, \theta, \phi\right) = -\left(\frac{r}{a}\right)^5 \rho(r, \theta, \phi).$$

Note that the combination should give an $\vec{E}$ field which satisfies the boundary condition

$$\vec{E} \times \hat{n}\Big|_{r=a} = 0.$$

(b) Show that the part (a) result is consistent with the expected image charge and location given a point charge located outside the sphere.

Exercise 3.3.4.

(a) Two small conducting spheres of radii a and b are separated by a distance d where $d \gg a, b$. Show that the coefficient of capacitance, C_{ab}, of the system is approximately equal to $-(ab)/d$.

(b) Find the approximate system capacitance, $C \equiv |Q/\Delta V|$ obtained by placing charges Q and $-Q$ on the two spheres, where ΔV is the difference in potentials. Show that

$$C \approx \frac{ab}{a + b - \frac{2ab}{d}}.$$

[You may use the result of Exercise 2.12.5, but only if you prove it first!]

Exercise 3.3.5. Given a hollow conducting cylinder of radius a, and a line charge of unit positive linear density which is parallel to and a distance ρ' from the cylinder's axis.

(a) By using the image method, find the location of a unit negative linear density line charge such that the resulting potential is constant on the cylinder. [*Ans.*: The image charge will be located at a distance $\rho'' = a^2/\rho'$ along the line connecting the real charge density to the cylinder's axis.]
(b) Using (a) and the fact that a potential plus a constant is still a potential, find the Green function inside the cylinder for line charges parallel to the cylinder's axis. [*Ans.*:

$$G_D(\rho, \phi; \rho', \phi') = \ln \left| \frac{a^4 + \rho^2 \rho'^2 - 2a^2 \rho\rho' \cos(\phi - \phi')}{a^2(\rho^2 + \rho'^2 - 2\rho\rho' \cos(\phi - \phi'))} \right|.]$$

(c) Assuming the Green function is continuous, present arguments that the Green function *outside* of the cylinder is identical to the inside form.

Exercise 3.3.6. A thin charged spherical shell of radius b is constructed outside and concentric to a grounded conducting sphere of radius a. (A thin wire through the shell connects the conducting sphere to the outside ground.) The thin charged shell has a uniform surface charge distribution and total charge Q. (It is not a conducting sphere!) Find the electric potential $\Phi(\vec{x})$ everywhere.

Exercise 3.3.7. A ring of charge of radius b surrounds a conducting sphere of radius a in its equator plane ($z = 0$), as shown in Figure 3.23. The coordinate origin at the center of the sphere. The center of the sphere

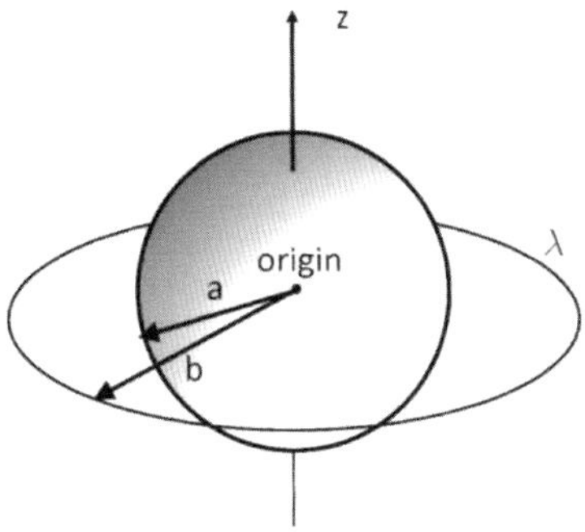

Fig. 3.23 Geometry for Exercise 3.3.7.

and ring coincide. The circular line charge has uniform line charge density λ (units: charge/length). The conducting sphere is maintained at zero potential (grounded).

(a) Find the induced charge on the sphere.
(b) Find the electric potential $\Phi(z)$ of the system measured along the z-axis, outside of the sphere. Confirm that $\Phi(z)|_{z=\pm a} = 0$.

Exercise 3.3.8.

(a) Consider the force on a point charge q originating from conducting sphere of radius a held at constant potential $V \neq 0$. The charge q is located a distance d away from the sphere's center ($d > a$). Show that one may always adjust the potential to make the force zero and determine the value of this potential in terms of q, a and d.
(b) Now find the force on the same point charge originating from a neutral, isolated conducting sphere of radius a.

Exercise 3.5.1. Solve the Laplace equation in two dimensions

$$\frac{\partial^2 \Phi}{\partial x^2} + \frac{\partial^2 \Phi}{\partial y^2} = 0,$$

by separation of variables, and find the potential (expressed as an infinite series) everywhere inside a two-dimensional box, of length a in the

x-direction and b in the y-direction, which has $\Phi = 0$ on its borders except along $y = b$ where $\Phi = V$ (V is a constant). [*Ans.*:

$$\Phi(x, y) = \frac{4V}{\pi} \sum_{n=1}^{\infty} \frac{1}{m} \frac{\sinh\left(\frac{m\pi y}{a}\right)}{\sinh\left(\frac{m\pi b}{a}\right)} \sin\left(\frac{m\pi x}{a}\right), m = 2n - 1.]$$

Exercise 3.5.2. Find the form of the potential in the two-dimensional space, $0 < x < a$, $y > 0$, as shown in Figure 3.24.

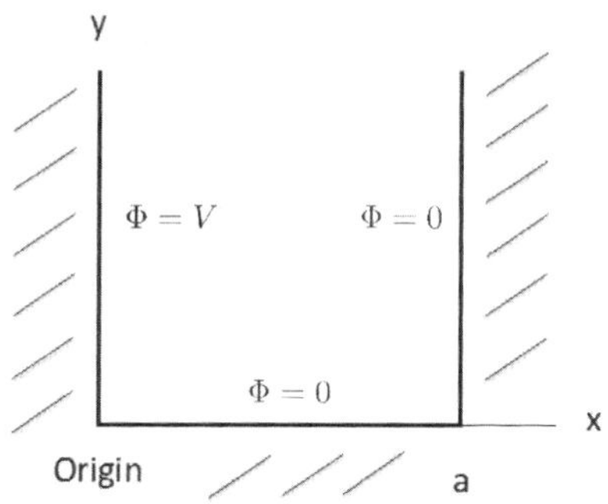

Fig. 3.24 Geometry for Exercise 3.5.2.

The vertical line at $x = 0$ is at potential V (a constant) whereas the horizontal line at $y = 0$ and the vertical line at $x = a$ are at potential zero. One also requires that the potential vanish as y approaches positive infinity. The solution in the region of interest can be represented as

$$\Phi = \Phi_1 + \Phi_2,$$

where

$$\Phi_1 = \sum_{n=1,2,3,\ldots} A_n \sin\left(\frac{n\pi x}{a}\right) \exp\left(-\frac{n\pi y}{a}\right),$$

$$\Phi_2 = V\left(1 - \frac{x}{a}\right).$$

Determine the coefficients A_n.

Exercise 3.5.3. Consider the same two-dimensional geometry as in Exercise 3.5.2, where conducting surfaces are located at $x = 0$ and a, as well as $y = 0$. The region of interest is thus: $0 \leq x \leq a$, $y \geq 0$.

(a) Show that the Dirichlet Green function, which satisfies

$$\nabla^2 G_D(\vec{x},\vec{x}\,') = -4\pi\delta(x-x')\delta(y-y'),$$

in the interior and

$$G_D(\vec{x},\vec{x}\,')|_{\vec{x}\in\text{walls}} = 0,$$

on the walls can be written as

$$G_D(\vec{x},\vec{x}\,') = 8\sum_{n=1}^{\infty}\frac{1}{n}\sin\left(\frac{n\pi x}{a}\right)\sin\left(\frac{n\pi x'}{a}\right)\sinh\left(\frac{n\pi y_<}{a}\right)e^{-\left(\frac{n\pi y_>}{a}\right)},$$

where $y_<$ ($y_>$) is the lesser (greater) of y and y'.

(b) Reconsider Exercise 3.5.2 and recover the previous result using the Green function from part (a). (You should see an interesting Fourier series for Φ_2!)

Exercise 3.5.4. Obtain the *Neumann* Green function $G_N(\vec{x},\vec{x}\,')$ in the same two-dimensional region, $0 < x < a$, $y > 0$, as in Exercises 3.5.2 and 3.5.3. The walls are now Neumann conductors, not normal conductors. The Green function satisfies

$$\nabla^2 G_N(\vec{x},\vec{x}\,') = -4\pi\delta(x-x')\delta(y-y'),$$

in the interior and will satisfy

$$\left.\frac{\partial G_N(\vec{x},\vec{x}\,')}{\partial n}\right|_{\vec{x}\in\text{walls}} = 0,$$

on the walls (n is a surface normal coordinate). Note that an alternate representation of a one-dimensional delta function is

$$\delta(x-x') = \frac{2}{a}\sum_{n=0}^{\infty}\delta_n\cos\left(\frac{n\pi x}{a}\right)\cos\left(\frac{n\pi x'}{a}\right),$$

where

$$\delta_n \equiv \begin{cases} 1/2, n=0 \\ 1, \ \ n>0 \end{cases}.$$

This delta function is built from even eigenfunctions whose *derivatives* (not *values*) vanish on the walls. Note the sum starts at $n = 0$.

[*Ans*: (up to an overall constant):

$$G_N(\vec{x},\vec{x}\,') = 8\sum_{n=1}^{\infty}\frac{1}{n}\cos\left(\frac{n\pi x}{a}\right)\cos\left(\frac{n\pi x'}{a}\right)\cosh\left(\frac{n\pi y_<}{a}\right)e^{-\left(\frac{n\pi y_>}{a}\right)} - 4\pi\frac{y_>}{a}.$$

Note the sum above starts at $n = 1$.]

Exercise 3.5.5. Find the two-dimensional Dirichlet Green function $G_D(\vec{x}, \vec{x}\,')$ representing a long line charge between two flat (infinite) parallel conductors which extend to $\pm\infty$ in the y-direction. The line charge is parallel to the plates. The surfaces are separated by the distance a, as shown in Figure 3.25. The potential is independent of the z-direction.

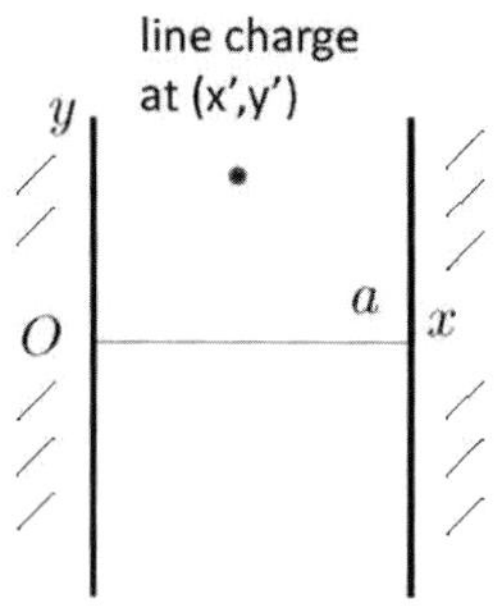

Fig. 3.25 Geometry for Exercise 3.5.5.

The two-dimensional Green function differential equation is

$$\left(\frac{\partial^2}{\partial x^2} + \frac{\partial^2}{\partial y^2}\right) G_D(\vec{x}, \vec{x}\,') = -4\pi\, \delta(x - x')\delta(y - y').$$

Assuming $G_D(\vec{x}, \vec{x}\,')$ may be written in the form

$$G_D(\vec{x}, \vec{x}\,') = \sum_{n=1,2,\ldots}^{\infty} \sin\left(\frac{n\pi x}{a}\right) \sin\left(\frac{n\pi x'}{a}\right) g_n(y, y'),$$

solve for the reduced Green function $g_n(y, y')$.

Exercise 3.5.6. Consider a two-dimensional box with side length a in the x-direction and side length b in the y-direction. You are given surface potentials $V = 0$ on the $y = 0$ boundary, $V = 0$ on the $x = 0$ and $x = a$ boundaries, and potential $V(x) = -Ex$ (E is a constant) on the $y = b$ boundary. Find the potential $\Phi(x, y)$ everywhere inside the box.

Exercise 3.6.1. Try to approximately reproduce the value of the Madelung constant for Na^+ as the reference atom in NaCl numerically from the Green function in Equation (3.98). Be aware that just putting $x = x' = y = y' = z = z' = a/2$ in this equation will result in a divergent expression because of the reference atom's self-energy. Can you remove this energy in an approximate way and get a result correct to at least three decimal places?

Exercise 3.7.1. Using separation of variables in cylindrical coordinates, evaluate the potential $\Phi(b, \phi)$ everywhere inside a charge-free conducting cylinder of radius b. The potential on the cylinder surface is given as

$$\Phi(b, \phi) = V_0\, \delta(\phi),$$

where $\delta(\phi)$ is a delta function and the azimuthal angle satisfies $-\pi < \phi < \pi$. (There are no other sources in this problem.) Find the Fourier series which gives the potential everywhere inside the cylinder. [*Ans*:

$$\Phi(\rho, \phi) = \frac{V_0}{2\pi} + \frac{V_0}{\pi} \sum_{m=1}^{\infty} \left(\frac{\rho}{b}\right) \cos(m\phi).]$$

Exercise 3.8.1. You are given the values of the potential on the interior surface of a conducting cylindrical pie with opening angle β, as shown; see Figure 3.26. The potentials V on the straight sides are constants, and the end potential $\Phi(b, \phi)$ at $\rho = b$ is a given function of the angle ϕ. Discontinuities in the surface potentials at $\phi = 0, \beta$ are allowed. Find the Fourier series representation of the potential $\Phi(\rho, \phi)$ in the interior.

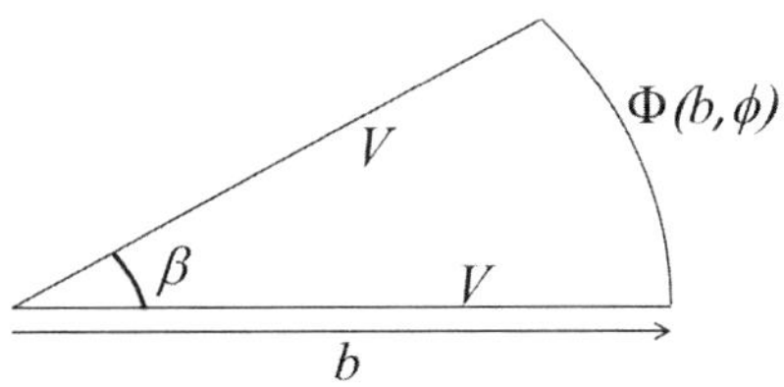

Fig. 3.26 Cylindrical pie geometry for Exercise 3.8.1.

Exercise 3.9.1. Using the Dirichlet Green function derived in Exercise 3.3.5, show that the potential inside a cylinder of radius b with surface potential $\Phi(b, \phi)$ is

$$\Phi(\rho, \phi) = \frac{1}{2\pi} \int_0^{2\pi} d\phi'\, \Phi(b, \phi') \frac{b^2 - \rho^2}{\rho^2 + b^2 - 2b\rho\cos(\phi - \phi')}.$$

This is called Poisson's integral.

Exercise 3.9.2. Using Exercise 3.9.1, rederive the expression (3.131) for the potential inside a long cylinder of radius b with potentials V_1 and V_2 on the two half-cylinders (see Figure 3.12).

Exercise 3.9.3. Fill in the details of the derivation of Equation (3.131) by summing the series in (3.130). [*Hint*: Define

$$Z \equiv \left(\frac{\rho}{b}\right) e^{i(\phi-\pi/2)},$$

and note that

$$\left(\frac{\rho}{b}\right)^{2n+1} \sin[(2n+1)(\phi-\pi/2)] = \text{Im}(Z^{2n+1}).]$$

Exercise 3.9.4.

(a) Using Fourier series methods, find the potential $\Phi^{\text{out}}(\rho,\phi)$ for the cylindrical halves problem, but this time in the region exterior to the cylinder $(\rho > b)$.

(b) By summing (see Exercise 3.9.3) or other means produce a closed form solution. Given the series solutions for the exterior and interior cases, is there an easy way of deriving one case from the other?

Exercise 3.9.5. We evaluated the potential $\Phi(\rho,\phi)$ for a point surface potential everywhere inside a charge-free conducting cylinder of radius b in Exercise 3.7.1 via separation of variables. Show that this series can be summed exactly in a simple form. (You may compare with the expected answer from Exercise 3.9.1.)

Exercise 3.11.1. (Adapted from Brown and Churchill, *op. cit.*) The two-dimensional Dirichlet problem described in Exercise 3.5.1 is conformally mapped to the space between the semicircular regions shown in Figure 3.27. All boundaries have zero potential except the boundary between $v=0$, $-r_0 < u < -1$, which has potential V.

(a) Show that the mapping $w = \exp(\pi z/b)$ takes the geometry and boundary conditions in Exercise 3.5.1 to the region shown in Figure 3.27.

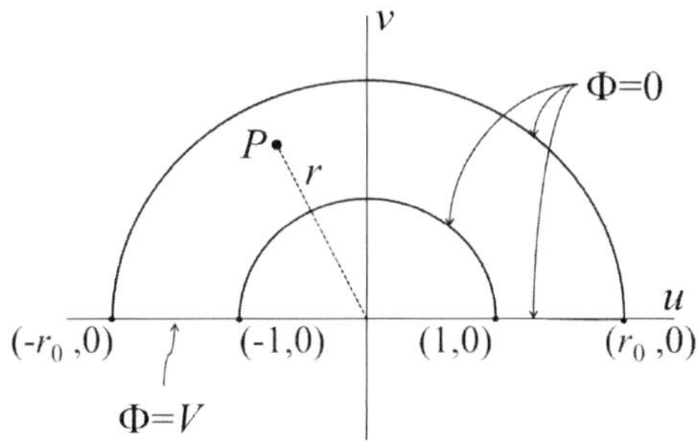

Fig. 3.27 Region for Exercise 3.11.1.

(b) Using this mapping, show that the potential everywhere inside the semicircular region is given by

$$\Phi(r,\theta) = \frac{4V}{\pi}\sum_{n=1}^{\infty}\frac{1}{m}\frac{\sinh\left(\frac{m\pi\theta}{\ln(r_0)}\right)}{\sinh\left(\frac{m\pi^2}{\ln(r_0)}\right)}\sin\left(m\pi\frac{\ln(r)}{\ln(r_0)}\right), \quad (m \equiv 2n-1),$$

where θ and r are cylindrical coordinates which locate a given point P in the semicircular region.

Exercise 3.11.2. Using the conformal mapping,

$$w = \frac{b^2}{z},$$

solve for the outside potential, $\Phi^{\text{out}}(\rho,\phi)$, of the cylinder of radius b in Section 3.9 given the inside solution, Equation (3.131).

Chapter 4

Electrostatics in Cylindrical and Spherical Coordinates

In Section 3.7 we applied the technique of *separation of variables* to find Green functions in geometries with rectangular symmetry. This technique was based on the fact that a complete set of solutions to the eigenfunction equation $\nabla^2\Phi(\vec{x}) = \lambda\Phi(\vec{x})$ (also called the *Helmholtz equation*) could be found separately for x, y, and z coordinates, thus enabling us to reduce three-dimensional problems with rectangular symmetry into three independent one-dimensional problems. The same technique can be applied to certain other symmetries as well. In particular, separation may also be applied to cylindrical and spherical coordinates, which we will discuss in detail in this chapter.[1]

4.1 Cylindrical coordinates and Bessel functions

Whenever there are fields expressed in cylindrical coordinates, Bessel functions almost always pop up, to the unending delight of physics students everywhere. Most textbooks introduce Bessel functions by applying separation of variables to Laplace's equation. We'll follow this approach in Section 4.5 — but in this section we'll use a generating function instead. This has the advantage of providing a powerful tool for proving properties of Bessel functions. We'll have occasion to use both approaches again when we treat time-dependent fields in spherical geometries (see Sections 10.10 and 10.11).

[1]In fact, there are 11 coordinate systems altogether in which the Helmholtz equation can be separated (see P. M. Morse and H. Feshbach, *Methods of Theoretical Physics, Part I*, McGraw-Hill (New York) 1953, Section 5.1). The other coordinate systems are of far less practical importance than rectangular, cylindrical, and spherical.

Before plunging into cylindrical coordinates, let's review our derivation of the free-space Green function $G_f(\vec{x}, \vec{x}\,')$ in Section 3.2. The function G_f is defined as the solution to

$$\nabla^2 G_f(\vec{x}, \vec{x}\,') = -4\pi\delta(\vec{x} - \vec{x}\,'), \tag{4.1}$$

in infinite space. Using the Fourier transform expression for $\delta(\vec{x} - \vec{x}\,')$, we can rewrite this as

$$\begin{aligned}\nabla^2 G_f(\vec{x}, \vec{x}\,') &= -\frac{4\pi}{(2\pi)^3}\int d^3k\, e^{i\vec{k}\cdot(\vec{x}-\vec{x}\,')} \\ &= -\frac{1}{2\pi^2}\int d^3k\, e^{ik_x(x-x')}e^{ik_y(y-y')}e^{ik_z(z-z')}.\end{aligned} \tag{4.2}$$

Since

$$\nabla^2 e^{ik_x(x-x')} = -k_x^2\, e^{ik_x(x-x')}$$

(and similarly for the y and z factors), by "inverting" the total Laplacian operator we obtain an expression for G_f:

$$G_f(\vec{x}, \vec{x}\,') = \frac{1}{2\pi^2}\int d^3k\, \frac{e^{ik_x(x-x')}e^{ik_y(y-y')}e^{ik_z(z-z')}}{k_x^2 + k_y^2 + k_z^2}. \tag{4.3}$$

Note that this derivation depended on our being able to write $e^{i\vec{k}\cdot(\vec{x}-\vec{x}\,')}$ in variable-separated form as the product of eigenfunctions of the Laplacian ∇^2. Unfortunately, this is not possible in cylindrical coordinates. However, we can do the next best thing, which is to write $e^{i\vec{k}\cdot(\vec{x}-\vec{x}\,')}$ as the *sum* of products of eigenfunctions. This is the approach we will take in the following discussion.

In the case of cylindrical coordinates, we may write

$$e^{i\vec{k}\cdot\vec{x}} = e^{i\vec{k}_\perp\cdot\vec{x}_\perp + k_z z} = (e^{ik_\perp\rho\cos(\phi-\alpha)})(e^{ik_z z}), \tag{4.4}$$

where $\vec{x} = (\rho, \phi, z)$ and $(k_\perp, \alpha, k_z)$ are the cylindrical coordinates for $\vec{x}$ and $\vec{k}$ respectively (as shown in Figure 4.1). The function $e^{ik_z z}$ is already an eigenfunction for the cylindrical Laplacian, so we may focus our attention on the ρ and ϕ dependence. Let us remind ourselves of the separated solution of the cylindrical Laplace equation we saw in Section 3.7, especially Equations (3.112) and (3.113). There we learned that the angular functions $e^{im\phi}$ for $m = 0, \pm1, \pm2, \ldots$ were solutions of the separated ϕ Laplace equation. Here, we simply assume that the complete cylindrical eigenfunctions are a product of this separated angular function and an unknown ρ dependent function. If we were to take a differential point of view at this point, we could deduce the differential equation for this function and try

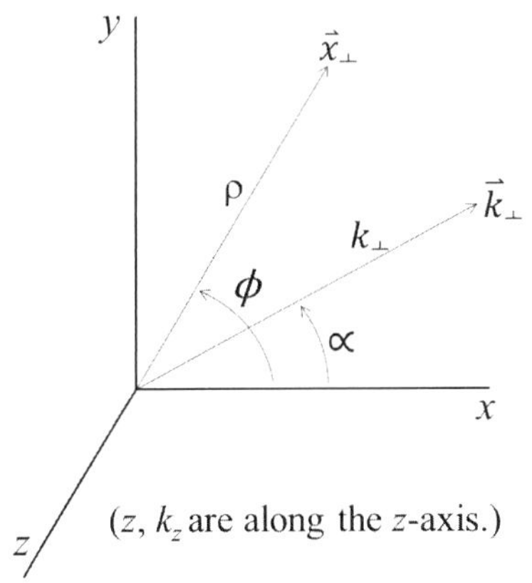

Fig. 4.1 Cylindrical coordinates.

to solve it. However, here we will keep our attention on the exponential in Equation (4.4), separating it and using it to *generate* the unknown ρ eigenfunction part.

Keeping this in mind we define

$$t \equiv k_\perp \rho; \qquad u \equiv ie^{i(\phi-\alpha)}, \tag{4.5}$$

from which it follows

$$e^{ik_\perp \rho \cos(\phi-\alpha)} = e^{\frac{t}{2}(u-\frac{1}{u})}. \tag{4.6}$$

Expanding the right-hand side of (4.6) in powers of u leads to the form,

$$e^{\frac{t}{2}(u-\frac{1}{u})} = \sum_{m=-\infty}^{\infty} u^m J_m(t). \tag{4.7}$$

The coefficients $J_m(t)$ turn out to be *Bessel functions* of integer order; and the left-hand side of (4.7) is the *generating function* for the Bessel functions. From (4.7) it is a straightforward exercise (Exercise 4.1.1) to obtain the Taylor series expansion for J_m (for $m \geq 0$):

$$J_m(t) = \sum_{n=0}^{\infty}(-1)^n \frac{(t/2)^{m+2n}}{n!(m+n)!}. \tag{4.8}$$

We may now investigate properties of the $J_m(t)$. Using the invariance of the generating function under the transformation $u \to -1/u$, we obtain that

$$\sum_{m=-\infty}^{\infty} u^m J_m(t) = \sum_{m=-\infty}^{\infty} (-1)^m u^{-m} J_m(t) = \sum_{m=-\infty}^{\infty} (-1)^m u^m J_{-m}(t). \tag{4.9}$$

Equating the left and right-hand sides of (4.9) term-by-term gives

$$J_{-m}(t) = (-1)^m J_m(t). \tag{4.10}$$

Combining Equations (4.6) and (4.7) and simplifying to $\alpha = 0$, we have

$$e^{it\cos\phi} = \sum_{m=-\infty}^{\infty} i^m e^{im\phi} J_m(t). \tag{4.11}$$

Multiplying both sides by $e^{-in\phi}/(2\pi)$ for n integer and integrating from 0 to 2π gives

$$\int_0^{2\pi} \frac{d\phi}{2\pi} e^{i(t\cos\phi - n\phi)} = \sum_{m=-\infty}^{\infty} i^m \int_0^{2\pi} \frac{d\phi}{2\pi} e^{i(m-n)\phi} J_m(t), \tag{4.12}$$

which leads to

$$J_n(t) = i^{-n} \int_0^{2\pi} \frac{d\phi}{2\pi} e^{i(t\cos\phi - n\phi)}, \tag{4.13}$$

where we have used the (easily verifiable) relation

$$\frac{1}{2\pi} \int_0^{2\pi} d\phi\, e^{i(m-n)\phi} = \delta_{mn}. \tag{4.14}$$

Equation (4.13) is called an *integral representation* for the Bessel function $J_n(t)$. Two additional equivalent forms are

$$J_n(t) = \frac{1}{2\pi} \int_{-\pi}^{\pi} d\phi\, e^{it\sin\phi - in\phi}, \tag{4.15}$$

$$J_n(t) = \frac{1}{\pi} \int_0^{\pi} d\phi\, \cos(t\sin\phi - n\phi). \tag{4.16}$$

We may develop a differential equation that $J_m(t)$ satisfies as follows. Consider

$$\frac{d}{dt} J_m(t) = \frac{1}{i^m} \int_0^{2\pi} \frac{d\phi}{2\pi} (i\cos\phi) e^{i(t\cos\phi - m\phi)}.$$

But

$$J_{m-1}(t) = \frac{1}{i^{m-1}} \int_0^{2\pi} \frac{d\phi}{2\pi} e^{i(t\cos\phi - (m-1)\phi)}$$
$$= \frac{1}{i^m} \int_0^{2\pi} \frac{d\phi}{2\pi} (ie^{i\phi}) e^{i(t\cos\phi - m\phi)},$$

and

$$J_{m+1}(t) = \frac{1}{i^m} \int_0^{2\pi} \frac{d\phi}{2\pi} \left(-ie^{-i\phi}\right) e^{i(t\cos\phi - m\phi)}.$$

Therefore

$$J_{m-1}(t) - J_{m+1}(t) = \frac{1}{i^m}\int_0^{2\pi}\frac{d\phi}{2\pi}i(e^{i\phi}+e^{-i\phi})e^{i(t\cos\phi - m\phi)}$$
$$= 2\frac{d}{dt}J_m(t). \tag{4.17}$$

Similarly we have

$$J_{m-1}(t) + J_{m+1}(t) = \frac{1}{i^m}\int_0^{2\pi}\frac{d\phi}{2\pi}(-2\sin\phi)e^{i(t\cos\phi - m\phi)}$$
$$= \frac{2}{i^m}\frac{1}{it}\int_0^{2\pi}\frac{d\phi}{2\pi}\frac{d}{d\phi}e^{it\cos\phi}e^{-im\phi}$$
$$= 0 - \frac{2}{i^m}\frac{1}{it}\int_0^{2\pi}\frac{d\phi}{2\pi}e^{it\cos\phi}\underbrace{\frac{d}{d\phi}e^{-im\phi}}_{-ime^{-im\phi}}$$
$$= \frac{2m}{t}J_m(t). \tag{4.18}$$

Equations (4.17) and (4.18) are called *recurrence relations*: they may be combined to obtain

$$J_{m+1}(t) = \left(\frac{m}{t} - \frac{d}{dt}\right)J_m(t), \tag{4.19}$$
$$J_{m-1}(t) = \left(\frac{m}{t} + \frac{d}{dt}\right)J_m(t). \tag{4.20}$$

Using first (4.19) then (4.20) gives

$$J_m(t) = \left(\frac{m-1}{t} - \frac{d}{dt}\right)J_{m-1}(t) = \left(\frac{m-1}{t} - \frac{d}{dt}\right)\left(\frac{m}{t} + \frac{d}{dt}\right)J_m(t)$$
$$\implies \left[\frac{d^2}{dt^2} + \frac{1}{t}\frac{d}{dt} - \frac{m^2}{t^2} + 1\right]J_m(t) = 0. \tag{4.21}$$

Equation (4.21) is the dimensionless form of *Bessel's equation*. Replacing the $k_\perp$ in (4.5) with k in order to streamline the notation gives us the more standard dimensional form:

$$\left[\frac{d^2}{d\rho^2} + \frac{1}{\rho}\frac{d}{d\rho} - \frac{m^2}{\rho^2} + k^2\right]J_m(k\rho) = 0. \tag{4.22}$$

Equation (4.22) gives our long-sought eigenfunctions of the ρ, ϕ part of the Laplacian operator:

$$\nabla^2 J_m(k\rho)e^{im\phi} = -k^2 J_m(k\rho)e^{im\phi}. \tag{4.23}$$

For the full Laplacian we have

$$\nabla^2 J_m(k\rho)e^{im\phi}e^{ik_z z} = -(k^2 + k_z^2)J_m(k\rho)e^{im\phi}e^{ik_z z}. \tag{4.24}$$

Notice that when $k_z^2 = -k^2$ we have a zero eigenvalue and therefore a solution to Laplace's equation. These eigenfunctions are fundamental to our construction of cylindrical Green functions.

Finally we mention that *Bessel functions of non-integer order* may be defined using various integral representations which generalize (4.15) to a real parameter ν.[2] The recurrence relations (4.17) and (4.18) are still valid in this case, and ν replaces m in Bessel's equation (4.22). A Taylor series expression for $J_\nu(t)$ may be derived either from the integral representation or most simply from a power series solution of the differential equation. One obtains

$$J_\nu(t) = \sum_{n=0}^{\infty}(-1)^n \frac{(t/2)^{\nu+2n}}{n!\,\Gamma(\nu+n+1)}, \quad \nu \neq -1, -2, -3, \ldots, \tag{4.25}$$

where $\Gamma(z)$ is the *gamma function*,[3]

$$\Gamma(z) \equiv \int_0^\infty dt\, t^{z-1}e^{-t}, \tag{4.26}$$

defined in general for complex z. From integration by parts for integers $n = 0, 1, 2, \ldots$ one may show that it satisfies

$$\Gamma(n+1) = n\,\Gamma(n) = n\,!. \tag{4.27}$$

Therefore (4.25) reduces to (4.8) when ν is an integer. The Bessel functions of order ν and $-\nu$ are linearly independent as long as ν is not an integer. Recall that when ν is an integer, we have the relationship (4.10) between $J_n(t)$ and $J_{-n}(t)$. The issue of finding another linearly independent solution to Bessel's equation when ν is an integer is resolved by the *Neumann function*, $N_\nu(t)$.[4] For general index ν it is given by

$$N_\nu(t) \equiv \left[\frac{J_\nu(t)\cos\nu\pi - J_{-\nu}(t)}{\sin\nu\pi}\right]. \tag{4.28}$$

When ν is an integer m, $N_m(t)$ is defined by

$$N_m(t) \equiv \lim_{\nu\to m} N_\nu(t), \tag{4.29}$$

where the limit in (4.29) is evaluated using the l'Hôpital rule. It turns out that the Neumann functions are singular at the origin, so are not valid free-space solutions in any domain that contains $t = 0$.

---

[2] J. T. Cushing, *Applied Analytical Methods for Physical Scientists*, John Wiley & Sons (New York) 1975, Ch. 7; P. M. Morse and H. Feshbach, *op. cit.*, Section 5.3.

[3] M. Abramowitz and I. A. Stegun, Eds., *Handbook of Mathematical Functions with Formulas, Graphs and Mathematical Tables*, National Bureau of Standards, 10th printing (Washington D. C.) 1972, Section 6.

[4] Alternate notation: $Y_\nu(t)$.

4.2 Completeness of Bessel functions

The generating function (4.7) gives us additional relations between the Bessel functions. The two-dimensional Fourier transform and inverse transform formulas

$$f(\vec{k}) = \int d^2x' f(\vec{x}\,')e^{-i\vec{k}\cdot\vec{x}\,'} \text{ and } f(\vec{x}) = \frac{1}{4\pi^2}\int d^2k f(\vec{k})e^{i\vec{k}\cdot\vec{x}} \tag{4.30}$$

can be combined to give

$$f(\vec{x}) = \int d^3x'\, f(\vec{x}\,')\left(\frac{1}{(2\pi)^2}\int d^2k\, e^{i\vec{k}\cdot(\vec{x}-\vec{x}\,')}\right), \tag{4.31}$$

which implies

$$\frac{1}{(2\pi)^2}\int d^2k\, e^{i\vec{k}\cdot(\vec{x}-\vec{x}\,')} = \delta^{(2)}(\vec{x}-\vec{x}\,'), \tag{4.32}$$

as expected from (2.34). Using polar coordinates ($\vec{x} = (\rho,\phi)$, $\vec{x}\,' = (\rho',\phi')$, $\vec{k} = (k,\alpha)$), we may rewrite (4.32) as

$$\delta^{(2)}(\vec{x}-\vec{x}\,') = \int \frac{dk\, kd\alpha}{(2\pi)^2} e^{ik\rho\cos(\phi-\alpha)}e^{-ik\rho'\cos(\phi'-\alpha)}. \tag{4.33}$$

We may re-express the exponentials in the integrand in terms of Bessel functions by using (4.7). Setting $t = k\rho$ and $u = ie^{i(\phi-\alpha)}$ in (4.7) gives

$$e^{ik\rho\cos(\phi-\alpha)} = \sum_{m=-\infty}^{\infty} i^m e^{im(\phi-\alpha)} J_m(k\rho), \tag{4.34}$$

while setting $t = k\rho'$ and $u = -ie^{-i(\phi'-\alpha)}$ in (4.7) gives

$$e^{-ik\rho'\cos(\phi'-\alpha)} = \sum_{m'=-\infty}^{\infty} (-i)^{m'} e^{-im'(\phi'-\alpha)} J_{m'}(k\rho'). \tag{4.35}$$

By substituting (4.34) and (4.35) into (4.33) and doing the α integral

$$\int_0^{2\pi} \frac{d\alpha}{2\pi} e^{-i\alpha(m-m')} = \delta_{mm'}, \tag{4.36}$$

we obtain

$$\delta^{(2)}(\vec{x}-\vec{x}\,') = \int_0^\infty \frac{dk}{2\pi}\, k \sum_{m=-\infty}^{\infty} e^{im(\phi-\phi')} J_m(k\rho')J_m(k\rho). \tag{4.37}$$

On the other hand, using (2.27) from Section 2.2 we have in polar coordinates

$$\delta^{(2)}(\vec{x}-\vec{x}\,') = \frac{1}{\rho}\delta(\rho-\rho')\delta(\phi-\phi'). \tag{4.38}$$

Combining (4.37) and (4.38), multiplying by $e^{-in(\phi-\phi')}$, and integrating over ϕ leads to:

$$\int_0^\infty dk\, kJ_n(k\rho)J_n(k\rho') = \frac{1}{\rho}\delta(\rho-\rho'). \tag{4.39}$$

Equation (4.39) is an example of a *completeness relation*: notice that it holds for *all* integers n. Plugging (4.39) back into (4.37) and equating with (4.38) gives another completeness relation, this time in the angular variable:

$$\sum_{m=-\infty}^{\infty} \frac{1}{2\pi} e^{im(\phi-\phi')} = \delta(\phi-\phi'). \tag{4.40}$$

Expression (4.39) leads to an analogue of the Fourier transform known as the *Hankel transform*:[5]

$$F(k) \equiv \int_0^\infty d\rho'\, \rho' f(\rho')J_n(k\rho') \tag{4.41}$$

$$\Longrightarrow\ f(\rho) = \int_0^\infty dk\, kF(k)J_n(k\rho). \tag{4.42}$$

Equation (4.40) also leads to discrete transform expressions for functions of ϕ. These transforms are only useful in problems with no boundaries in ϕ or ρ.

4.3 Zeros and orthogonality properties of Bessel functions

The differential equation (4.22) satisfied by $J_m(k\rho)$ can be written more simply as:

$$\left[\frac{d^2}{d\rho^2} + \frac{1}{\rho}\frac{d}{d\rho} - \frac{m^2}{\rho^2} + k^2\right] J_m(k\rho) = 0$$

$$\Longrightarrow \frac{1}{\rho}\frac{d}{d\rho}\left(\rho\frac{d}{d\rho}J_m(k\rho)\right) + \left(k^2 - \frac{m^2}{\rho^2}\right) J_m(k\rho) = 0. \tag{4.43}$$

Multiply by $\rho J_m(k'\rho)$ and integrate to obtain:

$$-\int_0^a d\rho\rho J_m(k'\rho)\frac{1}{\rho}\frac{d}{d\rho}\left(\rho\frac{d}{d\rho}J_m(k\rho)\right) = \int_0^a d\rho\,\rho\left(k^2 - \frac{m^2}{\rho^2}\right) J_m(k'\rho)J_m(k\rho). \tag{4.44}$$

[5]The completeness relation (4.39) and the Hankel transform equations are also valid for J_ν for real non-integer $\nu \geq -1/2$: see R. Piessens, "The Hankel Transform" in A. D. Poularikas, Ed., *The Transform and Applications Handbook*, 2nd ed., CRC Press (Boca Raton) 2011. When $\nu = -n$, these relations also hold because of (4.10).

Integrating the left-hand side of (4.44) by parts yields:

$$\int_0^a d\rho\, J_m(k\rho)\frac{d}{d\rho}\left(\rho\frac{dJ_m(k'\rho)}{d\rho}\right)$$
$$= \rho J_m(k\rho)\,\frac{dJ_m(k'\rho)}{d\rho}\bigg|_0^a - \int_0^a d\rho\,\rho\frac{dJ_m(k\rho)}{d\rho}\frac{dJ_m(k'\rho)}{d\rho}. \qquad (4.45)$$

The power series expansion of J_m in (4.8) shows that the $\rho = 0$ limit of the first term in (4.45) vanishes. The $\rho = a$ limit will also vanish if we choose k such that $J_m(ka) = 0$. Thus we define

$$k_{mn} \equiv \frac{x_{mn}}{a} \qquad \text{where } J_m(x_{mn}) = 0, \quad n = 1, 2, 3\ldots. \qquad (4.46)$$

Combining (4.44) and (4.45) gives

$$\int_0^a d\rho\,\rho\left(k_{mn}^2 - \frac{m^2}{\rho^2}\right) J_m(k_{mn}\rho)J_m(k_{mn'}\rho)$$
$$= \int_0^a d\rho\rho\frac{dJ_m(k_{mn}\rho)}{d\rho}\frac{dJ_m(k_{mn'}\rho)}{d\rho}. \qquad (4.47)$$

Note the right-hand side of (4.47) is symmetric in k_{mn} and $k_{mn'}$. If we interchange k_{mn} and $k_{mn'}$ in (4.47) and subtract this new equation from (4.47), we obtain

$$(k_{mn}^2 - k_{mn'}^2)\int_0^a d\rho\rho J_m(k_{mn}\rho)J_m(k_{mn'}\rho) = 0$$
$$\Longrightarrow \int_0^a d\rho\,\rho J_m(k_{mn}\rho)J_m(k_{mn'}\rho) = 0, \quad n \neq n'. \qquad (4.48)$$

These orthogonality relations will be useful when we compute the Dirchlet Green function for a cylindrical region.

When computing the Neumann Green function, we make use of similar relations. First, we define

$$\bar{k}_{mn} \equiv \frac{y_{mn}}{a} \qquad \text{where } \frac{d}{dy}J_m(y_{mn}) = 0, \quad n = 1, 2, 3\ldots. \qquad (4.49)$$

We may then use an analogous argument to obtain

$$\int_0^a d\rho\rho J_m(\bar{k}_{mn}\rho)J_m(\bar{k}_{mn'}\rho) = 0, \quad n \neq n'. \qquad (4.50)$$

Bessel functions satisfy the following normalization equations:

$$\int_0^a d\rho\rho[J_m(k_{mn}\rho)]^2 = \frac{a^2}{2}[J_{m+1}(x_{mn})]^2, \qquad (4.51)$$
$$\int_0^a d\rho\rho[J_m(\bar{k}_{mn}\rho)]^2 = \frac{a^2}{2}\left(1 - \frac{m^2}{y_{mn}^2}\right)[J_m(y_{mn})]^2. \qquad (4.52)$$

The derivation of (4.51) follows, while (4.52) is left as an exercise. Multiply both sides of Bessel's differential equation (4.43) by $2\rho^2 \mathrm{d}J_m(k\rho)/\mathrm{d}\rho$ to obtain

$$\left(2\rho^2 \frac{d}{d\rho} J_m(k\rho)\right)\left[\frac{1}{\rho}\frac{d}{d\rho}\left(\rho\frac{d}{d\rho}J_m(k\rho)\right) + \left(k^2 - \frac{m^2}{\rho^2}\right)J_m(k\rho)\right] = 0$$

$$\Longrightarrow \frac{d}{d\rho}\left(\rho\frac{d}{d\rho}J_m(k\rho)\right)^2 = \left(m^2 - k^2\rho^2\right)\frac{d}{d\rho}\left(J_m(k\rho)^2\right). \quad (4.53)$$

Set $k \equiv k_{mn}$ and integrate both sides from $\rho = 0$ to $\rho = a$, obtaining

$$\left(\rho\frac{d}{d\rho}J_m(k_{mn}\rho)\right)^2\Bigg|_0^a = \int_0^a d\rho\,\left(m^2 - k_{mn}^2\rho^2\right)\frac{d}{d\rho}[J_m(k_{mn}\rho)]^2. \quad (4.54)$$

The left-hand side of (4.54) can be evaluated directly while the right-hand side can be integrated by parts, giving

$$\begin{aligned} a^2\left(\frac{d}{d\rho}J_m(k_{mn}\rho)\right)^2\Bigg|_a &= \left(m^2 - k_{mn}^2\rho^2\right)J_m(k_{mn}\rho)^2\Big|_0^a + 2k_{mn}^2\int_0^a d\rho\,\rho[J_m(k_{mn}\rho)]^2 \\ &= 2k_{mn}^2\int_0^a d\rho\,\rho[J_m(k_{mn}\rho)]^2. \quad (4.55)\end{aligned}$$

In Section 4.1 we derived an expression for the Bessel function derivative (see (4.19)):

$$J_{m+1}(t) = \left(\frac{m}{t} - \frac{d}{dt}\right)J_m(t)$$

$$\Longrightarrow \frac{d}{d(k_{mn}\rho)}J_m(k_{mn}\rho)\Bigg|_{\rho=a} = \frac{m}{k_{mn}a}J_m(k_{mn}a) - J_{m+1}(k_{mn}a)$$

$$\Longrightarrow \frac{d}{d\rho}J_m(k_{mn}\rho)\Bigg|_{\rho=a} = -k_{mn}J_{m+1}(k_{mn}a). \quad (4.56)$$

Combining (4.54), (4.55) and (4.56) gives the result (4.51), completing the proof.

The above orthogonality and normalization relations may be summarized as:

$$\frac{2}{a^2[J_{m+1}(x_{mn})]^2}\int_0^a d\rho\rho J_m(k_{mn}\rho)J_m(k_{mn'}\rho) = \delta_{nn'}, \quad (4.57)$$

$$\frac{2}{a^2\left(1 - \frac{m^2}{y_{mn}^2}\right)[J_m(y_{mn})]^2}\int_0^a d\rho\,\rho J_m(\bar{k}_{mn}\rho)J_m(\bar{k}_{mn'}\rho) = \delta_{nn'}. \quad (4.58)$$

Using (4.57) and (4.58) we may directly compute the coefficients A_{mn}, B_{mn} in the following functional series expansions

$$f(\rho) = \sum_{n=1}^{\infty} A_{mn} J_m(k_{mn}\rho), \tag{4.59}$$

$$f(\rho) = \sum_{n=1}^{\infty} B_{mn} J_m(\bar{k}_{mn}\rho). \tag{4.60}$$

Note that sums are taken only over the index n: each fixed value of m is associated with a different expansion. Equation (4.59) is referred to as a Fourier-Bessel series or expansion. Inversion yields

$$A_{mn} = \frac{2}{a^2[J_{m+1}(x_{mn})]^2} \int_0^a d\rho' \rho' J_m\left(\frac{x_{mn}}{a}\rho'\right) f(\rho'), \tag{4.61}$$

$$B_{mn} = \frac{2}{a^2\left(1 - \frac{m^2}{y_{mn}^2}\right)[J_m(y_{mn})]^2} \int_0^a d\rho' \rho' J_m\left(\frac{y_{mn}}{a}\rho'\right) f(\rho'). \tag{4.62}$$

Substituting (4.61) into (4.59) and (4.62) into (4.60) gives:

$$f(\rho) = \int_0^a d\rho' \rho' f(\rho') \sum_{n=1}^{\infty} \frac{2}{a^2[J_{m+1}(x_{mn})]^2} J_m(k_{mn}\rho') J_m(k_{mn}\rho), \tag{4.63}$$

$$f(\rho) = \int_0^a d\rho' \rho' f(\rho') \sum_{n=1}^{\infty} \frac{2}{a^2\left(1 - \frac{m^2}{y_{mn}^2}\right)[J_m(y_{mn})]^2} J_m(\bar{k}_{mn}\rho') J_m(\bar{k}_{mn}\rho). \tag{4.64}$$

We shall assume without proof[6] the *completeness property* that all "reasonable" functions $f(\rho)$ can be written as a convergent series of the form (4.59) or (4.60). It follows from (4.63) and (4.64) that

$$\frac{1}{\rho}\delta(\rho - \rho') = \sum_{n=1}^{\infty} \frac{2}{a^2[J_{m+1}(x_{mn})]^2} J_m\left(\frac{x_{mn}\rho'}{a}\right) J_m\left(\frac{x_{mn}\rho}{a}\right); \tag{4.65}$$

$$\frac{1}{\rho}\delta(\rho - \rho') = \sum_{n=1}^{\infty} \frac{2}{a^2\left(1 - \frac{m^2}{y_{mn}^2}\right)[J_m(y_{mn})]^2} J_m\left(\frac{y_{mn}\rho'}{a}\right) J_m\left(\frac{y_{mn}\rho}{a}\right). \tag{4.66}$$

These expressions may be somewhat simplified by introducing the notation

$$J_{1m}(t) \equiv \frac{\sqrt{2}}{a} \frac{J_m(t)}{J_{m+1}(x_{mn})}, \tag{4.67}$$

$$J_{2m}(t) \equiv \frac{\sqrt{2}}{a\sqrt{1 - \frac{m^2}{y_{mn^2}}}} \frac{J_m(t)}{J_m(y_{mn})}. \tag{4.68}$$

[6]Readers who are interested in the rigorous mathematical details may consult R. Courant and D. Hilbert, *Methods of Mathematical Physics, Vol. I*, Wiley-VCH (Weinheim) 1989, Chapter 6.

The orthogonality and completeness relations may then be expressed as:

$$\int_0^a d\rho\, \rho J_{1m}(k_{mn}\rho) J_{1m}(k_{mn'}\rho) = \delta_{nn'}, \tag{4.69}$$

$$\int_0^a d\rho\, \rho J_{2m}(\bar{k}_{mn}\rho) J_{2m}(\bar{k}_{mn'}\rho) = \delta_{nn'}, \tag{4.70}$$

$$\sum_{n=1}^{\infty} J_{1m}(k_{mn}\rho') J_{1m}(k_{mn}\rho) = \frac{1}{\rho}\delta(\rho - \rho'), \tag{4.71}$$

$$\sum_{n=1}^{\infty} J_{2m}(\bar{k}_{mn}\rho') J_{2m}(\bar{k}_{mn}\rho) = \frac{1}{\rho}\delta(\rho - \rho'). \tag{4.72}$$

We emphasize that (4.69)–(4.72) are valid for *each* integer m.[7] Actual zeros of the various integer Bessel functions may be found using many types of computer applications, and are not listed here.

Note the apparent similarity between (4.71) and our earlier completeness relation (4.39): the latter is actually a limiting case of the former; see Exercise 4.6.1. This is analogous to the relationship between Fourier series on a finite interval and the continuous Fourier transform over an infinite interval.

4.4 Reduced Green function for the conducting cylinder

We now have the mathematical tools to solve potential problems for the region inside a cylinder with caps (see Figure 4.2). We suppose the curved surface and circular base are grounded conductors ($V = 0$), while the potential on the top surface is specified as $V(\rho, \phi)$. In this section we find the general solution using the reduced Green function, and in Section 4.5 we will give an alternative solution using separation of variables.

To begin, recall that the Green function is defined as a solution to Poisson's equation:

$$-\nabla^2 G_D(\vec{x}, \vec{x}\,') = 4\pi\delta(\vec{x} - \vec{x}\,'). \tag{4.73}$$

In Section 3.6 we introduced the strategy of building the Green function out of eigenfunctions of the Laplacian in two of the dimensions, leading to a

[7]These orthogonality and completeness relations can actually be shown to hold for J_ν for real non-integer $\nu \geq -1$: see G. N. Watson, *A Treatise on the Theory of Bessel Functions*, 2nd ed., reprinted in the Cambridge Mathematical Library series (Cambridge) 1996, Section 18.1.

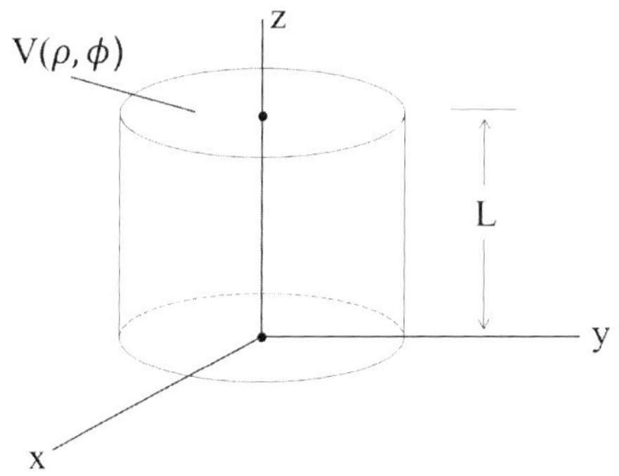

Fig. 4.2 Cylindrical region with caps.

reduced Green function in the third dimension. According to this strategy, the eigenfunctions are determined by the geometry and the boundary conditions. In this case we may use the eigenfunction expansion from Section 4.3 to write

$$\begin{aligned} 4\pi\delta(\vec{x}-\vec{x}\,') &= \frac{4\pi}{\rho}\delta(\rho-\rho')\delta(\phi-\phi')\delta(z-z') \\ &= 4\pi\sum_{n=1}^{\infty}\sum_{m=-\infty}^{\infty} J_{1m}(k_{mn}\rho')J_{1m}(k_{mn}\rho)\frac{e^{im(\phi-\phi')}}{2\pi}\delta(z-z'), \end{aligned} \tag{4.74}$$

where each term in the sum includes the factors $J_{1m}(k_{mn}\rho)e^{im\phi}$, which we have shown previously is an eigenfunction of the Laplacian with eigenvalue $-k_{mn}^2$. This suggests the following form for the Green function:

$$G_D(\vec{x},\vec{x}\,') = 4\pi\sum_{m=-\infty}^{\infty}\sum_{n=1}^{\infty} J_{1m}(k_{mn}\rho')J_{1m}(k_{mn}\rho)\frac{e^{im(\phi-\phi')}}{2\pi}g_{mn}(z,z'). \tag{4.75}$$

Plugging this expansion into Poisson's equation then leads to

$$\left[k_{mn}^2 - \frac{\partial^2}{\partial z^2}\right] g_{mn}(z,z') = \delta(z-z'). \tag{4.76}$$

Furthermore, the boundary conditions at the caps of the cylinder lead to

$$g_{mn}(z,z')|_{z=0,L} = 0. \tag{4.77}$$

We have already computed the solution to (4.76) in Section 3.6, when finding the reduced Green function for the conducting box; see Equation (3.97):

$$g_{mn}(z,z') = \frac{\sinh(k_{mn}z_<)\sinh[k_{mn}(L-z_>)]}{k_{mn}\sinh(k_{mn}L)}. \tag{4.78}$$

The Green function for the cylindrical region is therefore

$$G_D(\vec{x}, \vec{x}\,') = 2\sum_{n=1}^{\infty}\sum_{m=-\infty}^{\infty} e^{im(\phi-\phi')} J_{1m}(k_{mn}\rho') J_{1m}(k_{mn}\rho) \times \frac{\sinh(k_{mn}z_<)\sinh[k_{mn}(L-z_>)]}{k_{mn}\sinh(k_{mn}L)}. \tag{4.79}$$

To solve the given boundary value problem, we must evaluate:

$$\left.\frac{\partial}{\partial z'} g_{mn}(z, z')\right|_{z'=L} = -\frac{\sinh(k_{mn}z)}{\sinh(k_{mn}L)}. \tag{4.80}$$

This leads to the final solution for the potential:

$$\Phi(\vec{x}) = \sum_{n,m} C_{mn} e^{im\phi} J_{1m}(k_{mn}\rho)\sinh(k_{mn}z), \tag{4.81}$$

where

$$C_{mn} = \frac{1}{2\pi}\frac{1}{\sinh(k_{mn}L)}\int_0^a d\rho'\rho' \int_0^{2\pi} d\phi' e^{-im\phi'} J_{1m}(k_{mn}\rho')V(\rho', \phi'). \tag{4.82}$$

4.5 Potential inside a cylinder as a boundary value problem

The result of the previous section may also be obtained using separation of variables, which we used in Section 3.5 to find the potential inside a conducting box. In cylindrical coordinates, the Laplace equation is:

$$\frac{\partial^2\Phi}{\partial\rho^2} + \frac{1}{\rho}\frac{\partial\Phi}{\partial\rho} + \frac{1}{\rho^2}\frac{\partial^2\Phi}{\partial\phi^2} + \frac{\partial^2\Phi}{\partial z^2} = 0. \tag{4.83}$$

The assumption that $\Phi(\rho, \phi, z) = P(\rho)Q(\phi)Z(z)$ leads to the three equations:

$$\frac{d^2Z}{dz^2} - k^2 Z = 0, \tag{4.84}$$

$$\frac{d^2Q}{d\phi^2} + \nu^2 Q = 0, \tag{4.85}$$

$$\frac{d^2P}{d\rho^2} + \frac{1}{\rho}\frac{dP}{d\rho} + \left(k^2 - \frac{\nu^2}{\rho^2}\right)P = 0, \tag{4.86}$$

where we may recognize (4.86) as Bessel's equation (4.43). It follows directly from these equations that $Z(z) \sim e^{\pm kz}, Q(\phi) \sim e^{\pm i\nu\phi}$. If the azimuthal range is restricted, then in general ν is a (real) non-integer and

$P(\rho) \sim J_{\pm\nu}(k\rho)$ are independent solutions, where J_ν is given by the power series (4.25). If the full azimuthal range is allowed for ϕ, as is the case here, then consistency requires that $\nu = m$ where m is an integer. In this case one solution is $P(\rho) \sim J_m(k\rho)$ and the other is the Neumann function of Section 4.1. As was pointed out, the Neumann functions are singular at the origin, so are excluded here.

Now we impose boundary conditions on our separated solutions $\Phi(\rho, \phi, z) = J_m(k\rho)e^{im\phi}Z(z)$:

$$\begin{aligned} \Phi\big|_{\rho=a} = 0 &\implies J_m(ka) = 0 \\ &\implies k_{mn} = \frac{x_{mn}}{a}; \end{aligned} \tag{4.87}$$

$$\begin{aligned} \Phi\big|_{z=0} = 0 &\implies Z(0) = 0 \\ &\implies Z(z) \sim \sinh(k_{mn}z). \end{aligned} \tag{4.88}$$

We therefore obtain the general solution:

$$\Phi(\rho, \phi, z) = \sum_{n,m} C_{mn} e^{im\phi} J_{1m}(k_{mn}\rho) \sinh(k_{mn}z). \tag{4.89}$$

The boundary condition at $z = L$ gives

$$V(\rho, \phi) = \sum_{n,m} C_{mn} e^{im\phi} J_{1m}(k_{mn}\rho) \sinh(k_{mn}L). \tag{4.90}$$

We make use of our orthogonality formulas:

$$\begin{aligned} \delta_{mm'} &= \int_0^{2\pi} \frac{d\phi}{2\pi} e^{i\phi(m-m')}, \quad \delta_{nn'} = \int_0^a d\rho\, \rho J_{1m}(k_{mn}\rho) J_{1m}(k_{mn'}\rho) \\ \implies C_{mn} &= \frac{1}{2\pi \sinh(k_{mn}L)} \int_0^a d\rho' \rho' \int_0^{2\pi} d\phi\, e^{-im\phi} J_{1m}(k_{mn}\rho') V(\rho', \phi'), \end{aligned} \tag{4.91}$$

which agrees with our previous result in (4.81) and (4.82). Although this method is quicker than the reduced Green function method, it is less general.

4.6 Bessel functions of imaginary argument; asymptotic forms of Bessel functions

In the previous section, we obtained Bessel's equation for the radial function $P(\rho)$ due to the choice of $Z''(z) = k^2 Z(z)$ and $Q''(\phi) = -\nu^2 Q(\phi)$ in (4.84) and (4.85) for the z- and ϕ-dependent parts of the separated solution. This choice led to the functional form $Z \sim e^{\pm kz}$. However, in some cases (for instance where $\Psi(z=0) = \Psi(z=L) = 0$) it is preferable to choose $Z''(z) = -k^2 Z$, which leads to cosine and sine solutions for the z-dependent portion of the solution. Instead of Bessel's equation, we then have

$$\left[\frac{1}{\rho}\frac{d}{d\rho}\left(\rho\frac{d}{d\rho}\right) - \frac{\nu^2}{\rho^2} - k^2\right] P(\rho) = 0. \tag{4.92}$$

The solutions to (4.92) are linear combinations of the *Bessel functions of imaginary argument*, which are defined as follows:

$$I_\nu(k\rho) \equiv i^{-\nu} J_\nu(ik\rho), \tag{4.93}$$

$$K_\nu(k\rho) \equiv \frac{\pi}{2} i^{\nu+1} H_\nu^{(1)}(ik\rho). \tag{4.94}$$

($I_\nu(k\rho)$ and $K_\nu(k\rho)$ are real functions for $k\rho$ real.) $H_\nu^{(1)}$ is one of the *Hankel functions* of order ν, which are defined by

$$H_\nu^{(1)}(k\rho) \equiv J_\nu(k\rho) + iN_\nu(k\rho), \tag{4.95}$$

$$H_\nu^{(2)}(k\rho) \equiv J_\nu(k\rho) - iN_\nu(k\rho). \tag{4.96}$$

Note $H_\nu^{(2)}(k\rho)$ is just the complex conjugate of $H_\nu^{(1)}(k\rho)$.

The qualitative behavior of I_ν and K_ν is quite different from that of J_ν and N_ν. In particular, consider the asymptotic behavior of $J_\nu(k\rho)$ and $N_\nu(k\rho)$ as $\rho \to \infty$. If we retain only the leading-order terms in Bessel's equation, we obtain

$$\left[\frac{1}{\rho}\frac{d}{d\rho}\left(\rho\frac{d}{d\rho}\right) - \frac{\nu^2}{\rho^2} + k^2\right] P(\rho) = 0 \xrightarrow[\rho\to\infty]{} \frac{d^2P}{d\rho^2} + k^2 P = 0, \tag{4.97}$$

which indicates sinusoidal dependence. A more exact treatment[8] leads to the results (ν real)

$$J_\nu(k\rho) \xrightarrow[k\rho\gg 1,|\nu|]{} \sqrt{\frac{2}{\pi k\rho}} \cos\left(k\rho - \frac{\nu\pi}{2} - \frac{\pi}{4}\right), \tag{4.98}$$

$$N_\nu(k\rho) \xrightarrow[k\rho\gg 1,|\nu|]{} \sqrt{\frac{2}{\pi k\rho}} \sin\left(k\rho - \frac{\nu\pi}{2} - \frac{\pi}{4}\right). \tag{4.99}$$

On the other hand, applying these results to (4.93)–(4.94) leads immediately to

$$I_\nu(k\rho) \xrightarrow[k\rho\gg 1,|\nu|]{} \frac{e^{k\rho}}{\sqrt{2\pi k\rho}}, \tag{4.100}$$

$$K_\nu(k\rho) \xrightarrow[k\rho\gg 1,|\nu|]{} \sqrt{\frac{\pi}{2k\rho}} e^{-k\rho}. \tag{4.101}$$

[8]G. B. Arfken, H. J. Weber, F. E. Harris, *Mathematical Methods for Physicists*, 7th ed., Academic Press (Waltham, MA) 2013, Section 14.6.

It follows that $I_\nu(k\rho) \to \infty$ and $K_\nu(k\rho) \to 0$ as $\rho \to \infty$. Expansions for $k\rho \ll 1$ yield

$$J_\nu(k\rho) \underset{k\rho \ll 1}{\longrightarrow} \frac{1}{\Gamma(\nu+1)} \left(\frac{k\rho}{2}\right)^\nu \quad (\nu \neq -1, -2, -3, \dots), \tag{4.102}$$

and

$$N_\nu(k\rho) \underset{k\rho \ll 1}{\longrightarrow} -\frac{\Gamma(\nu)}{\pi} \left(\frac{2}{k\rho}\right)^\nu \quad (\nu > 0), \tag{4.103}$$

$$N_0(k\rho) \underset{k\rho \ll 1}{\longrightarrow} \frac{2}{\pi} \left(\ln\left(\frac{k\rho}{2}\right) + \gamma \right). \tag{4.104}$$

In addition

$$I_\nu(k\rho) \underset{k\rho \ll 1}{\longrightarrow} \frac{1}{\Gamma(\nu+1)} \left(\frac{k\rho}{2}\right)^\nu \quad (\nu \neq -1, -2, -3, \dots), \tag{4.105}$$

identical to (4.102), and

$$K_\nu(k\rho) \underset{k\rho \ll 1}{\longrightarrow} \frac{\Gamma(\nu)}{2} \left(\frac{2}{k\rho}\right)^\nu \quad (\nu > 0), \tag{4.106}$$

$$K_0(k\rho) \underset{k\rho \ll 1}{\longrightarrow} -\left(\ln\left(\frac{k\rho}{2}\right) + \gamma \right). \tag{4.107}$$

$\gamma = 0.57721\dots$ is known as Euler's constant and $\Gamma(z)$ is the gamma function.

Unlike the J_ν, the I_ν are not complete: there is no analog to the Hankel transform (4.39) for the functions I_ν. Furthermore, there are no orthogonality relations for I_ν, K_ν on finite intervals. Note the tradeoff between the ρ and z-dependent portions of the separated solution for Laplace's equation: if the ρ-dependent solutions are oscillatory and complete, then the z-dependent solutions are not, and vice versa. The reason for this is the $k_z^2 = -k^2$ condition for solutions to Laplace's equation mentioned after Equation (4.24), which is built into the separation equations (4.84) and (4.86).

4.7 Cylindrical free-space Green function using the Wronskian technique

In this section we will make use of the Bessel functions J_m, N_m, I_m, K_m introduced in the preceding sections to construct two different expansions for the Green function in free space in cylindrical coordinates. One expansion will utilize $\{J_m\}$, while the other will use $\{I_m, K_m\}$. The mathematical techniques we will introduce are quite general and can be applied to many other situations.

For the expansion in $\{J_m\}$, we shall follow the methodology of Section 4.4, and expand the delta function as

$$4\pi\delta(\vec{x}-\vec{x}\,') = \frac{4\pi}{\rho}\delta(\rho-\rho')\delta(\phi-\phi')\delta(z-z'), \tag{4.108}$$
$$= 2\int_0^\infty dk\, k \sum_{m=-\infty}^{\infty} e^{im(\phi-\phi')} J_m(k\rho)J_m(k\rho')\delta(z-z').$$

In light of the results of Section 4.4, we may expect a Green function solution of the form

$$G(\vec{x},\vec{x}\,') = 2\int_0^\infty dk\, k \sum_{m=-\infty}^{\infty} e^{im(\phi-\phi')} J_m(k\rho)J_m(k\rho')g_m(z,z'), \tag{4.109}$$

and taking the Laplacian gives (after some simplification)

$$-\nabla^2 G(\vec{x},\vec{x}\,') = 2\int_0^\infty dk\, k \sum_{m=-\infty}^{\infty} e^{im(\phi-\phi')} J_m(k\rho)J_m(k\rho')$$
$$\times \left[k^2 - \frac{\partial^2}{\partial z^2}\right] g_m(z,z'). \tag{4.110}$$

Comparison with (4.108) gives

$$\left[k^2 - \frac{\partial^2}{\partial z^2}\right] g_m(z,z') = \delta(z-z'). \tag{4.111}$$

In view of the discontinuity at $z = z'$, a variable-separated solution should have the form

$$g_m(z,z') = \begin{cases} f_1(z')\psi_1(z) & z \le z' \\ f_2(z')\psi_2(z) & z \ge z' \end{cases}, \tag{4.112}$$

where ψ_1 and ψ_2 are solutions to the homogeneous equation

$$k^2\psi_j - \psi_j'' = 0, \qquad j = 1, 2. \tag{4.113}$$

The symmetry relation $g_m(z,z') = g_m(z',z)$ then immediately tells us that $f_1 = \psi_2, f_2 = \psi_1$, so that

$$g_m(z,z') = \begin{cases} \psi_2(z')\psi_1(z) & z \le z' \\ \psi_1(z')\psi_2(z) & z \ge z' \end{cases}. \tag{4.114}$$

This can be written in abbreviated form as

$$g_m(z,z') = \psi_1(z_<)\psi_2(z_>). \tag{4.115}$$

Integrating (4.111) in the variable z over an infinitesimal interval containing z' yields

$$-\left.\frac{dg_m}{dz}\right|_{z=z'+} + \left.\frac{dg_m}{dz}\right|_{z=z'-} = 1,$$
$$\Longrightarrow -\psi_2'(z')\psi_1(z') + \psi_1'(z')\psi_2(z') = 1. \tag{4.116}$$

At this point we define the *Wronskian* of two functions as

$$W_z[\psi_1, \psi_2] \equiv \psi_1 \frac{d}{dz}\psi_2 - \psi_2 \frac{d}{dz}\psi_1, \tag{4.117}$$

where the subscript z on W_z indicates the variable of differentiation on the right-hand side. Using this notation and replacing $z' \to z$ we may rewrite (4.116) as

$$W_z[\psi_1(z), \psi_2(z)] = -1. \tag{4.118}$$

Now we already know that the homogeneous solutions have the form $e^{\pm kz}$. The finite boundary conditions at $z = \pm\infty$ require that $\psi_1(z) = C_1 e^{kz}$, $\psi_2(z) = C_2 e^{-kz}$, and without loss of generality one can choose $C_1 = C$ and $C_2 = 1$. The equation (4.118) then leads immediately to $2kC = 1$, which after some algebra gives the result

$$g_m(z, z') = \frac{1}{2k} e^{-k(z_> - z_<)}. \tag{4.119}$$

This leads to the following expression for the free-space Green function:

$$\frac{1}{|\vec{x} - \vec{x}\,'|} = \int_0^\infty dk \sum_{m=-\infty}^{\infty} e^{im(\phi-\phi')} J_m(k\rho) J_m(k\rho') e^{-k(z_> - z_<)}. \tag{4.120}$$

In (4.120), the basis functions in ρ are oscillatory, while those in z are non-oscillatory. We may derive an alternative expression to (4.120) that reverses this situation by starting with the following delta function expansion (compare (4.108)):

$$\begin{aligned}
4\pi\delta(\vec{x} - \vec{x}\,') &= \frac{4\pi}{\rho}\delta(\rho - \rho')\delta(\phi - \phi')\delta(z - z'), \\
&= \int_{-\infty}^{\infty} dk\, \frac{4\pi}{\rho}\delta(\rho - \rho') \sum_{m=-\infty}^{\infty} \frac{e^{im(\phi-\phi')}}{2\pi} \frac{e^{ik(z-z')}}{2\pi} \\
&= \frac{2}{\pi\rho}\delta(\rho - \rho') \int_0^\infty dk\, \cos[k(z - z')] \sum_{m=-\infty}^{\infty} e^{im(\phi-\phi')}.
\end{aligned} \tag{4.121}$$

We assume a Green function similar to (4.121), only this time with a ρ-dependent reduced Green function:

$$G(\vec{x}, \vec{x}\,') = \frac{2}{\pi} \int_0^\infty dk \cos[k(z - z')] \sum_{m=-\infty}^{\infty} e^{im(\phi - \phi')} g_m(\rho, \rho'). \tag{4.122}$$

The Green function defining equation gives

$$\frac{1}{\rho} \frac{d}{d\rho} \left(\rho \frac{dg_m}{d\rho} \right) - \left(k^2 + \frac{m^2}{\rho^2} \right) g_m = -\frac{1}{\rho} \delta(\rho - \rho'). \tag{4.123}$$

Solutions to the homogeneous form of (4.123) are the Bessel functions of imaginary argument $I_m(k\rho)$ and $K_m(k\rho)$.

Using the symmetry of the Green function as before, we have in analogy to (4.115)

$$g_m(\rho, \rho') = \psi_1(k\rho_<)\psi_2(k\rho_>). \tag{4.124}$$

Since $\rho = 0$ is within the region we must have $g_m(\rho, \rho')$ finite at $\rho = 0$, which implies that $\psi_1(k\rho) \propto I_m(k\rho)$. Similarly, the requirement that $g_m(\rho, \rho') \to 0$ as $\rho \to \infty$ leads to $\psi_2(k\rho) \propto K_m(k\rho)$: so we have

$$g_m(\rho, \rho') = A I_m(k\rho_<) K_m(k\rho_>). \tag{4.125}$$

Integrating (4.123) in the variable ρ over an infinitesimal interval containing ρ' gives

$$\begin{aligned}
-\left.\frac{dg_m}{d\rho}\right|_{\rho=\rho'+} + \left.\frac{dg_m}{d\rho}\right|_{\rho=\rho'-} &= \frac{1}{\rho'} \\
\Longrightarrow A\left[-\frac{dK_m(k\rho')}{d\rho'} I_m(k\rho') + K_m(k\rho') \frac{dI_m(k\rho')}{d\rho'} \right] &= \frac{1}{\rho'} \\
\Longrightarrow A \cdot W_{\rho'}[I_m(k\rho'), K_m(k\rho')] &= -\frac{1}{\rho'}.
\end{aligned} \tag{4.126}$$

By using the asymptotic forms for $I_m(k\rho), K_m(k\rho)$ as $\rho \to 0$ (or $\rho \to \infty$), it is straightforward to show that $A = 1$ in (4.126). Putting all the pieces together we have finally

$$\frac{1}{|\vec{x} - \vec{x}\,'|} = \frac{2}{\pi} \int_0^\infty dk \cos[k(z - z')] \sum_{m=-\infty}^{\infty} e^{im(\phi - \phi')} I_m(k\rho_<) K_m(k\rho_>). \tag{4.127}$$

In the preceding two examples, the Wronskian played a key role in determining the reduced Green function's overall coefficient from the discontinuity condition. In fact, this same technique can be generally applied to determine reduced Green function coefficients in other geometries. Typically, the reduced Green function in some variable x will have the form

$$g_m(x, x') = C\psi_1(x_<)\psi_2(x_>), \tag{4.128}$$

where $\psi_1(x)$ and $\psi_2(x)$ are homogeneous solutions. This form guarantees continuity of the Green function at $x = x'$. The discontinuity condition is expressible as

$$\left.\frac{dg_m}{dx'}\right|_{x'=x^-}^{x'=x^+} = D(x), \tag{4.129}$$

which simplifies to

$$C = \frac{D(x)}{W_x[\psi_1(x), \psi_2(x)]}. \tag{4.130}$$

Explicitly evaluating the Wronskian then gives the constant C.

4.8 Reduced Green function method for the conducting wedge

Not all cylindrically symmetric problems involve Bessel functions. For instance, let us consider the finite wedge Green function shown in Figure 4.3. Note that we have already examined an associated boundary value problem with $a \to \infty$ in Section 3.8. It is appropriate to use a line-charge source in three dimensions, so that two-dimensional symmetry is preserved. To find the Green function, we must solve:

$$\nabla^2 G(\vec{x}, \vec{x}\,') = -4\pi\frac{1}{\rho}\delta(\rho - \rho')\delta(\phi - \phi'). \tag{4.131}$$

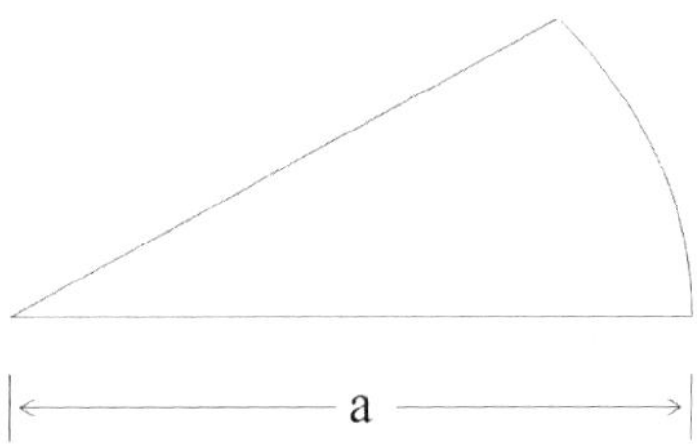

Fig. 4.3 Finite wedge.

The restricted range in ϕ is reminiscent of the conducting box problem that was solved in Section 3.6. We therefore take a leaf from our previous discussion and expand in Fourier sine series (see Equation (3.102)):

$$\frac{1}{2}\delta(\phi-\phi') = \frac{1}{\beta}\sum_{n=1}^{\infty}\sin\left(\frac{n\pi\phi}{\beta}\right)\sin\left(\frac{n\pi\phi'}{\beta}\right), \tag{4.132}$$

which is valid on the interval $0 \le \phi \le \beta$. We similarly expand the Green function

$$G(\vec{x},\vec{x}\,') = \frac{8\pi}{\beta}\sum_{n=1}^{\infty}\sin\left(\frac{n\pi\phi}{\beta}\right)\sin\left(\frac{n\pi\phi'}{\beta}\right) g_n(\rho,\rho'), \tag{4.133}$$

and obtain from Poisson's equation that

$$\left[\frac{1}{\rho}\frac{d}{d\rho}\left(\rho\frac{d}{d\rho}\right) - \frac{n^2\pi^2}{\rho^2\beta^2}\right] g_n = -\frac{1}{\rho}\delta(\rho-\rho'). \tag{4.134}$$

This could still be considered Bessel's function equation for $k = 0$. If we attempt to find homogeneous solutions of the form ρ^γ, we end up with the condition

$$\left[\frac{\gamma^2}{\rho^2} - \frac{n^2\pi^2}{\rho^2\beta^2}\right] g_n = 0, \tag{4.135}$$

which leads immediately to

$$g_n \sim \rho^{\pm\gamma} \quad \text{where } \gamma \equiv \frac{n\pi}{\beta}. \tag{4.136}$$

To satisfy the Green function boundary conditions we take

$$\psi_1(\rho') \equiv \rho'^\gamma; \quad \psi_2(\rho') \equiv \left(\rho'^\gamma - \frac{a^{2\gamma}}{\rho'^\gamma}\right), \tag{4.137}$$

and just as in Section 4.7 we have

$$g_n(\rho,\rho') = C\psi_1(\rho_<)\psi_2(\rho_>). \tag{4.138}$$

The discontinuity condition is

$$\left.\frac{dg_n}{d\rho}\right|_+ - \left.\frac{dg_n}{d\rho}\right|_- = -\frac{1}{\rho'}. \tag{4.139}$$

We may also compute

$$\begin{aligned} W_{\rho'}[\psi_1,\psi_2] &= \gamma\rho'^{\,\gamma}\left(\rho'^{\,\gamma-1} + \frac{a^{2\gamma}}{\rho'^{\,\gamma+1}}\right) - \gamma\left(\rho'^{\,\gamma} - \frac{a^{2\gamma}}{\rho'^{\,\gamma}}\right)\rho'^{\,\gamma-1} \\ &= \frac{2\gamma a^{2\gamma}}{\rho'}. \end{aligned} \tag{4.140}$$

It follows from (4.130) that $C = -(2\gamma a^{2\gamma})^{-1}$, so

$$g_n(\rho, \rho') = \frac{1}{2\gamma}\rho_<^\gamma\left(-\frac{\rho_>^\gamma}{a^{2\gamma}} + \frac{1}{\rho_>^\gamma}\right) \tag{4.141}$$

and

$$G(\vec{x}, \vec{x}\,') = \sum_{n=1}^{\infty} \frac{4}{n}\rho_<^\gamma\left(\frac{1}{\rho_>^\gamma} - \frac{\rho_>^\gamma}{a^{2\gamma}}\right)\sin\left(\frac{n\pi\phi}{\beta}\right)\sin\left(\frac{n\pi\phi'}{\beta}\right). \tag{4.142}$$

In the $a \to \infty$ limit, we have that (recalling that $\gamma = n\pi/\beta$)

$$G(\vec{x}, \vec{x}\,') = \sum_{n=1}^{\infty} \frac{4}{n}\left(\frac{\rho_<}{\rho_>}\right)^{n\pi/\beta}\sin\left(\frac{n\pi\phi}{\beta}\right)\sin\left(\frac{n\pi\phi'}{\beta}\right), \tag{4.143}$$

which gives the Green function for the wedge problem of Section 3.8. On the other hand, near the origin the lowest power of ρ dominates and

$$G(\vec{x}, \vec{x}\,') \approx 4\left(\frac{\rho}{\rho'}\right)^{\pi/\beta}\sin\left(\frac{\pi\phi}{\beta}\right)\sin\left(\frac{\pi\phi'}{\beta}\right). \tag{4.144}$$

This agrees with the $\rho \to 0$ behavior of the potential found in (3.120):

$$\Phi(\rho, \phi) = V + a_1\rho^{\pi/\beta}\sin\left(\frac{\pi\phi}{\beta}\right).$$

The three-dimensional version of this Green function is considered in Exercise 4.8.4.

4.9 Schwinger's construction of spherical harmonics

Now that we've beaten cylindrical symmetry to death, let us turn to spherical symmetry. Fortunately, we can make use of some of the techniques we have learned. Recall that our analysis of cylindrical symmetry in Section 4.1 began with the expansion of $e^{i\vec{k}\cdot\vec{x}_\perp}$ into a series whose terms were separated in the variables ρ and ϕ, such that the Laplacian operator acted on these terms in a particularly simple way. We then used these variable-separated terms to build an expansion for $\delta(\vec{x}_\perp - \vec{x}\,'_\perp)$, which we used to find expansions for Green functions. In tackling spherical symmetry, we take a similar but slightly different approach.[9] We will begin by finding variable-separated solutions to the Laplace equation, and we will make use of these solutions to build Green function expansions.

[9] J. Schwinger *et al.*, *Classical Electrodynamics*, Perseus Books (Reading, MA) 1998, Sec. 20.2.

We begin with the observation that $\nabla^2 e^{i\vec{k}\cdot\vec{r}} = -\vec{k}\cdot\vec{k}\, e^{i\vec{k}\cdot\vec{r}}$ and note that if we can find a complex vector $\vec{a}$ such that $\vec{a}\cdot\vec{a} = 0$, then

$$\vec{a}\cdot\vec{a} = 0 \implies \nabla^2 e^{i\vec{a}\cdot\vec{r}} = 0, \tag{4.145}$$

so that $e^{i\vec{a}\cdot\vec{r}}$ satisfies Laplace's equation. The conditions $\vec{a} \neq 0$ and $\vec{a}\cdot\vec{a} = 0$ require that $\vec{a}$ be complex, so we may write

$$a_1 = \frac{k_x + iC_1}{2}; \quad a_2 = \frac{k_y + iC_2}{2}; \quad a_3 = \frac{k_z + iC_3}{2}, \tag{4.146}$$

where $k_x, k_y, k_z, C_1, C_2, C_3$ are real constants; the factors of $1/2$ are for later convenience. That is, we are requiring

$$\vec{k} = 2\,\mathrm{Re}\,[\vec{a}]\,, \tag{4.147}$$

so that

$$e^{i\vec{k}\cdot\vec{r}} = e^{(\vec{a}+\vec{a}^*)\cdot\vec{r}}. \tag{4.148}$$

The requirement that $\vec{a}\cdot\vec{a} = 0$ yields two real equations; setting $C_3 = 0$ for simplicity leads directly to the solution:

$$\vec{a} = \left(\frac{1}{2}\left(k_x - i\frac{k}{k_\perp}k_y\right), \frac{1}{2}\left(k_y + i\frac{k}{k_\perp}k_x\right), \frac{k_z}{2}\right), \tag{4.149}$$

where $k^2 \equiv k_x^2 + k_y^2 + k_z^2$ and $k_\perp^2 \equiv k_x^2 + k_y^2$.

We are now ready to expand in power series:

$$e^{i\vec{a}\cdot\vec{r}} = \sum_{\ell=0}^{\infty} \frac{(i\vec{a}\cdot\vec{r})^\ell}{\ell!}. \tag{4.150}$$

It is easy to verify (for example, using Cartesian coordinates) that each monomial $(\vec{a}\cdot\vec{r})^\ell$ solves Laplace's equation: so we proceed to separate out the r, θ, ϕ dependence of each monomial. Re-expressing in spherical coordinates gives

$$\begin{aligned}
\frac{\vec{a}\cdot\vec{r}}{r} &= \frac{a_1 x}{r} + \frac{a_2 y}{r} + \frac{a_3 z}{r} \\
&= a_1 \sin\theta\left(\frac{e^{i\phi} + e^{-i\phi}}{2}\right) + a_2 \sin\theta\left(\frac{e^{i\phi} - e^{-i\phi}}{2i}\right) + a_3\cos\theta \\
&= \frac{(a_1 - ia_2)}{2}\sin\theta e^{i\phi} + \frac{(a_1 + ia_2)}{2}\sin\theta e^{-i\phi} + a_3.\cos\theta.
\end{aligned} \tag{4.151}$$

However, a_1, a_2, and a_3 are not independent: they are related by the condition that $a_1^2 + a_2^2 + a_3^2 = 0$, which may be rewritten as

$$(-a_1 + ia_2)(a_1 + ia_2) = a_3^2. \tag{4.152}$$

We may simplify (4.151) by defining a new set of variables (ξ_-, ξ_+) such that

$$\xi_-^2 \equiv \frac{a_1 + ia_2}{2}, \tag{4.153}$$

$$\xi_+^2 \equiv \frac{-a_1 + ia_2}{2}. \tag{4.154}$$

In light of (4.152) it follows that $a_3^2 = 4\xi_-^2\xi_+^2$ and we may choose ξ_- and ξ_+ so that

$$a_3 = 2\xi_-\xi_+. \tag{4.155}$$

Replacing a_1, a_2, a_3 with ξ_-, ξ_+ in (4.151) gives

$$\begin{aligned}\frac{\vec{a}\cdot\vec{r}}{r} &= -\xi_+^2 \sin\theta e^{i\phi} + \xi_-^2 \sin\theta e^{-i\phi} + 2\xi_+\xi_- \cos\theta \\ &= \xi_+^2 \left(-\sin\theta e^{i\phi} + \left(\frac{\xi_-}{\xi_+}\right)^2 \sin\theta e^{-i\phi} + 2\frac{\xi_-}{\xi_+}\cos\theta \right) \qquad (4.156) \\ &= \xi_+^2 \left(-\sin\theta e^{i\phi} + \frac{1}{\sin\theta e^{-i\phi}} \left[\left(\frac{\xi_-}{\xi_+}\sin\theta e^{-i\phi} + \cos\theta \right)^2 - \cos^2\theta \right] \right) \\ & \qquad (4.157) \\ &= \frac{\xi_+^2 e^{i\phi}}{\sin\theta} \left[\left(\frac{\xi_-}{\xi_+}\sin\theta e^{-i\phi} + \cos\theta \right)^2 - 1 \right], \qquad (4.158)\end{aligned}$$

where (4.157) is obtained from (4.156) by completing the square. Defining $\xi \equiv \xi_-/\xi_+$, it follows that

$$\begin{aligned}(\vec{a}\cdot\vec{r})^\ell &= r^\ell(\vec{a}\cdot\frac{\vec{r}}{r})^\ell \\ &= r^\ell \left(\frac{\xi_+^2 e^{i\phi}}{\sin\theta} \right)^\ell \left[(\xi \sin\theta e^{-i\phi} + \cos\theta)^2 - 1 \right]^\ell. \qquad (4.159)\end{aligned}$$

The expression $f(\xi) \equiv \left[(\xi \sin\theta e^{-i\phi} + \cos\theta)^2 - 1 \right]^\ell$ may be evaluated via Taylor expansion. First we introduce the variable

$$z \equiv \xi \sin\theta e^{-i\phi} + \cos\theta, \tag{4.160}$$

so that $f(z) = [z^2 - 1]^\ell$. Next, recall the Taylor series formula from basic calculus:

$$f(z) = \sum_{n=0}^{\infty} \frac{(z-b)^n}{n!} \left. \frac{d^n f(z)}{dz^n} \right|_{z=b} \tag{4.161}$$

and expand $f(z)$ around $b \equiv \cos\theta$. In this case, the highest power in the expansion is 2ℓ, so we have:

$$f(z) = \sum_{n=0}^{2\ell} \frac{(z-\cos\theta)^n}{n!} \frac{d^n}{dz^n}[z^2-1]^\ell \bigg|_{z=\cos\theta}. \tag{4.162}$$

We first evaluate the derivative at $z = \cos\theta$, then change variable from z to ξ using (4.160) to obtain:

$$f(\xi) = \sum_{n=0}^{2\ell} \frac{\left(\xi \sin\theta e^{-i\phi}\right)^n}{n!} \left(\frac{d}{d\cos\theta}\right)^n \left[\cos^2\theta - 1\right]^\ell. \tag{4.163}$$

Plugging (4.163) into (4.159) gives

$$(\vec{a}\cdot\vec{r})^\ell = r^\ell \left(\frac{\xi_+^2 e^{i\phi}}{\sin\theta}\right)^\ell \sum_{n=0}^{2\ell} \frac{\left(\xi \sin\theta e^{-i\phi}\right)^n}{n!} \left(\frac{d}{d\cos\theta}\right)^n \left[\cos^2\theta - 1\right]^\ell. \tag{4.164}$$

We redefine the summation index as $m \equiv \ell - n$ and recall that $\xi = \xi_-/\xi_+$ to obtain

$$(\vec{a}\cdot\vec{r})^\ell = r^\ell \left(\frac{\xi_+^2 e^{i\phi}}{\sin\theta}\right)^\ell \sum_{m=-\ell}^{\ell} \frac{((\xi_-/\xi_+)\sin\theta e^{-i\phi})^{\ell-m}}{(\ell-m)!} \left(\frac{d}{d\cos\theta}\right)^{\ell-m} [\cos^2\theta - 1]^\ell, \tag{4.165}$$

which may also be rewritten as

$$(\vec{a}\cdot\vec{r})^\ell = r^\ell 2^\ell \ell! \sum_{m=-\ell}^{\ell} \frac{\xi_+^{\ell+m}\xi_-^{\ell-m}}{\sqrt{(\ell+m)!(\ell-m)!}} \sqrt{\frac{(\ell+m)!}{(\ell-m)!}} (\sin\theta)^{-m} e^{im\phi} \times \left(\frac{d}{d\cos\theta}\right)^{\ell-m} \frac{(\cos^2\theta - 1)^\ell}{2^\ell \ell!}. \tag{4.166}$$

In light of (4.166), we define the (ℓ, m)'th *spherical harmonic* as

$$Y_{\ell m}(\theta,\phi) \equiv \sqrt{\frac{(2\ell+1)(\ell+m)!}{4\pi(\ell-m)!}} e^{im\phi} (\sin\theta)^{-m} \left(\frac{d}{d\cos\theta}\right)^{\ell-m} \frac{(\cos^2\theta - 1)^\ell}{2^\ell \ell!}, \tag{4.167}$$

so we may simplify (4.166) to

$$(\vec{a}\cdot\vec{r})^\ell = r^\ell 2^\ell \ell! \sum_{m=-\ell}^{\ell} \frac{\xi_+^{\ell+m}\xi_-^{\ell-m}}{\sqrt{(\ell+m)!(\ell-m)!}} \sqrt{\frac{4\pi}{2\ell+1}} Y_{\ell m}(\theta,\phi), \tag{4.168}$$

$$\Longrightarrow e^{i\vec{a}\cdot\vec{r}} = \sum_{\ell=0}^{\infty} \sum_{m=-\ell}^{\ell} i^\ell r^\ell 2^\ell \frac{\xi_+^{\ell+m}\xi_-^{\ell-m}}{\sqrt{(\ell+m)!(\ell-m)!}} \sqrt{\frac{4\pi}{2\ell+1}} Y_{\ell m}. \tag{4.169}$$

We have noted previously that $(\vec{a}\cdot\vec{r})^\ell$ satisfies the Laplace equation, so applying the Laplacian to (4.168) yields

$$0 = \sum_{m=-\ell}^{\ell} \frac{\xi_+^{\ell+m}\xi_-^{\ell-m}}{\sqrt{(\ell+m)!(\ell-m)!}}\sqrt{\frac{4\pi}{2\ell+1}}\nabla^2\left[r^\ell Y_{\ell m}(\theta,\phi)\right]. \tag{4.170}$$

Since ξ_+ and ξ_- are independent variables, all coefficients of $\xi_+^{\ell+m}\xi_-^{\ell-m}$ for $m=-\ell,\ldots,\ell$ must vanish separately, and we conclude

$$\nabla^2[r^\ell Y_{\ell m}(\theta,\phi)] = 0. \tag{4.171}$$

We have therefore achieved our original goal of finding variable-separated solutions to the Laplace equation in the variables r,θ,ϕ.

Let us now derive some basic properties of the spherical harmonics. First, applying the Laplacian in spherical coordinates to (4.171) gives

$$\left[\ell(\ell+1)r^{\ell-2}Y_{\ell m} + r^{\ell-2}\left(\frac{1}{\sin\theta}\frac{\partial}{\partial\theta}(\sin\theta\frac{\partial Y_{\ell m}}{\partial\theta}) + \frac{1}{\sin^2\theta}\frac{\partial^2}{\partial\phi^2}Y_{\ell m}\right)\right] = 0. \tag{4.172}$$

Furthermore we have

$$\frac{\partial^2}{\partial\phi^2}Y_{\ell m} = -m^2 Y_{\ell m}, \tag{4.173}$$

which leads to

$$\left[\frac{1}{\sin\theta}\frac{\partial}{\partial\theta}\left(\sin\theta\frac{\partial}{\partial\theta}Y_{\ell m}\right) - \frac{m^2}{\sin^2\theta}Y_{\ell m}\right] = -\ell(\ell+1)Y_{\ell m}. \tag{4.174}$$

In addition we may use functional invariance to derive some important identities. Equation (4.159)

$$(\vec{a}\cdot\vec{r})^\ell = r^\ell[-\xi_+^2\sin\theta e^{i\phi} + \xi_-^2\sin\theta e^{-i\phi} + 2\xi_+\xi_-\cos\theta]^\ell$$

is unchanged under the replacements

$$\xi_+ \leftrightarrow \xi_-; \quad \theta\to-\theta; \quad \phi\to-\phi. \tag{4.175}$$

Since (4.166) is derived from (4.159), this implies that we may make the same replacements without changing the equality, so that

$$\sum_{m=-\ell}^{\ell}\frac{\xi_+^{\ell+m}\xi_-^{\ell-m}}{\sqrt{(\ell+m)!(\ell-m)!}}\sqrt{\frac{4\pi}{2\ell+1}}Y_{\ell m}(\theta,\phi)$$
$$= \sum_{m'=-\ell}^{\ell}\frac{\xi_-^{\ell+m'}\xi_+^{\ell-m'}}{\sqrt{(\ell+m')!(\ell-m')!}}\sqrt{\frac{4\pi}{2\ell+1}}Y_{\ell m'}(-\theta,-\phi). \tag{4.176}$$

As before, since ξ_+ and ξ_- are independent variables, coefficients of terms of the form $\xi_+^a \xi_-^b$ must be the same in both expansions. Replacing m' with $-m$ in (4.176) yields the following identities (valid for all ℓ, m):

$$Y_{\ell,m}(\theta,\phi) = Y_{\ell,-m}(-\theta,-\phi). \tag{4.177}$$

Plugging (4.177) into (4.167) gives an alternate expression for the spherical harmonics:

$$Y_{\ell m}(\theta,\phi) = \sqrt{\frac{2\ell+1}{4\pi}}\sqrt{\frac{(\ell-m)!}{(\ell+m)!}}e^{im\phi}(-\sin\theta)^m \left(\frac{d}{d\cos\theta}\right)^{\ell+m} \frac{(\cos^2\theta-1)^\ell}{2^\ell \ell!}. \tag{4.178}$$

At this point we may define *Legendre polynomials* of order n,

$$P_n(x) \equiv \frac{1}{2^n n!}\left(\frac{d}{dx}\right)^n (x^2-1)^n, \tag{4.179}$$

for $n = 0, 1, 2, \ldots$. The first few Legendre polynomials are listed in Table 4.1. Some of the useful identities satisfied by the $P_n(x)$ are:

$$x\frac{dP_{n+1}(x)}{dx} - \frac{dP_n(x)}{dx} = (n+1)P_{n+1}(x), \tag{4.180}$$

$$(2n+1)xP_n(x) = (n+1)P_{n+1}(x) + nP_{n-1}(x), \tag{4.181}$$

$$\frac{dP_{n+1}(x)}{dx} = (2n+1)P_n(x) + \frac{dP_{n-1}(x)}{dx}, \tag{4.182}$$

$$(1-x^2)\frac{dP_n(x)}{dx} = -nx\,P_n(x) + n\,P_{n-1}(x). \tag{4.183}$$

Table 4.1 Legendre polynomials $P_n(x)$.

$P_0(x) = 1$
$P_1(x) = x$
$P_2(x) = \frac{1}{2}(3x^2-1)$
$P_3(x) = \frac{1}{2}(5x^3-3x)$
$P_4(x) = \frac{1}{8}(35x^4-30x^2+3)$

Table 4.2 Associated Legendre functions $P_n^m(\cos\theta)$.

$n=0$	$P_0^0 = 1$
$n=1$	$P_1^1 = -\sin\theta$
	$P_1^0 = \cos\theta$
$n=2$	$P_2^2 = 3\sin^2\theta$
	$P_2^1 = -3\sin\theta\cos\theta$
	$P_2^0 = \frac{1}{2}\left(3\cos^2\theta - 1\right)$
$n=3$	$P_3^3 = -15\sin^3\theta$
	$P_3^2 = 15\sin^2\theta\cos\theta$
	$P_3^1 = -\frac{1}{2}\sin\theta\left(15\cos^2\theta - 3\right)$
	$P_3^0 = \frac{1}{2}\left(5\cos^3\theta - 3\cos\theta\right)$

These simplify the expression for spherical harmonics:

$$Y_{\ell m}(\theta,\phi) = \sqrt{\frac{2\ell+1}{4\pi}}\, e^{im\phi} \times \begin{cases} \sqrt{\frac{(\ell-m)!}{(\ell+m)!}}(-\sin\theta)^m \left(\frac{d}{d\cos\theta}\right)^m P_\ell(\cos\theta) & \text{if } m \geq 0, \\ \sqrt{\frac{(\ell+m)!}{(\ell-m)!}}(\sin\theta)^{-m} \left(\frac{d}{d\cos\theta}\right)^{-m} P_\ell(\cos\theta) & \text{if } m \leq 0. \end{cases} \tag{4.184}$$

This may also be combined into the single form

$$Y_{\ell m}(\theta,\phi) = (-1)^{(m-|m|)/2}\sqrt{\frac{(2\ell+1)}{4\pi}}\sqrt{\frac{(\ell-|m|)!}{(\ell+|m|)!}}\, e^{im\phi} P_\ell^{|m|}(\cos\theta), \tag{4.185}$$

where $P_\ell^m(x)$ are *associated Legendre functions*,

$$P_n^m(x) \equiv (-1)^m (1-x^2)^{m/2} \left(\frac{d}{dx}\right)^m P_n(x), \tag{4.186}$$

Table 4.3 Spherical Harmonics $Y_{\ell m}(\theta, \phi)$.

$l = 0$	$Y_{00} = \dfrac{1}{\sqrt{4\pi}}$
$l = 1$	$Y_{11} = -\sqrt{\dfrac{3}{8\pi}} \sin\theta\, e^{i\phi}$ $Y_{10} = \sqrt{\dfrac{3}{4\pi}} \cos\theta$
$l = 2$	$Y_{22} = \dfrac{1}{4}\sqrt{\dfrac{15}{2\pi}} \sin^2\theta\, e^{2i\phi}$ $Y_{21} = -\sqrt{\dfrac{15}{8\pi}} \sin\theta\cos\theta\, e^{i\phi}$ $Y_{20} = \sqrt{\dfrac{5}{4\pi}} \left(\dfrac{3}{2}\cos^2\theta - \dfrac{1}{2}\right)$
$l = 3$	$Y_{33} = -\dfrac{1}{4}\sqrt{\dfrac{35}{4\pi}} \sin^3\theta\, e^{3i\phi}$ $Y_{32} = \dfrac{1}{4}\sqrt{\dfrac{105}{2\pi}} \sin^2\theta\cos\theta\, e^{2i\phi}$ $Y_{31} = -\dfrac{1}{4}\sqrt{\dfrac{21}{4\pi}} \sin\theta\left(5\cos^2\theta - 1\right) e^{i\phi}$ $Y_{30} = \sqrt{\dfrac{7}{4\pi}} \left(\dfrac{5}{2}\cos^3\theta - \dfrac{3}{2}\cos\theta\right)$

or

$$P_n^m(\cos\theta) \equiv (-1)^m (\sin\theta)^m \left(\frac{d}{d\cos\theta}\right)^m P_n(\cos\theta), \tag{4.187}$$

for $0 \le m \le n$. Note that $P_n^0(x) = P_n(x)$. The associated Legendre functions P_n^m for $n = 0, 1, 2, 3$ are listed in Table 4.2, and spherical harmonics $Y_{\ell m}$ for $l = 0, 1, 2, 3$ are listed in Table 4.3 for $m \ge 0$. By examination of (4.185) we may infer the following identities:

$$Y_{\ell m}(\theta, \phi) = (-1)^m Y_{\ell,-m}(\theta, -\phi), \tag{4.188}$$

$$Y^*_{\ell m}(\theta, \phi) = Y_{\ell m}(\theta, -\phi). \tag{4.189}$$

By setting $m = 0$ in (4.185) we also obtain

$$Y_{\ell 0}(\theta, \phi) = \sqrt{\frac{(2\ell + 1)}{4\pi}} P_\ell(\cos\theta), \tag{4.190}$$

so that (4.174) with $m = 0$ yields a differential equation that is satisfied by $P_\ell(\cos\theta)$.

4.10 Orthogonality properties of spherical harmonics

Recall that we began the previous section by defining a complex vector $\vec{a}$ such that $\vec{a}\cdot\vec{a} = 0$: following this definition, spherical harmonics arose from consideration of $e^{i\vec{a}\cdot\vec{x}}$, which satisfies Laplace's equation. We chose vector $\vec{a}$ to have the form (c.f. (4.146))

$$a_1 = \frac{k_x + iC_1}{2}; \quad a_2 = \frac{k_y + iC_2}{2}; \quad a_3 = \frac{k_z + iC_3}{2}.$$

It follows that

$$\vec{k} = \vec{a}^* + \vec{a}. \tag{4.191}$$

We may thus compute

$$\int d\Omega\, e^{i\vec{k}\cdot\vec{r}} = \int d\Omega\, e^{i\vec{a}^*\cdot\vec{r}+\vec{a}\cdot\vec{r}} = \sum_{\ell_1,\ell_2=0}^{\infty} i^{\ell_1} i^{\ell_2} \int d\Omega \frac{(\vec{a}^*\cdot\vec{r})^{\ell_1}}{\ell_1!}\frac{(\vec{a}\cdot\vec{r})^{\ell_2}}{\ell_2!}. \tag{4.192}$$

Plugging in the expansion (4.169) yields

$$\begin{aligned}\int d\Omega\, e^{i\vec{k}\cdot\vec{r}} = &\sum_{\ell_1,\ell_2=0} i^{\ell_1+\ell_2} \sum_{m_1=-\ell_1}^{\ell_1}\sum_{m_2=-\ell_2}^{\ell_2} (2r)^{\ell_1+\ell_2} \\ &\times \frac{\xi_+^{*\ell_1+m_1}\xi_+^{\ell_2+m_2}\xi_-^{*\ell_1-m_1}\xi_-^{\ell_2-m_2}}{\sqrt{(\ell_1+m_1)!(\ell_1-m_1)!(\ell_2+m_2)!(\ell_2-m_2)!}} \\ &\times\sqrt{\frac{4\pi}{2\ell_1+1}}\sqrt{\frac{4\pi}{2\ell_2+1}}\int d\Omega\, Y^*_{\ell_1 m_1}(\theta,\phi) Y_{\ell_2 m_2}(\theta,\phi).\end{aligned} \tag{4.193}$$

On the other hand, we may evaluate the integral directly as

$$\begin{aligned}\int d\Omega\, e^{i\vec{k}\cdot\vec{r}} &= 2\pi\int_0^{\pi} d\theta\, \sin\theta e^{ikr\cos\theta} \\ &= \frac{4\pi}{kr}\sin(kr) \\ &= 4\pi\sum_{\ell=0}^{\infty}(-1)^\ell \frac{r^{2\ell}k^{2\ell}}{(2\ell+1)!}.\end{aligned} \tag{4.194}$$

However, by the definition of $\vec{a}$ we have $k^2 = 2\vec{a}^* \cdot \vec{a}$, so that

$$\begin{aligned} k^{2\ell} &= (2\vec{a}^* \cdot \vec{a})^\ell \\ &= 2^{2\ell}(\xi_+^*\xi_+ + \xi_-^*\xi_-)^{2\ell} \\ &= 2^{2\ell}\sum_{m'=0}^{2\ell}\frac{2\ell!}{(2\ell-m')!m'!}(\xi_+^*\xi_+)^{2\ell-m'}(\xi_-^*\xi_-)^{m'}. \end{aligned} \tag{4.195}$$

Letting $m = \ell - m'$ and replacing $k^{2\ell}$ in (4.194) with (4.195) gives

$$\int d\Omega\, e^{i\vec{k}\cdot\vec{r}} = \sum_{\ell=0}^{\infty}\sum_{m=-\ell}^{\ell}\frac{4\pi}{2\ell+1}(-1)^\ell(2r)^{2\ell}\frac{(\xi_+^*\xi_+)^{\ell+m}(\xi_-^*\xi_-)^{\ell-m}}{(\ell+m)!(\ell-m)!}. \tag{4.196}$$

By comparing coefficients of like terms in (4.193) and (4.196), we may obtain a great deal of information about the integrals in (4.193). In particular, the coefficient of $\xi_+^{*\ell_1+m_1}\xi_+^{\ell_2+m_2}\xi_-^{*\ell_1-m_1}\xi_-^{\ell_2-m_2}$ in (4.193) must be zero unless $\ell_1 + m_1 = \ell_2 + m_2$ and $\ell_1 - m_1 = \ell_2 - m_2$. It follows that nonzero terms must have $\ell_1 = \ell_2$ and $m_1 = m_2$, so that

$$\int d\Omega\, Y_{\ell_1 m_1}^*(\theta,\phi)Y_{\ell_2 m_2}(\theta,\phi) = \delta_{\ell_1\ell_2}\delta_{m_1 m_2}. \tag{4.197}$$

Since we know the relationship between $Y_{\ell 0}(\theta,\phi)$ and $P_\ell(\cos\theta)$ from (4.190), by setting $(m_1 = m_2 = 0)$ we obtain

$$\int_{-1}^{1} d(\cos\theta)P_\ell(\cos\theta)P_{\ell'}(\cos\theta) = \frac{2}{2\ell+1}\delta_{\ell\ell'}. \tag{4.198}$$

4.11 The Coulomb expansion, completeness of spherical harmonics and the "Addition Theorem"

We now turn to consider the relation between Legendre polynomials and the free-space Green function. Letting γ denote the angle between $\vec{r}$ and $\vec{r}'$, we may express the Green function as:

$$\begin{aligned} \frac{1}{|\vec{r}-\vec{r}'|} &= \frac{1}{\sqrt{r^2+r'^2-2rr'\cos\gamma}} \\ &= \frac{1}{r\sqrt{1+(\frac{r'}{r})^2-2(\frac{r'}{r})\cos\gamma}} \qquad (4.199) \\ &= \frac{1}{r'\sqrt{1+(\frac{r}{r'})^2-2(\frac{r}{r'})\cos\gamma}}. \qquad (4.200) \end{aligned}$$

Defining $r_> \equiv \max(r, r')$, $r_< \equiv \min(r, r')$, and $x \equiv (r_<)/(r_>)$, we may expand either (4.199) or (4.200) in Taylor series to obtain:

$$\frac{1}{|\vec{r} - \vec{r}'|} = \sum_{\ell=0}^{\infty} \frac{r_<^\ell}{r_>^{\ell+1}} \frac{1}{\ell!} \frac{d^\ell}{dx^\ell} \frac{1}{\sqrt{1 + x^2 - 2x\cos\gamma}}\bigg|_{x=0}. \tag{4.201}$$

Let us define:

$$Q_\ell(\cos\gamma) \equiv \frac{1}{\ell!} \frac{d^\ell}{dx^\ell} \frac{1}{\sqrt{1 + x^2 - 2x\cos\gamma}}\bigg|_{x=0}. \tag{4.202}$$

It is not hard to show that $Q_\ell(\cos\gamma)$ is a polynomial in $\cos\gamma$ of degree ℓ. Actually, it turns out that $Q_\ell(\cos\gamma) = P_\ell(\cos\gamma)$, as we shall now show.[10] It is possible (but tedious) to show directly[11] that

$$\int_{-1}^{1} \frac{dx}{\sqrt{1 - 2\mu x + \mu^2}\sqrt{1 - 2\nu x + \nu^2}} = \frac{1}{\sqrt{\mu\nu}} \ln\left(\frac{1 + \sqrt{\mu\nu}}{1 - \sqrt{\mu\nu}}\right) \quad (|\mu|, |\nu| < 1)$$
$$= 2\sum_{\ell=0}^{\infty} \frac{(\mu\nu)^\ell}{(2\ell + 1)}. \tag{4.203}$$

On the other hand,

$$\frac{1}{\sqrt{1 - 2\mu x + \mu^2}} = \sum_{\ell'=0}^{\infty} \mu^{\ell'} Q_{\ell'}(x), \tag{4.204}$$

$$\frac{1}{\sqrt{1 - 2\nu x + \nu^2}} = \sum_{\ell=0}^{\infty} \nu^\ell Q_\ell(x). \tag{4.205}$$

Comparison shows that

$$\int_{-1}^{1} dx\, Q_\ell(x) Q_{\ell'}(x) = \frac{2}{2\ell + 1}\delta_{\ell\ell'}. \tag{4.206}$$

Now the n'th order polynomial $P_n(x)$ can be expressed as a linear combination of the $n + 1$ polynomials $Q_0(x), \ldots Q_n(x)$:

$$P_n(x) = \sum_{m=0}^{n} \alpha_{nm} Q_m(x). \tag{4.207}$$

It follows that

$$\int_{-1}^{1} dx\, P_n(x) Q_r(x) = \frac{2}{2r + 1}\alpha_{nr}, \qquad r = 0, 1, 2, \ldots, n. \tag{4.208}$$

[10]This argument is taken from Cushing, *op. cit.*, p. 172.

[11]See for example I. S. Gradshteyn and I. M. Ryzhik, *Table of Integrals, Series and Products*, A. Jeffrey and D. Zwillinger, Eds., 7th ed., Elsevier Academic Press (Burlington, MA) 2007, 2.261.

However, using integration by parts we can show

$$\int_{-1}^{1} dx\, P_n(x)x^m = \frac{1}{2^n n!}\int_{-1}^{1} dx\, \frac{d^n}{dx^n}[(-1+x^2)^n]x^m$$
$$= 0, \qquad m < n. \tag{4.209}$$

From this we conclude

$$\alpha_{nr} = 0, \qquad r < n, \tag{4.210}$$
$$\alpha_{nn}Q_n(x) = P_n(x). \tag{4.211}$$

In order to find the constants α_{nn}, we set $x = 1$ on both sides. One can show (see Exercise 4.11.2)

$$P_\ell(1) = 1. \tag{4.212}$$

On the other hand, if $\cos\gamma = 1$ then $\vec{r}$ and $\vec{r}\,'$ are parallel and

$$\frac{1}{|\vec{r}-\vec{r}\,'|} = \frac{1}{|r-r'|}. \tag{4.213}$$

Plugging into (4.201) and using definition (4.202) gives

$$\frac{1}{|r-r'|} = \sum_{\ell=0}^{\infty} \frac{r_<^\ell}{r_>^{\ell+1}} Q_\ell(1). \tag{4.214}$$

Comparison with the direct Taylor expansion of $1/|r-r'|$ gives $Q_\ell(1) = 1$. Since we know that $P_\ell(1) = 1$ and $P_\ell(x) \propto Q_\ell(x)$, it follows that $P_\ell(x) = Q_\ell(x)$ and

$$\frac{1}{|\vec{r}-\vec{r}\,'|} = \sum_{\ell=0}^{\infty} \frac{r_<^\ell}{r_>^{\ell+1}} P_\ell(\cos\gamma). \tag{4.215}$$

Equation (4.215) is known as the *Coulomb expansion*.

We may use the Coulomb expansion to derive a completeness relation for spherical harmonics. We start with the Poisson equation

$$\nabla^2 \frac{1}{|\vec{r}-\vec{r}\,'|} = -4\pi\delta(\vec{r}-\vec{r}\,'). \tag{4.216}$$

Substituting (4.215) in (4.216) and integrating across $r = r'$ gives

$$\left[-r^2\frac{\partial}{\partial r}\sum_{\ell=0}^{\infty} \frac{r_<^\ell}{r_>^{\ell+1}} P_\ell(\cos\gamma)\right]\Bigg|_{r'-\epsilon}^{r'+\epsilon} = 4\pi\delta(\cos\theta-\cos\theta')\delta(\phi-\phi'), \tag{4.217}$$

which results in

$$\sum_{\ell=0}^{\infty} \frac{2\ell+1}{4\pi} P_\ell(\cos\gamma) = \delta(\cos\theta-\cos\theta')\delta(\phi-\phi'). \tag{4.218}$$

At this point, we note that $P_\ell(\cos\gamma)$ can be written as a linear combination of the spherical harmonics $\{Y_{\ell m}(\theta,\phi)\}$:

$$P_\ell(\cos\gamma) = \sum_{m=-\ell}^{\ell} A_{\ell m} Y_{\ell m}(\theta,\phi). \tag{4.219}$$

The rigorous proof of this fact is somewhat technical (see Exercise 4.11.5), and requires showing that the $(2\ell+1)$-dimensional vector space spanned by $\{Y_{\ell m}(\theta,\phi)\}$ is invariant under rotations.

Substituting (4.219) into (4.218) gives

$$\sum_{\ell,m} \frac{(2\ell+1)}{4\pi} A_{\ell m} Y_{\ell m}(\theta,\phi) = \delta(\cos\theta - \cos\theta')\delta(\phi-\phi'). \tag{4.220}$$

Using the orthonormality relation (4.197) for spherical harmonics

$$\int d\Omega\, Y_{\ell m}(\theta,\phi) Y^*_{\ell' m'}(\theta,\phi) = \delta_{\ell\ell'}\delta_{mm'},$$

we obtain

$$A_{\ell m} = \frac{4\pi}{(2\ell+1)} Y^*_{\ell m}(\theta',\phi'). \tag{4.221}$$

Now using (4.221) in (4.220) gives us finally the completeness relation for spherical harmonics:

$$\sum_{\ell,m} Y_{\ell m}(\theta,\phi) Y^*_{\ell m}(\theta',\phi') = \delta(\cos\theta - \cos\theta')\delta(\phi-\phi'). \tag{4.222}$$

Plugging (4.221) into (4.219) gives the *addition theorem*:

$$P_\ell(\cos\gamma) = \frac{4\pi}{2\ell+1} \sum_{m=-\ell}^{\ell} Y_{\ell m}(\theta,\phi) Y^*_{\ell m}(\theta',\phi'), \tag{4.223}$$

$$= \frac{4\pi}{2\ell+1} \sum_{m=-\ell}^{\ell} Y^*_{\ell m}(\theta,\phi) Y_{\ell m}(\theta',\phi'). \tag{4.224}$$

Showing Equation (4.224) from (4.223) requires the use of (4.188) and (4.189) and is consistent with $P_\ell(\cos\theta)$ being real. Finally, plugging (4.223) and (4.224) into (4.215) gives a variable separated form of the Coulomb expansion,

$$\frac{1}{|\vec{r}-\vec{r}\,'|} = \sum_{\ell,m} \frac{r_<^\ell}{r_>^{\ell+1}} \frac{4\pi}{2\ell+1} Y_{\ell m}(\theta,\phi) Y^*_{\ell m}(\theta',\phi'), \tag{4.225}$$

$$= \sum_{\ell,m} \frac{r_<^\ell}{r_>^{\ell+1}} \frac{4\pi}{2\ell+1} Y^*_{\ell m}(\theta,\phi) Y_{\ell m}(\theta',\phi'). \tag{4.226}$$

4.12 Green function for concentric spheres

We will now compute Green functions for geometries with spherical symmetry. We will begin with the spherical cavity shown in Figure 4.4: the $a \to 0, b \to \infty$ limits will be treated as special cases. We have from (4.222) that

$$4\pi\delta(\vec{r}-\vec{r}\,') = 4\pi \sum_{\ell,m} Y^*_{\ell m}(\theta',\phi')Y_{\ell m}(\theta,\phi)\frac{1}{r^2}\delta(r-r'). \tag{4.227}$$

Since the Laplacian operator acts diagonally on spherical harmonics, we expect a Green function of the form

$$G(\vec{r},\vec{r}\,') = 4\pi \sum_{\ell,m} Y^*_{\ell m}(\theta',\phi')Y_{\ell m}(\theta,\phi)g_\ell(r,r') \tag{4.228}$$

$$\Longrightarrow\ -\nabla^2 G(\vec{r},\vec{r}\,') = 4\pi \sum_{\ell,m} Y^*_{\ell m}(\theta',\phi')Y_{\ell m}(\theta,\phi) \times \left(-\frac{1}{r^2}\frac{\partial}{\partial r}(r^2\frac{\partial g_\ell}{\partial r}) + \frac{\ell(\ell+1)}{r^2}g_\ell\right). \tag{4.229}$$

Comparison with (4.227) gives

$$-\frac{\partial}{\partial r}\left(r^2\frac{\partial g_\ell}{\partial r}\right) + \ell(\ell+1)g_\ell = \delta(r-r'). \tag{4.230}$$

The general homogeneous solution is $Ar^\ell + Br^{-\ell-1}$. The solutions that satisfy null boundary conditions at $r=a$ and $r=b$ are proportional to $(r^\ell - a^{2\ell+1}/r^{\ell+1})$ and $(r^\ell - b^{2\ell+1}/r^{\ell+1})$, respectively. Based on our previous experience with Green functions, we may conclude that the Green function has the form

$$g_\ell(r,r') = C\left(r_<^\ell - \frac{a^{2\ell+1}}{r_<^{\ell+1}}\right)\left(r_>^\ell - \frac{b^{2\ell+1}}{r_>^{\ell+1}}\right). \tag{4.231}$$

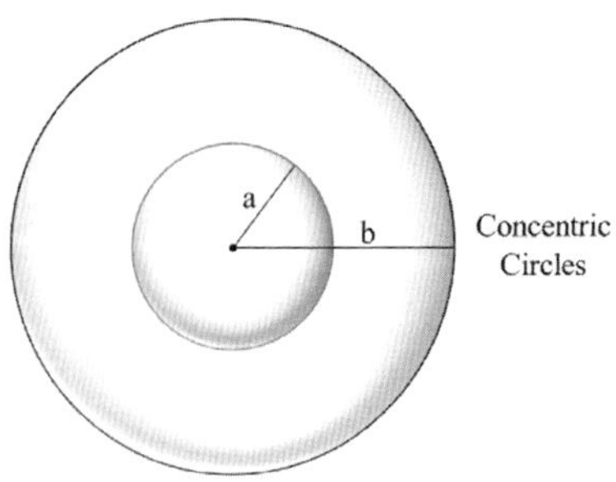

Fig. 4.4 Spherical cavity (for computation of Green function).

Using the fact that

$$-r^2 \frac{\partial g_\ell(r,r')}{\partial r}\bigg|_{r=r'^-}^{r=r'^+} = 1, \tag{4.232}$$

we may obtain

$$C = \frac{1}{(2\ell+1)\left(a^{2\ell+1} - b^{2\ell+1}\right)}. \tag{4.233}$$

In summary,

$$g_\ell(r,r') = \frac{1}{(2\ell+1)\left[1-(\frac{a}{b})^{2\ell+1}\right]}\left(r_<^\ell - \frac{a^{2\ell+1}}{r_<^{\ell+1}}\right)\left(\frac{1}{r_>^{\ell+1}} - \frac{r_>^\ell}{b^{2\ell+1}}\right). \tag{4.234}$$

Let us now consider some limiting cases of equation (4.234):

(1) $a \to 0, b \to \infty$ (free Green function). If we set $a = 0, b = \infty$ in (4.234), we obtain

$$g_\ell(r,r') = \frac{1}{2\ell+1}\frac{r_<^\ell}{r_>^{\ell+1}}. \tag{4.235}$$

Note that for $\ell = 0$ this function does not actually satisfy the Green function boundary condition at $r = 0$, since $g_0(0,r') \neq 0$. Instead, the condition at $a = 0$ ensures finiteness of the Green function at the origin. Plugging (4.235) into (4.228) recovers the free-space Green function (4.225).

(2) $a \to 0$, b finite (interior of a sphere). Substituting $a = 0$ in (4.234) gives

$$G(\vec{r},\vec{r}\,') = \sum_{\ell,m}\left(\frac{r_<^\ell}{r_>^{\ell+1}} - \frac{r^\ell r'^\ell}{b^{2\ell+1}}\right)\frac{4\pi}{2\ell+1}Y_{\ell m}^*(\theta',\phi')Y_{\ell m}(\theta,\phi). \tag{4.236}$$

Let us define $r_b \equiv b^2/r'$. Then we have

$$\frac{r^\ell r'^\ell}{b^{2\ell+1}} = \frac{b}{r'}\frac{r^\ell}{r_b^{\ell+1}}. \tag{4.237}$$

Note that $r_b > r$, so we may substitute (4.237) into (4.236) to obtain

$$G(\vec{r},\vec{r}\,') = \frac{1}{|\vec{r}-\vec{r}\,'|} - \frac{b}{r'}\frac{1}{|\vec{r}-\vec{r}_b|}, \tag{4.238}$$

where

$$\vec{r}_b \equiv \left(\frac{b^2}{r'},\theta',\phi'\right). \tag{4.239}$$

We thus obtain a negative image charge of magnitude b/r' located outside the sphere at position $\vec{r}_b$.

(3) $b \to \infty$, a finite (exterior of a sphere). When we take $b \to \infty$ in (4.234) we obtain

$$G(\vec{r}, \vec{r}\,') \to \sum_{\ell,m} \left(\frac{r_<^\ell}{r_>^{\ell+1}} - \frac{a^{2\ell+1}}{r^{\ell+1} r'^{\ell+1}} \right) \frac{4\pi}{2\ell+1} Y_{\ell m}^*(\theta', \phi') Y_{\ell m}(\theta, \phi). \tag{4.240}$$

We may define $r_a \equiv a^2/r'$ and obtain

$$\frac{a^{2\ell+1}}{r^{\ell+1} r'^{\ell+1}} = \frac{a}{r'} \frac{r_a^\ell}{r^{\ell+1}}. \tag{4.241}$$

Noting that $r_a < r$, it follows that

$$G(\vec{r}, \vec{r}\,') \to \frac{1}{|\vec{r} - \vec{r}\,'|} - \frac{a}{r'} \frac{1}{|\vec{r} - \vec{r_a}|}, \tag{4.242}$$

where

$$\vec{r_a} \equiv \left(\frac{a^2}{r'}, \theta', \phi' \right) \tag{4.243}$$

gives the location of the image charge inside the sphere.

In Section 3.3 we demonstrated the existence of a surface delta function in $\partial G_D / \partial n$. We can show this rigorously in the current geometry by differentiating the expansion (4.240):

$$\left. \frac{\partial G}{\partial n} \right|_{r=a} = -\left. \frac{\partial G}{\partial r} \right|_{r=a}$$
$$= -\sum_{\ell,m} \left(\frac{\ell a^{\ell-1}}{r'^{\ell+1}} + (\ell+1) \frac{a^{2\ell+1}}{r'^{\ell+1} a^{\ell+2}} \right) \frac{4\pi}{2\ell+1} Y_{\ell m}^*(\theta', \phi') Y_{\ell m}(\theta, \phi) \tag{4.244}$$

$$= -4\pi \sum_{\ell,m} \frac{a^{\ell-1}}{r'^{\ell+1}} Y_{\ell m}^*(\theta', \phi') Y_{\ell m}(\theta, \phi). \tag{4.245}$$

Taking limits of both sides, we obtain

$$\lim_{r' \to a^+} -\frac{1}{4\pi} \left. \frac{\partial G}{\partial n} \right|_{r=a} = a^{-2} \delta(\cos\theta - \cos\theta') \delta(\phi - \phi'), \tag{4.246}$$

which is consistent with (2.123).

4.13 Potential of conducting sphere in a uniform field via separation of variables

Let us now apply separation of variables in spherical coordinates to solve the potential outside a conducting sphere in a uniform field. We begin with Laplace's equation in spherical coordinates:

$$\frac{1}{r^2}\frac{\partial}{\partial r}\left(r^2\frac{\partial \Phi}{\partial r}\right)+\frac{1}{r^2\sin\theta}\frac{\partial}{\partial\theta}\left(\sin\theta\frac{\partial\Phi}{\partial\theta}\right)+\frac{1}{r^2\sin^2\theta}\frac{\partial^2\Phi}{\partial\phi^2}=0. \tag{4.247}$$

Variable-separated solutions have the form

$$\Phi = R(r)P(\theta)Q(\phi). \tag{4.248}$$

Substituting (4.248) into (4.247) gives

$$PQ\frac{d}{dr}\left(r^2\frac{dR}{dr}\right)+\frac{RQ}{\sin\theta}\frac{d}{d\theta}\left(\sin\theta\frac{dP}{d\theta}\right)+\frac{RP}{\sin^2\theta}\frac{d^2Q}{d\phi^2}=0, \tag{4.249}$$

and multiplying by $\sin^2\theta/\Phi$ gives

$$\frac{\sin^2\theta}{R}\left[\frac{d}{dr}\left(r^2\frac{dR}{dr}\right)\right]+\frac{\sin\theta}{P}\frac{d}{d\theta}\left(\sin\theta\frac{dP}{d\theta}\right)+\frac{1}{Q}\frac{d^2Q}{d\phi^2}=0. \tag{4.250}$$

The ϕ-dependent term in (4.250) must be a constant (denoted $-m^2$), so that

$$\frac{1}{Q}\frac{d^2Q}{d\phi^2}=-m^2 \implies Q\sim e^{\pm im\phi}. \tag{4.251}$$

If the full range of ϕ from 0 to 2π is allowed, then m must be an integer. Equation (4.250) now reduces to

$$\underbrace{\frac{1}{R}\frac{d}{dr}\left(r^2\frac{dR}{dr}\right)}_{r\text{ only}}+\underbrace{\frac{1}{P\sin\theta}\frac{d}{d\theta}\left(\sin\theta\frac{dP}{d\theta}\right)-\frac{m^2}{\sin^2\theta}}_{\theta\text{ only}}=0. \tag{4.252}$$

The θ-dependent terms of (4.252) must be equal to a constant, which with admirable foresight we denote by $-\ell(\ell+1)$. This leads (after some algebra) to the equations

$$\frac{d}{dr}\left(r^2\frac{dR}{dr}\right)-\ell(\ell+1)R=0 \tag{4.253}$$

and

$$\frac{1}{\sin\theta}\frac{d}{d\theta}\left(\sin\theta\frac{dP}{d\theta}\right)+\left[\ell(\ell+1)-\frac{m^2}{\sin^2\theta}\right]P=0. \tag{4.254}$$

The independent solutions to (4.253) are $R(r) \sim r^\ell,\ r^{-\ell-1}$. Equation (4.254) agrees with (4.174), and in Section 4.9 we found one solution (in the case where $\ell = 0, 1, 2, \ldots$ and $m = -\ell, \ldots, \ell$) to be the Legendre function of the first kind $P \sim P_\ell^{|m|}(\cos\theta)$. Another solution is given by $Q_\ell^{|m|}(\cos\theta)$, where $(m \geq 0)$

$$Q_\ell^m(x) \equiv (-1)^m (1-x^2)^{m/2} \frac{d^m}{dx^m} Q_\ell(x), \tag{4.255}$$

$$Q_\ell(x) \equiv P_\ell(x) \int^x \frac{dx'}{(1-x'^2)[P_\ell(x')]^2}. \tag{4.256}$$

The $Q_\ell(x)(Q_\ell^m(x))$ are called the *Legendre functions of the second kind* (add *associated*): they are infinite at the endpoints $x = \pm 1$. Their use in electrostatics is indicated in Exercise 4.13.2.

Let us now apply this machinery to find the potential for a conducting sphere in a uniform field; see Figure 4.5. We already saw one approach using the spherical conducting Green function for this situation in Section 3.3. The azimuthal symmetry implies that only $m = 0$ terms contribute, which reduces the variable-separated solution to the form

$$\Phi(r,\theta) = \sum_{\ell=0}^{\infty} A_\ell r^\ell P_\ell(\cos\theta) + \sum_{\ell=0}^{\infty} B_\ell r^{-\ell-1} P_\ell(\cos\theta). \tag{4.257}$$

The uniform electric field in the $+z$ direction of magnitude E_0 leads to the following boundary condition at ∞:

$$\Phi(r,\theta)\big|_{r\to\infty} = -E_0 z, \tag{4.258}$$

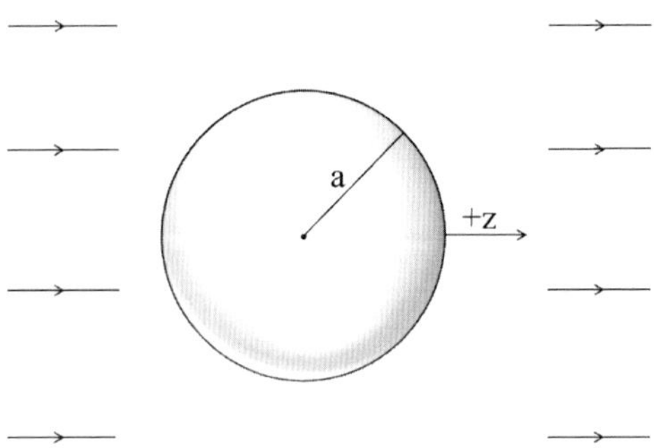

Fig. 4.5 Conducting sphere in uniform field.

which tells us that $A_1 = -E_0$ and $A_\ell = 0$ for $\ell > 1$. On the other hand, the conducting sphere has a constant potential V_0 at $r = a$ ($V_0 = 0$ if the sphere is grounded):

$$V_0 = \Phi(r,\theta)|_{r=a} \tag{4.259}$$

$$\implies V_0 = A_0 + A_1 a\cos\theta + \frac{B_0}{a} + \frac{B_1}{a^2}\cos\theta + \sum_{\ell=2}^{\infty} B_\ell a^{-\ell-1} P_\ell(\cos\theta). \tag{4.260}$$

It follows that $A_0 + B_0/a = V_0$, $A_1 a + B_1/a^2 = 0$, and $B_\ell = 0$ for $\ell > 1$. We may relate the constants A_0, B_0 to the total charge Q of the sphere by noting that

$$\begin{aligned} Q &= \int a^2 d\Omega\,\sigma = -\frac{a^2}{4\pi}\oint d\Omega\,\frac{\partial\Phi}{\partial r}\bigg|_{r=a} \\ &= -\frac{a^2}{4\pi}\oint d\Omega\,\frac{\partial}{\partial r}\left(A_0 + A_1 r\cos\theta + \frac{B_0}{r} + \frac{B_1}{r^2}\cos\theta\right)\bigg|_{r=a}. \\ &= B_0. \end{aligned} \tag{4.261}$$

From the above we may deduce that $A_0 = V_0 - Q/a$, $A_1 = -E_0$, $B_0 = Q$, and $B_1 = E_0 a^3$ so that

$$\Phi(r,\theta) = V_0 - \frac{Q}{a} + \frac{Q}{r} - E_0\left(r - \frac{a^3}{r^2}\right)\cos\theta. \tag{4.262}$$

Notice that V_0 and Q are independent parameters here. A grounded sphere ($V_0 = 0$) in this case does not necessarily have zero charge because the potential is not required to vanish at infinity.

4.14 Method of last resort: eigenfunction expansions

We have solved several problems using the reduced Green function strategy, which involves expanding the Green function in eigenfunctions in two of the dimensions leading to a reduced Green function equation in the third dimension. It is also possible to go whole hog and expand in all three dimensions. The solutions thus obtained have a more complicated form, but can be useful for perturbative calculations (see Sections 10.8 and 10.11). They also lead to useful identities for reduced Green functions.

Let us consider the eigenvalue equation corresponding to Laplace's equation (Helmholtz equation once again)

$$\nabla^2\psi(\vec{x}) + k^2\psi(\vec{x}) = 0, \tag{4.263}$$

where $\psi(\vec{x})$ is defined in a finite region of space (denoted V) and is required to satisfy Dirichlet, Neumann, or mixed boundary conditions. In general we will need *complex* solutions ψ, although our Green functions themselves will remain real. In a system with a confined geometry we have seen that nonzero solutions occur only for a set of real discrete values $k_1, k_2, \ldots$, so we obtain a discrete set of solutions ψ_n satisfying

$$\nabla^2 \psi_n(\vec{x}) + k_n^2 \psi_n(\vec{x}) = 0. \tag{4.264}$$

It is possible to show (using Green's second identity) that eigenfunctions corresponding to different eigenvalues are orthogonal:

$$\int_V d^3x\, \psi_n^*(\vec{x})\psi_m(\vec{x}) = 0 \qquad \text{when } n \neq m. \tag{4.265}$$

Furthermore, it is possible to orthogonalize different eigenfunctions corresponding to the same eigenvalue (using the Gram-Schmidt algorithm). Because V is finite, we may normalize the solutions as

$$\int_V d^3x\, \psi_n^*(\vec{x})\psi_m(\vec{x}) = \delta_{nm}. \tag{4.266}$$

We suppose that $\{\psi_n\}$ is a complete set of functions, so that any "reasonable" function $f(\vec{x})$ can be expressed as $f(\vec{x}) = \sum_m c_m \psi_m(\vec{x})$. Then the relation

$$\delta(\vec{x}\,' - \vec{x}) = \sum_n \psi_n^*(\vec{x}\,')\psi_n(\vec{x}), \tag{4.267}$$

may be established by integration; see Exercise 4.14.1.

The Dirichlet Green function $G_D(\vec{x}, \vec{x}\,')$ solves

$$\nabla^2 G_D(\vec{x}, \vec{x}\,') = -4\pi\delta(\vec{x} - \vec{x}\,'). \tag{4.268}$$

In view of (4.267), it is reasonable to expect a Green function of the form

$$G_D(\vec{x}, \vec{x}\,') = \sum_n a_n \psi_n^*(\vec{x}\,')\psi_n(\vec{x}), \tag{4.269}$$

with eigenvalues which obey the boundary condition on the surface S:

$$\psi_n(\vec{x})|_S = 0. \tag{4.270}$$

Plugging our eigenfunction expansions for $\delta(\vec{x}\,' - \vec{x})$ and $G_D(\vec{x}, \vec{x}\,')$ into (4.268) yields

$$\begin{aligned}
\nabla^2 \left(\sum_n a_n \psi_n^*(\vec{x}\,')\psi_n(\vec{x}) \right) &= -4\pi \sum_n \psi_n^*(\vec{x}\,')\psi_n(\vec{x}) \\
\Longrightarrow \sum_n k_n^2 a_n \psi_n^*(\vec{x}\,')\psi_n(\vec{x}) &= 4\pi \sum_n \psi_n^*(\vec{x}\,')\psi_n(\vec{x}) \\
\Longrightarrow a_n &= \frac{4\pi}{k_n^2},
\end{aligned} \tag{4.271}$$

which leads finally to the result

$$G_D(\vec{x}, \vec{x}\,') = 4\pi \sum_n \frac{\psi_n^*(\vec{x'})\psi_n(\vec{x})}{k_n^2}. \tag{4.272}$$

The symmetry condition $G_D(\vec{x}, \vec{x}\,') = G_D(\vec{x}\,', \vec{x})$ shows that this Green function is real.

We shall illustrate this process first with a simple example, namely the two-dimensional rectangle with Dirichlet boundary conditions. The box length is restricted to the range $0 \leq x \leq a$ in the x-direction and to the range $0 \leq y \leq b$ in the y-direction. The eigenvalue equation is:

$$\left[\frac{\partial^2}{\partial x^2} + \frac{\partial^2}{\partial y^2}\right]\psi + k^2\psi = 0. \tag{4.273}$$

The boundary conditions are:

$$\psi(0, y) = \psi(a, y) = \psi(x, 0) = \psi(x, b) = 0, \quad 0 \leq x \leq a, 0 \leq y \leq b. \tag{4.274}$$

Naturally, we seek variable-separated solutions: $(\psi(x, y) = \psi_x(x)\psi_y(y))$, which leads to

$$\frac{1}{\psi_x}\frac{d^2\psi_x}{dx^2} + \frac{1}{\psi_y}\frac{d^2\psi_y}{dy^2} = -k^2, \tag{4.275}$$

$$\Longrightarrow \quad \frac{d^2\psi_x}{dx^2} = -k_x^2\psi_x, \frac{d^2\psi_y}{dy^2} = -k_y^2\psi_y, k_x^2 + k_y^2 = k^2. \tag{4.276}$$

Using techniques that we have seen before, we arrive at the following normalized, variable-separated solutions:

$$\psi_{nm}(\vec{x}) = \sqrt{\frac{4}{ab}}\sin\left(\frac{n\pi x}{a}\right)\sin\left(\frac{m\pi y}{b}\right), \tag{4.277}$$

for n, m positive integers. From the theory of Fourier series, we know that these functions form a complete set. Using (4.272) we thus obtain the Green function

$$G_D(\vec{x}, \vec{x}\,') = \frac{16}{\pi ab}\sum_{n,m=1}^{\infty}\frac{\sin\left(\frac{n\pi x}{a}\right)\sin\left(\frac{n\pi x'}{a}\right)\sin\left(\frac{m\pi y}{b}\right)\sin\left(\frac{m\pi y'}{b}\right)}{(n/a)^2 + (m/b)^2}. \tag{4.278}$$

The expression in (4.278) is a double infinite series, whereas the equivalent reduced Green function solution would have only a single infinite sum. So, such solutions are not always well suited for practical computations (which is why this method is a "last resort").

As a final example, we compute the Dirichlet Green function in a cylinder of radius a of infinite length. In this case the eigenvalue equation is

$$\frac{1}{\rho}\frac{d}{d\rho}\left(\rho\frac{\partial\psi}{\partial\rho}\right)+\frac{1}{\rho^2}\frac{\partial^2\psi}{\partial\phi^2}+k^2\psi=0, \tag{4.279}$$

(where $\psi=\psi(\rho,\phi)$), with boundary conditions

$$\psi(a,\phi)=0, \psi(a,\phi)=\psi(a,\phi+2\pi). \tag{4.280}$$

(The second condition in (4.280) ensures that $\psi(\rho,\phi)$ is single-valued.) Assuming variable-separated solutions of the form $\psi=\psi_\rho(\rho)\psi_\phi(\phi)$ and plugging into (4.279) leads to

$$\rho^2\left[\frac{1}{\psi_\rho\rho}\frac{d}{d\rho}\left(\rho\frac{d\psi_\rho}{d\rho}\right)+k^2\right]+\frac{1}{\psi_\phi}\frac{d^2\psi_\phi}{d\phi^2}=0. \tag{4.281}$$

The ρ-dependent and ϕ-dependent terms must both be equal to the same constant, which we denote by $-m^2$. The ϕ-dependent term produces solutions

$$\begin{aligned}\frac{d^2\psi_\phi}{d\phi^2}&=-m^2\psi_\phi;\\ \Longrightarrow \psi_\phi&\sim e^{\pm im\phi},\end{aligned} \tag{4.282}$$

while the ρ-dependent term gives

$$\begin{aligned}&\frac{1}{\rho}\frac{d}{d\rho}\left(\rho\frac{d\psi_\rho}{d\rho}\right)+\left(k^2-\frac{m^2}{\rho^2}\right)\psi_\rho=0\\ &\Longrightarrow \psi_\rho\sim J_m(k\rho), N_m(k\rho).\end{aligned} \tag{4.283}$$

Applying the boundary conditions (4.280) and introducing the proper normalization leads to the variable-separated solutions

$$\psi_{mn}(\vec{x})=J_{1m}(k_{mn}\rho)\frac{1}{\sqrt{2\pi}}e^{im\phi}, \quad (n=1,2,\ldots,\ m=0,\pm1,\pm2,\ldots) \tag{4.284}$$

where the $\{k_{mn}\}$ are defined in terms of Bessel function roots $\{x_{mn}\}$:

$$k_{mn}\equiv\frac{x_{mn}}{a} \quad \text{where} \quad J_m(x_{mn})=0, \tag{4.285}$$

and J_{1m} is normalized and proportional to J_m:

$$J_{1m}(k_{mn}\rho)\equiv\frac{\sqrt{2}}{a}\frac{J_m(k_{mn}\rho)}{J_{m+1}(k_{mn}a)}. \tag{4.286}$$

Note that the solutions are oscillatory in both ρ and ϕ, and the oscillatory/exponential tradeoff observed in Section 4.6 is not present.

In view of previous results concerning the completeness of Bessel functions, we obtain the following Green function:

$$G_D(\vec{x}, \vec{x}\,') = 2a^2 \sum_{n=1}^{\infty} \sum_{m=-\infty}^{\infty} \frac{J_{1m}(k_{mn}\rho) J_{1m}(k_{mn}\rho') e^{im(\phi-\phi')}}{x_{mn}^2}. \qquad (4.287)$$

Notice that although the eigenfunctions are complex, the Green function is real.

Although eigenfunction solutions are too unwieldy for practical computations, they may be used to derive useful summation formulas by comparing with analytical solutions. For example, (4.278) can be compared with the reduced Green function solution to give a formula for one of the sums; while (4.287) can be compared with a single series form (Exercise 4.8.3) or ultimately to the image solution for the same problem (Exercise 3.3.5).

4.15 Going Deeper

4.15.1 *Special functions*

See the references in Section 1.6.1. In addition, see:

(1) R. Beals and R. Wong, *Special Functions: A Graduate Text*, Cambridge University Press (Cambridge) 2010.

(2) A. Gill, J. Segura and N. M. Temme, *Numerical Methods for Special Functions*, SIAM Publishing (Philadelphia) 2007.

(3) E. W. Weisstein, *Special Function*, MathWorld–A Wolfram Web Resource:
`http://mathworld.wolfram.com/SpecialFunction.html`.

4.15.2 *Eigenfunction expansions*

(1) P. M. Morse and H. Feshbach, *Methods of Theoretical Physics, Parts I and II*, McGraw-Hill (New York) 1953; Chapters 6 and 7.

(2) J. Matthews and R. L. Walker, *Mathematical Methods of Physics*, 2nd ed., Addison-Wesley (Redwood City)1970; Chapter 9.

(3) J. Van Bladel, *Electromagnetic Fields*, 2nd ed., IEEE Press (Piscataway) 2007; Sections 1.8, 1.10, 5.8.1.

4.16 Exercises

Exercise 4.1.1. Obtain the Taylor series expansion (4.8) for the Bessel function J_m from (a) the generating function expression (4.7) and (b) the integral representation (4.13).

Exercise 4.1.2. Using the generating function for Bessel functions,

$$e^{it\cos\phi} = \sum_{m=-\infty}^{\infty} i^m e^{im\phi} J_m(t),$$

prove the sum rules:

$$(a)\ \sum_{m=-\infty}^{\infty} (J_m(t))^2 = 1, \qquad (b)\ J_0(2t) = \sum_{m=-\infty}^{\infty} (-1)^m (J_m(t))^2 .$$

Exercise 4.1.3. Show that

$$\int_0^a d\rho\, \rho^{m+1} J_m(k\rho) = \frac{1}{k} a^{m+1} \left(J_{m+1}(ka) - \delta_{m,-1}\right).$$

Exercise 4.1.4. Show the integral representations (4.15) and (4.16) are correct from the preceding characterizations of the Bessel function $J_n(t)$.

Exercise 4.2.1.

(a) By evaluating the expression (the $\perp$ notation indicates the x and y components of a vector, and α is the cylindrical angle associated with $\vec{k}$, as in Figure 4.6),

$$\int_0^{2\pi} \frac{d\alpha}{(2\pi)^2} e^{i\vec{k}_\perp \cdot (\vec{x}-\vec{x}\,')_\perp}$$

two different ways (or any other way you can think of), show the Bessel function addition theorem,

$$J_0(kD) = \sum_{m=-\infty}^{\infty} J_m(k\rho) J_m(k\rho') e^{im(\phi-\phi')},$$

where

$$D \equiv |(\vec{x} - \vec{x}\,')_\perp|, \quad \rho = |\vec{x}_\perp|, \quad \rho' = |\vec{x}_\perp|.$$

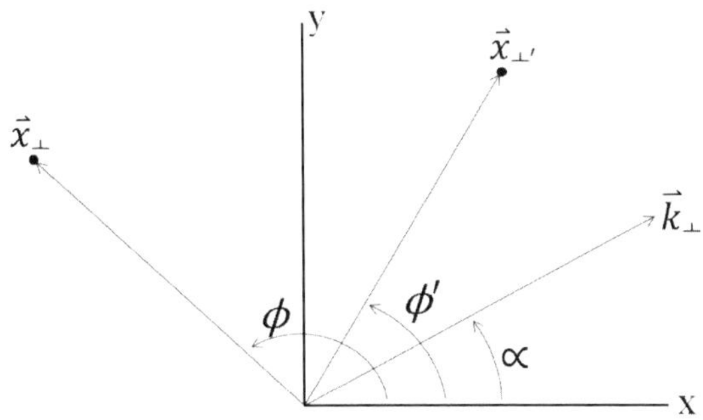

Fig. 4.6 Notation for Exercise 4.2.1.

(b) Using (a), show that

$$\int_0^\infty \frac{dk\, k}{2\pi} J_0(kD) = \frac{1}{\rho}\delta(\rho - \rho')\delta(\phi - \phi').$$

Exercise 4.3.1. Show that for zeros of derivatives of Bessel functions ($y_{mn} = \bar{k}_{mn} a$)

$$\int_0^a d\rho\, \rho \left[J_m(\bar{k}_{mn}\rho)\right]^2 = \frac{a^2}{2}\left(1 - \frac{m^2}{y_{mn}^2}\right)\left[J_m(y_{mn})\right]^2.$$

You may use the method described in the text after Equation (4.52).

Exercise 4.4.1. Using the reduced Green function technique, show that the Dirichlet Green function for the infinite conducting wall problem (see Figure 4.7) may be solved in cylindrical coordinates to give

$$\begin{aligned} G_D(\vec{x}, \vec{x}\,') &= 4\int_0^\infty dk\, k\, g(k; z, z') \left[\frac{1}{2} J_0(k\rho) J_0(k\rho') \right. \\ &\quad \left. + \sum_{m=1}^{\infty} J_m(k\rho) J_m(k\rho') \cos(m(\phi - \phi'))\right] \\ &= 2\int_0^\infty dk k\, g(k; z, z') J_0(kD). \end{aligned}$$

(For the second equality, see Exercise 4.2.1.) Find $g(k; z, z')$.

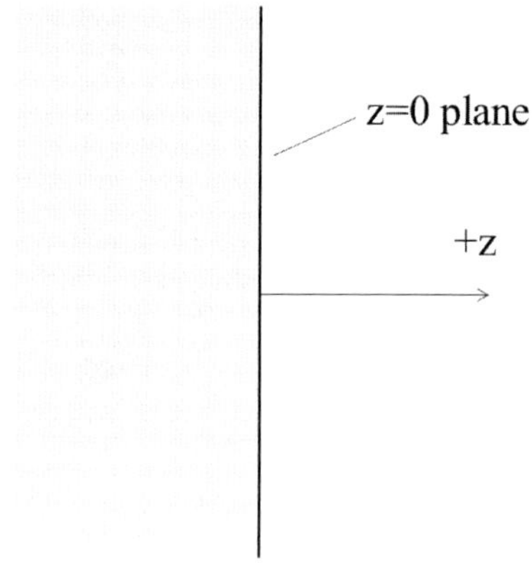

Fig. 4.7 Geometry for Exercise 4.4.1.

Exercise 4.4.2. Once again, using the reduced Green function technique, find the Dirchlet Green's function for an infinite, flat conducting plate in infinite space. *Take the surface of the conductor as the $z-x$ plane* and free space as the $y > 0$ region. Show that the Green function may be written in cylindrical coordinates (ρ, z, ϕ) as

$$G_D(\vec{x}, \vec{x}\,') = 4\int_0^\infty dk \sum_{m=1}^{\infty} J_m(k\rho)J_m(k\rho')\sin(m\phi)\sin(m\phi')e^{k(z_< - z_>)},$$

where $z_<$ ($z_>$) is the lesser (greater) of z and z'.

Exercise 4.4.3. Consider the parallel plate situation shown in Figure 4.8, with *Neumann* boundary conditions on the top and bottom plates. The bottom plate at $z = 0$ has $E_z|_{z=0} = 0$, whereas the top plate has $E_z|_{z=d} = E_z^0$ (a constant) for the circular disk region $\rho < a$, and $E_z|_{z=d} = 0$ otherwise.

(a) Show that the potential between the plates can be expressed as (up to an overall constant)

$$\Phi(\rho, z) = -aE_z^0 \int_0^\infty \frac{dk}{k} J_0(k\rho)\, J_1(ka)\frac{\cosh(kz)}{\sinh(kd)}.$$

[*Hint*: See results in Exercises 4.2.1 and 4.4.1 for this azimuthally symmetric problem.]

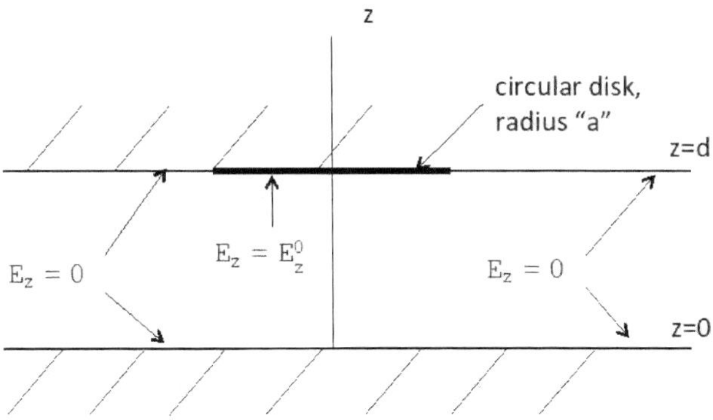

Fig. 4.8 Geometry for Exercise 4.4.3.

(b) Consider the same geometry as in (a) with the same Neumann boundary conditions as before on the top plate. However, now replace the Neumann condition $E_z|_{z=0} = 0$ with the Dirichlet condition $\Phi|_{z=0} = 0$ on the bottom plate. Show that the new potential can be expressed as

$$\Phi(\rho, z) = -aE_z^0 \int_0^\infty \frac{dk}{k} J_0(k\rho)\, J_1(ka) \frac{\sinh(kz)}{\cosh(kd)}.$$

Exercise 4.4.4. Consider the interior of a half-infinite conducting cylinder of radius a and infinite length ($z > 0$) bounded by a conducting plane at $z = 0$. Using cylindrical coordinates, find the Dirichlet Green function for this geometry using the reduced Green function technique.

Exercise 4.6.1. At the end of Section 4.3 it was stated that there was a limiting connection between the completeness relation (4.39), holding for ρ in the infinite domain $[0, \infty]$, and (4.71), which holds in the finite domain $[0, a]$. Using the asymptotic form for zeros of $J_\nu(k\rho)$ from (4.98), show that the two forms are consistent with one another for $a \to \infty$.

Exercise 4.6.2. Consider the volume inside of a right circular cylinder of radius a with conducting caps held a zero potential (grounded), as shown in Figure 4.9. The distance between the caps is L. The side of the cylinder is slightly separated from the caps and is raised to constant potential V.

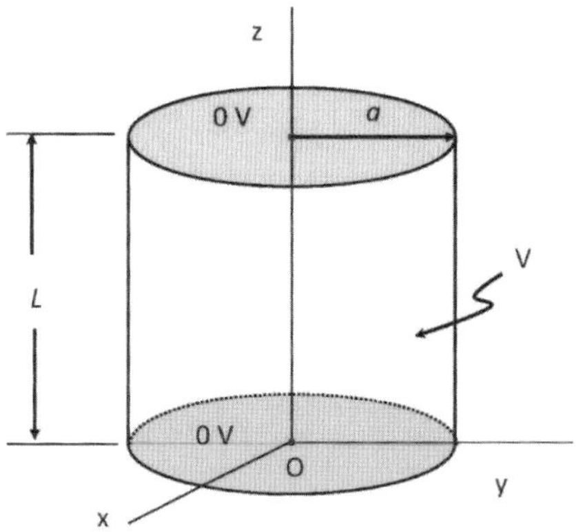

Fig. 4.9 Geometry for Exercise 4.6.2.

Using cylindrical coordinates, determine the potential $\Phi(\rho, z, \phi)$ everywhere inside the cylindrical volume.

Exercise 4.7.1.

(a) Recalling the Section 3.10 on variational techniques, use Equation (3.144) and the Bessel function expansion (4.120) to establish the starting point for calculating the capacitance of a thin, flat circular conductor of radius a with surface charge density $\sigma(\rho)$ carrying a total charge Q:

$$C^{-1}[\sigma] = \frac{4\pi^2}{Q^2}\int_0^\infty dk \left[\int_0^a d\rho\rho\, J_0(k\rho)\sigma(\rho)\right]^2 .$$

(b) Assuming the charge density form

$$\sigma(\rho) = A + B\left(\frac{\rho}{a}\right)^2 ,$$

where A and B are parameters, establish the lower limit on the capacitance: $C > 0.6213\ldots a$. The exact result from Exercise 2.6.1 is $C = \frac{2}{\pi}a = 0.6366\ldots a$. [*Hint*: A symbolic manipulation program can help with the integrals, which can also be looked up.]

Exercise 4.7.2. The half-space $z > 0$ in three dimensions is bounded by a conductor with *Neumann* boundary conditions. A circular patch S of radius a has

$$-\frac{\partial \Phi(\rho, z)}{\partial z} = E_z^0,$$

where $\Phi(\rho, z)$ is the potential and where E_z^0 is a constant, the rest of the surface has the electric field $E_z = 0$. Take the z-axis perpendicular to the conducting plane through the center of the circular patch. Using the Neumann Green function for this geometry and the Bessel function expansion (4.120), show that the potential may be written in integral form,

$$\Phi(\rho, z) = E_z^0 a \int_0^\infty \frac{dk}{k} J_0(k\rho) J_1(ka) e^{-kz}.$$

Exercise 4.7.3.

(a) Using the differential equations satisfied by both $J_\nu(t)$ and $J_{-\nu}(t)$,

$$\left[\frac{d^2}{dt^2} + \frac{1}{t}\frac{d}{dt} - \frac{\nu^2}{t^2} + 1\right] J_{\nu,-\nu}(t) = 0,$$

show that

$$W_t[J_\nu(t), J_{-\nu}(t)] = \frac{C}{t},$$

where C is a constant.

(b) Using (a) and the asymptotic form for $J_\nu(t)$, show

$$W_t[J_\nu(t), J_{-\nu}(t)] = -\frac{2\sin(\nu\pi)}{\pi t},$$

as well as

$$W_t[J_m(t), N_m(t)] = \frac{2}{\pi t},$$

for m an integer, where $N_m(t)$ is the Neumann Bessel function defined in Section 4.1.

Exercise 4.7.4. Consider an infinitely long conducting cylinder of radius a, with the z-axis of coordinates along the axis of symmetry of the cylinder (see Figure 4.10).

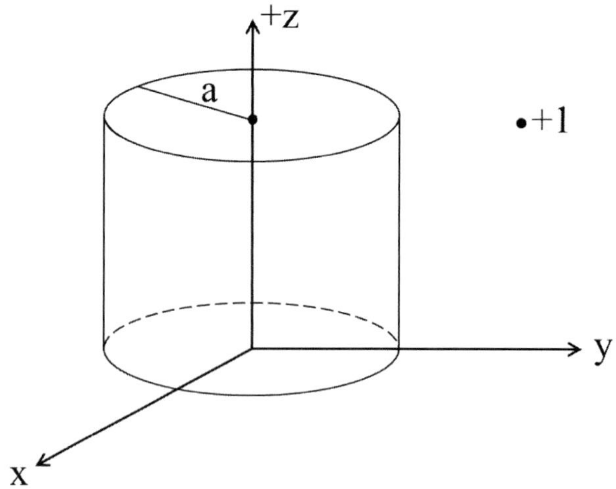

Fig. 4.10 Geometry for problem 4.7.4.

(a) Assume the Dirichlet Green function has the form

$$G_D(\vec{x}, \vec{x}\,') = 4\pi \sum_{m=-\infty}^{\infty} \frac{e^{im(\phi-\phi')}}{2\pi} \frac{1}{\pi} \int_0^\infty dk \cos[k(z-z')] g_m^{\text{out}}(k; \rho, \rho'),$$

and use

$$4\pi\delta(\vec{x} - \vec{x}\,') = 4\pi \sum_{m=-\infty}^{\infty} \frac{e^{im(\phi-\phi')}}{2\pi} \frac{1}{\pi} \int_0^\infty dk \cos[k(z-z')] \frac{1}{\rho}\delta(\rho - \rho'),$$

to find the reduced Green function $g_m^{\text{out}}(k; \rho, \rho')$ for the volume exterior to the cylinder.

(b) Assuming the same form for the Dirichlet Green function and delta function as in part (a), find the reduced Green function $g_m^{\text{in}}(k; \rho, \rho')$ for the interior solution.

Exercise 4.7.5. Consider the interior region of a cylindrical toroid of rectangular cross-section (see Figure 4.11). The height is L, and the inner and outer radii are a and b, respectively. Construct the Dirichlet Green function in cylindrical coordinates (ρ, z, ϕ) for the interior region of the cylinder. [*Hint*: Choosing ρ as the non-oscillatory part of the solution is easiest.]

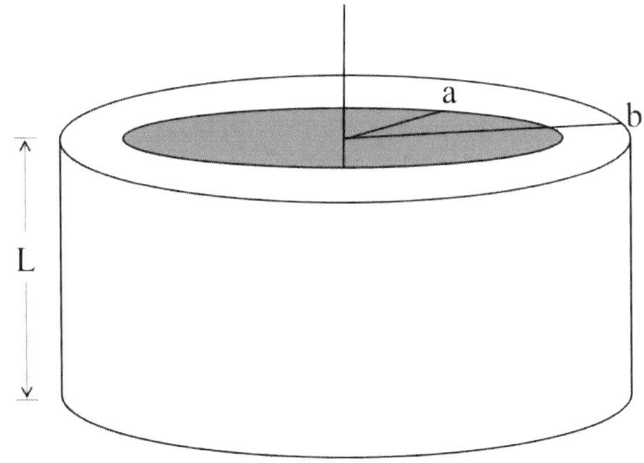

Fig. 4.11 Geometry for Exercise 4.7.5.

Exercise 4.7.6.

(a) Given

$$\frac{1}{|\vec{x}-\vec{x}\,'|} = \int_0^\infty dk \sum_{m=-\infty}^{\infty} e^{im(\phi-\phi')} J_m(k\rho) J_m(k\rho') e^{-k(z_>-z_<)},$$

in cylindrical coordinates, show *explicitly* that

$$\nabla^2 \frac{1}{|\vec{x}-\vec{x}\,'|} = -4\pi\,\delta(\vec{x}-\vec{x}\,'),$$

by carrying out the differential operations and using Equation (4.108).

(b) Using a cylindrical free-space Green function expansion, show that

$$\int_0^a \rho' d\rho' \int_0^{2\pi} \frac{d\phi'}{|\,\vec{x}-\vec{x}\,'\,|} = 2\pi\, a \int_0^\infty \frac{dk}{k}\, J_0(k\rho)\, J_1(ka)\, e^{-k|z-z'|},$$

where $J_0(k\rho)$ and $J_1(ka)$ are Bessel functions.

Exercise 4.8.1. In Equation (4.143) we found that the Dirichlet Green function for a line charge in the volume bounded by conducting walls intersecting at arbitrary angle β (see Figure 4.12) could be written as the infinite sum,

$$G_D(\vec{x},\vec{x}\,') = \sum_{n=1}^{\infty} \frac{4}{n} \left(\frac{\rho_<}{\rho_>}\right)^{\gamma} \sin\left(\frac{n\pi\phi}{\beta}\right) \sin\left(\frac{n\pi\phi'}{\beta}\right),$$

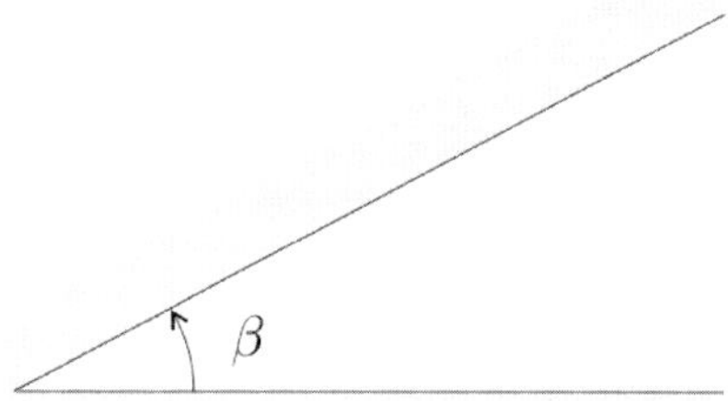

Fig. 4.12 Geometry for Exercise 4.8.1.

where $\rho_<$ ($\rho_>$) is the lesser (greater) of ρ and ρ' and $\gamma = n\pi/\beta$. Using the summation method of Section 3.9, show that this expression can be summed exactly into the closed form

$$G_D(\vec{x}, \vec{x}\,') = \ln \left[\frac{1 + (\rho_</\rho_>)^{2\pi/\beta} - 2\,(\rho_</\rho_>)^{\pi/\beta} \cos\left(\frac{\pi}{\beta}(\phi + \phi')\right)}{1 + (\rho_</\rho_>)^{2\pi/\beta} - 2\,(\rho_</\rho_>)^{\pi/\beta} \cos\left(\frac{\pi}{\beta}(\phi - \phi')\right)} \right].$$

Notice this expression agrees with Exercise 3.1.3 when $\beta = \pi/2$.

Exercise 4.8.2. Consider the closed, vacuum-filled wedge shown in Figure 4.13. It opens with angle β and is bounded by circular conducting surfaces at radii a and b ($a < b$). Find the Dirichlet Green function for a line charge of unit charge density in this geometry by the reduced Green function method.

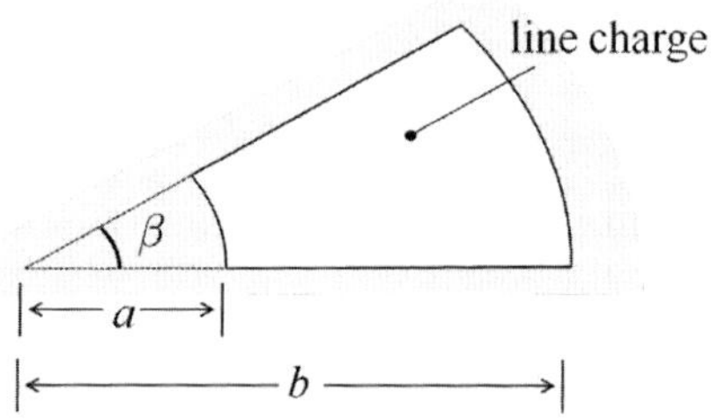

Fig. 4.13 Geometry for Exercise 4.8.2.

Exercise 4.8.3. By using a reduced Green function technique or other means, show that the Dirichlet Green function for a line-charge source in the cylindrical region $0 \leq \rho \leq a$ has the expansion,

$$G_D(\vec{x}, \vec{x}\,') = -\ln(\rho_>^2/a^2) + 2\sum_{m=1}^{\infty} \frac{\cos[m(\phi - \phi')]}{m}(\rho_<^m)\left(\frac{1}{\rho_>^m} - \frac{\rho_>^m}{a^{2m}}\right),$$

where $\rho_>(\rho_<)$ is the greater (lesser) of ρ and ρ'.

Extra: Using the summation method of Section 3.9, explicitly sum the Dirichlet Green function to get the image solution for a cylinder, Exercise 3.3.5(b).

Exercise 4.8.4. Consider an infinitely long vacuum-filled wedge-shaped hole (wedge radius a) inside a conductor in three dimensions. (See Figure 4.14: note the +1 charge is a *point charge*, not a line charge.) Use the reduced Green function method to derive a form for the Dirichlet Green function, $G_D(\rho, \phi, z; \rho', \phi', z')$, in cylindrical coordinates. It is possible to derive a differential equation for the reduced Green function in either the ρ or z variables. [*Hint*: Starting out with a correct form of the three-dimensional delta function is the key to doing this problem.]

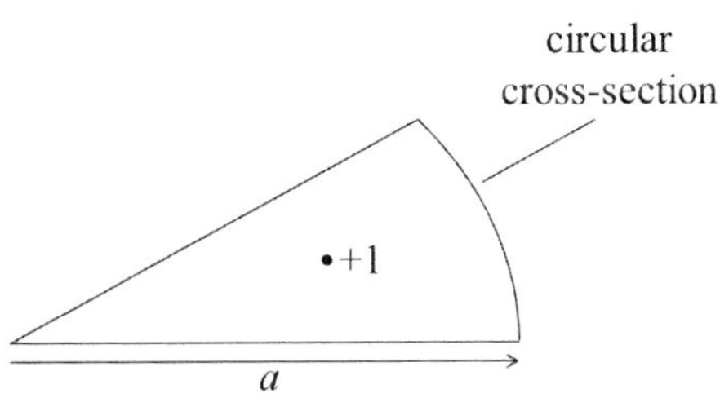

Fig. 4.14 Geometry for Exercise 4.8.4.

Exercise 4.8.5. Using a reduced Green function technique, construct the Dirichlet Green function $G_D(\rho, \phi; \rho', \phi')$ for a unit-density line-charge source positioned between grounded concentric conducting cylinders ($a > \rho' > b$); the inner radius is b and the outer radius is a.

Exercise 4.9.1.

(a) Using the binomial theorem, show that

$$P_\ell(x) = \frac{1}{2^\ell}\sum_{r=0}^{[\ell/2]} \frac{(-1)^r}{r!(\ell-r)!}\frac{(2\ell-2r)!}{(\ell-2r)!}x^{\ell-2r}$$
$$= \sum_{r=0}^{[\ell/2]} \frac{(-1)^r(2\ell-2r-1)!!}{(2r)!!\,(\ell-2r)!}x^{\ell-2r}.$$

Here the upper limit $[\ell/2]$ denotes the greatest integer in $\ell/2$. Note that $(0)! \equiv 1$. The definition of the double factorial quantity is

$$(N)!! \equiv \begin{cases} N(N-2)(N-4)\cdots 3\cdot 1, & N \text{ odd}, \\ N(N-2)(N-4)\cdots 2, & N \text{ even}. \end{cases}$$

$((0)!! = (-1)!! \equiv 1)$

(b) Using part (a) show that

$$P_\ell(0) = \begin{cases} 0, & \ell \text{ odd}, \\ (-1)^{\ell/2}\dfrac{(\ell-1)!!}{(\ell)!!}, & \ell \text{ even}. \end{cases}$$

(c) Also show that

$$\left(\frac{d}{dx}\right)^m P_\ell(x)\bigg|_{x=0} = \begin{cases} 0, & \ell-m \text{ odd}, \\ (-1)^{(\ell-m)/2}\dfrac{(\ell+m-1)!!}{(\ell-m)!!}, & \ell-m \text{ even}, \end{cases}$$

for $m \leq \ell$.

Exercise 4.9.2. The replacements

$$\theta \longrightarrow \pi-\theta, \phi \longrightarrow \phi+\pi, \quad \text{if } \phi < \pi$$
$$\theta \longrightarrow \pi-\theta, \phi \longrightarrow \phi-\pi, \quad \text{if } \phi \geq \pi$$

describe a parity inversion of the radius vector in spherical coordinates with polar angle θ and azimuthal angle ϕ. What effect does this change have on the spherical harmonics, $Y_{\ell,m}(\theta,\phi)$? [*Hint*: Think overall phase factor.]

Exercise 4.9.3.

(a) Using the formula (4.179), and given that the Legendre polynomials $P_n(x)$ are of order n in their argument$x$, show that they are orthogonal:

$$\int_{-1}^{1} dx\, P_n(x)P_m(x) = 0, \quad m \neq n.$$

(b) Using part (a), the recursion relation Equation (4.181), plus *mathematical induction*, show that

$$\int_{-1}^{1} dx\, P_n(x)P_n(x) = \frac{2}{2n+1}.$$

[*Induction*: Show the statement is true for $n = 0$; then assuming the induction hypothesis is true for n, show it is also true for the $n + 1$ case.]

Exercise 4.11.1.

(a) With the help of the *generating function* from Equation (4.204) or (4.205),

$$G(x,t) \equiv \frac{1}{\sqrt{1-2xt+t^2}} = \sum_{n=0}^{\infty} t^n P_n(x),$$

for the Legendre polynomials $P_n(x)$, obtain the relation Equation (4.180).

(b) Again, using the generating function show Equation (4.181).

[*Hint*: Consider derivatives with respect to t and x in (a) and (b).]

Exercise 4.11.2.

(a) Show by direct substitution that the Legendre polynomials $P_n(x)$ defined by Equation (4.179) satisfy Equation (4.182).

(b) Use the definition Equation (4.179) to also show that $P_n(1) = 1$.

Exercise 4.11.3. Evaluate the following integral as concisely as possible:

$$\int d\Omega' \, P_\ell(\cos\gamma_1) P_\ell(\cos\gamma_2),$$

where

$$\cos\gamma_1 \equiv \frac{\vec{r}\cdot\vec{r}\,'}{rr'},$$
$$\cos\gamma_2 \equiv \frac{\vec{r}\,'\cdot\vec{r}\,''}{r'r''}.$$

Exercise 4.11.4.

(a) Recalling the Section 3.10 on variational techniques, use Equation (3.144) and the Coulomb expansion (4.225) to establish the starting point for calculating the capacitance of a hemispherical conductor of radius a with surface charge density $\sigma(\cos\theta)$ carrying a total charge Q:

$$C^{-1}[\sigma] = \frac{4\pi^2 a^3}{Q^2} \sum_{\ell=0}^{\infty} \left[\int_0^1 dx \, P_\ell(x)\sigma(x) \right]^2 .$$

Note that $x \equiv \cos\theta$ and $P_\ell(x)$ is a Legendre polynomial.

(b) Assuming the charge density form

$$\sigma(x) = A + Bx,$$

where A and B are parameters, establish the lower limit on the capacitance: $C > 0.8052\ldots a$. Do you find the charge density increasing or decreasing in strength as the edge at $\theta = \pi/2$ is approached? [*Hint*: Integral lookup and numerical work is necessary to approximately evaluate the ℓ sums. The exact result is $C = (\frac{1}{2} + \frac{1}{\pi})a \approx 0.8183\ldots a$.[12]]

Exercise 4.11.5. The proof of the completeness relation (4.220) for spherical harmonics requires showing that $P_\ell(\cos\gamma)$ can be written as a linear combination of the spherical harmonics as in (4.219),

$$P_\ell(\cos\gamma) = \sum_m A_{\ell m} Y_{\ell m}(\theta, \phi),$$

[12]S. C. Loh, "Calculation of the Electric Field and the Capacitance of a Charged Spherical Bowl by Means of Toroidal Co-ordinates", Proc. Inst. E. E. **117**, 641 (1970).

where γ is the angle between $\vec{r}$ and $\vec{r}\,'$, and (r, θ, ϕ) are the spherical coordinates for $\vec{r}$. In this exercise you will show a more general statement,

$$Y_{\ell m}(\theta', \phi') = \sum_{m'=-\ell}^{\ell} C^{(\ell)}_{m\,m'} Y_{\ell\,m'}(\theta, \phi), \tag{4.288}$$

and use this to show (4.219).

(a) By expressing $(\vec{a} \cdot \vec{r})$ in Equation (4.168) in two different spherical coordinate systems centered on the same origin, first show that (4.168) gives rise to a linear relation of the form

$$(\xi_-)^{2\ell} \sum_{m=-\ell}^{\ell} \left(\frac{\xi_+}{\xi_-}\right)^{\ell+m} d_{\ell m} Y_{\ell m}(\theta, \phi)$$
$$= (\xi'_-)^{2\ell} \sum_{m=-\ell}^{\ell} \left(\frac{\xi'_+}{\xi'_-}\right)^{\ell+m} c_{\ell m} Y_{\ell m}(\theta', \phi'), \tag{4.289}$$

where the $\{c_{\ell m}\}$ and $\{d_{\ell m}\}$ are constants that are independent of $\xi_\pm$ and $\xi'_\pm$.

(b) Note that each different choice of $\vec{a}$ in (4.168) produces a different linear relation of the form (4.289) with a different value of $\left(\xi'_+/\xi'_-\right)$ on the left-hand side. Show that by choosing $2\ell + 1$ different values for $\vec{a}$, the resulting system of $2\ell + 1$ equations can be solved for $Y_{\ell m}(\theta', \phi')$ to obtain an equation of the form (4.288). [*Hint*: You will need to use invertibility properties of the *Vandermonde matrix*, which can be found on the web or in many standard references on linear algebra.]

(c) The final step in showing (4.219) is accomplished by noting that $P_\ell(\cos\gamma)$ is proportional to $Y_{\ell 0}(\theta', \phi')$ for any spherical coordinate system in which the z'-axis is chosen along $\vec{r}$.

Exercise 4.11.6.

(a) Using a spherical free-space Green function expansion, show that ($d\Omega = \sin\theta\ d\theta\, d\phi$)

$$\int \frac{d\Omega\,'}{|\,\vec{r} - \vec{r}\,'\,|} = \frac{4\pi}{r_>},$$

where $r_>$ is the greater of $r$and r'.

(b) Using the same expansion, evaluate as completely as possible the integral,

$$\int \frac{\cos\theta'\, d\Omega'}{|\vec{r}-\vec{r}\,'|}.$$

(Note that the Legendre polynomial $P_1(x) = x$.)

Exercise 4.12.1. A capacitor consists of two thin hemispherical conducting caps at potentials V and $-V$ separated by a thin gap, as shown in Figure 4.15. The radius of the sphere is a.

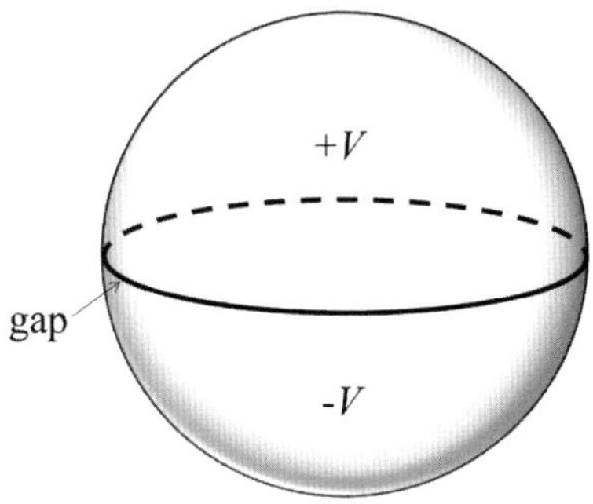

Fig. 4.15 Geometry for Exercise 4.12.1.

(a) Define the capacitance as

$$C \equiv \frac{Q}{V},$$

where Q is the charge on the upper cap. Treating this either as a boundary value problem or using the Green function for this geometry, show that

$$C = \frac{a}{2} \sum_{n=0,1,2,\ldots}^{\infty} [P_{2n}(0) - P_{2n+2}(0)]^2,$$

where $P_j(0)$ is the value of the Legendre polynomial at $x = 0$. [*Hint*: See Equation (4.182) and use the Exercise 4.11.2(b) result.]

(b) Using the result of Exercise 4.9.1, show that

$$C = \frac{a}{2} \sum_{n=0,1,2,\ldots}^{\infty} \left[(4n+3)\frac{(2n-1)!!}{(2n+2)!!}\right]^2.$$

Exercise 4.12.2. Find the electrostatic Neumann Green function, $G_N(\vec{x}, \vec{x}\,')$, for the outside region of a sphere of radius a. For simplicity, take the $+1$ charge on the $+z$ axis, as shown in Figure 4.16. The Coulomb expansion for $|\vec{x} - \vec{x}\,'|^{-1}$ should be helpful here. [*Ans*: Assuming $G_N \to 0$ for $r \to \infty$,

$$G_N(\vec{x}, \vec{x}\,') = \frac{1}{r_>} + \sum_{\ell=1}^{\infty} \left(\frac{r_<^\ell}{r_>^{\ell+1}} + \frac{\ell}{\ell+1} \frac{a^{2\ell+1}}{(rr')^{\ell+1}} \right) P_\ell(\cos\theta).]$$

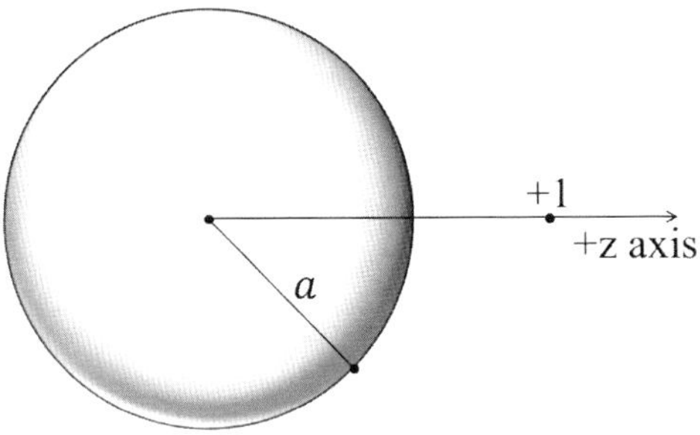

Fig. 4.16 Geometry for Exercise 4.12.2.

Exercise 4.12.3.

(a) Solve the Neumann interior problem for the region inside two concentric spheres of interior radius a and exterior radius b. Show that the $\vec{x}, \vec{x}\,'$ symmetric form of the solution is

$$\nabla^2 G_N(\vec{x}, \vec{x}\,') = -4\pi\delta(\vec{x} - \vec{x}\,'),$$
$$G_N(\vec{x}, \vec{x}\,') = \sum_\ell (2\ell + 1) g_\ell(r, r') P_\ell(\cos\gamma),$$

where γ is the angle between $\vec{x}$ and $\vec{x}\,'$ and (for $\ell > 0$ only)

$$g_\ell(r, r') = \frac{\ell(\ell+1)}{(2\ell+1)(b^{2\ell+1} - a^{2\ell+1})} \left(\frac{r_<^\ell}{\ell} + \frac{a^{2\ell+1}}{(\ell+1) r_<^{\ell+1}} \right) \times \left(\frac{r_>^\ell}{\ell} + \frac{b^{2\ell+1}}{(\ell+1) r_>^{\ell+1}} \right),$$

where $r_<$ ($r_>$) is the lesser (greater) of r and r'. Notice that $g_0(r, r')$ is not specified: can you determine it? Is it unique? Be sure to refresh yourself on the unusual properties of the interior Neumann Green function from Section 2.8. When $b \to \infty$ show your results agree with Exercise 4.12.2 above.

(b) Verify that the surface delta functions discussed in Section 2.8, Equation (2.115), are present in part (a).

Exercise 4.12.4.

(a) Using the Green function for a sphere of radius a or other means, show that the outside potential, $V(\theta, \phi)$, from a given potential, $\Phi(\vec{x})$, on the surface of a sphere can be written as

$$\Phi(\vec{x}) = \sum_{\ell,m} \left(\frac{a}{r}\right)^{\ell+1} Y_{lm}(\theta, \phi) B_{lm},$$

where

$$B_{lm} = \oint_S d\Omega' \, V(\theta', \phi') Y^*_{lm}(\theta', \phi').$$

What is the inside solution for $\Phi(\vec{x})$?

(b) Show that this is also equivalent to (+ for inside, − for outside)

$$\Phi(\vec{x}) = \pm \frac{a(a^2 - r^2)}{4\pi} \oint d\Omega' \frac{V(\theta', \phi')}{(a^2 + r^2 - 2ar\cos\gamma)^{3/2}},$$

where

$$\cos\gamma \equiv \frac{\vec{r} \cdot \vec{r}'}{rr'}.$$

Exercise 4.12.5. Using spherical coordinates obtain the interior solution for a hemisphere with potential 0 on the upper hemispherical surface at $r = a$ and potential V on the base at $z = 0$ (see Figure 4.17). Determine all coefficients necessary for the solution as an infinite series.

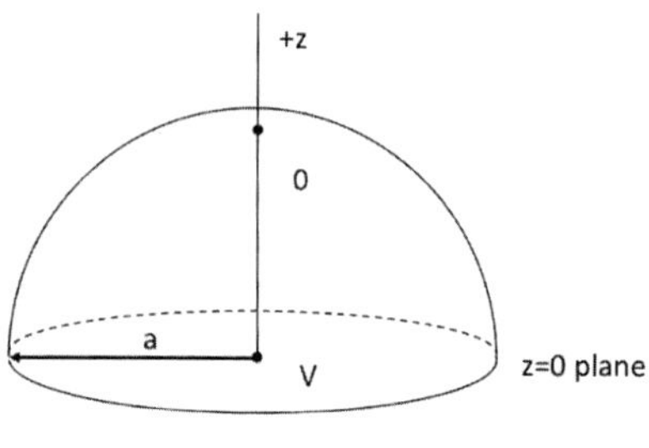

Fig. 4.17 Geometry for Exercise 4.12.5.

Exercise 4.12.6. Deduce the capacitances, C_{aa} and C_{bb} as well as the induction capacitance C_{ab} for concentric conducting spheres from the Dirichlet Green function of Section 4.12. Note that Section 2.12 gives the relevant starting point.

Exercise 4.13.1. A conducting sphere of radius a is split into two unequal pieces which are insulated from one another (see Figure 4.18). The top piece subtends a uniform polar angle, θ_0, as measured from the center of the sphere and is raised to potential V. The rest of the sphere has potential $-V$. Find the electric field, $\vec{E}$, at the center of the sphere, O.

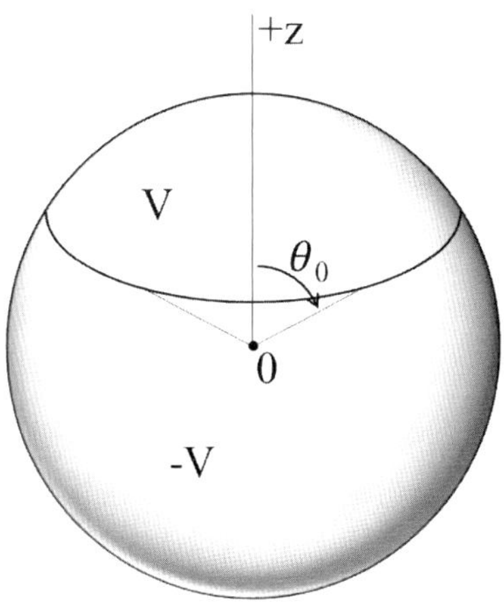

Fig. 4.18 Geometry for Exercise 4.13.1.

Exercise 4.13.2.

(a) Do point electrostatics on a sphere using the techniques of this chapter. That is, solve the scalar Green's function equation

$$\nabla^2_{\theta,\phi} G(\theta, \phi; \theta', \phi') = -2\pi\delta(\cos\theta - \cos\theta')\delta(\phi - \phi') + 2\pi\delta(\cos\theta + \cos\theta')\delta(\phi - \text{mod}_{2\pi}(\phi' + \pi)),$$

when there are unit + and − point charges present at angular location θ', ϕ' and the antipodal location $\pi - \theta'$, $\text{mod}_{2\pi}(\phi' + \pi)$, respectively. $\nabla^2_{\theta,\phi}$ is just the angular part of the ∇^2 operator for $r = 1$:

$$\nabla^2_{\theta,\phi}\Phi \equiv (\vec{r} \times \vec{\nabla})^2\Phi = \frac{1}{\sin\theta}\frac{\partial}{\partial\theta}\left(\sin\theta\frac{\partial\Phi}{\partial\theta}\right) + \frac{1}{\sin^2\theta}\frac{\partial^2\Phi}{\partial\phi^2}.$$

Show that

$$G(\theta,\phi;\theta',\phi') = Q_0(x) = \frac{1}{2}\ln\left(\frac{1+x}{1-x}\right)$$

where $Q_0(x)$ is the 0th order Legendre function of the second kind, and where $x = \cos\gamma = \hat{r}\cdot\hat{r}'$ is the cosine of the angle between the source and field.

(b) Also show one may write

$$G(\theta,\phi;\theta',\phi') = 2\sum_{\ell=\text{odd}} \frac{2\ell+1}{\ell(\ell+1)} P_\ell(\cos\gamma).$$

Exercise 4.13.3. The top part and bottom parts of the conducting spherical shell of radius a are separated at the constant polar angle θ_0, as shown in Figure 4.19.

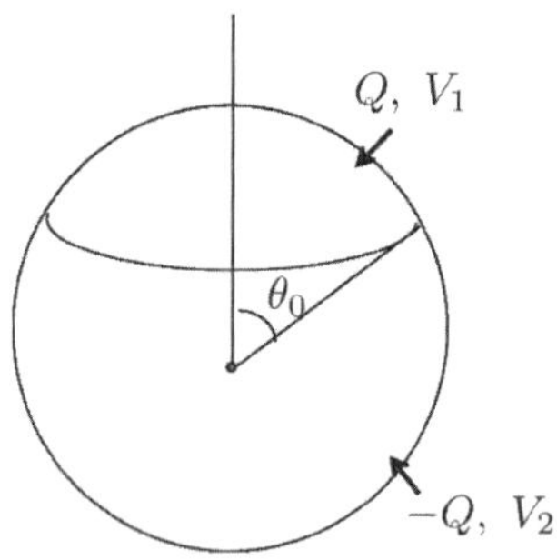

Fig. 4.19 Geometry for Exercise 4.13.3.

The top portion has charge Q and potential V_1, the bottom has charge $-Q$ and potential V_2.

(a) Show that the potentials V_1 and V_2 are related by

$$V_2 = -\left(\frac{1-\cos\theta_0}{1+\cos\theta_0}\right) V_1.$$

(b) Show that the capacitance of the system,

$$C \equiv \frac{Q}{V_1 - V_2},$$

is given by (use variable $x \equiv \cos\theta$ with $x_0 \equiv \cos\theta_0$)

$$C = \frac{a}{8} \sum_{\ell=1,2,3\ldots} (2\ell+1)^2 \left(\int_{x_0}^{1} dx\; P_\ell(x) \right)$$
$$\times \left[(1+x_0) \int_{x_0}^{1} dx\; P_\ell(x) + (x_0 - 1) \int_{-1}^{x_0} dx\; P_\ell(x) \right].$$

Explain a way the integrals can be done in summation form. (You don't have to do them!)

(c) Reconnect to Exercise 4.12.1 ($\theta_0 = \frac{\pi}{2}$) and recover the answer for the capacitance there.

Exercise 4.13.4. Consider a thin spherical conductor of radius a split in the middle. The two halves are slightly separated. The conducting surfaces are labelled 1 and 2 (see Figure 4.20).

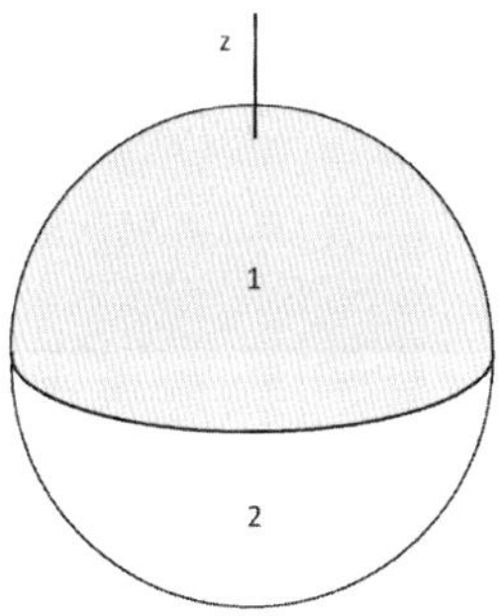

Fig. 4.20 Geometry for Exercise 4.13.4.

Show that the self-capacitance C_{11} $(= C_{22})$ of surface 1 and the inductive capacitance C_{12} $(= C_{21})$ of the two surfaces are given by

$$C_{11} = \frac{a}{4} \sum_{\ell=0,1,2,\ldots} (2\ell+1)^2 \left(\int_{0}^{1} dx\, P_\ell(x) \right)^2,$$

$$C_{12} = \frac{a}{4} \sum_{\ell=0,1,2,\ldots} (-1)^\ell (2\ell+1)^2 \left(\int_{0}^{1} dx\, P_\ell(x) \right)^2.$$

$x \equiv \cos\theta$ and $P_\ell(x)$ is a Legendre polynomial. (*Comment*: For this symmetrical circumstance, the *system* capacitance is given by $C = \frac{1}{2}(C_{11} - C_{12})$. The integrals can be evaluated using the same techniques as in Exercise 4.12.1.)

Exercise 4.13.5. Find the Dirichlet Green function $G_D(\vec{r}, \vec{r}')$ in spherical coordinates for the free half-space $0 < \theta < \frac{\pi}{2}$, $0 < \phi < 2\pi$ given a conducting plane at $\theta = \frac{\pi}{2}$, $0 < \phi < 2\pi$ (see Figure 4.21) in terms of a sum over the spherical harmonics, $Y_{\ell m}(\theta, \phi)$.

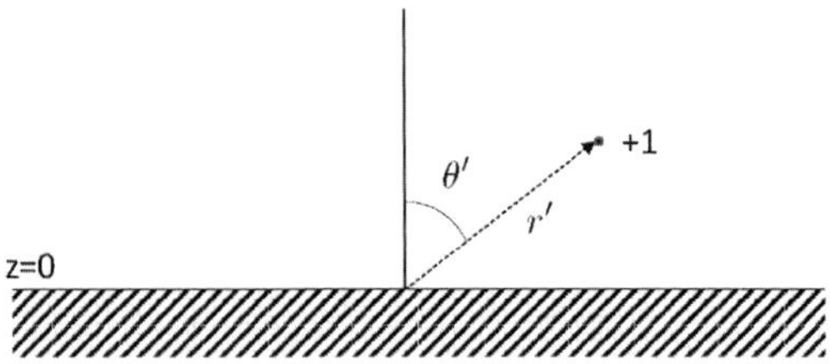

Fig. 4.21 Geometry for Exercise 4.13.5.

Exercise 4.14.1. Prove relation (4.267) for a complete set of orthonormal functions $\{\psi_n\}$.

Exercise 4.14.2.

Using the method of eigenfunction expansions or other means, show that another expression for the three-dimensional Dirichlet Green function for the infinite parallel plate capacitor problem (see Figure 4.22 and Exercises 3.1.1 and 3.2.3), is

$$G_D(\vec{x}, \vec{x}\,') = \frac{4}{a} \int_0^\infty dk\, k J_0(kD) \sum_{n=1}^{\infty} \frac{\sin(\frac{n\pi z}{a}) \sin(\frac{n\pi z'}{a})}{k^2 + (\frac{n\pi}{a})^2},$$

where $D \equiv |(\vec{x} - \vec{x}\,')_\perp|$ and J_0 is the Bessel function of order zero. (The k integral may be done; see Schwinger *et al.*, *op. cit.*, Chap. 18.)

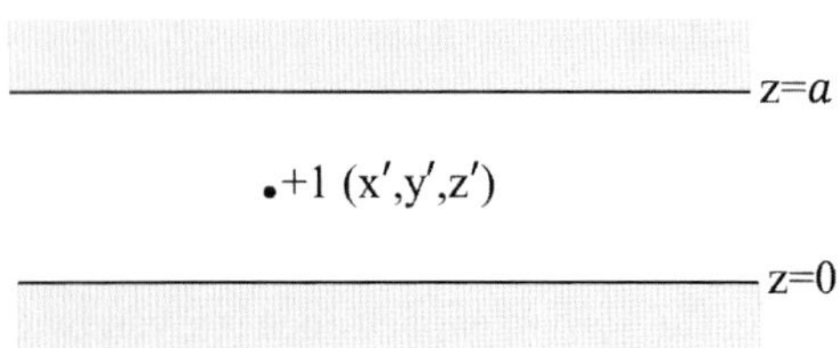

Fig. 4.22 Geometry for Exercise 4.14.2.

Exercise 4.14.3. Returning to Exercise 4.7.4(b), demonstrate that an eigenvalue expansion for the reduced Dirichlet Green function, $g_m^{\text{in}}(k;\rho,\rho')$, for the interior of a conducting cylinder of radius a is

$$g_m^{\text{in}}(k;\rho,\rho') = \sum_{n=1}^{\infty} \frac{J_{1m}(k_{mn}\rho)J_{1m}(k_{mn}\rho')}{k^2 + k_{mn}^2},$$

where the k_{mn} produce zeros of $J_m(x)$ through $k_{mn} = x_{mn}/a$.

Exercise 4.14.4. Do an eigenfunction expansion of the solution of the one-dimensional Dirichlet Green function with conducting boundary conditions at $x = 0, L$. Refresh yourself on its properties and interpretations in Section 2.9. The differential equation is

$$\frac{d^2 G_D(x',x)}{dx^2} = -\delta(x - x'),$$

and the boundary conditions are

$$G_D(0,x') = 0, \quad G_D(L,x') = 0.$$

Exercise 4.14.5. Solve for the three-dimensional Dirichlet Green function $G_D(\vec{x},\vec{x}\,')$ for the finite interior region of a conducting cylinder of radius a and z-direction height L by doing an eigenfunction expansion. When the z-direction eigensum is done, it will give Equation (4.79). Deduce therefore the sum rule

$$\sum_{n=1}^{\infty} \frac{\sin(k_n z)\sin(k_n z')}{k^2 + k_n^2} = \frac{L}{2}\frac{\sinh(kz_<)\sinh[k(L - z_>)]}{k\sinh(kL)},$$

where $k_n = \dfrac{n\pi}{L}$ and k is a real parameter. (This identity helps simplify the Exercise 4.14.2 answer as well.)

Exercise 4.14.6.

(a) Show that the eigenfunctions, $\psi(r,\theta,\phi)$, for the Dirichlet Green function $G_D(\vec{r},\vec{r}\,')$ for the interior of a spherical conductor are proportional to

$$\psi(r,\theta,\phi) \sim j_\ell(kr)Y_{\ell m}(\theta,\phi),$$

where $j_\ell(kr)$ is a special function (called a spherical Bessel function) which satisfies

$$\frac{1}{r^2}\frac{d}{dr}\left(r^2\frac{dj_\ell(kr)}{dr}\right) + \left(k^2 - \frac{\ell(\ell+1)}{r^2}\right)j_\ell(kr) = 0.$$

(b) How are the eigenvalues, k, determined inside a conducting sphere of radius a?

Chapter 5

Multipoles, Electrostatics of Macroscopic Media, Dielectrics

This chapter begins the introduction of a macroscopic model of matter, limited at this point to electric field interactions. The model will be built up from various field expansions in terms of idealized sources called multipole moments. We will also extend the concept of Green functions and examine the energy change implications of the model. Finally, we will be especially interested in how forces can be evaluated in different ways for various systems and situations.

5.1 Cartesian and spherical multipole expansions

Let us consider a localized charge with density $\rho(\vec{x}\,')$ in three-dimensional space, as shown in Figure 5.1. The potential is given by

$$\Phi(\vec{x}) = \int d^3x' \, \frac{\rho(\vec{x}\,')}{|\vec{x} - \vec{x}\,'|}. \tag{5.1}$$

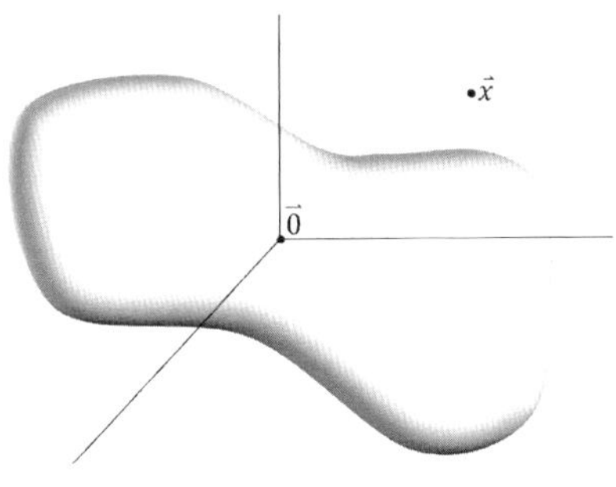

Fig. 5.1 Localized charge distribution and notation.

Note the origin need not lie within the charge distribution. If $r \equiv |\vec{x}|$ is large compared to the characteristic dimensions of the distribution, then a three-dimensional Taylor series expansion may be used to obtain

$$\frac{1}{|\vec{x}-\vec{x}\,'|} = \sum_{\ell=0}^{\infty} \frac{(\vec{x}\,'\cdot\vec{\nabla}'')^{\ell}}{\ell!}\left(\frac{1}{|\vec{x}-\vec{x}\,''|}\right)\Bigg|_{\vec{x}\,''=0}. \tag{5.2}$$

Using the fact that

$$\vec{\nabla}''\frac{1}{|\vec{x}-\vec{x}\,''|} = -\vec{\nabla}\frac{1}{|\vec{x}-\vec{x}\,''|}, \tag{5.3}$$

it follows that

$$\frac{(\vec{x}\,'\cdot\vec{\nabla}'')^{\ell}}{\ell!}\frac{1}{|\vec{x}-\vec{x}\,''|}\Bigg|_{\vec{x}\,''=0} = (-1)^{\ell}\frac{(\vec{x}\,'\cdot\vec{\nabla})^{\ell}}{\ell!}\frac{1}{r}, \tag{5.4}$$

which implies

$$\frac{1}{|\vec{x}-\vec{x}\,'|} = \frac{1}{r} - \vec{x}\,'\cdot\vec{\nabla}\frac{1}{r} + \frac{1}{2}(\vec{x}\,'\cdot\vec{\nabla})^2\frac{1}{r} - \ldots \qquad (r > r'). \tag{5.5}$$

The first and second derivative terms may be worked out explicitly:

$$(\vec{x}\,'\cdot\vec{\nabla})\frac{1}{r} = \sum_i x_i'\nabla_i\frac{1}{\sqrt{x^2+y^2+z^2}} = -\sum_i\frac{x_i'x_i}{r^3} = -\frac{\vec{x}\,'\cdot\vec{x}}{r^3}, \tag{5.6}$$

$$(\vec{x}\,'\cdot\vec{\nabla})^2\frac{1}{r} = \sum_{i,j} x_i'x_j'\nabla_i\nabla_j\frac{1}{\sqrt{x^2+y^2+z^2}} = -\sum_{i,j} x_i'x_j'\nabla_i\frac{x_j}{r^3}. \tag{5.7}$$

Equation (5.7) can be further simplified as

$$(\vec{x}\,'\cdot\vec{\nabla})^2\frac{1}{r} = \sum_{i,j} x_i'x_j'\frac{(3x_ix_j-\delta_{ij}r^2)}{r^5}. \tag{5.8}$$

These derivative expressions may be substituted into the Taylor expansion to obtain

$$\frac{1}{|\vec{x}-\vec{x}\,'|} = \frac{1}{r} + \frac{\vec{x}\,'\cdot\vec{x}}{r^3} + \frac{1}{2}\sum_{i,j}\frac{x_i'x_j'}{r^5}(3x_ix_j-\delta_{ij}r^2) + \ldots \tag{5.9}$$

The summation in (5.9) may be re-expressed as

$$\begin{aligned}\sum_{i,j}\frac{x_i'x_j'}{r^5}(3x_ix_j-\delta_{ij}r^2) &= \sum_{i,j}\frac{3x_i'x_j'x_ix_j - r'^2r^2}{r^5}\\ &= \sum_{i,j}\frac{x_ix_j}{r^5}(3x_i'x_j'-r'^2\delta_{ij}).\end{aligned} \tag{5.10}$$

We have therefore an approximate expression for the potential $\Phi(\vec{x})$ which is valid to second order in r'/r:

$$\Phi(\vec{x}) = \int d^3x' \rho(\vec{x}\,') \left\{ \frac{1}{r} + \frac{\vec{x}\,' \cdot \vec{x}}{r^3} + \frac{1}{2} \sum_{i,j} \frac{x_i x_j}{r^5} (3x'_i x'_j - r'^2 \delta_{ij}) + \ldots \right\}. \tag{5.11}$$

We introduce the notation

$$\Phi(\vec{x}) = \frac{q}{r} + \frac{\vec{x} \cdot \vec{p}}{r^3} + \frac{1}{2} \sum_{i,j} Q_{ij} \frac{x_i x_j}{r^5} + \ldots, \tag{5.12}$$

where

$$q \equiv \int d^3x' \rho(\vec{x}\,') \text{ is the charge (a scalar),}$$

$$\vec{p} \equiv \int d^3x' \vec{x}\,' \rho(\vec{x}\,') \text{ is the } \textit{electric dipole vector,}$$

$$Q_{ij} \equiv \int d^3x' (3x'_i x'_j - \delta_{ij} r'^2) \rho(\vec{x}\,') \text{ is the } \textit{quadrupole tensor.}$$

Note that in the case of a simple dipole consisting of two equal and opposite point charges $\pm q$ (q either sign), the dipole vector is $\vec{p} = q\vec{x}\,'$, where the vector $\vec{x}\,'$ points from the $-q$ to the $+q$ charge.

Although the quadrupole tensor Q_{ij} has $3 \times 3 = 9$ components, not all are independent: the tensor is symmetric ($Q_{ij} = Q_{ji}$) and *traceless*, since

$$\sum_i Q_{ii} = \int d^3x' (3r'^2 - 3r'^2) \rho(\vec{x}\,') = 0. \tag{5.13}$$

The symmetry and traceless conditions impose 4 conditions on Q_{ij}, reducing the number of independent components to 5. It follows that the number of independent components in the charge, dipole and quadrupole expressions are $1, 3, 5$ respectively. The pattern implied is indeed representative of a general trend, as we shall see below.

The fields corresponding to ideal point charge, dipole, and quadrupole charge distributions have different r-dependence as follows ($r \neq 0$):

$$\text{Point charge} \sim \frac{1}{r^2}: \quad \vec{E} = -\vec{\nabla}\left(\frac{q}{r}\right) = \frac{q\,\hat{r}}{r^2}, \tag{5.14}$$

$$\text{Point dipole} \sim \frac{1}{r^3}: \quad \vec{E}^d = -\vec{\nabla}\left(\frac{\vec{x}\cdot\vec{p}}{r^3}\right) = \frac{3(\vec{x}\cdot\vec{p})\vec{x} - \vec{p}\,r^2}{r^5}, \tag{5.15}$$

$$\begin{aligned}\text{Point quadrupole} \sim \frac{1}{r^4}: \quad E_i^Q &= -\nabla_i\left(\frac{1}{2}\sum_{j,k} Q_{jk}\frac{x_j x_k}{r^5}\right), \\ &= \frac{1}{2}\left\{\frac{5\sum_{j,k} Q_{jk}x_j x_k x_i - 2r^2\sum_k Q_{ik}x_k}{r^7}\right\}.\end{aligned} \tag{5.16}$$

Figure 5.2 shows field lines for point charge and dipole fields.

Since the matrix Q_{ij} is symmetric, it can be diagonalized via an appropriate rotation. In the case where the charge distribution has an axis of symmetry (which we choose as the z-axis, so that $Q_{11} = Q_{22}$) then there are two possible cases: either $Q_{33} > 0$ (the "prolate" case) or $Q_{33} < 0$ (the "oblate" case). The two cases are shown in Figure 5.2 (note that the charge distributions depicted will also have nonzero monopole moments). Clearly both the location and orientation of our axes will (in general) affect the moments.

The higher-order terms in the multipole expansion (5.12) become increasingly complicated, because the number of indices increases with each term. Fortunately this complication may be avoided by expanding the potential in spherical harmonics instead. Using the Coulomb expansion

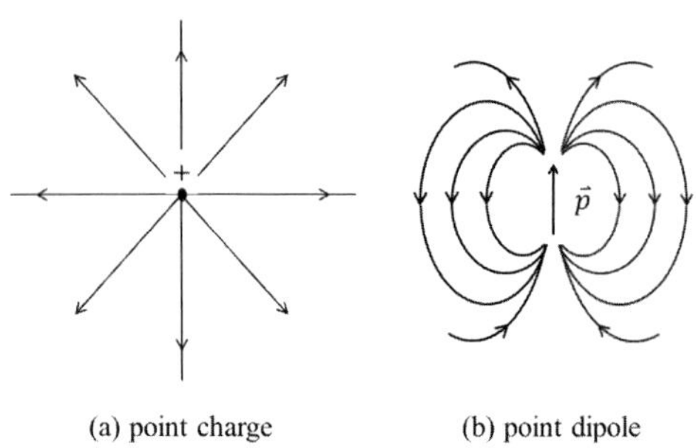

Fig. 5.2 Field lines for point charge and point dipole.

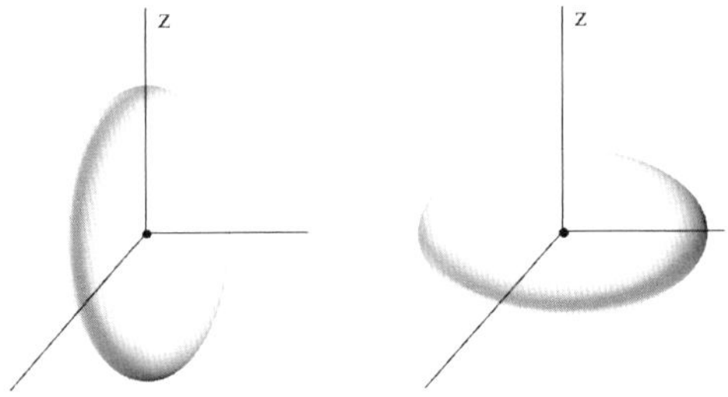

(a) $Q_{33}>0$ (prolate) (a) $Q_{33}<0$ (oblate)

Fig. 5.3 Prolate and oblate quadrupole charge distributions.

derived in Section 4.11 we have (for $r > r'$)

$$\frac{1}{|\vec{x}-\vec{x}\,'|} = \sum_{\ell,m} \frac{r'^{\ell}}{r^{\ell+1}} \sqrt{\frac{4\pi}{2\ell+1}} Y_{\ell m}(\theta,\phi) \sqrt{\frac{4\pi}{2\ell+1}} Y^{*}_{\ell m}(\theta',\phi'). \tag{5.17}$$

For field positions $r > |\vec{r}\,'_m|$, where $|\vec{r}\,'_m|$ is maximal for $\rho(\vec{r}\,'_m) \neq 0$, it follows that

$$\Phi(\vec{x}) = \sum_{\ell,m} \frac{1}{r^{\ell+1}} \sqrt{\frac{4\pi}{2\ell+1}} Y_{\ell m}(\theta,\phi) \int d^3x' r'^{\ell} \sqrt{\frac{4\pi}{2\ell+1}} Y^{*}_{\ell m}(\theta',\phi')\rho(\vec{x}\,'). \tag{5.18}$$

We introduce the notation

$$\rho_{\ell m} \equiv \int d^3x' r'^{\ell} \sqrt{\frac{4\pi}{2\ell+1}} Y^{*}_{\ell m}(\theta',\phi')\rho(\vec{x}\,'), \tag{5.19}$$

to obtain the following expansion for the potential:

$$\Phi(\vec{x}) = \sum_{\ell,m} \frac{1}{r^{\ell+1}} \sqrt{\frac{4\pi}{2\ell+1}} Y_{\ell m}(\theta,\phi)\rho_{\ell m}. \tag{5.20}$$

The number of multipoles for a given ℓ is therefore $2\ell+1$.

5.2 Multipole energy expansions

Consider a localized charge distribution and a point charge q_1 external to (and far from) the charge distribution (see Figure 5.4). The energy of interaction between q_1 and the localized distribution is given by

$$W = q_1\Phi(\vec{x}) = \frac{q_1 q}{r} + q_1 \frac{\vec{x}\cdot\vec{p}}{r^3} + \frac{q_1}{2}\sum_{i,j} Q_{ij}\frac{x_i x_j}{r^5} + \dots. \tag{5.21}$$

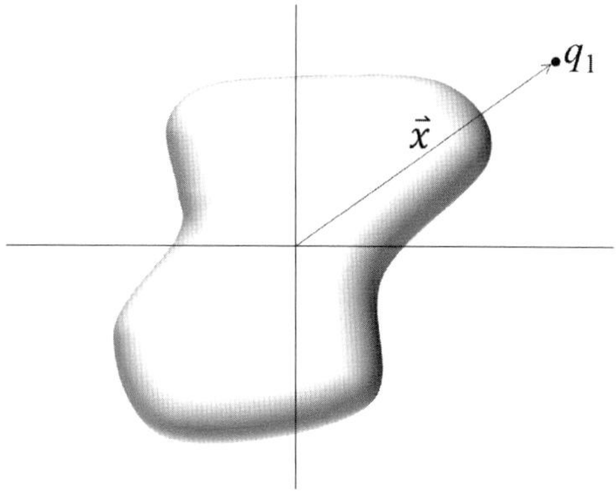

Fig. 5.4 Geometry for interaction energy calculation.

The electric field at the origin due to the charge q_1 at $\vec{x}$ is

$$\vec{E}^{(0)}(\vec{x}) \equiv -\frac{q_1 \vec{x}}{r^3}. \tag{5.22}$$

Keep in mind that $\vec{x}$ is the vector from the *multipole origin* to the *charge*.

A straightforward computation shows that

$$\frac{\partial E_j^{(0)}}{\partial x_i} = q_1 \left[\frac{3x_i x_j - \delta_{ij} r^2}{r^5} \right]. \tag{5.23}$$

It follows that

$$W = \frac{q_1 q}{r} - \vec{p} \cdot \vec{E}^{(0)} + \frac{q_1}{6} \sum_{i,j} Q_{ij} \left[\frac{3x_i x_j - \delta_{ij} r^2}{r^5} \right], \tag{5.24}$$

$$= \frac{q_1 q}{r} - \vec{p} \cdot \vec{E}^{(0)} + \frac{1}{6} \sum_{i,j} Q_{ij} \frac{\partial E_j^{(0)}}{\partial x_i} + \ldots. \tag{5.25}$$

We can now add the energy contributions from many q_1's. However, we must be careful because our origin is associated with the charge distribution, not the point charge. With this caveat we can read off the interaction energies for different cases:

Dipole - charge:

$$W_{dq} = -\vec{p} \cdot \vec{E}^{(0)} = q_1 \frac{\vec{x} \cdot \vec{p}}{r^3}. \tag{5.26}$$

The vector $\vec{x}$ points from the charge distribution to the charge, q_1.

Dipole - dipole:

$$W_{dd} = -\vec{p}_2 \cdot \vec{E}_1^d = \frac{-3(\vec{x} \cdot \vec{p}_1)(\vec{x} \cdot \vec{p}_2) + (\vec{p}_1 \cdot \vec{p}_2)r^2}{r^5}. \tag{5.27}$$

Equation (5.27) uses expression (5.15) for $\vec{E}^d$.

Dipole - quadrupole:

$$W_{dQ} = -\vec{p} \cdot \vec{E}^Q = -\frac{1}{2} \sum_{i,j} Q_{ij} \left\{ \frac{2r^2 p_i x_j - 5x_i x_j (\vec{p} \cdot \vec{x})}{r^7} \right\}. \tag{5.28}$$

Equation (5.28) uses expression (5.16) for $\vec{E}^Q$ with the replacement $\vec{x} \to -\vec{x}$. This is because the quadrupole is at $\vec{x}$ instead of the origin.

We compute also the quadrupole - dipole interaction energy to compare with (5.28).

$$\begin{aligned} W_{Qd} &= \frac{1}{6} \sum_{i,j} Q_{ij} \frac{\partial E_j^{(0)}}{\partial x_i} = \frac{1}{6} \sum_{i,j} Q_{ij} \frac{\partial}{\partial x_i} \left\{ \frac{3(\vec{x} \cdot \vec{p})x_j - p_j r^2}{r^5} \right\} \\ &= \frac{1}{2} \sum_{i,j} Q_{ij} \left\{ \frac{2r^2 p_i x_j - 5x_i x_j (\vec{p} \cdot \vec{x})}{r^7} \right\} \\ &= -W_{dQ}. \end{aligned} \tag{5.29}$$

In (5.29) we have reused the dipole electric field expression (5.15). The difference in sign between (5.28) and (5.29) occurs because of a difference in relative orientation between $\vec{p}$ and Q: In the first case, $\vec{x}$ points from dipole to quadrupole while in the second, $\vec{x}$ points from quadrupole to dipole. From the dipole-dipole interaction energy expressions, we may deduce the directions of the forces of interaction between two dipoles in various configurations (see Figure 5.5). The interaction energy expressions (5.26)–(5.28) explain a wealth of data having to do with forces between atoms in solids.

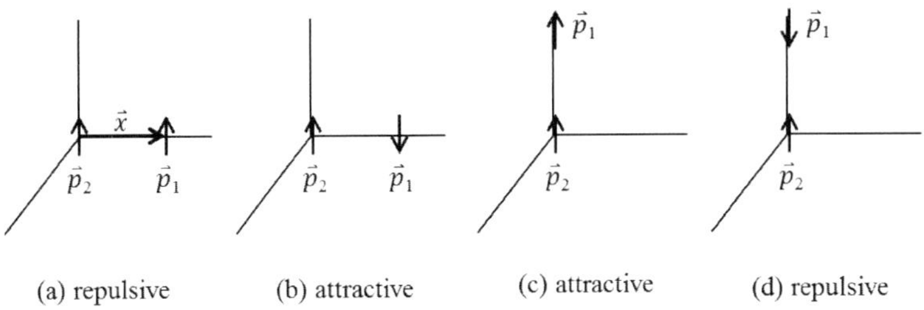

Fig. 5.5 Forces of interaction between dipoles.

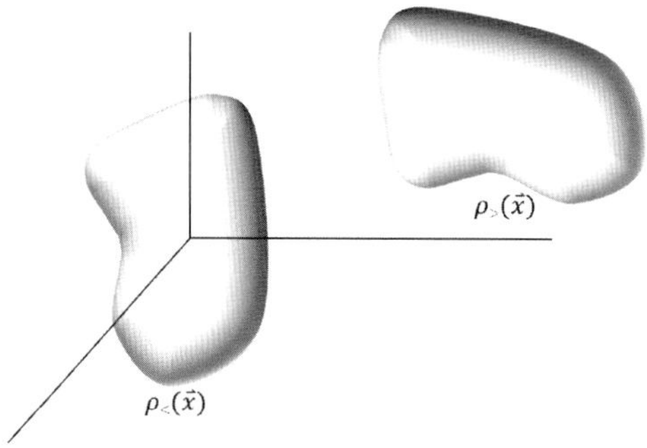

Fig. 5.6 An example of a separable charge configuration.

The situation becomes much more complicated for higher-order interactions. As mentioned above, it is simpler in this case to expand in spherical harmonics. In the case of charge densities we have for the interaction energy

$$W = \int d^3x\, d^3x' \frac{\rho_<(\vec{x})\rho_>(\vec{x}\,')}{|\vec{x} - \vec{x}\,'|}. \tag{5.30}$$

We assume these have no positions where both $\rho_<(\vec{x}) \neq 0$ and $\rho_>(\vec{x}\,') \neq 0$ for $r' < r$, as in Figure 5.6; we will call this a "separable configuration". Expanding in spherical harmonics gives

$$\frac{1}{|\vec{x} - \vec{x}\,'|} = \sum_{\ell,m} \frac{4\pi}{2\ell + 1} \frac{r_<^\ell}{r_>^{\ell+m}} Y^*_{\ell m}(\theta', \phi') Y_{\ell m}(\theta, \phi). \tag{5.31}$$

Separability implies one can uniquely associate $r_> = r'$ and $r_< = r$ in this context. It follows that

$$W = \sum_{\ell,m} \int d^3x' d^3x\, \rho_<(\vec{x}) \frac{4\pi}{2\ell + 1} Y_{\ell m}(\theta, \phi) \frac{r^\ell}{r'^{\ell+1}} Y^*_{\ell m}(\theta', \phi') \rho_>(\vec{x}\,'). \tag{5.32}$$

We define

$$(\rho_<)_{\ell m} \equiv \int d^3x \sqrt{\frac{4\pi}{2\ell + 1}} r^\ell Y^*_{\ell m}(\theta, \phi) \rho_<(\vec{x}), \tag{5.33}$$

$$(\rho_>)_{\ell m} \equiv \int d^3x \sqrt{\frac{4\pi}{2\ell + 1}} r^{-\ell-1} Y^*_{\ell m}(\theta, \phi) \rho_>(\vec{x}). \tag{5.34}$$

Using (4.188) and (4.189) these satisfy

$$(\rho_<)^*_{\ell m} = (-1)^m (\rho_<)_{\ell,-m}, \tag{5.35}$$

$$(\rho_>)^*_{\ell m} = (-1)^m (\rho_>)_{\ell,-m}. \tag{5.36}$$

We may now rewrite (5.32) as

$$W = \sum_{\ell,m} (\rho_<)^*_{\ell m} (\rho_>)_{\ell m}. \tag{5.37}$$

Of course for a distant source $\rho_>$ only a few of the lowest moments will contribute significantly. Equations (5.35) and (5.36) now verify that W is real.

It is straightforward to express multipole moments in terms of the $(\rho_<)_{\ell m}$. For example

$$\begin{aligned}\vec{p} &= \int d^3x' \rho(\vec{x}\,')r' \left[\sin\theta' \cos\phi\,\hat{\imath} + \sin\theta' \sin\phi'\,\hat{\jmath} + \cos\theta'\,\hat{k}\right], \\ &= \sqrt{\frac{8\pi}{3}} \int d^3x'\, \rho(\vec{x}\,')r' \left[\frac{1}{2}\left(-Y^*_{11} + Y^*_{1-1}\right)\hat{\imath} - \frac{i}{2}\left(Y^*_{11} + Y^*_{1-1}\right)\hat{\jmath} + \frac{1}{\sqrt{2}} Y^*_{10}\,\hat{k}\right], \\ &= \frac{1}{\sqrt{2}}\left(-(\rho_<)_{11} + (\rho_<)_{1-1}\right)\hat{\imath} - \frac{i}{\sqrt{2}}\left((\rho_<)_{11} + (\rho_<)_{1-1}\right)\hat{\jmath} + (\rho_<)_{10}\,\hat{k}.\end{aligned} \tag{5.38}$$

This gives

$$(\rho_<)_{11} = \frac{1}{\sqrt{2}}(-p_x + ip_y), \quad (\rho_<)_{1-1} = \frac{1}{\sqrt{2}}(p_x + ip_y), \quad (\rho_<)_{10} = p_z. \tag{5.39}$$

It follows from the definition of the electric dipole vector that if $q = 0$, then $\vec{p}$ is independent of the choice of origin for the coordinate system. This is also true for higher moments, if all lower-order ones vanish.[1]

5.3 External fields and forces on multipole distributions

We shall now compute the force on an arbitrary charge distribution in an external field (see Figure 5.7).

$$d\vec{F}(\vec{x}) = dq(\vec{E}^{(0)}(\vec{x}) + \vec{E}^r(\vec{x})), \tag{5.40}$$

$$\Longrightarrow \vec{F} = \int dq\vec{E}^{(0)}(\vec{x}) = \int d^3x\, \rho(\vec{x})\vec{E}^{(0)}(\vec{x}). \tag{5.41}$$

($\vec{E}^r(\vec{x})$ denotes the remainder field of the rest of $\rho(\vec{x})$, and does not contribute to the total force; a more complete explanation is presented in

[1]For an interesting alternative multipole expansion, see J. Schwinger *et al.*, *Classical Electrodynamics*, Perseus Books (Reading, MA) 1998, Chapter 22. See also Wikipedia: Multipole expansion, `http://en.wikipedia.org/wiki/Multipole_expansion`, for an equivalent interaction energy expression.

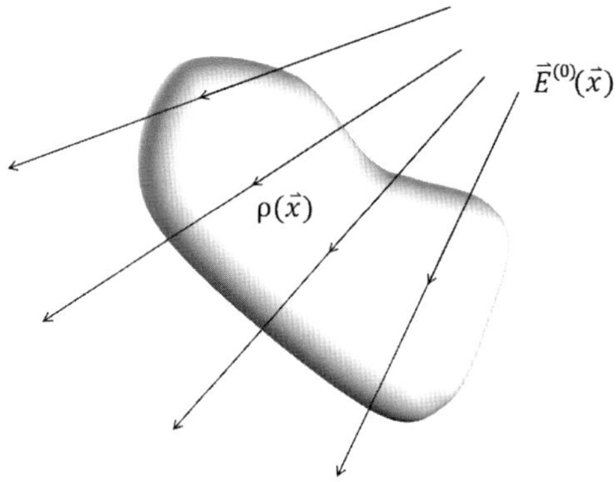

Fig. 5.7 Charge distribution in an external field.

Section 5.9.) We expand each component of the electric field in Taylor series:

$$\vec{E}^{(0)}(\vec{x}) = \vec{E}^{(0)}(0) + (\vec{x}\cdot\vec{\nabla}')\vec{E}^{(0)}(\vec{x}\,')\Big|_{\vec{x}\,'=0} + \frac{1}{2}(\vec{x}\cdot\vec{\nabla}')^2\vec{E}^{(0)}(\vec{x}\,')\Big|_{\vec{x}\,'=0} + \ldots \tag{5.42}$$

In Cartesian coordinates the second derivative term becomes

$$\begin{aligned}(\vec{x}\cdot\vec{\nabla}')^2\vec{E}^{(0)}(\vec{x}\,')\Big|_{\vec{x}\,'=0} &= \sum_{i,j} x_i \frac{\partial}{\partial x_i'} x_j \frac{\partial}{\partial x_j'} \vec{E}^{(0)}(\vec{x}\,')\Bigg|_{\vec{x}\,'=0} \\ &= \sum_{i,j} x_i x_j \left.\frac{\partial^2 \vec{E}^{(0)}(\vec{x}\,')}{\partial x_i' \partial x_j'}\right|_{\vec{x}\,'=0}.\end{aligned}$$

At this point we may make use of the following useful vector identity from Section 1.8:

$$\vec{\nabla}'(\vec{a}\cdot\vec{b}) = (\vec{a}\cdot\vec{\nabla}')\vec{b} + (\vec{b}\cdot\vec{\nabla}')\vec{a} + \vec{a}\times(\vec{\nabla}'\times\vec{b}) + \vec{b}\times(\vec{\nabla}'\times\vec{a}). \tag{5.43}$$

Applying this identity with $\vec{a}\equiv\vec{x}$, $\vec{b}\equiv\vec{E}^{(0)}(\vec{x}\,')$ (and noting that $\vec{\nabla}'\times\vec{E}^{(0)}(\vec{x}\,')=0$) yields

$$\vec{\nabla}'\left(\vec{x}\cdot\vec{E}^{(0)}(\vec{x}\,')\right) = (\vec{x}\cdot\vec{\nabla}')\vec{E}^{(0)}(\vec{x}\,'). \tag{5.44}$$

A second application of the same identity with $\vec{a}\equiv\vec{x}$ and $\vec{b}\equiv$ (right-hand side of (5.44)) gives

$$\vec{\nabla}'\left(\vec{x}\cdot(\vec{x}\cdot\vec{\nabla}')\vec{E}^{(0)}(\vec{x}\,')\right) = (\vec{x}\cdot\vec{\nabla}')(\vec{x}\cdot\vec{\nabla}')\vec{E}^{(0)}(\vec{x}\,'). \tag{5.45}$$

In the above we have to use:

$$\vec{\nabla}'\times\left[(\vec{x}\cdot\vec{\nabla}')\vec{E}^{(0)}(\vec{x}\,')\right] = \vec{\nabla}'\times\left[\vec{\nabla}'(\vec{x}\cdot\vec{E}^{(0)}(\vec{x}\,'))\right] = 0. \tag{5.46}$$

We may thus modify the quadrupole term in the multipole expansion of $\vec{E}^{(0}(\vec{x})$ to obtain:

$$\vec{E}^{(0)}(\vec{x}) = \vec{E}^{(0)}(0) + \vec{\nabla}' \left(\vec{x} \cdot \vec{E}^{(0)}(\vec{x}\,')\right)\Big|_{\vec{x}\,'=0} + \frac{1}{2}\,\vec{\nabla}' \left\{\vec{x} \cdot \left[(\vec{x} \cdot \vec{\nabla}')\vec{E}^{(0)}(\vec{x}\,')\right]\right\}\Big|_{\vec{x}\,'=0} + ... \tag{5.47}$$

Using index notation in the last term gives

$$\vec{E}^{(0)}(\vec{x}) = \vec{E}^{(0)}(0) + \vec{\nabla}' \left(\vec{x} \cdot \vec{E}^{(0)}(\vec{x}\,')\right)\Big|_{\vec{x}\,'=0} + \frac{1}{2}\,\vec{\nabla}' \sum_{i,j} x_i x_j \frac{\partial}{\partial x_j'} E_i^{(0)}(\vec{x}\,')\Bigg|_{\vec{x}\,'=0} + ... \tag{5.48}$$

$\vec{\nabla}' \cdot \vec{E}^{(0)} = 0$ for the external field in the region of interest, so we may add in the term

$$-\frac{1}{6} r^2 \vec{\nabla}' \cdot \vec{E}^{(0)}(\vec{x}\,')\Big|_{\vec{x}\,'=0} = -\frac{1}{6}\sum_i r^2 \frac{\partial E_i^{(0)}}{\partial x_i'}\Bigg|_{\vec{x}\,'=0} = -\frac{1}{6}\sum_{i,j} r^2 \delta_{ij} \frac{\partial E_i^{(0)}}{\partial x_j'}\Bigg|_{\vec{x}\,'=0}. \tag{5.49}$$

We now find that

$$\vec{E}^{(0)}(\vec{x}) = \vec{E}^{(0)}(0) + \vec{\nabla}' \left(\vec{x} \cdot \vec{E}^{(0)}(\vec{x}\,')\right)\Big|_{\vec{x}\,'=0} + \frac{1}{6}\,\vec{\nabla}' \sum_{i,j}(3x_i x_j - r^2\delta_{ij}) \frac{\partial E_i^{(0)}}{\partial x_j'}\Bigg|_{\vec{x}\,'=0} + ... \tag{5.50}$$

Thus for the force $\vec{F}$ from (5.41) we obtain

$$\vec{F} = \int d^3x\, \rho(\vec{x}) \left[\vec{E}^{(0)}(0) + \vec{\nabla}' \left(\vec{x} \cdot \vec{E}^{(0)}(\vec{x}\,')\right)\Big|_{\vec{x}\,'=0} + \frac{1}{6}\,\vec{\nabla}' \sum_{i,j}(3x_i x_j - r^2\delta_{ij})\, \frac{\partial E_i^{(0)}}{\partial x_j'}\Bigg|_{\vec{x}\,'=0} + \dots\right]. \tag{5.51}$$

Using our definitions of q, $\vec{p}$ and Q_{ij}, this then shows that (replacing $\vec{x}\,'$ with $\vec{x}$)

$$\vec{F} = q\vec{E}^{(0)}(0) + \vec{\nabla}(\vec{p} \cdot \vec{E}^{(0)}(\vec{x}))\Big|_{\vec{x}=0} + \vec{\nabla}\left[\frac{1}{6}\sum_{i,j} Q_{ij} \frac{\partial E_i^{(0)}}{\partial x_j}\right]\Bigg|_{\vec{x}=0} + ... \tag{5.52}$$

The vector character of this expression is associated with the gradient operator $\vec{\nabla}$. An alternative expression for the force may also be derived:

$$\vec{F} = q\vec{E}^{(0)}(0) + (\vec{p}\cdot\vec{\nabla})\vec{E}^{(0)}(\vec{x})\Big|_{\vec{x}=0} + \frac{1}{6}\sum_{i,j} Q_{ij}\left.\frac{\partial^2\vec{E}^{(0)}}{\partial x_i \partial x_j}\right|_{\vec{x}=0} + ... \quad (5.53)$$

Here, the vector character of this expression is associated with the electric field $\vec{E}^{(0)}$. Different assumptions are necessary to reach these two force expressions; see Exercise 5.3.1(a).

5.4 Electric polarization and the displacement field

In the preceding sections we have been talking about localized charge distributions that are small compared to the distances under consideration. Our results can be applied to the fields produced by individual atoms within a large collection of atoms: this will lead us to a characterization of the fields produced by macroscopic objects.

Consider a sample volume V from a material consisting of neutral atoms ($q = 0$) where atom i has dipole moment $\vec{p}_i$. In electrostatics (5.52) and (5.53) are equivalent. However, starting with (5.53) leads a little more directly to the final result. Summing the first nonzero term over atoms gives

$$\vec{F}_{\text{bulk}} \equiv \sum_{i=\text{atoms}} \vec{F}_i = \sum_i (\vec{p}_i\cdot\vec{\nabla})\vec{E}^{(0)}(\vec{x})\Big|_{\vec{x}=\vec{x}_i}. \quad (5.54)$$

Denote by $n(\vec{x})$ the density of atoms in the material. Then we can convert the sum (5.54) into an integral:

$$\vec{F}_{\text{bulk}} = \int_V d^3x \left(\vec{P}(\vec{x})\cdot\vec{\nabla}\right)\vec{E}^{(0)}(\vec{x}), \quad (5.55)$$

where the *electric polarization* $\vec{P}(\vec{x})$ is defined as

$$\vec{P}(\vec{x}) \equiv n(\vec{x})\vec{p}(\vec{x}). \quad (5.56)$$

The units of $\vec{P}(\vec{x})$ are (dipole strength)/(volume) $\sim$ (charge)/(distance)^2, the same as electric field. It is clear from (5.56) that the electric polarization can vary from location to location due either to density variations or to intrinsic changes in $\vec{p}$ from atom to atom.

Limiting the integration of (5.55) to the sample volume V gives

$$\vec{F}_{\text{bulk}} = \int_V d^3x \left(-\vec{\nabla}\cdot\vec{P}(\vec{x})\right)\vec{E}^{(0)}(\vec{x}) + \int_S da\,(\vec{P}\cdot\hat{n})\vec{E}^{(0)}(\vec{x}), \quad (5.57)$$

where an integration by parts has been performed, S is the surface of the sample volume and $\hat{n}$ is the outward unit normal from the sample. By comparison with $\vec{F} = \int d^3x\, \rho(\vec{x})\vec{E}^{(0)}(\vec{x})$ we may identify

$$\rho^d_{\text{eff}}(\vec{x}) \equiv -\vec{\nabla}\cdot\vec{P}, \tag{5.58}$$

$$\sigma^d_{\text{eff}}(\vec{x}) \equiv \vec{P}\cdot\hat{n}, \tag{5.59}$$

where ρ^d_{eff} denotes the *effective volume charge density* and σ^d_{eff} represents the *effective surface charge density* due to atomic dipoles. If the atoms $1, 2, \ldots$ also have quadrupole moments $[Q_{ij}]_n, n = 1, 2, \ldots$ then we may similarly define $q_{ij}(\vec{x}) \equiv n(\vec{x})Q_{ij}(\vec{x})$ and obtain a quadrupole contribution to the effective charge,

$$\rho^Q_{\text{eff}}(\vec{x}) \equiv \frac{1}{6}\sum_{i,j}\frac{\partial^2 q_{ij}(\vec{x})}{\partial x_i \partial x_j}, \tag{5.60}$$

as well as surface charge and surface electric polarization densities; see Exercise 5.4.1.

To summarize the preceding discussion, we have shown that neutral atoms within a bulk material give rise to an effective charge. We denote the effective charge due to all charges that are "bound up" in the atoms as ρ_{bound}. Altogether for bulk material we have

$$\vec{\nabla}\cdot\vec{E} = 4\pi[\rho_{\text{free}} + \rho_{\text{bound}}]. \tag{5.61}$$

If we assume that the bound charge is entirely due to the atomic dipole contributions

$$\rho_{\text{bound}} = \rho^d_{\text{eff}} = -\vec{\nabla}\cdot\vec{P}, \tag{5.62}$$

then

$$\vec{\nabla}\cdot\vec{E} = 4\pi[\rho_{\text{free}} - \vec{\nabla}\cdot\vec{P}]. \tag{5.63}$$

This motivates the definition of *displacement field*, which we also will refer to simply as the *D-field*, as

$$\vec{D} \equiv \vec{E} + 4\pi\vec{P}, \tag{5.64}$$

which implies the field equation

$$\vec{\nabla}\cdot\vec{D} = 4\pi\rho_{\text{free}}. \tag{5.65}$$

Usually we will assume that the macroscopic medium is *linear* and *isotropic*

$$\vec{P} = \chi(\vec{x})\vec{E}, \tag{5.66}$$

where $\chi(\vec{x})$ is called the *susceptibility*. We define the *dielectric constant* (alternatively "relative permittivity") $\epsilon(\vec{x})$ as

$$\epsilon(\vec{x}) \equiv 1 + 4\pi\chi(\vec{x}), \tag{5.67}$$

to obtain

$$\vec{D} = (1 + 4\pi\chi(\vec{x}))\vec{E} = \epsilon(\vec{x})\vec{E}, \tag{5.68}$$

$$\Longrightarrow \vec{P} = \frac{\epsilon(\vec{x}) - 1}{4\pi}\vec{E}. \tag{5.69}$$

If ϵ = constant throughout the material, then

$$\vec{\nabla} \cdot \vec{E} = 4\pi\frac{\rho_{\text{free}}}{\epsilon}. \tag{5.70}$$

Typically the electric fields are *reduced* by the effects of atoms in the material, so that $\epsilon > 1$. This is understandable in that dipoles tend to shield charges. One possible charge-shielding is due to *induced polarization*, and is depicted in Figure 5.8. Here the charge densities are distorted by the introduced field, leading to induced charge dipoles and a reduced effective charge in the interior. Of course, charge is conserved, and what seemingly disappears in the interior must reappear on the surface. Thus, Equations (5.58) and (5.59) arise together. Another mechanism is *orientation polarization* seen in Figure 5.9, which occurs in "polar" substances such as water (H_2O). This comes about because the negatively charged electron cloud is unequally shared, leading to more negative charge on the oxygen side and to more positive charge on the hydrogen side. However, higher temperatures tend to disorient the directions of such molecules so we would expect this mechanism to be less effective as the temperature is raised.

Note that in this section we have considered the dielectric constant in purely static circumstances, but usually it is measured at some non-zero

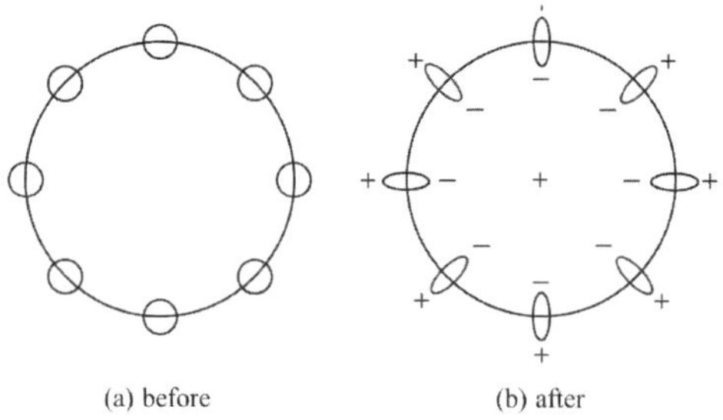

Fig. 5.8 Induced polarization.

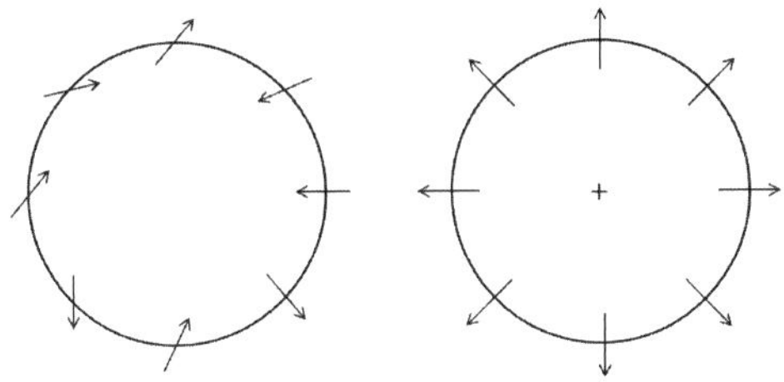

Fig. 5.9 Orientation polarization.

electromagnetic wave angular frequency ω. The response of the electrons and ions to incoming electromagnetic waves introduces frequency dependence: $\epsilon(\omega)$. There are an amazing variety of values which substances may attain. Water has a dielectric constant that ranges from 80.1 at 20°C to 34.5 at 200°C at extremely low frequencies (under 1 kHz), whereas for visible light it has $\epsilon \approx 1.77$ at room temperatures. A large variety of solid materials have dielectric constants which can exceed $\epsilon \sim 10^4$, which can have applications in the semiconductor industry. We will examine the frequency dependence of $\epsilon(\omega)$ more completely in Chapter 9.

5.5 Green functions in the presence of linear dielectrics

In this section, we show how to extend the Green function approach to accommodate dielectric media.

From (5.65) and (5.68) we have $\vec{\nabla} \cdot \left[\epsilon(\vec{x})\vec{E}\right] = 4\pi\rho_{\text{free}}$, and supposing that $\vec{E}$ is the gradient of a potential Φ we have

$$\vec{\nabla} \cdot \left[\epsilon(\vec{x})\vec{\nabla}\Phi(\vec{x})\right] = -4\pi\rho(\vec{x}), \tag{5.71}$$

where ρ denotes the free charge. Now let us assume that the Green function solves

$$\vec{\nabla} \cdot \left[\epsilon(\vec{x})\vec{\nabla}G(\vec{x},\vec{x}\,')\right] = -4\pi\delta(\vec{x}-\vec{x}\,'). \tag{5.72}$$

This still represents the electric field of a positive unit charge at $\vec{x}\,'$. We can now proceed in a manner similar to our original derivation using Green's second identity, Section 2.8:

$$\int d^3x' \left\{G(\vec{x},\vec{x}\,')\vec{\nabla}' \cdot \left[\epsilon(\vec{x}\,')\vec{\nabla}'\Phi(\vec{x}\,')\right] - \Phi(\vec{x}\,')\vec{\nabla}' \cdot \left[\epsilon(\vec{x}\,')\vec{\nabla}'G(\vec{x},\vec{x}\,')\right]\right\}$$

$$= -4\pi \int d^3x' \left\{ G(\vec{x}, \vec{x}\,')\rho(\vec{x}\,') - \Phi(\vec{x}\,')\delta(\vec{x} - \vec{x}\,') \right\}. \tag{5.73}$$

The two sides of (5.73) simplify to:

$$\text{RHS} = 4\pi\Phi(\vec{x}) - 4\pi \int d^3x'\, G(\vec{x}, \vec{x}\,')\rho(\vec{x}\,'). \tag{5.74}$$

$$\begin{aligned}\text{LHS} &= \int d^3x' \vec{\nabla}' \cdot \left[G(\vec{x}, \vec{x}\,')\epsilon(\vec{x}\,')\vec{\nabla}'\Phi(\vec{x}\,') - \vec{\nabla}' G(\vec{x}, \vec{x}\,')\epsilon(\vec{x}\,')\Phi(\vec{x}\,') \right] \\ &= \oint da' \left[\epsilon(\vec{x}\,')G(\vec{x}, \vec{x}\,')\frac{\partial \Phi}{\partial n'} - \epsilon(\vec{x}\,')\Phi(\vec{x}\,')\frac{\partial G}{\partial n'}(\vec{x}, \vec{x}\,') \right].\end{aligned} \tag{5.75}$$

We may choose Dirichlet boundary conditions on $G(\vec{x}, \vec{x}\,')$,

$$G_D(\vec{x}, \vec{x}\,')|_{\vec{x} \text{ on } S} = 0. \tag{5.76}$$

Note that these boundary conditions refer to conductor surfaces, not dielectric ones! If there are no conductor surfaces, the Green function is actually a free Green function. It follows that

$$\Phi(\vec{x}) = \int d^3x'\, G_D(\vec{x}, \vec{x}\,')\rho(\vec{x}\,') - \frac{1}{4\pi} \oint da' \epsilon(\vec{x}\,')\Phi(\vec{x}\,')\frac{\partial G_D}{\partial n'}. \tag{5.77}$$

It is possible to show as before that the dielectric Dirichlet Green function is symmetric in its arguments:

$$G_D(\vec{x}, \vec{x}\,') = G_D(\vec{x}\,', \vec{x}). \tag{5.78}$$

Dielectric Neumann Green functions can be defined as well.

5.6 Green function for the dielectric slab

Let us apply our knowledge to find the fields associated with a dielectric slab; see Figure 5.10. We obtain the free Green function $G(\vec{x}, \vec{x}\,')$ for $z' > 0$ by solving

$$z > 0: \quad \nabla^2 G(\vec{x}, \vec{x}\,') = -4\pi\delta(\vec{x} - \vec{x}\,'), \tag{5.79}$$

$$z < 0: \quad \nabla^2 G(\vec{x}, \vec{x}\,') = 0. \tag{5.80}$$

As before, we use Fourier expansions for the Green and delta functions:

$$4\pi\delta(\vec{x} - \vec{x}\,') = 4\pi \int \frac{d^2k}{(2\pi)^2} e^{i\vec{k}\cdot(\vec{x} - \vec{x}\,')_\perp} \delta(z - z'), \tag{5.81}$$

$$G(\vec{x}, \vec{x}\,') = 4\pi \int \frac{d^2k}{(2\pi)^2} e^{i\vec{k}\cdot(\vec{x} - \vec{x}\,')_\perp} g(z, z'), \tag{5.82}$$

$$-\nabla^2 G(\vec{x}, \vec{x}\,') = 4\pi \int \frac{d^2k}{(2\pi)^2} e^{i\vec{k}\cdot(\vec{x} - \vec{x}\,')_\perp} \left[k^2 - \frac{\partial^2}{\partial z^2} \right] g(z, z'). \tag{5.83}$$

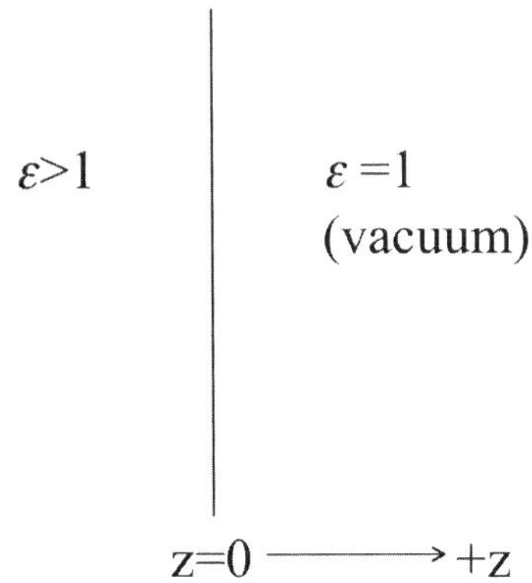

Fig. 5.10 Dielectric slab in a vacuum.

Equations (5.79) and (5.80) give

$$z > 0 : \left[-\frac{\partial^2}{\partial z^2} + k^2\right] g(z, z') = \delta(z - z'), \tag{5.84}$$

$$z < 0 : \left[-\frac{\partial^2}{\partial z^2} + k^2\right] g(z, z') = 0. \tag{5.85}$$

Our boundary conditions are

$$g\big|_{0^-}^{0^+} = 0, \quad (\Phi \text{ is continuous} \implies \vec{E}_{||} \text{ is continuous}) \tag{5.86}$$

$$\epsilon \frac{\partial}{\partial z} g\bigg|_{0^-} = \frac{\partial}{\partial z} g\bigg|_{0^+}, \qquad (D_n \text{ is continuous}) \tag{5.87}$$

The solutions in the various regions are:

$$z < 0 : \quad g = Ae^{kz}, \text{ (finite as } z \to -\infty) \tag{5.88}$$

$$0 < z < z' : \quad g = Be^{kz} + Ce^{-kz}, \tag{5.89}$$

$$z' < z : \quad g = De^{-kz}, \text{ (finite as } z \to +\infty) \tag{5.90}$$

From the above boundary conditions we obtain that

$$A = B + C, \quad \text{and} \quad \epsilon k A = k(B - C), \tag{5.91}$$

which lead to

$$B = \frac{\epsilon + 1}{2} A, \qquad C = \frac{1 - \epsilon}{2} A. \tag{5.92}$$

As z approaches z', we have

$$g\big|_{z'^-}^{z'^+} = 0, \tag{5.93}$$

$$-\frac{\partial}{\partial z}g\bigg|_{z'-}^{z'+} = 1, \tag{5.94}$$

which imply

$$De^{-kz'} = Be^{kz'} + Ce^{-kz'}, \tag{5.95}$$

$$kDe^{-kz'} + k(Be^{kz'} - Ce^{-kz'}) = 1. \tag{5.96}$$

Altogether we have four equations for the four unknowns A, B, C, D, which have the solution:

$$A = \frac{2}{\epsilon+1}\frac{1}{2k}e^{-kz'}, \tag{5.97}$$

$$B = \frac{1}{2k}e^{-kz'}, \tag{5.98}$$

$$C = -\frac{\epsilon-1}{\epsilon+1}\frac{1}{2k}e^{-kz'}, \tag{5.99}$$

$$D = -\frac{\epsilon-1}{\epsilon+1}\frac{1}{2k}e^{-kz'} + \frac{1}{2k}e^{kz'}. \tag{5.100}$$

In summary, the Green function for a point source at $z' > 0$ is

$$z < 0: \quad g = \frac{2}{\epsilon+1}\frac{1}{2k}e^{-k(z'-z)}, \tag{5.101}$$

$$0 < z < z': \quad g = \frac{1}{2k}\left[e^{-k(z'-z)} - \frac{\epsilon-1}{\epsilon+1}e^{-k(z+z')}\right], \tag{5.102}$$

$$z' < z: \quad g = \frac{1}{2k}\left[e^{-k(z-z')} - \frac{\epsilon-1}{\epsilon+1}e^{-k(z+z')}\right]. \tag{5.103}$$

Equations (5.102) and (5.103) can be combined as

$$z > 0 : g = \frac{1}{2k}\left[e^{-k|z-z'|} - \frac{\epsilon-1}{\epsilon+1}e^{-k(z+z')}\right]. \tag{5.104}$$

We may use our previous result from the free-space Green function from Section 3.2

$$4\pi\int\frac{d^2k}{(2\pi)^2}e^{i\vec{k}\cdot(\vec{x}-\vec{x}\,')_\perp}\frac{1}{2k}e^{-k|z-z'|} = \frac{1}{|\vec{x}-\vec{x}\,'|}, \tag{5.105}$$

to evaluate the integral (5.82), again where $z' > 0$:

$$z < 0: \quad G(\vec{x},\vec{x}\,') = \frac{1}{\epsilon}\frac{2\epsilon}{\epsilon+1}\frac{1}{|\vec{x}-\vec{x}\,'|}, \tag{5.106}$$

$$z > 0: \quad G(\vec{x},\vec{x}\,') = \frac{1}{|\vec{x}-\vec{x}\,'|} - \frac{\epsilon-1}{\epsilon+1}\frac{1}{|\vec{x}-\vec{x}\,''|}. \tag{5.107}$$

We have defined $\vec{x}'' \equiv (x', y', -z')$. Equation (5.106) also gives the $z' < 0$, $z > 0$ solution from the symmetry of G : $G(\vec{x}, \vec{x}') = G(\vec{x}', \vec{x})$. By comparing (5.106) with (5.70), we may identify $2\epsilon/(\epsilon + 1)$ as the effective "free" image charge. The fields associated with the Green functions for $z < 0$ and $z > 0$ are shown in Figures 5.11 and 5.12 respectively. The same results may be obtained directly with the image method, as we will show in Section 6.12; also see Exercise 5.6.5 for an image solution between dielectrics.

Recall that we found a surface charge when we solved for the Green function exterior to a conductor. This leads to the question of whether

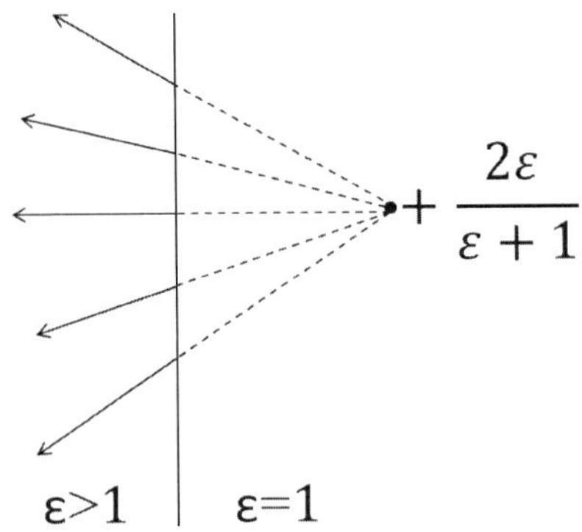

Fig. 5.11 Plane Green function electric field, $z < 0$ solution.

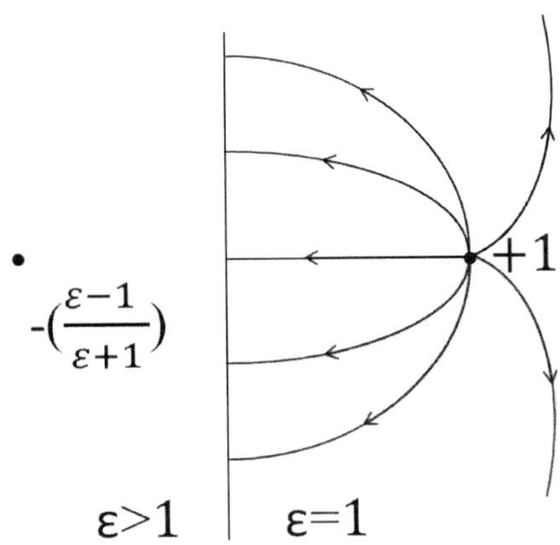

Fig. 5.12 Plane Green function electric field, $z > 0$ solution.

there is any charge on the surface of the dielectric. In fact, we have

$$-\vec{\nabla} \cdot \vec{P} = \rho_{\text{bound}}, \tag{5.108}$$

which implies

$$-(\vec{P}_2 - \vec{P}_1) \cdot \hat{n}_{21} = \sigma_{\text{bound}}. \tag{5.109}$$

$$\vec{P}_1 = \frac{\epsilon - 1}{4\pi}\vec{E}_1, \; \vec{P}_2 = 0, \tag{5.110}$$

from which it follows that

$$\sigma_{\text{bound}} = -\frac{1}{2\pi}\frac{\epsilon - 1}{\epsilon + 1}\frac{z'}{(\rho^2 + z'^2)^{3/2}}, \tag{5.111}$$

where $\rho^2 \equiv |\vec{x}_\perp - \vec{x}'_\perp|^2 = ((x - x')^2 + (y - y')^2)$. This gives rise to a surface delta function as $z' \to 0$, just as we saw before for a conductor in Section 3.2:

$$\lim_{z' \to 0^+} \frac{-2z'}{[\rho^2 + z'^2]^{3/2}} \to -4\pi\delta^{(2)}(\vec{x}_\perp - \vec{x}'_\perp), \tag{5.112}$$

$$\Longrightarrow \sigma_{\text{bound}} \to -\frac{\epsilon - 1}{\epsilon + 1}\delta^{(2)}(\vec{x}_\perp - \vec{x}'_\perp). \tag{5.113}$$

We close this section with a look at three special cases:

- The case $\epsilon \to \infty$ corresponds to a perfect conductor:

$$\begin{aligned} z < 0: &\quad G(\vec{x}, \vec{x}') = 0, \\ z > 0: &\quad G(\vec{x}, \vec{x}') = \frac{1}{|\vec{x} - \vec{x}'|} - \frac{1}{|\vec{x} - \vec{x}''|}. \end{aligned}$$

This agrees with our previous observation that conductors have no internal electric field and that conducting planes gives rise to image charges.

- The case $\epsilon \to 0$, corresponds to Neumann boundary conditions ($\vec{D}$ vanishes for $z < 0$ instead of $\vec{E}$.)

- Finally, in the trivial case ($\epsilon \to 1$), we may verify that we recover the free-space Green function.

5.7 Green function for the dielectric sphere

In this section we consider the free Green function exterior to a dielectric sphere, as shown in Figure 5.13. The defining equations for the Green function are

$$r > a : \nabla^2 G(\vec{x}, \vec{x}\,') = -4\pi\delta(\vec{x} - \vec{x}\,'), \tag{5.114}$$

$$r < a : -\nabla^2 G(\vec{x}, \vec{x}\,') = 0. \tag{5.115}$$

If we assume the functional form

$$G(\vec{x}, \vec{x}\,') = 4\pi \sum_{\ell,m} Y^*_{\ell m}(\theta', \phi') Y_{\ell m}(\theta, \phi) g_\ell(r, r'), \tag{5.116}$$

and we use a delta function expression of the form in (4.227), then we obtain the differential statements

$$-\frac{\partial}{\partial r}\left(r^2 \frac{\partial g_\ell}{\partial r}\right) + \ell(\ell+1) g_\ell = \delta(r - r'),\ r > a, \tag{5.117}$$

$$-\frac{\partial}{\partial r}\left(r^2 \frac{\partial g_\ell}{\partial r}\right) + \ell(\ell+1) g_\ell = 0, \qquad r < a. \tag{5.118}$$

From the continuity of Φ (hence $\vec{E}_{||}$) and D_n we obtain the boundary conditions

$$g_\ell\big|_{a^-}^{a^+} = 0, \tag{5.119}$$

$$\epsilon \frac{\partial g_\ell}{\partial r}\bigg|_{a^-} = \frac{\partial g_\ell}{\partial r}\bigg|_{a^+}. \tag{5.120}$$

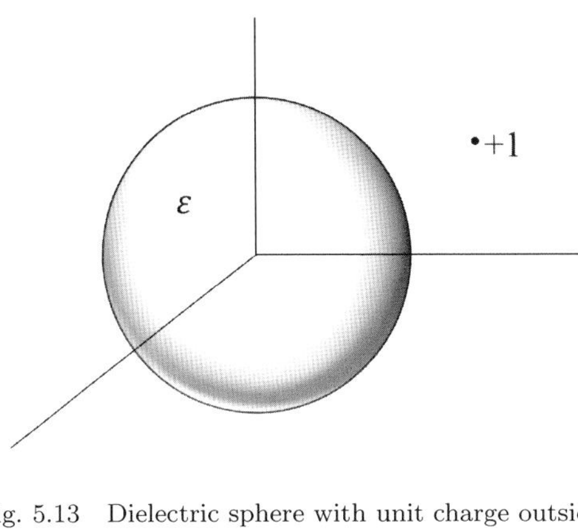

Fig. 5.13 Dielectric sphere with unit charge outside.

For $r' > a$ the solutions are

$$r < a: \quad g_\ell = A_\ell r^\ell, \tag{5.121}$$

$$a < r < r': \quad g_\ell = B_\ell r^\ell + C_\ell r^{-\ell-1}, \tag{5.122}$$

$$r' < r: \quad g_\ell = D_\ell r^{-\ell-1}. \tag{5.123}$$

The boundary conditions lead to

$$A_\ell a^\ell = B_\ell a^\ell + C_\ell a^{-\ell-1}, \tag{5.124}$$

$$\epsilon A_\ell \ell a^{\ell-1} = \ell B_\ell a^{\ell-1} - (\ell+1) C_\ell a^{-\ell-2}, \tag{5.125}$$

which can be rearranged to give

$$B_\ell = \frac{\ell(1+\epsilon)+1}{2\ell+1} A_\ell, \tag{5.126}$$

$$C_\ell = \frac{\ell(1-\epsilon)}{2\ell+1} a^{2\ell+1} A_\ell. \tag{5.127}$$

Other conditions at $r = r'$ are:

$$g_\ell \big|_{r'-}^{r'+} = 0, \tag{5.128}$$

$$-r'^2 \frac{\partial}{\partial r} g_\ell \bigg|_{r'-}^{r'+} = 1, \tag{5.129}$$

which give

$$B_\ell r'^\ell + C_\ell r'^{-\ell-1} = D_\ell r'^{-\ell-1}, \tag{5.130}$$

$$-r'^2 \left[-(\ell+1) D_\ell r'^{-\ell-2} - (\ell B_\ell r'^{\ell-1} - (\ell+1) C_\ell r'^{-\ell-2}) \right] = 1. \tag{5.131}$$

The four equations for the constants $A_\ell, B_\ell, C_\ell, D_\ell$ lead to the solution (as the reader may verify)

$$A_\ell = \frac{1}{\ell(1+\epsilon)+1} \frac{1}{r'^{\ell+1}}, \tag{5.132}$$

$$B_\ell = \frac{1}{2\ell+1} \frac{1}{r'^{\ell+1}}, \tag{5.133}$$

$$C_\ell = -\frac{(\epsilon-1)\ell}{\ell(1+\epsilon)+1} \frac{1}{2\ell+1} \frac{a^{2\ell+1}}{r'^{\ell+1}}, \tag{5.134}$$

$$D_\ell = C_\ell + \frac{r'^\ell}{2\ell+1}. \tag{5.135}$$

The Green function is now given by: $(r' > a)$

$$r < a: \quad G(\vec{x}, \vec{x}\,') = \sum_{\ell,m} Y^*_{\ell m}(\theta', \phi') Y_{\ell m}(\theta, \phi) \frac{4\pi}{\ell(1+\epsilon)+1} \frac{r^\ell}{r'^{\ell+1}}, \tag{5.136}$$

$$r > a: \quad G(\vec{x}, \vec{x}\,') = \sum_{\ell,m} Y^*_{\ell m}(\theta', \phi') Y_{\ell m}(\theta, \phi) \frac{4\pi}{2\ell+1} \frac{r^\ell_<}{r^{\ell+1}_>} \tag{5.137}$$

$$- \sum_{\ell,m} Y^*_{\ell m}(\theta', \phi') Y_{\ell m}(\theta, \phi) \frac{4\pi}{2\ell+1} \frac{\ell(\epsilon-1)}{\ell(1+\epsilon)+1} \frac{a^{2\ell+1}}{(rr')^{\ell+1}}. \tag{5.138}$$

At this point we recall the Addition Theorem for spherical harmonics:

$$P_\ell(\cos\gamma) = \sum_m \frac{4\pi}{2\ell+1} Y^*_{\ell m}(\theta', \phi') Y_{\ell m}(\theta, \phi). \tag{5.139}$$

For $r < a$ and $r > a$ this gives:

$$r < a: \; G(\vec{x}, \vec{x}\,') = \sum_{\ell=0}^{\infty} \frac{2\ell+1}{\ell(1+\epsilon)+1} \frac{r^\ell}{r'^{\ell+1}} P_\ell(\cos\gamma), \tag{5.140}$$

$$r > a: \; G(\vec{x}, \vec{x}\,') = \frac{1}{|\vec{x} - \vec{x}\,'|} - \sum_{\ell=1}^{\infty} \frac{(\epsilon-1)\ell}{\ell(1+\epsilon)+1} \frac{a^{2\ell+1}}{(rr')^{\ell+1}} P_\ell(\cos\gamma). \tag{5.141}$$

Equation (5.140) also gives the $r' < a, r > a$ form by interchanging r and r' due to the symmetry of the Green function.

There is no longer an image charge interpretation of these results except when $\epsilon \to \infty$. In this limit we almost recover (4.240) from (5.138) or (5.141), except for the $\ell = 0$ term. This is an explicit realization of the comment regarding this limit, in a footnote, in Section 1.4. (The $\epsilon \to 0$ limit for $r > a$ connects to Exercise 4.12.2.)

Looking at the $r > a$ solution when $r' \gg a$, we have:

$$G(\vec{x}, \vec{x}\,') \approx \frac{1}{|\vec{x} - \vec{x}\,'|} - \frac{\epsilon-1}{\epsilon+2} \frac{a^3}{r^2 r'^2} \cos\gamma. \tag{5.142}$$

Substituting $\cos\gamma = (\vec{x} \cdot \vec{x}\,')/(rr')$ in the previous equation yields

$$G(\vec{x}, \vec{x}\,') \approx \frac{1}{|\vec{x} - \vec{x}\,'|} + \frac{\vec{x} \cdot \vec{p}}{r^3}, \quad \text{where } \vec{p} = \frac{\epsilon-1}{\epsilon+2} a^3 \vec{E}'(0), \tag{5.143}$$

where $\vec{E}'(0) = -\vec{x}\,'/r'^3$ is the point charge's electric field evaluated at the origin. In this case $\vec{p}$ is the induced dipole moment of the sphere due to the positive charge. Therefore, the dipole moment induced in a sphere of radius a by a *uniform* electric field is

$$\vec{p} = \frac{\epsilon-1}{\epsilon+2} a^3 \vec{E}_{\text{const}}. \tag{5.144}$$

The other case for $r' \gg a$ is when $r < a$:

$$G(\vec{x}, \vec{x}\,') \approx \frac{1}{r'} - \frac{3}{\epsilon + 2}\vec{x} \cdot \vec{E}'(0), \tag{5.145}$$

where $\vec{E}'(0)$ is again the point charge's electric field. It follows that for the positive charge very far away, the electric field in the sphere is approximately:

$$\vec{E} = -\vec{\nabla} G(\vec{x}, \vec{x}\,') = \frac{3}{\epsilon + 2}\vec{E}'(0), \tag{5.146}$$

which is less than $\vec{E}'(0)$ when $\epsilon > 1$. $\vec{E}'(0)$ can again be replaced by any approximately constant field $\vec{E}_{\text{const.}}$ in which the sphere is immersed.

5.8 Field energy and dielectrics

In the absence of any constitutive relation between $\vec{D}$ and $\vec{E}$, all that we know for a given material is

$$\vec{\nabla} \cdot \vec{D} = 4\pi\rho, \tag{5.147}$$

$$\vec{\nabla} \times \vec{E} = 0, \tag{5.148}$$

where ρ refers to the free charge density. The second equation implies we may still take

$$\vec{E} = -\vec{\nabla}\Phi. \tag{5.149}$$

We continue to require $\vec{F} = q\vec{E}$, so that $q\Phi$ has the usual meaning of potential energy. The energy to move an infinitesimal charge is:

$$\underbrace{\delta W_1}_{\text{work } on \text{ the charge}} = \int_A^B (-\vec{F}_1) \cdot d\vec{\ell} = -\delta q_1 \int_A^B \vec{E} \cdot d\vec{\ell}, \tag{5.150}$$

$$= \delta q_1 (\Phi_B - \Phi_A). \tag{5.151}$$

Let A be our reference point (which can be at ∞ or any other point):

$$\Phi_A = 0, \Phi_B \to \Phi(\vec{x}_1). \tag{5.152}$$

Now suppose we move another charge from A to B':

$$\delta W_2 = \delta q_2 \Phi(\vec{x}_2) + O(\delta q_1 \delta q_2). \tag{5.153}$$

Summing, we obtain:

$$\delta W = \sum_i \delta W_i = \sum_i \delta q_i \Phi(\vec{x}_i), \tag{5.154}$$

and taking limits by replacing $\delta q_i \to d^3x\,\delta\rho(\vec{x})$ gives

$$\delta W = \int d^3x\,\delta\rho(\vec{x})\Phi(\vec{x}). \tag{5.155}$$

Equation (5.155) is always true: however, we can not integrate it in this form and further assumptions are necessary. Now, using the expression

$$\rho = \frac{1}{4\pi}\vec{\nabla}\cdot\vec{D}, \tag{5.156}$$

with $\rho \to \delta\rho$ and inserting into the expression for δW, we find

$$\begin{aligned}\delta W &= \frac{1}{4\pi}\int d^3x\,\vec{\nabla}\cdot\delta\vec{D}\Phi(\vec{x})\\ &= \frac{1}{4\pi}\int d^3x\,\left[\vec{\nabla}\cdot(\delta\vec{D}\Phi) - \delta\vec{D}\cdot\vec{\nabla}\Phi\right]\\ &= \frac{1}{4\pi}\int d^3x\,\vec{E}\cdot\delta\vec{D} + \frac{1}{4\pi}\oint_S da\,\Phi\,\delta\vec{D}\cdot\hat{n}.\end{aligned} \tag{5.157}$$

Given a localized charge distribution, we have $\Phi \to Q_{\text{tot}}/R$ as $|\vec{x}| = R \to \infty$, and the surface term in (5.157) vanishes:

$$\begin{aligned}\frac{1}{4\pi}\oint_{|\vec{x}|=R} da\,\Phi\,\delta\vec{D}\cdot\hat{n} &\underset{R\to\infty}{\longrightarrow} \frac{1}{4\pi}\oint_{|\vec{x}|=R} da\,\frac{Q_{\text{tot}}}{R}\delta\vec{D}\cdot\hat{n}\\ &\to \frac{Q_{\text{tot}}}{R}\delta q \underset{R\to\infty}{\longrightarrow} 0.\end{aligned} \tag{5.158}$$

We are left with the volume term:

$$\delta W = \frac{1}{4\pi}\int d^3x\vec{E}\cdot\delta\vec{D}. \tag{5.159}$$

This is as far as we can go unless we can write δW as a perfect differential. Assuming relations of the form:

$$D_i = \sum_j \epsilon_{ij}E_j, \tag{5.160}$$

where the coefficients ϵ_{ij} are assumed to be independent of variations in the charge density, then we have:

$$\begin{aligned}\delta D_i &= \sum_j \epsilon_{ij}\delta E_j,\\ \Longrightarrow \sum_i E_i\delta D_i &= \sum_{i,j}\epsilon_{ij}\delta E_j E_i.\end{aligned} \tag{5.161}$$

Also

$$\sum_i D_i\delta E_i = \sum_{i,j}\epsilon_{ij}E_j\delta E_i,$$

$$\Longrightarrow \vec{D}\cdot\delta\vec{E} = \vec{E}\cdot\delta\vec{D} \text{ if } \epsilon_{ij} = \epsilon_{ji}, \tag{5.162}$$

$$\Longrightarrow \vec{E}\cdot\delta\vec{D} = \frac{1}{2}\delta(\vec{E}\cdot\vec{D}). \tag{5.163}$$

The above certainly includes the isotropic case where $\vec{D} = \epsilon\vec{E}, \epsilon = \epsilon(\vec{x})$. However, it excludes for example a situation where

$$\vec{D} = \epsilon(E^2)\vec{E},$$

as we will see in Section 5.10, for then

$$\vec{E}\cdot\delta\vec{D} = \vec{D}\cdot\delta\vec{E} + \delta\epsilon(E^2)E^2.$$

So for a certain class of constitutive relations, we have the expression

$$W = \frac{1}{8\pi}\int d^3x\, \vec{E}\cdot\vec{D}, \tag{5.164}$$

for the field energy. This of course is consistent with the earlier electrostatic expression, (2.147), when $\vec{D} = \vec{E}$. We have followed the same logic in developing this expression as the earlier case: we have calculated the work done on small charge carriers and used energy conservation to deduce the field energy.

Other expressions for W can be developed. In particular, since we know all the static properties are determined by the Green function, then we should be able to express W in terms of $G(\vec{x},\vec{x}\,')$. Since $\vec{E} = -\vec{\nabla}\Phi$, the above gives for a localized charge distribution

$$W = \frac{1}{2}\int d^3x\, \rho(\vec{x})\Phi(\vec{x}), \tag{5.165}$$

where $\rho(\vec{x})$ represents the free charge density; self-energies are included. For Dirichlet boundary conditions, from (5.77) we have the expression

$$\Phi(\vec{x}) = \int d^3x'\, \rho(\vec{x}\,')G_D(\vec{x},\vec{x}\,') - \frac{1}{4\pi}\oint_S da'\Phi(\vec{x}\,')\epsilon(\vec{x}\,')\frac{\partial G_D}{\partial n'}. \tag{5.166}$$

As we noted before, the surface S in (5.166) pertains to the entire volume where fields are defined — it does not refer to the surfaces of dielectrics. If we suppose that $\Phi|_S = 0$ (which is certainly true for free space), then we obtain another expression for the energy:

$$W = \frac{1}{2}\int d^3x\rho(\vec{x})\Phi(\vec{x}) = \frac{1}{2}\int d^3xd^3x'\rho(\vec{x})G_D(\vec{x},\vec{x}\,')\rho(\vec{x}\,'). \tag{5.167}$$

This is just a generalization of the expression we derived in Section 2.10:

$$W = \frac{1}{2}\int d^3xd^3x'\,\frac{\rho(\vec{x})\rho(\vec{x}\,')}{|\vec{x}-\vec{x}\,'|}. \tag{5.168}$$

Now instead of introducing charge, think of introducing a dielectric. What is the amount of energy required to do this? We can calculate the finite field energy difference in the two configurations with the use of the Green functions for the two cases. Given that there may be grounded conductors present, we use (5.167) with G_D^0 when the dielectric is absent, and G_D when the dielectric is present. We therefore have the energy difference:

$$\begin{aligned}\Delta W &\equiv W - W_0 \\ &= \frac{1}{2}\int d^3x\, d^3x'\, \rho(\vec{x})\left[G_D(\vec{x},\vec{x}\,') - G_D^0(\vec{x},\vec{x}\,')\right]\rho(\vec{x}\,') \\ &= \frac{1}{2}\int d^3x\, \rho(\vec{x})[\Phi(\vec{x}) - \Phi_0(\vec{x})] \\ &= -\frac{1}{8\pi}\int d^3x\, \vec{D}\cdot\vec{\nabla}[\Phi(\vec{x}) - \Phi_0(\vec{x})] \\ &= \frac{1}{8\pi}\int d^3x\, \vec{D}\cdot(\vec{E} - \vec{E}_0).\end{aligned} \tag{5.169}$$

Now consider

$$\int d^3x\, \vec{D}\cdot\vec{E} = \int d^3x\, \vec{D}_0\cdot\vec{E} + \int d^3x\, (\vec{D} - \vec{D}_0)\cdot\vec{E}. \tag{5.170}$$

The second term in (5.170) integrates to zero by parts (using $\vec{E} = -\vec{\nabla}\Phi$ and $\vec{\nabla}\cdot\vec{D} = \vec{\nabla}\cdot\vec{D}_0$). Therefore

$$\Delta W = \frac{1}{8\pi}\int d^3x\, [\vec{E}\cdot\vec{D}_0 - \vec{D}\cdot\vec{E}_0]. \tag{5.171}$$

Now if we assume the isotropic case ($\vec{D} = \epsilon(\vec{x})\vec{E}$) and that some dielectric is initially present ($\vec{D}_0 = \epsilon_0(\vec{x})\vec{E}_0$), we have that

$$\Longrightarrow \Delta W = \frac{1}{8\pi}\int d^3x\, (\epsilon_0(\vec{x}) - \epsilon(\vec{x}))\vec{E}\cdot\vec{E}_0. \tag{5.172}$$

The integration in (5.172) is effectively over the volume of the dielectric. In the case where no dielectrics are initially present ($\epsilon_0 = 1$), we have

$$\Delta W = -\frac{1}{2}\int d^3x\, \vec{P}\cdot\vec{E}_0, \tag{5.173}$$

where we have used (5.69).

5.9 Bulk forces on dielectrics: theory

By using (5.173) or other means of finding an appropriate energy expression, we may determine the total force on a dielectric from

$$\vec{F}\cdot\delta\vec{x} = -(\delta W)_{\text{fixed charge}}$$
$$\Longrightarrow \vec{F} = -\left(\frac{\partial W}{\partial \vec{x}}\right)_Q, \tag{5.174}$$

where the notation indicates the derivative is carried out at fixed charge Q.

On the other hand, consider the movement of a dielectric in the presence of conductors kept at *fixed voltage* connected to idealized batteries. That is, assume all free charges are on the surfaces of conductors attached to the batteries. If W is the field energy, and W_b associated with the battery energy, we would expect:

$$\vec{F} = -\frac{\delta}{\delta\vec{x}}(W + W_b)_V. \tag{5.175}$$

But from (5.165) specialized to conductor surfaces,

$$W = \frac{1}{2}\sum_i \int da\, \sigma_i(\vec{x})V_i$$
$$= \frac{1}{2}\sum_i Q_i V_i, \tag{5.176}$$

so the change of field energy is given by:

$$\delta_V W = \frac{1}{2}\sum_i \delta Q_i V_i. \tag{5.177}$$

On the other hand, the battery's change in energy is

$$\delta_V W_b = \sum_i \delta\bar{Q}_i V_i. \tag{5.178}$$

However from conservation of charge

$$\delta\bar{Q}_i = -\delta Q_i$$
$$\Longrightarrow \delta_V W_b = -2\delta_V W$$
$$\Longrightarrow \vec{F} = +\left(\frac{\partial W}{\partial \vec{x}}\right)_V. \tag{5.179}$$

It is important to realize that in a given static (or instantaneous) situation, (5.174) and (5.179) give the same force in spite of the minus sign differences; see Exercise 5.9.5 in this context.

Energy methods as discussed above are helpful and simpler than other methods, but they give you no idea of *where* the forces originate. Going back to our earlier expression for bulk force, $\vec{F}_{\text{bulk}}$, from (5.57) we have

$$\vec{F}_{\text{bulk}} = \int_S da\, (\vec{P} \cdot \hat{n}) \vec{E}^{(0)}(\vec{x}), \tag{5.180}$$

which is true when $\nabla \cdot \vec{P} = 0$ in the interior volume. If $\epsilon =$ constant, we in fact have

$$\nabla \cdot \vec{P} = \frac{\epsilon - 1}{\epsilon} \rho_f, \tag{5.181}$$

from (5.69) and (5.70), which vanishes when there is no interior free charge. Note that

$$\vec{P} \cdot \hat{n} = \sigma_{\text{eff}} \tag{5.182}$$

on the surface from (5.59). Now consider a small surface element, da, as shown in Figure 5.14. Near the surface of each da, the field is given by:

$$\vec{E} = \vec{E}^{(0)} + \vec{E}^r + \vec{E}^s, \tag{5.183}$$

where $\vec{E}^s$ is the self-field ($\vec{E}^s = \pm 2\pi\sigma_{\text{eff}}\hat{n}$), $\vec{E}^r$ is from the rest of the surface and $\vec{E}^{(0)}$ is external. Therefore the average field at the interface is

$$\frac{\vec{E}_1 + \vec{E}_2}{2} = \vec{E}^{(0)} + \vec{E}^r. \tag{5.184}$$

From (5.180), (5.182) and (5.184) we can write the bulk force as:

$$\vec{F}_{\text{bulk}} = \int_S da\, \sigma_{\text{eff}} \left[\frac{\vec{E}_1 + \vec{E}_2}{2} - \vec{E}^r(\vec{x}) \right]. \tag{5.185}$$

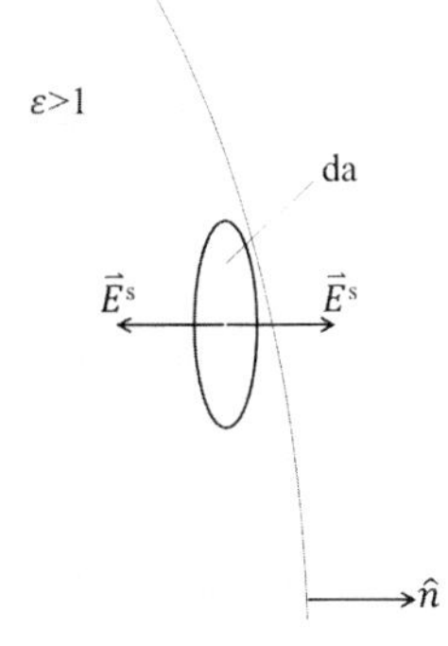

Fig. 5.14 Surface element on the surface.

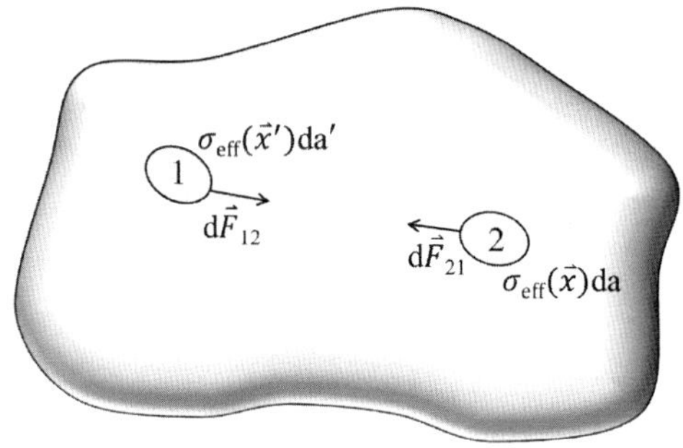

Fig. 5.15 Newton's 3rd law for surface charge elements.

Now consider the forces arising between charges $\sigma_{\rm eff}(\vec{x})da$ and all the other $\sigma_{\rm eff}(\vec{x}\,')da'$ in the rest of the surface, as shown in Figure 5.15. Newton's third law tells us that $d\vec{F}_{12} = -d\vec{F}_{21}$. This implies that the second term in (5.185) is zero when the integration is over the entire surface. This means we can always use either $\vec{E}^{(0)}$ or $(\vec{E}_1 + \vec{E}_2)/2$ in such expressions. However, this is not to say that there are not self-forces or stresses within a given material; for instance, in Section 2.11 we found that the outward pressure on the surface of a conductor is $2\pi\sigma^2$.

5.10 Nonlinear dielectric example: leading logarithm model

It turns out that the methods of electrostatics turn up in other areas of physics as well, as the following example shows.

The *leading logarithm model*[2] is a phenomenological (that is, not strictly rigorous) model for quark interactions. The model is based on quantum chromodynamics (QCD), to which various approximations and heuristic arguments are applied. We will sweep all the complicated details of the derivation under the rug and simply present the outcome, which is surprisingly simple. According to the model, quark-antiquark systems (these are called mesons) behave approximately like a pair of classical point charges of opposite sign, located within a dielectric medium with *nonlinear* dielectric

[2]S. Adler and T. Piran, "Flux Confinement in the Leading Logarithm Model", Phys. Lett. B **113**, 405 (1982); Erratum, *ibid*, B **121**, 455 (1983).

constant which satisfies:

$$\vec{E} = -\vec{\nabla}\Phi, \tag{5.186}$$

$$\vec{D} = \epsilon\vec{E}, \tag{5.187}$$

where ϵ is given by

$$\epsilon = 2\alpha \ln\left(\frac{E^2}{K^2}\right) \quad \text{where } E^2 = \vec{E}\cdot\vec{E}, \alpha > 0. \tag{5.188}$$

The formula gives a negative dielectric constant when $E^2 < K^2$. To avoid unphysical effects, one modifies this by saying that $D = 0$ (or $\epsilon = 0$) unless $E^2 > K^2$. This discontinuity in the form of the dielectric constant gives rise to two distinct regions in space corresponding to $\epsilon = 0$ and $\epsilon > 0$, respectively. The $\epsilon > 0$ region is called the *confinement region*: outside the confinement region, all fields are 0. The boundary of the confinement region is *dynamical*, that is it must be found by actually solving the equations. It turns out that the confinement region for two separated point charges is tube-shaped, as shown in Figure 5.16. Notice that the electric field lines flow everywhere tangential to the surface so that charge will not be lost, giving rise to effective homogeneous Neumann boundary conditions.

Let's consider the field energy for such a system, which according to (5.159) is given by:

$$\delta W = \frac{1}{4\pi}\int d^3x\, \vec{E}\cdot\delta\vec{D}. \tag{5.189}$$

Our only hope for evaluating W is to rewrite the integrand as a perfect differential. Making an "obvious" choice, we consider

$$\delta\left[\frac{1}{2}\vec{E}\cdot\vec{D} + \alpha E^2\right] = \frac{1}{2}\vec{D}\cdot\delta\vec{E} + \frac{1}{2}\vec{E}\cdot\delta\vec{D} + 2\alpha\vec{E}\cdot\delta\vec{E}. \tag{5.190}$$

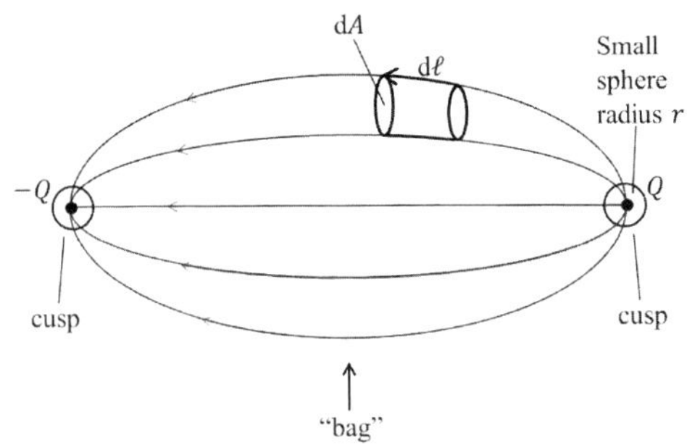

Fig. 5.16 Flux from Q to $-Q$ incorporating $d\ell\, dA$.

Now

$$\delta\vec{D} = \delta(\epsilon\vec{E}) = \left(2\alpha\ln\frac{E^2}{K^2}\right)\delta\vec{E} + \frac{4\alpha}{E^2}(\vec{E}\cdot\delta\vec{E})\vec{E}, \tag{5.191}$$

so that

$$\begin{aligned}\vec{E}\cdot\delta\vec{D} &= \left(2\alpha\ln\frac{E^2}{K^2}\vec{E}\right)\cdot\delta\vec{E} + 4\alpha\vec{E}\cdot\delta\vec{E}\\ &= \vec{D}\cdot\delta\vec{E} + 4\alpha\vec{E}\cdot\delta\vec{E}.\end{aligned} \tag{5.192}$$

Substituting this back into (5.190) yields:

$$\begin{aligned}\delta\left[\frac{1}{2}\vec{E}\cdot\vec{D} + \alpha E^2\right] &= \frac{1}{2}\vec{E}\cdot\delta\vec{D} - 2\alpha\vec{E}\cdot\delta\vec{E} + 2\alpha\vec{E}\cdot\delta\vec{E} + \frac{1}{2}\vec{E}\cdot\delta\vec{D}\\ &= \vec{E}\cdot\delta\vec{D}.\end{aligned} \tag{5.193}$$

We have succeeded in our quest, and we may write

$$W = \frac{1}{4\pi}\int d^3x\left[\frac{1}{2}\vec{E}\cdot\vec{D} + \alpha E^2\right]. \tag{5.194}$$

By assumption we have $|\vec{E}| > K$ whenever $|\vec{D}| > 0$. Using this fact and (5.194), we may give the following lower bound for W:

$$\begin{aligned}&W > \frac{1}{8\pi}\int d^3x\,\vec{E}\cdot\vec{D} = \frac{1}{8\pi}\int d^3x\,|\vec{E}||\vec{D}|\\ \Longrightarrow\ &W > \frac{K}{8\pi}\int d^3x\,|\vec{D}|.\end{aligned} \tag{5.195}$$

Equation (5.195) may be used to give a lower bound on the energy of the quark configuration shown in Figure 5.16. To estimate the integral in (5.195), we break the confinement region up into infinitesimally-thin tubes of constant flux, as shown in Figure 5.16. If we consider a small sphere of radius r centered at the $+Q$ charge, we have from Gauss's law that the total flux exiting the sphere is $4\pi Q$. This same flux enters the sphere of radius r centered at the $-Q$ charge. We may estimate the contribution to the field energy of a particular flux tube joining the two spheres as:

$$\delta W_{\text{tube}} > \frac{K}{8\pi}\int_{\text{tube}} d\ell\, dA\, D > \frac{K}{8\pi}(R-2r)\int_{\text{end of tube}} dA\, D. \tag{5.196}$$

The second inequality is simply due to the fact that $(R-2r)$ is the smallest tube length. The integral on the right-hand side of (5.196) is simply the portion of the flux exiting the small sphere around $+Q$ that enters the particular flux tube. If we sum over all flux tubes joining the two spheres, we obtain the total flux, and hence

$$\sum_{\text{tubes}} \delta W_{\text{tube}} > \frac{1}{2} K(R - 2r)|Q|. \tag{5.197}$$

Thus the field energy grows at least linearly at large R. Computer simulations confirm linear dependence on R for large R.

The physical significance of this result is staggering. Since energy grows with separation, the charges (which represent quarks) are not allowed to separate! This establishes, within the context of this particular model, one of the key results of QCD, namely that quarks are *confined* and never appear in isolation.

The above results are for two-quark systems: the three quark (baryon) equations, which have an effective $U(1) \times U(1)$ symmetry, may also be derived using similar methods.[3]

5.11 Bulk forces on dielectrics: examples

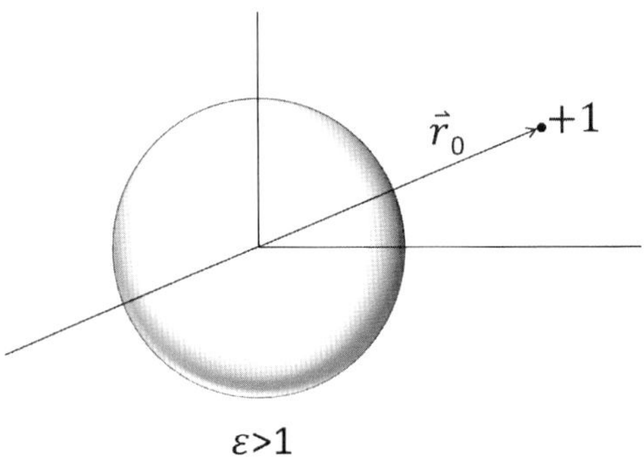

Fig. 5.17 Dielectric sphere, the easy way.

Since practice makes perfect, let us wrap up this chapter with two examples involving forces on dielectrics. For our first example we will consider the force between a dielectric sphere and a point charge, as shown in Figure 5.17. So that no one will ever suspect us of not being thorough, we will

[3] K. A. Milton, W. Wilcox and S. S. Pinsky, "Bag Formation In Three Quark Chromostatics", Phys. Rev. D **27**, 958 (1983); S. L. Adler, "Generalized Bag Models as Mean-Field Approximations to QCD", Phys. Lett. **110B**, 302 (1982).

solve this problem three different ways. The first and easiest method is to take the difference between the Green function expressions for energy with and without the sphere:

$$\Delta W = \frac{1}{2}\int d^3x\, d^3x'\, \rho(\vec{x})\left[G_D(\vec{x},\vec{x}\,') - G_D^0(\vec{x},\vec{x}\,')\right]\rho(\vec{x}\,'). \tag{5.198}$$

In this case ρ is a unit point charge, $\rho(\vec{x}) = \delta(\vec{x}-\vec{r}_0)$, so we obtain simply

$$\Delta W = \frac{1}{2}\left[G_D(\vec{x},\vec{x}\,') - G_D^0(\vec{x},\vec{x}\,')\right]\Big|_{\vec{x},\vec{x}\,'=\vec{r}_0}. \tag{5.199}$$

It turns out this is an $\infty - \infty$ limit, but since we are physicists this should not deter us in the slightest. We know where the ∞ comes from: it is the point charge self-energy, and the same energy contribution is present with or without the dielectric sphere. The G_D^0 term in (5.199) effectively cancels out the self-energy term in the expansion (5.141) with $r > a$:

$$\Delta W = -\frac{1}{2}\sum_{\ell=1}^{\infty}\frac{(\epsilon-1)\ell}{\ell(\epsilon+1)+1}\frac{a^{2\ell+1}}{r_0^{2\ell+2}}P_\ell(1), \tag{5.200}$$

which simplifies to (note $P_\ell(1) = 1$):

$$\Delta W = -\frac{\epsilon-1}{2r_0}\sum_{\ell=1}^{\infty}\frac{\ell}{\ell(1+\epsilon)+1}\left(\frac{a}{r_0}\right)^{2\ell+1}. \tag{5.201}$$

At large r_0 the leading behavior is

$$\Delta W \approx -\left(\frac{\epsilon-1}{\epsilon+2}\right)\frac{a^3}{2r_0^4} \implies F_r = -\left(\frac{\epsilon-1}{\epsilon+2}\right)\frac{2a^3}{r_0^5}, \tag{5.202}$$

where F_r denotes the force on the charge; the origin is at the center of the sphere. The force is attractive if $\epsilon > 1$, which is the physically relevant case. The r_0-dependence is different from the case of a grounded conducting sphere, which is inverse cube at large r_0. (Compare the Green function in (4.240), which has an induced $\ell = 0$ term, with (5.138), which does not.) Of course, the force on a neutral, isolated conducting sphere is also $1/r_0^5$, corresponding to the $\epsilon \to \infty$ limit of (5.202).

A considerably harder way to do the same problem is to use the explicit force expression from Section 5.9:

$$F_z = \int da\,(\vec{P}\cdot\hat{n})\left(\frac{E_{2z}+E_{1z}}{2}\right). \tag{5.203}$$

Region 1 is the dielectric sphere; Region 2 is the outside vacuum. We are taking the z-axis along $\vec{r}\,' = \vec{r}_0$. We start with

$$(\vec{P}\cdot\hat{n})\Big|_a = \frac{1}{4\pi}\left(E_{2r}\big|_a - E_{1r}\big|_a\right). \tag{5.204}$$

Expanding the Green function from Section 5.7 in the two regions up to $\sim 1/r'^3$, we have

$$G_1 \approx \frac{1}{r'} + \frac{3}{2+\epsilon}\frac{r}{r'^2}\underbrace{\cos\theta}_{P_1} + \frac{5}{3+2\epsilon}\frac{r^2}{r'^3}\frac{1}{2}\underbrace{(3\cos^2\theta - 1)}_{P_2}, \tag{5.205}$$

$$G_2 \approx \frac{1}{r'} + \frac{r}{r'^2}\cos\theta + \frac{r^2}{r'^3}\frac{1}{2}(3\cos^2\theta - 1) - \frac{(\epsilon-1)}{2+\epsilon}\frac{a^3}{r^2 r'^2}\cos\theta$$
$$- \frac{2(\epsilon-1)}{3+2\epsilon}\frac{a^5}{r^3 r'^3}\frac{1}{2}(3\cos^2\theta - 1). \tag{5.206}$$

We have:

$$E_{1r} = -\frac{\partial}{\partial r}G_1, \qquad E_{2r} = -\frac{\partial}{\partial r}G_2,$$
$$\Longrightarrow\; E_{1r}\big|_a \approx -\frac{3}{2+\epsilon}\frac{1}{r'^2}\cos\theta - \frac{5}{3+2\epsilon}\frac{a}{r'^3}(3\cos^2\theta - 1), \tag{5.207}$$
$$\Longrightarrow\; E_{2r}\big|_a \approx -\frac{3\epsilon}{2+\epsilon}\frac{1}{r'^2}\cos\theta - \frac{5\epsilon}{3+2\epsilon}\frac{a}{r'^3}(3\cos^2\theta - 1). \tag{5.208}$$

Plugging into (5.204) gives

$$\vec{P}\cdot\hat{n}\Big|_a \approx \frac{1}{4\pi}\left(\frac{3(1-\epsilon)}{2+\epsilon}\frac{1}{r'^2}\cos\theta + \frac{5(1-\epsilon)}{3+2\epsilon}\frac{a}{r'^3}(3\cos^2\theta - 1)\right). \tag{5.209}$$

Likewise

$$E_{1z} = -\frac{\partial}{\partial z}\left(\frac{3}{2+\epsilon}\frac{z}{r'^2} + \frac{5\epsilon}{3+2\epsilon}\frac{1}{r'^3}\frac{1}{2}(3z^2 - r^2)\right). \tag{5.210}$$

The expressions

$$\frac{\partial}{\partial z}\left(\frac{1}{r}\right) = -\frac{z}{r^3}, \qquad \frac{\partial r}{\partial z} = \frac{z}{r}, \tag{5.211}$$

now give

$$E_{1z}\big|_a = -\frac{3}{2+\epsilon}\frac{1}{r'^2} - \frac{10}{3+2\epsilon}\frac{a}{r'^3}\cos\theta, \tag{5.212}$$

$$E_{2z}\big|_a = -\frac{3}{2+\epsilon}\frac{1}{r'^2} - \frac{3(\epsilon-1)}{2+\epsilon}\frac{1}{r'^2}\cos^2\theta - \frac{2a}{r'^3}\cos\theta$$
$$+ \frac{(\epsilon-1)}{3+2\epsilon}\frac{a}{r'^3}\left(-15\cos^3\theta + 9\cos\theta\right). \tag{5.213}$$

It follows that

$$\left.\frac{E_{2z}+E_{1z}}{2}\right|_a = -\frac{3}{2+\epsilon}\frac{1}{r'^2} - \frac{3}{2}\frac{(\epsilon-1)}{2+\epsilon}\frac{1}{r'^2}\cos^2\theta - \left(\frac{8+2\epsilon}{3+2\epsilon}\right)\frac{a}{r'^3}\cos\theta$$

$$+\frac{(\epsilon-1)}{3+2\epsilon}\frac{a}{r'^3}\left(-\frac{15}{2}\cos^3\theta + \frac{9}{2}\cos\theta\right). \tag{5.214}$$

Now consider the product of (5.209) and (5.214). Notice that $1/r'^4$ terms go like $\int_{-1}^{1} d\cos\theta \begin{pmatrix}\cos\theta \\ \cos^3\theta\end{pmatrix} = 0$. The lowest order terms are $1/r'^5$. Defining $x \equiv \cos\theta$, we have

$$F_z \approx \frac{2\pi a^3}{4\pi r'^5}\int_{-1}^{1} dx \left\{ \frac{3(1-\epsilon)}{2+\epsilon}\frac{(\epsilon-1)}{3+2\epsilon}\left(-\frac{15}{2}x^4 + \frac{9}{2}x^2\right)\right.$$

$$\left. -\frac{3(1-\epsilon)}{2+\epsilon}\left(\frac{8+2\epsilon}{3+2\epsilon}\right)x^2 + \frac{5(1-\epsilon)}{3+2\epsilon}(3x^2-1)\left(-\frac{3}{2+\epsilon} - \frac{3}{2}\frac{(\epsilon-1)}{2+\epsilon}x^2\right)\right\}. \tag{5.215}$$

After more tedious algebra we have

$$F_z \approx \frac{2a^3}{r'^5}\left(\frac{\epsilon-1}{2+\epsilon}\right), \tag{5.216}$$

which means the sphere is attracted to the charge for $\epsilon > 1$.

Finally, we can also calculate the force explicitly using the *external* field:

$$F_z = \int da(\vec{P}\cdot\hat{n})E_z^0. \tag{5.217}$$

With notation as shown in Figure 5.18 we have

$$\Phi_0 = \frac{1}{\sqrt{r^2 + r'^2 - 2rr'\cos\theta}} = \sum_{\ell=0}^{\infty}\frac{r^\ell}{r'^{\ell+1}}P_\ell(\cos\theta) \tag{5.218}$$

$$= \frac{1}{r'} + \frac{r}{r'^2}\cos\theta + \frac{r^2}{r'^3}\frac{1}{2}(3\cos^2\theta - 1) + \ldots, \tag{5.219}$$

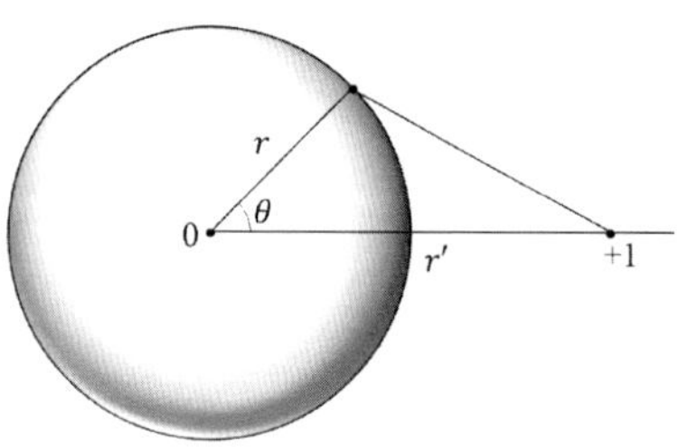

Fig. 5.18 Notation for force computation.

so that

$$E_z^0\Big|_a = -\frac{\partial}{\partial z}\Phi_0\Big|_a \approx -\frac{1}{r'^2} - 2\frac{a}{r'^3}\cos\theta. \tag{5.220}$$

Compare this with the complicated expression for $(E_{2z} + E_{1z})/2$ above. Then

$$F_z \approx 2\pi a^2 \frac{1}{4\pi}\frac{a}{r'^5}\int_{-1}^{1} dx\left[-\frac{6(1-\epsilon)}{2+\epsilon}x^2 - \frac{5(1-\epsilon)}{3+2\epsilon}(3x^2-1)\right]$$

$$\Longrightarrow F_z \approx \frac{2a^3}{r'^5}\frac{\epsilon-1}{2+\epsilon}. \tag{5.221}$$

Let us do another problem using the energy method. The problem is to find the fluid level of a dielectric fluid inside a cylindrically shaped capacitor which is fed by a large fluid reservoir, as shown in Figure 5.19. Initially the fluid and reservoir fluid levels are the same; after connection to the battery, the interior level is raised. Figure 5.20 shows the complicated field lines of the capacitor in cross-section. The outer radius of the capacitor is b and the inner radius is a. Inside or outside the fluid we have the cylindrically radial electric field:

$$E_\rho \approx -\frac{V}{\ln(b/a)}\frac{1}{\rho}, \tag{5.222}$$

(The $-$ sign indicates that the field points inward if $V > 0$.) Then for the free charge, Q_{plates}, supplied to the inner and outer cylindrical plates,

$$4\pi Q_{\text{plates}} = \int da\,\vec{D}\cdot\hat{n} \tag{5.223}$$

$$\Longrightarrow \Delta Q_{\text{plates}} = \frac{V\Delta z}{2\ln(b/a)}(\epsilon - 1), \tag{5.224}$$

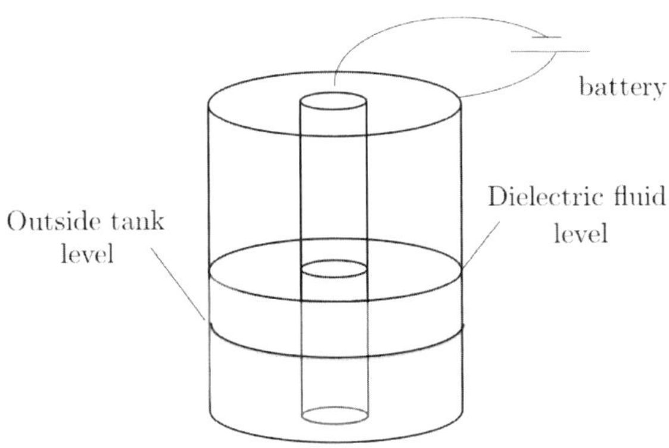

Fig. 5.19 Cylindrical capacitor.

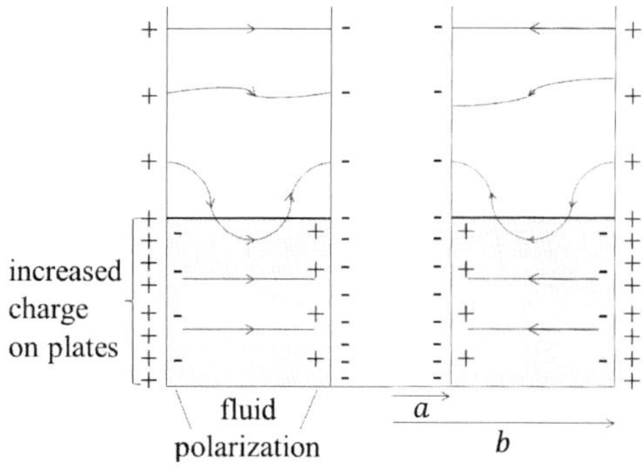

Fig. 5.20 Cross-section of a cylindrical tank.

where Δz is the overall fluid level change in the capacitor. The battery supplies an amount of energy given by

$$\Delta W^{\text{battery}} = V\Delta Q^{\text{battery}} = -V\Delta Q^{\text{plates}} = -\frac{\Delta z V^2}{2\ln(b/a)}(\epsilon - 1). \quad (5.225)$$

By energy conservation the change in energies must equal zero,

$$\Delta W^{\text{tot}} = \Delta W^{\text{gravity}} + \Delta W^{\text{field}} + \Delta W^{\text{battery}} = 0, \quad (5.226)$$

where

$$\Delta W^{\text{gravity}} = \frac{1}{2}\rho g(\Delta z)^2\pi(b^2 - a^2), \quad (5.227)$$

$$\Delta W^{\text{field}} = -\frac{1}{2}\Delta W^{\text{battery}}. \quad (5.228)$$

Here g is the gravitational acceleration and ρ is the density of the fluid. The factor of $1/2$ in Equation (5.227) refers to the center of mass of the disk of raised water. This equation neglects a negative gravitational contribution from the lowering of the outside tank level for a large tank. Equation (5.228) follows from the Section 5.9 discussion regarding the relationship between field and battery energy changes at fixed voltage. It follows that

$$0 = \frac{1}{2}\rho g(\Delta z)^2\pi(b^2 - a^2) - \frac{\Delta z V^2}{4\ln(b/a)}(\epsilon - 1)$$

$$\Longrightarrow (\epsilon - 1) \approx \frac{2\pi}{V^2}(b^2 - a^2)\rho g\Delta z\ln(b/a). \quad (5.229)$$

This allows evaluation of the dielectric constant for the fluid from a measurement of the height change in the fluid. This problem illustrates the relative simplicity of the energy method to find the overall force compared to the much harder task of finding the forces on the liquid.

5.12 Going Deeper

5.12.1 *Dielectric materials*

(1) O. Gallot-Lavallee, *Dielectric Materials and Electrostatics*, Wiley-ISTE (Hoboken, NJ) 2013.

(2) C. Kittel, *Introduction to Solid State Physics*, 8th ed., John Wiley & Sons (Hoboken, NJ) 2005; Chapter 16.

(3) L. D. Landau, L. P. Pitaevskii and E. M. Lifshitz, *Electrodynamics of Continuous Media*, 2nd ed. with corrections, Elsevier (New York) 1993; Chapter II.

(4) J. Martinez-Vega, Ed., *Dielectric Materials for Electrical Engineering*, John Wiley & Sons (Hoboken, NJ) 2010.

(5) B. K. P. Scaife, *Principles of Dielectrics*, revised ed., Oxford University Press (Oxford) 1998.

(6) A. R. Von Hippel, Ed., *Dielectric Materials and Applications*, Cambridge: Technology Press of MIT, reprinted by Artech House (Boston) 1995.

5.13 Exercises

Exercise 5.1.1. Either prove or give a counterexample to the following "Theorem": It is always possible to find an origin such that the dipole moment, $\vec{p}$, vanishes for a charge distribution whose total charge, q, is nonvanishing.

Exercise 5.1.2. A charge distribution has multipole moments q, $\vec{p}$, Q_{ij} with respect to one set of coordinate axes, and moments $q', \vec{p}\,'$ and Q'_{ij} with respect to another set whose origin is located at the point $\vec{R} = (X, Y, Z)$ relative to the first. (The axes are parallel.) Determine explicitly the connections between the monopole, dipole and quadrupole moments in the two coordinate frames.

Exercise 5.1.3. The center of a cubical volume with sides L is placed at the coordinate origin. Its sides are aligned perpendicularly with the x, y, z axes. Within the volume is a charge density $\rho(x, y, z) = Kx$, where K is a constant.

(a) Calculate the monopole, dipole and all the quadrupole moments of this charge distribution.
(b) What is the leading form of the electric field far away from the cube?

Exercise 5.1.4. Work out the form of the integral for the next pole element, R_{ijk}, in the expansion (see Equation (5.12)),

$$\Phi(\vec{x}) = \frac{q}{r} + \frac{\vec{x}\cdot\vec{p}}{r^3} + \frac{1}{2}\sum_{i,j} Q_{ij}\frac{x_i x_j}{r^5} + \frac{1}{6}\sum_{i,j,k} R_{ijk}\frac{x_i x_j x_k}{r^7} + \dots.$$

Show that R_{ijk} has only 7 independent elements.

Exercise 5.1.5. Consider the point charge distribution shown in Figure 5.21. The three charges are on the 3-axis and the distance of the two $+1$ charges from the origin is a. The -2 charge is located at the origin.

(a) Using Cartesian multipole expansion involving dipole ($\vec{p}$), and quadrupole terms (Q_{ij}), find the leading form of the potential far away from the charges, $r \gg a$.

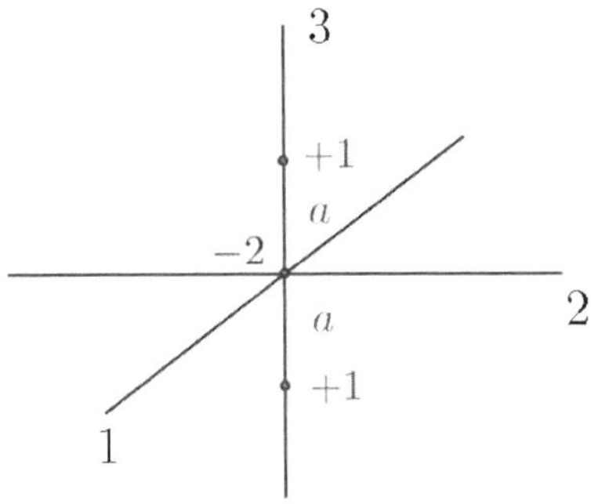

Fig. 5.21 Charge distribution for Exercise 5.1.5.

(b) Confirm the (a) result by a Taylor series expansion of the exact potential.

Exercise 5.1.6.

(a) A small neutral atom with an electric dipole moment $\vec{p} = (p_x, p_y, p_z)$ is located a distance d from a flat conducting surface. Take the conducting plane to be located on the $z = 0$ plane and the dipole to be on the positive z-axis. Find the electric potential $\Phi(\vec{x})$ of this system for $z > 0$. [*Hint*: Use the image method applied to an electric dipole.]

(b) For this part assume the atom's dipole points only along z: $\vec{p} = p_z \hat{k}$. Find the surface charge density $\sigma(\vec{x})$ induced on the conducting plane surface.

Exercise 5.2.1.

(a) For a cylindrically symmetric quadrupole ($Q_{11} = Q_{22} = -\frac{1}{2}Q_{33}$, all other $Q_{ij} = 0$) in an external potential, $\Phi(\vec{x})$, show that the energy of interaction between the field and the quadrupole is

$$W = \frac{1}{4} Q_{33} \frac{\partial^2 \Phi}{\partial z^2},$$

and the resulting force on the quadrupole is

$$\vec{F} = \frac{1}{4} Q_{33} \frac{\partial^2 \vec{E}}{\partial z^2}.$$

(b) The cylindrically symmetric point quadrupole is located at the origin and a point electric dipole $\vec{p} = (p_x, p_y, p_z)$ is located at position $\vec{x}_1 = (x_1, 0, 0)$, as shown in Figure 5.22.

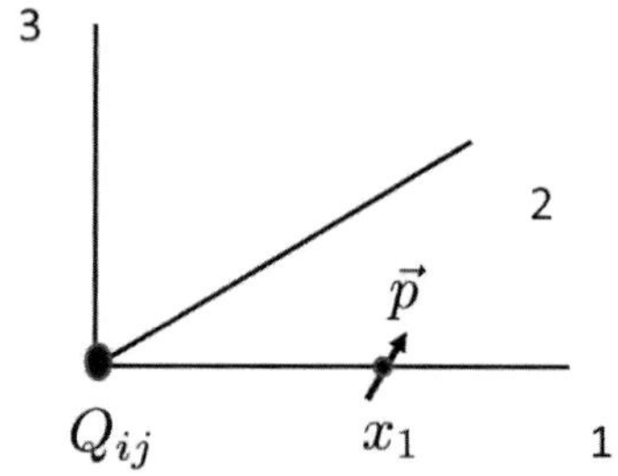

Fig. 5.22 Charge distribution for Exercise 5.2.1(b).

Using part (a), find the force on the quadrupole in the 1-direction due to the electric dipole.

Exercise 5.2.2. Two cylindrically symmetric quadrupoles $Q_{ij}^{(1)}$ and $Q_{ij}^{(2)}$ ($Q_{11}^{(i)} = Q_{22}^{(i)} = -\frac{1}{2}Q_{33}^{(i)}$, $i = \{1, 2\}$, all other components zero) are located a distance $r = |x_1|$ apart from one another with $Q_{ij}^{(1)}$ at the origin, as shown in Figure 5.23. Show that the energy of interaction of these two quadrupoles is

$$W_{QQ} = -\frac{9}{16}\frac{Q_{33}^{(1)}Q_{33}^{(2)}}{r^5}.$$

The result of Exercise 5.2.1 could be of help.

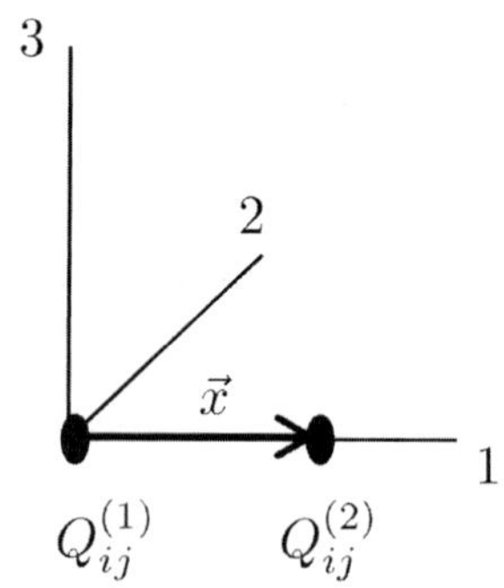

Fig. 5.23 Quadrupole orientation for Exercise 5.2.2.

Extra: Consider the two cylindrically symmetric quadrupoles again, arranged now so that their symmetry axes and separation are all along the 3 (z) axis. Find the energy of interaction of these two quadrupoles, W_{QQ}^{33}, in this configuration.

Exercise 5.2.3. A charge q is located a distance D from the center of a grounded conducting sphere of radius R ($D > R$). Using the image charge solution, the energy of interaction can be written as the product of the charge and image charge, divided by the distance between them:

$$W = q\left(-q\frac{R}{D}\right)\frac{1}{\left(D - \frac{R^2}{D}\right)} = -q^2\frac{R}{D^2}\frac{1}{\left(1 - \frac{R^2}{D^2}\right)}.$$

Given that this represents a separable configuration and using the sphere's center as the origin of coordinates, sum the infinite terms in the spherical multipole interaction energy from Equation (5.32) and recover this result.

Exercise 5.3.1.

(a) Show that an alternate form for the force we found in (5.51) is

$$\vec{F} = q\vec{E}^{(0)}(0) + (\vec{p}\cdot\vec{\nabla})\vec{E}^{(0)}(\vec{x})\Big|_{\vec{x}=0} + \frac{1}{6}\sum_{i,j} Q_{ij}\left.\frac{\partial^2 \vec{E}^{(0)}(\vec{x})}{\partial x_i \partial x_j}\right|_{\vec{x}=0} + \ldots.$$

Spell out the assumptions in your derivation.

(b) Show the i'th component of the torque N_i on an arbitrary charge distribution in an external field $\vec{E}^{(0)}(\vec{x})$ can be written as (first two terms)

$$N_i = (\vec{p}\times\vec{E}^{(0)}(0))_i + \frac{1}{3}\sum_{j,k,m} \epsilon_{ijk} Q_{jm}\left.\frac{\partial E_k^{(0)}(\vec{x})}{\partial x_m}\right|_{\vec{x}=0},$$

or

$$N_i = (\vec{p}\times\vec{E}^{(0)}(0))_i + \frac{1}{3}\sum_{j,k,m} \epsilon_{ijk} Q_{jm}\left.\frac{\partial E_m^{(0)}(\vec{x})}{\partial x_k}\right|_{\vec{x}=0},$$

where $\vec{p}$ is the dipole moment and Q_{jm} are the quadrupole moments. (Showing one form is sufficient.)

Exercise 5.4.1. Use the third term on the right in the force expression in Exercise 5.3.1(a) to argue that atoms with a nonzero quadrupole moment density, $q_{ij}(\vec{x}) \equiv n(\vec{x})Q_{ij}(\vec{x})$, contribute a bulk effective charge density,

$$\rho_{\text{eff}}^{Q}(\vec{x}) = \frac{1}{6}\sum_{i,j}\frac{\partial^2 q_{ij}(\vec{x})}{\partial x_i \partial x_j},$$

as well as effective surface charge density σ_{eff}^{Q} and surface electric polarization densities $\vec{\mathcal{P}}_{\text{eff}}^{Q}$:

$$\sigma_{\text{eff}}^{Q}(\vec{x}) = -\frac{1}{6}\sum_{i,j}\hat{n}_j\frac{\partial q_{ij}(\vec{x})}{\partial x_i},$$

$$(\mathcal{P}_{\text{eff}}^{Q})_j(\vec{x}) = \frac{1}{6}\sum_{i}\hat{n}_i q_{ij}(\vec{x}),$$

where $\hat{n}$ is the unit outward normal. For $\vec{\mathcal{P}}_{\text{eff}}^{Q}$ compare to Equation (5.55) restricted to a surface integral. [*Note*: This term may also be written as an effective volume contribution to the polarization, $\vec{P}_{\text{eff}}^{Q}$, plus $\vec{\mathcal{P}}_{\text{eff}}^{Q}$.]

Exercise 5.4.2.

(a) Show that in the volume of a linear, isotropic material with dielectric constant ϵ, the free charge and bound charge densities are related by

$$\rho_{\text{bound}} = \left(\frac{1-\epsilon}{\epsilon}\right)\rho_{\text{free}}.$$

(b) Show that (a) implies that the total bound surface charge for arbitrary geometry is given as

$$\int da\, \sigma_{\text{bound}} = \left(\frac{\epsilon - 1}{\epsilon}\right) Q_{\text{free}},$$

where Q_{free} is the total free charge in the volume.

Exercise 5.4.3. An infinitely long cylinder of dielectric material is placed in an initially uniform electric field of magnitude E_0 pointing in the $+y$ direction as shown in Figure 5.24. Treating this as a boundary value problem, show that the potentials inside and outside the cylinder are given by

$$\Phi_{\text{in}} = -\frac{2E_0}{\epsilon+1}\rho\sin\phi,$$

$$\Phi_{\text{out}} = -E_0\rho\sin\phi + E_0\frac{\epsilon-1}{\epsilon+1}a^2\rho^{-1}\sin\phi.$$

[*Hint*: See the form of the solutions allowed in Section 3.7.]

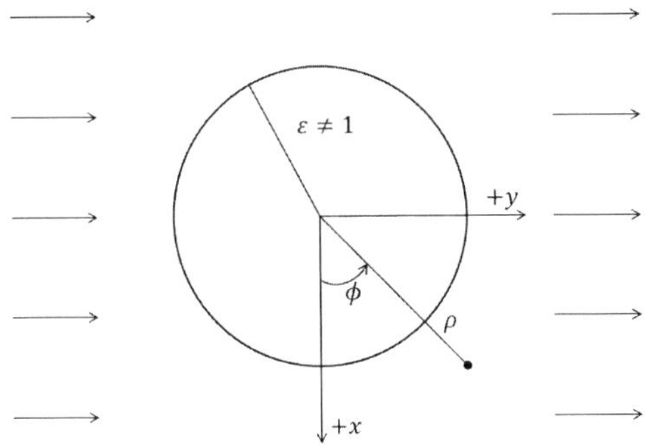

Fig. 5.24 Infinitely long cylinder in a uniform magnetic field.

Exercise 5.4.4.

(a) Argue that the potential $d\Phi$ due to a small dipole $d\vec{p}$ located at $\vec{x}\,'$,

$$d\Phi = \frac{d\vec{p}\cdot(\vec{x}-\vec{x}\,')}{|\vec{x}-\vec{x}\,'|^3},$$

gives rise to the potential

$$\Phi(\vec{x}) = \oint ds'\,\frac{\vec{P}\cdot\hat{n}'}{|\vec{x}-\vec{x}\,'|} - \int d^3x'\,\frac{\vec{\nabla}'\cdot\vec{P}(\vec{x}\,')}{|\vec{x}-\vec{x}\,'|},$$

when a space-dependent polarization $\vec{P}(\vec{x}\,')$ is introduced. Note that the surface and volume charge densities found, $\vec{P}\cdot\hat{n}$ and $-\vec{\nabla}\,'\cdot\vec{P}$ respectively, are consistent with (5.58) and (5.59).

(b) Apply part (a) to a spherical bubble of vacuum with radius a enclosed in a semi-infinite dielectric slab, as shown in Figure 5.25. Assume the slab has a uniform polarization, $\vec{P} = P_0\hat{z}$, outside the sphere. Show that the electric field induced by $\vec{P}$ everywhere inside the sphere is given by

$$\vec{E}_{\mathrm{P}} = \frac{4\pi}{3}P_0\hat{z}.$$

Exercise 5.4.5. A cubical-shaped empty hole (vacuum) exists within a piece of material which has a uniform electric polarization, $\vec{P} = P_0\hat{z}$, as shown in Figure 5.26.

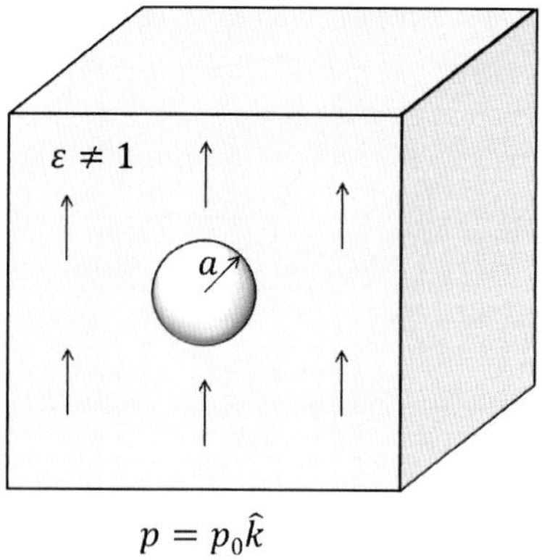

Fig. 5.25 Vacuum bubble in a semi-infinite slab.

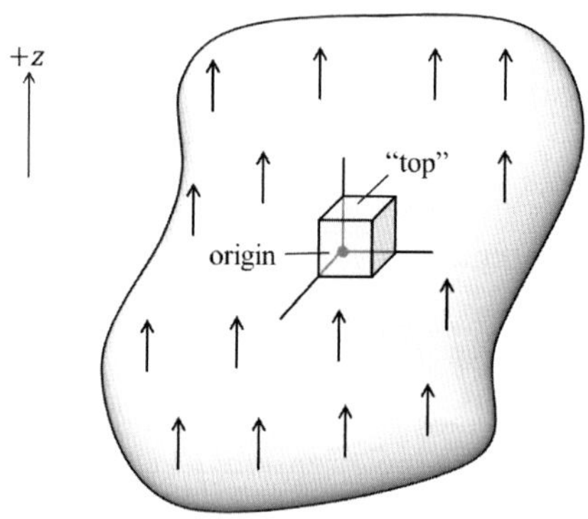

Fig. 5.26 Cubical vacuum within a uniformly polarized material.

(a) Taking the origin of coordinates at the center of the cubic hole, show that the electric field induced by $\vec{P}$ at the origin can be expressed as

$$\vec{E}_{\mathrm{P}}(0) = 2P_0 \int_{\text{“top”}} \frac{\vec{x}\,'da'}{r'^3},$$

where the integration is over the "top" of the cube. The Exercise 5.4.4(a) result is useful here.

(b) Using symmetry and the concept of solid angle, argue that this expression reduces to

$$\vec{E}_{\mathrm{P}}(0) = \frac{4\pi}{3} P_0 \hat{z}.$$

(c) Consider a model of induced polarization $\vec{P}$ in a uniform non-polar substance given by

$$\vec{P} = N\alpha_{\text{mol}}\vec{E}_{\text{T}},$$

where N is the atomic density, α_{mol} is the polarizability of an individual molecule or atom and $\vec{E}_{\text{T}}$ the total electric field $\vec{E}_{\text{T}} \equiv \vec{E} + \vec{E}_{\text{P}}$, where $\vec{E}$ is the applied field and $\vec{E}_{\text{P}}$ is the induced field due to the polarization of other atoms as modeled in Exercises 5.4.4(b) and 5.4.5(b). Using Equation (5.66), show that this implies

$$\alpha_{\text{mol}} = \frac{1}{N}\frac{\chi}{1 + \frac{4\pi}{3}\chi},$$

for the molecular polarizability. This is called the *Clausius-Mossotti* relation. Note that χ may also be related to the dielectric constant ϵ through (5.67).

Exercise 5.6.1. Two infinite perpendicular dielectric planes ($x - z$ plane and $y - z$ plane) meet at origin O. (The diagram shows a cut through the $x-y$ plane.) The dielectric variable $\epsilon \neq 1$ is a constant in the combined region $x < 0, y < 0$ for all z. Find the Green function in Cartesian coordinates for a unit point charge in the *vacuum* region $x' > 0, y' > 0$ located *outside* the dielectric.

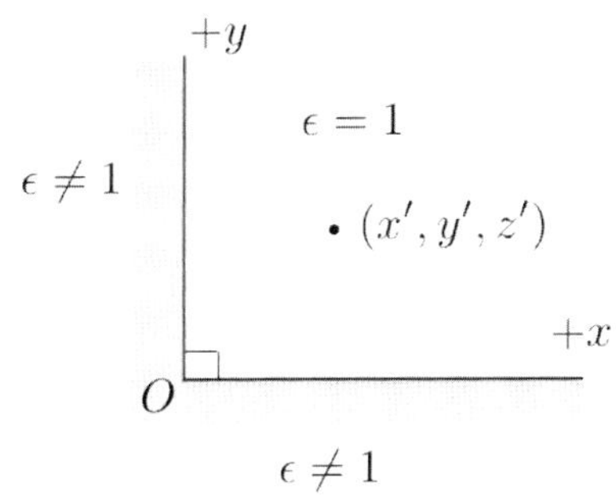

Fig. 5.27 Geometry for Exercise 5.6.1.

Exercise 5.6.2. A semi-infinite planar dielectric slab, with dielectric constant ϵ, is placed parallel to and a distance d above the surface of a perfectly

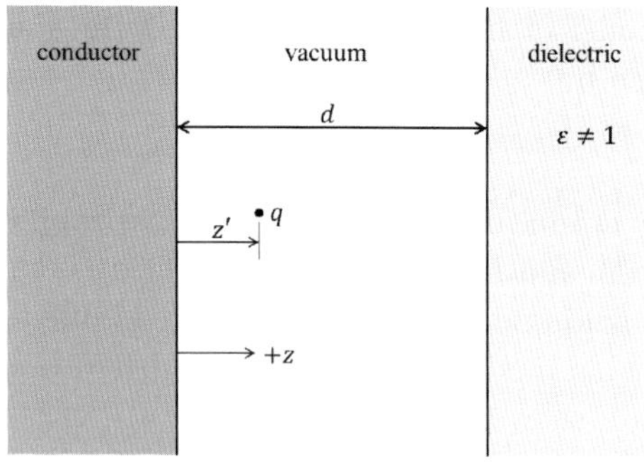

Fig. 5.28 Geometry for Exercise 5.6.2.

conducting plane (see Figure 5.28). Take the surface of the conducting plane to be $z = 0$ and the surface of the dielectric slab to be at $z = d > 0$. The Green function $G_D(\vec{x}, \vec{x}\,')$ for a positive unit charge in the region between the plate and slab will satisfy the differential equation,

$$\nabla^2 G_D(\vec{x}, \vec{x}\,') = -4\pi\delta(\vec{x} - \vec{x}\,'),$$

where $\delta(\vec{x} - \vec{x}\,')$ is a Dirac delta function. Given the Bessel function expansions (the m sums may be done using Exercise 4.2.1),

$$\delta(\vec{x} - \vec{x}\,') = \sum_{m=-\infty}^{\infty} \frac{e^{im(\phi-\phi')}}{2\pi} \int_0^\infty dk\, k J_m(k\rho) J_m(k\rho')\delta(z - z'),$$

$$G_D(\vec{x}, \vec{x}\,') = 4\pi \sum_{m=-\infty}^{\infty} \frac{e^{im(\phi-\phi')}}{2\pi} \int_0^\infty dk\, k J_m(k\rho) J_m(k\rho') g(z, z'),$$

show that, when boundary conditions at $z = 0$, d are supplied, this gives

$$g(z, z') = \begin{cases} f(z_>) \sinh(kz_<), & z < d \\ K(z') e^{k(2d-z)}, & z > d \end{cases}$$

where $z_<(_>)$ is the lesser (greater) of z and z' and

$$f(z_>) = \frac{e^{kz_>}}{k} \left(\frac{1 + \left(\frac{1+\epsilon}{1-\epsilon}\right) e^{2k(d-z_>)}}{1 + \left(\frac{1+\epsilon}{1-\epsilon}\right) e^{2kd}} \right),$$

$$K(z') = \left(\frac{2}{1-\epsilon}\right) \frac{\sinh(kz')}{k\left(1 + \left(\frac{1+\epsilon}{1-\epsilon}\right) e^{2kd}\right)}.$$

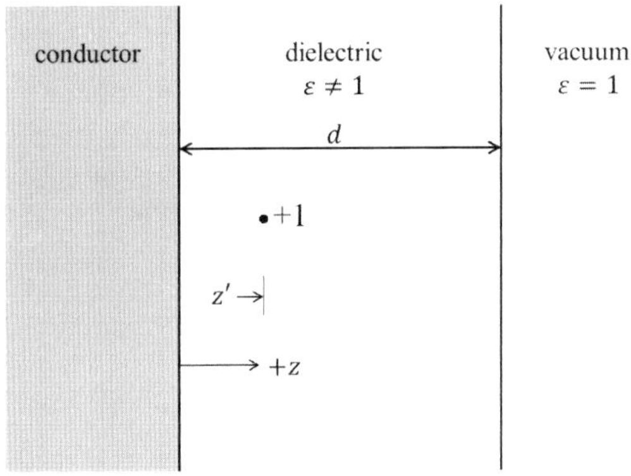

Fig. 5.29 Geometry for Exercise 5.6.3.

Exercise 5.6.3. Exercise 5.6.2 involved a unit point charge between parallel conductor and dielectric surfaces, as shown in Figure 5.28. Find a simple substitution of parameters which changes the above Dirichlet Green function to the Dirichlet Green function for the situation shown in Figure 5.29 where the vacuum and dielectric areas are interchanged.

Exercise 5.6.4. Consider a one-dimensional enclosed geometry as in Section 2.9. Conducting walls exist at $x = 0$ and $x = L$, and a dielectric slab with dielectric constant ϵ, is attached to one side of the metallic plates, extending from $x = 0$ to $x = d$ $(d < L)$. Find the one-dimensional Dirichlet Green function, $G_D(x', x)$, solving

$$-\frac{d^2 G_D(x', x)}{dx^2} = \delta(x - x'),$$

for a charge in the vacuum region, $d < x' < L$. (As a check, the form you find must reduce to the answer in Section 2.9 when $\epsilon = 1$ or $d = 0$.)

Exercise 5.6.5. A plane interface exists between two regions of unequal dielectric constant, ϵ_1 in the region $z < 0$ and ϵ_2 in the region $z > 0$, as shown in Figure 5.30. A point free charge q is located at $z' > 0$. Using the

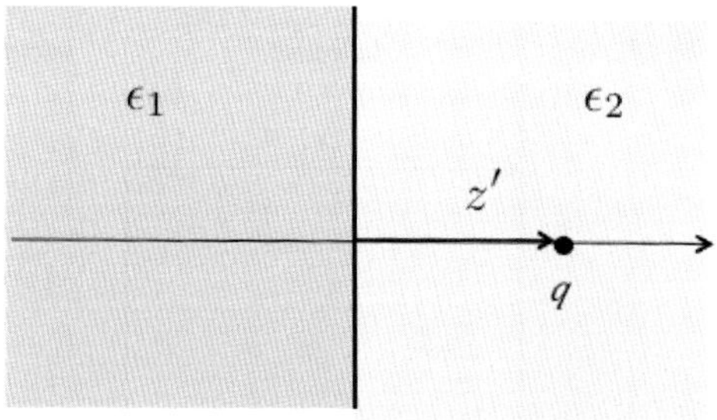

Fig. 5.30 Geometry for Exercise 5.6.5.

image method find the electric potential everywhere. [*Ans*:

$$\Phi = \frac{1}{\epsilon_2}\left(\frac{q}{R_1} + \frac{q'}{R_2}\right), \qquad z > 0,$$
$$\Phi = \frac{1}{\epsilon_1}\left(\frac{q''}{R_1}\right), \qquad z < 0,$$
$$q' = q\left(\frac{\epsilon_2 - \epsilon_1}{\epsilon_2 + \epsilon_1}\right), \quad q'' = 2q\left(\frac{\epsilon_1}{\epsilon_2 + \epsilon_1}\right),$$

where $R_1 \equiv |\vec{x} - \vec{x}'|$, $R_2 = |\vec{x} - \vec{x}''|$, for $\vec{x}'' = x'\hat{i} + y'\hat{j} - z'\hat{k}$.]

Exercise 5.6.6. Consider a uniform dielectric slab of thickness d that has infinite extent in the x, y directions (see Figure 5.31).

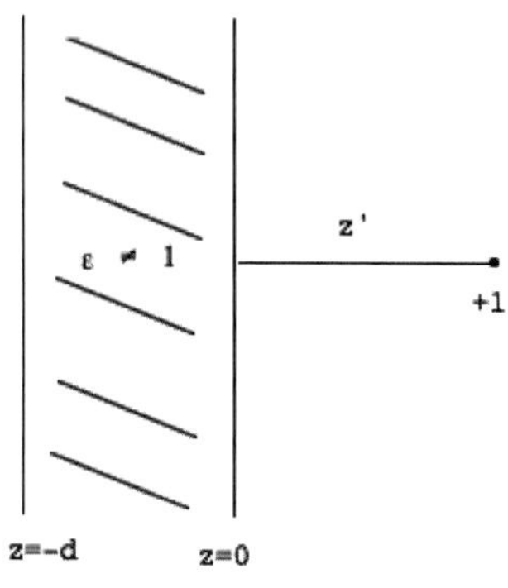

Fig. 5.31 Geometry for Exercise 5.6.6.

Assuming a Green function solution of the form of Equation (5.82) with source position $z' > 0$, give the *functional forms* of the reduced Green function, $g(z)$, in the various regions and the conditions on the coefficients. There is no need to solve for the coefficients, but argue that you have enough information to completely solve the problem.

Exercise 5.7.1. Using a reduced Green function technique and a line-charge source as for example in Exercise 4.8.3, find the Green function for a long cylinder of dielectric material with dielectric constant ϵ and radius a. Consider the case of the charge *outside* the cylinder as in Figure 5.32. [*Hint*: The answer is:

$$G(\vec{x}, \vec{x}') = 4 \sum_{m=0,1,2,\ldots}^{\infty} \cos[m(\phi - \phi')]\, g_m(\rho, \rho'),$$

$$g_m(\rho, \rho') = \begin{cases} \dfrac{1}{m(1+\epsilon)} \left(\dfrac{\rho}{\rho'}\right)^m, & \rho < a \\ \dfrac{1}{2m} \left(\dfrac{\rho_<}{\rho_>}\right)^m \left[1 + \left(\dfrac{1-\epsilon}{1+\epsilon}\right)\left(\dfrac{a}{\rho_<}\right)^{2m}\right], & \rho > a \end{cases}$$

for $m \neq 0$, and where we may write

$$g_0(\rho, \rho') = -\frac{1}{2} \ln\left(\frac{\rho_>}{K}\right).$$

A symbolic solver can help here. The positive constant K can be considered the length of the line charge; see the Section 3.1 discussion.]

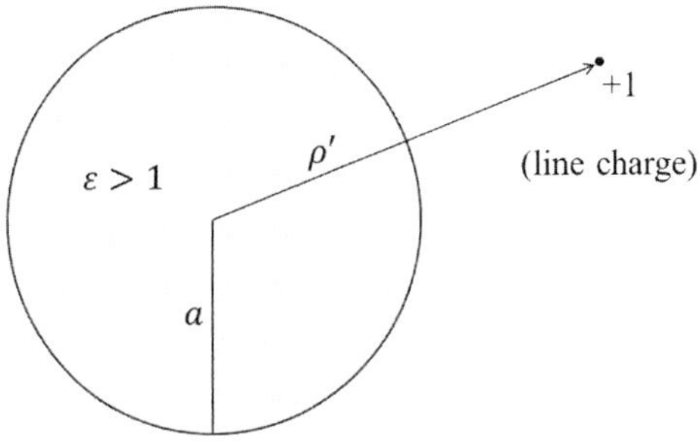

Fig. 5.32 Geometry for Exercise 5.9.10.

Exercise 5.7.2. Consider an infinitely long dielectric cylinder with $\epsilon > 1$ of radius a, with the z-axis of coordinates along the axis of symmetry of

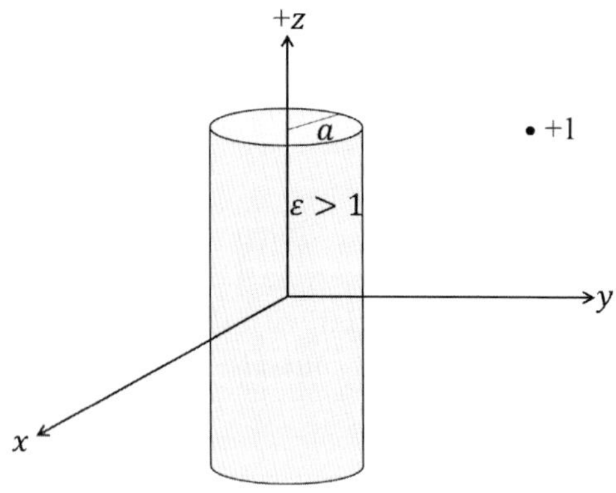

Fig. 5.33 Infinitely long cylinder for Exercise 5.7.2.

the cylinder (see Figure 5.33). We will calculate the Green function as in Exercise 5.7.1 but this time for a *point* charge outside of the cylinder. Assume the reduced form for the Green function (same as for Exercise 4.7.4),

$$G(\vec{x},\vec{x}\,') = \frac{2}{\pi}\sum_{m=-\infty}^{\infty} e^{im(\phi-\phi')}\int_0^\infty dk\cos[k(z-z')]g_m(\rho,\rho'),$$

and with

$$4\pi\delta(\vec{x}-\vec{x}\,') = \frac{2}{\pi}\sum_{m=-\infty}^{\infty} e^{im(\phi-\phi')}\int_0^\infty dk\cos[k(z-z')]\frac{1}{\rho}\delta(\rho-\rho').$$

Motivate the general form of solutions in the three regions, $\rho < a$, $a < \rho < \rho'$ and $\rho' < \rho$ in terms of linearly independent Bessel functions of imaginary argument. Apply the continuity and boundary conditions to obtain the equations which determine the coefficients. (It is not necessary to actually solve for these coefficients! Just show you have appropriate equations to uniquely determine them.)

Exercise 5.7.3. From the Green function solution for a positive unit charge in the presence of a spherical dielectric of radius a, generate the potential, both inside and outside, for a spherical dielectric placed in an initially uniform electric field; see Figure 5.34. Do this by the image technique used in Section 3.3. [*Ans.*:

$$\Phi_{\text{out}} = -E_0 z + \frac{\vec{p}\cdot\vec{x}}{r^3}; \qquad \vec{p} \equiv \frac{\epsilon-1}{\epsilon+2}a^3 E_0\hat{z},$$

$$\Phi_{\text{in}} = -E_0 z\left(\frac{3}{\epsilon+2}\right).]$$

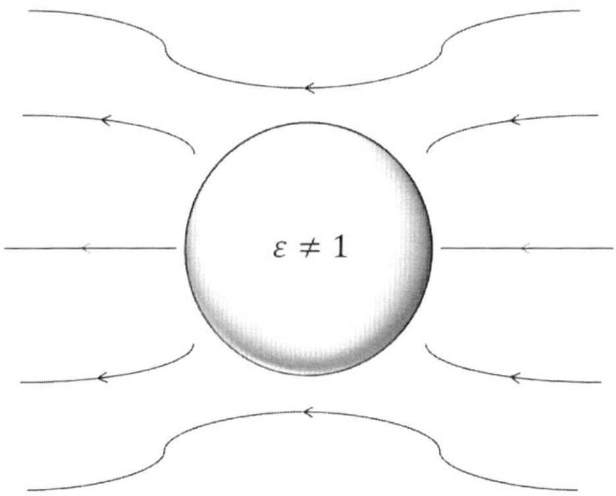

Fig. 5.34 Geometry for Exercise 5.7.3.

Exercise 5.7.4. Determine the Green function for a dielectric sphere of radius a when the source position is *inside* the sphere: $r' < a$. In the limit $r \gg a$, identify an *effective dipole moment* of the system and evaluate the total polarization charge on the surface of the sphere. Compare your answer with Exercise 5.4.2 above to verify your answer. [*Hint*: Part of the solution can be obtained directly by using the symmetry $G(\vec{x}\,', \vec{x}) = G(\vec{x}, \vec{x}\,')$ and the $r' > a$ solution given in Section 5.7.]

Exercise 5.7.5. Find the Green function for a spherical bubble of vacuum of radius a, embedded in a dielectric medium with a dielectric constant, ϵ. Consider the case of the free unit charge outside the bubble. [*Hint*: You do not have to start from scratch here. Compare the boundary conditions with that of a dielectric sphere in vacuum as given in Section 5.7.]

Exercise 5.7.6. An electric dipole of moment $\vec{p}_0$ (pointing in an arbitrary direction) is located at a distance $\vec{x}\,'$ away from the center of a dielectric sphere of radius a (see Figure 5.35). Find the leading (nontrivial) form of the electric potential, $\Phi(\vec{x})$:

(a) far away from the sphere and dipole ($r \gg r', a$)
(b) inside the sphere ($r < a, r' \gg a$).

Exercise 5.7.7. Consider a solid dielectric sphere of radius a located inside a hollow conducting sphere of inner radius b. The centers of the

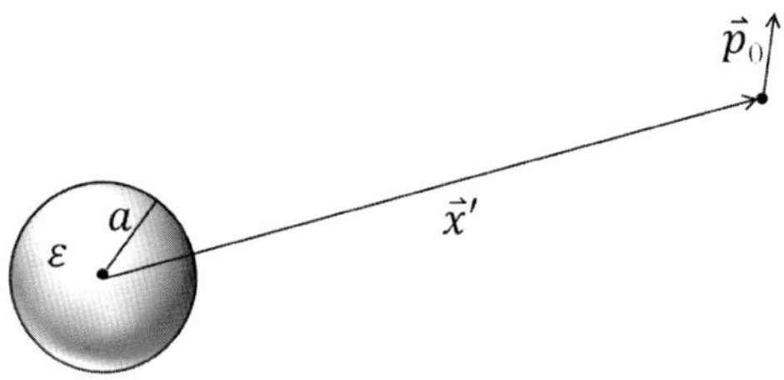

Fig. 5.35 Geometry for Exercise 5.7.6.

two spheres coincide. Assuming a Green function solution of the form of Equation (5.116) of the text with the point charge located at $a < r' < b$, give the functional forms of the function $g_\ell(r)$ for $r < b$ and the boundary conditions which determine the unknown coefficients. (You don't need to solve for the coefficients.)

Exercise 5.7.8. Use the Green function solution,

$$\Phi(\vec{x}) = \int d^3x'\, G(\vec{x}, \vec{x}\,')\rho(\vec{x}\,'),$$

to generate the inside potential ($r < a$) for a spherical dielectric of radius a placed in the field of a charged plane with uniform surface charge density, σ, located at $z = d$ on the z-axis (see Figure 5.36). Compare your answer for the electric field with that implied by Exercise 5.7.3.

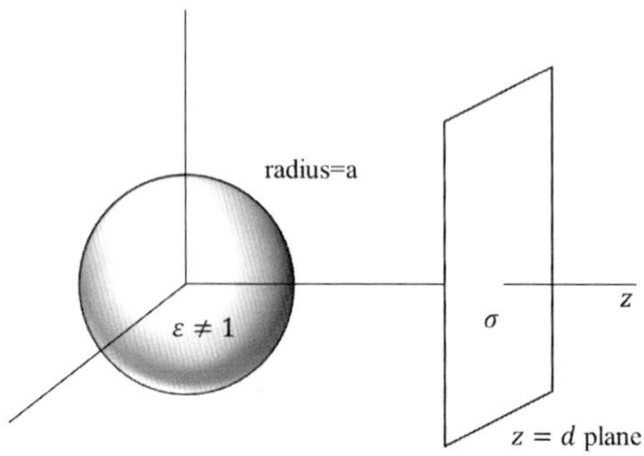

Fig. 5.36 Geometry for Exercise 5.7.8.

Exercise 5.7.9. A dielectric sphere with permittivity constant ε and radius a is surrounded by a circular line charge of radius b and total charge Q. The line charge is uniformly distributed, concentric with the sphere and located in the sphere's equator plane. Using the appropriate Green function, show that the electric potential *within* the sphere can be expressed as

$$\Phi(r,\theta) = Q \sum_{n=0,1,2,\ldots}^{\infty} \frac{(-1)^n(4n+1)}{2n(1+\varepsilon)+1} \frac{(2n-1)!!}{(2n)!!} \frac{r^{2n}}{b^{2n+1}} P_{2n}(\cos\theta).$$

The $P_{2n}(\cos\theta)$ is a Legendre polynomial. A result from Exercise 4.9.1 is useful.

Exercise 5.7.10. A uniform dielectric with permittivity ε is placed between the two halves of the thin spherical conductors of radius a from Exercise 4.13.4. Find the modified forms of the self-capacitance, $C_{11}(\varepsilon)$, and the inductive capacitance, $C_{12}(\varepsilon)$. Show that

$$C_{11}(\varepsilon) = C_{11} + \frac{a(\varepsilon-1)}{4} \sum_{\ell=1,2,3,\ldots} \ell(2\ell+1)\left(\int_0^1 dx\, P_\ell(x)\right)^2,$$

$$C_{12}(\varepsilon) = C_{12} + \frac{a(\varepsilon-1)}{4} \sum_{\ell=1,2,3,\ldots} (-1)^\ell \ell(2\ell+1)\left(\int_0^1 dx\, P_\ell(x)\right)^2,$$

where C_{11} and C_{12} are the forms from Exercise 4.13.4.

Exercise 5.7.11.

(a) Using a reduced Green function technique, find the Dirichlet Green function $G_D(\rho,\phi;\rho',\phi')$ for the region inside a conducting cylinder of radius a, inside of which is a solid dielectric cylinder of radius b. The axes of the two cylinders coincide. The unit-density line charge is positioned *outside* of the dielectric ($a > \rho' > b$), as shown in Figure 5.37. Find all the conditions (equations) which determine the unknown coefficients in your assumed reduced Green function. There is no need to solve for the coefficients, but argue that you have enough information to completely solve the problem.

(b) Explain how to obtain the $\rho' < b$, $a > \rho > b$ branch of the Dirichlet Green function for this problem directly from the part (a) solution.

(c) Explain how to obtain the Dirichlet Green function of the problem with the dielectric and vacuum regions reversed directly from the part (a) solution.

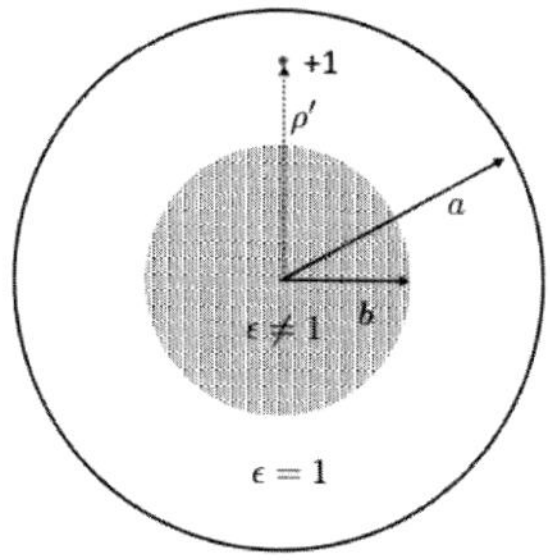

Fig. 5.37 Geometry for Exercise 5.7.11.

Exercise 5.7.12. A dielectric sphere with constant permittivity ϵ and radius a is centered at the origin. A unit point charge is located at a position $\vec{x}\,'$ outside the sphere ($|\vec{x}\,'| = r' > a$). Show that the point charge induces a quadrupole moment of the sphere,

$$Q_{ij} = \frac{2a^3(\epsilon - 1)}{3 + 2\epsilon} - \frac{\partial E'_j(0)}{\partial x'_i},$$

where

$$\vec{E}'(0) = -\frac{\vec{x}\,'}{r'^3},$$

is the electric field of the point charge evaluated at the origin. (Incidentally, this shows that a uniform field does not induce a quadrupole term.)

Exercise 5.8.1.

(a) Show for a material with a permanent polarization $\vec{P}(\vec{x})$ and no free charge, that

$$\int d^3x\, \vec{E} \cdot \vec{D} = 0,$$

where the integration is over all space.

(b) From Equation (5.27) the energy associated with an infinitesimal permanent electric dipole $\delta\vec{p}_i$ at $\vec{x}_i$ in an electric field $\delta\vec{E}_j(\vec{x}_i)$ due to a dipole $\delta\vec{p}_j$, can be written

$$\delta W_{ij}^{\text{int}} = -\delta\vec{p}_i \cdot \delta\vec{E}_j(\vec{x}_i), \qquad i \neq j.$$

From this expression, the connection $\vec{D} = \vec{E} + 4\pi\vec{P}$, and the result of part (a), argue that the interaction energy of a permanently electrically polarized piece or pieces of matter is

$$W^{\text{int}} = -\frac{1}{2}\int d^3x\, \vec{P} \cdot \vec{E} = \frac{1}{8\pi}\int d^3x\, \vec{E}^2,$$

where the integrations are over all space. (*Comment*: Even though we may exclude the $i = j$ self-energy term in $\delta W_{ij}^{\text{int}}$, there are actually self-energy contributions in W^{int} which arise in going from the discrete sum to the continuum integral.)

Exercise 5.9.1.

(a) Show that the normal force per unit area $\vec{\mathcal{F}} \cdot \hat{n}$ ($\hat{n}$ directed outward from the dielectric) on an arbitrary dielectric surface is given by

$$\vec{\mathcal{F}} \cdot \hat{n} = \frac{1}{8\pi}(E_{2n}^2 - E_{1n}^2),$$

where E_{2n}, E_{1n} are surface normal components of $\vec{E}_2$ and $\vec{E}_1$, as shown in Figure 5.38. (The Section 2.11 discussion may be helpful.)

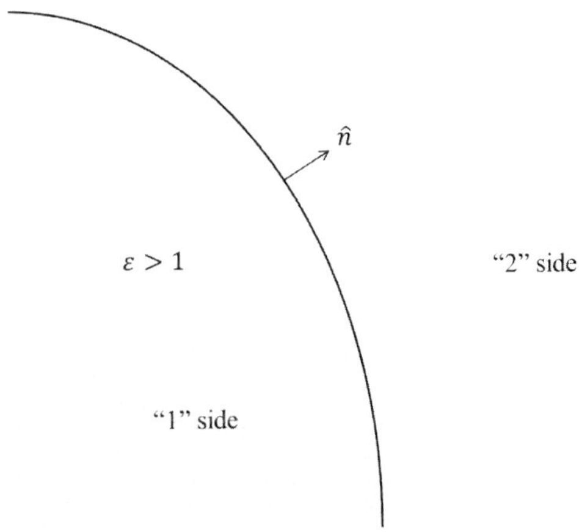

Fig. 5.38 Geometry for Exercise 5.9.1.

(b) When only bound charge, σ_b, is present show that

$$\vec{\mathcal{F}} \cdot \hat{n} = 2\pi\sigma_b^2 \left(\frac{1+\epsilon}{\epsilon - 1}\right).$$

Extra: When both free surface charge, σ_f, and bound surface charge, σ_b, are present show that the normal force per area is

$$\vec{F} \cdot \hat{n} = 2\pi \left[\left(\frac{\sigma_b \varepsilon}{\varepsilon - 1} + \sigma_f\right)^2 - \frac{\sigma_b^2}{(1-\varepsilon)^2}\right].$$

Exercise 5.9.2. Calculate the force on the half-infinite dielectric *plane*, due to the presence of the unit charge at $z' > 0$ (see Figure 5.39). Do it:

(a) By evaluating the force between the unit charge and the image charge.
(b) By using an energy method explained in the text:
$$\Delta W = \frac{1}{2}\int d^3x d^3x' \rho(\vec{x})[G_D(\vec{x},\vec{x}\,') - G_D^0(\vec{x},\vec{x}\,')]\rho(\vec{x}\,').$$
(c) By using the result of Exercise 5.9.1.

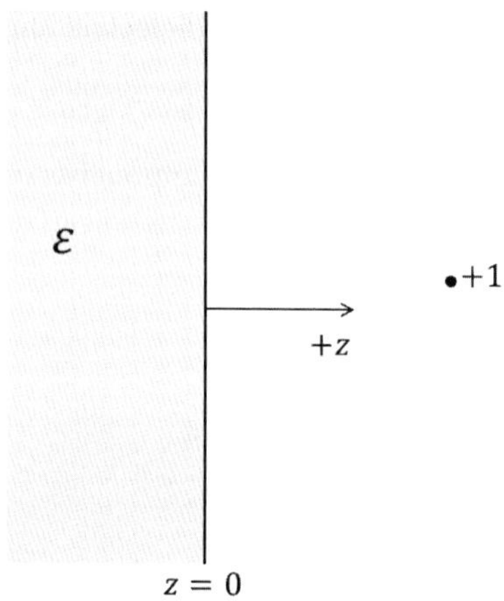

Fig. 5.39 Geometry for Exercise 5.9.2.

Exercise 5.9.3. Find the approximate force (attraction, repulsion?) between a dielectric rod with dielectric constant $\epsilon > 1$ length L and radius a ($L \gg a$), and a positive unit point charge located a distance $z \gg L$ from the rod. The point charge is located perpendicular to the rod's axis on the rod's midpoint plane, as shown in Figure 5.40. [*Hint*: The point charge's electric field at the rod's position will be approximately uniform. The result of Exercise 5.4.3 may be useful.]

Exercise 5.9.4. The middle of a long, thin cylinder of radius a and length L, with dielectric constant ϵ, is located a large distance z from a positive point charge ($z \gg L \gg a$). It is oriented with its lengthwise dimension pointed toward the charge, as shown in Figure 5.41. Find the force of the charge on the dielectric. Are they attracted or repelled for $\epsilon > 1$?

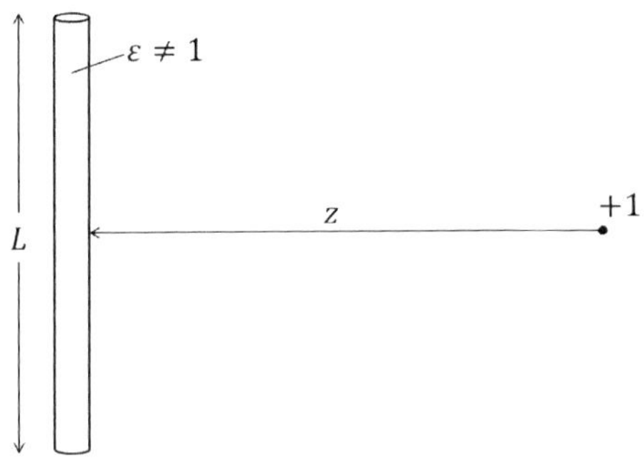

Fig. 5.40 Geometry for Exercise 5.9.3.

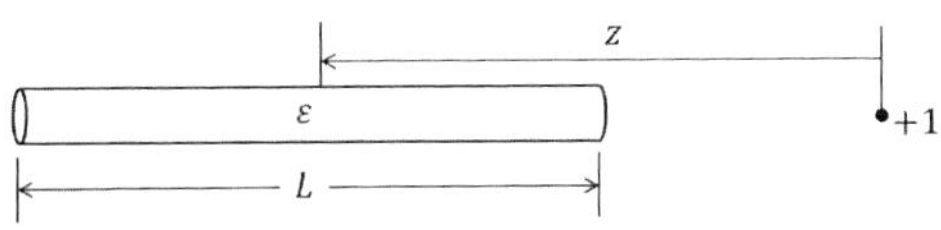

Fig. 5.41 Geometry for Exercise 5.9.4.

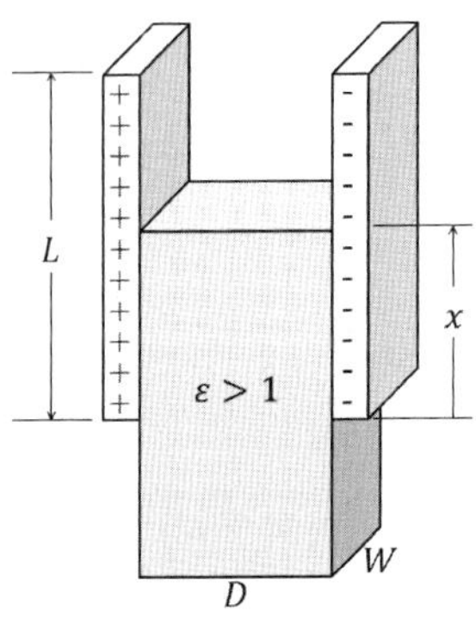

Fig. 5.42 Geometry for Exercise 5.9.5.

Exercise 5.9.5.

(a) Two parallel conducting plates of a capacitor of length L and width W are separated by a distance D. The region between the plates is filled to a distance x with a dielectric material with constant ϵ, as shown in Figure 5.42. If the plates are maintained at a constant potential V by connection to a battery, calculate the force F_x on the dielectric block. Neglect dielectric leading edge effects. Is the block pulled in or pushed out of the capacitor?

(b) The dielectric block has been withdrawn and a fixed charge $\pm Q$ has been placed on the plates. The magnitude of this charge is determined by $V/D = 4\pi Q/(LW)$ so that the potential in (a) is established when the dielectric block is not yet inserted. The block is reinserted a distance x so that it partially fills the space between the plates. Again neglecting edge effects, calculate the force F_x on the block. (*Soapbox comments*: You should find in part (a) that the charge on the conducting plates changes to keep V constant, whereas the voltage changes in part (b) for constant Q. Disconnecting the battery at any point in part (a) will not change the instantaneous force! Re-connecting the battery after the initial insertion in part (b) *will* lead to a change in the force, but only because the instantaneous value of the voltage has changed from it's initially established value!)

Exercise 5.9.6. Referring to Exercise 5.6.2 above, show that the z-direction force due to the unit point charge at z' on the dielectric surface may be written exactly as

$$F_z = 4\left(\frac{1+\epsilon}{1-\epsilon}\right)\int_0^\infty dk\, k\frac{e^{2kd}(\sinh(kz'))^2}{\left(1+\left(\frac{1+\epsilon}{1-\epsilon}\right)e^{2kd}\right)^2}.$$

Exercise 5.9.7.

(a) As a special case of Exercise 5.9.6, show that this force may be evaluated as

$$F_z \approx \frac{1}{4}\left(\frac{1-\epsilon}{1+\epsilon}\right)\left(\frac{1}{(d+z')^2}+\frac{1}{(d-z')^2}-\frac{2}{d^2}\right),$$

for weak dielectrics, $\epsilon - 1 \ll 1$.

(b) Using an image method, confirm this result.

Exercise 5.9.8.

(a) One may use the result of Exercise 5.6.2 in the $\epsilon \to \infty$ limit, producing a second conducting surface in place of the dielectric one. Using this Green function, show that the surface charge density on the formerly dielectric surface at $z = d$ due to the unit point charge at z' is

$$\sigma(\rho) = -\frac{1}{2\pi}\int_0^\infty dkk\left(\frac{\sinh kz'}{\sinh kd}\right)J_0(k\rho).$$

(b) Using part (a), now show that the force on the $z = d$ plate of a parallel plate capacitor dielectric is given by (the minus sign means attraction)

$$F_z = -\int_0^\infty dkk \left(\frac{\sinh kz'}{\sinh kd}\right)^2 .$$

Confirm this result as a special case of Exercise 5.9.6.

Exercise 5.9.9. A capacitor, which consists of a dielectric layer ($\epsilon \neq 1$) of width d between conducting plates, is in the process of being constructed. The dielectric is attached to the left plate, as shown in Figure 5.39. The initial distance between the metallic plates is $L > d$. The instantaneous distance, x, is given by $d < x < L$. A battery initially establishes a constant potential difference, ΔV, between the plates.

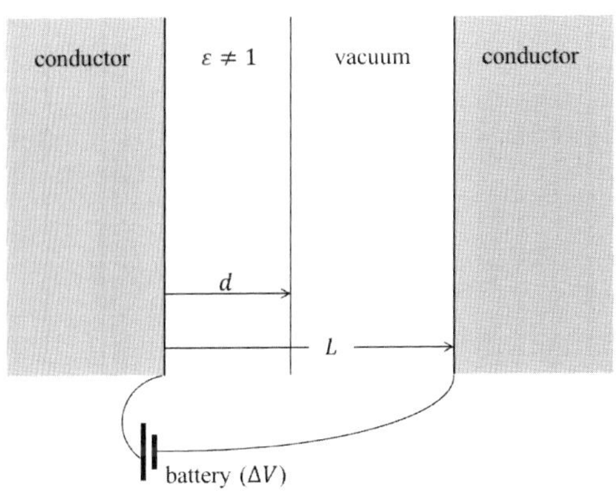

Fig. 5.43 Geometry for Exercise 5.9.9.

(a) Show that the force/area, $\mathcal{F}$, on the right metallic surface when the battery remains connected and the distance between the plates is reduced to x, is

$$\mathcal{F} = -\frac{1}{8\pi}\frac{\Delta V^2}{\left[d/\epsilon + (x-d)\right]^2}.$$

(Note that it is attractive, i.e., $\mathcal{F} < 0$).

(b) The battery is disconnected immediately after establishing the same initial voltage difference, ΔV, as in part (a). Now find the force per unit area, $\mathcal{F}$, on the right metallic plate when the distance between the plates is reduced to x, in terms of $\Delta V, \epsilon, d, L$, and x.

Exercise 5.9.10. In Exercise 5.7.1 we calculated the Green function for a line charge for the region outside of a long cylinder of dielectric material with radius a and dielectric constant $\epsilon > 1$. Define the force per unit length $\mathcal{F}$ as

$$\mathcal{F} \equiv -\frac{\partial \Delta \mathcal{W}}{\partial \rho'}, \qquad \Delta \mathcal{W} \equiv \frac{1}{2}[G_D(\vec{x},\vec{x}\,') - G_D^0(\vec{x},\vec{x}\,')]\Big|_{\vec{x}=\vec{x}\,'}$$

on the unit charge per length line charge. Given the above, find the induced force per unit length between the cylindrical post and the line charge. Sum the resulting series. Is it attractive or repulsive?

Exercise 5.10.1. In Section 5.10 it was shown for the dielectric quark confinement model that with

$$\vec{E} = -\vec{\nabla}\Phi, \vec{D} = \epsilon\vec{E},$$

$$\epsilon = 2\alpha \ln\left(\frac{E^2}{K^2}\right) \ (\alpha > 0),$$

$$W = \frac{1}{4\pi}\int d^3x \left[\frac{1}{2}\vec{E}\cdot\vec{D} + \alpha E^2\right],$$

under a first variation

$$\vec{E} \implies \vec{E} + \delta\vec{E},$$

we obtain

$$\delta W^{(1)} = \frac{1}{4\pi}\int d^3x \vec{E}\cdot\delta\vec{D}.$$

(a) If we require that

$$\vec{\nabla}\cdot\delta\vec{D} = 0,$$

(equivalent to $\delta\rho = 0$ since $\vec{\nabla}\cdot\vec{D} = 4\pi\rho$) argue that $\delta W^{(1)} = 0$. This shows that the variation gives an extremum in W.

(b) Consider a second variation of W. Keeping terms of second order in $\delta\vec{E}$, show that

$$\delta W^{(2)} > 0,$$

implying that the extremum found is a stable minimum.

Exercise 5.11.1. The bubble in Exercise 5.7.5 exists in a dielectric fluid. Find the force between the bubble and a positive unit free charge in the fluid for $r_0 \gg a$. Is it attractive or repulsive for $\epsilon > 1$? [*Hint*: Here is a shortcut. Think about what one would have to do to Equation (5.202) in order to adapt to this new situation.]

Exercise 5.11.2. A piece of linear dielectric material is brought slowly into a region where an initial electric field $\vec{E}_0(\vec{x})$ has already been established. The change in energy of the system is

$$W - W_0 = -\frac{1}{2}\int_V d^3x\, \vec{P}(\vec{x}) \cdot \vec{E}_0(\vec{x}),$$

from Equation (5.173), where $\vec{P}(\vec{x})$ is the polarization vector and the integration is only over the volume, V, of the introduced dielectric. Using the energy expression above and Equation (5.146), once again find the force between a dielectric sphere of radius a and a unit point charge located a distance r' from the center of the sphere when $r' \gg a$.

Exercise 5.11.3. A +1 charge is located at position $\vec{r}_0$ *inside* a spherical bubble of dielectric fluid of radius a with dielectric constant $\varepsilon > 1$ as shown in Figure 5.44.

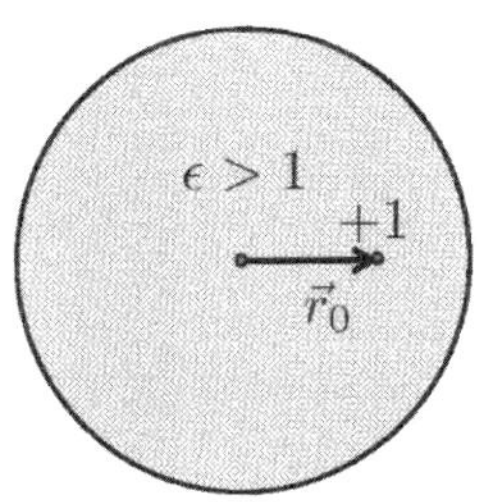

Fig. 5.44 System for Exercise 5.11.3.

Using a Green function approach, show that the radial force on the positive unit charge is

$$F_r = -\frac{\varepsilon - 1}{\varepsilon}\sum_{\ell=1}^{\infty}\frac{\ell(\ell+1)}{[\ell(1+\varepsilon)+1]}\frac{r_0^{2\ell-1}}{a^{2\ell+1}}.$$

Exercise 5.11.4. Find the approximate force between a dielectric sphere of radius a and a distant point dipole $\vec{p}$ located at $z = d > 0$ $(d \gg a)$. The origin of coordinates is taken at the center of the dielectric sphere. The dipole points in an arbitrary direction. Show that the components of the force F_i on the dipole are

$$F_i \approx -3\left(\frac{\varepsilon - 1}{\varepsilon + 2}\right)\frac{a^3}{d^7} \times \begin{cases} -p_3 p_i, & i = 1, 2 \\ 3p_3^2 + \vec{p}^2, & i = 3. \end{cases}$$

Chapter 6

Magnetostatics

Our mathematical development of magnetostatics will proceed analogously to our development of electrostatics in the preceding chapters. We will see how either a scalar or vector potential can be employed to fully describe fixed current situations. In the course of our investigations we will develop analogous methods and concepts, including the magnetic dipole moment, the macroscopic magnetization field, image method, and appropriate boundary condition considerations.

6.1 Analogy to electrostatics

In electrostatics, we started out with point charges; while in magnetostatics, we will start out with (infinitely thin) current-carrying wire elements represented as infinitesimal vectors $d\vec{\ell}$. Note that both point charges and wire elements are idealizations: in particular, the wire elements cannot exist in isolation, but must be part of a circuit.

Table 6.1 compares the fundamental force and field equations for electrostatics (with point charges) and magnetostatics (with wire elements). I_j denotes the magnitude of the current along $d\vec{\ell_j}$. Note that $\vec{F}_{12}$ refers to the force on element 1 at $\vec{x}_1$ by element 2 at $\vec{x}_2$. Also note that just as the definition of the electric field follows from comparison of the first and second lines of the table, the definition of the infinitesimal magnetic field $d\vec{B}(\vec{x})$ follows from comparison of the same lines.

Figure 6.1 shows possible relative orientations of infinitesimal parallel current-carrying elements, and the resulting forces. The conclusion is opposite from electrostatics: like current elements attract, while opposites

Table 6.1 Electrostatics versus magnetostatics.

Electrostatics of point charges	**Magnetostatics of wire elements**
$\vec{F} = q_1 \vec{E}_2$	$d\vec{F} = \frac{I_1}{c} d\vec{\ell}_1 \times \vec{B}_2$
$\vec{F}_{12} = q_1 q_2 \frac{(\vec{x}_1 - \vec{x}_2)}{\lvert\vec{x}_1 - \vec{x}_2\rvert^3}$	$d\vec{F}_{12} = \frac{I_1 I_2}{c^2} \frac{d\vec{\ell}_1 \times \left(d\vec{\ell}_2 \times (\vec{x}_1 - \vec{x}_2)\right)}{\lvert\vec{x}_1 - \vec{x}_2\rvert^3}$
$\vec{E}(\vec{x}) = q \frac{(\vec{x} - \vec{x}\,')}{\lvert\vec{x} - \vec{x}\,'\rvert^3}$	$d\vec{B}(\vec{x}) = \frac{I}{c} \frac{d\vec{\ell}\,' \times (\vec{x} - \vec{x}\,')}{\lvert\vec{x} - \vec{x}\,'\rvert^3}$

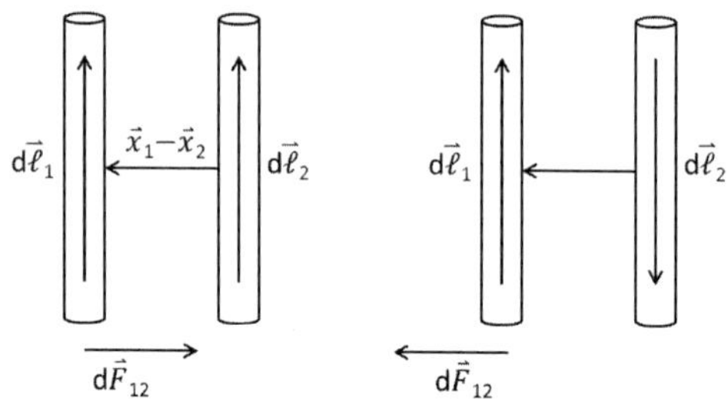

Fig. 6.1 Likes attract, opposites repel!

repel. Let us check whether or not the magnetostatic force laws in Table 6.1 satisfy Newton's third law, that is, does $d\vec{F}_{12} = -d\vec{F}_{21}$? We can use the $BAC - CAB$ rule to rewrite $d\vec{F}_{12}$ from Table 6.1 as

$$d\vec{F}_{12} = \frac{I_1 I_2}{c^2} \frac{[d\vec{\ell}_1 \cdot (\vec{x}_1 - \vec{x}_2)] d\vec{\ell}_2 - [d\vec{\ell}_1 \cdot d\vec{\ell}_2](\vec{x}_1 - \vec{x}_2)}{\lvert\vec{x}_1 - \vec{x}_2\rvert^3}. \tag{6.1}$$

The same simplification with $1 \leftrightarrow 2$ gives

$$-d\vec{F}_{21} = \frac{I_1 I_2}{c^2} \frac{[d\vec{\ell}_2 \cdot (\vec{x}_1 - \vec{x}_2)] d\vec{\ell}_1 - [d\vec{\ell}_1 \cdot d\vec{\ell}_2](\vec{x}_1 - \vec{x}_2)}{\lvert\vec{x}_1 - \vec{x}_2\rvert^3}. \tag{6.2}$$

Evidently $d\vec{F}_{12} \neq -d\vec{F}_{21}$. Figure 6.2 shows one situation where one current element feels a nonzero force, while the other experiences no force at all! Fortunately this is not a dealbreaker because $d\vec{F}_{12}$ and $d\vec{F}_{21}$ have no physical significances in themselves, so the inequality is a mere mathematical artifact. The real question is whether Newton's third law holds when forces between complete circuits are considered. Using the notation of Figure 6.3 where we identify $d\vec{\ell}_1 = d\vec{x}_1$, $d\vec{\ell}_2 = d\vec{x}_2$ for integration purposes,

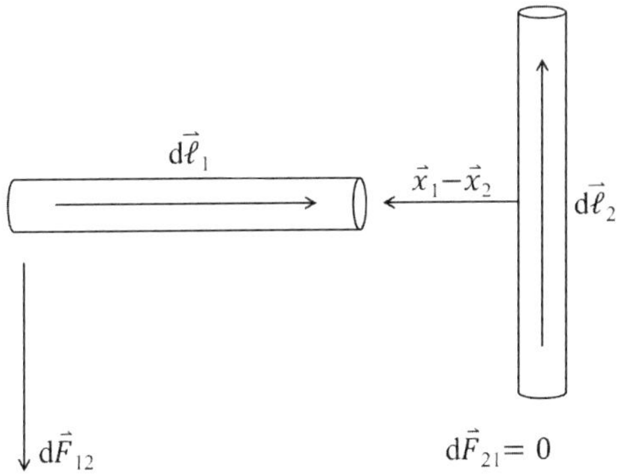

Fig. 6.2 Failure of Newton's third law for current elements.

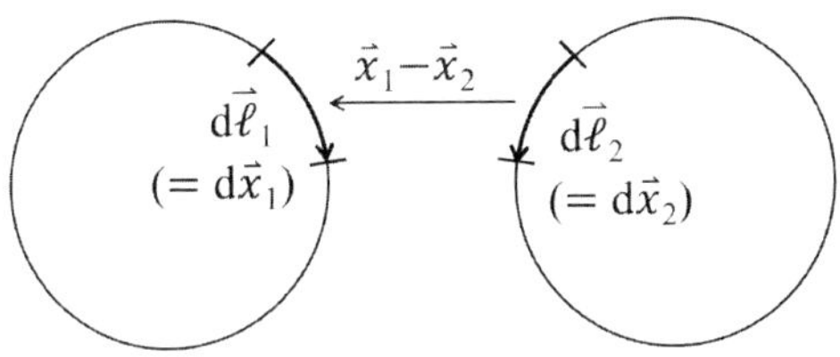

Fig. 6.3 Notation for current loop force calculation.

we compute

$$\begin{aligned}\vec{F}_{12} &= \frac{I_1 I_2}{c^2}\oint_1\oint_2 \frac{d\vec{x}_1 \times (d\vec{x}_2 \times (\vec{x}_1 - \vec{x}_2))}{|\vec{x}_1 - \vec{x}_2|^3} \\ &= \frac{I_1 I_2}{c^2}\oint_1\oint_2 \frac{[d\vec{x}_1 \cdot (\vec{x}_1 - \vec{x}_2)]d\vec{x}_2 - (d\vec{x}_1 \cdot d\vec{x}_2)(\vec{x}_1 - \vec{x}_2)]}{|\vec{x}_1 - \vec{x}_2|^3}.\end{aligned} \tag{6.3}$$

We may make use of the following identity

$$\begin{aligned}d_1\frac{1}{|\vec{x}_1 - \vec{x}_2|} &= -\frac{1}{|\vec{x}_1 - \vec{x}_2|^2}d_1|\vec{x}_1 - \vec{x}_2| \\ &= -\frac{1}{|\vec{x}_1 - \vec{x}_2|^2}d_1\sqrt{\sum_i (x_{1i} - x_{2i})^2} \\ &= -\frac{d\vec{x}_1 \cdot (\vec{x}_1 - \vec{x}_2)}{|\vec{x}_1 - \vec{x}_2|^3},\end{aligned} \tag{6.4}$$

to evaluate the first term in the integrand of (6.3):

$$\oint_1 \oint_2 \frac{[d\vec{x}_1 \cdot (\vec{x}_1 - \vec{x}_2)]d\vec{x}_2}{|\vec{x}_1 - \vec{x}_2|^3} = \oint_2 \oint_1 \left(d_1 \frac{1}{|\vec{x}_1 - \vec{x}_2|} \right) d\vec{x}_2$$
$$= \oint_2 d\vec{x}_2 \frac{1}{|\vec{x}_1 - \vec{x}_2|} \Bigg|_{\vec{x}_1{}'}^{\vec{x}_1{}'} = 0. \quad (6.5)$$

It follows that

$$\vec{F}_{12} = -\frac{I_1 I_2}{c^2} \oint_1 \oint_2 \frac{(\vec{x}_1 - \vec{x}_2)(d\vec{x}_1 \cdot d\vec{x}_2)}{|\vec{x}_1 - \vec{x}_2|^3}. \quad (6.6)$$

We note that (6.6) is antisymmetric under the exchange $1 \leftrightarrow 2$, so $\vec{F}_{12} = -\vec{F}_{21}$ for current loops.

From Table 6.1 we may integrate the equation for $d\vec{B}(\vec{x})$ to obtain the magnetic field for loops:

$$\vec{B}(\vec{x}) = \frac{I}{c} \oint d\vec{x}\,' \times \frac{(\vec{x} - \vec{x}\,')}{|\vec{x} - \vec{x}\,'|^3}. \quad (6.7)$$

The result is known as the *Biot-Savart law*. Alternative expressions for $\vec{B}(\vec{x})$ include

$$\vec{B}(\vec{x}) = \frac{I}{c} \oint \left(\vec{\nabla} \frac{1}{|\vec{x} - \vec{x}\,'|} \right) \times d\vec{x}\,', \quad (6.8)$$

where we integrate a cross-product of a gradient, or

$$\vec{B}(\vec{x}) = \frac{I}{c} \vec{\nabla} \times \oint \frac{d\vec{x}\,'}{|\vec{x} - \vec{x}\,'|}, \quad (6.9)$$

where the field is constructed as the curl of an integral. Going back to Table 6.1, we complete the expression for the force on a current loop as

$$\vec{F} = \frac{I}{c} \oint d\vec{x} \times \vec{B}(\vec{x}), \quad (6.10)$$

where the preceding expressions for the magnetic field $\vec{B}$ may be used.

Let us consider now the simple situation of two parallel wires, as shown in Figure 6.4. This is a simplification of a complete circuit, neglecting end and return wires which are assumed to be distant. We will obtain the force per unit length in two different ways. First, we begin with the magnetic field due to wire 1:

$$\vec{B}_1 \approx \frac{I_1}{c} \int_{-L}^{L} dz' \left(\vec{\nabla} \frac{1}{\sqrt{\rho^2 + z'^2}} \right) \times \hat{n}_1. \quad (6.11)$$

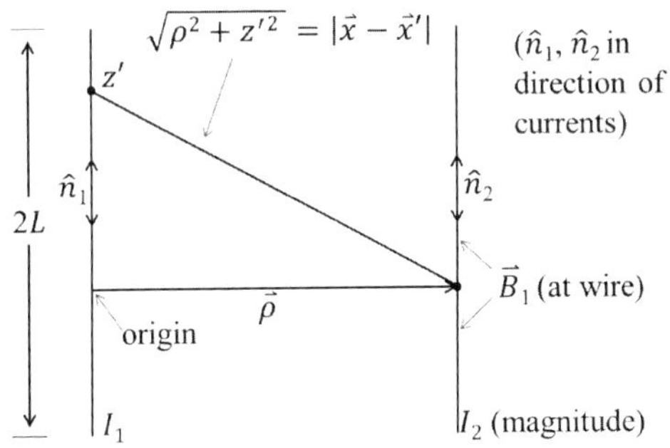

Fig. 6.4 Notation for parallel wires.

For $L \gg \rho$ we may approximate the integral as

$$\begin{aligned}\int_{-L}^{L} dz' \frac{1}{\sqrt{\rho^2 + z'^2}} &= 2\int_{0}^{L} dz' \frac{1}{\sqrt{\rho^2 + z'^2}} \\ &= 2\ln\left(z' + \sqrt{z'^2 + \rho^2}\right)\Big|_0^L \\ &= 2\ln\left(\frac{L}{\rho} + \sqrt{1 + L^2/\rho^2}\right) \\ &\approx 2\ln\frac{2L}{\rho} + \frac{\rho^2}{2L^2},\end{aligned} \tag{6.12}$$

so that

$$\begin{aligned}\vec{B}_1 &\approx \frac{I_1}{c}\vec{\nabla}\left(2\ln\frac{2L}{\rho} + \frac{\rho^2}{2L^2}\right) \times \hat{n}_1 \\ &\approx \frac{I_1}{c}\left(-2\frac{\hat{\rho}}{\rho}\left(1 - \frac{\rho^2}{2L^2}\right)\right) \times \hat{n}_1 \\ &\approx \frac{2I_1}{c\rho}(\hat{n}_1 \times \hat{\rho}).\end{aligned} \tag{6.13}$$

The force on wire 2 may then be computed as

$$\vec{F}_{21} = \frac{I_2}{c}\int dz\,\hat{n}_2 \times \vec{B}_1. \tag{6.14}$$

The total force is infinite if $L = \infty$, so instead we compute the force per length:

$$\begin{aligned}\frac{\vec{F}_{21}}{2L} = \frac{I_2}{c}\hat{n}_2 \times \vec{B}_1 &\approx \frac{2I_1 I_2}{c^2\rho}\hat{n}_2 \times (\hat{n}_1 \times \hat{\rho}) \\ &\approx -\frac{2I_1 I_2}{c^2}(\hat{n}_1 \cdot \hat{n}_2)\frac{\hat{\rho}}{\rho}.\end{aligned} \tag{6.15}$$

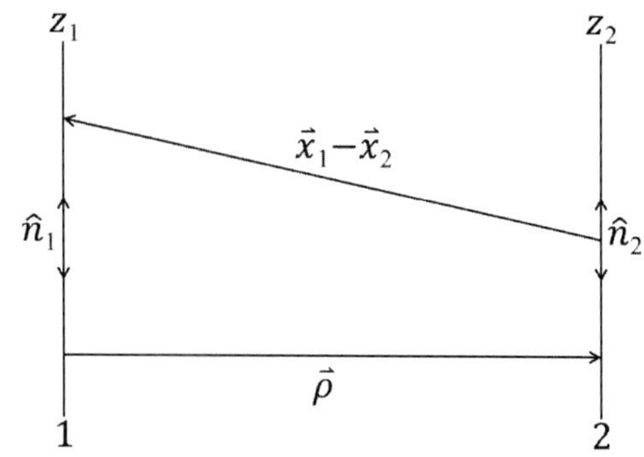

Fig. 6.5 Notation for direct force computation.

This confirms the rule that in magnetostatics likes attract, while opposites repel.

Just to be careful, we compute the force also directly via (6.6). We first introduce cylindrical coordinates as shown in Figure 6.5, so that

$$(\vec{x}_1 - \vec{x}_2) = -\vec{\rho} + (z_1 - z_2)\hat{k}; \quad d\vec{x}_1 \cdot d\vec{x}_2 = \hat{n}_1 \cdot \hat{n}_2 \, dz_1 dz_2, \tag{6.16}$$

and (6.6) becomes

$$\vec{F}_{21} = \frac{I_1 I_2}{c^2} \hat{n}_1 \cdot \hat{n}_2 \int_{-L}^{L} \int_{-L}^{L} \frac{dz_1 dz_2}{(\rho^2 + (z_1 - z_2)^2)^{3/2}} \left[-\vec{\rho} + \hat{k}(z_1 - z_2) \right]. \tag{6.17}$$

The term proportional to $\hat{k}$ vanishes by symmetry. The remaining integral may be evaluated exactly, giving

$$\vec{F}_{21} = \frac{I_1 I_2}{c^2} \hat{n}_1 \cdot \hat{n}_2 \, \vec{\rho} \left[\frac{2}{\rho} - \frac{2}{\rho^2} \sqrt{\rho^2 + 4L^2} \right]. \tag{6.18}$$

In order to compare with the previous result, we need the $L \gg \rho$ limit:

$$\sqrt{\rho^2 + 4L^2} \approx 2L + \frac{\rho^2}{4L}, \tag{6.19}$$

which gives

$$-\frac{2}{\rho} + \frac{2}{\rho^2} \sqrt{\rho^2 + 4L^2} \approx \frac{4L}{\rho^2} - \frac{2}{\rho} + \frac{1}{2L}. \tag{6.20}$$

The result for the force per length in the $L \gg \rho$ limit is thus

$$\frac{\vec{F}_{21}}{2L} \approx -\frac{2 I_1 I_2}{c^2} \hat{n}_1 \cdot \hat{n}_2 \frac{\hat{\rho}}{\rho}, \tag{6.21}$$

which agrees with our previous computation.

6.2 General equations for magnetostatics

Let us now generalize and introduce a volume current density:

$$I d\vec{x} \to \vec{J}(\vec{x}) d^3x. \tag{6.22}$$

When $\vec{J}_1$ and $\vec{J}_2$ are non-overlapping, our new force expression becomes

$$\vec{F}_{21} = -\frac{1}{c^2} \int d^3x_1 d^3x_2 \frac{(\vec{x}_1 - \vec{x}_2)\vec{J}_1(\vec{x}_1) \cdot \vec{J}_2(\vec{x}_2)}{|\vec{x}_1 - \vec{x}_2|^3}. \tag{6.23}$$

The magnetic field equations corresponding to (6.7)–(6.9) are

$$\vec{B}(\vec{x}) = \frac{1}{c} \int d^3x' \frac{\vec{J}(\vec{x}\,') \times (\vec{x} - \vec{x}\,')}{|\vec{x} - \vec{x}\,'|^3}, \tag{6.24}$$

$$= \frac{1}{c} \int d^3x' \left(\vec{\nabla} \frac{1}{|\vec{x} - \vec{x}\,'|} \right) \times \vec{J}(\vec{x}\,'), \tag{6.25}$$

$$= \frac{1}{c} \vec{\nabla} \times \int d^3x' \frac{\vec{J}(\vec{x}\,')}{|\vec{x} - \vec{x}\,'|}. \tag{6.26}$$

We may use (6.26) to immediately obtain

$$\vec{\nabla} \cdot \vec{B} = 0. \tag{6.27}$$

Using (6.22) in (6.10), we also obtain an expression for the magnetic force on a volume current:

$$\vec{F} = \frac{1}{c} \int d^3x\, \vec{J}(\vec{x}) \times \vec{B}(\vec{x}). \tag{6.28}$$

For a point charge q moving on a path $\vec{R}(t)$, the current is given by

$$\vec{J}(\vec{x}) = q \frac{d\vec{R}}{dt} \delta(\vec{x} - \vec{R}(t)), \tag{6.29}$$

and we obtain for the force

$$\vec{F} = \frac{q}{c} \frac{d\vec{R}}{dt} \times \vec{B}\left(\vec{R}(t)\right), \tag{6.30}$$

which is a statement of the magnetic part of the Lorentz force law.

Next let us compute $\vec{\nabla} \times \vec{B}$ using (6.26):

$$\vec{\nabla} \times \vec{B} = \frac{1}{c} \vec{\nabla} \times \left(\vec{\nabla} \times \int d^3x' \frac{\vec{J}(\vec{x}\,')}{|\vec{x} - \vec{x}\,'|} \right). \tag{6.31}$$

Using a second-order differential identity from Section 1.8, we have

$$\vec{\nabla} \times \vec{B} = \frac{1}{c}\vec{\nabla}\int d^3x' \vec{J}(\vec{x}\,') \cdot \vec{\nabla}\frac{1}{|\vec{x}-\vec{x}\,'|} - \frac{1}{c}\int d^3x' \vec{J}(\vec{x}\,')\underbrace{\nabla^2\frac{1}{|\vec{x}-\vec{x}\,'|}}_{-4\pi\delta(\vec{x}-\vec{x}\,')}$$

$$= -\frac{1}{c}\vec{\nabla}\int d^3x' \vec{J}(\vec{x}\,') \cdot \vec{\nabla}'\frac{1}{|\vec{x}-\vec{x}\,'|} + \frac{4\pi}{c}\vec{J}(\vec{x}). \tag{6.32}$$

We integrate the first term by parts, assuming $\vec{J}(\vec{x}\,')$ is localized, to obtain

$$\vec{\nabla} \times \vec{B} = \frac{4\pi}{c}\vec{J}(\vec{x}) + \frac{1}{c}\vec{\nabla}\int d^3x' \frac{\vec{\nabla}' \cdot \vec{J}(\vec{x}\,')}{|\vec{x}-\vec{x}\,'|}. \tag{6.33}$$

In the magnetostatic case we have that $\vec{J}(\vec{x},t) = \vec{J}(\vec{x})$, so the continuity equation (1.2) implies that $\partial\rho/\partial t$ is independent of time. If $\vec{\nabla} \cdot \vec{J} \neq 0$ at point $\vec{x}$, then $\rho(\vec{x})$ will either increase or decrease without bound at $\vec{x}$. It follows that only

$$\vec{\nabla} \cdot \vec{J} = 0, \tag{6.34}$$

is consistent with charge densities that are bounded for all time. We conclude that (6.34) is a necessary condition for magnetostatics. Note this condition is less restrictive than the electrostatic condition $\vec{J} = 0$. Under these conditions

$$\vec{\nabla} \times \vec{B} = \frac{4\pi}{c}\vec{J}(\vec{x}). \tag{6.35}$$

From this we see that a representation of $\vec{B}$ as $\vec{B} = -\vec{\nabla}\Phi_m$ cannot hold *everywhere*. However, in Section 6.5 we will see a very useful representation for $\vec{B}$ which is *almost* of this form in limited circumstances.

6.3 Ampère's law; vector potentials

In this section we derive two important mathematical consequences of the magnetostatic equations established in the previous section. Both have counterparts in electrostatics.

Our first consequence is *Ampère's law*, which is closely related to Gauss' law. Given any loop C and associated open surface S for which C is the boundary (see Figure 6.6), we may integrate the magnetic field equation (6.35) by dotting into the local normal vector $\hat{n}$ determined by the right

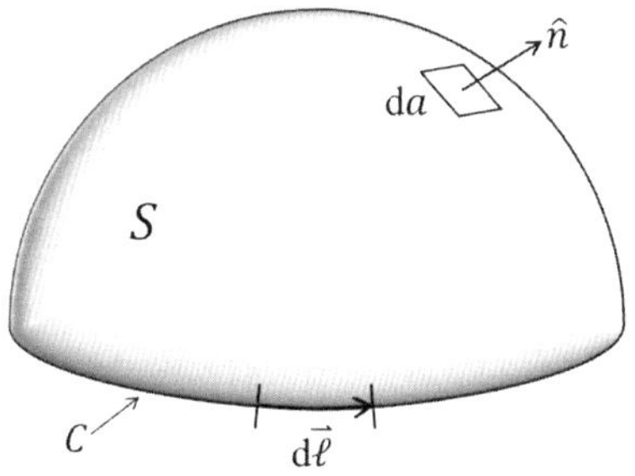

Fig. 6.6 Current loop C and surface S for Ampère's law calculation.

hand rule and integrating over the surface. We apply Stokes' theorem to obtain

$$\begin{aligned}\int_S (\vec{\nabla} \times \vec{B}) \cdot \hat{n}\, da &= \frac{4\pi}{c} \int_S \vec{J} \cdot \hat{n}\, da \\ \implies \oint_C \vec{B} \cdot d\vec{\ell} &= \frac{4\pi}{c} I. \end{aligned} \tag{6.36}$$

Equation (6.36) is called Ampère's law. It can be used to determine the $\vec{B}$ field in integration situations with simple symmetries; see for example Exercise 6.3.2. Note that the current I in this equation can be positive or negative.

Our second consequence is the *magnetostatic vector potential*, which is analogous to the scalar potential in electrostatics. Since $\vec{\nabla} \cdot \vec{B} = 0$, it is always possible to write (see Exercise 1.1.2)

$$\vec{B} = \vec{\nabla} \times \vec{A}, \tag{6.37}$$

which implies

$$\begin{aligned}\frac{4\pi}{c} \vec{J} &= \vec{\nabla} \times (\vec{\nabla} \times \vec{A}) \\ &= \vec{\nabla}(\vec{\nabla} \cdot \vec{A}) - \nabla^2 \vec{A}. \end{aligned} \tag{6.38}$$

If we modify $\vec{A}$ via a *gauge transformation* of the form

$$\vec{A} \to \vec{A} + \vec{\nabla}\Lambda \tag{6.39}$$

where Λ is an arbitrary scalar function, then $\vec{B}$ is unchanged. In particular, given $\vec{A}$ we may choose Λ such that $\nabla^2 \Lambda = -\vec{\nabla} \cdot \vec{A}$, which implies

$$\vec{\nabla} \cdot (\vec{A} + \vec{\nabla}\Lambda) = 0. \tag{6.40}$$

This choice of $\vec{A}$ with $\vec{\nabla}\cdot\vec{A}=0$ is called *Coulomb* or *radiation gauge.* With Coulomb gauge we have

$$-\nabla^2\vec{A}=\frac{4\pi}{c}\vec{J}. \tag{6.41}$$

Since each component of $\vec{A}$ independently satisfies Poisson's equation, we can apply the free Green function to find solutions for given sources in unbounded space. When boundaries or interfaces are present, one can develop a more complicated version of the theory of Green functions described in Chapters 2–4, but adapted to a vector field, to find solutions for various geometries.[1]

In unbounded space, in light of these comments and our earlier expression (6.26) for $\vec{B}$, we may choose

$$\vec{A}(\vec{x})=\frac{1}{c}\int d^3x'\,\frac{\vec{J}(\vec{x}\,')}{|\vec{x}-\vec{x}\,'|}. \tag{6.42}$$

We verify that this expression corresponds to Coulomb gauge for magnetostatics:

$$\begin{aligned}\vec{\nabla}\cdot\vec{A}&=\frac{1}{c}\int d^3x'\left(\vec{\nabla}\frac{1}{|\vec{x}-\vec{x}\,'|}\right)\cdot\vec{J}(\vec{x}\,')\\&=-\frac{1}{c}\int d^3x'\left(\vec{\nabla}'\frac{1}{|\vec{x}-\vec{x}\,'|}\right)\cdot\vec{J}(\vec{x}\,')\\&=0,\end{aligned}$$

via integration by parts and the condition (6.34).

6.4 Surface current considerations

We investigated the consistency of the discontinuous boundary condition (2.68) and the continuous field expression (2.69) in Section 2.6. In Chapter 1 we also derived the boundary condition

$$\hat{n}\times(\vec{B}_2-\vec{B}_1)=\frac{4\pi}{c}\vec{K}, \tag{6.43}$$

in free space for a surface current, $\vec{K}$. This surface current produces a magnetic field of course. We can modify (6.24) to express the magnetic field $\vec{B}$ due to this surface current, $\vec{K}$:

$$\vec{B}=\frac{1}{c}\int_S da'\vec{K}(\vec{x}\,')\times\frac{(\vec{x}-\vec{x}\,')}{|\vec{x}-\vec{x}\,'|^3}. \tag{6.44}$$

[1]For a more complete treatment of this topic see P. M. Morse and H. Feshbach, *Methods of Theoretical Physics, Part II*, McGraw-Hill (New York) 1953, Section 13.1.

We now ask: Is the boundary condition (6.43) consistent with the integral expression (6.44)? For a flat surface with normal vector $\hat{n}$, we compute

$$\hat{n} \times \vec{B} = \frac{1}{c} \int_S da' \frac{1}{|\vec{x} - \vec{x}\,'|^3} \left[\vec{K}(\vec{x}\,') \left(\hat{n} \cdot (\vec{x} - \vec{x}\,') \right) - (\vec{x} - \vec{x}\,')(\vec{K}(\vec{x}\,') \cdot \hat{n}) \right]. \tag{6.45}$$

The second term on the right in (6.45) vanishes for a current constrained to a surface. In Section 2.6 we developed the differential solid angle (as seen from location $\vec{x}$) as

$$d\Omega' = -da'\hat{n} \cdot \frac{(\vec{x} - \vec{x}\,')}{|\vec{x} - \vec{x}\,'|^3}. \tag{6.46}$$

Note $d\Omega'$ can be positive or negative, depending on the sign of $\hat{n} \cdot (\vec{x} - \vec{x}\,')$. It follows that

$$\hat{n} \times \vec{B} = -\frac{1}{c} \int_S \vec{K}(\vec{x}\,') d\Omega'. \tag{6.47}$$

The reader should note the similarity between (6.47) and the equation for the electric field due to a surface charge density σ on a flat surface

$$\vec{E} \cdot \hat{n} = -\int_S \sigma(\vec{x}\,') d\Omega'.$$

Now consider the case where $\vec{x}$ is very close to the surface, so that we can specify a (nearly) flat disc on the surface that subtends a solid angle $\pm 2\pi$ as seen from $\vec{x}$ (see Figure 6.7). Integrating over the disk gives the magnetic field due to the disk:

$$\hat{n} \times \vec{B}_2^d \approx \frac{2\pi}{c} \vec{K}, \tag{6.48}$$

$$\hat{n} \times \vec{B}_1^d \approx -\frac{2\pi}{c} \vec{K}. \tag{6.49}$$

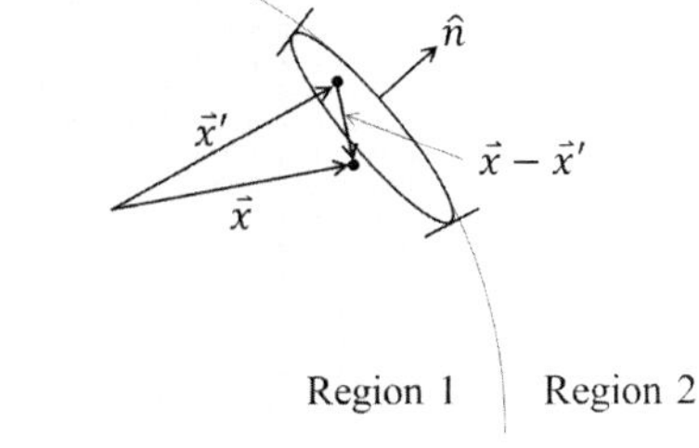

Fig. 6.7 Point $\vec{x}$ near surface with surface charge.

The magnetic field due to the remainder of the surface is continuous near $\vec{x}$, so there is a discontinuity in the total magnetic field given by

$$\hat{n} \times (\vec{B}_2 - \vec{B}_1) = \frac{4\pi}{c}\vec{K}. \tag{6.50}$$

Thus (6.43) and (6.44) are consistent with one another.

6.5 Solid angle formula for $\vec{B}$

We now return to our original expression for the magnetic field due to a closed loop, Equation (6.7), to develop a simple method for solving loop problems using the concept of solid angle. We use the notation shown in Figure 6.8. In Exercise 6.5.1 you will prove the following general result (valid for an arbitrary vector function $\vec{A}$)

$$\begin{aligned} \oint_{C'} d\vec{\ell}\,' \times \vec{A} &= \int_{S'} (d\vec{s}\,' \times \vec{\nabla}') \times \vec{A}, \\ &= \int_{S'} \sum_i ds'_i \vec{\nabla}' A_i - \int_{S'} d\vec{s}\,'(\vec{\nabla}' \cdot \vec{A}). \end{aligned} \tag{6.51}$$

We can integrate over the closed loop with current I in Figure 6.8 to form the magnetic field as:

$$\vec{B}(\vec{x}) = \frac{I}{c} \oint_{C'} d\vec{x}\,' \times \frac{(\vec{x} - \vec{x}\,')}{|\vec{x} - \vec{x}\,'|^3}. \tag{6.52}$$

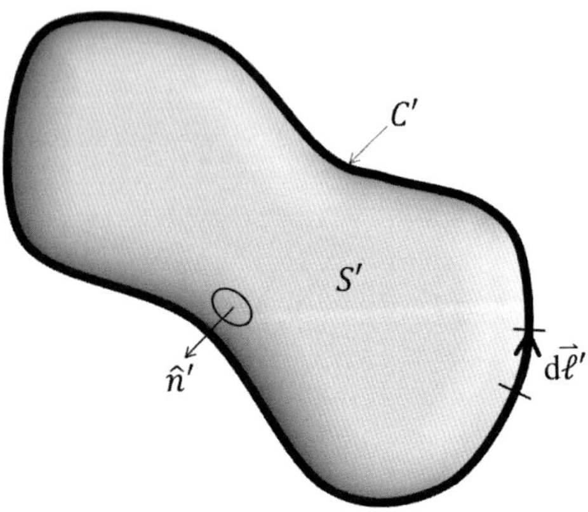

Fig. 6.8 Notation for wire loop: solid angle calculation.

Applying the above identity gives:

$$\vec{B}(\vec{x}) = \frac{I}{c}\left[\int_{S'}\sum_i ds_i'\vec{\nabla}'\frac{(\vec{x}-\vec{x}\,')_i}{|\vec{x}-\vec{x}\,'|^3} - \int_{S'} d\vec{s}\,'\vec{\nabla}'\cdot\left(\frac{(\vec{x}-\vec{x}\,')}{|\vec{x}-\vec{x}\,'|^3}\right)\right]$$
$$= \frac{I}{c}\left[-\vec{\nabla}\int_{S'} d\vec{s}\,'\cdot\frac{(\vec{x}-\vec{x}\,')}{|\vec{x}-\vec{x}\,'|^3} - \int_{S'} d\vec{s}\,'\vec{\nabla}'\cdot\left(\frac{(\vec{x}-\vec{x}\,')}{|\vec{x}-\vec{x}\,'|^3}\right)\right]. \tag{6.53}$$

The first term on the right in (6.53) may be simplified using expression (2.71) for differential solid angle

$$-\vec{\nabla}\int_{S'} d\vec{s}\,'\cdot\frac{(\vec{x}-\vec{x}\,')}{|\vec{x}-\vec{x}\,'|^3} = -\vec{\nabla}\int da'\hat{n}'\cdot\frac{(\vec{x}-\vec{x}\,')}{|\vec{x}-\vec{x}\,'|^3}$$
$$= \vec{\nabla}\Omega'(\vec{x}), \tag{6.54}$$

where $\Omega'(\vec{x})$ is the solid angle subtended by S' as viewed from $\vec{x}$. At first sight it might appear that this solid angle is dependent on the particular choice of S'; however, we will later argue that this is not the case.

The second term (6.53) can be simplified as

$$\int_{S'} d\vec{s}\,'\left(\vec{\nabla}'\cdot\frac{(\vec{x}-\vec{x}\,')}{|\vec{x}-\vec{x}\,'|^3}\right) = \int_{S'} d\vec{s}\,'\left(\nabla'^2\frac{1}{|\vec{x}-\vec{x}\,'|}\right)$$
$$= -4\pi\int_{S'} d\vec{s}\,'\delta(\vec{x}-\vec{x}\,'). \tag{6.55}$$

The right-hand side of (6.55) represents a "membrane" δ-function. This may look new, but we have actually encountered these previously in the Section 2.3 discussion. Figure 2.3 and Equation (2.39) reminds us of the "smearing" procedure that produces them and emphasizes how they can be consistently employed.

Putting the two terms in (6.53) together, our simplified expression for $\vec{B}$ for an arbitrary current loop becomes

$$\vec{B} = \frac{I}{c}\vec{\nabla}\Omega'(\vec{x}) + \frac{4\pi I}{c}\int_{S'} d\vec{s}\,'\delta(\vec{x}-\vec{x}\,'). \tag{6.56}$$

Equation (6.56) indicates that the characterization $\vec{B} = \frac{I}{c}\vec{\nabla}\Omega'$ will break down near the surface S'. This appears to introduce a discontinuity in the magnetic field, but we will see shortly that this expression actually gives a continuous form for $\vec{B}$ everywhere outside the wire.

If we drop the membrane δ-function, we immediately have $\vec{\nabla}\times\vec{B} = \vec{\nabla}\times\left(\frac{I}{c}\vec{\nabla}\Omega'(\vec{x})\right) = 0$ everywhere, which is incorrect. On the other hand, if we include the membrane term we obtain

$$\begin{aligned}\vec{\nabla} \times \vec{B} &= \frac{4\pi I}{c} \int_{S'} \left(\vec{\nabla}\delta(\vec{x} - \vec{x}\,')\right) \times d\vec{s}\,' \\ &= \frac{4\pi I}{c} \int_{S'} da'\, \hat{n}' \times \vec{\nabla}'\delta(\vec{x} - \vec{x}\,') \\ &= \frac{4\pi I}{c} \oint_{C'} \delta(\vec{x} - \vec{x}\,')d\vec{x}\,' \\ &= \frac{4\pi \vec{J}}{c},\end{aligned} \tag{6.57}$$

where a version of Stokes theorem has been used in the second to last step and where the final equality in (6.57) follows from the identification

$$\vec{J} = \sum_i I_i d\vec{x}\,'\delta(\vec{x} - \vec{x}\,'_i) \to I \oint_{C'} \delta(\vec{x} - \vec{x}\,')d\vec{x}\,'.$$

Ampère's law now follows immediately from Stokes' theorem and expression (6.56) for the magnetic field (see Figure 6.9):

$$\begin{aligned}\oint_C \vec{B} \cdot d\vec{\ell} &= \frac{I}{c} \oint_C d\vec{\ell} \cdot \left(\vec{\nabla}\Omega'(\vec{x}) + 4\pi \int_{S'} d\vec{s}\,'\delta(\vec{x} - \vec{x}\,')\right) \\ &= \frac{4\pi I}{c} \oint_C \int_{S'} d\vec{\ell} \cdot d\vec{s}\,'\delta(\vec{x} - \vec{x}\,') \\ &= \frac{4\pi}{c} \begin{cases} +I, \text{flow “in”} \\ -I, \text{flow “out”}. \end{cases}\end{aligned} \tag{6.58}$$

The first term on the right of the top form of (6.58) vanishes because it is a perfect differential evaluated on a closed contour. The second term uses the integration rule Equation (2.39) for membrane delta functions developed

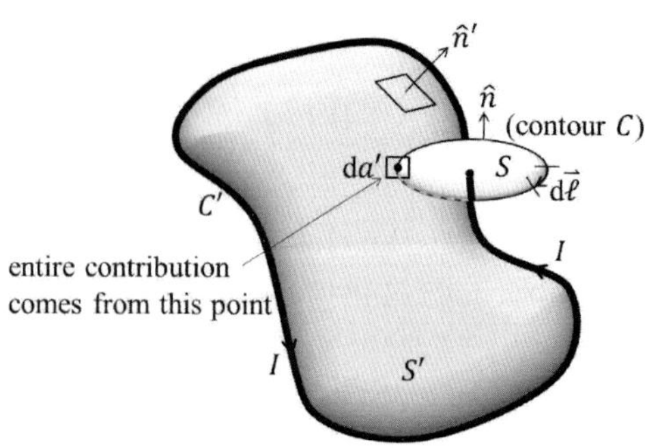

Fig. 6.9 Ampère's law with current loop.

in Section 2.3. When one is dealing with current loops it is possible to avoid using the membrane δ-function by allowing $\Omega'(\vec{x})$ to become multi-valued, similar to multi-valued functions in complex analysis. However, we shall find the single-valued characterization useful in our later treatment of magnetic monopoles in Section 7.8. Another separate overall motivation for our treatment is that evaluating a magnetic field from a scalar potential is easier than using the three components of the vector potential.

6.6 Circular current loop: solution using solid angle result for $\vec{B}$

We now apply formula (6.56) to compute exactly the magnetic field due to a circular current loop; this is the simplest complete circuit problem in magnetostatics (see Figure 6.10).

First, consider an arbitrary field point P as shown in Figure 6.10. This point is "outside" since $\hat{n}' \cdot \vec{x} > 0$; hence $\Omega(\vec{x}) < 0$. If P is anywhere on the z-axis and the opening angle of the solid angle cone is θ_0, which is continuous and assigned to be less than $\pi/2$ for $z > 0$ and greater than $\pi/2$

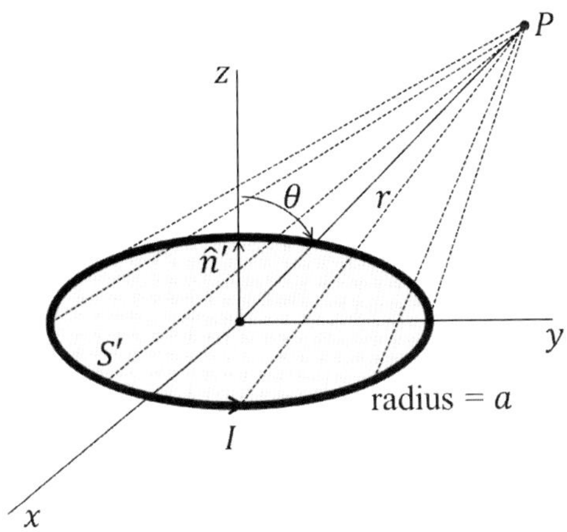

Fig. 6.10 Notation for circular current loop.

for $z < 0$, then Ω is given exactly by

$$\Omega_{\text{on}} = \begin{cases} -\displaystyle\int_0^{\theta_0} 2\pi \sin\theta d\theta, & z > 0, \\ -\displaystyle\int_\pi^{\theta_0} 2\pi \sin\theta d\theta, & z < 0, \end{cases} \tag{6.59}$$

$$= 2\pi \left(\cos\theta_0 - \text{sign}(z)\right)$$

$$= 2\pi \left(\frac{z}{\sqrt{z^2 + a^2}} - \text{sign}(z)\right). \tag{6.60}$$

Here we see the discontinuity in the solid angle at the position of the flat disk S' at $z' = 0$ that is bounded by the current loop. When approached from just above the surface we obtain $\Omega_{\text{on}}|_{z=0^+} = -2\pi$, when approached just below we obtain $\Omega_{\text{on}}|_{z=0^-} = 2\pi$. It follows that (replacing Ω' with Ω to simplify the notation)

$$\vec{B} = \frac{I}{c}\vec{\nabla}\Omega(\vec{x}). \tag{6.61}$$

Taking the gradient on either side gives

$$B_z = \frac{2\pi I}{c}\frac{a^2}{(a^2 + z^2)^{3/2}}. \tag{6.62}$$

Equation (6.60) is consistent with our known expression for solid angle when $|z| \gg a$, that is

$$\frac{z}{\sqrt{z^2 + a^2}} \approx \text{sign}(z)\left(1 - \frac{a^2}{2z^2}\right)$$

$$\implies \Omega_{\text{on}}(z) \approx -\,\text{sign}(z)\frac{\pi a^2}{z^2}. \tag{6.63}$$

By considering the change in solid angle when we displace the point P off-axis (which effectively rotates the loop relative to P's line of sight), we obtain (for $r \gg a$) that

$$\Omega \approx -\frac{\pi a^2}{r^2}\cos\theta, \tag{6.64}$$

where θ is the standard polar angle. Equation (6.61) gives us

$$B_r \approx \frac{\partial}{\partial r}\left(-\frac{I\pi a^2}{c}\frac{\cos\theta}{r^2}\right) = 2\left(\frac{I\pi a^2}{c}\right)\frac{\cos\theta}{r^3}, \tag{6.65}$$

$$B_\theta \approx \frac{1}{r}\frac{\partial}{\partial\theta}\left(-\frac{I\pi a^2}{c}\frac{\cos\theta}{r^2}\right) = \left(\frac{I\pi a^2}{c}\right)\frac{\sin\theta}{r^3}. \tag{6.66}$$

Let us now do the problem exactly. If we relate all variables to the primed surface S', we have

$$d\Omega'(\vec{x},\vec{x}\,') = da\,'\hat{n}' \cdot \frac{(\vec{x}\,' - \vec{x})}{|\vec{x} - \vec{x}\,'|^3} = -da\,'\frac{\partial}{\partial n'}\frac{1}{|\vec{x} - \vec{x}\,'|}, \tag{6.67}$$

where $\dfrac{\partial}{\partial n'} \equiv \hat{n}' \cdot \vec{\nabla}'$ is a primed normal derivative on the surface. In general we therefore have that

$$\Omega(\vec{x}) \equiv \int_{S'} d\Omega'(\vec{x},\vec{x}\,') = -\int_{S'} da\,'\frac{\partial}{\partial n'}\frac{1}{|\vec{x} - \vec{x}\,'|}. \tag{6.68}$$

Interchanging the $\vec{x}$ and $\vec{x}\,'$ derivatives, we have the simpler form

$$\Omega(\vec{x}) = \hat{n}' \cdot \vec{\nabla} \int_{S'} da' \left.\frac{1}{|\vec{x} - \vec{x}\,'|}\right|_{z'=0} = \frac{\partial}{\partial z}\int_{S'} da' \left.\frac{1}{|\vec{x} - \vec{x}\,'|}\right|_{z'=0}. \tag{6.69}$$

At this point the reader may be wondering whether we get a different solution if we choose a different surface S'. Fortunately, this is not the case. All surfaces with the same boundary subtend the same net solid angle. We can see this by noting that any two surfaces S'_1, S'_2 with the same boundary can be combined into a single closed surface S (with the caveat that the orientations of the two surfaces must be chosen in "opposite" directions). Then we may use the result from our discussion of Gauss' law that $\oint_S d\Omega_j = 0$ for any field point outside a closed surface S. It follows that $\oint_{S'_1} d\Omega_j - \oint_{S'_2} d\Omega_j = 0$, or $\oint_{S'_1} d\Omega_j = \oint_{S'_1} d\Omega_j$. The only thing that changes when a different S' is chosen is the location of the discontinuity in $\Omega(\vec{x})$. This still will not affect the final result for $\vec{B}$ since the membrane δ-function term will come in to cancel this discontinuity.

Having dealt with this issue, let us complete the exact calculation of $\vec{B}(\vec{x})$. Equation (4.120) for the free-space Green function in cylindrical coordinates gives

$$\frac{1}{|\vec{x} - \vec{x}\,'|} = \sum_{m=-\infty}^{\infty} \int_0^{\infty} dk\, e^{im(\phi-\phi')} J_m(k\rho) J_m(k\rho')e^{-k|z|}, \tag{6.70}$$

where we have used the fact that $e^{-k(z_> - z_<)} = e^{-k|z|}$ since $z' = 0$. Evaluation of (6.69) requires integrating (6.70) over ϕ' and ρ'. The ϕ' integral gives

$$\int_0^{2\pi} d\phi' e^{-im\phi'} = 2\pi\delta_{m0}. \tag{6.71}$$

To evaluate the ρ' integral, we use the Bessel function identity (4.20) to obtain (see also Exercise 4.1.3)

$$\begin{aligned}\int_0^a d\rho'\rho' J_0(k\rho') &= \int_0^a d\rho'\rho'\left(\frac{1}{k\rho'}+\frac{1}{k}\frac{d}{d\rho'}\right)J_1(k\rho')\\ &= \frac{1}{k}\int_0^a d\rho'\,\frac{d}{d\rho'}\left(\rho' J_1(k\rho')\right)\\ &= \frac{a}{k}J_1(ka). \end{aligned} \tag{6.72}$$

Expression (6.69) for the solid angle then becomes

$$\begin{aligned}\Omega(\vec{x}) &= 2\pi a\frac{\partial}{\partial z}\int_0^\infty dk\frac{1}{k}J_1(ka)J_0(k\rho)e^{-k|z|}\\ &= -2\pi a\,\mathrm{sign}(z)\int_0^\infty dk\, J_1(ka)J_0(k\rho)e^{-k|z|}, \end{aligned} \tag{6.73}$$

where sign(z) is $+1$ if $z>0$ and -1 if $z<0$.

As a check, expression (6.73) for the solid angle subtended by the loop should give us $\Omega(\vec{x})=0$ when $\rho>a, z=0$, and $\Omega=\mp 2\pi$ when $\rho<a, z=0^{\pm}$. Let us verify this. First we define the step function

$$H(\rho) \equiv a\int_0^\infty dkJ_1(ka)J_0(k\rho). \tag{6.74}$$

Using the relation $dJ_0(k\rho)/d\rho = -kJ_1(k\rho)$, we obtain the differential equation,

$$\begin{aligned}\frac{dH}{d\rho} &= a\int_0^\infty dkJ_1(ka)\frac{d}{d\rho}J_0(k\rho)\\ &= -a\int_0^\infty dkkJ_1(ka)J_1(k\rho)\\ &= -\delta(\rho-a), \end{aligned} \tag{6.75}$$

where we have used

$$\int_0^\infty dk\,kJ_m(k\rho)J_m(k\rho') = \frac{1}{\rho}\delta(\rho-\rho'), \tag{6.76}$$

which was derived in Section 4.2. Integration of (6.75) yields

$$H(\rho) = \begin{cases} C, & \rho<a,\\ C-1, & \rho>a, \end{cases} \tag{6.77}$$

where the constant C may be determined from the special case $\rho = 0$:

$$\begin{aligned} H(0) &= a\int_0^\infty dk J_1(ka)J_0(0) \\ &= \int_0^\infty dk \underbrace{aJ_1(ka)}_{-dJ_0(ka)/dk} \\ &= -J_0(ka)\big|_0^\infty = 1. \end{aligned} \tag{6.78}$$

Therefore

$$H(\rho) = \begin{cases} 1, \rho < a, \\ 0, \rho > a, \end{cases} \tag{6.79}$$

and

$$\lim_{z\to 0^+} \Omega(\vec{x}) = -2\pi H(\rho) = \begin{cases} -2\pi, \rho < a, \\ 0, \rho > a, \end{cases} \tag{6.80}$$

$$\lim_{z\to 0^-} \Omega(\vec{x}) = 2\pi H(\rho) = \begin{cases} 2\pi, \rho < a, \\ 0, \rho > a. \end{cases} \tag{6.81}$$

Now we find the nonzero magnetic field components. From Section 1.8 for cylindrical coordinates we have

$$\vec{\nabla}\psi = \hat{\rho}\frac{\partial \psi}{\partial \rho} + \hat{\phi}\frac{1}{\rho}\frac{\partial \psi}{\partial \phi} + \hat{z}\frac{\partial \psi}{\partial z}. \tag{6.82}$$

Using once again the relation $kJ_1(k\rho) = -dJ_0(k\rho)/d\rho$, we then find

$$B_\rho = \frac{2\pi a I}{c}\mathrm{sign}(z)\int_0^\infty dk k J_1(ka)J_1(k\rho)e^{-k|z|}, \tag{6.83}$$

$$B_z = \frac{2\pi a I}{c}\int_0^\infty dk k J_1(ka)J_0(k\rho)e^{-k|z|}. \tag{6.84}$$

Notice that $B_\rho\big|_{z=0} = 0$ (as we would expect from continuous differentiability) since

$$\int_0^\infty dk k J_1(ka)J_1(k\rho) = 0, \qquad \rho \neq a. \tag{6.85}$$

In the expression (6.84) for B_z we have actually omitted a term due to the discontinuity in Ω for $\rho < a$ coming from the derivative of the sign(z) term in (6.73). We note that with the assignment $z = a - \rho$ in (6.75), we have

$$\frac{dH(a-z)}{dz} = \delta(-z) = \delta(z), \tag{6.86}$$

$$H(a-z) = \begin{cases} 0, z < 0, \\ 1, z > 0. \end{cases} \tag{6.87}$$

This then gives

$$\frac{\partial}{\partial z}\text{sign}(z) = \frac{\partial}{\partial z}\left(H(a-z) - H(z-a)\right) = 2\delta(z) \tag{6.88}$$

$$\Longrightarrow \vec{B}_{\text{extra}} = \begin{cases} -\dfrac{4\pi I}{c}\delta(z)\hat{k}, \rho < a, \\ 0, \rho > a, \end{cases} \tag{6.89}$$

$$= -\frac{4\pi I}{c}\delta(z)\hat{k}H(\rho). \tag{6.90}$$

However, we have also left off the extra term in the representation of $\vec{B}$ that allows it to satisfy Ampère's law. Recalling that

$$\int_0^\infty \frac{dk}{2\pi}k \sum_{m=-\infty}^{\infty} e^{im(\phi-\phi')}J_m(k\rho)J_m(k\rho') = \frac{1}{\rho}\delta(\rho-\rho')\delta(\phi-\phi'), \tag{6.91}$$

from (4.37) and (4.38), we obtain on the $z'=0$ plane:

$$\begin{aligned} \frac{4\pi I}{c}\int d\vec{s}\,'\delta(\vec{x}-\vec{x}\,') &= \frac{4\pi I}{c}\hat{k}\,\delta(z)\int_0^a d\rho'\rho' \int_0^{2\pi} d\phi' \int_0^\infty \frac{dk}{2\pi}k \\ &\quad\times \sum_{m=-\infty}^{\infty} e^{im(\phi-\phi')}J_m(k\rho)J_m(k\rho'), \\ &= \frac{4\pi I}{c}\hat{k}\,\delta(z)\int_0^a d\rho'\rho' \int_0^\infty dk\, kJ_0(k\rho)J_0(k\rho'), \qquad (6.92) \\ &= \frac{4\pi aI}{c}\hat{k}\,\delta(z)\int_0^\infty dk\, J_0(k\rho)J_1(ka), \qquad (6.93) \\ &= \frac{4\pi I}{c}\delta(z)\hat{k}H(\rho), \qquad (6.94) \end{aligned}$$

where we have used (6.72) to go from (6.92) to (6.93). The two extra terms cancel, as they should. The expression (6.56) actually gives a $\vec{B}$ field continuous everywhere outside the wire.

6.7 Circular current loop: direct solution

In this section we solve the circular wire loop problem again, from a more standard point of view. This time we will use spherical coordinates (see Figure 6.11). We begin with the equation $\vec{B} = \vec{\nabla}\times\vec{A}$, and use the expression for $\vec{A}$ in Coulomb gauge:

$$\vec{A} = \frac{1}{c}\int d^3x' \frac{\vec{J}(\vec{x}\,')}{|\vec{x}-\vec{x}\,'|}. \tag{6.95}$$

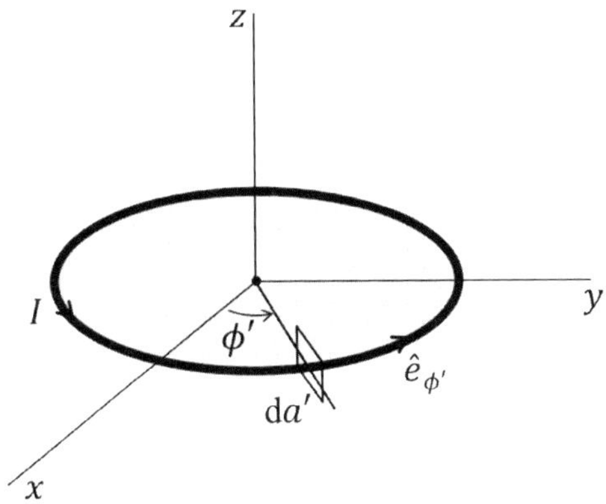

Fig. 6.11 Current loop in spherical coordinates.

In spherical coordinates we may write

$$\vec{J} = \frac{I}{a}\hat{e}_{\phi'}\delta(\cos\theta')\delta(r'-a), \tag{6.96}$$

where $\hat{e}_{\phi'} = -\hat{\imath}\sin\phi' + \hat{\jmath}\cos\phi'$. Let's confirm this:

$$\int \vec{J}\cdot\underbrace{\hat{n}}_{\hat{e}_{\phi'}}da' = \frac{I}{a}\int dr'\, r' d\cos\theta'\ \delta(\cos\theta')\delta(r'-a) = I. \tag{6.97}$$

Therefore

$$\vec{A}(\vec{x}) = \frac{I}{ca}\int dr' r'^2 d\Omega' \frac{(-\hat{\imath}\sin\phi' + \hat{\jmath}\cos\phi')}{|\vec{x}-\vec{x}\,'|}\delta(\cos\theta')\delta(r'-a), \tag{6.98}$$

where

$$|\vec{x}-\vec{x}\,'| = \left[r^2 + r'^2 - 2rr'(\cos\theta\cos\theta' + \sin\theta\sin\theta'\cos(\phi-\phi')\right]^{1/2}. \tag{6.99}$$

Since the geometry is azimuthally symmetric, the result will be independent of ϕ; so we may as well choose $\phi = 0$. Then since $|\vec{x}-\vec{x}\,'|$ is an even function of ϕ', it follows that the $\hat{\imath}$ term in (6.98) vanishes. Therefore

$$A_y(r,\theta,\phi=0) = A_\phi(r,\theta), \tag{6.100}$$

$$\Longrightarrow A_\phi = \frac{Ia}{c}\int_0^{2\pi} d\phi' \cos\phi' \left.\frac{1}{|\vec{x}-\vec{x}\,'|}\right|_{\theta'=\pi/2, r'=a, \phi=0}. \tag{6.101}$$

This result may actually be written directly in terms of elliptic integrals. However, we wish to find a form better suited to displaying the far field

$(r \gg a)$ and near field $(r \ll a)$ results. We use the Coulomb expansion

$$\frac{1}{|\vec{x}-\vec{x}\,'|} = \sum_{\ell,m} \frac{4\pi}{2\ell+1} \frac{r_<^\ell}{r_>^{\ell+1}} Y_{\ell m}^*(\theta',\phi') Y_{\ell m}(\theta,\phi), \tag{6.102}$$

$$\Longrightarrow \left.\frac{1}{|\vec{x}-\vec{x}\,'|}\right|_{\theta'=\pi/2, r'=a, \phi=0} = \sum_{\ell,m} \frac{4\pi}{2\ell+1} \frac{r_<^\ell}{r_>^{\ell+1}} Y_{\ell m}^*(\pi/2,\phi') Y_{\ell m}(\theta,0), \tag{6.103}$$

where $r_> = \max(r,a)$ and $r_< = \min(r,a)$. Using $\cos\phi' = (e^{i\phi'} + e^{-i\phi'})/2$, it follows that

$$\int_0^{2\pi} d\phi' \cos\phi' Y_{\ell m}^*(\pi/2,\phi') = \pi Y_{\ell m}^*(\pi/2,0)(\delta_{m,1} + \delta_{m,-1}), \tag{6.104}$$

$$\Longrightarrow A_\phi = \frac{2\pi I a}{c} \sum_\ell \frac{4\pi}{2\ell+1} \frac{r_<^\ell}{r_>^{\ell+1}} Y_{\ell 1}^*(\pi/2,0) Y_{\ell 1}(\theta,0). \tag{6.105}$$

We have explicitly

$$Y_{\ell m}(\theta,\phi) = \sqrt{\frac{2\ell+1}{4\pi}} \sqrt{\frac{(\ell-m)!}{(\ell+m)!}} e^{im\phi} (-\sin\theta)^m \left(\frac{d}{d\cos\theta}\right)^m P_\ell(\cos\theta). \tag{6.106}$$

Therefore

$$Y_{\ell 1}(\pi/2,0) = -\sqrt{\frac{2\ell+1}{4\pi\ell(\ell+1)}} \left(\frac{d}{d\cos\theta}\right) P_\ell(\cos\theta)\bigg|_{\cos\theta=0}. \tag{6.107}$$

The special value needed in (6.107) can be worked out; see the result of Exercise 4.9.1(c). Specializing to the single derivative case, we have

$$Y_{\ell 1}(\pi/2,0) = \begin{cases} 0, & \ell \text{ even}, \\ -\sqrt{\dfrac{2\ell+1}{4\pi\ell(\ell+1)}} \dfrac{(-1)^{(\ell-1)/2}\ell!!}{(\ell-1)!!}, & \ell \text{ odd}. \end{cases} \tag{6.108}$$

We finally put everything together to obtain

$$A_\phi = -\frac{2\pi I a}{c} \sum_{\ell \text{ odd}} \frac{4\pi}{2\ell+1} \frac{r_<^\ell}{r_>^{\ell+1}} \sqrt{\frac{2\ell+1}{4\pi\ell(\ell+1)}} \frac{(-1)^{(\ell-1)/2}\ell!!}{(\ell-1)!!} Y_{\ell 1}(\theta,0). \tag{6.109}$$

Alternatively, replacing ℓ with $2n+1$ we have

$$A_\phi = -\frac{\pi I a}{c} \sum_{n=0}^{\infty} \left(\frac{4\pi(2n+2)(2n+1)}{(4n+3)}\right)^{1/2} \times \frac{r_<^{2n+1}}{r_>^{2n+2}} \frac{(-1)^n(2n-1)!!}{(n+1)\,(2n)!!} Y_{2n+1,1}(\theta,0). \tag{6.110}$$

We may simplify this using the following expression from Section 4.9 for $m \geq 0$,

$$Y_{\ell m}(\theta,\phi) = \sqrt{\frac{2\ell+1}{4\pi}\frac{(\ell-m)!}{(\ell+m)!}}P_\ell^m(\cos\theta)e^{im\phi}, \tag{6.111}$$

$$\Longrightarrow Y_{2n+1,1}(\theta,0) = \left(\frac{4n+3}{4\pi(2n+2)(2n+1)}\right)^{1/2} P^1_{2n+1}(\cos\theta). \tag{6.112}$$

Therefore

$$A_\phi = -\frac{\pi I a}{c}\sum_{n=0}^{\infty}\frac{r_<^{2n+1}}{r_>^{2n+2}}\frac{(-1)^n(2n-1)!!}{(n+1)(2n)!!}P^1_{2n+1}(\cos\theta). \tag{6.113}$$

The non-zero magnetic field components are

$$B_r = \frac{1}{r\sin\theta}\frac{\partial}{\partial\theta}(\sin\theta A_\phi) = -\frac{1}{r}\frac{d}{d\cos\theta}(\sin\theta A_\phi), \tag{6.114}$$

$$B_\theta = -\frac{1}{r}\frac{\partial}{\partial r}(rA_\phi). \tag{6.115}$$

An alternative expression for $Y_{\ell m}(\theta,\phi)$ is (4.167),

$$Y_{\ell m}(\theta,\phi) = \sqrt{\frac{2\ell+1}{4\pi}\frac{(\ell+m)!}{(\ell-m)!}}e^{im\phi}(\sin\theta)^{-m}\left(\frac{d}{d\cos\theta}\right)^{\ell-m}\frac{(\cos^2\theta-1)^\ell}{2^\ell \ell!}, \tag{6.116}$$

which implies

$$\left(\frac{d}{d\cos\theta}\right)^m[(\sin\theta)^m Y_{\ell m}(\theta,\phi)] = \sqrt{\frac{2\ell+1}{4\pi}\frac{(\ell+m)!}{(\ell-m)!}}e^{im\phi}P_\ell(\cos\theta). \tag{6.117}$$

When $m = 1$ this gives

$$\frac{d}{d\cos\theta}[\sin\theta\, Y_{2n+1,1}(\theta,0)] = \sqrt{\frac{(4n+3)(2n+2)(2n+1)}{4\pi}}P_{2n+1}(\cos\theta), \tag{6.118}$$

so that from (6.114) and (6.110) we have

$$B_r = \frac{2\pi I a}{cr}\sum_{n=0}^{\infty}\frac{(-1)^n(2n+1)!!}{(2n)!!}\frac{r_<^{2n+1}}{r_>^{2n+2}}P_{2n+1}(\cos\theta). \tag{6.119}$$

Also, since

$$-\frac{1}{r}\frac{\partial}{\partial r}\left(r\frac{r_<^{2n+1}}{r_>^{2n+2}}\right) = \begin{cases}(2n+1)\dfrac{a^{2n+1}}{r^{2n+3}}, & r > a,\\ -(2n+2)\dfrac{r^{2n}}{a^{2n+2}}, & r < a,\end{cases} \tag{6.120}$$

we have from (6.115),

$$B_\theta = -\frac{\pi I a^2}{c}\sum_{n=0}^{\infty}\frac{(-1)^n(2n+1)!!}{(n+1)(2n)!!}P^1_{2n+1}(\cos\theta)$$
$$\times\begin{cases}\dfrac{1}{r^3}\left(\dfrac{a}{r}\right)^{2n}, & r>a,\\ -\left(\dfrac{2n+2}{2n+1}\right)\dfrac{1}{a^3}\left(\dfrac{r}{a}\right)^{2n}, & r<a.\end{cases} \tag{6.121}$$

Notice the apparent discontinuity in the B_θ field across $r = a$ in (6.121). Actually, everything is completely continuous except at the location of the current ($r = a, \theta = \pi/2$). This can be shown using the solid angle approach, where a step function representing the solid angle just above (-2π) or just below ($+2\pi$) the $\theta = \pi/2$ plane appears in the $r < a$ branch result of the solid angle $\Omega(\vec{x})$ expression; see Exercises 6.7.1 and 6.7.2.

The vector potential approach that we have used in this section may also be applied in cylindrical coordinates to derive the same expressions for B_ρ and B_z that we obtained in the previous section using the solid angle approach. We will make use of the vector potential in cylindrical coordinates for a current loop calculation in Section 6.12.

6.8 Current distributions and magnetic moments

In the previous sections, we have found exact solutions for a specific current loop geometry. We now switch gears, and find approximate (far-field) expressions for the vector potential for more general currents. This is just the analog of the multipole formalism for electrostatics (Section 5.1).

We begin with the Taylor expansion derived in Section 5.1,

$$\frac{1}{|\vec{x}-\vec{x}\,'|} = \sum_{n=0}^{\infty}\frac{(-1)^\ell(\vec{x}\,'\cdot\vec{\nabla})^\ell}{\ell!}\left(\frac{1}{r}\right) = \frac{1}{r} + \frac{\vec{x}\,'\cdot\vec{x}}{r^3} + \dots. \tag{6.122}$$

The vector potential in Coulomb gauge expands as

$$\vec{A}(\vec{x}) = \frac{1}{c}\int d^3x'\frac{\vec{J}(\vec{x})}{|\vec{x}-\vec{x}\,'|}$$
$$= \frac{1}{cr}\int d^3x'\vec{J}(\vec{x}\,') + \frac{1}{cr^3}\int d^3x'\vec{x}\cdot\vec{x}\,'\vec{J}(\vec{x}\,') + \dots. \tag{6.123}$$

It follows from the local magnetostatic identity $\vec{\nabla}' \cdot \vec{J}(\vec{x}\,') = 0$ that

$$\begin{aligned} 0 &= \int d^3x' x'_i \vec{\nabla}' \cdot \vec{J}(\vec{x}\,') = \int d^3x' x'_i \sum_j \nabla'_j J_j(\vec{x}\,') \\ &= \int d^3x' \left[\sum_j \nabla'_j \left(x'_i J_j(\vec{x}\,') \right) - J_i(\vec{x}\,') \right] \\ &= \int_S da' \hat{n}' \cdot \vec{J}(\vec{x}\,') x'_i - \int d^3x' J_i(\vec{x}\,'). \end{aligned} \tag{6.124}$$

Taking the surface S in (6.124) to be the surface at infinity, it follows that the first term on the right in (6.123) vanishes. It is to be understood throughout that we are dealing only with currents localized in a finite spatial region.

As to the second term in (6.123), we may write it in component form, and split the $\vec{x}\,'$ dependence into symmetric and antisymmetric terms:

$$\begin{aligned} \int d^3x' \, \vec{x} \cdot \vec{x}\,' \left(\vec{J}(\vec{x}\,') \right)_j &= \int d^3x' \sum_i x_i x'_i J_j(\vec{x}\,') \\ &= \frac{1}{2} \int d^3x' \sum_i x_i \left[(x'_i J_j + x'_j J_i) + (x'_i J_j - J_i x'_j) \right]. \end{aligned} \tag{6.125}$$

The symmetric term vanishes, as may be seen via the identity

$$\begin{aligned} 0 &= \int d^3x' x'_i x'_j \sum_k \nabla'_k J_k(\vec{x}\,') \\ &= - \int d^3x' \left(x'_i J_j + x'_j J_i \right). \end{aligned} \tag{6.126}$$

where we have integrated by parts. As to the antisymmetric term, we may use the identity

$$\left(-\vec{x} \times (\vec{x}\,' \times \vec{J}) \right)_j = \sum_i x_i \left(x'_i J_j - J_i x'_j \right), \tag{6.127}$$

to obtain

$$\int d^3x' \, \vec{x} \cdot \vec{x}\,' \vec{J}(\vec{x}\,') = -\frac{1}{2} \vec{x} \times \int d^3x' \left(\vec{x}\,' \times \vec{J}(\vec{x}\,') \right). \tag{6.128}$$

We define the *magnetic moment* (or *magnetic dipole moment*) as

$$\vec{m} \equiv \frac{1}{2c} \int d^3x \left(\vec{x} \times \vec{J}(\vec{x}) \right). \tag{6.129}$$

Note that $\vec{m}$ is independent of origin, as may be seen by replacing $\vec{x}$ with $\vec{x}+\vec{a}$ in (6.129).

The leading term in $\vec{A}(\vec{x})$ is

$$\vec{A}(\vec{x}) = \frac{\vec{m}\times\vec{x}}{r^3}. \tag{6.130}$$

For current loops

$$\vec{m} = \frac{I}{2c}\oint \vec{x}\times d\vec{\ell}. \tag{6.131}$$

We can employ our vector identity (6.51) again:

$$\vec{m} = -\frac{I}{2c}\left[\int_S \sum_i ds_i \vec{\nabla} x_i - \int_S d\vec{s}\,\vec{\nabla}\cdot\vec{x}\right]. \tag{6.132}$$

Noting that $\vec{\nabla} x_i = \hat{e}_i$ and $\vec{\nabla}\cdot\vec{x} = 3$ we have

$$\vec{m} = \frac{I}{c}\int d\vec{s} = \frac{I}{c}\vec{S}, \tag{6.133}$$

where $\vec{S} = (S_x, S_y, S_z)$, called the "directed area", denotes the vector areas of projections of the surface S onto the yz, xz, and xy-planes, respectively. Figure 6.12 illustrates $\vec{S}$ for a symmetric curved surface whose lower boundary is parallel to the xy-plane ; here $\vec{m} = \frac{I}{c}S'\hat{k}$, where S' is rectangular. At first sight it might seem that there would also be an x-component to $\vec{m}$ in Figure 6.12. However, there are both positive and negative components of the directed area on the yz-plane which cancel out so that $S_x = 0$. It also might seem that (6.133) would depend on the particular choice of $\vec{S}$. However, this turns out not to be the case. Consider for instance the example

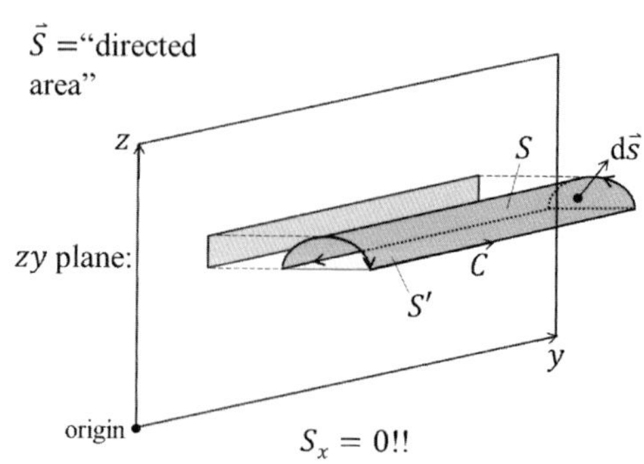

Fig. 6.12 Interpretation of $\vec{S}$.

where the boundary of S is a planar loop C. Even if S itself is not planar, the projected area onto any plane is independent of the choice of S so that

$$\vec{m} = \frac{I}{c} S\,\hat{n}, \tag{6.134}$$

where S now stands for the area of the planar loop and $\hat{n}$ is normal to the plane of C; the particular direction of $\hat{n}$ is given by the right-hand rule.

We will now compute the magnetic field due to a magnetic dipole from the vector potential (6.130). But before we do, we shall consider the analogous computation for an electric dipole to pave the way. In electrostatics the analogous expansion for the potential Φ is (cf. Equation (5.12)):

$$\Phi(\vec{x}) = \frac{q}{r} + \frac{\vec{x}\cdot\vec{p}}{r^3} + \dots. \tag{6.135}$$

The electric field due to the dipole term is

$$E_j^d(\vec{x}) = -\nabla_j \left(\frac{\vec{x}\cdot\vec{p}}{r^3}\right) = -\sum_i p_i \nabla_j \frac{x_i}{r^3}. \tag{6.136}$$

Now since

$$\nabla_i \frac{1}{r} = -\frac{x_i}{r^3}, \tag{6.137}$$

it follows that

$$E_j^d(\vec{x}) = \sum_i p_i \nabla_j \nabla_i \frac{1}{r}. \tag{6.138}$$

We will now show that

$$\nabla_j \nabla_i \frac{1}{r} = \frac{3x_i x_j - r^2 \delta_{ij}}{r^5} - \frac{4\pi}{3}\delta_{ij}\delta(\vec{x}). \tag{6.139}$$

This is a generalization of the free-space Poisson equation solution

$$\nabla^2 \frac{1}{r} = -4\pi\delta(\vec{x}),$$

which was first considered in Section 2.4.

We proceed via direct argument. Consider a small square box of volume $(2a)^3$ centered on the origin. For the $i = j = 1$ component, we have:

$$\nabla_x \nabla_x \frac{1}{r} = -\frac{\partial}{\partial x}\left(\frac{x}{r^3}\right). \tag{6.140}$$

Integrate this over the small box to obtain

$$\begin{aligned}\int d^3x\left(-\frac{\partial}{\partial x}\frac{x}{r^3}\right) &= -2a\int_{-a}^{a}dy\int_{-a}^{a}dz\frac{1}{(a^2+y^2+z^2)^{3/2}}\\ &= -8a\int_0^a dy\frac{z}{(a^2+y^2)\sqrt{a^2+y^2+z^2}}\Bigg|_0^a\\ &= -8a^2\int_0^a dy\frac{1}{(a^2+y^2)\sqrt{2a^2+y^2}}\\ &= -8\int_0^1 dx\frac{1}{(1+x^2)\sqrt{2+x^2}},\end{aligned} \tag{6.141}$$

where the last equality in (6.141) is obtained via the change of variable $x \equiv y/a$. An additional change of variable $x \equiv \tan u$ leads to

$$\begin{aligned}\int d^3x\left(-\frac{\partial}{\partial x}\frac{x}{r^3}\right) &= -\frac{8}{\sqrt{2}}\int_0^{\pi/4}\frac{\cos u du}{\sqrt{1-\frac{1}{2}\sin^2 u}}\\ &= -\frac{8}{\sqrt{2}}\cdot\sqrt{2}\sin^{-1}\left(\frac{1}{\sqrt{2}}\sin u\right)\Bigg|_0^{\pi/4} \qquad (6.142)\\ &= -4\pi/3, \qquad (6.143)\end{aligned}$$

where the evaluation in (6.142) is obtained from an integral table or trigonometric substitution. The result (6.143) is independent of the size of the box. It follows that

$$\nabla_x\nabla_x\frac{1}{r} = (\text{continuous piece}) - \frac{4\pi}{3}\delta(\vec{x}), \tag{6.144}$$

and similar results hold for the (y, y) and (z, z) components. To investigate the behavior of off-diagonal components, we look at

$$\nabla_x\nabla_y\frac{1}{r} = -\frac{\partial}{\partial x}\frac{y}{r^3}, \tag{6.145}$$

and integrate over the same small box centered at the origin:

$$\iiint dx\,dy\,dz\left(-\frac{\partial}{\partial x}\frac{y}{r^3}\right) = -\iint\frac{y}{r^3}dydz\Bigg|_{x=-a}^{x=a} = 0. \tag{6.146}$$

It follows that the only singular contribution is from the diagonal components.

Returning to (6.138) we find the field of a point dipole is

$$\vec{E}^d(\vec{x}) = \frac{3(\vec{p}\cdot\vec{x})\vec{x} - r^2\vec{p}}{r^5} - \frac{4\pi}{3}\vec{p}\,\delta(\vec{x}). \tag{6.147}$$

The $\delta(\vec{x})$ term has the effect of giving the correct volume integral of the electric field when dealing with a point electric dipole. We may also identify the effective source for point dipoles from taking the divergence of the field in (6.138):

$$\sum_j \nabla_j E_j^d(\vec{x}) = p_i \nabla_i \nabla^2 \frac{1}{r}$$
$$= -4\pi \vec{p} \cdot \vec{\nabla}\delta(\vec{x}). \tag{6.148}$$

However,

$$\vec{\nabla} \cdot \vec{E}^d(\vec{x}) = 4\pi\rho(\vec{x})$$
$$\Longrightarrow \rho(\vec{x}) = -\vec{p} \cdot \vec{\nabla}\delta(\vec{x}), \tag{6.149}$$

for a point electric dipole at the origin, or

$$\rho(\vec{x}) = -\vec{p} \cdot \vec{\nabla}\delta(\vec{x} - \vec{x}\,'), \tag{6.150}$$

for a dipole at source location $\vec{x}\,'$. Thus, for example, for grounded conductors in a given geometry, the implied scalar potential of a point dipole is

$$\Phi^d(\vec{x}) = \int d^3x''\, G_D(\vec{x}'', \vec{x})\rho(\vec{x}'')$$
$$= \vec{p} \cdot \vec{\nabla}' G_D(\vec{x}\,', \vec{x}). \tag{6.151}$$

We may obtain similar results in magnetostatics (Exercise 6.8.4). For a magnetic dipole, the corresponding magnetic field is

$$\vec{B}^m(\vec{x}) = \vec{\nabla} \times \left(\frac{\vec{m} \times \vec{x}}{r^3}\right)$$
$$= \frac{3(\vec{m} \cdot \vec{x})\vec{x} - r^2\vec{m}}{r^5} + \frac{8\pi}{3}\vec{m}\delta(\vec{x}). \tag{6.152}$$

In addition we have the implied magnetic dipole point current,

$$\vec{J}^m(\vec{x}) = -c\,\vec{m} \times \vec{\nabla}\delta(\vec{x}), \tag{6.153}$$

for a source at the origin. For a source at $\vec{x}\,'$, simply shift the argument of the delta function to $\vec{x} - \vec{x}\,'$.

In spherical coordinates when $\vec{x} \neq 0$ we find from (6.152) for a dipole oriented along the z-axis ($\vec{m} = m\hat{k}$),

$$B_r^m = \vec{B}^m \cdot \hat{e}_r = \frac{3r(\vec{m} \cdot \vec{x}) - \vec{m} \cdot \hat{e}_r r^2}{r^5} = \frac{2m\cos\theta}{r^3}, \tag{6.154}$$

$$B_\theta^m = \vec{B}^m \cdot \hat{e}_\theta = -\frac{\vec{m} \cdot \hat{e}_\theta}{r^3} = \frac{m\sin\theta}{r^3}, \tag{6.155}$$

$$B_\phi^m = 0. \tag{6.156}$$

These are just the forms we saw for the current loop in Section 6.6 (when $r \gg a$) because for a circular current loop of radius a,

$$m = |\vec{m}| = \frac{I}{c}\pi a^2. \tag{6.157}$$

6.9 External fields and forces on magnetic multipole distributions

In this section, we compute forces on localized current distributions. We shall proceed by analogy to our treatment of forces on localized charge distributions in electrostatics. As a reminder to the reader we began with

$$\vec{F} = \int d^3x\, \rho(\vec{x})\vec{E}(\vec{x}\,'), \tag{6.158}$$

in Section 5.3, and used the Taylor series expansion

$$E_i(\vec{x}) = E_i(0) + (\vec{x}\cdot\vec{\nabla}')E_i(\vec{x}\,')\Big|_{\vec{x}\,'=0} + \ldots. \tag{6.159}$$

Treating the electric field as external (previously denoted as $\vec{E}^{(0)}(\vec{x})$), we found in regions where the external charge distribution vanished that

$$\vec{F} = q\vec{E}(0) + (\vec{p}\cdot\vec{\nabla})\vec{E}(\vec{x})\Big|_{\vec{x}=0} + \ldots, \tag{6.160}$$

where $\vec{p}$ was the electric dipole moment. Since the curl of the external field vanishes in electrostatics, we also have

$$\vec{F} = q\vec{E}(0) + \vec{\nabla}(\vec{p}\cdot\vec{E}(\vec{x}))\Big|_{\vec{x}=0} + \ldots, \tag{6.161}$$

by a vector identity.

We seek the corresponding force expression(s) for magnetostatics. We start in similar fashion with the force equation

$$\vec{F} = \frac{1}{c}\int d^3x\, \vec{J}(\vec{x}) \times \vec{B}(\vec{x}), \tag{6.162}$$

and expand the external magnetic field

$$B_i(\vec{x}) = B_i(0) + \vec{x}\cdot\vec{\nabla}'B_i(\vec{x}\,')\Big|_{\vec{x}\,'=0} + \ldots, \tag{6.163}$$

to obtain

$$F_i = \frac{1}{c}\sum_{j,k}\epsilon_{ijk}\Big[B_k(0)\int d^3x\, J_j(\vec{x}) \\ + \sum_{\ell}\nabla'_\ell B_k(\vec{x}\,')|_{\vec{x}\,'=0}\int d^3x\, J_j(\vec{x})x_\ell + \ldots\Big]. \tag{6.164}$$

We now apply the identity (6.128), namely

$$\int d^3x\,(\vec{x}\,'\cdot\vec{x})\vec{J}(\vec{x}) = -\frac{1}{2}\vec{x}\,'\times\int d^3x\,\vec{x}\times\vec{J}(\vec{x})$$
$$= -\vec{x}\,'\times\vec{m}c. \tag{6.165}$$

Replacing $\vec{x}\,'$ in (6.165) with the constant vector $\vec{\nabla}'B_k(\vec{x}\,')\Big|_{\vec{x}\,'=0}$, we obtain

$$\int d^3x\left(\vec{\nabla}'B_k(\vec{x}\,')\Big|_{\vec{x}\,'=0}\cdot\vec{x}\right)\vec{J}(\vec{x}) = -c\vec{\nabla}'B_k(\vec{x}\,')\Big|_{\vec{x}\,'=0}\times\vec{m}$$
$$= c(\vec{m}\times\vec{\nabla}')B_k(\vec{x}\,')\Big|_{\vec{x}\,'=0}. \tag{6.166}$$

After putting this into (6.164) (and recalling $\int d^3x J_j(\vec{x}) = 0$) we have

$$F_i = \sum_{j,k}\epsilon_{ijk}\,(\vec{m}\times\vec{\nabla}')_j B_k(\vec{x}\,')\Big|_{\vec{x}\,'=0} + \ldots. \tag{6.167}$$

We now use index notation to write

$$(\vec{m}\times\vec{\nabla}')_j = \sum_{\ell,m}\epsilon_{j\ell m}m_\ell\nabla'_m, \tag{6.168}$$

which allows us to obtain

$$F_i = \sum_{\ell,m,j,k}\epsilon_{ijk}\epsilon_{j\ell m}m_\ell\,\nabla'_m B_k(\vec{x}\,')|_{\vec{x}\,'=0} + \cdots$$
$$= \left[\nabla'_i\left(\vec{m}\cdot\vec{B}(\vec{x}\,')\right) - m_i\left(\vec{\nabla}\cdot\vec{B}(\vec{x}\,')\right)\right]\Big|_{\vec{x}\,'=0} + \ldots, \tag{6.169}$$

where a summation identity for the epsilon products has been used; see Section 1.8. Finally, using $\vec{\nabla}\cdot\vec{B} = 0$ from Section 6.2, we have (replacing $\vec{x}\,'$ with $\vec{x}$ for simplicity)

$$\vec{F} = \vec{\nabla}\left(\vec{m}\cdot\vec{B}(\vec{x})\right)\Big|_{\vec{x}=0} + \ldots. \tag{6.170}$$

If in addition $\vec{\nabla}\times\vec{B} = 0$ for the external field in the region where $\vec{J}(\vec{x}) \neq 0$, we have the alternate expression

$$\vec{F} = (\vec{m}\cdot\vec{\nabla})\vec{B}(\vec{x})\Big|_{\vec{x}=0} + \ldots. \tag{6.171}$$

Equation (6.170) will be re-derived from a different point of view in Exercise 7.6.2. In this context if we define a scalar potential U by

$$U \equiv -\vec{m}\cdot\vec{B}, \tag{6.172}$$

and keep only the first term in (6.170), we have

$$\vec{F} = -\vec{\nabla}U. \tag{6.173}$$

If we take the magnetic moment as fixed, the form (6.171) indicates that the force on a magnetic particle will only be nonzero if the field is inhomogeneous along $\vec{m}$. This formula forms the beginning point in the understanding of the famous *Stern-Gerlach experiment* of 1922. Their measurement of the deflection of a beam of magnetic particles in such a $\vec{B}$ field gave a result at variance with classical ideas and helped lead modern science toward the ideas of *quantum mechanics*. The magnetic moment of a particle implies a quantity known as *spin*; spin systems form natural entryway for an explanation of quantum mechanical principles and systems.[2]

If the magnetic moment $\vec{m}$ is not fixed, the potential (6.172) tells us that it will rotate. In order to minimize U, it will tend to point along B. This implies a torque. Using $\vec{\mathcal{F}} \equiv \vec{J} \times \vec{B}$ to denote force density, we have

$$\vec{N} = \frac{1}{c}\int d^3x\, \vec{x} \times \vec{\mathcal{F}} = \frac{1}{c}\int d^3x\, \vec{x} \times (\vec{J} \times \vec{B}). \tag{6.174}$$

The leading order term in this expansion (0th order in $\vec{B}$) is

$$\vec{N} = \frac{1}{c}\int d^3x \left[\left(\vec{x} \cdot \vec{B}(0)\right)\vec{J}(\vec{x}) - \left(\vec{x} \cdot \vec{J}(\vec{x})\right)\vec{B}(0)\right] + \ldots. \tag{6.175}$$

We rely once again on the familiar identity (6.128) with $\vec{B}(0) = \vec{x}\,'$. The result is

$$\int d^3x \left(\vec{x} \cdot \vec{B}(0)\right)\vec{J}(\vec{x}) = -\vec{B}(0) \times \vec{m}c. \tag{6.176}$$

Using (6.126) with $i = j$ and summing over j, we can show that the second term in (6.175) vanishes. Therefore

$$\vec{N} = \vec{m} \times \vec{B}(0) + \ldots. \tag{6.177}$$

This is similar to the result in electrostatics, $\vec{N} = \vec{p} \times \vec{E}(0)$. Notice that the leading-order expressions for $\vec{F}$ and $\vec{N}$ for electric and magnetic dipoles are interchangeable via the substitution $\vec{p} \leftrightarrow \vec{m}$ in these static situations.

6.10 Introduction of "magnetization" and the $\vec{H}$ field

In parallel with our treatment of electrostatics, we are now ready to derive macroscopic Maxwell equations for magnetostatics.

[2]For an introductory treatment of quantum mechanics along these lines, see W. Wilcox, *Quantum Principles and Particles*, CRC Press (Boca Raton) 2012.

We briefly summarize our previous argument in Section 5.4 for the reader. Given a collection of atoms located in a volume V with boundary S, the force on each individual neutral atom was given by

$$\vec{F}_{\text{atom}} = \left(\vec{p}\cdot\vec{\nabla}\right)\vec{E}(\vec{x})\Big|_{\vec{x}=0} + \dots, \tag{6.178}$$

where the $\vec{E}$ considered was external. Denoting the atoms' number density by $n(\vec{x})$ we obtained

$$\vec{F}_{\text{bulk}} = \int d^3x\, n(\vec{x})\vec{F}_{\text{atom}}(\vec{x}). \tag{6.179}$$

We further defined the electric polarization as

$$\vec{P}(\vec{x}) \equiv n(\vec{x})\vec{p}(x), \tag{6.180}$$

and integrated by parts to obtain

$$\vec{F}_{\text{bulk}} = \int_V d^3x(-\vec{\nabla}\cdot\vec{P})\vec{E}(\vec{x}) + \int_S da(\vec{P}\cdot\hat{n})\vec{E}(\vec{x}), \tag{6.181}$$

giving rise to effective volume and surface charge densities

$$\rho_{\text{eff}} = -\vec{\nabla}\cdot\vec{P}, \tag{6.182}$$

$$\sigma^d_{\text{eff}}(\vec{x}) = \vec{P}\cdot\hat{n}. \tag{6.183}$$

The volume and surface contributions were thus separated. We also defined the displacement field

$$\vec{D} \equiv \vec{E} + 4\pi\vec{P}, \tag{6.184}$$

where $\vec{E}$ was the total electric field, and obtained finally the differential equations

$$\vec{\nabla}\cdot\vec{D} = 4\pi\rho_{\text{free}}, \tag{6.185}$$

$$\vec{\nabla}\times\vec{E} = 0. \tag{6.186}$$

The material constitutive relation specifying $\vec{D}$ as a function of $\vec{E}$ is necessary to solve these equations.

For magnetostatics, we can follow very similar steps. Unlike electrostatics where the starting points given by Equations (5.52) and (5.53) are equivalent because of the identity $\vec{\nabla}\times\vec{E} = 0$, here we must start with Equation (6.170) to compute the approximate bulk force. This is because the less general Equation (6.171) requires the non-overlap source condition

$\vec{\nabla} \times \vec{B} = 0$ at the position of $\vec{m}$. (We will reconsider the non-overlapping source situation in Section 7.7.) Thus, we start with

$$\vec{F}_{\text{atom}} = \vec{\nabla}\left(\vec{m} \cdot \vec{B}\right) = \sum_i m_i \vec{\nabla} B_i, \tag{6.187}$$

$$\vec{F}_{\text{bulk}} = \int d^3x\, n(\vec{x}) \vec{F}_{\text{atom}}. \tag{6.188}$$

We define the *magnetization* as

$$\vec{M}(\vec{x}) \equiv n(\vec{x}) \vec{m}(\vec{x}), \tag{6.189}$$

so that we may rewrite (6.188) as

$$\vec{F}_{\text{bulk}} = \int d^3x \sum_i M_i(\vec{x}) \vec{\nabla} B_i(\vec{x}). \tag{6.190}$$

At this point we make use of the identity

$$\int d^3x\, \vec{B} \times \left(\vec{\nabla} \times \vec{M}\right) = -\int d^3x \left(\vec{B} \cdot \vec{\nabla}\right) \vec{M} + \int d^3x \sum_i B_i \vec{\nabla} M_i, \tag{6.191}$$

which is straightforwardly proven using a differential form of the $BAC - CAB$ rule seen in Section 1.8. Note that the $\vec{\nabla}$ operator works only on $\vec{M}$. We integrate both terms on the right by parts to obtain

$$\int d^3x\, \vec{B} \times \left(\vec{\nabla} \times \vec{M}\right) = \int d^3x \left(\vec{\nabla} \cdot \vec{B}\right) \vec{M} - \int d^3x \sum_i M_i \vec{\nabla} B_i + \int da\, \hat{n}(\vec{B} \cdot \vec{M}) - \int da\, \vec{M}(\vec{B} \cdot \hat{n}). \tag{6.192}$$

The first term on the right in (6.192) vanishes because $\vec{\nabla} \cdot \vec{B} = 0$. The last two terms in (6.192) may be combined:

$$\int da\, \hat{n}(\vec{B} \cdot \vec{M}) - \int da\, \vec{M}(\vec{B} \cdot \hat{n}) = -\int da\, (\hat{n} \times \vec{M}) \times \vec{B}, \tag{6.193}$$

with the final result that

$$\vec{F}_{\text{bulk}} = \int_V d^3x \left(\vec{\nabla} \times \vec{M}\right) \times \vec{B} + \int_S da\, (\vec{M} \times \hat{n}) \times \vec{B}, \tag{6.194}$$

where volume and surface contributions are now separated. Comparison with the standard force equation $\vec{F} = \frac{1}{c} \int d^3x\, \vec{J}(\vec{x}) \times \vec{B}(\vec{x})$ leads to the definition of effective volume and surface currents

$$\vec{J}_{\text{eff}} \equiv c\vec{\nabla} \times \vec{M}, \tag{6.195}$$

$$\vec{K}_{\text{eff}} \equiv c\vec{M} \times \hat{n}. \tag{6.196}$$

We assume a model where the atoms in the material give rise to effective currents. We denote all currents which are "bound up" in the atoms as $\vec{J}_{\text{bound}}$. Thus we take

$$\vec{\nabla} \times \vec{B} = \frac{4\pi}{c} \left(\vec{J}_{\text{free}} + \vec{J}_{\text{bound}} \right). \tag{6.197}$$

If all the bound currents are due to the just calculated magnetic dipole forces, then

$$\vec{J}_{\text{bound}} = \vec{J}_{\text{eff}} = c\vec{\nabla} \times \vec{M}. \tag{6.198}$$

Defining the *H-field*

$$\vec{H} \equiv \vec{B} - 4\pi\vec{M}, \tag{6.199}$$

we obtain the equations

$$\vec{\nabla} \times \vec{H} = \frac{4\pi}{c} \vec{J}_{\text{free}}, \tag{6.200}$$

$$\vec{\nabla} \cdot \vec{B} = 0. \tag{6.201}$$

It is necessary to have a constitutive relation giving $\vec{H}$ as a function of $\vec{B}$ for the material. For linear, isotropic materials we have

$$\vec{B} = \mu\vec{H}, \tag{6.202}$$

where μ is the *magnetic permeability* of the material. If μ is constant in the material, it follows that

$$\vec{\nabla} \times \vec{B} = \frac{4\pi\mu}{c} \vec{J}_{\text{free}}. \tag{6.203}$$

Some materials have $\mu > 1$, enhancing the free current contribution, while others have $\mu < 1$, shielding the free current. The case $\mu > 1$ is called *paramagnetism*, and results from the lining up of internal magnetic fields in substances with non-zero atomic electron spins (and thus net angular momentum) along $\vec{B}$ since $U = -\vec{m} \cdot \vec{B}$ from Equation (6.172). The case $\mu < 1$ is called *diamagnetism*, and occurs when substances with no net atomic angular momentum react against the applied magnetic field; this is a consequence of Lenz's Law, which involves the time variation of magnetic flux applied to electronic orbits; we will study this in Section 7.3. Paramagnetic and diamagnetic materials typically have permeability values in the range $\mu - 1 \sim \pm 10^{-4,-5}$ from these mechanisms.

Ferromagnetic materials are those that possess permanent magnetic moments $\vec{M}$. Ferromagnetism arises from the long-range ordering of electron spins in a material. This happens within macroscopically small but microscopically large regions called *magnetic domains*. The μ in this case is defined as the incremental *change* of $\vec{B}$ with respect to $\vec{H}$ in (6.202). In such cases it is possible to attain $\mu \sim 10^5$, but in general the relationship between $\vec{H}$ and $\vec{B}$ is no longer linear. Ferromagnetic materials display a phenomenon known as *hysteresis*. As an external magnetic field is applied to such materials, there are a number of physical mechanisms which change the internal magnetic structure. Magnetic domains can change in size or rotate. In addition, the domain walls can move and new domains can be created or "nucleate". All these mechanisms can be present at the same time, and modeling the changes can be difficult; references 5, 8 and 10 of the *Going Deeper Magnetic materials* section give some of the models which are used to study and characterize magnetic materials. These domain changes do not go back to their original values when the external field is removed. In such cases the constitutive relation for $\vec{H}$ is no longer single-valued, but depends upon the prior history of preparation of the material.

An idealized hysteresis *limiting curve* of a magnetic substance is shown in Figure 6.13. The limiting curve is the saturation loop of a material where any value inside is allowed. Note that the measured curves themselves are not smooth; the substance displays small jumps in magnetization called *Barkhausen jumps* due to crystallographic defects. In describing Figure 6.13 we will refer to SI units, and define the "intrinsic induction" as $B_i \equiv M/\mu_0$, in which case we have $B_i = B - \mu_0 H$ for unidirectional fields B_i, B and H. The figure indicates that B_i goes asymptotically to the value B_s for large values of H; this is called magnetic "saturation". In addition, the value B_r

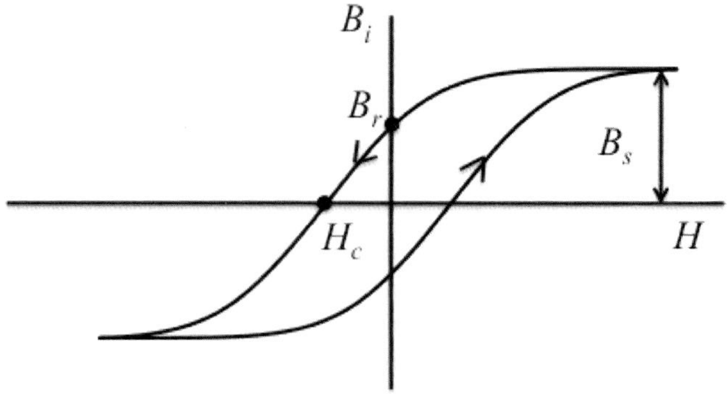

Fig. 6.13 The limiting curve for a hysteresis loop.

indicates the "remanence" value when the magnetic field, H, is turned off. Also, the magnetic field H_c associated with the removal of the B_i field on the limiting curve is termed the "coercivity force". All these concepts and more are necessary in modeling the behavior of magnetic materials.

6.11 Boundary conditions at material interfaces

We summarize our magnetostatic boundary conditions as

$$\left(\vec{B}_2 - \vec{B}_1\right) \cdot \hat{n} = 0, \tag{6.204}$$

$$\hat{n} \times \left(\vec{H}_2 - \vec{H}_1\right) = \frac{4\pi}{c}\vec{K}_{\text{free}}. \tag{6.205}$$

Consider now a material/vacuum interface where region 1 corresponds to the material ($\mu_1 \neq 1$) and region 2 corresponds to the vacuum ($\mu_2 = 1$). We are assuming $\vec{B} = \mu\vec{H}$ and that there are no free currents in the volume or surface of the material. Given the free surface current $\vec{K}_{\text{free}} = 0$, then

$$\vec{B}_2 \cdot \hat{n} = \vec{B}_1 \cdot \hat{n}, \tag{6.206}$$

$$\vec{H}_2 \times \hat{n} = \vec{H}_1 \times \hat{n}. \tag{6.207}$$

The situation we are discussing may be compared to the corresponding electrostatic situation with free volume and surface charge densities vanishing and $\vec{D} = \epsilon\vec{E}$ when $\epsilon_1 \neq 1$ and $\epsilon_2 = 1$:

$$\vec{D}_2 \cdot \hat{n} = \vec{D}_1 \cdot \hat{n}, \tag{6.208}$$

$$\vec{E}_2 \times \hat{n} = \vec{E}_1 \times \hat{n}. \tag{6.209}$$

The magnetostatic (6.200), (6.201) and electrostatic field equations (6.186), (6.185) and above boundary conditions are mathematically identical, with the replacements

$$\vec{B} \leftrightarrow \vec{D}, \vec{H} \leftrightarrow \vec{E}, \tag{6.210}$$

and

$$\mu \leftrightarrow \epsilon. \tag{6.211}$$

Given the mathematical equivalence of the field equations and boundary conditions of the two cases, the flat interface $\vec{E}, \vec{D}$ problem from Section 5.6 and the electrostatic interface problem of Exercise 5.6.5 (with appropriate changes in notation) give us insights into what to expect for $\vec{H}, \vec{B}$ fields near a flat interface with permeable materials. In the case of a highly permeable material ($\mu_1 \gg 1$) with magnetic sources outside the material, we expect suppression of the H_1 field *inside*. The relation (6.207) then helps us realize that the fields in the vacuum region must be nearly normal to the surface like electric fields near a conductor. In particular this implies that in this limit interfaces where $\vec{K}_{\text{free}} = 0$ behave as "equipotential" surfaces for H-field lines. On the other hand, for an inside source we expect suppression of the *outside* B_2, H_2 fields from shielding, and internal $\vec{B}$ fields nearly tangential to the inside surface from the boundary condition (6.206). Thus, the internal B-field lines are essentially "piped" through such materials.

Other field behaviors near interfaces can be determined from this analogy. For example, for $\mu_1 \approx 0$ we would expect outside fields from sources outside the material to be mainly tangential to the surface, just like the Neumann conductor discussed in Sections 1.4 and 2.7 for $\epsilon = 0$. Thus, the material seems to expel or repel magnetic field lines, exhibiting perfect diamagnetism. We will examine a situation for a flat interface in the next section where we will see that image currents enforce this condition. This is the situation found in Type 1 *superconductors*. Superconducting materials have the property that they conduct electricity without resistance below a certain critical temperature. Superconductivity is essentially a macroscopic manifestation of a quantum phenomenon involving the coupling of conductance electron spins in a state called a *Cooper pair*. Such a pair encounters much less internal resistance to motion.

In summary, if we are considering magnetostatics in a region where $\vec{J}_{\text{free}} = 0$, $\vec{K}_{\text{free}} = 0$, then from (6.200) we may always write $\vec{H}$ as the gradient of a scalar potential $\vec{H} = -\vec{\nabla}\Phi_m$ just as we may always write $\vec{E} = -\vec{\nabla}\Phi$ for electrostatics. Such situations are mathematically identical to the corresponding $\rho_{\text{free}} = 0$, $\sigma_{\text{free}} = 0$ electrostatic situation obtained via the replacements $\vec{B} \leftrightarrow \vec{D}$, $\vec{H} \leftrightarrow \vec{E}$, $\mu \leftrightarrow \epsilon$ described above.

6.12 Image method for magnetostatics

The *image method* that we used in electrostatics also applies to magnetostatics. Before applying the method to magnetostatics, we consider the general electrostatic case with a charge distribution in the vacuum with $z > 0$ located near a plane interface with a dielectric, as shown in Figure 6.14. We guess that the fields in the $z > 0$ region will be equivalent to the sum of fields produced by the actual charge density $\rho(x, y, z)$ and an

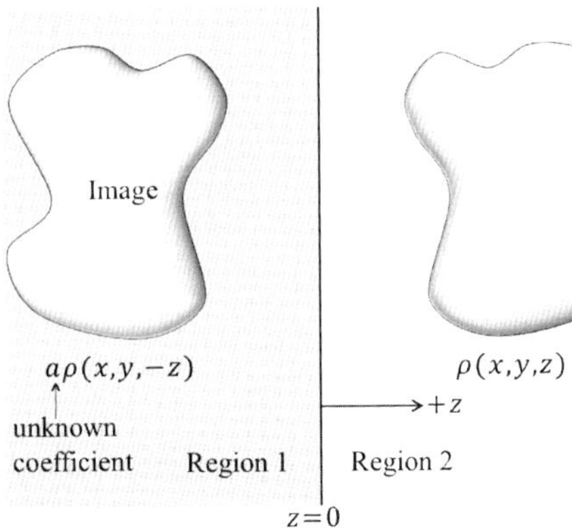

Fig. 6.14 Image method: electrostatic case, plane interface, field for $z > 0$. Image assumed to be in the Region 2 medium.

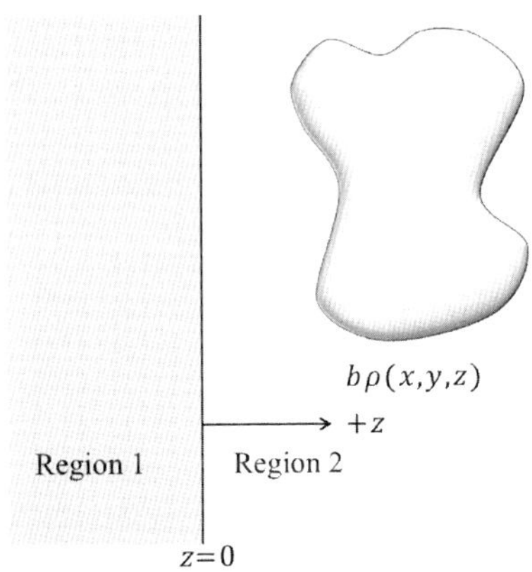

Fig. 6.15 Image method: electrostatic case, plane interface, field for $z < 0$. Source assumed to be in the Region 1 medium.

image charge $a \cdot \rho(x, y, -z)$, where a is to be determined. We also guess that the fields for $z < 0$ agree with the field due to a charge density $b \cdot \rho(x, y, z)$, where b is a constant to be determined (see Figure 6.15). In order to solve for a and b, we use the boundary conditions:

$$\epsilon E_{1z}\big|_{z=0} = E_{2z}\big|_{z=0}, \qquad \text{(normal } \vec{D}) \tag{6.212}$$

$$\vec{E}_{1\perp}\Big|_{z=0} = \vec{E}_{2\perp}\Big|_{z=0}. \qquad \text{(tangential } \vec{E}) \tag{6.213}$$

Given the above guesses, we have for the electric fields in regions 1 and 2:

$$\vec{E}_1 = b \int d^3x' \rho(x', y', z') \frac{(\vec{x} - \vec{x}\,')}{|\vec{x} - \vec{x}\,'|^3}, \tag{6.214}$$

$$\vec{E}_2 = a \int d^3x' \rho(x', y', -z') \frac{(\vec{x} - \vec{x}\,')}{|\vec{x} - \vec{x}\,'|^3} + \int d^3x' \rho(x', y', z') \frac{(\vec{x} - \vec{x}\,')}{|\vec{x} - \vec{x}\,'|^3}. \tag{6.215}$$

It follows that in region 2

$$\begin{aligned} E_{2z}|_{z=0} &= a \int_{-\infty}^{0} dz' \int da' \rho(x', y', -z') \frac{-z'}{|\vec{x} - \vec{x}\,'|^3} \\ &\quad + \int_{0}^{\infty} dz' \int da' \rho(x', y', z') \frac{-z'}{|\vec{x} - \vec{x}\,'|^3} \\ &= (-a + 1) \int_{0}^{\infty} dz' \int da' \rho(x', y', z') \frac{-z'}{|\vec{x} - \vec{x}\,'|^3}, \end{aligned} \tag{6.216}$$

while in region 1

$$\epsilon E_{1z}|_{z=0} = \epsilon b \int_{0}^{\infty} dz' \int da' \rho(x', y', z') \frac{-z'}{|\vec{x} - \vec{x}\,'|^3}. \tag{6.217}$$

Equating (6.216) and (6.217) via the boundary condition (6.212) leads to

$$-a + 1 = \epsilon b. \tag{6.218}$$

In similar fashion, we evaluate

$$\begin{aligned} \vec{E}_{2\perp}\,|_{z=0} &= a \int_{-\infty}^{0} dz' \int da' \rho(x', y', -z') \frac{(\vec{x} - \vec{x}\,')_\perp}{|\vec{x} - \vec{x}\,'|^3} \\ &\quad + \int_{0}^{\infty} dz' \int da' \rho(x', y', z') \frac{(\vec{x} - \vec{x}\,')_\perp}{|\vec{x} - \vec{x}\,'|^3} \\ &= (a + 1) \int_{0}^{\infty} dz' \int da' \rho(x', y', z') \frac{(\vec{x} - \vec{x}\,')_\perp}{|\vec{x} - \vec{x}\,'|^3}, \end{aligned} \tag{6.219}$$

$$\vec{E}_{1\perp}\Big|_{z=0} = b \int_{0}^{\infty} dz' \int da' \rho(x', y', z') \frac{(\vec{x} - \vec{x}\,')_\perp}{|\vec{x} - \vec{x}\,'|^3}. \tag{6.220}$$

It follows from the tangential boundary condition (6.213) that $a + 1 = b$, which together with (6.218) leads to

$$b = \frac{2}{\epsilon + 1}, \qquad a = -\frac{(\epsilon - 1)}{\epsilon + 1}, \tag{6.221}$$

in agreement with our results in Chapter 5 (cf. (5.106), (5.107)).

Doing the electrostatic case gives us confidence in the basic method. A similar approach works in magnetostatics. Consider a current distribution

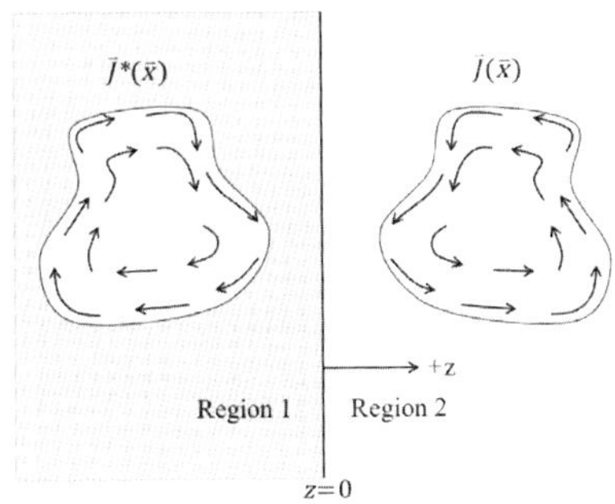

Fig. 6.16 Image method: magnetostatic case, plane interface, field for $z > 0$. Image assumed to be in the Region 2 medium.

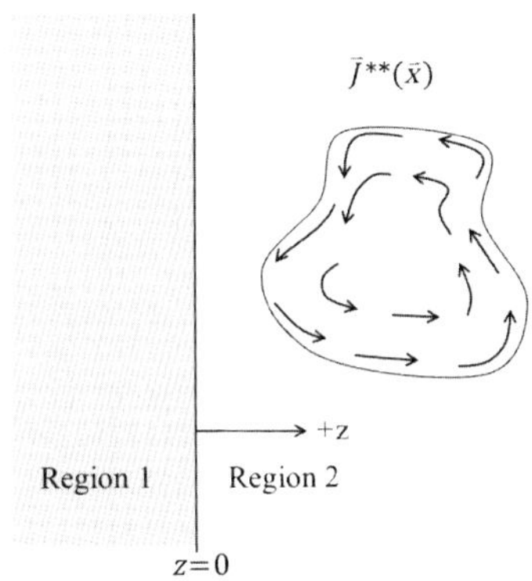

Fig. 6.17 Image method: magnetostatic case, plane interface, field for $z < 0$. Source assumed to be in the Region 1 medium.

$\vec{J}(\vec{x}) = (J_x(x, y, z), J_y(x, y, z), J_z(x, y, z))$ near a plane interface, as shown in Figure 6.16. For the magnetic field in the $z > 0$ region, we guess an image current J^* of the form

$$\vec{J}^*(\vec{x}) = (aJ_x(x, y, -z), aJ_y(x, y, -z), bJ_z(x, y, -z)) \tag{6.222}$$

while for $z < 0$ we guess a current $\vec{J}^{**}$ of the form (see Figure 6.17)

$$\vec{J}^{**}(\vec{x}) = c\vec{J}(\vec{x}). \tag{6.223}$$

Applying the boundary conditions,

$$B_{1z}\big|_{z=0} = B_{2z}\big|_{z=0}, \qquad \text{(normal } \vec{B}\text{)} \tag{6.224}$$

$$\frac{1}{\mu}\vec{B}_{1\perp}\bigg|_{z=0} = \vec{B}_{2\perp}\bigg|_{z=0}, \qquad \text{(tangential } \vec{H}\text{)} \tag{6.225}$$

leads to three equations in the three unknowns a, b, c. The final solution is (Exercise 6.12.1):

$$a = \left(\frac{\mu - 1}{\mu + 1}\right); \qquad b = -\left(\frac{\mu - 1}{\mu + 1}\right); \qquad c = \frac{2\mu}{\mu + 1}. \tag{6.226}$$

We demonstrate these results by solving exactly the problem of a circular current loop of radius a for $z > 0$ that is parallel to and a distance d away from the surface of a material with $\mu \neq 1$, as shown in Figure 6.18. The exact solution for the single component vector potential A_ϕ for a circular current loop in free space with geometry as in Figure 6.19 is

$$A_\phi(\rho, z) = \frac{2\pi I a}{c} \int_0^\infty dk e^{-k|z|} J_1(ka) J_1(k\rho). \tag{6.227}$$

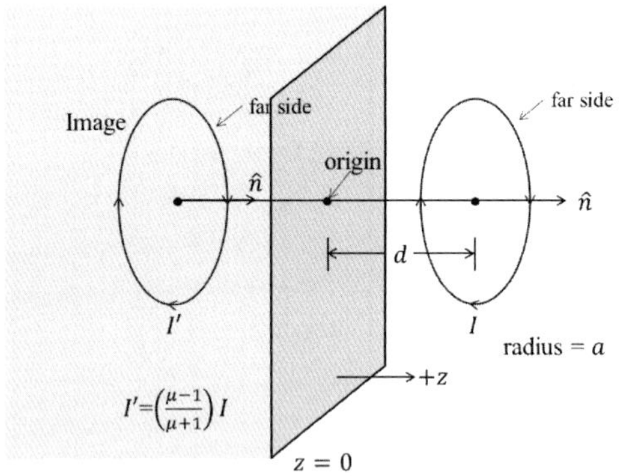

Fig. 6.18 Current loop near material interface (image current direction shown for $\mu > 1$); fields in $z > 0$ region.

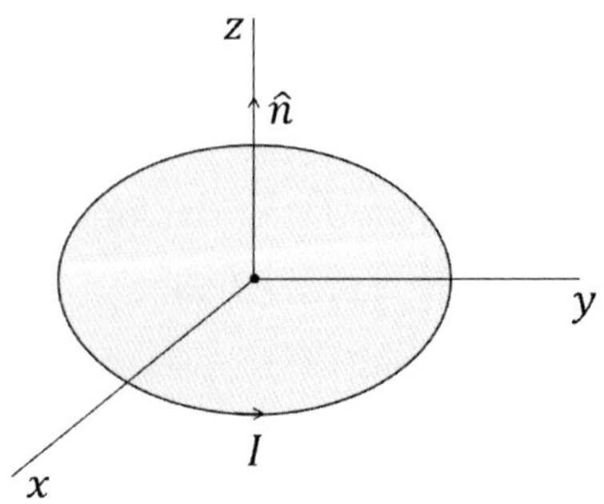

Fig. 6.19 Current loop geometry.

This can be established by starting with Equation (6.101) in cylindrical coordinates and using the Bessel function expansion (6.70). One can demonstrate that this expression gives the same result for B_ρ and B_z as in (6.83) and (6.84). Given this, the field due to the image loop will be given by shifting $z \to z + d$,

$$A'_\phi = \frac{2\pi I' a}{c} \int_0^\infty dk e^{-k|z+d|} J_1(ka) J_1(k\rho) \tag{6.228}$$

in the $z > 0$ region where

$$I' = I\left(\frac{\mu - 1}{\mu + 1}\right). \tag{6.229}$$

The force of the image field $\vec{B}\,'$ on the loop with current density $\vec{J}$ is

$$\vec{F} = \frac{1}{c} \int d^3x\, \vec{J}(\vec{x}) \times \vec{B}\,'(\vec{x}). \tag{6.230}$$

In cylindrical coordinates we obtain

$$\vec{J}(\vec{x}) = I\, \hat{e}_\phi \delta(z - d)\delta(\rho - a), \tag{6.231}$$

$$\vec{B}\,'(\vec{x}) = \vec{\nabla} \times \vec{A}\,'(\vec{x}). \tag{6.232}$$

$\vec{B}\,'$ only has components in the $\hat{e}_\rho$ and $\hat{k}$ directions:

$$B'_\rho = -\frac{\partial}{\partial z} A'_\phi, \tag{6.233}$$

$$B'_z = \frac{1}{\rho}\frac{\partial}{\partial \rho}\left(\rho A'_\phi\right). \tag{6.234}$$

We can then write

$$\begin{aligned} \vec{J} \times \vec{B}\,' &= I\,\delta(z - d)\delta(\rho - a)\hat{e}_\phi \times \left(B'_\rho \hat{e}_\rho + B'_z \hat{k}\right) \\ &= I\,\delta(z - d)\delta(\rho - a)\left[-B'_\rho \hat{k} + B'_z \hat{e}_\rho\right], \end{aligned} \tag{6.235}$$

where we have used $\hat{e}_\phi \times \hat{e}_\rho = -\hat{k}$ and $\hat{e}_\phi \times \hat{k} = \hat{e}_\rho$. Since $\hat{e}_\rho = \cos\phi\,\hat{\imath} + \sin\phi\,\hat{\jmath}$ it follows that the $\hat{e}_\rho$ term in (6.235) vanishes when integrated over ϕ; only the $\hat{k}$ term remains. To evaluate this term, we compute (note $z > 0$)

$$B'_\rho = -\frac{\partial}{\partial z} A'_\phi = \frac{2\pi I a}{c}\left(\frac{\mu - 1}{\mu + 1}\right)\int_0^\infty dk k e^{-k(z+d)} J_1(ka) J_1(k\rho), \tag{6.236}$$

which is just the appropriate form of (6.83). Plugging (6.236) into (6.230), we find (using cylindrical coordinates)

$$\begin{aligned} \vec{F} &= -\frac{2\pi I^2 a}{c^2}\hat{k}\left(\frac{\mu - 1}{\mu + 1}\right)\int \rho d\rho dz\, d\phi \\ &\quad \times \int_0^\infty dk\, k\, e^{-k(z+d)} J_1(ka) J_1(k\rho)\delta(z - d)\delta(\rho - a) \\ &= -\frac{4\pi^2 I^2 a^2}{c^2}\hat{k}\left(\frac{\mu - 1}{\mu + 1}\right)\int_0^\infty dk k\, e^{-2kd}(J_1(ka))^2. \end{aligned} \tag{6.237}$$

The force on the loop is towards the interface if $\mu > 1$ (paramagnetic) and away from the interface if $\mu < 1$ (diamagnetic). This agrees with our previous finding that likes attract and opposites repel in magnetostatics. If $d \gg a$, then

$$J_1(ka) \approx \frac{ka}{2}, \tag{6.238}$$

and we have

$$\begin{aligned} F_z &\approx -\frac{4\pi^2 I^2 a^2}{c^2}\left(\frac{\mu-1}{\mu+1}\right)\int_0^\infty dk k e^{-2kd}\frac{k^2a^2}{4} \\ &\approx -\frac{6\pi^2 I^2 a^4}{c^2(2d)^4}\left(\frac{\mu-1}{\mu+1}\right). \end{aligned} \tag{6.239}$$

There is an easy way to check this. For $d \gg a$ we may write the force on the current loop as $\vec{F} \approx \vec{\nabla}(\vec{m}\cdot\vec{B}\,')$ (cf. (6.170)), where $\vec{m}$ is the magnetic moment of the actual loop and $\vec{B}\,'$ refers to the magnetic field due to the image loop. From our earlier discussion of magnetic dipoles, we know that

$$\vec{m} = \frac{I}{c}\pi a^2 \hat{k}, \tag{6.240}$$

$$B'_r \approx 2\left(\frac{I'\pi a^2}{c}\right)\frac{\cos\theta'}{r'^3}, \tag{6.241}$$

$$B'_\theta \approx \left(\frac{I'\pi a^2}{c}\right)\frac{\sin\theta'}{r'^3}, \tag{6.242}$$

where (r', θ', ϕ') are spherical coordinates with origin at the image loop. On the z-axis ($\theta' = 0$) we have $\hat{r} = \hat{k}$, and

$$(B'_z)_{\theta'=0} = B'_r(\theta' = 0, r' = z) = 2\left(\frac{I'\pi a^2}{c}\right)\frac{1}{z^3}. \tag{6.243}$$

Therefore

$$F_z \approx \frac{2\pi^2 a^4 I I'}{c^2}\left.\frac{\partial}{\partial z}\frac{1}{z^3}\right|_{z=2d} \tag{6.244}$$

$$\approx -\frac{6\pi^2 I^2 a^4}{c^2(2d)^4}\left(\frac{\mu-1}{\mu+1}\right), \tag{6.245}$$

in agreement with (6.239).

6.13 Intrinsic and induced magnetization: theory and example

In this section we consider materials that possess an intrinsic magnetization $\vec{M}(\vec{x})$ even in the absence of an applied field. Such a material is depicted in Figure 6.20. There are no free currents, so $\vec{\nabla} \times \vec{H} = 0$ which implies that $\vec{H}$ arises from a scalar potential: $\vec{H} = -\vec{\nabla}\Phi_m$. We suppose that the overall magnetization is due to the sum of contributions $d\vec{m}$ from microscopic current loops within the material (see Figure 6.20). The vector potential associated with $d\vec{m}$ is (with origin located at $d\vec{m}$)

$$d\vec{A}^m = \frac{d\vec{m} \times \vec{x}}{r^3}, \tag{6.246}$$

and therefore

$$d\vec{B}^m = \vec{\nabla} \times \left(\frac{d\vec{m} \times \vec{x}}{r^3}\right). \tag{6.247}$$

From Equation (6.152),

$$\begin{aligned} d\vec{B}^m &= \frac{8\pi}{3} d\vec{m}\,\delta(\vec{x}) + \frac{3\vec{x}(d\vec{m}\cdot\vec{x}) - r^2 d\vec{m}}{r^5} \\ \Longrightarrow\ d\vec{H}^m &= d\vec{B}^m - 4\pi\, d\vec{m}\,\delta(\vec{x}) = -\frac{4\pi}{3} d\vec{m}\,\delta(\vec{x}) + \frac{3\vec{x}(d\vec{m}\cdot\vec{x}) - r^2 d\vec{m}}{r^5}. \end{aligned} \tag{6.248}$$

On the other hand from (6.147) with $\vec{p} \to d\vec{m}$,

$$\begin{aligned} -\vec{\nabla}\left(\frac{d\vec{m}\cdot\vec{x}}{r^3}\right) &= \sum_i dm_i \nabla_i \vec{\nabla}\left(\frac{1}{r}\right) \\ &= -\frac{4\pi}{3} d\vec{m}\,\delta(\vec{x}) + \frac{3\vec{x}\,(d\vec{m}\cdot\vec{x}) - d\vec{m}\,r^2}{r^5}. \end{aligned} \tag{6.249}$$

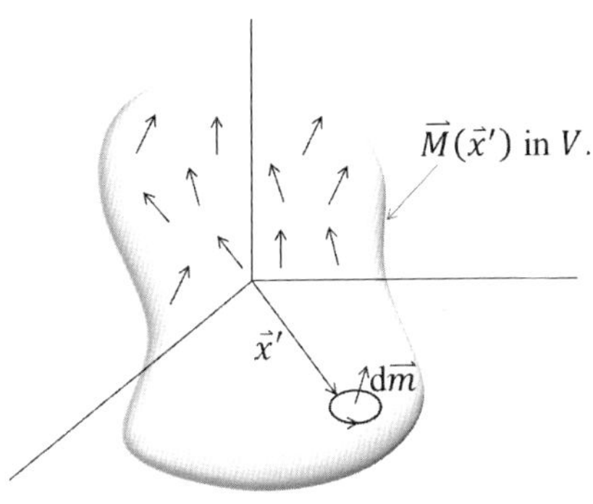

Fig. 6.20 Material with permanent magnetization.

This is the same as above for the $d\vec{H}^m$ field. Therefore,

$$d\Phi_m(\vec{x}) = \frac{d\vec{m}\cdot\vec{x}}{r^3}. \tag{6.250}$$

We have essentially changed the magnetostatic problem into an electrostatic one! When the dipole is at $\vec{x}\,'$ this becomes

$$d\Phi_m(\vec{x}) = \frac{d\vec{m}\cdot(\vec{x}-\vec{x}\,')}{|\vec{x}-\vec{x}\,'|^3}. \tag{6.251}$$

We get the same results by introducing *magnetic* dipole sources

$$d\rho_m(\vec{x}) = -d\vec{m}\cdot\vec{\nabla}\delta(\vec{x}-\vec{x}\,'), \tag{6.252}$$

with

$$\vec{\nabla}\cdot d\vec{H}^m = 4\pi\, d\rho_m, \tag{6.253}$$

in analogy to Equation (6.150).

If we now presume a continuous distribution of magnetic dipoles, then we may replace $d\vec{m}$ with $\vec{M}(\vec{x}\,')d^3x'$ and integrate, leading to

$$\begin{aligned}\Phi_m(\vec{x}) &= \int d^3x'\frac{\vec{M}(\vec{x}\,')\cdot(\vec{x}-\vec{x}\,')}{|\vec{x}-\vec{x}\,'|^3} &(6.254)\\ &= \int d^3x'\vec{M}(\vec{x}\,')\cdot\vec{\nabla}'\frac{1}{|\vec{x}-\vec{x}\,'|}\\ &= \int d^3x'\vec{\nabla}'\cdot\left(\frac{\vec{M}(\vec{x})}{|\vec{x}-\vec{x}\,'|}\right) - \int d^3x'\frac{\vec{\nabla}'\cdot\vec{M}(\vec{x}\,')}{|\vec{x}-\vec{x}\,'|}\\ &= \oint_S da'\frac{\vec{M}(\vec{x}\,')\cdot\hat{n}'}{|\vec{x}-\vec{x}\,'|} - \int_V d^3x'\frac{\vec{\nabla}'\cdot\vec{M}(\vec{x}\,')}{|\vec{x}-\vec{x}\,'|}, &(6.255)\end{aligned}$$

where S is the boundary of volume V which contains the material. The $\vec{H}$ field is given by $\vec{H} = -\vec{\nabla}\Phi_m$. This expression may be used when $\vec{M}$ is either intrinsic (permanent magnetization) or when it is induced. In the latter case the total field is $\vec{H} = \vec{H}^{\text{ext}} + \vec{H}^{\text{ind}}$, where $\vec{H}^{\text{ind}} = -\vec{\nabla}\Phi_m^{\text{ind}}$.

As an example consider the case of a uniformly magnetized sphere, as shown in Figure 6.21. To start off, we begin with the relatively simple problem of evaluating $\vec{H}$ at the origin. Since $\vec{\nabla}'\cdot\vec{M}(\vec{x}\,') = 0$ in V, we get

$$\vec{H}(\vec{x}) = -\vec{\nabla}\Phi_m = \oint da'\frac{\vec{M}(\vec{x}\,')\cdot\hat{n}'(\vec{x}-\vec{x}\,')}{|\vec{x}-\vec{x}\,'|^3}, \tag{6.256}$$

$$\Longrightarrow \vec{H}(0) = -\frac{M_0}{a^3}\oint da'\left(\hat{k}\cdot\hat{e}_{r'}\right)\vec{x}\,'. \tag{6.257}$$

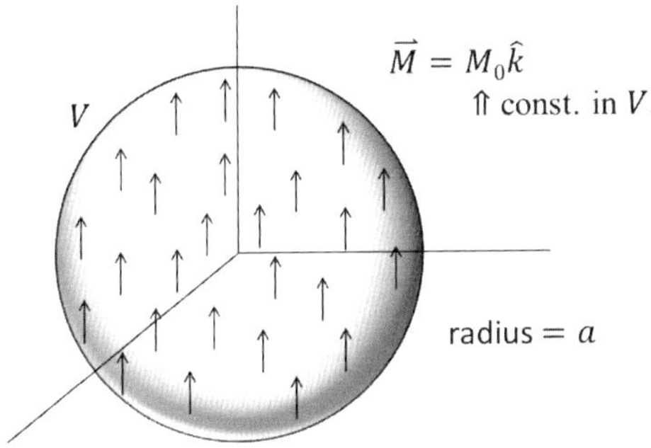

Fig. 6.21 Magnetized sphere.

In spherical coordinates, $da' = -a^2 d\cos\theta' d\phi'$ and $\vec{x}\,' = a[\sin\theta' cos\phi'\hat{\imath} + \sin\theta'\sin\phi'\hat{\jmath} + \cos\theta'\hat{k}]$; the ϕ-dependent terms in $\vec{x}\,'$ integrate to zero, and we are left with $(x \equiv \cos\theta')$

$$\vec{H}(0) = -2\pi M_0\hat{k}\int_{-1}^{1} dx\, x^2 = -\frac{4\pi}{3}\vec{M}, \tag{6.258}$$

$$\vec{B}(0) = \vec{H}(0) + 4\pi\vec{M} = \frac{8\pi}{3}\vec{M}. \tag{6.259}$$

Now we tackle the more ambitious goal of finding the field everywhere. Using (6.254) we have

$$\begin{aligned}\Phi_m(\vec{x}) &= M_0\int_{\text{sphere}} d^3x'\,\frac{(z-z')}{|\vec{x}-\vec{x}\,'|^3}\\ &= -M_0\frac{\partial}{\partial z}\int_{\text{sphere}} d^3x'\,\frac{1}{|\vec{x}-\vec{x}\,'|}.\end{aligned} \tag{6.260}$$

In the case where $\vec{x}$ is inside the sphere, then we may evaluate the integral as (take z' along $\vec{x}$):

$$\begin{aligned}\int_{\vec{x}\text{ in}} d^3x'\,\frac{1}{|\vec{x}-\vec{x}\,'|} &= \int\frac{r'^2\sin\theta' d\theta' d\phi' dr'}{\sqrt{r'^2+r^2-2r'r\cos\theta'}}\\ &= 2\pi\int_0^a dr'\frac{r'}{r}\,\sqrt{r'^2+r^2-2r'r\cos\theta}\Big|_0^{\pi}\\ &= 2\pi\int_0^a dr'\frac{r'}{r}(|r'+r|-|r'-r|)\\ &= \frac{2\pi}{r}\left[\int_0^r dr'\,r'(2r') + \int_r^a dr' r'(2r)\right]\\ &= 2\pi\left[a^2-\frac{1}{3}r^2\right].\end{aligned} \tag{6.261}$$

Therefore

$$\Phi_m = -2\pi M_0 \frac{\partial}{\partial z}\left[a^2 - \frac{1}{3}(x^2+y^2+z^2)\right]$$
$$= \frac{4\pi}{3} M_0 z. \tag{6.262}$$

We have finally for the $\vec{H}$- and $\vec{B}$-fields,

$$\vec{H} = -\frac{4\pi}{3}\vec{M}, \tag{6.263}$$
$$\vec{B} = \frac{8\pi}{3}\vec{M}, \tag{6.264}$$

everywhere inside. (Note the minus sign for $\vec{H}$ relative to $\vec{B}$.) When $\vec{x}$ is outside the sphere, then direct integration still works as above, with minor changes. An alternative evaluation method uses an expansion in Legendre polynomials (z' along $\vec{x}$ again):

$$\int_{\vec{x}\text{ out}} d^3x' \frac{1}{|\vec{x}-\vec{x}\,'|} = \int_{\vec{x}\text{ out}} d^3x' \sum_{\ell=0}^{\infty} \frac{r'^{\ell}}{r^{\ell+1}} P_\ell(\cos\theta'),$$
$$= \frac{4\pi}{r}\int_0^a r'^2 dr' = \frac{4\pi}{3}\frac{a^3}{r}. \tag{6.265}$$

This gives

$$\Phi_m(\vec{x}) = -M_0 \frac{\partial}{\partial z}\left(\frac{4\pi}{3}\frac{a^3}{r}\right) = \frac{4\pi}{3} M_0 a^3 \frac{\cos\theta}{r^2} = \frac{\vec{m}\cdot\vec{x}}{r^3}, \tag{6.266}$$

where

$$\vec{m} \equiv \frac{4\pi a^3}{3}\vec{M}. \tag{6.267}$$

Equations (6.266) and (6.267) show that the field external to the sphere is identical to that of a perfect magnetic dipole located at the origin with strength $4\pi a^3|M_0|/3$. Note that Equation (6.196) gives an effective surface current for this situation as

$$\vec{K}_{\text{eff}} = c\vec{M}\times\hat{n} = cM_0 \sin\theta\, \hat{e}_\phi, \tag{6.268}$$

where θ is the polar angle and $\hat{e}_\phi$ is the unit vector in the azimuthal direction. This connects to another treatment using the vector potential approach; see Exercise 6.7.3.

6.14 Going Deeper

6.14.1 *Magnetic materials*

(1) S. Blundell, *Magnetism in Condensed Matter*, Oxford University Press (New York) 2001.

(2) C. A. Brau, *Modern Problems in Classical Electrodynamics*, Oxford University Press (Oxford) 2004; Section 6.2.2.

(3) J. M. D. Coey, *Magnetism and Magnetic Materials*, Cambridge University Press (Cambridge) 2009.

(4) B. D. Cullity and C. D. Graham, *Introduction to Magnetic Materials*, 2nd ed., John Wiley & Sons, IEEE Press (Hoboken, NJ) 2009.

(5) D. Jiles and D. Atherton, "Theory of Magnetic Hysteresis", Journal of Magnetism and Magnetic Materials **61**, 48 (1986).

(6) C. Kittel, *Introduction to Solid State Physics*, 8th ed., John Wiley & Sons (Hoboken, NJ) 2005; Chapters 11, 12.

(7) L. D. Landau, L. P. Pitaevskii and E. M. Lifshitz, *Electrodynamics of Continuous Media*, 2nd ed. with corrections, Elsevier (New York) 1993; Chapters V and VI.

(8) F. Preisach, "Über die Magnetische Nachwirkung", Zeits. f. Physik A **94**, 277 (1935).

(9) N. A. Spaldin, *Magnetic Materials: Fundamentals and Applications*, 2nd ed., Cambridge University Press (Cambridge) 2011.

(10) J. Tellinen, "A Simple Scalar Model for Magnetic Hysteresis", IEEE Transactions on Magnetics, **24**, 2200 (1998).

6.15 Exercises

Exercise 6.1.1. An irregularly shaped planar circuit carrying current I is partly immersed in a uniform magnetic field $\vec{B}$ as shown. The $\vec{B}$ field points perpendicularly to the plane of the circuit and is nonzero only in the crosshatched region shown in Figure 6.22. Show that the force on the circuit is given by ($I > 0$)

$$\vec{F} = \frac{I}{c}\vec{L} \times \vec{B},$$

where $\vec{L}$ is the vector pointing from one side of the circuit to the other where the magnetic field vanishes.

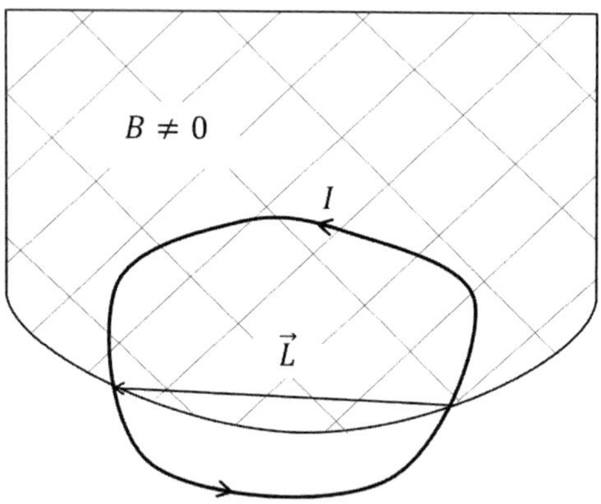

Fig. 6.22 Geometry for Exercise 6.1.1.

Exercise 6.3.1. A long straight wire with circular cross-section and radius a carries a uniform current density flowing along its length. Its cross-section is shown in Figure 6.23. The total current is I, a constant in time. The differential radial pressure exerted on a tube of radius ρ and arbitrary length along the wire by the current can be calculated as

$$\frac{dP(\rho)}{d\rho} = \frac{1}{A(\rho)}\frac{dF_\rho}{d\rho},$$

where F_ρ is the infinitesimal radial force and $A(\rho)$ is the cylindrical tube surface area. Integrate your expression to find $P(\rho)$ at all distances. Is the pressure inward or outward?

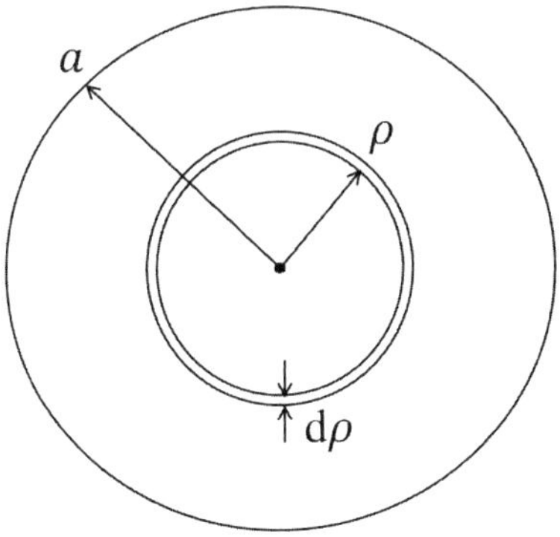

Fig. 6.23 Geometry for Exercise 6.3.1.

Exercise 6.3.2. An infinitely long, hollow cylinder of radius R has a uniform surface charge density, σ, on its surface (see Figure 6.24). It is spinning at angular velocity, ω, about the z-axis (symmetry axis). In addition, it is given a velocity, v_0, along the z-axis. Using Ampère's law, find the magnetic field $\vec{B}$ both inside and outside the cylinder.

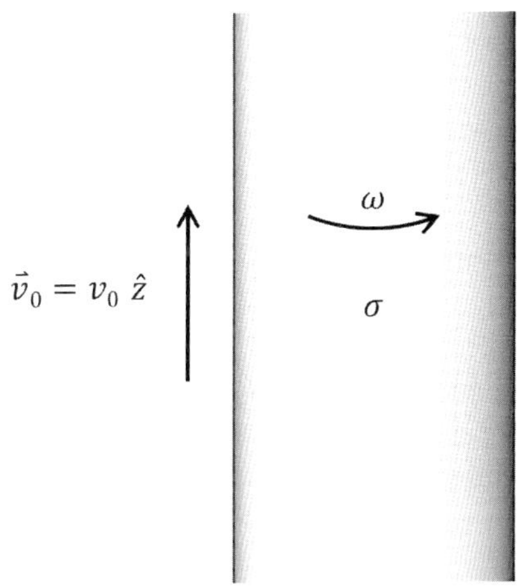

Fig. 6.24 Geometry for Exercise 6.3.2.

Exercise 6.4.1. A thin wall of constant surface current density $\vec{K} = K_0\hat{z}$ extends to infinity in the z- and y-directions (see Figure 6.25). Find the components of the resultant magnetic field on both sides of the sheet.

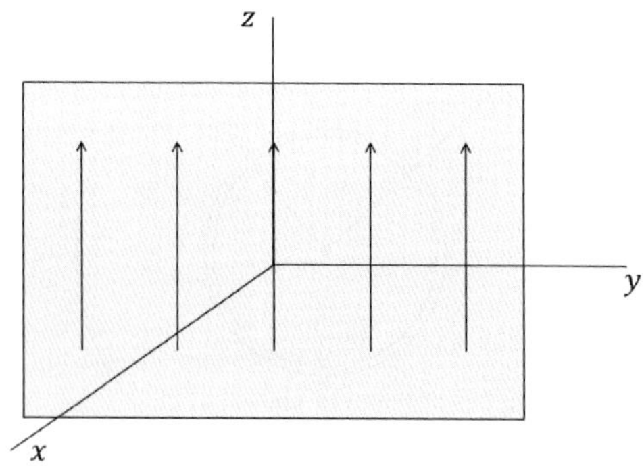

Fig. 6.25 Geometry for Exercise 6.4.1.

Exercise 6.5.1. Transform Stokes' theorem,

$$\oint_C \vec{A} \cdot d\vec{\ell} = \int_S da\, \hat{n} \cdot (\vec{\nabla} \times \vec{A})$$

into ($d\vec{s} = da\, \hat{n}$)

$$\oint_C d\vec{\ell} \times \vec{A} = \int_S (d\vec{s} \times \vec{\nabla}) \times \vec{A}.$$

Exercise 6.6.1.

(a) We found expressions for the z and ρ components of the magnetic field for a current loop of radius a carrying a current I in Equations (6.83) and (6.84). Build on these results by finding the z and ρ components of the magnetic field of a tightly wound solenoid of length d and radius a with n turns per unit length of insulated wire carrying current I; see Figure 6.26. Taking your origin at the geometrical center of the rod, show that in the region $-d/2 < z < d/2$ we have

$$B_\rho(\rho, z) = \frac{4\pi a n I}{c} \int_0^\infty dk\, J_1(ka) J_1(k\rho) e^{-kd/2} \sinh(kz),$$

$$B_z(\rho, z) = \frac{4\pi n I}{c} H(\rho) - \frac{4\pi a n I}{c} \int_0^\infty dk\, J_1(ka) J_0(k\rho) e^{-kd/2} \cosh(kz),$$

where $H(\rho)$ is the step function, which is 1 for $\rho < a$ and 0 for $\rho > a$. (The first term in B_z is the solution for an infinitely long solenoid; the second is a correction term.)

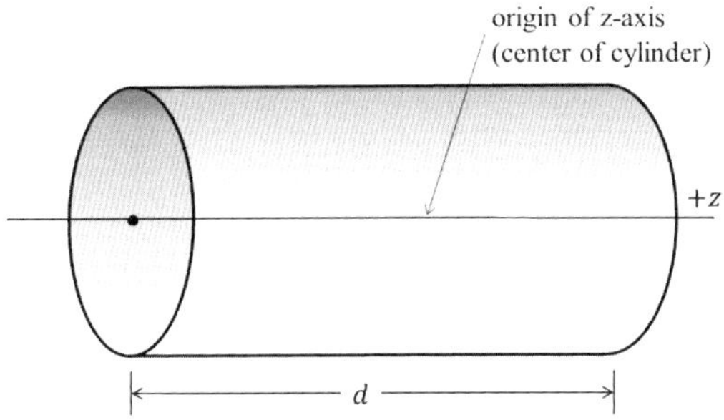

Fig. 6.26 Geometry for Exercise 6.6.1.

(b) Find some way of approximating the second term in the B_z expression and show that this term falls like $1/d^2$ for $d \gg \rho$, a and $z = 0$ (the middle). (This shows how the outside field falls off for longer solenoids.)

Exercise 6.6.2. A thin disk of radius R with a constant surface charge density σ is spinning with constant angular velocity ω about a line through the center of the disk, perpendicular to the disk's plane. The disk is in the $z = 0$ plane. Show that the magnetic field components B_ρ and B_z are given in cylindrical coordinates ρ and z by:

$$B_\rho = \text{sign}(z)\frac{2\pi\omega\sigma R^2}{c}\int_0^\infty dk\; e^{-k|z|}J_2(kR)J_1(k\rho),$$

$$B_z = \frac{2\pi\omega\sigma R^2}{c}\int_0^\infty dk\; e^{-k|z|}J_2(kR)J_0(k\rho).$$

Extra: In the limit $|z| >> \rho, a, b$, show that the B_z field becomes

$$B_z \approx \frac{\pi\sigma\omega}{2c}\frac{R^4}{|z|^3}.$$

From this result deduce the magnetic moment m_z of the spinning disk.

Exercise 6.6.3. A solid cylinder with radius R and length d and a constant internal volume charge density ρ_0 is set spinning at angular velocity ω around its symmetry axis. Taking your origin of coordinates at the center of the cylinder and your z-axis along the cylinder axis, show that the B_z field in the region $|z| < d/2$ is given by

$$B_z(\rho, z) = \frac{2\pi\rho_0\omega}{c}\left(R^2 - \rho^2\right) H(R) - \frac{4\pi\rho_0\omega R^2}{c}\int_0^\infty \frac{dk}{k} e^{-kd/2}\cosh(k\,z) J_2(k\,R) J_0(k\rho),$$

where $H(R)$ is a step function which is 0 for $\rho > R$ and 1 for $\rho < R$. [*Hint*: Use the result of either Exercise 6.6.1 or Exercise 6.6.2.]

Exercise 6.6.4. Given the volume current for a thin circular wire of radius a in the $z' = 0$ plane which carries current I,

$$\vec{J}(\vec{x}\,') = I\,a\frac{\delta(\rho' - a)}{\rho'}\delta(z')\,\hat{e}_{\phi'},$$

(cylindrical coordinates) where the x, y components of the azimuthal unit vector $\hat{e}_{\phi'}$ are

$$\hat{e}_{\phi'} = -\sin\phi'\,\hat{x} + \cos\phi'\,\hat{y},$$

show that an alternate form of the vector potential may be evaluated as

$$\vec{A}(\rho, \phi) = \frac{4Ia}{c}\hat{e}_\phi \int_0^\infty dk\cos(kz) I_1(k\rho_<) K_1(k\rho_>).$$

See Section 4.6 for the definition of the modified Bessel functions $I_m(x)$ and $K_m(x)$. $I_1(k\rho_<)$ is evaluated for the lesser of ρ and a, while $K_1(k\rho_>)$ is evaluated for the greater of those same quantities.

Exercise 6.6.5. A truncated circular cone with opening angle α has its symmetry axis oriented along the z-axis. The origin is taken at the theoretical tip of the cone. The end of the cone is shaved off at a distance L_1 from the origin along the z-axis from the origin, and the distance from the origin to the top of the cone along z is L_2. The cone carries a uniform surface charge density, σ (see Exercise 2.4.4), and is rotated about the z-axis at angular velocity ω, so that the surface current is given by $\vec{K} = \sigma\omega\rho\,\hat{e}_\phi$ in cylindrical coordinates; see Figure 6.27.

The magnetic field along z from a single circular current loop is known from Equation (6.62). Sum the currents and evaluate B_z at the origin (theoretical tip).

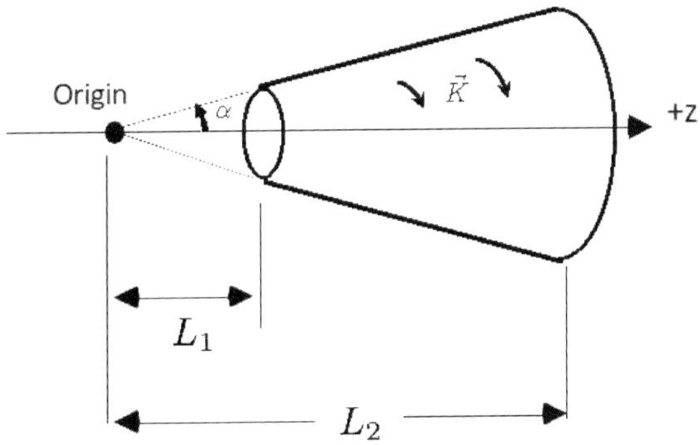

Fig. 6.27 Geometry for Exercise 6.6.5.

Exercise 6.7.1. Use the Coulomb expansion (4.225) in (6.69) to evaluate the solid angle of a flat circular surface of radius a in spherical rather than cylindrical coordinates. Orient the surface in the $x-y$ plane with the z axis through the middle, like the current loop in Sections 6.6 and 6.7. Show that

$$\Omega(\vec{x}) = -2\pi \sum_{n=0,1,2,\ldots}^{\infty} \left(\frac{a}{r}\right)^{2n+2} \frac{(-1)^n(2n+1)!!}{(2n+2)!!} P_{2n+1}(\cos\theta),$$

when $r \geq a$ and that

$$\Omega(\vec{x}) = -2\pi \sum_{n=0,1,2,\ldots}^{\infty} \frac{(-1)^n(2n-1)!!}{(2n)!!} \left[\frac{4n+3}{2n+2} - \left(\frac{r}{a}\right)^{2n+1}\right] P_{2n+1}(\cos\theta),$$

when $r \leq a$. Notice the two results agree when $r = a$. For the latter case show that we may separate out a sign function to give the form,

$$\Omega(\vec{x}) = -2\pi\,\mathrm{sign}(z) + 2\pi \sum_{n=0,1,2,\ldots}^{\infty} \frac{(-1)^n(2n-1)!!}{(2n)!!} \left(\frac{r}{a}\right)^{2n+1} P_{2n+1}(\cos\theta).$$

[*Hints*: Equations (4.181)–(4.183) as well as Exercise 4.12.1. Also note that

$$\hat{z} = \cos\theta\,\hat{r} - \sin\theta\,\hat{\theta} \implies \frac{\partial}{\partial z} = \cos\theta\frac{\partial}{\partial r} + \frac{\sin^2\theta}{r}\frac{\partial}{\partial\cos\theta}.]$$

Exercise 6.7.2. Using the expressions for the solid angle derived in the last exercise, recover the results for the B_r and B_θ fields in Equations (6.119) and (6.121) for both the $r > a$ and $r < a$ cases.

Exercise 6.7.3. A uniform surface charge density, σ, is smeared over a thin spherical shell of radius a which is set spinning at a constant angular velocity ω; see Figure 6.28. Using the vector potential approach to calculating the magnetic field, show that the magnetic fields inside and outside the sphere are:

$$\vec{B}(\vec{x}) = \begin{cases} \dfrac{8\pi a\sigma\omega}{3c}\hat{z}, & r < a, \\ \dfrac{3\vec{x}\,(\vec{m}\cdot\vec{x}) - r^2\vec{m}}{r^5}, & r > a, \end{cases}$$

with

$$\vec{m} = \frac{4\pi\sigma\omega a^4}{3c}\hat{z}.$$

This shows that the magnetic field outside is that of a perfect magnetic dipole. (This situation has the same $\vec{B}$ field as the uniformly magnetized sphere discussed in Section 6.13, if one chooses $M_0 = \sigma\omega a/c$.)

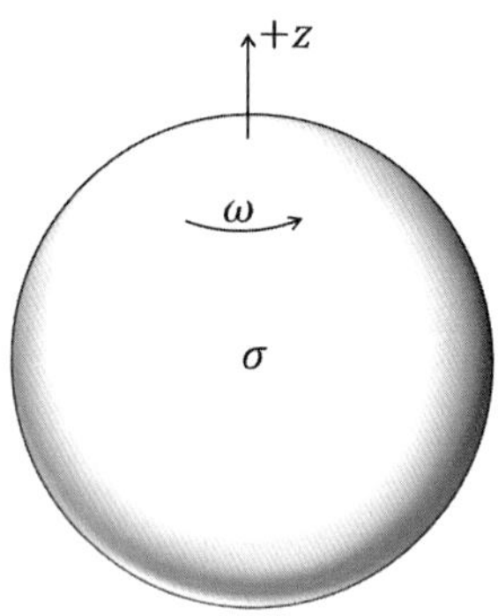

Fig. 6.28 Geometry for Exercise 6.7.3.

Exercise 6.7.4. A *solid* sphere of radius a with a constant volume charge density ρ is spinning with constant angular velocity ω about a line through the center of the disk (the z axis).

(a) Show that the field outside the sphere is a perfect magnetic dipole field with a magnetic moment, m, given by $m = 4\pi a^5\omega\rho/(15c)$.
(b) Show that the field inside the sphere is a combination of an r-dependent dipole field ($m = 4\pi r^5\omega\rho/(15c)$) and an additional $\hat{z}$ directed field, which vanishes at $r = a$.

Exercise 6.7.5. A rotating charged sphere as in Exercise 6.7.3 (radius a, angular frequency ω) has a non-uniform surface charge density given by $\sigma(\theta) = \sigma_0 \cos\theta$, where θ is the polar angle in spherical coordinates. The rotating sphere generates a surface current, $\vec{K}(\theta) = \omega\, a\, \sigma(\theta) \sin\theta\, \hat{e}_\phi$, giving rise to a vector potential. Show that the magnetic field generated inside $(r < a)$ and outside $(r > a)$ the sphere in Cartesian coordinates is

$$\vec{B} = \begin{cases} \dfrac{4\pi\sigma_0\omega}{5c}\left(-x\,\hat{x} - y\,\hat{y} + 2z\,\hat{z}\right), r < a \\ \dfrac{4\pi\sigma_0 a^5\omega}{5c\,r^7}\left(x(5z^2 - r^2)\hat{x} + y(5z^2 - r^2)\hat{y} - z(5(x^2+y^2) - 2r^2)\hat{z}\right), r > a. \end{cases}$$

Exercise 6.8.1.

(a) Consider again the B_z field from a current loop for a current I from Equation (6.84). Similar to Exercise 6.6.1, show that B_z field of a solenoid with current I, radius a, length d, and n insulated turns per unit length for $|z| > d/2$ is now

$$B_z = \frac{4\pi a n I}{c} \int_0^\infty dk\; J_1(ka) J_0(k\rho) e^{-k|z|} \sinh(kd/2),$$

for an origin at the geometrical center of the rod.

(b) For $|z| \gg d, \rho, a$ show that the field is approximately

$$B_z \approx \frac{2\pi a^2 I N}{c} \frac{1}{|z|^3},$$

where N is the total number of loops $(dn \equiv N)$.

(c) Treating the solenoid to lowest order as a point magnetic dipole, show the result in (b) from a multipole point of view.

Exercise 6.8.2. An extremely thin, flat piece of material is shaved off the end of a permanent magnet; see Figure 6.29. The magnetization in the material is uniform and perpendicular to its plane. ($\vec{M} = M_0\,\hat{n}$, *inside*) Show that the field is equivalent to a circuit element of the same shape as the outside boundary, C, of the material, with effective current $I = cM_0 d$, where d is the thickness of the material. [*Hint*: The form for the $\vec{A}$ field you want to end up with is

$$\vec{A} = M_0 d \oint_C \frac{d\vec{\ell}\,'}{|\vec{x} - \vec{x}\,'|}.$$

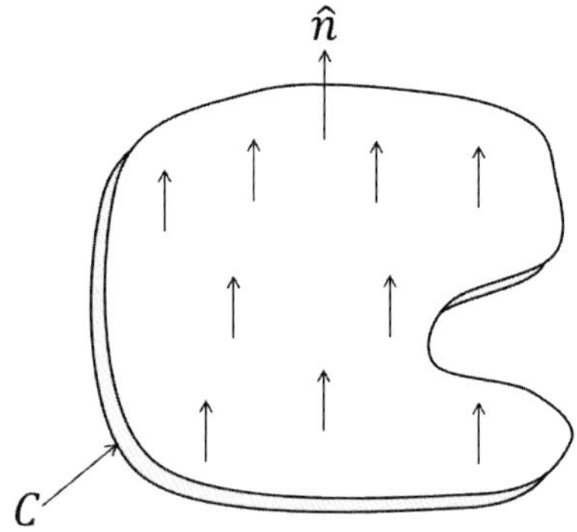

Fig. 6.29 Geometry for Exercise 6.8.2.

See Equations (6.42) and (6.22). An alternate approach uses the Exercise 6.13.1 result. Actually, the assumption above that the thin slab must be flat may be removed if the magnetization remains perpendicular to the surface.]

Exercise 6.8.3.

(a) Consider the circuit shown in Figure 6.30(a). Find its magnetic moment.

(b) A circular current loop with radius R is bent 90° along the x-diagonal, as shown in Figure 6.30(b). Again, find the magnetic moment of the loop.

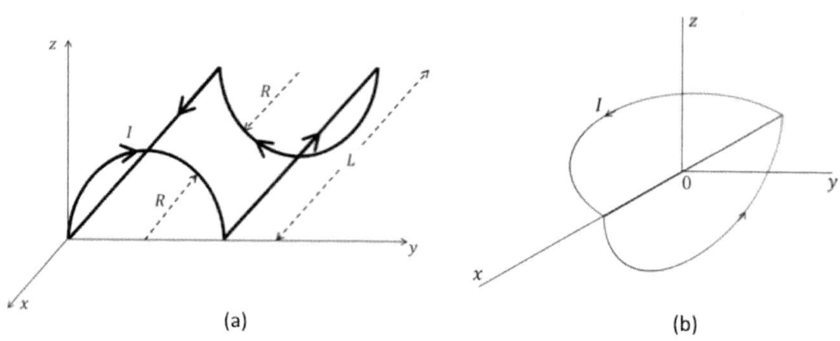

Fig. 6.30 Geometry for Exercise 6.8.3.

Exercise 6.8.4.

(a) Show that the magnetic field implied by

$$\vec{A}^m(\vec{x}) = \frac{\vec{m} \times \vec{r}}{r^3}$$

is

$$\vec{B}^m(\vec{x}) = \frac{8\pi}{3}\vec{m}\delta(\vec{x}) + \frac{3\vec{r}(\vec{m} \cdot \vec{r}) - mr^2}{r^5}.$$

(b) Also show that

$$\vec{\nabla} \times \vec{B}^m = \frac{4\pi}{c}\vec{J}^m(\vec{x}),$$

where

$$\vec{J}^m(\vec{x}) = -c\vec{m} \times \vec{\nabla}\delta(\vec{x}).$$

[*Hint*: The identity derived for $\nabla_i \nabla_j \frac{1}{r}$ should be of use.]

Exercise 6.8.5.

(a) Show the following cyclic identity in magnetostatics for a given spatial current $J_i(x')$ (i, j and k are free indices):

$$0 = \int d^3x' \, \left(x'_i x'_j J_k(\vec{x}\,') + x'_i J_j(\vec{x}\,') x'_k + J_i(\vec{x}\,') x'_j x'_k\right).$$

(b) In addition using part (a) show that

$$\frac{1}{c}\int d^3x' \, x'_i x'_j J_k(\vec{x}\,') = -\frac{1}{2}\sum_r (\varepsilon_{kir} m_{rj} + \varepsilon_{kjr} m_{ri}),$$

where the ε_{ijk} are the antisymmetric permutation symbols, and the "magnetic quadrupole moments" m_{ij} are given by

$$m_{ij} \equiv \frac{2}{3c}\int d^3x' \, \left(\left(\vec{x}\,' \times \vec{J}(\vec{x}\,')\right)_i x'_j\right).$$

(c) The first three terms of the Taylor series expansion of the Coulomb potential are

$$\frac{1}{|\vec{x} - \vec{x}\,'|} = \frac{1}{r} + \frac{\vec{x} \cdot \vec{x}\,'}{r^3} + \frac{1}{2}\sum_{i,j} \frac{x'_i x'_j}{r^5}(3x_i x_j - \delta_{ij} r^2) + \ldots.$$

Show that the next term in the Taylor series expansion for the vector potential in (6.123), termed "magnetic quadrupole" and denoted as $A_k^{(q)}(\vec{x})$, can be written

$$A_k^{(q)}(\vec{x}) = -\frac{1}{2cr^5}\sum_{i,j,r} \varepsilon_{kir} Q_{ij}(\vec{x})\, m_{rj}$$

where

$$Q_{ij}(\vec{x}) \equiv (3x_i x_j - r^2\,\delta_{ij}),$$

Exercise 6.8.6.

(a) Using the result of Exercise 6.8.5, show that the magnetic quadrupole field can be expressed as

$$B_k^{(q)} = \frac{3}{2\,r^7}\sum_{i,j} s_{ij}(5x_i x_j x_k - \delta_{ik} r^2 x_j - \delta_{jk} r^2 x_i),$$

where

$$s_{ij} \equiv \frac{1}{2}(m_{ij} + m_{ji}).$$

(b) Show that there are only 5 components of the symmetric magnetic quadrupole tensor s_{ij} just as for the electric quadrupole tensor.

(c) By direct computation or by comparing to the result in Exercise 6.7.5, deduce the values of the s_{ij} which reproduce the magnetic field in this exercise for $r > a$, showing this is a pure magnetic quadrupole field.

Exercise 6.9.1. Consider a point magnetic dipole with arbitrary magnetic moment $\vec{m}$ located next to a long straight thin wire carrying current I along the z-axis. With the dipole located at $x = x_0, y = 0$ in Cartesian coordinates, find the components of the force $\vec{F}$ and torque $\vec{N}$ on the dipole.

Exercise 6.9.2. The expression for the force on an arbitrary current loop in a magnetic field $\vec{B}$ is given by

$$\vec{F} = \frac{I}{c}\oint d\vec{x}\, \times \vec{B}(\vec{x}),$$

where $d\vec{x}$ is taken along the current. By doing a Taylor series expansion of the ith component of the magnetic field $B_i(\vec{x})$ in the vicinity of the loop,

$$B_i(\vec{x}) \approx B_i(0) + \vec{x} \cdot \vec{\nabla}' \, B_i(\vec{x}\,')|_{\vec{x}\,'=0},$$

re-derive the expression for the force on a magnetic dipole,

$$\vec{F} \approx \vec{\nabla}(\vec{m} \cdot \vec{B}),$$

where the magnetic moment, $\vec{m}$, is defined by

$$\vec{m} \equiv \frac{I}{2c} \oint \vec{x} \times d\vec{x}.$$

Exercise 6.10.1.

(a) Use the expression for a bound surface current,

$$\vec{K}_{\text{eff}} = c\vec{M} \times \hat{n},$$

($\hat{n}$ is the outward normal) and the equation for the surface torque on a magnet placed in a uniform external magnetic field $\vec{B}_0$,

$$\vec{N} = \oint_s da \, \vec{x} \times \left[\frac{1}{c} \vec{K}_{\text{eff}} \times \vec{B}_0 \right],$$

to show that

$$\vec{N} = \vec{m} \times \vec{B}_0,$$

where $\vec{m} \equiv \int d^3x \, \vec{M}$, and $\vec{M}$ and $\vec{B}_0$ are constants in V. (The answer here is not surprising given (6.177); however, this shows *where* the torque is arising.)

(b) A hemisphere of radius a is immersed in an external uniform magnetic field oriented along the x-axis, as shown in Figure 6.31.

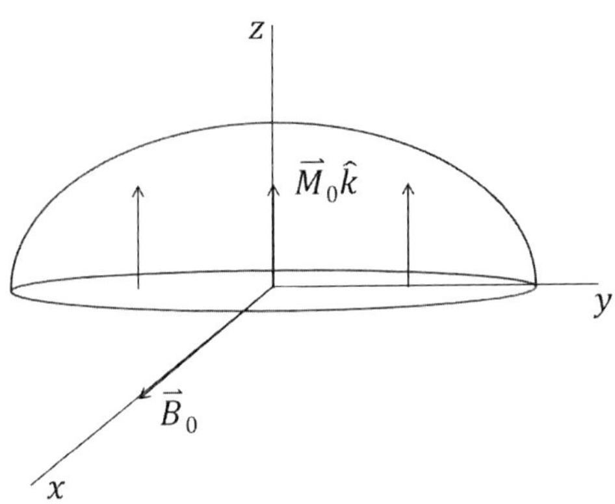

Fig. 6.31 Geometry for Exercise 6.10.1.

The magnetization within the hemisphere is also uniform, and is given by $\vec{M} = M_0\hat{k}$. Use part (a) to evaluate the torque on the hemisphere.

Exercise 6.10.2. A point magnetic dipole is located at the center of a spherical cavity (radius a) of vacuum embedded in a region of constant magnetic permeability μ; see Figure 6.32.

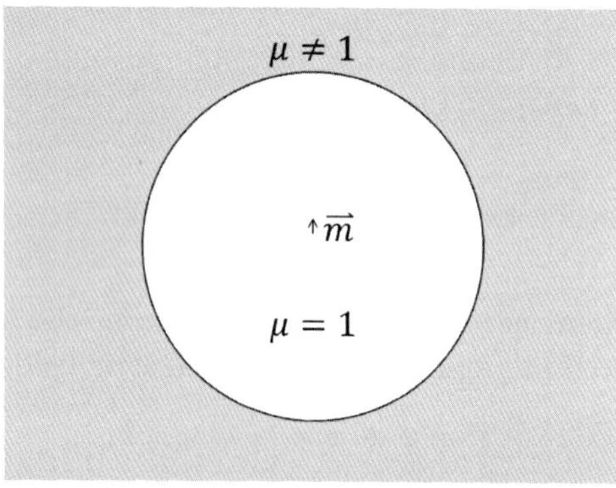

Fig. 6.32 Geometry for Exercise 6.10.2.

(a) Assume that the vector potentials inside and outside the sphere can be written as

$$\vec{A}_{in} = \frac{\vec{m} \times \vec{r}}{r^3} + \frac{C_1}{a^3}\vec{m} \times \vec{r},$$
$$\vec{A}_{out} = C_2\frac{\vec{m} \times \vec{r}}{r^3}.$$

Find the values of $C_{1,2}$ and thus the values of the magnetic fields, $\vec{B}_{in,out}$.

(b) The same magnetic dipole is now put at the center of a sphere of magnetic permeability μ surrounded by vacuum. Find the $\vec{B}$ field inside and outside the sphere. [*Hint*: There is an easy way of getting (b) from (a).]

(c) Find the bound surface current $\vec{K}_b$ for part (a) and show it has the same form as for the free current $\vec{K}_f$ for the rotating sphere in Exercise 6.7.3. Do this by showing that the two problems are connected by the substitution

$$\sigma\omega a \leftrightarrow \frac{3c}{4\pi a^3}\left(\frac{\mu - 1}{2\mu + 1}\right).$$

Exercise 6.11.1.

(a) Find the magnetic scalar potential Φ_m and thus the $\vec{B}, \vec{H}$ fields for a spherical piece of permeable material with magnetic permeability constant μ and radius a in a uniform external magnetic field $\vec{B}_0$. The easy way to do this is to use electric/magnetic interchangeability in the context of the answer given to Exercise 5.7.3.

(b) A magnetization $\vec{M}$ is induced in the sphere from the external field. Use the part (a) result to deduce that

$$\vec{M} = \frac{3}{4\pi}\left(\frac{\mu - 1}{\mu + 2}\right)\vec{B}_0.$$

Exercise 6.12.1. A current distribution $\vec{J}(\vec{x})$ exists in vacuum adjacent to a semi-infinite plane slab of material having permeability $\mu \neq 1$ and filling the half-space $z < 0$; see Figure 6.33. Show that for $z > 0$ the magnetic field can be calculated by replacing the medium of permeability μ by an image current $\vec{J}^*$ with components,

$$(a\ J_x(x, y, -z), a\ J_y(x, y, -z), b\ J_z(x, y, -z))$$

where

$$a = -b = \left(\frac{\mu - 1}{\mu + 1}\right).$$

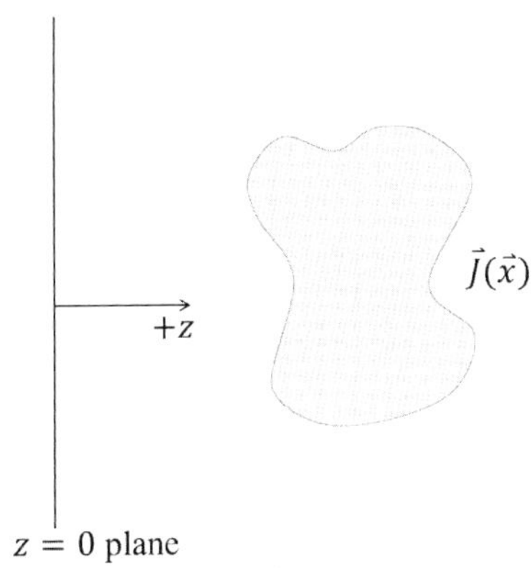

Fig. 6.33 Geometry for Exercise 6.12.1.

In addition, show that for $z < 0$ the magnetic field appears to be due to a current distribution $\vec{J}^{**} = c\vec{J}$, where $c = 2\mu/(\mu + 1)$.

Exercise 6.12.2. In contrast to Exercise 6.12.1, the free current distribution $\vec{J}(\vec{x})$ is embedded in the semi-infinite plane slab of material with magnetic susceptibility $\mu \neq 1$, filling the half-space $z > 0$; the other side is free space. The magnetic field for $z > 0$ is assumed to be given by the field of the real current, $\vec{J}(\vec{x})$, plus an image current distribution $\vec{J}^*(\vec{x})$ embedded in the same material. Likewise, the situation for $z < 0$ is assumed to be due to a different current, $\vec{J}^{**}(\vec{x})$, at the location of the real current, but in a vacuum. Find how the components of the currents, $\vec{J}^*(\vec{x})$ and $\vec{J}^{**}(\vec{x})$, are related to the real current, $\vec{J}(\vec{x})$, for $z > 0$ and $z < 0$. [*Hint*: An easy approach is to just use the previous solution in Exercise 6.12.1 with appropriate changes.]

Exercise 6.12.3. Two flat, circular surfaces with constant surface charge densities σ and $-\sigma$ are located a very small distance Δd apart forming a flat electric dipole surface with dipole constant D; see Figure 6.34. The dipole plane is parallel to and a distance d away from a flat semi-infinite interface of dielectric constant ϵ. Using the appropriate Green function, find the $\vec{E}$ and $\vec{D}$ fields in the $z > 0$ and $z < 0$ regions (but not on or between the circular surfaces). Then show that the substitutions,

$$\epsilon \Longrightarrow \mu, D(= \sigma\Delta d) \Longrightarrow \frac{I}{c},$$

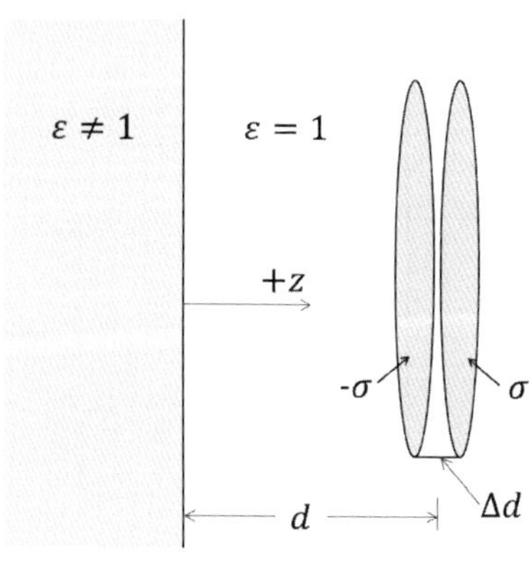

Fig. 6.34 Geometry for Exercise 6.12.3.

give the $\vec{B}$ field from a thin current loop at the edge of the circular surfaces outside a flat interface of permeability μ. [*Ans.:* For the $\vec{B}$ field we have

$$\vec{B} = \frac{I}{c}\left(\vec{\nabla}\Omega_{\text{source}} + \frac{\mu - 1}{\mu + 1}\vec{\nabla}\Omega_{\text{image}}\right), z > 0,$$

$$\vec{B} = \frac{I}{c}\frac{2\mu}{1+\mu}\vec{\nabla}\Omega_{\text{source}}, z < 0,$$

where Ω_{source} and Ω_{image} represent the solid angles subtended by the source and image circular loop surfaces. This problem illustrates the interchangeability of electro- and magnetostatics in source-free regions.]

Exercise 6.12.4. Let us build on the results of Exercise 6.12.3. In open space (with no excluded volumes) it is given that $G(\vec{x}, \vec{x}'; \epsilon)$ represents the *electrostatic* Green function for a point charge in the presence of dielectrics with dielectric constant ϵ. The connection between the source $\rho(\vec{x}\,')$ and the potential $\Phi(\vec{x})$ is given by

$$\Phi(\vec{x}) = \int d^3x' G(\vec{x}, \vec{x}'; \epsilon)\rho(\vec{x}\,').$$

Now introduce an electric dipole surface S' and appropriate dipole sources, starting with Equation (6.150) integrated over S'. Use interchangeability between electrostatics and magnetostatics as we already saw in Exercise 6.12.3,

$$\epsilon \implies \mu, D \implies \frac{I}{c},$$

to argue that the analogous magnetic scalar potential $\Phi_m(\vec{x})$ is given by

$$\Phi_m(\vec{x}) = \frac{I}{c}\int_{S'} da' \frac{\partial}{\partial n'} G(\vec{x}, \vec{x}'; \mu),$$

now representing a current loop carrying current I at the edge of the surface S'. A right-hand rule applies to the $\hat{n}'$ and I directions. We are taking the gradient normal to the surface as in Equation (6.68). Using the $\vec{E} \to \vec{H}$ interchangeability, this means we have

$$\vec{H} = -\vec{\nabla}\Phi_m.$$

This representation holds in source free-regions, which in this case means outside the electric dipole surface S'.

Exercise 6.12.5. Let's define a perfect electric conductor as a material in which $\vec{E}$ and $\vec{B}$ fields are both exactly zero (limits: $\epsilon \implies \infty, \mu \implies 0$).

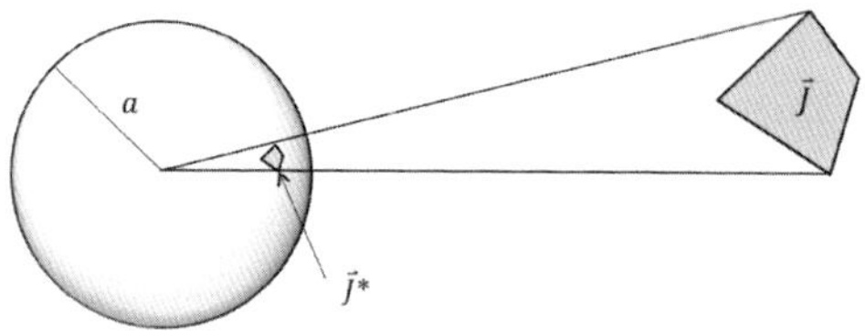

Fig. 6.35 Geometry for Exercise 6.12.5.

Consider a spherical region of radius a which acts as a perfect conductor. There is a magnetostatic current density located outside the conductor as shown in Figure 6.35. Given the real outside current as $\vec{J}(r,\theta,\phi)$ (spherical coordinates), show that the real current plus an image current $\vec{J}^*(r,\theta,\phi)$ with the functional form $(r < a)$

$$\vec{J}^*(r,\theta,\phi) = -\left(\frac{a}{r}\right)^5 \vec{J}\left(\frac{a^2}{r},\theta,\phi\right),$$

or

$$\vec{J}^*\left(\frac{a^2}{r},\theta,\phi\right) = -\left(\frac{r}{a}\right)^5 \vec{J}(r,\theta,\phi),$$

gives a magnetic field $\vec{B}$ which can be made to satisfy the boundary condition,

$$\vec{B}\cdot\hat{n}|_{r=a} = 0,$$

at the sphere's surface. (This is the magnetostatic analog of the spherical image problem in electrostatics we considered in Exercise 3.3.3.)

Exercise 6.12.6. An infinite plane slab with an interface at $z = 0$ has a magnetic susceptibility μ. A small magnetic dipole $\vec{m}$ is located a distance d from the plane (see Figure 6.36). The direction of $\vec{m}$ is located with the azimuthal and polar angles ϕ and θ, respectively, measured with respect to the perpendicular to the plane (the z axis). Find the force on the magnetic dipole. Is it attractive or repulsive for $\mu > 1$?

Exercise 6.13.1. A small magnetic dipole, $d\vec{m}$, located at $\vec{x}'$, gives rise to a vector potential given by

$$d\vec{A}(\vec{x}) = \frac{d\vec{m}\times(\vec{x}-\vec{x}')}{|\vec{x}-\vec{x}'|^3}.$$

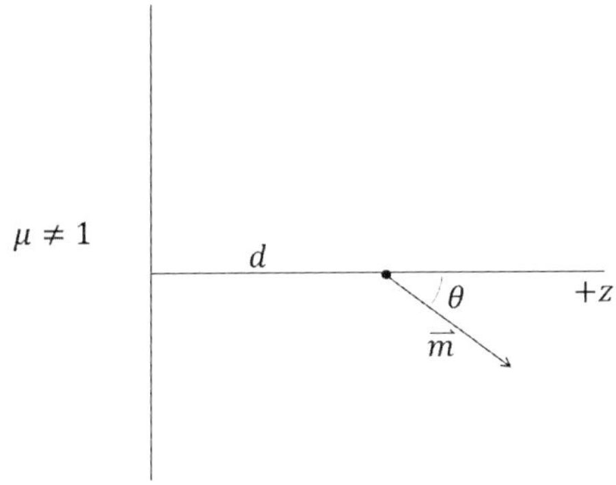

Fig. 6.36 Geometry for Exercise 6.12.6.

Show that, for a volume magnetization, this leads to the expression

$$\vec{A}(\vec{x}) = \int_V d^3x' \frac{\vec{\nabla}' \times \vec{M}(\vec{x}')}{|\vec{x} - \vec{x}'|} - \int_S da' \frac{\hat{n}' \times \vec{M}(\vec{x}')}{|\vec{x} - \vec{x}'|},$$

where the volume and surface contributions are explicitly separated. (These results are consistent with volume and surface currents given by $\vec{J}_{\text{eff}} = c\vec{\nabla} \times \vec{M}, \vec{K}_{\text{eff}} = c\vec{M} \times \hat{n}$; see Equations (6.195), (6.196).)

Exercise 6.13.2. A spherical magnet of radius a has a split magnetic profile as shown in Figure 6.37. The top half of the magnet has a uniform magnetization $\vec{M}$ in the positive z-direction, and the bottom half has a magnetization in the $-\vec{M}$ direction.

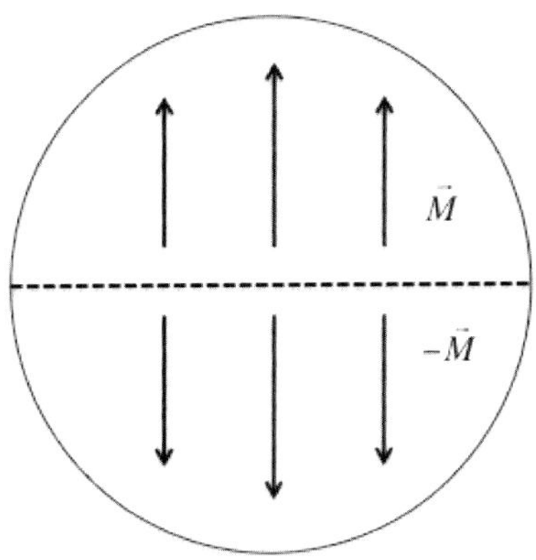

Fig. 6.37 Exercise 6.13.2 magnet.

(a) Using the magnetic scalar potential expression Equation (6.255), first find the magnetic field $\vec{B}$ far away from the magnet ($r \gg a$).

(b) Now find a form for the magnetic scalar potential everywhere as an expansion over Legendre polynomials $P_n(\cos\theta)$ using the variable separated Coulomb expansion (4.225) and the explicit expression for the Legendre polynomials given in Exercise 4.9.1. Show that the final form may be written as

$$\Phi_m(r,\theta) = 4\pi M_0 a^2 \sum_{n=0,1,2,\ldots} \frac{r_<^{2n}}{r_>^{2n+1}} C_n P_{2n}(\cos\theta),$$

where $r_>$ ($r_<$) is the greater (lesser) of r and a. Find an expression for the coefficients C_n. Make sure the potential at long-range agrees with your considerations in part (a).

Exercise 6.13.3.

(a) A long, straight rod of arbitrary but uniform cross-section has a constant volume magnetization, $\vec{M} = M_0\hat{k}$. The rod ends with a flat surface in the xy-plane with surface area S (see Figure 6.38). Choosing the coordinate origin O arbitrarily, show directly that far away from the end we can write

$$\vec{H}(\vec{x}) \approx (M_0 S)\frac{(\vec{x} - \vec{x}\,'')}{|\vec{x} - \vec{x}\,''|^3},$$

Where $\vec{x}\,''$ is a "typical" point on S.

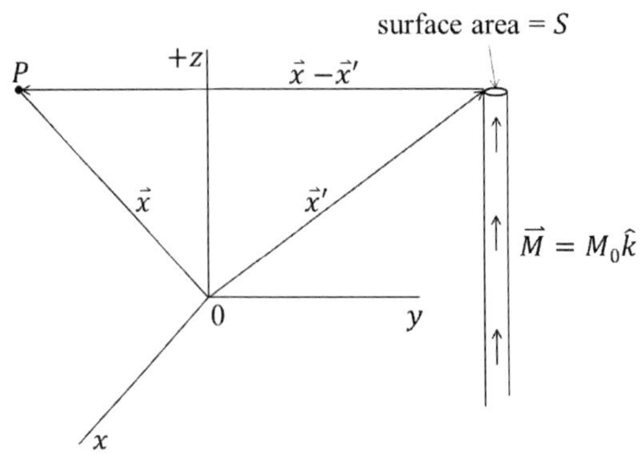

Fig. 6.38 Geometry for Exercise 6.13.3.

(b) Assuming $\vec{x}$ is far enough from the ends so that the result in (a) holds, find $\vec{H}$ for a thin bar magnet of circular cross-section with radius R, length $L \gg R$ and uniform magnetization $\vec{M} = M_0\hat{k}$.

Exercise 6.13.4.

(a) Consider a very long piece of magnetized material of arbitrary cross-section with a permanent, uniform magnetization, as shown in Figure 6.39: The upper surface S is flat and lies in the xy-plane. Show that H_z is given by

$$H_z(\vec{x}) = -M_0\Omega(\vec{x}),$$

both inside and outside the material, where $\Omega(\vec{x})$ is the solid angle subtended by S at $\vec{x}$.

(b) Apply (a) to find H_z and B_z at the center of a cube of side b with uniform magnetization, $\vec{M} = M_0\hat{k}$.

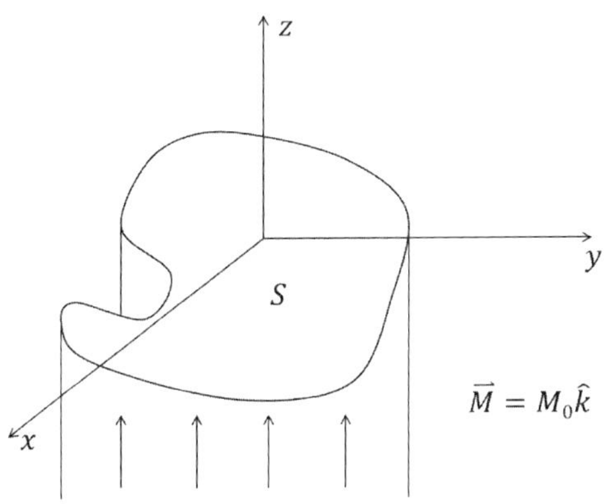

Fig. 6.39 Geometry for Exercise 6.13.4.

Exercise 6.13.5. A magnet in the shape of a rod has a uniform magnetization $\vec{M}$ along the z-axis. It is of length d along z and has the cross-section of a circle of radius a. Use an origin located in the geometric center of the rod.

(a) Using the concept of magnetic scalar potential, show that the $\vec{H}$ field cylindrical coordinate components H_ρ and H_z in the region $-d/2 < z < d/2$ are given by the integral expressions,

$$H_\rho(\rho, z) = 4\pi M_0 a \int_0^\infty dk\, J_1(ka) J_1(k\rho) e^{-kd/2} \sinh(kz),$$

$$H_z(\rho, z) = -4\pi M_0 a \int_0^\infty dk\, J_1(ka) J_0(k\rho) e^{-kd/2} \cosh(kz).$$

(b) Harmonize these results with Exercise 6.6.1 for the $\vec{B}$ field from a solenoid of length d and circular radius a. Use the connection between the effective surface current $\vec{K}_{\text{eff}} = c\vec{M} \times \hat{n}$ and the actual current in Exercise 6.6.1.

Exercise 6.13.6. An extremely thin (but uniformly thick) hollow shell has a constant area magnetization, M_A ($M_A = M\, d$, where M is the volume magnetization and d is the shell thickness) which points in the direction of the object's outward surface normal, $\hat{n}'$. Show that the magnetic scalar potential for this situation is simply given by:

$$\Phi_m(\vec{x}) = \begin{cases} 0 & \text{outside} \\ -4\pi M_A & \text{inside.} \end{cases}$$

(If you cannot show the above, try arguing that the $\vec{B}$ field must vanish both inside and outside the shell by looking at the $\vec{A}$ field.)

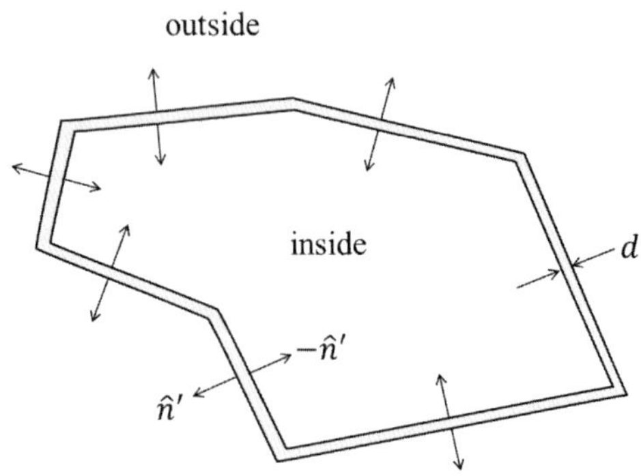

Fig. 6.40 Geometry for Exercise 6.13.6.

Chapter 7

Time Varying Fields I

In this chapter we begin to consider situations where charges and fields can vary with time. We begin by giving an argument that makes plausible the Maxwell equations that we wrote down at the beginning of Chapter 1. After generalizing these equations to macroscopic materials using two independent formalisms, we show how the static scalar and vector potentials can be generalized to yield a second-order formulation for the time-dependent case. We will also fill in some gaps in the discussion of magnetostatics by deriving expressions for energies and forces associated with static current distributions and bulk materials. Finally we address the puzzling asymmetry between electric and magnetic charge in the Maxwell equations.

7.1 Plausibility argument leading to Maxwell's equations

Up to this point we have been considering the static electromagnetic equations. In the vacuum, these equations are

$$\text{electrostatics:} \begin{cases} \vec{\nabla} \cdot \vec{E} = 4\pi\rho, \\ \vec{\nabla} \times \vec{E} = 0. \end{cases} \tag{7.1}$$

$$\text{magnetostatics:} \begin{cases} \vec{\nabla} \cdot \vec{B} = 0, \\ \vec{\nabla} \times \vec{B} = \dfrac{4\pi}{c}\vec{J}. \end{cases} \tag{7.2}$$

We have also used the Lorentz force law

$$\vec{F} = q\left(\vec{E} + \frac{\vec{v}}{c} \times \vec{B}\right), \tag{7.3}$$

to calculate forces in static situations. Current conservation also implies the continuity equation

$$\frac{\partial \rho}{\partial t} + \vec{\nabla} \cdot \vec{J} = 0. \tag{7.4}$$

In electrostatics $\partial\rho/\partial t = 0$ by definition, and in Section 6.2 we proved that $\partial\rho/\partial t$ also vanishes in magnetostatic situations. We now relax this assumption, and investigate the form of the field equations when time variation of the sources and fields is considered. Although we cannot rigorously derive these equations from the static case, nonetheless we can give plausibility arguments that motivate their form. The following discussion (largely due to Schwinger[1]) uses the basic strategy of looking at various static situations from "non-stationary" reference frames.

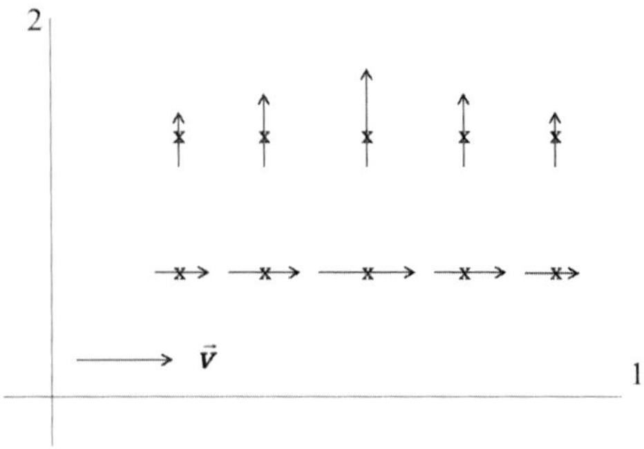

Fig. 7.1 Static electric field components with spatial gradients.

First we choose a particular frame as the "stationary" frame and suppose that only static electric fields are present in this frame. For later convenience, we denote spacetime coordinates and fields in this frame by $(\vec{x}\,', t')$ and $\vec{E}\,'$, respectively. Then we have

$$\frac{\partial \vec{E}\,'(\vec{x}\,', t')}{\partial t'} = 0. \qquad \text{(static form)} \tag{7.5}$$

We may describe the same situation from the point of view of a "slow-moving" frame with constant velocity $\vec{v} = v\hat{e}_1$ by applying standard (non-relativistic) physics. For simplicity we first consider the two-dimensional situation shown in Figure 7.1. The moving observer will see a changing electric field due to spatial variations in the field. In the following discussion we will use $(\vec{x}, t)$ to denote the space-time coordinates of the *moving*

[1] J. Schwinger *et al.*, *Classical Electrodynamics*, Perseus Books (Reading, MA) 1998, Section 1.2.

observer. Since $\vec{v}$ is non-relativistic, we may choose $t = t'$. In the situation shown in Figure 7.1, it is not difficult to see that for example

$$\frac{\partial E_1'}{\partial t} = v_1 \frac{\partial E_1'}{\partial x_1}, \qquad \frac{\partial E_2'}{\partial t} = v_1 \frac{\partial E_2'}{\partial x_1}.$$

These equations illustrate the interesting (and potentially confusing) fact that $\partial/\partial t \neq \partial/\partial t'$ even though $t = t'$. The reason is that the coordinate $\vec{x}$ depends not only on $\vec{x}\,'$, but also t'.

We may generalize to three dimensions by making use of the Galilean transformation

$$\vec{x} = \vec{x}\,' - \vec{v}t', \quad t = t',$$

and the chain rule to obtain

$$\frac{\partial \vec{E}\,'(\vec{x}\,', t')}{\partial t'} = \left(\frac{\partial}{\partial t} - \vec{v} \cdot \vec{\nabla}\right) \vec{E}\,'(\vec{x}\,'(\vec{x}, t,), t), \tag{7.6}$$

$$\frac{\partial \vec{E}\,'(\vec{x}\,', t')}{\partial x_j'} = \frac{\partial \vec{E}\,'(\vec{x}\,'(\vec{x}, t,), t)}{\partial x_j}, \qquad (j = 1, 2, 3) \tag{7.7}$$

which together with the electrostatic equations in the stationary frame imply that (arguments of the fields understood)

$$\left(\frac{\partial}{\partial t} - \vec{v} \cdot \vec{\nabla}\right) \vec{E}\,' = 0, \quad \vec{\nabla} \cdot \vec{E}\,' = 4\pi\rho.$$

We shall refer to the expression $\partial/\partial t - \vec{v} \cdot \vec{\nabla}$ as a *connective derivative*. Note that $-\vec{v}$ may be identified with the velocity of the charge distribution with respect to the moving axes.

Now since $\vec{v}$ is constant, we have from the $BAC - CAB$ rule

$$\vec{\nabla} \times \left(\vec{v} \times \vec{E}\,'\right) = \vec{v}\left(\vec{\nabla} \cdot \vec{E}\,'\right) - \left(\vec{v} \cdot \vec{\nabla}\right) \vec{E}\,',$$
$$= 4\pi\rho\vec{v} - \left(\vec{v} \cdot \vec{\nabla}\right) \vec{E}\,'. \tag{7.8}$$

By combining these equations with (7.6) and (7.7), we obtain in the moving frame

$$\frac{1}{c}\frac{\partial \vec{E}\,'}{\partial t} - 4\pi\rho\frac{\vec{v}}{c} + \vec{\nabla} \times \left(\frac{\vec{v}}{c} \times \vec{E}\,'\right) = 0. \tag{7.9}$$

The second term in (7.9) may be related to current as follows. In the stationary frame we have

$$\frac{\partial \rho}{\partial t'} = 0, \tag{7.10}$$

which according to the chain rule (7.6) becomes in the moving frame (since $\vec{v}$ is a constant)

$$\frac{\partial \rho}{\partial t} - \left(\vec{v} \cdot \vec{\nabla}\right) \rho = 0 \Longrightarrow \frac{\partial \rho}{\partial t} + \vec{\nabla} \cdot (-\rho \vec{v}) = 0. \tag{7.11}$$

From this we identify (from the moving observer's point of view)

$$\vec{J} = -\vec{v}\rho, \tag{7.12}$$

which makes sense since the moving observer sees charge moving in the $-\vec{v}$ direction. Putting $\vec{J} = -\vec{v}\rho$ into (7.9) gives

$$-\vec{\nabla} \times \left(\frac{\vec{v}}{c} \times \vec{E}\,'\right) - \frac{1}{c}\frac{\partial \vec{E}\,'}{\partial t} = \frac{4\pi}{c}\vec{J}. \tag{7.13}$$

Note that although the derivatives in this equation are in terms of unprimed coordinates, the field $\vec{E}\,'$ is that observed by the *primed* ("stationary") observer. There is no guarantee that the electric and magnetic fields in the unprimed frame should be equal to the corresponding fields in the primed frame. In fact, we may infer that (7.13) is a time-dependent generalization of one of the static equations (7.1)–(7.2). Because of the presence of $\vec{J}$, the corresponding static equation must be the second magnetostatic equation (7.2). However, the correspondence is only possible if we identify

$$\vec{B} = -\frac{\vec{v}}{c} \times \vec{E}\,'. \tag{7.14}$$

This is consistent with the static equation $\vec{\nabla} \cdot \vec{B} = 0$ because (again using that $\vec{v}$ is a constant)

$$\vec{\nabla} \cdot \vec{B} = \frac{\vec{v}}{c} \cdot \vec{\nabla} \times \vec{E}\,' = 0. \tag{7.15}$$

Furthermore, it is plausible to identify $\vec{E}$ with $\vec{E}\,'$, since then we have by the chain rule (7.7)

$$\vec{\nabla} \times \vec{E} = \vec{\nabla} \times \vec{E}\,' = \vec{\nabla}\,' \times \vec{E}\,' = 0.$$

In terms of unprimed fields then, (7.13) becomes

$$\vec{\nabla} \times \vec{B} - \frac{1}{c}\frac{\partial \vec{E}}{\partial t} = \frac{4\pi}{c}\vec{J}. \tag{7.16}$$

The $(1/c)\partial\vec{E}/\partial t$ term in (7.16) is called the *displacement current*, and was introduced by Maxwell. We view the arguments of these newly defined fields to be simply $(\vec{x}, t)$, appropriate for independently measured fields in the unprimed frame.

It appears that a static pure electric field, as seen by a moving observer, is no longer purely electric! This is actually not so surprising, if we consider that the charge density that is a source of $\vec{E}$ in the stationary frame becomes a current when seen by the moving observer.

Although the above argument was restricted to a particular case, we conjecture that (7.16) is a general result. The added term does not spoil the linearity of the equations. Moreover it is consistent with the continuity equation (7.4), as we already showed in Section 1.1; see the discussion leading to (1.2).

To complete the full set of time-dependent Maxwell equations, we now imagine a situation with static magnetic fields only. Actually the situation cannot be as simple as this, since we assume in classical electrodynamics that moving charges are the source of all magnetic fields. We can however consider magnetic fields in a region where the electric field is zero, such as the region outside a current loop. We mimic the previous argument and choose a "stationary" frame in which $\partial \vec{B}'/\partial t' = 0$. Just as before, in the "moving" frame we replace $\partial/\partial t'$ with the connective derivative to obtain

$$\frac{\partial \vec{B}'}{\partial t} - \left(\vec{v} \cdot \vec{\nabla}\right) \vec{B}' = 0. \tag{7.17}$$

From the static equations we know $\vec{\nabla}' \cdot \vec{B}' = 0$, and the chain rule gives us $\vec{\nabla} \cdot \vec{B}' = 0$. The $BAC - CAB$ identity yields

$$\begin{aligned} &\vec{\nabla} \times \left(\vec{v} \times \vec{B}'\right) = \vec{v}\left(\vec{\nabla} \cdot \vec{B}'\right) - \left(\vec{v} \cdot \vec{\nabla}\right) \vec{B}' \\ \Longrightarrow \quad &\frac{1}{c}\frac{\partial \vec{B}'}{\partial t} + \vec{\nabla} \times \left(\frac{\vec{v}}{c} \times \vec{B}'\right) = 0. \end{aligned} \tag{7.18}$$

If we again suppose that this represents a generalization of the static Maxwell equations, the only possibility is the second electrostatic equation, $\vec{\nabla} \times \vec{E} = 0$. This correspondence is complete if we identify the electric field in the moving frame as

$$\vec{E} = k \cdot \frac{\vec{v}}{c} \times \vec{B}'. \tag{7.19}$$

for some constant k. To determine k, we compute the force on a unit charge that is stationary in the unprimed frame, using Equation (7.3). An observer in the unprimed frame will calculate the force as $\vec{E}$, but an observer in the primed frame will calculate the force as $(\vec{v}/c) \times \vec{B}'$. These two observations are consistent only if we set $k = 1$. Finally, it is reasonable that $\vec{B} = \vec{B}'$

since we have already noted that $\vec{\nabla} \cdot \vec{B}\,' = 0$. The final resulting equation is seen below in the "Faraday law", Equation (7.21).

In summary, we have given plausible justifications for all four Maxwell equations:

"Coulomb" $$\vec{\nabla} \cdot \vec{E} = 4\pi\rho, \tag{7.20}$$

"Faraday" $$\vec{\nabla} \times \vec{E} + \frac{1}{c}\frac{\partial \vec{B}}{\partial t} = 0, \tag{7.21}$$

"Ampère"(modified by Maxwell) $$\vec{\nabla} \times \vec{B} - \frac{1}{c}\frac{\partial \vec{E}}{\partial t} = \frac{4\pi}{c}\vec{J}, \tag{7.22}$$

"No-monopole" $$\vec{\nabla} \cdot \vec{B} = 0. \tag{7.23}$$

It seems that we have tied everything neatly together by conjecturing that $\vec{E}$ and $\vec{B}$ are not absolute fields, but rather depend upon the motion of observers. Unfortunately we still have a consistency problem because (7.20)–(7.23) are not *covariant* (that is, form-invariant) under Galilean transformations. To show this, we consider two frames (primed and unprimed) with relative velocity $\vec{v}$ as in Figure 7.1, and express the *unprimed* Maxwell equations in terms of *primed* variables (which no longer denotes a "stationary" frame). We have seen already that the derivatives in the two frames are related by

$$\vec{\nabla}\,' = \vec{\nabla}; \qquad \frac{\partial}{\partial t'} = \frac{\partial}{\partial t} - \vec{v} \cdot \vec{\nabla}, \tag{7.24}$$

which can be inverted as

$$\vec{\nabla} = \vec{\nabla}\,'; \qquad \frac{\partial}{\partial t} = \frac{\partial}{\partial t'} + \vec{v} \cdot \vec{\nabla}\,'. \tag{7.25}$$

The charges and currents transform as

$$\rho = \rho', \qquad \vec{J} = \vec{J}\,' - \vec{v}\rho', \tag{7.26}$$

while the fields in primed and unprimed frames are related by

$$\vec{E} = \vec{E}\,' + \frac{\vec{v}}{c} \times \vec{B}\,'; \qquad \vec{B} = \vec{B}\,' - \frac{\vec{v}}{c} \times \vec{E}\,'. \tag{7.27}$$

If we write the Maxwell equations in the unprimed frame, and replace each unprimed quantity with the corresponding primed expression, then the equations do not remain the same. We can however obtain covariant equations to first order in v/c by making some slight adjustments. If we replace (7.25) and (7.26) with

$$\vec{\nabla} = \vec{\nabla}\,' + \frac{\vec{v}}{c^2}\frac{\partial}{\partial t'}; \qquad \frac{\partial}{\partial t} = \frac{\partial}{\partial t'} + \vec{v} \cdot \vec{\nabla}', \tag{7.28}$$

$$\rho = \rho' - \frac{\vec{v}}{c^2} \cdot \vec{J}\,'; \qquad \vec{J} = \vec{J}\,' - \vec{v}\rho', \tag{7.29}$$

then we get form invariance to first order in v/c (as you will show in Exercise 7.1.1). What is the physical significance of this? Also, what is the *general* transformation that Maxwell's equations are covariant under? We will discuss these issues further in Chapter 13.

7.2 Faraday's law

Now that our equations include time dependence, we may use them to explain a wealth of phenomena. In this section we investigate the connection between magnetic and electric fields in an arbitrary spatial loop, as shown in Figure 7.2. The connection we will find has important consequences for the behavior of a current-carrying loop (such as a thin conducting wire) in the presence of a changing magnetic field.

We first integrate the Faraday equation (7.21) over a surface S bounded by the loop, then apply Stokes' theorem on the $\vec{\nabla} \times \vec{E}$ term to obtain

$$\int_S da\,\hat{n} \cdot \left(\vec{\nabla} \times \vec{E} + \frac{1}{c}\frac{\partial \vec{B}}{\partial t} \right) = 0$$

$$\Longrightarrow \oint_C \vec{E} \cdot d\vec{\ell} = -\frac{1}{c}\frac{d}{dt}\int_S da\,\hat{n} \cdot \vec{B}. \tag{7.30}$$

We define the *electromotive force* or *emf* $\mathcal{E}$ associated with loop C as follows:

$$\mathcal{E} \equiv \oint_C \vec{E} \cdot d\vec{\ell}. \tag{7.31}$$

The reader should take note of two important details. First, $\mathcal{E}$ is not really a force, since it has the units of voltage; this unfortunate terminology dates

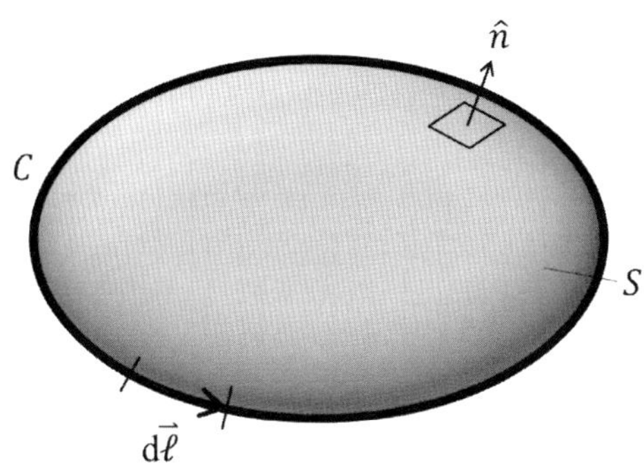

Fig. 7.2 Closed loop C in space (and surface S bounded by C).

back to Maxwell, and has become standard. Second, (7.31) shows that the electromotive force vanishes whenever $\vec{E}$ is the gradient of a potential (as in the electrostatic case).

We define the *magnetic flux* through surface S as follows:

$$\mathcal{F} \equiv \int_S da\, \hat{n} \cdot \vec{B}. \tag{7.32}$$

These definitions enable us to state *Faraday's Law*

$$\mathcal{E} = -\frac{1}{c}\frac{d\mathcal{F}}{dt}. \tag{7.33}$$

Note that Equation (7.33) is true for any surface S which has bounding contour C.

Now let us apply this to a current loop at C. The all-important minus sign in (7.33) indicates that the induced current in a conducting loop flows in a direction so as to produce a magnetic field that *opposes* the change in flux in the circuit. This important fact is known as *Lenz' Law*, and is the key to understanding the phenomenon of diamagnetism that was described in Section 6.10. In this case, electron orbits play the role of permanently modified current loops, opposing the change in conditions and reducing the interior $\vec{B}$ field for fixed external sources.

7.3 Derivations of macroscopic Maxwell equations

Next we consider the time-dependent macroscopic Maxwell equations. In this section we will present two derivations of these equations based on different simplified phenomenological models. The first is from Schwinger,[2] and emphasizes the direct calculation of forces on atoms in the material. The second is explained in Jackson's textbook,[3] and is based on effective sources.

The following considerations assume a stationary frame of reference, relative to which all atoms in the material are moving at the same constant velocity $\vec{v}$. We consider a single atom whose center of mass is located at $\vec{R}(t)$, so $\vec{v} \equiv d\vec{R}/dt$.

[2] J. Schwinger *et al.*, *Classical Electrodynamics*, Perseus Books (Reading, MA) 1998, Chapter 4.

[3] J. D. Jackson, *Classical Electrodynamics*, 3rd ed., John Wiley & Sons (New York) 1999, Section 6.6. Jackson's treatment follows the method in Ref. 5 of the upcoming *Going Deeper* Section 7.9.1.

7.3.1 *Schwinger model of macroscopic medium*

Our approach here will be similar to that used in Chapters 5 and 6 to derive macroscopic equations for electrostatics and magnetostatics respectively. As before, we shall begin with summations over fields due to individual atoms, and then approximate the sums by integrals over density functions. We emphasize that this procedure is entirely heuristic: nonetheless, it leads to equations which are extremely useful in situations where the classical and non-relativistic approximations are valid.

We model a typical atom in the material as a system of K point particles ($K-1$ electrons surrounding a heavy nucleus) with charges, locations and velocities (relative to the stationary frame) equal to e_k, $x_k(t)$ and $v_k(t)$ respectively, $k = 1 \ldots K$. We suppose atoms have zero net charge, so that

$$\sum_{k=1}^{K} e_k = 0. \tag{7.34}$$

The net force on the atom due to electric and magnetic fields is

$$\vec{F}_{\text{atom}} = \sum_{k=1}^{K} \left[e_k \vec{E}(\vec{x}_k) + e_k \frac{\vec{v}_k}{c} \times \vec{B}(\vec{x}_k) \right]. \tag{7.35}$$

Let's make sure the form of the atomic force is consistent with a simple model of the charge density and current. We postulate a classical point charge model given by

$$\rho_k(\vec{x}\,', t) = e_k \delta(\vec{x}\,' - \vec{x}_k(t)), \qquad \vec{J}_k(\vec{x}\,', t) = e_k \frac{d\vec{x}_k}{dt} \delta(\vec{x}\,' - \vec{x}_k(t)), \tag{7.36}$$

which satisfies charge and current continuity. The electromagnetic force is given by

$$\vec{F}_k = \int d^3x' \left(\rho_k(\vec{x}\,', t) \vec{E}(\vec{x}\,') + \frac{1}{c} \vec{J}_k(\vec{x}\,', t) \times \vec{B}(\vec{x}\,') \right), \tag{7.37}$$

which with (7.36) implies (7.35) when the sum over k is done. Note the integral is over all space.

We assume that the atom is small enough so that $\vec{E}$ and $\vec{B}$ vary slowly over the extent of the atom. In this case, we can expand $\vec{E}(\vec{x}_k)$ and $\vec{B}(\vec{x}_k)$ about the center of mass $\vec{R}$ of the charge distribution:

$$\vec{E}(\vec{x}_k) = \vec{E}(\vec{R}) + [(\vec{x}_k - \vec{R}) \cdot \vec{\nabla}] \vec{E}(\vec{R}) + \ldots, \tag{7.38}$$

$$\vec{B}(\vec{x}_k) = \vec{B}(\vec{R}) + [(\vec{x}_k - \vec{R}) \cdot \vec{\nabla}] \vec{B}(\vec{R}) + \ldots, \tag{7.39}$$

so that the force on the atom may be approximated by

$$\vec{F}_{\text{atom}} = \sum_k \left\{ \overbrace{[e_k\vec{E}(\vec{R}) + e_k\frac{\vec{v}_k}{c} \times \vec{B}(\vec{R})]}^{\text{term (1)}} + \overbrace{e_k[(\vec{x}_k - \vec{R}) \cdot \vec{\nabla}]\vec{E}(\vec{R})}^{\text{term (2)}} \right.$$

$$\left. + \overbrace{e_k\frac{\vec{v}_k}{c} \times \left([(\vec{x}_k - \vec{R}) \cdot \vec{\nabla}]\vec{B}(\vec{R})\right)}^{\text{term (3)}} + \dots \right\}. \tag{7.40}$$

All "terms" in (7.40) are actually vectors. In term (2) we identify the electric dipole moment of the atom,

$$\vec{p}(t) \equiv \sum_k e_k(\vec{x}_k - \vec{R}) = \sum_k e_k\vec{x}_k, \tag{7.41}$$

which results from

$$\rho(\vec{x}\,', t) \equiv \sum_k e_k\delta(\vec{x}\,' - \vec{x}_k + \vec{R}); \qquad \vec{p} = \int d^3x'\,\vec{x}\,'\rho(\vec{x}\,', t), \tag{7.42}$$

where it is understood that $\vec{x}_k$ actually depends on t. In term (1) we may then recognize

$$\sum_k e_k\vec{v}_k = \frac{d\vec{p}}{dt}, \tag{7.43}$$

so that (7.40) simplifies to

$$\vec{F}_{\text{atom}} = (\vec{p} \cdot \vec{\nabla})\vec{E}(\vec{R}) + \frac{1}{c}\frac{d\vec{p}}{dt} \times \vec{B}(\vec{R}) + \text{term (3)} + \dots. \tag{7.44}$$

The notation $\vec{E}(\vec{R}), \vec{B}(\vec{R})$ is shorthand: more precisely, these fields should be written as $\vec{E}(\vec{R}(t), t), \vec{B}(\vec{R(t)}, t)$ since they may depend on time both explicitly and through the motion of the center of mass. In the following argument, for simplicity we shall further abbreviate $\vec{E}(\vec{R})$ and $\vec{B}(\vec{R})$ as $\vec{E}$ and $\vec{B}$, respectively.

The second term in (7.44) may be rewritten as

$$\frac{1}{c}\frac{d\vec{p}}{dt} \times \vec{B} = \frac{1}{c}\frac{d}{dt}(\vec{p} \times \vec{B}) - \vec{p} \times \frac{1}{c}\frac{d\vec{B}}{dt}, \tag{7.45}$$

and by the chain rule,

$$\frac{d\vec{B}}{dt} = \frac{\partial\vec{B}}{\partial t} + (\vec{v} \cdot \vec{\nabla})\vec{B}, \tag{7.46}$$

where $\vec{v} = d\vec{R}/dt$ is the velocity of the center of mass of the atom relative to the laboratory. Note that $\vec{v}$ can be broken up as bulk velocity plus the relative velocity of the atom with respect to the material. In this treatment we will assume that the relative atomic velocities are small, so that $\vec{v}$ can be taken equal to the bulk velocity for all the atoms. Of course, this is really not true because atoms have relative thermal and vibrational velocities, but for the most part these can be neglected. (This approximation will reappear in the upcoming (7.88) also.) Under this condition we have

$$\begin{aligned}\frac{1}{c}\frac{d\vec{p}}{dt}\times\vec{B} &= \frac{1}{c}\frac{d}{dt}(\vec{p}\times\vec{B}) - \vec{p}\times\underbrace{\frac{1}{c}\frac{\partial\vec{B}}{\partial t}}_{-\vec{\nabla}\times\vec{E}} - \vec{p}\times\left[\left(\frac{\vec{v}}{c}\cdot\vec{\nabla}\right)\vec{B}\right] \\ &= \frac{1}{c}\frac{d}{dt}(\vec{p}\times\vec{B}) + \vec{\nabla}(\vec{E}\cdot\vec{p}) - (\vec{p}\cdot\vec{\nabla})\vec{E} - \left(\frac{\vec{v}}{c}\cdot\vec{\nabla}\right)(\vec{p}\times\vec{B}).\end{aligned} \tag{7.47}$$

By comparison with (7.44), we now have

$$\vec{F}_{\text{atom}} = \vec{\nabla}(\vec{p}\cdot\vec{E}) + \frac{1}{c}\frac{d}{dt}(\vec{p}\times\vec{B}) - \left(\frac{\vec{v}}{c}\cdot\vec{\nabla}\right)(\vec{p}\times\vec{B}) + \text{term (3)} + \ldots. \tag{7.48}$$

Let us go to work on term (3) now. We start by adding and subtracting a term proportional to $\vec{v}$,

$$\begin{aligned}\text{term (3)} &= \left(\frac{1}{2}+\frac{1}{2}\right)\frac{1}{c}\sum_k e_k(\vec{v}_k - \vec{v})\times[(\vec{x}_k - \vec{R})\cdot\vec{\nabla}]\vec{B} \\ &\quad + \frac{1}{c}\sum_k e_k\,\vec{v}\times[(\vec{x}_k - \vec{R})\cdot\vec{\nabla}]\vec{B},\end{aligned} \tag{7.49}$$

where we are indicating a break-up of the first term into two equal pieces. One piece we leave alone; in the other we use the $BAC-CAB$ rule identity:

$$[(\vec{x}_k-\vec{R})\times(\vec{v}_k-\vec{v})]\times\vec{\nabla} = -(\vec{x}_k-\vec{R})[(\vec{v}_k-\vec{v})\cdot\vec{\nabla}] + (\vec{v}_k-\vec{v})[(\vec{x}_k-\vec{R})\cdot\vec{\nabla}]. \tag{7.50}$$

This now allows us to write (7.49) in the intermediate form,

$$\begin{aligned}\text{term (3)} &= \frac{1}{2c}\sum_k e_k(\vec{v}_k - \vec{v})\times\left[(\vec{x}_k - \vec{R})\cdot\vec{\nabla}\right]\vec{B} \\ &\quad + \frac{1}{2c}\sum_k e_k(\vec{x}_k - \vec{R})\times\left[(\vec{v}_k - \vec{v})\cdot\vec{\nabla}\right]\vec{B} \\ &\quad + \frac{1}{2c}\sum_k e_k\left[[(\vec{x}_k - \vec{R})\times(\vec{v}_k - \vec{v})]\times\vec{\nabla}\right]\times\vec{B} \\ &\quad + \frac{1}{c}\sum_k e_k\vec{v}\times\left[(\vec{x}_k - \vec{R})\cdot\vec{\nabla}\right]\vec{B}.\end{aligned} \tag{7.51}$$

We need to simplify (7.51)! Two types of multipole moments will allow us to do this. The first two terms on the right in (7.51) may be combined using the time derivative of the *non-traceless* form of the quadrupole tensor,

$$Q'_{ij} \equiv 3\sum_k e_k(x_k - R)_i(x_k - R)_j \qquad (i,j = 1,2,3). \tag{7.52}$$

The other quantity we need for the third term on the right of (7.51) is the atomic magnetic moment definition (relative to the center of mass of the atom),

$$\vec{m} \equiv \frac{1}{2c}\sum_k e_k(\vec{x}_k - \vec{R}) \times (\vec{v}_k - \vec{v}), \tag{7.53}$$

which displays the classical connection between magnetic moment and particle angular momentum,

$$\vec{m}_k = \frac{e_k}{2mc}\vec{L}_k, \tag{7.54}$$

for the kth particle, where m is the mass of the particle and $\vec{L}_k$ is the angular momentum. Equation (7.53) follows from our previous definition of magnetic moment (see (6.129))

$$\vec{m} = \frac{1}{2c}\int d^3x'\, \vec{x}\,' \times \vec{J}(\vec{x}\,'),$$

using the current density

$$\vec{J}(\vec{x}\,') = \sum_k e_k(\vec{v}_k - \vec{v})\delta(\vec{x}\,' - \vec{x}_k + \vec{R}).$$

Notice however that

$$\int d^3x'\, \vec{J}(\vec{x}\,') = \sum_k e_k\vec{v}_k = \frac{d\vec{p}}{dt} \neq 0, \tag{7.55}$$

so the lowest moment does not vanish. Therefore $\vec{m}$ as defined above is dependent upon our choice of the origin; this is a failure of the classical treatment of particle paths. We assume from the dynamics of heavy nuclei and light electrons, that the center of mass in our classical model is the most appropriate place to center our coordinates.

Introducing the time derivative of Q'_{ij} and the magnetic moment $\vec{m}$ in (7.51), we now achieve the much simpler form,

$$\text{term (3)} = \frac{1}{6c}\sum_{i,j,\ell,m} \hat{e}_i\epsilon_{ij\ell}\dot{Q}'_{jm}\nabla_m B_\ell + (\vec{m}\times\vec{\nabla})\times\vec{B} + \frac{\vec{v}}{c}\times\left[\vec{p}\cdot\vec{\nabla}\right]\vec{B}, \tag{7.56}$$

where the sum on k in the last term in (7.51) has produced the electric dipole moment $\vec{p}$ from the definition (7.41). The first term on the right involves the time-derivative of the non-traceless quadrupole moments Q'_{jm}. This term has a partner from the electric side if we Taylor expand the electric field to the next order in (7.38). (An additional quadrupole term emerges from the expansion of (7.39) to the next order.) These higher order terms imply an effective current and charge density which will be examined in Exercise 7.3.3. For purposes of simplicity we will not consider these terms in the rest of the discussion.

At long last, we obtain a complete expression for the force on an atom in the material by plugging (7.56) minus the quadrupole term into (7.48). Using a vector identity and the Maxwell equation (7.23) to simplify the second term on the right in (7.56), we can write

$$\begin{aligned}\vec{F}_{\text{atom}} &= \vec{\nabla}(\vec{p}\cdot\vec{E} + \vec{m}\cdot\vec{B}) + \frac{1}{c}\frac{d}{dt}(\vec{p}\times\vec{B}) \\ &\quad \underbrace{-\vec{p}\times\left[\frac{\vec{v}}{c}\cdot\vec{\nabla}\right]\vec{B} + \frac{\vec{v}}{c}\times\left[\vec{p}\cdot\vec{\nabla}\right]\vec{B}}_{[(\vec{p}\times\frac{\vec{v}}{c})\times\vec{\nabla}]\times\vec{B}} + \ldots,\end{aligned} \tag{7.57}$$

which simplifies to

$$\vec{F}_{\text{atom}} = \vec{\nabla}\left(\vec{p}\cdot\vec{E} + \left(\vec{m} + \vec{p}\times\frac{\vec{v}}{c}\right)\cdot\vec{B}\right) + \frac{1}{c}\frac{d}{dt}(\vec{p}\times\vec{B}) + \ldots. \tag{7.58}$$

Since the atomic moments are constants, we should understand that the quantities $\vec{p}$ and $\left(\vec{m} + \vec{p}\times\frac{\vec{v}}{c}\right)$ can be pulled through the gradients in the first two terms. A very similar argument to the atomic force gives the microscopic torque as

$$\vec{N}_{\text{atom}} = \vec{p}\times\vec{E} + \vec{m}\times\vec{B}, \tag{7.59}$$

which is reasonable in light of results on electric and magnetic moments from previous chapters. The student is invited to consider this in Exercise 7.3.1.

To obtain the bulk force on the material, we introduce a density of atoms and integrate the force per atom times the density. The notation for the density function must reflect the fact that the material is experiencing an overall bulk velocity. Previously, $\vec{R}(t)$ was pointing at an atom's center of mass. Now, we take it to locate an origin which is at rest with the material, which is moving at a uniform velocity $\vec{v}$. The variable $\vec{x}$ is a spatial

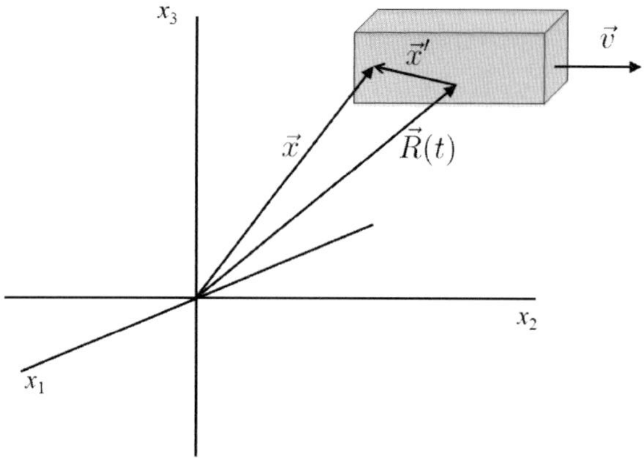

Fig. 7.3 Location variables for integrating over atoms in the Schwinger model.

integration variable which locates each atom as we integrate; see Figure 7.3. The variable $\vec{x}\,'$ points from the moving origin to the atom; $\vec{x} = \vec{x}\,' + \vec{R}(t)$. We sum over the contributions of the atoms. The arguments of the fields now locate each atom at $\vec{x}$: $\vec{E}(\vec{x}, t), \vec{B}(\vec{x}, t)$. Denoting the atomic density function as $n(\vec{x}\,', t')$, we have as in (7.5) and (7.6):

$$\frac{\partial n(\vec{x}\,', t')}{\partial t'} = 0 \implies \frac{\partial n(\vec{x} - \vec{R}(t), t)}{\partial t} + \vec{v} \cdot \vec{\nabla} n(\vec{x} - \vec{R}(t), t) = 0. \tag{7.60}$$

In addition, the atom's moments are made space and time dependent:

$$\vec{p} \longrightarrow \vec{p}(\vec{x}, t), \quad \vec{m} \longrightarrow \vec{m}(\vec{x}, t).$$

We define the fields

$$\vec{P}(\vec{x}, t) \equiv n(\vec{x} - \vec{R}(t), t)\, \vec{p}(\vec{x}, t), \quad \vec{M}(\vec{x}, t) \equiv n(\vec{x} - \vec{R}(t), t)\, \vec{m}(\vec{x}, t). \tag{7.61}$$

In this way we obtain the bulk force:

$$\begin{aligned} \vec{F}_{\text{bulk}} &= \int d^3x\, n(\vec{x} - \vec{R}(t), t) \vec{F}_{\text{atom}}(\vec{x}, t), \\ &= \int d^3x \sum_{i=1}^{3} \left[P_i \vec{\nabla} E_i + \left(M_i + \left(\vec{P} \times \frac{\vec{v}}{c} \right)_i \right) \vec{\nabla} B_i \right] + \frac{1}{c} \int d^3x \frac{\partial}{\partial t} (\vec{P} \times \vec{B}). \end{aligned} \tag{7.62}$$

The spatial integrals cover all space, including surfaces. Note also that a careful treatment shows that the total time derivative in the last term in (7.58) has changed to a partial time derivative in (7.62) under spatial averaging; see Exercise 7.3.2.

From (7.62) we see the material has an effective magnetization given by

$$\vec{M}_{\text{eff}} = \vec{M} + \vec{P} \times \frac{\vec{v}}{c}. \tag{7.63}$$

In view of the definitions of $\vec{H}$ and $\vec{P}$ in (6.199) and (5.64) respectively, we thus have effectively

$$\vec{B} - \vec{H} = 4\pi\vec{M} + (\vec{D} - \vec{E}) \times \frac{\vec{v}}{c}, \tag{7.64}$$

for moving macroscopic materials, in non-relativistic approximation. In Chapter 13 we will give a faster, more elegant derivation of (7.63) based on transformation properties of the fields; see Exercise 13.3.3.

We take $\vec{v} = 0$, so the material is now at rest in the lab ($n(\vec{x} - \vec{R}(t)) \to n(\vec{x})$). Now consider

$$\begin{aligned} \frac{1}{c}\vec{P} \times \frac{\partial \vec{B}}{\partial t} &= -\vec{P} \times (\vec{\nabla} \times \vec{E}), \\ &= -\sum_i P_i \vec{\nabla} E_i + (\vec{P} \cdot \vec{\nabla})\vec{E}, \\ &= -\sum_i P_i \vec{\nabla} E_i - \vec{E}(\vec{\nabla} \cdot \vec{P}) + \sum_i \nabla_i (P_i \vec{E}). \end{aligned} \tag{7.65}$$

The first term on the last line of (7.65) cancels with a term in (7.62), while the third term leads to a surface integral which vanishes on the surface at infinity. This gives

$$\vec{F}_{\text{bulk}} = \int d^3x \left[-(\vec{\nabla} \cdot \vec{P})\vec{E} + \frac{1}{c}\frac{\partial \vec{P}}{\partial t} \times \vec{B} \right] + \int d^3x \sum_i [M_i \vec{\nabla} B_i]. \tag{7.66}$$

Using the result in (6.192) again maintaining integrations over all space, we have finally

$$\vec{F}_{\text{bulk}} = \int d^3x \left[-(\vec{\nabla} \cdot \vec{P})\vec{E} + \frac{1}{c}\frac{\partial \vec{P}}{\partial t} \times \vec{B} + (\vec{\nabla} \times \vec{M}) \times \vec{B} \right]. \tag{7.67}$$

By comparing this with the usual continuous force expression, we may identify effective charge and current densities that are "bound" within the material:

$$\rho_{\text{b}} = -\vec{\nabla} \cdot \vec{P}, \tag{7.68}$$

$$\vec{J}_{\text{b}} = c\vec{\nabla} \times \vec{M} + \frac{\partial \vec{P}}{\partial t}. \tag{7.69}$$

(The subscript "b" corresponds to "bound" as used in Chapters 5 and 6.) These identifications are consistent since we have identically that

$$\frac{\partial \rho_{\rm b}}{\partial t} + \vec{\nabla} \cdot \vec{J}_{\rm b} = 0. \tag{7.70}$$

The Schwinger procedure for defining the macroscopic fields is to average the microscopic fundamental fields and sources over macroscopically small but microscopically large regions in space and time to yield macroscopic fields and effective sources. In this way, one transitions from a fundamental description to one involving only averaged fields and the free current and charge densities. (Note, however, our use of "bound" corresponds to his "effective" sources!) These give the standard macroscopic Maxwell equations, which we will present in a moment. However, first let us get a different perspective on the situation, one involving an emphasis on the sources rather than the forces.

7.3.2 *Jackson model of macroscopic medium*

Jackson considers a microscopic model which allows for charged atoms and additional currents. However, in order to compare Jackson's approach with Schwinger's, here we will restrict ourselves to the case of neutral atoms. We will see that many of the steps are directly analogous.

In this development one first computes the average current density of the $n^{\rm th}$ atom (or molecule). To do this, let $\vec{x}_n(t)$ denote the center of mass of the $n^{\rm th}$ molecule (with associated velocity $\vec{v}_n(t)$), and let $\vec{x}_{jn}(t)$ represent the location of the $j^{\rm th}$ point charge within of the $n^{\rm th}$ molecule relative to $\vec{x}_n(t)$ (with associated relative velocity $\vec{v}_{jn}(t)$). The $\vec{x}_n(t)$ and $\vec{x}_{jn}(t)$ are assumed to be classical time-dependent trajectories. (The time dependence of these quantities will be suppressed for simplicity in the following.) We may then express the current due to one molecule as

$$\vec{j}_n(\vec{x}, t) = \sum_{j(n)} e_j \left(\vec{v}_n + \vec{v}_{jn}\right) \delta\left(\vec{x} - \vec{x}_n - \vec{x}_{jn}\right). \tag{7.71}$$

Next, we average this current with a test function $f(\vec{x})$ whose spatial integral is one, and which varies slowly over the dimensions of the molecule:

$$\begin{aligned} \langle \vec{j}_n(\vec{x}, t)\rangle &= \int d^3x'\, f(\vec{x} - \vec{x}\,')\vec{j}_n(\vec{x}\,', t) \\ &= \sum_{j(n)} e_j \left(\overbrace{\vec{v}_n}^{\text{term 1}} + \overbrace{\vec{v}_{jn}}^{\text{term 2}} \right) f\left(\vec{x} - \vec{x}_n - \vec{x}_{jn}\right). \end{aligned} \tag{7.72}$$

Keeping only the first-order Taylor expansion term leads to

$$
\begin{aligned}
[\text{term } 1]_n &= \sum_{j(n)} e_j \vec{v}_n [f(\vec{x}-\vec{x}_n) - \vec{x}_{jn}\cdot\vec{\nabla} f(\vec{x}-\vec{x}_n)], \\
&= \underbrace{0}_{\text{neutral atom}} \quad - \sum_{j(n)} e_j \vec{v}_n \left(\vec{x}_{jn}\cdot\vec{\nabla}\right) f(\vec{x}-\vec{x}_n), \\
&= -\vec{v}_n \left(\vec{p}_n \cdot \vec{\nabla} f(\vec{x}-\vec{x}_n)\right), \qquad (7.73)
\end{aligned}
$$

where

$$
\vec{p}_n \equiv \sum_{j(n)} e_j \vec{x}_{jn}, \qquad (7.74)
$$

is the atomic dipole moment in this development.

The second term in (7.72) is more challenging to deal with. Again to first order in the Taylor expansion, we have

$$
\begin{aligned}
[\text{term } 2]_n &= \sum_{j(n)} e_j \vec{v}_{jn} [f(\vec{x}-\vec{x}_n) - \vec{x}_{jn}\cdot\vec{\nabla} f(\vec{x}-\vec{x}_n)] \\
&= \sum_{j(n)} e_j \vec{v}_{jn} f(\vec{x}-\vec{x}_n) - \sum_{j(n)} e_j \vec{v}_{jn} (\vec{x}_{jn}\cdot\vec{\nabla}) f(\vec{x}-\vec{x}_n), \\
&= \frac{d\vec{p}_n}{dt} f(\vec{x}-\vec{x}_n) - \left(\frac{1}{2}+\frac{1}{2}\right) \sum_{j(n)} e_j \vec{v}_{jn} (\vec{x}_{jn}\cdot\vec{\nabla}) f(\vec{x}-\vec{x}_n). \quad (7.75)
\end{aligned}
$$

We are indicating a breakup of the second term in (7.75) into equal parts. In one term we use the vector identity,

$$
(\vec{x}\times\vec{v})\times\vec{\nabla} f = -\vec{x}(\vec{v}\cdot\vec{\nabla}) f + \vec{v}(\vec{x}\cdot\vec{\nabla}) f. \qquad (7.76)
$$

We now have

$$
\begin{aligned}
[\text{term } 2]_n &= \frac{d\vec{p}_n}{dt} f(\vec{x}-\vec{x}_n) - \frac{1}{2} \sum_{j(n)} e_j \left(\vec{v}_{jn}\vec{x}_{jn} + \vec{x}_{jn}\vec{v}_{jn}\right)\cdot\vec{\nabla} f(\vec{x}-\vec{x}_n) \\
&\quad - \frac{1}{2} \sum_{j(n)} e_j (\vec{x}_{jn}\times\vec{v}_{jn}) \times \vec{\nabla} f(\vec{x}-\vec{x}_n). \qquad (7.77)
\end{aligned}
$$

We again need moments to simplify this equation! In this context

$$
\vec{m}_n \equiv \frac{1}{2c} \sum_{j(n)} e_j \vec{x}_{jn} \times \vec{v}_{jn}, \qquad (7.78)
$$

is the molecular magnetic moment for the nth atom, and

$$
(Q'_n)_{ik} \equiv 3 \sum_{j(n)} e_j (x_{jn})_i (x_{jn})_k, \qquad (7.79)
$$

is the non-traceless quadrupole tensor for the atom. This results in

$$[\text{term } 2]_n = \frac{d\vec{p}_n}{dt} f(\vec{x} - \vec{x}_n) - \frac{1}{6} \sum_{i,k} \hat{e}_i (\dot{Q}'_n)_{ik} \nabla_k f(\vec{x} - \vec{x}_n)$$
$$- c\vec{m}_n \times \vec{\nabla} f(\vec{x} - \vec{x}_n). \tag{7.80}$$

The second term in (7.80) involves the time derivative of the non-traceless quadrupole tensor. For purposes of simplicity of the discussion, we will not consider this term further (although as an exercise the student may!) We then obtain

$$[\text{term } 2]_n = c\vec{\nabla} \times (\vec{m}_n f(\vec{x} - \vec{x}_n)) + \frac{\partial}{\partial t} [\vec{p}_n f(\vec{x} - \vec{x}_n)] - \vec{p}_n \frac{\partial}{\partial t} f(\vec{x} - \vec{x}_n). \tag{7.81}$$

The partial time derivative is the appropriate object here in a context in which the variable $\vec{x}$ also appears and where $\partial \vec{p}_n / \partial t = d\vec{p}_n / dt$. To obtain the macroscopic fields, according to (7.72) we must sum $[\text{term } 1]_n$ plus $[\text{term } 2]_n$ over the n atomic contributions. As a reminder, for $[\text{term } 1]_n$ from (7.73) we have

$$\sum_n [\text{term } 1]_n = -\sum_n \vec{v}_n \left(\vec{p}_n \cdot \vec{\nabla} f(\vec{x} - \vec{x}_n) \right). \tag{7.82}$$

Let us now consider $[\text{term } 2]_n$:

$$\sum_n [\text{term } 2]_n = \sum_n c\vec{\nabla} \times (\vec{m}_n f(\vec{x} - \vec{x}_n)) + \frac{\partial}{\partial t} \sum_n \vec{p}_n f(\vec{x} - \vec{x}_n)$$
$$- \sum_n \vec{p}_n \frac{\partial}{\partial t} f(\vec{x} - \vec{x}_n),$$
$$= c\vec{\nabla} \times \vec{M}(\vec{x}, t) + \frac{\partial}{\partial t} \vec{P}(\vec{x}, t) - \sum_n \vec{p}_n \frac{\partial}{\partial t} f(\vec{x} - \vec{x}_n), \tag{7.83}$$

where we have defined macroscopic magnetization and polarization fields as

$$\vec{M}(\vec{x}, t) \equiv \sum_n \vec{m}_n f(\vec{x} - \vec{x}_n), \tag{7.84}$$

$$\vec{P}(\vec{x}, t) \equiv \sum_n \vec{p}_n f(\vec{x} - \vec{x}_n). \tag{7.85}$$

Furthermore,

$$\frac{\partial}{\partial t} f(\vec{x} - \vec{x}_n) = -\vec{v}_n \cdot \vec{\nabla} f(\vec{x} - \vec{x}_n), \tag{7.86}$$

by the chain rule. We obtain finally

$$\sum_n < \vec{j}_n(\vec{x},t) >= \sum_n ([\text{term } 1]_n + [\text{term } 2]_n),$$

$$= c\vec{\nabla} \times \vec{M}(\vec{x},t) + \frac{\partial}{\partial t}\vec{P}(\vec{x},t) + \sum_n \left(\vec{p}_n(\vec{v}_n \cdot \vec{\nabla})f - \vec{v}_n(\vec{p}_n \cdot \vec{\nabla})f\right). \quad (7.87)$$

The last two terms in (7.87) involve the atomic velocities, $\vec{v}_n$. As in the previous discussion, we may assume that the atoms are essentially fixed in position; then $\vec{v} = \vec{v}_n$ for all n. Then we have

$$\sum_n < \vec{j}_n(\vec{x},t) >= c\vec{\nabla} \times \left[\vec{M}(\vec{x},t) + \vec{P}(\vec{x},t) \times \frac{\vec{v}}{c}\right] + \frac{\partial}{\partial t}\vec{P}(\vec{x},t), \quad (7.88)$$

which is in agreement with Equation (7.69) for the velocity independent terms, and which also implies an effective magnetization of $\vec{M}_{\text{eff}} = \vec{M} + \vec{P} \times \frac{\vec{v}}{c}$ as in (7.63).

Although Schwinger's derivation deals with forces and Jackson's with currents, the steps in the two developments are seen to be exactly parallel. The above derivation gave the average current: a similar argument applied to charge density yields the result

$$\sum_n \langle \rho_n(\vec{x},t) \rangle = -\vec{\nabla} \cdot \vec{P}, \quad (7.89)$$

to lowest order in the Taylor expansion, in agreement with (7.68); see Exercise 7.3.4. These results give the same $\vec{F}_{\text{bulk}}$ as in (7.67).

7.3.3 *Summary of macroscopic equations*

Both Schwinger's and Jackson's arguments lead to the same set of effective field equations that take the average electronic properties of the material into account. To obtain these equations, first we recognize that there may be additional charges and currents within the material that are externally produced: we identify these as the *free charge* and *free current* as previously introduced in Sections 5.4 and 6.10. This leads to the expressions

$$\rho = \rho_{\text{free}} + \rho_{\text{b}}; \qquad \vec{J} = \vec{J}_{\text{free}} + \vec{J}_{\text{b}}. \quad (7.90)$$

With the addition of these free currents, our field equations become

$$\vec{\nabla} \times \vec{B} - \frac{1}{c}\frac{\partial \vec{E}}{\partial t} = \frac{4\pi}{c}\left(\vec{J}_{\text{free}} + \frac{\partial \vec{P}}{\partial t} + c\vec{\nabla} \times \vec{M}\right), \quad (7.91)$$

$$\vec{\nabla} \cdot \vec{E} = 4\pi(\rho_{\text{free}} - \vec{\nabla} \cdot \vec{P}), \quad (7.92)$$

$$\vec{\nabla} \times \vec{E} + \frac{1}{c}\frac{\partial \vec{B}}{\partial t} = 0, \tag{7.93}$$

$$\vec{\nabla} \cdot \vec{B} = 0. \tag{7.94}$$

As in our previous discussions of macroscopic media (in Sections 5.4 and 6.10), we may define

$$\vec{D} \equiv \vec{E} + 4\pi\vec{P}, \tag{7.95}$$

$$\vec{H} \equiv \vec{B} - 4\pi\vec{M}. \tag{7.96}$$

Then, our macroscopic Maxwell equations simplify to

$$\vec{\nabla} \times \vec{H} - \frac{1}{c}\frac{\partial \vec{D}}{\partial t} = \frac{4\pi}{c}\vec{J}_{\text{free}}, \tag{7.97}$$

$$\vec{\nabla} \cdot \vec{D} = 4\pi\rho_{\text{free}}, \tag{7.98}$$

$$\vec{\nabla} \times \vec{E} + \frac{1}{c}\frac{\partial \vec{B}}{\partial t} = 0, \tag{7.99}$$

$$\vec{\nabla} \cdot \vec{B} = 0. \tag{7.100}$$

We close this section with a few important observations. In the above all fields are considered averaged macroscopic fields, even $\vec{E}$ and $\vec{B}$. Note also that the last two equations (7.99) and (7.100) did not enter our macroscopic considerations except as dynamical equations satisfied by the fundamental fields. The macroscopic form is the same as the fundamental form and should be considered as constraints. In the laboratory, the fields $\vec{D}$ and $\vec{H}$ are easier to measure than $\vec{E}$ and $\vec{B}$ because the charges and currents in (7.97) and (7.98) can be considered under experimental control. Although we have gone some way in defining the averaging processes involved, we still have not precisely specified the averaged fields. In practical situations the measurement apparatus actually performs the averaging. In quite a different sense, these equations also form a basis for our description of magnetic charge, as explained in Section 7.8.

7.4 Second-order formulation of the vacuum Maxwell equations

In Section 2.4 we reformulated electrostatics in terms of potentials, and in Section 6.3 we did the same for magnetostatics. In this section we will tie these together and present a potential formulation for the vacuum Maxwell equations that includes both the electrostatic scalar potential and the magnetostatic vector potential as special cases. In keeping with the fact that the electromagnetic field has four source components (charge and the three components of current density), our potential will have four independent components.[4]

We begin with the vector potential $\vec{A}$ defined in Section 6.3 by

$$\vec{\nabla} \times \vec{A} = \vec{B}. \tag{7.101}$$

Equation (7.101) correctly incorporates the condition $\vec{\nabla} \cdot \vec{B} = 0$. Faraday's equation may then be rewritten as

$$\vec{\nabla} \times \left(\vec{E} + \frac{1}{c}\frac{\partial \vec{A}}{\partial t} \right) = 0. \tag{7.102}$$

From vector calculus it follows that every function with zero curl can be written as a gradient, so there exists a function Φ such that

$$\begin{aligned} -\vec{\nabla}\Phi &= \vec{E} + \frac{1}{c}\frac{\partial \vec{A}}{\partial t} \\ \Longrightarrow \vec{E} &= -\vec{\nabla}\Phi - \frac{1}{c}\frac{\partial \vec{A}}{\partial t}. \end{aligned} \tag{7.103}$$

The other Maxwell equations may now be written in terms of the potentials $\vec{A}$ and Φ (recall Exercise 1.1.2):

$$\vec{\nabla} \cdot \vec{E} \Longrightarrow \nabla^2\Phi + \frac{1}{c}\frac{\partial}{\partial t}(\vec{\nabla} \cdot \vec{A}) = -4\pi\rho. \tag{7.104}$$

$$\vec{\nabla} \times \vec{B} - \frac{1}{c}\frac{\partial \vec{E}}{\partial t} \Longrightarrow \vec{\nabla}^2\vec{A} - \frac{1}{c^2}\frac{\partial^2 \vec{A}}{\partial t^2} - \vec{\nabla}\left(\vec{\nabla} \cdot \vec{A} + \frac{1}{c}\frac{\partial \Phi}{\partial t} \right) = -\frac{4\pi}{c}\vec{J}. \tag{7.105}$$

The choice of $\vec{A}$ and Φ is not unique, because the transformation

$$\vec{A} \Longrightarrow \vec{A} + \vec{\nabla}\Lambda, \tag{7.106}$$

$$\Phi \Longrightarrow \Phi - \frac{1}{c}\frac{\partial \Lambda}{\partial t}, \tag{7.107}$$

leaves $\vec{B}$ and $\vec{E}$ unchanged. In order to simplify the equations, we may impose the additional condition

$$\vec{\nabla} \cdot \vec{A} + \frac{1}{c}\frac{\partial \Phi}{\partial t} = 0, \tag{7.108}$$

[4]We shall see in Chapter 13 that these are in fact the components of a relativistically covariant 4-vector.

which is called the *Lorenz gauge.*[5] It is always possible to satisfy the Lorenz gauge, since given any $\vec{A}_0$ and Φ_0 we may define

$$F \equiv \vec{\nabla} \cdot \vec{A}_0 + \frac{1}{c}\frac{\partial \Phi_0}{\partial t}, \tag{7.109}$$

and then solve

$$\nabla^2 \Lambda - \frac{1}{c^2}\frac{\partial^2 \Lambda}{\partial t^2} = -F. \tag{7.110}$$

The resulting function Λ may be used to redefine $\vec{A}$ and Φ according to (7.106) and (7.107). In this case, then (7.104) and (7.105) simplify to

$$\nabla^2 \Phi - \frac{1}{c^2}\frac{\partial^2 \Phi}{\partial t^2} = -4\pi\rho, \tag{7.111}$$

$$\nabla^2 \vec{A} - \frac{1}{c^2}\frac{\partial^2 \vec{A}}{\partial t^2} = -\frac{4\pi}{c}\vec{J}, \tag{7.112}$$

which are *wave equations* (with sources) in $\vec{A}$ and Φ.

Even with the Lorenz gauge, the specification of $\vec{A}$ and Φ is still not unique. Given any solution of the homogeneous wave equation

$$\nabla^2 \Lambda - \frac{1}{c^2}\frac{\partial^2 \Lambda}{\partial t^2} = 0, \tag{7.113}$$

then we may make the replacements (7.106), (7.107) and still obtain solutions to (7.106) and (7.107). The Lorenz gauge is only consistent if currents are conserved, as we see in

$$\begin{aligned} -\left(\nabla^2 - \frac{1}{c^2}\frac{\partial^2}{\partial t^2}\right)\left(\vec{\nabla}\cdot\vec{A} + \frac{1}{c}\frac{\partial \Phi}{\partial t}\right) &= 0, \\ \Longrightarrow \frac{4\pi}{c}\left(\vec{\nabla}\cdot\vec{J} + \frac{\partial \rho}{\partial t}\right) &= 0, \end{aligned} \tag{7.114}$$

where we have interchanged commuting differential operators to obtain the result.

An alternative to the Lorenz gauge that is useful in many situations is the *radiation gauge* (also known as *Coulomb gauge* or *transverse gauge*), given by

$$\vec{\nabla} \cdot \vec{A} = 0. \tag{7.115}$$

We have already seen this gauge in Section 6.3, where we showed that it always exists in magnetostatic situations by choosing Λ appropriately. The same Λ works in non-static situations, if we also modify Φ according

[5] Note "Lorenz" (after Ludwig Lorenz) rather than "Lorentz" (after Hendrik Lorentz).

to (7.107). In this case, the second order potential form of the Maxwell equations becomes

$$\nabla^2\Phi = -4\pi\rho, \tag{7.116}$$

$$\vec{\nabla}^2\vec{A} - \frac{1}{c^2}\frac{\partial^2\vec{A}}{\partial t^2} = -\frac{4\pi}{c}\vec{J} + \frac{1}{c}\frac{\partial\vec{\nabla}\Phi}{\partial t}. \tag{7.117}$$

Like the Lorenz gauge, this gauge is only consistent if charge is conserved as we see once more in

$$-\left(\nabla^2 - \frac{1}{c^2}\frac{\partial^2}{\partial t^2}\right)\vec{\nabla}\cdot\vec{A} = 0$$

$$\Longrightarrow \frac{4\pi}{c}\left(\vec{\nabla}\cdot\vec{J} - \frac{1}{4\pi}\underbrace{\frac{\partial}{\partial t}\nabla^2\Phi}_{-4\pi\frac{\partial\rho}{\partial t}}\right) = 0. \tag{7.118}$$

Both the Lorenz gauge and Coulomb gauge give rise to wave equations. In Section 8.6 we will discuss the solution of these equations.

7.5 Magnetostatic field energy

In this section we derive an expression for the energy due to currents in magnetostatic systems. It may seem strange that we did not treat this concept in the magnetostatics chapter. Although it is not absolutely necessary to wait, it helps to have the time-dependent macroscopic Maxwell equations in place to put part of our argument in context. It will also help to have this section fresh and handy for the considerations leading to forces which are derived for quasi-static circumstances in the next section.

Recall that in electrostatics we computed the energy of a static charge distribution by assembling the charges. We may use a similar procedure to find the energy of a static current distribution. Specifically, we assemble macroscopic current distributions (either circuits or volume distributions of currents) out of infinitesimal, resistanceless current loops. We imagine that each tiny circuit comes equipped with its own battery whose mission in life is to keep the current flowing at a constant rate. In terms of this picture, we may compute the energy necessary to establish a given magnetic field configuration W by calculating the total energy expended by the loops' batteries. Macroscopic currents in a circuit are constructed by "tiling" the circuit with infinitesimal loops, such that the currents in the interior of the circuit cancel and the boundary currents comprise the observed macroscopic current.

Now consider a single infinitesimal current loop. When the magnetic field through the loop is changed, the current in the loop is given by Ohm's law

$$\mathcal{E}_0 + \mathcal{E} = IR \implies \mathcal{E}_0 = -\mathcal{E} \quad \text{for } R = 0, \tag{7.119}$$

where $\mathcal{E}_0$ is the emf due to the battery, and $\mathcal{E}$ is the induced emf due to changing magnetic flux. During the time interval δt, an amount of charge $\delta q \equiv I\delta t$ is moved around the circuit. The work done by the battery during the time interval δt is thus

$$\delta W \equiv \mathcal{E}_0 \delta q = \mathcal{E}_0 I \delta t = -\mathcal{E} I \delta t, \tag{7.120}$$

so that Faraday's law gives

$$\delta W = \frac{1}{c} I \delta \mathcal{F}. \tag{7.121}$$

We do not include the energy required to assemble the infinitesimal circuits themselves, which turns out to be infinite (see Section 7.6). This infinite contribution is the *magnetostatic self-energy*, which is analogous to electrostatic self-energy encountered in Section 2.10. Since the infinitesimal circuits are neither created nor destroyed in any physical process, we can neglect this term.

Using Stokes' theorem we may rewrite $\delta\mathcal{F}$ as

$$\delta \mathcal{F} = \int_S da\, \hat{n} \cdot \underbrace{\delta \vec{B}}_{\vec{\nabla} \times \delta \vec{A}} = \oint_C d\vec{\ell} \cdot \delta \vec{A}. \tag{7.122}$$

To get the total work done to assemble the macroscopic distribution, we sum over all infinitesimal loops:

$$\delta W \equiv \sum_i \delta W_i = \frac{1}{c} \sum_i I_i \delta \mathcal{F}_i = \frac{1}{c} \sum_i \oint_{C_i} I_i d\vec{\ell}_i \cdot \delta \vec{A}. \tag{7.123}$$

In the continuum limit, the discrete sum in (7.123) becomes a continuous integral:

$$\sum_i \oint_{C_i} \to \int_V, \qquad I_i d\vec{\ell} \to \vec{J}(\vec{x}) d^3x$$

$$\implies \delta W = \frac{1}{c} \int d^3x\, \vec{J}(\vec{x}) \cdot \delta \vec{A}(\vec{x}). \tag{7.124}$$

If we assemble the circuits slowly, then we can neglect the displacement current term in the macroscopic Ampère-Maxwell Equation (7.97) and rewrite (7.124) as

$$\delta W = \frac{1}{4\pi}\int d^3x \left(\vec{\nabla}\times\vec{H}\right)\cdot\delta\vec{A}. \tag{7.125}$$

Using the component form to integrate by parts, we find

$$\begin{aligned}\left(\vec{\nabla}\times\vec{H}\right)\cdot\delta\vec{A} &= \sum_{i,j,k}\epsilon_{ijk}\delta A_i\nabla_j H_k \\ &= \sum_{i,j,k}\epsilon_{ijk}\nabla_j(\delta A_i H_k) - \sum_{i,j,k}\epsilon_{ijk}(\nabla_j\delta A_i)H_k \\ &= [\text{surface term}] + \delta\vec{B}\cdot\vec{H}\end{aligned}$$

$$\Longrightarrow \delta W = \frac{1}{4\pi}\int d^3x\,\vec{H}\cdot\delta\vec{B}. \tag{7.126}$$

This is identical to the electrostatic expression (5.159) with replacements $\vec{E}\to\vec{H}$, $\vec{D}\to\vec{B}$. (We have seen this correspondence before in (6.210).)

Now during the assembly process the free currents will change as more circuit elements are brought in, and thus $\vec{H}$ will change. So we can't go any further without specific assumptions about the medium. If we assume our medium is linear and that the permeability tensor is symmetric ($\mu_{ij} = \mu_{ji}$), then

$$\begin{aligned}B_i &= \sum_j \mu_{ij}H_j \\ \Longrightarrow \vec{B}\cdot\vec{H} &= \sum_{i,j}\mu_{ij}H_iH_j \\ \Longrightarrow \delta\vec{B}\cdot\vec{H} &= \vec{B}\cdot\delta\vec{H} = \frac{1}{2}\delta\left(\vec{B}\cdot\vec{H}\right).\end{aligned} \tag{7.127}$$

It follows that

$$W = \frac{1}{8\pi}\int d^3x\,\vec{B}\cdot\vec{H}. \tag{7.128}$$

This can be integrated by parts using $\vec{B} = \vec{\nabla}\times\vec{A}$ to obtain the general expression

$$W = \frac{1}{2c}\int d^3x\,\vec{J}\cdot\vec{A} = \frac{1}{2c^2}\int d^3x d^3x'\,\frac{\vec{J}(\vec{x})\cdot\vec{J}(\vec{x}\,')}{|\vec{x}-\vec{x}\,'|}. \tag{7.129}$$

If the current consists of a number of closed circuits, then the expression simplifies to

$$W = \sum_{i,j}\frac{I_iI_j}{2c^2}\oint_i\oint_j\frac{d\vec{x}_i\cdot d\vec{x}_j}{|\vec{x}_i-\vec{x}_j|}. \tag{7.130}$$

It is possible to show (Exercise 7.5.3) that this is equivalent to

$$W = \frac{1}{2c}\sum_i I_i \mathcal{F}_i, \tag{7.131}$$

where $\mathcal{F}_i$ is total magnetic flux through circuit i:

$$\mathcal{F}_i = \int_{S_i} da_i\, \hat{n}_i \cdot \vec{B}. \tag{7.132}$$

In the above discussion we have computed the field energy W via a conceptual process of assembling current distributions from infinitesimal resistanceless circuits, where each circuit has a battery that maintains constant current. Alternatively, (7.131) may be derived by assembling the infinitesimal resistanceless circuits at fixed flux rather than fixed current. In this case, the field energy is computed as the work done by external sources to maintain constant flux. We encountered similar alternatives in electrostatics, where electrostatic energy was computed either by bringing in conductors at fixed voltage (analogous to fixed current case) or fixed charge (analogous to fixing the magnetic flux).

Yet another derivation of (7.131) is as follows. (It does not refer to the medium.) We assume a linear relationship between I_i and $\mathcal{F}_i$, and we parametrize the process of building up the currents by the variable α, where α varies from 0 to 1.

$$\begin{aligned} I_i(\alpha) &\equiv \alpha I_i, \\ \mathcal{F}_i(\alpha) &= \mathcal{F}_i \alpha \implies \delta\mathcal{F}_i \equiv \mathcal{F}_i \delta\alpha. \end{aligned}$$

Then using the expression for δW in (7.121) we obtain

$$\begin{aligned} \delta_\alpha W &\equiv \frac{1}{c}\sum_i I_i \mathcal{F}_i \alpha\, d\alpha, \\ \implies W &= \int_0^1 d\alpha\, \delta_\alpha W = \frac{1}{2c}\sum_i I_i \mathcal{F}_i, \end{aligned} \tag{7.133}$$

which leads immediately to (7.131). (A variation of this argument in the electrostatic case is not possible since building up charge this way would violate charge conservation!)

The *self-inductance* and *mutual-inductance* of current loops are related to the field energies calculated. For a model of thin-wired circuits the definition of these quantities is given by comparing (7.130) to the definition,

$$W \equiv \frac{1}{2}\sum_i L_i I_i^2 + \sum_{i<j} M_{ij} I_i I_j \tag{7.134}$$

for $i = 1, ..., N$ circuits, each with current I_i. This gives us

$$M_{ij} = M_{ji} = \frac{1}{c^2} \oint \oint \frac{d\vec{x}_i \cdot d\vec{x}_j}{|\vec{x}_i - \vec{x}_j|}, \tag{7.135}$$

for the mutual-inductance, M_{ij} $(i \neq j)$, and

$$L_i = \frac{1}{c^2} \oint \oint \frac{d\vec{x}_i \cdot d\vec{x}\,'_i}{|\vec{x}\,'_i - \vec{x}_i|}, \tag{7.136}$$

for the self-inductance, L_i. We can also show that

$$M_{ij} = \frac{\mathcal{F}_{ij}}{cI_j}, \quad L_i = \frac{\mathcal{F}_{ii}}{cI_i}, \tag{7.137}$$

where $\mathcal{F}_{ij}$ is magnetic flux flowing through circuit i due to circuit j,

$$\mathcal{F}_{ij} = \int_{s_i} da_i\, \hat{n}_i \cdot \vec{B}_j. \tag{7.138}$$

The total flux through circuit i is then $\mathcal{F}_i \equiv \sum_j \mathcal{F}_{ij}$. Self and mutual-inductance are fundamental concepts in the construction of electronic circuits and devices.

7.6 Magnetostatic field energy and forces on current loops

It is also possible to directly compute the field energy of two interacting circuits by integrating the force as the two circuits are brought together from infinity. This alternative method is confirmatory rather than independent; however, we will learn something important that we haven't seen already.

We begin with two circuits, as shown in Figure 7.4. In the magnetostatics chapter we computed the force of loop 2 on loop 1 in Equation (6.6) as

$$\vec{F}_{12} = -\frac{I_1 I_2}{c^2} \oint_1 \oint_2 \frac{(\vec{x}_1 - \vec{x}_2) d\vec{x}_1 \cdot d\vec{x}_2}{|\vec{x}_1 - \vec{x}_2|^3}. \tag{7.139}$$

If loop (1) is translated by $\vec{R}$ as shown in Figure 7.4, then the force can be expressed as

$$\vec{F}_{12}(\vec{R}) = -\frac{I_1 I_2}{c^2} \oint_1 \oint_2 \frac{(\vec{x}_1 - \vec{x}_2 + \vec{R}) d\vec{x}_1 \cdot d\vec{x}_2}{|\vec{x}_1 - \vec{x}_2 + \vec{R}|^3}, \tag{7.140}$$

which can also be written

$$\vec{F}_{12}(\vec{R}) = I_1 I_2 \vec{\nabla}_{\vec{R}} M_{12}(\vec{R}), \tag{7.141}$$

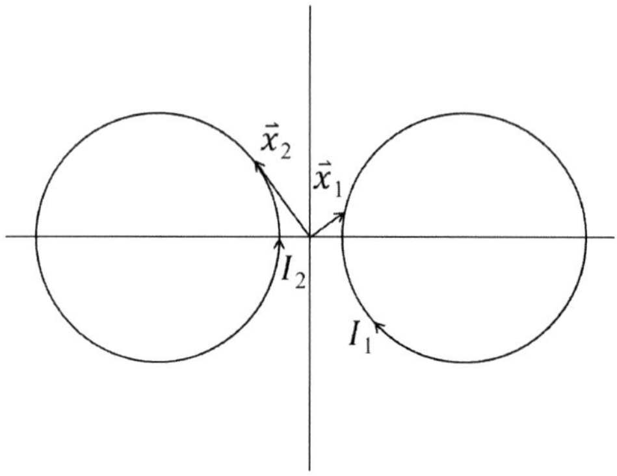

Fig. 7.4 Two interacting circuits: notation.

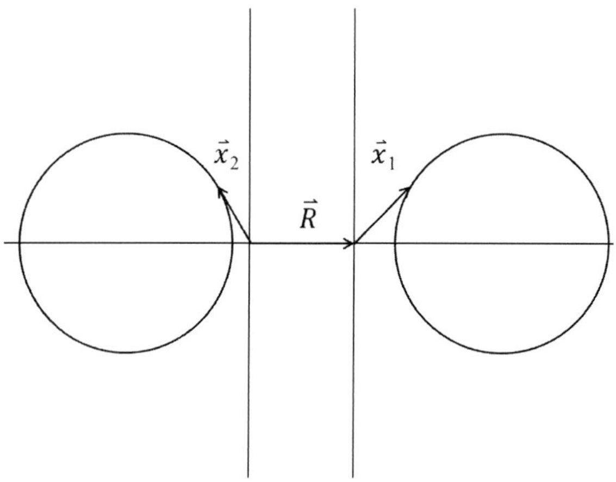

Fig. 7.5 New coordinate system.

where M_{12} is the mutual inductance, given here by

$$M_{12}(\vec{R}) = \frac{1}{c^2} \oint_1 \oint_2 \frac{d\vec{x}_1 \cdot d\vec{x}_2}{|\vec{x}_1 - \vec{x}_2 + \vec{R}|}. \tag{7.142}$$

As an aside, we remark that the mutual inductance satisfies the Laplace equation

$$\nabla_R^2 M_{12}(\vec{R}) = 0, \tag{7.143}$$

as long as the circuits don't overlap. This makes it possible to compute $M_{12}(\vec{R})$ for all $\vec{R}$, using multipole expansions of the homogeneous Poisson equation, together with the computation of $M_{12}(R\hat{n})$ for a fixed direction $\hat{n}$ as a boundary condition; see Exercise 7.6.4.

It now follows that the mechanical energy of assembly is

$$W_{\text{mech}} = -\int_{\infty}^{0} \vec{F}_{12}(\vec{R}) \cdot d\vec{R},$$

$$\Longrightarrow W_{\text{mech}} = -\frac{I_1 I_2}{c^2} \oint_1 \oint_2 \frac{d\vec{x}_1 \cdot d\vec{x}_2}{|\vec{x}_1 - \vec{x}_2|}. \tag{7.144}$$

Surprisingly, this result is the negative of what we found previously in (7.130). The difference is due to the fact that in the mechanical process of bringing the loops together, we have changed the energy of the batteries as well as the field,

$$\delta_I W_{\text{mech}} = \delta_I W + \delta_I W_{\text{b}}, \tag{7.145}$$

where W denotes the field energy and W_{b} is the battery energy. The notation δ_I indicates that I is kept constant during the process. Note that W includes the self-energies of the two circuits; however, these do not change since I is constant. We have

$$W = \frac{1}{2c} \sum_i I_i \mathcal{F}_i, \Longrightarrow \delta_I W = \frac{1}{2c} \sum_i I_i \delta\mathcal{F}_i. \tag{7.146}$$

Now the change in the ith battery energy is the negative of the work done by the batteries to move a small amount of charge δq_i. The change in the battery energy is therefore

$$\delta_I W_{\text{b}} = -\sum_i \mathcal{E}_{0i} \delta q_i = \sum_i \mathcal{E}_i \delta q_i. \tag{7.147}$$

The ith battery's emf ($\mathcal{E}_{0i}$) opposes the induced emf ($\mathcal{E}_i$) in order to keep the current constant. It follows from Faraday's law that

$$\delta_I W_{\text{b}} = -\sum_i \left(\frac{1}{c} \frac{\delta \mathcal{F}_i}{\delta t} \right) \delta q_i, \tag{7.148}$$

which can be rearranged to give

$$\delta_I W_{\text{b}} = -\sum_i \frac{1}{c} \delta\mathcal{F}_i \frac{\delta q_i}{\delta t} = -\sum_i \frac{1}{c} \delta\mathcal{F}_i I_i = -2\delta_I W. \tag{7.149}$$

(We saw a similar relation for electrostatics in Section 5.9: $\delta_V W_{\text{b}} = -2\delta_V W$.) We finally obtain justification for the minus sign in (7.144):

$$\delta_I W_{\text{mech}} = \delta_I W + \delta_I W_{\text{b}} = \delta_I W - 2\delta_I W = -\delta_I W. \tag{7.150}$$

An additional consequence of the above argument is an alternate expression for the magnetostatic force at constant current:

$$\vec{F} = -\frac{\partial}{\partial \vec{x}}(W + W_b)_I = +\left(\frac{\partial W}{\partial \vec{x}}\right)_I, \tag{7.151}$$

which is analogous to (5.179) and gives the electrostatic force at constant voltage. On the other hand, when calculating the magnetostatic force at fixed flux there is no battery energy, and we have

$$\vec{F} = -\left(\frac{\partial W}{\partial \vec{x}}\right)_F. \tag{7.152}$$

This same sign was encountered in (5.174), which gives the electrostatic force on a charge distribution at fixed charge. We see that the situation of constant charge (voltage) in electrostatics is analogous to constant flux (current) in magnetostatics. We will see relevant examples in the upcoming Section 7.7. The point bears repeating that although the signs differ, the two expressions will give the same result for the magnetostatic force in instantaneous circumstances, just as we saw in the electrostatic case.

We may verify (7.151) in the case of parallel wires, shown in Figure 7.6. Recall that the force between wires was explicitly calculated two ways in Section 6.1. The field energy for this situation is

$$W = \int dz_1 dz_2 \frac{I_1 I_2}{c^2} \frac{\hat{n}_1 \cdot \hat{n}_2}{|\vec{x}_1 - \vec{x}_2|}, \quad \left(|\vec{x}_1 - \vec{x}_2| = \sqrt{\rho^2 + (z_1 - z_2)^2}\right)$$

$$\Longrightarrow W = \frac{I_1 I_2}{c^2} \hat{n}_1 \cdot \hat{n}_2 \int_{-L}^{L} \int_{-L}^{L} \frac{dz_1 dz_2}{\sqrt{\rho^2 + (z_1 - z_2)^2}}. \tag{7.153}$$

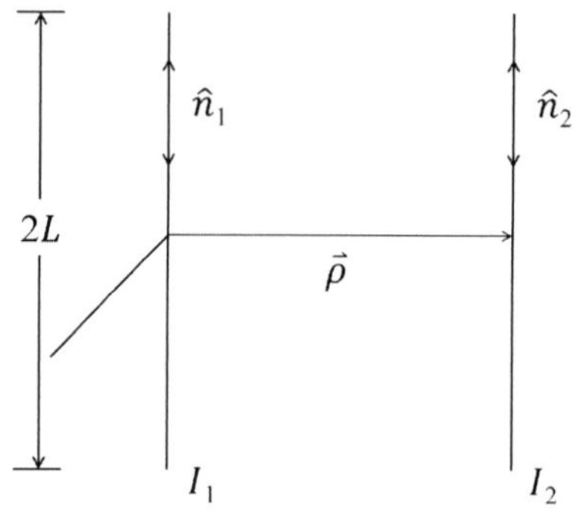

Fig. 7.6 Force between parallel wires.

The integral may be evaluated as follows:

$$z' \equiv \frac{z_1 + z_2}{2}, \quad z'' \equiv z_1 - z_2 \quad (dz_1 dz_2 = dz' dz'')$$

$$\Longrightarrow \int\int \frac{dz_1 dz_2}{\sqrt{\rho^2 + (z_1 - z_2)^2}} = \int_{-L}^{L} dz' \int_{-2L}^{2L} \frac{dz''}{\sqrt{\rho^2 + z''^2}}$$
$$\underbrace{\approx}_{L \gg \rho} \int_{-L}^{L} dz' 2\ln\left(\frac{4L}{\rho}\right)$$

$$\Longrightarrow \frac{W}{2L} \approx \frac{2I_1 I_2}{c^2} \hat{n}_1 \cdot \hat{n}_2 \ln\left(\frac{4L}{\rho}\right), \tag{7.154}$$

so that the force of wire 1 on 2 per length is

$$\frac{F_\rho}{2L} = +\frac{\partial}{\partial\rho}\left(\frac{W}{2L}\right) \approx -\frac{2I_1 I_2}{c^2} \frac{\hat{n}_1 \cdot \hat{n}_2}{\rho}. \tag{7.155}$$

This agrees with (6.15) and (6.21).

The electrostatic energy of a point charge was found to be infinite (see (5.118)). What about the magnetostatic analog? For a single circular current loop of radius a we have

$$W_{\text{self}} = \frac{I^2 a^2}{2c^2} \int d\phi \, d\phi' \left. \frac{\hat{e}_\phi \cdot \hat{e}_{\phi'}}{|\vec{x} - \vec{x}\,'|} \right|_{\substack{\rho=\rho'=a \\ z=z'=0}}, \tag{7.156}$$

from (7.130). Now

$$|\vec{x} - \vec{x}\,'||_{z=z'=0, \rho=\rho'=a} = a\left[(\cos\phi - \cos\phi')^2 + (\sin\phi - \sin\phi')^2\right]^{1/2}$$
$$= \sqrt{2} a [1 - \cos(\phi - \phi')]^{1/2}, \tag{7.157}$$

and

$$\hat{e}_\phi \cdot \hat{e}_{\phi'} = (-\hat{\imath}\sin\phi + \hat{\jmath}\cos\phi) \cdot (-\hat{\imath}\sin\phi + \hat{\jmath}\cos\phi')$$
$$= \cos(\phi - \phi'). \tag{7.158}$$

In the integration,

$$\phi_1 \equiv \frac{1}{2}(\phi' + \phi'), \text{ limits} : (0, 2\pi)$$
$$\phi_2 \equiv \phi - \phi', \text{ limits} : (-2\pi, 2\pi)$$

so that

$$W_{\text{self}} = \frac{\pi I^2 a}{\sqrt{2} c^2} \int_{-2\pi}^{2\pi} \frac{d\phi_2 \cos\phi_2}{\sqrt{1 - \cos\phi_2}}. \tag{7.159}$$

When $\phi_2 \approx 0, 2\pi, -2\pi$ we must be careful in our evaluations because of the singularity in the integrand. Let us introduce a symmetric form of the integral which integrates from $-2\pi+\epsilon$ to $-\epsilon$, then continues the integration from ϵ to $2\pi - \epsilon$ for $\epsilon \to 0^+$. One can show this is equivalent to evaluating the integral:

$$W_{\text{self}}(\epsilon) \equiv \frac{4\pi I^2 a}{\sqrt{2}c^2} \int_{\epsilon}^{\pi} \frac{d\phi \cos\phi}{\sqrt{1-\cos\phi}}, \tag{7.160}$$

in the limit $\epsilon \to 0^+$. The integral may be looked up or evaluated by computer to yield:

$$W_{\text{self}}(\epsilon) = \frac{4\pi I^2 a}{c^2} \left[\left(\ln\left(\frac{4}{\epsilon}\right) - 2 \right) + O(\varepsilon^2) \right]. \tag{7.161}$$

This implies that (7.159) is logarithmically divergent. This self-energy result implies the self-inductance of the loop also is divergent. Of course, the self-energy of an actual current loop is not infinite. A more realistic model would use a nonsingular current density $\vec{J}(\vec{x})$, in which case we would in general get a finite result. This is similar to the situation in electrostatics, where the self-energy is infinite (linearly) for a point charge but finite for continuous distributions.

7.7 Bulk forces on magnetic materials: theory and examples

We may compute the bulk forces on magnetic materials in the same way that we computed forces on dielectrics in Section 5.9. Just as before, there is an energy method and a direct force computation: we will demonstrate both below.

7.7.1 *Energy method*

We begin with

$$\Delta W = W - W_0, \tag{7.162}$$

where W_0 and W are respectively the initial and final magnetostatic field energies (initially, the material is placed at spatial infinity or some convenient location).

We suppose that $\vec{J}$ is fixed during the process. It is then possible to show (see Exercise 7.7.1)

$$\Delta W = \frac{1}{8\pi}\int_{\Delta V} d^3x \left(\frac{1}{\mu_0} - \frac{1}{\mu}\right) \vec{B}\cdot\vec{B}_0, \tag{7.163}$$

where ΔV is the introduced bulk volume. If $\mu_0 = 1$ we may rewrite this as

$$\Delta W = \frac{1}{2}\int_{\Delta V} d^3x\, \vec{M}\cdot\vec{B}_0, \tag{7.164}$$

where the factor of $1/2$ corresponds to the fact that this is not a permanent dipole (see (5.173) for the electrostatic analog). This expression and

$$F_\xi = +\left(\frac{\partial \Delta W}{\partial \xi}\right)_I \tag{7.165}$$

can be used together to find the force.

This is usually the easiest method for finding forces. Often approximations are made to simplify calculations. Here we provide some examples that illustrate the method.

Energy method, Example 1: magnetized rod in solenoid

A cylindrical solenoid of length L with constant current I produces an approximately uniform magnetic field $\vec{B}_0$ within the solenoid. A thick cylindrical rod, which nearly fills the solenoidal cavity, with cross-section A and magnetic permeability $\mu \neq 1$ is partially inserted into the solenoid parallel to its axis, as shown in Figure 7.7. The magnetization within the rod is given by

$$\vec{M} = \frac{1}{4\pi}(\vec{B} - \vec{H}) = \frac{\vec{B}}{4\pi}\left(1 - \frac{1}{\mu}\right), \tag{7.166}$$

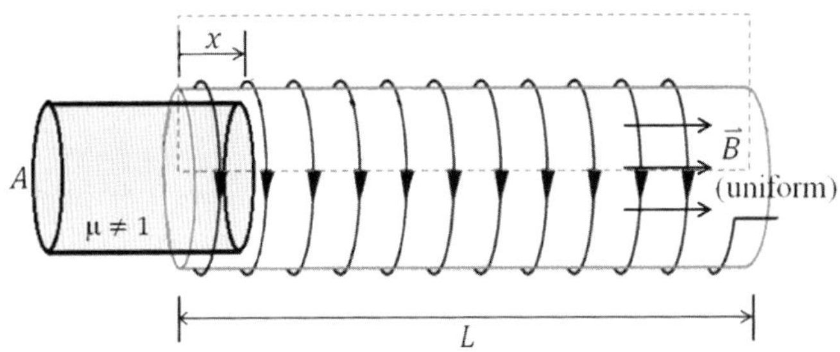

Fig. 7.7 Force on magnetically permeable rod inserted in a solenoid.

where $\vec{B}$ is nearly constant inside the solenoid. We may use Ampère's law on the dashed loop in Figure 7.7 to compute the magnetic field $\vec{B}$. Neglecting edge effects, we obtain

$$\oint \vec{H} \cdot d\vec{\ell} = \frac{4\pi}{c} NI, \tag{7.167}$$

where N is the number of turns in the solenoid. When $x = 0$ the left-hand side of (7.167) is equal to $B_0 L$, and since the value is independent of x we have

$$\frac{1}{\mu} Bx + B(L - x) = B_0 L \implies B = \frac{B_0 L}{\left[L + x\left(\frac{1}{\mu} - 1\right)\right]}. \tag{7.168}$$

Using this together with (7.164) and (7.166) gives

$$\Delta W \approx \frac{1}{2} \int d^3x \frac{\vec{B} \cdot \vec{B}_0}{4\pi} \left(1 - \frac{1}{\mu}\right) = \frac{\vec{B}_0^2 L A x (1 - \frac{1}{\mu})}{8\pi [L + x(\frac{1}{\mu} - 1)]},$$

$$\implies F_x = + \left(\frac{\partial \Delta W}{\partial x}\right)_I = \frac{\vec{B}_0^2 A L^2}{8\pi} \frac{(1 - \frac{1}{\mu})}{[L + x(\frac{1}{\mu} - 1)]^2}. \tag{7.169}$$

The rod is pulled inward or pushed outward according to whether $\mu > 1$ (the paramagnetic case) or $\mu < 1$ (the diamagnetic case).

Actually, the problem above is oversimplified in that we are applying the normal $\vec{B}$ boundary condition at the end of the rod and ignoring the sides. We would expect this to hold for the case where the permeable material has not been inserted too far into the solenoid. In Exercise 7.7.2 we also consider a variant associated with a far insertion; then the boundary condition is more appropriately applied to the sides using tangential $\vec{H}$. We also consider a case where the flux (rather than the current) is held constant.

Energy method, Example 2: current loop and magnetic material

A current loop is located outside of and parallel to a large block of material with magnetic permeability $\mu \neq 1$ as shown in Figure 7.8. Since the field energy is the volume integral of $\vec{M} \cdot \vec{B}_0$, we would expect that the force between the loop and the material will be attractive in the para-magnetic case ($\mu > 1$), and repulsive in the diamagnetic case ($\mu < 1$). This is actually what we found in (6.237) using the image method. In Exercise 7.7.3 you will use the energy method to arrive at the same result.

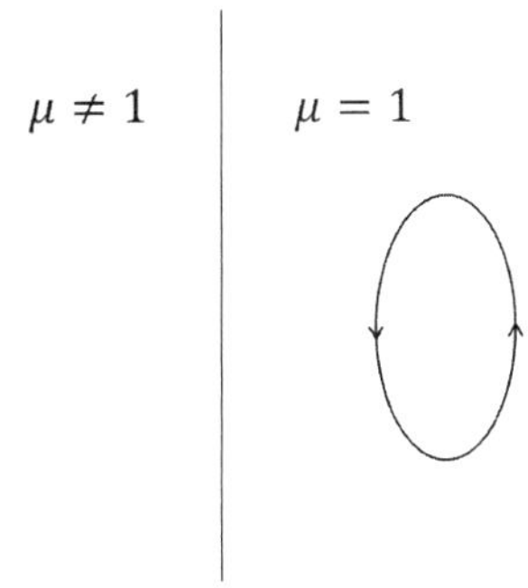

Fig. 7.8 Example 2.

7.7.2 *Force method*

We may also use explicit integral formulas to compute the magnetostatic force, as we did in Section 6.9 for electrostatics. If we have an external field $\vec{B}$, then

$$\vec{F}_{\text{bulk}} = \frac{1}{c}\int d^3x\, \vec{J}_{\text{b}} \times \vec{B}, \tag{7.170}$$

where the bound current density within the volume is

$$\vec{J}_{\text{b}} = c\vec{\nabla} \times \vec{M}. \tag{7.171}$$

Previously our integrals were over all space, and did not have explicit material surface contributions. In this case, when $\vec{J}_{\text{b}}$ is integrated over all space, the discontinuity in $\vec{M}$ at the boundary of the medium leads to a surface term. Thus we may break the force up into volume and surface contributions as

$$\vec{F}_{\text{bulk}} = \int_V d^3x(\vec{\nabla} \times \vec{M}) \times \vec{B} + \frac{1}{c}\int_S da\, \vec{K}_{\text{b}} \times \vec{B}. \tag{7.172}$$

The surface current $\vec{K}_{\text{b}}$ was computed in (6.196), and also may be deduced from the following argument. Using the notation of Figure 7.9, we integrate both sides of (7.171) over the area A within the infinitesimal loop to obtain

$$\begin{aligned}
&\int_A da\, \hat{t} \cdot \vec{J}_{\text{b}} = \int_A da\, \hat{t} \cdot \left(c\vec{\nabla} \times \vec{M}\right) \\
&\implies \vec{K}_{\text{b}} \cdot \hat{t}\Delta\ell = -c\vec{M} \cdot \vec{\ell}_{21}\Delta\ell \\
&\implies \vec{K}_{\text{b}} \cdot \hat{t} = \left(c\vec{M} \times \hat{n}_{21}\right) \cdot \hat{t} \\
&\implies \vec{K}_{\text{b}} = c\vec{M} \times \hat{n}.
\end{aligned} \tag{7.173}$$

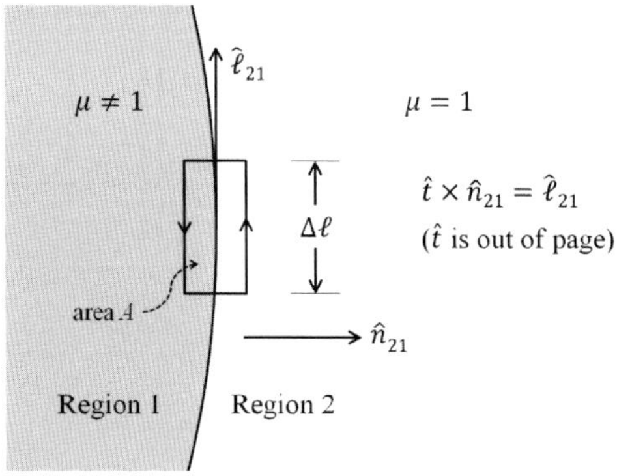

Fig. 7.9 Notation for surface current derivation.

Yet another way to obtain the surface current is to use the magnetic boundary conditions on the loop in Figure 7.9:

$$\hat{n} \times (\vec{B}_2 - \vec{B}_1) = \frac{4\pi}{c}\left(\vec{K}_{\text{free}} + \vec{K}_{\text{b}}\right). \tag{7.174}$$

Plugging in the relations

$$\vec{B}_2 = \vec{H}_2; \quad \vec{B}_1 = \vec{H}_1 + 4\pi\vec{M}; \quad \hat{n} \times (\vec{H}_2 - \vec{H}_1) = \frac{4\pi}{c}\vec{K}_{\text{free}}$$

$$\Longrightarrow \vec{K}_{\text{b}} = \frac{c}{4\pi}\hat{n} \times -4\pi\vec{M} = c\vec{M} \times \hat{n}. \tag{7.175}$$

Having established the surface current, we return now to the bulk force and consider some special cases. For a linear isotropic medium,

$$\vec{M} = \frac{\vec{B}}{4\pi}\left(1 - \frac{1}{\mu(\vec{x})}\right), \tag{7.176}$$

and when μ is constant then

$$\begin{aligned}\vec{J}_{\text{b}} = c\vec{\nabla} \times \vec{M} &= \frac{c}{4\pi}\left(1 - \frac{1}{\mu}\right)\underbrace{\vec{\nabla} \times \vec{B}}_{\frac{4\pi}{c}\vec{J}_{\text{tot}}}\\ &= \left(1 - \frac{1}{\mu}\right)\left(\vec{J}_{\text{free}} + \vec{J}_{\text{b}}\right)\\ \Longrightarrow \vec{J}_{\text{b}} &= (\mu - 1)\vec{J}_{\text{free}},\end{aligned} \tag{7.177}$$

which is appropriate for volume currents. In this case, if $\vec{J}_{\text{free}} = 0$ in V then only the surface term is left and

$$\begin{aligned}d\vec{F}|_{\text{surface}} &= \frac{da}{c}\left(\vec{K}_{\text{free}} + \vec{K}_{\text{b}}\right) \times \vec{B},\\ &= \frac{1}{4\pi}da\left[\hat{n} \times (\vec{B}_2 - \vec{B}_1)\right] \times \vec{B}.\end{aligned} \tag{7.178}$$

For $\vec{B}$ in (7.178) we may use either the external field or the average at the surface, $(\vec{B}_1 + \vec{B}_2)/2$, just like in the electrostatic case. (In this case $\vec{B}_{1\text{self}}$, $\vec{B}_{2\text{self}}$ are tangent to surface and will average to zero). In the latter case:

$$\begin{aligned} d\vec{F}_{\text{bulk}} &= \frac{1}{8\pi} da[\hat{n} \times (\vec{B}_2 - \vec{B}_1)] \times \left(\vec{B}_1 + \vec{B}_2\right), \\ &= \frac{1}{8\pi} da[-\hat{n}(\vec{B}_2^2 - \vec{B}_1^2) + (\vec{B}_2 - \vec{B}_1)(B_{1n} + B_{2n})] \end{aligned} \tag{7.179}$$

$$\begin{aligned} \Longrightarrow \mathcal{F}_n &\equiv \frac{d\vec{F}_{\text{bulk}} \cdot \hat{n}}{da} = \frac{1}{8\pi}[-\vec{B}_2^2 + \vec{B}_1^2 + ((B_{2n})^2 - (B_{1n})^2)], \\ \mathcal{F}_n &= \frac{1}{8\pi}[(B_{1||})^2 - (B_{2||})^2]. \end{aligned} \tag{7.180}$$

In the case where region 2 is vacuum, $\vec{B}_1 = \mu \vec{H}_1$, and no free currents are present, we obtain from the tangential H boundary conditions that

$$\mathcal{F}_n \to \frac{(B_{2||})^2}{8\pi}(\mu^2 - 1). \tag{7.181}$$

We may apply this result to Example 2 of the energy method (depicted in Figure 7.8) to show once again that the force is attractive or repulsive depending on whether $\mu > 1$ or $\mu < 1$ respectively; see Exercise 7.7.4.

In the case where there are no free currents or sources within the material being studied, then the force can be recast in an essentially "electrostatic" form, as follows. Here we denote the magnetic field by $\vec{B}_e$ to emphasize that we assume it is entirely due to external sources. We begin with the identity

$$\vec{\nabla}\left(\vec{M} \cdot \vec{B}_e\right) = \left(\vec{M} \cdot \vec{\nabla}\right)\vec{B}_e + \left(\vec{B}_e \cdot \vec{\nabla}\right)\vec{M} + \vec{B}_e \times \left(\vec{\nabla} \times \vec{M}\right) + \vec{M} \times \left(\vec{\nabla} \times \vec{B}_e\right). \tag{7.182}$$

Since $\vec{\nabla} \times \vec{B}_e = 0$ in the region of interest, we have

$$\left(\vec{\nabla} \times \vec{M}\right) \times \vec{B}_e = \left(\vec{M} \cdot \vec{\nabla}\right)\vec{B}_e + \left(\vec{B}_e \cdot \vec{\nabla}\right)\vec{M} - \vec{\nabla}\left(\vec{M} \cdot \vec{B}_e\right). \tag{7.183}$$

We may apply this now to our force expression (7.170) and (7.171):

$$\int d^3x \left(\vec{\nabla} \times \vec{M}\right) \times \vec{B}_e = \underbrace{\int d^3x \left(\vec{M} \cdot \vec{\nabla}\right)\vec{B}_e}_{(1)} + \underbrace{\int d^3x \left(\vec{B}_e \cdot \vec{\nabla}\right)\vec{M}}_{(2)} \tag{7.184}$$

$$- \underbrace{\int d^3x \vec{\nabla}\left(\vec{M} \cdot \vec{B}_e\right)}_{(3)}. \tag{7.185}$$

Some hard work is necessary to simplify these terms:

$$(1) = \int d^3x \sum_i \nabla_i(M_i\vec{B}_e) - \int d^3x \left(\vec{\nabla}\cdot\vec{M}\right)\vec{B}_e$$

$$= \int da\left(\hat{n}\cdot\vec{M}\right)\vec{B}_e - \int d^3x\left(\vec{\nabla}\cdot\vec{M}\right)\vec{B}_e, \tag{7.186}$$

$$(2) = \int d^3x \sum_i \nabla_i\left(B_{ei}\vec{M}\right) - \int d^3x\left(\vec{\nabla}\cdot\vec{B}_e\right)\vec{M}$$

$$= \int da\left(\hat{n}\cdot\vec{B}_e\right)\vec{M}, \tag{7.187}$$

$$(3) = -\int d^3x\vec{\nabla}\left(\vec{M}\cdot\vec{B}_e\right) = -\int da\left(\vec{M}\cdot\vec{B}_e\right)\hat{n}. \tag{7.188}$$

From the original surface term in $\vec{F}_{\text{bulk}}$ we get

$$\int da\,(\vec{M}\times\hat{n})\times\vec{B}_e = -\int da\,(\hat{n}\cdot\vec{B}_e)\vec{M} + \int da\,(\vec{M}\cdot\vec{B}_e)\hat{n}. \tag{7.189}$$

Putting all the pieces together yields

$$\vec{F}_{\text{bulk}} = \int d^3x\;(-\vec{\nabla}\cdot\vec{M})\vec{B}_e + \int da\;(\hat{n}\cdot\vec{M})\vec{B}_e, \tag{7.190}$$

where it is understood the first term is restricted to the volume V of the material. The effective interaction behaves as if the magnetic material possessed a magnetic "charge density" of $-\vec{\nabla}\cdot\vec{M}$ and "surface charge" of $\vec{M}\cdot\hat{n}$. We could have inferred this also from (6.255).

7.8 Magnetic charge and the macroscopic Maxwell equations

We now return from our excursion into magnetostatics, and start another topic. Instead of assuming that moving electric charge is the source of all magnetic fields, we will examine the implications of supposing that magnetic charges also exist. To do this, we add magnetic charge and current to the fundamental Maxwell equations in the obvious, symmetrical manner:

$$\vec{\nabla}\cdot\vec{E} = 4\pi\rho_e, \tag{7.191}$$

$$\vec{\nabla}\times\vec{B} - \frac{1}{c}\frac{\partial\vec{E}}{\partial t} = \frac{4\pi}{c}\vec{J}_e, \tag{7.192}$$

$$\vec{\nabla}\cdot\vec{B} = 4\pi\rho_m, \tag{7.193}$$

$$\vec{\nabla}\times\vec{E} + \frac{1}{c}\frac{\partial\vec{B}}{\partial t} = -\frac{4\pi}{c}\vec{J}_m. \tag{7.194}$$

Now there are two types of conserved quantities and associated currents, $(\rho_e, \vec{J}_e)$ and $(\rho_m, \vec{J}_m)$. In addition, the generalized Lorentz force on particles with both electrical charge e and magnetic charge g is given by

$$\vec{F} = e\left(\vec{E} + \frac{\vec{v}}{c} \times \vec{B}\right) + g\left(\vec{B} - \frac{\vec{v}}{c} \times \vec{E}\right). \tag{7.195}$$

The question is, does this change introduce new physics? The surprising answer is: Not necessarily! Consider the following *duality transformation*:

$$x_e = x'_e \cos\xi + x'_m \sin\xi, \tag{7.196}$$

$$x_m = -x'_e \sin\xi + x'_m \cos\xi, \tag{7.197}$$

where

$$x_{e,m} = \rho_{e,m} \quad \text{or} \quad \vec{J}_{e,m}, \tag{7.198}$$

and similarly redefine fields in the same way,

$$\vec{E} = \vec{E}' \cos\xi + \vec{B}' \sin\xi, \tag{7.199}$$

$$\vec{B} = -\vec{E}' \sin\xi + \vec{B}' \cos\xi. \tag{7.200}$$

Then we obtain a new, primed set of equations which have the same form as the original (unprimed) set. This implies that electric or magnetic charges or currents are not absolutes, but can be rotated to different values: just like gauge fields, their value depends on some convention.

There are some invariants under the duality transformation. For example, we have

$$\rho_m^2 + \rho_e^2 = {\rho'_m}^2 + {\rho'_e}^2, \tag{7.201}$$

$$\vec{J}_m^2 + \vec{J}_e^2 = \vec{J}_m'^2 + \vec{J}_e'^2, \tag{7.202}$$

so that the sum of squares cannot change. Figure 7.10 shows that if all the particles in the universe have the same ratio (ρ_m/ρ_e), we can always find a rotation angle ξ such that the duality transformation (7.196), (7.197) produces $\rho'_m = 0$. However, if different particles have different ratios there is no single value of ξ that can produce $\rho'_m = 0$ for all particles. This is the non-trivial case we consider here.

Let us consider the case of a purely magnetic particle interacting with a purely electric one. Then we cannot use the Maxwell equations with $\rho_m, \vec{J}_m = 0$. This is inconvenient. However, there is a formulation of magnetic charge that uses the standard macroscopic Maxwell equations with $\rho_m, \vec{J}_m = 0$. To begin to understand it, let's go back to the discussion

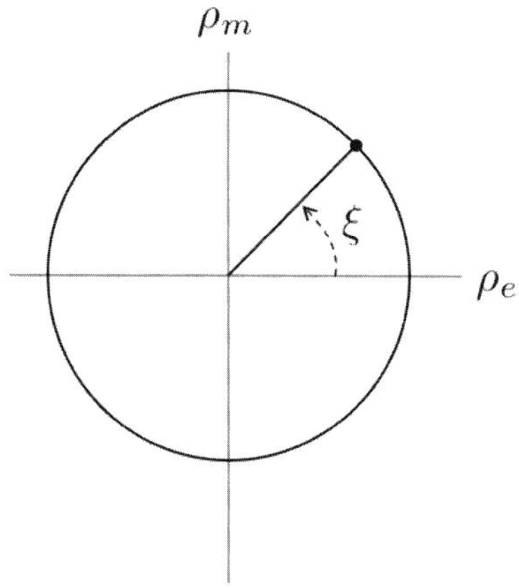

Fig. 7.10 ρ_m, ρ_e charge definition diagram.

in Section 6.13 involving space-distributed infinitesimal magnetic dipoles, $d\vec{m}$. We learned that

$$\begin{aligned}
&\vec{\nabla}\cdot d\vec{H}_m = 4\pi\, d\rho_m,\\
&d\rho_m = -d\vec{m}\cdot\vec{\nabla}\delta(\vec{x}-\vec{x}\,'),\\
\Longrightarrow\ &d\vec{H}_m = -\vec{\nabla}d\Phi_m = -\vec{\nabla}\left(\frac{d\vec{m}\cdot(\vec{x}-\vec{x}\,')}{|\vec{x}-\vec{x}\,'|^3}\right),
\end{aligned} \tag{7.203}$$

as well as

$$\begin{aligned}
&\vec{\nabla}\times d\vec{B}_m = \frac{4\pi}{c}d\vec{J}_m,\\
&d\vec{J}_m = -c\,d\vec{m}\times\vec{\nabla}\delta(\vec{x}-\vec{x}\,'),\\
\Longrightarrow\ &d\vec{B}_m = \vec{\nabla}\times d\vec{A}_m = \vec{\nabla}\times\left(\frac{d\vec{m}\times(\vec{x}-\vec{x}\,')}{|\vec{x}-\vec{x}\,'|^3}\right),
\end{aligned} \tag{7.204}$$

where we have used the current from Equation (6.153) in obtaining Equation (7.204).

Now instead of a volume distribution of magnetic dipoles as we have considered previously, we consider a line (or "string") of magnetization made of magnetic dipoles as shown in Figure 7.11. (A similar situation was considered in Exercise 6.13.3.) We write $d\vec{m} \equiv -g\,d\vec{x}\,'$ (where $g > 0$ to be consistent with the figure), and obtain

$$\begin{aligned}
&\vec{A}(\vec{x}) = -g\int_L d\vec{x}\,'\times\frac{(\vec{x}-\vec{x}\,')}{|\vec{x}-\vec{x}\,'|^3} = g\int_L d\vec{x}\,'\times\vec{\nabla}\frac{1}{|\vec{x}-\vec{x}\,'|}\\
\Longrightarrow\ &\vec{B}(\vec{x}) = g\int_L \vec{\nabla}\times\left(d\vec{x}\,'\times\vec{\nabla}\frac{1}{|\vec{x}-\vec{x}\,'|}\right).
\end{aligned} \tag{7.205}$$

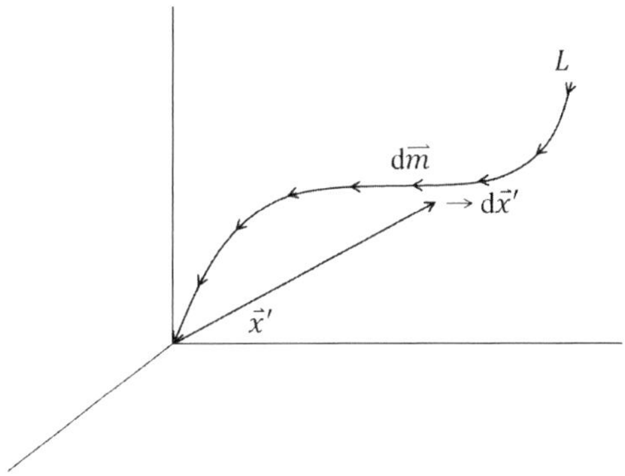

Fig. 7.11 Line of magnetization.

(This definition of $\vec{A}$ is consistent with the Coulomb gauge, since it can be shown that $\vec{\nabla}\cdot\vec{A}=0$.) Rearrangement via the $BAC-CAB$ rule gives

$$\begin{aligned}
&\vec{\nabla}\times\left(d\vec{x}\,'\times\vec{\nabla}\frac{1}{|\vec{x}-\vec{x}\,'|}\right)\\
&=\sum_i\left[\nabla_i\left(d\vec{x}\,'\nabla_i\frac{1}{|\vec{x}-\vec{x}\,'|}\right)-\nabla_i\left(dx_i'\vec{\nabla}\frac{1}{|\vec{x}-\vec{x}\,'|}\right)\right]\\
&=d\vec{x}\,'\nabla^2\frac{1}{|\vec{x}-\vec{x}\,'|}-(d\vec{x}\,'\cdot\vec{\nabla})\left(\vec{\nabla}\frac{1}{|\vec{x}-\vec{x}\,'|}\right)\\
&=-4\pi d\vec{x}\,'\delta(\vec{x}-\vec{x}\,')+d\vec{x}\,'\cdot\vec{\nabla}'\left(\vec{\nabla}\frac{1}{|\vec{x}-\vec{x}\,'|}\right). \qquad (7.206)
\end{aligned}$$

Integrating, we find:

$$\begin{aligned}
&\int_L d\vec{x}\,'\cdot\vec{\nabla}'\left(\vec{\nabla}\frac{1}{|\vec{x}-\vec{x}\,'|}\right)=\vec{\nabla}\frac{1}{|\vec{x}-\vec{x}\,'|}\Big|_{x'=0}^{x'=\infty}=\frac{\vec{x}}{|\vec{x}|^3}\\
&\Longrightarrow\ \vec{B}(\vec{x})=g\frac{\vec{x}}{|\vec{x}|^3}-4\pi g\int_L d\vec{x}\,'\delta(\vec{x}-\vec{x}\,'), \qquad (7.207)
\end{aligned}$$

where the last term in (7.207) represents a directed line delta function. We verify that the magnetic field satisfies $\vec{\nabla}\cdot\vec{B}=0$:

$$\begin{aligned}
\vec{\nabla}\cdot\vec{B}&=g\nabla^2\frac{1}{|\vec{x}|}-4\pi g\int_L(d\vec{x}\,'\cdot\vec{\nabla})\delta(\vec{x}-\vec{x}\,')\\
&=4\pi g\delta(\vec{x})+4\pi g\int_L(d\vec{x}\,'\cdot\vec{\nabla}')\delta(\vec{x}-\vec{x}\,')\\
&=4\pi g\,\delta(\vec{x})+4\pi g\delta(\vec{x}-\vec{x}\,')\Big|_{x'=0}^{x'=\infty}=0. \qquad (7.208)
\end{aligned}$$

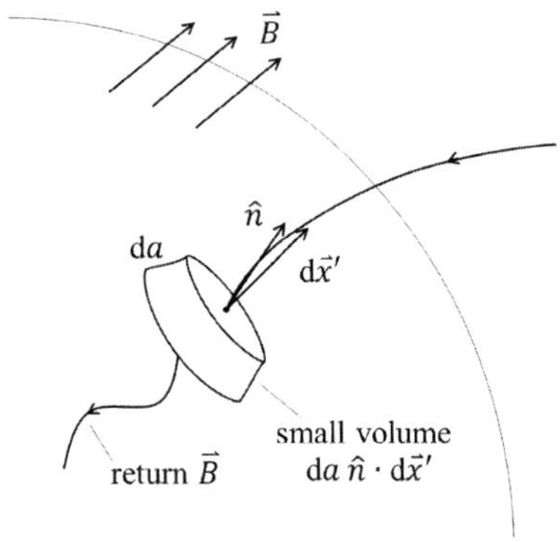

Fig. 7.12 Flux through closed surface.

(The integration of the delta function in (7.208) is formal as far as the upper limit is concerned: we will say more about this later.) The delta-function term in (7.207) represents an intense inward line of $\vec{B}$ flux that balances the outgoing flux so that the total flux on any closed surface enclosing the origin is zero. We may verify this by integrating (see Figure 7.12):

$$\mathcal{F} = \int_S da\,\hat{n}\cdot\vec{B} = g\int_S da\,\frac{\vec{x}\cdot\hat{n}}{|\vec{x}|^3} - 4\pi g\int_L\int_S da\,\hat{n}\cdot d\vec{x}\,'\delta(\vec{x}-\vec{x}\,'). \quad (7.209)$$

The first integral in (7.209) gives the total solid angle subtended by the closed surface (viewed from the origin), so the result is 4π. In the second integral the integration element $da\,\hat{n}\cdot d\vec{x}\,'$ is a volume element including the point $\vec{x}\,'$, so the delta function integrates to 1. The two integrals thus cancel to give a net flux of zero.

We may obtain the magnetization $\vec{M}(x)$ by summing, that is, integrating the contributions from all the infinitesimal dipoles along L. Therefore

$$\vec{M}(\vec{x}) = \int_L d\vec{m}(\vec{x}\,')\delta(\vec{x}-\vec{x}\,') = -g\int_L d\vec{x}\,'\,\delta(\vec{x}-\vec{x}\,') \quad (7.210)$$

$$\Longrightarrow \vec{H}(\vec{x}) = \vec{B}(\vec{x}) - 4\pi\vec{M}(\vec{x}) = g\frac{\vec{x}}{|\vec{x}|^3}, \quad (7.211)$$

which is a pure monopole field everywhere! There is no singularity in $\vec{H}$ along L (except at the endpoint). The connection we have found between $\vec{H}$, $\vec{B}$ and $\vec{M}$ very much is in the spirit of Equation (6.56) for the $\vec{B}$ field from a loop of current. Here there are singularities which cancel in $\vec{B}$ and $\vec{M}$ giving a continuous $\vec{H}$; there, singularities in $\vec{\nabla}\Omega$ and the membrane term cancel giving a continuous $\vec{B}$ field.

We are trying to represent a localized magnetic charge, and the string we have introduced should not be observable. This is accomplished if we take $\vec{H}$ as the observable field; the field $\vec{B}$ is simply an auxiliary field necessary within the framework of the macroscopic Maxwell equations. To confirm the consistency of this picture, we may compute $\vec{H}$ directly:

$$\begin{aligned}\Phi_m(\vec{x}) &= -g\int_L \frac{d\vec{x}\,'\cdot(\vec{x}-\vec{x}\,')}{|\vec{x}-\vec{x}\,'|^3} = -g\int_L d\vec{x}\,'\cdot\vec{\nabla}'\frac{1}{|\vec{x}-\vec{x}\,'|},\\ &= \frac{-g}{|\vec{x}-\vec{x}\,'|}\Big|_{x'=0}^{x'=\infty} = \frac{g}{|\vec{x}|},\end{aligned}$$

which implies

$$\vec{H}(\vec{x}) = g\frac{\vec{x}}{|\vec{x}|^3}. \tag{7.212}$$

Let's say the string is moved, as shown in Figure 7.13. The strings must come together again before or at spatial infinity, otherwise one can show that there is an effect of moving the string (namely, infinite radiation!) We have

$$\begin{aligned}\vec{A}_{L_2}-\vec{A}_{L_1} &= -g\int_{L_2} d\vec{x}\,'\times\frac{(\vec{x}-\vec{x}\,')}{|\vec{x}-\vec{x}\,'|^3} + g\int_{L_1} d\vec{x}\,'\times\frac{(\vec{x}-\vec{x}\,')}{|\vec{x}-\vec{x}\,'|^3},\\ &= -g\oint_{L_2-L_1} d\vec{x}\,'\times\frac{(\vec{x}-\vec{x}\,')}{|\vec{x}-\vec{x}\,'|^3}.\end{aligned} \tag{7.213}$$

(cf. (6.52)). Using (6.51), we may then compute

$$\begin{aligned}&\oint d\vec{\ell}\,'\times\vec{A} = \int\sum_i ds_i'\vec{\nabla}'A_i - \int d\vec{s}\,'(\vec{\nabla}'\cdot\vec{A})\\ \Longrightarrow\ &\vec{A}_{L_2}-\vec{A}_{L_1} = -g\vec{\nabla}\Omega(\vec{x}) - 4\pi g\int d\vec{s}\,'\delta(\vec{x}-\vec{x}\,').\end{aligned} \tag{7.214}$$

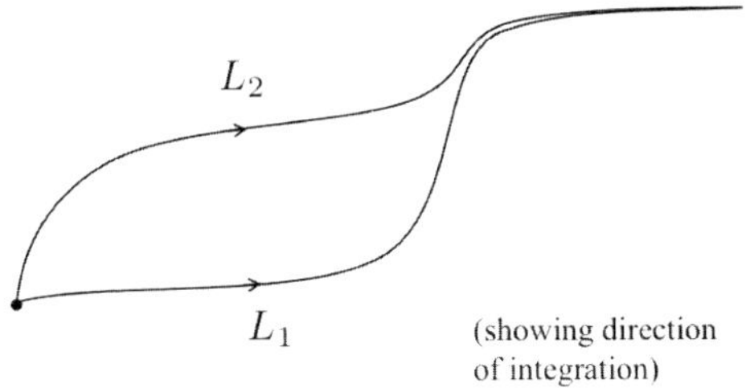

Fig. 7.13 The string is moved.

The last term in (7.214) is a membrane δ-function. Again, we saw the very same form in Section 6.5, when we computed $\vec{B}$ from a loop integral. Just as before, this term makes $\vec{A}_{L_2} - \vec{A}_{L_1}$ continuous across the surface. Equation (7.214) can be viewed as a generalization of the gauge transformation $\vec{A} \Longrightarrow \vec{A} + \vec{\nabla}\Lambda$ that we saw before in (7.106); this connection is further explored in Exercises 7.8.3 and 7.8.4.

Let us make sure that this expression for $\vec{A}_{L_2} - \vec{A}_{L_1}$ is consistent with earlier expression for $\vec{B}$. We will need the identity:

$$\vec{\nabla} \times \int d\vec{s}\,'\delta(\vec{x} - \vec{x}\,') = \int \left(\vec{\nabla}\delta(\vec{x} - \vec{x}\,')\right) \times d\vec{s}\,'$$
$$= \int da\,\hat{n}' \times \vec{\nabla}'\delta(\vec{x} - \vec{x}\,') = \oint_{L_2 - L_1} d\vec{x}\,'\delta(\vec{x} - \vec{x}\,'), \qquad (7.215)$$

which together with (7.214) gives

$$\vec{\nabla} \times \left(\vec{A}_{L_2} - \vec{A}_{L_1}\right) = -4\pi g \oint_{L_2 - L_1} d\vec{x}\,'\delta(\vec{x} - \vec{x}\,'). \qquad (7.216)$$

But from the earlier expression:

$$\vec{B}_{1,2}(\vec{x}) = g\frac{\vec{x}}{|\vec{x}|^3} - 4\pi g \int_{L_{1,2}} d\vec{x}\,'\delta(\vec{x} - \vec{x}\,'),$$
$$\Longrightarrow \vec{B}_2 - \vec{B}_1 = -4\pi g \int_{L_2} d\vec{x}\,'\delta(\vec{x} - \vec{x}\,') - 4\pi g \int_{L_1} (-d\vec{x}\,')\delta(\vec{x} - \vec{x}\,'),$$
$$= -4\pi g \oint_{L_2 - L_1} d\vec{x}\,'\,\delta(\vec{x} - \vec{x}\,'). \qquad (7.217)$$

This is the same as above.

The usual macroscopic Maxwell equations (7.97)–(7.100), if properly interpreted, can replace the Maxwell equations (7.191)–(7.194) with explicit $\rho_m, \vec{J}_m$ sources. Essentially, we are treating electric charge as "free" and magnetic charge as "bound". We can extend our static results to the dynamical case; see in this context Exercise 7.8.2 where the two sourceless Maxwell equations, (7.193) and (7.194), are transformed into dynamical equations for $\vec{H}$ and $\vec{D}$ with explicit point magnetic sources. Thus, when we are dealing with the pure *magnetic* case, we can use the usual classical point particles. It is only when we must deal with electric and magnetic charge *simultaneously* that one of them must be given a string. The fields $\vec{H}$ and $\vec{D}$ in this formalism are considered the observable (measurable) ones; the $\vec{B}$ and $\vec{E}$ are considered auxiliary fields which are needed in this macroscopic Maxwell approach.

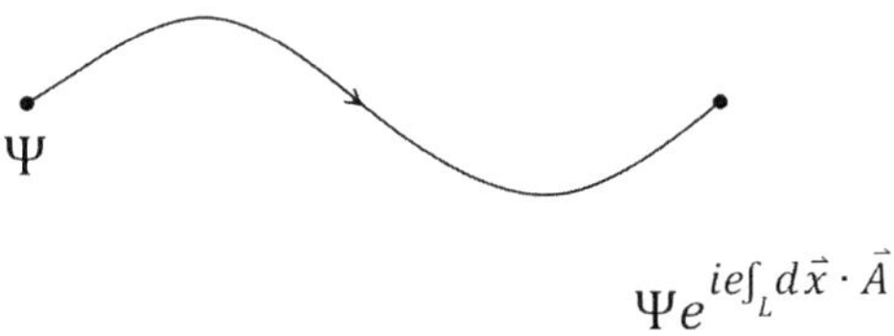

Fig. 7.14 Path-dependent phase.

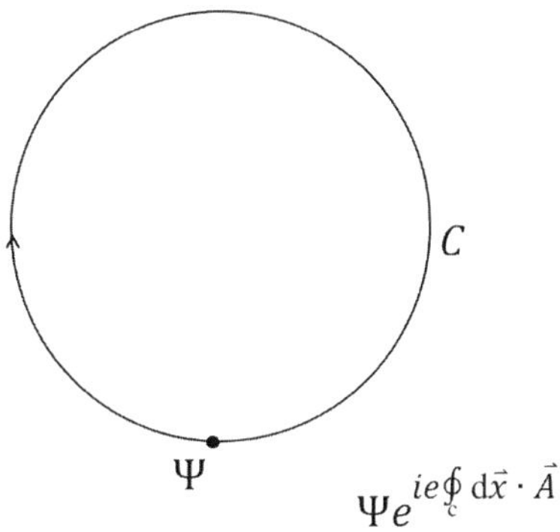

Fig. 7.15 Transporting a charged particle in a closed loop.

This construction leads to an important consequence for electric charge; to see this we will need a smidgen of quantum mechanics. A particle with a general charge e in an electromagnetic field $A(\vec{x})$ picks up a path-dependent phase which in "natural units" ($\hbar = c = 1$) is given by $ie\int_L d\vec{x}\cdot\vec{A}(\vec{x})$, as shown in Figure 7.14. In particular, we can transport the particle in a *closed* loop, as in Figure 7.15. So let's imagine transporting our electrically charged particle in a closed loop when the string from a magnetic monopole is either at L_1 or L_2; see Figure 7.16. We insist that the string is not real, and therefore there should be no observable effects of moving it. Uniqueness of the wave function requires (this gives the magnetic flux through C'):

$$e\oint_{C'} d\vec{x}\,'\cdot\left\{\vec{A}_{L_2}-\vec{A}_{L_1}\right\} = 2\pi n, \quad n = 0, \pm 1, \pm 2, \ldots. \tag{7.218}$$

Using (7.214), the left-hand side gives

$$\begin{aligned} e\oint_{C'} d\vec{x}\,'\cdot&\left\{-g\vec{\nabla}'\Omega(\vec{x}\,') - 4\pi g\int_S d\vec{s}\,\delta(\vec{x}-\vec{x}\,')\right\}, \\ &= -4\pi e\,g\oint_{C'}\int_S \underbrace{d\vec{s}\cdot d\vec{x}\,'}_{\pm dV}\,\delta(\vec{x}-\vec{x}\,') = \pm 4\pi e\,g. \end{aligned} \tag{7.219}$$

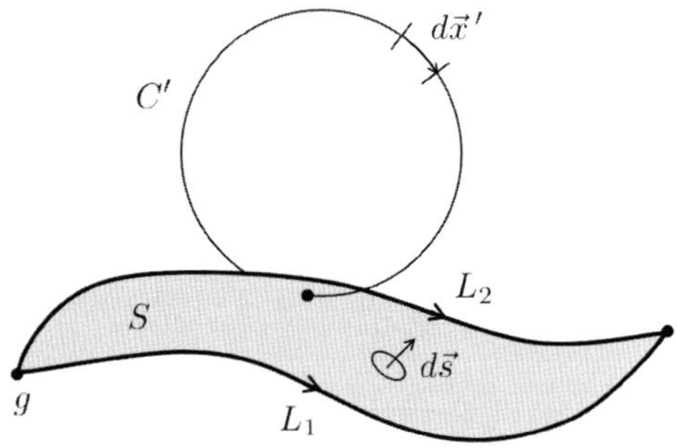

Fig. 7.16 Transporting the charged particle in the vicinity of a string.

This gives rise to the famous Dirac quantization condition:

$$e_a g_b = \frac{n}{2}. \quad \left(\frac{e_a g_b}{\hbar c} = \frac{n}{2} \text{ Gaussian units}\right) \tag{7.220}$$

We are putting a, b subscripts on the charges above to emphasize that the charges are on separate particles. The original quantization argument was put forward by P. A. M. Dirac in 1931.[6] Such a condition would explain the quantization of electric charge if there is only one magnetic monopole in the universe. Actually, this requires either e or g to be quantized. In particular since in natural units for e the magnitude of the charge on an electron,

$$e^2 \approx \frac{1}{137}, \quad \left(\frac{e^2}{\hbar c} \approx \frac{1}{137} \text{ Gaussian}\right) \tag{7.221}$$

if we let

$$\begin{aligned} e^s &= \text{magnitude of smallest nonzero electric charge} \\ &\approx \frac{1}{3\sqrt{137}} \quad \text{(down quark)} \\ g^s &= \text{magnitude of smallest nonzero magnetic charge} \\ &\Longrightarrow (g^s)^2 \approx \frac{9}{4} \times 137. \end{aligned} \tag{7.222}$$

There is an alternative formulation of magnetic charge due to Schwinger.[7] The "Schwinger string" looks like Figure 7.17 (compare to

[6]P. A. M. Dirac, "Quantised Singularities in the Electromagnetic Field", Proc. Roy. Soc. A **133**, 60 (1931).
[7]See Ref. 5 citations in the *Going Deeper* Section 7.9.2.

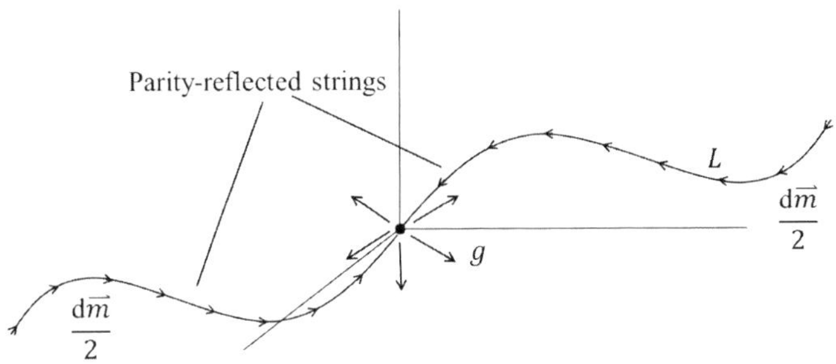

Fig. 7.17 Schwinger string.

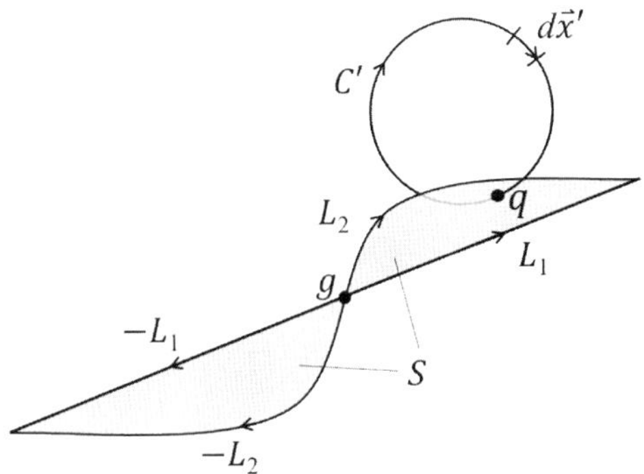

Fig. 7.18 Separate electric and magnetic charges.

Figure 7.11.) The magnetization is now an odd function of its argument, bringing magnetic charge in from two mirror directions:

$$\vec{M}(\vec{x}) = -\frac{1}{2} g \int_L d\vec{x}' \left[\delta(\vec{x}-\vec{x}') - \delta(\vec{x}+\vec{x}')\right]. \tag{7.223}$$

Now we take the electric charge in a loop as before; see Figure 7.18. For separate electric and magnetic charges we still require the flux to be quantized as in (7.218), but now the left-hand side gives

$$e \oint_{C'} d\vec{x}' \cdot \left\{ -g\vec{\nabla}'\Omega(\vec{x}) - 2\pi g \int d\vec{s}\left[\delta(\vec{x}-\vec{x}') - \delta(\vec{x}+\vec{x}')\right] \right\} = \pm 2\pi e\, g \tag{7.224}$$

$$\Longrightarrow \; e_a g_b = n. \tag{7.225}$$

The factor of two has gone away from the Dirac condition because each half of the Schwinger string carries only half of the entire magnetic flux.

Actually, if one assumes each particle can have both magnetic and electric charge, then Schwinger's condition is

$$e_a g_b - e_b g_a = n. \tag{7.226}$$

Such an object is called a *dyon*. Notice this condition it is invariant under our original duality transformation,

$$e_{a,b} = e'_{a,b} \cos\xi + g'_{a,b} \sin\xi, \tag{7.227}$$

$$g_{a,b} = -e'_{a,b} \sin\xi + g'_{a,b} \cos\xi, \tag{7.228}$$

whereas Dirac's condition, which makes a real invariant distinction between electric and magnetic charge, does not. Using Schwinger's condition, the magnitude of the smallest magnetic charge (with no accompanying electric charge) is determined by

$$(g^s)^2 \approx 9 \times 137. \tag{7.229}$$

The interaction between monopoles is very strong, approximately $(137)^2 \approx 18,000$ times stronger than between two electrons. Note that there are theoretical reasons to believe that one should actually use the electron charge, not the down quark charge, in calculating the smallest magnetic charge.

As we have seen, the difference between Dirac's and Schwinger's strings is the number of flux lines attached to the magnetic charge (as well as the way they are attached for the Schwinger string). One could also postulate alternative theories with 3, 4, 5,... strings attached. (See also Exercise 7.8.6.) Such theories produce different flux quantization conditions. Thus, there is an element of arbitrariness in the classical models presented above. Only a truly quantum theory of magnetic charge could resolve this ambiguity. In modern field theories, so-called anomaly cancellations enforce similar quantization conditions on charges, and the charges which emerge in these theories are always quantized.[8]

We need to tie up a few loose strings before we finish... (sorry!). In (7.208) there was a reference to an integration that extended to spatial infinity where the string ends. This is not entirely satisfactory. We can remove reference to infinity if we only allow an equal number of magnetic and anti-magnetic charges. The strings then extend from one type of charge to the opposite and there is no need to involve integrations to spatial infinity.

[8]For more in-depth explanation of magnetic charge in the context of field theories, see Section 7.9.2, Ref. 7.

We need to mention another elegant approach to including magnetic charge in Maxwell's equations that is topological in nature. Given a string along either the positive or negative z-axis, one defines two vector potentials: one excludes the region near the positive z-axis when the string is along $+z$, the other includes this axis but excludes the region near the negative z-axis when the string is along $-z$. Taking the two potentials together leads to the correct expression for the magnetic field everywhere.[9]

Many searches for magnetic monopoles have been performed, and as a result upper limits have been set on the fluxes of such particles in cosmic rays and particle colliders.[10] There are no confirmed observations.

Finally, as far as the authors know, no clearly recognized consistent quantum theory of point magnetic charge exists. However, a consistent basis for a quantum theory of extended magnetic charge certainly exists. In modern "grand unified" theories, magnetic monopoles emerge as "soliton" (finite energy particle-like states) solutions and have an electromagnetic size determined by the Compton wavelength of the associated gauge bosons of the theory. The magnetic charge on these solitons is quantized and they have masses between the early universe symmetry breaking scale $\sim 10^{16}$ GeV/c^2 and the Planck mass scale, $\sqrt{\hbar c/G} \sim 10^{19}$ GeV/c^2, where G is Newton's gravitational constant. See the *Going Deeper* references for more information on the cosmological implications and other theoretical aspects.

7.9 Going Deeper

7.9.1 *Macroscopic Maxwell equations*

(1) K. Cho, *Reconstruction of Macroscopic Maxwell Equations*, Springer Tracts in Modern Physics, Vol. 237 (Berlin) 2010.

(2) S. R. De Groot, *The Maxwell Equations: Nonrelativistic Multipole Expansion to All Orders*, Physica **31**, 953 (1965).

---

[9]See for example the explanation in: J. J. Sakurai and J. Napolitano, *Modern Quantum Mechanics*, 2nd ed., Addison-Wesley (Boston) 2011.

[10]See the particle physics review in Ref. 6, Section 7.9.2.

(3) Yu. A. Il'inskii and L. V. Keldysk, *Electromagnetic Response of Material Media*, Plenum Press (New York) 1994; Chapter 1.

(4) F. N. H. Robinson, *Macroscopic Electromagnetism*, Pergamon Press (Oxford) 1973.

(5) G. Russakoff, *A Derivation of the Macroscopic Maxwell Equations*, Am. J. Phys. **38**, 1188 (1970).

(6) K. Schram, *Quantum Statistical Derivation of the Macroscopic Maxwell Equations*, Physica **26**, 1080 (1960).

7.9.2 *Magnetic charge*

(1) M. F. Atiyah and N. J. Hitchin, *The Geometry and Dynamics of Magnetic Monopoles*, Princeton University Press (Princeton) 1988.

(2) S. Balestra *et al.*, *Magnetic Monopole Bibliography-II*, arXiv: 1105.5587.

(3) P. D. B. Collins, A. D. Martin and E. J. Squires, *Particle Physics and Cosmology*, John Wiley & Sons (New York) 1989; Chapter 17.

(4) N. S. Manton and P. Sutcliffe, *Topological Solitons*, Cambridge University Press (Cambridge) 2004; Chapter 8.

(5) K. A. Milton, *Theoretical and Experimental Status of Magnetic Monopoles*, Rep. Prog. Phys. **69**, 1637 (2006).

(6) K. Olive *et al.* (Particle Data Group), Chin. Phys. C **38**, 090001 (2014); `http://pdg.lbl.gov`; Review, "Magnetic Monopoles".

(7) J. Preskill, Ann. Rev. Nucl. Part. Sci., *Magnetic Monopoles*, **34**, 461 (1984).

(8) A. Rajantie, *Introduction to Magnetic Monopoles*, Contem. Phys. **53**, 195 (2012).

7.10 Exercises

Exercise 7.1.1. Given two frames, primed and unprimed, such that the primed frame is moving with velocity $\vec{v}$ with respect to the unprimed frame. Show, to first order in $\vec{v}$ only, that the free-space Maxwell equations given by (7.20)–(7.23) are form invariant under the following transformations. [Note that $\vec{v}$ in this problem corresponds to $-\vec{v}$ in (7.28)–(7.29); in Sections 13.1–13.3 we will need the equations in this form.]

$$\begin{cases} \vec{\nabla}\,' = \vec{\nabla} + \dfrac{\vec{v}}{c^2}\dfrac{\partial}{\partial t} \\ \dfrac{\partial}{\partial t'} = \dfrac{\partial}{\partial t} + \vec{v}\cdot\vec{\nabla} \end{cases} \qquad \begin{cases} \vec{E}\,' = \vec{E} + \dfrac{\vec{v}}{c}\times\vec{B} \\ \vec{B}\,' = \vec{B} - \dfrac{\vec{v}}{c}\times\vec{E} \end{cases} \qquad \begin{cases} \rho' = \rho - \dfrac{\vec{v}}{c^2}\cdot\vec{J} \\ \vec{J}\,' = \vec{J} - \vec{v}\rho \end{cases}$$

Exercise 7.2.1. The very long flat rectangular circuit in Figure 7.19 has a mass M, a source of emf $\mathcal{E}_0$, and a resistance R. The long wires carrying current I have a distance D between them. The battery produces a current following Ohm's law. Gravity plays no role in this problem and the circuit moves frictionlessly. Given that the end of the circuit is immersed in a uniform magnetic field, $\vec{B}$, directed out of the page, find the equation of the circuit that follows from the Lorentz force, Ohm's law and Faraday's law. This will give the differential equation that determines $x(t)$. Show the

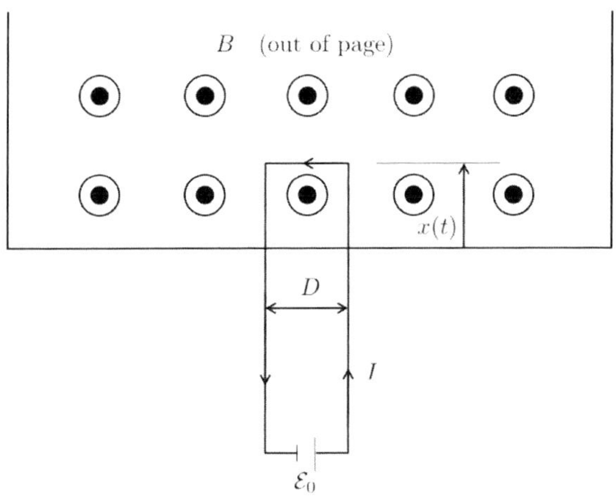

Fig. 7.19 Geometry for Exercise 7.2.1.

motion exponentially approaches a terminal velocity. [*Note*: For a similar circuit with a self-inductance term, see Exercise 7.6.5.]

Exercise 7.3.1.

(a) Starting with ($\vec{E}$ and $\vec{B}$ are external fields and the material is at rest in the lab)

$$\vec{N}_{\text{atom}} = \sum_k e_k(\vec{x}_k - \vec{R}) \times \left[\vec{E} + \frac{\vec{v}_k}{c} \times \vec{B}\right],$$

show that

$$\vec{N}_{\text{atom}} = \vec{p} \times \vec{E} + \vec{m} \times \vec{B},$$

where

$$\vec{p} = \sum_k e_k(\vec{x}_k - \vec{R}) = \sum_k e_k\vec{x}_k, \quad \vec{m} = \frac{1}{2c}\sum_k e_k(\vec{x}_k - \vec{R}) \times \vec{v}_k.$$

(b) Given the torque expression ($n(\vec{x})$ is the density of neutral atoms),

$$\vec{N}_{\text{bulk}} = \int d^3x\, n(\vec{x}) \left[\vec{x} \times \vec{F}_{\text{atom}} + \vec{N}_{\text{atom}}\right],$$

where from (a)

$$\vec{N}_{\text{atom}} = \vec{p} \times \vec{E} + \vec{m} \times \vec{B},$$

and from (7.58) with $\vec{v} = 0$

$$\vec{F}_{\text{atom}} = \sum_i [p_i\vec{\nabla}E_i + m_i\vec{\nabla}B_i] + \frac{1}{c}\frac{\partial}{\partial t}(\vec{p} \times \vec{B}),$$

show that

$$\vec{N}_{\text{bulk}} = \int d^3x\, \vec{x} \times \left[\rho_{\text{b}}\vec{E} + \frac{1}{c}\vec{J}_{\text{b}} \times \vec{B}\right],$$

where $\rho_{\text{b}} = -\vec{\nabla} \cdot \vec{P}$, and $\vec{J}_{\text{b}} = \partial\vec{P}/\partial t + c\vec{\nabla} \times \vec{M}$.

Exercise 7.3.2. Can you justify in a more detailed manner the change to a partial time derivative in the last term in (7.62)?

Exercise 7.3.3. Show that the effective charge density and current associated with the non-traceless quadrupole contribution in the Schwinger model for a piece of matter at rest are

$$\rho^Q_{\text{eff}}(\vec{x},t) = \frac{1}{6}\sum_{i,j}\nabla_i\nabla_j q'_{ij}(\vec{x},t),$$

$$\left(\vec{j}^{\,Q}_{\text{eff}}\right)_j = -\frac{1}{6}\sum_i \nabla_i \frac{\partial q'_{ij}(\vec{x},t)}{\partial t},$$

where $q'_{ij}(\vec{x},t) \equiv n(\vec{x})Q'_{ij}(\vec{x},t)$. Note that the combination satisfies the continuity equation, signifying a conserved current. (We saw an earlier version of the quadrupole charge density for an electrostatic external field application in Exercise 5.4.1.)

Exercise 7.3.4. Using the Jackson semi-classical formalism assuming neutral atoms, show that the time-averaged bound charge density gives

$$\sum_n \langle \rho_n(\vec{x},t)\rangle = -\vec{\nabla}\cdot\vec{P},$$

where $\vec{P}$ is the model electric polarization field. The charge density associated with an atom labelled by n, $\rho_n(\vec{x},t)$, is defined by

$$\rho_n(\vec{x},t) = \sum_{j(n)} e_j \delta(\vec{x}-\vec{x}_n-\vec{x}_{jn}),$$

where $\vec{x}_n$ locates the *nth* atom relative to the origin, $\vec{x}_{jn}$ locates the *jth* particle relative to the *nth* atom, and $\delta(\vec{x}-\vec{x}_n-\vec{x}_{jn})$ is a three-dimensional Dirac delta function.

Exercise 7.3.5. Verify the form (cf. Section 7.8) of the continuous duality symmetry

$$\vec{X} = \vec{X}\,'\cos\xi + \vec{Y}\,'\sin\xi,$$
$$\vec{Y} = -\vec{X}\,'\sin\xi + \vec{Y}\,'\sin\xi,$$

where

$$\{\vec{X},\vec{Y}\} = \{\vec{E},\vec{H}\} \quad \text{or} \quad \{\vec{D},\vec{B}\},$$

present in the macroscopic Maxwell equations (7.97)–(7.100) in regions where $\rho_e^{\text{free}} = \vec{J}_e^{\text{free}} = 0$. (This generalizes the discrete form (6.210) which we originally saw connected electrostatic and magnetostatic cases.)

Exercise 7.3.6. We dropped terms proportional to the bulk velocity, $\vec{v}$, in the characterization of the effective quadrupole current (see Exercise 7.3.3) in both the Schwinger and Jackson macroscopic models. Now keeping such terms, using *both* models, show that there are additional terms in the effective quadrupole current given by

$$\left(j^Q_{eff}\right)_j = \frac{1}{6}\sum_{i,k}\left(v_j\nabla_i\nabla_k {q'}^{(S)}_{ik} - v_i\nabla_i\nabla_k {q'}^{(S)}_{jk}\right), \quad \text{(Schwinger)}$$

$$\sum_n\left\langle\left(j^Q_n\right)_j\right\rangle = \frac{1}{6}\sum_{i,k}\left(v_j\nabla_i\nabla_k {q'}^{(J)}_{ik} - v_i\nabla_i\nabla_k {q'}^{(J)}_{jk}\right), \quad \text{(Jackson)}$$

where ${q'}^{(J,S)}_{ik} = {q'}^{(J,S)}_{ik}(\vec{x},t)$ is the non-traceless quadrupole density, given by

$${q'}^{(S)}_{ik} = n(x - R(t))Q'_{ik}, \quad \text{(Schwinger)}$$

$${q'}^{(J)}_{ik} = \sum_n (Q'_n)_{ik} f(\vec{x} - \vec{x}_n), \quad \text{(Jackson)}$$

[*Hints*: In the Schwinger model, consider the quadrupole term of (7.56) and the next higher-order term in (7.39). In the Jackson model, consider the quadrupole term in (7.80) and the next higher-order term in (7.73). In both models, assume there is no atomic thermal motion: $\vec{v} = \vec{v}_n$ for all atoms or molecules.]

Exercise 7.3.7. Expanding the magnetic field to second order in a Taylor series as (take the atomic center of mass as the origin, $\vec{R} = 0$) one has

$$B_i(\vec{x}) = B_i(0) + (\vec{x}\cdot\vec{\nabla}')B_i(\vec{x}\,')|_{x'=0} + \frac{1}{2}(\vec{x}\cdot\vec{\nabla}')(\vec{x}\cdot\vec{\nabla}')B_i(\vec{x}\,')|_{x'=0},$$

or using index notation,

$$B_i(\vec{x}) = B_i(0) + \sum_j x_j\nabla'_j B_i(\vec{x}\,')|_{x'=0} + \frac{1}{2}\sum_{j,k} x_j x_k \nabla'_j\nabla'_k B_i(\vec{x}\,')|_{x'=0}.$$

(a) From the second term on the right, and using the starting point Equation (7.37) in the Schwinger model, show that the new implied magnetic quadrupole ("m.q.") atomic force term is

$$F^{\text{m.q.}}_\ell = \frac{1}{2c}\sum_{m,n,j,k}\varepsilon_{\ell mn}\nabla_j\nabla_k B_n\int d^3x\, x_j x_k J_m(x),$$

where we are defining $\nabla_j\nabla_k B_n \equiv \nabla'_j\nabla'_k B_n(\vec{x}\,')|_{x'=0}$.

(b) Referring back to Exercise 6.8.5, show the magnetic quadrupole atomic force term can be written as

$$F_\ell^{\text{m.q.}} = \frac{1}{2}\sum_{i,j} m_{ij}\nabla_\ell\nabla_j B_i,$$

or alternately as

$$F_\ell^{\text{m.q.}} = \frac{1}{2}\sum_{i,j} m_{ij}\nabla_i\nabla_j B_\ell,$$

in source-free regions where $\vec{\nabla}\times\vec{B} = 0$ holds.

Exercise 7.3.8. Using the result of the force on an atom from Exercise 7.3.7, find the implied effective current in the macroscopic Maxwell equations using the Schwinger formalism from Section 7.3.1. The new quadrupole force was

$$F_\ell^{m.q.} = \frac{1}{2}\sum_{i,j} m_{ij}\nabla_\ell\nabla_j B_i.$$

Show that the implied bound current $\vec{J}_b^{m.q.}$ can be written like the first term of (7.69),

$$\vec{J}_b^{m.q.} = c\vec{\nabla}\times\vec{M}^{m.q.},$$

where the jth component of $\vec{M}^{m.q.}$ is given by

$$M_j^{new} \equiv -\frac{1}{2}\sum_i \nabla_i M_{ji},$$

where

$$M_{ij}(\vec{x},t) \equiv n(\vec{x},t)m_{ij}(\vec{x},t).$$

$n(\vec{x},t)$ is the density of material as in (7.61) with $\vec{v} = 0$. This quadrupole contribution can thus be considered another mechanism contributing to the *total* magnetization field of the material.

Exercise 7.3.9. Given the results in Exercise 6.8.5 and starting with the expression for the torque on a localized current distribution, $\vec{J}(\vec{x})$,

$$\vec{N} = \frac{1}{c}\int d^3x\, \vec{x}\times\left(\vec{J}\times\vec{B}\right),$$

where

$$\vec{B}(\vec{x}) \simeq \vec{B}(0) + \left(\vec{x} \cdot \vec{\nabla}\right) \vec{B}(0),$$

(vector notation) or

$$B_i(\vec{x}) \simeq B_i(0) + \sum_j x_j \nabla_j B_i(0),$$

(index notation; $\nabla_j B_i(0)$ designates the jth derivative of the ith component of $\vec{B}$ evaluated at the origin) derive an expression for the quadrupole term in the magnetostatic torque:

$$N_i = \left(\vec{m} \times \vec{B}(0)\right)_i + ??.$$

The unknown term will involve the m_{ij} from Exercise 7.3.7 and the first gradients of the magnetic field, $\nabla_j B_i(0)$.

Exercise 7.4.1. The current on the right hand side of Equation (7.117),

$$\vec{J}_T \equiv \vec{J} - \frac{1}{4\pi} \frac{\partial \vec{\nabla} \Phi}{\partial t},$$

is called the *transverse current*. Show that it may be written as

$$\vec{J}_T = \vec{\nabla} \times \left(\vec{\nabla} \times \int \frac{d^3x'}{4\pi} \frac{\vec{J}(\vec{x}\,', t)}{|\vec{x} - \vec{x}\,'|}\right).$$

Exercise 7.4.2. In the Coulomb gauge, the electromagnetic field equations are given by (7.116) and (7.116), whereas in the Lorenz gauge they are given by (7.111) and (7.112). The Coulomb gauge condition is given by (7.115) and the Lorenz condition is given by (7.110). The Lorenz (L) and Coulomb (C) vector and scalar potentials are related by

$$\vec{A}_L = \vec{A}_C + \vec{\nabla}\Lambda,$$

$$\Phi_L = \Phi_C - \frac{1}{c} \frac{\partial \Lambda}{\partial t}.$$

Show that the gauge function Λ satisfies the equation

$$\nabla^2 \Lambda - \frac{1}{c^2} \frac{\partial^2}{\partial t^2} \Lambda = F.$$

Find the explicit form of the function F for this gauge transformation. (Note we are in free space and all fields and sources have space and time arguments $(\vec{x}, t)$.)

Exercise 7.5.1.

(a) Show for a piece of matter with intrinsic magnetization $\vec{M}$ (and no free currents), that

$$\int d^3x\, \vec{B} \cdot \vec{H} = 0,$$

when the integration is over all space. (Recall the Section 6.13 discussion.)

(b) From Equation (6.172) the energy associated with an infinitesimal permanent magnetic dipole $\delta\vec{m}_i$ at $\vec{x}_i$ in the magnetic field $\delta\vec{B}_j(\vec{x}_i)$ due to a dipole $\delta\vec{m}_j$, can be written

$$\delta U_{ij}^{\text{int}} = -\delta\vec{m}_i \cdot \delta\vec{B}_j(\vec{x}_i), \qquad i \neq j.$$

From this expression, the connection $\vec{B} = \vec{H} + 4\pi\vec{M}$, and the result of part (a), argue that the interaction energy of a permanently electrically polarized piece or pieces of matter is

$$U^{\text{int}} = -\frac{1}{2}\int d^3x\, \vec{M} \cdot \vec{B} = -\frac{1}{8\pi}\int d^3x\, \vec{B}^2,$$

where the integrations are over all space. [*Notes*: As in the electrostatic analog Exercise 5.8.1 there are actually self-energy contributions in U^{int} which arise in going from the discrete sum to the continuum integral. The sign of the result is the opposite of the expected positive field energy because it includes the battery energy necessary to maintain the values of the magnetic moments $\delta\vec{m}_i$ in the accumulated fields. Also note that part (a) is the "reason" there is a minus sign difference between the internal $\vec{B}$ and $\vec{H}$ fields in Equations (6.264) and (6.263).]

Exercise 7.5.2. A spherically shaped permanent magnet with radius a and uniform volume magnetization $\vec{M}_1$ is fixed in position while another magnet, with the same radius but different uniform magnetization $\vec{M}_2$ is introduced, as shown in Figure 7.20. Find the work done ΔU^{int} in bringing the magnet's centers a distance $\vec{R}$ apart. The result from Exercise 7.5.1(b) above should be useful. The result should look quite familiar.

Exercise 7.5.3. We saw mutual and self-inductance arise briefly in Section 7.5.

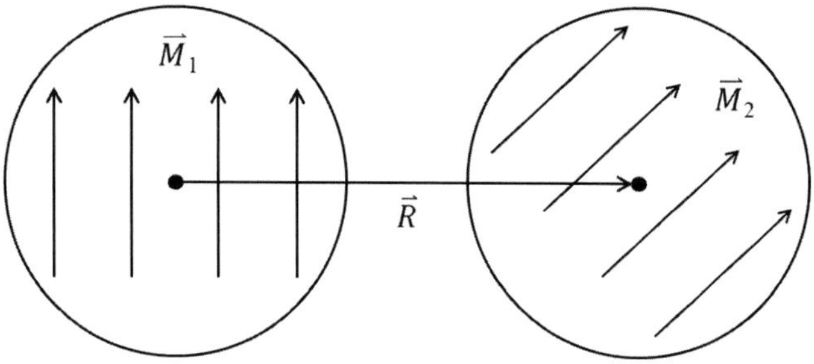

Fig. 7.20 Geometry for Exercise 7.5.2.

(a) Show that the relations in Equation (7.137) are true, and thus that Equation (7.131) holds.
(b) Consider two separate circuits with (finite) self-inductances L_1, L_2 and mutual inductance M_{12}. Show that $L_1 L_2 \geq (M_{12})^2$ from energy positivity.

Exercise 7.5.4. A small planar circuit of arbitrary shape and area A is located at the center of a much larger circular circuit of radius R. The planes of the two circuits coincide. The smaller circuit carries a steady current I_1 and the larger one a steady current I_2. Find the mutual inductance, M_{12}, of this system.

Exercise 7.5.5. Two small, planar circuits with areas a_1 and a_2 and associated current magnitudes I_1 and I_2 are located a large distance, R, apart. ($R \gg$ sizes of the circuits.) $\hat{n}_1$ and $\hat{n}_2$ are unit vectors pointing in the direction of their magnetic moments. Find their mutual inductance and demonstrate explicitly that $M_{12} = M_{21}$. [*Ans.*: $M_{12}(\vec{R}) = \dfrac{a_1 a_2}{c^2}\left[\dfrac{3(\hat{R}\cdot\hat{n}_1)(\hat{R}\cdot\hat{n}_2) - \hat{n}_1\cdot\hat{n}_2}{R^3}\right]$.]

Exercise 7.5.6. Comparing (7.129) to the first term of (7.134) gives the form for the self-inductance L for a space-distributed current loop, $\vec{J}(\vec{x})$, as

$$L = \frac{1}{I^2 c^2}\iint d^3x\, d^3x\,'\frac{\vec{J}(\vec{x})\cdot\vec{J}(\vec{x}\,')}{|\vec{x}-\vec{x}\,'|},$$

where I is the total current flowing in the circuit.

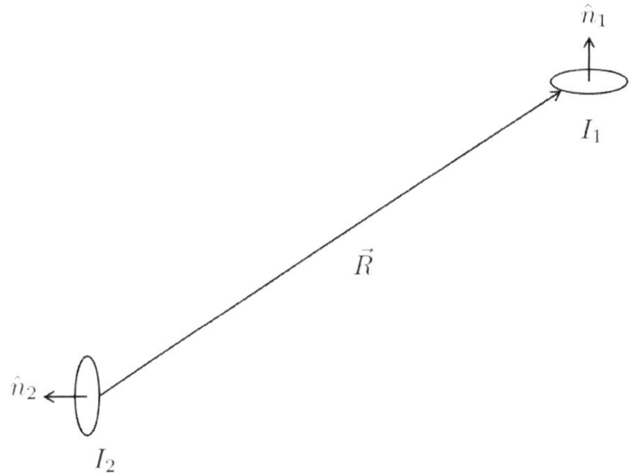

Fig. 7.21 Geometry for Exercise 7.5.5.

Consider a thin cylinder of radius a and height L with a constant surface current, $\vec{K} = K_0\,\hat{e}_\phi$, where $\hat{e}_\phi$ is the cylindrical unit vector in the azimuthal direction. Take your origin O at one end of the cylinder, as shown in Figure 7.22.

Using the formula above show that the self-inductance of the cylinder is given by

$$L_{cyl.} = \frac{2\pi^2 a^2}{L^2 c^2} \int_0^\infty \frac{dk}{k}\ (J_1(ka))^2 \left(L + \frac{1}{k}(e^{-kL} - 1) \right).$$

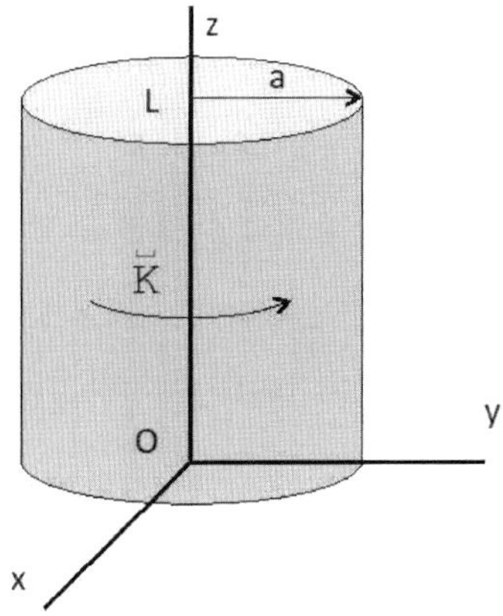

Fig. 7.22 Geometry for Exercise 7.5.6.

[*Hints*: Use cylindrical coordinates and the Coulomb expansion Equation (4.120).]

Exercise 7.5.7. A thin circular disk of radius R with a constant surface charge density σ is spinning with constant angular velocity ω about a line through the center of the disk, perpendicular to the disk's plane. The disk is in the $z = 0$ plane. The magnetic field components are given in cylindrical coordinates by (see Exercise 6.6.2)

$$B_\rho = \frac{2\pi\omega\sigma R^2}{c}\operatorname{sign}(z)\int_0^\infty dk\, e^{-k|z|}\, J_1(k\rho)J_2(kR),$$

$$B_z = \frac{2\pi\omega\sigma R^2}{c}\int_0^\infty dk\, e^{-k|z|} J_0(k\rho)J_2(kR).$$

Using

$$L_{disk} = \frac{2W_{disk}}{(I_{disk})^2},$$

show that the self-inductance, L, of the spinning disk is given by

$$L_{disk} = \frac{64\pi R}{15c^2}.$$

Note that the total current, I, flowing past a fixed line along the plate in this case will be

$$I = \frac{\text{total charge}}{\text{period of motion}} = \frac{\omega\,\sigma\,R^2}{2}.$$

Also note: A Bessel function integral that is necessary is

$$\int_0^\infty \frac{dx}{x^2}(\,J_2(x)\,)^2 = \frac{4}{15\pi}.$$

Exercise 7.5.8. This exercise gives an alternate derivation of L_{disk} in Exercise 7.5.7 from a flux, rather than an explicit energy, point of view.

(a) Using the connection between magnetic flux and interaction energy, show that the interaction energy of two co-planar circular current loops of radii a and b is

$$W_{ab} = \left(\frac{2\pi I_a a}{c}\right)\left(\frac{2\pi I_b b}{c}\right)\int_0^\infty dk\, J_1(ka)J_1(kb).$$

(b) Based on part (a), the interaction energy for infinitesimal circular coplanar currents is

$$dW_{ab} = \frac{dI_a\, dI_b}{2}\left(\frac{2\pi a}{c}\right)\left(\frac{2\pi b}{c}\right)\int_0^\infty dk\; J_1(ka)J_1(kb).$$

We may now use the infinitesimal currents associated with the spinning, charged disk:

$$dI_a = \omega\sigma a\, da,\; dI_b = \omega\sigma b\, db.$$

Now integrate both currents over the circular surface of radius R, avoiding double counting of the interactions. This will give the self-energy of the disk, W_{disk}, from which the self-inductance immediately follows.

Exercise 7.6.1.

(a) Find the general expression for the force of circuit 2 on 1, $\vec{F}_{12}$, for the small planar circuits in Exercise 7.5.5. [*Hint*: Equation (7.141).]
(b) Using your expression, give the particular forces $\vec{F}_{12}$ for the (A) and (B) configurations of distant circuits of areas a_1 and a_2 arranged as in Figure 7.23, where vector $\vec{R}$ locates circuit 1 relative to circuit 2.

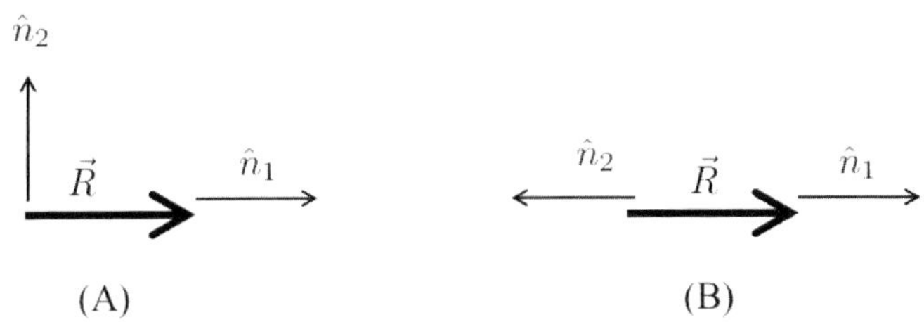

Fig. 7.23 Geometry for Exercise 7.6.1.

Exercise 7.6.2. Assuming localized, non-overlapping current distributions $\vec{J}_1$ (giving rise to the field $\vec{A}_1$, treated here as "external") and $\vec{J}_2$, show that the field interaction energy

$$W_{12} = \frac{1}{c}\int d^3x\, \vec{A}_1 \cdot \vec{J}_2,$$

when $\vec{A}_1(x)$ is expanded in a Taylor series about the origin gives rise to the lowest order form,

$$W_{12} \approx \vec{m}_2 \cdot \vec{B}_1,$$

which shows that Equations (6.170) and (7.151) are consistent:

$$\vec{F} = \left(\frac{\partial W_{12}}{\partial \vec{x}}\right)_I \implies \vec{F} = \vec{\nabla}(\vec{m}_2 \cdot \vec{B}_1).$$

Exercise 7.6.3.

(a) Given two current loops as shown in Figure 7.24, show that the mutual inductance, M_{12}, of the two loops is given by ($J_1(x)$ is a Bessel function)

$$M_{12} = \frac{4\pi^2 a^2}{c^2} \int_0^\infty dk e^{-kR} (J_1(ka))^2.$$

[*Hint*: Equation (6.227) and the connection between field energy and mutual inductance, Equation (7.134).]

(b) Using Equation (7.141), find the force between these current loops. (You may compare your result with the force of an image loop computed in (6.237) by making the appropriate identifications.)

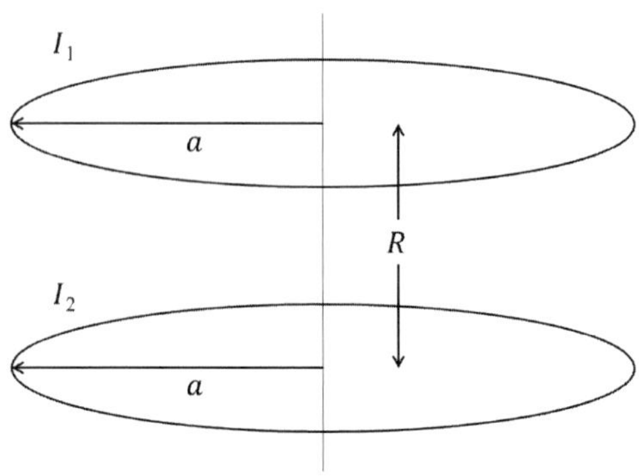

Fig. 7.24 Geometry for Exercise 7.6.3.

Exercise 7.6.4. Referring to Exercise 7.6.3(a), investigate the mutual inductance $M_{12}(\theta)$ when the loop carrying current I_1 is located at a polar angle θ relative to the z axis passing through the center of the loop carrying current I_2. The distance between the center of the loops is R and the planes of the current loop planes are parallel and do not intersect. Given the fact that the mutual inductance satisfies Equation (7.143), we may assume the azimuthally symmetric form ($P_\ell(x)$ is a Legendre polynomial)

$$M_{12}(\theta) = \sum_{\ell=0,1,\ldots} \frac{A_\ell}{R^{\ell+1}} P_\ell(\cos\theta).$$

Comparing to the known solution above for $\theta = 0$, find the first three coefficients, $A_{0,1,2}$, then predict the far-field ($R \gg a$) mutual inductance when $\theta = \pi/2$.

Exercise 7.6.5. The very long flat rectangular circuit in Figure 7.25 has a mass M a resistance R and a self-inductance L. It produces a current by Ohm's law. Gravity plays no role in this problem and the circuit moves frictionlessly. There is no battery in the circuit. Given that the end of the circuit is immersed in a uniform magnetic field $\vec{B}$ directed out of the page, find the equation for the position of the circuit $x(t)$ that follows from the Lorentz force, Ohm's law, and Faraday's law for arbitrary initial values $\ddot{x}(0), \dot{x}(0), x(0)$. Show that a damped harmonic oscillator equation subject to constant force results. Compare to

$$\ddot{x}(t) + 2\beta\,\dot{x}(t) + \omega_0^2\,x(t) + const. = 0,$$

and identify the natural frequency ω_0 and damping constant β. [*Notes*: This problem is similar to Exercise 7.2.1 but with no battery and a self-inductance term in the circuit. A circuit element with a self-inductance L

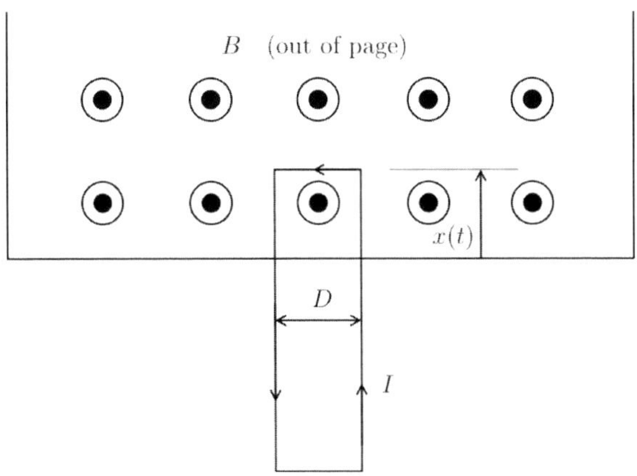

Fig. 7.25 Geometry for Exercise 7.6.5.

is associated with a electromotive force change $-LdI(t)/dt$ where $I(t)$ is the current, when the result for the self-inductance in Equation (7.137) is combined with Faraday's law.]

Exercise 7.6.6. A loop of radius a and current I_2 is situated parallel to and a distance R from the rotating circular disk of Exercise 7.5.7, as shown in Figure 7.26.

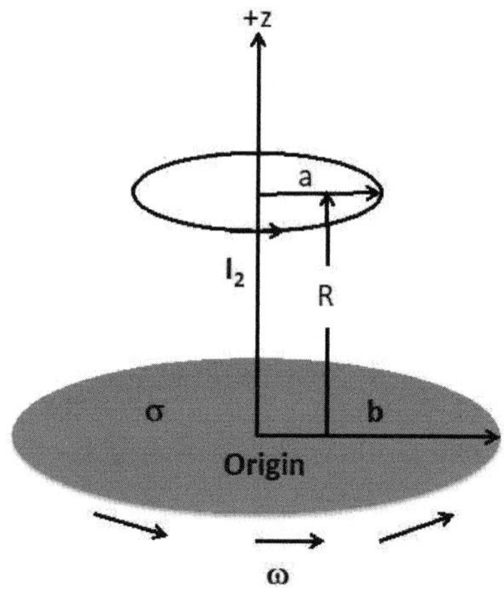

Fig. 7.26 Geometry for Exercise 7.6.6.

(a) Using the fields from Exercise 7.5.7, or other means, find the mutual inductance of the system, M_{12}:

$$M_{12} = \frac{8\pi^2 a}{c^2} \int_0^\infty \frac{dk}{k}\, e^{-kR} J_2(kb) J_1(ka).$$

(b) From part (a), find the magnetic force of the disk on the loop.

Extra: Confirm your result in (b) by a *direct* force calculation.

Exercise 7.6.7.

(a) Calculate the mutual inductance per length, M_{12}/L, of two long identically shaped flat rectangular circuits with parallel sides, L, which are much longer than the end sides, d. Neglect edge effects. The distance

R is measured between the near sides of circuits 2 and 1, as shown in Figure 7.27.

(b) Using the connection between mutual inductance and force, find the force of circuit 2 on 1, F_{12}. With currents in the direction shown (clockwise), is the overall force repulsive or attractive?

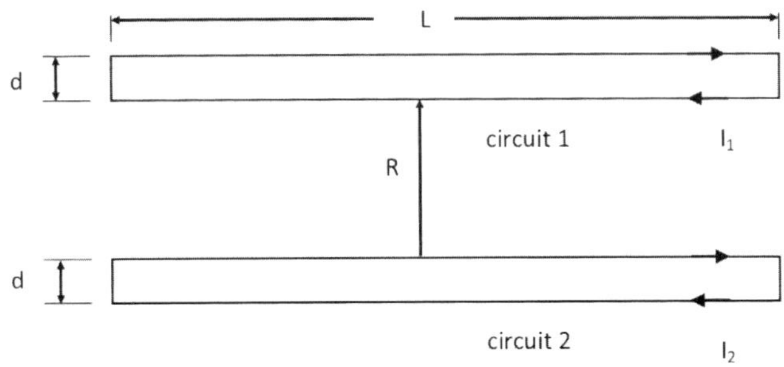

Fig. 7.27 Geometry for Exercise 7.6.7.

Exercise 7.6.8. A circular cylinder of length D and radius a has a uniform azimuthal surface current $\vec{K}$ flowing on it. Its symmetry axis is coincident to the axis of a circular loop of wire of radius a carrying current I_2 as shown in Figure 7.28. The wire loop is in the $z = 0$ plane. The distance between the bottom of the cylinder and the wire loop is R. The total current flowing around the cylinder is $I_1 = K\,D$, where $\vec{K} = K\,\hat{e}_\phi$.

Show that the force in the z-direction of loop 2 on the cylinder is

$$F_z = \frac{I_1 I_2}{D}\left[M_{12}(R+D) - M_{12}(R)\right],$$

where

$$M_{12}(R') = \frac{4\pi^2 a^2}{c^2}\int_0^\infty dk\, e^{-kR'}\,(J_1(ka))^2,$$

is the mutual inductance between two concentric single-wire current loops of radius a positioned a distance R' apart.

Exercise 7.7.1. An object of permeability μ is placed in a magnetic field whose current sources are fixed. Show that the change in energy,

$$\Delta W \equiv W - W_0,$$

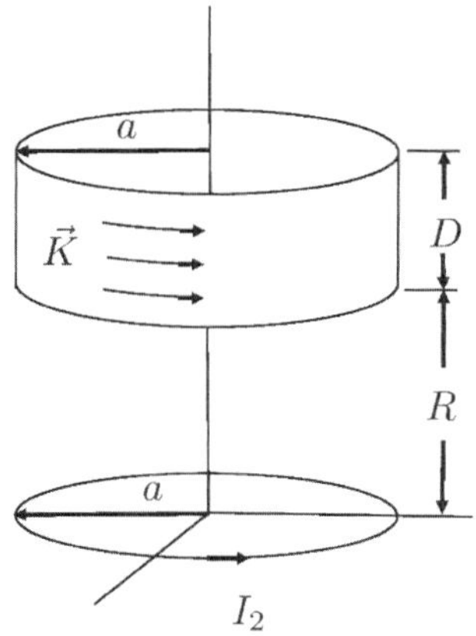

Fig. 7.28 Geometry for Exercise 7.6.8.

where $W = \frac{1}{2c}\int d^3x\, \vec{J}\cdot\vec{A}$, gives

$$\Delta W = \frac{1}{2}\int_V d^3x\, \vec{M}\cdot\vec{B}_0,$$

where V is the volume occupied by the object, as given in Equation (7.164).

Exercise 7.7.2.

(a) Consider the situation described in the text with a thick permeable rod inserted a long distance x into a solenoid of length L with an approximate uniform field. The current in the solenoid is fixed at I producing the initial field $\vec{B}_0$. The appropriate boundary condition is now on the solenoid's side rather than end, requiring that $\vec{H}$ is a constant everywhere. Find the approximate force on the rod, F_x.

(b) An approximately uniform magnetic field $\vec{B}_0$ is produced in a superconducting solenoid of length L, which maintains constant magnetic flux within the solenoid. A thick rod of material of permeability μ and cross-section A is inserted a short distance into the solenoid. Considering the boundary condition at the rod's end and neglecting magnet side effects, find the force F_x on the rod.

Exercise 7.7.3. A circular current loop of radius a and current I in vacuum is oriented with its plane parallel to and a distance d form a flat interface of constant permeability $\mu \neq 1$ in the region $z < 0$ as shown in Figure 7.8. Take the interface plane surface to be given by $z = 0$. Using the field energy expression $\Delta W = \frac{1}{2}\int d^3x\, \vec{M} \cdot \vec{B}_0$, show that the force of the current loop on the interface is given by

$$F_z = \frac{4\pi^2 I^2 a^2}{c^2}\left(\frac{\mu - 1}{\mu + 1}\right)\int_0^\infty dk\, k e^{-2kd}(J_1(ka))^2,$$

which agrees with (6.237) (but differs in sign because it gives the force of the interface on the current).

Exercise 7.7.4. Calculate the force on the plane interface in Figure 7.8 again using the explicit force/area expression (7.181). The answer is given in Exercise 7.7.3.

Exercise 7.7.5. The middle of a long, thin cylinder of radius a and length L, with permeability μ is located a large distance z from a point magnetic dipole, $\vec{m}$, where $z \gg L \gg a$. It is oriented with its axis oriented perpendicular to the direction to the dipole, but along $\vec{m}$, as shown in Figure 7.29. The interaction field energy of the configuration is (see Exercise 7.7.1)

$$\Delta W = \frac{1}{2}\int d^3x\, \vec{M} \cdot \vec{B}_0,$$

where $\vec{M}$ is the magnetization and $\vec{B}_0$ is the field before the introduction of the cylinder. Find the force of the dipole on the cylinder. Are they attracted or repelled for $\mu > 1$?

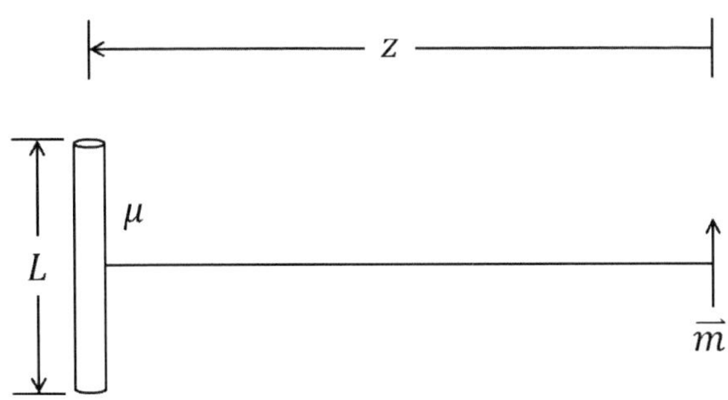

Fig. 7.29 Geometry for Exercise 7.7.5.

Exercise 7.7.6. Let's return to the spinning uniformly charged spherical shell of radius a with surface charge density σ of Exercise 6.7.3. Again, the magnetic fields inside and outside the sphere are:

$$\vec{B}(\vec{x}) = \begin{cases} \dfrac{8\pi a\sigma\omega}{3c}\hat{z} & r < a, \\ \dfrac{3\vec{x}(\vec{m}\cdot\vec{x}) - r^2\vec{m}}{r^5} & r > a, \end{cases}$$

with

$$\vec{m} = \frac{4\pi\sigma\omega a^4}{3c}\hat{z}.$$

Show that the total force on the top hemisphere is (attractive),

$$F_z = -\frac{\pi^2}{c^2}a^4\omega^2\sigma^2.$$

Exercise 7.7.7. A long, straight solenoid (circular cross-section) of radius a carrying a current I and having n insulated wire turns per length, is situated such that its end is touching the surface of a large, flat piece of magnetic material with magnetic permeability μ.

(a) Argue that the B field of the solenoid is the same as from a magnet with uniform lengthwise magnetization M, where $M = nI/c$.
(b) Using (a) and an image source, show that the solenoid sticks to the flat surface with a force of attraction given by

$$F = 2\pi AMM',$$

where $M' = (\mu - 1)/(\mu + 1)M$ and $A = \pi a^2$ (repulsive if $\mu < 1$). [*Hint*: Equation (7.190) and Exercise 6.13.4(a).]

Exercise 7.7.8. As a generalized variation on the previous problem, find an exact, closed form integral expression involving Bessel functions for the force on the rightmost of two cylindrical magnets of radius a and length L (uniform volume magnetization $\vec{M}$) which are co-axial and whose ends are a distance $2D$ apart, as seen in Figure 7.30. Check that in the appropriate limit, your result agrees with the previous result above in Exercise 7.7.7(b). [*Hints*: You can start with expressions for the force between co-planar current carrying wires in Section 6.12 or use a force expression in Section 7.7

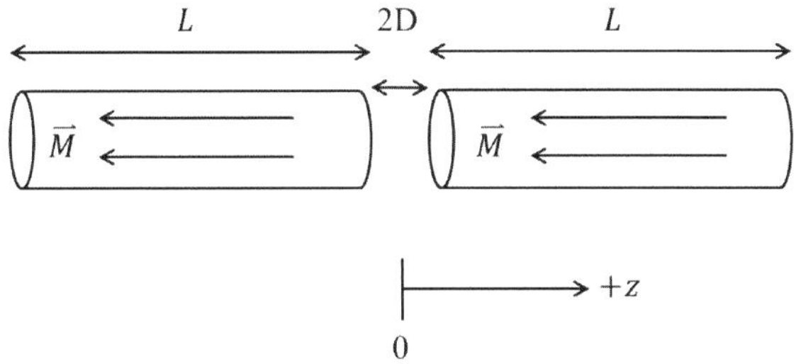

Fig. 7.30 Geometry for Exercise 7.7.8.

and the connection to the solid angle expression for $\vec{B}$ fields from Exercise 6.13.4(a).]

Exercise 7.7.9. A small sphere of magnetic permeability μ and radius b is situated a large distance, z, from a point magnetic dipole, $\vec{m}$, as seen in Figure 7.31. Starting with the scalar potential solution inside a dielectic sphere in an initially uniform electric field (Exercise 5.7.3), the electric-magnetic analogy of Section 6.11, and the energy method of Section 7.7, find the force on the sphere; it should be proportional to z^{-7}. When $\mu > 1$ is the sphere attracted to or repelled from the magnetic dipole?

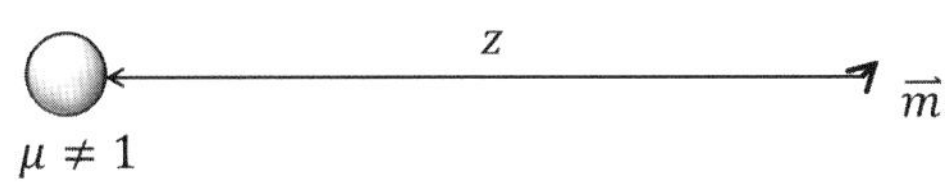

Fig. 7.31 Geometry for Exercise 7.7.9.

Exercise 7.7.10. We saw in Exercise 5.9.1 that the normal force per area $\vec{\mathcal{F}} \cdot \hat{n}$ ($\hat{n}$ directed outward from the material) on a dielectric interface is given by

$$\vec{\mathcal{F}} \cdot \hat{n} = 2\pi\sigma_b^2 \left(\frac{\epsilon + 1}{\epsilon - 1}\right).$$

Similarly, show that the normal force per area $\vec{\mathcal{F}} \cdot \hat{n}$ on a magnetic interface is given by

$$\vec{\mathcal{F}} \cdot \hat{n} = \frac{2\pi}{c^2} \vec{K}_b^2 \left(\frac{\mu + 1}{\mu - 1}\right),$$

where $\vec{K}_b$ is the bound surface current.

Extra: When both bound and free currents ($\vec{K}_f$) are present, show that

$$\vec{\mathcal{F}} \cdot \hat{n} = \frac{2\pi}{c^2}\left(\frac{\mu}{\mu - 1}\right)^2 \left[\vec{K}_b^2 - \frac{1}{\mu^2}\left(\vec{K}_b + (1-\mu)\vec{K}_f\right)^2\right].$$

Exercise 7.7.11. A thin glass rod of radius R and length L has a uniform electric charge density, σ, smeared on its outer cylindrical surface. We have $L \gg R$. The rod is non-permeable; $\mu = 1$. It is set spinning counterclockwise from above about the z-axis with an angular velocity, ω. Take the origin of coordinates at the center of the rod. The magnetic field generated at an arbitrary point, $\vec{r}$, is the same as for a set of magnetic charges, g and $-g$, located at the ends of the rod; namely

$$\vec{H} = g\left(\frac{\hat{r}_g}{r_g^2} - \frac{\hat{r}_{-g}}{r_{-g}^2}\right),$$

where $\hat{r}_g$ is a unit vector pointing from the g charges to the field point, and r_g is the associated distance; similar notation for the charge $-g$. Find the effective value of the magnetic charge g for the spinning rod.

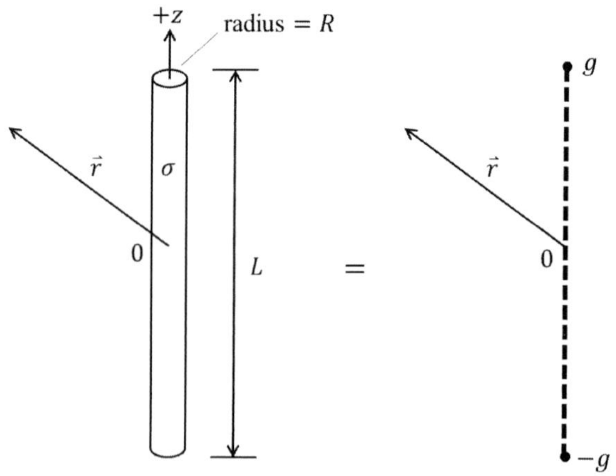

Fig. 7.32 Geometry for Exercise 7.7.11.

Exercise 7.7.12. Refer to Figure 7.7. Consider the insertion of thick rod of magnetic material with permeability μ into a solenoid at fixed current I with a boundary condition applied at the end of the rod (initial insertion).

Find the force on the rod in the x-direction *using an explicit force method.* Identify the location(s) of the forces. Does your result agree with the result Equation (7.168) of the text using the volume method? [*Hint*: Model the magnetic field as ending abruptly at the end of the solenoid; note this is *inside* the volume of the inserted rod.]

Exercise 7.8.1. We consider an arbitrary (truncated) line of magnetization in the text; this turns out to be a description of a magnetic point charge. Consider now a uniform layer of magnetization, $\vec{M}_s$, on a flat surface, S, extending to infinity in one direction, as shown in Figure 7.33. The magnetization is pointed along the direction which extends to infinity. The sheet ends on an arbitrary line whose length is L. The sheet is bordered by lines L_1 and L_2, also extending to infinity. Show that this describes a string of magnetic charge. Where is the magnetic charge located? What is the total charge in the system? (Carefully explain the meanings of the symbols you use.)

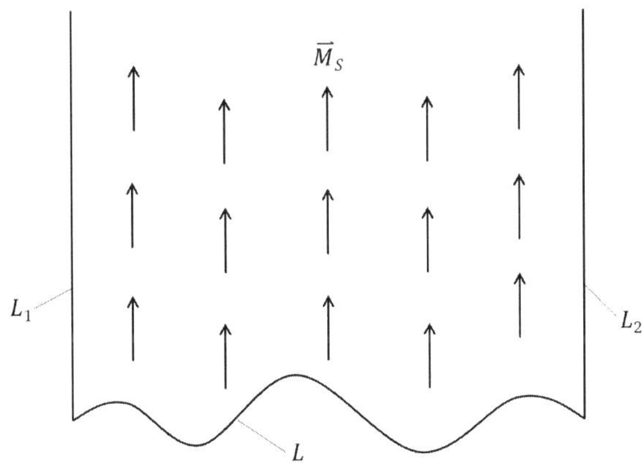

Fig. 7.33 Geometry for Exercise 7.8.1.

Exercise 7.8.2. Given the magnetization and polarization of a rigid, moving magnetic string (the magnetic charge, g, is located at $\vec{R}(t)$; the integration limits on $\vec{x}''$ are 0 to ∞ along some given path)

$$\vec{M}(\vec{x}) = -g \int d\vec{x}''\, \delta(\vec{x} - \vec{R}(t) - \vec{x}''),$$

$$\vec{P}(\vec{x}) = \frac{\vec{v}(t)}{c} \times \vec{M}(\vec{x}),$$

where Equation (7.210) has been generalized and $\vec{v}(t) = d\vec{R}/dt$, show that the sourceless Maxwell equations,

$$\vec{\nabla} \cdot \vec{B} = 0, \quad \vec{\nabla} \times \vec{E} + \frac{1}{c}\frac{\partial \vec{B}}{\partial t} = 0,$$

are transformed into ($\vec{B} = \vec{H} + 4\pi\vec{M}, \vec{E} = \vec{D} - 4\pi\vec{P}$)

$$\vec{\nabla} \cdot \vec{H} = 4\pi g\, \delta(\vec{x} - \vec{R}(t)), \quad \vec{\nabla} \times \vec{D} + \frac{1}{c}\frac{\partial \vec{H}}{\partial t} = -\frac{4\pi}{c} g\, \vec{v}(t)\delta(\vec{x} - \vec{R}(t)).$$

[*Notes*: This shows that the $\vec{D}, \vec{H}$ fields are just the nonsingular fields of a moving magnetic point charge located at $\vec{R}(t)$; all reference to the string has disappeared! The general statement,

$$\frac{\partial}{\partial t} f(\vec{x} - \vec{R}(t)) + \vec{\nabla} \cdot \left(\vec{v}(t) f(\vec{x} - \vec{R}(t))\right) = 0,$$

where f is an arbitrary function of its argument, may be useful.]

Exercise 7.8.3. Show that the (second-order) Maxwell equations,

$$\nabla^2\Phi + \frac{1}{c}\frac{\partial}{\partial t}(\vec{\nabla} \cdot \vec{A}) = -4\pi(\rho_{\text{free}} - \vec{\nabla} \cdot \vec{P}),$$

$$\nabla^2\vec{A} - \frac{1}{c^2}\frac{\partial^2 \vec{A}}{\partial t^2} - \vec{\nabla}\left(\vec{\nabla} \cdot \vec{A} + \frac{1}{c}\frac{\partial \Phi}{\partial t}\right) = -\frac{4\pi}{c}\left(\vec{J}_{\text{free}} + c\vec{\nabla} \times \vec{M} + \frac{\partial \vec{P}}{\partial t}\right),$$

are invariant under the time-independent generalized gauge transformation,

$$\vec{A} \to \vec{A} - g\vec{\nabla}\Omega_s(\vec{x}) - 4\pi g \int_S d\vec{s}\,'\delta(\vec{x} - \vec{x}\,'),$$

$$\vec{M} \to \vec{M} - g\oint_C d\vec{x}\,'\delta(\vec{x} - \vec{x}\,'),$$

$$\Phi \to \Phi, \quad \vec{P} \to \vec{P},$$

where $\Omega_s(\vec{x})$ is the solid angle subtended by a finite open surface, S, the contour of which, C, is given by two possible locations of a magnetic string. It is given by

$$\Omega_s(\vec{x}) = \int_S d\vec{s}\,' \cdot \vec{\nabla}\left(\frac{1}{|\vec{x} - \vec{x}\,'|}\right).$$

[*Notes*: This exercise generalizes the usual continuous gauge functions $\Lambda(\vec{x})$ in $\vec{A} \to \vec{A} + \vec{\nabla}\Lambda$ to functions which are only piecewise continuous. The discontinuity is the string change surface as in Figure 7.13.]

Exercise 7.8.4. In reference to Exercise 7.8.3, show the Maxwell equations are also invariant under the more general time dependent transformations, describing a moving magnetic charge,

$$\vec{A} \to \vec{A} - g\vec{\nabla}\Omega_s(\vec{x},t) - 4\pi g \int_S d\vec{s}\,''\delta(\vec{x} - \vec{R}(t) - \vec{x}\,''),$$

$$\vec{M} \to \vec{M} - g \oint_C d\vec{x}\,''\delta(\vec{x} - \vec{R}(t) - \vec{x}\,''),$$

$$\Phi \to \Phi - \frac{4\pi g}{c}\vec{v}(t) \cdot \int_S d\vec{s}\,''\delta(\vec{x} - \vec{R}(t) - \vec{x}\,'') + \frac{g}{c}\frac{\partial\Omega_s(\vec{x},t)}{\partial t},$$

$$\vec{P} \to \vec{P} - \frac{g}{c}\vec{v}(t) \times \oint_C d\vec{x}\,''\delta(\vec{x} - \vec{R}(t) - \vec{x}\,''),$$

where $\vec{R}(t)$ describes the given path of the magnetically charged particle and $\vec{v}(t) = d\vec{R}/dt$.

Exercise 7.8.5. Confirm that the transformation equations (7.196) and (7.197) for sources and fields leaves the form of the Maxwell equations (7.191)–(7.194) (pick two of them, say) as well as the Lorentz force expression (7.195) unchanged.

Exercise 7.8.6. Consider the following sheet of magnetization in the $x-y$ plane: In each small area, $da = \rho\, d\rho\, d\phi$ in the $z = 0$ plane, the infinitesimal magnetic moment $d\vec{m}$ is given by

$$d\vec{m} = -\frac{g}{2\pi}\frac{\hat{\rho}}{\rho}da = -\frac{g}{2\pi}\hat{\rho}\, d\rho\, d\phi,$$

where ρ is the distance from $d\vec{m}$ to the origin and $\hat{\rho}$ is the unit vector pointing radially outward in this plane.

(a) Find the $\vec{B}$ and $\vec{H}$ fields everywhere, including the $z = 0$ plane.

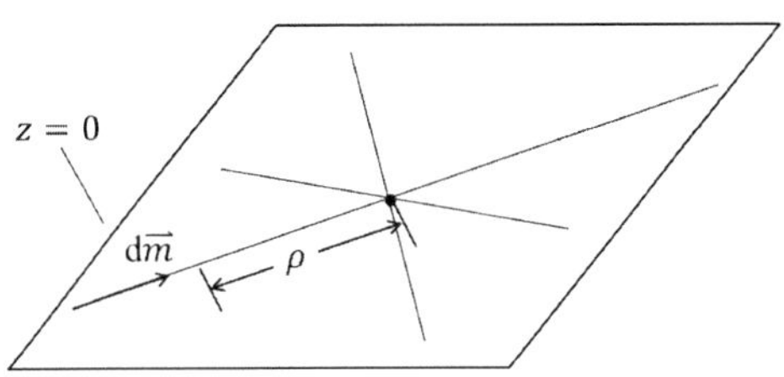

Fig. 7.34 Geometry for Exercise 7.8.6.

(b) Find the total amount of magnetic charge implied by this magnetized sheet construction.

(c) Can you generalize this construction to represent any azimuthally symmetric sheet of magnetization? Considering the argument for quantization of electric/magnetic charge given in the text, does this way of representing magnetic charge have any implications for the charge quantization condition? Discuss.

Chapter 8

Time Varying Fields II

Our discussion in Section 7.1 should have brought an awareness that electric and magnetic fields cannot be separate and independent objects. In this chapter we will deepen our understanding of these as interrelated aspects of a single entity, namely the *electromagnetic field*. All subsequent material depends on this new perspective, beginning with Chapter 9 where we introduce electromagnetic waves.

We will begin this chapter with the fundamental conservation laws for electromagnetic fields: energy, momentum, and angular momentum. Next, we construct time-dependent Green functions which we use to derive expressions for vector and scalar potentials: these expressions will be fundamental to our treatment of radiation in Chapter 11. Finally, we explore transformation properties of classical fields and quantities under orthogonal transformations and time reversal: this is preliminary to the more complete (relativistic) discussion in Section 13.1 and the presentation in Section 14.2 of aspects of the more fundamental quantum formalism of electromagnetism.

8.1 Conservation of energy; energy flux

In this section we discuss conservation laws in classical electromagnetism. We begin with the microscopic Maxwell equations (without magnetic charges or currents). The rate at which work is done by the fields acting on one particle is

$$\frac{dE_m}{dt} = \vec{F} \cdot \vec{v}, \tag{8.1}$$

where the subscript m refers to "mechanical" work. The right-hand side of (8.1) can be re-expressed using the Lorentz force law, as follows:

$$\vec{F} = q\left(\vec{E} + \frac{\vec{v}}{c} \times \vec{B}\right), \tag{8.2}$$

$$\Longrightarrow \vec{F} \cdot \vec{v} = q\vec{v} \cdot \vec{E}. \tag{8.3}$$

For a system with many charges we obtain

$$\sum_i \vec{F}_i \cdot \vec{v}_i = \sum_i q_i\, \vec{v}_i \cdot \vec{E}_i. \tag{8.4}$$

Imagine now a continuous distribution of current given by current density $\vec{J}(\vec{x})$. Making the replacements

$$q_i\, \vec{v}_i \rightarrow \vec{J}(\vec{x})\, d^3x; \qquad \sum_i \rightarrow \int_V \tag{8.5}$$

we obtain the equation

$$\frac{dE_m}{dt} = \int_V d^3x\, \vec{J} \cdot \vec{E}, \tag{8.6}$$

which presumes that no current leaves the volume V.

Next we manipulate (8.6) so that it becomes a statement about fields. From the Ampère-Maxwell equation

$$\vec{J} = \frac{c}{4\pi}\left(\vec{\nabla} \times \vec{B} - \frac{1}{c}\frac{\partial \vec{E}}{\partial t}\right), \tag{8.7}$$

it follows that

$$\vec{J} \cdot \vec{E} = \frac{c}{4\pi}\left((\vec{\nabla} \times \vec{B}) \cdot \vec{E} - \frac{1}{c}\frac{\partial \vec{E}}{\partial t} \cdot \vec{E}\right). \tag{8.8}$$

Using the differential identity

$$\vec{\nabla} \cdot (\vec{E} \times \vec{B}) = \vec{B} \cdot \underbrace{(\vec{\nabla} \times \vec{E})}_{-\frac{1}{c}\frac{\partial \vec{B}}{\partial t}} - \vec{E} \cdot (\vec{\nabla} \times \vec{B}), \tag{8.9}$$

we solve for $(\vec{\nabla} \times \vec{B}) \cdot \vec{E}$ and plug into (8.8) to obtain

$$\begin{aligned} \vec{J} \cdot \vec{E} &= \frac{c}{4\pi}\left[-\frac{1}{c}\frac{\partial \vec{B}}{\partial t} \cdot \vec{B} - \frac{1}{c}\frac{\partial \vec{E}}{\partial t} \cdot \vec{E} - \vec{\nabla} \cdot (\vec{E} \times \vec{B})\right] \\ &= -\frac{\partial}{\partial t}\left[\frac{\vec{E}^2 + \vec{B}^2}{8\pi}\right] - \vec{\nabla} \cdot \left[\frac{c}{4\pi}\vec{E} \times \vec{B}\right]. \end{aligned} \tag{8.10}$$

Equation (8.10) motivates the definitions of *energy density* u and *energy flux* (also known as *Poynting vector*) $\vec{S}$ as

$$u \equiv \frac{\vec{E}^2 + \vec{B}^2}{8\pi}, \tag{8.11}$$

$$\vec{S} \equiv \frac{c}{4\pi}\vec{E} \times \vec{B}. \tag{8.12}$$

Note that $\vec{S}$ has units of energy per area per time, as we would expect for an energy flux. If the volume V is current-free (that is, $\vec{J} = 0$), from (8.10) it follows that we have

$$\frac{\partial u}{\partial t} + \vec{\nabla} \cdot \vec{S} = 0, \tag{8.13}$$

which has the same form as the continuity equation.

When $\vec{J} \neq 0$ we have

$$\frac{\partial u}{\partial t} + \vec{\nabla} \cdot \vec{S} + \vec{J} \cdot \vec{E} = 0, \tag{8.14}$$

which integrates to

$$\frac{d}{dt}\int_V d^3x\, u + \oint_S da\, \hat{n} \cdot \vec{S} + \frac{dE_m}{dt} = 0. \tag{8.15}$$

The three terms in (8.15) have physical interpretations as follows:

$\frac{d}{dt}\int_V d^3x\, u$: Rate of change of field energy in V;

$\oint_S da\, \hat{n} \cdot \vec{S}$: Rate of flow of field energy across S;

$\frac{dE_m}{dt}$: Rate of transfer of energy to matter.

The transfer of energy to matter is often in the form of heat.

This energy conservation law has important consequences for oscillating fields (which we examine in later chapters). A sinusoidally-varying current density $\vec{J}_c(\vec{x})\cos\omega t + \vec{J}_s(\vec{x})\sin\omega t$ can be conveniently expressed as the real part of a complex function:

$$\vec{J}(\vec{x}, t) \equiv \vec{J}(\vec{x})e^{-i\omega t}, \quad \text{where} \quad \vec{J}(\vec{x}) \equiv \vec{J}_c(\vec{x}) + i\vec{J}_s(\vec{x}). \tag{8.16}$$

We shall see in Section 11.1 that the real part of the current $\vec{J}(\vec{x}, t)$ in (8.16) gives rise to oscillating electric and magnetic fields which are the real parts of complex functions:

$$\begin{cases} \vec{E}(\vec{x}, t) = \vec{E}(\vec{x})e^{-i\omega t}, \\ \vec{B}(\vec{x}, t) = \vec{B}(\vec{x})e^{-i\omega t}. \end{cases} \tag{8.17}$$

The vector functions $\vec{E}(\vec{x})$, $\vec{B}(\vec{x})$ are in general complex, and reflect the fields' phases at each spatial location $\vec{x}$. They may also be functions of the oscillation angular frequency ω. The dependence here will be implicit, but we consider harmonic fields with explicit ω dependence beginning in Section 9.5.

Although only the real parts of $\vec{E}(\vec{x},t)$, $\vec{B}(\vec{x},t)$ and $\vec{J}(\vec{x},t)$ are physically significant, it turns out the complex fields and currents (including imaginary parts) also satisfy Maxwell's equations. This can be seen by writing

$$\begin{aligned}\mathrm{Im}[\vec{E}(\vec{x},t)] &= \mathrm{Re}[-i\vec{E}(\vec{x},t)] = \mathrm{Re}\left[\vec{E}(\vec{x})e^{-i\omega t - i\pi/2}\right] \\ &= \mathrm{Re}\left[\vec{E}\left(\vec{x}, t+\frac{\pi}{2\omega}\right)\right], \end{aligned} \tag{8.18}$$

and similar results hold for $\vec{B}$ and $\vec{J}$. Since Maxwell's equations are invariant under time translation $t \to t + \pi/(2\omega)$, it follows that the imaginary parts (as well as the real parts) satisfy Maxwell's equations.

The complex fields enable us to express time-averaged energy flow in a particularly simple way. We compute

$$\begin{aligned}\mathrm{Re}\left(\vec{J}(\vec{x},t)\right)\cdot\mathrm{Re}\left(\vec{E}(\vec{x},t)\right) &= \frac{\vec{J}(\vec{x},t)+\vec{J}^*(\vec{x},t)}{2}\cdot\frac{\vec{E}(\vec{x},t)+\vec{E}^*(\vec{x},t)}{2} \\ &\xrightarrow{\text{time average}} \frac{1}{2}\mathrm{Re}\left[\vec{J}(\vec{x})^*\cdot\vec{E}(\vec{x})\right], \end{aligned} \tag{8.19}$$

where the other terms in (8.19) vanish under time averaging because they multiply either $e^{2i\omega t}$ or $e^{-2i\omega t}$.

To find the time-averaged version of the conservation law (8.15) for oscillating fields, we first compute:

$$\begin{aligned}\frac{1}{2}\vec{J}^*\cdot\vec{E} &= \frac{c}{8\pi}\left[-\frac{1}{c}\frac{\partial\vec{B}}{\partial t}\cdot\vec{B}^* - \frac{1}{c}\frac{\partial\vec{E}^*}{\partial t}\cdot\vec{E} - \vec{\nabla}\cdot(\vec{E}\times\vec{B}^*)\right] \\ &= -\frac{i\omega}{8\pi}\left[|\vec{E}(\vec{x})|^2 - |\vec{B}(\vec{x})|^2\right] - \frac{c}{8\pi}\vec{\nabla}\cdot(\vec{E}\times\vec{B}^*). \end{aligned} \tag{8.20}$$

The corresponding integral form is

$$\frac{1}{2}\int d^3x\,\vec{J}^*(\vec{x})\cdot\vec{E}(\vec{x}) + 2i\omega\int d^3x\,(w_e - w_m) + \oint_S da\,\vec{S}\cdot\hat{n} = 0, \tag{8.21}$$

where the time-averaged quantities w_e, w_m and $\vec{S}$ are defined as:

$$w_e \equiv \frac{1}{16\pi}|\vec{E}(\vec{x})|^2, \tag{8.22}$$

$$w_m \equiv \frac{1}{16\pi}|\vec{B}(\vec{x})|^2, \tag{8.23}$$

$$\vec{S} \equiv \frac{c}{8\pi}\vec{E}(\vec{x})\times\vec{B}^*(\vec{x}). \tag{8.24}$$

(This definition of $\vec{S}$ differs from (8.12) but is closely related; this is the definition that is typically used for oscillating fields.) If we consider a portion of the closed surface S to be the "input" surface S_i, then we may rewrite (8.21) as follows:

$$\oint_{S_i} da\,\vec{S}\cdot(-\hat{n}) = \frac{1}{2}\int_V d^3x\,\vec{J}^*(\vec{x})\cdot\vec{E}(\vec{x}) + 2i\omega\int d^3x\,(w_e - w_m) + \oint_{S-S_i} da\,\vec{S}\cdot\hat{n}. \tag{8.25}$$

To get time-averaged energy conservation, we take real parts of all terms in (8.25). The volume integral of $w_e - w_m$ drops out; and the remaining terms have natural physical interpretations:

$\mathrm{Re}\oint_{S_i} da\,\vec{S}\cdot(-\hat{n})$: Power input through surface S_i;

$\mathrm{Re}\,\frac{1}{2}\int_V d^3x\,\vec{J}^*(\vec{x})\cdot\vec{E}(\vec{x})$: Rate of work done by fields on matter in V;

$\mathrm{Re}\oint_{S-S_i} da\,\vec{S}\cdot\hat{n}$: Rate of outward flow of field energy through $S - S_i$.

The imaginary parts of the terms in (8.25) gives the balance of *stored* energies, as we will see in a moment.

It is enlightening to apply these equations to electrical components within an electrical circuit. Consider an electrical component (with resistance, capacitance, and inductance) connected via two terminals to an external voltage source, as shown in Figure 8.1. We apply (8.25) first to a surface S' surrounding the voltage source, which we suppose has negligible internal electric and magnetic field energy so that $w_e = w_m = 0$. We further suppose that the Poynting vector $\vec{S}$ is negligible except at the terminals: this supposition is accurate at low frequencies, when radiation effects are negligible.[1] Taking the "input surface" as the two terminals, we have

$$\begin{aligned}\oint_{S'} da\,\vec{S}\cdot(-\hat{n}') &= \frac{1}{2}\int d^3x\,\vec{J}^*(\vec{x})\cdot\vec{E}(\vec{x})\\ &= \frac{1}{2}\int \left(\vec{J}^*(\vec{x})dA\right)\cdot\left(\vec{E}(\vec{x})dx\right)\\ &= \frac{1}{2}I_i^* V,\end{aligned} \tag{8.26}$$

where I_i refers to the internal current within the voltage source, which is the *negative* of the current supplied to the electrical component.

[1] Radiation effects are contained in the last term of (8.25) and will be considered in Section 11.1 when we study antennas.

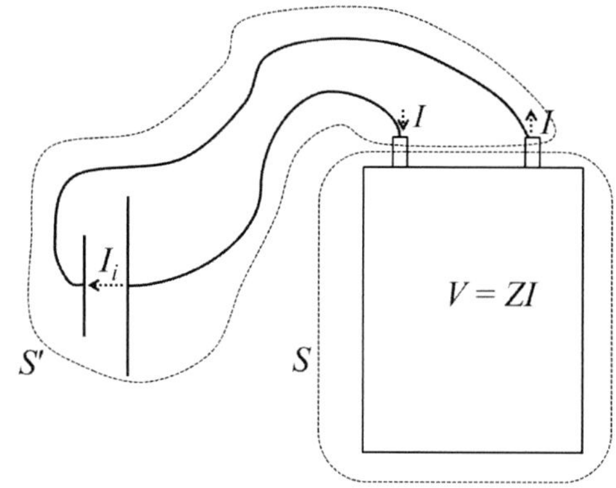

Fig. 8.1 Electrical component in circuit.

Turning now to the electrical component, we consider the surface S surrounding the component shown in Figure 8.1. Once again we take the input surface as the terminals, only this time the sign of the normal vector is reversed: the power input to the electrical component is the power output from the voltage source. Combining (8.25) and (8.26) gives (using $-\hat{n}' = \hat{n}$ and $I = -I_i$)

$$\frac{1}{2}I^* V = \frac{1}{2}\int d^3x\, \vec{J}^*(\vec{x}) \cdot \vec{E}(\vec{x}) + 2i\omega \int d^3x\, w_e - 2i\omega \int d^3x\, w_m. \tag{8.27}$$

Typically there is a linear relationship between V and I, and we define the (complex) *impedance* Z by

$$Z \equiv V/I; \qquad R \equiv \mathrm{Re}[Z]; \quad X \equiv -\mathrm{Im}[Z]. \tag{8.28}$$

where R and X are the *resistance* and *reactance*, respectively.[2] Also, the quantity $Y \equiv 1/Z$ is called the *admittance*.

The last two terms in (8.27) represent stored electrical and magnetic energies. For capacitors we have $Q = CV$, and it follows from (8.28) and $I = dQ/dt$ that $X_C \equiv -1/(\omega C)$ when $Q = Q_0 e^{-i\omega t}$. Similarly, for inductors we have $V = L\, dI/dt$, which leads to reactance $X_L \equiv \omega L$ when $I = I_0 e^{-i\omega t}$. The three energy terms on the right-hand side of (8.27) are associated with resistance, capacitance and inductance, respectively (see Exercise 8.1.1).

[2]Electrical engineers define X with a '+' instead of a '−'.

8.2 Conservation of momentum; Maxwell stress tensor

In the previous section we dealt with conservation of energy; we turn now to the conservation of momentum using the microscopic Maxwell equations. The situation is somewhat more complicated, since momentum has three components as opposed to energy which has only one. We begin as before with a single particle:

$$\frac{d\vec{P}_m}{dt} = q\left(\vec{E} + \frac{\vec{v}}{c} \times \vec{B}\right). \tag{8.29}$$

For n particles we have

$$\frac{d\vec{P}_m}{dt} = \sum_i q_i \left(\vec{E}_i + \frac{\vec{v}_i}{c} \times \vec{B}_i\right), \tag{8.30}$$

and passing to the continuum limit gives us

$$\frac{d\vec{P}_m}{dt} = \int d^3x \left(\rho\vec{E} + \frac{1}{c}\vec{J} \times \vec{B}\right). \tag{8.31}$$

We may now express ρ and $\vec{J}$ in terms of fields, using Maxwell's equations:

$$\rho\vec{E} + \frac{1}{c}\vec{J} \times \vec{B} = \frac{1}{4\pi}\left[\vec{E}(\nabla \cdot \vec{E}) + \frac{1}{c}\vec{B} \times \frac{\partial \vec{E}}{\partial t} - \vec{B} \times (\vec{\nabla} \times \vec{B})\right]. \tag{8.32}$$

Using the fact that

$$\frac{1}{c}\vec{B} \times \frac{\partial \vec{E}}{\partial t} = -\frac{1}{c}\frac{\partial}{\partial t}\left(\vec{E} \times \vec{B}\right) + \frac{1}{c}\vec{E} \times \underbrace{\frac{\partial \vec{B}}{\partial t}}_{-c\vec{\nabla}\times\vec{E}}, \tag{8.33}$$

we may rewrite the right-hand side of (8.32) as

$$\begin{aligned}\rho\vec{E} + \frac{1}{c}\vec{J} \times \vec{B} = {} & \frac{1}{4\pi}\left[\vec{E}(\vec{\nabla} \cdot \vec{E}) - \vec{E} \times (\vec{\nabla} \times \vec{E}) + \vec{B}(\vec{\nabla} \cdot \vec{B}) - \vec{B} \times (\vec{\nabla} \times \vec{B})\right] \\ & - \frac{1}{4\pi c} \cdot \frac{\partial}{\partial t}(\vec{E} \times \vec{B}),\end{aligned} \tag{8.34}$$

where we have added the term $\vec{B}(\vec{\nabla} \cdot \vec{B})$ (which is equal to 0) to make the expression symmetric in $\vec{E}$ and $\vec{B}$. In component form we have ($i, j = 1, 2, 3$)

$$\begin{aligned}\left[\vec{E}(\vec{\nabla} \cdot \vec{E}) - \vec{E} \times (\vec{\nabla} \times \vec{E})\right]_i &= \sum_j \left[E_i \nabla_j E_j - E_j \nabla_i E_j + E_j \nabla_j E_i\right] \\ &= \sum_j \nabla_j \left(E_i E_j - \frac{1}{2}\vec{E} \cdot \vec{E}\, \delta_{ij}\right),\end{aligned} \tag{8.35}$$

and similarly for the $\vec{B}$ terms in (8.34). The component form of (8.34) is thus

$$\left[\rho\vec{E} + \frac{1}{c}\vec{J}\times\vec{B}\right]_i = \frac{1}{4\pi}\sum_j \nabla_j \left[E_iE_j + B_iB_j - \frac{1}{2}(E^2+B^2)\delta_{ij}\right] - \frac{1}{4\pi c}\frac{\partial}{\partial t}(\vec{E}\times\vec{B})_i. \tag{8.36}$$

In light of (8.36), we define the *Maxwell stress tensor*[3] as

$$T_{ij} \equiv -\frac{1}{4\pi}\left[E_iE_j + B_iB_j - \frac{1}{2}(E^2+B^2)\delta_{ij}\right]. \tag{8.37}$$

From the definition it is evident that T_{ij} is symmetric:

$$T_{ij} = T_{ji}. \tag{8.38}$$

We may also define the *momentum density* $\vec{g}$ as

$$\vec{g} \equiv \frac{1}{4\pi c}(\vec{E}\times\vec{B}), \tag{8.39}$$

where $\vec{g}$ has the dimensions of momentum per volume. Notice that the energy flux (cf. Equation (8.12)) and momentum density are closely related:

$$\vec{S} = c^2\vec{g}, \tag{8.40}$$

so that

$$\text{energy flux} = c^2 \times (\text{momentum density}).$$

In Section 11.1 we shall see that electromagnetic fields and their associated momentum density travel at velocity c in the vacuum, so that $c \times$ (momentum density) can be interpreted as a *momentum flux*. From (8.40) we thus obtain

$$\text{energy flux} = c \times (\text{momentum flux}),$$

which points to the relativistic relationship $E = pc$ for massless particles.

When $\rho = \vec{J} = 0$ the momentum conservation equation (8.36) simplifies to

$$\frac{\partial g_i}{\partial t} + \sum_j \nabla_j T_{ij} = 0, \tag{8.41}$$

which once again has the form of the continuity equation, except that the conserved quantity here is a 3-component vector rather than a single number as with the energy and charge conservation equations.

[3] In some references T_{ij} is defined as the negative of (8.37).

In the general case where charges and currents are present, we have

$$\frac{\partial g_i}{\partial t} + \sum_j \nabla_j T_{ij} + f_i = 0, \tag{8.42}$$

where the *electromagnetic force density* f_i is defined by

$$f_i \equiv \left[\rho \vec{E} + \frac{1}{c}\vec{J} \times \vec{B}\right]_i . \tag{8.43}$$

The integrated form of (8.43) is

$$\frac{d}{dt}\int_V d^3x\, g_i + \sum_j \oint_S da\, n_j T_{ij} + F_i = 0, \tag{8.44}$$

where n_j is the jth component of the outward unit normal and F_i is the ith component of the total force. The terms in (8.44) have natural physical interpretations:

$\frac{d}{dt}\int_V d^3x\, g_i$: Rate of change of field momentum in V;

$\sum_j \oint_S da\, n_j T_{ij}$: Rate of field momentum flow across S;

$F_i = \int_V d^3x\, f_i$: Rate of momentum transfer to matter within V.

Let us consider the simple static example shown in Figure 8.2. According to (8.44), we may express the static force on q_2 as

$$F_i = -\sum_j \oint_S da\, n_j T_{ij}. \tag{8.45}$$

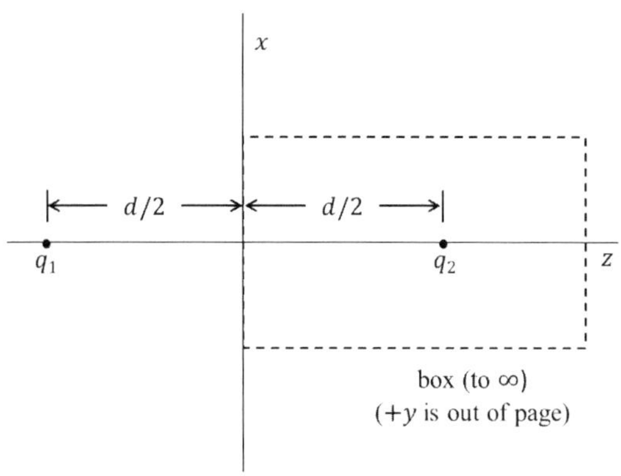

Fig. 8.2 Field stresses and forces on charges.

We may enlarge the box shown in Figure 8.2 to include the entire half-space $z > 0$, so that the surface becomes the $z = 0$ plane (the stress tensor $T_{ij} \to 0$ as $r \to \infty$ fast enough so that the surface integrals on the box's other faces will vanish in the limit). In this case, the outward normal at the surface $z = 0$ is $\hat{n} = (0, 0, -1)$, and (8.44) becomes

$$F_i = \int_{z=0} da\, T_{iz}. \tag{8.46}$$

It is natural to break the electric field into fields due to q_1 and q_2 (denoted by $\vec{E}_1$ and $\vec{E}_2$ respectively), so that

$$\vec{E}_1\Big|_{z=0} = q_1 \frac{(\vec{x} - \vec{x}'_1)}{|\vec{x} - \vec{x}'_1|^3} = q_1 \frac{[(x, y, 0) - (0, 0, -d/2)]}{[x^2 + y^2 + d^2/4]^{3/2}},$$

and

$$\vec{E}_2\Big|_{z=0} = q_2 \frac{(\vec{x} - \vec{x}'_2)}{|\vec{x} - \vec{x}'_2|^3} = q_1 \frac{[(x, y, 0) - (0, 0, d/2)]}{[x^2 + y^2 + d^2/4]^{3/2}}.$$

By combining $\vec{E}_1$ and $\vec{E}_2$ we obtain ($r^2 \equiv x^2 + y^2 + d^2/4$)

$$E_x\big|_{z=0} = (q_1 + q_2)\frac{x}{r^3},$$

$$E_y\big|_{z=0} = (q_1 + q_2)\frac{y}{r^3},$$

$$E_z\big|_{z=0} = (q_1 - q_2)\frac{d/2}{r^3},$$

which in turn gives

$$\begin{aligned}
T_{xz} &= -\frac{1}{4\pi} E_x E_z = -\frac{1}{4\pi}(q_1^2 - q_2^2)\frac{x \cdot d/2}{r^6}, \\
T_{yz} &= -\frac{1}{4\pi} E_y E_z = -\frac{1}{4\pi}(q_1^2 - q_2^2)\frac{y \cdot d/2}{r^6}, \\
T_{zz} &= \frac{1}{8\pi}\left(E_x^2 + E_y^2 - E_z^2\right) \\
&= \frac{(q_1 + q_2)^2}{8\pi r^6}(x^2 + y^2) - \frac{(q_1 - q_2)^2}{8\pi r^6}\frac{d^2}{4}.
\end{aligned}$$

We see immediately from the $-\infty$ to $+\infty$ integration on x and y that

$$F_x = F_y = 0,$$

as expected. We may rewrite the integral for F_z in polar coordinates (using $\rho^2 \equiv x^2 + y^2$) as

$$\begin{aligned}
F_z &= \int_0^{2\pi} d\phi \int_0^{\infty} \rho d\rho \left[\frac{(q_1 + q_2)^2 \rho^2}{8\pi[\rho^2 + d^2/4]^3} - \frac{(q_1 - q_2)^2 d^2/4}{8\pi[\rho^2 + d^2/4]^3}\right], \\
&= \frac{(q_1 + q_2)^2}{4} \int_0^{\infty} \frac{\rho^3 d\rho}{[\rho^2 + d^2/4]^3} - \frac{(q_1 - q_2)^2 d^2}{32} \int_0^{\infty} \frac{2\rho d\rho}{[\rho^2 + d^2/4]^3}.
\end{aligned} \tag{8.47}$$

The first integral in (8.47) evaluates to (see Gradshteyn and Rhyzik, *Tables of Integrals, Series, and Products*, 7th ed., Elsevier Academic Press (Burlington, MA) 2007, 2.114 (2)):

$$\int_0^\infty \frac{\rho^3 d\rho}{[d^2/4+\rho^2]^3} = -\left(\rho^2 + \frac{d^2}{8}\right)\frac{1}{2(d^2/4+\rho^2)^2}\bigg|_0^\infty = \frac{1}{d^2}, \tag{8.48}$$

while the second integral works out to

$$\int_0^\infty \frac{2\rho d\rho}{[\rho^2+d^2/4]^3} = -\frac{1}{2[\rho^2+d^2/4]^2}\bigg|_0^\infty = \frac{8}{d^4}. \tag{8.49}$$

Altogether, the force on charge q_2 is

$$F_z = \frac{(q_1+q_2)^2}{4d^2} - \frac{(q_1-q_2)^2}{4d^2} = \frac{q_1 q_2}{d^2}. \tag{8.50}$$

The preceding calculation provides yet another way of finding the force on a dielectric slab in the presence of a point charge, using the image charge shown in Figure 8.3. The force on the z' charge (the negative of force on slab) is mathematically equivalent to the problem we just did using the image charge

$$q_1 = -\left(\frac{\epsilon-1}{\epsilon+1}\right), \qquad q_2 = 1. \tag{8.51}$$

Then (8.50) gives

$$F_z^{\text{dielectric}} = -F_z^{\text{charge}} = \left(\frac{\epsilon-1}{\epsilon+1}\right)\frac{1}{4z'^2}, \tag{8.52}$$

which agrees with our previous computation in Exercise 5.9.2.

In general we have to be careful when using the stress tensor if interfaces between materials are present. We will return to this point in Section 8.5.

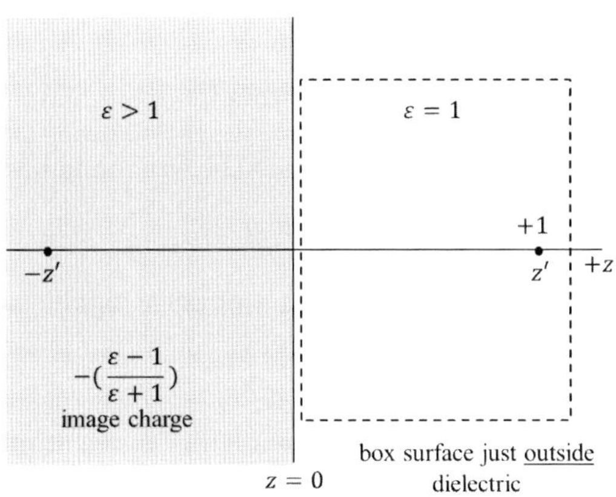

Fig. 8.3 Dielectric ($z < 0$) and point charge.

8.3 Conservation of angular momentum; shear tensor

Next we consider the conservation of angular momentum when electromagnetic fields are present. Summing over particles and taking the continuum limit, we have

$$\vec{N} = \sum_i \vec{x}_i \times \vec{F}_i \to \int d^3x\, \vec{x} \times \vec{f}, \tag{8.53}$$

where f is the force density. In component form, we have

$$N_i = \sum_{j,k} \int d^3x\, \epsilon_{ijk} x_j f_k. \tag{8.54}$$

If we cross $\vec{x}$ into the local momentum conservation equation (8.42), we obtain the local angular momentum conservation law:

$$\sum_{j,k} \epsilon_{ijk} x_j \left[\frac{\partial g_k}{\partial t} + \sum_l \nabla_l T_{kl} + f_k \right] = 0. \tag{8.55}$$

Let's define the *torque density* as

$$n_i \equiv \sum_{j,k} \epsilon_{ijk} x_j f_k \qquad (\text{or } \vec{n} = \vec{x} \times \vec{f}), \tag{8.56}$$

(please don't mistake $\vec{n}$ for a normal vector!) and the *field angular momentum density* as

$$\mathcal{L}_i \equiv \sum_{j,k} \epsilon_{ijk} x_j g_k \qquad (\text{or } \vec{\mathcal{L}} = \vec{x} \times \vec{g}). \tag{8.57}$$

You will show in Exercise 8.3.3 that we may rewrite the angular momentum conservation equation (8.55) as

$$\frac{\partial \mathcal{L}_i}{\partial t} + \sum_j \nabla_j M_{ij} + n_i = 0, \tag{8.58}$$

where the *shear tensor* M_{ij} is given by

$$M_{ij} \equiv \sum_{k,l} \epsilon_{ikl} x_k T_{lj}, \tag{8.59}$$

in index notation. Equation (8.58) can also be written in integrated form as

$$\frac{d}{dt} \int d^3x\, \mathcal{L}_i + \sum_j \oint_S da\, n_j M_{ij} + N_i = 0, \tag{8.60}$$

where as before each term in (8.60) has a physical significance:

$\frac{d}{dt} \int d^3x\, \mathcal{L}_i$: Rate of change of the field angular momentum in V;

$\sum_j \oint_S da\, n_j M_{ij}$: Rate of change of the angular momentum flow across S;

$N_i = \int d^3x\, n_i$: Rate of transfer of the angular momentum to matter (torque).

8.4 Effective conservation laws for macroscopic media

In this section we consider two alternative approaches to conservation laws for macroscopic media.

Recall that in Section 8.2 we derived the momentum conservation equation in the vacuum by starting with the force density equation

$$f_i \equiv \left[\rho\vec{E} + \frac{1}{c}\vec{J}\times\vec{B}\right]_i, \tag{8.61}$$

and using the Maxwell equations to obtain the result (see (8.37))

$$f_i = -\sum_j \nabla_j T_{ij} - \frac{\partial}{\partial t}\left(\frac{\vec{E}\times\vec{B}}{4\pi c}\right)_i. \tag{8.62}$$

Let us now consider momentum conservation when macroscopic media are present. One possible approach is to include the bound charge as part of the total charge (and similarly for current density). In effect, we may replace ρ with $\rho + \rho_{\mathrm{b}}$ and $\vec{J}$ with $\vec{J} + \vec{J}_{\mathrm{b}}$ in (8.61) and use to obtain

$$4\pi(\rho + \rho_b) = \vec{\nabla}\cdot\vec{E}, \tag{8.63}$$

$$\frac{4\pi}{c}(\vec{J} + \vec{J}_b) = \vec{\nabla}\times\vec{B} - \frac{1}{c}\frac{\partial\vec{E}}{\partial t}, \tag{8.64}$$

which gives us

$$\begin{aligned} f_i &= \left[(\rho + \rho_{\mathrm{b}})\vec{E} + \frac{1}{c}(\vec{J} + \vec{J}_{\mathrm{b}})\times\vec{B}\right]_i \\ &= -\sum_j \nabla_j T_{ij} - \frac{\partial}{\partial t}\left(\frac{\vec{E}\times\vec{B}}{4\pi c}\right)_i. \end{aligned} \tag{8.65}$$

In this equation, ρ and $\vec{J}$ now represent ρ_{free} and $\vec{J}_{\mathrm{free}}$, respectively.

In summary: if we place ρ_{b} and $\vec{J}_{\mathrm{b}}$ on the same footing as ρ and $\vec{J}$, then there is no need to refer to macroscopic fields at all. For example, the energy conservation equation (8.14) when $\rho_{\mathrm{b}}, \vec{J}_{\mathrm{b}} \neq 0$, becomes

$$(\vec{J} + \vec{J}_{\mathrm{b}})\cdot\vec{E} = -\frac{\partial u}{\partial t} - \vec{\nabla}\cdot\vec{S}, \tag{8.66}$$

with u and $\vec{S}$ as before. When we are given constitutive relations, we know how to find ρ_{b} and $\vec{J}_{\mathrm{b}}$ from the various discontinuities and gradients of the fields (see for example (5.58), (5.59), (6.195), and (6.196)).

Alternatively, it is possible to derive energy and momentum conservation laws in the presence of macroscopic media by using the macroscopic Maxwell equations, so that the macroscopic fields $\vec{H}$ and $\vec{D}$ account for the effects of $\rho_{\rm b}$ and $\vec{J}_{\rm b}$. We used this approach in Sections 5.8 and 7.5 to derive the macroscopic electric and magnetic energy density expressions $\vec{E}\cdot\vec{D}/(8\pi)$ and $\vec{B}\cdot\vec{H}/(8\pi)$, respectively. We can similarly derive an energy conservation equation by starting with the macroscopic Ampère-Maxwell equation:

$$\vec{J}\cdot\vec{E} = \frac{1}{4\pi}\vec{E}\cdot\left[c\vec{\nabla}\times\vec{H} - \frac{\partial\vec{D}}{\partial t}\right], \tag{8.67}$$

and using the vector identity

$$\vec{\nabla}\cdot(\vec{E}\times\vec{H}) = \vec{H}\cdot(\vec{\nabla}\times\vec{E}) - \vec{E}\cdot(\vec{\nabla}\times\vec{H}) \tag{8.68}$$

to obtain

$$\vec{J}\cdot\vec{E} = -\frac{1}{4\pi}\left[c\vec{\nabla}\cdot(\vec{E}\times\vec{H}) + \vec{E}\cdot\frac{\partial\vec{D}}{\partial t} + \vec{H}\cdot\frac{\partial\vec{B}}{\partial t}\right]. \tag{8.69}$$

If we have specific constitutive relations, we can simplify this further. For example, in the case of linear, isotropic homogeneous materials with $\vec{D} = \epsilon\vec{E}$ and $\vec{B} = \mu\vec{H}$, we have

$$\vec{J}\cdot\vec{E} = -\frac{\partial}{\partial t}\left[\frac{\vec{E}\cdot\vec{D} + \vec{B}\cdot\vec{H}}{8\pi}\right] - \vec{\nabla}\cdot\left(\frac{c}{4\pi}\vec{E}\times\vec{H}\right). \tag{8.70}$$

This leads naturally to the definitions for energy density u and energy flux $\vec{S}$:

$$u \equiv \frac{1}{8\pi}(\vec{E}\cdot\vec{D} + \vec{B}\cdot\vec{H}), \tag{8.71}$$

$$\vec{S} \equiv \frac{c}{4\pi}\vec{E}\times\vec{H}. \tag{8.72}$$

Note that these definitions are specific to the case of linear, isotropic homogeneous materials with non-harmonic fields. In other contexts, different expressions for u and $\vec{S}$ are used; the reader should be careful to observe the context. In particular, definitions (8.71), (8.72) will be used when we discuss waveguides in Chapter 10.

Given these understandings, a modified form of the harmonic energy conservation law in Section 8.1 holds for fields given by Equation (8.17). For linear, isotropic materials, one may derive that

$$\frac{1}{2}I^*V = \frac{1}{2}\int d^3x\, \vec{J}_f^*(\vec{x})\cdot\vec{E}(\vec{x}) + 2i\omega\int d^3x\,(w_e - w_m) + \oint_{S-S_i} da\,\vec{S}\cdot\hat{n}. \tag{8.73}$$

Note that $\vec{J}_f$ represents the free current and the bounding surfaces S_i and $S - S_i$ are taken to be in free space. We define as usual $V = IZ$, where Z is the impedance. In this context the effective time-averaged energy densities are defined as

$$w_e \equiv \frac{1}{16\pi}\vec{E}(\vec{x}) \cdot \vec{D}^*(\vec{x}), \tag{8.74}$$

$$w_m \equiv \frac{1}{16\pi}\vec{B}(\vec{x}) \cdot \vec{H}^*(\vec{x}). \tag{8.75}$$

Note that w_e and w_m can be complex in phenomenological applications. The effective form of the time-averaged Poynting vector inside the material is

$$\vec{S} \equiv \frac{c}{8\pi}\vec{E}(\vec{x}) \times \vec{H}^*(\vec{x}). \tag{8.76}$$

We again caution the reader to be alert to the context for the correct employment of these quantities. Of course, these expressions for w_e, w_m and $\vec{S}$ reduce to those in Equations (8.22)–(8.24) in free space.

8.5 Maxwell stress tensor example: dielectric slab

We mentioned in the previous section that we can use the Maxwell stress tensor even when macroscopic media are present. Here we illustrate this using the same example we discussed in Section 8.2, namely a point charge outside of a dielectric slab. This example will also illustrate the comment at the end of Section 8.2 that discretion is required when employing the stress tensor when material interfaces are present.

We compute the force on the slab using the integrated form of the momentum conservation equation (8.65) with all time derivatives set equal to zero (here ρ is ρ_{free} and $\vec{J}$ is $\vec{J}_{\text{free}}$):

$$f_i \equiv \left[(\rho + \rho_{\text{b}})\vec{E} + \frac{1}{c}(\vec{J} + \vec{J}_{\text{b}}) \times \vec{B}\right]_i = -\sum_j \nabla_j T_{ij} \tag{8.77}$$

$$\Longrightarrow F_i \equiv \int_V d^3x(\rho + \rho_{\text{b}})E_i = -\sum_j \oint_S da\, n_j T_{ij}, \tag{8.78}$$

where we integrate over the surface shown in Figure 8.4. In Section 5.6 we derived the Green function for this geometry as

$$G(\vec{x}, \vec{x}\,') = \begin{cases} \dfrac{2}{\epsilon + 1}\dfrac{1}{|\vec{x} - \vec{x}\,'|}, & z < 0, \\[2ex] \dfrac{1}{|\vec{x} - \vec{x}\,'|} - \left(\dfrac{\epsilon - 1}{\epsilon + 1}\right)\dfrac{1}{|\vec{x} - \vec{x}\,''|}, & z > 0, \end{cases} \tag{8.79}$$

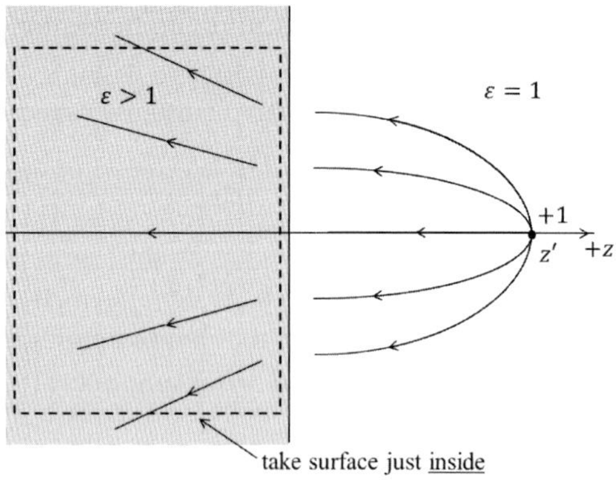

Fig. 8.4 Fields in dielectric in the presence of a charge.

where $\vec{x}'' \equiv (x', y', -z')$. This implies that

$$\vec{E}|_{z=0^-} = \frac{2}{\epsilon+1}\frac{\vec{x}-\vec{x}'}{|\vec{x}-\vec{x}'|^3}, \tag{8.80}$$

which gives (with $r^2 \equiv x^2 + y^2 + z'^2$)

$$\begin{aligned} E_x|_{z=0^-} &= \frac{2}{\epsilon+1}\frac{x}{r^3}, \\ E_y|_{z=0^-} &= \frac{2}{\epsilon+1}\frac{y}{r^3}, \\ E_z|_{z=0^-} &= -\frac{2}{\epsilon+1}\frac{z'}{r^3}. \end{aligned}$$

As before, $F_x = F_y = 0$ because of symmetry. To find F_z, we compute T_{zz}:

$$T_{zz} = \frac{1}{8\pi}[E_x^2 + E_y^2 - E_z^2] = \frac{1}{8\pi}\left(\frac{2}{\epsilon+1}\right)^2 \frac{(x^2+y^2-z'^2)}{r^6},$$

so that

$$\begin{aligned} F_z = -\int_{z=0^-} da\, T_{zz} &= -\frac{1}{8\pi}\left(\frac{2}{\epsilon+1}\right)^2 2\pi \int_0^\infty d\rho\, \rho \left[\frac{\rho^2 - z'^2}{(\rho^2+z'^2)^3}\right], \\ &= -\frac{1}{4}\left(\frac{2}{\epsilon+1}\right)^2 \left\{\frac{1}{4z'^2} - \frac{1}{4z'^2}\right\} = 0. \end{aligned} \tag{8.81}$$

It seems at first sight that the net force on the dielectric is zero. Recall however the definition in (8.78) of F_z as the volume integral of the force density:

$$F_i = \int_V d^3x\, (\rho + \rho_{\mathrm{b}}) E_i. \tag{8.82}$$

Since we have chosen a surface S within the dielectric, the force F_i does not include the forces on the *surface* of the dielectric. This calculation therefore confirms our previous statements that the force on the dielectric comes from the surface only.

In summary, in computing forces using T_{ij}, one must be careful to include appropriate physical surfaces within the volume of integration. In fact, if we include the interface within the volume of integration in our force calculation then we would just repeat the work at the end of Section 8.2.

8.6 Retarded potentials

In electrostatics, we saw that the potential of a stationary unit point charge gave the Green function in free space. In electrodynamics, the potentials of a moving unit point charge are similarly related to Green functions. Here we say "potentials" rather than "potential" because the moving particle has both scalar and vector potentials. In this section we will both derive these potentials and demonstrate their connection to Green functions.

With reference to the microscopic Maxwell equations, the equations for the potentials Φ and $\vec{A}$ given the Lorenz condition (7.108)

$$\vec{\nabla} \cdot \vec{A} + \frac{1}{c}\frac{\partial \Phi}{\partial t} = 0, \tag{8.83}$$

are

$$\Box \Phi = 4\pi\rho, \tag{8.84}$$

$$\Box \vec{A} = \frac{4\pi}{c}\vec{J}, \tag{8.85}$$

where $\Box$ is the *D'Alembertian*

$$\Box \equiv \frac{1}{c^2}\frac{\partial^2}{\partial t^2} - \nabla^2. \tag{8.86}$$

Equations (8.84) and (8.85) are wave equations, as seen in Section 7.4. To solve these for simple geometries, we can take a Green function approach which includes time development as well as spatial variation. The generic form of the wave equation is

$$\Box \psi(\vec{x}, t) = -4\pi f(\vec{x}, t). \tag{8.87}$$

In the Green function approach, we solve instead

$$\Box G(\vec{x}, t; \vec{x}\,', t') = 4\pi\delta(\vec{x} - \vec{x}\,')\delta(t - t'). \tag{8.88}$$

Just as with spatial Green functions, we will use the freedom in the specification of boundary conditions on G to simplify the resulting solution for $\psi(\vec{x},t)$. First, let us find the analog of the Dirichlet Green function that we used in electrostatics. We assume Dirichlet boundary conditions on G

$$G(\vec{x},t;\vec{x}\,',t')|_{\vec{x}\text{ on S}} = 0, \tag{8.89}$$

and in addition, the *causality* condition

$$G(\vec{x},t;\vec{x}\,',t') = 0,\ t < t'. \tag{8.90}$$

Before we actually solve for $\psi(\vec{x},t)$, we derive an important and physically interesting property of the Green function. Let us define

$$G_1 \equiv G(\vec{x},t;\vec{x}\,',t'); \qquad G_2 \equiv G(\vec{x},-t;\vec{x}\,'',-t''). \tag{8.91}$$

In analogy with Green's second identity (2.95), we then compute:

$$-G_2\times\left(\Box G_1 = 4\pi\delta(\vec{x}-\vec{x}\,')\delta(t-t')\right), \tag{8.92}$$

$$-G_1\times\left(\Box G_2 = 4\pi\delta(\vec{x}-\vec{x}\,'')\delta(t-t'')\right). \tag{8.93}$$

Taking the difference and integrating over all $\vec{x}$ and t, we obtain

$$\begin{aligned}
&-4\pi[G(\vec{x}\,',-t';\vec{x}\,'',-t'') - G(\vec{x}\,'',t'';\vec{x}\,',t')]\\
&= \int_{-\infty}^{\infty} dt \int d^3x \left\{ \left[G_2\nabla^2 G_1 - G_1\nabla^2 G_2\right] - \frac{1}{c^2}\left[G_2\frac{\partial^2 G_1}{\partial t^2} - G_1\frac{\partial^2 G_2}{\partial t^2}\right]\right\}.
\end{aligned} \tag{8.94}$$

Now we use the vector identity

$$\vec{\nabla}\cdot\left[G_2\vec{\nabla}G_1 - G_1\vec{\nabla}G_2\right] = G_2\nabla^2 G_1 - G_1\nabla^2 G_2, \tag{8.95}$$

to convert the first term in (8.94) into a surface term, which vanishes because of the Dirichlet condition. The second term can be written

$$\begin{aligned}
&-\frac{1}{c^2}\int d^3x \int_{-\infty}^{\infty} dt\left[G_2\frac{\partial^2 G_1}{\partial t^2} - G_1\frac{\partial^2 G_2}{\partial t^2}\right]\\
&\quad = -\frac{1}{c^2}\int d^3x \int_{-\infty}^{\infty} dt\frac{\partial}{\partial t}\left[G_2\frac{\partial G_1}{\partial t} - G_1\frac{\partial G_2}{\partial t}\right],
\end{aligned} \tag{8.96}$$

$$\begin{aligned}
&\quad = -\frac{1}{c^2}\int d^3x \left[G(\vec{x},-t;\vec{x}\,'',-t'')\frac{\partial}{\partial t}G(\vec{x},t;\vec{x}\,',t')\right.\\
&\qquad \left.\left. -G(\vec{x},t;\vec{x}\,',t')\frac{\partial}{\partial t}G(\vec{x},-t;\vec{x}\,'',-t'')\right]\right|_{t=-\infty}^{t=\infty},
\end{aligned} \tag{8.97}$$

which now vanishes because of the causality condition. As a result we obtain *time reversal* symmetry

$$G(\vec{x}\,',t';\vec{x}\,'',t'') = G(\vec{x}\,'',-t'';\vec{x}\,',-t'), \tag{8.98}$$

so that (unlike time-independent Green functions) the Green function is not symmetric under argument interchange $(\vec{x}\,', t') \leftrightarrow (\vec{x}\,'', t'')$. Note however we can still replace $\Box$ with $\Box'$ in the Green function equation (8.88).

Let us return to the Green function solution for $\psi(\vec{x}, t)$. We follow a similar algebraic procedure as in the static case, and compute:

$$\begin{aligned} &- G(\vec{x}, t; \vec{x}\,', t')\,\Box'\psi(\vec{x}\,', t') + \psi(\vec{x}\,', t')\,\Box' G(\vec{x}, t; \vec{x}\,', t') \\ &\quad = G(\vec{x}, t; \vec{x}\,', t')\,[-4\pi f(\vec{x}\,', t')] + \psi(\vec{x}\,', t')\,[-4\pi\delta(\vec{x} - \vec{x}\,')\delta(t - t')]\,. \end{aligned} \tag{8.99}$$

Integrating over all space and times $t_0 < t' < \infty$ (where t_0 is a fixed time less than t) gives

$$\begin{aligned} &\int_{t_0}^{\infty} dt' \int d^3x' \left\{[G\nabla'^2\psi - \psi\nabla'^2 G] - \frac{1}{c^2}\left[G\frac{\partial^2\psi}{\partial t'^2} - \psi\frac{\partial^2 G}{\partial t'^2}\right]\right\}, \\ &\qquad = -4\pi\int_{t_0}^{\infty} dt' \int d^3x'[Gf] + 4\pi\psi(\vec{x}, t). \end{aligned} \tag{8.100}$$

The upper limits of the time integrals $\int_{t_0}^{\infty} dt'[\ldots]$ can be replaced by $t + \epsilon$, because $G(\vec{x}, t; \vec{x}\,', t') = 0$ for $t' > t$. Solving for $\psi(\vec{x}, t)$, and using vector differential identities to change volume integrals into surface integrals we find

$$\begin{aligned} \psi(\vec{x}, t) = &\int_{t_0}^{t+\epsilon} dt' \int d^3x' G(\vec{x}, t; \vec{x}\,', t') f(\vec{x}\,', t') \\ &- \frac{1}{4\pi}\int_{t_0}^{t+\epsilon} dt' \oint_S da'\psi(\vec{x}\,', t')\frac{\partial}{\partial n'}G(\vec{x}, t; \vec{x}\,', t') \\ &+ \frac{1}{4\pi c^2}\int d^3x' \left[G(\vec{x}, t; \vec{x}\,', t')\frac{\partial}{\partial t'}\psi(\vec{x}\,', t')\right. \\ &\left.\left. - \psi(\vec{x}\,', t')\frac{\partial}{\partial t'}G(\vec{x}, t; \vec{x}\,', t')\right]\right|_{t'=t_0}. \end{aligned} \tag{8.101}$$

Equation (8.101) gives us $\psi(\vec{x}, t)$ for $t > t_0$, as long as we know:

$$\psi(\vec{x}\,', t')|_{t_0},\ \frac{\partial}{\partial t'}\psi(\vec{x}\,', t')|_{t_0} \text{ for all } \vec{x}\,' \text{ in } V;$$

$$\psi(\vec{x}\,', t')|_S \text{ for all } t' \text{ in } t_0 < t' < t \text{ and for all } \vec{x}\,' \text{ on } S.$$

If $\psi|_{t'=t_0} = \frac{\partial}{\partial t'}\psi\big|_{t'=t_0} = 0$, then (8.101) reduces to the Dirichlet form we are familiar with from electrostatics, except with an extra integral over time.

We mention in passing that there is also a Green function treatment of the wave equation (8.85) for the vector potential. However, practical boundary conditions lead to a situation where there J_k influences A_i for $i \neq k$. As a result, the Green function takes the form of a second-rank tensor (or dyad). For more details, we refer the reader to Morse and Feshbach's herculean opus.[4]

Just as in the static case, the Green function depends only on the geometry; and once we have it, we can in principle calculate the solution $\psi(\vec{x}, t)$ for any source in that particular geometry. We demonstrate this first in the easiest case, namely unbounded space and time. Once again drawing on our experience with the static case, we express the delta functions as integrals over eigenfunctions:

$$\delta(\vec{x} - \vec{x}\,') = \int \frac{d^3k}{(2\pi)^3} e^{i\vec{k}\cdot(\vec{x}-\vec{x}\,')}, \quad \delta(t - t') = \int_{-\infty}^{\infty} \frac{d\omega}{2\pi} e^{-i\omega(t-t')}. \tag{8.102}$$

Next we assume that the Green function has the form

$$G(\vec{x}, t; \vec{x}\,', t') = \int \frac{d^3k}{(2\pi)^3} \int_{-\infty}^{\infty} \frac{d\omega}{2\pi} e^{i\vec{k}\cdot(\vec{x}-\vec{x}\,')-i\omega(t-t')} G(k, \omega), \tag{8.103}$$

where k as usual denotes $|\vec{k}|$. Plugging into the Green function defining relation, we find

$$\begin{aligned} \Box G &= \int \frac{d^3k}{(2\pi)^3} \int_{-\infty}^{\infty} \frac{d\omega}{2\pi} e^{i\vec{k}\cdot(\vec{x}-\vec{x}\,')-i\omega(t-t')} \left(k^2 - \frac{\omega^2}{c^2}\right) G(k, \omega) \\ &= 4\pi \int \frac{d^3k}{(2\pi)^3} \int_{-\infty}^{\infty} \frac{d\omega}{2\pi} e^{i\vec{k}\cdot(\vec{x}-\vec{x}\,')-i\omega(t-t')} \\ \Longrightarrow G(k, \omega) &= \frac{4\pi}{k^2 - \frac{\omega^2}{c^2}}. \end{aligned} \tag{8.104}$$

Formally then we have the expression

$$G(\vec{x}, t; \vec{x}\,', t') = 4\pi \int \frac{d^3k}{(2\pi)^3} \int_{-\infty}^{\infty} \frac{d\omega}{2\pi} \frac{e^{i\vec{k}\cdot(\vec{x}-\vec{x}\,')-i\omega(t-t')}}{k^2 - \frac{\omega^2}{c^2}}. \tag{8.105}$$

Unfortunately this expression is not well defined because the denominator has singularities when $\omega = \pm kc$. We can interpret the singular integral in terms of contour integrals; but we need some prescription (preferably one with a physical interpretation) for choosing the right contour. To do this, we choose the k_z-axis along $\vec{x} - \vec{x}\,'$, so that

$$d^3k \to 2\pi k^2 dk \sin\theta d\theta; \qquad \vec{k}\cdot(\vec{x} - \vec{x}\,') \to k|\vec{x} - \vec{x}\,'| \cos\theta.$$

[4]P. M. Morse and H. Feshbach, *Methods of Theoretical Physics, Part I*, McGraw-Hill (New York) p. 1789 *et seq.*

With the definitions $x \equiv \cos\theta$ and $R \equiv |\vec{x} - \vec{x}\,'|$, the $d\theta$ integral in (8.105) can be evaluated via change of variable:

$$\begin{aligned}\int_0^\pi d\theta\, \sin\theta e^{k|\vec{x}-\vec{x}\,'|\cos\theta} &= \int_{-1}^1 dx\, e^{ikRx}\\ &= \frac{2i}{ikR}\frac{(e^{ikR} - e^{-ikR})}{2i}\\ &= \frac{2}{kR}\sin kR, \end{aligned} \tag{8.106}$$

so that the Green function simplifies to

$$\begin{aligned} G(\vec{x}, t; \vec{x}\,', t') &= \frac{1}{\pi^2 R}\int_{-\infty}^{\infty} d\omega\, e^{-i\omega(t-t')}\int_0^\infty dk \frac{k}{k^2 - \frac{\omega^2}{c^2}} \sin kR.\\ &= \frac{1}{\pi^2 R}\int_{-\infty}^{\infty} d\omega\, e^{-i\omega(t-t')}\int_0^\infty dk \frac{k}{k^2 - \frac{\omega^2}{c^2}} \frac{e^{ikR} - e^{-ikR}}{2i}\\ &= \frac{1}{2i\pi^2 R}\int_{-\infty}^{\infty} d\omega\, e^{-i\omega(t-t')}\int_{-\infty}^\infty dk \frac{k\, e^{ikR}}{k^2 - \frac{\omega^2}{c^2}}. \end{aligned} \tag{8.107}$$

At this point, in order to avoid singularities we make the prescription (later we will provide the justification)

$$\frac{\omega}{c} \to \frac{\omega}{c} + i\epsilon \quad (\epsilon > 0). \tag{8.108}$$

In other words, we suppose that ω has a vanishingly small, positive imaginary part. This shifts the poles of the integrand in (8.107) as shown in Figure 8.5(a) where the complex variable z is replacing k. Furthermore, because of the exponential suppression of the factor e^{izR} when $|z|R \gg 1$ and $\mathrm{Im}(z) > 0$, the dk integral in (8.107) can be identified with the contour integral shown in Figure 8.5(a), so that

$$\int_{-\infty}^{\infty} dk \frac{ke^{ikR}}{k^2 - \frac{\omega^2}{c^2}} \to \oint_C dz \frac{ze^{izR}}{(z - \frac{\omega}{c} - i\epsilon)(z + \frac{\omega}{c} + i\epsilon)}. \tag{8.109}$$

Figure 8.5(b) shows an alternative prescription which yields the same result. Rather than shifting the poles, instead we modify the contour to avoid the poles. The two small semicircles in Figure 8.5(b) have radius ϵ where $\epsilon \to 0^+$. Note the crucial point that in both (a) and (b), the pole with $\mathrm{Re}\, z > 0$ is inside the contour while the pole with $\mathrm{Re}\, z < 0$ is outside. Later when we encounter similar integrals, we will either shift the pole or modify the contour, whichever is more convenient.

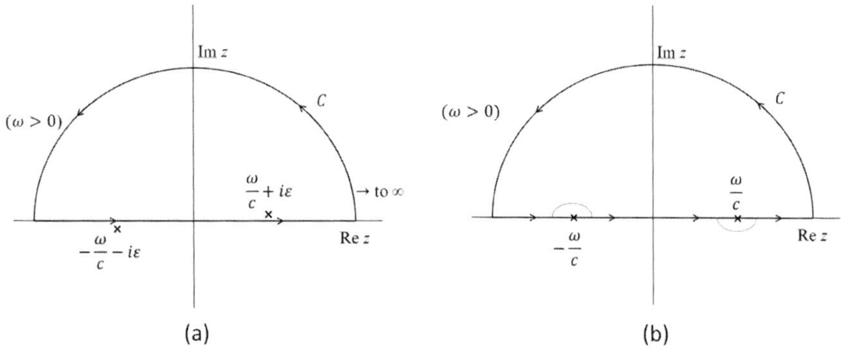

Fig. 8.5 Alternative contour integrals for retarded Green function.

Using the fact that the clockwise integral over a simple closed contour is $2\pi i$ times the sum of enclosed poles, we obtain

$$\oint_C dz \frac{ze^{izR}}{(z - \frac{\omega}{c} - i\epsilon)(z + \frac{\omega}{c} + i\epsilon)} = 2\pi i \left(\frac{\frac{\omega}{c} e^{i(\omega/c)R}}{2\frac{\omega}{c}} \right)$$
$$= i\pi e^{i(\omega/c)R}. \tag{8.110}$$

This gives us what is known as the *retarded* free Green function

$$G^{\text{ret}}(\vec{x}, t; \vec{x}\,', t') = \frac{1}{R} \int_{-\infty}^{\infty} \frac{d\omega}{2\pi} e^{-i\omega[(t-t')-R/c]}$$
$$= \frac{\delta((t - t') - \frac{R}{c})}{R}, \tag{8.111}$$

where $R \equiv |\vec{x} - \vec{x}\,'|$ as mentioned above.

Notice the following important properties of G^{ret}:

(1) $G^{\text{ret}}(\vec{x}, t; \vec{x}\,', t') = G^{\text{ret}}(\vec{x}\,', -t'; \vec{x}, -t)$ (so that time-reversal symmetry is obeyed); and
(2) $G^{\text{ret}}(\vec{x}, t; \vec{x}\,', t')$ vanishes except when $t = t' + \frac{R}{c}$ (which justifies the name "retarded").

By causality we only require $G^{\text{ret}}(\vec{x}, t; \vec{x}\,', t') = 0$ for $t < t' + \frac{R}{c}$. The δ-function form of G^{ret} is particular to three dimensions: the retarded Green function in one or two space dimensions has a "tail" and vanishes only for $t < t' + \frac{R}{c}$ in general.

The causal form for $G^{\text{ret}}(\vec{x}, t; \vec{x}\,', t')$ above is a result of our choice of contour. To show that a different choice results in a physically different

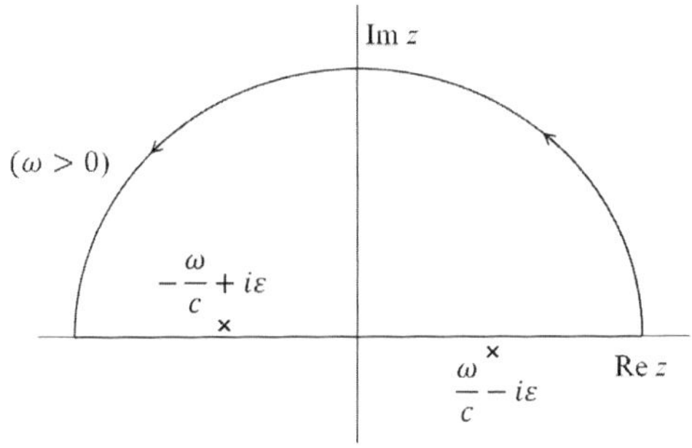

Fig. 8.6 Pole locations for advanced Green function.

Green function, we may use instead the prescription

$$\frac{\omega}{c} \to \frac{\omega}{c} - i\epsilon, \quad \epsilon > 0.$$

The pole locations then change, as shown in Figure 8.6. We may carry out the same analysis for this new case. Instead of (8.110), we obtain

$$\begin{aligned}\oint_C dz \frac{ze^{izR}}{(z - \frac{\omega}{c} + i\epsilon)(z + \frac{\omega}{c} - i\epsilon)} &= 2\pi i \left(\frac{\frac{\omega}{c} e^{-i\omega/cR}}{2\frac{\omega}{c}} \right) \\ &= i\pi e^{-i\omega/cR}. \end{aligned} \tag{8.112}$$

which leads to the so-called *advanced* form of the Green function

$$G^{\text{adv}}(\vec{x}, t; \vec{x}\,', t') = \frac{\delta((t - t') + \frac{R}{c})}{R}. \tag{8.113}$$

If we choose the retarded solution, then the potentials in unbounded space may be written (still assuming the Lorenz condition):

$$\begin{aligned}\Phi(\vec{x}, t) &= \int d^3x' dt' \frac{\delta((t - t') - \frac{R}{c})}{|\vec{x} - \vec{x}\,'|} \rho(\vec{x}\,', t') \\ &= \int d^3x' \frac{\rho(\vec{x}\,', t - \frac{R}{c})}{|\vec{x} - \vec{x}\,'|} \end{aligned} \tag{8.114}$$

and

$$\begin{aligned}\vec{A}(\vec{x}, t) &= \int d^3x' dt' \frac{\delta((t - t') - \frac{R}{c})}{|\vec{x} - \vec{x}\,'|} \frac{1}{c} \vec{J}(\vec{x}\,', t') \\ &= \int d^3x' \frac{1}{c} \frac{\vec{J}(\vec{x}\,', t - \frac{R}{c})}{|\vec{x} - \vec{x}\,'|}. \end{aligned} \tag{8.115}$$

8.7 Green function for half-infinite geometry

Next, we derive the Green function for the half-infinite geometry. Since image charges worked so well in the static case, we may reasonably expect that the time-dependent Green function also has an image charge interpretation.

The Green function equation is

$$\left(\nabla^2 - \frac{1}{c^2}\frac{\partial}{\partial t^2}\right) G(\vec{x},t;\vec{x}\,',t') = -4\pi\delta(\vec{x}-\vec{x}\,')\delta(t-t'), \tag{8.116}$$

subject to initial conditions

$$G = \frac{\partial G}{\partial t} \equiv 0, \;\; t < t', \tag{8.117}$$

and boundary conditions

$$G(\vec{x},t;\vec{x}\,',t')|_{z=0} = 0. \tag{8.118}$$

Before solving this equation, we first develop an alternate integral representation of the free Green function, which will enable us to interpret the integrals we encounter in solving (8.116)–(8.118). We may rewrite (8.105) as

$$G_{\text{free}} = 4\pi \int_{-\infty}^{\infty} \frac{d\omega}{2\pi} e^{-i\omega(t-t')} \int \frac{d^2k_\perp}{(2\pi)^2} e^{i\vec{k}_\perp\cdot(\vec{x}-\vec{x}\,')} \int_{-\infty}^{\infty} \frac{dk_z}{2\pi} \frac{e^{ik_z(z-z')}}{k_z^2+k_\perp^2-\omega^2/c^2}. \tag{8.119}$$

Let us first evaluate the rightmost integral in the case where $k_\perp^2 > \omega^2/c^2$. We have:

$$k_z^2 + k_\perp^2 - \omega^2/c^2 = (k_z+i\gamma)(k_z-i\gamma), \tag{8.120}$$

where

$$\gamma \equiv \sqrt{k_\perp^2 - \omega^2/c^2}. \tag{8.121}$$

In the case where $z - z' > 0$, the integrand $\to 0$ as $\text{Im}[k_z] \to \infty$ and we complete the integration contour using the upper semicircle (see Figure 8.7) to obtain

$$\int_{-\infty}^{\infty} \frac{dk_z}{2\pi} \frac{e^{ik_z(z-z')}}{k_z^2+k_\perp^2-\omega^2/c^2} = \frac{2\pi i}{2\pi}\frac{e^{-\gamma(z-z')}}{2i\gamma} = \frac{e^{-\gamma(z-z')}}{2\gamma}. \tag{8.122}$$

In the case where $z - z' < 0$, we follow the lower semicircle with the result:

$$\int_{-\infty}^{\infty} \frac{dk_z}{2\pi} \frac{e^{ik_z(z-z')}}{k_z^2+k_\perp^2-\omega^2/c^2} = -\frac{2\pi i}{2\pi}\frac{e^{-\gamma(z-z')}}{-2i\gamma} = \frac{e^{-\gamma(z-z')}}{2\gamma}. \tag{8.123}$$

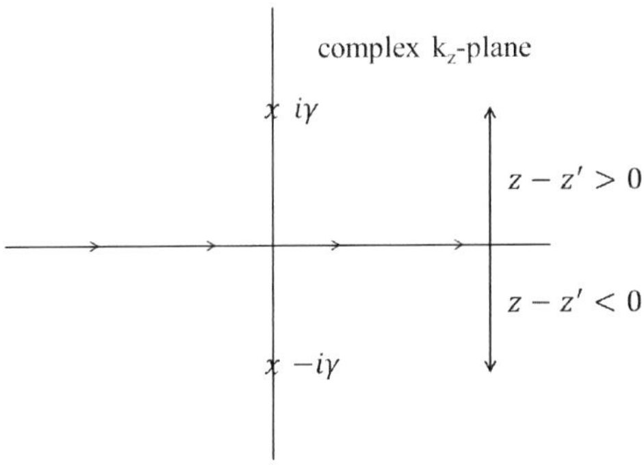

Fig. 8.7 Integration plane for (8.119).

We must also evaluate the integral when $\frac{\omega^2}{c^2} > k_\perp^2$. We re-express the integrand's denominator as

$$k_z^2 - \frac{\omega^2}{c^2} + k_\perp^2 = (k_z + \theta)(k_z - \theta), \tag{8.124}$$

where

$$\theta \equiv \sqrt{\frac{\omega^2}{c^2} - k_\perp^2}. \tag{8.125}$$

As with the free-space Green function, we replace $\frac{\omega}{c}$ with $\frac{\omega}{c} + i\epsilon$ with $\epsilon > 0$, and perform a Taylor expansion with the result that

$$\begin{aligned} \sqrt{\frac{\omega^2}{c^2} - k_\perp^2} &\to \sqrt{\frac{\omega^2}{c^2} - k_\perp^2 + 2i\epsilon\frac{\omega}{c}} \\ &\approx \theta\left(1 + \frac{i\epsilon\frac{\omega}{c}}{\theta^2}\right) && (8.126) \\ &\approx \theta + i\epsilon S_\omega, && (8.127) \end{aligned}$$

where we have rescaled ϵ to absorb other constant factors, and $S_\omega \equiv \text{sign}(\omega)$. The poles are now located at $\pm(\theta + i\epsilon S_\omega)$.

When $S_\omega > 0$, the $z - z' > 0$ and $z - z' < 0$ cases can be combined (see Figure 8.8) to obtain

$$\int_{-\infty}^{\infty} \frac{dk_z}{2\pi} \frac{e^{ik_z(z-z')}}{(k_z^2 - \frac{\omega^2}{c^2} + k_\perp^2)} = i\frac{e^{i\theta|z-z'|}}{2\theta} \qquad (S_\omega > 0). \tag{8.128}$$

When $S_\omega < 0$, we similarly obtain

$$\int_{-\infty}^{\infty} \frac{e^{ik_z(z-z')}}{k_z^2 - \frac{\omega^2}{c^2} + k_\perp^2} = \frac{ie^{-i\theta|z-z'|}}{-2\theta}. \tag{8.129}$$

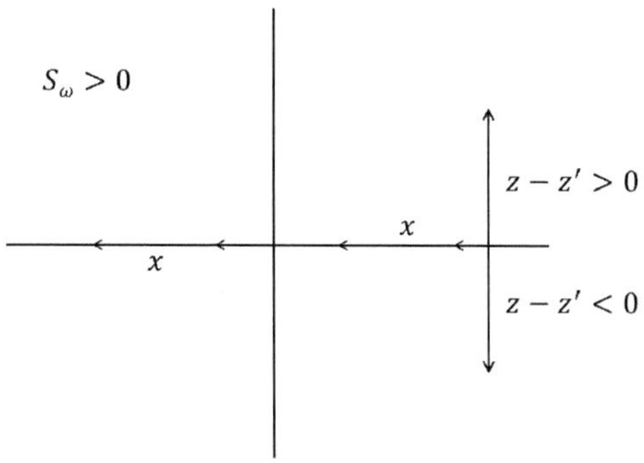

Fig. 8.8 Poles for Green function integration.

Equations (8.128) and (8.129) can be incorporated into the single equation:

$$\int_{-\infty}^{\infty} \frac{dk_z}{2\pi} \frac{e^{ik_z(z-z')}}{k_z^2 - \frac{\omega^2}{c^2} + k_\perp^2} = \frac{i}{2S_\omega\theta} e^{iS_\omega\theta|z-z'|}. \tag{8.130}$$

In summary, we have obtained the following alternative integral representation of the free-space Green function:

$$G_{\text{free}} = \int \frac{d^2k_\perp}{(2\pi)^2} \int_{-\infty}^{\infty} d\omega e^{i\vec{k}_\perp\cdot(\vec{x}-\vec{x}\,')} e^{-i\omega(t-t')} \times \begin{cases} \frac{1}{\gamma} e^{-\gamma|z-z'|}, \ k_\perp^2 > \frac{\omega^2}{c^2}, \\ \frac{i}{S_\omega\theta} e^{iS_\omega\theta|z-z'|}, k_\perp^2 < \frac{\omega^2}{c^2}. \end{cases} \tag{8.131}$$

Further simplifications of $G^{\text{ret,adv}}$ are examined in Exercise 8.7.2.

Let us now return to the problem at hand, namely the Green function for the half-infinite geometry $z > 0$. We assume the form

$$G = 4\pi \int \frac{d^2k_\perp}{(2\pi)^2} \int_{-\infty}^{\infty} \frac{d\omega}{2\pi} e^{i\vec{k}_\perp\cdot(\vec{x}-\vec{x}\,')-i\omega(t-t')} g(z, z', k_\perp, \omega), \tag{8.132}$$

and plugging into $\Box G(\cdots) = -4\pi\delta(\cdots)$ gives

$$\begin{aligned} \Box G &= 4\pi \int \frac{d^2k_\perp}{(2\pi)^2} \int_{-\infty}^{\infty} \frac{d\omega}{2\pi} e^{i\vec{k}_\perp\cdot(\vec{x}-\vec{x}\,')-i\omega(t-t')} \left[-k_\perp^2 + \frac{\omega^2}{c^2} - \frac{\partial^2}{\partial z^2}\right] g \\ &= -4\pi \int \frac{d^2k_\perp}{(2\pi)^2} \int_{-\infty}^{\infty} \frac{d\omega}{2\pi} e^{i\vec{k}_\perp\cdot(\vec{x}-\vec{x}\,')-i\omega(t-t')} \delta(z - z'), \end{aligned} \tag{8.133}$$

which yields

$$\left[k_\perp^2 - \frac{\omega^2}{c^2} - \frac{\partial^2}{\partial z^2}\right] g = \delta(z - z'). \tag{8.134}$$

The Dirichlet boundary conditions lead to the condition

$$g(z,z',k_\perp,\omega)|_{z=0} = 0. \tag{8.135}$$

We already know the solution when $\omega = 0$:

$$g = \frac{1}{2k_\perp}\left(e^{-k_\perp|z-z'|} - e^{-k_\perp|z+z'|}\right). \tag{8.136}$$

When $k_\perp^2 > \omega^2/c^2$ we may simply replace $k_\perp$ in (8.136) with

$$\gamma = \sqrt{k_\perp^2 - \omega^2/c^2},$$

so that

$$g = \frac{1}{2\gamma}\left(e^{-\gamma|z-z'|} - e^{-\gamma|z+z'|}\right), \qquad k_\perp^2 > \frac{\omega^2}{c^2}. \tag{8.137}$$

It remains to solve (8.134) when $k_\perp^2 < \omega^2/c^2$. In the extreme case where $k_\perp = 0$ we have

$$\left(-\frac{\omega^2}{c^2} - \frac{\partial^2}{\partial z^2}\right) g = \delta(z - z'). \tag{8.138}$$

Using well-established techniques we obtain:

$$g = \begin{cases} C_1 e^{i(\omega/c)z} + C_2 e^{-i(\omega/c)z}, & (z > z') \\ C_3 e^{i(\omega/c)z} + C_4 e^{-i(\omega/c)z}, & (z < z'). \end{cases} \tag{8.139}$$

To determine the constants we must use boundary conditions at 0 and ∞, as well as the interface conditions at $z = z'$. The rule for the advanced Green function ($\omega/c \to \omega/c + i\epsilon, \epsilon > 0$) implies that $C_2 = 0$ to avoid exponential growth at ∞. From the boundary condition $g|_{z=0} = 0$ we obtain $C_3 = -C_4$. The two remaining conditions are

$$g\Big|_{z=z'-}^{z=z'+} = 0, \text{ and } -\frac{\partial}{\partial z} g\Big|_{z=z'-}^{z=z'+} = 1, \tag{8.140}$$

which lead to the results

$$C_1 = \frac{ic}{2\omega}\left(e^{-i(\omega/c)z'} - e^{i(\omega/c)z'}\right), \quad C_3 = \frac{ic}{2\omega} e^{i(\omega/c)z'}. \tag{8.141}$$

Putting everything together, we have for the case $k_\perp = 0$:

$$g(z,z',0,\omega) = \frac{ic}{2\omega}\left(e^{i(\omega/c)|z-z'|} - e^{i(\omega/c)|z+z'|}\right). \tag{8.142}$$

When $|\omega/c| > k_\perp > 0$, it appears that all we have to do is replace (ω/c) in (8.138) with $[(\omega/c)^2 - k_\perp^2]^{1/2}$ and crank through the same solution. This is indeed possible; however, there is a subtlety involving the boundary condition at $z = \infty$. In analogy to (8.139), for $z > z'$ we obtain

$$C_1 e^{i\theta z} + C_2 e^{-i\theta z}, \quad (z > z'), \tag{8.143}$$

where $\theta \equiv [(\omega/c)^2 - k_\perp^2]^{1/2}$. If we use the prescription $\omega/c \to \omega/c + i\epsilon$, we obtain

$$C_1 e^{i(\theta + iS_\omega \epsilon)z} + C_2 e^{-i(\theta + iS_\omega \epsilon)z}, \quad (z > z'), \tag{8.144}$$

where S_ω denotes the sign of ω (as in (8.127). It follows that $C_1 = 0$ or $C_2 = 0$ depending on whether S_ω is negative or positive, respectively. The upshot is that S_ω enters into the solution, which works out to

$$g(z, z', k_\perp, \omega) = \begin{cases} \frac{1}{2\gamma}\left(e^{-\gamma|z-z'|} - e^{-\gamma|z+z'|}\right), & k_\perp^2 > \dfrac{\omega^2}{c^2}, \\ \frac{i}{2S_\omega\theta}\left(e^{iS_\omega\theta|z-z'|} - e^{iS_\omega\theta|z+z'|}\right), & k_\perp^2 < \dfrac{\omega^2}{c^2}. \end{cases} \tag{8.145}$$

At this point we may compare our result with the alternative representation of the free Green function given in (8.131). We obtain finally,

$$G(\vec{x}, t; \vec{x}\,', t') = \frac{\delta\left((t-t') - \frac{|\vec{x}-\vec{x}\,'|}{c}\right)}{|\vec{x}-\vec{x}\,'|} - \frac{\delta\left((t-t') - \frac{|\vec{x}-\vec{x}\,''|}{c}\right)}{|\vec{x}-\vec{x}\,''|}, \tag{8.146}$$

where $\vec{x}\,'' \equiv (x', y', -z')$.

The physical significance of this result is shown in Figures 8.9 and 8.10. Before time $t = t' + z'/c$, the Green function is identical to the free-space Green function, but after that time the image charge field effectively reproduces reflection from the boundary. In fact, we could have solved directly for the Green function for this geometry by an obvious image method (for a similar application of images, see Exercise 8.7.3). Our treatment above

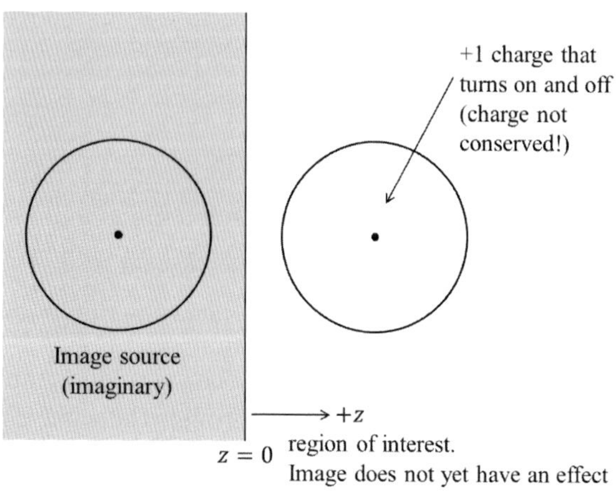

Fig. 8.9 Image charge field, $t < t' + z'/c$.

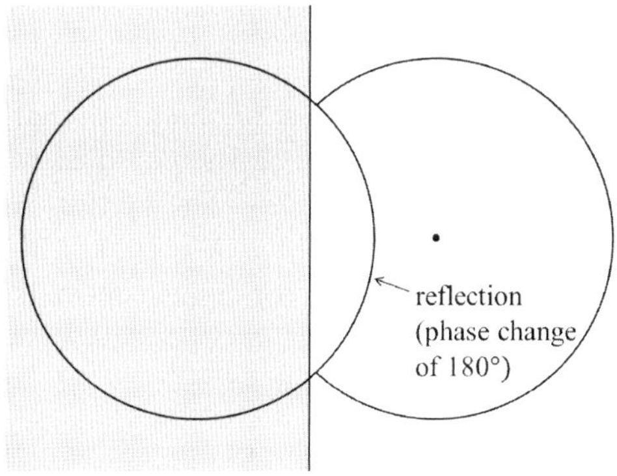

Fig. 8.10 Image charge field, $t > t' + z'/c$.

was intended to draw parallels with the reduced Green function method that was used in Chapter 3 for the same geometry (compare (3.31) with (8.145)). Furthermore, the techniques used will be useful in Section 10.2, where we discuss reflection and absorption of electromagnetic waves at dielectric boundaries.

8.8 Transformation properties of electromagnetic quantities

The electromagnetic field solutions involve various quantities that transform differently under change of reference frame. In this section we classify different types of quantities and their transformation properties. In particular, we will define and characterize vectors, scalars, and tensors.

Note that the specification of vectors, scalars, and tensors depends on the set of transformations under consideration. In this section we shall consider *only* the set of three-dimensional rotations: hence our classification is "non-relativistic". Nonetheless, most of the results in this section carry over to the relativistic case, which (as far as the mathematics goes) amounts to enlarging the group of transformations to include Lorentz transformations.

Physically we may distinguish between *active* rotations (rotations of the physical system) and *passive* rotations (rotations of the reference frame). Mathematically there is no difference between the two (the mathematics can't tell what's moving), but physically they represent opposite relative rotations of object and coordinate system. Our results below are equally valid for passive and active rotations.

We begin by defining a *vector* as a set of quantities that transform under 3-dimensional rotations the same way as the coordinates $x_i, i = 1, 2, 3$ of a point in space (typically we will use $i, j, k \in \{1, 2, 3\}$ as spatial indices; when we get to relativity, we will use $\mu, \nu, \lambda = \{0, 1, 2, 3\}$). We may write the vector transformation equation in component form as

$$x'_i = \sum_{j=1}^{3} a_{ij} x_j, \tag{8.147}$$

or in matrix form

$$\begin{pmatrix} x'_1 \\ x'_2 \\ x'_3 \end{pmatrix} = \begin{pmatrix} a_{11} & a_{12} & a_{13} \\ a_{21} & a_{22} & a_{23} \\ a_{31} & a_{32} & a_{33} \end{pmatrix} \begin{pmatrix} x_1 \\ x_2 \\ x_3 \end{pmatrix}, \quad \text{or} \quad x' = ax. \tag{8.148}$$

The a_{ij} have a geometrical interpretation as *direction cosines*: that is,

$$a_{ij} = \hat{e}'_i \cdot \hat{e}_j = \cos(\hat{e}'_i, \hat{e}_j), \tag{8.149}$$

where $\hat{e}_j$ and $\hat{e}'_j$ are the unit vectors along the coordinate axes in the untransformed and transformed frames, respectively.

At this point, we consider the mathematical properties of rotation matrices $\{a_{ij}\}$. The condition that rotations preserve the length of vectors leads to the equation

$$\sum_i a_{ij} a_{ik} = \delta_{jk}, \tag{8.150}$$

or

$$a^T a = 1 \implies a^T = a^{-1}. \tag{8.151}$$

Matrices satisfying (8.151) are called *orthogonal* matrices. The orthogonality condition (8.151) represents six equations in nine unknowns, which implies that any three of the a_{ij} uniquely specify the matrix.

In addition to the vector transformation law, we will also need the transformation of the gradient operator $\vec{\nabla}$. According to the chain rule,

$$\frac{\partial}{\partial x'_i} = \sum_j \left(\frac{\partial x_j}{\partial x'_i}\right) \frac{\partial}{\partial x_j}, \tag{8.152}$$

and also

$$\sum_i \left(\frac{\partial x_j}{\partial x'_i}\right) \left(\frac{\partial x'_i}{\partial x_k}\right) = \delta_{jk}. \tag{8.153}$$

From (8.147) we have

$$\left(\frac{\partial x'_i}{\partial x_k}\right) = a_{ik}, \tag{8.154}$$

and comparison of (8.150) and (8.153) gives

$$\left(\frac{\partial x_j}{\partial x'_i}\right) = a_{ij}. \tag{8.155}$$

Plugging into (8.152) gives

$$\frac{\partial}{\partial x'_i} = \sum_j a_{ij} \frac{\partial}{\partial x_j}, \tag{8.156}$$

which has the same form as (8.147).

We have seen that all rotation matrices are orthogonal — but is the converse true? It turns out that the matrix determinant function is the key to answering this question. Recall the following properties of matrix determinants:

$$\det AB = \det A \det B, \tag{8.157}$$

$$\det A^T = \det A, \tag{8.158}$$

$$\det A^{-1} = (\det A)^{-1}, \tag{8.159}$$

and, if $\lambda_i (i = 1, \ldots, n)$ are the eigenvalues,

$$\det A = \prod_{i=1}^{n} \lambda_i. \tag{8.160}$$

In three dimensions, we also have the familiar formula

$$\det A = \sum_{i,j,k} \epsilon_{ijk} A_{1i} A_{2j} A_{3k} = \sum_{i,j,k} \epsilon_{ijk} A_{i1} A_{j2} A_{k3}, \tag{8.161}$$

where A_{ij} are the elements of the matrix A. This can be extended to the $n \times n$ case using the *completely antisymmetric symbol* (also called the *Levi-Civita symbol*), defined as

$$\epsilon_{i_1 i_2 \ldots i_n} = \begin{cases} 0, & \text{if any two indices are equal} \\ 1, & \text{if } (i_1 i_2 \ldots i_n) \text{ is an even permutation of } (1, 2, \ldots n) \\ -1, & \text{if } (i_1 i_2 \ldots i_n) \text{ is an odd permutation of } (1, 2, \ldots n.) \end{cases} \tag{8.162}$$

For orthogonal matrices $a^T a = 1$, so we have:

$$\begin{aligned} &\Longrightarrow (\det a)^2 = 1, \\ &\Longrightarrow \det a = \pm 1 \qquad \text{only}. \end{aligned} \tag{8.163}$$

Now $\det 1 = 1$; and any rotation a is continuously connected to 1, in the sense that there is a continuous series of rotations that brings us from 1 to a. Furthermore, det is a continuous function of the matrix entries. It follows that it must be the case that $\det a = 1$ for any rotation, since there is no way that the determinant could "jump" from 1 to -1 within the continuous series of rotations connecting 1 to a.

It remains to characterize the orthogonal matrices with $\det a = -1$. One example (in three dimensions) is the 3×3 matrix -1, which produces the transformation shown in Figure 8.11. It should be clear that the transformation shown in Figure 8.11 does not correspond to any manipulation of the object. An orthogonal transformation a with $\det a = -1$ is called a (spatial) *inversion*. The transformation $a = -1$ is called a *complete inversion*, *point inversion*, or *parity reversal*. Plane reflections are also inversions. Vectors change sign under inversion, as shown in Figure 8.12.

There is another type of vectorial quantity which behaves differently under inversion. Given vectors $\vec{A}$ and $\vec{B}$, consider the vector product

$$(\vec{A} \times \vec{B})_i \equiv \sum_{j,k} \epsilon_{ijk} A_j B_k. \tag{8.164}$$

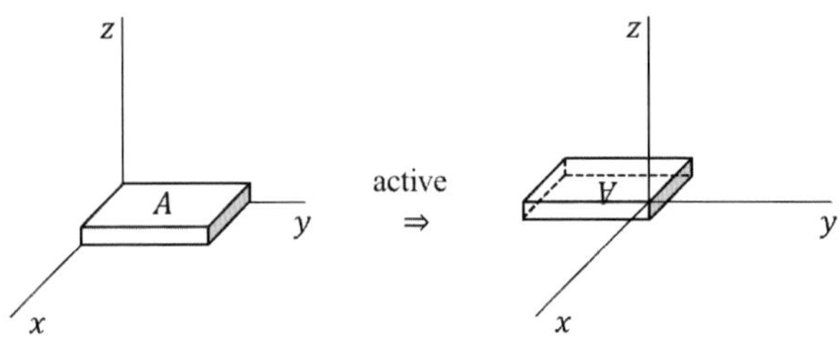

Fig. 8.11 Transformation produced by the matrix -1 in 3 dimensions.

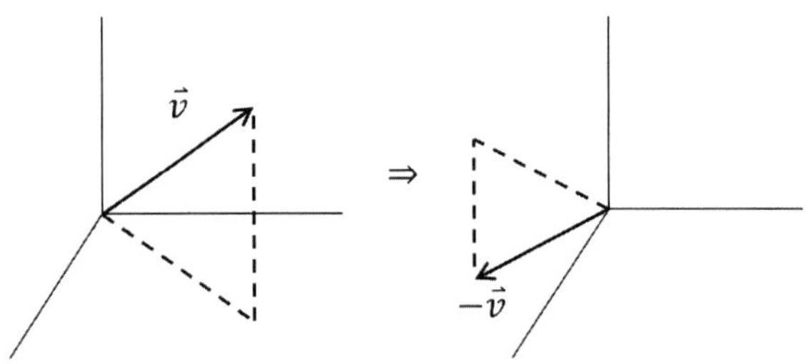

Fig. 8.12 Complete inversion of a vector.

In Exercise 8.8.1 you will show that for any orthogonal matrix a,

$$\sum_{j,k} \epsilon_{ijk} a_{jl} a_{km} = (\det a) \sum_n \epsilon_{nlm} a_{in}. \tag{8.165}$$

Thus

$$\begin{aligned}
(\vec{A}\,' \times \vec{B}\,')_i &= \sum_{j,k} \epsilon_{ijk} A'_j B'_k \\
&= \sum_{j,k,l,m} \epsilon_{ijk} a_{jl} A_l a_{km} B_m \\
&= (\det a) \sum_{n,l,m} \epsilon_{nlm} a_{in} A_l B_m \\
&= (\det a) \sum_n a_{in} (\vec{A} \times \vec{B})_n.
\end{aligned} \tag{8.166}$$

Thus $\vec{A} \times \vec{B}$ does *not* change sign under inversion; such quantities are called *pseudovectors*. Angular momentum of a particle, $\vec{L} = \vec{x} \times \vec{p}$, is one example of a pseudovector. Consider for instance the effect of reflection on a particle undergoing circular motion. A reflection in the xy-plane is given by the matrix

$$a = \begin{pmatrix} 1 & 0 & 0 \\ 0 & 1 & 0 \\ 0 & 0 & -1 \end{pmatrix}.$$

We find that (see Figure 8.13)

$$L'_x = -L_x, \quad L'_y = -L_y, \quad L'_z = +L_z.$$

The signs are exactly opposite from a reflected vector but are consistent with (8.166) since $\det a = -1$.

A quantity is a *scalar* if it remains invariant under orthogonal transformation. For example, we may verify that the dot product of two vectors is a scalar: we have

$$\begin{aligned}
\vec{A}' \cdot \vec{B}' &= \sum_i A'_i B'_i = \sum_{i,j,k} a_{ij} A_j a_{ik} B_k \\
&= \sum_{j,k} \underbrace{\left(\sum_i a_{ij} a_{ik} \right)}_{\delta_{jk}} A_j B_k = \vec{A} \cdot \vec{B}.
\end{aligned} \tag{8.167}$$

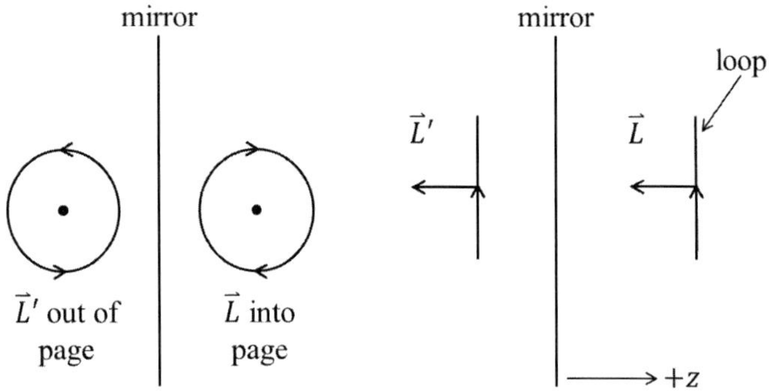

Fig. 8.13 Transformation of angular momentum under reflection.

The dot product of a vector and a pseudovector is a *pseudoscalar*. For example, if $\vec{A}, \vec{B}, \vec{C}$ are vectors we may have

$$\vec{A}' \cdot (\vec{B}' \times \vec{C}') = \sum_{i,j,k} \epsilon_{ijk} A'_i B'_j C'_k. \tag{8.168}$$

We may evaluate this *scalar triple product* as follows: using the pseudovector transformation rule (8.166) for $\vec{B} \times \vec{C}$ and vector transformation for $\vec{A}$, we obtain

$$\begin{aligned} \sum_i A'_i (\vec{B}' \times \vec{C}')_i &= \sum_{i,j,k} (a_{ik} A_k)(\det a) a_{ij} (\vec{B} \times \vec{C})_j, \\ &= (\det a) \sum_{j,k} \delta_{jk} A_k (\vec{B} \times \vec{C})_j \\ &= (\det a) \sum_k A_k (\vec{B} \times \vec{C})_k. \end{aligned} \tag{8.169}$$

Physical fields which are a function of space and time may be classified according to their transformation properties. For instance, a *scalar field* is a field whose values at each space-time point transform under orthogonal change of coordinates as scalars (here we are taking the passive transformation point of view). If Φ and Φ' are used to denote the field expressed in original and transformed coordinates respectively, then a scalar field must satisfy:

$$\Phi'(\vec{x}\,') = \Phi(\vec{x}). \tag{8.170}$$

In accordance with (8.148) the 3-component field

$$\vec{A}(\vec{x}) = (A_1(\vec{x}), A_2(\vec{x}), A_3(\vec{x})),$$

is called a *vector field* if it transforms under rotations as

$$A'_i(\vec{x}\,') = \sum_j a_{ij} A_j(\vec{x}). \tag{8.171}$$

Similarly, pseudoscalar and pseudovector fields also exist and transform with an extra factor of $\det a$:

$$\Phi'(\vec{x}\,') = (\det a)\Phi(\vec{x}); \qquad A'_i(\vec{x}\,') = (\det a)\sum_j a_{ij} A_j(\vec{x}). \tag{8.172}$$

Tensors transform as direct products of vectors. For example, a rank two tensor T_{ij} transforms under orthogonal transformation as

$$T'_{ij} = \sum_{k,l} a_{ik} a_{jl} T_{kl}. \tag{8.173}$$

Let us verify that the Maxwell stress tensor introduced in Section 8.2 does indeed obey this rule. From (8.37) we have in primed coordinates:

$$T'_{ij} \equiv -\frac{1}{4\pi}\left[E'_i E'_j + B'_i B'_j - \frac{1}{2}\delta_{ij}(E'^2 + B'^2)\right]. \tag{8.174}$$

Since $\vec{E}$ is the gradient of a scalar field, thus $\vec{E}$ is a vector field: therefore

$$E'_i = \sum_j a_{ij} E_j \implies E'_i E'_j = \sum_{k,l} a_{ik} a_{jl} E_k E_l. \tag{8.175}$$

When considering current loops, one can infer that $\vec{B}$ is a pseudovector field as in the similar situation in Figure 8.13. This is confirmed by the expression for $d\vec{B}$ in Table 6.1. It follows from $(\det a)^2 = 1$ that

$$B'_j = \det a \sum_k a_{ik} B_k \implies B'_i B'_j = \sum_{k,l} a_{ik} a_{jl} B_k B_l. \tag{8.176}$$

The Kronecker delta δ_{ij} also transforms as a tensor:

$$\delta_{ij} = \sum_k a_{ik} a_{jk} = \sum_{k,l} a_{ik} a_{jl} \delta_{kl}, \tag{8.177}$$

while $\vec{E}^2$ and $\vec{B}^2$ transform as scalars. Putting (8.175)–(8.177) into (8.174) yields the tensor transformation rule (8.173), as expected.

We close this section with a discussion of *time reversal*, which is of key importance in quantum mechanics and relativistic theories (as we shall see in Sections 13.5). From the passive point of view, time reversal simply means that we measure negative times as positive, and vice versa. For example, if we have a particle trajectory $\vec{x}(t)$, then under time reversal the trajectory is

$$\vec{x}\,'(t') = \vec{x}(t), \quad \text{where } t' = -t. \tag{8.178}$$

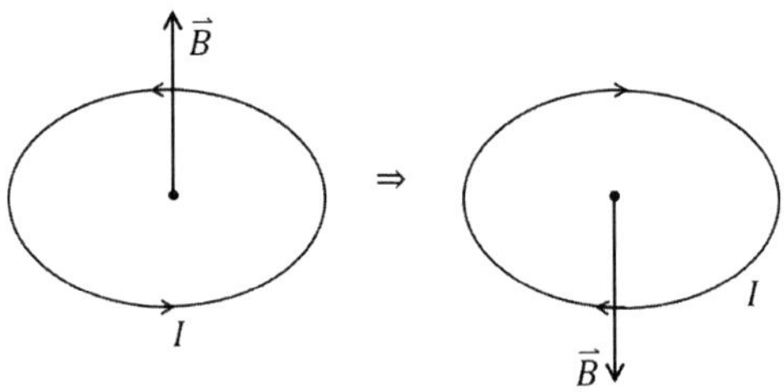

Fig. 8.14 Magnetic field under time reversal.

Velocities change sign under time reversal:

$$\frac{d\vec{x}\,'(t')}{dt'} = \frac{d\vec{x}(t)}{d(-t)} = -\frac{d\vec{x}(t)}{dt}. \tag{8.179}$$

Accelerations however are invariant: this can also be seen geometrically, since the direction of curvature of a trajectory is the same regardless which direction it is parametrized. It follows that Newton's force law is time-reversal invariant for forces that are given by time-reversal invariant vector fields; such is the case for example with conservative forces, which are given by the gradient of a (possibly time-dependent) potential. One such example is electrostatics where the electric field transforms as $\vec{E}\,'(\vec{x}\,',t') = \vec{E}(\vec{x},-t)$. On the other hand, we see from Figure 8.14 that if we perform a time-reversal transformation on a steady current loop, then the $\vec{B}$ field changes direction (where we are assuming that charge is invariant under time-reversal). This sign change together with the sign change of the velocity ensures that the Lorentz force is time-reversal invariant. Note $\vec{B}\,'(\vec{x}\,',t') = -\vec{B}(\vec{x},-t)$ does not mean $\vec{B}$ is an odd function of t. Rather, it means that if we reverse the motion, the corresponding fields will also be reversed. The general time reversal transformation for fields and sources in Maxwell's equations, as well as the Lorentz force law, is given by:

$$t' = -t, \quad \vec{v}\,' = -\vec{v}, \quad \rho'(\vec{x}\,',t') = \rho(\vec{x},-t), \quad \vec{J}\,'(\vec{x}\,',t') = -\vec{J}(\vec{x},-t), \tag{8.180}$$

for kinematical and source quantities and sources, as well as

$$\vec{E}\,'(\vec{x}\,',t') = \vec{E}(\vec{x},-t), \quad \vec{B}\,'(\vec{x}\,',t') = -\vec{B}(\vec{x},-t), \tag{8.181}$$

for fields. This transformation is checked in Exercise 8.8.2. The time reversal situation for Green functions following from the identity (8.98) is explored in Exercise 10.12.9.

8.9 Going Deeper

8.9.1 *Green functions*

(1) G. B. Arfken, H. J. Weber, F. E. Harris, *Mathematical Methods for Physicists*, 7th ed., Academic Press (Waltham, MA) 2013; Chapter 10.

(2) D. M. Cannell, *George Green: Mathematician and Physicist 1793–1841*, Athlone Press (London) 1993.

(3) D. G. Duffy, *Green's Functions with Applications*, Chapman & Hall/CRC (Boca Raton) 2001.

(4) P. M. Morse and H. Feshbach, *Methods of Theoretical Physics, Part I*, McGraw-Hill (New York) 1953; Chapter 7.

(5) G. F. Roach, *Green's Functions*, 2nd ed., Cambridge University Press (Cambridge) 1982.

(6) I. Stakgold and M. J. Holst, *Green's Functions and Boundary Value Problems*, 3rd ed., John Wiley & Sons (New York) 2011.

8.10 Exercises

Exercise 8.1.1.

(a) Apply (8.27) in the case of Ohm's law ($\vec{J} = \sigma\vec{E}$) at low frequencies where the conductivity, σ, is real, to the case of a wire of length l and cross-sectional area A to show that $R = \dfrac{l}{\sigma A}$.

(b) Given a small elementary capacitor of area A and plate separation d, show the electric field energy term in (8.27) gives rise to a capacitive impedance $\left(V = i\dfrac{I}{\omega C}\right)$ with $C = \dfrac{A}{4\pi d}$.

(c) Given an inductor with area A , length l, and N turns of wire, show the magnetic field energy term in (8.27) gives rise to an inductive impedance $(V = -i\omega LI)$ with $L = \dfrac{4\pi N^2 A}{lc^2}$.

Exercise 8.1.2. Consider a parallel plate capacitor subject to an harmonic input current, $I_i = Ie^{-i\omega t}$. The thin conducting plates are circular in shape with radius a and the plate separation is d. The region between the plates is vacuum. The z-axis goes through the center of the capacitor plates. At low frequencies ($\omega a/c \ll 1$) we have

$$\vec{E} = E_0 e^{-i\omega t}\,\hat{z},$$

where E_0 is a given constant. We are neglecting fringing fields.

(a) As the frequency is increased the fields inside the capacitor change. Now calculate the electric and magnetic fields inside the capacitor in powers of the angular frequency ω. Using the Maxwell equations, to lowest order in correction, show that inside we have

$$\vec{E} = \left(1 - \frac{\omega^2\rho^2}{4c^2}\right) E_0\, e^{-i\omega t}\,\hat{z},$$

$$\vec{B} = -\frac{i\omega\rho}{2c} E_0\, e^{-i\omega t}\,\hat{e}_\phi.$$

ρ is the radial distance variable in cylindrical coordinates.

(b) Assuming cylindrical symmetry construct the field equation that the harmonic electric field satisfies in source-free regions. Show that the solution in cylindrical coordinates gives rise to the Bessel equation. Then argue from the known zero frequency field $\vec{E}|_{\omega=0} = E_0\,\hat{z}$ that the electric and magnetic fields inside the capacitor are (take E_0 real)

$$\vec{E} = E_0\,\hat{z}\,e^{-i\omega t}J_0(k\rho), \vec{B} = -iE_0\,\hat{\phi}\,e^{-i\omega t}J_1(k\rho),$$

where $k = \omega/c$, and $J_0(k\rho)$, $J_1(k\rho)$ are the 0th and 1st order regular Bessel functions, respectively.

(c) Evaluate the input charge $Q_i(t) = Qe^{-i\omega t}$ where $I_i = \frac{dQ_i(t)}{dt}$ on the top plate of the capacitor. Show that the magnitude of the charge coefficient Q is related to the magnitude of the electric field by

$$|Q| = \frac{|E_0|}{2}\frac{a^2}{x}\,|J_1(x)|\,,$$

where $x \equiv ka$.

Exercise 8.1.3. In a situation where there is no free current is present, Equation (8.27) and the definition

$$Z(\omega) \equiv i\left(\frac{1}{\omega\,C(\omega)} - \omega L(\omega)\right),$$

for the impedance may be satisfied by

$$C(\omega) = \frac{|I|^2}{4\omega^2}\frac{1}{\displaystyle\int d^3x\,w_e}, \quad L(\omega) = \frac{4}{|I|^2}\int d^3x\,w_m.$$

Now using results for the capacitor from Exercise 8.1.2, show that the frequency-dependent capacitance and inductance are

$$C(\omega) = \frac{a^2}{d}\frac{J_1(x)^2}{x^2\left(J_0(x)^2 + J_1(x)^2\right)},$$

$$L(\omega) = \frac{d}{2c^2}\frac{2\left(J_0(x)^2 - \frac{2}{x}J_0(x)J_1(x) + J_1(x)^2\right)}{J_1(x)^2},$$

where $x = \dfrac{\omega a}{c}$. Also show the impedance is given very simply by

$$Z(\omega) = i\frac{2d}{a\,c}\frac{J_0(x)}{J_1(x)}.$$

Do a Taylor series in x to verify $C \simeq a^2/4d$ and $L \simeq d/2c^2$ to lowest order.

Extra: By expanding to the next highest power for these quantities, show that

$$C(\omega) \approx \frac{a^2}{4d\left(1 + \dfrac{x^4}{192}\right)}, \quad L(\omega) \approx \frac{d}{2c^2}\left(1 + \frac{x^2}{12}\right).$$

Completely unfair question that may be ignored: Prove that $C(\omega) < C(0)$ and $L(\omega) > L(0)$.

Exercise 8.1.4. Consider a long circular inductor of radius a, length $L \gg a$ with N turns of wire. The inductor is oriented along the z-axis. An harmonic current source sends a current $I_i(t) = Ie^{-i\omega t}$ through the closely spaced wires, giving rise to an approximately uniform magnetic field inside the inductor. At very low frequencies ($\omega a/c \ll 1$), the harmonic field is given approximately by

$$\vec{B} = B_0\, e^{-i\omega t}\,\hat{z}.$$

Note that B_0 is a function of I. We are neglecting fringing fields.

(a) As the frequency is increased the fields inside change. Now calculate the electric and magnetic fields inside the inductor in powers of the angular frequency ω. To lowest order in correction, show that inside the inductor we have

$$\vec{B} = \left(1 - \frac{\omega^2\rho^2}{4c^2}\right) B_0 e^{-i\omega t}\,\hat{z},$$
$$\vec{E} = \frac{i\omega\rho}{2c} B_0 e^{-i\omega t}\,\hat{e}_\phi.$$

(b) In Exercise 8.1.2 we learned that the $\vec{E}$ and $\vec{B}$ fields satisfy the Bessel function differential equation in free space. Given the setup of the problem, argue that the solution inside the solenoid is given in cylindrical coordinates as

$$\vec{B} = B_0 e^{-i\omega t}\,\hat{z}\, J_0(k\rho),$$
$$\vec{E} = iB_0 e^{-i\omega t}\,\hat{e}_\phi J_1(k\rho).$$

$J_0(k\rho)$ and $J_1(k\rho)$ are Bessel functions and $k = \omega/c$.

Exercise 8.2.1. For the free-space stress tenor, show that

$$\text{(a)}\ \sum_{k=1}^{3} T_{kk} = u, \qquad \text{(b)}\ \sum_{k,l} (T_{kl})^2 = 3u^2 - 2(c\vec{g})^2,$$

where $\vec{g}$ is the field momentum density (8.39) and u is the energy density (8.11).

Exercise 8.2.2. With $\vec{g}$ and u as in Exercise 8.2.1, show that

$$\det(T_{kl}) = u[(c\vec{g})^2 - u^2].$$

Write down as general a form as possible using a few unknown numerical coefficients. Use dimensional analysis assuming Lorentz and $\vec{E}, \vec{B}$ sign symmetries to determine which types of terms are present. Evaluate the coefficients by looking at special cases. Brute force is discouraged!

Exercise 8.2.3. In Section 3.4 we considered a neutral conducting sphere in a uniform electric field with the two halves of the sphere separated by a small air gap (see Figure 3.7). We found that the potential was

$$\Phi = -E_0\left(r - \frac{a^3}{r^2}\right)\cos\theta.$$

Use the stress tensor expression for the force

$$F_i = -\sum_j \oint_S da\, n_j T_{ij},$$

and adopt an enclosing volume around the top half of the sphere to calculate the force on the top half of the sphere.

Extra: Include a total charge Q on the sphere in the calculation and repeat the calculation.

Exercise 8.2.4. Reconsider the inductor from Exercise 8.1.4. It has a radius, a, number of turns of wire, N, and length $L \gg a$ and is subjected to an harmonic current $I_i(t) = I\, e^{-i\omega t}$. The solutions *outside* the inductor are (one actually requires outgoing wave boundary conditions for these types of Bessel functions; look ahead to Section 12.1)

$$\vec{B} = B_1 e^{-i\omega t}\hat{z}\, H_0^{(1)}(k\rho),$$
$$\vec{E} = iB_1 e^{-i\omega t}\hat{e}_\phi H_1^{(1)}(k\rho),$$

where B_1 is a constant and $H_0^{(1)}(k\rho)$ and $H_1^{(1)}(k\rho)$ are Hankel functions (see Equation (4.95)).

(a) Employing the force expression coming from the energy momentum tensor in Equation (8.44), show that the time-averaged force per unit area P (pressure) on the coil of wires which make up the inductor is

$$P^{avg} = \frac{1}{16\pi}\left(|B_z|^2\Big|_{a^-} - |B_z|^2\Big|_{a^+}\right).$$

The $a^-(a^+)$ notation indicates the evaluation is done slightly inside (outside) the surface.

(b) Using the fields, this evaluates to

$$P^{avg} = \frac{1}{16\pi}\left(|B_0|^2 J_0(x)^2 - |B_1|^2|H_0^{(1)}(x)|^2\right),$$

where $x \equiv ka$. Determine B_0 and B_1 using the electric and magnetic field boundary conditions at the surface of the inductor. Note that as $\omega \to 0$ one should find

$$B_1(\omega) \to 0, \;\; B_0(\omega) \to \frac{4\pi NI}{Lc}.$$

Note also that the *total* force on the wires of course is zero by symmetry.

Exercise 8.3.1. The static electromagnetic force and torque (with respect to the origin) on an object are given by

$$F_i = -\sum_j \oint_S da\, n_j T_{ij}, \qquad N_i = -\sum_j \oint_S da\, n_j M_{ij},$$

where $M_{ij} \equiv \sum_{k,l} \epsilon_{ikl} x_k T_{lj}$.

(a) Prove that if the magnetic and/or electric field is constant (in space and time) at all points on the surface of an object, then the net force on the object is zero.

(b) Consider the same question with torque instead of force: that is, given constant electric, magnetic fields on an arbitrarily shaped closed surface, find out whether or not the net torque on the object vanishes.

Exercise 8.3.2. Using the expressions for $\vec{F}$ and $\vec{N}$ in Exercise 8.3.1, compute (a) the force and (b) the torque on a sphere of radius a centered at the origin if the magnetic field just outside its surface is given by

$$\vec{B} = \frac{3\hat{n}(\hat{n}\cdot\vec{m}) - \vec{m}}{a^3} + \vec{B}_0,$$

where $\vec{B}_0$ is a constant vector, $\hat{n}$ is the unit vector along $\vec{r}$, and $\vec{m}$ is another (arbitrary) constant vector. Try to avoid brute force. Show that $\vec{F} = 0$ but that $\vec{N} = \vec{m} \times \vec{B}_0$. [*Hints*: Physical intuition is useful, as are the results from Exercise 8.3.1. Some surface integrals involving combinations of normal unit vectors such as n_i, $n_i n_j$, $n_i n_j n_k$, and so on can be worked out with Gauss's theorem, but many such terms are zero.]

Exercise 8.3.3.

(a) Given that the Maxwell stress tensor T_{ij} is symmetric, show that the angular momentum conservation equation (8.55) may be written in the form (8.58), where the shear tensor M_{ij} is given by (8.59).

(b) The momentum conservation equation (8.42) remains true when T_{ij} is replaced by $T_{ij} + \tau_{ij}$ as long as $\sum_j \nabla_j \tau_{ij} = 0$. Let $\tau_{ij}^{(\mathrm{a})}$ be the antisymmetric part of τ_{ij}:

$$\tau_{ij}^{(\mathrm{a})} \equiv \frac{\tau_{ij} - \tau_{ij}^T}{2}.$$

Suppose that $\tau_{ij}^{(\mathrm{a})}$ can be expressed as a pure divergence, that is:

$$\tau_{ij}^{(\mathrm{a})} \equiv \sum_k \nabla_k \mathcal{F}_{kij}.$$

Show that a conservation of angular momentum equation analogous to (8.58) holds using the modified shear tensor M'_{ij} where

$$M'_{ij} = \sum_{k,l} \epsilon_{ikl} \left(x_k (T_{lj} + \tau_{lj}) - \mathcal{F}_{jlk} \right).$$

Thus there is a freedom in the choice of T_{ij} and M_{ij} which preserves the physical conservation equations.

Exercise 8.4.1. Using the macroscopic Maxwell equations show that the modified form of the harmonic energy law for linear, isotropic materials is given by Equation (8.73) with definitions (8.74)–(8.76).

Exercise 8.6.1. Find the solution of the free, one-dimensional wave equation Green function,

$$\left(\frac{\partial^2}{\partial x^2} - \frac{1}{c^2} \frac{\partial^2}{\partial t^2} \right) G(x,t;x',t') = -4\pi \delta(x - x') \delta(t - t').$$

(the 4π is optional) with retarded time boundary conditions,

$$G(x,t;x',t') = 0 \quad \text{for } t < t'.$$

[*Hint*: A dynamical step function will emerge.]

Exercise 8.6.2. Find and solve for the one-dimensional *massive* retarded Green function. The Green function wave equation is

$$\left(\frac{\partial^2}{\partial x^2} - \frac{1}{c^2}\frac{\partial^2}{\partial t^2} - \mu^2\right) G_\mu(x,t;x',t') = -4\pi\delta(t-t')\delta(x-x'),$$

($\mu^2 = \dfrac{m^2c^2}{h^2}$) subject to retarded time boundary conditions. (An integral table may be useful. A step function and a Bessel function emerge this time.)

Exercise 8.6.3.

(a) By considering wave equations, show that the two dimensional Green function is given by

$$G^{2D}(\vec{x}_\perp,t;\vec{x}'_\perp,t') = \int_{-\infty}^{\infty} dz\, G^{3D}(\vec{x},t;\vec{x}',t'),$$

up to a possible overall multiplicative constant.

(b) Given that

$$G^{3D}(\vec{x},t;\vec{x}',t') = \frac{\delta(\tau - \frac{R}{c})}{R}$$

where $R \equiv |\vec{x} - \vec{x}'|$, $\tau \equiv t - t'$, carry out the integration in (a) to find $G^{2D}(\vec{x}_\perp,t;\vec{x}'_\perp,t')$.

Exercise 8.6.4.

(a) Show that the retarded three dimensional Green function,

$$G^{3D}(\vec{x},t;\vec{x}',t') = \frac{\delta(\tau - \frac{R}{c})}{R},$$

($R = |\vec{x} - \vec{x}'|$, $\tau = t - t'$) satisfies

$$\left(\vec{\nabla}^2 - \frac{1}{c^2}\frac{\partial^2}{\partial t^2}\right) G^{3D}(\vec{x},t;\vec{x}',t') = -4\pi\delta^{(3)}(\vec{x} - \vec{x}')\delta(t-t'),$$

directly by carrying out the specified derivatives. (You may choose $\vec{x}' = t' = 0$ for simplicity.)

(b) By using the integral form or by other means, prove (or argue) that

$$\lim_{(t-t')\to 0^+} G^{3D}(\vec{x},t;\vec{x}',t') = 0,$$

$$\lim_{(t-t')\to 0^+} \frac{\partial}{\partial t'} G^{3D}(\vec{x},t;\vec{x}',t') = -4\pi c^2 \delta^{(3)}(\vec{x}-\vec{x}'),$$

where $\delta^{(3)}(\vec{x}-\vec{x}')$ is the three-dimensional spatial function. (The second result is the analog of the surface delta function we talked about in Section 2.8 for the electrostatic case.)

Exercise 8.6.5. Consider the Green function for the one-dimensional Schrödinger equation

$$\left(\frac{\partial^2}{\partial x^2} + ia\frac{\partial}{\partial t}\right) G_S(x,t;x',t') = -4\pi\delta(x-x')\delta(t-t'),$$

where a is a positive constant.

(a) Construct the explicit form of the Green function by contour integration. Obtain the retarded solution (given in many books):

$$G_S^{\text{ret}}(x,t;x',t') = 2i\sqrt{\frac{\pi}{ia(t-t')}}\exp\left(\frac{ia(x-x')^2}{4(t-t')}\right) H(t-t').$$

(b) Show (in analogy to Exercise 8.6.4(b) above) that

$$\lim_{(t-t')\to 0^+} G_S^{\text{ret}}(x,t;x',t') = \frac{4\pi i}{a}\delta(x-x').$$

Extra: Similarly consider the retarded Green function for the *three-dimensional* Schrödinger equation

$$\left(\vec{\nabla}^2 + ia\frac{\partial}{\partial t}\right) G_{3S}^{\text{ret}}(\vec{x},t;\vec{x}',t') = -4\pi\,\delta(\vec{x}-\vec{x}')\delta(t-t'),$$

where a is a positive constant. Show that the solution is

$$G_{3S}^{\text{ret}}(\vec{x},t;\vec{x}',t') = \frac{1}{2}\sqrt{\frac{a}{i\pi(t-t')^3}}\exp\left(\frac{ia}{4}\frac{|\vec{x}-\vec{x}'|^2}{(t-t')}\right) H(t-t').$$

Exercise 8.6.6. Show that the free Green functions for the wave equation may also be written ($R = |\vec{x}-\vec{x}\,'|$)

$$G^{\text{ret}}(\vec{x},t;\vec{x}',t') = \frac{2}{c}H(t-t')\,\delta\left(\frac{R^2}{c^2} - (t-t')^2\right),$$

$$G^{\text{adv}}(\vec{x},t;\vec{x}',t') = \frac{2}{c}H(t'-t)\,\delta\left(\frac{R^2}{c^2} - (t-t')^2\right),$$

where $H(x)$ is the *Heaviside step function*

$$H(x) = \begin{cases} 1, & x > 0, \\ 0, & x < 0. \end{cases}$$

Exercise 8.6.7. The charge and current densities given by ($\vec{v} \equiv d\vec{x}(t)/dt$)

$$\begin{aligned} \rho(\vec{x}, t) =& e\, \delta(\vec{x} - \vec{x}(t)), \\ \vec{J}(\vec{x}, t) =& e\, \vec{v}(t)\delta(\vec{x} - \vec{x}(t)), \end{aligned}$$

represent a moving point charge. Derive the form of the corresponding *Liénard-Wiechert potentials*,

$$\begin{aligned} \Phi(\vec{x}, t) &= \frac{e}{|\vec{x} - \vec{x}(t_R)| - (\vec{x} - \vec{x}(t_R)) \cdot \dfrac{\vec{v}(t_R)}{c}}, \\ \vec{A}(\vec{x}, t) &= \frac{\vec{v}(t_R)}{c} \Phi(\vec{x}, t), \end{aligned}$$

where the *retarded time* t_R is defined recursively as

$$t_R \equiv t - \frac{|\vec{x} - \vec{x}(t_R)|}{c}.$$

Exercise 8.6.8.

(a) Given a charge e in uniform motion along a straight line, so that the charge and current density are given as in the previous problem with $\vec{v}(t)$ replaced by a constant vector $\vec{v}$. Show that the scalar and vector potentials are given by

$$\Phi(\vec{x}, t) = \frac{e}{\sqrt{(\vec{x} - \vec{v}t)^2 + ((\vec{x} \cdot \vec{v})^2 - x^2 v^2)/c^2}},$$

$$\vec{A}(\vec{x}, t) = \frac{\vec{v}}{c} \Phi(\vec{x}, t).$$

(b) Obtain exact expressions for both $\vec{E}(\vec{x}, t)$ and $\vec{B}(\vec{x}, t)$, and show that

$$\vec{B}(\vec{x}, t) = \frac{\vec{v}}{c} \times \vec{E}(\vec{x}, t).$$

Exercise 8.6.9.

(a) Consider a time-dependent point electric dipole, $\vec{p}(t)$ located at the origin. Write down the dipole's charge density $\rho^p(\vec{x},t)$ and current density, $\vec{J}^p(\vec{x},t)$. Evaluate the electric and magnetic fields, $\vec{E}^p$ and $\vec{B}^p$, from the retarded propagator. [*Ans.*:

$$\vec{E}^p = \frac{3\vec{x}\,(\vec{p}\cdot\vec{x})}{|\vec{x}|^5} + \frac{3\vec{x}\,(\dot{\vec{p}}\cdot\vec{x})}{c|\vec{x}|^4} + \frac{\vec{x}\,(\ddot{\vec{p}}\cdot\vec{x})}{c^2|\vec{x}|^3} - \frac{\vec{p}}{|\vec{x}|^3} - \frac{\dot{\vec{p}}}{c|\vec{x}|^2} - \frac{\ddot{\vec{p}}}{c^2|\vec{x}|} - \frac{4\pi}{3}\vec{p}\,\delta(\vec{x}),$$

$$\vec{B}^p = \frac{\ddot{\vec{p}}\times\vec{x}}{c^2|\vec{x}|^2} + \frac{\dot{\vec{p}}\times\vec{x}}{c|\vec{x}|^3},$$

where $\vec{p}$ means $\vec{p}(t - \frac{|\vec{x}|}{c})$.]

(b) Consider a time-dependent point magnetic dipole, $\vec{m}(t)$ located at the origin. Write down the source's charge density, $\rho^m(\vec{x},t)$, and current density, $\vec{J}^m(\vec{x},t)$. Explicitly evaluate the scalar and vector potentials from the retarded propagator. [*Note*: although this part doesn't require that you find the electric and magnetic fields, there is actually an easy way to find them by applying duality to the electric dipole solutions.]

Exercise 8.7.1. Referring to Exercise 8.6.1, find the appropriate retarded one-dimensional solution for the $x > 0$ region when there is a "wall" with Dirichlet boundary conditions at $x = 0$.

Exercise 8.7.2. The three-dimensional free Green function satisfies,

$$\left(\vec{\nabla}^2 - \frac{1}{c^2}\frac{\partial^2}{\partial t^2}\right) G(\vec{x},t;\vec{x}\,',t') = -4\pi\delta(\vec{x}-\vec{x}\,')\delta(t-t').$$

(a) Show that the free Green function (either retarded or advanced), may be written using positive ω's only as

$$G^{\text{ret,adv}}(\vec{x},t;\vec{x}\,',t') = 8\pi\text{Re}\left(\int_0^\infty \frac{d\omega}{2\pi} e^{-i\omega(t-t')} G^{\text{ret,adv}}(\omega)\right),$$

$$G^{\text{ret,adv}}(\omega) = \int \frac{d^3k}{(2\pi)^3} \frac{e^{i\vec{k}\cdot(\vec{x}-\vec{x}\,')}}{\vec{k}^2 - (\frac{\omega}{c} \pm i\epsilon)^2},$$

where ϵ is a small positive quantity, and the +(-) sign is for retarded (advanced).

(b) Assuming a solution of the form,

$$G^{\text{ret,adv}}(\vec{x},t;\vec{x}\,',t') = 4\pi \int \frac{d^3k}{(2\pi)^3} e^{i\vec{k}\cdot(\vec{x}-\vec{x}\,')} g^{\text{ret,adv}}(k;t,t'),$$

($k \equiv |k|$) for the free Green functions, solve for $g^{\text{ret,adv}}(k;t,t')$.

Exercise 8.7.3. A unit point source acts at a time, t', and at a distance, $\vec{x}' = (0,0,z')$, from one side of a pair of perfectly conducting parallel plates, as shown in Figure 8.15. Find the solution of the Dirichlet wave equation Green function, $G_{||}(\vec{x},t;\vec{x}\,',t')$, for this geometry as a series of image sources. Characterize where and when the image sources act.

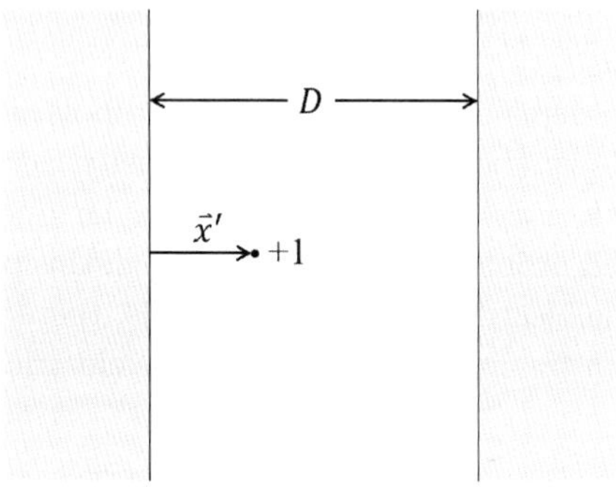

Fig. 8.15 Geometry for Exercise 8.7.3.

Exercise 8.7.4. We applied Dirichlet boundary conditions, Equation (8.118), in deriving the three-dimensional Green function $G_D(\vec{x},t;\vec{x}',t')$, Equation (8.146), for infinite plane geometry. Now apply the Neumann boundary condition,

$$\left.\frac{\partial G_N(\vec{x},t;\vec{x}',t')}{\partial z}\right|_{z=0} = 0,$$

for the same plane geometry to find the three-dimensional Neumann Green function, $G_N(\vec{x},t;\vec{x}',t')$.

Exercise 8.7.5.

(a) Solve for the one-dimensional Green function, $G^{2m}(x,t;x',t')$, which satisfies the wave equation

$$\left(\frac{d^2}{dx^2} - \frac{1}{c^2}\frac{\partial^2}{\partial t^2}\right) G^{2m}(x,t;x',t') = -\delta(x-x')\delta(t-t'),$$

where

$$G^{2m}(x,t;x',t') = \int_{-\infty}^{\infty} \frac{d\omega}{2\pi}\, e^{-i\omega(t-t')} g^{2m}(x,x';\omega),$$

for the region between two mirrors at $x = 0$ and $x = d$. From the differential equation for $g^{2m}(x,x';\omega)$ and the Dirichlet boundary conditions $g^{2m}(0,x';\omega) = 0$, $g^{2m}(d,x';\omega) = 0$ on the mirrors, show that

$$g^{2m}(x,x';\omega) = \frac{\sin(kx_<)\sin(k(d-x_>))}{k\sin(kd)},$$

where $x_<$ ($x_>$) is the lesser (greater) of x and x' and $k \equiv \omega/c$.

(b) Using the Green function rule $k \to k + i\epsilon$ allows one to do a Taylor series expansion of $1/\sin(kd)$. Doing the ω integrals, show the multiple reflection expansion result

$$\begin{aligned} &G^{2m}(x,t;x',t') = \\ &\frac{c}{2}\sum_{n=0}^{\infty}\Bigg[H\left(t-t'-\frac{1}{c}\left(|x-x'|+2dn\right)\right) - H\left(t-t'-\frac{1}{c}\left(x+x'+2dn\right)\right) \\ &\qquad +H\left(t-t'-\frac{1}{c}\left(-|x-x'|+2d(n+1)\right)\right) \\ &\qquad -H\left(t-t'-\frac{1}{c}\left(-x-x'+2d(n+1)\right)\right)\Bigg], \end{aligned}$$

where $H(x)$ is the Heaviside step function. A single step function is seen in the one-dimensional free Green function in Exercise 8.6.1. The implied image sources giving rise to the infinite additional terms are the same as in Exercise 8.7.3.

Exercise 8.8.1.

(a) Starting with the definition of the determinant of a 3×3 matrix,

$$\det a = \sum_{l,m,n} \epsilon_{lmn} a_{1l} a_{2m} a_{3n}$$

show that

$$\sum_{j,k} \epsilon_{ijk} a_{jl} a_{km} = \det a \sum_{n} \epsilon_{nlm} a_{in}$$

where a_{ij} are elements of a general orthogonal transformation.

(b) Use (a) to prove that $\vec{\nabla} \times \vec{A}(\vec{x})$ produces a pseudovector if $\vec{A}(\vec{x})$ transforms as a vector field.

(c) Show that $\vec{\nabla} \cdot \vec{A}(\vec{x})$ transforms as a scalar.

(d) Show that $\vec{\nabla}\Phi(\vec{x})$ transforms as a vector, if Φ is a scalar field.

Exercise 8.8.2. Check the form invariance of Maxwell's equations under the time-reversal transformations (8.180) and (8.181). If magnetic charges and currents are added, as in Equations (7.193) and (7.194), how do these quantities transform?

Chapter 9

Plane Electromagnetic Waves and Propagation in Matter

Now we begin to get to the heart of the matter, literally. We use the formalism in the previous chapters to build up an understanding of electromagnetic interactions in matter. We will depart in many instances from exact approaches in order to simplify and understand, although some results, such as the Kramers-Kronig relations do form a rigorous background. The internal model introduced will be summarized in macroscopic constitutive relations, which will be remarkably simple, but effective. Applications are many: surface interactions, plasmas, metals, wave propagation and energy loss. Our basic tool for this investigation will be the plane wave, which we will now study.

9.1 Plane waves in dielectric media

We will now consider plane waves in a nonconducting (dielectric) medium. For now, we will assume that ϵ and μ are true constants. Since we have no sources, the relevant Maxwell equations are:

$$\vec{\nabla} \cdot \vec{E} = 0, \tag{9.1}$$

$$\vec{\nabla} \times \vec{E} + \frac{1}{c}\frac{\partial \vec{B}}{\partial t} = 0, \tag{9.2}$$

$$\vec{\nabla} \cdot \vec{B} = 0, \tag{9.3}$$

$$\vec{\nabla} \times \vec{B} - \frac{\mu\epsilon}{c}\frac{\partial \vec{E}}{\partial t} = 0, \tag{9.4}$$

where (9.4) is a modified version of the macroscopic Ampère-Maxwell equation (7.97) in a uniform medium.

As before, we decouple the $\vec{E}$ and $\vec{B}$ equations by taking derivatives:

$$\vec{\nabla} \times \left(\vec{\nabla} \times \vec{E} + \frac{1}{c}\frac{\partial \vec{B}}{\partial t} \right) = 0$$

$$\Longrightarrow \vec{\nabla}(\underbrace{\vec{\nabla} \cdot \vec{E}}_{=0}) - \nabla^2 \vec{E} + \frac{1}{c}\frac{\partial}{\partial t} \underbrace{\vec{\nabla} \times \vec{B}}_{\frac{\mu\epsilon}{c}\frac{\partial \vec{E}}{\partial t}} = 0$$

$$\Longrightarrow \left(\nabla^2 - \frac{\mu\epsilon}{c^2}\frac{\partial^2}{\partial t^2} \right) \vec{E} = 0. \tag{9.5}$$

In the same way, we may derive

$$\left(\nabla^2 - \frac{\mu\epsilon}{c^2}\frac{\partial^2}{\partial t^2} \right) \vec{B} = 0. \tag{9.6}$$

Equations (9.5) and (9.6) are modified wave equations for each component of the electric and magnetic fields, respectively. The speed of propagation is $c/\sqrt{\mu\epsilon}$, which may be confirmed by computing retarded Green function solutions analogous to (8.111). Properties associated with these fields such as energy and momentum density also propagate with the same speed. However, the assumption that ϵ and μ are constants independent of wave frequency (and thus with no time dependence) is unrealistic. The upcoming chapter will give a more practical interpretation of these equations with less restrictive assumptions.

Equations (9.5) and (9.6) have plane wave solutions,

$$\vec{E} = \mathrm{Re}\left(\vec{\mathcal{E}}\, e^{i\vec{k}\cdot\vec{x} - i\omega t} \right); \quad \vec{B} = \mathrm{Re}\left(\vec{\mathcal{B}}\, e^{i\vec{k}\cdot\vec{x} - i\omega t} \right), \tag{9.7}$$

where $\vec{\mathcal{E}}, \vec{\mathcal{B}}$ are complex vectors. In the following, we will see that new wave equations replacing (9.5) and (9.6) apply to plane waves where μ and ϵ are allowed to vary with angular frequency, ω. These waves travel in the $\vec{k}$ direction, and the magnitude of $\vec{k}$ or *wave number* k is related to ω via the *dispersion relation*

$$k(\omega) = \sqrt{\mu(\omega)\epsilon(\omega)}\,\frac{\omega}{c}. \tag{9.8}$$

We identify surfaces of constant phase through the equation

$$\vec{k} \cdot \vec{x} - \omega t = \text{constant},$$

$$\Longrightarrow \vec{k} \cdot \frac{d\vec{x}}{dt} = \omega. \tag{9.9}$$

Defining the unit vector $\hat{k} \equiv \vec{k}/k$, and letting $\vec{x}$ move along $\hat{k}$ while maintaining constant phase so that $\vec{x} \equiv v_{\text{phase}} t\,\hat{k}$, we have the *phase velocity*

$$v_{\text{phase}} \equiv \frac{\omega}{k(\omega)} = \frac{c}{n(\omega)}, \tag{9.10}$$

where $n(\omega)$ is the (frequency-dependent) *index of refraction*,

$$n(\omega) \equiv \sqrt{\mu(\omega)\epsilon(\omega)}. \tag{9.11}$$

The phase velocity appears to be a natural generalization of the propagation speed described earlier. However, we will see in Section 9.8 that it can be different from signal propagation speed. In fact, the phase velocity can be greater than c if $\mu = 1$ and $\epsilon < 1$, as we will see in Section 9.6.

From $\vec{\nabla} \cdot \vec{E} = 0$ we get

$$\vec{\nabla} \cdot \vec{E} = \text{Re}\left(i\vec{\mathcal{E}} \cdot \vec{k}\, e^{i\vec{k}\cdot\vec{x} - i\omega t}\right) = 0 \quad \Longrightarrow \vec{\mathcal{E}} \cdot \hat{k} = 0. \tag{9.12}$$

From $\vec{\nabla} \times \vec{E} = -\frac{1}{c}(\partial \vec{B}/\partial t)$ we also get

$$\vec{\mathcal{B}} = n(\omega)\,\hat{k} \times \vec{\mathcal{E}} \quad \Longrightarrow \vec{\mathcal{B}} \cdot \hat{k} = 0. \tag{9.13}$$

We find that the oscillations of the $\vec{E}$ and $\vec{B}$ fields are *transverse* to the direction of motion. Since there are two directions perpendicular to a given $\vec{k}$, there can be two linearly independent states of $\vec{E}$ or $\vec{B}$.

To construct a general plane wave, we first choose any two unit vectors $\hat{\epsilon}_{1,2}$ satisfying $\hat{\epsilon}_1 \cdot \hat{\epsilon}_2 = 0$. Defining $\hat{k} \equiv \hat{\epsilon}_1 \times \hat{\epsilon}_2$, we may write the following plane wave solution:

$$\vec{E}_j = \text{Re}\left(\mathcal{E}_j \hat{\epsilon}_j e^{i\vec{k}\cdot\vec{x} - i\omega t}\right) \qquad (j = 1, 2), \tag{9.14}$$

$$\vec{B}_j = \sqrt{\mu\epsilon}\,\hat{k} \times \vec{E}_j, \tag{9.15}$$

with arbitrary complex numbers $\mathcal{E}_1, \mathcal{E}_2$. Then the general form for a plane wave with propagation direction $\hat{k}$ can be written

$$\vec{E}(\vec{x}, t) = \sum_{j=1,2} \vec{E}_j(\vec{x}, t), \tag{9.16}$$

$$\vec{B}(\vec{x}, t) = \sum_{j=1,2} \vec{B}_j(\vec{x}, t). \tag{9.17}$$

In the case where $\mathcal{E}_{1,2}$ are real, then $\vec{E}(\vec{x}, t)$ oscillates linearly along the direction $\mathcal{E}_1 \hat{\epsilon}_1 + \mathcal{E}_2 \hat{\epsilon}_2$, as shown in Figure 9.1.

On the other hand, a complex phase difference between $\mathcal{E}_1$ and $\mathcal{E}_2$ translates into a phase difference in time between the two components, and the

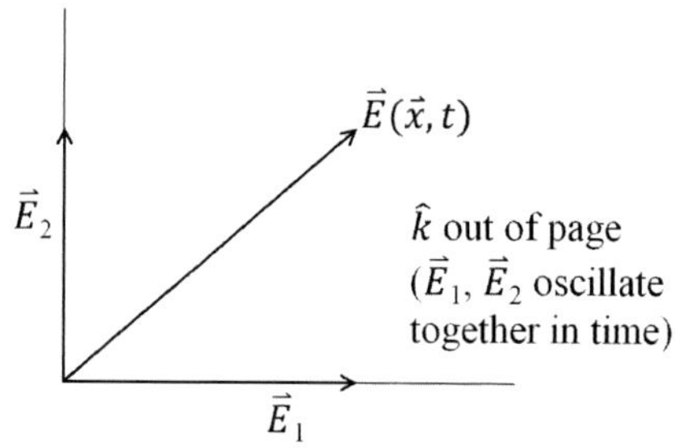

Fig. 9.1 Linearly-polarized electromagnetic wave.

resulting $\vec{E}(\vec{x},t)$ can change in both magnitude and direction at a given $\vec{x}$, or along a given line in space at fixed t. To understand this variation more clearly, we introduce another orthonormal basis with the same span as $\hat{\epsilon}_1, \hat{\epsilon}_2$:

$$\hat{\epsilon}_\pm \equiv \frac{1}{\sqrt{2}} \left(\hat{\epsilon_1} \pm i\hat{\epsilon}_2\right). \tag{9.18}$$

The vectors $\hat{\epsilon}_+, \hat{\epsilon}_-$ are orthonormal in a complex sense:

$$\hat{\epsilon}_+^* \cdot \hat{\epsilon}_- = \hat{\epsilon}_-^* \cdot \hat{\epsilon}_+ = 0, \tag{9.19}$$

$$\hat{\epsilon}_+^* \cdot \hat{\epsilon}_+ = \hat{\epsilon}_-^* \cdot \hat{\epsilon}_- = 1. \tag{9.20}$$

We may rewrite the plane wave electric field $\vec{E}$ in terms of this new orthonormal basis as

$$\vec{E}(\vec{x},t) = \vec{E}_+ + \vec{E}_-, \tag{9.21}$$

where

$$\vec{E}_\pm = \mathrm{Re}\left(\mathcal{E}_\pm \hat{\epsilon}_\pm e^{ik\cdot\vec{x}-i\omega t}\right). \tag{9.22}$$

To clarify this new representation, we consider the case where $\mathcal{E}_\pm$ are both real. Then we have

$$\vec{E}_+ = \frac{\mathcal{E}_+}{\sqrt{2}}(\hat{\epsilon}_1 \cos(\vec{k}\cdot\vec{x}-\omega t) - \hat{\epsilon}_2 \sin(\vec{k}\cdot\vec{x}-\omega t)), \tag{9.23}$$

$$\vec{E}_- = \frac{\mathcal{E}_-}{\sqrt{2}}(\hat{\epsilon}_1 \cos(\vec{k}\cdot\vec{x}-\omega t) + \hat{\epsilon}_2 \sin(\vec{k}\cdot\vec{x}-\omega t)). \tag{9.24}$$

The time variations of $\vec{E}_+$ and $\vec{E}_-$ at a fixed spatial location are shown in Figure 9.2.

From the quantum-mechanical point of view, the fact that only two independent polarizations of the electric (and magnetic) fields are possible

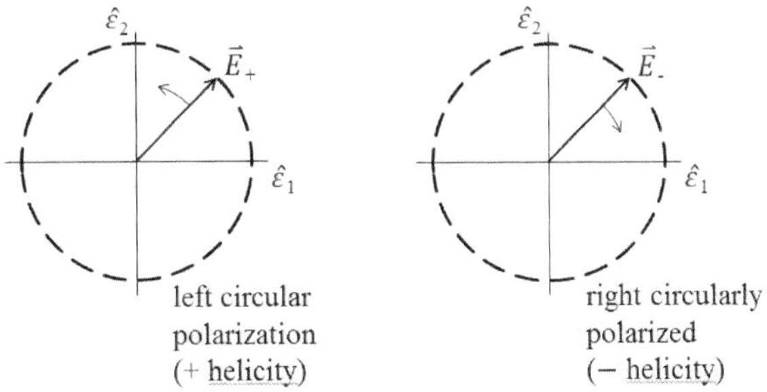

Fig. 9.2 Counterclockwise and clockwise polarizations of plane electromagnetic waves. The direction of propagation $\hat{k}$ points out of the page.

is due to the photon nature of light. It turns out that photon spin can only point parallel ($\hat{\epsilon}_+$ polarization) or antiparallel ($\hat{\epsilon}_-$ polarization) to $\hat{k}$; in particle physics language, these are referred to as positive and negative helicity states, respectively. In optics these two possibilities are referred to as left- and right-circular polarization, again respectively. The general state of polarization when the $\mathcal{E}_\pm$ are complex is known as elliptical polarization because the tips of the electric and magnetic field vectors describe fixed elliptical forms.

9.2 Reflection and refraction of plane waves from dielectric interfaces I: $E_\perp$ polarization

We now consider the reflection and refraction of plane waves at a plane interface between dielectrics.[1] We will use parts of this model again to find the fields inside an imperfect conductor in the next chapter.

Figure 9.3 shows the interface between the two dielectrics, and the notation we'll be using. In particular, the normal vector $\hat{n}$ is along the z-axis, and $\vec{k}_\perp$ is the component of $\vec{k}$ which is perpendicular to $\hat{n}$. We will assume that the sources for waves are in region 2 ($z > 0$), and that dielectric constants satisfy

$$\epsilon(z,\omega) = \begin{cases} \epsilon_2(\omega), \, z > 0 \\ \epsilon_1(\omega), \, z < 0. \end{cases} \tag{9.25}$$

[1]This section is adapted from J. Schwinger *et al.*, *Classical Electrodynamics*, Perseus Books (Reading, MA) 1998, Chapter 41.

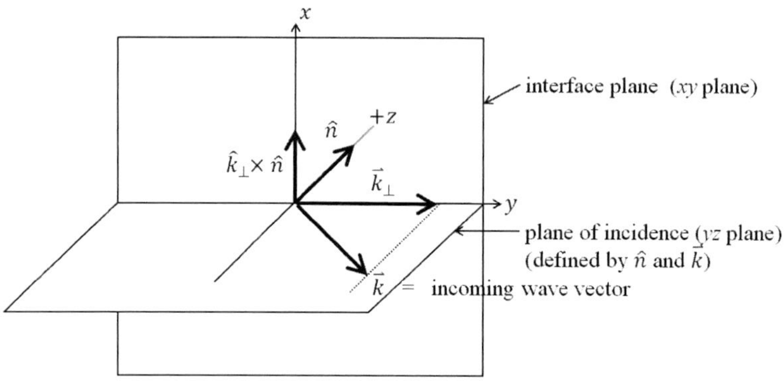

Fig. 9.3 Interface between dielectrics: notation.

Thus we are assuming linear, homogeneous media whose dielectric properties depend on frequency. Such media are called *dispersive*.

We may express any vector field as a linear combination of the vectors $\hat{n}, \hat{k}_\perp$, and $\hat{k}_\perp \times \hat{n}$ ($\vec{k}_\perp$ is perpendicular to $\hat{n}$). For example, we may write the electric field $\vec{E}$ as

$$\vec{E} = E_z\hat{n} + E_{||}\hat{k}_\perp + E_\perp(\hat{k}_\perp \times \hat{n}). \tag{9.26}$$

The notation here is tricky but logical: $E_\perp$ is the component of $\vec{E}$ perpendicular to the plane of incidence (defined by $\vec{k}$ and $\hat{n}$); while $E_{||}$ is the component parallel to the same plane. Note that E_z, $E_{||}$, and $E_\perp$ are not independent components, because the three are related via $\vec{E} \cdot \vec{k} = 0$.

The relevant Maxwell equations are:

$$\vec{\nabla} \times \vec{H} = \frac{1}{c}\frac{\partial \vec{D}}{\partial t} + \frac{4\pi}{c}\vec{J}, \tag{9.27}$$

$$-\vec{\nabla} \times \vec{E} = \frac{1}{c}\frac{\partial \vec{B}}{\partial t}. \tag{9.28}$$

We will concentrate on examining the behavior of electromagnetic waves of each specific frequency near the interface (assuming steady-state conditions).

Taking Fourier transforms of (9.27) and (9.28) using the formula

$$\int_{-\infty}^{\infty} dt\, e^{i\omega t} F(\vec{x}, t) \equiv F(\vec{x}, \omega), \tag{9.29}$$

(note $F^*(\vec{x},-\omega)=F(\vec{x},\omega)$ when ω is real), we obtain

$$\vec{\nabla}\times\vec{H}=-\frac{i\omega}{c}\vec{D}+\frac{4\pi}{c}\vec{J}, \tag{9.30}$$

$$\vec{\nabla}\times\vec{E}=\frac{i\omega}{c}\vec{B}, \tag{9.31}$$

where all fields and the current density in (9.30) and (9.31) are functions of $(\vec{x},\omega)$. We also assume $\mu=1$ (though in Exercise 9.3.1 and 9.3.2 we remove this assumption). Assuming that $\epsilon(z,\omega)$ is real, we obtain

$$\vec{D}(\vec{x},\omega)=\epsilon(z,\omega)\vec{E}(\vec{x},\omega), \tag{9.32}$$

$$\vec{H}(\vec{x},\omega)=\vec{B}(\vec{x},\omega). \tag{9.33}$$

At this point we apply Fourier transforms in coordinate space, making use of the translational independence of the fields in the x- and y-directions:

$$\int d^2x_\perp e^{-i\vec{k}_\perp\cdot\vec{x}_\perp}F(\vec{x},\omega)\equiv F(z,\vec{k}_\perp,\omega). \tag{9.34}$$

In the Maxwell equations, this means that in (9.30) and (9.31) the $\vec{\nabla}$ operator may be replaced everywhere with

$$\vec{\nabla}\to i\vec{k}_\perp+\hat{n}\frac{\partial}{\partial z}, \tag{9.35}$$

leading to

$$\begin{aligned}\left(i\vec{k}_\perp+\hat{n}\frac{\partial}{\partial z}\right)&\times\left(B_z\hat{n}+B_{||}\hat{k}_\perp+B_\perp(\hat{k}_\perp\times\hat{n})\right)\\ &=\frac{-i\omega\epsilon}{c}\left(E_z\hat{n}+E_{||}\hat{k}_\perp+E_\perp(\hat{k}_\perp\times\hat{n})\right)\\ &\quad+\frac{4\pi}{c}\left(J_z\hat{n}+J_{||}\hat{k}_\perp+J_\perp(\hat{k}_\perp\times\hat{n})\right),\end{aligned} \tag{9.36}$$

$$\begin{aligned}-\left(i\vec{k}_\perp+\hat{n}\frac{\partial}{\partial z}\right)&\times\left(E_z\hat{n}+E_{||}\hat{k}_\perp+E_\perp(\hat{k}_\perp\times\hat{n})\right)\\ &=\frac{-i\omega}{c}\left(B_z\hat{n}+B_{||}\hat{k}_\perp+B_\perp(\hat{k}_\perp\times\hat{n})\right).\end{aligned} \tag{9.37}$$

Upon matching components, (9.36) breaks down to:

$$\hat{n}:\ k_\perp B_\perp=\frac{\omega\epsilon}{c}E_z+\frac{4\pi i}{c}J_z, \tag{9.38}$$

$$\hat{k}_\perp:\ \frac{\partial}{\partial z}B_\perp=-\frac{i\omega\epsilon}{c}E_{||}+\frac{4\pi}{c}J_{||}, \tag{9.39}$$

$$\hat{k}_\perp\times\hat{n}:\ \frac{\partial}{\partial z}B_{||}-iB_zk_\perp=\frac{i\omega\epsilon}{c}E_\perp-\frac{4\pi}{c}J_\perp, \tag{9.40}$$

which are sufficient to specify $\vec{E}$ if $\vec{B}$ and $\vec{J}$ are known. Similarly, (9.37) gives:

$$\hat{n}: \; k_\perp E_\perp = -\frac{\omega}{c} B_z, \tag{9.41}$$

$$\hat{k}_\perp: \; \frac{\partial}{\partial z} E_\perp = \frac{i\omega}{c} B_{||}, \tag{9.42}$$

$$\hat{k}_\perp \times \hat{n}: \; \frac{\partial}{\partial z} E_{||} - iE_z k_\perp = -\frac{i\omega}{c} B_\perp, \tag{9.43}$$

which specify $\vec{B}$ if $\vec{E}$ is known. We decouple the equations by taking z derivatives and substituting, thus obtaining second-order equations. For instance, using (9.40),(9.41), and (9.42) we obtain an equation for $E_\perp$:

$$\begin{aligned}\frac{\partial^2}{\partial z^2} E_\perp &= \frac{i\omega}{c} \frac{\partial}{\partial z} B_{||} \\ &= \frac{i\omega}{c} \left(iB_z k_\perp + \frac{i\omega\epsilon}{c} E_\perp - \frac{4\pi}{c} J_\perp \right), \\ &= -\frac{\omega}{c} k_\perp \left(-\frac{c}{\omega} k_\perp E_\perp \right) - \frac{\omega^2 \epsilon}{c^2} E_\perp - i\frac{4\pi\omega}{c^2} J_\perp,\end{aligned} \tag{9.44}$$

which after rearrangement gives

$$\left[\frac{\partial^2}{\partial z^2} - k_\perp^2 + \frac{\omega^2 \epsilon}{c^2} \right] E_\perp = -i\frac{4\pi\omega}{c^2} J_\perp. \tag{9.45}$$

In dual fashion, (9.38), (9.39) and (9.43) yield (note ϵ is a function of z and ω)

$$\begin{aligned}\frac{\partial}{\partial z}\frac{1}{\epsilon}\frac{\partial}{\partial z} B_\perp &= -\frac{i\omega}{c}\frac{\partial}{\partial z} E_{||} + \frac{4\pi}{c}\frac{\partial}{\partial z}\left(\frac{1}{\epsilon} J_{||}\right), \\ &= -\frac{i\omega}{c}(iE_z k_\perp - i\frac{\omega}{c} B_\perp) + \frac{4\pi}{c}\frac{\partial}{\partial z}\left(\frac{1}{\epsilon} J_{||}\right), \\ &= \frac{\omega}{c} k_\perp \left(\frac{k_\perp c}{\omega\epsilon} B_\perp - \frac{4\pi i}{\omega\epsilon} J_z \right) - \frac{\omega^2}{c^2} B_\perp + \frac{4\pi}{c}\frac{\partial}{\partial z}\left(\frac{1}{\epsilon} J_{||}\right)\end{aligned}$$

$$\Longrightarrow \left[\frac{\partial}{\partial z}\frac{1}{\epsilon}\frac{\partial}{\partial z} - \frac{k_\perp^2}{\epsilon} + \frac{\omega^2}{c^2} \right] B_\perp = -\frac{4\pi i}{c\epsilon} k_\perp J_z + \frac{4\pi}{c}\frac{\partial}{\partial z}\left(\frac{1}{\epsilon} J_{||}\right). \tag{9.46}$$

Note that both (9.45) and (9.46) reduce to the wave equation (with sources) if $\epsilon =$ constant. In this case, if the sources $J_\perp, J_{||}$ and J_z are given then we can solve these equations for $E_\perp, B_\perp$. From these we may find $E_{||}, B_{||}, E_z$ and B_z from the relations (9.38)–(9.43).

We shall construct and solve the reduced Green function for (9.45) (and later for (9.46)) in the case of a plane interface between media whose dielectric constants depend only on ω. If we consider first a single medium, the Green function for $E_\perp$ has the same reduced wave equation form as (8.134):

$$-\left(\frac{\partial^2}{\partial z^2}+\frac{\omega^2\epsilon(\omega)}{c^2}-k_\perp^2\right)g_{E_\perp}(z,z')=\delta(z-z'). \tag{9.47}$$

Let's put this in perspective. The space-time field $E_\perp(\vec{x},t)$ in free space is given by

$$E_\perp(\vec{x},t)=\int\frac{d^2k_\perp}{(2\pi)^2}\int\frac{d\omega}{2\pi}e^{i\vec{k}_\perp\cdot\vec{x}_\perp-i\omega t}E_\perp(z,\vec{k}_\perp,\omega), \tag{9.48}$$

where

$$E_\perp(z,\vec{k}_\perp,\omega)=\frac{4\pi i\omega}{c^2}\int_{-\infty}^{\infty}dz'\,g_{E_\perp}(z,z')J_\perp(z',\vec{k}_\perp,\omega). \tag{9.49}$$

When $z\neq z'$, the solutions to $g_{E_\perp}(z,z')$ have the same form as (8.131), except for an additional factor of $\epsilon(\omega)$ (which we abbreviate as ϵ for convenience):

$$g_{E_\perp}(z,z')\sim\begin{cases}\exp\left(\pm\sqrt{k_\perp^2-\frac{\omega^2\epsilon}{c^2}}\,z\right), & k_\perp^2-\frac{\omega^2\epsilon}{c^2}>0,\\ \exp\left(\pm iS_\omega\sqrt{\frac{\omega^2\epsilon}{c^2}-k_\perp^2}\,z\right), & \frac{\omega^2\epsilon}{c^2}-k_\perp^2>0.\end{cases} \tag{9.50}$$

The upper solution in (9.50) describes a damped static solution while the lower solution gives traveling waves. Using the results of Exercise 8.7.2, we may assume that $\omega>0$ without loss of generality. We will work with the traveling wave branch; if we need to reconstruct the full Green function, the other branch follows by analytic continuation. Let us define (note this corresponds to θ defined in Section 8.7)

$$k_{1,2}\equiv\left(\frac{\omega^2\epsilon_{1,2}}{c^2}-k_\perp^2\right)^{1/2}>0. \tag{9.51}$$

Putting the $(\vec{x}_\perp,t)$ dependence in (9.48) together with the factor of $e^{ik_{1,2}z}$ in $g_{E_\perp}(z,z')$, we see that $E_\perp(\vec{x},t)$ is a superposition of waves of the form

$$e^{i\vec{k}_{1,2}\cdot\vec{x}-i\omega t}\equiv e^{i\vec{k}_\perp\cdot\vec{x}_\perp}e^{\pm ik_{1,2}z}e^{-i\omega t}, \tag{9.52}$$

where the vectors

$$\vec{k}_{1,2}\equiv(\vec{k}_\perp,\pm k_{1,2}), \tag{9.53}$$

represent directions of propagation in the two dielectrics. The $+$ sign in (9.53) corresponds to plane waves with a positive component of propagation along z (or $\hat{n}$); and similarly for the $-$ sign. From (9.51) we have the dispersion relation

$$(\vec{k}_{1,2})^2=\frac{\omega^2\epsilon_{1,2}}{c^2}, \tag{9.54}$$

where the $\epsilon_{1,2}$ are unknown (but positive) functions of ω. (Note that we will denote the magnitude of $\vec{k}_{1,2}$ as $|\vec{k}_{1,2}|$, to distinguish it from $k_{1,2}$ in (9.53).)

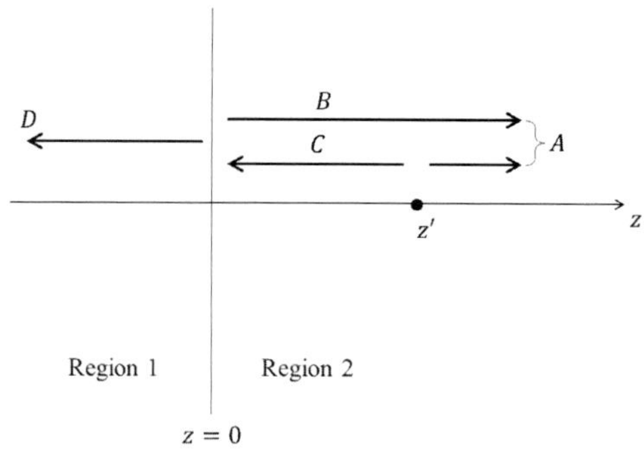

Fig. 9.4 Propagating waves at the interface between dielectrics with source at z'.

Let's compute the Green function for the case where sources are in region 2. (A similar computation covers the case where sources are in region 1.) In this case the solution for $g_{E_\perp}(z,z')$ must have the form:

$$g_{E_\perp}(z,z') = \begin{cases} Ae^{ik_2 z} & z > z', \\ Be^{ik_2 z} + Ce^{-ik_2 z} & z' > z > 0, \\ De^{-ik_1 z} & z < 0, \end{cases} \tag{9.55}$$

which corresponds to propagating waves as shown in Figure 9.4. The other boundary conditions on $g_{E_\perp}$ are

$$g_{E_\perp}(z,z')\Big|_{z=0^-}^{z=0^+} = 0 \quad (E_\perp \text{ is continuous across interface}), \tag{9.56}$$

$$\frac{\partial}{\partial z} g_{E_\perp}(z,z')\Big|_{z=0^-}^{z=0^+} = 0 \quad \left(\text{from } \frac{\partial E_\perp}{\partial z} = \frac{i\omega}{c} B_{||}\right). \tag{9.57}$$

($B_{||}$ is continuous across the interface if the free surface current is zero; remember $\mu = 1$ on both sides.) At $z = z'$, the usual conditions (from the differential equation) are

$$g_{E_\perp}(z,z')\Big|_{z=z'^-}^{z=z'^+} = 0, \tag{9.58}$$

$$-\frac{\partial}{\partial z} g_{E_\perp}(z,z')\Big|_{z=z'^-}^{z=z'^+} = 1. \tag{9.59}$$

We obtain equations for A, B, C, D:

$$B + C = D, \tag{9.60}$$

$$B - C = -\frac{k_1}{k_2} D, \tag{9.61}$$

$$Ae^{ik_2 z'} = Be^{ik_2 z'} + Ce^{-ik_2 z'}, \tag{9.62}$$

$$ik_2\left(-Ae^{ik_2z'} + Be^{ik_2z'} - Ce^{-ik_2z'}\right) = 1, \tag{9.63}$$

which lead to the solution

$$A = \frac{k_2 - k_1}{k_2 + k_1}\frac{i}{2k_2}e^{ik_2z'} + \frac{i}{2k_2}e^{-ik_2z'}, \tag{9.64}$$

$$B = \frac{k_2 - k_1}{k_2 + k_1}\frac{i}{2k_2}e^{ik_2z'}, \tag{9.65}$$

$$C = \frac{i}{2k_2}e^{ik_2z'}, \tag{9.66}$$

$$D = \frac{2k_2}{k_2 + k_1}\frac{i}{2k_2}e^{ik_2z'}. \tag{9.67}$$

Using these coefficients, we may combine the wave expressions for $z > z'$ and $z' > z > 0$ into a single form for $z > 0$:

$$g_{E_\perp}(z, z') = \begin{cases} \dfrac{k_2 - k_1}{k_2 + k_1}\dfrac{i}{2k_2}e^{ik_2(z+z')} + \dfrac{i}{2k_2}e^{ik_2|z-z'|}, & z > 0, \\ \dfrac{2k_2}{k_2 + k_1}\dfrac{i}{2k_2}e^{-ik_1z}e^{ik_2z'}, & z < 0, \end{cases} \tag{9.68}$$

which is valid for $\omega^2\epsilon/c^2 - k_\perp^2 > 0$.

We may check consistency with Section 8.7 by setting $\epsilon_{1,2} = 1$: In this case, $k_1 = k_2 = \theta$ and we obtain

$$g_{E_\perp}(z, z') \to \frac{i}{2\theta}e^{i\theta|z-z'|} \quad \text{(valid for all } z\text{)}, \tag{9.69}$$

which is the same form as the $g(z, z')$ found in Section 8.7 for traveling waves when $\omega^2/c^2 > k_\perp^2, \omega > 0$. Also note that for $\epsilon_2 = 1$ and $\epsilon_1 \to \infty$ or $k_1 \to \infty$ (corresponding to a conductor in region 1),

$$g_{E_\perp}(z, z') \to \begin{cases} \dfrac{i}{2\theta}\left(e^{i\theta|z-z'|} - e^{i\theta(z+z')}\right), & z > 0, \\ 0, & z < 0, \end{cases} \tag{9.70}$$

which is of the same form for the reduced Green function for the conducting plane also for $\omega^2/c^2 > k_\perp^2$ and $\omega > 0$ (see (8.145)).

All of the well-known electromagnetic properties at a dielectric interface follow directly from the equations we have derived above. For example, the angle of incidence is equal to the angle of reflection because the $\vec{k}$ vectors for transmitted and reflected waves are $\vec{k}_\perp - k_2\hat{z}$ and $\vec{k}_\perp + k_2\hat{z}$ respectively. Snell's law also follows from the fact that $|\vec{k}_\perp| =$ constant: defining angles i and t as in Figure 9.5, we have

$$\sqrt{k_\perp^2 + (k_2)^2}\sin i = \sqrt{k_\perp^2 + (k_1)^2}\sin t,$$

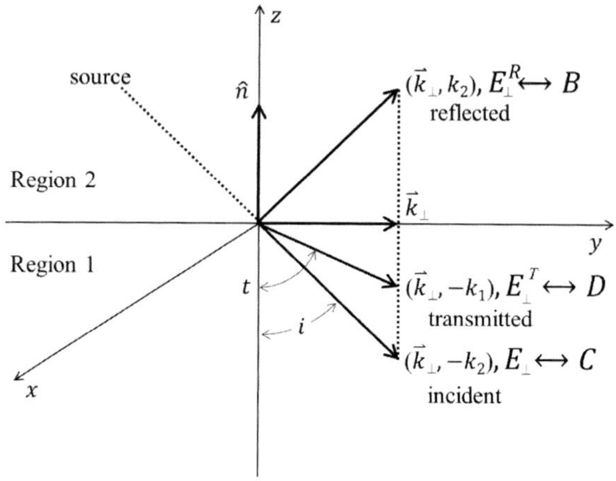

Fig. 9.5 Three-dimensional depiction of the transmitted and reflected waves at the dielectric interface. The yz plane is the plane of incidence.

and defining the *indices of refraction* n_1, n_2 as

$$n_{1,2}^2 \equiv \epsilon_{1,2}\mu_{1,2}, \tag{9.71}$$

(here $\mu_{1,2} = 1$) it follows that

$$k_\perp^2 + (k_{1,2})^2 = \frac{\omega^2}{c^2}\epsilon_{1,2} = \frac{\omega^2}{c^2}n_{1,2}^2,$$

which implies

$$n_2 \sin i = n_1 \sin t. \tag{9.72}$$

The ratios

$$\frac{D}{C} = \frac{2k_2}{k_1 + k_2}, \tag{9.73}$$

$$\frac{B}{C} = \frac{k_2 - k_1}{k_1 + k_2}, \tag{9.74}$$

are identified as the *transmission* and *reflection* coefficients, respectively, from the directional flow in Figure 9.5. Note that for this polarization the $E_\perp$ components are measured with respect to the x-axis ($\hat{k}_\perp \times \hat{n}$ direction). The physical significance of the coefficients in our solution is as follows: C is proportional to the incident incoming amplitude, which we shall call $E_\perp$; D is proportional to the transmitted amplitude, denoted as $E_\perp^T$; and finally

B is proportional to the reflected amplitude, designated as $E_\perp^R$. Thus, we have

$$\frac{D}{C} = \frac{E_\perp^T}{E_\perp}, \tag{9.75}$$

$$\frac{B}{C} = \frac{E_\perp^R}{E_\perp}. \tag{9.76}$$

Since the wave numbers k_1 and k_2 are

$$k_1 = |\vec{k_1}| \cos t = \frac{\omega}{c} n_1 \cos t, \tag{9.77}$$

$$k_2 = |\vec{k_2}| \cos i = \frac{\omega}{c} n_2 \cos i, \tag{9.78}$$

we have

$$\frac{E_\perp^T}{E_\perp} = \frac{2k_2}{k_1 + k_2} = \frac{2n_2 \cos i}{n_1 \cos t + n_2 \cos i}, \tag{9.79}$$

$$\frac{E_\perp^R}{E_\perp} = \frac{k_2 - k_1}{k_1 + k_2} = \frac{n_2 \cos i - n_1 \cos t}{n_1 \cos t + n_2 \cos i}. \tag{9.80}$$

For normal incidence ($i = t = 0$) these become

$$\frac{E_\perp^T}{E_\perp} \to \frac{2n_2}{n_1 + n_2} \implies \text{pure reflection if } n_1/n_2 \implies \infty, \tag{9.81}$$

$$\frac{E_\perp^R}{E_\perp} \to \frac{n_2 - n_1}{n_1 + n_2} \implies \text{phase reversal if } n_1 > n_2. \tag{9.82}$$

9.3 Reflection and refraction of plane waves from dielectric interfaces II: $B_\perp$ polarization

So far we have solved for the field component $E_\perp$ which arises from currents $\vec{J}_\perp$ that are perpendicular to the interface. In this section we solve for $B_\perp$, which is due to currents $\vec{J}_{||}$ and $\vec{J}_z$: this corresponds to linearly polarized waves whose electric fields are perpendicular to $E_\perp$.

The reduced Green function for $B_\perp$ is

$$-\left[\frac{\partial}{\partial z}\frac{1}{\epsilon}\frac{\partial}{\partial z} - \frac{k_\perp^2}{\epsilon} + \frac{\omega^2}{c^2}\right] g_{B_\perp}(z, z') = \delta(z - z'), \tag{9.83}$$

or

$$-\left[\frac{\partial}{\partial z}\frac{1}{\epsilon}\frac{\partial}{\partial z} + \frac{k_{1,2}^2}{\epsilon}\right] g_{B_\perp}(z, z') = \delta(z - z'), \tag{9.84}$$

with $k_{1,2}$ defined as before (see (9.53)). Given the plane at $z = 0$, the traveling wave solutions in the various regions when the source is at $z' > 0$ are (assuming sources in Region 2, as before)

$$g_{B_\perp}(z,z') = \begin{cases} Ae^{ik_2 z}, \ z > z', \\ Be^{ik_2 z} + Ce^{-ik_2 z}, \ z' > z > 0, \\ De^{-ik_1 z}, \ 0 > z. \end{cases} \tag{9.85}$$

However, the boundary conditions are now

$$g_{B_\perp}(z,z')\Big|_{z=0^-}^{z=0^+} = 0, \quad (B_\perp \text{ continuous}) \tag{9.86}$$

$$\frac{1}{\epsilon}\frac{\partial}{\partial z} g_{B_\perp}(z,z')\Big|_{z=0^-}^{z=0^+} = 0. \quad \left(\frac{ic}{\omega}\frac{1}{\epsilon}\frac{\partial}{\partial z}B_\perp = E_{||}\right) \tag{9.87}$$

($E_{||}$ = continuous.) In addition

$$g_{B_\perp}(z,z')\Big|_{z=z'^-}^{z=z'^+} = 0, \tag{9.88}$$

$$-\frac{1}{\epsilon_2}\frac{\partial}{\partial z} g_{B_\perp}(z,z')\Big|_{z=z'^-}^{z=z'^+} = 1. \tag{9.89}$$

The resulting equations for A, B, C, D are

$$B + C = D, \tag{9.90}$$

$$B - C = -\frac{\epsilon_2}{\epsilon_1}\frac{k_1}{k_2} D, \tag{9.91}$$

$$Ae^{ik_2 z'} = Be^{ik_2 z'} + Ce^{-ik_2 z'}, \tag{9.92}$$

$$ik_2(-Ae^{ik_2 z'} + Be^{ik_2 z'} - Ce^{-ik_2 z'}) = \epsilon_2. \tag{9.93}$$

The solution is

$$A = \frac{\epsilon_1 k_2 - \epsilon_2 k_1}{\epsilon_1 k_2 + \epsilon_2 k_1}\frac{i\epsilon_2}{2k_2} e^{ik_2 z'} + \frac{i\epsilon_2}{2k_2} e^{-ik_2 z'}, \tag{9.94}$$

$$B = \frac{\epsilon_1 k_2 - \epsilon_2 k_1}{\epsilon_1 k_2 + \epsilon_2 k_1}\frac{i\epsilon_2}{2k_2} e^{ik_2 z'}, \tag{9.95}$$

$$C = \frac{i\epsilon_2}{2k_2} e^{ik_2 z'}, \tag{9.96}$$

$$D = \frac{2\epsilon_1 k_2}{\epsilon_1 k_2 + \epsilon_2 k_1}\frac{i\epsilon_2}{2k_2} e^{ik_2 z'}. \tag{9.97}$$

The result is the same as for $E_\perp$ (cf. Equations (9.64)–(9.67)) except k_1 or k_2 is replaced with k_1/ϵ_1 or k_2/ϵ_2, respectively, everywhere in the coefficients of the exponents. The transmission and reflection coefficients are:

$$\frac{D}{C} = \frac{B_\perp^T}{B_\perp} = \frac{2k_2/\epsilon_2}{k_1/\epsilon_1 + k_2/\epsilon_2}, \tag{9.98}$$

$$\frac{B}{C} = \frac{B_\perp^R}{B_\perp} = \frac{k_2/\epsilon_2 - k_1/\epsilon_1}{k_1/\epsilon_1 + k_2/\epsilon_2}, \tag{9.99}$$

or using Equations (9.77) and (9.78),

$$\frac{B_\perp^T}{B_\perp} = \frac{2n_1 \cos i}{n_1 \cos i + n_2 \cos t}, \tag{9.100}$$

$$\frac{B_\perp^R}{B_\perp} = \frac{n_1 \cos i - n_2 \cos t}{n_1 \cos i + n_2 \cos t}. \tag{9.101}$$

From the field equations (9.38)–(9.43) and our results for $E_\perp$ and $B_\perp$, we can also find results for the other components of $\vec{E}$ and $\vec{B}$ (see Exercise 9.3.2 in the $\mu \neq 1$ case).

For normal incidence we have

$$\frac{B_\perp^T}{B_\perp} = \frac{2n_1}{n_1 + n_2}, \quad \left(\text{compare: } \frac{E_\perp^T}{E_\perp} = \frac{2n_2}{n_1 + n_2}\right), \tag{9.102}$$

$$\frac{B_\perp^R}{B_\perp} = \frac{n_1 - n_2}{n_1 + n_2}. \quad \left(\text{compare: } \frac{E_\perp^R}{E_\perp} = \frac{n_2 - n_1}{n_1 + n_2}\right). \tag{9.103}$$

Note that when $|n_1| \gg |n_2|$, the transmitted fields inside the material are mainly magnetic. We shall see in Section 9.6 that this situation applies to conductors, where $|n_1| = \sqrt{\epsilon(\omega)} \sim \omega^{-1/2}$ at low frequencies.

9.4 Brewster's angle and total internal reflection

It is possible for the reflection coefficient to vanish. In the $E_\perp$ case, this happens only when $k_1 = k_2$ or $\epsilon_1 = \epsilon_2$: that is, there is no interface. On the other hand, in the $B_\perp$ case we obtain from (9.99) the condition

$$\frac{k_1}{\epsilon_1} = \frac{k_2}{\epsilon_2}. \tag{9.104}$$

Using the fact that

$$\frac{k_{1,2}^2}{\epsilon_{1,2}^2} = \left(\frac{\omega^2}{c^2}\frac{1}{\epsilon_{1,2}} - \frac{k_\perp^2}{\epsilon_{1,2}^2}\right), \tag{9.105}$$

we obtain the condition

$$\frac{\epsilon_2 - \epsilon_1}{\epsilon_2 \epsilon_1} \left[\frac{\omega^2}{c^2} - k_\perp^2 \frac{\epsilon_1 + \epsilon_2}{\epsilon_1 \epsilon_2} \right] = 0, \tag{9.106}$$

which occurs in general when

$$\frac{\omega}{c} = \sqrt{\frac{\epsilon_1 + \epsilon_2}{\epsilon_1 \epsilon_2}} |\vec{k}_\perp|. \tag{9.107}$$

But $|\vec{k}_\perp| = (\omega n_2/c) \sin i$, so this defines a particular angle i_B (known as *Brewster's angle*), where

$$\sin i_B = \sqrt{\frac{\epsilon_1}{\epsilon_1 + \epsilon_2}}, \qquad or \qquad \tan i_B = \sqrt{\frac{\epsilon_1}{\epsilon_2}} = \frac{n_1}{n_2}. \tag{9.108}$$

At this angle $B_\perp$ is completely transmitted. It follows that for an unpolarized incident wave, the reflected wave will be linearly polarized with $\vec{E}$ perpendicular to the plane of incidence (i.e. parallel to the interface).

Although the transmission coefficient never vanishes, when $n_2 > n_1$ for large incidence angle the transmitted wave essentially disappears. This follows from Snell's law (9.72):

$$n_1 \sin t = n_2 \sin i, \quad \text{so that} \quad \frac{n_2}{n_1} > 1 \implies t > i. \tag{9.109}$$

Thus there are values of i such that

$$\frac{n_2}{n_1} \sin i > 1. \tag{9.110}$$

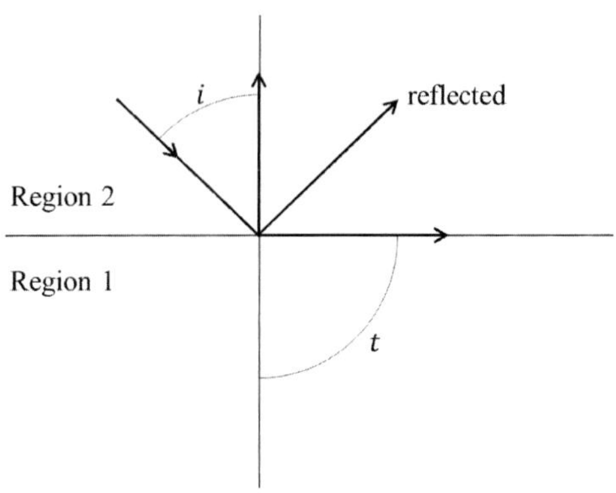

Fig. 9.6 Transmitted and reflected waves at dielectric interface.

When $\sin i = n_1/n_2$, then $t = 90^\circ$ and the transmitted light ray grazes along the interface as shown in Figure 9.6. For angles i greater than this, the transmitted wave number becomes complex:

$$k_1 = \sqrt{\frac{\omega^2}{c^2}\epsilon_1 - k_\perp^2} = i\frac{\omega}{c}\sqrt{(n_2 \sin i)^2 - n_1^2} \equiv i\kappa. \tag{9.111}$$

Then the form of the wave in region 1 has the form

$$e^{i\vec{k}_\perp \cdot \vec{x}_\perp} e^{-\kappa z} e^{-i\omega t},$$

which means the wave penetrates a distance $\sim 1/\kappa$ into region 1, but attenuates exponentially so that there is no net flow of energy into region 1 (under steady state conditions). This is the "damped static solution" possibility mentioned after Equation (9.50).

Of course, we don't have to use the above Green function formalism to get reflection and transmission coefficients. We could have simply enforced the correct boundary conditions (normal $\vec{D}$, normal $\vec{B}$, tangential $\vec{E}$, tangential $\vec{H}$) at the interface on a given plane wave. However our treatment exhibits more closely the connection with the previous Green functions for free space and the conducting plane, and emphasizes that all the relevant information is in these functions.

9.5 Simple model for constitutive relations

Let us now discuss some simple models for constitutive relations in metals, dielectrics and plasmas. These models lead to a qualitative understanding of the electromagnetic properties of these materials, especially their response to external fields. Keep in mind that these are only *phenomenological* descriptions, and do not presume to provide a rigorous physical explanation for the observed properties.

In our development of the macroscopic Maxwell equations, we made a distinction between bound and free charges and currents. In actual materials, these charges and currents are almost always electrons. Consider such an electron, either bound or free, moving under the influence of an external electric field, but slowed by a resistive force. The equation of motion is

$$m\left[\ddot{\vec{x}} + \gamma\dot{\vec{x}} + \omega_0^2\vec{x}\right] = e\,\vec{E}(t), \tag{9.112}$$

where $e < 0$ is the electron charge. This is simply the damped, driven oscillator equation from basic physics. ω_0 is the natural frequency of the electron in equation, which corresponds to the energy level for a bound electron and is 0 for a free or conductive (valence) electron. γ is a damping constant primarily due to radiation for bound electrons, whereas for free electrons the origin is in collisions with other electrons, lattice imperfections, impurities, etc. For bound electrons, usually $\gamma \ll \omega_0$.

We will not discuss the solutions to this equation here in detail, but only point out the properties that we will need. (In Exercise 9.5.2 you will derive the Green function, and use it to obtain the general solution.)

First, consider the model for conductors. In this case $\omega_0 = 0$ and we have

$$m(\ddot{\vec{x}} + \gamma\dot{\vec{x}}) = e\vec{E}(t). \tag{9.113}$$

We define $\vec{v}(t) \equiv \dot{\vec{x}}$, and rewrite (9.113) as

$$\frac{d}{dt}(e^{\gamma t}\vec{v}(t)) = \frac{e}{m}e^{\gamma t}\vec{E}(t). \tag{9.114}$$

Integrate from $t' = -\infty$ (or from whatever time the initial conditions are given) to obtain:

$$\begin{aligned}\int_{-\infty}^{t} dt'\,\frac{d}{dt'}\left(e^{\gamma t'}\vec{v}(t')\right) &= \frac{e}{m}\int_{-\infty}^{t} dt'\,e^{\gamma t'}\vec{E}(t')\\ \Longrightarrow \vec{v}(t) &= \frac{e}{m}\int_{-\infty}^{t} dt'\,e^{-\gamma(t-t')}\vec{E}(t'). \end{aligned}\tag{9.115}$$

Notice in this model the response is nonlocal in time: that is, $\vec{v}(t)$ depends upon $\vec{E}(t')$ at earlier times. The main contribution to the integral comes from time differences of order γ^{-1}. Assuming that conduction electrons have a constant density n_c, the current $\vec{J}$ is given by

$$\vec{J} = n_c e\vec{v} = \frac{n_c e^2}{m}\int_{-\infty}^{t} dt'\,e^{-\gamma(t-t')}\vec{E}(t'). \tag{9.116}$$

For the specific case of a constant electric field, this becomes

$$\vec{J} = \frac{n_c e^2}{m\gamma}\vec{E} \equiv \sigma\vec{E}, \tag{9.117}$$

giving Ohm's law with σ being interpreted as the static conductivity. Note that σ is inversely proportional to γ, which makes sense physically.

A more general situation is where the electric field exhibits harmonic time variation, as in the case of electromagnetic waves. We set

$$\vec{E}(t) = \mathrm{Re}\left(\vec{\mathcal{E}}(\omega)e^{-i\omega t}\right). \tag{9.118}$$

(where as usual $\vec{\mathcal{E}}(\omega)$ may be complex). Then, the current density becomes

$$\begin{aligned}\vec{J}(t) &= \mathrm{Re}\left(\frac{n_c e^2}{m}\vec{\mathcal{E}}(\omega)\int_{-\infty}^{t} dt'\, e^{-\gamma(t-t')}e^{-i\omega t'}\right),\\ &= \mathrm{Re}\left(\frac{n_c e^2}{m}\frac{1}{\gamma - i\omega}\vec{\mathcal{E}}(\omega)e^{-i\omega t}\right),\\ &= \mathrm{Re}\left(\sigma(\omega)\vec{\mathcal{E}}(\omega)e^{-i\omega t}\right). \end{aligned}\tag{9.119}$$

Equation (9.119) generalizes Ohm's law with a frequency-dependent conductivity

$$\sigma(\omega) = \frac{n_c e^2}{m}\frac{1}{\gamma - i\omega},\tag{9.120}$$

and agrees with the static value (9.117) for $\omega = 0$. The real and imaginary parts of $\sigma(\omega)$ are

$$\mathrm{Re}\,\sigma(\omega) = \frac{n_c e^2}{m}\frac{\gamma}{\gamma^2 + \omega^2},\tag{9.121}$$

$$\mathrm{Im}\,\sigma(\omega) = \frac{n_c e^2}{m}\frac{\omega}{\gamma^2 + \omega^2},\tag{9.122}$$

which are even and odd functions of ω respectively. In addition, we get the amplitude and phase of $\sigma(\omega)$ as

$$\sigma(\omega) = |\sigma(\omega)|e^{i\delta},\tag{9.123}$$

$$|\sigma(\omega)| = \frac{n_c e^2}{m}\frac{1}{\sqrt{\gamma^2 + \omega^2}},\tag{9.124}$$

$$\tan\delta = \frac{\omega}{\gamma}.\tag{9.125}$$

The complex conductivity (9.123) produces a phase shift of δ in the current relative to the electric field:

$$\begin{aligned}\vec{J}(t) &= \mathrm{Re}(\sigma(\omega)\vec{\mathcal{E}}(\omega)e^{-i\omega t}),\\ &= \frac{n_c e^2}{m}\frac{1}{\sqrt{\gamma^2 + \omega^2}}\mathrm{Re}(\vec{\mathcal{E}}(\omega)e^{-i\omega t + i\delta}).\end{aligned}\tag{9.126}$$

For copper, σ is essentially real well beyond the microwave region ($\omega \lesssim 10^{11}\ \mathrm{sec}^{-1}$). If the real part of the conductivity is integrated over all positive angular frequencies, we find

$$\int_0^\infty d\omega\,\mathrm{Re}\,\sigma(\omega) = \frac{n_c e^2}{2m}\int_{-\infty}^{\infty}\frac{d\omega}{\gamma}\frac{1}{1+\omega^2/\gamma^2} = \frac{n_c e^2}{2m}\pi,\tag{9.127}$$

which is independent of γ. Equation (9.127) can be used to determine the density of conduction electrons $n_{\rm c}$.

If we are dealing with bound electrons, we must include the $\omega_0^2\vec{x}$ term in the differential equation. Let us re-examine the situation of harmonic time dependence with this additional term. Consider

$$m\left[\ddot{\vec{x}}+\gamma\dot{\vec{x}}+\omega_0^2\vec{x}\right]=e\,\mathrm{Re}\left[\vec{\mathcal{E}}(\omega)e^{-i\omega t}\right]. \tag{9.128}$$

The steady state solution to this is given by assuming

$$\vec{x}(t)=\mathrm{Re}\left[\vec{x}(\omega)e^{-i\omega t}\right], \tag{9.129}$$

from which we find

$$\vec{x}(t)=\frac{e}{m}\mathrm{Re}\left[\frac{\vec{\mathcal{E}}(\omega)e^{-i\omega t}}{-\omega^2+\omega_0^2-i\gamma\omega}\right]. \tag{9.130}$$

Within this model we are trying to calculate the effective macroscopic properties (that is, the constitutive relations). If the electrons are in their "equilibrium positions" (defined as $\vec{x}_i(t)\equiv 0$ for all electrons i) then we assume that the polarization, $\vec{P}$, of the sample vanishes. The polarization as a function of electron displacements $\{\vec{x}_i(t)\}$ is defined as

$$\vec{P}(t)\equiv\frac{e}{V}\sum_{i=\mathrm{atoms}}\vec{x}_i(t). \tag{9.131}$$

Supposing that all electrons have the same displacement ($\vec{x}_i=\vec{x}$), then

$$\vec{P}(t)=n_{\rm b}e\vec{x}(t)\equiv\mathrm{Re}\left[\chi(\omega)\vec{\mathcal{E}}(\omega)e^{-i\omega t}\right], \tag{9.132}$$

where $n_{\rm b}$ is the density of bound electrons, and the frequency-dependent susceptibility $\chi(\omega)$ is given by

$$\chi(\omega)=\frac{n_{\rm b}e^2}{m}\frac{1}{-\omega^2+\omega_0^2-i\omega\gamma}. \tag{9.133}$$

In the static case

$$\chi(0)=\frac{n_{\rm b}e^2}{m\omega_0^2}>0, \tag{9.134}$$

where the m here is not necessarily the actual electron mass, but can be an *effective mass* that partially includes the effect of binding on the bound electrons. We see that $\epsilon(0)\equiv 1+4\pi\chi(0)>1$ as expected. For general ω it is natural to define a frequency-dependent dielectric constant $\epsilon(\omega)$ by $\epsilon(\omega)=1+4\pi\chi(\omega)$, so that

$$\vec{D}(t)=\mathrm{Re}(\epsilon(\omega)\vec{\mathcal{E}}(\omega)e^{-i\omega t}). \tag{9.135}$$

From (9.133) we have

$$\epsilon(\omega) = 1 + \frac{4\pi n_{\rm b} e^2}{m} \frac{1}{\omega_0^2 - \omega^2 - i\omega\gamma} \tag{9.136}$$

$$\Longrightarrow \text{Re}\ \epsilon(\omega) = 1 + \frac{4\pi n_{\rm b} e^2}{m} \frac{(\omega_0^2 - \omega^2)}{(\omega_0^2 - \omega^2)^2 + \omega^2\gamma^2} \tag{9.137}$$

$$\Longrightarrow \text{Im}\ \epsilon(\omega) = 4\pi \frac{n_{\rm b} e^2}{m} \frac{\omega\gamma}{(\omega_0^2 - \omega^2)^2 + \omega^2\gamma^2}. \tag{9.138}$$

We will study the dependence of ϵ on ω (that is, *dispersion*) in more detail later.

We may derive an integral rule for bound electrons which is analogous to the conductivity integral rule in (9.127). This time we evaluate (see Exercise 9.5.4)

$$\begin{aligned} \int_0^\infty d\omega\, \omega\, \text{Im}\ \epsilon(\omega) &= \frac{2\pi n_{\rm b} e^2}{m} \int_{-\infty}^\infty d\omega\, \frac{\omega^2\gamma}{(\omega_0^2 - \omega^2)^2 + \omega^2\gamma^2}, \\ &= \frac{2\pi^2 n_{\rm b} e^2}{m}. \end{aligned} \tag{9.139}$$

Equations (9.127) and (9.139) are closely related, and may actually be combined into a single equation. We may write

$$\vec{\nabla} \times \vec{H} = \frac{4\pi}{c}\vec{J} + \frac{1}{c}\frac{\partial \vec{D}}{\partial t}, \tag{9.140}$$

where the two terms on the right-hand side are the *conduction current* and *displacement current* respectively. We have

$$\vec{D} = \text{Re}(\epsilon(\omega)\vec{\mathcal{E}}e^{-i\omega t}), \tag{9.141}$$

$$\vec{J} = \text{Re}(\sigma(\omega)\vec{\mathcal{E}}e^{-i\omega t}) \tag{9.142}$$

so that

$$\frac{4\pi}{c}\vec{J} + \frac{1}{c}\frac{\partial \vec{D}}{\partial t} = \text{Re}\left[\left(\frac{4\pi}{c}\sigma(\omega) - \frac{i\omega}{c}\epsilon(\omega)\right)\vec{\mathcal{E}}e^{-i\omega t}\right]. \tag{9.143}$$

This suggests the definition of an *effective conductivity* as

$$\sigma_{\rm eff}(\omega) \equiv \sigma(\omega) - \frac{i\omega}{4\pi}\epsilon(\omega). \tag{9.144}$$

Then integrating the real part of $\sigma_{\rm eff}$ over positive ω values yields

$$\begin{aligned} \int_0^\infty d\omega\, \text{Re}\, \sigma_{\rm eff}(\omega) &= \int_0^\infty d\omega\, \text{Re}\, \sigma(\omega) + \frac{1}{4\pi}\int_0^\infty d\omega\, \omega\, \text{Im}\, \epsilon(\omega) \\ &= \frac{e^2\pi}{2m}(n_{\rm c} + n_{\rm b}), \end{aligned} \tag{9.145}$$

where n_c and n_b represent conduction and displacement electrons, respectively. Alternatively we could define an effective dielectric constant:

$$\epsilon_{\text{eff}}(\omega) \equiv \epsilon(\omega) + i\frac{4\pi\sigma(\omega)}{\omega}, \tag{9.146}$$

from which it follows that

$$k^2 = \frac{\mu\omega^2}{c^2}\left(\epsilon(\omega) + i\frac{4\pi\sigma(\omega)}{\omega}\right). \tag{9.147}$$

Letting $k \equiv \alpha + i\beta$ it is possible to solve explicitly for α and β; see Exercise 9.5.5 at the end of the chapter for the case of normally incident waves on a surface. The factor of β causes damping and energy loss to the medium; Chapter 10 will deal with the forms induced for β when the waves we are dealing with propagate in a metallic waveguide.

9.6 Modeling of plasmas, metals and dielectrics

In this section, we apply the phenomenological formulas we have derived to various physical situations.

First, we consider the general case of a material with both conductance ($\omega_0 = 0$) electrons and bound electrons with multiple resonant frequencies $\{\omega_{0,i}\}, i = 1, 2, \ldots$. We adopt the effective dielectric point of view represented by (9.146), and in the following we will use $\epsilon(\omega)$ to denote $\epsilon_{\text{eff}}(\omega)$ in this section for convenience. In this case our former $\epsilon(\omega)$ in (9.136) generalizes to

$$\epsilon(\omega) = 1 + \frac{4\pi e^2}{m}\sum_i \frac{n_{b,i}}{\omega_{0,i}^2 - \omega^2 - i\omega\gamma_i} + i\frac{4\pi}{\omega}\left(\frac{e^2 n_c}{m}\frac{1}{\gamma - i\omega}\right). \tag{9.148}$$

The final term in (9.148) is the contribution from conductance electrons.

Assuming we have a partial conductor, then at sufficiently low frequencies (that is, $\omega \ll \omega_{0,\min}, \gamma$) we have

$$\epsilon(\omega) \approx \epsilon_0 + i\frac{4\pi\sigma(0)}{\omega}, \tag{9.149}$$

where

$$\epsilon_0 = 1 + \frac{4\pi e^2}{m}\sum_i \frac{n_{b,i}}{\omega_{0,i}^2}. \tag{9.150}$$

Notice $\epsilon(\omega) \to +\infty$ as $\omega \to 0^+$. In our earlier discussions of dielectric Green functions, we recovered perfect conductor Green functions by taking the $\epsilon \to +\infty$ limit. Now we know why this works, although we see it is more appropriate to take this limit through positive imaginary numbers.

If we are working with a very good conductor ($\sigma(0)/\omega \gg 1$), then at low frequencies

$$k^2 = \frac{\omega^2}{c^2}\epsilon \approx i\frac{4\pi\mu\sigma(0)\omega}{c^2} \tag{9.151}$$

$$\Longrightarrow k \approx (1+i)\frac{\sqrt{2\pi\omega\mu\sigma(0)}}{c}. \tag{9.152}$$

Thus a plane wave solution inside the conductor is damped:

$$e^{i\vec{k}\cdot\vec{x}} \to e^{-\hat{k}\cdot\vec{x}/\delta}e^{i\hat{k}\cdot\vec{x}/\delta}, \tag{9.153}$$

where δ is the *skin depth*, given by

$$\delta = \frac{c}{\sqrt{2\pi\mu\omega\sigma(0)}}. \tag{9.154}$$

Notice that at high frequencies for either dielectrics ($\omega \gg \omega_{0,\max}$) or conductors ($\omega \gg \gamma$), we have

$$\epsilon(\omega) \approx 1 - \frac{4\pi e^2 N}{m\omega^2} \equiv 1 - \frac{\omega_p^2}{\omega^2}, \tag{9.155}$$

where $N \equiv N_{\mathrm{b}} + n_{\mathrm{c}}$, $\sum_i n_{b,i} \equiv N_{\mathrm{b}}$ is the total density of bound electrons, and ω_p is called the *plasma frequency*. ($\omega_p \lesssim \omega_{0,\max}$ is assured in plasmas as well as most materials.) In dielectrics, this turns out to apply only for $\omega \gg \omega_p$. In this region the dielectric constant is slightly less than unity. For metals and plasmas, (9.155) can hold for $\omega < \omega_p$ as well. Then the dispersion relation $k^2 = (\omega/c)^2\epsilon$ gives damping:

$$k = \frac{i}{c}\sqrt{\omega_p^2 - \omega^2}, \quad \omega^2 < \omega_p^2, \tag{9.156}$$

while the case $\omega^2 > \omega_p^2$ gives oscillating solutions:

$$k = \frac{1}{c}\sqrt{\omega^2 - \omega_p^2}, \quad \omega^2 > \omega_p^2. \tag{9.157}$$

Such relations can be used to describe the response of plasmas, metals or dielectrics to electromagnetic waves. For plasmas, $\gamma \approx 0$ and $n_{\mathrm{b},i} = 0$, and the response is damped for frequencies down to $\omega \approx 0$. For metals, this leads to the so-called *ultraviolet transparency* as one increases ω through ω_p. The same thing happens for plasmas, except at much lower frequencies (in the microwave range).

We can get a different perspective on the physical situation by looking at the time behavior of fields rather than the spatial behavior (that is, the k values) discussed above. First, let us consider the motion of free charges in the simple case where Ohm's law holds and the conductivity is nearly independent of frequency (which is characteristic of highly resistive materials and some semiconductors). In this case we have the frequency-independent equation

$$\vec{J} = \sigma(0)\vec{E} = \frac{\sigma(0)}{\epsilon_0}\vec{D}, \tag{9.158}$$

where we are assuming that $\sigma(0) \ll \omega$ in (9.149). This may be combined with charge conservation

$$\frac{\partial \rho(\vec{x},t)}{\partial t} + \vec{\nabla}\cdot\vec{J}(\vec{x},t) = 0, \tag{9.159}$$

to obtain

$$\frac{\partial \rho}{\partial t} + \vec{\nabla}\cdot\left(\frac{\sigma(0)}{\epsilon_0}\vec{D}\right) = 0. \tag{9.160}$$

However, because

$$\vec{\nabla}\cdot\vec{D} = 4\pi\rho, \tag{9.161}$$

we find that

$$\left[\frac{\partial}{\partial t} + \frac{4\pi\sigma(0)}{\epsilon_0}\right]\rho(\vec{x},t) = 0. \tag{9.162}$$

For an initial charge density, $\rho(\vec{x},0)$, this gives rise to the solution:

$$\rho(\vec{x},t) = \rho(\vec{x},0)\exp\left(-\frac{4\pi\sigma(0)}{\epsilon_0}t\right). \tag{9.163}$$

Physically, (9.163) reflects the fact that free charges are expelled from the interior to the surface.

On the other hand, in the case of conductors the frequency dependence of $\sigma(\omega)$ plays an important role. Again we assume that ϵ_0 varies slowly with ω. We replace $\rho(\vec{x},t)$ in (9.162) with the Fourier transform expression

$$\rho(\vec{x},t) = \int_{-\infty}^{\infty}\frac{d\omega}{2\pi}e^{-i\omega t}\rho(\vec{x},\omega), \tag{9.164}$$

to obtain (also replacing $\sigma(0)$ with $\sigma(\omega)$):

$$\int_{-\infty}^{\infty}\frac{d\omega}{2\pi}e^{-i\omega t}\left[-i\omega + \frac{4\pi\sigma(\omega)}{\epsilon_0}\right]\rho(\vec{x},\omega) = 0. \tag{9.165}$$

Applying $\partial/\partial t + \gamma$ to both sides gives

$$\left(\frac{\partial}{\partial t} + \gamma\right) \int_{-\infty}^{\infty} \frac{d\omega}{2\pi} e^{-i\omega t} \left[-i\omega + \frac{4\pi n_c e^2}{m\epsilon_0} \frac{1}{\gamma - i\omega}\right] \rho(\vec{x}, \omega) = 0$$

$$\Longrightarrow \int_{-\infty}^{\infty} \frac{d\omega}{2\pi} e^{-i\omega t} \left[-\omega^2 - i\gamma\omega + \frac{4\pi n_c e^2}{m\epsilon_0}\right] \rho(\vec{x}, \omega) = 0, \qquad (9.166)$$

which corresponds to the differential equation

$$\left[\frac{\partial^2}{\partial t^2} + \gamma \frac{\partial}{\partial t} + \frac{(\omega_p^c)^2}{\epsilon_0}\right] \rho(\vec{x}, t) = 0, \qquad (9.167)$$

where ω_p^c is the conductance part of the plasma frequency.

Equation (9.167) completely characterizes the behavior of free charges in both conductive and resistive materials. The underdamped case ($\gamma/2 < \omega_p^c/\sqrt{\epsilon_0}$) corresponds to conductors: the charge within the material varies as $\rho(t) \sim e^{-\gamma t/2} \cos(\omega_1 t + \delta)$, where $\omega_1 = \sqrt{(\omega_p^c)^2/\epsilon_0 - \gamma^2/4}$ and δ is a phase factor. The interpretation is that the charge disturbance moves to the surface as a damped oscillation in time. The severely overdamped case ($\gamma/2 \gg \omega_p^c/\sqrt{\epsilon_0}$) corresponds to highly resistive materials: the charge decays as $\rho(t) \sim e^{-(4\pi\sigma(0)/\epsilon_0)t}$ for large enough times ($t \gg 1/\gamma$) showing that the charges are also expelled but with the exponential decay constant of (9.163).

9.7 Kramers-Kronig relations

The Kramers-Kronig relations are equations that relate the real and imaginary parts of the frequency-dependent dielectric constant $\epsilon(\omega)$. Before proving these relations, we first introduce the electromagnetic *response function*, which enables us to translate frequency dependent dispersion into the time-domain. Using the Fourier transform on the $\vec{D}$ field gives

$$\vec{D}(\vec{x}, t) = \frac{1}{2\pi} \int_{-\infty}^{\infty} d\omega \, e^{-i\omega t} \vec{D}(\vec{x}, \omega), \qquad (9.168)$$

and using the inverse Fourier transform on the $\vec{E}$ field gives

$$\vec{E}(\vec{x}, \omega) = \int_{-\infty}^{\infty} dt' \, e^{i\omega t'} \vec{E}(\vec{x}, t'). \qquad (9.169)$$

Assuming that $\vec{D}(\vec{x}, \omega) = \epsilon(\omega)\vec{E}(\vec{x}, \omega)$, we may combine (9.168) and (9.169) to obtain

$$\vec{D}(\vec{x}, t) = \int_{-\infty}^{\infty} dt' \, \vec{E}(\vec{x}, t') \int_{-\infty}^{\infty} \frac{d\omega}{2\pi} e^{-i\omega(t-t')} \epsilon(\omega), \qquad (9.170)$$

where the order of integration has been switched. If we let $\tau \equiv t - t'$, then can write this as

$$\vec{D}(\vec{x},t) = \vec{E}(\vec{x},t) + \int_{-\infty}^{\infty} d\tau\, \vec{E}(\vec{x},t-\tau)G(\tau), \tag{9.171}$$

where $G(\tau)$ is the response function, defined as

$$G(\tau) \equiv \int_{-\infty}^{\infty} \frac{d\omega}{2\pi} e^{-i\omega\tau}[\epsilon(\omega) - 1]. \quad \left(\int_{0^-}^{\infty} d\tau\, G(\tau)e^{i\omega\tau} = \epsilon(\omega) - 1\right). \tag{9.172}$$

Physically, we would expect that $\vec{D}(\vec{x},t)$ should not depend on $\vec{E}(\vec{x},t-\tau)$ at future times ($\tau < 0$). In view of (9.171) this implies that $G(\tau) = 0$ for $\tau < 0$. If we make this assumption, then we may infer that the poles of $\epsilon(\omega)$ must lie in the lower half plane. We have actually seen this before: recall our expression for G^{ret},

$$G^{\text{ret}} = 4\pi \int \frac{d^3k}{(2\pi)^3} e^{i\vec{k}\cdot(\vec{x}-\vec{x}\,')} \int_{-\infty}^{\infty} \frac{d\omega}{2\pi} \frac{e^{-i\omega(t-t')}}{k^2 - \left(\frac{\omega}{c} + i\epsilon\right)^2}, \tag{9.173}$$

which is just (8.105) with the substitution $\omega/c \to \omega/c + i\epsilon$. This has the pole structure shown in Figure 9.7. If we do the ω integral we find

$$G^{\text{ret}} = 4\pi c\, H(t-t') \int \frac{d^3k}{(2\pi)^3} \frac{1}{k} e^{i\vec{k}\cdot(\vec{x}-\vec{x}\,')} \sin[kc(t-t')]. \tag{9.174}$$

This causal form followed from the absence of poles in the upper half ω−plane. Similarly, the poles of $\epsilon(\omega)$ should be in the lower half plane in order to guarantee the causality of $G(\tau)$.

The representation in (9.172) and the hypothesis of no poles in the upper half plane only ensure causality: $G(\tau) = 0$, $\tau < 0^-$. There may be other physical requirements for $G(\tau)$ as well:

- Time locality: $G(\tau) \to 0$ as $\tau \to \infty$;

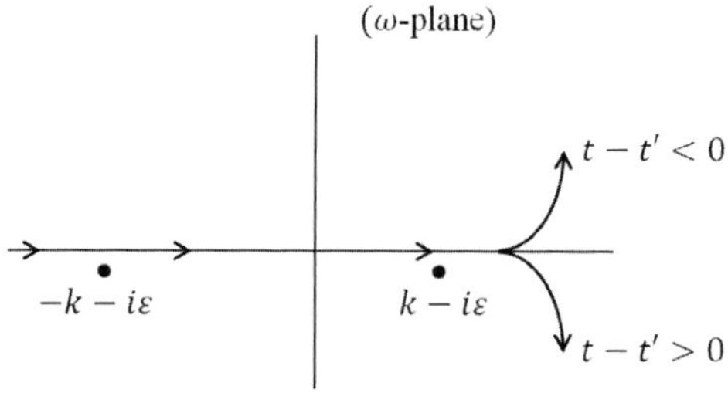

Fig. 9.7 Poles for retarded Green function.

- Response damping: $G(\tau)$ finite $\forall\tau$;
- Time continuity at $\tau = 0$: $G(0^-) = G(0^+)$.

For example, poles at real nonzero ω would violate time locality; on the other hand power functions such as ω^n, $n = 0, 1, 2, \ldots$ produce time-localized infinite responses. In addition, a simple pole at $\omega = 0$ would imply a material with a *static* conductivity, $\sigma(\omega) =$ constant, for all ω and give a discontinuous function at $\tau = 0$. Elimination of such behaviors will result in a relation between $\vec{E}$ and $\vec{D}$ much like our simple model connecting $\vec{J}$ and $\vec{E}$.

Using these ideas, we can derive a relationship between the real and imaginary parts of $\epsilon(\omega)$. We begin with the obvious evaluation

$$0 = \frac{1}{2\pi}\oint dz\,\frac{\epsilon(z) - 1}{z - \omega + i\epsilon}, \tag{9.175}$$

with poles and closed contour path as shown in Figure 9.8. We have that $\lim_{\omega\to\infty}(\epsilon(\omega) - 1) \sim \omega^{-2}$, as follows from replacing ω with $\omega + i\epsilon$:

$$\begin{aligned} \epsilon(\omega) &= 1 + \int_{0^-}^{\infty} d\tau G(\tau) e^{+i\omega\tau}, \\ G(\tau) &= G(0^+) + \tau G'(0^+) + \ldots \\ \implies \epsilon(\omega) &= 1 + i\frac{G(0^+)}{\omega} - \frac{G'(0^+)}{\omega^2} + \ldots \end{aligned} \tag{9.176}$$

We know that $G(0^-) = 0$, so continuity implies that $G(0^+) = 0$ and

$$\lim_{\omega\to\infty}(\epsilon(\omega) - 1) \sim \omega^{-2}, \tag{9.177}$$

which we also observed in our simple models in the previous chapter. The integral around the semicircular part of the contour in Figure 9.8 thus vanishes so that

$$0 = \int_{-\infty}^{\infty}\frac{d\omega'}{2\pi}\frac{\epsilon(\omega') - 1}{\omega' - \omega + i\epsilon}. \tag{9.178}$$

However, we may write

$$\frac{1}{\omega' - \omega + i\epsilon} = P\frac{1}{\omega' - \omega} - i\pi\delta(\omega' - \omega), \tag{9.179}$$

where the P denotes the *principal part* of the integrand (see Figure 9.9).

Thus we may rewrite (9.178) as

$$-\frac{i}{2}(\epsilon(\omega) - 1) + P\int_{-\infty}^{\infty}\frac{d\omega'}{2\pi}\frac{\epsilon(\omega') - 1}{\omega' - \omega} = 0, \tag{9.180}$$

$$\implies \epsilon(\omega) - 1 = -\frac{i}{\pi}P\int_{-\infty}^{\infty} d\omega'\,\frac{\epsilon(\omega') - 1}{\omega' - \omega}. \tag{9.181}$$

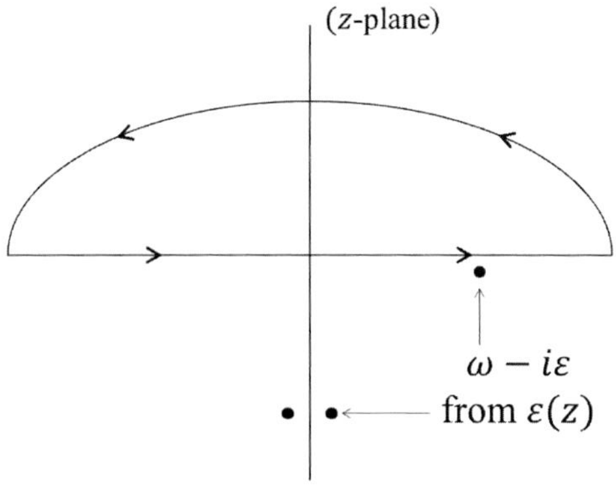

Fig. 9.8 Poles of the integrand in (9.175).

Fig. 9.9 Principal part of integrand, represented as the sum of two contour integrals.

Taking real and imaginary parts leads to

$$\mathrm{Re}\,\epsilon(\omega) = 1 + \frac{1}{\pi} P \int_{-\infty}^{\infty} d\omega' \, \frac{\mathrm{Im}\,\epsilon(\omega')}{\omega' - \omega}, \tag{9.182}$$

$$\mathrm{Im}\,\epsilon(\omega) = -\frac{1}{\pi} P \int_{-\infty}^{\infty} d\omega' \, \frac{\mathrm{Re}\,\epsilon(\omega') - 1}{\omega' - \omega}, \tag{9.183}$$

which are the *Kramers-Kronig relations*.

As a mathematical aside, note that we may write

$$\frac{1}{\omega' - \omega - i\epsilon} = P \frac{1}{\omega' - \omega} + i\pi\delta(\omega' - \omega), \tag{9.184}$$

as depicted in Figure 9.10. So, for example, if $f(z)$ is nonsingular on C, then using either (9.179) or (9.184) we have (see Figure 9.11)

$$P \oint_C \frac{f(z)dz}{z - \alpha} = 0 + i\pi f(\alpha) = 2\pi i f(\alpha) - i\pi f(\alpha).$$

The response function $G(\tau)$ in (9.172) must be real, which implies the relation $\epsilon(-\omega) = \epsilon^*(\omega^*)$ in the complex ω-plane. This further implies $\mathrm{Re}\,\epsilon(\omega)$

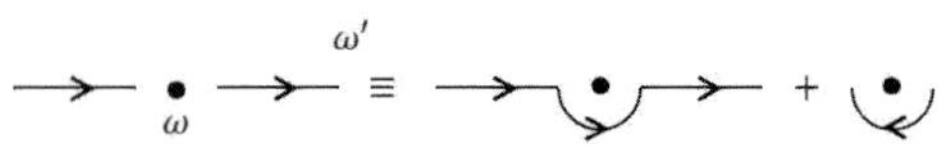

Fig. 9.10 Principal part of integrand (alternative representation).

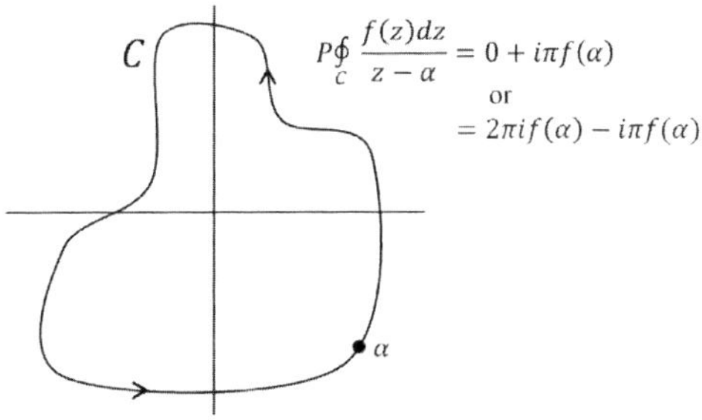

Fig. 9.11 Principle part of closed contour integral.

is even and $\mathrm{Im}\,\epsilon(\omega)$ is odd in ω for ω on the real axis. This means we may write

$$\mathrm{Re}\,\epsilon(\omega) - 1 = \frac{2}{\pi} P \int_0^\infty d\omega' \frac{\omega' \,\mathrm{Im}\,\epsilon(\omega')}{\omega'^2 - \omega^2}, \tag{9.185}$$

$$\mathrm{Im}\,\epsilon(\omega) = -\frac{2\omega}{\pi} P \int_0^\infty d\omega' \frac{\mathrm{Re}\,\epsilon(\omega') - 1}{\omega'^2 - \omega^2}. \tag{9.186}$$

Let's now go back and look at an important consequence of the Kramers-Kronig dispersion relations, namely *sum rules* for the dielectric constant ϵ (which we saw already in our simple model in the last chapter). We saw that $\epsilon(\omega) = 1 - G'(0^+)/\omega^2 + \ldots$, and we may therefore define

$$\omega_{\mathrm{p}}^2 \equiv \lim_{\omega\to\infty} \omega^2(1 - \epsilon(\omega)). \tag{9.187}$$

In Exercise 9.7.2 you will show that the Kramers-Kronig dispersion relations applied to (9.187) give

$$\omega_p^2 = \frac{2}{\pi} \int_0^\infty d\omega\, \omega \mathrm{Im}\,\epsilon(\omega). \tag{9.188}$$

In our simple model from Section 9.6 (with no conductance electrons),

$$\epsilon(\omega) = 1 + \frac{4\pi e^2}{m} \sum_i \frac{n_{\mathrm{b},i}}{\omega_{0,i}^2 - \omega^2 - i\omega\gamma_i}, \tag{9.189}$$

then either (9.188) or (9.189) gives

$$\omega_p^2 = \frac{4\pi e^2}{m} \sum_i n_{\mathrm{b}}^i = \frac{4\pi e^2}{m} N_{\mathrm{b}}, \tag{9.190}$$

which recovers the model form of the square of the plasma frequency.

There is another sum rule for the real part of $\epsilon(\omega)$. Consider our model in the case of a single resonance

$$\mathrm{Re}\,(\epsilon(\omega)-1)=\frac{4\pi n_{\mathrm{b}}e^2}{m}\frac{(\omega_0^2-\omega^2)}{(\omega_0^2-\omega^2)^2+\omega^2\gamma^2}. \tag{9.191}$$

The integral from 0 to ∞ of the right-hand side of (9.191) may be evaluated using

$$\int_0^\infty d\omega\,\frac{\omega^2}{(\omega_0^2-\omega^2)^2+\omega^2\gamma^2}=\frac{\pi}{2\gamma} \tag{9.192}$$

(This integral was used previously in (9.139)), and

$$\int_0^\infty d\omega\,\frac{1}{(\omega_0^2-\omega^2)^2+\omega^2\gamma^2}=\frac{\pi}{2\gamma\omega_0^2}. \tag{9.193}$$

Altogether, we obtain from (9.191) that

$$\int_0^\infty d\omega\,[\mathrm{Re}\,\epsilon(\omega)-1]=0. \tag{9.194}$$

In fact, in Exercise 9.7.2 you will show using just the Kramers-Kronig relations that in general

$$\int_0^N d\omega\,[\mathrm{Re}\,\epsilon(\omega)-1]=\frac{\omega_p^2}{N}+\mathcal{O}\left(\frac{1}{N^3}\right). \tag{9.195}$$

How do these sum rules for $\mathrm{Re}\,\epsilon(\omega)$ and $\mathrm{Im}\,\epsilon(\omega)$ change if there is a pole at $\omega=0$, as we saw in (9.149) for conductors? Suppose the effective dielectric function is

$$\epsilon_{\mathrm{eff}}(\omega)=\epsilon(\omega)+\frac{i4\pi\sigma(\omega)}{\omega}, \tag{9.196}$$

where $\epsilon(\omega)$ satisfies the usual Kramers-Kronig relations and

$$\sigma(\omega)=\frac{\gamma\sigma(0)}{\gamma-i\omega}. \tag{9.197}$$

In Exercise 9.7.3 you will show that

$$\mathrm{Re}\,(\epsilon_{\mathrm{eff}}(\omega)-1)=\frac{2}{\pi}P\int_0^\infty d\omega'\,\frac{\omega'\,\mathrm{Im}\,\epsilon_{\mathrm{eff}}(\omega')}{\omega'^2-\omega^2}, \tag{9.198}$$

$$\mathrm{Im}\,\epsilon_{\mathrm{eff}}(\omega)=\frac{4\pi\sigma(0)}{\omega}-\frac{2\omega}{\pi}P\int_0^\infty d\omega'\,\frac{\mathrm{Re}\,\epsilon_{\mathrm{eff}}(\omega')-1}{\omega'^2-\omega^2}, \tag{9.199}$$

and

$$\omega_{\mathrm{p,eff}}^2=\omega_{\mathrm{p}}^2+4\pi\gamma\sigma(0), \tag{9.200}$$

$$\int_0^N d\omega'\,[\mathrm{Re}\,\epsilon_{\mathrm{eff}}(\omega')-1]=\frac{\omega_{\mathrm{p,eff}}^2}{N}-2\pi^2\sigma(0)+\mathcal{O}\left(\frac{1}{N^3}\right). \tag{9.201}$$

9.8 Dispersion in one-dimension: theory and example

Let's discuss some additional consequences of dispersion relations. We shall limit the discussion to some background and an illustrative example which is characteristic of the dispersive character of materials in general.

We begin with recalling Equation (9.5) for plane waves, where μ and ϵ can be considered angular frequency dependent:

$$\left(\nabla^2 - \frac{\mu\epsilon}{c^2}\frac{\partial^2}{\partial t^2}\right)\vec{E} = 0. \tag{9.202}$$

The phase velocity is

$$v_p = \frac{\omega}{k} = \frac{c}{\sqrt{\mu\epsilon}}. \tag{9.203}$$

In regions where $\epsilon > 1$, then $v_p < c$. However, we have already seen cases (such as (9.155)) where ϵ has the form

$$\epsilon(\omega) = 1 - \frac{\omega_p^2}{\omega^2},$$

from which it follows that $v_p > c$. But special relativity says that energy cannot be transported faster than c. How may we resolve this apparent contradiction?

To answer this, we need some additional background for the wave equation. We begin with the Maxwell equations

$$\vec{\nabla} \times \vec{H} = \frac{1}{c}\dot{\vec{D}} + \frac{4\pi}{c}\vec{J}, \tag{9.204}$$

$$\vec{\nabla} \times \vec{E} = -\frac{1}{c}\dot{\vec{B}}. \tag{9.205}$$

We consider the nonconducting case where $\vec{J} = 0$, so that

$$\vec{D}(\vec{x},t) = \vec{E}(\vec{x},t) + \int_{0^-}^{\infty} d\tau\, \vec{E}(\vec{x},t-\tau)\overbrace{\int_{-\infty}^{\infty}\frac{d\omega'}{2\pi}e^{-i\omega'\tau}\,[\epsilon(\omega')-1]}^{G(\tau)}, \tag{9.206}$$

where the $d\tau$ integral in (9.206) includes $\tau = 0$ (which turns out to be important later). In the case where $\mu = 1$ we have

$$\begin{cases} \vec{\nabla} \times \vec{B} = \frac{1}{c}\dot{\vec{E}} + \frac{1}{c}\int_{0^-}^{\infty} d\tau\, \frac{\partial}{\partial t}\vec{E}(\vec{x},t-\tau)G(\tau) \\ \vec{\nabla} \times \vec{E} = -\frac{1}{c}\dot{\vec{B}}. \end{cases} \tag{9.207}$$

$$\implies \nabla^2 \vec{E} = \frac{1}{c^2}\frac{\partial^2 \vec{E}}{\partial t^2} + \frac{1}{c^2}\int_{0-}^{\infty} d\tau\, \frac{\partial^2}{\partial t^2}\vec{E}(\vec{x}, t-\tau)G(\tau). \tag{9.208}$$

This is our new wave equation, which is nonlocal in time. For simplicity we replace $\vec{E}(\vec{x}, t)$ with a scalar function in one dimension $u(x, t)$. We want to solve

$$\frac{\partial^2 u(x,t)}{\partial x^2} = \frac{1}{c^2}\frac{\partial^2 u}{\partial t^2} + \frac{1}{c^2}\int_{0-}^{\infty} d\tau\, \frac{\partial^2}{\partial t^2}u(x, t-\tau)G(\tau). \tag{9.209}$$

We infer a solution of the form

$$u(x,t) = \frac{1}{\sqrt{2\pi}}\int_{-\infty}^{\infty} d\omega\, e^{-i\omega t}[A(\omega)e^{+ik(\omega)x} + B(\omega)e^{-ik(\omega)x}], \tag{9.210}$$

where two terms are included in the integral because a wave with a given ω can go in either the positive or negative direction. It follows that the first two terms in (9.209) can be re-expressed as integrals:

$$\frac{\partial^2 u}{\partial x^2} = \frac{1}{\sqrt{2\pi}}\int_{-\infty}^{\infty} d\omega\, e^{-i\omega t}(-k(\omega)^2)[\ldots], \tag{9.211}$$

$$\frac{1}{c}\frac{\partial^2 u}{\partial t^2} = \frac{1}{\sqrt{2\pi}}\int_{-\infty}^{\infty} d\omega\, e^{-i\omega t}\left(\frac{-\omega^2}{c^2}\right)[\ldots]. \tag{9.212}$$

The third term in (9.209) may be simplified using the definition of $G(\tau)$ in (9.172):

$$\frac{1}{c^2}\int_{0-}^{\infty} d\tau\, \frac{\partial^2}{\partial t^2}u(x, t-\tau)G(\tau)$$
$$= -\frac{1}{c^2}\frac{\partial^2 u}{\partial t^2} + \frac{1}{\sqrt{2\pi}}\int_{-\infty}^{\infty} d\omega\, e^{-i\omega t}\left(-\epsilon(\omega)\frac{\omega^2}{c^2}\right)[\ldots]. \tag{9.213}$$

Plugging (9.213) into (9.209) and canceling the common term leads to

$$k(\omega)^2 = \epsilon(\omega)\frac{\omega^2}{c^2} \implies k(\omega) = \pm\, n(\omega)\frac{\omega}{c}. \tag{9.214}$$

Either root may work as long as we are consistent. We may obtain A, B from the initial conditions on $u(x, t)$:

$$u(0,t) = \frac{1}{\sqrt{2\pi}}\int_{-\infty}^{\infty} d\omega\, e^{-i\omega t}[A(\omega) + B(\omega)], \tag{9.215}$$

$$\frac{\partial u(0,t)}{\partial x} = \frac{1}{\sqrt{2\pi}}\int_{-\infty}^{\infty} d\omega\, (ik)e^{-i\omega t}[A(\omega) - B(\omega)], \tag{9.216}$$

which may be inverted to obtain:

$$A(\omega) = \frac{1}{2\sqrt{2\pi}}\int_{-\infty}^{\infty} dt\, e^{+i\omega t}\left[u(0,t) + \frac{1}{ik(\omega)}\frac{\partial u(0,t)}{\partial x}\right], \tag{9.217}$$

$$B(\omega) = \frac{1}{2\sqrt{2\pi}}\int_{-\infty}^{\infty} dt\, e^{+i\omega t}\left[u(0,t) - \frac{1}{ik(\omega)}\frac{\partial u(0,t)}{\partial x}\right]. \tag{9.218}$$

Equations (9.217) and (9.218) show that $A^*(-\omega) = A(\omega), B^*(-\omega) = B(\omega)$ as long as $k^*(-\omega) = -k(\omega)$, which in turn follows from

$$n^*(-\omega) = n(\omega) \tag{9.219}$$

(and a similar property holds for $\epsilon(\omega)$). Using (9.210) it follows that $u(x,t)$ is real, and the integration in (9.210) can be restricted to positive ω values (by taking twice the real part).

Alternatively, instead of the ω-integral (9.210) we may postulate a solution in the form of an integral over k:

$$u(x,t) = \mathrm{Re}\,\frac{1}{\sqrt{2\pi}}\int_{-\infty}^{\infty} dk\, e^{-i\omega(k)t}[A(k)e^{ikx}]. \tag{9.220}$$

We then obtain

$$\frac{\partial^2 u}{\partial x^2} = \mathrm{Re}\,\frac{1}{\sqrt{2\pi}}\int_{-\infty}^{\infty} dk\, e^{-i\omega t}(-k^2)A(k)e^{ikx}, \tag{9.221}$$

$$\frac{1}{c}\frac{\partial^2 u}{\partial t^2} = \mathrm{Re}\frac{1}{\sqrt{2\pi}}\int_{-\infty}^{\infty} dk\, e^{-i\omega t}\left(\frac{-\omega^2}{c^2}\right)A(k)e^{ikx}, \tag{9.222}$$

$$\begin{aligned}\frac{1}{c^2}\int_{0^-}^{\infty} d\tau\,\frac{\partial^2}{\partial t^2}u(x,t-\tau)G(\tau) = {}&-\frac{1}{c^2}\frac{\partial^2 u}{\partial t^2}\\ &+\mathrm{Re}\frac{1}{\sqrt{2\pi}}\int_{-\infty}^{\infty} dk\, e^{-i\omega t}\left(-\epsilon\frac{\omega^2}{c^2}\right)A(k)e^{ikx}.\end{aligned} \tag{9.223}$$

Hence $u(x,t)$ is a solution of the wave equation as long as

$$k^2 = \epsilon(\omega(k))\frac{\omega^2(k)}{c^2}, \tag{9.224}$$

and $\epsilon(-\omega^*(k)) = \epsilon^*(\omega(k))$. Since the two solutions for $\omega(k)$ are negatives of each other, we may arbitrarily pick one. We may re-express the initial conditions as

$$u(x,0) = \frac{1}{2\sqrt{2\pi}}\int_{-\infty}^{\infty} dk\,[Ae^{+ikx} + A^*e^{-ikx}], \tag{9.225}$$

$$\frac{\partial u(x,0)}{\partial t} = \frac{1}{2\sqrt{2\pi}}\int_{-\infty}^{\infty} dk\,(-i\omega(k))[Ae^{+ikx} - A^*e^{-ikx}], \tag{9.226}$$

which may then be inverted to obtain:

$$A(k) = \frac{1}{\sqrt{2\pi}}\int_{-\infty}^{\infty} dx\, e^{-ikx}\left[u(x,0) + \frac{i}{\omega(k)}\frac{\partial u(x,0)}{\partial t}\right], \tag{9.227}$$

$$A^*(-k) = \frac{1}{\sqrt{2\pi}}\int_{-\infty}^{\infty} dx\, e^{-ikx}\left[u(x,0) - \frac{i}{\omega(k)}\frac{\partial u(x,0)}{\partial t}\right]. \tag{9.228}$$

These equations show that for consistency, we must have

$$\omega^*(k) = \omega(-k). \tag{9.229}$$

Now that we know something about the equation we are solving, let's consider a particular example of a dispersion relation:

$$\omega(k) = \nu\left(1 + \frac{a^2k^2}{2}\right). \tag{9.230}$$

This can be considered an approximation to Equations (9.156), (9.157) when $|k| \ll \omega_p/c$. The relation $k^2 = \epsilon\,\omega^2/c^2$ for $\mu = 1$ implies

$$\epsilon(\omega) = \frac{2c^2}{a^2\omega}\left(\frac{1}{\nu} - \frac{1}{\omega}\right). \tag{9.231}$$

Notice that

$$\omega > \nu \implies k \text{ real} \implies \text{solution oscillates in space,}$$
$$\omega < \nu \implies k \text{ imaginary} \implies \text{solution is damped in space.}$$

Our equation is

$$\frac{\partial^2 u}{\partial x^2} = \frac{1}{c^2}\frac{\partial^2 u}{\partial t^2} + \frac{1}{c^2}\int_{0-}^{\infty} d\tau\, \frac{\partial^2}{\partial t^2}u(x, t-\tau)G(\tau), \tag{9.232}$$

$$G(\tau) = \frac{1}{2\pi}\int_{-\infty}^{\infty} d\omega\, [\epsilon(\omega) - 1]e^{-i\omega\tau}. \tag{9.233}$$

We may replace $\partial^2 u/\partial t^2$ with $\partial^2 u/\partial\tau^2$, and integrate by parts:

$$\frac{\partial^2 u}{\partial x^2} = \frac{1}{c^2}\frac{\partial^2 u}{\partial t^2} + \frac{1}{c^2}\int_{0-}^{\infty} d\tau\, u(x, t-\tau)\frac{\partial^2 G(\tau)}{\partial\tau^2}. \tag{9.234}$$

Unfortunately $\epsilon(\omega) - 1$ has pieces proportional to a constant and $\sim 1/\omega$ which imply a discontinuous response function, so strictly speaking our previous model (and the K-K relations) do not apply. In addition, the real $\epsilon(\omega)$ form in Equation (9.231) is not even in ω, which means the response function is not real. Nonetheless, let's just plow ahead and see what we get, as we will see it illustrates some important aspects of wave behavior. The second time derivative of the response function is:

$$\begin{aligned}\frac{\partial^2}{\partial\tau^2}G(\tau) &= \frac{1}{2\pi}\int_{-\infty}^{\infty} d\omega\, e^{-i\omega\tau}\left[\frac{2c^2}{a^2}\left(1 - \frac{i\omega}{i\nu}\right) + \omega^2\right], \\ &= \frac{2c^2}{a^2}\left[\delta(\tau) + \frac{1}{i\nu}\delta'(\tau)\right] - \delta''(\tau).\end{aligned} \tag{9.235}$$

Substituting into (9.234) gives

$$\frac{\partial^2 u}{\partial x^2} = \frac{1}{c^2}\int_{0^-}^{\infty} d\tau\, u(x,t-\tau)\left[\frac{2c^2}{a^2}(\delta(\tau)+\frac{1}{i\nu}\delta'(\tau))-\delta''(\tau)\right]+\frac{1}{c^2}\frac{\partial^2 u}{\partial t^2}$$
$$= \frac{2}{a^2}\left[u-\frac{i}{\nu}\frac{\partial u}{\partial t}\right]. \tag{9.236}$$

Let $y(x,t) \equiv u(x,t)e^{i\nu t}$. Then

$$-\frac{a^2\nu}{2}\frac{\partial^2 y}{\partial x^2} = i\frac{\partial y}{\partial t}. \tag{9.237}$$

This is essentially the Schrödinger equation. The wave packet in this case is behaving like a nonrelativistic particle of mass

$$m_{\text{eff}} = \frac{\hbar}{a^2\nu}. \tag{9.238}$$

The general solution is

$$y(x,t) = \frac{1}{\sqrt{2\pi}}\int_{-\infty}^{\infty} dk\, e^{i(kx-\omega_s(k)t)}g(k), \tag{9.239}$$

$(\omega_s(k) \equiv a^2\nu k^2/2)$ where

$$g(k) = \frac{1}{\sqrt{2\pi}}\int_{-\infty}^{\infty} dx\, e^{-ikx}y(x,0). \tag{9.240}$$

We suppose that $y(x,0)$ has the form of a Gaussian wave packet:

$$y(x,0) = \frac{c}{\sqrt{\delta_x}}\exp\left(i\bar{k}x-\frac{x^2}{2\delta_x^2}\right), \tag{9.241}$$

where δ_x, $\bar{k}$ and c are real constants. Notice that $\langle k\rangle = \bar{k}$, where $\langle\ldots\rangle$ has the usual quantum-mechanical meaning. The real part of $y(x,0)$ is shown in Figure 9.12. Notice the oscillations within the Gaussian envelope. If we substitute (9.241) into (9.240) and plug the result into (9.239), after evaluating the Gaussian integrals we end up with

$$u(x,t) = y(x,t)e^{-i\nu t}$$
$$= \frac{c}{\sqrt{\delta_x + i\dfrac{a^2\nu t}{\delta_x}}}\exp\left(\frac{-(x-a^2\nu\bar{k}t)^2}{2(\delta_x^2+ia^2\nu t)}\right)\exp(i\bar{k}x-i\omega(\bar{k})t), \tag{9.242}$$

where the first exponential is interpreted as the new Gaussian envelope and the second as a phase. In this example $u(x,t)$ is not real; this is an inevitable consequence not of the complex initial state (9.241), but of the odd ω term in $\epsilon(\omega)$. Note that simply taking the real part of solutions

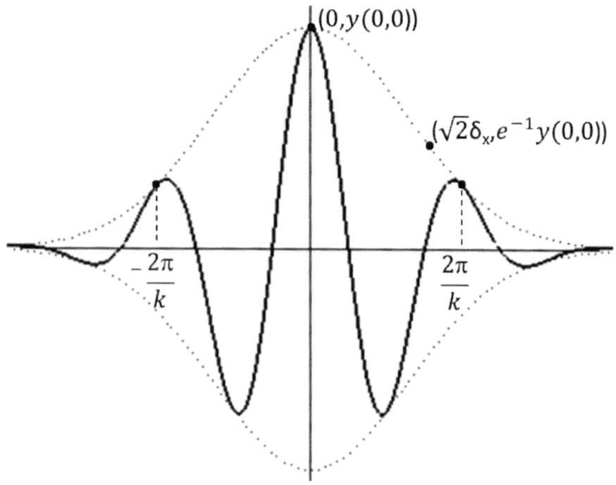

Fig. 9.12 Real part of $y(x,0)$. The "envelope" has the form $y(0,0)\exp(-x^2/(2\delta_x^2))$.

such as (9.242) is unacceptable, as there is no associated wave equation for this quantity. (This treacherous situation does not occur for the original unapproximated dispersion relation (9.155), as seen in Exercise 9.8.1.) We may compute the phase speed and envelope speed for this example as

$$\text{phase speed} = \frac{\nu}{\bar{k}}\left(1+\frac{a^2\bar{k}^2}{2}\right) = \frac{\omega(\bar{k})}{\bar{k}}, \tag{9.243}$$

$$\text{envelope speed} = a^2\nu\bar{k}\left(=\frac{\hbar\bar{k}}{m_{\text{eff}}}\right) = \frac{d\omega(k)}{dk}|_{k=\bar{k}}. \tag{9.244}$$

The quantity on the right in Equation (9.244) is called the *group velocity* and is (for materials that are weakly dispersive) the approximate speed of energy transmission. This corresponds to making the *stationary phase approximation*,

$$\frac{\partial(kx-\omega(|k|)t)}{\partial k} \approx 0,$$

in the argument of the exponent in (9.220). We will encounter this approximation again in the scattering considerations of Section 12.3.

9.9 Charged particle energy loss in materials

The results we have found for the effective dielectric constant from our simple model of constitutive relations are sufficient to give us a rough understanding of dissipative energy loss processes for charged particles traveling through materials. There are two types of interactions possible, both electromagnetic of course. At nonrelativistic energies, we expect we should be able to use our semiclassical energy loss model, which parameterized free and bound electron energy interactions in terms of the damping constant γ and resonance values $\omega_{0,i}$. This is possible because all the electron properties, both bound and free, should be summed up in the effective dielectric function, $\epsilon_{\text{eff}}(\omega)$. Experimentally, these interactions would be characterized as resulting in electronic excitation and ionization of the atoms in the material. At higher energies the next layer in physical structure, the nucleus, begins to play a role in energy losses as incoming charged particles will be accelerated by these fixed scattering centers, resulting in radiative energy losses. We will defer discussion of these types of losses until more theory has been developed in Chapter 11 leading up to Section 11.4, where we will eventually compare the rates of losses from these two mechanisms.

A straight-line path of a particle such as an electron or a light ion at sufficiently high energies is characterized by

$$\rho(\vec{x},t) = q\,\delta(\vec{x}-\vec{v}t), \tag{9.245}$$

$$\vec{J}(\vec{x},t) = q\vec{v}\,\delta(\vec{x}-\vec{v}t), \tag{9.246}$$

where q is the charge of the incoming particle and where $\vec{v}$ is a constant vector. Notice that the incoming particle mass does not enter this characterization, just its velocity. However, when we need the value of this mass below we will assign it to be M, which can distinguish it from the electron's mass m in this discussion. We define the momentum and angular frequency Fourier transforms of these as

$$\rho(\vec{k},\omega) \equiv \frac{1}{(2\pi)^2}\int_{-\infty}^{\infty} dt\, e^{i\omega t}\int d^3x\, e^{-i\vec{k}\cdot\vec{x}}\rho(\vec{x},t), \tag{9.247}$$

$$\vec{J}(\vec{k},\omega) \equiv \frac{1}{(2\pi)^2}\int_{-\infty}^{\infty} dt\, e^{i\omega t}\int d^3x\, e^{-i\vec{k}\cdot\vec{x}}\vec{J}(\vec{x},t), \tag{9.248}$$

resulting in the simple forms

$$\rho(\vec{k},\omega) = \frac{q}{2\pi}\,\delta(\omega-\vec{k}\cdot\vec{v}), \tag{9.249}$$

$$\vec{J}(\vec{k},\omega) = \frac{q\vec{v}}{2\pi}\,\delta(\omega-\vec{k}\cdot\vec{v}). \tag{9.250}$$

If we Fourier transform the source equation for charge density,

$$\vec{\nabla}\cdot\vec{D}(\vec{x},t) = 4\pi\rho(\vec{x},t), \tag{9.251}$$

we obtain

$$i\vec{k}\cdot\vec{D}(\vec{k},\omega) = 4\pi\rho(\vec{k},\omega). \tag{9.252}$$

We will introduce the effects of dispersion through the relationship

$$\vec{D}(\vec{k},\omega) = \epsilon_{\text{eff}}(\omega)\vec{E}(\vec{k},\omega), \tag{9.253}$$

where $\epsilon_{\text{eff}}(\omega)$ includes the contribution of conductance electrons as in Exercise 9.7.3.

We now need to relate the electric field to the underlying potentials to produce a formal solution in momentum space. We use the nonrelativistic approximation

$$\vec{E}(\vec{x},t) = -\vec{\nabla}\Phi(\vec{x},t), \tag{9.254}$$

leaving out the dependence on the vector potential. At nonrelativistic velocities, we expect the trajectories of particles to be determined mainly by the interaction of the classical charge densities, as for example in Rutherford scattering. This assumption neglects the effect of the associated magnetic field, which we expect will become comparable with the electric contribution for relativistic particle velocities.

The momentum and frequency Fourier transform of (9.254) is

$$\vec{E}(\vec{k},\omega) = -i\vec{k}\,\Phi(\vec{k},\omega). \tag{9.255}$$

Now solving for $\Phi(\vec{k},\omega)$, and therefore the electric field, using (9.249), (9.252), (9.253) and (9.255) gives

$$\vec{E}(\vec{k},\omega) = -\frac{2iq\vec{k}}{\vec{k}^2\epsilon_{\text{eff}}(\omega)}\delta(\omega - \vec{k}\cdot\vec{v}). \tag{9.256}$$

We would like to understand the effect of energy loss on an electron traveling at nonrelativistic velocities through a medium. The tool for this is the *stopping power*, which is defined as the energy loss per path length,

$$\frac{dE}{dx} \equiv \frac{1}{v}\frac{dE}{dt} = -\frac{1}{v}\int d^3x\,\vec{J}(\vec{x},t)\cdot\vec{E}(\vec{x},t), \tag{9.257}$$

where we have used (8.6). Note the negative sign in the definition: this represents work done *on* the system. Doing the spatial integral gives us the intermediate form

$$\begin{aligned}\frac{dE}{dx} &= -\frac{1}{v}\int d^3k\,\vec{J}(-\vec{k},t)\cdot\vec{E}(\vec{k},t)\\ &= -\frac{1}{2\pi v}\int d^3k\int_{-\infty}^{\infty}d\omega\int_{-\infty}^{\infty}d\omega'\,e^{-i(\omega+\omega')t}\,\vec{J}(-\vec{k},\omega')\cdot\vec{E}(\vec{k},\omega).\end{aligned} \tag{9.258}$$

Using (9.250) and (9.256) above now tells us

$$\begin{aligned}\frac{dE}{dx} &= \frac{iq^2}{2\pi^2 v}\int \frac{d^3k}{\vec{k}^2}\vec{k}\cdot\vec{v}\int_{-\infty}^{\infty} d\omega\, \frac{\delta(\omega-\vec{k}\cdot\vec{v})}{\epsilon_{\rm eff}(\omega)},\\ &= \frac{iq^2}{2\pi^2 v}\int \frac{d^3k}{\vec{k}^2}\frac{\vec{k}\cdot\vec{v}}{\epsilon_{\rm eff}(\vec{k}\cdot\vec{v})}. \end{aligned} \tag{9.259}$$

We can replace $1/\epsilon_{\rm eff}(\omega)$ with its odd part, as seen in Section 9.7:

$$\text{Odd}\left(\frac{1}{\epsilon_{\rm eff}(\omega)}\right) = i\,\text{Im}\left(\frac{1}{\epsilon_{\rm eff}(\omega)}\right). \tag{9.260}$$

This gives

$$\frac{dE}{dx} = -\frac{q^2}{2\pi^2 v}\int \frac{d^3k}{\vec{k}^2}\,\vec{k}\cdot\vec{v}\;\text{Im}\left(\frac{1}{\epsilon_{\rm eff}(\vec{k}\cdot\vec{v})}\right). \tag{9.261}$$

It is convenient to define $\omega \equiv \vec{k}\cdot\vec{v}$. Then, assuming cylindrical symmetry about $\vec{v}$ and using cylindrical coordinates, we may write the integration measure divided by $\vec{k}^2 = k_\perp^2 + k_z^2$ as

$$\frac{d^3k}{\vec{k}^2} = 2\pi\frac{k_\perp dk_\perp d\left(\frac{\omega}{v}\right)}{k_\perp^2 + \left(\frac{\omega}{v}\right)^2}, \tag{9.262}$$

where we have done the angular integration and $k_z = \omega/v$. The perpendicular part of the integral, however, requires an upper limit:

$$\int_0^{k_0} \frac{k_\perp dk_\perp}{k_\perp^2 + \left(\frac{\omega}{v}\right)^2} = \frac{1}{2}\ln\left(1 + \left(\frac{vk_0}{\omega}\right)^2\right), \tag{9.263}$$

where k_0 is the maximum value of $k_\perp$. We need an understanding of the relationship between this quantity and the other kinematical quantities in the interactions to proceed. Drawing on the quantum mechanical concept of particle momentum, we interpret $\hbar k_0$ as the maximum transverse momentum transfer from M to m in the scattering event. This transferred momentum is just twice the magnitude of the momentum of either particle in the center of mass (or momentum) frame, $p^{\rm cm}$, at a scattering angle of $90°$. In this frame we have from nonrelativistic dynamics that

$$p^{\rm cm} = \frac{Mm}{M+m}v. \tag{9.264}$$

Thus, we set

$$\hbar k_0 = 2p^{\rm cm} = \zeta m v, \tag{9.265}$$

where

$$\zeta \equiv \frac{2}{1 + m/M}. \tag{9.266}$$

The two extremes of this function are $\zeta = 1$ for $M = m$ and $\zeta = 2$ for $M \gg m$.

We still have the remaining longitudinal part of the integration to finish. Making the above replacement means that inside the ω integral we have the factor

$$\ln\left(1 + \left(\frac{vk_0}{\omega}\right)^2\right) \longrightarrow \ln\left(1 + \left(\frac{\zeta mv^2}{\hbar\omega}\right)^2\right), \tag{9.267}$$

to integrate over. Since this factor is a slowly varying logarithm, it makes sense to try to pull it outside the integral. We replace ω in (9.267) with an average effective value, $\bar{\omega}$. $\hbar\bar{\omega}$ is viewed as an average ionization energy for the particle along its path, which can be measured and tabulated for various substances. The theory predicts this parameter is approximately independent of energy. Within this picture, the average ionization energy satisfies $mv^2/(\hbar\bar{\omega}) \gg 1$. Given this, the modified integral now reads:

$$\frac{dE}{dx} = -\frac{q^2}{\pi v^2}\ln\left(\frac{\zeta mv^2}{\hbar\bar{\omega}}\right)\int_{-\infty}^{\infty} d\omega\, \omega\, \mathrm{Im}\left(\frac{1}{\epsilon_{\text{eff}}(\omega)}\right). \tag{9.268}$$

The integral above reminds us of the sum rule in (9.188), which also has an extension in Exercise 9.7.3. In fact, making the same analytic assumptions for $1/\epsilon_{\text{eff}}(\omega)$ as for $\epsilon_{\text{eff}}(\omega)$ (this is a nontrivial assumption) allows us to derive a similar rule. The connection between the two cases arises from

$$\lim_{\omega\to\infty} \frac{1}{\epsilon_{\text{eff}}(\omega)} \to 1 + \frac{(\omega_p)^2_{\text{eff}}}{\omega^2}, \tag{9.269}$$

where

$$(\omega_p)^2_{\text{eff}} \equiv \frac{4\pi(n_{\text{b}} + n_{\text{c}})e^2}{m} = \frac{4\pi n_{\text{b}} e^2}{m} + 4\pi\gamma\sigma(0), \tag{9.270}$$

includes the conductance density, and e denotes the electron charge. The sum rule is

$$(\omega_p)^2_{\text{eff}} = -\frac{2}{\pi}\int_0^{\infty} d\omega\, \omega\, \mathrm{Im}\left(\frac{1}{\epsilon_{\text{eff}}(\omega)}\right), \tag{9.271}$$

which differs only by a sign from the result in Exercise 9.7.3. Finally, this gives

$$\begin{aligned}\frac{dE}{dx} &= \frac{q^2(\omega_p)^2_{\text{eff}}}{v^2}\ln\left(\frac{\zeta mv^2}{\hbar\bar{\omega}}\right)\\ &= \frac{4\pi N Z_1^2 e^4}{mv^2}\ln\left(\frac{\zeta mv^2}{\hbar\bar{\omega}}\right),\end{aligned} \tag{9.272}$$

where N is the total electron charge density, and we are using $|q| = Z_1|e|$ for the projectile. This remarkably simple nonrelativistic equation and its higher-energy generalizations are fundamental in characterizing the penetration of charged particles in both experimental and medical applications. Some of the most prominent quantum physicists of the last century including Bohr, Bethe and Fermi have contributed to the extended theory of stopping power. An important organization which has existed since 1928, the International Commission on Radiation Units and Measurements (ICRU), has studied and compiled results from many investigations, and their reports are extremely useful to programs of study involving radiation diagnosis, protection and measurement.

9.10 Going Deeper

9.10.1 *Plasma physics*

(1) P. M. Bellan, *Fundamentals of Plasma Physics*, Cambridge University Press (Cambridge) 2008.

(2) A. J. Bittencourt, *Fundamentals of Plasma Physics*, 3rd ed., Springer (New York) 2004.

(3) F. F. Chen, *Plasma Physics and Controlled Fusion, Vol. 1*, 2nd ed., Springer (New York) 2006.

(4) D. R. Nicholson, *Introduction to Plasma Theory*, John Wiley & Sons (New York) 1983.

(5) R. B. White, *The Theory of Toroidally Confined Plasmas*, 3rd ed., Imperial College Press (London) 2014.

9.10.2 *Dissipative charged particle energy loss*

(1) K. A. Olive *et al.* (Particle Data Group), Chin. Phys. C, 38, 090001 (2014); `http://pdg.lbl.gov`; Review, "Passage of Particles Through Matter."

(2) C. A. Brau, *Modern Problems in Classical Electrodynamics*, Oxford University Press (Oxford) 2004; Sections 7.3.1, 7.3.2.

(3) J. D. Jackson, *Classical Electrodynamics*, 3rd ed., John Wiley & Sons (New York) 1999; Sections 13.1–13.3.

(4) "Stopping Powers for Electrons and Positrons", ICRU Report 37, International Commission on Radiation Units and Measurements (Bethesda, MD) 1984.

(5) "Stopping Powers and Ranges for Protons and Alpha Particles", ICRU Report 49, International Commission on Radiation Units and Measurements (Bethesda, MD) 1993.

(6) "Stopping of Ions Heavier than Helium", ICRU Report 73, Journal of the ICRU **5**, Oxford University Press, 2005.

9.11 Exercises

Exercise 9.1.1.

(a) For plane wave solutions to Maxwell's equations,

$$\vec{E}(\vec{x},t) = \mathrm{Re}\left(\vec{\mathcal{E}}\, e^{i\vec{k}\cdot\vec{x}-i\omega t}\right),$$
$$\vec{B}(\vec{x},t) = \mathrm{Re}\left(\vec{\mathcal{B}}\, e^{i\vec{k}\cdot\vec{x}-i\omega t}\right),$$

in a linear isotropic material for which $\vec{\mathcal{D}} = \epsilon\vec{\mathcal{E}}$ and $\vec{\mathcal{B}} = \mu\vec{\mathcal{H}}$, where $\epsilon \equiv \epsilon(\omega)$ and $\mu \equiv \mu(\omega)$ are real constants, show that the time averaged electric and magnetic energy densities are equal: $w_e = w_m$. Recall that (Section 8.4)

$$w_e = \frac{1}{16\pi}\vec{\mathcal{E}}\cdot\vec{\mathcal{D}}^*, \quad w_m = \frac{1}{16\pi}\vec{\mathcal{B}}\cdot\vec{\mathcal{H}}^*.$$

(b) Show that the harmonic Poynting vector,

$$\vec{S} = \frac{c}{8\pi}\vec{\mathcal{E}}\times\vec{\mathcal{H}}^*,$$

is given by

$$\vec{S} = v_{\mathrm{phase}} w_{em}\hat{k},$$

where w_{em} is the total energy density.

(c) Given the Cartesian coordinates of the unit vector $\hat{k}$ as

$$\hat{k} = (\sin\theta\cos\phi, \sin\theta\sin\phi, \cos\theta),$$

using spherical coordinate angles, find the forms for the linear polarization vectors $\hat{\epsilon}_1$ and $\hat{\epsilon}_2$. Require $\hat{\epsilon}_1$ to be in the xy-plane and $\hat{\epsilon}_1\times\hat{\epsilon}_2 = \hat{k}$.

Exercise 9.1.2. The most general plane wave electric field $\vec{E} = \vec{E}_+ + \vec{E}_-$ can be written in terms of

$$\vec{E}_\pm = \mathrm{Re}\left(\mathcal{E}_\pm\hat{\epsilon}_\pm e^{i\vec{k}\cdot\vec{x}-i\omega t}\right),$$
$$\hat{\epsilon}_\pm = \frac{1}{\sqrt{2}}(\hat{\epsilon}_1 \pm i\hat{\epsilon}_2).$$

(a) Suppose that

$$\mathcal{E}_+ = r\mathcal{E}_-,$$

where r is a real constant (consider $r \neq 1$), but the $\mathcal{E}_\pm$ are both complex in general. Show that this results in elliptical polarization of the wave. That is, show that the 1, 2 field components form an ellipse:

$$\frac{(E_1)^2}{(1+r)^2} + \frac{(E_2)^2}{(1-r)^2} = \frac{|\mathcal{E}_-|^2}{2},$$

where $E_1 = (\vec{E}_+ + \vec{E}_-)_1$, and similarly for E_2.

(b) Now suppose that

$$\mathcal{E}_+ = r\, e^{i\delta}\mathcal{E}_-,$$

where δ is a real phase. Show that

$$\frac{(E_1')^2}{(1+r)^2} + \frac{(E_2')^2}{(1-r)^2} = \frac{|\mathcal{E}_-|^2}{2},$$

where

$$E_1' = \cos(\delta/2)E_1 - \sin(\delta/2)E_2,$$
$$E_2' = \cos(\delta/2)E_2 + \sin(\delta/2)E_1$$

giving a rotated ellipse.

Exercise 9.1.3. (Adapted from Jackson, *op. cit.*, problem 7.28.)

(a) Show that a wave traveling in the $\hat{z}$ (or $\hat{e}_3$) direction which has a finite extent in the transverse directions (large compared to its wavelength) has an electric field given approximately by (real part understood)

$$\vec{E}(\vec{x},t) \approx \left(\vec{\mathcal{E}}_0(\vec{x}_T) + \frac{i\hat{e}_3}{k}\vec{\nabla}\cdot\vec{\mathcal{E}}_0(\vec{x}_T)\right)e^{i(kz-\omega t)},$$

in source-free regions ($\vec{J} = \rho = 0$), where $\vec{x}_T = (x, y)$. Notice, the "transverse" wave has developed a z-component! [*Hint*: Apply $\vec{\nabla}\cdot\vec{E} = 0$.]

(b) Show that the associated magnetic field (to the same order) is given by

$$\vec{B}(\vec{x},t) \approx \frac{kc}{\omega}\left(\hat{e}_3\times\vec{\mathcal{E}}_0(\vec{x}_T) + \frac{i\hat{e}_3}{k}\vec{\nabla}\cdot(\hat{e}_3\times\vec{\mathcal{E}}_0(\vec{x}_T))\right)e^{i(kz-\omega t)}.$$

Once again, $\vec{B}(\vec{x},t)$ develops a small z-component. Also notice that no dispersion relation between k and ω has been specified, so that the above can be used in vacuum or non-vacuum regions.

Exercise 9.1.4. (Adapted from Jackson, *op. cit.*, problem 7.29.)

(a) As an extension of the last exercise, consider circularly polarized plane waves where

$$\vec{\mathcal{E}}_0(\vec{x}_T) = E_0(\hat{e}_1 \pm i\hat{e}_2),$$

holds everywhere (E_0 is a complex constant in space). Show that all components of the time-averaged angular momentum (real part understood)

$$\vec{L} = \frac{1}{8\pi c}\int d^3x\, \vec{x} \times (\vec{E} \times \vec{B}^*),$$

vanish.

(b) Using the space-dependent fields from Exercise 9.1.3 with

$$\vec{\mathcal{E}}_0(\vec{x}_T) = E_0(\vec{x}_T)(\hat{e}_1 \pm i\hat{e}_2),$$

and assuming that $E_0(\vec{x}_T)$ is an even function of x and y and its phase is independent of position, show that L_1 and L_2 still vanish but that L_3 is given by

$$L_3 \approx \pm\frac{1}{4\pi\omega}\int d^3x\, |E_0|^2.$$

Show also to lowest order for slowly varying $E_0(\vec{x}_T)$,

$$U = \frac{1}{16\pi}\int d^3x\, (|\vec{E}|^2 + |\vec{B}|^2),$$
$$\approx \frac{1}{4\pi}\int d^3x\, |E_0|^2,$$

so that

$$\frac{L_3}{U} \approx \pm\frac{1}{\omega},$$

where throughout the upper sign is for positive helicity, the lower sign is for negative helicity.

(c) Explain the physics behind the result that the wave in (a) has $L_3 = 0$, while the (b) wave has $L_3 \neq 0$.

Exercise 9.2.1. In one space dimension a discontinuity in dielectric constant exists between two semi-infinite regions so that

$$\epsilon = \begin{cases} \epsilon_2, & x > 0, \\ \epsilon_1, & x < 0, \end{cases}$$

with $\epsilon_{1,2}$ true constants, independent of ω. The origin of coordinates is taken at the interface and the source point has $x' > 0$. Find the Green function for the one-dimensional wave equation,

$$\left(\frac{\partial^2}{\partial x^2} - \frac{\epsilon}{c^2}\frac{\partial^2}{\partial t^2}\right) G(x,t;x',t') = -\delta(t-t')\delta(x-x'),$$

subject to the usual boundary conditions for tangential electric field components. Be sure to solve for both the $x > 0$ and $x < 0$ regions. Give an image source interpretation of your results, identifying reflection and transmission coefficients and propagation speeds in the two regions. [*Hints*: This problem is very close to the reduced Green function exercise solved in Section 9.2 except for the final step, which is the reconstruction of $G(x,t;x',t')$ itself for which Exercises 8.6.1 and 8.7.1 are relevant.]

Exercise 9.2.2. Consider a plane electromagnetic wave of angular frequency ω incident normally on a planar nonconducting material of depth D and index of refraction n, from vacuum. The backing material is a perfect conductor (see Figure 9.13). Assume the incoming wave has the initial form $\mathrm{Re}\,(\vec{E}_0 e^{i(kx-\omega t)})$, where $\vec{E}_0$ is linearly polarized. Write down the boundary conditions on the electric, magnetic field components at the two interfaces. Solve for the fields everywhere and show that the phase ϕ of the reflected wave relative to the incoming along the vacuum-dielectric interface is given

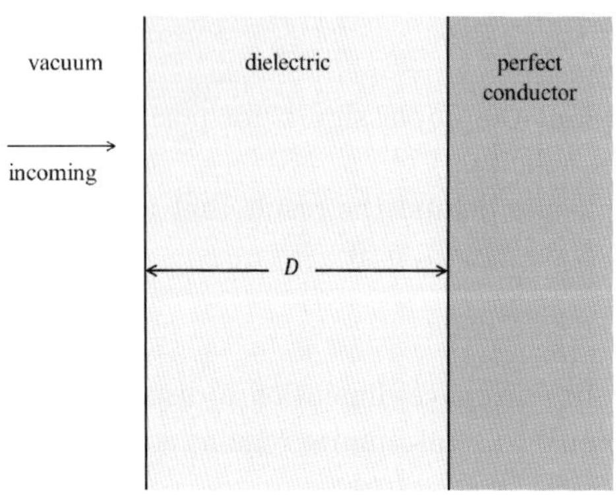

Fig. 9.13 Situation for Exercise 9.2.2.

by

$$\phi = \tan^{-1}\left(\frac{2}{n\cot(k_1 D) - \left(\frac{1}{n}\right)\tan(k_1 D)}\right),$$

where $k_1 = \omega n/c$ and $\mu = 1$ in the dielectric.

Exercise 9.2.3. Consider a dielectric sphere of radius a with dielectric constant ϵ_1 immersed in a semi-infinite background of dielectric material ϵ_2. A reduced Green function $g(\omega, r)$ for a source located at the center of the sphere is defined by

$$G(r, t; t') \equiv \int_{-\infty}^{\infty} \frac{d\omega}{2\pi} e^{-i\omega(t-t')} g(\omega, r),$$

where

$$\left(\nabla^2 - \frac{\epsilon}{c^2}\frac{\partial^2}{\partial t^2}\right) G(r, t; t') = -\frac{\delta(r)}{r^2}\delta(t - t'),$$

with

$$\epsilon = \begin{cases} \epsilon_1, & r < a, \\ \epsilon_2, & r > a. \end{cases}$$

Let

$$g(\omega, r) = \begin{cases} A\dfrac{e^{i\omega n_1 r/c}}{r} + B\dfrac{e^{-i\omega n_1 r/c}}{r}, & r < a, \\ C\dfrac{e^{i\omega n_2 r/c}}{r}, & r > a, \end{cases}$$

($n_{1,2} = \sqrt{\epsilon_{1,2}}$) with spatial boundary conditions

$$g\Big|_{a^-}^{a^+} = 0, \quad \frac{d}{dr}(r\,g)\Big|_{a^-}^{a^+} = 0,$$

appropriate for transverse electric fields. Find exact solutions for A, B and C. Show they may be put into the form ($A + B = 1$)

$$B = -\sum_{j=1}^{\infty} R_j e^{2ij\omega n_1 a/c}, \quad C = e^{-i\omega n_2 a/c}\sum_{j=1}^{\infty} T_j e^{i(2j-1)\omega n_1 a/c},$$

where the first two reflection and transmission coefficients are given by

$$T_1 = \frac{2n_1}{n_1 + n_2}, R_1 = \frac{n_2 - n_1}{n_1 + n_2}, T_2 = R_1 T_1, R_2 = R_1^2.$$

Extra: The above gives the lowest order solution for the first 2 transmission and reflection coefficients, $T_{1,2}$ and $R_{1,2}$. Now using boundary conditions at the surface of the sphere, find all of the transmission and reflection coefficients, T_i and R_i, for arbitrary i.

Exercise 9.2.4. From the reduced Green function for $E_\perp$ polarization $g_{E_\perp}(z,z')$, Equation (9.68), reconstruct the real space Green functions $G_{E_\perp}(\vec{x},t;\vec{x}',t')$ defined through

$$G_{E_\perp}(\vec{x},t;\vec{x}',t') = 4\pi \int \frac{d^2k_\perp}{(2\pi)^2} \int \frac{d\omega}{(2\pi)} e^{i\vec{k}_\perp\cdot(\vec{x}-\vec{x}')} e^{-i\omega(t-t')} g_{E_\perp}(z,z'),$$

in the case where the permittivities $\epsilon_{1,2}$ are real constants (not functions of angular frequency, ω). Interpret your answer physically. [*Hints*: Note that the reflection, transmission coefficients are functions only of angle, not $k_\perp$ or ω. You may compare to the reflection Green function expression, Equation (8.145), for evaluation of integrals.]

Exercise 9.2.5.

(a) Consider the Green function which solves the differential equations in Exercise 9.2.3 of the text but for three spherical regions $r < a$, $a < r < b$ and $r > b$. The dielectric constant takes on the values

$$\epsilon = \begin{cases} \epsilon_1, r < a \\ \epsilon_2, a < r < b \\ \epsilon_3, r > b \end{cases}$$

in the 3 regions. (The indices of refraction are $n_i = \sqrt{\epsilon_i}$. Assume $n_2(b-a) < an_1$.) Apply boundary conditions and solve for the reduced Green function $g(\omega,r)$ in the three regions. (A symbolic manipulator is helpful.) Using the coefficient in $g(\omega,r)$ which describes transition to the outside ϵ_3 region, show that the first 4 waves to reach the outside region have transition amplitudes ($T_{1,2,3,4}$) and times ($t_{1,2,3,4}$) necessary to travel to $r = b^+$ from the origin given by $\left(T_0 \equiv \left(\frac{4n_1n_2}{(n_1+n_2)(n_2+n_3)}\right)\right)$:

First wave:

$$T_1 = T_0,\ t_1 = (an_1 + (b-a)n_2)/c;$$

Second wave:

$$T_2 = T_0\left(\frac{(n_1-n_2)(n_3-n_2)}{(n_1+n_2)(n_2+n_3)}\right),\ t_2 = (an_1 + 3(b-a)n_2)/c;$$

Third wave:

$$T_3 = T_0\left(\frac{n_2-n_1}{n_1+n_2}\right),\ t_3 = (3an_1 + (b-a)n_2)/c;$$

Fourth wave:

$$T_4 = T_0\left(\frac{n_3-n_2}{n_2+n_3}\right)\left(\frac{4n_1n_2}{(n_1+n_2)^2} - \left(\frac{n_2-n_1}{n_1+n_2}\right)^2\right),$$
$$t_4 = (3an_1 + 3(b-a)n_2)/c.$$

An appropriate Taylor series expansion of the amplitude for transition into the $r > b$ region identifies these coefficients. Can you draw diagrams which show the paths these waves take and the associated amplitude factors associated with internal transitions and reflections?

(b) All the transition coefficients can be identified by using the *trinomial theorem* on the transition amplitude. Show that this gives a set of transition coefficients given by

$$T_{mjk} \equiv T_0\frac{m!}{j!k!(m-j-k)!}R_1^jR_2^kR_3^{m-j-k},$$

where

$$R_1 = \frac{(n_2-n_1)(n_2-n_3)}{(n_1+n_2)(n_2+n_3)},\ R_2 = \frac{(n_3-n_2)}{(n_2+n_3)},\ R_3 = \frac{(n_2-n_1)}{(n_1+n_2)}.$$

Confirm that the sum of all such coefficients adds to one:

$$\sum_{m=0}^{\infty}\sum_{j=0}^{m}\sum_{k=0}^{m-j} T_{mjk} = 1.$$

Exercise 9.2.6. Consider Equation (9.49) in the context of $E_\perp$ waves scattering from a plane interface ($\mu_1 = \mu_2 = 1$ case) at $z = 0$. Given the two surface currents at $z' = 0$ and $z' = a$ ($a > 0$),

$$J_\perp(z',\vec{k}_\perp,\omega) = \delta(z')\,K_\perp^1(\vec{k}_\perp,\omega) + \delta(z'-a)\,K_\perp^2(\vec{k}_\perp,\omega),$$

show that the ratio of left-moving ($\sim E_\perp^{\text{left}}e^{-ik_1z}$) to right-moving ($\sim E_\perp^{\text{right}}e^{ik_2z}$) amplitudes for $|z| > a$, is given by

$$\frac{E_\perp^{\text{left}}}{E_\perp^{\text{right}}} = \frac{K_\perp^1 + K_\perp^2\,e^{ik_2a}}{K_\perp^1 + K_\perp^2\left(\cos(k_2a) - i\dfrac{k_1}{k_2}\sin(k_2a)\right)}.$$

All quantities have argument $(\vec{k}_\perp,\omega)$.

Exercise 9.3.1. Consider the situation of plane waves incident on a flat interface, but this time don't set $\mu = 1$. Derive the equations satisfied by $E_\perp$ and $H_\perp$ and show that their Green functions are interchanged under $\epsilon(z) \leftrightarrow \mu(z)$. Use these facts and the previous solution for $B_\perp$ in Section 9.3 to write down the transmission and reflection coefficients for $H_\perp$ and $E_\perp$. The answers are:

$$\frac{H_\perp^T}{H_\perp} = \frac{2k_2/\epsilon_2}{k_1/\epsilon_1 + k_2/\epsilon_2}, \quad \frac{H_\perp^R}{H_\perp} = \frac{k_2/\epsilon_2 - k_1/\epsilon_1}{k_1/\epsilon_1 + k_2/\epsilon_2},$$

$$\frac{E_\perp^T}{E_\perp} = \frac{2k_2/\mu_2}{k_1/\mu_1 + k_2/\mu_2}, \quad \frac{E_\perp^R}{E_\perp} = \frac{k_2/\mu_2 - k_1/\mu_1}{k_1/\mu_1 + k_2/\mu_2}.$$

Note that

$$k_1 = \frac{\omega}{c} n_1 \cos t, \quad k_2 = \frac{\omega}{c} n_2 \cos i,$$

where $n = \sqrt{\epsilon\mu}$ for each region, and t and i are the angles of transmission and incidence, respectively.

Exercise 9.3.2. Continue the investigation of the various amplitudes when $\mu_{1,2} \neq 1$. From the results derived for the reflection and transmission coefficients for $E_\perp$ and $H_\perp$ in Exercise 9.3.1, find similar expressions for $E_{||}$ and $H_{||}$. The answers are:

$$\frac{E_{||}^T}{E_{||}} = \frac{2k_1/\epsilon_1}{k_1/\epsilon_1 + k_2/\epsilon_2}, \quad \frac{E_{||}^R}{E_{||}} = \frac{k_1/\epsilon_1 - k_2/\epsilon_2}{k_1/\epsilon_1 + k_2/\epsilon_2},$$

$$\frac{H_{||}^T}{H_{||}} = \frac{2k_1/\mu_1}{k_1/\mu_1 + k_2/\mu_2}, \quad \frac{H_{||}^R}{H_{||}} = \frac{k_1/\mu_1 - k_2/\mu_2}{k_1/\mu_1 + k_2/\mu_2}.$$

Exercise 9.3.3. Establish that the electromagnetic energy flow across a plane interface is conserved for the $E_\perp$ polarization. Use the complex Poynting vector to define energy flows across the plane interface ($z = 0$) for initial, transmitted and reflected fields:

$$P^I \equiv -\frac{c}{8\pi}\left(\vec{E} \times \vec{H}^*\right) \cdot \hat{e}_3\big|_{z=0},$$

$$P^T \equiv -\frac{c}{8\pi}\left(\vec{E}^T \times \vec{H}^{T*}\right) \cdot \hat{e}_3\big|_{z=0},$$

$$P^R \equiv \frac{c}{8\pi}\left(\vec{E}^R \times \vec{H}^{R*}\right) \cdot \hat{e}_3\big|_{z=0}.$$

($\vec{E}$ and $\vec{H}^*$ denote incident fields.) Defining the *energy* transmission and reflection coefficients,

$$T \equiv \frac{P^T}{P^I}, \quad R \equiv \frac{P^R}{P^I},$$

and using expressions in Exercises 9.3.1 and 9.3.2 for the various amplitudes, show that

$$T = \frac{4k_1 k_2/(\mu_1 \mu_2)}{(k_1/\mu_1 + k_2/\mu_2)^2}, \quad R = \frac{(k_1/\mu_1 - k_2/\mu_2)^2}{(k_1/\mu_1 + k_2/\mu_2)^2},$$

giving

$$T + R = 1.$$

Extra: Find the energy reflection and transmission coefficients for $H_\perp$ polarization. Of course, make sure that energy is being conserved in the process: $T + R = 1$.

The next three exercises build on the transfer matrix formalism introduced in Problem 7.8 of Jackson's textbook.

Exercise 9.3.4. A single frequency electromagnetic plane wave is normally incident from vacuum on a stack of materials of different thicknesses, t_i, and indices of refraction, n_i, in the ith layer. Given that an electric field amplitude may be written as

$$E = E_+ e^{ikx} + E_- e^{-ikx},$$

in each layer, the 2×2 transfer matrix, T, associated with the transition can be formulated as:

$$E' = TE,$$

$$\Rightarrow \begin{pmatrix} E'_+ \\ E'_- \end{pmatrix} = \begin{pmatrix} t_{22}^* & t_{21}^* \\ t_{21} & t_{22} \end{pmatrix} \begin{pmatrix} E_+ \\ E_- \end{pmatrix}.$$

Note the complex conjugates and the assumption that all indices of refraction are real. Also note this form for T is maintained across multiple transitions.

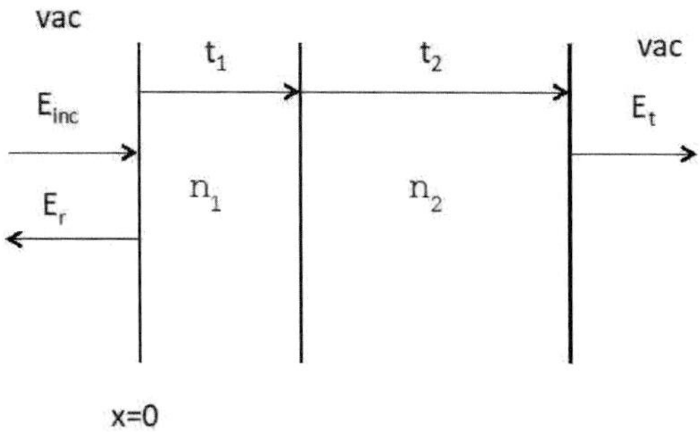

Fig. 9.14 Layers in Exercise 9.3.4.

(a) Using the transfer matrix formalism, show that energy is conserved for any T matrix: $|E_i|^2 = |E_r|^2 + |E_t|^2$, where E_i is the incident electric field, E_r is the reflected field, and E_t is the transmitted electric field. Note the *given* reflection and transmission amplitude results:

$$r \equiv \frac{E_r}{E_i} = -\frac{t_{21}}{t_{22}}, \; t \equiv \frac{E_t}{E_i} = \frac{Det\ (T)}{t_{22}}.$$

You may assume $Det\ (T) = 1$, although this may be proven for any vacuum-to-vacuum set of layer transitions. (Try your hand at it!)

(b) Find the reflection, $R \equiv |r|^2$, and transmission, $T \equiv |t|^2$, coefficients for a two-material interface with indices of refraction n_1 and n_2, as shown in Figure 9.14. Plot R and T for the values (units can be appropriately assigned)

$$n_1 = 4, \quad n_2 = 2, \quad t_1 = 1, \quad t_2 = 2, \quad k_1 = 6,$$

as a function of k_2 for $0 < k_2 < \pi$. Make sure that $T + R = 1$.

Exercise 9.3.5. Imagine sending the incident wave in the *opposite* direction on the interfaces using the transfer formalism of Exercise 9.3.4. Define the opposite direction refection and transmission amplitudes

$$r^{op} \equiv \frac{E_r^{op}}{E_i^{op}}, \; t^{op} \equiv \frac{E_t^{op}}{E_i^{op}}.$$

Show the relations

$$r^{op} \equiv -r^* \frac{t}{t^*}, \; t^{op} \equiv t,$$

where t and r are the *forward* transmission and reflection amplitudes. [*Hint*: Reverse directions and think inverse!]

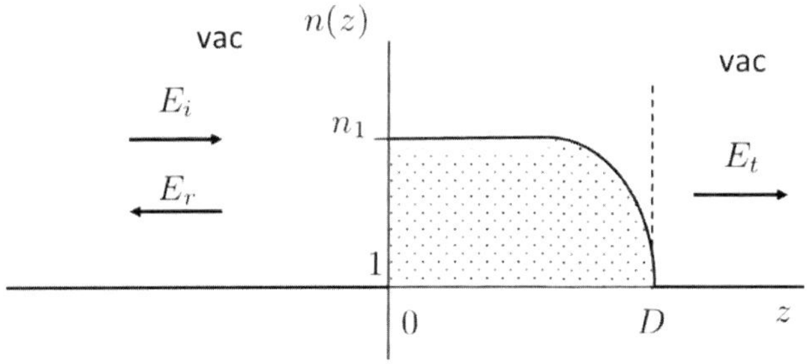

Fig. 9.15 Situation for Exercise 9.3.6.

Exercise 9.3.6. An electromagnetic plane wave moving from vacuum is normally incident on a uniform sheet of material which has a real index of refraction $n_1 > 1$. The index of refraction of the material is initially constant and then continuously goes to 1 within a finite distance, D, as shown in Figure 9.15. The E_i and E_r fields define incident and reflected electric plane waves, respectively, in the region $z < 0$, and E_t defines the transmitted electric plane wave in the region $z > D$. Evaluate the reflection $t \equiv E_t/E_i$ and refraction $r \equiv E_r/E_i$ amplitudes (up to an arbitrary phase) and *require* that energy is conserved: $|r|^2 + |t|^2 = 1$. [*Hints*: You will have to establish that no scattering takes place *within* the dielectric, only at the discontinuous interface. Next match E, B boundary conditions at the interface at $z = 0$, then use the Poynting vector and energy conservation. The transfer matrix formalism of Exercise 9.3.4 can be helpful.]

Exercise 9.4.1.

(a) Using the results of Exercise 9.3.1, show that the Brewster angle θ_B for $H_\perp$ polarization may be expressed as ($x \equiv \mu_2/\mu_1$, $n \equiv \sqrt{\mu\epsilon}$; source in the "2" region)

$$\tan^2\theta_B = \left(\frac{1 - x^2\left(\dfrac{n_1}{n_2}\right)^2}{\left(\dfrac{n_2}{n_1}\right)^2 - 1}\right).$$

(b) For $0 < x < 1$, show that the Brewster angle does not exist for

$$\frac{1}{x} > \left(\frac{n_1}{n_2}\right) > 1,$$

whereas for $x > 1$ it does not exist for

$$\frac{1}{x} < \left(\frac{n_1}{n_2}\right) < 1.$$

(c) Again using the result of Exercise 9.3.1, can you tell if there a Brewster angle $\bar{\theta}_B$ for $E_\perp$ polarization for $\mu_1 \neq \mu_2$? [*Hint*: You should find solutions for $E_\perp$ in the gaps for $H_\perp$!]

Exercise 9.5.1. When an alternating current is sent into a uniform conducting wire for which Ohm's law holds, Equation (9.117), we have learned that the conductivity turns complex. This means there is a phase difference between source field $\vec{E}$ and the response current $\vec{J}$, as seen in the simple model in the text.

(a) Is this an inductive or capacitive phase? Equation (8.73) should help you answer this.
(b) If the wire has a length l and cross-section A, and a uniform $\vec{E}$ field inside, find the implied reactance X_C and capacitance C (if capacitive) or the reactance X_L and inductance L (if inductive) from the simple model.

Exercise 9.5.2. The retarded Green function for the linear differential equation

$$\left[\frac{d^2}{dt^2} + \gamma\frac{d}{dt} + \omega_0^2\right]\vec{x}(t) = \frac{e}{m}\vec{E}(t)$$

satisfies

$$\left[\frac{d^2}{dt^2} + \gamma\frac{d}{dt} + \omega_0^2\right]G^{\text{ret}}(t,t') = \delta(t-t').$$

Causality requires

$$G^{\text{ret}}(t,t') = 0, t < t'.$$

(a) Show that in general

$$G^{\text{ret}}(t',t'') = G^{\text{ret}}(-t'',-t').$$

(b) Show that the particular solution is ($\vec{x}, d\vec{x}/dt = 0$ at $t = -\infty$)

$$\vec{x}(t) = \frac{e}{m}\int_{-\infty}^{t^+} dt'\, G^{\text{ret}}(t,t')\vec{E}(t'),$$

where $t^+ = t + \epsilon$, $\epsilon > 0$.

(c) Show explicitly that ($H(\tau)$ is the Heaviside step function, $H(\tau) = 1$, $\tau > 0$; $H(\tau) = 0$, $\tau < 0$)

$$G^{\text{ret}}(t,t') = \begin{cases} \dfrac{H(t-t')}{\sqrt{\omega_0^2 - \gamma^2/4}} e^{-(\gamma/2)(t-t')} \sin\left[\sqrt{\omega_0^2 - \gamma^2/4}\,(t-t')\right], & \omega_0 > \gamma/2, \\ \dfrac{H(t-t')}{\sqrt{\gamma^2/4 - \omega_0^2}} e^{-(\gamma/2)(t-t')} \sinh\left[\sqrt{\gamma^2/4 - \omega_0^2}\,(t-t')\right], & \omega_0 < \gamma/2. \end{cases}$$

(d) Set $\omega_0 = 0$ and show that we recover the result

$$\vec{v}(t) = \frac{e}{m}\int_{-\infty}^{t} dt' e^{-\gamma(t-t')}\vec{E}(t').$$

Exercise 9.5.3. Given the time equation for a field $\psi(t)$,

$$\left(\frac{d^2}{dt^2} + \gamma\frac{d}{dt} + \omega_0^2\right)\psi(t) = f(t),$$

and its retarded Green function,

$$\left(\frac{d^2}{dt^2} + \gamma\frac{d}{dt} + \omega_0^2\right)G^{\text{ret}}(t,t') = \delta(t-t'),$$

where $G^{\text{ret}}(t,t') = G^{\text{ret}}(-t',-t)$ (time reversal) and $G^{\text{ret}}(t,t') = 0$ for $t < t'$ (causality), show that the complete solution (particular plus homogeneous) for $\psi(t)$ is given by ($G^{\text{ret}} \equiv G^{\text{ret}}(t,t')$ everywhere; t_i is the given initial time)

$$\psi(t) = \int_{t_i}^{\infty} dt' G^{\text{ret}} f(t') - \left(\psi(t')\frac{dG^{\text{ret}}}{dt'} - G^{\text{ret}}\frac{d\psi(t')}{dt'} - \gamma\psi(t')G^{\text{ret}}\right)\Big|_{t'=t_i}.$$

Exercise 9.5.4. Show (by complex integration or otherwise) that

$$\int_{-\infty}^{\infty} d\omega\, \frac{\omega^2\gamma}{(\omega^2 - \omega_0^2)^2 + \omega^2\gamma^2} = \pi.$$

Exercise 9.5.5.

(a) Using (9.147)

$$k^2 = \frac{\mu\omega^2}{c^2}\left(\epsilon + i\frac{4\pi\sigma}{\omega}\right),$$

write $k \equiv \alpha + i\beta$ and solve explicitly for α and β. (Consider ϵ, σ real and frequency independent.)

(b) A plane polarized electromagnetic wave of frequency ω in free space is incident normally on a flat surface with dielectric constant ϵ, permittivity μ and conductivity σ. Obtain an exact expression in terms of α and β for the phase $\phi(\omega)$ of the reflected wave relative to the incident wave. Find the high ($\sigma/\omega \ll 1$) and low ($\sigma/\omega \gg 1$) frequency limits of $\phi(\omega)$ in terms of ϵ, μ and σ. Show that the phase is close to π in either limit, increasing as a function of ω in one case, decreasing as a function of ω in the other, if $\epsilon/\mu > 1$.

(c) Find a simple formula for the critical frequency $\omega_{\rm cr}$ at which this phase is maximized. In addition, get an expression for this maximum phase, $\phi_{\rm max}$. (Note, however, that the assumption that ϵ and σ are frequency independent is unrealistic.)

Exercise 9.5.6. An electromagnetic plane polarized wave of angular frequency ω, $\vec{E} = E_0\,\hat{x}\,e^{i(kz-\omega t)}$, is normally incident on a uniform sheet of material with index of refraction n and width d. The waves begin and end in vacuum; for definiteness, take the origin of your z-coordinate at the location of the initial interface and the location of the second interface at $z = d$.

(a) Given the result for k^2 in Equation (9.147), with $\epsilon(\omega) \equiv \epsilon$ and $\sigma(\omega) \equiv \sigma$ frequency independent constants as in Exercise 9.5.5, show that the expression for the transmission amplitude $t \equiv E_t/E_i$ is given by (we used a symbolic manipulator)

$$t = \frac{4n\,e^{i(n-1)kd}}{(n+1)^2 - e^{2ikd}(n-1)^2}$$

where

$$ik = \frac{1}{\delta_{new}}\left(-1 + i\frac{Z^2}{z}\right), \quad n = \frac{ck}{\omega},$$

$$Z \equiv \left(1 + \sqrt{1+z^2}\right)^{1/2}, \quad z \equiv \frac{2}{\epsilon\mu\gamma^2}, \quad \gamma \equiv \frac{\omega\delta}{c},$$

$$\delta \equiv \frac{c}{\sqrt{2\pi\mu\sigma\omega}}, \quad \delta_{new} = \frac{Z}{\sqrt{z}}\delta.$$

Note that for $\gamma \ll 1.0$ (a good conductor or sufficiently low frequencies) we have $\delta_{new} \approx \delta$, whereas for $\gamma \gg 1.0$ (a poor conductor or sufficiently high frequencies) we have $\delta_{new} \approx \sqrt{\epsilon\mu}\,\gamma\delta$.

(b) Set $\epsilon = 1$, $\mu = 1$, and plot the transmission coefficient $|t|^2$ as a function of $D \equiv d/\delta_{new}$ for three cases: $\gamma = .01$ in the range $0 < D < 0.1$; $\gamma = 1.0$ in the range $0 < D < 4.0$; and $\gamma = 10.0$ in the range $0 < D < 4.0$. Interpret and/or comment on the results plotted in this fashion.

Exercise 9.5.7.

(a) Let's go back to Exercise 9.5.6. There we had normal incidence on a material with index of refraction n. When n turns complex and the thickness d becomes large, the reflection amplitude results for very large d implies that energy is not conserved. Energy is being lost to the surface of the material, with skin depth δ_{new}, through heating. For normal incidence show that the ratio of reflection to incident amplitudes gives

$$\frac{E_r}{E_i} \overset{d\to\infty}{\rightarrow} \frac{(1-n)}{(1+n)}.$$

(b) Going back to the same model as in Exercise 9.5.6, assuming $\epsilon(\omega) = \epsilon$ and $\sigma(\omega) = \sigma$ are constants, one has (Equation (9.147))

$$k^2 = \frac{\mu\omega^2}{c^2}\left(\epsilon + i\frac{4\pi\sigma}{\omega}\right), \text{ where } k = n\frac{\omega}{c}.$$

Using part (a), show that the fractional energy loss

$$\mathbb{F}(z, \epsilon\mu) \equiv 1 - \left|\frac{E_r}{E_i}\right|^2,$$

is given by the expression

$$\mathbb{F}(z, \epsilon\mu) = \frac{4\sqrt{\frac{\epsilon\mu}{2}}Z(z)}{\left(1 + \sqrt{\frac{\epsilon\mu}{2}}Z(z)\right)^2 + \frac{\epsilon\mu}{2}\left(\frac{z}{Z(z)}\right)^2},$$

where

$$Z(z) \equiv \left(1+\sqrt{1+z^2}\right)^{1/2}, z \equiv \frac{4\pi\sigma}{\epsilon\omega}.$$

$z \gg 1$ and $z \ll 1$ are the good conductor and poor conductor limits, respectively.

(c) Show the following statements:

(1) $\mathbb{F}(z,\epsilon\mu) \approx \sqrt{\frac{8}{\epsilon\mu z} - \frac{4}{\epsilon\mu z}}, \quad z \gg 1.$

(2) $\mathbb{F}(z,\epsilon\mu) \approx \frac{4\sqrt{\epsilon\mu}}{(1+\sqrt{\epsilon\mu})^2} + \frac{\sqrt{\epsilon\mu}\,(1-3\epsilon\mu)}{2(1+\sqrt{\epsilon\mu})^4} z^2, \quad z \ll 1.$

(3) Normally, one has $\epsilon\mu > 1$, which renders $\mathbb{F}(0,\epsilon\mu)$ a maximum in z. However, numerically show that for $\epsilon\mu = 1/6$, $\mathbb{F}(z,\epsilon\mu)$ actually maximizes near $z \approx 3.873$.

Exercise 9.6.1. Apply the w_e term in Equation (8.73) to model a parallel plate capacitor with an inserted resistive dielectric. The dielectric has cross-section A and width d. A uniform electric field is assumed in this model. In this case the Z factor, $V = IZ$, is

$$Z = \frac{1}{1/R - i\omega C},$$

describing the addition of the impedances in parallel. Now, using the low frequency model form of the dielectric constant given in Equation (9.149),

$$\epsilon(\omega) \approx \epsilon_0 + i\frac{4\pi\sigma(0)}{\omega},$$

show that the implied form for the capacitance C and resistance R are

$$C = \frac{\epsilon_0 A}{4\pi d}, \qquad R = \frac{d}{A\sigma(0)},$$

coming from the real and imaginary parts of $\epsilon(\omega)$, respectively.

Exercise 9.6.2. Equation (9.167) is the differential equation for free charge in many materials. The plasma frequency in this case is due to the conductance electrons present:

$$(\omega_p^c)^2 = \frac{4\pi e^2 n_c}{m}.$$

Write down the general solution to Equation (9.167) in both the underdamped ($\gamma < 2\omega_p^c/\sqrt{\epsilon_0}$) and overdamped ($\gamma > 2\omega_p^c/\sqrt{\epsilon_0}$) cases. Verify the behavior

$$\rho(t) \sim e^{-(\gamma/2)t} \cos(\omega_1 t + \delta)$$

($\omega_1 = \sqrt{(\omega_p^c)^2/\epsilon_0 - \gamma^2/4}$) in the underdamped case, and the behavior

$$\rho(t) \sim e^{-(4\pi\sigma(0)/\epsilon_0)t},$$

in the severely overdamped case for times $t \gg 1/\gamma$, where the static conductivity is given as

$$\sigma(0) = \frac{n_c e^2}{m\gamma}.$$

Exercise 9.6.3. A radio enthusiast in Waco listening on a certain frequency, f, finds that one can detect signals from a short wave radio transmitter in Bozeman, Montana, which is 1230 miles away, but transmitters significantly closer or further become weaker. Given that the signal is reflecting from an atmospheric layer approximately 300 miles above the Earth where essentially all atoms are ionized, that the free electron density in this layer is 10^6 cm^{-3} (bound electrons cannot contribute), and that the index of refraction is given by Equation (9.155), find the approximate frequency of the enthusiast's radio signal.

Exercise 9.6.4.

(a) Using the expression for harmonic power conservation, Equation (8.73), show that for a material of length d and cross-sectional area A containing both free and bound electrons in a uniform electric field the general impedance, $Z(\omega)$, is given by

$$Z(\omega) = \frac{d}{A\,\sigma_{\text{eff}}(\omega)},$$

where $\sigma_{\text{eff}}(\omega)$ is given by Equation (9.144).

(b) An electric field $\vec{E}$ is established in a capacitor whose plates are separated by the material described in part (a). The equivalent circuit for this system adds the resistance and inductance in series, which is in turn added to the capacitance in parallel, as shown in Figure 9.16. Use the low frequency expression

$$\sigma_{\rm eff}(\omega) = \frac{n_c e^2}{m}\frac{1}{\gamma - i\omega} - i\frac{\epsilon_0\omega}{4\pi},$$

for $\sigma_{\rm eff}(\omega)$, where ϵ_0 is a constant. Evaluate the resistance, R, capacitance, C, and inductance, L of this system. (Exercises 9.5.1 and 9.6.1 of the text are similar problems.)

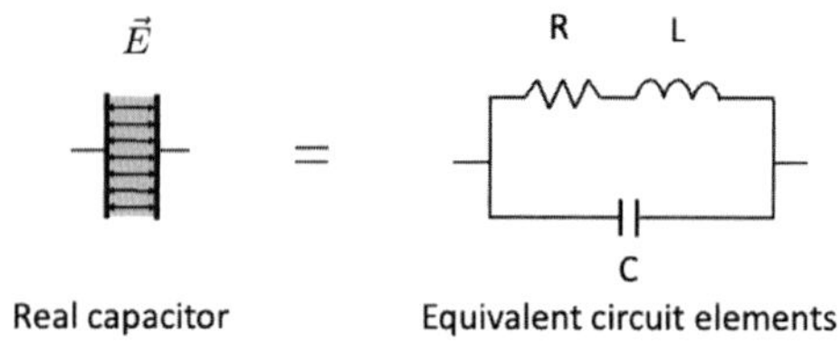

Fig. 9.16 Real and equivalent circuits for Exercise 9.6.4.

(c) The equivalent circuit just forms a series LRC circuit. Show that its resonance frequency, $1/\sqrt{LC}$, is simply related to the plasma frequency of the material from which it is composed.

Exercise 9.7.1. Establish that the single resonance response dielectric function, Equation (9.136), which has a single pole in the lower half complex plane, gives rise to a response function, Equation (9.172), which is just the retarded Green function from Exercise 9.5.2 multiplied by an overall constant:

$$G(\tau) = \frac{4\pi n_b e^2}{m} G^{\rm ret}(\tau, 0).$$

Exercise 9.7.2.

(a) Using the definition

$$\omega_p^2 \equiv \lim_{\omega\to\infty} \omega^2(1 - \epsilon(\omega))$$

and the Kramers-Kronig dispersion relations, show that

$$\omega_p^2 = \frac{2}{\pi}\int_0^\infty d\omega'\, \omega' {\rm Im}\,\epsilon(\omega').$$

(b) Likewise show that

$$\int_0^N d\omega' \, [\mathrm{Re}\, \epsilon(\omega') - 1] = \frac{\omega_p^2}{N} + \mathcal{O}\left(\frac{1}{N^3}\right).$$

Exercise 9.7.3.

(a) In (9.148) we modeled a conductor using an effective (dynamic) dielectric constant of the form

$$\epsilon_{\mathrm{eff}}(\omega) = \epsilon(\omega) + \frac{4\pi i \sigma(\omega)}{\omega},$$

where

$$\sigma(\omega) = \frac{\gamma \sigma(0)}{\gamma - i\omega}.$$

Suppose that $\epsilon(\omega)$ satisfies the usual Kramers-Kronig dispersion relations. Show that $\epsilon_{\mathrm{eff}}(\omega)$ satisfies the modified relations:

$$\mathrm{Re}\, \epsilon_{\mathrm{eff}}(\omega) - 1 = \frac{2}{\pi} P \int_0^\infty d\omega' \, \frac{\omega' \mathrm{Im}\, \epsilon_{\mathrm{eff}}(\omega')}{\omega'^2 - \omega^2},$$

$$\mathrm{Im}\, \epsilon_{\mathrm{eff}}(\omega) = \frac{4\pi\sigma(0)}{\omega} - \frac{2\omega}{\pi} P \int_0^\infty d\omega' \, \frac{\mathrm{Re}\, \epsilon_{\mathrm{eff}}(\omega') - 1}{\omega'^2 - \omega^2}.$$

(b) In this situation, show that the modified forms of the results in Exercise 9.7.2(a) and (b) are

$$(\omega_p)^2_{\mathrm{eff}} = \omega_p^2 + 4\pi\gamma\sigma(0) = \frac{2}{\pi} \int_0^\infty d\omega' \, \omega' \mathrm{Im}\, \epsilon_{\mathrm{eff}}(\omega')$$

and

$$\int_0^N d\omega' [\mathrm{Re}\, \epsilon_{\mathrm{eff}}(\omega') - 1] = \frac{(\omega_p)^2_{\mathrm{eff}}}{N} - 2\pi^2 \sigma(0) + \mathcal{O}\left(\frac{1}{N^3}\right).$$

Exercise 9.7.4. Use the constraint

$$\int_{-\infty}^\infty dt \, \frac{\partial P(\vec{x}, t)}{\partial t} \cdot \vec{E}(\vec{x}, t) > 0,$$

where $\vec{E}(\vec{x}', t')$ is an arbitrary electric field to show that $\mathrm{Im}\, \epsilon(\omega) > 0$ for real $\omega > 0$. Assume that

$$\vec{E}(\vec{x}, \omega) = \int_{-\infty}^\infty dt \, e^{i\omega t} \vec{E}(\vec{x}, t), \; \epsilon(\omega) = \int_{-\infty}^\infty dt \, e^{i\omega t} \epsilon(t),$$

$$\vec{P}(\vec{x}, \omega) = \frac{(\epsilon(\omega) - 1)}{4\pi} \vec{E}(\vec{x}, \omega).$$

(Comment: Note that ($\mu = 1$)

$$\vec{J}_{\rm b} = \frac{\partial \vec{P}(\vec{x},t)}{\partial t},$$

so the above is just a statement that the total work done by the field on the bound charges is positive.)

Exercise 9.7.5.

(a) Given the representation for $\epsilon(z)$ in the upper half complex plane,

$$\epsilon(z) = 1 + \frac{1}{2\pi i}\int_{-\infty}^{\infty} d\omega' \, \frac{\epsilon(\omega') - 1}{\omega' - z},$$

show that

$$\text{Im}(\epsilon(z)) = \frac{\text{Im}(z)}{\pi}\int_{-\infty}^{\infty} d\omega' \frac{\text{Im}(\epsilon(\omega'))}{|\omega' - z|^2}.$$

(b) Using (a) or any other method for the case where

$$\epsilon(-z) = \epsilon^*(z^*) \qquad \text{and} \qquad \text{Im}\,(\epsilon(\omega)) > 0 \text{ for } \omega > 0,$$

show that $\text{Im}(\epsilon(z)) > 0$ for $\text{Re}(z) > 0$ and $\text{Im}(\epsilon(z)) < 0$ for $\text{Re}(z) < 0$ in the upper half complex plane.

Exercise 9.7.6. We may write the causal connection between $\vec{D}$ and $\vec{E}$, Equation (9.171) of the notes, as

$$\vec{D}(\vec{x},t) = \vec{E}(\vec{x},t) + \int_{0^-}^{\infty} d\tau \, H(\tau) g(\tau) \vec{E}(\vec{x}, t - \tau),$$

where $H(\tau)$ is the Heaviside step function,

$$H(\tau) = i\int_{-\infty}^{\infty} \frac{d\nu}{2\pi}\frac{e^{-i\nu\tau}}{\nu + i\epsilon}.$$

($H(\tau) = 1$, $\tau > 0$; $H(\tau) = 0$, $\tau < 0$; $\epsilon > 0$) $g(\tau)$ has the Fourier transform

$$g(\tau) = \int_{-\infty}^{\infty} \frac{d\omega'}{2\pi} e^{-i\omega'\tau} \tilde{g}(\omega'),$$

and is a real function ($\tau > 0$).

(a) Using the above connections, and given that $\vec{D}(\omega) = \epsilon(\omega)\vec{E}(\omega)$, show that one has

$$\epsilon(\omega) - 1 = i\int_{-\infty}^{\infty} \frac{d\omega'}{2\pi}\frac{\tilde{g}(\omega')}{\omega - \omega' + i\epsilon}.$$

(b) Choosing $g(-\tau) = -g(\tau)$, show that $\widetilde{g}(\omega')$ is odd in ω' and purely imaginary. Then, show that the result in part (a) gives

$$\operatorname{Re}\epsilon(\omega) - 1 = \frac{1}{\pi} P \int_{-\infty}^{\infty} d\omega' \, \frac{\operatorname{Im}\epsilon(\omega')}{\omega' - \omega}.$$

What equation do we obtain if we choose $g(-\tau) = g(\tau)$?

Exercise 9.7.7. Given (as derived for example in Exercise 9.7.6)

$$\operatorname{Re}\epsilon(\omega) - 1 = \frac{1}{\pi} P \int_{-\infty}^{\infty} d\omega' \, \frac{\operatorname{Im}\epsilon(\omega')}{\omega' - \omega},$$

show that the other Kramers-Kronig relation,

$$\operatorname{Im}\epsilon(\omega) = -\frac{1}{\pi} P \int_{-\infty}^{\infty} d\omega' \, \frac{\operatorname{Re}\epsilon(\omega') - 1}{\omega' - \omega},$$

can be obtained by an appropriate contour integration. [*Hint*: Work out the double principle part product,

$$P\frac{1}{\omega' - \omega} P\frac{1}{\omega'' - \omega'},$$

using forms such as (9.179) or (9.184).]

Exercise 9.7.8.

(a) (Compare to Exercise 9.7.5) Given the representation for the complex permittivity $\epsilon(z)$ in the upper half complex plane,

$$\epsilon(z) = 1 + \frac{1}{2\pi i} \int_{-\infty}^{\infty} d\omega' \frac{\epsilon(\omega') - 1}{\omega' - z},$$

show that

$$\operatorname{Re}(\epsilon(z) - 1) = \frac{1}{\pi} \int_{-\infty}^{\infty} d\omega' \frac{\operatorname{Im}(\epsilon(\omega'))\operatorname{Re}(\omega' - z)}{|\omega' - z|^2}.$$

[*Hint*: In order to use the residue theorem, make sure your integrand functions are analytic. For example, even if $g(z)$ is analytic, $\operatorname{Im}(g(z))$ and $\operatorname{Re}(g(z))$ are not.]

(b) Using (a) for the case where

$$\epsilon(-z) = \epsilon^*(z^*) \text{ and } \operatorname{Im}(\epsilon(\omega)) > 0 \text{ for } \omega > 0,$$

show that $\text{Re}(\epsilon(z) - 1) > 0$ everywhere in the upper half plane for $\text{Im}(z) \geq |\text{Re}(z)|$. In particular, this establishes that $\text{Re}(\epsilon(z)) - 1 > 0$ for the positive imaginary axis. [Note: Putting the Exercise 9.7.5 result for $\text{Im}(\epsilon(z))$ when $\text{Re}(z) \neq 0$ together with this result establishes that $\epsilon(z)$ has no zeros in upper half plane.]

Exercise 9.8.1. We briefly discussed the behavior of the dielectric constant at high frequencies in metals or plasmas where it was given approximately by ($\mu = 1$)

$$\epsilon(\omega) = 1 - \frac{\omega_p^2}{\omega^2},$$

where ω_p is the plasma frequency.

(a) Derive the explicit form of the response function, $G(\tau)$, for this $\epsilon(\omega)$. Find the group, v_g, and phase, v_p, velocities implied by this dielectric relation, and demonstrate that $v_p > c > v_g$.

(b) In one dimension, show that the electric field ($\vec{E}(\vec{x}, t) \to u(x, t)$) satisfies the modified wave equation,

$$\frac{\partial^2 u}{\partial x^2} = \frac{1}{c^2}\frac{\partial^2 u}{\partial t^2} + \frac{\omega_p^2}{c^2}u.$$

(c) Solve this equation (called the Klein-Gordon equation) using separation of variables for a "box" of length a and boundary conditions $u|_{x=0,a} = 0$. Show that the resonant angular frequencies of the separated time function are given by

$$\omega_n^2 = \omega_p^2 + \left(\frac{cn\pi}{a}\right)^2, \quad n = 1, 2, 3, \ldots$$

(d) Solve for the time behavior of the one-dimensional Klein-Gordon equation for $u(x, t)$, subject to the initial conditions,

$$u(x, 0) = \begin{cases} s\,x, & 0 \leq x \leq a/2 \\ s(a - x), & a/2 \leq x \leq a \end{cases}$$

$$\frac{du(x, 0)}{dt} = 0$$

in a box of length a. [*Hint*: There are two linearly independent solutions, $e^{\pm i\omega_n t}\sin\left(\frac{n\pi x}{a}\right)$, with two sets of coefficients.]

Exercise 9.8.2. Consider a dielectric function given by

$$\epsilon(\omega) = 1 + \frac{\omega_p^2}{\omega_0^2}\left(1 + \frac{\omega^2}{\omega_0^2}\right),$$

as a function of angular frequency, ω. You can consider ω_p the plasma frequency and ω_0 a resonance frequency. Given the sourced wave equation for a unit pulse emitted at the origin (in space and time) in one dimension x (see Equation (9.209)),

$$\frac{\partial^2 u(x,t)}{\partial x^2} - \frac{1}{c^2}\frac{\partial^2 u(x,t)}{\partial t^2} - \frac{1}{c^2}\int_{0^-}^{\infty} d\tau \, \frac{\partial^2 u(x,t-\tau)}{\partial t^2} G(\tau) = -\delta(x)\delta(t),$$

($G(\tau)$ is the response function) solve for $u(x,t)$ and display the results of a numerical evaluation for the given parameters:

$$\omega_p = 8, \quad \omega_0 = 10, \quad c = 1, \quad t = 10.$$

Display $u(x, t = 10)$ as a function of x for the range $0 < x < 15$ for these parameters.

Exercise 9.9.1.

(a) Assuming the absence of poles for $1/\epsilon_{\text{eff}}(\omega)$ in the upper half plane and using the connection (9.269), write appropriately modified K-K relations for $1/\epsilon_{\text{eff}}(\omega)$ and show the sum rule (9.271).

(b) Taking a one resonance model form for $\epsilon_{\text{eff}}(\omega)$ or $\epsilon(\omega)$ from (9.148), show explicitly the result in part (a) holds for the appropriate form of ω_p^2. What is one crucial assumption one must make for the result to hold for the multi-resonance form? Provide some numerical evidence that the assumption holds.

Exercise 9.9.2. Using numerical means, calculate and plot the expected stopping range $R(W_0)$ of an electron with an initial kinetic energy $W_0 = \frac{1}{2}mv_0^2$ for aluminum (atomic number $= 13$, atomic mass ≈ 27 gm/mol, density ≈ 2.7 gm/cm^3, $\hbar\bar{\omega} \approx 166$ eV). Very roughly, use $\hbar\bar{\omega}$ as the lowest energy scattering event in your energy integral. Show the range as a function of W_0 for 10 KeV $< W_0 <$ 100 KeV. (Careful evaluations in Ref. 4, Section 9.10.2, yield $R(W_0) = 1.31 \times 10^{-4}$ cm at 10 KeV and 6.94×10^{-3} cm at 100 KeV.)

Exercise 9.9.3. In Sections 9.5, 9.7 and 9.9 we encountered some sum rules for the effective dielectric permittivity $\epsilon(\omega)$ in dispersive materials. An effective model of $\epsilon(\omega)$ involving n pole contributions is given by (see Equation (9.189))

$$\epsilon(\omega) = 1 + \omega_p^2 \sum_{i=1}^{n} \frac{f_i}{\omega_{0,i}^2 - \omega^2 - i\omega\,\gamma_i},$$

where ω_p^2 is the square of the plasma frequency, as given in (9.190), and the $f_i \geq 0$ are numbers called the *oscillator strengths* which satisfy

$$\sum_{i=1}^{n} f_i = 1.$$

(a) Build a two-resonance model with $n = 2$, $\gamma_1 = 1$, $\gamma_2 = 2$, $\omega_{0,1} = 2$, $\omega_{0,2} = 1$, $\omega_p = 5$, $f_1 = 0.5$, $f_2 = 0.5$ (arbitrary units). Numerically verify the sum rules (see (9.188) and (9.271))

$$\int_0^\infty d\omega \left[\omega \frac{2}{\pi\,\omega_p^2} \operatorname{Im}(\epsilon(\omega))\right] = 1,$$

$$\int_0^\infty d\omega \left[\omega \frac{2}{\pi\,\omega_p^2} \operatorname{Im}(1/\epsilon(\omega))\right] = -1.$$

Using a plotting program, show the value of the *integrands* (in square brackets above) in the region $0 < \omega < 10$.

(b) With the same parameters as above, supplemented by $N = 10$, now numerically verify (see Equation (9.195))

$$\int_0^N d\omega \left[\frac{N}{\omega_p^2} \operatorname{Re}(\epsilon(\omega) - 1)\right] \approx 1$$

(find *actual* number) and numerically determine the result,

$$\int_0^N d\omega \left[\frac{N}{\omega_p^2} \operatorname{Re}(1/\epsilon(\omega) - 1)\right] \approx ??.$$

Show the *integrands* of both integrals in the $0 < \omega < N$ region.

Chapter 10

Waveguides and Resonant Cavities

Electromagnetism is not just a theory for physicists to play with: it is an eminently practical theory which is widely used in communication and power transmission. In this chapter we'll consider some important aspects of these practical applications which have helped to shape our world.

10.1 Boundary conditions near an imperfect conductor

In this section we derive boundary conditions between a dielectric medium and an imperfect conductor. In later sections, we'll apply these conditions to boundaries of various shapes.

As long as a boundary is not too sharply curved (we'll explain the precise meaning of "sharp" later), it can be modeled locally as a plane as shown in Figure 10.1. As indicated in the figure, the normal vector $\hat{n}$ points in the $+z$ direction.

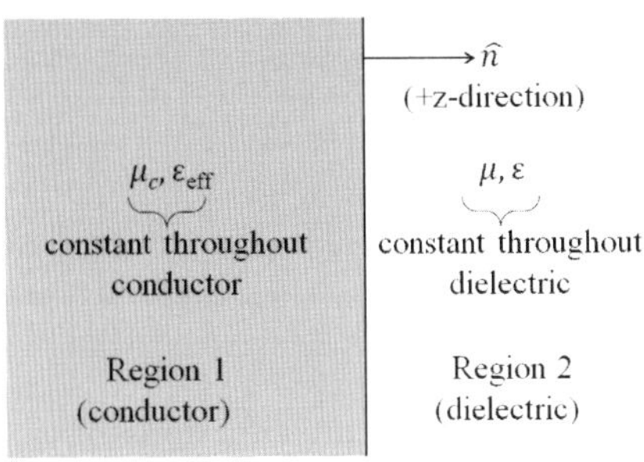

Fig. 10.1 Local representation of general conductor-dielectric boundary.

The dielectric region obeys the macroscopic Maxwell equations with dielectric constant ϵ and permeability μ. The conducting medium also obeys the macroscopic Maxwell equations with constants ϵ_c and μ_c, and is "Ohmic" so that

$$\vec{J}(\vec{x},\omega) = \sigma(\omega)\vec{E}(\vec{x},\omega), \tag{10.1}$$

as in (9.119), where the (harmonic) current density and electric field are given by $\mathrm{Re}\left(\vec{J}(\vec{x},\omega)e^{-i\omega t}\right)$ and $\mathrm{Re}\left(\vec{E}(\vec{x},\omega)e^{-i\omega t}\right)$ respectively. In Section 9.5 we showed that in this case, the medium can be treated as a dielectric with a frequency-dependent effective dielectric constant defined by (9.146):

$$\epsilon_{\mathrm{eff}} \equiv \epsilon_c + i\frac{4\pi\sigma(\omega)}{\omega}. \tag{10.2}$$

Our starting-point is the boundary conditions derived in Chapter 1, which are generally valid for either static or time-varying fields. Using Σ and $\vec{K}$ to represent the (free) surface charge and current densities respectively, we then have:

$$(\vec{D}_2 - \vec{D}_1)\cdot\hat{n} = 4\pi\Sigma, \tag{10.3}$$

$$(\vec{B}_2 - \vec{B}_1)\cdot\hat{n} = 0, \tag{10.4}$$

$$\hat{n}\times(\vec{E}_2 - \vec{E}_1) = 0, \tag{10.5}$$

$$\hat{n}\times(\vec{H}_2 - \vec{H}_1) = \frac{4\pi}{c}\vec{K}. \tag{10.6}$$

Following our "Ohmic" assumptions, the Fourier transforms in time of (10.3)–(10.6) are

$$\epsilon E_{2z} - \epsilon_{\mathrm{eff}}E_{1z} = 4\pi\Sigma, \tag{10.7}$$

$$B_{2z} - B_{1z} = 0, \tag{10.8}$$

$$E_{2\perp} - E_{1\perp} = 0, \tag{10.9}$$

$$E_{2||} - E_{1||} = 0, \tag{10.10}$$

$$\frac{B_{2\perp}}{\mu} - \frac{B_{1\perp}}{\mu_c} = \frac{4\pi}{c}(\vec{K}\times\hat{n})_\perp, \tag{10.11}$$

$$\frac{B_{2||}}{\mu} - \frac{B_{1||}}{\mu_c} = \frac{4\pi}{c}(\vec{K}\times\hat{n})_{||}, \tag{10.12}$$

where all fields in these equations are functions of ω. Figure 10.2 shows the geometrical relationships between the various field components. The notation is the same as in Section 9.2: the $\|$ and $\perp$ subscripts on the fields

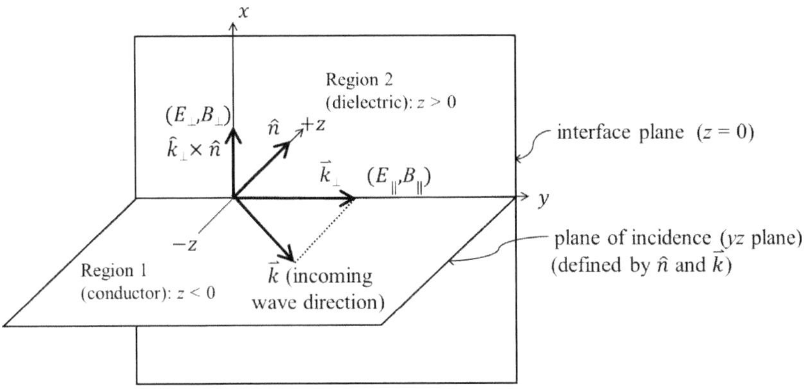

Fig. 10.2 Reference figure for Equations (10.7)–(10.12).

denote fields parallel and perpendicular to the plane of incidence, which is spanned by the incoming wave vector $\vec{k}$ and surface normal $\hat{n}$; while $\vec{k}_\perp$ denotes the component of $\vec{k}$ perpendicular to $\hat{n}$ (so that $|k_\perp| = |k|\sin i = n_2(\omega/c)\sin i$).

Equations (10.7)–(10.12) enable us to express the fields in the dielectric (Region 2) in terms of the fields inside the conductor (Region 1). So if the Region 1 fields can be determined, the Region 2 fields will follow as well. Once the ideal limiting case is established, we will correct the connection between the fields in our upcoming considerations, preparatory to introducing sources into Region 2. We'll consider only the case where there are no free charges or currents, so $\Sigma = \vec{K} = 0$.

Before taking on the general case, let's warm up with two important limiting situations. First, the *perfect electric conductor* corresponds to $|\epsilon_{\text{eff}}| \to \infty$ and $\mu_c \to 0$, so that $\vec{E}_1 = \vec{B}_1 = 0$ inside the conductor. Therefore just outside the conductor we have

$$\hat{n} \times \vec{E}_1 = 0 \implies \hat{n} \times \vec{E}_2 = 0, \tag{10.13}$$

$$\vec{B}_1 \cdot \hat{n} = 0 \implies \vec{B}_2 \cdot \hat{n} = 0. \tag{10.14}$$

Note that even though free currents and charges are zero, nonetheless the *effective* surface charge Σ_{eff} and surface current $\vec{K}_{\text{eff}}$ can be nonzero in response to the applied fields. We shall soon find that these charges and currents are distributed near the surface for good, but not perfect conductors.

Our second limiting case is the *perfect magnetic conductor*, where $|\mu_c| \to \infty$ and $\epsilon_{\text{eff}} \to 0$. In this case $\vec{D}_1 = \vec{H}_1 = 0$, so that just outside the conductor we have

$$\hat{n} \times \vec{H}_1 = 0 \implies \hat{n} \times \vec{H}_2 = 0, \tag{10.15}$$

$$\vec{D}_1 \cdot \hat{n} = 0 \implies \vec{D}_2 \cdot \hat{n} = 0. \tag{10.16}$$

These equations describe the leading logarithm quark model discussed in Section 5.10. In that particular case we also had $\hat{n} \times \vec{D} = 0$ at the surface since $\epsilon \to 0$ in the confinement region as well.

In the remainder of this section we will consider "good" electrical conductors, which for our purposes are Ohmic media which additionally satisfy:

$$\epsilon_c \approx 1; \quad \sigma(\omega) \approx \sigma(0) \equiv \sigma; \quad \frac{\sigma}{\omega} \gg 1. \tag{10.17}$$

The relevant Maxwell equations within the conductor are (considering fields as functions of $\vec{x}$ and t):

$$\vec{\nabla} \times \vec{H}_1 = \frac{1}{c}\frac{\partial}{\partial t}\vec{D}_1 + \frac{4\pi}{c}\vec{J}, \tag{10.18}$$

$$-\vec{\nabla} \times \vec{E}_1 = \frac{1}{c}\frac{\partial}{\partial t}\vec{B}_1. \tag{10.19}$$

Taking Fourier transforms in the time variable, we obtain

$$\vec{\nabla} \times \vec{H}_1 = -\frac{i\omega\epsilon_{\text{eff}}}{c}\vec{E}_1, \tag{10.20}$$

$$\vec{\nabla} \times \vec{E}_1 = \frac{i\omega}{c}\vec{B}_1, \tag{10.21}$$

where the fields in (10.20) and (10.21) are functions of $(\vec{x}, \omega)$. We may also Fourier transform in the spatial variables x and y, so that the independent variables are $(z, \vec{k}_\perp, \omega)$, as in (9.35):

$$\vec{\nabla} \to i\vec{k}_\perp + \hat{n}\frac{\partial}{\partial z}. \tag{10.22}$$

At this point we return to pesky issue of boundary curvature, which we alluded to earlier. We'd like to identify the boundary with the mathematical surface $z = 0$, because then the boundary conditions are independent of $\vec{k}_\perp$ and the situation is simplified enormously. This identification is a good approximation as long as (a) the variation in z due to curvature makes a negligible contribution to $\vec{\nabla}$ in (10.22); and (b) we only consider field points that are close enough that the boundary "looks flat" (see Figure 10.3). Condition (a) holds if

$$|\vec{k}| \gg 1/R_{\min}, \tag{10.23}$$

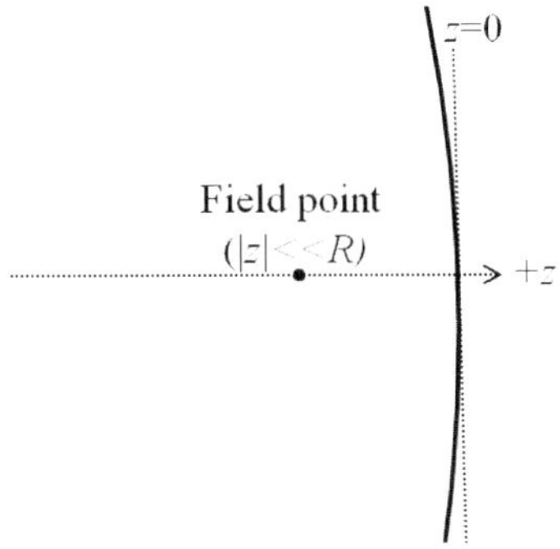

Fig. 10.3 The field point says, "That boundary looks pretty flat to me!"

where $R_{\rm min}$ is the minimum radius of curvature of the surface. On the other hand, condition (b) is valid as long as we only consider field points with $|z| \ll R_{\rm min}$. We shall see shortly that fields are only appreciable for $|z| \lesssim \delta$, where δ is the skin depth defined in (9.154) as

$$\delta \equiv \frac{c}{\sqrt{2\pi\sigma\mu_c\omega}}. \tag{10.24}$$

Thus, condition (b) is tantamount to assuming that

$$\delta \ll R_{\rm min}. \tag{10.25}$$

Relations (10.17), (10.23) and (10.25) summarize the physical assumptions required for the following discussion. In the usual good-conductor case with $\mu_c, \epsilon_c \approx 1$ then (10.25) follows from (10.23), and is thus redundant.

From the vector components of the $\vec{\nabla} \times \vec{H}$ equation (10.20) we find

$$k_\perp H_{1\perp} = \frac{\omega\epsilon_{\rm eff}}{c} E_{1z}, \tag{10.26}$$

$$\frac{\partial}{\partial z} H_{1\perp} = -\frac{i\omega\epsilon_{\rm eff}}{c} E_{1||}, \tag{10.27}$$

$$-\frac{\partial}{\partial z} H_{1||} + ik_\perp H_{1z} = -\frac{i\omega\epsilon_{\rm eff}}{c} E_{1\perp}. \tag{10.28}$$

Similarly, from the $\vec{\nabla} \times \vec{E}$ equation (10.21), we obtain:

$$k_\perp E_{1\perp} = -\frac{\omega}{c} B_{1z}, \tag{10.29}$$

$$\frac{\partial}{\partial z} E_{1\perp} = \frac{i\omega}{c} B_{1||}, \tag{10.30}$$

$$\frac{\partial}{\partial z} E_{1||} - ik_\perp E_{1z} = -\frac{i\omega}{c} B_{1\perp}. \tag{10.31}$$

Equations (10.26)–(10.31) correspond to (9.38)–(9.43) with $\vec{J} = \sigma(\omega)\vec{E}$ and $\vec{B} \to \vec{H}$ in the first three equations. Since by assumption $\epsilon_{\rm eff}$ and μ_c are independent of z throughout the conductor, we may derive equations for the $||$, $\perp$, and z components of $\vec{E}_1$ and $\vec{H}_1$ that are analogous to (9.45) and (9.46):

$$\left[\frac{\partial^2}{\partial z^2} + k_1^2\right] E_{1||,\perp,z} = 0, \tag{10.32}$$

$$\left[\frac{\partial^2}{\partial z^2} + k_1^2\right] H_{1||,\perp,z} = 0, \tag{10.33}$$

where as in (9.51) we define

$$k_1 \equiv \sqrt{\frac{\omega^2 \mu_c \epsilon_{\rm eff}}{c^2} - k_\perp^2}. \tag{10.34}$$

As indicated above, we take the perfect electrical conductor case as our starting-point. Then just outside the conductor ($z = 0^+$), we find

$$E_{2||,\perp}(0^+) = 0; \qquad B_{2z}(0^+) = 0; \qquad E_{2z}(0^+), B_{2||,\perp}(0^+) \neq 0. \tag{10.35}$$

The continuity of tangential $\vec{H}$ together with (10.33) gives

$$H_{1||,\perp}(z) = e^{-ik_1 z} H_{2||,\perp}(0^+). \tag{10.36}$$

We can now find $E_{1||}$ and $E_{1\perp}$ (as shown in Figure 10.2) using (10.27)–(10.29) as well as (10.34) and (10.36):

$$E_{1||}(z) = \frac{ic}{\omega\epsilon_{\rm eff}} \frac{\partial}{\partial z} H_{1\perp} = \left(\frac{ck_1}{\omega\epsilon_{\rm eff}}\right) e^{-ik_1 z} H_{2\perp}(0^+), \tag{10.37}$$

$$E_{1\perp}(z) = \frac{-i\omega\mu_c}{ck_1^2} \frac{\partial}{\partial z} H_{1||} = -\left(\frac{\omega\mu_c}{ck_1}\right) e^{-ik_1 z} H_{2||}(0^+). \tag{10.38}$$

Under our "good conductor" assumptions (10.17), we may approximate

$$k_1 = \left(\frac{\omega^2\mu_c}{c^2}\left(\epsilon_c + i\frac{4\pi\sigma}{\omega}\right) - k_\perp^2\right)^{1/2} \approx \frac{\sqrt{2\pi\sigma\mu_c\omega}}{c}(1+i) = \frac{1}{\delta}(1+i), \tag{10.39}$$

where δ is the skin depth (10.24). In this case the constants on the right-hand sides of (10.37) and (10.38) are equal (to lowest order):

$$\frac{ck_1}{\omega\epsilon_{\rm eff}} \approx \frac{\omega\mu_c}{ck_1} \approx \sqrt{\frac{\mu_c\omega}{8\pi\sigma}}(1-i). \tag{10.40}$$

Since z is negative inside the conductor we may write $z = -|z|$ so that

$$H_{1||,\perp}(z) \approx H_{2||,\perp}(0^+) e^{-|z|/\delta} e^{i|z|/\delta}. \tag{10.41}$$

Notice the characteristic exponential decay, with the rate of decay determined by the skin depth.

Again referring to Figure 10.2, we have

$$\hat{n} \times \vec{H}_2^{\rm tang}(0^+) = \hat{n} \times (H_{2\perp}\hat{k}_\perp \times \hat{n} + H_{2||}\hat{k}_\perp)$$
$$= H_{2\perp}\hat{k}_\perp - H_{2||}\hat{k}_\perp \times \hat{n}, \tag{10.42}$$
$$\vec{E}_1^{\rm tang} = E_{1||}\hat{k}_\perp + E_{1\perp}\hat{k}_\perp \times \hat{n}. \tag{10.43}$$

Applying (10.37) and (10.38) to the $\hat{k}_\perp$ and $\hat{k}_\perp \times \hat{n}$ components of (10.42)–(10.43) respectively, and using the approximation (10.40) we obtain

$$\vec{E}_1^{\rm tang}(z) \approx \sqrt{\frac{\omega\mu_c}{8\pi\sigma}}(1-i)\hat{n} \times \vec{H}_2^{\rm tang}(0^+)e^{-|z|/\delta}e^{i|z|/\delta}. \tag{10.44}$$

Equation (10.44) is polarization independent. Note the phase of $\vec{E}_1^{\rm tang}$ differs from $\vec{H}_2^{\rm tang}(0^+)$, and its magnitude is small compared to the $\vec{B}_1$ field (as we indicated in Section 9.3):

$$\frac{\left|\vec{E}_1^{\rm tang}\right|}{\left|\vec{B}_1^{\rm tang}\right|} \approx \frac{1}{\mu_c}\sqrt{\frac{\omega\mu_c}{4\pi\sigma}}. \tag{10.45}$$

The normal components of $\vec{E}_1$ and $\vec{H}_1$ may also be solved for in this limit. Recalling that $k_\perp = n_2(\omega/c)\sin i$ where i is the (vacuum) incidence angle, we have:

$$E_{1z}(z) = \frac{ck_\perp}{\omega\epsilon_{\rm eff}}H_{1\perp}(z) = \frac{ck_\perp}{\omega\epsilon_{\rm eff}}H_{2\perp}(0^+)e^{-ik_1 z}$$
$$\approx -i\frac{n_2\omega\sin i}{4\pi\sigma}H_{2\perp}(0^+)e^{-|z|/\delta}e^{i|z|/\delta}, \tag{10.46}$$
$$H_{1z}(z) = -\frac{ck_\perp}{\omega\mu_c}E_{1\perp}(z) = \frac{k_\perp}{k_1}H_{2||}(0^+)e^{-ik_1 z}$$
$$\approx n_2\sin i\sqrt{\frac{\omega}{8\pi\sigma\mu_c}}(1-i)H_{2||}(0^+)e^{-|z|/\delta}e^{i|z|/\delta}. \tag{10.47}$$

Here the normal components are polarization-dependent, unlike $\vec{E}_1^{\rm tang}$. Both normal components are small, but as with the tangential components the electric component is generally much smaller than the magnetic:

$$\frac{|E_{1z}|}{|B_{1z}|} \approx \frac{|B_{2\perp}|}{|B_{2||}|}\frac{1}{\mu_c}\sqrt{\frac{\omega\mu_c}{4\pi\sigma}}. \tag{10.48}$$

Equations (10.46) and (10.47) provide some nice consistency checks. From (10.26) applied to Region 2 we have (note $\epsilon_{\rm eff} \to \epsilon$ in Region 2)

$$\epsilon E_{2z}(0^+) = k_\perp\frac{c}{\omega}H_{2\perp}(0^+). \tag{10.49}$$

Combining this with the first line of (10.46) gives

$$E_{1z}(z) = \frac{\epsilon E_{2z}(0^+)}{\epsilon_{\text{eff}}} e^{-ik_1 z}, \tag{10.50}$$

or

$$D_{1z}(z) = D_{2z}(0^+) e^{-ik_1 z}, \tag{10.51}$$

which also follows directly from the wave equation for E_z and the fact that $D_{1z}(0^-) = \epsilon E_{2z}(0^+)$. Similarly, we know from (10.29) applied to Region 2 that

$$B_{2z}(0^+) = -\frac{ck_\perp}{\omega} \overbrace{E_{2\perp}(0^+)}^{E_{1\perp}(0^-)}, \tag{10.52}$$

which together with the first line of (10.47) gives (note $E_{1\perp}(z) = E_{1\perp}(0^-)e^{-ik_1 z}$)

$$\begin{aligned} H_{1z}(z) &= \frac{B_{2z}(0^+)}{\mu_c} e^{-ik_1 z} \\ \Longrightarrow\ B_{1z}(z) &= B_{2z}(0^+) e^{-ik_1 z}, \end{aligned} \tag{10.53}$$

which also follows directly from the wave equation for $B_{1z}(z)$.

Note that now $\vec{E} \cdot \vec{B} = E_z B_z \neq 0$ on both sides of the interface, so our initial assumptions (10.35) are shown to be approximations. One consequence of nonzero $\vec{E}_1^{\text{tang}}$ is a current inside Region 1: denoting $\vec{H}_2^{\text{tang}}(0^+)$ by $\vec{H}_2^{\text{tang}}$, we have

$$\vec{J}(z) = \sigma \vec{E}_1^{\text{tang}} = \sqrt{\frac{\omega \mu_c \sigma}{8\pi}} (1-i) \hat{n} \times \vec{H}_2^{\text{tang}} e^{-|z|/\delta} e^{i|z|/\delta}. \tag{10.54}$$

This leads to

$$\begin{aligned} \vec{K}_{\text{eff}} \equiv \int_{-\infty}^{0} dz\, \vec{J}(z) &= \sqrt{\frac{\omega \mu_c \sigma}{8\pi}} (1-i) \frac{\delta}{1-i} \hat{n} \times \vec{H}_2^{\text{tang}}, \\ &= \frac{c}{4\pi} \hat{n} \times \vec{H}_2^{\text{tang}}. \end{aligned} \tag{10.55}$$

This is consistent with our model for a perfect conductor because on the boundary

$$\hat{n} \times (\vec{H}_2 - \underbrace{\vec{H}_1}_{0})\big|_S = \frac{4\pi}{c} \vec{K}_{\text{free}} \Longrightarrow \vec{K}_{\text{eff}} = \vec{K}_{\text{free}}. \tag{10.56}$$

We close this section with some general observations about power loss, an important topic that we shall revisit in Sections 10.7 and 10.8. Recalling (8.72), the energy flow across the $z = 0$ boundary may be expressed as

$$S_z = \frac{c}{4\pi} \hat{n} \cdot (\vec{E} \times \vec{H})\Big|_{z=0}, \tag{10.57}$$

where in this case $\hat{n} = \hat{e}_z$. We may approach the $z = 0$ boundary from either side, because the boundary conditions guarantee continuity. For harmonic fields, we write (as in Chapter 8)

$$\vec{E}(\vec{x}, t) = \frac{1}{2}[\vec{E}e^{-i\omega t} + \vec{E}^* e^{i\omega t}], \tag{10.58}$$

where $\vec{E}$ and $\vec{E}^*$ in (10.58) are understood to be functions of $\vec{x}$. We also have

$$\begin{aligned} \vec{E} \times \vec{H} &= \frac{1}{4}[\vec{E}e^{-i\omega t} + \vec{E}^* e^{i\omega t}] \times [\vec{H}e^{-i\omega t} + \vec{H}^* e^{i\omega t}] \\ &= \frac{1}{2}\mathrm{Re}[\vec{E}^* \times \vec{H} + \vec{E} \times \vec{H}e^{-2i\omega t}]. \end{aligned} \tag{10.59}$$

It follows that the time-averaged energy flux in the z-direction may be computed as

$$\begin{aligned} \langle S_z \rangle &= \frac{c}{8\pi}\mathrm{Re}[\hat{n} \cdot (\vec{E}_2^* \times \vec{H}_2)] \\ &= \frac{c}{8\pi}\mathrm{Re}[\vec{E}_2^* \cdot (\vec{H}_2 \times \hat{n})], \end{aligned} \tag{10.60}$$

where $\vec{E}^*$ denotes $\vec{E}^*(0^+)$ and $\langle \ldots \rangle$ denotes time average. Since only the tangential part of the electric field contributes to this expression, from (10.44) we have

$$\begin{aligned} \langle S_z \rangle &\approx \frac{c}{8\pi}\mathrm{Re}\left[\sqrt{\frac{\omega\mu_c}{8\pi\sigma}}(1+i)(\hat{n} \times \vec{H}_2^*) \cdot (\vec{H}_2 \times \hat{n})\right] \\ &= -\frac{\omega\mu_c\delta}{16\pi}|\hat{n} \times \vec{H}_2|^2. \end{aligned} \tag{10.61}$$

Equation (10.61) relates $\langle S_z \rangle$ to fields outside the conductor. In Sections 10.7 and 10.8, we'll use this to find approximate expressions for power loss.

10.2 General considerations for waveguides of arbitrary cross-section

Waveguides are cavities with z-independent geometry. Cylindrical waveguides are most common, but rectangular and other shapes are also used. In this section, we will consider general properties of waveguides with arbitrary cross-section.

Based on our previous experience with boundary conditions, we expect fields to be plane waves in the z-direction but standing waves in the x and

y directions. As in the previous section we assume that ϵ and μ inside the waveguide are either constant or depend only on frequency, and that no free currents or charges are present (so in particular the conductivity inside the cavity is zero). The fields inside the cavity are (expressed as functions of $\vec{x}$ and ω)

$$\vec{\nabla} \times \vec{H} = -\frac{i\omega\epsilon}{c}\vec{E}, \tag{10.62}$$

$$\vec{\nabla} \times \vec{E} = \frac{i\omega\mu}{c}\vec{H}. \tag{10.63}$$

We assume plane wave solutions for $\vec{H}$ and $\vec{E}$ of the form $\vec{A}(x,y)e^{i(kz-\omega t)}$, where k can be positive or negative. We may then replace

$$\vec{\nabla} \to \vec{\nabla}_\perp + ik\,\hat{e}_3, \tag{10.64}$$

where "$\perp$" means perpendicular to the z-axis. We then have

$$(\vec{\nabla}_\perp + ik\hat{e}_3) \times (\vec{H}_\perp + H_z\hat{e}_3) = -\frac{i\omega\epsilon}{c}(\vec{E}_\perp + E_z\hat{e}_3), \tag{10.65}$$

$$(\vec{\nabla}_\perp + ik\hat{e}_3) \times (\vec{E}_\perp + E_z\hat{e}_3) = \frac{i\omega\mu}{c}(\vec{H}_\perp + H_z\hat{e}_3). \tag{10.66}$$

It follows upon taking components parallel and perpendicular to z that

$$\vec{\nabla}_\perp \times \vec{H}_\perp = -\frac{i\omega\epsilon}{c}E_z\hat{e}_3, \tag{10.67}$$

$$\vec{\nabla}_\perp H_z - ik\vec{H}_\perp = -\frac{i\omega\epsilon}{c}\hat{e}_3 \times \vec{E}_\perp, \tag{10.68}$$

$$\vec{\nabla}_\perp \times \vec{E}_\perp = i\frac{\omega\mu}{c}H_z\hat{e}_3, \tag{10.69}$$

$$\vec{\nabla}_\perp E_z - ik\vec{E}_\perp = i\frac{\omega\mu}{c}\hat{e}_3 \times \vec{H}_\perp. \tag{10.70}$$

We may simplify the problem by solving for $\vec{H}_\perp, \vec{E}_\perp$ in terms of H_z, E_z. Using (10.67)–(10.70) and defining the *cutoff wavenumber* squared

$$\gamma^2 \equiv \frac{\omega^2\mu\epsilon}{c^2} - k^2, \tag{10.71}$$

we have (assuming that $\gamma \neq 0$)

$$\vec{E}_\perp = \frac{i}{\gamma^2}\left[k\vec{\nabla}_\perp E_z - \frac{\omega\mu}{c}(\hat{e}_3 \times \vec{\nabla}_\perp)H_z\right], \tag{10.72}$$

$$\vec{H}_\perp = \frac{i}{\gamma^2}\left[k\vec{\nabla}_\perp H_z + \frac{\omega\epsilon}{c}(\hat{e}_3 \times \vec{\nabla}_\perp)E_z\right]. \tag{10.73}$$

If we can now find differential equations for E_z and H_z, we will have solved the problem in principle. It follows from (10.62) and (10.63) that

$$\left(\nabla^2 + \frac{\omega^2\mu\epsilon}{c^2}\right)\{E_z, H_z\} = 0, \tag{10.74}$$

or using (10.64)

$$(\nabla_\perp^2 + \gamma^2)\{E_z, H_z\} = 0. \tag{10.75}$$

This is just the two-dimensional Helmholtz equation. It remains to give boundary conditions for E_z and H_z. We may write (10.72) and (10.73) in the neighborhood of a point on the boundary in terms of a local Cartesian coordinate system, as shown in Figure 10.4. We refer to $\hat{n}$ in this case as an inwardly directed normal.

$$E_x = \frac{i}{\gamma^2}\left[k\frac{\partial E_z}{\partial x} + \frac{\omega\mu}{c}\frac{\partial H_z}{\partial y}\right], \tag{10.76}$$

$$E_y = \frac{i}{\gamma^2}\left[k\frac{\partial E_z}{\partial y} - \frac{\omega\mu}{c}\frac{\partial H_z}{\partial x}\right], \tag{10.77}$$

$$H_x = \frac{i}{\gamma^2}\left[k\frac{\partial H_z}{\partial x} - \frac{\omega\epsilon}{c}\frac{\partial E_z}{\partial y}\right], \tag{10.78}$$

$$H_y = \frac{i}{\gamma^2}\left[k\frac{\partial H_z}{\partial y} + \frac{\omega\epsilon}{c}\frac{\partial E_z}{\partial x}\right]. \tag{10.79}$$

In the rest of this section we consider the ideal case of a perfectly conducting boundary. Then the boundary condition $\hat{n} \times \vec{E}|_s = 0$ implies

$$E_z|_s = 0, \tag{10.80}$$

which is independent of our local frame. In addition, we have the local statements

$$E_x|_s = 0, \tag{10.81}$$

$$H_y|_s = 0, \tag{10.82}$$

where (10.82) follows from $\hat{n} \cdot \vec{B}|_s = 0$. We also have that

$$\frac{\partial E_z}{\partial x}|_s = 0 \text{ (since } x \text{ is parallel to the surface).} \tag{10.83}$$

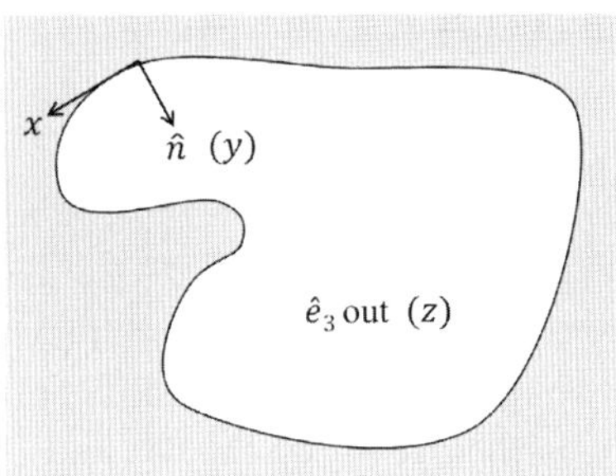

Fig. 10.4 Local Cartesian coordinate system.

Then, from (10.76) or (10.79) we get

$$\frac{\partial H_z}{\partial y}|_s = 0. \tag{10.84}$$

The conditions on E_z and H_z at the boundary may be summarized in coordinate-free form as

$$E_z|_s = 0, \tag{10.85}$$

$$(\hat{n} \cdot \vec{\nabla}) H_z|_s = 0. \tag{10.86}$$

The equations and boundary conditions for E_z and H_z are completely decoupled, so they may be chosen independently. We choose to characterize solutions as follows:

- $H_z = 0$ everywhere: *transverse magnetic* (TM) waves
- $E_z = 0$ everywhere: *transverse electric* (TE) waves
- $H_z = E_z = 0$ everywhere: *transverse electric magnetic* (TEM) waves

It can be shown that TE, TM and TEM waves form a complete set of solutions to (10.67)–(10.70) above; in other words, an arbitrary solution can be expressed as a linear combination of these three basic types. We will assume this fact without proof. In addition, since we are solving a real equation with real boundary conditions, then the E_z, H_z fields can always be chosen as functions of a single complex phase, *i.e.*, real functions times a complex constant, for non-degenerate cutoff wavenumbers. (We will examine some degenerate cases in the exercises.) This fact will become important later.

TEM waves require $\gamma = 0$, so (10.76)–(10.79) cannot be used. Instead, from $H_z = E_z = 0$ and (10.67)–(10.70) we have

$$\vec{\nabla}_\perp \times \vec{E}_\perp = 0, \tag{10.87}$$

$$\vec{\nabla}_\perp \cdot \vec{E}_\perp = 0. \tag{10.88}$$

Note that (10.87) and (10.88) together give $\nabla_\perp^2 \vec{E}_\perp = 0$ (which is consistent with (10.75) with $\gamma = 0$). As a result, to find TEM solutions for $\vec{E}_\perp$ we just need to solve a two-dimensional electrostatic problem with Dirichlet conditions for $\vec{E}_\perp$. $\vec{H}_\perp$ is then given by (10.68) and (10.71) with $\gamma = 0$:

$$\vec{H}_\perp = \text{sign}(k)\sqrt{\frac{\epsilon}{\mu}}\,\hat{e}_3 \times \vec{E}_\perp. \tag{10.89}$$

TEM modes are only possible for certain geometries. In particular, if the boundary of the cavity is connected, then the scalar potential Φ is constant on the entire boundary, so the uniqueness theorem for Dirichlet problems implies that $\vec{E}_\perp = 0$ everywhere inside. But this is not necessarily true if the boundary is disconnected, such as the case of the cavity between two coaxial cylinders.

In the TM case, the relation between $\vec{E}_\perp$ and $\vec{H}_\perp$ may be obtained as follows (using (m) to indicate TM mode):

$$\vec{E}_\perp^{(m)} = \frac{ik}{\gamma^2}\vec{\nabla}_\perp E_z^{(m)}, \tag{10.90}$$

$$\vec{H}_\perp^{(m)} = \frac{i\omega\epsilon}{\gamma^2 c}(\hat{e}_3 \times \vec{\nabla}_\perp)E_z^{(m)} \tag{10.91}$$

$$\implies \vec{H}_\perp^{(m)} = \frac{\omega\epsilon}{kc}\hat{e}_3 \times \vec{E}_\perp^{(m)}. \tag{10.92}$$

In the TE case, we have similarly (using (e) to indicate TE modes):

$$\vec{E}_\perp^{(e)} = -\frac{i\omega\mu}{\gamma^2 c}(\hat{e}_3 \times \vec{\nabla}_\perp)H_z^{(e)}, \tag{10.93}$$

$$\vec{H}_\perp^{(e)} = \frac{ik}{\gamma^2}\vec{\nabla}_\perp H_z^{(e)} \tag{10.94}$$

$$\implies \vec{H}_\perp^{(e)} = \frac{kc}{\omega\mu}\hat{e}_3 \times \vec{E}_\perp^{(e)}. \tag{10.95}$$

Both TE and TM cases can be summarized in the single equation

$$\vec{H}_\perp = \frac{1}{Z}\hat{e}_3 \times \vec{E}_\perp, \tag{10.96}$$

where the *wave impedance* Z is defined as

$$Z \equiv \begin{cases} \dfrac{ck}{\epsilon\omega} & \text{TM}, \\ \dfrac{\mu\omega}{ck} & \text{TE}. \end{cases} \tag{10.97}$$

The wave impedance can be positive, negative, or complex. In upcoming Section 10.4 we will see that the power flow in the waveguide is related to its inverse. The wave impedance Z is actually dimensionless, but Z/c forms a resistance in Gaussian units. There is no special name for this; the units are simply s/cm.

10.3 TE and TM modes for rectangular waveguides with perfectly-conducting walls

In this section we find explicitly the TE and TM modes for an ideal (perfectly-conducting) rectangular waveguide with dimensions $a \times b$ as shown in Figure 10.5. (There are no TEM modes, since the boundary is connected.) For the TE modes (with $E_z = 0$), we have from (10.75) the fundamental wave equation for the (x, y) dependence

$$\left(\frac{\partial^2}{\partial x^2} + \frac{\partial^2}{\partial y^2} + \gamma^2\right) H_z = 0. \tag{10.98}$$

A variable-separated solution of the form $H_z = I(x)J(y)e^{i(kx-\omega t)}$ must then satisfy

$$\frac{\partial^2 I}{\partial x^2} = -\kappa^2 I, \qquad \frac{\partial^2 J}{\partial y^2} + (\gamma^2 - \kappa^2)J = 0. \tag{10.99}$$

It follows that $I(x)$ has the form $I(x) = C_1 \cos(\kappa x + \alpha)$, and the boundary conditions lead to

$$\frac{\partial I(0)}{\partial x} = \frac{\partial I(a)}{\partial x} = 0 \implies \alpha = 0, \kappa = \frac{m\pi}{a}, \; m = 0, 1, 2, \ldots \tag{10.100}$$

$J(y)$ has a similar form, and the boundary conditions lead to

$$J(y) = C_2 \cos\left(\frac{n\pi}{b} y\right), \; n = 0, 1, 2, \ldots. \tag{10.101}$$

From (10.99) it follows for the (m, n) TE mode that

$$\gamma_{mn}^2 = \pi^2 \left(\frac{m^2}{a^2} + \frac{n^2}{b^2}\right), \tag{10.102}$$

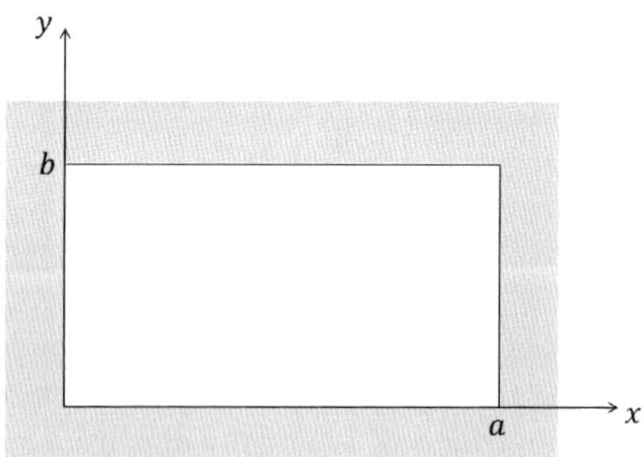

Fig. 10.5 Rectangular Waveguide.

so from (10.71) we obtain

$$k = \pm\frac{\sqrt{\mu\epsilon}}{c}\sqrt{\omega^2 - \omega_{mn}^2}, \tag{10.103}$$

where

$$\omega_{mn} \equiv \frac{c}{\sqrt{\mu\epsilon}}\gamma_{mn} \equiv \frac{c\pi}{\sqrt{\mu\epsilon}}\left(\frac{m^2}{a^2} + \frac{n^2}{b^2}\right)^{1/2}, \tag{10.104}$$

is called the *cutoff frequency* for TE modes. If $\omega > \omega_{mn}$, then k is real and the solution is a traveling wave; but if $\omega < \omega_{mn}$ then k is imaginary which leads to exponential attenuation in the z-direction. In either case we have the (x, y) dependence (as long as m and n are not both zero)

$$H_z^{(e)}(\vec{x}_\perp, k, \omega) = H_0 \cos\left(\frac{m\pi x}{a}\right)\cos\left(\frac{n\pi y}{b}\right), \tag{10.105}$$

where H_0 can be complex, and the corresponding physical field is given by the real part of $H_z(\vec{x}_\perp, k, \omega)e^{i(kx-\omega t)}$ where k is given by (10.103).

A similar derivation gives the TM modes for the same waveguide:

$$E_z^{(m)}(\vec{x}_\perp, k, \omega) = E_0 \sin\left(\frac{m'\pi x}{a}\right)\sin\left(\frac{n'\pi y}{b}\right). \tag{10.106}$$

For TM modes m' and n' must both be nonzero, so the lowest mode has m', $n' = 1$. Except for the $m = 0$ or $n = 0$ TE modes (one of which is the lowest mode for the system), all other TE and TM modes are degenerate in the sense that for every γ_{mn} there exists both TE and TM modes that satisfy the same dispersion relation (10.103).

10.4 Orthogonality properties for modes in ideal waveguides

In this section we will establish orthogonality properties of the different modes (eigenfunctions) in waveguides of arbitrary cross-section with perfectly-conducting walls. These properties have important implications for the physics: for instance, two orthogonal modes may be present simultaneously without interfering with each other, and the total power is the sum of power contributions from each individual mode. Orthogonality properties will also be needed in our next physical application: the effect of thin, transverse obstructions on transmission of modes in waveguides (see Sections 10.5 and 10.6).

We will use λ and μ as generic mode indices. Each mode λ corresponds to a unique value of γ^2 for which (10.75) has a solution with the appropriate boundary conditions; to clarify this relationship, we write γ^2 as ${\gamma_\lambda}^2$. However, k is determined only up to a $\pm$ sign from ${\gamma_\lambda}^2$ via (10.71), which means there are actually two modes present. These two modes appear implicitly in our earlier expressions, but it is now time to make the notation explicit. We will henceforth denote these modes as λ_+ and λ_-. We will adopt the convention that $k_{\lambda_+} (= -k_{\lambda_-})$ is positive real (when k_λ^2 is positive) or positive imaginary (when k_λ^2 is negative). Using the freedom to choose phases, we require that there will be no $\pm$-mode dependence in the z-fields. To simplify the notation in the following discussion we'll use $E_{z\lambda}$ and $H_{z\lambda}$ to represent $E_z(\vec{x}_\perp, k_\lambda, \omega)$ and $H_z(\vec{x}_\perp, k_\lambda, \omega)$. Given this, the k_{λ_s} ($s = \pm$) and ω-dependent waveguide fields are expressed as:

$$\vec{E}_{\lambda_s}(\vec{x}) = \left[E_{z\lambda}\hat{e}_3 + \vec{E}_{\perp\lambda_s}\right] e^{ik_{\lambda_s} z}, \tag{10.107}$$

$$\vec{H}_{\lambda_s}(\vec{x}) = \left[H_{z\lambda}\hat{e}_3 + \vec{H}_{\perp\lambda_s}\right] e^{ik_{\lambda_s} z}. \tag{10.108}$$

The fields $\vec{E}_{\perp\lambda_s}$ are related to the z-fields via (10.90) and (10.93), which we now write as

$$\vec{E}^{(m)}_{\perp\lambda_s} = \frac{ik_{\lambda_s}}{\gamma_\lambda^2}\vec{\nabla}_\perp E^{(m)}_{z\lambda}, \tag{10.109}$$

$$\vec{E}^{(e)}_{\perp\lambda_s} = -\frac{i\omega\mu}{\gamma_\lambda^2 c}\hat{e}_3 \times \vec{\nabla}_\perp H^{(e)}_{z\lambda}, \tag{10.110}$$

where (e) and (m) again denote TE and TM cases, respectively. (The extension of this notation to the z-fields is unnecessary but quickens understanding.) The relations (10.109) and (10.110) give us immediately:

$$\vec{E}^{(m)}_{\perp\lambda_-} = -\vec{E}^{(m)}_{\perp\lambda_+} \text{ and } \vec{E}^{(e)}_{\perp\lambda_-} = \vec{E}^{(e)}_{\perp\lambda_+}. \tag{10.111}$$

As we pointed out in Section 10.2, the solutions to the Helmholtz equation $E^{(m)}_{z\lambda}, H^{(e)}_{z\lambda}$ are functions of a single phase in space. Again, using the freedom to choose such phases, we will find it most useful in the upcoming calculations to choose the $\vec{E}^{(e,m)}_{\perp\lambda_s}$ fields *real* for both propagating and non-propagating modes. As far as $\vec{H}^{(e,m)}_{\perp\lambda_s}$ fields are concerned, (10.96) and (10.97) give

$$\vec{H}_{\perp\lambda_s} = \frac{1}{Z_{\lambda_s}}\hat{e}_3 \times \vec{E}_{\perp\lambda_s}, \tag{10.112}$$

where from (10.97)

$$Z_{\lambda_s} \equiv \begin{cases} \dfrac{ck_{\lambda_s}}{\epsilon\omega} & \text{TM}, \\ \dfrac{\mu\omega}{ck_{\lambda_s}} & \text{TE}. \end{cases} \tag{10.113}$$

For cutoff modes, in the engineering convention for electrical circuits, positive imaginary Z_{λ_s} is considered capacitive and negative imaginary Z_{λ_s} is considered inductive. Using these relations and (10.111), we find that

$$\vec{H}^{(m)}_{\perp\lambda_-} = \vec{H}^{(m)}_{\perp\lambda_+} \quad \text{and} \quad \vec{H}^{(e)}_{\perp\lambda_-} = -\vec{H}^{(e)}_{\perp\lambda_+}. \tag{10.114}$$

Let's first consider the orthogonality of E_z fields corresponding to two different TM modes, which we will distinguish with labels λ and μ. From Green's second formula we have

$$\begin{aligned} &\int da \left(E^{(m)*}_{z\mu} \nabla^2_\perp E^{(m)}_{z\lambda} - E^{(m)}_{z\lambda} \nabla^2_\perp E^{(m)*}_{z\mu} \right) \\ &\quad = \oint_C d\ell \left(E^{(m)*}_{z\mu} \frac{\partial}{\partial n} E^{(m)}_{z\lambda} - E^{(m)}_{z\lambda} \frac{\partial}{\partial n} E^{(m)*}_{z\mu} \right). \end{aligned} \tag{10.115}$$

The right-hand side refers to the contour C which bounds the waveguide cross-section S. But we also know that the modes satisfy:

$$(\nabla^2_\perp + \gamma^2_\lambda) E^{(m)}_{z\lambda} = (\nabla^2_\perp + \gamma^2_\mu) E^{(m)}_{z\mu} = 0; \quad E^{(m)}_{z\lambda}\big|_C = E^{(m)}_{z\mu}\big|_C = 0. \tag{10.116}$$

The boundary conditions at C imply the right-hand side of (10.115) is zero, while the derivative expressions applied to the left-hand side lead to

$$\begin{aligned} &\left(-\gamma^2_\lambda + \gamma^2_\mu\right) \int da\, E^{(m)*}_{z\mu} E^{(m)}_{z\lambda} = 0 \\ &\qquad \Longrightarrow \int da\, E^{(m)*}_{z\mu} E^{(m)}_{z\lambda} = 0 \qquad \text{if } \gamma^2_\lambda \neq \gamma^2_\mu. \end{aligned} \tag{10.117}$$

The results for H_z for TE modes are very similar:

$$\int da\, H^{(e)*}_{z\mu} H^{(e)}_{z\lambda} = 0 \qquad \text{if } \gamma^2_\lambda \neq \gamma^2_\mu. \tag{10.118}$$

In the case of degenerate TE or TM modes, using standard orthogonalization techniques it's always possible to choose an orthogonal basis for the vector subspace of degenerate modes. With our usual clairvoyance, we choose normalizations that lead to nicer expressions later:

$$\int da\, E^{(m)*}_{z\lambda} E^{(m)}_{z\mu} = \frac{\gamma^2_\lambda}{|k_\lambda|^2} \delta_{\lambda\mu}, \tag{10.119}$$

$$\int da\, H^{(e)*}_{z\lambda} H^{(e)}_{z\mu} = \frac{\gamma^2_\lambda c^2}{\omega^2 \mu^2} \delta_{\mu\lambda}. \tag{10.120}$$

The relations (10.109) and (10.110) also make it possible for us to compute orthogonality relations for $\vec{E}_\perp$. In the TM case we have (the complex conjugate on the $\vec{E}^{(e),(m)}_{\perp\lambda_s}$ fields is purely window dressing; $s, s' = \pm$):

$$\begin{aligned}
\int da\, \vec{E}^{(m)*}_{\perp\lambda_s} \cdot \vec{E}^{(m)}_{\perp\mu_{s'}} &= \frac{k^*_{\lambda_s} k_{\mu_{s'}}}{\gamma^2_\lambda \gamma^2_\mu} \int da\, \vec{\nabla}_\perp E^{(m)*}_{z\lambda} \cdot \vec{\nabla}_\perp E^{(m)}_{z\mu} \\
&= \frac{k^*_{\lambda_s} k_{\mu_{s'}}}{\gamma^2_\lambda \gamma^2_\mu} \left\{ \oint dl\, E^{(m)*}_{z\lambda} \frac{\partial}{\partial n} E^{(m)}_{z\mu} - \int da\, E^{(m)*}_{z\lambda} \nabla^2_\perp E^{(m)}_{z\mu} \right\} \\
&= \frac{k^*_{\lambda_s} k_{\mu_{s'}}}{\gamma^2_\lambda} \int da\, E^{(m)*}_{z\lambda} E^{(m)}_{z\mu} \\
&= \begin{cases} \delta_{\lambda\mu} & s = s', \\ -\delta_{\lambda\mu} & s \neq s'. \end{cases}
\end{aligned} \tag{10.121}$$

The TE case ends up with a similar result:

$$\begin{aligned}
\int da\, \vec{E}^{(e)*}_{\perp\lambda_s} \cdot \vec{E}^{(e)}_{\perp\mu_{s'}} &= -\frac{\omega^2\mu^2}{\gamma^2_\lambda \gamma^2_\mu c^2} \int da\, (\hat{e}_3 \times \vec{\nabla}_\perp H^{(e)*}_{z\lambda}) \cdot (\hat{e}_3 \times \vec{\nabla}_\perp H^{(e)}_{z\mu}) \\
&= -\frac{\omega^2\mu^2}{\gamma^2_\lambda \gamma^2_\mu c^2} \int da\, \left[(\hat{e}_3 \times \vec{\nabla}_\perp H^{(e)*}_{z\lambda}) \times \hat{e}_3 \right] \cdot \vec{\nabla}_\perp H^{(e)}_{z\mu}) \\
&= \frac{\omega^2\mu^2}{\gamma^2_\lambda \gamma^2_\mu c^2} \int da\, \vec{\nabla}_\perp H^{(e)*}_{z\lambda} \cdot \vec{\nabla}_\perp H^{(e)}_{z\mu} \\
&= \frac{\omega^2\mu^2}{\gamma^2_\lambda \gamma^2_\mu c^2} \gamma^2_\mu \int da\, H^{(e)*}_{z\lambda} H^{(e)}_{z\mu} \\
&= \delta_{\lambda\mu}.
\end{aligned} \tag{10.122}$$

In the mixed TM, TE case, where λ is TM and μ is TE we find

$$\begin{aligned}
\int da\, \vec{E}^{(m)*}_{\perp\lambda_s} \cdot \vec{E}^{(e)}_{\perp\mu_{s'}} &= -\frac{k^*_{\lambda_s} \omega\mu}{\gamma^2_\lambda \gamma^2_\mu c} \int da\, \vec{\nabla}_\perp E^{(m)*}_{z\lambda} \cdot (\hat{e}_3 \times \vec{\nabla}_\perp H^{(e)}_{z\mu}) \\
&= -\frac{k^*_{\lambda_s} \omega\mu}{\gamma^2_\lambda \gamma^2_\mu c} \left\{ \oint d\ell\, E^{(m)*}_{z\lambda} \hat{n} \cdot (\hat{e}_3 \times \vec{\nabla}_\perp H^{(e)}_{z\mu}) \right. \\
&\qquad \left. - \int da\, E^{(m)*}_{z\lambda} \underbrace{\vec{\nabla}_\perp \cdot (\hat{e}_3 \times \vec{\nabla}_\perp H^{(e)}_{z\mu})}_{-\hat{e}_3 \cdot (\vec{\nabla}_\perp \times \vec{\nabla}_\perp H^{(e)}_{z\mu})} \right\} \\
&= 0.
\end{aligned} \tag{10.123}$$

The relation (10.112) in same mode case (both TM or both TE) also leads to

$$\int da\, \vec{H}^*_{\perp\lambda_s} \cdot \vec{H}_{\perp\mu_{s'}} = \frac{1}{Z^*_{\lambda_s} Z_{\mu'_s}} \int da\, \underbrace{(\hat{e}_3 \times \vec{E}^*_{\perp\lambda_s}) \cdot (\hat{e}_3 \times \vec{E}_{\perp\mu_{s'}})}_{\vec{E}^*_{\perp\lambda_s} \cdot \vec{E}_{\perp\mu_{s'}}},$$

$$= \frac{1}{|Z_\lambda|^2} \begin{cases} \delta_{\lambda\mu} & (\text{TM; TE, } s = s') \\ -\delta_{\lambda\mu} & (\text{TE, } s \neq s'). \end{cases} \tag{10.124}$$

In the mixed case we get

$$\int da\, \vec{H}^{(m)*}_{\perp\lambda_s} \cdot \vec{H}^{(e)}_{\perp\mu_{s'}} = \int da\, \vec{E}^{(m)*}_{\perp\lambda_s} \cdot \vec{E}^{(e)}_{\perp\mu_{s'}} = 0. \tag{10.125}$$

For handy reference, here we summarize the orthogonality relations we've found so far:

$$\int da\, E^{(m)*}_{z\lambda} E^{(m)}_{z\mu} = \frac{\gamma_\lambda^2}{|k_\lambda|^2} \delta_{\lambda\mu}, \tag{10.126}$$

$$\int da\, H^{(e)*}_{z\lambda} H^{(e)}_{z\mu} = \frac{\gamma_\lambda^2 c^2}{\omega^2 \mu^2} \delta_{\lambda\mu}, \tag{10.127}$$

$$\int da\, \vec{E}^{(m)*}_{\perp\lambda_s} \cdot \vec{E}^{(m)}_{\perp\mu_s} = \int da\, \vec{E}^{(e)*}_{\perp\lambda_s} \cdot \vec{E}^{(e)}_{\perp\mu_s} = \delta_{\lambda\mu}, \tag{10.128}$$

$$\int da\, \vec{H}^{(m)*}_{\perp\lambda_s} \cdot \vec{H}^{(m)}_{\perp\mu_s} = \int da\, \vec{H}^{(e)*}_{\perp\lambda_s} \cdot \vec{H}^{(e)}_{\perp\mu_s} = \frac{\delta_{\lambda\mu}}{|Z_\lambda|^2}. \tag{10.129}$$

In addition to these relations, we have that TM fields are always orthogonal to the corresponding TE fields (as in (10.125), for example). The reader should also be careful when computing inner products of fields with different sign indices: there may be a sign change in the result, according to (10.111) and (10.114).

The power flow for the λ_+ and λ_- modes are in opposite directions for propagating modes. We can see this by computing the power flows at $z = 0$:

$$\frac{c}{8\pi}\text{Re}\left(\int da\, \hat{e}_3 \cdot \left(\vec{E}^*_{\lambda_s}\Big|_{z=0} \times \vec{H}_{\lambda_s}\Big|_{z=0}\right)\right) = \frac{c}{8\pi}\text{Re}\left(\int da\, \hat{e}_3 \cdot \left(\vec{E}^*_{\perp\lambda_s} \times \vec{H}_{\perp\lambda_s}\right)\right)$$

$$= \frac{c}{8\pi}\text{Re}\left(\frac{1}{Z_{\lambda_s}} \int da\, \underbrace{\hat{e}_3 \cdot \left(\vec{E}^*_{\perp\lambda_s} \times (\hat{e}_3 \times \vec{E}_{\perp\lambda_s})\right)}_{|\vec{E}_{\perp\lambda_s}|^2}\right)$$

$$= \frac{c}{8\pi}\text{Re}\left(\frac{1}{Z_{\lambda_s}}\right). \tag{10.130}$$

Note that Z_{λ_s} shares the sign of s. This expression gives us another practical interpretation for the wave impedance Z_{λ_s} and confirms that there is no power flow from non-propagating modes.

In the general case we may expand the E and H fields as (remember, we are looking at a fixed frequency ω)

$$\vec{E}(\vec{x}) = \sum_\lambda \left(A_{\lambda_+} \vec{E}_{\lambda_+}(\vec{x}) + A_{\lambda_-} \vec{E}_{\lambda_-}(\vec{x}) \right), \tag{10.131}$$

$$\vec{H}(\vec{x}) = \sum_\lambda \left(A_{\lambda_+} \vec{H}_{\lambda_+}(\vec{x}) + A_{\lambda_-} \vec{H}_{\lambda_-}(\vec{x}) \right), \tag{10.132}$$

where the mode index λ ranges over both TE and TM modes. It follows from our previous orthogonality relations that the coefficient A_{λ_s} for mode λ_s can be recovered using

$$A_{\lambda_s} = \frac{1}{2} \int da \left(\vec{E}_\perp \big|_{z=0} \cdot \vec{E}^*_{\perp\lambda_s} + |Z_\lambda|^2 \vec{H}_\perp \big|_{z=0} \cdot \vec{H}^*_{\perp\lambda_s} \right), \tag{10.133}$$

which is valid regardless of whether λ is TE or TM (see Exercise 10.4.1).

Usually one is most interested only in the *lowest* mode in an arbitrary waveguide. Recall that TM and TE modes are both determined by the equation

$$(\nabla_\perp^2 + \gamma^2)\psi = 0, \text{ where } \psi = \begin{cases} E_z & \text{TM}, \\ H_z & \text{TE}, \end{cases} \tag{10.134}$$

and boundary conditions are

$$\psi|_C = 0, \text{ TM}; \qquad \frac{\partial \psi_z}{\partial n}\bigg|_C = 0, \text{ TE}. \tag{10.135}$$

It follows that TM and TE modes correspond to Dirichlet and Neumann solutions (respectively) on the same domain. The lowest Neumann solution is the (trivial) constant solution, corresponding to $\gamma = 0$; but mathematicians have shown that under very general conditions, apart from this trivial solution the $n+1$'st smallest (positive) Neumann eigenvalue is strictly smaller than the n'th smallest Dirichlet eigenvalue.[1] It follows from this that the lowest nontrivial TE mode is always the lowest mode in the waveguide (as long as no TEM modes are present).

In Section 10.13 (Appendix) we give the fields for waveguide modes in rectangular and cylindrical waveguides.

[1] E. T. Kornhauser and I. Stakgold, "A variational Theorem for $\nabla^2 u + \lambda u = 0$ and its applications", J. Math. and Phys. **31**, 45–54 (1952); N. Filonov, "On an inequality between Dirichlet and Neumann eigenvalues for the Laplace operator", St. Petersburg Mathematics Journal, **16**, 413–416 (2005).

10.5 Reflection and impedance properties of thin obstructions in waveguides

Now we apply the results of the previous to the following practical situation: a waveguide of arbitrary cross-section, into which is introduced a thin, transverse, conducting obstruction of arbitrary shape in the $z = 0$ plane. (Two particular examples are shown in Figures 10.6 and 10.7.) As before, ω is taken as a fixed parameter; and we'll consider the case where ω is in the correct window of values to insure that only the lowest mode is propagating. We'll treat this situation as a scattering problem: we send in a continuous propagating wave from a source located at $z = -\infty$, and assume a steady-state situation has been reached.

Given our incoming wave, on the source side of the obstruction ($z < 0$) we can express the steady-state electric field as an expansion,

$$\vec{E}_{\rm s}(\vec{x}) = \vec{E}_{1_+}(\vec{x}) + R\vec{E}_{1_-}(\vec{x}) + \sum_{\lambda\neq 1} A_{\lambda_-}\vec{E}_{\lambda_-}(\vec{x}), \tag{10.136}$$

where the subscript 's' refers to the "source" side of the obstruction, and the subscript "1" refers to the lowest mode (which according to the previous section is *always* a TE mode). Note that we could multiply the right-hand side of (10.136) by an overall complex amplitude, but this would only complicate things and wouldn't change our conclusions. Note also that no λ_+ terms are included in the sum, since for non-propagating λ_+ modes we have $k_{\lambda_+} = i|k_{\lambda_+}|$ which leads to (unphysical) exponential growth as $z \to -\infty$.

On the non-source side ($z > 0$, indicated by the subscript 'ns'), we have

$$\vec{E}_{\rm ns}(\vec{x}) = T\vec{E}_{1_+}(\vec{x}) + \sum_{\lambda\neq 1} A_{\lambda_+}\vec{E}_{\lambda_+}(\vec{x}). \tag{10.137}$$

Tangential $\vec{E}$ vanishes at the surface of the conducting obstruction; and $\vec{E}$ is continuous at $z = 0$ within the gaps in the obstruction. In either case, it follows that $\vec{E}_\perp(\vec{x})$ is continuous at $z = 0$ for all (x, y) in the waveguide. By using orthogonality of the $\vec{E}_{\perp\lambda}(\vec{x})$ components in (10.136) and (10.137) at $z = 0$ with $\lambda = 1$, and employing (10.111), we get

$$1 + R = \int da\, \vec{E}\big|_{z=0} \cdot \vec{E}^*_{\perp 1_+} = T, \tag{10.138}$$

where the integrand in (10.138) is only nonzero in the gaps in the obstruction. If we do the same thing for each of the higher modes λ, we find

$$A_{\lambda_+} = \int da\, \vec{E}\big|_{z=0} \cdot \vec{E}^*_{\perp\lambda_+} = \begin{cases} A_{\lambda_-} & \text{TE,} \\ -A_{\lambda_-} & \text{TM,} \end{cases} \qquad (\lambda \neq 1). \tag{10.139}$$

Note that (10.139) implies the excitation of the λ_+ and λ_- modes by the obstruction are not independent. We can also expand $\vec{H}$ just as we've done for $\vec{E}$. The expansions for $\vec{H}_s$ and $\vec{H}_{ns}$ (analogous to (10.136) and (10.137)) are

$$\vec{H}_s(\vec{x}) = \vec{H}_{1_+}(\vec{x}) + R\vec{H}_{1_-}(\vec{x}) + \sum_{\lambda\neq 1} A_{\lambda_-}\vec{H}_{\lambda_-}(\vec{x}), \tag{10.140}$$

$$\vec{H}_{ns}(\vec{x}) = T\vec{H}_{1_+}(\vec{x}) + \sum_{\lambda\neq 1} A_{\lambda_+}\vec{H}_{\lambda_+}(\vec{x}). \tag{10.141}$$

$\vec{H}$ is not continuous across the entire cross-sectional area at $z = 0$, but it will be continuous across the gaps. Using this continuity in Equations (10.140) and (10.141), and employing (10.112) as well as (10.114), it is possible to show that

$$-2RY_{1_+}\vec{E}_{\perp 1_+} = 2\sum_{\lambda\neq 1} Y_{\lambda_+}A_{\lambda_+}\vec{E}_{\perp\lambda_+}, \quad \text{(valid in gaps } \textit{only}\text{)} \tag{10.142}$$

where

$$Y_{\lambda_+} \equiv 1/Z_{\lambda_+}, \tag{10.143}$$

is the admittance, as defined in Section 8.1. Plugging in the formula for A_{λ_+} from (10.139), we have

$$-2RY_{1_+}\vec{E}_{\perp 1_+} = 2\sum_{\lambda\neq 1} Y_{\lambda_+}\vec{E}_{\perp\lambda_+}\int da\,\vec{E}\big|_{z=0}\cdot\vec{E}^*_{\perp\lambda_+}. \tag{10.144}$$

This can be regarded as an integral equation which determines $\vec{E}|_{z=0}$. After dotting both sides with $\vec{E}^*|_{z=0}$ and integrating over da we find

$$-2R(1+R^*)Y_{1_+} = 2\sum_{\lambda\neq 1} Y_{\lambda_+}\left|\int_{\text{gaps}} da\,\vec{E}^*\big|_{z=0}\cdot\vec{E}_{\perp\lambda_+}\right|^2, \tag{10.145}$$

where (10.138) has been used. Dividing both sides by $Y_{1+}|1+R|^2$ leads to

$$Y_d \equiv -\frac{2R}{T} = \frac{2\sum_{\lambda\neq 1} Y_{\lambda_+}\left|\int_{\text{gaps}} da\,\vec{E}\big|_{z=0}\cdot\vec{E}_{\perp\lambda_+}\right|^2}{Y_{1_+}\left|\int_{\text{gaps}} da\,\vec{E}\big|_{z=0}\cdot\vec{E}_{\perp 1_+}\right|^2}, \tag{10.146}$$

($T = 1 + R$) where we are using the fact that the $\vec{E}_{\perp\lambda_+}$ field is real. Note that we may write this as

$$Y_d = \frac{2\sum_{\lambda\neq 1} Y_{\lambda_+}|A_{\lambda_+}|^2}{Y_{1_+}|T|^2}, \tag{10.147}$$

where the A_{λ_+} appear as weight factors. Also note the integrals in (10.146) can be taken either over the entire waveguide cross-section at $z = 0$ or just over the gaps, since $\vec{E}_\perp = 0$ at the surface of the obstruction. The dimensionless quantity Y_d is called the *relative shunt admittance*, and is an important concept in transmission-line theory. Exercise 10.5.1(b) helps to give an additional physical interpretation for Y_d. We will see below that (10.146) sometimes has helpful calculational properties.

Before we talk about this, let us point out that there is an alternate approach which can yield a useful expression for $Z_d = 1/Y_d$ as well. To see this, we use (10.140) at $z = 0^-$ as well as (10.114) and (10.139) for TE modes to obtain:

$$\vec{H}_{\perp \mathrm{s}} = \vec{H}_{\perp 1_+}(1 - R) - \sum_{\lambda \neq 1} A_{\lambda_+} \vec{H}_{\perp \lambda_+}. \tag{10.148}$$

The orthogonality relation (10.129) gives

$$1 - R = |Z_1|^2 \int da\, \vec{H}_\mathrm{s}\big|_{z=0} \cdot \vec{H}^*_{\perp 1_+}, \tag{10.149}$$

$$A_{\lambda_+} = -|Z_\lambda|^2 \int da\, \vec{H}_\mathrm{s}\big|_{z=0} \cdot \vec{H}^*_{\perp \lambda_+}. \tag{10.150}$$

Similarly, from Equation (10.141) at $z = 0^+$ we find

$$\vec{H}_{\perp \mathrm{ns}} = T\vec{H}_{\perp 1_+} + \sum_{\lambda \neq 1} A_{\lambda_+} \vec{H}_{\perp \lambda_+}, \tag{10.151}$$

and the same orthogonality relation gives

$$T = 1 + R = |Z_1|^2 \int da\, \vec{H}_\mathrm{ns}\big|_{z=0} \cdot \vec{H}^*_{\perp 1_+}, \tag{10.152}$$

$$A_{\lambda_+} = |Z_\lambda|^2 \int da\, \vec{H}_\mathrm{ns}\big|_{z=0} \cdot \vec{H}_{\perp \lambda_+}, \tag{10.153}$$

(N.B. $\lambda \neq 1$ in (10.150) and (10.153)). To simplify the notation we define

$$\vec{H}_\mathrm{diff} \equiv (\vec{H}_\mathrm{ns} - \vec{H}_\mathrm{s})\big|_{z=0}, \tag{10.154}$$

(note that $\vec{H}_\mathrm{diff} = 0$ in the gaps) and from combining the above we find

$$R = \frac{1}{2}|Z_1|^2 \int_\mathrm{obs.} da\, \vec{H}_\mathrm{diff} \cdot \vec{H}^*_{\perp 1_+} = -\frac{1}{2} Z_{1_+} \int_\mathrm{obs.} da\, \vec{E}_{\perp 1_+} \cdot (\hat{e}_3 \times \vec{H}_\mathrm{diff}), \tag{10.155}$$

$$A_{\lambda_+} = \frac{1}{2}|Z_\lambda|^2 \int_\mathrm{obs.} da\, \vec{H}_\mathrm{diff} \cdot \vec{H}^*_{\perp \lambda_+} = -\frac{1}{2} Z_{\lambda_+} \int_\mathrm{obs.} da\, \vec{E}_{\perp \lambda_+} \cdot (\hat{e}_3 \times \vec{H}_\mathrm{diff}), \tag{10.156}$$

where we are integrating only over the obstruction. We have used the connection (10.112) and the fact that the $\vec{E}_{\perp\lambda_+}$ fields are real. On the other hand, from (10.137) we have at all points on the surface of the conducting obstruction:

$$0 = (1+R)\vec{E}_{\perp 1_+} + \sum_{\lambda\neq 1} A_{\lambda_+}\vec{E}_{\perp\lambda_+}. \quad \text{(valid on obstruction } \textit{only}\text{)}. \tag{10.157}$$

Using the rightmost form of (10.156) in (10.157) gives

$$(1+R)\vec{E}_{\perp 1_+} = \frac{1}{2}\sum_{\lambda\neq 1} Z_{\lambda_+}\vec{E}_{\perp\lambda_+}\int_{\text{obs.}} da\,\vec{E}^*_{\perp\lambda_+}\cdot(\hat{e}_3\times\vec{H}_{\text{diff}}). \tag{10.158}$$

This equation is analogous to (10.144) above. Dotting both sides into $(\hat{e}_3 \times \vec{H}^*_{\text{diff}})$ and integrating over the obstruction gives

$$-2R^*(1+R)Y^*_{1_+} = \frac{1}{2}\sum_{\lambda\neq 1} Z_{\lambda_+}\left|\int_{\text{obs.}} da\,\vec{E}_{\perp\lambda_+}\cdot(\hat{e}_3\times\vec{H}_{\text{diff}})\right|^2, \tag{10.159}$$

where the rightmost form of (10.155) has been used. Dividing both sides by $4|R|^2Y^*_{1_+}$ gives our result:

$$Z_d = -\frac{T}{2R} = \frac{1}{2}\frac{\sum_{\lambda\neq 1} Z_{\lambda_+}\left|\int_{\text{obs.}} da\,\vec{K}\cdot\vec{E}_{\perp\lambda_+}\right|^2}{Z_{1_+}\left|\int_{\text{obs.}} da\,\vec{K}\cdot\vec{E}_{\perp 1_+}\right|^2}, \tag{10.160}$$

where

$$\vec{K} = \frac{c}{4\pi}\hat{e}_3\times\vec{H}_{\text{diff}}, \tag{10.161}$$

is the surface current on the obstruction, understood to be at $z = 0$. More compactly,

$$Z_d = \frac{1}{2}\frac{\sum_{\lambda\neq 1} Z_{\lambda_+}|Y_{\lambda_+}A_{\lambda_+}|^2}{Z_{1_+}|Y_{1_+}R|^2}, \tag{10.162}$$

from (10.155) and (10.156), which is consistent with (10.147).

The punch line to all this is that for certain types of obstructions (10.146) and (10.160) can be used to form upper limits on Y_d and Z_d. This only happens, however, in cases where the admittances Y_{λ_+} (or impedances Z_{λ_+}) enter with a single complex phase. Consider a model where $\vec{V}_0$ is given by ($\vec{\Phi}_\lambda$ real)

$$X\vec{\Phi}_1\int_D da\,\vec{V}_0\cdot\vec{\Phi}_1 = \sum_{\lambda\neq 1}\vec{\Phi}_\lambda\int_D da\,\vec{V}_0\cdot\vec{\Phi}_\lambda. \tag{10.163}$$

Table 10.1 Substitutions for application of model (10.163).

Quantity	Y_d case (10.144)	Z_d case (10.158)
X	$\frac{1}{2}Y_d e^{-i\alpha}$	$2Z_d e^{i\alpha}$
$\vec{\Phi}_\lambda$	$\sqrt{\lvert Y_\lambda \rvert}\, \vec{E}_{\perp\lambda_+}$	$\sqrt{\lvert Z_\lambda \rvert}\, \vec{E}_{\perp\lambda_+}$
$\vec{V}_0$	$\vec{E}\big\rvert_{z=0}$	$\vec{K} = \frac{c}{4\pi}\hat{e}_3 \times \vec{H}_{\text{diff}}$
D	gaps	obstruction

Solve for X and substitute $\vec{V}$ for $\vec{V}_0$:

$$X(\vec{V}) \equiv \frac{\sum_{\lambda\neq 1}[\int_D da\, \vec{V}\cdot\vec{\Phi}_\lambda]^2}{[\int_D da\, \vec{V}\cdot\vec{\Phi}_1]^2} = \sum_{\lambda\neq 1}\left[\frac{\int_D da\, \vec{V}\cdot\vec{\Phi}_\lambda}{\int_D da\, \vec{V}\cdot\vec{\Phi}_1}\right]^2. \tag{10.164}$$

It turns out both (10.144) and (10.158) can be brought into the form of (10.163) with the substitutions shown in Table 10.1, where the single phase α is defined by ($\lambda \neq 1$)

$$Y_{\lambda_+} = e^{i\alpha}|Y_\lambda|. \tag{10.165}$$

One can show that the choice $\vec{V} = \vec{V}_0$ minimizes $X(\vec{V})$ (see Exercise 10.5.2); thus, an arbitrary $\vec{V}$ produces an *upper* limit to the true $X(\vec{V}_0)$ value. The key to this fact is that each and every term in the numerator is real and non-negative. The astute reader will note the replacement of the absolute value squared terms in the summands of (10.146) and (10.160) with the squared summands in (10.164). Consider the following: the common form (10.163) holds identically for both the real and imaginary parts of $\vec{V}_0$. Denoting these as $(\vec{V}_0)_R$ and $(\vec{V}_0)_I$, one valid non-trivial solution is $(\vec{V}_0)_R \neq 0, (\vec{V}_0)_I = 0$. What's more, the completeness of the modes assumed here assures us that such a choice results in a unique solution up to an inessential position-independent phase factor. Thus, it is always possible in the single-phase case to make such a replacement, which certainly simplifies the situation.

In the next section we will simply investigate two special cases in which the above variational method can be carried out.

10.6 Variational examples: parallel rectangular strips in rectangular waveguides

In this section, we'll apply our results from the previous section to the particular situation shown in Figure 10.6, where the waveguide is rectangular ($a > b$) and the obstruction consists of parallel strips in the $z = 0$ plane. Using Section 10.13, we obtain the (normalized) lowest TE mode for rectangular waveguides, $m = 1$, $n = 0$:

$$(H_z^{(e)})_{10} = -i\frac{c\pi}{\omega\mu a}\sqrt{\frac{2}{ab}}\cos\frac{\pi x}{a}, \tag{10.166}$$

$$(H_{x+}^{(e)})_{10} = -\frac{ck_{1_+}}{\omega\mu}\sqrt{\frac{2}{ab}}\sin\frac{\pi x}{a}, \tag{10.167}$$

$$(E_y^{(e)})_{10} = \sqrt{\frac{2}{ab}}\sin\frac{\pi x}{a}, \tag{10.168}$$

$$(H_{y+}^{(e)})_{10} = (E_x^{(e)})_{10} = (E_z^{(e)})_{10} = 0, \tag{10.169}$$

(k_{10_+} is shortened to k_{1_+}) where

$$k_{1_+} = \sqrt{\frac{\omega^2\mu\epsilon}{c^2} - \frac{\pi^2}{a^2}}. \tag{10.170}$$

We may deduce several characteristics of the solution as follows. The incident field $\vec{E}_{\perp 1_+}$ from (10.168) and (10.169) is y-independent, as is the

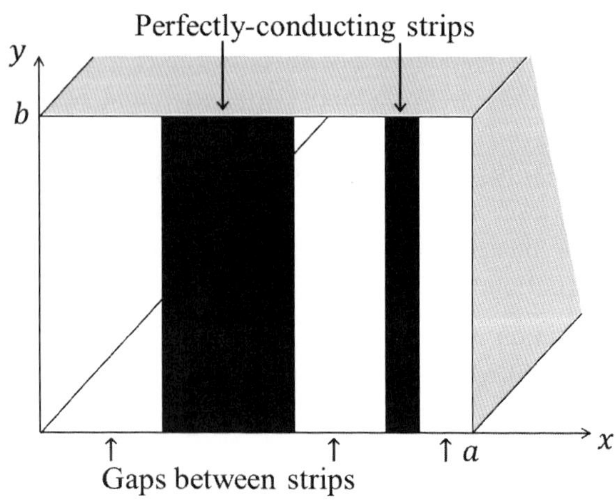

Fig. 10.6 Thin, parallel rectangular obstructions in a rectangular waveguide ($a > b$).

geometry shown in Figure 10.6. That is, the strips (and gaps) are parallel to the y-axis, so they do not introduce any y-dependence into the equations. It follows that we should only include in the solution those modes $\vec{E}_{\lambda_+}$ that are y-independent. Since E_x and E_z vanish along the walls at $y = 0$ and $y = b$, this means that $E_x = E_z = 0$ remains true everywhere. Of course, the TM modes are eliminated because $E_z \neq 0$. Furthermore, only TE modes with $n = 0$ have both $E_z = 0$ and $E_x = 0$, so these are the only modes that contribute to the sum. What's more, for the cutoff λ_+ modes of the incident wave, we have

$$Y_{\lambda_+} = \frac{1}{Z_{\lambda_+}} = \begin{cases} -\dfrac{i\epsilon\omega}{c\kappa_{\lambda_+}} & \text{TM}, \\ \dfrac{ic\kappa_{\lambda_+}}{\mu\omega} & \text{TE}, \end{cases} \tag{10.171}$$

where

$$\kappa_{\lambda_+} = \frac{\sqrt{\mu\epsilon}}{c}\sqrt{\omega_\lambda^2 - \omega^2}. \tag{10.172}$$

We therefore have in this case ($\lambda \to m$)

$$\frac{2Y_{m_+}}{Y_{1_+}} = \frac{2i}{|k_1|}\frac{\sqrt{\mu\epsilon}}{c}\left(\omega_m^2 - \omega^2\right)^{1/2},$$

$$\left(\omega_m^2 = \omega_1^2 m^2, \quad \omega_1^2 = \frac{c^2\pi^2}{\mu\epsilon}\frac{1}{a^2}\right)$$

$$\Longrightarrow \frac{2Y_{m_+}}{Y_{1_+}} = \frac{2i\pi}{a|k_1|}\sqrt{m^2 - \frac{\omega^2}{\omega_1^2}}. \tag{10.173}$$

Thus, the cutoff modes enter with the single phase $\alpha = \pi/2$ and we may expect to set variational principle limits on Y_d and Z_d.

Now, let us apply these ideas. In this case we recognize that only the field $E_y \neq 0$. From Section 10.13 we find

$$(E_y^{(e)})_{m0} = \sqrt{\frac{2}{ab}}\sin\left(\frac{m\pi x}{a}\right). \tag{10.174}$$

Then using Equation (10.146) for Y_d we find (remember, field phases do not matter)

$$Y_d = 2\sum_{m=2}^{\infty}\frac{Y_{m_+}}{Y_{1_+}}\left[\frac{\int_{\text{gaps}} dx\, E_y(x)\sin\left(\frac{m\pi x}{a}\right)}{\int_{\text{gaps}} dx\, E_y(x)\sin\left(\frac{\pi x}{a}\right)}\right]^2. \tag{10.175}$$

In obtaining (10.175), the trivial y integration cancelled on top and bottom. Then employing (10.173) gives

$$Y_d = \frac{2i\pi}{a|k_1|} \sum_{m=2} \sqrt{m^2 - \frac{\omega^2}{\omega_1^2}} \left[\frac{\int_{\text{gaps}} dx\, E_y(x) \sin\left(\frac{m\pi x}{a}\right)}{\int_{\text{gaps}} dx\, E_y(x) \sin\left(\frac{\pi x}{a}\right)} \right]^2 . \tag{10.176}$$

In the Z_d case all fields are again independent of y. From Equation (10.160) and the previous expression for Y_{m_+}/Y_{1_+}, we obtain

$$Z_d = \frac{-i|k_1|a}{2\pi} \sum_{m=2} \frac{1}{\sqrt{m^2 - \frac{\omega^2}{\omega_1^2}}} \left[\frac{\int_{\text{strips}} dx\, K_y(x) \sin\left(\frac{m\pi x}{a}\right)}{\int_{\text{strips}} dx\, K_y(x) \sin\left(\frac{\pi x}{a}\right)} \right]^2 , \tag{10.177}$$

where again the y-integration is trivial. This now provides an upper limit to both iZ_d and $-iY_d$, which brackets iZ_d, for example, between upper and lower limits.

Let us now consider a slightly different physical situation as shown in Figure 10.7 where the conducting strips are now parallel to the longer cross-sectional dimension ($a > b$). As in the previous case, we assume that only the lowest TE mode is propagating, and all other modes are attenuated. Once again, we use symmetry considerations to infer properties of the solution. The lowest TE mode has $m = 1$, $n = 0$, and (10.169) has $(E_x^{(e)})_{10} = 0$ for the incident field. The obstructions introduce no new dependence of E_x on x; this is consistent with $E_x = 0$ for the solution. It can similarly be argued that the x-dependence of E_y will be unaffected, so $E_y \sim \sin\left(\frac{\pi x}{a}\right)$. This is basically the same argument as in the other

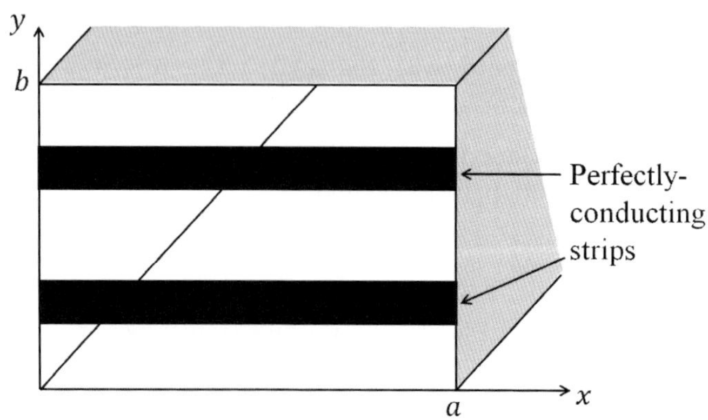

Fig. 10.7 Thin obstructions in waveguide: alternate geometry ($a > b$).

situation except that there we saw that the y-independence of *all* fields was unaffected.

The question is: How can we meet these conditions on E_x and E_y with a linear combination of modes? From Section 10.13 we have (non-propagating modes)

$$(E_x^{(e)})_{mn} = \frac{-2\pi n}{\gamma_{mn} b\sqrt{ab}} \cos\left(\frac{m\pi x}{a}\right) \sin\left(\frac{n\pi y}{b}\right) \cdot \begin{cases} \frac{1}{\sqrt{2}} & m = 0 \text{ or } n = 0, \\ 1 & \text{otherwise}. \end{cases} \tag{10.178}$$

$$(E_{x+}^{(m)})_{mn} = \frac{2\pi m}{\gamma_{mn} a\sqrt{ab}} \cos\left(\frac{m\pi x}{a}\right) \sin\left(\frac{n\pi y}{b}\right) \quad (m, n > 0). \tag{10.179}$$

If we use TE modes, the condition $E_x = 0$ implies that we can only use modes with $n = 0$. However, from (10.174) we have $(E_y^{(e)})_{m0} \sim \sin\left(\frac{m\pi x}{a}\right)$, so only the $m = 1$ mode is allowed. For $m, n > 0$, TE and TM modes with the same m, n are degenerate. To meet the conditions $E_x = 0$, $E_y \sim \sin\left(\frac{\pi x}{a}\right)$, we may form new combinations of (unnormalized) degenerate modes as follows:

$$(\vec{E}^{\text{new}})_{1n} \equiv \frac{(\vec{E}^{(m)})_{1n}}{b} + \frac{(\vec{E}^{(e)})_{1n}}{an}. \tag{10.180}$$

In fact, we have

$$\frac{(E_{x+}^{(m)})_{1n}}{b} + \frac{(E_x^{(e)})_{1n}}{an} = 0, \tag{10.181}$$

for $n = 1, 2, 3, \ldots$, and also from Section 10.13

$$(E_y^{\text{new}})_{1n} \sim \sin\left(\frac{\pi x}{a}\right), \tag{10.182}$$

as desired. Therefore, these are the modes compatible with the boundary conditions. It is not clear whether this obstruction will give a positive imaginary Y_d (as before) or a negative imaginary value. We will have to just work it out. Using the same techniques as in the previous example, the following expression for Y_d can be derived (see Exercise 10.6.2):

$$Y_d = -4i|k_1| \sum_{n=1}^{\infty} \frac{1}{\sqrt{\frac{n^2\pi^2}{b^2} - k_1^2}} \left[\frac{\int_{\text{gaps}} dy\, E_y(y) \cos\left(\frac{n\pi y}{b}\right)}{\int_{\text{gaps}} dy\, E_y(y)}\right]^2. \tag{10.183}$$

As before, we may find upper limits for iY_d by replacing $E_y(y)$ in (10.183) with arbitrary real functions. It is also possible to derive an expression for Z_d similar to (10.177) (see Exercise 10.6.3), which makes it possible to find lower limits for iY_d.

10.7 Power loss in non-ideal waveguides: power method

Until now we've been looking at situations in the ideal case where the waveguide's walls are perfectly conducting. Unfortunately real life is not so simple. We shall see in this section that imperfect conductivity leads to power loss for an electromagnetic wave propagating down a waveguide. Not surprisingly, the rate of power loss is of vital practical importance. In this section we discuss one method for approximating this power loss; and in the next section we present a more accurate (albeit more difficult) approach.

From (10.60) we have that

$$\vec{S} = \frac{c}{8\pi}\mathrm{Re}(\vec{E}^* \times \vec{H}) = \frac{c}{8\pi}\mathrm{Re}(\vec{E} \times \vec{H}^*), \tag{10.184}$$

is the appropriate form of the energy flux vector for harmonic fields. We shall use (10.184) to compute the energy flux for TM and TE modes for propagating modes in a waveguide.

In the case of TM modes, we have $H_z = 0$ and restate Equations (10.90) and (10.91):

$$\vec{E}_\perp = \frac{ik}{\gamma^2}\vec{\nabla}_\perp E_z, \tag{10.185}$$

$$\vec{H}_\perp = \frac{i\omega\epsilon}{\gamma^2 c}\hat{e}_3 \times (\vec{\nabla}_\perp E_z). \tag{10.186}$$

We take the wavenumber k to be real (positive or negative) and temporarily leave off mode labels. By computing the appropriate cross-products, we find

$$\begin{aligned}\frac{c}{8\pi}(\vec{E}\times\vec{H}^*) &= \frac{c}{8\pi}\left(\frac{-i\omega\epsilon}{\gamma^2 c}\right)\left[E_z\hat{e}_3 + \frac{ik}{\gamma^2}\vec{\nabla}_\perp E_z\right] \times \left[\hat{e}_3 \times (\vec{\nabla}_\perp E_z^*)\right] \\ &= \frac{\omega k\epsilon}{8\pi\gamma^4}\left[\hat{e}_3|\vec{\nabla}_\perp E_z|^2 + i\frac{\gamma^2}{k}E_z\vec{\nabla}_\perp E_z^*)\right].\end{aligned} \tag{10.187}$$

For TE modes, we have $E_z = 0$, and a similar calculation leads to

$$\frac{c}{8\pi}(\vec{E}\times\vec{H}^*) = \frac{\omega k\mu}{8\pi\gamma^4}\left[\hat{e}_3|\vec{\nabla}_\perp H_z|^2 - i\frac{\gamma^2}{k}H_z^*\vec{\nabla}_\perp H_z\right]. \tag{10.188}$$

At this point we may unify our treatment of TM and TE modes by defining

$$\psi \equiv \begin{cases} E_z & \text{TM modes} \\ H_z & \text{TE modes}\end{cases}; \qquad \zeta \equiv \begin{cases} \epsilon & \text{TM modes} \\ \mu & \text{TE modes}\end{cases}, \tag{10.189}$$

so that the following equation for $\vec{S}$ is valid for both TM and TE modes:

$$\vec{S} = \text{Re}\left(\frac{\omega k\zeta}{8\pi\gamma^4}\left[\hat{e}_3|\vec{\nabla}_\perp\psi|^2 - i\frac{\gamma^2}{k}\psi^*\vec{\nabla}_\perp\psi\right]\right). \tag{10.190}$$

For pure (*i.e.*, nondegenerate) modes, we have seen in Section 10.2 that ψ can be taken as real (or if given a complex phase, it has the same phase everywhere in space). In this case, $\psi\vec{\nabla}_\perp\psi^*$ is real, so we have from (10.190)

$$\vec{S} = \frac{\omega k\zeta}{8\pi\gamma^4}\hat{e}_3|\vec{\nabla}_\perp\psi|^2. \tag{10.191}$$

The energy flow in (10.191) is entirely in the $\hat{e}_3$ direction. Tangential flow may occur in the case of degenerate modes since then $\psi\vec{\nabla}_\perp\psi^*$ in (10.190) may not be purely real. Although not shown here, it may also occur for mixed TE, TM modes, even if not degenerate. This is a type of stored energy. Since we're primarily interested in power transmission along the $\hat{e}_3$ direction, we won't consider tangential flow any further.

By integrating across a cross-section of the waveguide we obtain the total transmitted power:

$$P = \int da\,\vec{S}\cdot\hat{e}_3 = \frac{k\omega\zeta}{8\pi\gamma^4}\int da\,|\vec{\nabla}_\perp\psi|^2. \tag{10.192}$$

For pure modes ψ can be considered as a function of x, y (independent of z), and (10.192) implies there are no energy losses. However, in non-ideal waveguides with imperfectly-conducting walls pure modes are no longer solutions. We may treat imperfect conductivity as having a perturbative effect on k, where in general k becomes complex. The effect on k will depend on the mode, so in order to keep track of modes we reintroduce λ as a mode label. The form of λ may vary depending on the situation: for example, in the case of rectangular modes λ has the form $\{m, n, s\}$ where m and n are integers and $s = \pm 1$ is a direction label. Letting k_λ (resp. k'_λ) denote the perfect-conductor (resp. imperfect-conductor) k-values for mode λ and frequency ω, we have

$$k'_\lambda = k_\lambda + \alpha_\lambda + i\beta_\lambda, \tag{10.193}$$

where $\alpha_\lambda + i\beta_\lambda$ is a small perturbation that also depends on λ and ω. (As usual, we are taking ω as a fixed parameter.) In this discussion we assume that $k_\lambda > 0$ above cutoff so that the waves are traveling in the $+z$ direction. From our previous discussion of good-conductor boundary conditions in Section 10.1 we should expect $\alpha_\lambda \approx \beta_\lambda$ (see e.g. (10.39), which holds inside the bounding conductor).

In order to approximate the power loss, we assume that the perturbed field ψ'_λ has the form

$$\psi'_\lambda \approx \psi_\lambda e^{i(\alpha_\lambda + i\beta_\lambda)z}, \tag{10.194}$$

in keeping with the shift in k given in (10.193). Note that this expression cannot be exact, since ψ'_λ still satisfies the unperturbed equation $(\nabla_\perp^2 + \gamma_\lambda^2)\psi'_\lambda = 0$ with perfect electric conductor boundary conditions. Nonetheless we trip merrily along and substitute the form (10.194) into (10.192), which yields

$$P(z) \approx P_0 e^{-2\beta_\lambda z}. \tag{10.195}$$

Thus the power decays exponentially with decay constant $2\beta_\lambda$, where β_λ can be estimated via

$$\beta_\lambda = -(2P_0)^{-1} \left.\frac{dP}{dz}\right|_{z=0}. \tag{10.196}$$

We should expect this approximation to break down near the mode's cutoff frequency where $k_\lambda \approx 0$, because then k'_λ in (10.193) can no longer be considered a perturbation of k_λ.

In the remainder of this section, we show how to evaluate β_λ. We already have an expression for P_0 in (10.192), which we may further simplify by using the two-dimensional versions of Green's first identity and Stokes' law:

$$\begin{aligned}
P_0 &= \frac{k_\lambda \omega \zeta}{8\pi\gamma_\lambda^4} \int da \left[\vec{\nabla}_\perp \cdot (\psi_\lambda^* \vec{\nabla}_\perp \psi_\lambda) - \psi_\lambda^* \nabla_\perp^2 \psi_\lambda\right] \\
&= \frac{k_\lambda \omega \zeta}{8\pi\gamma_\lambda^4} \left[\oint_C d\ell\, (\psi_\lambda^* \vec{\nabla}_\perp \psi_\lambda) \cdot \hat{n} - \int da\, \psi_\lambda^* \nabla_\perp^2 \psi_\lambda\right] \\
&= \frac{k_\lambda \omega \zeta}{8\pi\gamma_\lambda^4} \left[\oint_C d\ell\, \psi_\lambda^* \frac{\partial \psi_\lambda}{\partial n} + \int da\, \psi_\lambda^* \gamma_\lambda^2 \psi_\lambda\right],
\end{aligned} \tag{10.197}$$

where we have also used $\nabla_\perp^2 \psi_\lambda = -\gamma_\lambda^2 \psi_\lambda$ which is valid in both TM and TE modes. The contour integral vanishes because of boundary conditions, leaving us with

$$P_0 = \frac{\omega k_\lambda \zeta}{8\pi\gamma_\lambda^2} \int da\, |\psi_\lambda|^2. \tag{10.198}$$

For the derivative in (10.196) we need to take a different approach. Here we specialize to TM modes: the TE case is left as an exercise. From (10.61) we find for imperfect conductors that

$$\langle S_{\text{surface}} \rangle = \frac{\omega \mu_c \delta}{16\pi} |\hat{n} \times \vec{H}|^2 = -\frac{dP}{da}, \tag{10.199}$$

where da is an infinitesimal element of area, δ is the skin depth, and μ_c is the magnetic permeability of the conductor. We may integrate (10.199) around the edge of the waveguide (right outside the conductor) to obtain

$$-\frac{dP}{dz} = \frac{\omega\mu_c\delta}{16\pi}\oint d\ell\,|\hat{n}\times\vec{H}|^2. \tag{10.200}$$

Since we're looking at the TM case, we use (10.186) to get approximately (using $\gamma'_\lambda \approx \gamma_\lambda$ to lowest order):

$$\begin{aligned}
\oint d\ell\,|\hat{n}\times\vec{H}|^2 &= \oint d\ell\,|\hat{n}\times\vec{H}_\perp|^2 \\
&\approx \frac{\omega^2\epsilon^2}{\gamma_\lambda^4 c^2}\oint d\ell\,\left|\hat{n}\times\left(\hat{e}_3\times\vec{\nabla}_\perp E_z\right)\right|^2 \\
&\approx \frac{\omega^2\epsilon^2}{\gamma_\lambda^4 c^2}\oint d\ell\,\left|\hat{e}_3\frac{\partial}{\partial n}E_z\right|^2 \\
\Longrightarrow\quad -\frac{dP}{dz} &\approx \frac{\omega\mu_c\delta}{16\pi}\frac{\omega^2\epsilon^2}{\gamma_\lambda^4 c^2}\oint d\ell\,\left|\frac{\partial E_z}{\partial n}\right|^2.
\end{aligned} \tag{10.201}$$

So from (10.196), (10.198) and (10.201) we obtain finally

$$\beta_\lambda \approx \tilde{\beta}_\lambda \equiv \frac{\mu_c\delta}{4k_\lambda\mu}\left(\frac{\omega}{\omega_\lambda}\right)^2\frac{\oint d\ell\,\left|\frac{\partial E_z}{\partial n}\right|^2}{\int da\,|E_z|^2},\quad \text{(TM modes)} \tag{10.202}$$

where E_z refers to the perfect-conductor fields, and ω_λ is the *cutoff angular frequency* for mode λ, defined through the relation

$$\gamma_\lambda^2 = \frac{\mu\epsilon}{c^2}\omega_\lambda^2, \tag{10.203}$$

which was already introduced in (10.104) for rectangular waveguide modes.

In Exercise 10.7.2 you will prove the corresponding expression for TE modes:

$$\beta_\lambda \approx \tilde{\beta}_\lambda \equiv \frac{\mu_c\delta}{4k_\lambda\mu}\frac{\oint d\ell\left[\gamma_\lambda^2|H_z|^2 + \frac{k_\lambda^2}{\gamma_\lambda^2}\left|\frac{\partial H_z}{\partial \ell}\right|^2\right]}{\int da\,|H_z|^2}.\quad \text{(TE modes)} \tag{10.204}$$

Note that the $\tilde{\beta}_\lambda$ in (10.202), (10.204) share the sign of $k_{\lambda_\pm}$ in the general case. In both TM and TE modes, the approximate attenuation constant $\tilde{\beta}_\lambda$ becomes infinite when $\omega \to \omega_\lambda$ since this means $k_\lambda \to 0$. In the next section we'll see that this qualitative behavior is not accurate and that a finite result can be obtained. As far as high frequencies are concerned, since $\delta \propto \omega^{-1/2}$ and $k_\lambda \sim \omega$ when $\omega \gg \omega_\lambda$, it can be shown for both TE and TM that $\tilde{\beta}_\lambda \sim \omega^{1/2}$ as $\omega \to \infty$.

10.8 Power loss in waveguides: perturbation of boundary conditions method

The method described in the previous section for estimating power loss in imperfect-conductor waveguides is limited both in accuracy and applicability. In this section we follow a more careful approach which leads to a more accurate estimate. Here we'll consider TM modes only, and leave the TE case (which is similar) as an exercise (see Exercise 10.8.2). As in the previous section, we will treat the imperfect-conductor case as a perturbation of the perfect-conductor case. Thus, for every perfect-conductor TM mode λ with cutoff wavenumber γ_λ and corresponding field E_z (and $H_z = 0$), we seek an imperfect-conductor TM mode with cutoff wavenumber $\gamma'_\lambda \approx \gamma_\lambda$ and corresponding field E'_z (and $H'_z = 0$). We will derive an expression for an appropriate perturbed boundary condition that includes the effects of finite conductivity. See Figure 10.8 for the notation: note in particular that $\hat{n}$ is again used to denote the inward-pointing normal.

From (10.44) we have within the (imperfect) conductor that

$$\vec{E}\,'^{\text{tang}}\Big|_S \approx \sqrt{\frac{\omega\mu_c}{8\pi\sigma}}(1-i)\hat{n}\times\vec{H}\,'^{\text{tang}}, \tag{10.205}$$

which also holds just inside the waveguide (right next to the conductor) since $\vec{E}\,'^{\text{tang}}$ is continuous. Since $H'_z|_S = 0$ (S specifies the waveguide surface in the section) for TM modes, we may replace $\vec{H}\,'^{\text{tang}}$ with $\vec{H}\,'_\perp$ (note $\hat{z}$ is tangent to the boundary), and expression (10.186) for $\vec{H}\,'_\perp$ enables us

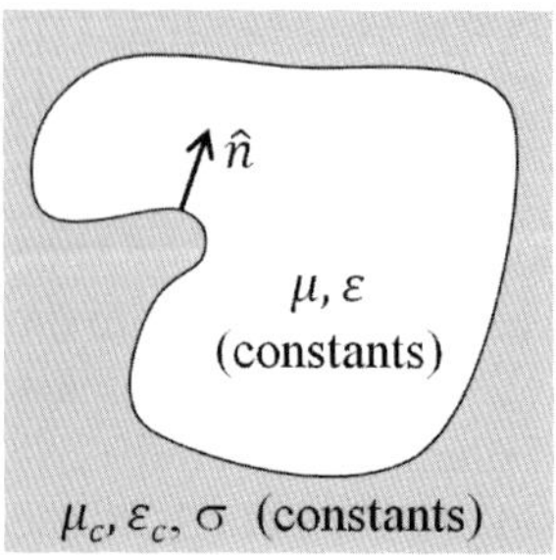

Fig. 10.8 Notation for waveguide with imperfectly-conducting walls.

to evaluate to lowest order (using $\gamma'_\lambda \approx \gamma_\lambda$ and $\partial E'_z/\partial n|_S \approx \partial E_z/\partial n|_S$):

$$\begin{aligned} E'_z|_S &\approx \sqrt{\frac{\omega\mu_c}{8\pi\sigma}}(1-i)\hat{e}_3\cdot(\hat{n}\times\vec{H}'_\perp)\Big|_S \\ &\approx \frac{i\omega\epsilon}{\gamma_\lambda^2 c}\sqrt{\frac{\omega\mu_c}{8\pi\sigma}}(1-i)\hat{e}_3\cdot(\hat{n}\times(\hat{e}_3\times\vec{\nabla}_\perp E'_z))\Big|_S \\ &\approx \frac{i\omega\epsilon}{\gamma_\lambda^2 c}\sqrt{\frac{\omega\mu_c}{8\pi\sigma}}(1-i)\left.\frac{\partial E_z}{\partial n}\right|_S . \end{aligned} \tag{10.206}$$

After some additional algebra we obtain

$$\begin{aligned} E'_z|_S &\approx \frac{(i+1)}{2}\delta\left(\frac{\omega}{\omega_\lambda}\right)^2\frac{\mu_c}{\mu}\left.\frac{\partial E_z}{\partial n}\right|_S \\ &\equiv f^{\mathrm{TM}}\left.\frac{\partial E_z}{\partial n}\right|_S , \end{aligned} \tag{10.207}$$

where ω_λ was defined in (10.203).

We have finally arrived at a boundary-value problem that specifies the (approximate) perturbed field E'_z:

$$\Big(\underbrace{\frac{\partial^2}{\partial x^2}+\frac{\partial^2}{\partial y^2}}_{\nabla_\perp^2}+{\gamma'_\lambda}^2\Big)E'_z = 0; \quad E'_z|_S = f^{\mathrm{TM}}\left.\frac{\partial E_z}{\partial n}\right|_S . \tag{10.208}$$

Just as in the perfectly-conducting case, the value of γ'_λ can now be determined by restrictions imposed by the boundary conditions. In the following, we'll show two different ways to do this.

One method for finding γ'_λ begins by applying the two-dimensional Green identity

$$\int da\,[\phi\nabla_\perp^2\psi - \psi\nabla_\perp^2\phi] = \oint d\ell\left[\psi\frac{\partial\phi}{\partial n}-\phi\frac{\partial\psi}{\partial n}\right], \tag{10.209}$$

with $\phi = E_z^*$ and $\psi = E'_z$ and boundary conditions $E_z^*|_S = 0$ and $E'_z|_S = f^{\mathrm{TM}}(\partial E_z/\partial n)$. This gives us

$$\begin{aligned} &\int da[E_z^*(-{\gamma'_\lambda}^2 E'_z) - E'_z(-\gamma_\lambda^2)E_z^*] = f^{\mathrm{TM}}\oint d\ell\,\frac{\partial E_z}{\partial n}\frac{\partial E_z^*}{\partial n} \\ &\quad\Longrightarrow (-{\gamma'_\lambda}^2+\gamma_\lambda^2)\int da\,E_z^*E'_z = f^{\mathrm{TM}}\oint d\ell\left|\frac{\partial E_z}{\partial n}\right|^2 \\ &\quad\Longrightarrow -{\gamma'_\lambda}^2+\gamma_\lambda^2 = {k'_\lambda}^2 - k_\lambda^2 \approx f^{\mathrm{TM}}\frac{\oint d\ell\,\left|\frac{\partial E_z}{\partial n}\right|^2}{\int da|E_z|^2}. \end{aligned} \tag{10.210}$$

Note that since γ'_λ is complex it no longer corresponds to a cutoff wavenumber as does γ_λ in (10.203): this has important physical implications, as we shall see shortly. The right-hand side of (10.210) may be re-expressed in terms of the $\tilde{\beta}_\lambda$ given by (10.202):

$$k'^{\,2}_\lambda - k_\lambda^2 \approx 2k_\lambda \tilde{\beta}_\lambda (1+i), \tag{10.211}$$

with k_λ^2 given by (10.71).

A second, more systematic method for finding γ'_λ from (10.208) is as follows. Let $\{\phi_\nu\}$ be a complete set of normalized solutions to the unperturbed eigen-problem, where ν is an index that indicates the mode (similar to λ in the previous discussion):

$$(\nabla_\perp^2 + \gamma_\nu{}^2)\phi_\nu(\vec{x}) = 0, \qquad \phi_\nu|_S = 0. \tag{10.212}$$

The Green function for the perturbed problem (with $\gamma'_\lambda \approx \gamma_\lambda$ for a particular *fixed* λ) satisfies

$$(\nabla_\perp^2 + \gamma'^{\,2}_\lambda)G_\lambda = -4\pi\delta(\vec{x} - \vec{x}\,'), \qquad G_\lambda|_S = 0. \tag{10.213}$$

We may use the Green function to solve the mixed boundary-value problem

$$(\nabla_\perp^2 + \gamma'^{\,2}_\lambda)\psi_\lambda = 0, \quad \psi_\lambda|_S = -f(\vec{x})\left.\frac{\partial \psi_\lambda}{\partial n}\right|_S, \tag{10.214}$$

where $\hat{n}$ now refers to the waveguide outward-pointing normal, and where we are considering a surface position dependent function, $f(\vec{x})$. Using the Green identity and the boundary conditions in (10.213) we obtain an expression for ψ_λ as follows:

$$\begin{aligned} &\int da' [G_\lambda \nabla'^2_\perp \psi_\lambda - \psi_\lambda \nabla'^2_\perp G_\lambda] = 4\pi\psi_\lambda(\vec{x}) \\ \Longrightarrow &\oint d\ell' \left[G_\lambda \frac{\partial \psi_\lambda}{\partial n'} - \frac{\partial G_\lambda}{\partial n'}\psi_\lambda\right] = 4\pi\psi_\lambda(\vec{x}) \\ \Longrightarrow &\psi_\lambda(\vec{x}) = \frac{1}{4\pi}\oint d\ell' \frac{\partial G_\lambda}{\partial n'}\frac{\partial \psi_\lambda}{\partial n'} f(\vec{x}\,'). \end{aligned} \tag{10.215}$$

We also have the eigenfunction expansion for the Green function (see Section 4.14)

$$G_\lambda(\vec{x}, \vec{x}\,') = 4\pi \sum_\nu \frac{\phi_\nu^*(\vec{x}\,')\phi_\nu(x)}{\gamma_\nu{}^2 - \gamma'^{\,2}_\lambda}, \tag{10.216}$$

which is valid as long as the $\{\phi_\nu\}$ are complete. Assuming that ψ_λ is a small perturbation of ϕ_λ, we may "bootstrap" using (10.215):

$$\psi_\lambda(\vec{x}) \approx \frac{1}{4\pi}\oint d\ell' \frac{\partial G_\lambda}{\partial n'}\frac{\partial \phi_\lambda}{\partial n'} f(\vec{x}\,'), \tag{10.217}$$

and replacing the Green function with its eigenfunction expansion gives

$$\psi_\lambda(\vec{x}) \approx \sum_\nu \phi_\nu(\vec{x}) \frac{\oint d\ell' \, \dfrac{\partial \phi_\nu^*}{\partial n'} \dfrac{\partial \phi_\lambda}{\partial n'} f(\vec{x}\,')}{{\gamma_\nu}^2 - {\gamma'_\lambda}^2}. \tag{10.218}$$

We can now obtain approximate expressions for γ'_λ. For example, to lowest order we have $\int da\, \psi_\lambda \phi_\lambda^* \approx 1$, so that

$$1 \approx \int da\, \phi_\lambda^* \phi_\lambda \frac{\oint d\ell' \left| \dfrac{\partial \phi_\lambda}{\partial n'} \right|^2 f(\vec{x}\,')}{{\gamma_\lambda}^2 - {\gamma'_\lambda}^2} \tag{10.219}$$

$$\Longrightarrow {\gamma_\lambda}^2 - {\gamma'_\lambda}^2 \approx \oint d\ell' \left| \frac{\partial \phi_\lambda}{\partial n'} \right|^2 f(\vec{x}\,'). \tag{10.220}$$

If instead of the normalized eigenfunction ϕ_λ we use an unnormalized version ϕ_λ^u, then we must divide by a normalizing factor:

$$\gamma_\lambda^2 - {\gamma'_\lambda}^2 \approx \frac{\oint d\ell' \left(\dfrac{\partial \phi_\lambda^{u*}}{\partial n'} f(x') \dfrac{\partial \phi_\lambda^u}{\partial n'} \right)}{\int da\, |\phi_\lambda^u|^2}. \tag{10.221}$$

When $f = f^{\rm TM}$, this gives the same TM results as in (10.210). Higher-order corrections can be obtained through extensions of this method: see Morse and Feshbach[2] for details.

We can go beyond perturbation theory at this point using (10.211). Let us recall the forms of $\tilde{\beta}_\lambda$ in (10.202) and (10.204), which inform us that the combination $k_\lambda \tilde{\beta}_\lambda$ is real and positive both above and below cutoff; for simplicity we will consider both as positive in the following for $\omega > \omega_\lambda$. Also recall that β_λ was originally defined in (10.193) as the imaginary part of k'_λ; the $\tilde{\beta}_\lambda$ which appears in (10.211) is only an estimate. But now we can solve (10.211) and take the imaginary part to find an improved estimate for β_λ (which we denote as $\tilde{\beta}_\lambda^{\rm new}$):

$$\tilde{\beta}_\lambda^{\rm new} \equiv \frac{1}{\sqrt{2}} \left(\sqrt{k_\lambda^2 (k_\lambda + 2\tilde{\beta}_\lambda)^2 + 4k_\lambda^2 \tilde{\beta}_\lambda^2} - k_\lambda (k_\lambda + 2\tilde{\beta}_\lambda) \right)^{1/2}. \tag{10.222}$$

In the case of slight attenuation ($\beta_\lambda / k_\lambda \ll 1$) the reader may verify that $\tilde{\beta}_\lambda^{\rm new} = \tilde{\beta}_\lambda$ to lowest order. However, at the cutoff frequency ($\omega = \omega_\lambda$) we

[2]P. M. Morse and H. Feshbach, *Methods of Theoretical Physics, Part II*, McGraw-Hill (New York) 1953, Section 9.2.

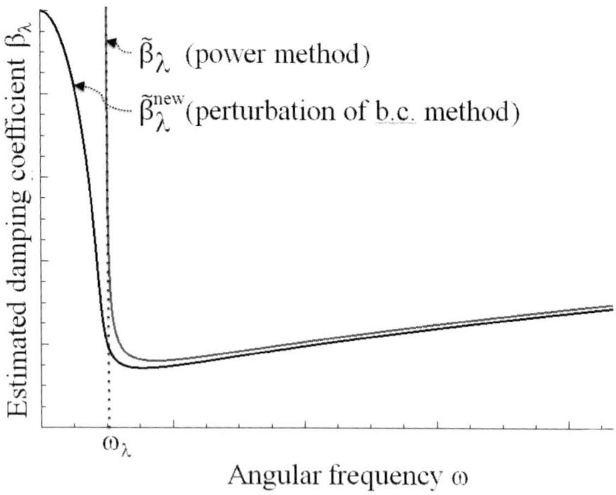

Fig. 10.9 ω-dependence of damping coefficient estimates $\tilde{\beta}_\lambda$ (power method) and $\tilde{\beta}_\lambda^{\rm new}$ (perturbation of b.c. method), for TM mode λ.

have

$$(\tilde{\beta}_\lambda^{\rm new})_{\omega=\omega_\lambda} = (\sqrt{2}-1)^{1/2}(k_\lambda\tilde{\beta}_\lambda)^{1/2}|_{\omega=\omega_\lambda}$$
$$= \left((\sqrt{2}-1)\frac{\mu_c\delta}{4\mu}\frac{\oint |d\ell\, \frac{\partial E_z}{\partial n}|^2}{\int da\,(E_z)^2}\right)^{1/2}, \qquad (10.223)$$

for TM modes, which is now finite, unlike $(\tilde{\beta}_\lambda)_{\omega=\omega_\lambda}$. This method may be extended to $\omega < \omega_\lambda$ by again solving for the imaginary part of k'_λ from the analytically continued version of (10.211) (see Exercise 10.8.5).

Figure 10.9 shows the dependence of $\tilde{\beta}_\lambda^{\rm new}$ on ω for TM mode λ above and below ω_λ. In the perfect-conductor case, $\omega < \omega_\lambda$ corresponds to attenuated modes; the finiteness of $\tilde{\beta}_\lambda^{\rm new}$ indicates that these attenuated modes still exist in the imperfect-conductor case (albeit with modified attenuation).

One may try to repeat these calculations in the TE case. In Exercise 10.8.2 the reader will find the corresponding estimate for $\gamma_\lambda^2 - {\gamma'_\lambda}^2 = {k'_\lambda}^2 - k_\lambda^2$. This turns out to be a more subtle problem due to the appearance of a second term in the lowest-order finite conductivity expression for $\partial H'_z/\partial n|_S$. Constructing $\tilde{\beta}_\lambda^{\rm new}$ from $\tilde{\beta}_\lambda$ in the TE case at angular frequencies at and above ω_λ is straightforward; however, below cutoff one encounters the issue that $f^{\rm TE} \sim \omega^{-1/2}$ at low angular frequencies, which implies in general $\tilde{\beta}_\lambda^{\rm new}$ diverges weakly like $\omega^{-1/4}$. This is not surprising given that $k_\lambda \tilde{\beta}_\lambda$ is now becoming large with respect to γ_λ^2 and the perturbation method is failing. Thus, the $\tilde{\beta}_\lambda^{\text{new}}$ has a more limited applicability in the TE case, but still confirms that the damping constant is finite near cutoff.

The methods above assume the mode λ is nondegenerate. The exercises indicate some mathematical approaches available when degeneracies are present.

10.9 TE and TM modes in an ideal rectangular resonant cavity

We have seen that waveguides may be used for *transmission* of radiation. Resonant cavities, on the other hand, are used for *tuning*, that is, obtaining radiation of a definite frequency. In the following sections, we will find resonant modes for cavities of different shapes.

We'll start with the simplest case, namely rectangular cavities. A natural approach to finding the TE and TM modes is to build on our results for rectangular waveguides, and impose the following additional boundary conditions (from (10.13) and (10.14)):

$$E_{x,y}\big|_{z=0,\,d} = 0, \tag{10.224}$$

$$H_z\big|_{z=0,\,d} = 0. \tag{10.225}$$

In the TM case we have $H_z = 0$, so (10.225) is satisfied automatically. In the rectangular waveguide we found traveling wave solutions:

$$E_z(\pm k) \equiv E_0 \sin\left(\frac{m\pi x}{a}\right) \sin\left(\frac{n\pi y}{b}\right) e^{\pm ikz}, \tag{10.226}$$

for $m, n = 1, 2, 3, \ldots$. We may obtain standing-wave solutions which also satisfy (10.224) by taking linear combinations of solutions. Consider for instance the linear combination

$$E_z \equiv \frac{E_z(k) + E_z(-k)}{2} = E_0 \sin\left(\frac{m\pi x}{a}\right) \sin\left(\frac{n\pi y}{b}\right) \cos(kz). \tag{10.227}$$

Then from $\vec{E}_\perp(k) = (ik/\gamma_{mn}^2)\vec{\nabla}_\perp E_z(k)$ (γ_{mn} given by (10.102)) we obtain

$$E_x = -\frac{k}{\gamma_{mn}^2}\left(\frac{m\pi}{a}\right) E_0 \cos\left(\frac{m\pi x}{a}\right) \sin\left(\frac{n\pi y}{b}\right) \sin(kz), \tag{10.228}$$

$$E_y = -\frac{k}{\gamma_{mn}^2}\left(\frac{n\pi}{b}\right) E_0 \sin\left(\frac{m\pi x}{a}\right) \cos\left(\frac{n\pi y}{b}\right) \sin(kz). \tag{10.229}$$

The condition (10.224) now gives

$$k = \frac{\pi p}{d}, \ p = 0, 1, 2, \ldots. \tag{10.230}$$

Likewise in the TE case, for non-negative integers m and n (not both zero) we have

$$H_z(\pm k) \equiv H_0 \cos\left(\frac{m\pi x}{a}\right) \cos\left(\frac{n\pi y}{b}\right) e^{\pm ikz}, \tag{10.231}$$

and defining

$$H_z \equiv \frac{H_z(k) - H_z(-k)}{2} = H_0 \cos\left(\frac{m\pi x}{a}\right) \cos\left(\frac{n\pi y}{b}\right) \sin(kz), \tag{10.232}$$

we may impose the boundary conditions (10.225) and obtain conditions on k:

$$H_z|_{z=d} = 0 \implies k = \frac{\pi p}{d}, \ p = 1, 2, 3, \ldots. \tag{10.233}$$

The boundary conditions (10.224) will also be satisfied since

$$\vec{E}_\perp = -\frac{i\omega\mu}{\gamma^2 c} \hat{e}_3 \times \vec{\nabla}_\perp H_z \implies \vec{E}_\perp\Big|_{z=0,\, d} = 0. \tag{10.234}$$

In both TE and TM cases, we have discrete frequencies:

$$\omega^2_{m,n,p} = \frac{c^2}{\mu\epsilon}\left(\gamma^2_{mn} + k^2\right) = \frac{\pi^2 c^2}{\mu\epsilon}\left[\frac{n^2}{a^2} + \frac{m^2}{b^2} + \frac{p^2}{d^2}\right] \tag{10.235}$$

where

$$\begin{aligned} \text{TM}: &\quad m, n > 0, p \geq 0; \\ \text{TE}: &\quad m > 0 \text{ or } n > 0, p > 0. \end{aligned}$$

The lowest TM mode has $m, n = 1, \ p = 0$ with frequency

$$\omega^2_{1,1,0} = \frac{\pi^2 c^2}{\mu\epsilon}\left[\frac{1}{a^2} + \frac{1}{b^2}\right], \tag{10.236}$$

whereas the lowest TE mode has either m or $n = 0$ and $p = 1$ with corresponding frequency

$$\omega^2_{1,0,1} = \frac{\pi^2 c^2}{\mu\epsilon}\left[\frac{1}{a^2} + \frac{1}{d^2}\right] \text{ or } \omega^2_{0,1,1} = \frac{\pi^2 c^2}{\mu\epsilon}\left[\frac{1}{b^2} + \frac{1}{d^2}\right]. \tag{10.237}$$

If $a \neq b \neq d$ and we arrange the dimensions such that $a < b < d$, then the two TE modes are always the lowest:

$$\frac{1}{b^2} + \frac{1}{d^2} < \frac{1}{a^2} + \frac{1}{d^2} < \frac{1}{a^2} + \frac{1}{b^2}.$$

If $m, n, p > 0$ then $\omega_{m,n,p}$ given by (10.235) is degenerate in the sense that both TE and TM modes exist with the same frequency.

10.10 Eigenmode expansion for ideal spherical resonant cavity (I): spherical Bessel functions

For our next case we'll look at a spherical resonant cavity (vacuum interior) with perfectly-conducting walls. Besides its practical importance, this example enables us to introduce two important classes of functions: spherical Bessel functions and vector spherical harmonics, both of which are widely useful in quantum mechanics as well as electromagnetism.

You may recall that in Chapter 4 we presented two alternative approaches to introduce Bessel functions: via generating function (Section 4.1), and using separation of variables (Section 4.5). Here we'll take a similar tack with spherical Bessel functions.

First we'll take a separation-of-variables approach. Beginning with the scalar wave equation

$$\left(\nabla^2 - \frac{1}{c^2}\frac{\partial^2}{\partial t^2}\right)\psi(\vec{x},t) = 0, \tag{10.238}$$

and write the solution as a Fourier transform

$$\psi(\vec{x},t) = \int_{-\infty}^{\infty} d\omega\, e^{-i\omega t}\psi(\vec{x},\omega), \tag{10.239}$$

so that the function $\psi(\vec{x},\omega)$ satisfies the Helmholtz equation

$$(\nabla^2 + k^2)\psi(\vec{x},\omega) = 0, \text{ (where } k \equiv |\omega|/c). \tag{10.240}$$

(The integral in (10.239) will actually be replaced with a sum over discrete frequencies in a confined geometry.)

Equation (10.240) closely resembles Laplace's equation, and we already know that the variable-separated solutions of Laplace's equation have the form $f_l(r)Y_{lm}(\theta,\phi)$. So we suppose that variable-separated solutions of (10.240) have the same form, and see if we can solve for $f_l(r)$.

We may rewrite the Laplacian operator as

$$\nabla^2 = \frac{1}{r}\frac{\partial^2}{\partial r^2}(r) - \frac{L^2}{r^2}, \tag{10.241}$$

where $\vec{L}$ is the *angular momentum operator*

$$\vec{L} \equiv \frac{1}{i}\vec{r}\times\vec{\nabla}. \tag{10.242}$$

In spherical coordinates L^2 is given explicitly by

$$L^2 = \vec{L}\cdot\vec{L} = -\left[\frac{1}{\sin\theta}\frac{\partial}{\partial\theta}\left(\sin\theta\frac{\partial}{\partial\theta}\right) + \frac{1}{\sin^2\theta}\frac{\partial^2}{\partial\phi^2}\right]. \tag{10.243}$$

From (4.173) and (4.174) we have

$$L^2 Y_{lm} = l(l+1)Y_{lm}, \tag{10.244}$$

which leads to

$$\frac{1}{r}\frac{d^2}{dr^2}(rf_l) + \left(k^2 - \frac{l(l+1)}{r^2}\right) f_l = 0. \tag{10.245}$$

Using the substitution

$$f_l(r) \equiv \frac{u_l(r)}{\sqrt{r}}, \tag{10.246}$$

we obtain

$$\left[\frac{d^2}{dr^2} + \frac{1}{r}\frac{d}{dr} + k^2 - \frac{(l+\frac{1}{2})^2}{r^2}\right] u_l(r) = 0, \tag{10.247}$$

which is identical to Bessel's equation (4.22) with $m = l + 1/2$. It follows that the solutions $u_l(r)$ to (10.247) may be identified as "Bessel functions of half-integer order".

Closed-form solutions for $f_l(r)$ in (10.245) can be found using induction on l. For $l = 0$, defining $x \equiv kr$ we have

$$\frac{d^2(xf_0)}{dx^2} + xf_0 = 0, \tag{10.248}$$

with two linearly independent solutions

$$j_0 \equiv \frac{\sin x}{x} \quad \text{and} \quad n_0 \equiv -\frac{\cos x}{x}. \tag{10.249}$$

Solutions for $l > 0$ may be found using induction. Assume that f_l is a solution of (10.245), and make the substitution

$$f_l = (-x)^l R_l, \tag{10.250}$$

which leads to

$$\frac{d^2 R_l}{dx^2} + \frac{2(l+1)}{x}\frac{dR_l}{dx} + R_l = 0. \tag{10.251}$$

Differentiating with respect to x, we find

$$\frac{d^3 R_l}{dx^3} + \frac{2(l+1)}{x}\frac{d^2 R_l}{dx^2} + \left[1 - \frac{2(l+1)}{x^2}\right]\frac{dR_l}{dx} = 0. \tag{10.252}$$

If we now define R_{l+1} by

$$R_{l+1} \equiv \frac{1}{x}\frac{dR_l}{dx}, \tag{10.253}$$

we may replace dR_l/dx by xR_{l+1} in (10.252) to obtain (after some algebra)

$$\frac{d^2R_{l+1}}{dx^2} + \frac{2(l+2)}{x}\frac{dR_{l+1}}{dx} + R_{l+1} = 0, \tag{10.254}$$

which implies that $f_{l+1} \equiv (-x)^{l+1}R_{l+1}$ also solves (10.245) with $l \to l+1$.

An inductive relation for the f_l's can be obtained by applying substitution (10.250) to the inductive relation (10.253):

$$\frac{df_l}{dx} + f_{l+1} = \frac{l}{x}f_l. \tag{10.255}$$

Alternatively, if we first iterate (10.253) before the substitution we obtain a closed-form expression for f_l:

$$f_l = (-x)^l R_l = (-x)^l \left(\frac{1}{x}\frac{d}{dx}\right)^l R_0 = (-x)^l \left(\frac{1}{x}\frac{d}{dx}\right)^l f_0. \tag{10.256}$$

We thus have two linearly independent solutions for each l:

$$j_l(x) \equiv (-x)^l \left(\frac{1}{x}\frac{d}{dx}\right)^l \left(\frac{\sin x}{x}\right), \tag{10.257}$$

$$n_l(x) \equiv -(-x)^l \left(\frac{1}{x}\frac{d}{dx}\right)^l \left(\frac{\cos x}{x}\right), \tag{10.258}$$

which are called the *spherical Bessel functions.*

Later we will need to know the limiting behavior of these functions when $x \to 0$ and $x \to \infty$. For $x \to 0$, we may expand $\sin x$, $\cos x$ in (10.257) and (10.258) in Taylor series and retain only the lowest-order terms to obtain

$$j_l(x) \underset{x\to 0}{\longrightarrow} \frac{x^l}{(2l+1)!!}, \tag{10.259}$$

$$n_l(x) \underset{x\to 0}{\longrightarrow} -x^{-(l+1)}(2l-1)!!. \tag{10.260}$$

Note that $j_l(x)$ is regular at $x = 0$, while $n_l(x)$ is singular.

On the other hand, for $x \gg l$ we may begin once again with (10.257) and (10.258) and retain only the lowest-order dependence on $1/x$ to obtain

$$j_l(x) \underset{x\gg l}{\longrightarrow} (-x)^l \left(\frac{1}{x}\right)^{l+1} \left(\frac{d}{dx}\right)^l \sin x = \frac{\sin(x - \pi l/2)}{x}, \tag{10.261}$$

and

$$n_l(x) \underset{x\gg l}{\longrightarrow} -(-x)^l \left(\frac{1}{x}\right)^{l+1} \left(\frac{d}{dx}\right)^l \cos x = -\frac{\cos(x - \pi l/2)}{x}. \tag{10.262}$$

These results also allow us to express spherical Bessel functions in terms of regular Bessel functions. Equations (10.246) and (10.247) imply that $j_l(x)$ and $n_l(x)$ are linear combinations of $J_{\ell+1}(x)/\sqrt{x}$ and $N_{\ell+1}(x)/\sqrt{x}$. However, we know already the asymptotic behavior for $J_{l+1/2}$ and $N_{l+1/2}$ from (4.98) and (4.99):

$$J_{l+1/2}(x) \underset{x \gg l}{\longrightarrow} \sqrt{\frac{2}{\pi x}} \cos\left(x - \frac{l\pi}{2} - \frac{\pi}{2}\right) = \sqrt{\frac{2}{\pi x}} \sin\left(x - \frac{l\pi}{2}\right), \tag{10.263}$$

$$N_{l+1/2}(x) \underset{x \gg l}{\longrightarrow} \sqrt{\frac{2}{\pi x}} \sin\left(x - \frac{l\pi}{2} - \frac{\pi}{2}\right) = -\sqrt{\frac{2}{\pi x}} \cos\left(x - \frac{l\pi}{2}\right). \tag{10.264}$$

Comparison of (10.261) and (10.262) with (10.263) and (10.264) yields

$$j_l(x) = \sqrt{\frac{\pi}{2x}} J_{l+1/2}(x), \tag{10.265}$$

$$n_l(x) = \sqrt{\frac{\pi}{2x}} N_{l+1/2}(x). \tag{10.266}$$

We will also make use of the *spherical Hankel functions*, defined by

$$h_l^{(1)}(x) \equiv j_l(x) + in_l(x), \tag{10.267}$$

$$h_l^{(2)}(x) \equiv j_l(x) - in_l(x). \tag{10.268}$$

From (10.261) and (10.262) we may derive the asymptotic behavior

$$h_l^{(1)}(x) \underset{x \gg l}{\longrightarrow} \frac{-ie^{i(x-l\pi/2)}}{x}, \tag{10.269}$$

$$h_l^{(2)}(x) \underset{x \gg l}{\longrightarrow} \frac{ie^{-i(x-l\pi/2)}}{x}. \tag{10.270}$$

The lowest order spherical Bessel and spherical Hankel functions of the first kind are listed for convenience in Tables 10.2 and 10.3.

As promised at the beginning of this section, we now start over from scratch and derive the same results using a generating function. Our argument parallels the discussion in Section 4.1. In that section, we obtained the δ-function as a sum of variable-separated terms in cylindrical coordinates by first expressing the δ-function as a Fourier integral in cylindrical coordinates, and then expanding the (ρ, ϕ)-dependent part of the integrand as a series in $e^{im\phi}$. The functions $e^{im\phi}$ were chosen because they correspond to the angular dependence of the variable-separated solutions to Laplace's equation in cylindrical coordinates. The analogous argument in spherical

Table 10.2 Spherical Bessel functions.

	$j_l(x)$	$n_l(x)$
$l=0$	$\dfrac{\sin x}{x}$	$-\dfrac{\cos x}{x}$
$l=1$	$\dfrac{\sin x}{x^2} - \dfrac{\cos x}{x}$	$-\dfrac{\cos x}{x^2} - \dfrac{\sin x}{x}$
$l=2$	$\left(\dfrac{3}{x^3} - \dfrac{1}{x}\right)\sin x - \dfrac{3}{x^2}\cos x$	$-\left(\dfrac{3}{x^3} - \dfrac{1}{x}\right)\cos x - \dfrac{3}{x^2}\sin x$
$l=3$	$\left(\dfrac{15}{x^4} - \dfrac{6}{x^2}\right)\sin x - \left(\dfrac{15}{x^3} - \dfrac{1}{x}\right)\cos x$	$-\left(\dfrac{15}{x^4} - \dfrac{6}{x^2}\right)\cos x - \left(\dfrac{15}{x^3} - \dfrac{1}{x}\right)\sin x$

Table 10.3 Spherical Hankel functions of the first kind.

	$h_l^{(1)}(x)$
$l=0$	$-\dfrac{ie^{ix}}{x}$
$l=1$	$-\dfrac{e^{ix}}{x}\left(1 + \dfrac{i}{x}\right)$
$l=2$	$\dfrac{ie^{ix}}{x}\left(1 + \dfrac{3i}{x} - \dfrac{3}{x^2}\right)$
$l=3$	$\dfrac{e^{ix}}{x}\left(1 + \dfrac{6i}{x^2} - \dfrac{15}{x^2} - \dfrac{15i}{x^3}\right)$

coordinates follows a similar procedure, but this time the expansion is in terms of spherical harmonics since they give the angular dependence of variable-separated solutions to Laplace's equation in spherical coordinates.

Let us begin the argument as in Section 4.1, with an integral expression for the delta function:

$$\delta(\vec{x} - \vec{x}\,') = \int \frac{d^3k}{(2\pi)^3} e^{i\vec{k}\cdot(\vec{x}-\vec{x}\,')}. \tag{10.271}$$

On the one hand, we may also write the delta-function in variable-separated form (using spherical coordinates) as

$$\delta(\vec{x}-\vec{x}\,') = \frac{1}{r^2}\delta(r-r')\delta(\cos\theta-\cos\theta')\delta(\phi-\phi'). \tag{10.272}$$

On the other hand, we may re-express the exponent in (10.271) as

$$\vec{k}\cdot(\vec{x}-\vec{x}\,') = kr\cos\gamma - kr'\cos\gamma', \tag{10.273}$$

and then expand the two complex exponentials in terms of the complete set of eigenfunctions $\{P_l\}$, for instance:

$$e^{ikr\cos\gamma} = \sum_l i^l(2l+1)P_l(\cos\gamma)j_l(kr), \tag{10.274}$$

where the constant factors $i^l(2l+1)$ are chosen by "clairvoyance". Of course at this point we don't know that the j_l's in (10.274) are actually spherical Bessel functions: the proof will come later.

In terms of the angles shown in Figure 10.10, we may rewrite (10.274) using the addition theorem (4.224) as

$$e^{ikr\cos\gamma} = \sum_{l,m} 4\pi i^l Y^*_{lm}(\alpha,\beta)Y_{lm}(\theta,\phi)j_l(kr). \tag{10.275}$$

Substitution yields

$$\frac{1}{r^2}\delta(r-r')\delta(\cos\theta-\cos\theta')\delta(\phi-\phi') = \frac{(4\pi)^2}{(2\pi)^3}\sum_{l,l'}\sum_{m,m'} i^{l-l'}$$

$$\int dk k^2 \sin\alpha d\alpha\, d\beta\, j_l(kr)j_{l'}(kr')Y^*_{lm}(\alpha,\beta)Y_{l'm'}(\alpha,\beta)Y_{lm}(\theta,\phi)Y^*_{l'm'}(\theta',\phi'). \tag{10.276}$$

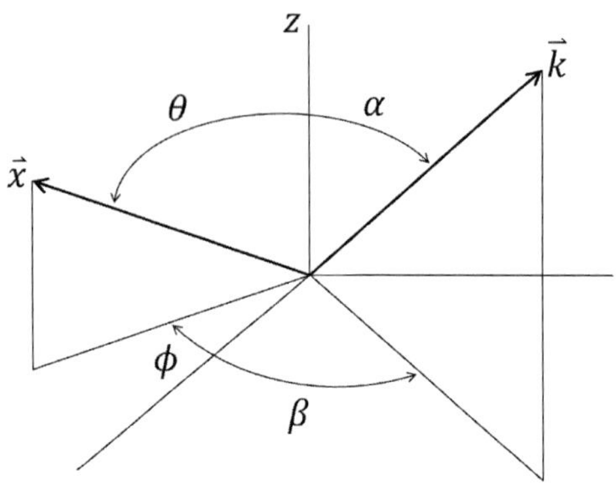

Fig. 10.10 Angle definitions.

The α and β integrals may be evaluated, resulting in:

$$\frac{1}{r^2}\delta(r-r')\delta(\cos\theta-\cos\theta')\delta(\phi-\phi') = \frac{2}{\pi}\sum_{l,m} Y^*_{lm}(\theta',\phi')Y_{lm}(\theta,\phi) \times \int_0^\infty dk k^2\, j_l(kr)j_l(kr'). \quad (10.277)$$

It follows from the completeness relation (4.222) that

$$\int_0^\infty dk k^2 j_l(kr)j_l(kr') = \frac{\pi}{2r^2}\delta(r-r'). \quad (10.278)$$

In our original discussion of Legendre polynomials we showed

$$\int_{-1}^{1} dx\, P_l(x)P_{l'}(x) = \frac{2}{2l+1}\delta_{ll'}, \quad (10.279)$$

which can be applied to the generating function (10.274) to give an integral representation for $j_l(x)$ ($z = \cos\gamma$):

$$j_l(x) = \frac{1}{2i^l}\int_{-1}^{1} dz\, e^{ixz} P_l(z). \quad (10.280)$$

From this we may compute directly

$$j_0(x) = \frac{\sin x}{x}, \quad (10.281)$$

which agrees with the j_0 found by separation of variables. Using the Legendre polynomial identities (4.180)–(4.183) we can also deduce recursion relations such as

$$\frac{d}{dx}j_l(x) - j_{l-1}(x) = -\frac{(l+1)}{x}j_l(x), \quad (10.282)$$

$$\frac{d}{dx}j_l(x) + j_{l+1}(x) = \frac{l}{x}j_l(x). \quad (10.283)$$

Recursion relation (10.283) is identical to (10.255), which together with (10.281) implies that the j_l's defined by (10.280) are identical to the j_l's which were derived using separation of variables.

10.11 Eigenmode expansion for ideal spherical resonant cavity (II): vector spherical harmonics

Now that we have armed ourselves with spherical Bessel functions, we are ready to finish our conquest of the spherical resonant cavity. Our series solution for the Helmholtz equation (10.240) is now ($k \equiv \omega/c > 0$)

$$\psi(\vec{x},\omega) = \sum_{l,m}[A_{lm}j_l(kr) + B_{lm}n_l(kr)]Y_{lm}(\theta,\phi). \quad (10.284)$$

We may apply this to the electric and magnetic fields, since in the vacuum we have

$$(\nabla^2 + k^2)\vec{E} = 0 \text{ and } (\nabla^2 + k^2)\vec{B} = 0. \tag{10.285}$$

For an arbitrary vector field $\vec{A}$ we have

$$\nabla^2(\vec{r}\cdot\vec{A}) = \vec{r}\cdot\nabla^2\vec{A} + 2\vec{\nabla}\cdot\vec{A}. \tag{10.286}$$

Since $\vec{\nabla}\cdot\vec{E} = \vec{\nabla}\cdot\vec{B} = 0$ in the vacuum, we have

$$(\nabla^2 + k^2)\vec{r}\cdot\vec{E} = (\nabla^2 + k^2)\vec{r}\cdot\vec{B} = 0. \tag{10.287}$$

We may thus treat the radial components of $\vec{E}$ and $\vec{B}$ in just the same way that we treated E_z and B_z in our discussion of waveguide modes in Section 10.2. In keeping with this analogy, it is thus natural to define *spherical transverse electric (TE) modes* (denoted by superscript (e); we might also call this "radial magnetic") via the conditions

$$\vec{r}\cdot\vec{E}^{(e)}_{lm} = 0, \tag{10.288}$$

$$\vec{r}\cdot\vec{B}^{(e)}_{lm} = \frac{l(l+1)}{k} g_{lm}(kr) Y_{lm}(\theta,\phi), \tag{10.289}$$

where

$$g_{lm}(kr) \equiv A_{lm} j_l(kr) + B_{lm} n_l(kr), \tag{10.290}$$

and the normalization factor in (10.289) is chosen for later convenience. To obtain the nonzero components of the electric field, we may use the Fourier transform of the Faraday equation:

$$\vec{B}(\vec{x},\omega) = -\frac{i}{k}\vec{\nabla}\times\vec{E}, \tag{10.291}$$

so that

$$k\vec{r}\cdot\vec{B} = \frac{1}{i}\vec{r}\cdot(\vec{\nabla}\times\vec{E}) = \frac{1}{i}(\vec{r}\times\vec{\nabla})\cdot\vec{E} = \vec{L}\cdot\vec{E}. \tag{10.292}$$

In view of (10.289), this gives

$$\vec{L}\cdot\vec{E}^{(e)}_{lm} = l(l+1) g_{lm}(kr) Y_{lm}(\theta,\phi). \tag{10.293}$$

In addition, we have

$$\begin{aligned}(\vec{r}\times\vec{L})\cdot\vec{E}^{(e)}_{lm} &= \frac{1}{i}\left[\vec{r}\times(\vec{r}\times\vec{\nabla})\right]\cdot\vec{E}^{(e)}_{lm} \\ &= \vec{r}\cdot\left[(\vec{r}\cdot\vec{\nabla})\vec{E}^{(e)}_{lm}\right] - r^2\vec{\nabla}\cdot\vec{E}^{(e)}_{lm} = 0.\end{aligned} \tag{10.294}$$

It turns out that (10.293) and (10.294) determine $\vec{E}^{(e)}_{lm}$ uniquely, as we shall now show. First we'll need the following theorem (which follows from the upcoming completeness and orthogonality relations):

Theorem. *Any vector field $\vec{A}$ may be represented (except perhaps at the origin) as follows:*

$$\vec{A}(r,\theta,\phi) = \sum_{l=1}^{\infty}\sum_{m=-l}^{l}\Big[C_{lm}(r)\vec{X}_{lm} + K_{lm}(r)\hat{r}\times\vec{X}_{lm}\Big] + \hat{r}(\hat{r}\cdot\vec{A}), \quad (10.295)$$

where the $\{C_{lm}\}$ and $\{K_{lm}\}$ are arbitrary functions of r, and the vector spherical harmonics *$\vec{X}_{lm}(\theta,\phi)$ are defined as*

$$\vec{X}_{lm} \equiv \frac{\vec{L}Y_{lm}(\theta,\phi)}{\sqrt{l(l+1)}}, \quad l = 1,2,3,\ldots;\ m=-l,\ldots,l. \quad (10.296)$$

The following properties of $\vec{X}_{lm}$ may be derived from (10.296) and the definition of $\vec{L}$ in (10.242):

$$\hat{r}\cdot\vec{X}_{lm} = 0, \quad (10.297)$$

$$\hat{r}\cdot(\hat{r}\times\vec{X}_{lm}) = 0, \quad (10.298)$$

$$\hat{r}\times(\hat{r}\times\vec{X}_{lm}) = -\vec{X}_{lm}, \quad (10.299)$$

$$\vec{X}_{lm}\cdot(\hat{r}\times\vec{X}_{lm}) = 0. \quad (10.300)$$

The following operator statements can be proved similarly:

$$\vec{L}\times\vec{L} = i\vec{L}, \quad (10.301)$$

$$(\hat{r}\times\vec{L})\cdot\vec{X}_{lm} = \vec{L}\cdot(\vec{X}_{lm}\times\hat{r}) = 0, \quad (10.302)$$

$$(\hat{r}\times\vec{L})\cdot(\hat{r}\times\vec{X}_{lm}) = \vec{L}\cdot\vec{X}_{lm}. \quad (10.303)$$

The following orthogonality relations are also satisfied by the $\vec{X}_{lm}$:

$$\int d\Omega\,\vec{X}^*_{l'm'}\cdot\vec{X}_{lm} = \delta_{ll'}\delta_{mm'}, \quad (10.304)$$

$$\int d\Omega\,\vec{X}^*_{l'm'}\cdot(\hat{r}\times\vec{X}_{lm}) = 0, \quad (10.305)$$

$$\int d\Omega(\hat{r}\times\vec{X}^*_{l'm'})\cdot(\hat{r}\times\vec{X}_{lm}) = \int d\Omega\,(\vec{X}^*_{l'm'}\times(\hat{r}\times\vec{X}_{lm}))\cdot\hat{r}$$
$$= \int d\Omega\,\vec{X}^*_{l'm'}\cdot\vec{X}_{lm}. \quad (10.306)$$

These relations can be applied to the vector field decomposition (10.295) to obtain expressions for the coefficient functions $C_{lm}(r)$ and $K_{lm}(r)$:

$$C_{lm}(r) = \int d\Omega\, \vec{X}^*_{lm} \cdot \vec{A}, \tag{10.307}$$

$$K_{lm}(r) = \int d\Omega\, (\hat{r} \times \vec{X}^*_{lm}) \cdot \vec{A}. \tag{10.308}$$

In addition there is a completeness relation given by:

$$\sum_{l,m} \left[\vec{X}^*_{lm}(\theta', \phi')\vec{X}_{lm}(\theta, \phi) + (\hat{r}' \times \vec{X}^*_{lm})(\hat{r} \times \vec{X}_{lm})\right]$$
$$= (\hat{e}_\theta \hat{e}_\theta + \hat{e}_\phi \hat{e}_\phi)\delta(\cos\theta - \cos\theta')\delta(\phi - \phi'). \tag{10.309}$$

This expression is formed out of "dyadics", which are matrices made out of products of vectors. We will see other uses of such structures in Chapter 12. Relations (10.304) and (10.309) are proved in Exercises 10.11.1 and 10.11.2, respectively.

Returning to the task at hand, we obtain $\vec{E}^{(e)}_{lm}$ explicitly by substituting expansion (10.295) for $\vec{E}^{(e)}_{lm}$ into (10.293) and (10.294) respectively. From (10.293) we get (also making use of (10.302))

$$\vec{L} \cdot \vec{E}^{(e)}_{lm} = \vec{L} \cdot \sum_{l',m'} [C_{l'm'}\vec{X}_{l'm'} + K_{l'm'}\hat{r}\times\vec{X}_{l'm'}]$$

$$\Longrightarrow l(l+1)g_{lm}(kr)Y_{lm} = \sum_{l',m'} \frac{l'(l'+1)C_{l'm'}}{\sqrt{l'(l'+1)}} Y_{l'm'}$$

$$\Longrightarrow C_{l'm'} = \sqrt{l(l+1)}g_{lm}(kr)\delta_{ll'}\delta_{mm'}, \tag{10.310}$$

and from (10.294) we find

$$(\hat{r} \times \vec{L}) \cdot \vec{E}^{(e)}_{lm} = (\hat{r} \times \vec{L}) \cdot \sum_{l',m'} \left[C_{l'm'}\vec{X}_{l'm'} + K_{l'm'}\hat{r} \times \vec{X}_{l'm'}\right]$$

$$\Longrightarrow 0 = \sum_{l',m'} \frac{K_{l'm'}l'(l'+1)}{\sqrt{l'(l'+1)}} Y_{l'm'} \Longrightarrow K_{l'm'} = 0. \tag{10.311}$$

At long last we have our explicit expressions for $\vec{E}$ and $\vec{B}$ fields for TE modes:

$$\vec{E}^{(e)}_{lm}(\vec{x}, \omega) = g_{lm}(kr)\vec{L}Y_{lm}(\theta, \phi), \tag{10.312}$$

$$\vec{B}^{(e)}_{lm}(\vec{x}, \omega) = -\frac{i}{k}\vec{\nabla} \times \vec{E}^{(e)}_{lm}(\vec{x}, \omega), \tag{10.313}$$

where (10.313) comes from (10.291).

It should be no surprise that the above argument for TE modes is paralleled by a similar argument for *spherical transverse magnetic (TM) modes* (denoted by superscript (m); we might also call this "radial electric") which we define by the conditions

$$\vec{r} \cdot \vec{B}_{lm}^{(m)} = 0, \tag{10.314}$$

$$\vec{r} \cdot \vec{E}_{lm}^{(m)} = -\frac{l(l+1)}{k} f_{lm}(kr) Y_{lm}(\theta, \phi), \tag{10.315}$$

where

$$f_{lm} \equiv A'_{lm} j_l(kr) + B'_{lm} n_l(kr), \tag{10.316}$$

and the odd-looking normalization factor in (10.315) is chosen for later convenience as usual. The explicit expressions for $\vec{E}$ and $\vec{B}$ in this case are

$$\vec{B}_{lm}^{(m)}(\vec{x}, \omega) = f_{lm}(kr) \vec{L} Y_{lm}(\theta, \phi), \tag{10.317}$$

$$\vec{E}_{lm}^{(m)}(\vec{x}, \omega) = \frac{i}{k} \vec{\nabla} \times \vec{B}_{lm}^{(m)}(\vec{x}, \omega). \tag{10.318}$$

We may now obtain the resonant frequencies for the spherical cavity by imposing the boundary conditions. For TE modes, regularity at $r = 0$ means that the coefficient of n_l in (10.290) must be equal to zero, so we can replace g_{lm} in (10.312) with j_l. Also, perfect electrical conductor boundary conditions at $r = a$ imply that

$$\hat{r} \times \vec{E}_{lm} \Big|_{r=a} = \hat{r} \cdot \vec{B}_{lm} \Big|_{r=a} = 0. \tag{10.319}$$

Applying these conditions to TE modes gives (from (10.289) and (10.312))

$$j_l(ka) = 0, \qquad \text{(TE modes)} \tag{10.320}$$

which determines the allowable values of k. (The roots of (10.320) can be found online, or in a mathematical handbook.)

For TM modes, regularity at $r = 0$ once again implies that we may replace f_{lm} in (10.315) with j_l. As far as satisfying the boundary conditions (10.319), the condition on $\vec{B}_{lm}$ holds trivially due to (10.314), but the condition on $\vec{E}_{lm}$ requires some computational gymnastics. To evaluate $\hat{r} \times \vec{E}_{lm}$, we expand

$$\begin{aligned} \vec{\nabla} \times (j_l \vec{L} Y_{lm}) &= \vec{\nabla} j_l \times \vec{L} Y_{lm} + j_l \vec{\nabla} \times \vec{L} Y_{lm} \\ &= \frac{dj_l}{dr} \hat{r} \times \vec{L} Y_{lm} + j_l \vec{\nabla} \times \vec{L} Y_{lm}. \end{aligned} \tag{10.321}$$

The second term on the right-hand side can be evaluated using components:

$$\begin{aligned}(\vec{\nabla}\times\vec{L})_j Y_{lm} &= \frac{1}{i}\left[\vec{\nabla}\times(\vec{r}\times\vec{\nabla})\right]_j Y_{lm}\\ &= \frac{1}{i}\sum_i\left[\nabla_i r_j\nabla_i - \nabla_i r_i\nabla_j\right]Y_{lm}\\ &= \frac{1}{i}\left[\nabla_j + r_j\nabla^2 - 3\nabla_j - (\vec{r}\cdot\vec{\nabla})\nabla_j\right]Y_{lm},\end{aligned}$$

so that

$$(\vec{\nabla}\times\vec{L})Y_{lm} = \frac{1}{i}\left[-2\vec{\nabla} + \vec{r}\nabla^2 - (\vec{r}\cdot\vec{\nabla})\vec{\nabla}\right]Y_{lm}. \tag{10.322}$$

Taking $\hat{r}\times$ on both sides of (10.321) and using (10.299) gives

$$\begin{aligned}\hat{r}\times(\vec{\nabla}\times(j_l\vec{L}Y_{lm})) &= \frac{dj_l}{dr}\hat{r}\times(\hat{r}\times\vec{L})Y_{lm} + j_l\frac{\vec{r}}{r}\times(\vec{\nabla}\times\vec{L})Y_{lm}\\ &= -\frac{dj_l}{dr}\vec{L}Y_{lm} + j_l\left[-\frac{2}{r}\vec{L}Y_{lm} - (\vec{r}\cdot\vec{\nabla})\frac{1}{r}\vec{L}Y_{lm}\right]\\ &= -\left(\frac{dj_l}{dr} + \frac{j_l}{r}\right)\vec{L}Y_{lm}\\ &= -\frac{1}{r}\frac{d(rj_l)}{dr}\vec{L}Y_{lm}.\end{aligned} \tag{10.323}$$

Multiplying by i/k gives finally

$$\hat{r}\times\vec{E}^{(m)}_{lm}(\vec{x},\omega) = -\frac{i}{kr}\frac{d(rj_l)}{dr}\vec{L}Y_{lm}, \tag{10.324}$$

so that the eigenvalue condition imposed by the boundary conditions is

$$\frac{d}{dr}(rj_l(kr))|_{r=a} = 0. \quad \text{(TM modes)} \tag{10.325}$$

The lowest-energy modes are of special practical importance. Notice that according to (10.296) there is no X_{00}, so the index l in (10.320) and (10.325) begins with $l = 1$. For TE modes with $l = 1$, the eigenvalue condition is

$$j_1(ka) = \frac{\sin ka}{(ka)^2} - \frac{\cos ka}{ka} = 0, \tag{10.326}$$

which can be solved numerically to yield

$$ka = 4.493409, 7.725252, 10.904132, \ldots \qquad \text{(TE modes, } l = 1\text{)}. \tag{10.327}$$

For TM modes with $l = 1$, the eigenvalue condition evaluates to:

$$\sin ka\left(1 - \frac{1}{(ka)^2}\right) + \frac{\cos ka}{ka} = 0, \tag{10.328}$$

which can be solved numerically to yield

$$ka = 2.74371, 6.11676, 9.31662, \ldots \qquad \text{(TM modes, } l = 1\text{)}. \tag{10.329}$$

The lowest mode is TM with $l = 1$.

In the exercises we will examine two applications of spherical modes. Exercise 10.11.4 deals with the "MIT bag model" of hadrons (which uses perfect *magnetic* conductor boundary conditions) and Exercise 10.12.2 deals with *Schumann resonances*, which are low-frequency electromagnetic resonances in the ionosphere. Cavity resonators are at the heart of many electronic devices which require high-frequency signal amplification, including e.g. radar, microwave communications, medical devices, and particle physics accelerators. An important application of resonant cavity technology is the radio-frequency devices called *klystrons* used in particle accelerators (see Exercise 13.7.5).

10.12 Energy loss and frequency shift in non-ideal resonant cavities

In Sections 10.7 and 10.8 we discussed the energy loss for different waveguide modes due to imperfectly-conducting walls. In this section we'll do the same thing for resonant cavities. Since a resonant cavity is an example of a damped, driven harmonic oscillator, we'll begin by looking at some general facts about harmonic oscillators. In particular, we will review the concept of an oscillator's *quality factor*, which is invariably denoted by Q. Actually there are several alternative definitions of Q, as we shall see shortly.

The standard equation for a harmonically driven oscillator is

$$\ddot{x} + 2\beta\dot{x} + \omega_0^2 x = A\cos\omega t, \tag{10.330}$$

which has a particular solution of the form

$$x_p(t) = D(\omega)\cos(\omega t - \delta). \tag{10.331}$$

By plugging (10.331) into (10.330) we find

$$D(\omega) = \frac{A}{\sqrt{(\omega_0^2 - \omega^2)^2 + 4\omega^2\beta^2}}, \tag{10.332}$$

$$\delta = \arctan\left(\frac{2\omega\beta}{\omega_0^2 - \omega^2}\right), \tag{10.333}$$

where the branches for the arctan must respect continuity. The driving frequency which produces the maximum (resonant) amplitude is denoted by ω_R, and is determined by the condition

$$\left.\frac{dD(\omega)}{d\omega}\right|_{\omega=\omega_R} = 0 \implies \omega_R^2 = \omega_0^2 - 2\beta^2. \tag{10.334}$$

One definition for the Q of an oscillator (which we will denote as Q_1) is

$$Q_1 \equiv \frac{\omega_R}{\Delta\omega}, \tag{10.335}$$

where $\Delta\omega$ is the peak width for the $|D(\omega)|^2$ versus ω curve at half the maximum height, as shown in Figure 10.11. Q_1 thus reflects the "sharpness" of the squared amplitude versus frequency curve. In the case where $\beta \ll \omega_0$, it can be shown that

$$\omega_R \approx \omega_0; \quad \Delta\omega \approx 2\beta \implies Q_1 \approx \frac{\omega_0}{2\beta} \qquad (\beta \ll \omega_0). \tag{10.336}$$

It's also possible to define the quality factor in similar fashion based on the $\omega D(\omega)$ versus ω curve, where $\omega D(\omega)$ represents the maximum value of $|\dot{x}_p(t)|$ for the oscillator when driven at frequency ω. It can be shown that this gives $\omega_R = \omega_0$ and $\Delta\omega = 2\beta$ exactly, and so we may also *define* $Q_2 \equiv \omega_0/(2\beta)$.

Yet another possible definition (which we refer to as Q_3) is in terms of energy loss when the oscillator is driven at resonance frequency:

$$Q_3 \equiv 2\pi \frac{\text{average total energy at } \omega = \omega_R}{\text{energy loss during one period}}. \tag{10.337}$$

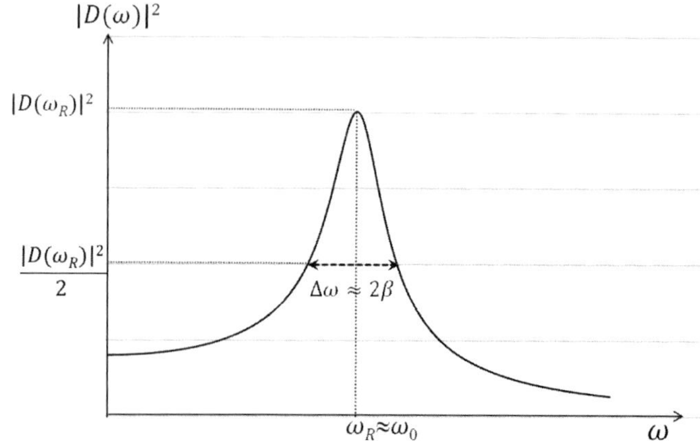

Fig. 10.11 Squared amplitude versus frequency of a damped oscillator, in the case where $\beta \ll \omega_0$.

It can be shown that all of these different definitions of Q agree in the limit where $\beta \ll \omega_0$. Since this is the case we're considering, we will drop the subscript and refer simply to Q.

Let us now evaluate the quality factor Q_λ for a cavity with imperfectly-conducting walls oscillating in mode λ. We will use two methods: the first starts with (10.337), while the second uses $Q_\lambda \approx \omega_\lambda/(2\beta_\lambda)$ and takes advantage of our previous results for energy loss in waveguides (Sections 10.7 and 10.8).

Here we go with the first method. Based on (10.337), we may evaluate the quality factor Q_λ for a cavity with imperfectly-conducting walls oscillating in mode λ as follows:

$$\begin{aligned} Q_\lambda &\equiv \omega_\lambda \frac{\text{average total energy}}{(\omega_\lambda/2\pi) \times \text{ energy loss during one period}} \\ &= \omega_\lambda \frac{\text{average total energy}}{\text{average power loss}} \\ &= \omega_\lambda \frac{U}{\int_{\text{walls}} \frac{dP}{da}\, da}, \end{aligned} \tag{10.338}$$

where ω_λ is the mode frequency for the same cavity with perfectly-conducting walls. From previous results we have

$$U = \frac{1}{16\pi} \int d^3x \left[\epsilon|\vec{E}|^2 + \mu|\vec{H}|^2\right], \tag{10.339}$$

$$\int_{\text{walls}} \frac{dP}{da}\, da = \frac{\mu_c \omega_\lambda \delta}{16\pi} \int_{\text{walls}} da\, |\vec{H}^{\text{tang}}|^2 \tag{10.340}$$

$$\Longrightarrow Q_\lambda = \frac{1}{\mu_c \delta} \frac{\int d^3x \left[\epsilon|\vec{E}|^2 + \mu|\vec{H}|^2\right]}{\int_{\text{walls}} da\, |\vec{H}^{\text{tang}}|^2}. \tag{10.341}$$

(We suppress a mode index on the skin depth δ for resonant cavities.) Let's compute Q_λ for a cylindrical cavity with arbitrary cross-section (not necessarily circular) that is also bounded at $z = 0$ and $z = d$ (so the cavity is essentially a cylindrical waveguide with caps). Then for TM modes the boundary conditions at $z = 0$ and $z = d$ lead to (in analogy to (10.227) and (10.230))

$$E_z = \psi_\lambda(x, y) \cos\left(\frac{p\pi z}{d}\right),\ p = 0, 1, 2, \ldots, \tag{10.342}$$

and just as in waveguides (see (10.75)) we have

$$\left[\nabla_\perp^2 + \gamma_\lambda^2\right] \psi_\lambda(x, y) = 0, \tag{10.343}$$

$$\psi_\lambda(x, y)|_S = 0, \tag{10.344}$$

where

$$\gamma_\lambda^2 \equiv \frac{{\omega_\lambda}^2\mu\epsilon}{c^2} - k^2 \quad \text{and} \quad k \equiv \frac{p\pi}{d}. \tag{10.345}$$

(Note that γ_λ depends also on p, but to avoid notational chaos we refrain from adding another subscript.) We may also conclude from our previous discussion of waveguides (see (10.72) and (10.73)) for TM modes that

$$\vec{E}_\perp = -\frac{p\pi}{d\gamma_\lambda^2}\sin\left(\frac{p\pi z}{d}\right)\vec{\nabla}_\perp\psi_\lambda, \tag{10.346}$$

$$\vec{H}_\perp = \frac{i\epsilon\omega_\lambda}{c\gamma_\lambda^2}\cos\left(\frac{p\pi z}{d}\right)\hat{e}_3 \times \vec{\nabla}_\perp\psi_\lambda. \tag{10.347}$$

The first term in the numerator of (10.341) is obtained as follows:

$$\begin{aligned}
\epsilon\int d^3x\,|\vec{E}_\perp|^2 &= \epsilon\left(\frac{p\pi}{d\gamma_\lambda^2}\right)^2\frac{d}{2}\int dxdy\,\vec{\nabla}_\perp\psi_\lambda^* \cdot \vec{\nabla}_\perp\psi_\lambda \\
&= \frac{1}{2}\epsilon d\left(\frac{p\pi}{d\gamma_\lambda}\right)^2\int da\,|\psi_\lambda|^2 \\
\Longrightarrow \epsilon\int d^3x\,|\vec{E}|^2 &= \frac{\epsilon d}{2}\left(1 + \left(\frac{p\pi}{d\gamma_\lambda}\right)^2\right)\int da\,|\psi_\lambda|^2.
\end{aligned} \tag{10.348}$$

Now

$$\begin{aligned}
\frac{{\omega_\lambda}^2\mu\epsilon}{c^2} &= \gamma_\lambda^2 + \left(\frac{p\pi}{d}\right)^2 \\
\Longrightarrow \frac{{\omega_\lambda}^2\mu\epsilon}{c^2\gamma_\lambda^2} &= 1 + \left(\frac{p\pi}{d\gamma_\lambda}\right)^2 \\
\Longrightarrow \epsilon\int d^3x|\vec{E}|^2 &= \left(\frac{\mu\epsilon^2{\omega_\lambda}^2 d}{2c^2\gamma_\lambda^2}\right)\int da\,|\psi_\lambda|^2.
\end{aligned} \tag{10.349}$$

For the second term in the numerator of (10.341), when $p \neq 0$ we have

$$\begin{aligned}
\mu\int d^3x|\vec{H}_\perp|^2 &= \left(\frac{\mu\epsilon^2{\omega_\lambda}^2}{c^2{\gamma_\lambda}^4}\right)\frac{d}{2}\int da\,\underbrace{|\hat{e}_3 \times \vec{\nabla}_\perp\psi_\lambda|^2}_{\vec{\nabla}_\perp\psi_\lambda^* \cdot \vec{\nabla}_\perp\psi_\lambda} \\
&= \left(\frac{\mu\epsilon^2{\omega_\lambda}^2 d}{2c^2\gamma_\lambda^2}\right)\int da\,|\psi_\lambda|^2 \\
&= \epsilon\int d^3x\,|\vec{E}|^2.
\end{aligned} \tag{10.350}$$

Next, computing the denominator in (10.341) requires integrating over the entire surface of the cavity. For the lateral portion of the boundary we get

(again for $p \neq 0$)

$$|\vec{H}^{\text{tang}}|^2 = |\hat{n} \times \vec{H}_\perp|^2 = \left(\frac{\epsilon^2 {\omega_\lambda}^2}{c^2 {\gamma_\lambda}^4}\right) \cos^2\left(\frac{p\pi z}{d}\right) |\hat{n} \times (\hat{e}_3 \times \vec{\nabla}_\perp \psi_\lambda)|^2$$
$$= \left(\frac{\epsilon^2 {\omega_\lambda}^2}{c^2 {\gamma_\lambda}^4}\right) \cos^2\left(\frac{p\pi z}{d}\right) \left|\frac{\partial \psi_\lambda}{\partial n}\right|^2. \tag{10.351}$$

For the caps (don't forget these!)

$$|\vec{H}^{\text{tang}}|^2 = |\hat{e}_3 \times \vec{H}_\perp|^2 = \left(\frac{\epsilon^2 {\omega_\lambda}^2}{c^2 {\gamma_\lambda}^4}\right) |\hat{e}_3 \times (\hat{e}_3 \times \vec{\nabla}_\perp \psi_\lambda)|^2$$
$$= \left(\frac{\epsilon^2 {\omega_\lambda}^2}{c^2 {\gamma_\lambda}^4}\right) |\vec{\nabla}_\perp \psi_\lambda|^2. \tag{10.352}$$

Altogether the entire boundary gives

$$\int_{\text{walls}} da\, |\vec{H}^{\text{tang}}|^2 = \frac{\epsilon^2 \omega_\lambda^2}{c^2 {\gamma_\lambda}^4} \left[\underbrace{2\gamma_\lambda^2 \int da\, |\psi_\lambda|^2}_{\text{2 caps}} + \underbrace{\frac{d}{2} \oint d\ell \left|\frac{\partial \psi_\lambda}{\partial n}\right|^2}_{\text{lateral}} \right], \tag{10.353}$$

so explicitly

$$Q^{\text{TM}}_{\lambda, p\neq 0} = \frac{\mu d}{2\mu_c \delta} \frac{\int da\, |\psi_\lambda|^2}{\int da\, |\psi_\lambda|^2 + \frac{d}{4\gamma_\lambda^2} \oint d\ell \left|\frac{\partial \psi_\lambda}{\partial n}\right|^2}. \tag{10.354}$$

For $p = 0$ the expression differs slightly:

$$Q^{\text{TM}}_{\lambda, p=0} = \frac{\mu d}{\mu_c \delta} \frac{\int da\, |\psi_\lambda|^2}{\int da\, |\psi_\lambda|^2 + \frac{d}{2\gamma_\lambda^2} \oint d\ell \left|\frac{\partial \psi_\lambda}{\partial n}\right|^2}. \tag{10.355}$$

In Exercise 10.12.3 you will show a corresponding result for TE modes:

$$Q^{\text{TE}}_{\lambda, p\neq 0} = \frac{\frac{\mu d}{2\mu_c \delta} \left(\frac{\omega_\lambda^2 \mu\epsilon}{c^2}\right) \left(\frac{d}{p\pi}\right)^2 \int da\, |\psi_\lambda|^2}{\int da\, |\psi_\lambda|^2 + \frac{d}{4\gamma_\lambda^2} \oint d\ell \left|\frac{\partial \psi_\lambda}{\partial \ell}\right|^2 + \frac{d\gamma_\lambda^2}{4} \left(\frac{d}{p\pi}\right)^2 \oint d\ell\, |\psi_\lambda|^2}. \tag{10.356}$$

In other exercises you will find Q's for other geometries, such as the spherical cavity (see Exercises 10.12.4, 10.12.5, and 10.12.6).

As mentioned above, it is also possible to obtain Q_λ from $Q_\lambda \approx \omega_\lambda/(2\beta_\lambda)$. Finding ω_λ is straightforward, but getting β_λ is a bit more tortuous. Just as in waveguides, the imperfect conductivity of the cavity's walls leads to complex frequency shifts for TM and TE modes. These shifts may be computed directly by changing the line (resp. surface) integrals in (10.202) and (10.204) to surface (resp. volume) integrals. Alternatively, we may use a modified version of the boundary conditions technique discussed in Section 10.8 for nondegenerate modes, or a generalization of Exercises 10.8.3 and 10.8.4 and similar developments for degenerate modes.

If mode λ experiences a frequency shift $\Delta\omega_\lambda$, then the time dependence of the fields changes:

$$\psi \sim e^{-i\omega_\lambda t} \to \psi \sim e^{-i(\omega_\lambda + \mathrm{Re}[\Delta\omega_\lambda])t} e^{\mathrm{Im}[\Delta\omega_\lambda]t}. \tag{10.357}$$

The characteristic equation for the harmonic oscillator equation (10.330) has roots $\omega_\pm \approx -i\beta \pm \omega_0$ which implies that $-\beta$ can be identified with the $\mathrm{Im}\,[\Delta\omega_\lambda]$ in (10.357). It follows that

$$Q_\lambda \approx \frac{\omega_\lambda}{2\beta_\lambda} = -\frac{\omega_\lambda}{2\mathrm{Im}\,[\Delta\omega_\lambda]}. \tag{10.358}$$

Thus (10.358) can be used to find $\Delta\omega_\lambda$ if Q_λ is already known (and vice versa). From the forms of f^{TM} and f^{TE} (Equation (10.207) and Exercise 10.2.1), we find that $\mathrm{Re}\,[\Delta\omega_\lambda] \approx \mathrm{Im}\,[\Delta\omega_\lambda]$. This shows that $\omega_0 \to \omega_0 - \beta$ in the harmonic oscillator equation (10.330) would actually more closely describe cavity oscillations, to lowest order in β.

10.13 Appendix: waveguide modes for rectangular and cylindrical waveguides

For propagating modes ($\omega > \omega_\lambda$) the dispersion relation is

$$k_{\lambda_\pm} = \pm\sqrt{\frac{\omega^2 \epsilon\mu}{c^2} - \gamma_\lambda^2} = \pm\frac{\sqrt{\epsilon\mu}}{c}\sqrt{\omega^2 - \omega_\lambda^2},$$

where

$$\omega_\lambda = \frac{c\gamma_\lambda}{\sqrt{\epsilon\mu}}.$$

For non-propagating modes ($\omega < \omega_\lambda$) $k_{\lambda_\pm} = \pm i|k_{\lambda_\pm}|$. Note that for this text the normalizations are chosen so that transverse electric fields are always *real*.

10.13.1 *Rectangular waveguide modes*

Rectangular modes are indexed by two nonnegative integers m, n such that

$$\gamma_{mn} \equiv \pi\sqrt{\frac{m^2}{a^2} + \frac{n^2}{b^2}},$$

plus the above sign index $s = \pm 1$ on k_{mn_s}.

TE rectangular modes ($n = 0$ or $m = 0$ allowed):

With the definition

$$H_0 \equiv -\frac{2i}{\sqrt{ab}}\left(\frac{\gamma_{mn}c}{\omega\mu}\right)\cdot\begin{cases}\frac{1}{\sqrt{2}} & m = 0 \text{ or } n = 0,\\ 1 & m, n \neq 0,\end{cases}$$

we have (suppressing the (m, n) index on the fields for compactness of notation)

$$\begin{aligned}
H_z^{(e)} &= H_0 \cos\left(\frac{m\pi x}{a}\right)\cos\left(\frac{n\pi y}{b}\right)\\
E_x^{(e)} &= -i\frac{\omega\mu}{c\gamma_{mn}^2}\left(\frac{n\pi}{b}\right)H_0\cos\left(\frac{m\pi x}{a}\right)\sin\left(\frac{n\pi y}{b}\right),\\
E_y^{(e)} &= i\frac{\omega\mu}{c\gamma_{mn}^2}\left(\frac{m\pi}{a}\right)H_0\sin\left(\frac{m\pi x}{a}\right)\cos\left(\frac{n\pi y}{b}\right),\\
H_{x\pm}^{(e)} &= -i\frac{k_{mn\pm}}{\gamma_{mn}^2}\left(\frac{m\pi}{a}\right)H_0\sin\left(\frac{m\pi x}{a}\right)\cos\left(\frac{n\pi y}{b}\right),\\
H_{y\pm}^{(e)} &= -i\frac{k_{mn\pm}}{\gamma_{mn}^2}\left(\frac{n\pi}{b}\right)H_0\cos\left(\frac{m\pi x}{a}\right)\sin\left(\frac{n\pi y}{b}\right).
\end{aligned}$$

TM rectangular modes ($n, m > 0$):

With the definition

$$E_0 \equiv -\frac{2i}{\sqrt{ab}}\left(\frac{\gamma_{mn}}{|k_{mn}|}\right)\cdot\begin{cases}1 & \text{if } k_{mn} \text{ is real},\\ -i & \text{if } k_{mn} \text{ is imaginary},\end{cases}$$

we have

$$\begin{aligned}
E_z^{(m)} &= E_0 \sin\left(\frac{m\pi x}{a}\right)\sin\left(\frac{n\pi y}{b}\right),\\
E_{x\pm}^{(m)} &= i\frac{k_{mn\pm}}{\gamma_{mn}^2}\left(\frac{m\pi}{a}\right)E_0\cos\left(\frac{m\pi x}{a}\right)\sin\left(\frac{n\pi y}{b}\right),\\
E_{y\pm}^{(m)} &= i\frac{k_{mn\pm}}{\gamma_{mn}^2}\left(\frac{n\pi}{b}\right)E_0\sin\left(\frac{m\pi x}{a}\right)\cos\left(\frac{n\pi y}{b}\right),\\
H_x^{(m)} &= -i\frac{\omega\epsilon}{c\gamma_{mn}^2}\left(\frac{n\pi}{b}\right)E_0\sin\left(\frac{m\pi x}{a}\right)\cos\left(\frac{n\pi y}{b}\right),\\
H_y^{(m)} &= i\frac{\omega\epsilon}{c\gamma_{mn}^2}\left(\frac{m\pi}{a}\right)E_0\cos\left(\frac{m\pi x}{a}\right)\sin\left(\frac{n\pi y}{b}\right).
\end{aligned}$$

10.13.2 *Cylindrical waveguide modes*

The mode labels for a cylindrical waveguide of radius R are integers m, n ($m = 0, 1, 2, \ldots$; $n = 1, 2, 3, \ldots$) where m refers to the Bessel function of integer order, $J_m(x)$, and n refers to the nth zero of either $J_m(x)$ (TM case) or its first derivative (TE case), excluding nodes at the origin. There is again a sign index $s = \pm 1$ on k_{mn_s}. Note the two-fold degeneracy in ϕ space for nonzero m and the fact that there are some additional TE, TM degeneracies coming from the Bessel function relation $dJ_0(x)/dx = -J_1(x)$.

TE cylindrical modes:

$$H_z = H_0 J_m(\gamma_{mn}\rho) \cdot \begin{cases} \sin(m\phi), \\ \cos(m\phi), \end{cases}$$

$$E_\rho = \frac{i\omega\mu}{c\gamma_{mn}^2}\frac{m}{\rho} H_0 J_m(\gamma_{mn}\rho) \cdot \begin{cases} \cos(m\phi), \\ -\sin(m\phi), \end{cases}$$

$$E_\phi = -\frac{i\omega\mu}{c\gamma_{mn}^2} H_0 \frac{\partial J_m(\gamma_{mn}\rho)}{\partial\rho} \cdot \begin{cases} \sin(m\phi), \\ \cos(m\phi), \end{cases}$$

$$H_{\rho\pm} = \frac{ik_{mn\pm}}{c\gamma_{mn}^2} H_0 \frac{\partial J_m(\gamma_{mn}\rho)}{\partial\rho} \cdot \begin{cases} \sin(m\phi), \\ \cos(m\phi), \end{cases}$$

$$H_{\phi\pm} = -\frac{ik_{mn\pm}}{\gamma_{mn}^2}\frac{m}{\rho} H_0 J_m(\gamma_{mn}\rho) \cdot \begin{cases} \cos(m\phi), \\ -\sin(m\phi). \end{cases}$$

γ_{mn} is determined by the root condition

$$J_m'(y_{mn}) = 0, \; y_{mn} = \gamma_{mn}R,$$

($y_{mn} \neq 0$) where $J_m'(y_{mn})$ is the derivative with respect to its argument of the Bessel function of order m.

TM cylindrical modes:

$$E_z = E_0 J_m(\gamma_{mn}\rho) \cdot \begin{cases} \sin(m\phi), \\ \cos(m\phi), \end{cases}$$

$$E_{\rho\pm} = \frac{ik_{mn\pm}}{\gamma_{mn}^2} E_0 \frac{\partial J_m(\gamma_{mn}\rho)}{\partial\rho} \cdot \begin{cases} \sin(m\phi), \\ \cos(m\phi), \end{cases}$$

$$E_{\phi\pm} = \frac{ik_{mn\pm}}{\gamma_{mn}^2}\frac{m}{\rho}E_0 J_m(\gamma_{mn}\rho)\cdot\begin{cases}\cos(m\phi),\\ -\sin(m\phi),\end{cases}$$

$$H_\rho = -\frac{i\omega\epsilon}{c\gamma_{mn}^2}\frac{m}{\rho}E_0 J_m(\gamma_{mn}\rho)\cdot\begin{cases}\cos(m\phi),\\ -\sin(m\phi),\end{cases}$$

$$H_\phi = \frac{i\omega\epsilon}{c\gamma_{mn}^2}E_0\frac{\partial J_m(\gamma_{mn}\rho)}{\partial\rho}\cdot\begin{cases}\sin(m\phi),\\ \cos(m\phi).\end{cases}$$

γ_{mn} is now determined by

$$J_m(x_{mn}) = 0,\ \ x_{mn} = \gamma_{mn}R,$$

$(x_{mn} \neq 0)$ where again $m = 0, 1, 2, \ldots$.

The roots to $J'_m(y)$ or $J_m(x)$ are given in, for example, M. Abramowitz and I. Stegun, *Handbook of Mathematical Functions*, or can be generated by numerical methods. The normalization constants H_0, E_0 can be determined using the results for integrating products of Bessel functions found in (4.51) and (4.52); note the useful combinations (4.67) and (4.68). These constants must be chosen to be independent of the sign factor in $k_{mn\pm}$, to produce real transverse electric fields and correctly account for the $m = 0$ case.

10.14 Going Deeper

10.14.1 *Electromagnetic waveguides and cavities*

(1) R. E. Collin, *Field Theory of Guided Waves*, 2nd ed., Wiley-IEEE Press (Piscataway, NJ) 1991.

(2) R. F. Harrington, *Time Harmonic Electromagnetic Waves*, 2nd ed., John Wiley-IEEE Press (New York) 2001; Chapters 4-7.

(3) D. A. Hill, *Electromagnetic Fields in Cavities*, John Wiley-IEEE Press (Hoboken, NJ) 2009.

(4) J. D. Jackson, *Classical Electrodynamics*, 2nd ed., John Wiley.

(5) S. F. Mahmoud, *Electromagnetic Waveguides: Theory and Applications*, Peter Peregrinus (London) 1991.

(6) P. M. Morse and H. Feshbach, *Methods of Theoretical Physics, Part II*, McGraw-Hill (New York) 1953; Chapter 13.

(7) J. Schwinger and D. S. Saxon, Discontinuities in Waveguides, Notes on Lectures by Julian Schwinger, Gordon and Breach (New York) 1968.

(8) W. R. Smythe, *Static and Dynamic Electricity*, 3rd ed., Hemisphere Publishing (New York) 1989; Chapter XIII.

(9) J. Van Bladel, *Electromagnetic Fields*, 2nd ed., John Wiley-IEEE Press (Piscataway, NJ) 2007; Chapters 10 and 15.

(10) T. Van Duzer, J. R. Whinnery and S. Ramo, *Fields And Waves in Communication Electronics*, 3rd ed., John Wiley & Sons (New York) 1994.

10.15 Exercises

Exercise 10.1.1. Show that a better representation of the skin depth for normal incidence, $\delta_n \equiv 1/\mathrm{Im}\,[k_1]|_{k_\perp=0}$, is

$$\delta_n = \frac{c}{4\pi\sigma}\sqrt{\frac{2\epsilon_c}{\mu_c}}\left(\sqrt{1+\left(\frac{4\pi\sigma}{\omega\epsilon_c}\right)^2}+1\right)^{1/2},$$

which for a good conductor agrees with (10.24), but for a poor one results in

$$\delta_n = \frac{c}{2\pi\sigma}\sqrt{\frac{\epsilon_c}{\mu_c}},$$

which is frequency independent.

Exercise 10.1.2. A consistent phenomenology can also be worked out in the good magnetic conductor limit, which we take to be $k_1 \approx \omega\sqrt{\mu_c\epsilon_{\text{eff}}}/c$, $|\mu_c/\epsilon_{\text{eff}}| \gg 1$, $\epsilon_c \ll 4\pi\sigma/\omega$. Starting with

$$H_{2||,\perp}(0^+) = 0; \qquad E_{2z}(0^+) = 0; \qquad H_{2z}(0^+), E_{2||,\perp}(0^+) \neq 0$$

just outside the magnetic conductor, derive

$$\vec{H}_1^{\text{tang}}(z) \approx -\sqrt{\frac{\epsilon_{\text{eff}}}{\mu_c}}e^{-ik_1 z}\,\hat{n}\times\vec{E}_2^{\text{tang}}(0^+),$$

$$\vec{K}_{\text{eff}} = i\frac{\sigma}{k_1}\vec{E}_2^{\text{tang}}(0^+),$$

$$\langle S_z\rangle \approx -\frac{c}{8\pi}\mathrm{Re}\left[\sqrt{\frac{\epsilon_{\text{eff}}}{\mu_c}}\right]|\vec{E}_2^{\text{tang}}(0^+)|^2$$

as the analogs of Equations (10.44), (10.55) and (10.61), respectively.

Exercise 10.1.3.

(a) Consider the electric and magnetic fields in an approximately flat conductor. The time-averaged magnetic force per unit area on the surface is

$$\left\langle\vec{f}\right\rangle_m = \frac{1}{2c}\mathrm{Re}\left[\int_{-\infty}^{0} dz\,\vec{J}^*(z)\times\vec{B}_1(z)\right].$$

Using the model presented in Section 10.1, evaluate the normal z-component as

$$\langle f_z\rangle_m = -\frac{\mu_c}{16\pi}\left|\vec{H}_2^{\text{tang}}\right|^2.$$

(b) Now calculate the electric component of the surface force,

$$\left\langle \vec{f} \right\rangle_e = \frac{1}{2c}\mathrm{Re}\left[\int_{-\infty}^{0} dz\, \rho^*(z)\vec{E}_1(z)\right].$$

Use the equation of continuity and the Ohmic current $\vec{J} = \sigma\, \vec{E}^{\mathrm{tang}}$, where σ is the conductivity, to get the expression for the charge density, $\rho^*(z)$. Show that the normal component evaluates to

$$\langle f_z \rangle = \frac{1}{16\pi}\frac{|\, D_{2z}\,|^2}{|\epsilon_{\mathrm{eff}}|},$$

where the effective dielectric constant in region 1 is

$$\epsilon_{\mathrm{eff}} = \frac{4\pi i\sigma}{\omega}.$$

For plane waves, note that D_{2z} may be related to the non-zero quantity $H_{2\perp}(0^+)$ via Equation (10.49).

Exercise 10.2.1. Using Equations (10.44), (10.76) and (10.78), show that the correct conducting boundary conditions on the H_z, E_z fields at finite conductivity are given by ($\hat{n}$ an outward normal):

$$E_z|_S = -f^{\mathrm{TM}}\left(\frac{\partial E_z}{\partial n} + \frac{k_\lambda c}{\omega\epsilon}\frac{\partial H_z}{\partial \ell}\right)\bigg|_S,$$

$$\frac{\partial H_z}{\partial n}\bigg|_S = \left(-f^{\mathrm{TE}}H_z + \frac{k_\lambda c}{\omega\mu}\frac{\partial E_z}{\partial \ell}\right)\bigg|_S,$$

where

$$f^{\mathrm{TE}} \equiv -\frac{(1+i)}{2}\delta\,\gamma_\lambda^2\frac{\mu_c}{\mu}, \quad f^{\mathrm{TM}} \equiv \frac{(1+i)}{2}\delta\,\frac{\omega^2}{\omega_\lambda^2}\frac{\mu_c}{\mu}.$$

The derivative with respect to ℓ is a tangential derivative along x for example in Equation (10.76); it is counterclockwise. Of course, the original TM, TE modes in Section 10.2 are just lowest order solutions of these equations for $f^{\mathrm{TM,TE}} = 0$.

Exercise 10.2.2.

(a) Show that the position-dependent surface charge density, Σ_{eff}, on a perfectly conducting waveguide of arbitrary cross-section, filled with a uniform medium with dielectric constant ϵ and permeability, μ, is given by (inward normal $\hat{n}$, and $\partial\ell$ a counterclockwise tangential derivative)

$$\Sigma_{\text{eff}} = \begin{cases} \dfrac{\mu\epsilon\omega}{4\pi i c\gamma^2} \left.\dfrac{\partial H_z}{\partial \ell}\right|_s, & \text{TE case} \\ \dfrac{ik}{4\pi\gamma^2} \left.\dfrac{\partial D_z}{\partial n}\right|_s, & \text{TM case.} \end{cases}$$

(b) Show that these imply that the current in (10.55) is conserved on the surface.

Exercise 10.2.3.

(a) Consider a waveguide consisting of coaxial metal cylinders with radii a and b ($b > a$). For TE and TM modes, find the eigenvalue condition which determines the cutoff wavenumbers γ_λ in terms of the Bessel functions $J_m(x)$ and $N_m(x)$.
(b) When $(b - a) \ll b$, one may write down explicit solutions in terms of elementary functions instead of Bessel functions. Use these simpler functions to solve for the explicit cutoff frequencies for TE and TM modes. Other than the TEM case $\gamma_0 = 0$, what is the lowest nonzero cutoff wavenumber of these modes?

Exercise 10.2.4. Let us assume the interior of our waveguides actually has a small static conductivity, σ_w, perhaps from air. Show that the wave impedances, Z, for all cases (TEM, TM, TE) acquire a small inductive imaginary part,

$$Z \longrightarrow Z\left(1 - i\frac{2\pi\sigma_w}{\epsilon\,\omega}\right),$$

for $\omega \gg 4\pi\sigma_w/\epsilon$ for TEM or $\omega \gg \gamma c/\sqrt{\epsilon\mu} \gg 4\pi\sigma_w/\epsilon$ for TM and TE, where γ is the cutoff wavenumber defined in (10.71), and ϵ and μ are the real static dielectric and permeability constants, respectively.

Exercise 10.3.1.

(a) (Inspired by J. Schwinger *et al.*, *op. cit.*, Ch. 44.) One can investigate variational principles for waveguides. Given

$$\left(\nabla_\perp^2 + \gamma^2\right)\psi = 0,$$

show that

$$I_\psi \equiv \frac{\int da|\vec{\nabla}_\perp \psi|^2}{\int da|\psi|^2},$$

forms a variational estimate for cutoff parameter γ^2 under the change $\psi \to \psi + \delta\psi$ for either TE ($\psi = H_z$) or TM ($\psi = E_z$) modes, where the integration is over the cross-section of the waveguide. Assume a position-independent phase for the amplitude ψ. Explain the restriction or lack of restriction on the types of variations allowed in the two cases.

(b) TM mode: Find an estimate for the lowest cutoff frequency γ of a circular waveguide of radius a with the ansatz

$$\psi(\rho) = A\cos\left(\frac{\pi\rho}{2a}\right),$$

where ρ is the cylindrical radial coordinate. We find

$$\gamma a \simeq \frac{\pi}{2}\sqrt{\frac{\pi^2+4}{\pi^2-4}} \simeq 2.4146.$$

The exact ground state, which has $E_z = E_0 J_0(\gamma\rho)$ (Section 10.13.2) is $\gamma a \simeq 2.4048$ from the first zero of $J_0(\gamma a)$.

(c) TE mode: Find an estimate for the lowest cutoff frequency γ again for the circular waveguide of radius a. Using the dimensionless variable $t \equiv \rho/a$, show that the explicit variational form is

$$(\gamma a)^2 = \frac{\displaystyle\int_0^1 dt\, t\left[\left(\frac{dP(t)}{dt}\right)^2 + \left(\frac{P(t)}{t}\right)^2\right]}{\displaystyle\int_0^1 dt\, t\,(P(t))^2},$$

for

$$\psi(\rho) = P(t)\sin(\varphi) \text{ or } P(t)\cos(\varphi).$$

Show that the assumed form

$$P(t) = t\left(1 - \frac{t^2}{3}\right),$$

which has the appropriate boundary conditions, when substituted above gives an estimate $\gamma a \approx 1.8423$. The exact ground state, which has $H_z = H_0 J_1(\gamma\rho)\cos\phi$ or $H_z = H_0 J_1(\gamma\rho)\sin\phi$ (Section 10.13.2) is $\gamma a = 1.8412$ from the first zero root of $J_1'(\gamma a) = 0$.

Exercise 10.3.2. Let us start with the variational expression for TE modes in a waveguide,

$$I_H \equiv \frac{\int da \left|\vec{\nabla}_\perp H_z\right|^2}{\int da \left|H_z\right|^2},$$

which equals γ^2 for the exact ground state field. By keeping *second order* terms in δH_z in the variation, show that under the change

$$H_z \to H_z + \delta H_z,$$

I_H gives an *upper limit* to γ^2. Note the field equation for H_z is given by Equation (10.75) and the boundary condition by (10.84). Assume a position-independent-phase for H_z. In addition, for the variational field δH_z assume

$$\hat{n} \cdot \vec{\nabla}_\perp \delta H_z \Big|_S = 0, \quad \delta H_z = \sum_{n=1}^{\infty} \delta a_n \, H_z^{(n)},$$

where δa_n are arbitrary coefficients, and for non-ground states assume

$$\left(\vec{\nabla}_\perp + \gamma_n^2\right) H_z^{(n)} = 0, \, \gamma_n^2 \geq \gamma^2 \text{ for } n \geq 2 \; (\gamma \equiv \gamma_1).$$

Exercise 10.4.1. Verify Equation (10.133) for all λ_s cases.

Exercise 10.4.2. Find an expression for the normalization constants H_0 for TE cylindrical modes and E_0 for TM cylindrical modes in Section 10.13.

Exercise 10.5.1. Consider a thin obstruction in an ideal waveguide of arbitrary cross-section, as described in Section 10.5.

(a) Suppose the obstruction has zero area. Show using (10.140) and (10.141) that $R = 0$ and $A_{\lambda_+} = 0$ for all modes $\lambda \neq 1$. Conclude that no scattering occurs in this situation.
(b) In view of definition (10.96) for wave impedance, show that

$$(\vec{H}^{\infty}_{\perp s} - \vec{H}^{\infty}_{\perp ns}) = \frac{1}{Z_d Z_{1_+}} \hat{e}_3 \times \vec{E}^{\infty}_{\perp ns},$$

for $Z_d = 1/Y_d$, where Y_d is the relative shunt admittance defined in (10.146). The fields $\vec{H}^{\infty}_{\perp s}, \vec{H}^{\infty}_{\perp ns}$ and $\vec{E}^{\infty}_{\perp ns}$ are measured far from the obstruction ($\implies |z| \gg$ cutoff length scales c/ω_m).

Exercise 10.5.2. Show that $X(\vec{V})$ given by (all functions real)

$$X(\vec{V}) \equiv \frac{\sum_{\lambda\neq 1}\left[\int_D da\,\vec{\Phi}_\lambda\cdot\vec{V}\right]^2}{\left[\int_D da\,\vec{\Phi}_1\cdot\vec{V}\right]^2}$$

is minimized by $\vec{V}=\vec{V}_0$ where $\vec{V}_0$ is determined by

$$X(\vec{V}_0)\vec{\Phi}_1\int_D da\,\vec{\Phi}_1\cdot\vec{V}_0=\sum_{\lambda\neq 1}\vec{\Phi}_\lambda\int_D da\,\vec{\Phi}_\lambda\cdot\vec{V}_0.$$

Exercise 10.5.3. For waveguide modes described in Section 10.5, we developed some formulas for relative shunt admittance, Y_d, or impedance, Z_d. We found a unified treatment for both cases. In this situation verify that the integral equations (10.144) and (10.158) can actually both be brought into the form of Equation (10.163) using the substitutions from Table 10.1.

Exercise 10.6.1. Either by using a modified expression for the A_{λ_+} from (10.141) in (10.147) or by integration by parts, derive the generally more convergent form for the waveguide obstacle shown in Figure 10.6:

$$Y_d=\frac{2i\pi}{a|k_1|}\sum_{m=2}\sqrt{m^2-\frac{\omega^2}{\omega_1^2}\frac{1}{m^2}}\left[\frac{\int_{\text{gaps}}dx\,H_z(x)\cos\left(\frac{m\pi x}{a}\right)}{\int_{\text{gaps}}dx\,H_z(x)\cos\left(\frac{\pi x}{a}\right)}\right]^2.$$

Exercise 10.6.2. Derive expression (10.183) for the rectangular waveguide with obstructions shown in Figure 10.7, in which only the lowest TE mode is propagated. Note that the solutions on the source and non-source side should be written as linear combinations of the modes $(\vec{E}^{\text{new}})_{1n}$ defined in (10.180).

Exercise 10.6.3. Derive an expression for Z_d (analogous to (10.177)) for the rectangular waveguide with obstructions shown in Figure 10.7, in which only the lowest TE mode is propagated. As in Exercise 10.6.2, the fields should be written as linear combinations of the modes defined in (10.180).

Exercise 10.6.4. Consider Equation (10.183) of the text for the rectangular waveguide with $a > b$. For this problem we will assume a rectangular *gap* in the waveguide of width d centered along the y-axis. In the limit $k_1 \to 0^+$ (that is, k_1 approaches zero from positive values) one has $Y_d \to Y_d^0$ where

$$Y_d^0 = -\frac{4i|k_1|b}{\pi}\sum_{n=1}^{\infty}\frac{1}{n}\frac{\left[\int_{\text{gap}} dy\, E_y(y)\cos\left(\frac{n\pi y}{b}\right)\right]^2}{\left[\int_{\text{gap}} dy\, E_y(y)\right]^2},$$

which is called the *static admittance*. This expression will give us an (upper limit) estimate to the real quantity iY_d in the static limit (i.e., near cutoff).

(a) Defining $\bar{y} \equiv y - b/2$ and assuming $E_y(y)$ is an even function of position relative to the center of the strip, show that

$$Y_d^0 = \frac{i|k_1|b}{\pi}\frac{\int_{-d/2}^{d/2} d\bar{y}\int_{-d/2}^{d/2} d\bar{y}'\; E_y(\bar{y})E_y(\bar{y}')\ln\left(2\left|\cos\left(\frac{2\pi\bar{y}}{b}\right) - \cos\left(\frac{2\pi\bar{y}'}{b}\right)\right|\right)}{\left[\int_{-d/2}^{d/2} d\bar{y}\, E_y(\bar{y})\right]^2},$$

when the n sum is performed.

(b) Now assuming that $E_y(y)$ =constant in the gap, numerically perform the integrals to show or verify that

1. $iY_d^0 \simeq 0.641\,\frac{|k_1|b}{\pi}$ when $d/b = 0.2$
2. $iY_d^0 \simeq 0.984\,\frac{|k_1|b}{\pi}$ when $d/b = 0.1$

(One can also do the explicit numerical sum to check these results.)

Exercise 10.7.1.

(a) Examine the propagating TE, TM, and TEM cases separately in an arbitrarily-shaped waveguide of length L filled with a uniform medium and calculate the ratio of electric to magnetic field energies, U_E/U_B:

$$\frac{U_E}{L} = \frac{1}{16\pi}\int da\,\vec{E}\cdot\vec{D}^*, \quad \frac{U_B}{L} = \frac{1}{16\pi}\int da\,\vec{B}\cdot\vec{H}^*.$$

(b) Adding up the energies for TE and TM modes from part (a),

$$U = U_E + U_B,$$

derive

$$\frac{U}{L} = \frac{1}{8\pi}\left(\frac{\omega}{\omega_\lambda}\right)^2 \zeta\int da|\psi|^2,$$

where ω_λ is the cutoff angular frequency for the mode and ψ and ζ are defined in (10.189).

(c) From the expression for group velocity,

$$v_g \equiv \left(\frac{\partial k(\omega)}{\partial \omega}\right)^{-1},$$

and Equation (10.198), show that

$$v_g = \frac{P_0 L}{U}.$$

Exercise 10.7.2. Using the power method, derive the expression (10.204) for the estimated attenuation constant $\tilde{\beta}_\lambda$ for nondegenerate TE modes in a waveguide with a uniform cross-section.

Exercise 10.7.3.

(a) Show that the attenuation constants $\tilde{\beta}_\lambda$ from the power method for the lowest TM and TE (assuming non-mixing of the degenerate modes) modes of a cylindrical waveguide of radius R which has been filled with a medium with dielectric constant ϵ and permeability μ are given by:

$$\tilde{\beta}_\lambda = \begin{cases} \left(\dfrac{\mu_c \epsilon\, \delta}{2c^2 R}\right) \dfrac{\omega^2}{k_{01}}, & \text{TM case } (m = 0, n = 1), \\ \left(\dfrac{\mu_c \epsilon\, \delta}{2c^2 R}\right) \dfrac{\omega^2}{k_{11}} \left[\dfrac{1}{y_{11}^2 - 1} + \left(\dfrac{\omega_{11}}{\omega}\right)^2\right], & \text{TE case } (m = 1, n = 1). \end{cases}$$

y_{11} is the first zero of $dJ_1(y)/dy$ and ω_{11} is the associated cutoff frequency.

(b) Find the angular frequency condition for minimum attenuation for these modes. Do these frequencies exceed the threshold frequencies of other modes?

Exercise 10.7.4. Some industrious 2-dimensional Flatlanders wish to construct a waveguide. (Remember the two-dimensional considerations in Chapter 1: only the fields E_x, E_y and B_z exist.) Their waveguide consists of the region between two parallel conducting strips, filled with a medium for which $\epsilon, \mu =$ constant. The distance between the strips is b. Take the direction of energy transmission to be along the x direction.

(a) Assuming perfectly conducting "walls", find the eigenmodes of the system. Write out the full expressions for the E_x, E_y and H_z fields. (The z direction may not exist, but the H_z field does.) Does an analog of a TEM mode exist?

(b) Find an expression for the average rate of energy transmission (per unit length) along x for the modes in (a):

$$\mathcal{P} \equiv \int_0^b dy\, \langle S_x \rangle.$$

(c) Assuming that Equation (10.199) or its equivalent here continues to hold, find an approximate expression for the beta coefficient of the traveling wave modes found in part (a) when the conducting boundaries are imperfect.

Exercise 10.7.5. Using the results from Exercise 10.2.2, show that the attenuation coefficients $\tilde{\beta}$ for TM and TE modes given by (10.202) and (10.204) may be written in unified fashion as (mode labels are left off)

$$\tilde{\beta} = \frac{\pi \mu_c \delta \omega}{2PZ^2\epsilon^2} \oint d\ell \left[|\Sigma_{\text{eff}}|^2 + \frac{\epsilon^2}{c^2} |Z(\vec{K}_{\text{eff}})_\ell|^2 \right],$$

where Σ_{eff} is the effective surface charge density, $(\vec{K}_{\text{eff}})_\ell$ is the effective tangential surface current along ℓ (corresponding to the x direction in Figure 10.4), Z is the wave impedance factor in Equation (10.97), and P is given by Equation (10.198).

Exercise 10.7.6. Consider the transmission of power in an arbitrary waveguide using a *mixture* of traveling TE, TM modes. (They may or may not be degenerate, that is, have the same cutoff frequency, γ.)

(a) Show the expression for the net power transmitted along the waveguide (in the z-direction) is given by

$$P = \frac{\omega}{8\pi} \int da \left\{ \frac{k_1 \epsilon}{\gamma_1^4} |\vec{\nabla}_\perp E_{1z}|^2 + \frac{k_2 \mu}{\gamma_2^4} |\vec{\nabla}_\perp H_{2z}|^2 \right\},$$

where the "1" subscripts specify TM mode values and "2" designates TE values.

(b) By examining the tangential part of the harmonic Poynting vector, designated as

$$\frac{c}{8\pi}\mathrm{Re}\left[(\vec{E}\times\vec{H}^*)_\perp\right],$$

show that a tangential flow of energy in the waveguide is induced. [*Hint*: There should be two such terms which emerge, involving both the H_{2z} and E_{1z} fields.]

Exercise 10.7.7.

(a) Derive a formula for the power loss, β^{TEM}, for waveguide TEM harmonic modes in a waveguide with inner and outer boundaries and arbitrary cross-section. The inner surface is held at potential V and the outer surface is grounded. The waveguide is filled with a nonconductive material with uniform dielectric and permeability constants ϵ and μ. Given that $\vec{E}_\perp = -\vec{\nabla}_\perp\Phi$ in the waveguide, show that:

$$\beta^{\mathrm{TEM}} = \frac{c}{8\pi\delta\sigma}\sqrt{\frac{\epsilon}{\mu}}s(k)\frac{\oint_{all} d\ell\left(\frac{\partial\Phi}{\partial n}\right)^2}{V\oint_{inner} d\ell\left(\frac{\partial\Phi}{\partial n}\right)}.$$

σ is waveguide conductivity, δ is skin depth, $s(k)$ is the sign of k, and n is an outer normal coordinate relative to the waveguide interior.

(b) Evaluate β^{TEM} for a waveguide with circular cross-sections with concentric inner and outer conducting radii a and b, respectively, as shown in Figure 10.12.

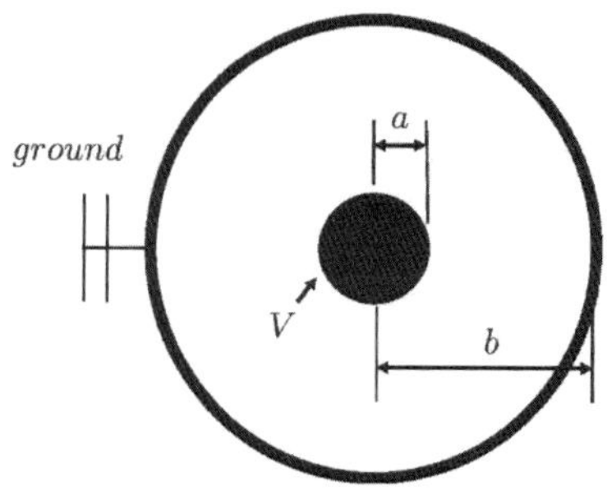

Fig. 10.12 Waveguide cross-section for Exercise 10.7.7.

Exercise 10.8.1. Consider the TM modes in a rectangular waveguide with long side $a > b$. Define the cutoff frequencies by

$$\omega_\lambda = \frac{c}{\sqrt{\mu\epsilon}}\gamma_\lambda, \quad \gamma_\lambda^2 = (\gamma_x)_\lambda^2 + (\gamma_y)_\lambda^2.$$

(a) Find the *exact* transcendental equations which determine the allowed $(\gamma_x)_\lambda$, $(\gamma_y)_\lambda$ values when the surface boundary condition is

$$\left(E_z + f^{\text{TM}}\frac{\partial E_z}{\partial n}\right)\bigg|_S = 0.$$

$\hat{n}$ is an *outward* normal and f^{TM} is some (possibly complex) number.

(b) Using the eigenvalue equations determined in (a), and assuming

$$f^{\text{TM}} = \frac{(1+i)}{2}\delta\left(\frac{\omega}{\omega_\lambda}\right)^2\frac{\mu_c}{\mu},$$

$$|\gamma_x f^{\text{TM}}| \ll 1, \quad |\gamma_y f^{\text{TM}}| \ll 1,$$

$$k'_\lambda = k_\lambda + \alpha_\lambda + i\beta_\lambda,$$

where k_λ is the k-value corresponding to $f^{\text{TM}} = 0$, find α_λ and β_λ for the lowest TM mode, to first non-vanishing order in f^{TM}. Assume $|\alpha_\lambda|$, $|\beta_\lambda| \ll |k_\lambda|$, so that you are not near cutoff.

(c) In (b), you should have found that the cutoff frequency is lowered by this perturbation ($\alpha_\lambda > 0$), which represents the effect of a good (but not perfect) conducting wall. Give a physicist's reason why this happens in general.

Exercise 10.8.2.

(a) The correct waveguide boundary conditions for finite conductivity are given in Exercise 10.2.1. Show that the lowest-order implementations of these boundary conditions for finite conductivity are ($\hat{n}$ outward normal)

$$E'_z|_S = -f^{\text{TM}}\left(\frac{\partial E_z}{\partial n}\right)\bigg|_S,$$

in the TM case, and

$$\frac{\partial H'_z}{\partial n}\bigg|_S = -f^{\text{TE}}\left(H_z - \left(\frac{k_\lambda}{\gamma_\lambda^2}\right)^2\frac{\partial^2 H_z}{\partial \ell^2}\right)\bigg|_S,$$

in the TE case. The perfect conductor fields are unprimed.

(b) Using part (a) and given that TE modes with $E_z = 0$ are solutions for $f^{\text{TM,TE}} = 0$, show that the perturbation of boundary conditions method described in Section 10.8 applied to nondegenerate TE modes in a waveguide gives

$$k_\lambda'^{\,2} - k_\lambda^2 = 2k_\lambda(1+i)\tilde{\beta}_\lambda,$$

where $\tilde{\beta}_\lambda$ is the power method value for the attenuation constant for nondegenerate TE modes given in (10.204) and computed in Exercise 10.7.2. Note this is equivalent to the corresponding result (10.211) for TM modes.

Exercise 10.8.3.

(a) Consider the case of propagating, degenerate TE waveguide modes ($E_z = 0$), subject to finite conductivity boundary conditions. Given the TE boundary condition (n represents an outward normal, $\partial\ell$ a counterclockwise derivative along the surface; $\psi = H_z$; see Exercise 10.8.2),

$$\left.\frac{\partial\psi}{\partial n}\right|_S = -f^{\text{TE}}\left(\psi - \frac{k_\lambda^2}{\gamma_\lambda^4}\frac{\partial^2\psi}{\partial\ell^2}\right)\bigg|_S,$$

($k_\lambda^2 = \mu\epsilon\omega^2/c^2 - \gamma_\lambda^2$) show that, assuming n-fold degeneracy and that the degenerate TE modes are orthogonal,

$$\sum_{i=1}^{n}\left[(\gamma_\lambda'^{\,2} - \gamma_\lambda^2)N_i\delta_{ji} + \Delta_{ji}\right]a_i = 0,$$

$$N_i = \int da\,|\phi^{(i)}|^2,\ \psi = \sum_{i=1}^{n} a_i\phi^{(i)},$$

$$\Delta_{ji} = -\oint d\ell\, f^{\text{TE}}\left[\phi^{(j)*}\phi^{(i)} + \frac{k_\lambda^2}{\gamma_\lambda^4}\frac{\partial\phi^{(j)*}}{\partial\ell}\frac{\partial\phi^{(i)}}{\partial\ell}\right],$$

where $\phi^{(i)}$ are the original degenerate eigenmodes:

$$(\nabla_\perp^2 + \gamma_\lambda^2)\phi^{(i)} = 0,\quad \left.\frac{\partial\phi^{(i)}}{\partial n}\right|_S = 0.$$

As in the treatment in Section 10.8 for TM modes, f^{TE} can be considered to be a surface position dependent function. (Challenge to the reader: Show that there are actually contributions from both $\partial f^{\text{TE}}/\partial\ell|_S$ and $\partial f^{\text{TM}}/\partial\ell|_S$ terms in Δ_{ij}, which cancel if the common variation arises for example from the conductivity, σ.)

(b) The two lowest TE modes of a cylindrical waveguide of radius R are specified by $m = n = 1$ for λ:

$$\psi^{(1,2)} = \frac{\sqrt{2}\, y_{11} J_1(\gamma_{11}\rho)}{\sqrt{\pi} R \sqrt{y_{11}^2 - 1}\, J_1(y_{11})} \cdot \begin{cases} \sin(m\phi), \\ \cos(m\phi). \end{cases}$$

The reader can check that this gives $N_i = 1$. Taking

$$f^{\mathrm{TE}}(\phi) = f_1 + f_2 \sin(2\phi),$$

($f_{1,2}$ constants) use the theory in (a) to find the splitting of the lowest modes (cf. Exercise 10.7.3(a) result):

$$\gamma_\lambda'^2 - \gamma_\lambda^2 = -\Delta_{11} \pm \Delta_{12},$$

$$\Delta_{11} = \Delta_{22} = -\frac{2 f_1 \mu\epsilon R\, \omega^2}{y_{11}^2 c^2} \left[\frac{1}{(y_{11}^2 - 1)} + \left(\frac{\omega_{11}}{\omega} \right)^2 \right],$$

$$\Delta_{12} = \Delta_{21} = \frac{f_2 \mu\epsilon R \omega^2}{y_{11}^2 c^2} \left[\frac{1}{(y_{11}^2 - 1)} - \left(\frac{y_{11}^2 + 1}{y_{11}^2 - 1} \right) \left(\frac{\omega_{11}}{\omega} \right)^2 \right].$$

Exercise 10.8.4.

(a) We have not yet treated the case of degenerate TE, TM waveguide modes, subject to finite conductivity boundary conditions, where both E_z and H_z are non-zero. A treatment follows from the discussion and boundary value results in Exercise 10.2.1. We are solving

$$(\nabla_\perp^2 + \gamma_\lambda'^2)\psi = 0, \text{ given that } (\nabla_\perp^2 + \gamma_\lambda^2)\phi_i = 0,$$

where $\psi = E_z, H_z$ and ϕ_1 and ϕ_2 are the normalized, unmixed solutions for TM and TE, respectively. Show the solution of this 2×2 system may be characterized as

$$\sum_{i=1}^{2} \left[(\gamma_\lambda'^2 - \gamma_\lambda^2)\delta_{ji} + \Delta_{ji} \right] a_i = 0,$$

as in Exercise 10.8.3 ($N_i = 1$), but with $E_z = a_1\phi_1$, $H_z = a_2\phi_2$. For the $\gamma_\lambda'^2$ elements, show that one has new eigenvalues,

$$\gamma_\lambda'^2 - \gamma_\lambda^2 = -\frac{1}{2}(\Delta_{11} + \Delta_{22}) \pm \frac{1}{2}\sqrt{(\Delta_{11} - \Delta_{22})^2 + 4\Delta_{12}\Delta_{21}},$$

where ($\hat{n}$ outward)

$$\Delta_{11} = 2k_\lambda(1+i)\tilde{\beta}^{\mathrm{TM}},$$

$$\Delta_{22} = 2k_\lambda(1+i)\tilde{\beta}^{\mathrm{TE}},$$

$$\Delta_{12} = f^{\mathrm{TM}} \left(\frac{k_\lambda c}{\omega\epsilon} \right) \oint d\ell \frac{\partial \phi_1}{\partial n} \frac{\partial \phi_2}{\partial \ell},$$

$$\Delta_{21} = \frac{\epsilon}{\mu}\Delta_{12}.$$

(b) For the rectangular waveguide with arbitrary sides a and b along x and y, respectively, these modes are twofold degenerate (in general), and the z-fields are given by Equations (10.105) and (10.106) of the text ($m' = m$ and $n' = n$; $m, n \neq 0$). The normalized, unmixed TM, TE eigenmodes can be taken as

$$\phi_1 = \frac{2}{\sqrt{ab}} \sin\left(\frac{m\pi x}{a}\right) \sin\left(\frac{n\pi y}{b}\right), \quad \text{TM}$$
$$\phi_2 = \frac{2}{\sqrt{ab}} \cos\left(\frac{m\pi x}{a}\right) \cos\left(\frac{n\pi y}{b}\right), \quad \text{TE.}$$

Obtain an explicit form for Δ_{12} for the rectangular waveguide and show that it vanishes when $a = b$, i.e., nondegenerate perturbation theory gives the correct result in a *square* waveguide.

Exercise 10.8.5.

(a) Given

$$k_\lambda'^{\,2} - k_\lambda^2 = 2k_\lambda(1+i)\tilde{\beta}_\lambda,$$

solve for the imaginary part of k_λ' ($\tilde{\beta}_\lambda^{\rm new}$) assuming $\omega > \omega_\lambda$ and k_λ and $\tilde{\beta}_\lambda$ are positive (*Ans.*: Equation (10.222)). Analytically continue the expression for the $\omega < \omega_\lambda$ case using $k_\lambda \to i\kappa_\lambda$, where

$$\kappa_\lambda \equiv \frac{\sqrt{\mu\epsilon}}{c}\sqrt{\omega_\lambda^2 - \omega^2}.$$

(b) Show that the $\tilde{\beta}_\lambda$ coefficient may be written

$$\tilde{\beta}_\lambda = N^{\rm TM} \frac{\mu\epsilon}{c^2} \frac{(\omega_\lambda \omega^3)^{1/2}}{k_\lambda},$$

in the TM case (Equation (10.202)) and

$$\tilde{\beta}_\lambda = N_1^{\rm TE} \frac{\mu\epsilon}{c^2} \frac{(\omega_\lambda^5 \, \omega^{-1})^{1/2}}{k_\lambda} + N_2^{\rm TE} k_\lambda \left(\frac{\omega_\lambda}{\omega}\right)^{1/2},$$

in the TE case (Equation (10.204)), where

$$N^{\rm TM} = \frac{c^3}{4\epsilon\mu^2}\sqrt{\frac{\mu_c}{2\pi\sigma\omega_\lambda^5}}\frac{\oint d\ell \left|\frac{\partial E_z}{\partial n}\right|^2}{\int da |E_z|^2},$$

$$N_1^{\rm TE} = \frac{c}{4\mu}\sqrt{\frac{\mu_c}{2\pi\sigma\omega_\lambda}}\frac{\oint d\ell \, |H_z|^2}{\int da |H_z|^2},$$

$$N_2^{\rm TE} = \frac{c^3}{4\epsilon\mu^2}\sqrt{\frac{\mu_c}{2\pi\sigma\omega_\lambda^5}}\frac{\oint d\ell \left|\frac{\partial H_z}{\partial \ell}\right|^2}{\int da |H_z|^2},$$

are dimensionless constants.

(c) For plotting purposes fix a length scale such that $\sqrt{\mu\epsilon}\,\omega_\lambda/c = 1$ and assume $N^{\rm TM} = 0.1$ for mode λ. Defining $x \equiv \sqrt{\mu\epsilon}\,\omega/c$, plot $\tilde{\beta}_\lambda$ for $1 < x < 10$ and $\tilde{\beta}_\lambda^{\rm new}$ for $0 < x < 10$. The result should be Figure 10.9!

Exercise 10.8.6.

(a) Derive the expression for the distortion of shape eigenmode equations for degenerate TE or TM modes for the eigenvector expansion coefficients, a_i. That is, show that for N degenerate eigenmodes we have

$$\sum_{i=1}^{N} \left[\left(\gamma^2 - \gamma_0^2\right) N_i \delta_{ji} + \Delta_{ji}\right] a_i = 0, \quad (j = 1, 2, .., N)$$

$$N_i = \int da\, |\psi_0^{(i)}|^2,$$

$$\Delta_{ji} = \oint_C dl\, \delta(x,y) \left[\frac{\partial \psi_0^{(j)*}}{\partial n}\frac{\partial \psi_0^{(i)}}{\partial n} - \psi_0^{(j)*}\frac{\partial^2 \psi_0^{(i)}}{\partial n^2}\right],$$

where $\delta(x,y)$ is the distortion function. The solution is then

$$\psi = \sum_{i=1}^{N} a_i \psi_0^{(i)}.$$

[*Hints*: Begin with Green's second identity, Equation (2.95), then do a Taylor expansion in the new TM or TE fields, producing the distortion function, $\delta(x,y)$. γ_0 is the degenerate cutoff frequency, which appears in

$$\left(\nabla_\perp^2 + \gamma_0^2\right)\psi_0^{(i)} = 0 \;\; (i = 1, 2, .., N),$$

and γ represent the new cutoff values. See Jackson, *op. cit.*, problems 8.12 and 8.13 for additional details.]

(b) Use this formalism to calculate the splitting of the two ($N = 2$) $m = 1$ TM modes of a waveguide with circular cross-section of radius R when the distortion function is given by

$$\delta(x,y) = \Delta R \cos(2\phi),$$

where ϕ is the azimuthal angle. Take (see Section 10.13.2)

$$\psi_0^{(1)} = AJ_1(\gamma_0\rho)\sin\phi, \quad \psi_0^{(2)} = AJ_1(\gamma_0\rho)\cos\phi,$$

where $\gamma_0 = 3.83171/R$ from the first root of $J_1(x) = 0$. Show the new cutoffs $\gamma^{(1)}$ and $\gamma^{(2)}$ satisfy

$$\left(\gamma^{(1)}\right)^2 - \gamma_0^2 = Q/R^2, \; \left(\gamma^{(2)}\right)^2 - \gamma_0^2 = -Q/R^2.$$

Find the form of Q (numerical value not necessary) and the form of the corresponding eigenvectors. (Note that this state is actually *3-fold* degenerate because there is a $m = 0$ TE state with the same γ_0 cutoff value. We will ignore this complication for this problem!)

Exercise 10.9.1. Consider the expressions I_E and II_E defined in an arbitrary resonator geometry of volume V:

$$I_E \equiv \frac{\int d^3x \left(\vec{\nabla} \times \vec{E}^*\right) \cdot \left(\vec{\nabla} \times \vec{E}\right)}{\int d^3x \, \vec{E}^* \cdot \vec{E}}, \quad II_E \equiv \frac{\int d^3x \, \vec{E}^* \cdot \left[\vec{\nabla} \times \left(\vec{\nabla} \times \vec{E}\right)\right]}{\int d^3x \, \vec{E}^* \cdot \vec{E}}.$$

These forms are useful for variational estimates of resonator squared wavenumbers, k^2, when exact fields are impossible to determine. Assume the ground state electric field amplitude $\vec{E}$ has a position-independent phase. Show that $I_E = II_E$ for any field $\vec{E}$ such that $\hat{n} \times \vec{E}\Big|_S = 0$ on the surface S of the resonator. Thus, the two forms are completely equivalent for variational purposes with conducting boundary conditions.

Exercise 10.9.2.

(a) Consider an arbitrary resonator geometry of volume V. Form the variational quantity

$$I_E \equiv \frac{\int d^3x |\vec{\nabla} \times \vec{E}\,|^2}{\int d^3x |\vec{E}\,|^2},$$

for the ground state. Assume the ground state has a single position-independent phase. Derive the result

$$\delta^{(2)} I_E \equiv \frac{1}{\int d^3x |\vec{E}\,|^2} \left(-k^2 \int d^3x |\delta\vec{E}\,|^2 + \int d^3x |\vec{\nabla} \times \delta\vec{E}\,|^2\right),$$

representing *second order* terms generated in II under $\vec{E} \to \vec{E} + \delta\vec{E}$. Note that the field equations are

$$\vec{\nabla} \cdot \vec{E} = 0, \quad \left(\nabla^2 + k^2\right) \vec{E} = 0,$$

and the boundary conditions are

$$\vec{E} \times \hat{n}|_s = 0, \quad \delta\vec{E} \times \hat{n}|_s = 0.$$

You may assume the first variation vanishes for these boundary conditions. The derivation is very similar to the one in Exercise 10.9.3.

(b) Use the (a) result to show for a complete orthogonal set of states $\vec{E}_n$ such that

$$\vec{\nabla} \cdot \vec{E}_n = 0, \quad \left(\nabla^2 + k_n^2\right) \vec{E}_n = 0,$$

for $n = 1, 2, 3 \ldots$ with $k_n > k$ when $n \geq 2$, and

$$\vec{E}_n \times \hat{n}|_s = 0,$$

$$\int d^3x \, \vec{E}_n \cdot \vec{E}_{n'} = 0, \;\; n \neq n',$$

that the general variation

$$\delta \vec{E} = \sum_{n=1}^{\infty} \delta a_n \vec{E}_n$$

(δa_n are arbitrary parameters) results in $\delta^{(2)} I_E > 0$, implying that a variational estimate using I_E provides an *upper limit* to the true ground state k value.

Exercise 10.9.3. Consider an arbitrary resonator geometry of volume V containing a linear isotropic medium (ϵ, μ). Assume the ground state has only a position-independent phase. The quantity ($\vec{\nabla} \cdot \vec{H} = 0$, $\vec{\nabla} \times \vec{H} = -ik\,\vec{D}$ and $\left(\nabla^2 + k^2\right) \vec{H} = 0$ are given)

$$I_H \equiv \frac{\int d^3x \, |\vec{\nabla} \times \vec{H}|^2}{\int d^3x \, |\vec{H}|^2},$$

for the exact lowest modes gives the associated k^2 value. Now show under a first order variation of the fields, $\vec{H} \to \vec{H} + \delta \vec{H}$, that

$$\delta I_H = \frac{-2ik}{\int d^3x \, |\vec{H}|^2} \oint da \left[(\delta \vec{H}^* - \delta F \, \vec{H}^*) \cdot (\vec{D} \times \hat{n}) \right],$$

where

$$\delta F \equiv \frac{\int d^3x \, \vec{H} \cdot \delta \vec{H}^*}{\int d^3x \, |\vec{H}|^2} \text{ and } \vec{\nabla} \times \vec{H}\,|_S = -ik\,\vec{D}\,|_S.$$

This shows that the variation vanishes if the associated $\vec{D}$ field satisfies $\vec{D} \times \hat{n}\,|_S = 0$.

Exercise 10.9.4. Using the variational form from Exercise 10.9.3, find an estimate for kR for the TM $p = 0$, $m = 0$ mode (see Equation (10.342) and Section 10.13.2) of a circular cylindrical cavity (vacuum) of radius R and length d using the trial $\vec{D}$ field assumption

$$\vec{D}(\rho) = D_0 \hat{z} \cos\left(\frac{\pi\rho}{2R}\right),$$

which has the correct surface boundary condition. The $\vec{H}$ field from $\vec{\nabla} \times \vec{H} = -ik\vec{D}$ is (verify!)

$$\vec{H}(\rho) = -i\frac{kD_0}{\rho}\hat{e}_\phi \int_0^\rho d\rho'\,\rho' \cos\left(\frac{\pi\rho'}{2R}\right).$$

Show that

$$kR \simeq \frac{\pi}{2}\sqrt{\frac{\pi^2 - 4}{2.5009}} \simeq 2.4064.$$

The exact field has (Section 10.13.2)

$$E_z = E_0 J_0(kR).$$

The exact answer to this accuracy is $kR = 2.4048$ (the lowest zero root of $J_0(kR)$).

Exercise 10.10.1.

(a) Using the Taylor series for $\sin x$ and $\cos x$, show the small argument limit of the spherical Bessel functions, Equations (10.259) and (10.260).
(b) Similarly, argue that the large argument limit ($x \gg l$) of the spherical Bessel functions is given by the middle part of Equations (10.261) and (10.262), which, when evaluated, leads to the right hand sides of these same two equations.

Exercise 10.10.2. Starting with the integral expression (10.280) for spherical Bessel functions, deduce the spherical Bessel function recurrence relation

$$\frac{dj_\ell(x)}{dx} = j_{\ell-1}(x) - \frac{(\ell+1)}{x}j_\ell(x).$$

You may use the Legendre identity (4.183),

$$zP_\ell(z) = P_{\ell-1}(z) + \frac{(z^2-1)}{\ell}\frac{dP_\ell(z)}{dz},$$

or any other convenient method.

Exercise 10.11.1.

(a) Using the definitions

$$\hat{\epsilon}_\pm \equiv \frac{\hat{e}_x \pm i\hat{e}_y}{\sqrt{2}}; \quad L_\pm = L_x \pm iL_y,$$

where $\vec{L}$ is defined by (10.242), show that

$$\vec{X}_{lm} = \frac{1}{\sqrt{l(l+1)}}\left[\frac{\hat{\epsilon}_- L_+}{\sqrt{2}} + \frac{\hat{\epsilon}_+ L_-}{\sqrt{2}} + \hat{e}_z L_z\right] Y_{lm}.$$

(b) It can be shown[3] that

$$L_\pm Y_{lm} = \sqrt{(l \mp m)(l \pm m + 1)}Y_{l,m\pm 1}, \quad L_z Y_{lm} = mY_{lm}.$$

Use these facts and part (a) to show the orthogonality relation (10.304),

$$\int d\Omega\, \vec{X}^*_{l'm'}(\theta,\phi)\cdot\vec{X}_{lm}(\theta,\phi) = \delta_{ll'}\delta_{mm'}.$$

Exercise 10.11.2.

(a) Show that (sum starts at $l = 1$)

$$\sum_{l,m} \vec{X}^*_{lm}(\theta',\phi')\cdot\vec{X}_{lm}(\theta,\phi) = \delta(\cos\theta - \cos\theta')\delta(\phi - \phi'),$$

by using standard properties of the spherical harmonics, Y_{lm}, where $\vec{X}_{lm}(\theta,\phi)$ is defined by (10.296).

(b) Show that (for each l value)

$$\sum_m \left(\hat{r}' \times \vec{X}^*_{lm}(\theta',\phi')\right)\cdot\vec{X}_{lm}(\theta,\phi) = 0.$$

(c) Using parts (a) and (b) above or other means, prove the completeness relation (10.309),

$$\sum_{l,m}\left[\vec{X}^*_{lm}(\theta',\phi')\vec{X}_{lm}(\theta,\phi) + (\hat{r}'\times\vec{X}'^*_{lm})(\hat{r}\times\vec{X}_{lm})\right]$$
$$= (\hat{e}_\theta\hat{e}_\theta + \hat{e}_\phi\hat{e}_\phi)\delta(\cos\theta - \cos\theta')\delta(\phi-\phi').$$

---

[3]G. Arfken, H. Weber, and F. Harris, *Mathematical Methods for Physicists*, 7th Academic Press (Amsterdam) 2013, Section 16.1.

Exercise 10.11.3. Reduce as indicated:

(a) $\vec{\nabla} Y_{lm} = A(r)\hat{r}\, Y_{lm} + B(r)\hat{r} \times \vec{X}_{lm} + C(r)\vec{X}_{lm}$
(b) $\vec{\nabla} \times \vec{X}_{lm} = D(r)\hat{r}\, Y_{lm} + E(r)\hat{r} \times \vec{X}_{lm} + F(r)\vec{X}_{lm}$
(c) $\vec{\nabla} \times \left(\hat{r} \times \vec{X}_{\ell m}\right) = G(r) Y_{\ell m}\hat{r} + H(r)\hat{r} \times \vec{X}_{\ell m} + I(r)\vec{X}_{\ell m}$.

The $\vec{X}_{lm}$ are vector spherical harmonics, defined in Equation (10.296). Find the unknown functions, $A(r) - I(r)$ above.

Exercise 10.11.4. The so-called MIT bag model of hadrons assumes that such particles can be represented as combinations of quark or gluon eigenmodes confined in a spherical resonant cavity of radius a with perfect *magnetic* conductor boundary conditions on the surface, S:

$$\hat{r} \times \vec{B}\Big|_S = 0,$$
$$\hat{r} \cdot \vec{E}\Big|_S = 0.$$

This model allows a very rough description of so-called *glueball* states as the TE or TM eigenmodes of a spherical cavity populated by two or three *gluons*. Gluons are massless, and the energy of a state with TE or TM eigenvalue $x = ka$ is given by $E = \hbar kc$ where $\hbar$ is the reduced Planck's constant.

(a) Find the eigenvalue conditions which determine the normal modes for TM and TE excitations. Give at least the first three ka values associated with $l = 1$ and 2.
(b) Determine the parity of the $\vec{A}$, $\vec{E}$ and $\vec{B}$ fields for both TM and TE modes. (See Exercise 4.9.2 and Section 8.8.)
(c) The best way to determine the energies of these states presently is a set of computational methods called "Lattice QCD". These types of calculations introduce a space-time set of lattice points to discretize the quark and gluon degrees of freedom of the field theory known as Quantum Chromodynamics. Look up the literature on such lattice QCD glueball calculations and make a report to the class on what you find. In particular, what are some of the computational issues involved in these calculations?

Exercise 10.11.5.

(a) Using results from Exercise 10.11.1, show the summation identity involving the vector spherical harmonics:

$$\sum_{m=-l}^{l} \vec{X}^{*}_{lm}(\theta,\phi) \cdot \vec{X}_{lm}(\theta,\phi) = \frac{2l+1}{4\pi}.$$

(b) Show that

$$\vec{X}_{lm}(\theta',\phi') = (-1)^l \vec{X}_{lm}(\theta,\phi),$$

for the coordinate inversion:

$$\theta' = \pi - \theta,\ \phi' = \phi + \pi \text{ if } \phi < \pi,$$
$$\theta' = \pi - \theta,\ \phi' = \phi - \pi \text{ if } \phi \geq \pi.$$

Exercise 10.11.6. Obtain the form (up to normalization constants) of the electric and magnetic fields for all modes of a resonator consisting of the space between concentric spheres of radii a and b ($a > b$). Obtain as well the eigenfunction conditions which determine the ka values for the various modes. (Note there is a $k = 0$ electric mode also.)

Exercise 10.12.1. Those industrious Flatlanders now wish to construct a rectangular resonant cavity in their two-dimensional space. It has sizes a along the x-axis and b along the y-axis.

(a) Find the resonant angular frequencies when the "cavity" walls are perfectly conducting.
(b) Find the Q_λ of the Flatlanders' cavity modes by using an appropriately modified Equation (10.341).

Exercise 10.12.2. Schumann resonances are low frequency electromagnetic energy peaks in the Earth-ionosphere cavity spectrum excited by lightning discharges; the spherical region of height $h = b - a$ between the Earth's surface and ionosphere forms the effective resonant cavity. For perfect conductor boundary conditions the predicted low frequency TM resonances are approximately given by the Schumann formula

$$\omega_l^S \equiv \sqrt{l(l+1)}\,\frac{c}{a},$$

which gives $\omega_l^S/2\pi = 10.6$, 18.4, 26.0, 33.5, and 41.1 Hz ($l = 1, 2, 3, 4$ and 5) for an approximate Earth radius $a = 6370$ km.

(a) Improve on the prediction above by incorporating a finite conductivity condition on the ionosphere. The model geometry is depicted in Figure 10.13 where the Earth's surface ($r = a$) is treated as a perfect conductor (primarily due to seawater) whereas the ionosphere ($r = b$) is considered an imperfect conductor. Using the TM modes (10.317) and (10.318) show in the limit $h \ll a$ that an approximate solution to (10.245) gives the new resonance formula

$$\omega_l \approx \omega_l^S \left(1 - \frac{\delta_l}{4h}\right),$$

when considering the real part of the new low eigenfrequencies. Also, from the imaginary part one finds

$$Q_l \approx \frac{2h}{\delta_l},$$

where Q_l is the quality factor Equation (10.358) and δ_l is the skin depth Equation (10.24) evaluated for the lth mode. (Higher angular frequency modes $\sim c/h$ also exist for both TM and TE cases. Note similarity with Exercise 10.2.3(b) above.)

(b) Prepare a table of modified angular resonance values by (unjustifiably) fitting $\omega_1/2\pi$ to the first *measured* resonance value, 7.8 instead of 10.6 Hz, and then computing the $\omega_l/2\pi$ and associated Q_l values

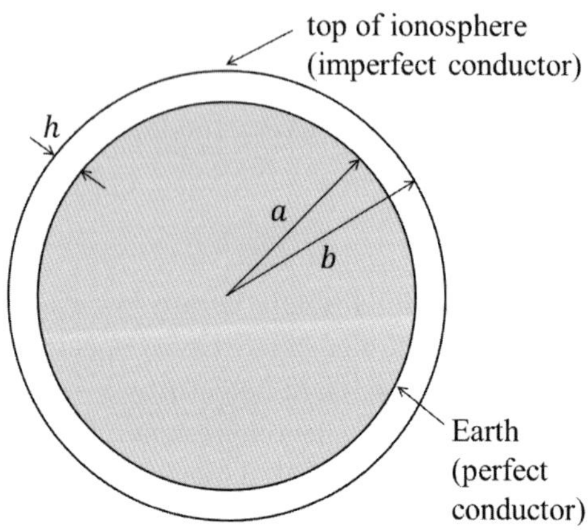

Fig. 10.13 Earth ionosphere model for Exercise 10.12.2.

for $l = 2$, 3, 4, and 5. Compare your results with experimental values in the literature. How good is this model? What improvements can be made? (A helpful reference on Schumann resonances with plentiful references is: A. Nickolaenko and M. Hayakawa, *Schumann Resonances for Tyros: Essentials of Global Electromagnetic Resonance in the Earth-Ionosphere Cavity*, Springer (Tokyo) 2014.)

Exercise 10.12.3. Using Equation (10.341), derive Equation (10.356) ($k^2 = \omega^2\mu\epsilon/c^2$):

$$Q^{\mathrm{TE}}_{\lambda,p\neq 0} = \frac{\mu d}{2\mu_c\delta}k^2\left(\frac{d}{p\pi}\right)^2$$
$$\times \frac{\int da\,|\psi_\lambda|^2}{\int da\,|\psi_\lambda|^2 + \frac{d}{4\gamma^2}\oint d\ell\left|\frac{\partial\psi_\lambda}{\partial\ell}\right|^2 + \frac{d\gamma^2}{4}\left(\frac{d}{p\pi}\right)^2\oint d\ell\,|\psi_\lambda|^2},$$

applicable to nondegenerate TE modes in a cylindrical cavity of arbitrary cross-section and length d. The quantity $\partial\ell$ denotes a tangential surface derivative, perpendicular to z.

Exercise 10.12.4. Using (10.341), show for TE modes in a spherical conducting cavity of radius a (vacuum interior) made from an imperfect conductor, that

$$Q^{\mathrm{TE}}_\lambda = \frac{a}{\mu_c\delta},$$

for all TE modes, independent of λ. (The necessary radial integral can be done using techniques from Chapter 4.)

Exercise 10.12.5. Consider again the spherical resonant cavity (vacuum) of radius a made from an imperfect conductor.

(a) Given the boundary condition (10.205), show that for TE modes the exact eigenmode condition is given by

$$j_l(x) = \frac{iC}{1 + \dfrac{iC(l+1)}{x}}j_{l+1}(x),$$

where $x = ka$, $j_l(x)$ is the spherical Bessel function, and

$$C = \frac{1-i}{2}\mu_c\delta k.$$

(b) Let λ denote the mode associated with the nth root of (a) and define the quality factor Q_λ as in (10.358). Using a Taylor series expansion and recursion relations, show that for $|C|(l+1) \ll x$ one recovers the Exercise 10.12.4 result:

$$Q_\lambda^{\mathrm{TE}} = \frac{a}{\mu_c \delta}.$$

Exercise 10.12.6.

(a) Consider (once again!) a spherical resonant cavity (vacuum) of radius a made from an imperfect conductor. Using (10.205) on the surface, show that the exact eigenmode condition for TM modes can be expressed as

$$\frac{d}{dx}(x j_l(x)) = iCx j_l(x),$$

where $x = ka$, $j_l(x)$ is the spherical Bessel function, and

$$C = \frac{1-i}{2}\mu_c \delta k.$$

(b) Using Taylor series and appropriate differential equations, and again using the general form for Q_λ from Exercise 10.12.5, show that the quality factor for mode $\lambda \equiv (l, n)$ is given approximately by

$$Q_\lambda^{\mathrm{TM}} \approx \frac{a}{\mu_c \delta}\left(1 - \frac{l(l+1)}{x^2}\right).$$

Exercise 10.12.7. Consider an arbitrarily shaped but hollow (no interior surfaces) resonant cavity filled with a uniform material with dielectric constant ϵ and permeability μ. Assuming perfect conducting boundary conditions on the surface, show that the electric and magnetic energies are equal regardless of the manner of excitation (any mode). That is, show that (using harmonic fields)

$$\epsilon \int d^3x\, |\vec{E}|^2 = \frac{1}{\mu}\int d^3x\, |\vec{B}|^2.$$

Exercise 10.7.1(a) will give a similar result in the waveguide case. This knowledge for example allows a simplification of the integrals in the formula for the Q of a cavity.

Exercise 10.12.8.

(a) Using (10.341), find the quality factors Q_λ of the TE modes of a rectangular cavity with sides a, b, and d along x, y, and z, respectively. Ignore possible mixing with degenerate TM modes.
(b) Repeat part (a), this time using (10.358) instead of (10.341). Obtain a transcendental eigenvalue equation from the Exercise 10.8.2(a) TE boundary condition. Of course your answers in (a) and (b) should agree!

Exercise 10.12.9.

(a) A retarded Green function for the scalar wave equation in an arbitrary resonant cavity of volume V satisfies:

$$\left(\nabla^2 - \frac{1}{c^2}\frac{\partial^2}{\partial t^2}\right) G^{\text{ret}}(\vec{x}, t; \vec{x}', t') = -4\pi\delta(\vec{x} - \vec{x}')\delta(t - t').$$

Assuming a complete set of orthogonal eigenmodes, $\phi_k(\vec{x})$, with the properties

$$\left(\nabla^2 + \gamma_k^2\right)\phi_k(\vec{x}) = 0, \ \ \phi_k(\vec{x})|_S = 0,$$

$$\sum_k \phi_k^*(\vec{x}\,')\phi_k(\vec{x}) = \delta(\vec{x} - \vec{x}\,'),$$

$$\int_V d^3x\, \phi_{k'}^*(\vec{x})\phi_k(\vec{x}) = \delta_{k'k},$$

and that

$$G^{\text{ret}}(\vec{x}, t; \vec{x}\,', t') = 0, \ \ t < t',$$
$$G^{\text{ret}}(\vec{x}, t; \vec{x}\,', t') = 4\pi \sum_k \phi_k(\vec{x})\phi_k^*(\vec{x}\,')\Gamma_k^{\text{ret}}(t - t'),$$

find the functional form for $\Gamma_k^{\text{ret}}(t - t')$.
(b) Use the identity (cf. Equation (8.98))

$$G^{\text{ret}}(\vec{x}, t; \vec{x}\,', t') = G^{\text{ret}}(\vec{x}\,', -t'; \vec{x}, -t),$$

and part (a) to argue that the $\phi_k(\vec{x})$, for nondegenerate modes can always be chosen real.

Exercise 10.12.10. Calculate the Q value for the *lowest* frequency mode of a thin cylindrical cavity bounded by planes at $z = 0$ and $z = d$ with a

(a) circular cross-section of radius a ($d \ll a$) [*Ans.*: $Q_{p=0}^{\text{TM, cir.}} \approx \dfrac{\mu d}{\mu_c \delta} \dfrac{1}{1 + d/a}$];

(b) rectangular cross-section of length a along x and b along y ($d \ll a, b$) [*Ans.*: $Q_{p=0}^{\text{TM, rec.}} \approx \dfrac{\mu d}{\mu_c \delta} \dfrac{1}{1 + 2d\left(\dfrac{a^3 + b^3}{a^3 b + b^3 a}\right)}$].

Chapter 11

Radiation of Systems and Point Particles

In this chapter we examine the phenomenon of radiation. The explanation of radiation in terms of Maxwell's equations was one of the supreme triumphs of electromagnetic theory, and indeed of all physics. We will begin by presenting two different formalisms for calculating electromagnetic radiation (for harmonic and nonharmonic currents). Afterwards, we will discuss various approximations (such as the multipole formalism), and some special radiating situations, including nuclear bremsstrahlung and synchrotron radiation.

11.1 Electromagnetic radiation of systems: harmonic formalism

The foundational equations for all of our radiation calculations in this chapter are the free-space retarded potentials in Lorenz gauge from (8.114) and (8.115). Recalling that $R \equiv |\vec{x} - \vec{x}\,'|$, we have:

$$\Phi(\vec{x}, t) = \int d^3x' \, \frac{\rho(\vec{x}', t - R/c)}{R}, \tag{11.1}$$

$$\vec{A}(\vec{x}, t) = \frac{1}{c} \int d^3x' \, \frac{\vec{J}(\vec{x}\,', t - R/c)}{R}. \tag{11.2}$$

In this section we develop the harmonic formalism, which is useful in situations where both charges and currents are varying harmonically:

$$\rho(\vec{x}, t) = \mathrm{Re}\left[\rho(\vec{x})e^{-i\omega t}\right], \tag{11.3}$$

$$\vec{J}(\vec{x}, t) = \mathrm{Re}\left[\vec{J}(\vec{x})e^{-i\omega t}\right], \tag{11.4}$$

where $\rho(\vec{x})$ and $\vec{J}(\vec{x})$ are complex-valued. In the following, we will do all computations with complex quantities, with the understanding that physical fields are obtained by taking the real parts. The reader should understand that the real parts of these quantities correspond to the physical fields. Of course, sources do not have to consist of a single frequency, but could be periodic with many components. There is nothing to limit us in generalizing the harmonic formalism to include a series (or integral in the general case) of terms with different frequencies $\omega_1, \omega_2, \ldots$ to replace (11.3) and (11.4):

$$\rho(\vec{x}, t) = \mathrm{Re}\left[\sum_n \rho_n(\vec{x}) e^{-i\omega_n t}\right], \tag{11.5}$$

$$\vec{J}(\vec{x}, t) = \mathrm{Re}\left[\sum_n \vec{J}_n(\vec{x}) e^{-i\omega_n t}\right]. \tag{11.6}$$

This would lead us to a formalism with periodic time signals originating from a Fourier-decomposable frequency spectrum. In the large time averaging limit the various frequency components would not mix, however. This means an *incoherent* addition in the upcoming radiative power expression (see Equation (11.23)). Thus our single frequency results are a lot more general than we might at first suspect!

If we plug the (complex) single-frequency charges and currents into (11.1)–(11.2), we find ($k \equiv \omega/c$)

$$\Phi(\vec{x}, t) = \Phi(\vec{x}) e^{-i\omega t}, \text{ where } \Phi(\vec{x}) = \int d^3x' \, \frac{\rho(\vec{x}\,') e^{ikR}}{R}, \tag{11.7}$$

$$\vec{A}(\vec{x}, t) = \vec{A}(\vec{x}) e^{-i\omega t}, \text{ where } \vec{A}(\vec{x}) = \frac{1}{c}\int d^3x' \, \frac{\vec{J}(\vec{x}\,') e^{ikR}}{R}. \tag{11.8}$$

We can now eliminate reference to the field Φ. Using the Lorenz gauge condition

$$\vec{\nabla} \cdot \vec{A} + \frac{1}{c}\frac{\partial \Phi}{\partial t} = 0, \tag{11.9}$$

with $\partial/\partial t \to -i\omega$ for harmonic fields, we find

$$\Phi(\vec{x}) = -\frac{i}{k} \vec{\nabla} \cdot \vec{A}(\vec{x}). \tag{11.10}$$

The same result can also be obtained via the continuity equation.

Of course, we are ultimately interested in the physical fields. The $\vec{B}$ field is given as usual by $\vec{B} = \vec{\nabla} \times \vec{A}$, while $\vec{E}$ can be computed from

$$\frac{1}{c}\frac{\partial \vec{E}(\vec{x}, t)}{\partial t} = \vec{\nabla} \times \vec{B}(\vec{x}, t) \implies \vec{E}(\vec{x}) = \frac{i}{k} \vec{\nabla} \times \vec{B}. \tag{11.11}$$

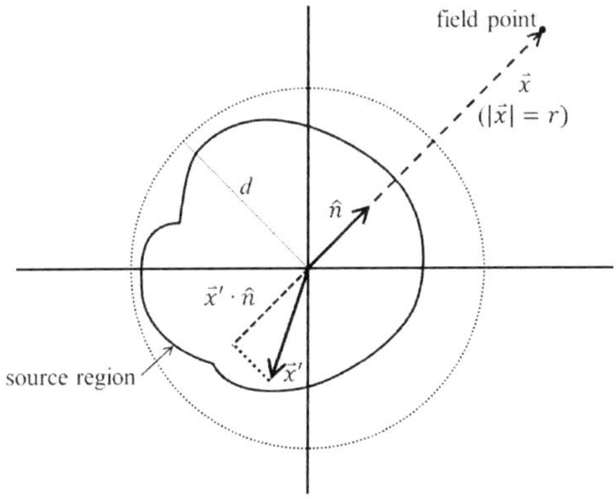

Fig. 11.1 Notation used in radiation calculations.

At this point we need to start making approximations. These approximations will take different forms in different regions in space. In order to clearly indicate these regions, we introduce the following notation, which is depicted in Figure 11.1. We suppose that the charges/currents are located in the vicinity of $\vec{x}\,' = 0$, and that all charges and currents vanish outside of a sphere of radius d. We let $r \equiv |\vec{x}|$ denote the distance between the field point and the origin, and we let $\hat{n}$ be the unit vector in the direction of $\vec{x}$. The fact that we're looking at a fixed frequency means that an additional length scale is also present, namely the wavelength $\lambda \equiv 2\pi/k$. We shall find that both the source size d and the wavelength must be considered in determining approximate forms.

Let us look at the case where $d, r \ll \lambda$, which is equivalent to $kd, kr \ll 1$. First, for $kd \ll 1$, we have $kR = k|\vec{x} - \vec{x}\,'| = kr + \mathcal{O}(kd)$. This leaves us with

$$\vec{A}(\vec{x}, t) \approx \left(\frac{1}{c} \int d^3x' \, \frac{\vec{J}(\vec{x}\,')}{|\vec{x} - \vec{x}\,'|} \right) e^{i(kr - \omega t)} \qquad (d \ll \lambda). \tag{11.12}$$

We identify an outgoing wave $e^{i(kr-\omega t)}$. The amplitude associated with this wave is given by the additional $kr \ll 1$ approximation, and corresponds to the *magnetostatic* vector potential

$$\vec{A}(\vec{x}) \approx \left(\frac{1}{c} \int d^3x' \, \frac{\vec{J}(\vec{x}\,')}{|\vec{x} - \vec{x}\,'|} \right) \qquad (d, r \ll \lambda). \tag{11.13}$$

We term this the *near zone* approximation.

Next we consider the situation $\lambda, d \ll r$, where the relative size of λ and d is not specified. The region of space where this condition is satisfied we term the *far zone*. The condition $d \ll r$ allows us to expand:

$$R = |\vec{x} - \vec{x}\,'| = \sqrt{x^2 + x'^2 - 2\vec{x}\cdot\vec{x}\,'} = r - \hat{n}\cdot\vec{x}\,' + \mathcal{O}\left(\frac{d^2}{r}\right), \qquad (11.14)$$

where $\hat{n} \equiv \hat{r}$. To first order we have

$$\begin{aligned} \vec{A}(\vec{x}) &\approx \frac{e^{ikr}}{c}\int d^3x'\,\frac{\vec{J}(\vec{x}')e^{-ik\hat{n}\cdot\vec{x}\,'}}{r - \hat{n}\cdot\vec{x}\,'} \\ &\approx \frac{e^{ikr}}{cr}\int d^3x'\,\vec{J}(\vec{x}')e^{-ik\hat{n}\cdot\vec{x}'}. \end{aligned} \qquad (11.15)$$

Let's use this approximation to calculate the angular distribution of radiated power due to a localized charge/current distribution. For this, we'll need to find $\vec{E}$ and $\vec{B}$. First we have

$$\vec{B}(\vec{x}) = \vec{\nabla}\times\vec{A}(\vec{x}) = \frac{1}{c}\int d^3x'\,\vec{\nabla}\left(\frac{e^{ikr}}{r}e^{-ik\hat{n}\cdot\vec{x}\,'}\right)\times\vec{J}(\vec{x}\,'). \qquad (11.16)$$

Just for reference purposes, we'll start off calculating the gradient above exactly, but then only keep relevant terms. We have

$$\vec{\nabla}\left(\frac{e^{ikr}}{r}e^{-ik\hat{n}\cdot\vec{x}\,'}\right) = ik\frac{e^{ikr}}{r}e^{-ik\hat{n}\cdot\vec{x}\,'}\left(\hat{n} - \vec{\nabla}(\hat{n}\cdot\vec{x}\,') - \frac{\hat{n}}{kr}\right). \qquad (11.17)$$

At this point we invoke the $r \gg \lambda$ approximation, which justifies dropping the $(kr)^{-1}$ term. As for the middle term in (11.17), we have

$$\nabla_i(n_j x'_j) = x'_j\nabla_i\left(\frac{x_j}{r}\right) = x'_j\left[\frac{r^2\delta_{ij} - x_i x_j}{r^3}\right] = \mathcal{O}\left(\frac{d}{r}\right), \qquad (11.18)$$

so this term also can be neglected because $d \ll r$. We are left with

$$\begin{aligned} \vec{B} &\approx ik\frac{e^{ikr}}{cr}\hat{n}\times\int d^3x'\,\vec{J}(\vec{x}\,')e^{-ik\hat{n}\cdot\vec{x}\,'} \\ &\approx ik\,\hat{n}\times\vec{A}(\vec{x}), \end{aligned} \qquad (11.19)$$

which shows that $\vec{B} \sim 1/r$ in the far zone where $r \gg d, \lambda$. For the same reasons we can now write down by inspection

$$\begin{aligned} \vec{E} = \frac{i}{k}\vec{\nabla}\times\vec{B} &\approx -\frac{ike^{ikr}}{rc}\hat{n}\times\left[\hat{n}\times\int d^3x'\,\vec{J}(\vec{x}')e^{-ik\hat{n}\cdot\vec{x}\,'}\right] \\ &= -ik\hat{n}\times\left(\hat{n}\times\vec{A}\right) \\ &= \vec{B}\times\hat{n}, \end{aligned} \qquad (11.20)$$

which implies that $\vec{E}^2 \approx \vec{B}^2$. The time-averaged angular distribution of radiated power from (8.24) is now (in the large r radiation limit)

$$\left(\frac{dP}{d\Omega}\right)_{\text{avg}} = \frac{da}{d\Omega}\left(\frac{dP}{da}\right)_{\text{avg}} = r^2 \frac{c}{8\pi}\text{Re}\left(\hat{n}\cdot(\vec{E}\times\vec{B}^*)\right)$$
$$= r^2 \frac{c}{8\pi}|\vec{B}|^2. \tag{11.21}$$

This can also be written as

$$\left(\frac{dP}{d\Omega}\right)_{\text{avg}} = \frac{k^2}{8\pi c}\left|\hat{n}\times\int d^3x'\,\vec{J}(\vec{x}\,')e^{-ik\hat{n}\cdot\vec{x}\,'}\right|^2, \tag{11.22}$$

which is independent of r, as we would expect from energy conservation. As mentioned previously, if frequency components $\omega_1, \omega_2, \ldots$ were present, we would simply recover the total angular power as

$$\left(\frac{dP}{d\Omega}\right)_{\text{avg}} = \sum_n \frac{k_n^2}{8\pi c}\left|\hat{n}\times\int d^3x'\,\vec{J}_n(\vec{x}\,')e^{-ik_n\hat{n}\cdot\vec{x}\,'}\right|^2, \tag{11.23}$$

where $k_n = \omega_n/c$. We will see an instance of such an incoherent addition of frequency components in the periodic trajectory results of Section 11.8; also see Exercise 11.2.4.

To illustrate the use of this formalism, we look at the particular example of the loop current shown in Figure 11.2. We have

$$\vec{J}(\vec{x}') = \frac{I}{a}\delta(r'-a)\delta(\cos\theta')\hat{e}_{\phi'} \quad (\hat{e}_{\phi'} = -\sin\phi'\hat{\imath} + \cos\phi'\hat{\jmath}), \tag{11.24}$$

$$\frac{dP}{d\Omega} = \frac{k^2 c}{8\pi}|\hat{n}\times\vec{\mathcal{I}}|^2, \tag{11.25}$$

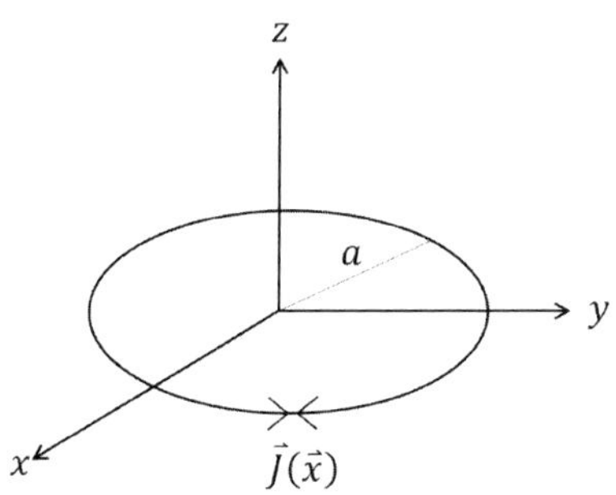

Fig. 11.2 Harmonic current loop in the xy-plane.

where

$$\vec{\mathcal{I}} \equiv \frac{1}{c}\int d^3x'\, \vec{J}(\vec{x}\,')e^{-ik\hat{n}\cdot\vec{x}\,'}$$
$$= \frac{I}{ac}a^2\int_0^{2\pi} d\phi'(-\sin\phi'\hat{\imath} + \cos\phi'\hat{\jmath})e^{-ik\hat{n}\cdot\vec{x}\,'}. \tag{11.26}$$

By azimuthal symmetry, we may choose our coordinate system so that the field point lies in the $\phi = 0$ half-plane. Then the exponent is

$$\vec{x}' \cdot \hat{n} = a(\cos\theta\underbrace{\cos\theta'}_{0} + \sin\theta\underbrace{\sin\theta'}_{1}\cos(\phi - \phi')),$$
$$= a\sin\theta\cos\phi'. \tag{11.27}$$

By symmetry, the first term in the integrand (11.26) vanishes, and the integral can be rewritten as

$$\vec{\mathcal{I}} = \frac{Ia}{2c}\hat{\jmath}\int_0^{2\pi} d\phi'\,(e^{i\phi'} + e^{-i\phi'})e^{-i(ka\sin\theta)\cos\phi'}. \tag{11.28}$$

Notice the similarity with the integral representation of $J_m(x)$ from (4.13):

$$J_m(x) = \frac{1}{i^m}\int_0^{2\pi}\frac{d\phi}{2\pi}e^{i(x\cos\phi - m\phi)}. \tag{11.29}$$

Since $J_m(x)$ is real, this may also be written

$$J_m(x) = J_m(x)^* = \frac{(-1)^m}{i^m}\int_0^{2\pi}\frac{d\phi}{2\pi}e^{-i(x\cos\phi - m\phi)}. \tag{11.30}$$

A bit of twiddling leads to

$$\vec{\mathcal{I}} = \frac{-\pi iIa\hat{\jmath}}{c}(J_1(ka\sin\theta) - J_{-1}(ka\sin\theta)) = \frac{-2\pi iIa\hat{\jmath}}{c}J_1(ka\sin\theta). \tag{11.31}$$

Since we have chosen our coordinate system such that $\vec{x}$ is in the $\phi = 0$ half-plane, $\hat{\jmath}$ actually corresponds to $\hat{\phi}$. Thus we may write

$$\vec{\mathcal{I}} = \frac{-2\pi iIa}{c}J_1(ka\sin\theta)\hat{\phi} \tag{11.32}$$
$$\implies \left(\frac{dP}{d\Omega}\right)_{\text{avg}} = \frac{k^2c}{8\pi}\left(\frac{2\pi Ia}{c}\right)^2 J_1(ka\sin\theta)^2\underbrace{|\hat{n}\times\hat{\phi}|^2}_{1},$$
$$= \frac{(Iak)^2\pi}{2c}J_1(ka\sin\theta)^2. \tag{11.33}$$

In the case where $\lambda \gg a$ (that is, $ka \ll 1$), we may further approximate

$$J_1(x) \approx \frac{x}{2} \implies \left(\frac{dP}{d\Omega}\right)_{\text{avg}} \approx \frac{I^2(ka)^4}{8c}\pi\sin^2\theta. \tag{11.34}$$

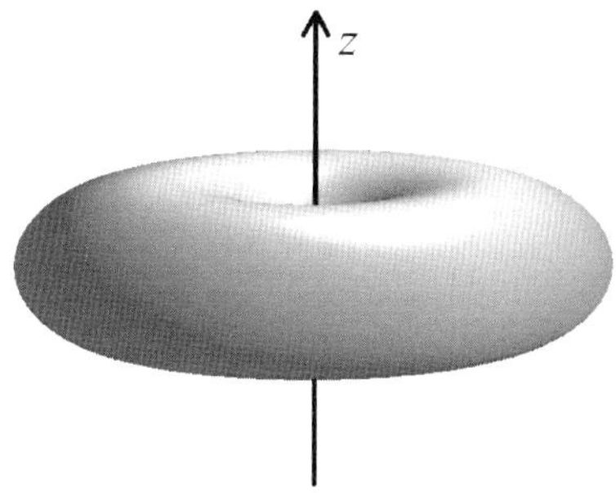

Fig. 11.3 Radiation pattern due to current loop of radius a, when $ka \ll 1$.

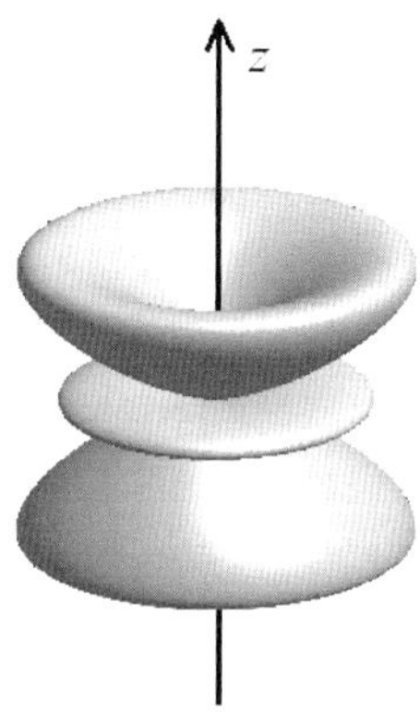

Fig. 11.4 Radiation pattern for $3.832 < ka < 7.016$.

This corresponds to the (azimuthally symmetric) radiation pattern shown in Figure 11.3.

When $ka > 1$, the zeros of J_1^2 produce additional lobes in the radiation pattern (11.33). In particular, $J_1(x)$ has zeros at

$$x = 3.832, 7.016, 10.173, \ldots$$

When $3.832 < ka < 7.016$ we expect three (azimuthally-symmetric) lobes in the radiation pattern, as shown in Figure 11.4. Every time ka increases past another zero, another pair of lobes is added.

As another more realistic example, consider the center-fed thin linear antenna of length d. The current density in the antenna is

$$J_z(x,y,z) = I \sin\left(\frac{kd}{2} - k|z|\right)\delta(x)\delta(y), \quad -d/2 \leq z \leq d/2. \tag{11.35}$$

Notice the current has nodes at both ends. Also notice the current is in phase at the point of excitation (implying the feeding current has a phase difference of π) consistent with being an even function of z. We calculate the angular power by using (11.35) in (11.22). The result is given by

$$\left(\frac{dP}{d\Omega}\right)_{\text{avg}} = \frac{I^2}{2\pi c}\left[\frac{\cos\left(\frac{kd}{2}\cos\theta\right) - \cos\left(\frac{kd}{2}\right)}{\sin\theta}\right]^2. \tag{11.36}$$

For $kd \ll 1$ this gives

$$\left(\frac{dP}{d\Omega}\right)_{\text{avg}} \approx \frac{I^2(kd)^4}{2\pi c}\sin^2\theta, \tag{11.37}$$

similar to (11.34). Again, additional lobes appear for larger antenna sizes compared to the wavelength. One may show that there are $2n$ zeros ($n = 0, 1, 2, \ldots$) of the radiation pattern in (11.36) (excluding $\theta = 0, \pi$) resulting in $2n + 1$ azimuthally symmetric radiation lobes for $2\pi n < kd < 2\pi(n + 1)$. For $kd \gg 1$ the radiation pattern shifts to deposit more energy in the lobes in the vicinity of $\theta \approx 0, \pi$.

In the exercises, you will derive similar results for other simple antenna arrays. Note that finding the actual current distribution in antenna array elements is a difficult problem similar to finding the surface charge distribution on conductors. Only for extremely thin wires can a sinusoidal form as in the last example be justified.

11.2 Electromagnetic radiation of systems: real source formalism

The harmonic formalism described in the previous section is good for antennas and other radiating systems, where the charge/current distributions vary harmonically in time. In this section, we will explain a complimentary formalism, which can be used for charges and currents with arbitrary time dependence.

As before, we begin with the potentials

$$\Phi(\vec{x}, t) = \int d^3x' \, \frac{\rho(\vec{x}\,', t - R/c)}{R}, \tag{11.38}$$

$$\vec{A}(\vec{x}, t) = \frac{1}{c}\int d^3x' \, \frac{\vec{J}(\vec{x}\,', t - R/c)}{R}. \tag{11.39}$$

Since we're assuming $r \gg d$, we may use the same approximation as before:

$$|\vec{x} - \vec{x}\,'| = r - \hat{n} \cdot \vec{x}\,' + \mathcal{O}\left(\frac{d^2}{r}\right).$$

To first order in d/r, we obtain

$$\Phi(\vec{x}, t) \approx \frac{1}{r} \int d^3x' \rho(\vec{x}\,', t_r), \tag{11.40}$$

$$\vec{A}(\vec{x}, t) \approx \frac{1}{cr} \int d^3x' \, \vec{J}(\vec{x}\,', t_r), \tag{11.41}$$

where t_r is the first-order approximation to the retarded time in the far zone,

$$t_r \equiv t - \frac{r}{c} + \frac{1}{c} \hat{n} \cdot \vec{x}\,'. \tag{11.42}$$

This time we'll need both Φ and $\vec{A}$ to calculate the $\vec{B}$ and $\vec{E}$ fields, via the usual equations

$$\vec{B} = \vec{\nabla} \times \vec{A}; \qquad \vec{E} = -\vec{\nabla}\Phi - \frac{1}{c}\frac{\partial \vec{A}}{\partial t}. \tag{11.43}$$

Both Φ and $\vec{A}$ are functions of t_r. By the chain rule, for an arbitrary function $f(t_r)$ we have

$$\vec{\nabla} f(t_r) = \frac{\partial f(t_r)}{\partial t_r} \vec{\nabla} t_r = \frac{\partial f(t_r)}{\partial t} \vec{\nabla} t_r, \tag{11.44}$$

where

$$\begin{aligned} \vec{\nabla} t_r &= \vec{\nabla}\left(t - \frac{r}{c} + \frac{1}{c}\hat{n} \cdot \vec{x}\,'\right), \\ &= -\frac{\hat{n}}{c} + \frac{1}{c} \underbrace{\left(\frac{\vec{x}\,'}{r} - \frac{(\vec{x} \cdot \vec{x}\,')\vec{x}}{r^3}\right)}_{\text{from (11.18)}}. \end{aligned} \tag{11.45}$$

The variable $\vec{x}\,'$ refers to the charge distribution, so $|\vec{x}\,'| \leq d$. Again using $r \gg d$, we get to lowest order

$$\vec{\nabla} t_r \approx -\frac{\hat{n}}{c}. \tag{11.46}$$

By assumption, r is much greater than any length scale in the problem (including λ). So we may discard all terms proportional to $(1/r^n)$ where

$n > 1$, to obtain the first-order approximation:

$$\vec{B} = \vec{\nabla} \times \vec{A} \approx \frac{1}{cr} \int d^3x' \underbrace{\vec{\nabla} \times \vec{J}(\vec{x}\,', t_r)}_{\vec{\nabla} t_r \times \frac{\partial \vec{J}(\vec{x}\,', t_r)}{\partial t}}$$

$$\approx -\frac{\hat{n}}{c} \times \left[\frac{1}{cr} \int d^3x' \frac{\partial \vec{J}(\vec{x}\,', t_r)}{\partial t} \right], \tag{11.47}$$

$$\approx -\frac{\hat{n}}{c} \times \frac{\partial \vec{A}}{\partial t}. \tag{11.48}$$

Likewise

$$\vec{E} = -\vec{\nabla}\Phi - \frac{1}{c}\frac{\partial \vec{A}}{\partial t} \approx -\frac{1}{r} \int d^3x' \; \underbrace{\vec{\nabla}\rho(\vec{x}\,', t_r)}_{-\frac{\hat{n}}{c}\frac{\partial \rho(\vec{x}\,', t_r)}{\partial t},} - \frac{1}{rc^2} \int d^3x' \; \frac{\partial \vec{J}(\vec{x}\,', t_r)}{\partial t}$$

$$\approx \frac{1}{rc} \left[\hat{n} \int d^3x' \frac{\partial \rho(\vec{x}\,', t_r)}{\partial t} - \frac{1}{c} \int d^3x' \frac{\partial \vec{J}(\vec{x}\,', t_r)}{\partial t} \right]. \tag{11.49}$$

We may re-express the $\partial\rho/\partial t$ term using the continuity equation as follows:

$$\begin{aligned} \frac{\partial \rho(\vec{x}\,', t_r)}{\partial t} &= - \left. \vec{\nabla}' \cdot \vec{J}(\vec{x}\,', t_r) \right|_{t_r = \text{const.}} \\ &= - \left(\vec{\nabla}' \cdot \vec{J}(\vec{x}\,', t_r(\vec{x}\,')) - \frac{\partial \vec{J}(\vec{x}\,', t_r)}{\partial t} \cdot \vec{\nabla}' t_r \right) \\ &\approx -\vec{\nabla}' \cdot \vec{J}(\vec{x}\,', t_r(\vec{x}\,')) + \frac{\hat{n}}{c} \cdot \frac{\partial \vec{J}(\vec{x}\,', t_r)}{\partial t}, \end{aligned} \tag{11.50}$$

where we have re-asserted the non-constant $\vec{x}\,'$ dependence of t_r in the middle step. Since the current distribution is bounded, we have

$$\int d^3x' \, \vec{\nabla}' \cdot \vec{J}(\vec{x}\,', t_r(\vec{x}\,')) = 0. \tag{11.51}$$

Therefore

$$\begin{aligned} \vec{E} &\approx \frac{1}{rc^2} \left[\hat{n} \int d^3x' \, \hat{n} \cdot \frac{\partial \vec{J}(\vec{x}\,', t_r)}{\partial t} - \int d^3x' \, \frac{\partial \vec{J}(\vec{x}\,', t_r)}{\partial t} \right] \\ &= \hat{n} \times \left[\frac{\hat{n}}{c} \times \frac{1}{cr} \int d^3x' \, \frac{\partial \vec{J}(\vec{x}\,', t_r)}{\partial t} \right], \end{aligned} \tag{11.52}$$

$$\approx \hat{n} \times \left[\frac{\hat{n}}{c} \times \frac{\partial \vec{A}}{\partial t} \right] \approx \vec{B} \times \hat{n}. \tag{11.53}$$

Note that $\vec{E} \approx \vec{B} \times \hat{n}$ and $\vec{E}^2 \approx \vec{B}^2$, just as in the harmonic formalism (see (11.15)).

The instantaneous angular distribution of power is now seen to be (note that in this case the fields are real and our equalities are in the large r radiation limit)

$$\begin{aligned}\frac{dP(t)}{d\Omega} &= \frac{cr^2}{4\pi}\hat{n}\cdot(\vec{E}\times\vec{B}) = \frac{cr^2}{4\pi}\vec{B}^2 \\ &= \frac{1}{4\pi c}\left[\hat{n}\times\frac{1}{c}\int d^3x'\,\frac{\partial\vec{J}(\vec{x}\,',t_r)}{\partial t}\right]^2, \qquad (11.54)\\ &= \frac{r^2}{4\pi c}\left[\hat{n}\times\frac{\partial\vec{A}(\vec{x},t)}{\partial t}\right]^2, \qquad (11.55)\end{aligned}$$

where we have used (11.47) and (11.48) to re-express $\vec{B}^2$. Note in particular that for steady currents there is no radiation, as expected.

It is instructive to compare (11.54) with the corresponding expression (11.22) for the harmonic formalism. The ω^2 in (11.22) has been replaced with a squared time derivative. Also, there is no averaging factor of 1/2. We can verify that (11.54) is consistent with (11.22) by plugging

$$\vec{J}(\vec{x}\,',t) = \vec{J}(\vec{x}\,')\sin(\omega t - \delta(\vec{x}\,')), \qquad (11.56)$$

into (11.54), where $\vec{J}(\vec{x}\,')$ is real and $\delta(\vec{x}\,')$ is a position dependent phase. This gives

$$\frac{dP(t)}{d\Omega} = \frac{1}{4\pi c}\left[\hat{n}\times\int d^3x'\,\vec{J}(\vec{x}\,')\frac{\omega}{c}\cos(\omega t_r - \delta(\vec{x}\,'))\right]^2. \qquad (11.57)$$

Using

$$\begin{aligned}\cos(\omega t_r - \delta(\vec{x}\,')) &= \cos(\omega t)\cos\left[\omega\left(-\frac{r}{c}+\frac{1}{c}\hat{n}\cdot\vec{x}\,'\right)-\delta(\vec{x}\,')\right]\\ &\quad - \sin(\omega t)\sin\left[(\omega\left(-\frac{r}{c}+\frac{1}{c}\hat{n}\cdot\vec{x}\,'\right)-\delta(\vec{x}\,')\right], \qquad (11.58)\end{aligned}$$

and averaging $\cos^2(\omega t)$, $\sin^2(\omega t)$, $\cos(\omega t)\sin(\omega t)$ over a single time period $T = 2\pi/\omega$, we find

$$\begin{aligned}\left(\frac{dP}{d\Omega}\right)_{\rm avg} &= \frac{k^2}{8\pi c}\left\{\left[\hat{n}\times\int d^3x'\,\vec{J}(\vec{x}\,')\cos\left[\omega\left(-\frac{r}{c}+\frac{1}{c}\hat{n}\cdot\vec{x}\,'\right)-\delta(\vec{x}\,')\right]\right]^2\right.\\ &\quad \left. + \left[\hat{n}\times\int d^3x'\vec{J}(\vec{x}\,')\sin\left[\omega\left(-\frac{r}{c}+\frac{1}{c}\hat{n}\cdot\vec{x}\,'\right)-\delta(\vec{x}\,')\right]\right]^2\right\}. \qquad (11.59)\end{aligned}$$

With the definitions

$$\vec{J}^{\text{har}}(\vec{x}\,') \equiv \vec{J}(\vec{x}\,')e^{i\delta(\vec{x}\,')}, \tag{11.60}$$

$$\vec{A}^{\text{har}} \equiv \frac{e^{ikr}}{cr}\int d^3x'\,\vec{J}^{\text{har}}(\vec{x}\,')e^{-ik\hat{n}\cdot\vec{x}\,'}, \tag{11.61}$$

and using that $\vec{J}(\vec{x}\,')$ is real, this gives

$$\left(\frac{dP}{d\Omega}\right)_{\text{avg}} = \frac{k^2}{8\pi c}\left|\hat{n}\times\int d^3x'\,\vec{J}^{\text{har}}(\vec{x}\,')e^{-ik\hat{n}\cdot\vec{x}'}\right|^2, \tag{11.62}$$

$$= \frac{k^2r^2c}{8\pi}\left|\hat{n}\times\vec{A}^{\text{har}}\right|^2. \tag{11.63}$$

which agrees with (11.22).

11.3 Frequency distribution of radiated energy; impulsive scattering

We have computed angular distributions for instantaneous radiated power, and for time-averaged radiated power from a periodic charge/current distribution. In this section, we consider the radiation as a sum (more exactly, an integral) of frequency components, and investigate the radiated energy as a function of *frequency*, as well as angle of emission.

In terms of the power distribution $dP(t)/d\Omega$, the energy distribution as a function of angle is

$$\frac{dE}{d\Omega} = \int_{-\infty}^{\infty} dt\,\frac{dP(t)}{d\Omega}, \tag{11.64}$$

where we use (11.54) or (11.55) to define the right-hand side. To simplify the notation, we define

$$\vec{\mathcal{J}}(\vec{x},t) \equiv \frac{1}{c}\int d^3x'\,\frac{\partial}{\partial t}\vec{J}(\vec{x}\,',t_r) = r\frac{\partial}{\partial t}\vec{A}(\vec{x},t), \tag{11.65}$$

where we are using (11.41) in the large r limit. Then (11.64) can be rewritten as

$$\frac{dE}{d\Omega} = \frac{1}{4\pi c}\int_{-\infty}^{\infty} dt\,[\hat{n}\times\vec{\mathcal{J}}(\vec{x},t)]^2. \tag{11.66}$$

Since we want to consider frequency components, naturally we'll need Fourier transforms, which were introduced in Section 2.2. But here we'll be using a *different* convention for the Fourier transform normalization, so that the Fourier transform of $f(\omega)$ will be defined as

$$f(t) \equiv \frac{1}{\sqrt{2\pi}} \int_{-\infty}^{\infty} d\omega\, f(\omega) e^{-i\omega t}, \tag{11.67}$$

and the corresponding inverse transform formula is

$$f(\omega) = \frac{1}{\sqrt{2\pi}} \int_{-\infty}^{\infty} dt\, f(t) e^{i\omega t}. \tag{11.68}$$

We will be using *Parseval's theorem*, which is a beautiful and general result from Fourier analysis. (You will prove this theorem in Exercise 11.3.1.) Using the Fourier transform conventions above, the theorem says that

$$\int_{-\infty}^{\infty} dt |f(t)|^2 = \int_{-\infty}^{\infty} d\omega |f(\omega)|^2. \tag{11.69}$$

Letting $\vec{\mathcal{J}}(\vec{x},\omega)$ be the Fourier inverse transform of $\vec{\mathcal{J}}(\vec{x},t)$, we may establish

$$\vec{\mathcal{J}}(\vec{x},\omega) = -i\omega r \vec{A}(\vec{x},\omega), \tag{11.70}$$

where

$$\begin{aligned} \vec{A}(\vec{x},\omega) &= \int_{-\infty}^{\infty} \frac{dt}{\sqrt{2\pi}}\, e^{i\omega t} \vec{A}(\vec{x},t), \\ &= \frac{1}{cr} \int_{-\infty}^{\infty} \frac{dt}{\sqrt{2\pi}}\, e^{i\omega t} \int d^3x' \vec{J}(\vec{x}', t_r), \\ &= \frac{e^{ikr}}{cr} \int d^3x' \vec{J}(\vec{x}',\omega) e^{-ik\hat{n}\cdot\vec{x}'}. \end{aligned} \tag{11.71}$$

We have interchanged integrals and used (11.41) and $t_r = t - r/c + \hat{n}\cdot\vec{x}'/c$ in this reduction. Parseval's theorem now gives us immediately that

$$\frac{dE}{d\Omega} = \frac{r^2}{4\pi c} \int_{-\infty}^{\infty} d\omega\, \omega^2 |\hat{n} \times \vec{A}(\vec{x},\omega)|^2. \tag{11.72}$$

Since $\vec{A}(\vec{x},t)$ is real, it follows from (11.68) that $|\vec{A}(\vec{x},-\omega)| = |\vec{A}(\vec{x},\omega)|$, and we may rewrite (11.72) as

$$\frac{dE}{d\Omega} = \frac{r^2}{2\pi c} \int_{0}^{\infty} d\omega\, \omega^2 \left|\hat{n} \times \vec{A}(\vec{x},\omega)\right|^2. \tag{11.73}$$

At this point let us pause to consider the physics. We have written the angular energy distribution as an integral involving the frequency components of the vector potential. It makes sense therefore to identify

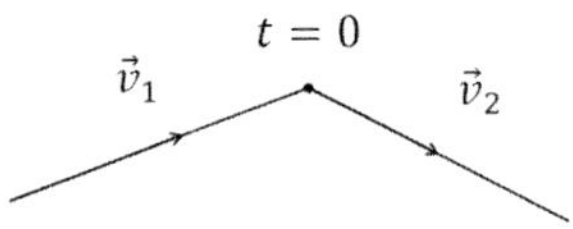

Fig. 11.5 Notation for impulsive scattering.

the integrand in (11.73) as a *differential energy density as a function of frequency*. We are thus motivated to define:

$$\frac{d^2E(\omega)}{d\omega d\Omega} \equiv \frac{k^2r^2c}{2\pi}\left|\hat{n}\times\vec{A}(\vec{x},\omega)\right|^2 \tag{11.74}$$

$$= \frac{k^2}{2\pi c}\left|\hat{n}\times\int d^3x'\,\vec{J}(\vec{x}\,',\omega)e^{-ik\hat{n}\cdot\vec{x}'}\right|^2. \tag{11.75}$$

Notice the close resemblance to (11.22), which is in the harmonic formalism. Equations (11.74) and (11.75) are foundational in developing expressions for field energy distribution expressions in many laboratory situations (see for example Sections 11.7 and 11.8).

As an application of (11.74), (11.75) we will compute $d^2E(\omega)/(d\omega d\Omega)$ for a particle of charge e undergoing impulsive scattering, as shown in Figure 11.5. In this case we have

$$\vec{J}(\vec{x},t) = \begin{cases} e\,\vec{v}_1\delta\left(\vec{x}-\vec{v}_1t\right), & t<0,\\ e\,\vec{v}_2\delta\left(\vec{x}-\vec{v}_2t\right), & t>0.\end{cases} \tag{11.76}$$

We have

$$\begin{aligned}\int d^3x'\,\vec{J}(\vec{x}\,',\omega)e^{-ik\hat{n}\cdot\vec{x}\,'} &= \frac{1}{\sqrt{2\pi}}\int_{-\infty}^{\infty}dt\,e^{i\omega t}\int d^3x'\,e^{-ik\hat{n}\cdot\vec{x}\,'}\,\vec{J}(\vec{x}\,',t)\\ &= \frac{1}{\sqrt{2\pi}}\left[e\,\vec{v}_1\int_{-\infty}^{0}dt\,e^{i\omega t(1-\frac{1}{c}\hat{n}\cdot\vec{v}_1)}\right.\\ &\qquad\left.+e\,\vec{v}_2\int_{0}^{\infty}dt\,e^{i\omega t(1-\frac{1}{c}\hat{n}\cdot\vec{v}_2)}\right].\end{aligned} \tag{11.77}$$

We now use

$$\int_0^{\infty}dt\,e^{ia\omega t} = \frac{i}{a\omega};\quad \int_{-\infty}^{0}dt\,e^{ia\omega t} = -\frac{i}{a\omega}, \tag{11.78}$$

for $a>0$. The first result is obtained by evaluating the integral with $\omega\to\omega+i\epsilon$ (see (8.108)) and letting $\epsilon\to 0$; the second follows by complex conjugation and $t\to -t$. Plugging into (11.77) gives

$$\int d^3x'\,\vec{J}(\vec{x}\,',\omega)e^{-ik\hat{n}\cdot\vec{x}\,'} = \frac{1}{\sqrt{2\pi}}\left[\frac{-ie\,\vec{v}_1}{\omega(1-\frac{1}{c}\hat{n}\cdot\vec{v}_1)}+\frac{ie\,\vec{v}_2}{\omega(1-\frac{1}{c}\hat{n}\cdot\vec{v}_2)}\right]. \tag{11.79}$$

Finally we have from (11.79) and (11.75) that

$$\frac{d^2E(\omega)}{d\omega d\Omega} = \frac{e^2}{4\pi^2c^3}\left|\hat{n}\times\left(\frac{\vec{v}_2}{(1-\frac{1}{c}\hat{n}\cdot\vec{v}_2)} - \frac{\vec{v}_1}{(1-\frac{1}{c}\hat{n}\cdot\vec{v}_1)}\right)\right|^2. \tag{11.80}$$

In the ultra-relativistic limit, the denominators in (11.80) produce strong radiation beams near the $\hat{n} \sim \hat{v}_1, \hat{v}_2$ directions; however, the numerator actually produces a near-zero in these exact directions. In the nonrelativistic case ($|\vec{v}_1|, |\vec{v}_2| \ll c$) this reduces to

$$\begin{aligned}\frac{d^2E(\omega)}{d\omega d\Omega} &\approx \frac{e^2}{4\pi^2c^3}\left|\hat{n}\times(\vec{v}_2-\vec{v}_1)\right|^2\\ &= \frac{e^2}{4\pi^2c^3}\left|\vec{v}_2-\vec{v}_1\right|^2\sin^2\theta.\end{aligned} \tag{11.81}$$

The energy density is independent of frequency, which implies that the emitted energy is formally infinite. If we integrate over solid angles we obtain

$$\frac{dE(\omega)}{d\omega} \approx \frac{2}{3}\frac{e^2}{\pi c^3}\left|\vec{v}_2-\vec{v}_1\right|^2. \tag{11.82}$$

The idealization to a sharp impulse is what leads to the infinite amount of radiation. Equation (11.80) is one of the basic ingredients needed to understand processes where a charged particle is sharply deflected by atomic forces, leading to radiation and energy loss. One of these situations occurs when a charged particle enters a different medium, leading to *transition radiation*. Then, the direction of the radiation generated is modified by the boundary conditions at the surface of the material, such as we studied for example in Section 9.2. We will not have time to discuses this subject, but the reader will find some very comprehensive references in the *Going Deeper* section.

11.4 Multipole expansion: physical interpretation

The real-source formalism introduced in Section 11.2 led to integral expressions for potentials and fields in the far zone where $r \gg d, \lambda$. In the case where the source size d is much smaller than the wavelength λ, it is possible to expand these expressions in Taylor series. These expansions give us a much clearer picture of the angular distribution of the radiation fields, and of their fall-off as distance from the source increases.

Let us begin by defining

$$t_0 \equiv t - r/c \qquad \text{(the “origin retarded time”)}. \tag{11.83}$$

Then we may rewrite (11.41) as

$$\vec{A}(\vec{x}, t) \approx \frac{1}{rc} \int d^3x' \, \vec{J}\left(\vec{x}\,', t_0 + \frac{\hat{n} \cdot \vec{x}\,'}{c}\right). \tag{11.84}$$

Then, expanding in Taylor series around $t = t_0$, we have

$$\vec{J}\left(\vec{x}\,', t_0 + \frac{\hat{n} \cdot \vec{x}\,'}{c}\right) = \vec{J}(\vec{x}\,', t_0) + \frac{\partial \vec{J}(\vec{x}\,', t_0)}{\partial t}\left(\frac{\hat{n} \cdot \vec{x}\,'}{c}\right) + \frac{1}{2}\frac{\partial^2 \vec{J}(\vec{x}\,', t_0)}{\partial t^2}\left(\frac{\hat{n} \cdot \vec{x}\,'}{c}\right)^2 + \dots \tag{11.85}$$

Expression (11.85) plugged into (11.84) gives us the multipole expansion for the vector potential $\vec{A}$. The first term in the expansion has the same form as the current density in the far-field region, except that it is evaluated at the “origin retarded time”. In the remainder of this section we provide some physical insight into the significance of the higher order terms in the expansion, which involve time derivatives of the current density evaluated at the same time.

Figure 11.1 shows that $\hat{n} \cdot \vec{x}\,'$ is the projection of the distance between origin and source point onto the line-of sight from origin to field point. It follows that $\hat{n} \cdot \vec{x}\,'/c$ is the approximate travel time difference for light going from $\vec{x}\,'$ to the origin for the given far zone observer. For harmonic sources we have $|\partial^m \vec{J}/\partial t^m| \sim \omega^m |\vec{J}|$: so in order for the higher-order terms in (11.85) to be small compared to the first, we must have

$$\omega \left|\frac{\hat{n} \cdot \vec{x}\,'}{c}\right| \ll 1 \implies \left|\frac{\hat{n} \cdot \vec{x}\,'}{c}\right| \ll \frac{1}{\omega}. \tag{11.86}$$

In other words, the time scale associated with light traveling from the various portions of the source is small compared with the time scale(s) associated with the local time fluctuation of the currents. This is guaranteed if

$$d \ll \frac{\lambda}{2\pi} \quad \left(d \ll \frac{\lambda}{2\pi} \ll r\right). \tag{11.87}$$

For nonharmonic sources with a continuous frequency spectrum, this condition only holds for low frequencies ($\omega \ll c/d$). On the other hand, for antenna systems at fixed frequency, this condition requires a small antenna size d relative to reduced wavelength ($kd \ll 1$).

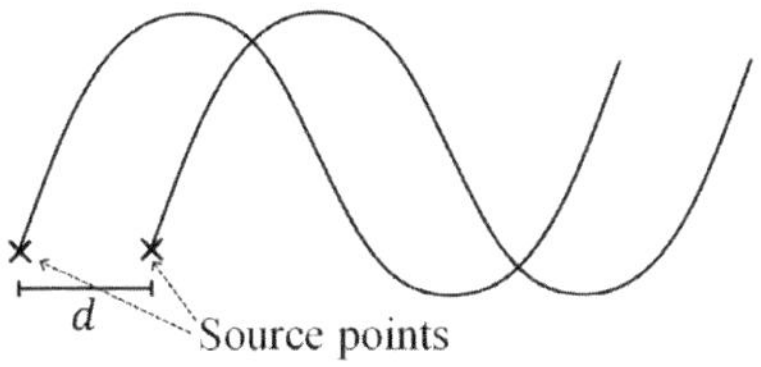

Fig. 11.6 Incoherency between radiation from different point sources, when $d \not\ll \lambda/2\pi$.

We may gain some intuition into this situation with the aid of Figure 11.6. If the source size is comparable to the wavelength, then even if different source points are in phase, the radiation originating from the different points will be out of phase. Interpreted along these lines, we see that (11.87) is a type of *coherency* requirement with respect to our origin.

Figure 11.6 also applies in the case of electromagnetic radiation arising from a moving point particle. To have coherency at a particular frequency ω, the distance travelled by the particle during one period $(2\pi/\omega)$ should be much smaller than the wavelength:

$$v\frac{2\pi}{\omega} \ll \lambda \implies v \ll c. \tag{11.88}$$

So for point-particle sources, coherency corresponds to nonrelativistic velocities rather than small source size.

In summary, the multipole expansion can be considered as a perturbative approach to deal with incoherency. In the next section we will consider the lowest-order terms in the expansion.

11.5 Dipole and quadrupole contributions to radiated power

Let's do the math. We'll begin by taking just the lowest-order term $\vec{A}(\vec{x}\,', t)$ in the multipole expansion. From (11.84) and (11.85) we have

$$\vec{A}(\vec{x}, t) \approx \frac{1}{rc}\int d^3x'\,\vec{J}(\vec{x}\,', t_0). \tag{11.89}$$

We may simplify the right-hand side using

$$\vec{J}(\vec{x}\,', t_0) = \sum_i \nabla_i' \left(\vec{x}\,' J_i(\vec{x}\,', t_0)\right) - \vec{x}\,'\vec{\nabla}' \cdot \vec{J}(\vec{x}\,', t_0)$$

$$\Longrightarrow \int d^3x' \vec{J}(\vec{x}\,', t_0) = \underbrace{\int d^3x' \sum_i \nabla_i' \left(\vec{x}\,' J_i(\vec{x}\,', t_0)\right)}_{0} - \int d^3x'\, \vec{x}\,'\vec{\nabla}' \cdot \vec{J}(\vec{x}\,', t_0)$$

$$= \int d^3x'\, \vec{x}\,' \frac{\partial \rho(\vec{x}\,', t_0)}{\partial t}, \tag{11.90}$$

where we have assumed a bounded charge distribution and used the continuity equation to obtain the last line. Now recall that the dipole moment is defined by

$$\vec{p}(t_0) = \int d^3x'\, \vec{x}\,' \rho(\vec{x}\,', t_0). \tag{11.91}$$

We obtain finally that to lowest order

$$\vec{A}(\vec{x}, t) \approx \frac{\dot{\vec{p}}(t_0)}{rc}. \tag{11.92}$$

From (11.54) we obtain the angular distribution of power emitted as

$$\frac{dP(t)}{d\Omega} = \frac{1}{4\pi c^3}\left[\hat{n} \times \ddot{\vec{p}}(t_0)\right]^2 = \frac{\left[\ddot{\vec{p}}(t_0)\right]^2}{4\pi c^3} \sin^2\theta(t_0). \tag{11.93}$$

This displays an instantaneous dipole form. The total (instantaneous) power is

$$P(t) = \frac{2[\ddot{\vec{p}}(t_0)]^2}{3c^3}. \tag{11.94}$$

In the case of harmonic sources note that in general the components of the vector $\vec{p}(t_0)$ may not oscillate with the same time phase. Thus, although (11.93) always displays an *instantaneous* dipole form, its *time average* in general will not. In the case where the components of $\vec{p}(t_0)$ oscillate in phase, we have

$$\vec{p}(t_0) = \vec{p}\sin(\omega t_0 + \alpha), \tag{11.95}$$

where $\vec{p}$ is a constant (real) vector and α is a phase factor. Using (11.93) we may then obtain the time-averaged power results

$$\left(\frac{dP}{d\Omega}\right)_{\text{avg}} = \frac{\omega^4}{8\pi c^3}\left[\hat{n} \times \vec{p}\right]^2, \tag{11.96}$$

$$P_{\text{avg}} = \frac{\omega^4[\vec{p}]^2}{3c^3} = \frac{ck^4[\vec{p}]^2}{3}. \tag{11.97}$$

In the case of a single nonrelativistic particle with $\vec{p}(t_0) = e\vec{x}(t_0)$, we get

$$\frac{dP(t)}{d\Omega} = \frac{e^2}{4\pi c^3}\left(\hat{n} \times \ddot{\vec{x}}(t_0)\right)^2, \tag{11.98}$$

$$P(t) = \frac{2e^2[\ddot{\vec{x}}(t_0)]^2}{3c^3}. \tag{11.99}$$

This expression for $P(t)$ (which resembles (11.82)) is called the *Larmor formula*. The instantaneous radiation pattern is identical in shape to Figure 11.3, where $\hat{z}$ now points along the direction of acceleration $\ddot{\vec{x}}$. The pattern is independent of the particle velocity in this nonrelativistic approximation. We will see this is *not* true for relativistic particles.

Now let's take the next step and include the first perturbative term in the multipole expansion of $\vec{A}(\vec{x},t)$. Until now, the lowest-order expansion has only produced electric dipole radiation: we'll see shortly that first-order adds magnetic dipole and electric quadrupole terms to the mix.

To first order, we have

$$\vec{J}(\vec{x}\,', t_0 + \frac{1}{c}\hat{n}\cdot\vec{x}\,') \approx \vec{J}(\vec{x}\,', t_0) + \frac{\partial \vec{J}(\vec{x}\,', t_0)}{\partial t}\left(\frac{\hat{n}\cdot\vec{x}\,'}{c}\right) \tag{11.100}$$

$$\begin{aligned}\implies \frac{\partial \vec{A}(\vec{x},t)}{\partial t} &\approx \frac{1}{r}\frac{d}{dt}\int d^3x'\,\frac{1}{c}\vec{J}(\vec{x}\,', t_0 + \frac{1}{c}\hat{n}\cdot\vec{x}\,'),\\ &\approx \frac{1}{r}\left(\frac{\ddot{\vec{p}}(t_0)}{c} + \frac{1}{c^2}\frac{d^2}{dt^2}\int d^3x'\,\vec{J}(\vec{x}\,', t_0)\hat{n}\cdot\vec{x}\,'\right). \end{aligned} \tag{11.101}$$

Using component notation, the new term can be decomposed into two separate terms, which we will label suggestively as (e.q.) and (m.d.):

$$\begin{aligned}&\int d^3x'\,J_i(\vec{x}\,', t_0)\hat{n}\cdot\vec{x}\,'\\ &\quad = \int d^3x' \sum_j n_j\left[\frac{1}{2}\underbrace{(x'_j J_i + x'_i J_j)}_{\text{(e.q.)}} + \frac{1}{2}\underbrace{(x'_j J_i - x'_i J_j)}_{\text{(m.d.)}}\right]. \end{aligned} \tag{11.102}$$

Beginning with (m.d.), we may rewrite this term in vector form and simplify it as follows:

$$\begin{aligned}\text{(m.d.)} &= -\frac{1}{2}\int d^3x'\,\hat{n}\times\left[\vec{x}\,'\times\vec{J}(\vec{x}\,', t_0)\right]\\ &= -c\hat{n}\times\vec{m}(t_0), \end{aligned} \tag{11.103}$$

where $\vec{m}(t_0)$ is the *magnetic dipole moment* defined in (6.129),

$$\vec{m}(t_0) \equiv \frac{1}{2c}\int d^3x'\, \vec{x}' \times \vec{J}(\vec{x}\,', t_0). \tag{11.104}$$

Having achieved our objective with the magnetic dipole term, we move on to the (e.q.) term. Using integration by parts, we find

$$\begin{aligned}
(\text{e.q.}) &= \frac{1}{2}\sum_j n_j \int d^3x'\,(x'_j J_i + x'_i J_j) \\
&= -\frac{1}{2}\sum_j n_j \int d^3x'\, x'_i x'_j\, \underbrace{\vec{\nabla}' \cdot \vec{J}(\vec{x}\,', t_0)}_{-\frac{\partial}{\partial t}\rho(\vec{x}\,', t_0)} \\
&= \frac{1}{2}\sum_j n_j \frac{d}{dt}\int d^3x'\, x'_i x'_j\, \rho(\vec{x}\,', t_0) \\
&= \frac{1}{6}\sum_j n_j \frac{d}{dt}\left(Q_{ij}(t_0) + \langle r^2(t_0)\rangle \delta_{ij}\right),
\end{aligned} \tag{11.105}$$

where Q_{ij} is the *electric quadrupole tensor* defined in Section 5.1,

$$Q_{ij}(t_0) \equiv \int d^3x'\,(3x'_i x'_j - r'^2\delta_{ij})\rho(\vec{x}\,', t_0), \tag{11.106}$$

and $\langle r^2(t_0)\rangle$ is the second moment of the charge density,

$$\langle r^2(t_0)\rangle \equiv \int d^3x'\, r'^2 \rho(\vec{x}\,', t_0), \tag{11.107}$$

which has dimensions charge $\times$ length^2. We define a vector (in component notation)

$$Q_i(\hat{n}, t_0) \equiv \sum_j Q_{ij}(t_0) n_j, \tag{11.108}$$

so that we can re-express the (e.q.) term as

$$(\text{e.q.}) = \frac{1}{6}\dot{\vec{Q}} + \frac{1}{6}\hat{n}\frac{d}{dt}\langle r^2(t_0)\rangle. \tag{11.109}$$

Putting (lowest order) + (m.d.) + (e.q.) terms together, we have

$$\frac{\partial \vec{A}(\vec{x}, t)}{\partial t} \approx \frac{1}{r}\left(\frac{\ddot{\vec{p}}(t_0)}{c} - \frac{\hat{n}}{c} \times \ddot{\vec{m}}(t_0) + \frac{1}{6c^2}\dddot{\vec{Q}}(\hat{n}, t_0) + \frac{1}{6c^2}\hat{n}\frac{d^3}{dt^3}\langle r^2(t_0)\rangle\right). \tag{11.110}$$

Thus the emitted power angular distribution is

$$\frac{dP(t)}{d\Omega} = r^2 \frac{1}{4\pi c}\left[\hat{n} \times \frac{\partial \vec{A}(\vec{x},t)}{\partial t}\right]^2$$
$$= \frac{1}{4\pi c^3}\left[\hat{n} \times \left(\ddot{\vec{p}}(t_0) - \hat{n} \times \ddot{\vec{m}}(t_0) + \frac{1}{6c}\dddot{\vec{Q}}(\hat{n},t_0) + \underbrace{\frac{1}{6c}\hat{n}\frac{d^3}{dt^3}\left\langle r^2(t_0)\right\rangle}_{0}\right)\right]^2, \tag{11.111}$$

where the final term vanishes because of the cross-product. This vanishing has the physically interesting result that radially-symmetric current distributions (also called *breathing modes*) emit no radiation, at least to first order. In fact, it is possible to show that radiation is *exactly* zero for such modes (see Exercise 11.5.5). The harmonic analog of (11.111),

$$\left(\frac{dP}{d\Omega}\right)_{\text{avg}} = \frac{ck^4}{8\pi}\left|\hat{n} \times \left(\vec{p} - \hat{n} \times \vec{m} - i\frac{k}{6}\vec{Q}(\hat{n})\right)\right|^2, \tag{11.112}$$

may also be derived (see Exercise 11.5.2). Note that the complex quantities $\vec{p}$, $\vec{m}$ and $\vec{Q}(\hat{n})$ in this formalism, defined by Equations (11.91), (11.104) and (11.106) with $\rho(\vec{x}\,', t_0) \to \rho(\vec{x}\,')$ and $\vec{J}(\vec{x}\,', t_0) \to \vec{J}(\vec{x}\,')$, are distinct from the similarly named objects in the real formalism.

In general, there is interference between the three moments (corresponding to the three nonzero terms in (11.111)) due to the squaring operation. It is instructive nonetheless to compute the power distribution for each individual term. In the pure magnetic dipole case, we have:

$$\frac{dP(t)}{d\Omega} = \frac{1}{4\pi c^3}\left[\hat{n} \times (\hat{n} \times \ddot{\vec{m}}(t_0))\right]^2 = \frac{1}{4\pi c^3}\left[\hat{n} \times \ddot{\vec{m}}(t_0)\right]^2$$
$$= \frac{[\ddot{\vec{m}}(t_0)]^2}{4\pi c^3}\sin^2\theta(t_0), \tag{11.113}$$

$$P(t) = \frac{2[\ddot{\vec{m}}(t_0)]^2}{3c^3}. \tag{11.114}$$

These are identical to the electric dipole expressions (11.93) and (11.94) except that $\vec{p}$ has been replaced by $\vec{m}$.

For harmonically-varying magnetic dipoles where the components have the same time phase, we may define a constant magnetic dipole vector analogous to the electric dipole vector $\vec{p}$ in (11.95) and obtain (in analogy to (11.97))

$$\left(\frac{dP}{d\Omega}\right)_{\text{avg}} = \frac{\omega^4}{8\pi c^3}\left[\hat{n}\times\vec{m}\right]^2, \tag{11.115}$$

$$P_{\text{avg}} = \frac{\omega^4\vec{m}^2}{3c^3} = \frac{ck^4\vec{m}^2}{3}. \tag{11.116}$$

We can also check that our magnetic dipole expressions are consistent with the current loop results computed in Section 11.1. For a loop in the xy-plane, we have

$$\vec{m} = \frac{1}{2c}\int d^3x'\,\vec{x}\,'\times\vec{J}(\vec{x}\,') = \frac{I}{c}\pi a^2\hat{k}$$

$$\Longrightarrow \left(\frac{dP}{d\Omega}\right)_{\text{avg}} = \frac{I^2(ka)^4\pi}{8c}\sin^2\theta, \tag{11.117}$$

in perfect agreement with the $ka \ll 1$ result (11.34).

Finally, we look at the electric quadrupole term. For a pure quadrupole,

$$\frac{dP(t)}{d\Omega} = \frac{1}{144\pi c^5}\left[\hat{n}\times\dddot{\vec{Q}}(\hat{n},t_0)\right]^2. \tag{11.118}$$

The instantaneous angular dependence is no longer proportional to $\sin^2\theta(t_0)$ because $\vec{Q}(\hat{n},t_0)$ is also angle-dependent. For harmonic sources whose components vary in phase we have

$$\vec{Q}(\hat{n},t_0) = \vec{Q}(\hat{n})\sin(\omega t_0+\alpha), \tag{11.119}$$

which leads to

$$\left(\frac{dP}{d\Omega}\right)_{\text{avg}} = \frac{c}{288\pi}k^6[\hat{n}\times\vec{Q}(\hat{n})]^2. \tag{11.120}$$

The angular dependence is complicated, so we will content ourselves with computing the total power. We have (letting $\vec{Q}$ represent either $\dddot{\vec{Q}}(\hat{n},t_0)$ or $\vec{Q}(\hat{n})$)

$$\begin{aligned}[\hat{n}\times\vec{Q}]\cdot[\hat{n}\times\vec{Q}] &= \hat{n}\cdot[\vec{Q}\times(\hat{n}\times\vec{Q})],\\ &= \hat{n}\cdot\left[\hat{n}\vec{Q}^2-\vec{Q}(\hat{n}\cdot\vec{Q})\right] = \vec{Q}^2-(\hat{n}\cdot\vec{Q})^2\\ &= \sum_{i,j,k}Q_{ki}Q_{jk}n_in_j - \sum_{i,j,k,\ell}Q_{ij}Q_{k\ell}n_in_jn_kn_\ell.\end{aligned} \tag{11.121}$$

It is possible to compute $P(t)$ and P_{avg} by integrating (11.121) over all angles. This is the approach taken in Exercise 11.5.12. There is however an elegant method for computing power that avoids these messy integrations, which we now present.

First, we compute $[\hat{n} \times \vec{Q}]^2$ in the special case where the current distribution is azimuthally symmetric. Let's choose the z-axis along the axis of symmetry. Then from (11.106) we find that the quadrupole tensor Q_{ij} is diagonal. (The same result holds true for the moment of inertia tensor in classical mechanics.) If we define $Q_0 \equiv Q_{33}$, then the tracelessness of Q_{ij} implies that

$$Q_{11} = Q_{22} = -\frac{Q_{33}}{2} = -\frac{Q_0}{2}. \tag{11.122}$$

Substituting these values into (11.121), we find:

$$\begin{aligned}
[\hat{n} \times \vec{Q}]^2 &= \sum_i (Q_{ii})^2 n_i^2 - \sum_i Q_{ii} n_i^2 \sum_k Q_{kk} n_k^2, \\
&= Q_0^2 \left(\cos^2\theta + \frac{1}{4}\sin^2\theta \right) - Q_0^2 \left(\cos^2\theta - \frac{1}{2}\sin^2\theta \right)^2, \\
&= \frac{9}{4} Q_0^2 \cos^2\theta \sin^2\theta.
\end{aligned} \tag{11.123}$$

Using $\int_0^\pi d\theta \, \sin\theta \sin^2\theta \cos^2\theta = \frac{4}{15}$ we find for the real harmonic case with azimuthal symmetry

$$P_{\text{avg}} = \frac{ck^6}{240}(Q_0)^2. \tag{11.124}$$

For instantaneous power, we may replace Q_0 with $\dddot{Q}_0$ everywhere in the above argument to obtain

$$P(t) = \frac{1}{120c^5}(\dddot{Q}_0(t_0))^2, \tag{11.125}$$

for the azimuthally-symmetric case.

Now comes some magic. We know that $P(t)$ is a rotationally invariant scalar, since total power doesn't depend on orientation. We also know from (11.121) that $P(t)$ is a polynomial of degree 2 in the entries of Q. There is a deep mathematical theorem that characterizes all rotational invariants that are polynomials in matrix entries.[1] The theorem states that all such invariants can be expressed as a sum of terms, where each term is a constant times a product of factors of the form

$$\mathrm{Tr}\left[Q^{(1)} \ldots Q^{(n)}\right], \tag{11.126}$$

where each $Q^{(j)}$ is equal to either Q or Q^T. The possible second-order terms that fit this description are:

$$\mathrm{Tr}\,[Q]^2; \;\; \mathrm{Tr}\,[QQ]; \;\; \mathrm{Tr}\left[Q^T Q\right]; \;\; \mathrm{Tr}\left[Q^T Q^T\right]. \tag{11.127}$$

[1] C. Procesi, "The invariant theory of $n \times n$ matrices", Adv. in Math. **19**: 306–381 (1976).

Since Q is symmetric and traceless, the reader may verify that each term in (11.127) is either equal to zero or $\sum_{ij}(Q_{ij})^2$. Thus $P(t)$ must be expressible as a constant times $\sum_{ij}(Q_{ij})^2$. To find this constant, all we have to do is look at the azimuthally symmetric case. From (11.122) we have $\sum_{ij}(Q_{ij})^2 = (3/2)Q_0^2$. Thus we may rewrite (11.124) as

$$P_{\text{avg}} = \frac{ck^6}{360}\sum_{i,j}(Q_{ij})^2, \tag{11.128}$$

and our invariance argument proves that this is the *general* equation for real harmonic sources. Similarly,

$$P(t) = \frac{1}{180c^5}\sum_{i,j}(\dddot{Q}_{ij}(t_0))^2, \tag{11.129}$$

is the general equation for instantaneous power from an arbitrarily-varying electric quadrupole. This miraculous technique of inferring general expressions using invariance or symmetry is tremendously powerful, and is widely used in physics.

The results in this section are limited to the lowest three moments $\vec{p}$, $\vec{m}$ and Q_{ij} from a dynamic charge distribution. For more complicated sources, many multipoles may be necessary. The situation is similar to the electrostatic case in Sections 5.1 and 5.2, where the potential, or equivalently the charge distribution, was expanded in the spherical harmonics Y_{lm}. A similar spherical multipole formalism exists for harmonic radiation based upon the vector spherical harmonics, $\vec{X}_{lm}$ and $\hat{n} \times \vec{X}_{lm}$, introduced in Section 10.11 for the spherical resonator.[2] However, unlike the $1/r^{l+1}$ field falloff in the electrostatic case for the lth multipole, all the radiation multipoles contribute to the $\sim 1/r$ radiation field. This advanced topic will not be covered here, but we will see a further use of the spherical harmonic multipole formalism in the scattering results of Section 12.5. We will also compare the low moments obtained from various formalisms, including a simpler spherical multipole formalism, for an antenna example in Exercise 11.5.13.

[2] J. D. Jackson, *Classical Electrodynamics*, 3rd ed., John Wiley & Sons (New York) 1999, Section 9.7; C. J. Bouwkamp and H. B. G. Casimir, "On Multipole Expansions in the Theory of Electromagnetic Radiation", Physica **20**, 539 (1954).

11.6 Point particle radiation: Larmor and Liénard results

Let's look at radiation from point particles from the most general point of view. Exercise 8.6.7 examines the case where (here e is a general charge):

$$\rho(\vec{x},t) = e\delta(\vec{x}-\vec{x}(t)); \qquad \vec{J}(\vec{x},t) = e\vec{v}(t)\delta\left(\vec{x}-\vec{x}(t)\right), \tag{11.130}$$

where $\vec{x}$ (without argument) denotes the field point, and $\vec{x}(t)$ (with argument) denotes the particle position. The exact expression for the vector potential turned out to be (the so-called Liénard-Wiechert potentials)

$$\Phi(\vec{x},t) = \frac{e}{R - \frac{1}{c}\vec{R}\cdot\vec{v}(t_R)}; \quad \vec{A}(\vec{x},t) = \frac{\vec{v}(t_R)}{c}\Phi(\vec{x},t), \tag{11.131}$$

where t_R is the *exact retarded time*, defined implicitly by:

$$t_R \equiv t - \frac{R}{c}; \quad \vec{R} \equiv \vec{x} - \vec{x}(t_R); \quad R \equiv |\vec{R}|, \tag{11.132}$$

and

$$\vec{v}(t_R) \equiv \left.\frac{d\vec{x}(t)}{dt}\right|_{t=t_R}. \tag{11.133}$$

We emphasize that (11.132) are *not* explicit equations for t_R and $\vec{R}$, since each is defined in terms of the other. Rather, these equations express the interrelationships between these quantities.

In order to compute the power via (11.55) we will need to compute $\partial A/\partial t$. For this we will need the partial time derivatives of $\vec{R}$, R, and t_R. So we compute (using (11.132) to obtain $\partial\vec{R}/\partial t$):

$$\begin{aligned}\frac{\partial R}{\partial t} &= \frac{1}{2R}\frac{\partial R^2}{\partial t} = \frac{\vec{R}}{R}\cdot\frac{\partial\vec{R}}{\partial t} \\ &= -\hat{R}\cdot\underbrace{\dot{\vec{x}}(t_R)}_{\vec{v}(t_R)}\frac{\partial t_R}{\partial t}.\end{aligned} \tag{11.134}$$

However

$$\frac{\partial t_R}{\partial t} = 1 - \frac{1}{c}\frac{\partial R}{\partial t}, \tag{11.135}$$

so we may plug this back into (11.134) to get

$$\frac{\partial R}{\partial t} = -\hat{R}\cdot\vec{v}(t_R)\left(1 - \frac{1}{c}\frac{\partial R}{\partial t}\right). \tag{11.136}$$

Solving this for $\partial R/\partial t$ gives

$$\frac{\partial R}{\partial t} = \frac{-\hat{R}\cdot\vec{v}(t_R)}{1-\hat{R}\cdot\frac{1}{c}\vec{v}(t_R)}, \tag{11.137}$$

and plugging this into (11.135) gives

$$\frac{\partial t_R}{\partial t} = 1 + \frac{\frac{1}{c}\hat{R}\cdot\vec{v}}{1-\hat{R}\cdot\vec{v}/c} = \frac{1}{1-\hat{R}\cdot\frac{1}{c}\vec{v}(t_R)}, \tag{11.138}$$

and therefore

$$\frac{\partial\vec{v}(t_R)}{\partial t} = \dot{\vec{v}}(t_R)\frac{\partial t_R}{\partial t} = \frac{\dot{\vec{v}}(t_R)}{1-\frac{1}{c}\hat{R}\cdot\vec{v}(t_R)}. \tag{11.139}$$

Now if we try to use these derivatives to compute $\partial\vec{A}/\partial t$ where $\vec{A}$ is given by (11.131), we'll end up with a fairly complicated expression. Fortunately, we are interested in the far-field region, where the equations simplify considerably. In this region r is much larger than any other length scale in the problem, so in particular we're assuming the particle remains in a bounded region. Therefore, we may throw out all terms that fall off faster than $1/r$. This means that we may replace t_R in our equations with the *approximate retarded time*, defined implicitly by

$$t_r \equiv t - \frac{1}{c}\left(r - \hat{n}\cdot\vec{x}(t_r)\right). \tag{11.140}$$

In the far-field limit we may also replace the expression (11.131) for $\vec{A}(\vec{x},t)$ with

$$\vec{A}(\vec{x},t) \approx \frac{e}{cr}\frac{\vec{v}(t_r)}{1-\hat{n}\cdot\frac{1}{c}\vec{v}(t_r)}. \tag{11.141}$$

(An alternative derivation of (11.141) is described in Exercise 11.6.1.) This equation is an *exact* statement in an *instantaneous* frame of reference with an origin given by $\vec{x}(t_R) = 0$, because then $t_R = t_r = t - r/c$.

If we apply (11.139) to the approximation (11.141), we get

$$\frac{\partial\vec{A}}{\partial t} \approx \frac{e}{cr}\left[\frac{\frac{1}{c}(\hat{n}\cdot\dot{\vec{v}})\vec{v} + \left(1-\frac{1}{c}(\hat{n}\cdot\vec{v})\right)\dot{\vec{v}}}{(1-\frac{1}{c}\hat{n}\cdot\vec{v})^3}\right]_{\text{ret}}, \tag{11.142}$$

where the subscript "ret" indicates that $\vec{v}$ and $\dot{\vec{v}}$ represent $\vec{v}(t_r)$ and $\dot{\vec{v}}(t_r)$, respectively.

To find the power distribution, we'll need to compute $\hat{n} \times \partial \vec{A}/\partial t$. The computation is simplified if we first rewrite the numerator in (11.142):

$$
\begin{aligned}
-\hat{n} \times \left(\left(\hat{n} - \frac{\vec{v}}{c} \right) \times \dot{\vec{v}} \right) &= -\hat{n} \times (\hat{n} \times \dot{\vec{v}}) + \hat{n} \times \left(\frac{\vec{v}}{c} \times \dot{\vec{v}} \right) \\
&= -\hat{n}(\hat{n} \cdot \dot{\vec{v}}) + \dot{\vec{v}} + \frac{\vec{v}}{c}(\hat{n} \cdot \dot{\vec{v}}) - \dot{\vec{v}} \left(\frac{\hat{n} \cdot \vec{v}}{c} \right) \qquad (11.143)
\end{aligned}
$$

$$
\begin{aligned}
\Longrightarrow \hat{n} \times \frac{\partial \vec{A}}{\partial t} &\approx \hat{n} \times \frac{e}{cr} \left[\frac{\hat{n}(\hat{n} \cdot \dot{\vec{v}}) - \hat{n} \times ((\hat{n} - \frac{\vec{v}}{c}) \times \dot{\vec{v}})}{(1 - \frac{1}{c}\hat{n} \cdot \vec{v})^3} \right]_{\text{ret}} \\
&\approx \frac{e}{cr} \hat{n} \times \left[\frac{-\hat{n} \times ((\hat{n} - \frac{\vec{v}}{c}) \times \dot{\vec{v}})}{(1 - \frac{1}{c}\hat{n} \cdot \vec{v})^3} \right]_{\text{ret}} . \qquad (11.144)
\end{aligned}
$$

For any vector $\vec{Q}$ we have:

$$
[\hat{n} \times (\hat{n} \times \vec{Q})]^2 = [\hat{n} \times \vec{Q}]^2, \qquad (11.145)
$$

so we obtain finally

$$
\frac{dP(t)}{d\Omega} = r^2 \frac{1}{4\pi c} \left[\hat{n} \times \frac{\partial \vec{A}}{\partial t} \right]^2_{\text{ret}} = \frac{e^2}{4\pi c^3} \left[\frac{[\hat{n} \times ((\hat{n} - \vec{\beta}) \times \dot{\vec{v}})]^2}{(1 - \hat{n} \cdot \vec{\beta})^6} \right]_{\text{ret}} . \qquad (11.146)
$$

($\vec{\beta} \equiv \vec{v}/c$). In deriving this result, we have not made any assumption on the velocity $\vec{v}$ (in contrast to our multipole results in the previous section, which required the assumption that $v \ll c$). In fact, (11.146) is an exact result in the far-field approximation.

We should mention a subtlety in this situation that was not present in our previous computations. In previous cases involving continuous $\vec{J}$ distributions, the source was taken as being at rest with respect to the chosen origin. Now, however, the source is in motion. This means that the instantaneous rate of energy emission at the source at time t_r is not equal in general to the instantaneous rate of absorption at radius r at time t. This difference in these rates has nothing to do with the relativistic transformations we will study in Chapter 13, since both t and t_r are measured in the same frame of reference, namely the absorption frame. Rather, it's a kinematical consequence of the finite speed of light, and is closely related to the Doppler effect. The energy absorbed in the infinitesimal time interval $[t, t + dt]$ was emitted in time interval $[t_r, t_r + dt_r]$. It follows that the angular distribution of power emission is given by

$$\frac{dP(t_r)}{d\Omega} \equiv \frac{d^2E(t_r)}{dt_r d\Omega} = \left.\frac{d^2E(t_r)}{dt\,d\Omega}\right|_{t_r\to t} \frac{dt}{dt_r} = \frac{dP(t)}{d\Omega}\left(1 - \frac{1}{c}\hat{n}\cdot\vec{v}\right)$$
$$= \frac{e^2}{4\pi c^3}\left[\frac{\left[\hat{n}\times\left((\hat{n}-\vec{\beta})\times\dot{\vec{v}}\right)\right]^2}{\left(1-\hat{n}\cdot\vec{\beta}\right)^5}\right]_{\text{ret}} . \qquad (11.147)$$

Although the *instantaneous* rates of emission and absorption differ, the *total* energy emitted by the particle in its complete history must be the same as the total energy absorbed. In particular, if the motion is periodic the energy emitted over a period is equal to the energy absorbed in the same period.

The kinematical factor dt/dt_r in (11.147) has a nice geometrical interpretation, which is shown in Figure 11.7.[3] The two wavefronts shown in the figure both pass through $\vec{x}$, but are emitted at different times (t_R and $t_R + \Delta t_R$, respectively) and arrive at $\vec{x}$ at different times (t and $t + \Delta t$, respectively). From the figure we have:

$$X = c[\overbrace{(t+\Delta t)}^{\text{absorbed}} - \overbrace{(t_R+\Delta t_R)}^{\text{emitted}}],$$
$$= R + c(\Delta t - \Delta t_R). \qquad (11.148)$$

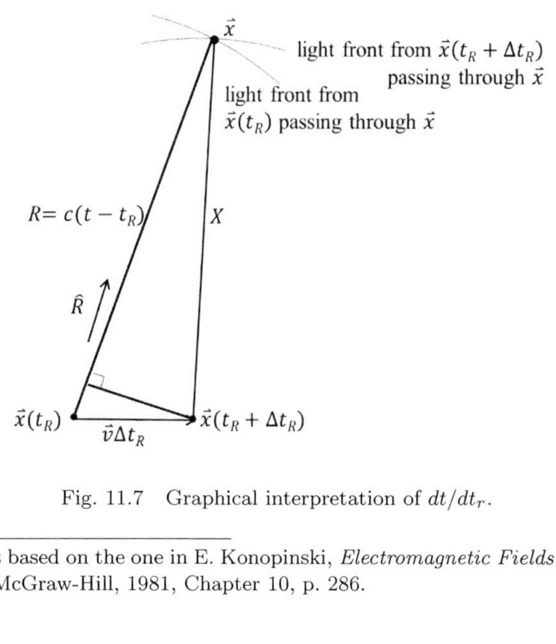

Fig. 11.7 Graphical interpretation of dt/dt_r.

[3]The figure is based on the one in E. Konopinski, *Electromagnetic Fields and Relativistic Particles*, McGraw-Hill, 1981, Chapter 10, p. 286.

On the other hand, the diagram also shows that (as $\Delta t_R \to 0$)

$$\begin{aligned} X &\approx R - \vec{v} \cdot \hat{R} \Delta t_R \\ \Longrightarrow \Delta t &\approx \Delta t_R \left(1 - \frac{1}{c} \vec{v} \cdot \hat{R}\right). \end{aligned} \tag{11.149}$$

In the small time interval limit, this says that

$$\text{absorption interval} = \text{emission interval} \times \left(1 - \frac{\text{line-of-sight velocity}}{c}\right),$$

which includes the familiar Doppler factor. In the far field limit, this then gives the dt/dt_r factor in Equation (11.147).

It's worth taking a little time to consider the differing roles of t and t_r in going forward from here. As far as the angular distribution of absorbed radiation is concerned, the quantity $dP(t)/d\Omega$ is what's actually measured in the laboratory. In contrast, the quantity $dP(t_r)/d\Omega$ is a sort of hybrid beast with an *emission* rate divided by an *absorption* (or laboratory) solid angle. However, because of the interdependence of time/angle in the $t \to t_r$ transformation, we realize that each absorption angular position at large r is associated with a unique transformation relative to the emission frame. If our goal is to uncover a general expression for the radiated power which is not tied to any particular absorption frame, it would be best to adopt the emission frame *before* the integration over angles above. This will result in an integrated total $P(t_r) = dE(t_r)/dt_r$ which has simple transformation properties when moving between reference frames. In fact in Section 13.7 we will see that the $P(t_r)$ we obtain (upcoming Equation (11.160)) is what we will call a Lorentz invariant. Once equipped with such a general expression, we can recover the particular form of the radiated power in any other inertial frame.

One reference frame of particular interest is the instantaneous rest frame of the emitting particle. To find $dP(t)/d\Omega$ in this frame, we set $\vec{v}(t_r) = 0$ in (11.147) which implies $dt/dt_r = 1$ so that

$$\frac{dP(t_r)}{d\Omega} = \frac{dP(t)}{d\Omega} = \frac{e^2}{4\pi c^3}(\hat{n} \times \dot{\vec{v}}(t_r))^2. \tag{11.150}$$

This equation differs slightly from (11.98) in that $\hat{n} \times \dot{\vec{v}}(t_r)$ replaces $\hat{n} \times \dot{\vec{v}}(t_0)$. Actually, if we keep the higher-order corrections characterizing the particle's trajectory in our multipole formalism, then we would just reform the $\dot{\vec{v}}(t_r)$ from the $\dot{\vec{v}}(t_0)$ that appears in the electric dipole result. In fact, if in addition we choose our spatial origin so that $\vec{x}(t_r) = 0$, then $t_r = t_0$, $dt_r/dt_0 = 1$ and we have

$$\frac{dP(t_0)}{d\Omega} = \frac{dP(t)}{d\Omega} = \frac{e^2}{4\pi c^3}(\hat{n}\times\dot{\vec{v}}(t_0))^2, \tag{11.151}$$

$$P(t) = \frac{2e^2}{3c^3}[\dot{\vec{v}}(t_0)]^2. \tag{11.152}$$

These agree exactly with (11.98) and the Larmor formula (11.99). Thus, the Larmor formula is an *exact* result in this particular frame of reference.

Since (11.152) is an exact result in a known frame of reference, we could directly find the power in any other frame *if* we knew the transformation laws connecting arbitrary frames of reference. We don't know these transformation laws yet, but we will come back to this point of view later (see Exercise 13.7.1). For now, we can manage to find the general result for $P(t_r)$ by integrating our expression for $dP(t_r)/d\Omega$ over all solid angles. To do this, let's take our instantaneous z-axis along $\vec{v}(t_r)$ and choose our origin so that $\vec{x}(t_r) = 0$, in which case $t_r = t_0$. Therefore the numerator in (11.147) becomes

$$\begin{aligned}\left[\hat{n}\times\left(\left(\hat{n}-\frac{\vec{v}}{c}\right)\times\dot{\vec{v}}\right)\right]^2 &= \left[\hat{n}(\hat{n}\cdot\dot{\vec{v}})-\dot{\vec{v}}-\frac{\vec{v}}{c}(\hat{n}\cdot\dot{\vec{v}})+\dot{\vec{v}}\left(\frac{\hat{n}\cdot\vec{v}}{c}\right)\right]^2\\ &= -(\hat{n}\cdot\dot{\vec{v}})^2\left(1-\frac{v^2}{c^2}\right)+\dot{\vec{v}}^2\left(1-\left(\frac{\hat{n}\cdot\vec{v}}{c}\right)\right)^2+2\left(\frac{\vec{v}\cdot\dot{\vec{v}}}{c}\right)(\hat{n}\cdot\dot{\vec{v}})\left(1-\left(\frac{\hat{n}\cdot\vec{v}}{c}\right)\right)\\ &= -\dot{v}^2\cos^2\theta_{\dot{v}}(1-\beta^2)+\dot{v}^2\,(1-\beta\cos\theta)^2+2\frac{v\dot{v}^2}{c}\cos\gamma\cos\theta_{\dot{v}}(1-\beta\cos\theta),\end{aligned} \tag{11.153}$$

where we have introduced the notation

$$\dot{v}\equiv|\dot{\vec{v}}|;\quad v\equiv|\vec{v}|;\quad \beta\equiv\frac{v}{c}, \tag{11.154}$$

and $\theta_{\dot{v}}, \theta$ and γ are as shown in Figure 11.8. From this figure we may also see that

$$\cos\theta_{\dot{v}} = \cos\theta\cos\gamma+\sin\theta\sin\gamma\cos\phi. \tag{11.155}$$

The denominator in (11.147) is somewhat easier to express:

$$\left(1-\frac{1}{c}\hat{n}\cdot\vec{v}\right) = (1-\beta\cos\theta), \tag{11.156}$$

The only ϕ-dependence in the expression is in the factors of $\cos^2\theta_{\dot{v}}$ and $\cos\theta_{\dot{v}}$ in the numerator. The necessary ϕ-integrals in integrating over all solid angles are given by

$$\int_0^{2\pi} d\phi\cos^2\theta_{\dot{v}} = 2\pi\left(\cos^2\theta\cos^2\gamma+\frac{1}{2}\sin^2\theta\sin^2\gamma\right), \tag{11.157}$$

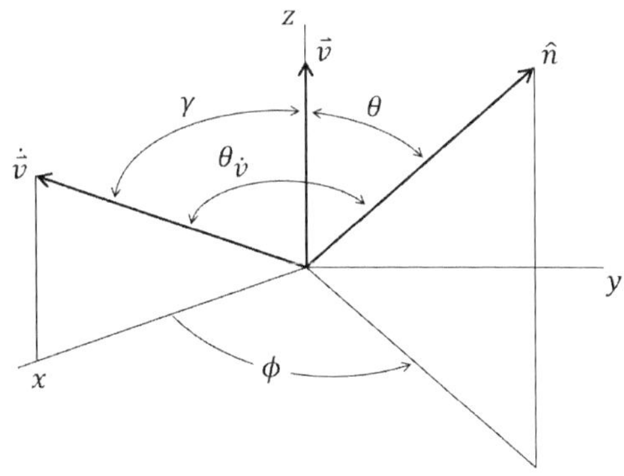

Fig. 11.8 Notation used in calculation of $dP(t_r)/d\Omega$.

$$\int_0^{2\pi} d\phi \cos\theta_{\dot{v}} = 2\pi\cos\theta\cos\gamma. \tag{11.158}$$

We now have

$$P(t_r) = \frac{e^2\dot{v}^2}{2c^3}\int_0^{\pi}\frac{d\theta\sin\theta}{(1-\beta\cos\theta)^5}\left[-(1-\beta^2)\left(\cos^2\theta\cos^2\gamma + \frac{1}{2}\sin^2\theta\sin^2\gamma\right)\right.$$
$$\left. + (1-\beta\cos\theta)^2 + 2\beta\cos\theta\cos^2\gamma(1-\beta\cos\theta)\right]. \tag{11.159}$$

The rest of the integration is straightforward (see Exercise 11.6.3), and the result is

$$\begin{aligned} P(t_r) &= \frac{e^2\dot{v}^2}{2c^3}\frac{\frac{4}{3} - \frac{4}{3}\beta^2\sin^2\gamma}{(1-\beta^2)^3} \\ &= \frac{2e^2}{3c^3}\frac{\dot{v}^2 - (\vec{\beta}\times\dot{\vec{v}})^2}{(1-\beta^2)^3}, \end{aligned} \tag{11.160}$$

where all $\vec{v}, \dot{\vec{v}}$ are evaluated at the approximate retarded time t_r in general. This is the *Liénard* result, a generalization of the Larmor formula to arbitrary frames.

To get a better feeling for the physics involved here, let us look at two special cases of the above general results. First let's take the case where $\vec{v}$ and $\dot{\vec{v}}$ are parallel, and consider the angular dependence of $dP(t)/d\Omega$. From (11.146) we have

$$\left.\frac{dP(t)}{d\Omega}\right|_{\vec{v}\parallel\dot{\vec{v}}} = \frac{e^2}{4\pi c^3}\frac{(\hat{n}\times\dot{\vec{v}})^2}{(1-\frac{1}{c}\hat{n}\cdot\vec{v})^6} = \frac{e^2\dot{v}^2}{4\pi c^3}\frac{\sin^2\theta}{(1-\beta\cos\theta)^6}. \tag{11.161}$$

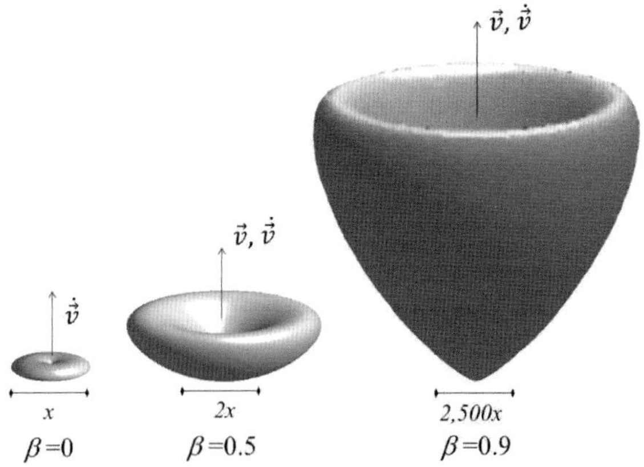

Fig. 11.9 Radiation patterns when $\dot{\vec{v}}||\vec{v}$, for three different values of β. The intervals below each pattern indicate the relative scales for the three distributions shown.

Figure 11.9 shows the radiation pattern for different values of β. The beam's angle relative to the direction of motion (denoted by θ_m) is determined by the condition

$$\begin{aligned} 0 &= \frac{d}{d\theta}\left(\frac{dP(t)}{d\Omega}\right)\bigg|_{\theta=\theta_m} \\ \Longrightarrow 0 &= 4\beta\cos^2\theta_m + 2\cos\theta_m - 6\beta = 0 \\ \Longrightarrow \cos\theta_m &= \frac{1}{4\beta}(\sqrt{1+24\beta^2}-1). \end{aligned} \tag{11.162}$$

For $\beta \approx 1$ we define $\delta \equiv 1-\beta$, so $\delta \ll 1$ and

$$\begin{aligned} \cos\theta_m &= \frac{1}{4(1-\delta)}(\sqrt{1+24(1-\delta)^2}-1) \\ &\approx \frac{1}{4}(1+\delta)\left(5\left(1-\frac{24}{25}\delta\right)-1\right) \\ &= 1-\frac{\delta}{5} \\ &= 1-\frac{1}{5}(1-\beta). \end{aligned} \tag{11.163}$$

Using $\cos\theta_m \approx 1-\theta_m^2/2$ for $\theta_m \ll 1$, we get finally in the extreme relativistic limit

$$\theta_m \approx \sqrt{\frac{2}{5}(1-\beta)} \approx \sqrt{\frac{1-\beta^2}{5}} = \frac{1}{\gamma\sqrt{5}}, \tag{11.164}$$

where γ is the relativistic γ-factor

$$\gamma^2 \equiv \frac{1}{1-\beta^2}. \tag{11.165}$$

The total emitted power is

$$P(t_r) = \frac{2}{3}\frac{e^2}{c^3}\frac{\dot{v}^2}{(1-\beta^2)^3} = \frac{2}{3}\frac{e^2}{c^3}\gamma^6\dot{v}^2, \tag{11.166}$$

which is minimized in the reference frame where $v = 0$ (in which case it reduces to Larmor's formula).

For our second case, we consider the situation where $\vec{v}$ and $\dot{\vec{v}}$ are perpendicular. In this case we find

$$\left[\hat{n} \times \left(\left(\hat{n} - \frac{\vec{v}}{c}\right) \times \dot{\vec{v}}\right)\right]^2 = -\dot{v}^2 \sin^2\theta \cos^2\phi(1-\beta^2) + \dot{v}^2(1-\beta\cos\theta)^2, \tag{11.167}$$

which leads to

$$\frac{dP(t)}{d\Omega} = \frac{e^2}{4\pi c^3}\frac{\dot{v}^2}{(1-\beta\cos\theta)^4}\left[1 - \frac{(1-\beta^2)\sin^2\theta\cos^2\phi}{(1-\beta\cos\theta)^2}\right]. \tag{11.168}$$

Figure 11.10 shows the radiation pattern given by (11.168), which is peaked in the forward direction and resembles a train headlight when $\beta \approx 1$. The total emitted power is

$$P(t_r) = \frac{2e^2}{3c^3}\frac{\dot{v}^2}{(1-\beta^2)^3}(1-\beta^2) = \frac{2e^2}{3c^3}\frac{\dot{v}^2}{(1-\beta^2)^2} = \frac{2}{3}\frac{e^2}{c^3}\dot{v}^2\gamma^4. \tag{11.169}$$

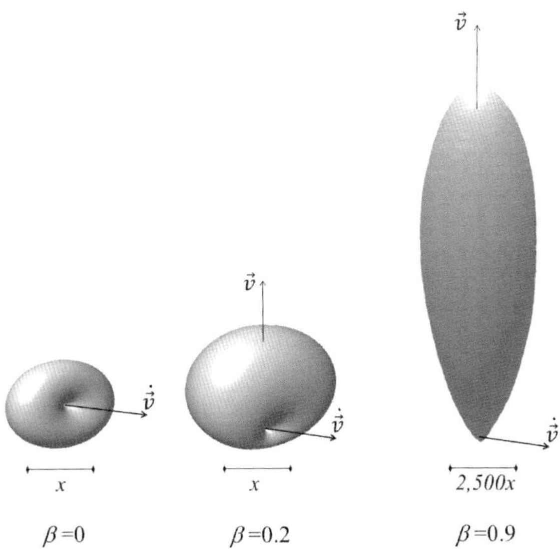

Fig. 11.10 Radiation patterns when $\dot{\vec{v}} \perp \vec{v}$, for three different values of β. The intervals below each pattern indicate the relative scales for the three distributions shown.

For a fixed $\dot{v}^2$ and β, $P_{\text{linear}} = \gamma^2 P_{\text{circular}}$ so that apparently the radiation loss is much larger in the case of linear acceleration. However, we will see in Section 13.7 that due to relativistic effects this conclusion is reversed if we fix the magnitude of the force (given by $|d\vec{p}/dt|$) instead of $\dot{v}^2$.

For periodic circular motion we have (radius $= R$)

$$\dot{v}^2 = \omega^2 v^2 = \omega \frac{v^3}{R} = \omega c^3 \frac{\beta^3}{R}, \tag{11.170}$$

and the energy lost in one period $T = 2\pi/\omega$ is inversely proportional to R:

$$\Delta E = PT = \frac{4\pi e^2}{3R} \frac{\beta^3}{(1-\beta^2)^2}. \tag{11.171}$$

We will re-derive this result in Section 11.9. Further engineering considerations for circular particle accelerators will be discussed in Section 13.7.

11.7 Nonrelativistic treatment of bremsstrahlung

An important application of point particle radiation occurs when charged particles are deflected as they interact with atomic nuclei. This type of radiation is called *bremsstrahlung* ("braking radiation" in German). This process was referred to in Section 9.9, where we discussed dissipative energy losses and introduced the idea of stopping power. Here we will treat only the nonrelativistic case, as in the earlier discussion. In this case, the energy loss due to dissipation dominates the radiative losses. However, it turns out that at ultrarelativistic energies the situation reverses, and nuclear bremsstrahlung becomes the dominant energy loss mechanism. We will start in a relativistic context in order to see where the various approximations come into play. As we proceed, we will see the limitations of the classical theory.

To lowest order we know the approximate forces involved from the assumed classical path. We will use the accelerations experienced along that path to approximately evaluate the vector potential and thus find the radiation generated. The kinematical situation is shown in Figure 11.11. The relativistic force law in the atomic rest frame is given by

$$\vec{F} = \frac{d\vec{p}}{dt}, \quad \vec{p} = M\gamma\vec{\beta}c, \tag{11.172}$$

where M is the mass of the incoming particle, which can be any type of charged particle. We will take the velocity to be along the z-direction, while

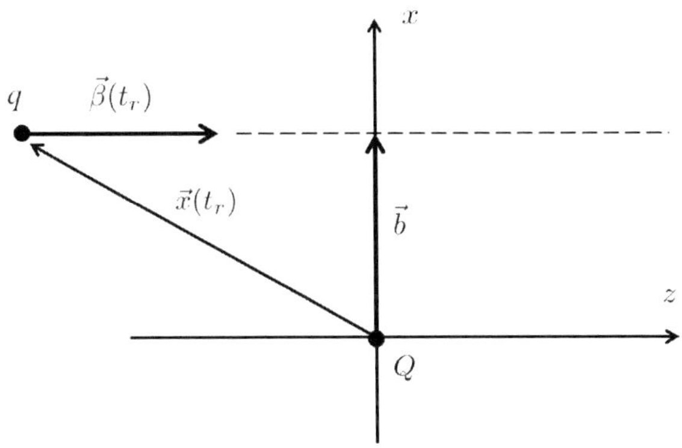

Fig. 11.11 Kinematical situation for bremsstrahlung calculation.

the impact parameter, which characterizes the particle's path, is taken along the x-direction:

$$\vec{\beta} = (0, 0, \beta), \quad \vec{b} = (b, 0, 0). \tag{11.173}$$

Note that $\vec{b} \cdot \vec{\beta} = 0$. The force is taken to be Coulombic:

$$\vec{F}_C = \frac{qQ\hat{r}}{r^2}. \tag{11.174}$$

Here q is the charge of the incoming particle and Q is the charge of the scattering center, which we assume to be fixed.

As a side note to our calculation, we remark that the assumption (11.174) significantly limits the applicability of the model. In fact, the incoming particle is influenced by atomic electrons as well as the nucleus. It's possible to make a more realistic calculation that includes the effect of atomic screening by using an effective force of the form

$$\vec{F}_C^S = \frac{qQ\hat{r}}{r^2} e^{-r/a}, \tag{11.175}$$

where the parameter a corresponds to the size of the atom. It turns out that the influence of screening increases for both $\beta \ll 1$ and $\beta \approx 1$. In both of these situations, the minimum momentum transfer to the scattering center defines length scales that are comparable to a in (11.175), so that screening becomes significant. If the incoming particles are heavier than electrons, low-velocity screening effects are less significant. The present calculation assumes small deflections of the incoming particles, which also means that velocities can't be too small.

We return to the main flow of our argument. Assuming that particle deflections are small, the maximum change in potential energy of the particle must be small compared to the incoming kinetic energy

$$\frac{|qQ|}{b} \ll \frac{1}{2}Mv^2, \tag{11.176}$$

which corresponds to a minimum impact parameter of

$$b_{\min} = \frac{2|qQ|}{Mv^2}. \tag{11.177}$$

For the particle's path, we will use the straight-path classical expression

$$\vec{x}(t) = \vec{b} + \vec{\beta}ct. \tag{11.178}$$

The relativistic expression

$$\begin{aligned} \frac{d\vec{p}}{dt} &= \frac{d}{dt}\left(\gamma\vec{\beta}Mc\right) \\ &= Mc\gamma\left(\dot{\vec{\beta}} + \gamma^2\vec{\beta}\,(\vec{\beta}\cdot\dot{\vec{\beta}})\right), \end{aligned} \tag{11.179}$$

can be used to compute the equation of motion (see Exercise 11.7.1), which in the nonrelativistic limit ($\gamma \approx 1$) becomes

$$\dot{\vec{\beta}}(t) \approx \left(\frac{qQ}{Mc}\right)\frac{\vec{b} + \vec{\beta}ct}{\left(b^2 + (\beta ct)^2\right)^{3/2}}. \tag{11.180}$$

We are now in a position to begin the electromagnetic part of the calculation. We'll need an expression for the vector potential $\vec{A}$ in order to compute the differential energy density from (11.74). Using the simplified form of (11.142) given in (11.144), we have the exact expression

$$\frac{\partial\vec{A}(\vec{x},t)}{\partial t} = \frac{q}{r}\left[\frac{-\hat{n}\times\left((\hat{n}-\vec{\beta})\times\dot{\vec{\beta}}\right)}{(1-\hat{n}\cdot\vec{\beta})^3}\right]_{\text{ret}}. \tag{11.181}$$

In the nonrelativistic limit we then find

$$\begin{aligned} \vec{A}(\vec{x},\omega) &= \frac{i}{\sqrt{2\pi}\,\omega}\int_{-\infty}^{\infty} dt\, e^{i\omega t}\frac{\partial\vec{A}(\vec{x},t)}{\partial t} \\ &\approx -\frac{iq}{\sqrt{2\pi}\,\omega r}\left[\hat{n}\times\left(\hat{n}\times\int_{-\infty}^{\infty} dt\, e^{i\omega t}\dot{\vec{\beta}}(t_r)\right)\right]. \end{aligned} \tag{11.182}$$

The relationship between the observer's time t and the approximate retarded time t_r in the nonrelativistic limit is given by

$$\begin{aligned} t &= t_r + \frac{r}{c} - \frac{\hat{n}\cdot\vec{x}(t_r)}{c} \\ &\approx t_r + \frac{r}{c} - \frac{\hat{n}\cdot\vec{b}}{c}. \end{aligned} \tag{11.183}$$

Plugging both (11.183) and (11.180) into (11.182) gives

$$\vec{A}(\vec{x},\omega) = -\frac{q}{\sqrt{2\pi}\,\omega\, r}\, e^{i\omega/c(r-\hat{n}\cdot\vec{b})}\,\hat{n}\times\left[\left(\hat{n}\times\vec{b}\right)\mathcal{I} + \left(\hat{n}\times\vec{\beta}c\right)\mathcal{J}\right], \quad (11.184)$$

where ($t_r \to t$ in the integrations)

$$\begin{aligned}\mathcal{I} &\equiv \int_{-\infty}^{\infty} dt\, \frac{e^{i\omega t}}{(b^2+(\beta c\, t)^2)^{3/2}} \\ &= \frac{1}{(\beta c)^3}\int_{-\infty}^{\infty} dt\,\frac{e^{i\omega t}}{((b/(\beta c))^2+t^2)^{3/2}},\end{aligned} \quad (11.185)$$

and

$$\begin{aligned}\mathcal{J} &\equiv \int_{-\infty}^{\infty} dt\, \frac{t\, e^{i\omega t}}{(b^2+(\beta c\, t)^2)^{3/2}} \\ &= \frac{1}{(\beta c)^3}\int_{-\infty}^{\infty} dt\,\frac{t\, e^{i\omega t}}{((b/(\beta c))^2+t^2)^{3/2}}.\end{aligned} \quad (11.186)$$

These integral expressions may be identified as Bessel functions of imaginary argument,[4] which we first encountered in Section 4.6:

$$K_1(\omega T) = \frac{T}{2\omega}\int_{-\infty}^{\infty} dt\,\frac{e^{i\omega t}}{(T^2+t^2)^{3/2}}, \quad (11.187)$$

$$K_0(\omega T) = \frac{1}{2i\omega}\int_{-\infty}^{\infty} dt\,\frac{t\, e^{i\omega t}}{(T^2+t^2)^{3/2}}. \quad (11.188)$$

We can then write our integrals as

$$\mathcal{I} = \frac{2\omega}{b\,(\beta c)^2}\, K_1\left(\frac{b\,\omega}{\beta c}\right), \quad (11.189)$$

$$\mathcal{J} = \frac{2i\omega}{(\beta c)^3}\, K_0\left(\frac{b\,\omega}{\beta c}\right). \quad (11.190)$$

These results may be plugged into expression (11.184) for $\vec{A}(\vec{x},\omega)$, so that we may then compute $\left|\hat{n}\times\vec{A}(\vec{x},\omega)\right|^2$ to find the differential energy density from (11.74). Helpfully, the $\mathcal{I}$ and $\mathcal{J}$ parts of the integral have opposite complex phases so that taking the absolute square is straightforward. The final result is

$$\frac{d^2E}{d\omega d\Omega} = \frac{\omega^2 q^2}{\pi^2\beta^4 c^5}\left(\frac{qQ}{Mc}\right)^2\left[\left(\hat{n}\times\hat{b}\right)^2 K_1^2\left(\frac{b\,\omega}{\beta c}\right) + \left(\hat{n}\times\hat{\beta}\right)^2 K_0^2\left(\frac{b\,\omega}{\beta c}\right)\right]. \quad (11.191)$$

[4]See M. Abramowitz and I. A. Stegun, Eds., *Handbook of Mathematical Functions with Formulas, Graphs and Mathematical Tables*, p. 376, 9.6.25, and I. S. Gradshteyn and I. M. Ryzhik, *Table of Integrals, Series and Products*, 3.754 (3).

Next, we integrate over angles. According to our initial definitions, we have

$$\left(\hat{n}\times\hat{b}\right)^2 = \cos^2\theta + \sin^2\theta\sin^2\phi, \tag{11.192}$$

$$\left(\hat{n}\times\hat{\beta}\right)^2 = \sin^2\theta, \tag{11.193}$$

where θ and ϕ are the scattering angles. In our final calculation we will give the energy loss for a beam of particles rather than a single particle, so that the classical impact parameter direction $\hat{b}$ must be averaged over to get the physical angular dependence. Essentially, this means that we should average our energy angular cross-section (11.191) over the angle ϕ, in which case we have

$$\frac{1}{2\pi}\int d\phi\,(\cos^2\theta + \sin^2\theta\sin^2\phi) = \frac{1}{2}(\cos^2\theta + 1), \tag{11.194}$$

which is an averaged dipole pattern. The other dipole pattern from (11.193) is azimuthally symmetric already. Completing the angular integration, and getting ready for the next integration step, we write the result as

$$\frac{dE}{d\omega} = \frac{8q^4Q^2}{3\pi\beta^2c^5M^2b^2}\left[x^2K_1^2(x) + x^2K_0^2(x)\right], \tag{11.195}$$

where

$$x \equiv \frac{b\omega}{v}. \tag{11.196}$$

Let's think about stopping power in this context. We are now in possession of an expression (11.195) for the frequency distribution of radiation from the path deflection of a single particle. We would like to turn this into an expression for the energy loss of a particle beam as it passes through a medium. The impact parameter b is a classical parameter which should be integrated over, as is typically done in classical cross-section computations. Assuming cylindrical symmetry, this implies an integration factor of $2\pi\int db\,b$. We also need to integrate over the angular frequency spectrum produced. In addition, assuming that sources are incoherent we can multiply the result by the density of scatterers: this gives a factor of N/Z_2, where N is the nuclear charge density, or magnitude of electron charge density for neutral materials, and Z_2 is the number of elementary charges per nucleus. The result of these steps gives the expression:

$$\begin{aligned}\frac{dE}{dx} &\equiv 2\pi\frac{N}{Z_2}\int_0^{\omega_{\max}} d\omega\int_{b_{\min}}^{\infty} db\,b\left(\frac{dE}{d\omega}\right)\\ &= \frac{16q^4Q^2N}{3Z_2\beta^2c^5M^2}\int_0^{\omega_{\max}} d\omega\int_{b_{\min}}^{\infty}\frac{db}{b}\left[x^2K_1^2(x) + x^2K_0^2(x)\right].\end{aligned} \tag{11.197}$$

Note the upper and lower limits supplied on the ω and b integrals as well as the order of the integrations. In fact, if we were to interchange the integrations and ignore the upper limit on ω, the resulting integral would be completely finite, but wrong. The upper limit in the ω integral is recognizing that the angular frequency certainly can not exceed the limit based on the incoming energy of the particle. Nonrelativistically, we thus set

$$\hbar\omega_{\text{max}} = \frac{1}{2}Mv^2, \tag{11.198}$$

where we are interfacing with the quantum mechanical connection between energy and angular frequency. This is very similar to the stopping power calculation in Section 9.9, where we recognized a kinematical upper limit to the integrated momentum.

Carrying on, we may replace db/b in (11.197) with dx/x. The final integral in (11.197) is

$$\int_{b_{\text{min}}}^{\infty} \frac{db}{b}\left[x^2K_1^2(x) + x^2K_0^2(x)\right] = \int_{x_{\text{min}}}^{\infty} dx\, x\left[K_1^2(x) + K_0^2(x)\right], \tag{11.199}$$

where

$$x_{\text{min}} = \frac{b_{\text{min}}\omega}{v}. \tag{11.200}$$

Both integrands are well-behaved at large values of their argument, as seen in the asymptotic statement

$$K_\nu(x) \underset{x\gg 1,|\nu|}{\longrightarrow} \sqrt{\frac{\pi}{2k\rho}}e^{-x}, \tag{11.201}$$

from (4.101). The $K_0(x)$ integral is actually finite even for $x_{\text{min}} \to 0$, but the $K_1(x)$ integral is more problematical. We recall that (see (4.106) and (4.107))

$$K_1(x) \underset{x\ll 1}{\longrightarrow} \frac{1}{x}, \quad K_0(x) \underset{x\ll 1}{\longrightarrow} -\left(\ln\left(\frac{x}{2}\right) + \gamma\right), \tag{11.202}$$

telling us the $K_1(x)$ integral in (11.199) has a logarithmic divergence. In the $x_{\text{min}} \ll 1$ limit, we find that we may parameterize

$$\int_{x_{\text{min}}}^{\infty} dx\, x\left[K_1^2(x) + K_0^2(x)\right] = \ln\left(\frac{\lambda}{x_{\text{min}}}\right), \tag{11.203}$$

with $\lambda = 1.12292$ to a good approximation (Exercise 11.7.2). We now have

$$\frac{dE}{dx} = \frac{16q^4Q^2N}{3Z_2\beta^2c^5M^2}\int_0^{\omega_{\text{max}}} d\omega \ln\left(\frac{\lambda v}{\omega b_{\text{min}}}\right). \tag{11.204}$$

Our next step is to write a dimensionless form of the last integral:

$$\int_0^{\omega_{\max}} d\omega \ln\left(\frac{\lambda v}{\omega b_{\min}}\right) = \left(\frac{Mv^2}{2\hbar}\right)\int_0^1 d\bar{x}\, \ln\left(\frac{\lambda}{a\bar{x}}\right), \tag{11.205}$$

where

$$\bar{x} \equiv \frac{2\hbar\omega}{Mv^2}, \quad a \equiv \frac{b_{\min} Mv}{2\hbar}. \tag{11.206}$$

There seem to be two possible choices for $b_{\min}$ in our calculation:

(1) Classical: $b_{\min} = \dfrac{2|qQ|}{Mv^2} \implies a = \dfrac{|qQ|}{\hbar v}$.

(2) Semiclassical: $b_{\min} = \dfrac{\hbar}{Mv} \implies a = \dfrac{1}{2}$.

Possibility (1) is motivated by (11.177) above. However, this equation is actually just a statement of the limitation of the applicability of the classical calculation, not a restriction on the approach distance to the scattering center. Possibility (2) is motivated from a semiclassical limit on z component of nonrelativistic angular momentum. It is definitely an improvement upon (1). However, it also has a problem for it does not naturally give a smooth frequency spectrum cutoff at $\omega = \omega_{\max}$. We could force this by taking instead $b_{\min} = 2\lambda\hbar/(Mv) \implies a = \lambda$, but as we said this isn't natural.

At this point let's not mince words: all of these possibilities are wrong! Semiclassical ideas are no substitute for actual quantum kinematics. The correct replacement consists of:

$$\ln\left(\frac{\lambda}{a\bar{x}}\right) \longrightarrow \ln\left(\frac{(1+\sqrt{1-\bar{x}}\,)^2}{\bar{x}}\right), \tag{11.207}$$

where $\bar{x}$ retains its meaning from (11.206). Obviously, the classical theory can't reproduce this behavior. This logarithm arises in the quantum calculation of bremsstrahlung as a result of energy and momentum conservation in the integration over scattering angles in the low energy limit (see Exercise 11.7.3). This particular result was included among many others in a quantum mechanical bremsstrahlung calculation for various types of stopping power published by Bethe and Heitler in 1934.[5]

We may now integrate:

$$\int_0^1 d\bar{x}\, \ln\left(\frac{(1+\sqrt{1-\bar{x}})^2}{\bar{x}}\right) = 2, \tag{11.208}$$

[5]H. Bethe and W. Heitler, "On the Stopping of Fast Particles and on the Creation of Positive Electrons", Proc. R. Soc. Lond. **A**, 146 (1934).

giving

$$\frac{dE}{dx} = \frac{16q^4Q^2N}{3Z_2\hbar Mc^3} = \frac{16\alpha Z_1^4 Z_2 e^4 N}{3Mc^2}, \tag{11.209}$$

as our nonrelativistic result for the stopping power from nuclear bremsstrahlung, with some significant help from quantum correspondence. Here $\alpha \equiv e^2/(\hbar c)$ is the *fine structure constant*, $|q| = Z_1|e|$, and $|Q| = Z_2|e|$, where e is the electron charge. In this approximation one obtains a constant value for the stopping power that is independent of the particle energy. The bremsstrahlung energy from (11.209) is smaller than the dissipative energy loss in (9.272) by a ratio of approximately

$$\left(\frac{dE}{dx}\right)_{\text{brem.}} \Big/ \left(\frac{dE}{dx}\right)_{\text{diss.}} \sim \alpha\beta^2 Z_1^2 Z_2 \frac{m}{M}, \tag{11.210}$$

aside from the dissipative logarithmic factor: here the mass m is the electron mass, while M is the mass of the incoming particle (which may also be an electron).

As is suggested by using an (unjustified!) upper limit of $\omega_{\text{max}} = E/\hbar = \gamma Mc^2/\hbar$ in (11.205), the relativistic form of bremsstrahlung stopping power increases approximately linearly with total particle energy, E, and eventually becomes the dominant energy loss mechanism at ultrarelativistic energies; this means for example $\gamma \gtrsim 1/\alpha$ for $Z_1 \approx 1, Z_2 \approx 1, M = m$ from the modified version of (11.210). Although it is tempting to go back and attempt to confirm this with a relativistic calculation, we can see that in the end it would again be hampered by an inexact knowledge of the energy spectrum of the radiation, which can only come from the correct relativistic kinematics. Thus, it is better to follow the quantum mechanical treatments using the appropriate cross-sections and kinematics from this point forward. We have reached the limits of our classical considerations in this regard.

11.8 Radiation from periodic trajectories: general considerations

We will now consider radiation due to a particle in general periodic motion. We will be using the "real source formalism" presented in Section 11.2. This approach requires that the current be bounded in *time* as well as space, so technically it cannot be applied to strictly periodic currents. However, we may consider currents that are periodic over several periods, and take the limit as the number of periods goes to infinity. We will see that although the energy distribution as a function of frequency tends towards infinity, we may still calculate an average power distribution with finite total power by averaging over periods.

Let's start by deriving an expression for $d^2E/(d\omega d\Omega)$ for radiation from a moving particle. Beginning with (11.74)

$$\frac{d^2E}{d\omega d\Omega} = \frac{\omega^2 r^2}{2\pi c}\left|\hat{n}\times\vec{A}(\vec{x},\omega)\right|^2, \tag{11.211}$$

and recalling the definition of $\vec{A}(\vec{x},\omega)$ from expression (11.71), we have

$$\begin{aligned}\vec{A}(\vec{x},\omega) &= \frac{1}{cr}\int d^3x'\int_{-\infty}^{\infty} dt\,\vec{J}(\vec{x}\,',\omega)e^{-ik\hat{n}\cdot\vec{x}\,'}\\ &= \frac{e^{ikr}}{cr}\int d^3x'\int_{-\infty}^{\infty}\frac{dt}{\sqrt{2\pi}}\,\vec{J}(\vec{x}\,',t)e^{i\omega(t-\frac{1}{c}\hat{n}\cdot\vec{x}\,')}.\end{aligned} \tag{11.212}$$

We may write the particle trajectory as

$$\vec{J}(\vec{x},t) = e\vec{v}(t)\delta\left(\vec{x}-\vec{x}(t)\right), \tag{11.213}$$

so that

$$\vec{A}(\vec{x},\omega) = \frac{e^{ikr}}{\sqrt{2\pi}r}\frac{e}{c}\underbrace{\int_{-\infty}^{\infty} dt\,\vec{v}(t)e^{i\omega(t-\frac{1}{c}\hat{n}\cdot\vec{x}(t))}}_{\equiv\vec{\chi}}, \tag{11.214}$$

which implies

$$\frac{d^2E}{d\omega d\Omega} = \frac{e^2\omega^2}{4\pi^2c^3}\left|\hat{n}\times\vec{\chi}\right|^2. \tag{11.215}$$

Let's say now that the motion is periodic in time for $-NT \le t \le NT$, where T is the period and N is a positive integer. Then we can break up the integral as

$$\vec{\chi} = \vec{\chi}_{\text{start}} + \vec{\chi}_{\text{stop}} + \underbrace{\sum_{n=-N}^{N-1}\int_{nT}^{(n+1)T} dt\,\vec{v}(t)e^{i\omega(t-\frac{1}{c}\hat{n}\cdot\vec{x}(t))}}_{\equiv\vec{\chi}_N}, \tag{11.216}$$

where $\vec{\chi}_{\text{start,stop}}$ are the integrals over the time periods $(-\infty,-NT)$ and (NT,∞) respectively. This equation reflects the fact any realistic motion of a classical particle involves an aperiodic start and stop. We could model $\vec{\chi}_{\text{start,stop}}$ in many different ways. For example, we can imagine the particle to be confined to the circular trajectory for all times, but its velocity to be increased over some finite time interval to its periodic value. For this discussion, all that will matter about $\vec{\chi}_{\text{start,stop}}$ is that they are finite, and are assumed to be independent of N.

For the periodic portion of the motion, we make a change of variable on each interval,

$$t_n \equiv t - nT, \tag{11.217}$$

so that

$$\vec{\chi}_N = \sum_{n=-N}^{N-1} e^{i\omega T n} \int_0^T dt_n\, \vec{v}(t_n + nT) e^{i\omega(t_n - \frac{1}{c}\hat{n}\cdot\vec{x}(t_n+nT))}. \tag{11.218}$$

But

$$\vec{x}(t+nT) = \vec{x}(t); \quad \vec{v}(t+nT) = \vec{v}(t), \tag{11.219}$$

so all the integrals in (11.218) are equal and

$$\vec{\chi}_N = \left(\sum_{n=-N}^{N-1} e^{i\omega T n}\right) \int_0^T dt\, \vec{v}(t) e^{i\omega(t - \frac{1}{c}\hat{n}\cdot\vec{x}(t))}. \tag{11.220}$$

Notice that the integral is now independent of n. We have

$$\frac{d^2E}{d\omega d\Omega} = \frac{e^2\omega^2}{4\pi^2c^3}\left|\hat{n}\times\left(\vec{\chi}_{\text{start}} + \vec{\chi}_{\text{stop}} + \sum_{n=-N}^{N-1} e^{i\omega T n}\int_0^T dt\, \vec{v}(t)\, e^{i\omega(t-\frac{1}{c}\hat{n}\cdot\vec{x}(t)}\right)\right|^2. \tag{11.221}$$

The limit

$$\frac{d^2P(\omega)}{d\omega d\Omega} \equiv \lim_{N\to\infty}\frac{1}{2NT}\left(\frac{d^2E}{d\omega d\Omega}\right), \tag{11.222}$$

defines a quantity which gives the frequency and angular dependence of the average power. Since $\vec{\chi}_{\text{start,stop}}$ are independent of N they drop out in the limit and we have

$$\frac{d^2P(\omega)}{d\omega d\Omega} = \frac{e^2\omega^2}{4\pi^2c^3}Q(\omega)\left|\hat{n}\times\int_0^T dt\, \vec{v}(t) e^{i\omega(t-\frac{1}{c}\hat{n}\cdot\vec{x}(t))}\right|^2, \tag{11.223}$$

where

$$Q(\omega) \equiv \lim_{N\to\infty}\frac{1}{2NT}\sum_{n,n'=-N}^{N-1} e^{i\omega T(n-n')}. \tag{11.224}$$

The sum in (11.224) may be evaluated using the geometric series summation formula as follows:

$$\begin{aligned}\sum_{n,n'=-N}^{N-1} e^{i\omega T(n-n')} &= \left|\sum_{n=-N}^{N-1} e^{i\omega Tn}\right|^2 = \left|e^{-i\omega TN}\sum_{n=0}^{2N-1} e^{i\omega Tn}\right|^2 \\ &= \left|\frac{1-e^{i\omega 2TN}}{1-e^{i\omega T}}\right|^2 = \left|\frac{e^{-i\omega TN}-e^{i\omega TN}}{e^{-i\omega T/2}-e^{i\omega T/2}}\right|^2 \\ &= \left|\frac{\sin(\omega TN)}{\sin(\omega T/2)}\right|^2 \\ \Longrightarrow Q(\omega) &= \lim_{N\to\infty}\frac{1}{2NT}\left[\frac{\sin(\omega TN)}{\sin(\omega T/2)}\right]^2. \end{aligned} \tag{11.225}$$

With the definition

$$\omega_0 \equiv \frac{2\pi}{T}, \tag{11.226}$$

we see that the ratio of sines is bounded and the limit $N \to \infty$ gives zero as long as $\omega \neq m\omega_0$ for any integer m. On the other hand, for *fixed* N then we have by l'Hôpital's rule

$$\lim_{\omega\to m\omega_0}\frac{1}{N}\left(\frac{\sin(\omega TN)}{\sin(\omega T/2)}\right)^2 = \frac{1}{N}\left(\frac{(NT\cos(m\omega_0 NT)}{(T/2)\cos(\omega_0 T/2)}\right)^2 = 4N. \tag{11.227}$$

Since the limit of (11.227) is infinite as $N \to \infty$ evidently $Q(\omega)$ cannot be a function. Rather, it is a distribution. We may reasonably suspect that $Q(\omega)$ is a sum of delta functions at $\omega = m\omega_0$, $m = \pm1, \pm2, \pm3, \ldots$: such a distribution is called a *periodic delta function*. Graphically, this resembles the evenly spaced positive impulses in Figure 3.9.

To confirm that $Q(\omega)$ is indeed a periodic delta function, we first need to simplify (11.225). (Another approach to finding $Q(\omega)$ is given in Exercise 11.8.1.) Note that only the singular behavior near $\omega = m\omega_0$ concerns us, since $Q = 0$ for other values of ω. The denominator in (11.225) corresponds to a second-order pole near $\omega = m\omega_0$, since $\omega \approx m\omega_0 \implies \sin^2(\omega T/2) \approx [(\omega - m\omega_0)T/2]^2$. It follows that

$$\left[\frac{1}{\sin(\omega T/2)}\right]^2 \text{ and } \sum_{m=1}^{\infty}\left[\frac{1}{(\omega T/2 - m\pi)^2}\right]$$

have the same singular behavior for $\omega > 0$, and thus

$$Q(\omega) = \lim_{N\to\infty}\frac{1}{2NT}\sum_{m=1}^{\infty}\left[\frac{\sin^2(\omega TN)}{(\omega T/2 - m\pi)^2}\right] \quad (\omega > 0). \tag{11.228}$$

Notice we do not make a replacement of the numerator, since the increasingly rapid oscillation as $N \to \infty$ strongly influences the limit's behavior.

To evaluate the limit, we consider a delta function form based upon the nonnegative function

$$f(x - x_0) \equiv \frac{1}{2\pi} \frac{\sin^2[2(x - x_0)]}{(x - x_0)^2}. \tag{11.229}$$

One can show that

$$\int_{-\infty}^{\infty} dx\, f(x - x_0) = 1, \tag{11.230}$$

using, for example, integration by parts and contour integration. Referring to Section 2.2, we may now construct a delta function using $f(x - x_0)$ and Equation (2.9) as (replacing $1/\epsilon \to N$, with N a positive integer)

$$\delta(x - x_0) = \lim_{N\to\infty} \frac{1}{2\pi N} \frac{\sin^2[2N(x - x_0)]}{(x - x_0)^2}. \tag{11.231}$$

From this, we now build a periodic delta function by summation:

$$\sum_{m=1}^{\infty} \delta(x - mx_0) = \lim_{N\to\infty} \frac{1}{2\pi N} \sum_{m=1}^{\infty} \frac{\sin^2[2N(x - mx_0)]}{(x - mx_0)^2}. \tag{11.232}$$

Comparing with $Q(\omega)$ we identify

$$x = \omega T/2, \;\; x_0 = \pi,$$

so that

$$Q(\omega) \to \frac{\pi}{T} \sum_{m=1}^{\infty} \delta(\omega T/2 - m\pi) = \frac{\omega_0^2}{2\pi} \sum_{m=1}^{\infty} \delta(\omega - m\omega_0). \tag{11.233}$$

Thus, the spectrum is discrete for any periodic particle trajectory and consists of harmonics of the fundamental angular frequency ω_0. This can be considered as an intermediate situation between impulsive scattering involving all frequencies and pure harmonic sources.

From (11.223) we have

$$\begin{aligned} \frac{d^2P(\omega)}{d\omega d\Omega} &= \frac{e^2\omega_0^4 m^2}{(2\pi c)^3} \sum_{m=1}^{\infty} \delta(\omega - m\omega_0) \left| \hat{n} \times \int_0^T dt\, \vec{v}(t) e^{i\omega_0 m(t - \frac{1}{c}\hat{n}\cdot\vec{x}(t))} \right|^2 \\ &= \sum_{m=1}^{\infty} \delta(\omega - m\omega_0) \frac{dP_m}{d\Omega}, \end{aligned} \tag{11.234}$$

where we define

$$\frac{dP_m}{d\Omega} \equiv \frac{e^2\omega_0^4 m^2}{(2\pi c)^3} \left| \hat{n} \times \int_0^T dt\, \vec{v}(t) e^{i\omega_0 m(t - \frac{1}{c}\hat{n}\cdot\vec{x}(t))} \right|^2. \qquad (11.235)$$

$dP_m/d\Omega$ may be interpreted as the angular distribution of power delivered to the mth harmonic.

It is instructive to connect this result with our previous formula (11.22) for $(dP/d\Omega)_{\rm avg}$ for harmonic sources. This can be done using the *Poisson sum rule:*[6]

$$\sum_{n=-\infty}^{\infty} f(an) = \frac{\sqrt{2\pi}}{a} \sum_{m=-\infty}^{\infty} F\left(\frac{2\pi m}{a}\right), \qquad (11.236)$$

where

$$F(\omega) = \frac{1}{\sqrt{2\pi}} \int_{-\infty}^{\infty} dt\, f(t) e^{i\omega t}. \qquad (11.237)$$

Based on (11.214), (11.216) and (11.220), we define $\vec{A}^P(\vec{x},\omega)$ as the periodic motion's contribution to $\vec{A}(\vec{x},\omega)$:

$$\vec{A}^P(\vec{x},\omega) \equiv \frac{e^{ikr}}{\sqrt{2\pi}\, r} \frac{e}{c} \left(\sum_{n=-N}^{N-1} e^{i\omega T n} \right) \int_0^T dt\, \vec{v}(t) e^{i\omega(t - \frac{1}{c}\hat{n}\cdot\vec{x}(t))}, \qquad (11.238)$$

where we have used expression (11.220) for $\vec{\chi}_N$. Taking the limit as $N \to \infty$ and using the Poisson sum rule, we find

$$\vec{A}^P(\vec{x},\omega) \to \frac{e\omega_0}{\sqrt{2\pi} c r} \sum_{m=-\infty}^{\infty} e^{i\frac{r}{c}(\omega_0 m)} \delta(\omega - m\omega_0) \int_0^T dt\, \vec{v}(t)\, e^{i\omega_0 m(t - \frac{1}{c}\hat{n}\cdot\vec{x}(t))}. \qquad (11.239)$$

The Fourier inverse transform is given by

$$\vec{A}^P(\vec{x},t) \equiv \frac{1}{\sqrt{2\pi}} \int_{-\infty}^{\infty} d\omega\, e^{-i\omega t} \vec{A}^P(\vec{x},\omega) = \frac{2}{\sqrt{2\pi}} \mathrm{Re} \int_0^{\infty} d\omega\, e^{-i\omega t} \vec{A}^P(\vec{x},\omega), \qquad (11.240)$$

which can be rewritten as

$$\vec{A}^P(\vec{x},t) = \mathrm{Re} \sum_{m=1}^{\infty} e^{-i\omega_0 m t} \vec{A}_m(\vec{x}), \qquad (11.241)$$

(much like the harmonic case), where

$$\vec{A}_m(\vec{x}) \equiv \frac{e^{i\frac{r}{c}(\omega_0 m)}}{r} \frac{e\omega_0}{\pi c} \int_0^T dt\, \vec{v}(t) e^{i(\omega_0 m)(t - \frac{1}{c}\hat{n}\cdot\vec{x}(t))}. \qquad (11.242)$$

[6] P. M. Morse and H. Feshbach, *Methods of Theoretical Physics, Part I*, McGraw-Hill Book Company (New York) 1953, p. 466.

(Note that $\vec{A}_0(\vec{x}) = 0$ because the motion is periodic.) Likewise, the frequency spectrum for the periodic part of the current density is

$$\begin{aligned}\vec{J}^P(\vec{x},\omega) &= \frac{1}{\sqrt{2\pi}}\int_{-NT}^{NT} dt\, e^{+i\omega t}\vec{J}(\vec{x},t)\\ &= \frac{e}{\sqrt{2\pi}}\int_{-NT}^{NT} dt\, e^{+i\omega t}\vec{v}(t)\delta(\vec{x}-\vec{x}(t))\\ &= \frac{e}{\sqrt{2\pi}}\sum_{n=-N}^{N-1}\int_{nT}^{(n+1)T} dt_n\, e^{+i\omega t_n}\vec{v}(t_n)\delta(\vec{x}-\vec{x}(t_n))\\ &= \frac{e}{\sqrt{2\pi}}\sum_{n=-N}^{N-1} e^{+i\omega nT}\int_0^T dt\,\vec{v}(t)\delta(\vec{x}-\vec{x}(t)). \end{aligned} \tag{11.243}$$

Once again, taking the limit as $N \to \infty$ and using the Poisson sum rule, we find

$$\vec{J}^P(\vec{x},\omega) = \frac{e\omega_0}{\sqrt{2\pi}}\sum_{m=-\infty}^{\infty}\delta(\omega - m\omega_0)\int_0^T dt\,\vec{v}(t)\delta(\vec{x}-\vec{x}(t))e^{im\omega_0 t}. \tag{11.244}$$

The Fourier inverse transform is

$$\vec{J}^P(\vec{x},t) \equiv \frac{1}{\sqrt{2\pi}}\int_{-\infty}^{\infty} d\omega\, e^{-i\omega t}\vec{J}^P(\vec{x},\omega) = \frac{2}{\sqrt{2\pi}}\mathrm{Re}\int_0^{\infty} d\omega\, e^{-i\omega t}\vec{J}^P(\vec{x},\omega), \tag{11.245}$$

or

$$\vec{J}^P(\vec{x},t) = \mathrm{Re}\sum_{m=1}^{\infty} e^{-i\omega_0 mt}\vec{J}_m(\vec{x}), \tag{11.246}$$

(which also resembles the harmonic case) where

$$\vec{J}_m(\vec{x}) = \frac{\omega_0 e}{\pi}\int_0^T dt\,\vec{v}(t)\delta(\vec{x}-\vec{x}(t))e^{i\omega_0 mt}. \tag{11.247}$$

Then the usual harmonic relationship between $\vec{A}_m(\vec{x})$ and $\vec{J}_m(\vec{x})$ holds for each m,

$$\vec{A}_m(\vec{x}) = \frac{e^{i\frac{r}{c}\omega_0 m}}{r}\int d^3x'\,\frac{1}{c}\vec{J}_m(\vec{x}\,')e^{-i\omega_0 m\frac{1}{c}\hat{n}\cdot\vec{x}\,'}. \tag{11.248}$$

Comparing this with expression (11.235) for $dP_m/d\Omega$, we have

$$\frac{dP_m}{d\Omega} = \frac{(\omega_0 m)^2 r^2}{8\pi c}|\hat{n}\times\vec{A}_m(\vec{x})|^2. \tag{11.249}$$

Thus, each m value behaves as an independent harmonic source for periodic motion. This gives us further insight into the discrete frequency contribution occurring in Equation (11.234).

11.9 Synchrotron radiation from circular periodic motion

We will now examine *synchrotron radiation*, which is defined as radiation due to a particle in circular periodic motion. We shall adopt the model shown in Figure 11.12.

From (11.235) we have the angular power distribution for the mth harmonic is

$$\frac{dP_m}{d\Omega} = \frac{e^2\omega_0^4 m^2}{(2\pi c)^3}\left|\hat{n}\times\int_0^T dt\,\vec{v}(t)e^{i\omega_0 m(t-\frac{1}{c}\hat{n}\cdot\vec{x}(t))}\right|^2. \quad (11.250)$$

Because of the azimuthal symmetry, we may assume that the field point is in the xz-plane so that:

$$\hat{n} = (\sin\theta, 0, \cos\theta), \quad (11.251)$$

$$\vec{x}(t) = R(\cos\omega_0 t, \sin\omega_0 t, 0), \quad (11.252)$$

$$\vec{v}(t) = v(-\sin\omega_0 t, \cos\omega_0 t, 0), \quad (11.253)$$

where $v \equiv R\omega_0$. The integral in (11.250) can be broken up as

$$\int_0^T dt\,\vec{v}(t)e^{i\omega_0 m(t-\frac{1}{c}\hat{n}\cdot\vec{x}(t))} \equiv v(-(\text{term 2})\hat{\imath} + (\text{term 1})\hat{\jmath}), \quad (11.254)$$

where

$$(\text{term 1}) \equiv \int_0^T dt\,\cos\omega_0 t\,\exp\left[i\omega_0 m\left(t - \frac{1}{c}R\sin\theta\cos\omega_0 t\right)\right], \quad (11.255)$$

$$(\text{term 2}) \equiv \int_0^T dt\,\sin\omega_0 t\,\exp\left[i\omega_0 m\left(t - \frac{1}{c}R\sin\theta\cos\omega_0 t\right)\right]. \quad (11.256)$$

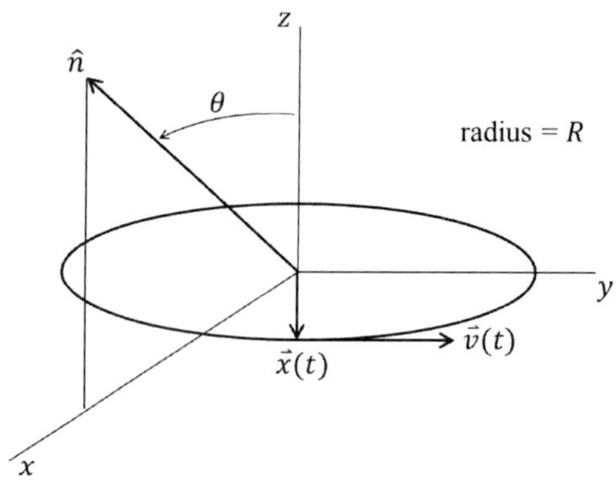

Fig. 11.12 Circular periodic motion (for synchrotron radiation calculation).

Both terms bear some resemblance to the integral expression for Bessel functions in (4.13):

$$J_m(z) = \frac{1}{i^m}\int_0^{2\pi}\frac{d\phi}{2\pi}\, e^{i(z\cos\phi - m\phi)}. \tag{11.257}$$

Since $J_m(z)$ is real, by taking complex conjugates and rearranging we get

$$\int_0^{2\pi}\frac{d\phi}{2\pi}\, e^{i(m\phi - z\cos\phi)} = i^m(-1)^m J_m(z). \tag{11.258}$$

Letting $\phi \equiv \omega_0 t$ and

$$z \equiv \frac{m\omega_0 R}{c}\sin\theta = m\beta\sin\theta, \quad \beta \equiv \frac{\omega_0 R}{c}, \tag{11.259}$$

we find

$$\begin{aligned}
(\text{term 1}) &= \frac{2\pi}{\omega_0}\int_0^{2\pi}\frac{d\phi}{2\pi}\underbrace{\cos\phi}_{\frac{1}{2}(e^{i\phi}+e^{-i\phi})} e^{i(m\phi - z\cos\phi)}, \\
&= \frac{\pi}{\omega_0}\left[\int_0^{2\pi}\frac{d\phi}{2\pi}\, e^{i((m+1)\phi - z\cos\phi)} + \int_0^{2\pi}\frac{d\phi}{2\pi}\, e^{i((m-1)\phi - z\cos\phi)}\right], \\
&= (-1)^{m+1} i^{m+1}\frac{\pi}{\omega_0}[J_{m+1}(z) - J_{m-1}(z)].
\end{aligned} \tag{11.260}$$

Using similar methods, we also find

$$\begin{aligned}
(\text{term 2}) &= \frac{\pi}{i\omega_0}\left[\int_0^{2\pi}\frac{d\phi}{2\pi}\, e^{i((m+1)\phi - z\cos\phi)} - \int_0^{2\pi}\frac{d\phi}{2\pi}\, e^{i((m-1)\phi - z\cos\phi)}\right] \\
&= (-1)^{m+1} i^{m}\frac{\pi}{\omega_0}[J_{m+1}(z) + J_{m-1}(z)].
\end{aligned} \tag{11.261}$$

Altogether we have

$$\begin{aligned}
&\int_0^T dt\,\vec{v}(t)e^{i\omega_0 m(t - \frac{1}{c}\hat{n}\cdot\vec{x}(t))} \\
&\quad = v(-1)^m i^m\frac{\pi}{\omega_0}[(J_{m+1}(z) + J_{m-1}(z))\hat{i} - i(J_{m+1}(z) - J_{m-1}(z))\hat{j}].
\end{aligned} \tag{11.262}$$

To complete our evaluation of $dP_m/d\Omega$, we compute

$$\begin{aligned}
\left|\hat{n}\times\int_0^T dt\,\vec{v}(t)e^{i\omega_0 m(t - \frac{1}{c}\hat{n}\cdot\vec{x}(t))}\right|^2 &= \frac{v^2\pi^2}{\omega_0^2}\Big|-i\hat{k}\sin\theta(J_{m+1} - J_{m-1}) \\
&\quad + \hat{j}\cos\theta(J_{m+1} + J_{m-1}) + i\,\hat{i}\cos\theta(J_{m+1} - J_{m-1})\Big|^2 \\
&= \frac{\pi^2 v^2}{\omega_0^2}\big[\sin^2\theta(J_{m+1} - J_{m-1})^2 \\
&\quad + \cos^2\theta[(J_{m+1} + J_{m-1})^2 + (J_{m+1} - J_{m-1})^2]\big] \\
&= \frac{\pi^2 v^2}{\omega_0^2}\left[2J_{m+1}^2 + 2J_{m-1}^2 - \sin^2\theta(J_{m+1} + J_{m-1})^2\right].
\end{aligned} \tag{11.263}$$

Using (11.259) and the Bessel function recurrence relation (4.18),

$$J_{m+1}(z) + J_{m-1}(z) = \frac{2m}{z} J_m(z),$$

we find

$$\left| \hat{n} \times \int dt\, \vec{v}(t) e^{i\omega_0 m(t - \frac{1}{c}\hat{n}\cdot\vec{x}(t))} \right|^2 = \frac{4\pi^2 c^2}{\omega_0^2} \left[\frac{\beta^2}{2} (J_{m+1}^2 + J_{m-1}^2) - J_m^2 \right], \tag{11.264}$$

which gives finally

$$\frac{dP_m}{d\Omega} = \frac{e^2 \omega_0^2 m^2}{2\pi c} \left[\beta^2 \frac{J_{m+1}^2(z) + J_{m-1}^2(z)}{2} - J_m^2(z) \right], \tag{11.265}$$

where z and β are defined in (11.259). We may further simplify this using the recurrence relation (4.17),

$$J_{m-1} - J_{m+1} = 2J'_m,$$

which enables us to write

$$\begin{aligned} \frac{J_{m+1}^2 + J_{m-1}^2}{2} &= \frac{1}{2} \left[\left(\frac{m}{z} J_m - J'_m \right)^2 + \left(\frac{m}{z} J_m + J'_m \right)^2 \right] = \frac{m^2}{z^2} J_m^2 + J_m'^2, \\ &= \frac{1}{\beta^2 \sin^2 \theta} J_m^2 + J_m'^2. \end{aligned} \tag{11.266}$$

Finally we have

$$\frac{dP_m}{d\Omega} = \frac{e^2 \omega_0 m^2}{2\pi} \frac{\beta^3}{R} \left[\left(\frac{J_m(z)}{\beta \tan \theta} \right)^2 + J_m'^2(z) \right]. \tag{11.267}$$

Another useful quantity is P_m, the *total* power emitted into the mth harmonic in one period. We could find P_m by integrating (11.267) over θ, but we don't know how to do this analytically. So instead we take an alternative approach, and integrate (11.250) over angles. Using the identity

$$|\hat{n} \times \vec{I}|^2 = |\vec{I}|^2 - |\hat{n} \cdot \vec{I}|^2, \tag{11.268}$$

we may rewrite (11.250) as

$$\begin{aligned} \frac{dP_m}{d\Omega} = \frac{e^2 \omega_0^4 m^2}{(2\pi c)^3} \Bigg[&\left| \int_0^T dt\, \vec{v}(t)\, e^{i\omega_0 m(t - \frac{1}{c}\hat{n}\cdot\vec{x}(t))} \right|^2 \\ &- \left| \int_0^T dt\, \hat{n} \cdot \vec{v}(t)\, e^{i\omega_0 m(t - \frac{1}{c}\hat{n}\cdot\vec{x}(t))} \right|^2 \Bigg]. \end{aligned} \tag{11.269}$$

The second integrand in (11.269) may be simplified by noting that

$$\frac{d}{dt}e^{i\omega_0 m(t-\frac{1}{c}\hat{n}\cdot\vec{x}(t))} = i\omega_0 m\left(1-\frac{1}{c}\hat{n}\cdot\vec{v}(t)\right)e^{i\omega_0 m(t-\frac{1}{c}\hat{n}\cdot\vec{x}(t))}, \qquad (11.270)$$

and thus

$$\hat{n}\cdot\vec{v}e^{i\omega_0 m(t-\frac{1}{c}\hat{n}\cdot\vec{x}(t))} = ce^{i\omega_0 m(t-\frac{1}{c}\hat{n}\cdot\vec{x}(t))} - \frac{c}{i\omega_0 m}\frac{d}{dt}e^{i\omega_0 m(t-\frac{1}{c}\hat{n}\cdot\vec{x}(t))}. \qquad (11.271)$$

The second term in (11.271) integrates to zero because of periodicity, and we're left with

$$\frac{dP_m}{d\Omega} = \frac{e^2\omega_0^4 m^2}{(2\pi c)^3}\left[\left|\int_0^T dt\,\vec{v}(t)\,e^{i\omega_0 m(t-\frac{1}{c}\hat{n}\cdot\vec{x}(t))}\right|^2 - c^2\left|\int_0^T dt\,e^{i\omega_0 m(t-\frac{1}{c}\hat{n}\cdot\vec{x}(t))}\right|^2\right]. \qquad (11.272)$$

Let's rewrite (11.272) by re-expressing the two squared terms as double integrals over two independent time variables:

$$\frac{dP_m}{d\Omega} = \frac{e^2\omega_0^4 m^2}{(2\pi c)^3}\int_0^T dt\int_0^T dt'[\vec{v}(t)\cdot\vec{v}(t')-c^2]e^{i\omega_0 m(t-t'-\frac{1}{c}\hat{n}\cdot(\vec{x}(t)-\vec{x}(t')))}. \qquad (11.273)$$

We will integrate over $d\Omega$ first, by exchanging the order of integration to bring the $d\Omega$ integral inside. Taking our z-axis along $|\vec{x}(t)-\vec{x}(t')|$, we may write $\hat{n}\cdot(\vec{x}(t)-\vec{x}(t'))$ as $|\vec{x}(t)-\vec{x}(t')|\cos\theta$. Using the integral

$$\begin{aligned}\int d\Omega e^{-i\omega_0 m(\frac{1}{c}\hat{n}\cdot(\vec{x}(t)-\vec{x}(t')))} &= 2\pi\int d\theta\,\sin\theta e^{-i\omega_0 m(\frac{1}{c}|\vec{x}(t)-\vec{x}(t')|\cos\theta)}\\ &= 2\pi\frac{2\sin\left(\frac{1}{c}\omega_0 m|\vec{x}(t)-\vec{x}(t')|\right)}{\frac{1}{c}\omega_0 m|\vec{x}(t)-\vec{x}(t')|},\end{aligned} \qquad (11.274)$$

we find

$$\begin{aligned}P_m = 2\pi\frac{e^2\omega_0^4 m^2}{(2\pi c)^3}&\int_0^T dt\int_0^T dt'[\vec{v}(t)\cdot\vec{v}(t')-c^2]e^{i\omega_0 m(t-t')}\\ &\times\frac{2\sin\left(\frac{1}{c}\omega_0 m|\vec{x}(t)-\vec{x}(t')|\right)}{\frac{1}{c}\omega_0 m|\vec{x}(t)-\vec{x}(t')|}.\end{aligned} \qquad (11.275)$$

Using geometrical arguments (see Figure 11.13) we may evaluate:

$$\vec{v}(t)\cdot\vec{v}(t') = v^2\cos(\omega_0(t-t')), \qquad (11.276)$$

$$|\vec{x}(t)-\vec{x}(t')| = 2R\left|\sin\left(\frac{\omega_0(t-t')}{2}\right)\right|. \qquad (11.277)$$

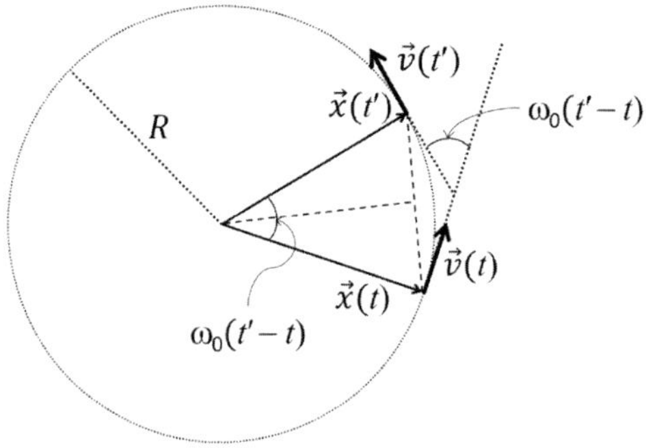

Fig. 11.13 Evaluation of $|\vec{x}(t) - \vec{x}(t')|$ and $\vec{v}(t) \cdot \vec{v}(t')$. Here $\vec{x}(t)$ and $\vec{x}(t')$ are equal sides of an isosceles triangle with vertex angle $\omega_0(t-t')$, and the angle between $\vec{v}(t)$ and $\vec{v}(t')$ is also $\omega_0(t-t')$.

As a result, we find that the integrand depends on t, t' via $t - t'$ *only*. So it is convenient to change variables to relative and average times, τ and τ', as in:

$$\tau \equiv t - t'; \; \tau' \equiv \frac{t+t'}{2}. \tag{11.278}$$

Since the integrand in (11.275) is periodic in both t and t' with period T, we are allowed to change the integration boundary and take our integration limits as $-T/2 < \tau, \tau' < T/2$. The Jacobian of the transformation $(t, t') \to (\tau, \tau')$ is equal to 1, and the integral over τ' may be performed immediately to obtain

$$\begin{aligned} P_m = {} & 2\pi T \frac{e^2 \omega_0^4 m^2}{(2\pi c)^3} \int_{-T/2}^{T/2} d\tau [v^2 \cos \omega_0 \tau) - c^2] e^{i\omega_0 m \tau} \\ & \times \frac{2 \sin\left(\frac{2R}{c} \omega_0 m |\sin(\omega_0 \tau/2)|\right)}{\frac{2R}{c} \omega_0 m |\sin(\omega_0 \tau/2)|}. \end{aligned} \tag{11.279}$$

The fraction in the integrand is an even function of $|\sin(\omega_0 \tau/2)|$, so we may remove the absolute values. Notice that the real part of the integrand is an even function of τ, while the imaginary part is odd and thus integrates to 0. We define $\phi \equiv \omega_0 \tau$, and use $\beta = v/c$, $T = 2\pi/\omega_0$ to get

$$P_m = \frac{e^2 \omega_0 m}{R} \int_0^{\pi} \frac{d\phi}{\pi} \cos m\phi (\beta^2 \cos\phi - 1) \frac{\sin(2m\beta \sin(\phi/2))}{\sin(\phi/2)}. \tag{11.280}$$

This somewhat resembles the Bessel function integral representation from (11.257), so we will try to massage this integral until we get something resembling our expression for P_m. From (11.280) it looks like we want a sine instead of cosine in the complex exponent, so we substitute $\phi \to \phi - \pi/2$ to get

$$i^m J_m(z) = \int_{\pi/2}^{5\pi/2} \frac{d\phi}{2\pi}\, e^{i(z\sin\phi - m\phi + m\pi/2)} = i^m \int_{-\pi}^{\pi} \frac{d\phi}{2\pi}\, e^{i(z\sin\phi - m\phi)}, \tag{11.281}$$

where we have also used periodicity of the integrand to shift the integral limits to $[-\pi, \pi]$. The real part of the integrand is even and the imaginary part is odd, so we may simplify:

$$J_m(z) = \int_0^{\pi} \frac{d\phi}{\pi} \cos(z\sin\phi - m\phi). \tag{11.282}$$

Next, separate the integral into two pieces on $[0, \pi/2]$ and $[\pi/2, \pi]$ respectively, and substitute $\phi \to \pi - \phi$ in the second integral:

$$J_m(z) = \int_0^{\pi/2} \frac{d\phi}{\pi} \left[\cos(z\sin\phi - m\phi) + \cos(z\sin\phi + m\phi - m\pi)\right].$$

Letting $m \to 2m$, this then gives

$$\begin{aligned} J_{2m}(z) &= \int_0^{\pi/2} \frac{d\phi}{\pi} \left[\cos(z\sin\phi - 2m\phi) + \cos(z\sin\phi + 2m\phi)\right] \\ &= \int_0^{\pi/2} \frac{d\phi}{\pi} \left[2\cos(z\sin\phi)\cos(2m\phi)\right]. \end{aligned} \tag{11.283}$$

Replacing $\phi \to \phi/2$, we find

$$J_{2m}(z) = \int_0^{\pi} \frac{d\phi}{\pi} \cos(z\sin(\phi/2))\cos(m\phi). \tag{11.284}$$

Straightforward calculations give

$$J'_{2m}(z) = -\int_0^{\pi} \frac{d\phi}{\pi} \sin(\phi/2)\cos(m\phi)\sin(z\sin(\phi/2)), \tag{11.285}$$

$$\int_0^x dz\, J_{2m}(z) = \int_0^{\pi} \frac{d\phi}{\pi} \cos(m\phi)\frac{\sin(x\sin(\phi/2))}{\sin(\phi/2)}. \tag{11.286}$$

Finally, we may express the total power in the mth harmonic as:

$$P_m = \frac{e^2}{R} m\omega_0 \left[2\beta^2 J'_{2m}(2m\beta) - (1-\beta^2)\int_0^{2m\beta} dz\, J_{2m}(z)\right]. \tag{11.287}$$

So far we have evaluated power expressions for the different harmonics separately. If we want the corresponding expressions for *total* power, one approach is to sum over harmonics; so for instance the total emitted power

$dP/d\Omega$ is the sum of $dP_m/d\Omega$ for $m = 1, 2, \ldots$. Unfortunately, summing the terms in (11.267) analytically is difficult, if not impossible. However, we may take a different approach and average the *instantaneous* emitted power over one period:

$$\left(\frac{dP}{d\Omega}\right)_{\text{avg}} = \frac{1}{T}\int_0^T dt_r \, \frac{dP(t_r)}{d\Omega}. \tag{11.288}$$

From (11.147) we have

$$\frac{dP(t_r)}{d\Omega} = \frac{e^2}{4\pi c^3}\left[\frac{[\hat{n}\times\left((\hat{n}-\vec{\beta})\times\dot{\vec{v}}\right)]^2}{\left(1-\hat{n}\cdot\vec{\beta}\right)^5}\right]_{t_r}. \tag{11.289}$$

Using azimuthal symmetry, and anticipating the high energy limit, this time we choose

$$\hat{n} = (\cos\theta, 0, \sin\theta), \tag{11.290}$$

as opposed to (11.251). Also, since we are integrating over an entire period we may choose our time origin so that $\vec{\beta}$ is parallel to the x-axis when $t_r = 0$, so that

$$\vec{\beta} \equiv \beta(\cos(\omega_0 t_r), \sin(\omega_0 t_r), 0). \tag{11.291}$$

It follows immediately that $\hat{n}\cdot\vec{\beta} = \beta\cos\theta\cos(\omega_0 t_r)$, and much more laboriously that

$$\begin{aligned}[\hat{n}\times\left((\hat{n}-\vec{\beta})\times\dot{\vec{v}}\right)]^2 = R^2{\omega_0}^4\big[&\sin^2\theta + \beta^2\cos^2\theta \\ &+ \cos^2\theta\cos^2(\omega_0 t_r) - 2\beta\cos\theta\cos(\omega_0 t_r)\big].\end{aligned} \tag{11.292}$$

Plugging this into (11.289) and defining $\phi \equiv \omega_0 t_r$ allows us to write (11.288) as

$$\left(\frac{dP}{d\Omega}\right)_{\text{avg}} = \frac{e^2\beta^3}{4\pi RT}\int_0^{2\pi} d\phi \frac{[\sin^2\theta + \beta^2\cos^2\theta + \cos^2\theta\cos^2\phi - 2\beta\cos\theta\cos\phi]}{(1-\beta\cos\theta\cos\phi)^5}. \tag{11.293}$$

This integral may be done exactly (see Exercise 11.9.2), and the result is

$$\left(\frac{dP}{d\Omega}\right)_{\text{avg}} = \frac{e^2\beta^3}{16RT}\left[\frac{3-3\cos^2\theta(1-\beta^2)}{(1-\beta^2\cos^2\theta)^{5/2}} + \frac{5-\cos^2\theta-4\beta^2\cos^4\theta}{(1-\beta^2\cos^2\theta)^{7/2}}\right]. \tag{11.294}$$

For $\beta \approx 1$, it is clear that the power is maximized at small angles $\theta \ll 1$, which in this case means the synchrotron plane. By integrating over $d\Omega$, it is possible to recover expression (11.171) for the total energy E emitted in one period (Exercise 11.9.3):

$$E = T\int d\Omega \left(\frac{dP}{d\Omega}\right)_{\text{avg}} = \frac{4\pi e^2}{3R}\frac{\beta^3}{(1-\beta^2)^2}. \tag{11.295}$$

11.10 Going Deeper

11.10.1 *Transition radiation*

(1) C. A. Brau, *Modern Problems in Classical Electrodynamics*, Oxford University Press (Oxford) 2004; Sections 5.2.4, 10.2.2.

(2) V. L. Ginzburg, V. N. Tsytovich, *Transition Radiation and Transition Scattering*, IOP Publishing (Bristol) 1990.

(3) J. D. Jackson, *Classical Electrodynamics*, 3rd ed., John Wiley & Sons (New York) 1999; Section 13.7.

(4) L. D. Landau, L. P. Pitaevskii and E. M. Lifshitz, *Electrodynamics of Continuous Media*, 2nd ed. with corrections, Elsevier (New York) 1993; Section 116.

(5) K. Olive *et al.* (Particle Data Group), Chin. Phys. C **38**, 090001 (2014): `http://pdg.lbl.gov`; Review, "Passage of Particles Through Matter", Section 32.7.3.

11.10.2 *Bremsstrahlung*

(1) C. A. Brau, *Modern Problems in Classical Electrodynamics*, Oxford University Press (Oxford) 2004; Sections 5.2.2, 10.2.1.

(2) E. Haug and W. Nakel, *The Elementary Process of Bremsstrahlung*, World Scientific Lecture Notes in Physics, Vol. 73, World Scientific (New Jersey) 2004.

(3) C. Itzykson and J.-B. Zuber, *Quantum Field Theory*, McGraw-Hill (New York) 1980; Section 5-2-4.

(4) J. D. Jackson, *Classical Electrodynamics*, 3rd ed., John Wiley & Sons (New York) 1999; Chapter 15.

(5) A. V. Korol and A. V. Solov'yov, *Polarization Bremsstrahlung*, Springer Series on Atomic, Optical and Plasma Physics, Springer (Heidelberg) 2014.

11.10.3 *Synchrotron radiation*

(1) P. Duke, *Synchrotron Radiation: Production and Properties*, Oxford University Press (Oxford) 2000.

(2) L. D. Landau and E. M. Lifshitz, *Classical Theory of Fields*, 4th ed. with corrections, Butterworth-Heinemann (Oxford) 2000; Section 74.

(3) S. Mobilio, F. Boscherini and C. Meneghini, Eds., *Synchrotron Radiation: Basics, Methods and Applications*, Springer (Heidelberg) 2014.

(4) J. Schwinger *et al.*, *Classical Electrodynamics*, Perseus Books (Reading, MA) 1998; Chapters 38, 39 and 40.

(5) H. Wiedemann, *Synchrotron Radiation*, Springer (Berlin) 2003.

(6) P. Willmott, *An Introduction to Synchrotron Radiation: Techniques and Applications*, John Wiley & Sons (Chichester) 2011.

11.11 Exercises

Exercise 11.1.1.

(a) Using a computer graphics package, investigate the radiation patterns for the center-fed linear antenna. Illustrate the radiation patterns for $kd = \pi$, 3π and 5π.

(b) Prove the statement in the text relating the number of radiation zeros, $2n$, to the kd value for this antenna. In this process, show they are located at

$$\cos\theta = \pm\left(1 - \frac{4\pi m}{kd}\right), \quad m = 1, 2, \ldots, n,$$

where n is the greatest integer less than $kd/(2\pi)$.

Exercise 11.1.2. A center-fed harmonic linear antenna of length d is oriented along the z-axis and centered at the coordinate origin. A phase angle δ is introduced between the two arms of the antenna. The harmonic sinusoidal current density J_z is given by

$$J_z(x,y,z) = \begin{cases} I\sin(\frac{kd}{2} - kz)\delta(x)\delta(y), & \frac{d}{2} > z > 0, \\ Ie^{i\delta}\sin(\frac{kd}{2} + kz)\delta(x)\delta(y), & -\frac{d}{2} < z < 0. \end{cases}$$

(a) Show that the average angular power radiated $(dP/d\Omega)_{\text{avg}}$ is given by:

$$\frac{I^2}{2\pi c}\left[\frac{\cos\left(\frac{kd}{2}\cos\theta + \delta/2\right) - \cos(\frac{kd}{2})\cos(\delta/2) + \cos\theta\sin(\frac{kd}{2})\sin(\delta/2)}{\sin\theta}\right]^2.$$

(b) Plot the intensity patterns for $kd = 2\pi$ with $\delta = \pi/2, \pi$ and $3\pi/2$.

(c) Show that the total average power $P_{\text{avg}}(kd,\delta)$ is periodic with period 2π in δ, and is an even function of δ.

(d) Using numerical means show that the ratio of total average powers $P_{\text{avg}}(kd,\pi)/P_{\text{avg}}(kd,0)$ exhibits both enhancements and suppressions as a function of kd. Try to characterize the kd values of these extrema.

Exercise 11.1.3.

(a) A long, straight current-carrying wire of length $2L$ between two cities (see Figure 11.14) has a harmonically driven current of the form

$$\vec{J}'(\vec{x}\,',t) = \vec{J}(\vec{x}\,')e^{-i\omega t}, \quad \vec{J}(\vec{x}\,') = I\delta(x')\delta(y')\hat{k},$$

(I is a constant current) along $-L < z' < L$. Show that the average angular radiated power is

$$\left(\frac{dP}{d\Omega}\right)_{\text{avg}} = \frac{I^2}{2\pi c}\frac{\sin^2(kL\cos\theta)}{\cos^2\theta}\sin^2\theta.$$

(b) The exact average angular distribution of radiated power for a transmission line, modeled as if the Earth were not present and with the origin at the middle of the span, can be taken as the answer to part (a). Given this, calculate the total average power loss per unit length,

$$\frac{P_{\text{avg}}}{2L} = \frac{1}{2}\mathcal{R}I^2,$$

associated with this transmission line in the $L \to \infty$ limit. Find the numerical result for $\mathcal{R}$ in the familiar SI units Ω/m (ohms per meter) for a typical 60 Hz signal. [*Hint*: A direct assault will probably be repulsed. Try an outflanking movement of integrals or an integration package.]

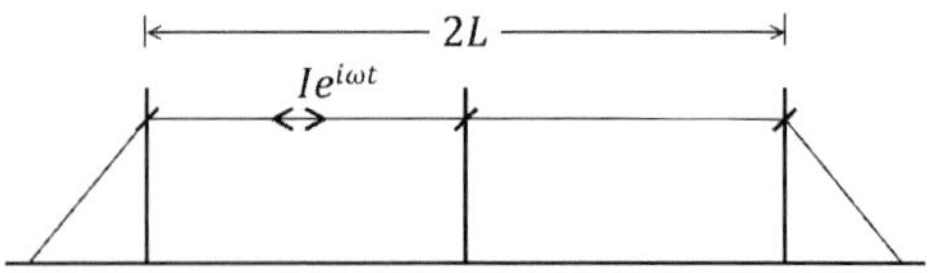

Fig. 11.14 Setup for Exercise 11.1.3.

Exercise 11.1.4. Consider an harmonic current on the surface of a sphere of radius a:

$$\vec{J}(\vec{x}') = K\sin\theta'\hat{e}_{\phi'}\delta(r'-a),$$

where θ' is the usual polar angle, $\hat{e}_{\phi'}$ is the unit azimuthal vector and r' is the distance from the sphere's center. (This current could be due to an harmonic rotation/oscillation of the sphere about the z-axis with angular frequency ω when a uniform surface charge is placed on the surface.)

(a) Show that
$$\left(\frac{dP}{d\Omega}\right)_{avg} = \frac{2\pi}{c}(Kka^2)^2\sin^2\theta(j_1(ka))^2.$$
($j_1(ka)$ is a spherical Bessel function.) Notice that for $j_1(ka) = 0$ there is zero radiation produced! (Result due to Arnold Sommerfeld in 1904.)

(b) Find the low frequency limit of your expression, $ka \ll 1$.

Exercise 11.1.5. Consider a phase-separated harmonic current on the surface of a sphere of radius a:
$$\vec{J}(\vec{x}') = \begin{cases} K\sin\theta'\hat{e}_{\phi'}\delta(r'-a), & 0<\theta<\frac{\pi}{2}, \\ e^{i\delta}K\sin\theta'\hat{e}_{\phi'}\delta(r'-a), & \frac{\pi}{2}<\theta<\pi, \end{cases}$$
where θ' is the usual polar angle, $\hat{e}_{\phi'}$ is the unit azimuthal vector and r' is the distance from the sphere's center.

(a) Numerically evaluate and present a plot for the average angular power, $\frac{dP}{d\Omega}_{avg}$. In addition, numerically evaluate the total average power $P_{avg,1}$ for $ka = 4.493409458$ (first zero of $j_1(ka)$), setting $K^2a^2/c = 1$. Use four values of δ: $\pi/32$, $\pi/16$, $\pi/8$, $\pi/4$. Show that for $\delta \ll 1$, $P_{avg,1} \approx N_1\delta^2$ for this first zero. Find N_1.

(b) Find the form of the low frequency limit, $ka \ll 1$, of the average angular power $\frac{dP}{d\Omega}_{avg,low}$ and total power, $P_{avg,low}$, for this system. What happens when $\delta = \pi$?

Exercise 11.1.6. Consider a phase-separated harmonic current on the surface of a cylinder of radius a and length L whose center is oriented along the z-axis:
$$\vec{J}(\vec{x}') = \begin{cases} K\,\hat{e}_{\phi'}\delta(\rho'-a), & 0<z'<L/2 \\ e^{i\delta}K\,\hat{e}_{\phi'}\delta(\rho'-a), & 0>z'>-L/2 \end{cases},$$
where ρ' is the distance from the z-axis in cylindrical coordinates and $\hat{e}_{\phi'}$ is the unit azimuthal vector.

(a) Evaluate the average angular power radiated. Show that
$$\frac{dP}{d\Omega}_{avg} =$$
$$\frac{4\pi K^2a^2}{c}\left(\frac{\sin\left(\frac{kL}{4}\cos\theta\right)J_1(ka\sin\theta)}{\cos\theta}\right)^2\left(1+\cos\left(\frac{kL}{2}\cos\theta+\delta\right)\right).$$
$J_1(x)$ is a regular Bessel function.

(b) Using numerical means, plot the angular distribution for $K^2a^2/c = 1$, $L = 4a$ given $ka = 1$ with $\delta = 0$ and $\delta = \pi$.

Exercise 11.2.1. You are given the time dependent sources (see Section 6.8 to see how these arise):

$$\text{electric dipole at origin: } \rho(\vec{x}, t) = -\vec{p}(t) \cdot \vec{\nabla}\delta(\vec{x}),$$

$$\text{magnetic dipole at origin: } \vec{J}(\vec{x}, t) = -c\vec{m}(t) \times \vec{\nabla}\delta(\vec{x}).$$

Starting with the exact formula in the real formalism, derive the results $(t_0 = t - r/c)$,

$$\text{electric dipole at origin: } \frac{dP(t)}{d\Omega} = \frac{1}{4\pi c^3}\left[\hat{n} \times \ddot{\vec{p}}(t_0)\right]^2,$$

$$\text{magnetic dipole at origin: } \frac{dP(t)}{d\Omega} = \frac{1}{4\pi c^3}\left[\hat{n} \times \ddot{\vec{m}}(t_0)\right]^2,$$

for the angular distribution of radiation from a pure electric dipole and magnetic dipole, respectively.

Exercise 11.2.2. Consider a thin insulating ring of radius a and charge density $\rho(\vec{x}\,') = \rho_0 a\delta(r' - a)\delta(\cos\theta')\sin\phi'$ in spherical coordinates, where ϕ is the usual azimuthal angle. (We may set $\rho_0 = \Lambda/a^2$, where Λ is a linear charge density.) It is spun through its symmetry axis at a constant angular velocity ω. The charge density function becomes $\rho(\vec{x}\,', t) = \rho_0 a\delta(r' - a)\delta(\cos\theta')\sin(\phi' - \omega t)$.

(a) Show that its exact power distribution is

$$\frac{dP(t)}{d\Omega} = \frac{\pi\rho_0^2\omega^4 a^8}{4c^3}\left[(J_0(x) + J_2(x))^2\sin^2\gamma(t_0) - 4J_0(x)J_2(x)\cos^2(\phi - \omega t_0)\right],$$

$(t_0 = t - r/c)$ where J_0, J_2 are Bessel functions, $x \equiv ka\sin\theta$ where $k = \omega/c$, and $\gamma(t_0)$ is defined as

$$\ddot{\vec{p}}(t_0) \cdot \hat{n} \equiv |\ddot{\vec{p}}(t_0)|\cos\gamma(t_0).$$

(b) Show that the time-averaged radiation rate is

$$\left(\frac{dP}{d\Omega}\right)_{\text{avg}} = \frac{\pi\rho_0^2\omega^4 a^8}{8c^3}\left[(J_0(x) - J_2(x))^2 + \cos^2\theta(J_0(x) + J_2(x))^2\right],$$

where θ is the usual polar angle. (Notice that this result is independent of ϕ, as it should be.)

Exercise 11.2.3. A real but harmonic current density, $\vec{J}(\vec{x}\,',t)$, with angular frequency ω on a single-wave circular antenna of radius a is given in spherical coordinates by

$$\vec{J}(\vec{x}\,',t) = \frac{I}{a}\delta(r'-a)\delta(\cos\theta')\sin(\phi')\sin(\omega t)\hat{e}_{\phi'},$$

where $\hat{e}_{\phi'}$ is the azimuthal unit vector ($\hat{e}_{\phi'} = -\hat{\imath}\sin\phi' + \hat{\jmath}\cos\phi'$). Find the radiated instantaneous angular power exactly. The answer is very simple and involves Bessel functions of order 0 and 2. It is of the general form ($x \equiv ka\sin\theta$):

$$\frac{dP(t)}{d\Omega} = \frac{(I\pi ka)^2}{4\pi c^3}\left[G(\theta,\phi,t_0)(J_2(x))^2 + H(\theta,\phi,t_0)(J_0(x))^2 \right. \\ \left. + I(\theta,\phi,t_0)J_0(x)J_2(x)\right].$$

Solve for the time and angle dependent functions $G(\theta,\phi,t_0)$, $H(\theta,\phi,t_0)$, $I(\theta,\phi,t_0)$.

Exercise 11.2.4. Assuming many frequency components of the real current,

$$\vec{J}(\vec{x}\,',t) = \sum_n \vec{J}_n(\vec{x}\,')\sin(\omega_n t - \delta_n(\vec{x}\,')),$$

in (11.54), argue that (11.62) is replaced with

$$\left(\frac{dP}{d\Omega}\right)_{\text{avg}} = \sum_n \frac{k_n^2}{8\pi c}\left|\hat{n}\times\int d^3x'\,\vec{J}_n^{\,\text{har}}(\vec{x}\,')e^{-ik_n\hat{n}\cdot\vec{x}\,'}\right|^2,$$

with $\vec{J}_n^{\,\text{har}}(\vec{x}\,') \equiv \vec{J}_n(\vec{x}\,')e^{i\delta_n(\vec{x}\,')}$, when the time averaging is over a large time period, $T\to\infty$.

Exercise 11.3.1. Prove Parseval's Theorem (11.69). Be careful to use the normalizations (11.67) and (11.68) in your proof.

Exercise 11.3.2.

(a) Using (11.75) with the given current ($Q_1 + Q_2 = Q$)

$$\vec{J}(\vec{x},t) = \begin{cases} Q_1\vec{v}_1\delta\left(\vec{x}-\vec{v}_1 t\right) + Q_2\vec{v}_2\delta\left(\vec{x}-\vec{v}_2 t\right), & t<0, \\ Q\vec{v}\,\delta\left(\vec{x}-\vec{v}t\right), & t>0, \end{cases}$$

and assuming relativistic momentum conservation,

$$\vec{p}_1 + \vec{p}_2 = \vec{P},$$

where $\vec{p}_1 = m_1\gamma_1\vec{v}_1$, $\vec{p}_2 = m_2\gamma_2\vec{v}_2, \vec{P} = M\gamma\vec{v}$, show that the frequency/angular cross-section is

$$\frac{d^2E(\omega)}{d\omega d\Omega} = \frac{1}{4\pi^2c^3}\left|\hat{n}\times\left(\frac{Q\vec{v}}{(1-\frac{1}{c}\hat{n}\cdot\vec{v})} - \frac{Q_1\vec{v}_1}{(1-\frac{1}{c}\hat{n}\cdot\vec{v}_1)} - \frac{Q_2\vec{v}_2}{(1-\frac{1}{c}\hat{n}\cdot\vec{v}_2)}\right)\right|^2.$$

(b) An "amplitude zero" is defined (in this case) by the condition (which can be made to appear relativistically invariant)

$$\frac{Q}{M\gamma(1-\frac{1}{c}\hat{n}\cdot\vec{v})} = \frac{Q_1}{m_1\gamma_1(1-\frac{1}{c}\hat{n}\cdot\vec{v}_1)} = \frac{Q_2}{m_2\gamma_2(1-\frac{1}{c}\hat{n}\cdot\vec{v}_2)}.$$

Show that the amplitude in (a) completely vanishes under these conditions when momentum conservation is taken into account. [*Note*: This result is applicable to "Standard Model" quark reactions such as $u\bar{d} \to W^+\gamma$, which has $Q_u = 2/3$, $Q_{\bar{d}} = 1/3$ and $Q_{W^+} = 1$, in units of the magnitude of the charge on an electron. See M. A. Samuel, "Amplitude Zeros", Phys. Rev. **D** 27, 2724 (1983).]

Exercise 11.5.1.

(a) Starting with the instantaneous nonrelativistic dipole radiation formula

$$\frac{dP(t)}{d\Omega} = \frac{e^2}{4\pi c^3}|\hat{n}\times\ddot{\vec{x}}(t_0)|^2,$$

($t_0 = t - r/c$) and using the results of Section 11.3, show that the frequency spectrum may be written as

$$\frac{d^2E}{d\Omega d\omega} = \frac{e^2}{2\pi c^3}|\hat{n}\times\ddot{\vec{x}}(\omega)|^2,$$

where

$$\ddot{\vec{x}}(\omega) \equiv \frac{1}{\sqrt{2\pi}}\int_{-\infty}^{+\infty} dt\, e^{i\omega t}\ddot{\vec{x}}(t_0).$$

(b) For uniform acceleration $\vec{a}$ along $\hat{k}$,

$$\ddot{x}(t_0) = \begin{cases} 0, & t_0 < 0, \\ a\,\hat{k}, & 0 \le t_0 \le T, \\ 0, & T < t_0, \end{cases}$$

and using part (a), find the angular/frequency energy cross-section, $d^2E/(d\Omega d\omega)$, for this process. Integrate in angle to find the frequency distribution, $dE/d\omega$, and plot as a function of ω.

(c) Find the total energy emitted in this process by integration. [See the interesting perspective on uniform acceleration in J. Schwinger *et al.*, *Classical Electrodynamics*, Perseus Books (Reading, MA) 1998, Chapter 37.]

Exercise 11.5.2. Starting with the expression Equation (11.19) of the text, show that the leading terms, proportional to $1/r$, in the magnetic field for a harmonic source ($\rho(\vec{x},t) = \rho(\vec{x})e^{-i\omega t}$, $\vec{J}(\vec{x},t) = \vec{J}(\vec{x})e^{-i\omega t}$) with electric dipole ($\vec{p}$), magnetic dipole ($\vec{m}$), and electric quadrupole multipoles ($Q_i(\hat{n}) \equiv \sum_j \hat{n}_j Q_{ij}$) is

$$\vec{B}(\vec{x}) = k^2 \frac{e^{ikr}}{r} \hat{n} \times \left(\vec{p} - \hat{n} \times \vec{m} - i\frac{k}{6}\vec{Q}(\hat{n}) \right).$$

Use this result to argue that the harmonic analog to Equation (11.111) for real sources is given by Equation (11.112).

Exercise 11.5.3.

(a) Go back to the rotating insulating ring of Exercise 11.2.2. Take the $ka \ll 1$ (nonrelativistic) limit of the exact Exercise 11.2.2(a) result for $dP(t)/d\Omega$.

(b) Find the approximate nonrelativistic differential instantaneous power emitted using Equation (11.93), and show that it agrees with the (a) part result.

(c) The insulating ring is set spinning on its axis with an initial kinetic energy E_0. Because of its accelerated motion, it radiates and slows down. Using conservation of energy, separating variables and integrating over an integral number of cycles, show that in nonrelativistic approximation the time behavior of the energy $E(t)$ is given by

$$\frac{1}{E(t)} - \frac{1}{E_0} = Kt,$$

where E_0 is the energy at $t = 0$ and K is a positive constant, given in terms of ρ_0, m, and a. Find the constant K.

Exercise 11.5.4. Let's go back to Exercise 11.5.3 (based on 11.2.2) and change it a bit. We now rotate the charged ring instead about an axis in the

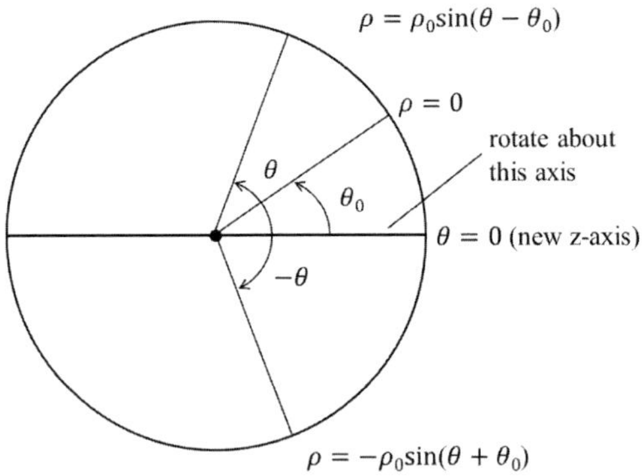

Fig. 11.15 Setup for Exercise 11.5.4.

original xy-plane at an arbitrary angle, θ_0, with respect to the line defining the charge density $\rho = 0$ point, as in Figure 11.15. In the long-wavelength limit, find the instantaneous power angular distribution, $dP(t)/d\Omega$.

Exercise 11.5.5. Given the expression for instantaneous angular power,

$$\frac{dP(t)}{d\Omega} = \frac{1}{4\pi c}\left[\hat{n} \times \int d^3x' \, \frac{1}{c}\frac{\partial \vec{J}(\vec{x}\,', t_r)}{\partial t}\right]^2,$$

where $t_r = t_0 + \hat{n}\cdot\vec{x}\,'/c$ ($t_0 = t - r/c$), show that for a spherically symmetric radial current,

$$\vec{J}(\vec{x}\,', t) = \hat{n}' f(r', t), \; (r' = |\vec{x}\,'|)$$

there is in fact no radiation, even though charges are being accelerated. (This is the "breathing mode" mentioned in Section 11.5, where we gave an approximate argument. Here, you are to prove it *exactly*.)

Exercise 11.5.6. Consider a thin spherical shell of radius a, slightly separated along its equator, with static equal but opposite potentials on the two halves (see Figure 11.16). The shell is rotated at a constant angular velocity ω about an axis. Find the time-averaged angular power, $(dP/d\Omega)_{\text{avg}}$, radiated from the system in the long-wavelength limit when the axis of rotation is:

(a) the z-axis;
(b) the x-axis.

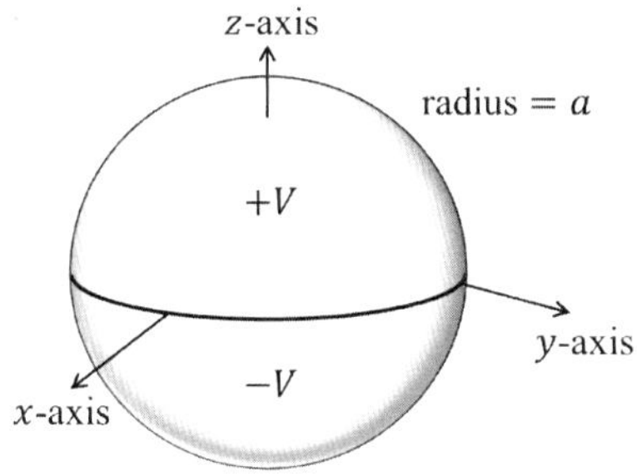

Fig. 11.16 Setup for Exercise 11.5.6.

Exercise 11.5.7.

(a) A pulsar is a rotating neutron star with a strong magnetic field. Often this magnetic field is not oriented along the rotation axis of the star, but rotates at a polar angle θ_r around the star; see Figure 11.17. Given a uniform angular rotation rate of the star, Ω, find an expression for the instantaneous radiation power loss from the lowest order multipole moment. [*Ans.*:

$$P(t) = \frac{2\vec{m}^2\Omega^4}{3c^3}\sin^2\theta_r.]$$

(b) Define the Earth to be in the direction $\hat{n} = (\sin\theta_E,\ 0,\ \cos\theta_E)$. Assuming the Earth and the pulsar star are at rest with respect to one another, find the time dependence of the instantaneous power signal received at the Earth.

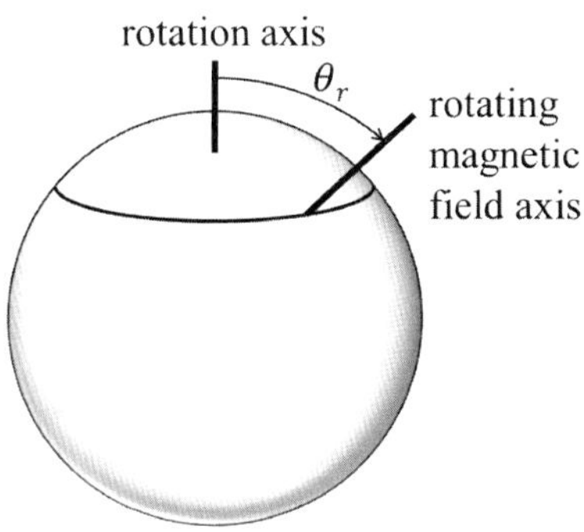

Fig. 11.17 Setup for Exercise 11.5.7.

Exercise 11.5.8.

(a) Starting with Equations (11.93)–(11.97), find $dP(t)/d\Omega$, $P(t)$ and $(dP/d\Omega)_{\text{avg}}$ for the nonrelativistic motion of a particle with charge e moving in a circle of radius R in a uniform, constant magnetic field, $\vec{B}$.
(b) As the particle radiates, it changes its orbit parameters and looses energy. Show that in nonrelativistic approximation the instantaneous energy $E(t)$ as a function of time is given by

$$E(t) = E_0 \, e^{-\lambda t},$$

where E_0 is the energy at $t = 0$. Find the positive constant λ.

Exercise 11.5.9. Find the instantaneous power radiated per unit solid angle $dP(t)/d\Omega$, as well as the time-averaged power $(dP/d\Omega)_{\text{avg}}$, for two nonrelativistic charges, e, moving with coordinates given by:

(a) $\vec{x}_1(t) = A\hat{\imath}\cos(\omega t)$, $\vec{x}_2(t) = A\hat{\jmath}\cos(\omega t)$;
(b) $\vec{x}_1(t) = A\hat{\imath}\sin(\omega t)$, $\vec{x}_2(t) = A\hat{\jmath}\cos(\omega t)$.

In what sense are these results equivalent to the motion of a single charge?

Exercise 11.5.10.

(a) A cylinder of height L and radius a has a harmonic source given everywhere on the cylinder by

$$\vec{J}(\vec{x}\,') = K\delta(\rho' - a)\hat{e}_{\phi'},$$

where $\hat{e}_{\phi'}$ is the azimuthal unit vector. In the long wavelength limit, find the average power angular distribution, $(dP/d\Omega)_{\text{avg}}$.
(b) Do this problem exactly, then show that you recover the long wavelength answer above when ka, $kL \ll 1$.

Exercise 11.5.11. (Referring to Exercise 3.1.4.) A particle of charge e is attached to a string of length L and oscillates above a grounded semi-infinite conducting plane. The distance from the point of attachment to the plane is D. (Gravity is not operating.) Exercise 3.1.4 computed the angular frequency of small oscillations, ω; assume ω is known here. Now compute the instantaneous power radiated from this system, $P(t)$, in lowest order dipole approximation assuming the oscillation angle $\theta(t)$ is given by

$$\theta(t) \approx \theta_0 \sin(\omega t), \quad |\theta_0| \ll 1.$$

Exercise 11.5.12. Using direct integration of expression (11.121) for $|\hat{n} \times \vec{Q}|^2$, give an alternate proof of formulas (11.128) and (11.129) for P_{avg} and $P(t)$ respectively. Note that by making appropriate choices of the function in Gauss' theorem (see Exercise 1.4.2 variants) one may evaluate the complete angular integral of two $(n_i n_j)$ and four $(n_i n_j n_k n_\ell)$ radial unit normals.

Exercise 11.5.13. In Section 10.10 we presented the expansion (Equation (10.275))

$$e^{i\vec{k}\cdot\vec{x}} = 4\pi \sum_{l,m} i^l j_l(kr) Y^*_{lm}(\theta', \phi') Y_{lm}(\theta, \phi),$$

where the spherical angles θ', ϕ' give the direction of $\vec{k}$, θ and ϕ give the $\vec{x}$ direction, and $j_l(kr)$ is the spherical Bessel function of order l.

(a) Use this expansion (the "Poor Man") to show that for harmonic sources ($k = \omega/c$; l sum starts at 0)

$$\left(\frac{dP}{d\Omega}\right)_{\text{avg}} = \frac{2\pi k^2}{c} \left| \hat{n} \times \sum_{l,m} \vec{J}(l,m) Y_{lm}(\theta, \phi) \right|^2, \tag{11.296}$$

where

$$\vec{J}(l,m) \equiv (-i)^l \int d^3x' \vec{J}(\vec{x}\,') j_l(kr') Y^*_{lm}(\theta', \phi').$$

(b) Given the expression for the current density in spherical coordinates for a linear center-fed antenna current oriented along the z-axis

$$\vec{J}(\vec{x}\,') = \hat{n}' \frac{I(r')}{2\pi r'^2} [\delta(\cos\theta' - 1) - \delta(\cos\theta' + 1)], \quad (r' < d/2)$$

obtain the expression for total averaged power

$$P_{\text{avg}} = \frac{2\pi k^2}{c} \sum_l \left(|\vec{J}(l,0)|^2 Q_l - P_l \operatorname{Re}(\vec{J}(l,0) \cdot \vec{J}^*(l+2,0)) \right), \tag{11.297}$$

where

$$l \text{ even:} \quad \vec{J}(l,m) = 2\hat{k}\delta_{m,0}(-i)^l\sqrt{\frac{2l+1}{4\pi}}\int_0^{d/2} dr'\, I(r')j_l(kr'),$$

$$l \text{ odd:} \quad \vec{J}(l,m) = 0,$$

$$Q_l = \frac{1}{2l+1}\left(\frac{(l+1)(l+2)}{(2l+3)} + \frac{l(l-1)}{(2l-1)}\right),$$

$$P_l = \frac{2}{2l+3}\frac{(l+1)(l+2)}{\sqrt{(2l+1)(2l+5)}}.$$

Some results on spherical harmonics in standard references were used. The result (11.297) shows why this expansion is not used in general: the various multipole terms are not orthogonal to one another when the total power is computed. The mixture is caused by the angular function $\sin^2\theta$ in this case.

(c) For the half-wave antenna ($kd = \pi$) with current

$$I(r') = I\sin\left(kd/2 - kr'\right),$$

compute the time-averaged angular power distribution (11.296) as well as the total averaged power (11.297) using $\vec{J}(0,0)$ and $\vec{J}(2,0)$. Compare to the exact result and the lowest spherical multipole field result.

1. Exact:

$$\left(\frac{dP}{d\Omega}\right)_{\text{avg}} = \frac{I^2}{2\pi c}\frac{\cos^2(\frac{\pi}{2}\cos\theta)}{\sin^2\theta}, \quad P_{\text{avg}} = \frac{1}{2}(\gamma + \text{ci}(2\pi) + \ln(2\pi)) \approx 1.21883\frac{I^2}{c}.$$

($\gamma = 0.57721566490$ is the Euler-Mascheroni constant, also seen in (4.107) and $\text{ci}(x)$ is the cosine integral.)

2. Spherical Multipole ($l = 1$):[7]

$$\left(\frac{dP}{d\Omega}\right)_{\text{avg}} \approx \frac{9I^2}{2\pi^3 c}\sin^2\theta, \quad P_{\text{avg}} \approx \frac{12}{\pi^2}\frac{I^2}{c}.$$

3. Poor Man ($l = 0$ in (11.296); $\vec{J}(0,0)$ only in (11.297)):

$$\left(\frac{dP}{d\Omega}\right)_{\text{avg}} \approx ?, \qquad P_{\text{avg}} \approx ?$$

4. Poor Man ($l = 0, 2$ in (11.296); $\vec{J}(0,0)$ and $\vec{J}(2,0)$ in (11.297)):

$$\left(\frac{dP}{d\Omega}\right)_{\text{avg}} \approx ?, \qquad P_{\text{avg}} \approx ?$$

What is the percentage error of categories 2, 3 and 4 compared with the exact P_{avg}?

[7]See J. D. Jackson, *op. cit.*, Tables 9.1 and 9.2.

Exercise 11.5.14.

(a) Let us reconnect to the harmonic formalism of Section 11.1 for multipoles. Consider a pure magnetic dipole source, $\vec{m}$, in harmonic formalism. Starting with Equation (11.15), and using the reduction in Equation (11.102), show that the leading far-field contribution in this formalism is

$$\vec{A}_m = ik\frac{e^{ikr}}{r}\hat{n} \times \vec{m}$$

giving

$$\vec{B}_m = ik\frac{e^{ikr}}{r}\left[ik\,\hat{n} \times (\hat{n} \times \vec{m}) - \frac{1}{r}\left(3\hat{n}\,(\hat{n}\cdot\vec{m}) - \vec{m}\right)\right].$$

(b) Use the harmonic formalism from Section 8.3 applied to radiation of angular momentum. Apply Equation (8.60) in a steady-state situation to obtain an equation for the torque,

$$N_i = \frac{1}{8\pi}\text{Re}\left[\sum_{j,k,\ell}\oint da\, n_j\, \epsilon_{ik\ell}\, x_k \left(E_\ell E_j^* + B_\ell B_j^* - \frac{1}{2}(|\vec{E}|^2 + |\vec{B}|^2)\delta_{\ell j}\right)\right],$$

where we have used the averaging procedure in Equation (8.19) as well as the Maxwell tensor Equation (8.37). Use the field from part (a) to show the torque on an harmonic magnetic dipole source is

$$\vec{N} = C\,\text{Im}\,(\vec{m}^* \times \vec{m}).$$

An integration over unit vectors in the sense of Exercise 8.3.2 was performed. Note that the leading radiation term in $\vec{B}_m$ does not actually contribute. Solve for the constant C.

Extra: Do the same calculation for an harmonic electric dipole source, $\vec{p}$ rather than a magnetic one. This is not so tough given the electric/magnetic duality $\vec{B} \to \vec{E}$, $\vec{E} \to -\vec{B}$ under $\vec{m} \to \vec{p}$.

Exercise 11.5.15.

(a) Consider an harmonic antenna of length d, but with a current,

$$J_z(z) = Ie^{-i\omega t}\sin(kz)\delta(x)\delta(y), \quad -d/2 \le z \le d/2.$$

Using Equation (11.22) in the limit $kd \ll 1$ show that

$$\frac{dP}{d\Omega}_{avg} = \frac{I^2(kd)^6}{1152\pi c}\cos^2\theta\sin^2\theta.$$

(b) Treat the same problem by the multipole expansion method. Evaluate the electric dipole, magnetic dipole and relevant electric quadrupole terms and show that you get the same result as in part (a). [*Hint/Warning*: There is charge build-up on the ends of the antenna when kd is not an integral multiple of 2π. That is, unlike the center-fed problem in Section 11.1, the ends of the antenna are not current nodes.]

Exercise 11.6.1. Prove (11.141) from the approximation

$$\vec{A}(x,t) \approx \frac{1}{r}\int d^3x'\,\frac{1}{c}\vec{J}(\vec{x}\,',t_r) = \frac{1}{r}\int d^3x'\,dt'\,\frac{1}{c}\vec{J}(\vec{x}\,',t')\delta(t'-t_r).$$

Exercise 11.6.2.

(a) Using the forms ($\vec{R} = \vec{x} - \vec{x}(t_R)$, $t_R = t - R/c$, $\vec{\beta} = \vec{v}/c$)

$$\Phi = \frac{e}{R(1-\hat{R}\cdot\vec{\beta})}, \quad \vec{A} = \frac{\vec{v}}{c}\Phi,$$

and the rules displayed in the text for taking derivatives of $\vec{x}(t_R)$, $\vec{v}(t_R)$, etc. for a moving point particle, show that,

$$\vec{E} = -\vec{\nabla}\Phi - \frac{1}{c}\frac{\partial\vec{A}}{\partial t},$$

is given explicitly by

$$\vec{E} = \vec{E}_v + \vec{E}_a,$$

where

$$\vec{E}_v = \frac{e}{R^2}\frac{1-\beta^2}{(1-\hat{R}\cdot\vec{\beta}\,)^3}(\hat{R}-\vec{\beta}),$$

$$\vec{E}_a = \frac{e}{cR}\frac{\hat{R}\times[(\hat{R}-\vec{\beta})\times\dot{\vec{\beta}}]}{(1-\hat{R}\cdot\vec{\beta}\,)^3},$$

are called the velocity and acceleration fields, respectively. Notice that $\vec{E}_v \sim 1/R^2$ and so never contributes to radiation.

(b) Show $\vec{B} = \hat{R}\times\vec{E}$ exactly. [*Hint*: Two useful identities for this purpose are:

$$-\hat{R}_i\frac{\partial\beta_j}{\partial t} = \nabla_i v_j; \quad \left(\hat{R}\times\frac{\vec{\beta}}{c}\right)\frac{\partial\Phi}{\partial t} = \vec{\nabla}\Phi\times(\hat{R}-\vec{\beta}).$$

Prove them, then use them.]

Exercise 11.6.3. Finish the integration leading to (11.160) of the text.

Exercise 11.6.4.

(a) Using the results of Exercise 11.6.2, find the time-averaged electric field on the symmetry axis of a circular synchrotron of radius a. You are given the position vector $\vec{x}(t)$ for the synchrotron particle as

$$\vec{x}(t) = a(\cos\omega t, \sin\omega t, 0).$$

You should find a very simple result.

(b) Now consider the electric field along the x-axis in the plane of the synchrotron, outside of the synchrotron radius. For these numerical evaluations, set $a = 1$.

1. Numerically calculate the ratio

$$\frac{\left\langle (E_{v,x})^2 \right\rangle}{\left\langle (E_{a,x})^2 \right\rangle},$$

where $\left\langle (E_{v,x})^2 \right\rangle$ is the time-average of the square of the x-component of the velocity field and $\left\langle (E_{a,x})^2 \right\rangle$ is the time-average of the square of the x-component of the acceleration field. Plot this ratio as a function of position, x, for the range $10 < x < 1000$ for $\beta = 0.1$ and 0.99.

2. Numerically calculate the ratio

$$\frac{\left\langle (E_{v,y})^2 \right\rangle}{\left\langle (E_{a,y})^2 \right\rangle},$$

where $\left\langle (E_{v,y})^2 \right\rangle$ is the time-average of the square of the y-component of the velocity field and $\left\langle (E_{a,y})^2 \right\rangle$ is the time-average of the square of the y-component of the acceleration field. Plot this ratio as a function of position, x, for the range $10 < x < 1000$ for $\beta = 0.1$ and 0.99. (A log-log plot is useful here.) Comment on your results relative to the expected behavior of this ratio in each case.

Exercise 11.6.5. A point particle which carries no electric charge but has a time-constant electric dipole moment, $\vec{p}$, follows a given path $\vec{x}(t)$. Starting with the charge density and current density expressions,

$$\rho(\vec{x}, t) = -\vec{p} \cdot \vec{\nabla}\, \delta(\vec{x} - \vec{x}(t)), \;\; \vec{J}(\vec{x}, t) = -c\vec{\beta}(t)\, \vec{p} \cdot \vec{\nabla}\, \delta(\vec{x} - \vec{x}(t)),$$

where $\vec{\beta}(t) = \dfrac{d\vec{x}(t)}{dt}$, solve for the *leading order* term in the vector potential, $\vec{A}(\vec{x},t)$, which dominates at large distances $r = |\vec{x}|$ and contributes to the power radiated. Simplify your expression as much as possible.

Exercise 11.7.1. Solve for the exact form of the particle acceleration for the semiclassical path. That is, using (11.179) and considering motion along the $\hat{b}$ and $\hat{\beta}$ directions, show that

$$\dot{\vec{\beta}}(t) = \left(\frac{eQ}{\gamma mc}\right) \frac{\vec{b} + \gamma^{-2}\vec{\beta}ct}{(b^2 + (\beta ct)^2)^{3/2}}.$$

Exercise 11.7.2. Confirm the parameterization of the integral in (11.203) by numerical means.

Exercise 11.7.3. In the work by Bethe and Heitler quoted in the text, a relativistic cross-section appears which is integrated over angles and frequencies to get stopping power. In our notation, the low energy term in their formula is proportional to the logarithmic quantity,

$$\mathcal{L} \equiv \frac{1}{2} \ln \left(\frac{p^2 + pp' - E\hbar\omega/c^2}{p^2 - pp' - E\hbar\omega/c^2}\right).$$

All symbols have their usual relativistic meaning and are magnitudes only: $p = |\vec{p}|$ is the initial particle momentum, $p' = |\vec{p}\,'|$ is the final momentum and E is the initial particle energy. Conserving energy with the relation $E = E' + \hbar\omega$ where E' is the final particle energy, show that one has the nonrelativistic reduction

$$\mathcal{L} \longrightarrow \ln \left(\frac{\left(1 + \sqrt{1 - \bar{x}}\right)^2}{\bar{x}}\right),$$

where $\bar{x} = 2\hbar\omega/mv^2$.

Exercise 11.8.1. Use the Poisson sum rule (11.236) to derive (11.233) for the expression $Q(\omega)$ which is defined in (11.224). [*Hint*: Write $Q(\omega)$ as the product of two sums, and apply the Poisson sum rule to one of them.]

Exercise 11.8.2. In the text we consider certain finite Fourier series for time-periodic systems. Such sums also occur for space-periodic systems

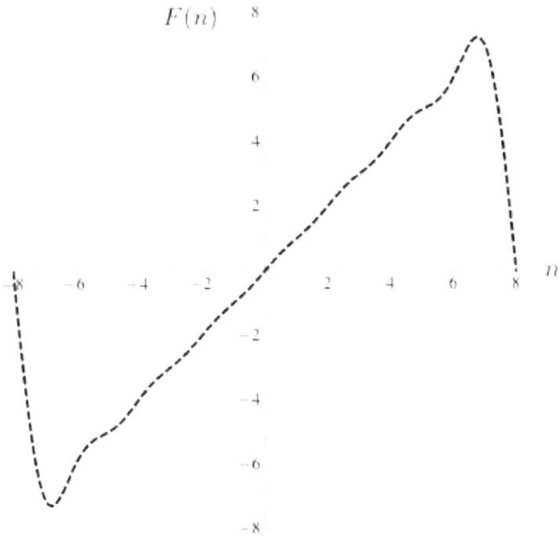

Fig. 11.18 Reference figure for Exercise 11.8.2.

defining a one-dimensional lattice. Define the "momentums", $q_l = \pi l/N$, $l = 0, \pm 1, \pm 2, \ldots, \pm(N-1), N$, for an even lattice of $2N$ points. These give periodic boundary conditions, $e^{-iq_l 2N} = 1$, for the momentums. Given the lattice "completeness" statement,

$$\frac{1}{2N} \sum_{n=-N+1}^{N} e^{-iq_l n} = \delta_{l,0},$$

find the finite Fourier transform of the following function ($n = -N+1, -N+2, \ldots, N-1, N$):

$$F(n) = \begin{cases} n, & n \neq N \\ 0, & n = N \end{cases}$$

($F(n)$ is the closest analog to a linear function on this periodic lattice.) That is, assume that one may write $F(n) = \sum_{l=-N+1}^{N} F_l e^{-iq_l n}$, and solve for the coefficients, F_l. Include graphs which shows your function for $N = 4, 8$; see Figure 11.18 which shows the function generated for the $N = 8$ case, plotting n as a continuous variable.

Extra: In a similar manner also solve for the series which yields function $G(n) = n^2$ on this lattice for $N = 8$. In this case the real part is needed when you plot as a function of continuous n.

Exercise 11.8.3. (A mathematical integration and graphics package is necessary for this problem.) Equations (11.211) and (11.214) may also be employed for any other complete particle trajectory. This problem gives an idea of the frequency distribution of radiation expected when the trajectory is neither impulsive, as in the sample at the end of Section 11.3, nor completely periodic. Consider particles approaching a circle of radius R with a quarter turn, half turn and full turn around the circle. Let $\omega_0 = v/R \implies R/c = \beta/\omega_0$ and the observation direction be $\hat{n} = (0, 1, 0)$:

(a) Quarter turn trajectory:

$$\vec{v}(t) = \begin{cases} v(-1, 0, 0), & t > \frac{\pi}{2\omega_0}, \\ v(-\sin(\omega_0 t), \cos(\omega_0 t), 0), & 0 < t < \frac{\pi}{2\omega_0}, \\ v(0, 1, 0), & t < 0. \end{cases}$$

(b) Half turn trajectory:

$$\vec{v}(t) = \begin{cases} v(0, -1, 0), & t > \frac{\pi}{\omega_0}, \\ v(-\sin(\omega_0 t), \cos(\omega_0 t), 0), & 0 < t < \frac{\pi}{\omega_0}, \\ v(0, 1, 0), & t < 0. \end{cases}$$

(c) Full turn trajectory:

$$\vec{v}(t) = \begin{cases} v(0, 1, 0), & t > \frac{2\pi}{\omega_0}, \\ v(-\sin(\omega_0 t), \cos(\omega_0 t), 0), & 0 < t < \frac{2\pi}{\omega_0}, \\ v(0, 1, 0), & t < 0. \end{cases}$$

For convenience in your evaluations choose a time and distance scale such that $\omega_0 = 1, c = 1$. In each case make qualitative and quantitative observations on the frequency spectrum. Make sure to check the low velocity ($\beta \ll 1$) and high velocity ($\beta \approx 1$) cases. Can you see a periodic spectrum begin to emerge?

Exercise 11.8.4. A particle with the charge of an electron, e, executes an impulsive path in the xy-plane enclosing a square of side L as shown in Figure 11.19. Take the speed of the particle to be v on all parts. Take your origin at the center of the box. Let the zero of time be when the particle is at position: $x = -L/2$, $y = -L/2$. The period of motion is $T = 4L/v \Rightarrow \omega_0 = \dfrac{\pi v}{2L}$ is the fundamental angular frequency.

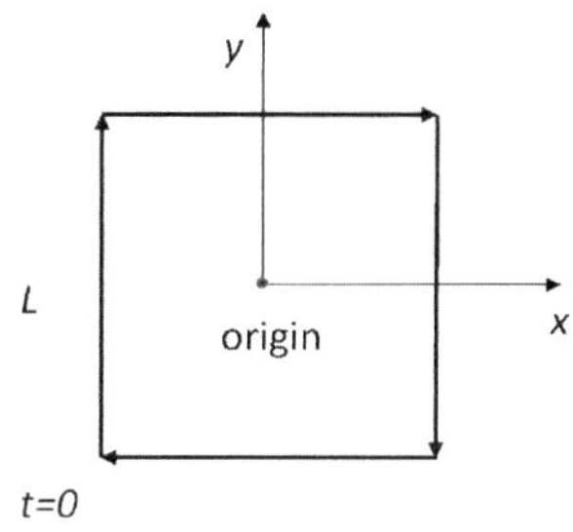

Fig. 11.19 Particle path for Exercise 11.8.4.

(a) Do a nonrelativistic reduction of the power radiated into the mth harmonic for m values such that $\beta m \ll 1$. Show that only $m =$ odd integers contribute, and one obtains simply

$$\frac{dP_m}{d\Omega} = \frac{8e^2\omega_0^2 v^2}{(2\pi c)^3}[\cos^2\theta + 1],$$

where θ is the polar angle. In spite of the non-azimuthally symmetric path, the answer is still independent of ϕ.

(b) Now do a numerical investigation of the general expression using an evaluation and graphics package. Consider the case of $L = 1$ cm, $e =$ electron charge, $v = c/100$ and $m = 25$. Show there is a suppression of the power near $\theta = \pi/2$ relative to the nonrelativistic expression. Now consider the case $m = 205$ with $\phi = 0$ and $\pi/4$. Show that the suppression becomes more pronounced and that one begins to see azimuthal dependence. Also demonstrate with $v = c/2$, $m = 2$, $\phi = 0$ that the suppression of even m begins to break down at relativistic velocities. (*Comment*: The sharp corners here are unrealistic. Although there is suppression of the harmonic power for $\beta m > 1$, the sum over m undoubtedly is not convergent.)

Exercise 11.8.5.

(a) A particle of charge e undergoes simple harmonic motion along the z-axis: $z(t) = a\cos(\omega_0 t)$. Show that the radiation frequency spectrum is discrete and that the power radiated into the mth harmonic is given by

$$\frac{dP_m}{d\Omega} = \frac{e^2\omega_0^2 m^2}{2\pi c}(J_m(m\frac{\omega_0 a}{c}\cos\theta))^2\tan^2\theta.$$

(b) Do a numerical investigation of the *integrated* power emitted into the mth harmonic, P_m, as a function of $\beta \equiv \omega_0 a/c$ for m values from 1 to 100. Show that the parametrization,

$$\ln\left(\frac{P_m}{P_1}\right) = C_1 + C_2 m,$$

(using the natural log) produces a reasonable fit for larger m values ($m \geq 20$) at each β value. Make a table of C_1 and C_2 values for $\beta = 0.1$, 0.5, 0.9 and 0.99. Does this simple fit improve or get worse as the β value is increased? Can you explain the value of C_2 in the nonrelativistic limit?

Exercise 11.9.1. Evaluate $dP_m/d\Omega$ using Equation (11.265) in the nonrelativistic limit $\beta \ll 1$, using the fact that

$$J_m(z) \approx \frac{1}{m!}\left(\frac{z}{2}\right)^m \qquad \text{for } z \ll 1.$$

Show that the $m = 1$ term predominates so that $dP/d\Omega \approx dP_1/d\Omega$, and that your result agrees with the nonrelativistic limit of expression (11.294) (which uses a different definition of θ; see (11.251) and (11.290)).

Exercise 11.9.2.

(a) Derive Equation (11.294) of this chapter either by a symbolic integration package or explicitly. If explicitly, the following integral formula is from Gradshteyn and Rhyzik, *op. cit.*, 3.661 (4):

$$I_n \equiv \int_0^{2\pi} \frac{dx}{(1 - A\cos x)^{n+1}} = \frac{2\pi}{(1 - A^2)^{(n+1)/2}} P_n\left(\frac{1}{\sqrt{1 - A^2}}\right)$$

where P_n is the nth Legendre polynomial. Rewrite expression (11.293) for $dP/d\Omega$ in terms of I_4, dI_3/dA, and d^2I_2/dA^2. Then, plug in explicit expressions for the Legendre polynomials and simplify to obtain the result (11.294).

(b) Now using Equation (11.294) show that in the extreme relativistic case ($\beta \approx 1$) the angular power distribution can be approximated as:

$$\frac{dP}{d\Omega} \underset{\beta\approx 1}{\approx} \frac{7}{16}\frac{e^2\omega_0}{2\pi c T}\frac{1}{(\gamma^{-2} + \theta^2)^{5/2}}\left[1 + \frac{5}{7}\frac{\theta^2}{\gamma^{-2} + \theta^2}\right].$$

Exercise 11.9.3. Integrate expression (11.294) for $dP/d\Omega$ over angles either by a symbolic integration package or explicitly. If explicitly, you may use the following integral formulas from Gradshteyn and Rhyzik, *op. cit.*, 2.271(6), 2.272(6), and 2.273(7):

$$\int \frac{dx}{u^{2n+1}} = \frac{1}{a^n} \sum_{k=0}^{n-1} \frac{(-1)^k}{2k+1} \binom{n-1}{k} \frac{c^k x^{2k+1}}{u^{2k+1}},$$

$$\int \frac{x^2 dx}{u^{2n+1}} = \frac{1}{a^{n-1}} \sum_{k=0}^{n-2} \frac{(-1)^k}{2k+3} \binom{n-2}{k} \frac{c^k x^{2k+3}}{u^{2k+3}},$$

$$\int \frac{x^4 dx}{u^{2n+1}} = \frac{1}{a^{n-2}} \sum_{k=0}^{n-3} \frac{(-1)^k}{2k+5} \binom{n-3}{k} \frac{c^k x^{2k+5}}{u^{2k+5}},$$

where

$$u \equiv \sqrt{a + cx^2},$$

$$\binom{n}{k} \equiv \frac{n(n-1)\cdots(n-k+1)}{1 \cdot 2 \cdots k}, \quad \binom{n}{0} \equiv 1.$$

Show that the result agrees with expression (11.171) for the total energy E emitted in one period.

Exercise 11.9.4. An antenna consists of a rotating rectangle of size $L \times 2R$ spinning at a constant angular velocity, ω_0. It has two straight, thin sides of length L with uniform charge density and total charges Q and $-Q$ a distance R from the rotation axis. See Figure 11.20

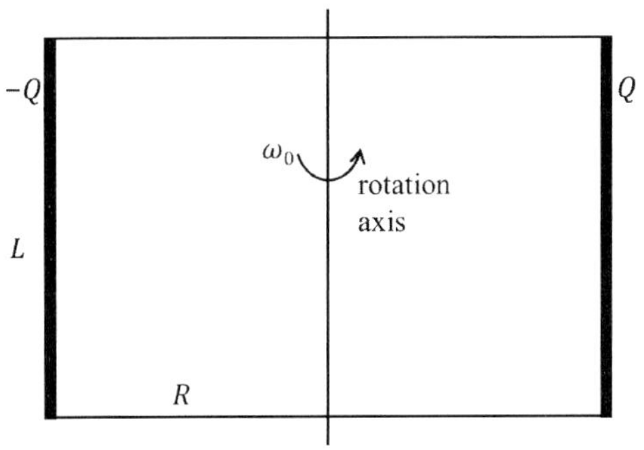

Fig. 11.20 Reference figure for Exercise 11.9.4.

Show that the angular power radiated into the mth harmonic is

$$\frac{dP_m}{d\Omega} = \frac{8Q^2 c}{\pi L^2} \frac{\sin^2\left(\frac{m\omega_0 L}{2c}\cos\theta\right)}{\cos^2\theta}\left[\frac{\beta^2}{2}(J_{m+1}^2(z) + J_{m-1}^2(z)) - J_m^2(z)\right],$$

($\beta = R\omega_0/c,\ z = m\beta\sin\theta$) when m is odd, and zero when m is even.

Exercise 11.9.5. A general, real charge, $\rho(\vec{x}\,',t)$, and current, $\vec{J}(\vec{x}\,',t)$, distribution which maintains its shape while rotating with angular frequency ω_0 in a circle of radius a, is given in cylindrical coordinates by (ρ_0 has units of charge/length^3)

$$\rho(\vec{x}\,',t) = \rho_0 a^2 \delta(\rho' - a)\delta(z')f(\phi'(t)),$$

$$\vec{J}(\vec{x}\,',t) = \omega_0 a \rho(\vec{x}\,',t)\hat{e}_{\phi'},$$

where $\phi'(t) = \phi' - \omega_0 t$. The arbitrary real function, $f(\phi'(t))$, may be expressed as

$$f(\phi'(t)) = \frac{a_0}{2} + \sum_{n=1}^{\infty}(a_n\cos(n\phi'(t)) + b_n\sin(n\phi'(t))),$$

in terms of the real Fourier series coefficients a_n and b_n. Show that the angular distribution of synchrotron radiation given off in the mth harmonic ($m = 1, 2, 3, \ldots$) by this system is given by

$$\frac{dP_m}{d\Omega} = \pi^2(a_m^2 + b_m^2)\left(\frac{\rho_0 a^3}{e}\right)^2\left(\frac{dP_m}{d\Omega}\right)_{\text{synchrotron}},$$

in terms of the synchrotron formula for a point particle, Equations (11.265) or (11.267), of the text. Apply to a point particle with charge e and $f(\phi'(t)) = \delta(\phi'(t))$ to verify the normalization of this result.

Exercise 11.9.6. A "shaker synchrotron" guides a particle with charge e on a circular path of radius R with constant velocity $v_1 = R\omega$ in the $x - y$ plane. In addition, the particle is oscillated along the z-direction with the same angular frequency ω but with amplitude D: $v_2 \equiv D\omega$. The components of the position and velocity are

$$\vec{x}(t) = (R\cos(\omega t), R\sin(\omega t), D\cos(\omega t)),$$

$$\vec{v}(t) = (-v_1\sin(\omega t), v_1\cos(\omega t), -v_2\sin(\omega t)).$$

We will observe in the $\phi = 0$ plane, which means the unit normal is

$$\hat{n} = (\sin\theta, 0, \cos\theta)\,,$$

where θ is the usual polar angle. Using numerical and/or analytical means, build the expression for the instantaneous emission angular power, $\dfrac{dP(t_r)}{d\Omega}$. Average this over one period, $T = 2\pi/\omega$, to get it in the form

$$\left\langle \frac{dP(t_r)}{d\Omega} \right\rangle_{avg.} \equiv \frac{1}{T}\int_0^T dt_r \frac{dP(t_r)}{d\Omega} \equiv \frac{e^2\omega^2}{2\pi c}\cdot\frac{d\Pi(\theta)}{d\Omega}.$$

Numerically plot the dimensionless quantity $\dfrac{d\Pi(\theta)}{d\Omega}$ for $0 < \theta < \pi$ for

(a) $v_1 = 0.3\,c,\ v_2 = 0.3\,c$;

(b) $v_1 = 0.9\,c,\ v_2 = 0.3\,c$;

(c) $v_1 = 0.3\,c,\ v_2 = 0.9\,c$.

Exercise 11.9.7.

(a) Find the angular power radiated in the mth harmonic for the shaker synchrotron from Exercise 11.9.6. The rather simple answer is

$$\frac{dP_m}{d\Omega} = \frac{e^2\omega^2 m^2}{2\pi c}\left[\frac{v_1^2}{c^2}\left(\frac{dJ_m(z')}{dz'}\right)^2 + \left(\frac{v_1\cos\theta - v_2\sin\theta}{v_1\sin\theta + v_2\cos\theta}\right)^2 (J_m(z'))^2\right],$$

where

$$z' \equiv \frac{m\omega}{c}(R\sin\theta + D\cos\theta) = \frac{m}{c}(v_1\sin\theta + v_2\cos\theta)\,.$$

(b) Do a numerical plot of the angular power radiated into the lowest harmonic for the same 3 cases listed in Exercise 11.9.6. Write

$$\frac{dP_m}{d\Omega} = \frac{e^2\omega^2}{2\pi c}\cdot\frac{d\Pi_m(\theta)}{d\Omega},$$

and do angular plots of $\dfrac{d\Pi_1(\theta)}{d\Omega}$, that is, for $m = 1$.

Extra: Try numerically summing these expressions to recover the total $\dfrac{d\Pi(\theta)}{d\Omega}$ from Exercise 11.9.6:

$$\frac{d\Pi(\theta)}{d\Omega} \equiv \sum_{m=1}^{m_{\max}} \frac{d\Pi_m(\theta)}{d\Omega}.$$

Exercise 11.9.8. A modified "shaker synchrotron" guides a particle with charge e on a circular path of radius R with constant velocity $v_1 = R\,\omega_1$ in the $x - y$ plane. The components of the velocity are the same as in Exercise 11.9.7, except that $\omega \to \omega_1$ in the x and y directions, and $\omega \to \omega_2$ in the z-direction. In addition, the particle is oscillated along the z-direction with a *different* angular velocity ω_2 and with amplitude D: $v_2 \equiv D\omega_2$. In nonrelativistic approximation derive the expression for the instantaneous angular power radiated at retarded time t_r, $\dfrac{dP(t_r)}{d\Omega}$. Then, show that one has

$$\left\langle \frac{dP(t_r)}{d\Omega} \right\rangle_\phi \equiv \frac{1}{2\pi}\int d\phi \frac{dP(t_r)}{d\Omega}$$
$$= \frac{e^2}{8\pi c^3}\left[R^2\omega_1^4(1+\cos^2\theta) + 2\,D^2\omega_2^4 \sin^2\theta\cos^2(\omega_2 t_r)\right],$$

when the instantaneous angular power is averaged over the azimuthal angle φ.

Exercise 11.9.9.

(a) A straight connecting bar of length $2R$ has charges e and $\pm e$ placed at its ends and spins at angular velocity ω_0 in the $x - y$ plane, producing a *coherent* radiation amplitude. Show that the harmonic angular power distributions for the cases with charges (e,e) and (e,-e) at the ends are given by

$$\frac{dP_m^{(e,e)}}{d\Omega} = \begin{cases} 4\dfrac{dP_m}{d\Omega}, & m \text{ even} \\ 0, & m \text{ odd} \end{cases}, \quad \frac{dP_m^{(e,-e)}}{d\Omega} = \begin{cases} 0, & m \text{ even} \\ 4\dfrac{dP_m}{d\Omega}, & m \text{ odd} \end{cases}$$

where $\dfrac{dP_m}{d\Omega}$ is given by Equations (11.265) or (11.267).

(b) Given the (a) result, numerically evaluate the total power ratio as a function of β:

$$R_{-+}(\beta) \equiv \frac{\sum\limits_{m=\text{odd}}^{m_{\max}} P_m^{e,-e}(\beta)}{\sum\limits_{m=\text{even}}^{m_{\max}} P_m^{e,e}(\beta)}.$$

(Note: One can determine good $m_{\max}$ values by making sure that

$$\frac{R}{e^2\omega_0}\sum_{m=1,2,3,\ldots}^{m_{\max}} P_m(\beta) = \frac{2}{3}\frac{\beta^3}{(1-\beta^2)^2},$$

in the synchrotron case.) Then make a table of $R_{-+}(\beta)$ values for $\beta = 0.1$ to $\beta = 0.9$ in unit steps of 0.1. Are your results surprising? Explain as best you can what is going on here.

Chapter 12

Scattering and Diffraction

One of the most important and practical subjects for technology purposes is the scattering and diffraction of electromagnetic waves. Following the ideas introduced in Chapter 11, we present a simplified discussion of these subjects, which nonetheless covers the fundamentals. First, we'll talk about some general considerations related to scattering and then address scattering at short wave lengths in the form of conducting sphere scattering. Next, we will look at partial wave techniques in the same context. Finally, we will discuss aspects of diffraction, or the transmission of plane waves through holes or gaps in flat conducting surfaces. Throughout the chapter we will assume incoming plane waves, which means we will be relying on the harmonic source formalism introduced in Section 11.1.

12.1 Definition of polarized scattering cross-section

As mentioned above we are dealing with harmonic sources, so the electric and magnetic fields have the form $\vec{E}(\vec{x})e^{-i\omega t}$ and $\vec{B}(\vec{x})e^{-i\omega t}$ respectively. The Maxwell equations for $\vec{E}(\vec{x})$ and $\vec{B}(\vec{x})$ (henceforth abbreviated as $\vec{E}$ and $\vec{B}$) are ($k \equiv \omega/c$)

$$\vec{\nabla} \times \vec{B} = -ik\vec{E} + \frac{4\pi}{c}\vec{J}, \tag{12.1}$$

$$\vec{\nabla} \times \vec{E} = ik\vec{B}, \tag{12.2}$$

which can be combined as

$$\vec{\nabla} \times (\vec{\nabla} \times \vec{E}) = k^2\vec{E} + ik\frac{4\pi}{c}\vec{J}. \tag{12.3}$$

We also have from the first equation

$$ik\vec{\nabla} \cdot \vec{E} = \frac{4\pi}{c}\vec{\nabla} \cdot \vec{J}, \tag{12.4}$$

from which it follows that

$$-(\vec{\nabla}^2 + k^2)\vec{E} = \frac{4\pi}{c} ik \left(\vec{J} + \frac{1}{k^2} \vec{\nabla}(\vec{\nabla} \cdot \vec{J}) \right). \tag{12.5}$$

We assume the existence of an incoming plane wave of the form

$$\vec{E}^{(i)}_{\text{inc}} = E_0 \hat{\epsilon}_{0i} e^{i\vec{k}_0 \cdot \vec{x}}, \tag{12.6}$$

where ($\vec{k}_0$ = initial wavenumber)

$$\hat{\epsilon}_{0i} \cdot \vec{k}_0 = 0, \quad (i = 1, 2). \tag{12.7}$$

Here we are allowing for two distinct polarizations corresponding to each incoming plane wave direction. In the following, we will leave off the index i unless we need to take the presence of multiple incoming polarizations explicitly into account. The most general incoming wave can be written as a linear combination of waves of the form (12.6).

We may solve (12.5) by finding the corresponding Green function. Recall that

$$\left(\nabla^2 - \frac{1}{c^2} \frac{\partial^2}{\partial t^2} \right) G^{\text{ret.}} = -4\pi \delta(\vec{x} - \vec{x}\,') \delta(t - t'), \tag{12.8}$$

where from (8.111) we have ($R \equiv |\vec{x} - \vec{x}\,'|$)

$$G^{\text{ret}} = \frac{1}{R} \delta \left(t - t' - \frac{R}{c} \right). \tag{12.9}$$

Taking Fourier inverse transforms of both sides of (12.8) with respect to $(t - t')$ results in

$$(\nabla^2 + k^2) G^{\text{out}}(\vec{x}, \vec{x}\,'; k) = -4\pi \delta(\vec{x} - \vec{x}\,'), \tag{12.10}$$

where

$$\begin{aligned} G^{\text{out}}(\vec{x}, \vec{x}\,'; k) &\equiv \int_{-\infty}^{\infty} d(t - t')\, e^{i\omega(t-t')} G^{\text{ret}} \\ &= \int_{-\infty}^{\infty} d(t - t')\, e^{i\omega(t-t')} \frac{1}{R} \delta \left(t - t' - \frac{R}{c} \right) \\ &= \frac{e^{ikR}}{R}. \end{aligned} \tag{12.11}$$

This Green function represents *outgoing* spherical waves. In terms of $G^{\text{out}}(\vec{x}, \vec{x}\,'; k)$, the solution to (12.5) can be written as

$$\vec{E} = \vec{E}_{\text{inc}} + \frac{ik}{c} \int d^3x' \, \frac{e^{ik|\vec{x} - \vec{x}\,'|}}{|\vec{x} - \vec{x}\,'|} \left(\vec{J} + \frac{1}{k^2} \vec{\nabla}\,' \vec{\nabla}\,' \cdot \vec{J} \right). \tag{12.12}$$

The above solution is for volume currents $\vec{J}$, but an analogous equation exists for surface currents which, together with boundary conditions, can be considered an integral equation for $\vec{E}$. We'll deal with a conductor surface presently. We can perform integration by parts on the above to obtain

$$\vec{E} = \vec{E}_{\text{inc}} + ik\left(\overleftrightarrow{1} + \frac{1}{k^2}\vec{\nabla}\vec{\nabla}\right)\cdot\vec{A}(\vec{x}), \tag{12.13}$$

where $\vec{A}(x)$ is the harmonic form of the vector potential (recall (11.8)),

$$\vec{A}(\vec{x}) = \frac{1}{c}\int d^3x'\,\frac{e^{ik|\vec{x}-\vec{x}\,'|}}{|\vec{x}-\vec{x}\,'|}\vec{J}. \tag{12.14}$$

Here the notation $\left(\overleftrightarrow{1} + k^{-2}\vec{\nabla}\vec{\nabla}\right)$ represents a matrix operator (also called a *dyadic*); we saw a dyadic angular form previously in (10.309). The expression $\overleftrightarrow{1}$ represents the identity matrix.

Expression (12.13) is more exact than necessary. In the far zone (large-r limit) we may make the replacements:

$$\vec{A}_{\text{sc}}(\vec{x}) = \frac{e^{ikr}}{r}\int d^3x'\,\frac{1}{c}\vec{J}(\vec{x}\,')e^{-ik\hat{n}\cdot\vec{x}\,'},\ \vec{\nabla}\to i\vec{k}. \tag{12.15}$$

Note that in this chapter for the final direction vector we use the notation $\hat{k}\equiv\hat{n}$, not to be confused with the $z-$component unit vector! Then we have

$$\vec{E}\to\vec{E}_{\text{inc}}+\vec{E}_{\text{sc}}, \tag{12.16}$$

where

$$\begin{aligned}\vec{E}_{\text{sc}} &\equiv ik\left(\overleftrightarrow{1} - \hat{k}\hat{k}\right)\cdot\vec{A}_{\text{sc}}\\ &= -ik\,\hat{k}\times\left(\hat{k}\times\vec{A}_{\text{sc}}\right).\end{aligned} \tag{12.17}$$

Here $\vec{E}_{\text{sc}}$ refers to the *asymptotic* scattered field, and is in fact identical to (11.20). From previous discussions we have also

$$\vec{B}\to\vec{B}_{\text{inc}}+\vec{B}_{\text{sc}}, \tag{12.18}$$

$$\vec{B}_{\text{sc}} = ik\,\hat{k}\times\vec{A}_{\text{sc}}. \tag{12.19}$$

Then, the time-averaged scattered power is

$$\frac{dP}{d\Omega} = \frac{k^2r^2c}{8\pi}\left|\hat{k}\times\vec{A}_{\text{sc}}\right|^2 = \frac{r^2c}{8\pi}|\vec{E}_{\text{sc}}|^2 = \frac{r^2c}{8\pi}|\vec{B}_{\text{sc}}|^2. \tag{12.20}$$

In order to express $\vec{E}_{\rm sc}$ in terms of its polarized components, we shall make use of the matrix identity

$$\sum_f \hat{\epsilon}_f \hat{\epsilon}_f^* + \hat{k}\hat{k} = \overleftrightarrow{1}, \tag{12.21}$$

where f takes on two values and where we always choose $\hat{\epsilon}_1$, $\hat{\epsilon}_2$ so that $\hat{\epsilon}_i^* \cdot \hat{\epsilon}_j = \delta_{ij}$. For example, in the case where $\hat{k} = \hat{z}$ we may choose linear polarizations:

$$\hat{\epsilon}_1 = (1,0,0),\ \hat{\epsilon}_2 = (0,1,0),\ \hat{k} = (0,0,1),$$
$$\Longrightarrow\ \hat{\epsilon}_1\hat{\epsilon}_1^* + \hat{\epsilon}_2\hat{\epsilon}_2^* + \hat{k}\hat{k} = \overleftrightarrow{1}.$$

If instead we choose circular polarization (as in Equation (9.18)), for example

$$\hat{\epsilon}_+ = \frac{1}{\sqrt{2}}(1,i,0),\ \hat{\epsilon}_- = \frac{1}{\sqrt{2}}(1,-i,0),\ \hat{k} = (0,0,1),$$

then we find

$$\hat{\epsilon}_+\hat{\epsilon}_+^* + \hat{\epsilon}_-\hat{\epsilon}_-^* + \hat{k}\hat{k} = \frac{1}{2}\begin{pmatrix} 1 & -i & 0 \\ i & 1 & 0 \\ 0 & 0 & 0 \end{pmatrix} + \frac{1}{2}\begin{pmatrix} 1 & i & 0 \\ -i & 1 & 0 \\ 0 & 0 & 0 \end{pmatrix} + \begin{pmatrix} 0 & 0 & 0 \\ 0 & 0 & 0 \\ 0 & 0 & 1 \end{pmatrix} = \begin{pmatrix} 1 & 0 & 0 \\ 0 & 1 & 0 \\ 0 & 0 & 1 \end{pmatrix}.$$

In general we have

$$\vec{E}_{\rm sc} = \sum_f (\hat{\epsilon}_f \hat{\epsilon}_f^* + \hat{k}\hat{k}) \cdot \vec{E}_{\rm sc} = \sum_f \hat{\epsilon}_f (\vec{E}_{\rm sc} \cdot \hat{\epsilon}_f^*), \tag{12.22}$$

where we have used $\hat{k} \cdot \vec{E}_{\rm sc} = 0$. Then we may express the total power distribution as the sum of polarized power distributions:

$$\frac{dP}{d\Omega} = \frac{r^2 c}{8\pi}\left|\sum_f \hat{\epsilon}_f(\vec{E}_{\rm sc} \cdot \hat{\epsilon}_f^*)\right|^2 = \frac{r^2 c}{8\pi}\sum_f \left|\hat{\epsilon}_f^* \cdot \vec{E}_{\rm sc}\right|^2 \equiv \sum_f \left(\frac{dP}{d\Omega}\right)_f. \tag{12.23}$$

This interpretation is consistent with previous expressions for power in terms of $\vec{E}_{\rm sc} \times \vec{B}_{\rm sc}^*$ (see (8.24) for example); with the definitions

$$(\vec{E}_{\rm sc})_f \equiv \hat{\epsilon}_f (\vec{E}_{\rm sc} \cdot \hat{\epsilon}_f^*), \tag{12.24}$$
$$(\vec{B}_{\rm sc}^*)_f \equiv \hat{k} \times \hat{\epsilon}_f^* (\vec{E}_{\rm sc}^* \cdot \hat{\epsilon}_f), \tag{12.25}$$

we have

$$\begin{aligned} \mathrm{Re}\left((\vec{E}_{\rm sc})_f \times (\vec{B}_{\rm sc}^*)_f\right) \cdot \hat{k} &= \left(\mathrm{Re}\big(\underbrace{\hat{\epsilon}_f \times (\hat{k} \times \hat{\epsilon}_f^*)}_{\hat{k}}\big) \cdot \hat{k}\right) \left|\vec{E}_{\rm sc} \cdot \hat{\epsilon}_f^*\right|^2 \\ &= \left|\hat{\epsilon}_f^* \cdot \vec{E}_{\rm sc}\right|^2, \end{aligned} \tag{12.26}$$

and thus

$$\left(\frac{dP}{d\Omega}\right)_f = \frac{r^2 c}{8\pi}\mathrm{Re}\left((\vec{E}_{\rm sc})_f \times (\vec{B}^*_{\rm sc})_f\right)\cdot\hat{k}. \tag{12.27}$$

Let us pause here for a moment to consider the larger meaning of the power expression (12.23). The specific fields being considered here, $\vec{E}_{\rm sc}$ and $\vec{B}_{\rm sc}$, are the outgoing radiation fields produced by a scattering event. In Chapter 11 we were also dealing with outgoing radiation fields $\vec{E}_{\rm out}, \vec{B}_{\rm out} \sim e^{ikr}/r$ originating from specific types of sources. It should be abundantly clear that expression (12.23) applies to both situations. This gives us a quick and easy way to convert our total power or energy expressions in Chapter 11 to polarized versions. As we can see from (12.17), (12.15), (12.20) and (12.23), we need only unambiguously identify the structures $|\hat{k}\times(\hat{k}\times\vec{A}_{\rm out})| = |\hat{k}\times\vec{A}_{\rm out}|$ in the unpolarized expressions. Then we have

$$\begin{aligned}\hat{\epsilon}^*_f\cdot\vec{E}_{\rm out} &= -ik\,\hat{\epsilon}^*_f\cdot(\hat{k}\times(\hat{k}\times\vec{A}_{\rm out}))\\ &= ik\,\hat{\epsilon}^*_f\cdot\vec{A}_{\rm out},\end{aligned} \tag{12.28}$$

and we may make the replacement

$$|\hat{k}\times(\hat{k}\times\vec{A}_{\rm out})| = |\hat{k}\times\vec{A}_{\rm out}| \longrightarrow |\hat{\epsilon}^*_f\cdot\vec{A}_{\rm out}|, \tag{12.29}$$

in order to identify a polarized power or energy distribution from any of the formulas in Chapter 11. Thus, for example from (11.22) we have

$$\left(\frac{dP}{d\Omega}\right)_f = \frac{k^2}{8\pi c}\left|\hat{\epsilon}^*_f\cdot\int d^3x'\,\vec{J}(\vec{x}\,')e^{-ik\hat{n}\cdot\vec{x}\,'}\right|^2. \tag{12.30}$$

As another example we have from (11.75)

$$\left(\frac{d^2E(\omega)}{d\omega d\Omega}\right)_f = \frac{k^2}{2\pi c}\left|\hat{\epsilon}^*_f\cdot\int d^3x'\,\vec{J}(\vec{x}\,',\omega)e^{-ik\hat{n}\cdot\vec{x}'}\right|^2. \tag{12.31}$$

See Exercise 12.1.2 for a further application of this idea to the harmonic synchrotron power expressions.

Getting back to the scattering situation at hand, we shall find it useful to write $\vec{E}_{\rm sc}$ in the form

$$\vec{E}_{\rm sc} = \frac{e^{ikr}}{r}E_0\vec{f}(\vec{k},\vec{k}_0), \tag{12.32}$$

where E_0 is the initial electric field amplitude. $\vec{f}(\vec{k},\vec{k}_0)$ is called the *scattering amplitude*. Then we may write

$$\left(\frac{dP}{d\Omega}\right)_f = \frac{c|E_0|^2}{8\pi}\left|\hat{\epsilon}^*_f\cdot\vec{f}(\vec{k},\vec{k}_0)\right|^2, \tag{12.33}$$

and thus from (12.28)

$$\hat{\epsilon}_f^* \cdot \vec{f}(\vec{k}, \vec{k}_0) = \frac{re^{-ikr}}{E_0} \hat{\epsilon}_f^* \cdot \vec{E}_{\rm sc} = \frac{ikre^{-ikr}}{E_0} \hat{\epsilon}_f^* \cdot \vec{A}_{\rm sc}$$
$$= \frac{ik}{cE_0} \hat{\epsilon}_f^* \cdot \int d^3x' \, \vec{J}(\vec{x}\,') e^{-i\vec{k}\cdot\vec{x}\,'}. \tag{12.34}$$

At this point the dependence of $\vec{E}_{\rm sc}$ and $\vec{f}(\vec{k}, \vec{k}_0)$ on $\vec{k}_0$ and $\hat{\epsilon}_0$, where the subscript indicates initial value, has not been brought out yet. As we shall see later, this dependence arises because $\vec{J}$ is really induced by $\vec{E}_{\rm inc}$ in scattering situations. In the harmonic formalism in free space, one has $|\vec{k}_0| = |\vec{k}|$.

Using expression (8.24) for $\vec{S}$ and the fact that $\vec{E}_0 \times \vec{B}_0^* = \hat{k}|E_0|^2$ for plane waves, we have for the initial energy flux

$$|\vec{S}_0| = \frac{c}{8\pi}|E_0|^2. \tag{12.35}$$

Taking both incoming and final polarizations into account, we define the *polarized cross-sections* as

$$\left(\frac{d\sigma}{d\Omega}\right)_{f,i} \equiv \left(\frac{dP}{d\Omega}\right)_{f,i} \Big/ \, |\vec{S}_0| \equiv |\hat{\epsilon}_f^* \cdot \vec{f}_i(\vec{k}, \vec{k}_0)|^2, \tag{12.36}$$

where $\vec{f}_i(\vec{k}, \vec{k}_0)$ are the scattering amplitudes corresponding to the initial polarizations $\hat{\epsilon}_{0i}$ $(i = 1, 2)$. The *unpolarized cross-section* is defined as

$$\left(\frac{d\sigma}{d\Omega}\right)_{\rm unpol} \equiv \frac{1}{2}\sum_{f,i} \left|\hat{\epsilon}_f^* \cdot \vec{f}_i(\vec{k}, \vec{k}_0)\right|^2, \tag{12.37}$$

where the factor of 1/2 is due to the averaging over the two initial polarization states.

12.2 Kirchhoff identity for scattering

In this section, we show how to find scattering amplitudes in terms of the fields immediately surrounding the region where scattering takes place. This can be very useful in cases where we don't have a detailed knowledge of $\vec{J}$. The expression we'll derive is actually exact in the $r \to \infty$ limit. Figure 12.1 shows our notation for the scattering system. Note that all currents are assumed to be situated *inside* the surface S.

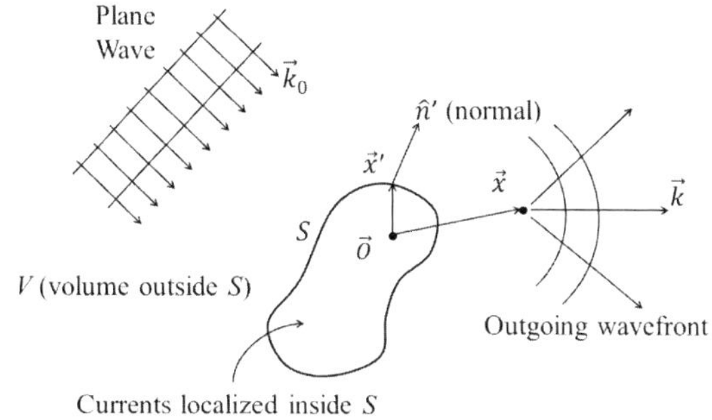

Fig. 12.1 Notation for derivation of Kirchhoff's identity.

First we use Green's second identity to obtain a new expression for the electric field. The Helmholtz equations for $\vec{E}_{\rm sc}$ and $G^{\rm out}$ are:

$$-(\vec{\nabla}'^2 + k^2)\vec{E}_{\rm sc}(\vec{x}\,') = \frac{4\pi}{c} ik\left(\vec{J}(\vec{x}\,') + \frac{1}{k^2}\vec{\nabla}'(\vec{\nabla}' \cdot \vec{J}(\vec{x}\,'))\right), \tag{12.38}$$

$$(\vec{\nabla}'^2 + k^2)G^{\rm out}(\vec{x}, \vec{x}\,', k) = -4\pi\delta(\vec{x} - \vec{x}\,'). \tag{12.39}$$

We multiply these equations by $G^{\rm out}(\vec{x}, \vec{x}\,', k)$ and $\vec{E}_{\rm sc}(\vec{x}\,',)$ respectively and add them, then integrate both sides over all space *outside* of S. After applying Green's second identity to each vector component of the left-hand side, we obtain ($\vec{J}$, $\vec{E}_{\rm sc}$, and $G^{\rm out}$ in the integrals denote $\vec{J}(\vec{x}\,')$, $\vec{E}_{\rm sc}(\vec{x}\,')$, and $G^{\rm out}(\vec{x}, \vec{x}\,', k)$ respectively)

$$\frac{1}{4\pi}\oint_S da' \left[\vec{E}_{\rm sc}(-\hat{n}' \cdot \vec{\nabla}' G^{\rm out} - G^{\rm out}(-\hat{n}' \cdot \vec{\nabla}')\vec{E}_{\rm sc}\right]$$
$$= -\vec{E}_{\rm sc}(\vec{x}) + \frac{ik}{c}\int_V d^3x'\, G^{\rm out}\left[(\vec{J} + \frac{1}{k^2}\vec{\nabla}'(\vec{\nabla}' \cdot \vec{J})\right]. \tag{12.40}$$

Notice that we have used $-\hat{n}'$ as the outward-pointing normal, which points *into* S.

There is an important point that we should mention here. To apply Green's second identity correctly, we should be integrating over a *bounded* volume. We solve this problem by including a bounding surface "at infinity" that is at a finite but large distance. It can be shown that the surface integrals "at infinity" vanish.[1] We will assume this result, and retain only the surface integral over S.

[1]For a rigorous demonstration of this, the reader may consult J. D. Jackson, *Classical Electrodynamics*, 3rd ed., John Wiley & Sons (New York) 1999, Section 10.6. The result follows from evaluating (12.48) on the surface at "infinity", and uses the transverse nature of $\vec{E}_{\rm sc}$ and $\vec{B}_{\rm sc}$ expressed in (11.20).

Continuing our argument, we solve (12.40) for $\vec{E}_{\rm sc}(\vec{x})$:

$$\vec{E}_{\rm sc}(\vec{x}) = \frac{ik}{c}\int_V d^3x'\, G^{\rm out}\left[\vec{J} + \frac{1}{k^2}\vec{\nabla}'(\vec{\nabla}'\cdot\vec{J})\right] + \frac{1}{4\pi}\oint_S da'\left[(\hat{n}'\cdot\vec{\nabla}'G^{\rm out})\vec{E}_{\rm sc} - G^{\rm out}(\hat{n}'\cdot\vec{\nabla}')\vec{E}_{\rm sc}\right]. \quad (12.41)$$

There are no currents in the region outside S, so the volume integral disappears:

$$\vec{E}_{\rm sc}(\vec{x}) = \frac{1}{4\pi}\oint_S da'\left[(\hat{n}'\cdot\vec{\nabla}'G^{\rm out})\vec{E}_{\rm sc} - G^{\rm out}(\hat{n}'\cdot\vec{\nabla}')\vec{E}_{\rm sc}\right]. \quad (12.42)$$

We're not quite done yet: we want to re-express the $(\hat{n}'\cdot\vec{\nabla}')\vec{E}_{\rm sc}$ to avoid having to take derivatives of the field at S. First we re-express the second term in the integrand:

$$\vec{E}_{\rm sc}(\vec{x}) = \frac{1}{4\pi}\oint_S da'\left[2(\hat{n}'\cdot\vec{\nabla}'G^{\rm out})\vec{E}_{\rm sc} - (\hat{n}'\cdot\vec{\nabla}')\left(G^{\rm out}\vec{E}_{\rm sc}\right)\right]. \quad (12.43)$$

We now invoke a general vector identity which we first used back in (6.51):

$$\oint_C d\vec{l}' \times \vec{A} = \int_{S'} da\left[\sum_i (n_i'\vec{\nabla}'A_i) - \hat{n}'\vec{\nabla}'\cdot\vec{A}\right], \quad (12.44)$$

where S' is an arbitrary open surface bounded by C. In our case we take $\vec{A} \equiv (G^{\rm out}\vec{E}_{\rm sc})$ and $S' \equiv S$ (so C vanishes). This gives

$$0 = \oint_S da\left[\sum_i \left(n_i'\vec{\nabla}'\left(G^{\rm out}(E_{\rm sc})_i\right)\right) - \hat{n}'\vec{\nabla}'\cdot(G^{\rm out}\vec{E}_{\rm sc})\right]. \quad (12.45)$$

We may add this integrand to the integrand in (12.43) without changing the equality:

$$\begin{aligned}
\vec{E}_{\rm sc}(\vec{x}) &= \frac{1}{4\pi}\oint_S da'\Bigg[2(\hat{n}'\cdot\vec{\nabla}'G^{\rm out})\vec{E}_{\rm sc} - \hat{n}'\vec{\nabla}'\cdot(G^{\rm out}\vec{E}_{\rm sc}) \\
&\quad + \sum_i n_i'\vec{\nabla}'\left(G^{\rm out}(E_{\rm sc})_i\right) - (\hat{n}'\cdot\vec{\nabla}')\left(G^{\rm out}\vec{E}_{\rm sc}\right)\Bigg] \\
&= \frac{1}{4\pi}\oint_S da'\Big[2(\hat{n}'\cdot\vec{\nabla}'G^{\rm out})\vec{E}_{\rm sc} - \hat{n}'\vec{\nabla}'\cdot(G^{\rm out}\vec{E}_{\rm sc}) \\
&\quad + \hat{n}'\times(\vec{\nabla}'\times(G^{\rm out}\vec{E}_{\rm sc}))\Big] \\
&= \frac{1}{4\pi}\oint_S da'\Big[2(\hat{n}'\cdot\vec{\nabla}'G^{\rm out})\vec{E}_{\rm sc} - \hat{n}'(\vec{E}_{\rm sc}\cdot\vec{\nabla}'G^{\rm out}) \\
&\quad + ik(\hat{n}'\times\vec{B}_{\rm sc})G^{\rm out} + \underbrace{\hat{n}'\times(\vec{\nabla}'G^{\rm out}\times\vec{E}_{\rm sc})}_{(\hat{n}'\cdot\vec{E}_{\rm sc})\vec{\nabla}'G^{\rm out} - \vec{E}_{\rm sc}\hat{n}'\cdot\vec{\nabla}'G^{\rm out}}\Big],
\end{aligned} \quad (12.46)$$

where we have used the $BAC-CAB$ rule twice, and the fact that $\vec{\nabla}\cdot\vec{E}_{\rm sc} = 0$ on S. The remaining terms collapse into:

$$\vec{E}_{\rm sc}(\vec{x}) = \frac{1}{4\pi}\oint_S da' \Big[(ik(\hat{n}' \times \vec{B}_{\rm sc})G^{\rm out} + (\hat{n}' \times \vec{E}_{\rm sc}) \times \vec{\nabla}' G^{\rm out} + (\hat{n}' \cdot \vec{E}_{\rm sc})\vec{\nabla}' G^{\rm out}\Big]. \tag{12.47}$$

We re-emphasize that the integrand depends only on the scattered fields evaluated at the surface S. At this point we may replace:

$$G^{\rm out} \to \frac{e^{ikr}}{r}e^{-i\vec{k}\cdot\vec{x}\,'}; \quad \vec{E}_{\rm sc} \to E_0\frac{e^{ikr}}{r}\vec{f}(\vec{k},\vec{k}_0); \quad \vec{\nabla}' G^{\rm out} \to -i\vec{k}G^{\rm out},$$

to obtain finally the *vector Kirchhoff identity*

$$\vec{f}(\vec{k},\vec{k}_0) = \frac{ik}{4\pi E_0}\oint_S da'\, e^{-i\vec{k}\cdot\vec{x}\,'} \Big[(\hat{n}' \times \vec{B}_{\rm sc}) + \hat{k} \times (\hat{n}' \times \vec{E}_{\rm sc}) - \hat{k}(\hat{n}' \cdot \vec{E}_{\rm sc})\Big]. \tag{12.48}$$

For a fixed polarization direction $\hat{\epsilon}^*$ we have

$$\hat{\epsilon}^* \cdot \vec{f}(\vec{k},\vec{k}_0) = \frac{ik}{4\pi E_0}\oint_S da'\, e^{-i\vec{k}\cdot\vec{x}\,'} \Big[\hat{\epsilon}^* \cdot (\hat{n}' \times \vec{B}_{\rm sc}) + \hat{\epsilon}^* \cdot (\hat{k} \times (\hat{n}' \times \vec{E}_{\rm sc}))\Big]. \tag{12.49}$$

Since we have made no approximations in our derivation other than $r \to \infty$, (12.48) may be considered as an "exact" asymptotic expression for the scattering amplitude.

12.3 Short wavelength scattering for a conducting sphere

The results of the previous section may be used to determine the scattering produced by a conducting object. In this case, the surface S is taken just outside the object. Here we treat the case of a conducting sphere, as shown in Figure 12.2. We will require some additional approximations (as explained below), so our results are no longer asymptotically exact as in the previous section.

To find the scattered fields $\vec{E}_{\rm sc}$, $\vec{B}_{\rm sc}$, we note first that they must give the correct boundary conditions at the conductor surface. From (10.13) and (10.14) we have the general perfect-conductor boundary conditions

$$\hat{n}' \cdot (\vec{B}_{\rm inc} + \vec{B}_{\rm sc})\Big|_s = 0, \tag{12.50}$$

$$\hat{n}' \times (\vec{E}_{\rm inc} + \vec{E}_{\rm sc})\Big|_s = 0. \tag{12.51}$$

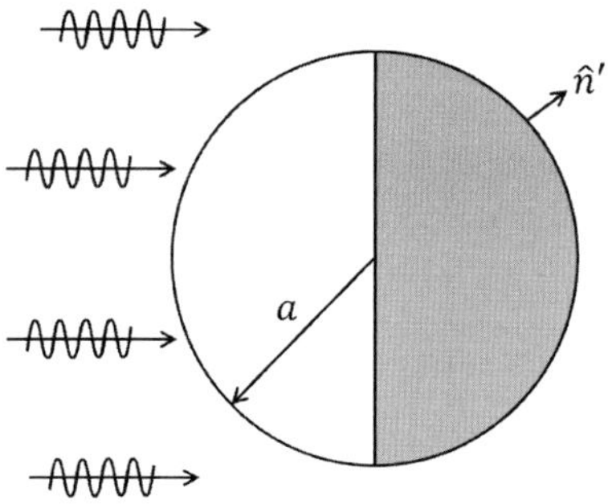

Fig. 12.2 Small-wavelength scattering from a conducting sphere, with shadow and illuminated side.

For the other boundary conditions we may distinguish between the illuminated side and the shadow side (see Figure 12.2). On the shadow side, from physical observations of light we expect the $\vec{E}$ and $\vec{B}$ fields to vanish at the boundary when the wavelength of the incoming radiation is small compared to the dimensions of the scattering object, or that $ka \gg 1$. On the illuminated side, when $ka \gg 1$ the surface appears locally flat, so we may use the boundary conditions corresponding to reflection from a flat, perfectly conducting surface. Referring to the geometry in Figure 9.3 and using Exercise 9.3.2 in the $n_1 \gg n_2$ limit, we have ("ill" denotes illuminated side)

$$\hat{n}' \cdot \vec{E}_{\text{sc}}|_{\text{ill}} \approx \hat{n}' \cdot \vec{E}_{\text{inc}}, \tag{12.52}$$

and from (9.101) with $n_1 \gg n_2$ we find

$$\hat{n}' \times \vec{B}_{\text{sc}}|_{\text{ill}} \approx \hat{n}' \times \vec{B}_{\text{inc}}. \tag{12.53}$$

We may summarize the shadow-side and illuminated-side boundary conditions as ("sh" denotes shadow side)

$$\vec{E}_{\text{sc}}|_{\text{sh}} \approx -\vec{E}_{\text{inc}}, \tag{12.54}$$

$$\vec{B}_{\text{sc}}|_{\text{sh}} \approx -\vec{B}_{\text{inc}}, \tag{12.55}$$

$$\vec{E}_{\text{sc}}|_{\text{ill}} \approx \hat{n}'(\hat{n}' \cdot \vec{E}_{\text{inc}}) + \hat{n}' \times (\hat{n}' \times \vec{E}_{\text{inc}}), \tag{12.56}$$

$$\vec{B}_{\text{sc}}|_{\text{ill}} \approx -\hat{n}'(\hat{n}' \cdot \vec{B}_{\text{inc}}) - \hat{n}' \times (\hat{n}' \times \vec{B}_{\text{inc}}). \tag{12.57}$$

In our derivation of the Kirchhoff identity we used the Green function G^{out} for free space (with no bounding surfaces), so we should verify that these boundary conditions are consistent with the physical fields inside as well as outside the conductor. Inside the conductor the physical fields vanish, so we must have $\vec{E}_{\rm sc} = -\vec{E}_{\rm inc}$ and $\vec{B}_{\rm sc} = -\vec{B}_{\rm inc}$. The standard boundary conditions (1.20) and (1.30) then give

$$\frac{1}{c}\vec{K} = \frac{1}{4\pi}\hat{n}' \times (\vec{B}_{\rm inc} + \vec{B}_{\rm sc}), \tag{12.58}$$

$$\sigma = \frac{\hat{n}'}{4\pi} \cdot (\vec{E}_{\rm inc} + \vec{E}_{\rm sc}), \tag{12.59}$$

as the surface current and charge density. In Exercise 12.3.1 you will use these expressions to give an alternative derivation of the forward scattering amplitude.

The scattering amplitude is the sum of contributions from the shadow and illuminated sides. We begin with the shadow-side contribution. Substituting (12.54) and (12.55) into (12.49) and using the plane wave expressions for $\vec{E}_{\rm inc}$ and $\vec{B}_{\rm inc}$, we find for initial wave number $\vec{k}_0$ and any polarization vector $\hat{\epsilon}^*$ that

$$\hat{\epsilon}^* \cdot \vec{f}_{\rm sh} = \frac{k}{4\pi i}\int_{\rm sh} da'\, \hat{\epsilon}^* \cdot \left[(\hat{k} + \hat{k}_0) \times (\hat{n}' \times \hat{\epsilon}_0) + (\hat{n}' \cdot \hat{\epsilon}_0)\hat{k}_0\right] e^{-i(\vec{k}-\vec{k}_0)\cdot\vec{x}\,'}. \tag{12.60}$$

Since $ka \gg 1$, the integrand is highly oscillatory and the integral nearly vanishes unless $\vec{k} \approx \vec{k}_0$. In this case we only need to consider the case where $\vec{k} \approx \vec{k}_0$, and we may approximate

$$\hat{\epsilon}^* \cdot \left[(\hat{k} + \hat{k}_0) \times (\hat{n}' \times \hat{\epsilon}_0) + (\hat{n}' \cdot \hat{\epsilon}_0)\hat{k}_0\right] \underset{\vec{k}\approx\vec{k}_0}{\approx} -2(\hat{k}_0 \cdot \hat{n}')(\hat{\epsilon}^* \cdot \hat{\epsilon}_0), \tag{12.61}$$

which gives

$$\hat{\epsilon}^* \cdot \vec{f}_{\rm sh}(\vec{k}, \vec{k}_0) \approx \frac{ik\hat{\epsilon}^* \cdot \hat{\epsilon}_0}{2\pi}\int_{\rm sh} da'\, e^{-i(\vec{k}-\vec{k}_0)\cdot\vec{x}\,'}(\hat{n}' \cdot \hat{k}_0). \tag{12.62}$$

As we said, for $ka \gg 1$ the exponential is highly oscillatory, implying the integral is appreciably different from zero only for $\theta \lesssim 1/(ka)$. Under these conditions (take z' along $\hat{k}_0$)

$$\begin{aligned} &(\vec{k}_0 - \vec{k}) \cdot \vec{x}\,' = -\vec{k}_\perp \cdot \vec{x}\,'_\perp + \mathcal{O}\left(\frac{1}{ka}\right), \\ \Longrightarrow &\int_{\rm sh} da'\, e^{-i(\vec{k}-\vec{k}_0)\cdot\vec{x}\,'}(\hat{n}' \cdot \hat{k}_0) \approx \int_{\rm sh} d^2x'_\perp\, e^{-i\vec{k}_\perp\cdot\vec{x}\,'_\perp}, \end{aligned} \tag{12.63}$$

where $d^2x'_\perp$ is the projected area along $\vec{k}_0$, $d^2x'_\perp \equiv da'\cos\theta'$. For the sphere (using $|\vec{k}_\perp| = k\sin\theta$)

$$\int_{\rm sh} da'\, e^{-i\vec{k}_\perp\cdot\vec{x}'_\perp}\cos\theta' = a^2\int_0^{\pi/2} d\theta'\,\sin\theta'\cos\theta'\int_0^{2\pi} d\phi'\, e^{-ika\sin\theta\sin\theta'\cos\phi'}$$
$$= 2\pi a^2\int_0^{\pi/2} d\theta'\,\sin\theta'\cos\theta' J_0(ka\sin\theta\sin\theta')$$
$$= \frac{2\pi a^2}{(ka\sin\theta)^2}\int_0^{ka\sin\theta} dx\, x\underbrace{J_0(x)}_{\frac{1}{x}\frac{d}{dx}(xJ_1(x))} = 2\pi a^2\frac{J_1(ka\sin\theta)}{ka\sin\theta}$$
$$\Longrightarrow \hat{\epsilon}^*\cdot\vec{f}(\vec{k},\vec{k}_0)\Big|_{\rm sh} \approx ika^2\,\hat{\epsilon}^*\cdot\hat{\epsilon}_0\frac{J_1(ka\sin\theta)}{ka\sin\theta}. \tag{12.64}$$

Ignoring possible interference with the illuminated-side contribution, we have for the shadow-side contribution to the polarized cross-section

$$\left(\frac{d\sigma}{d\Omega}\right)_{f,i}\Bigg|_{\rm sh} = k^2a^4\left|\frac{J_1(ka\sin\theta)}{ka\sin\theta}\right|^2|\hat{\epsilon}_f^*\cdot\hat{\epsilon}_{0i}|^2, \tag{12.65}$$

where as before we take into account the possibility of multiple incoming and outgoing polarizations. To obtain the unpolarized cross-section, we choose linear polarizations and specific directions for $\hat{\epsilon}_{0i}$ and $\hat{\epsilon}_f^*$. A convenient choice is to take $\hat{\epsilon}_{01}$, $\hat{\epsilon}_1^*$ as perpendicular to the $(\vec{k}_0,\vec{k})$ plane, and $\hat{\epsilon}_{02}$, $\hat{\epsilon}_2^*$ as parallel to the $(\vec{k}_0,\vec{k})$ plane. In this case we have:

$$\hat{\epsilon}_{01}\cdot\hat{\epsilon}_1^* = 1;\quad \hat{\epsilon}_{02}\cdot\hat{\epsilon}_2^* = \cos\theta;\quad \hat{\epsilon}_{01}\cdot\hat{\epsilon}_2^* = \hat{\epsilon}_{02}\cdot\hat{\epsilon}_1^* = 0.$$

It follows directly that

$$\frac{1}{2}\sum_{f,i}|\hat{\epsilon}_f^*\cdot\hat{\epsilon}_{0i}|^2 = \frac{1}{2}(1+\cos^2\theta) = 1+\mathcal{O}\left((ka)^{-2}\right)\approx 1,$$

so that

$$\frac{d\sigma}{d\Omega}\Bigg|_{\rm sh} \approx k^2a^4\left|\frac{J_1(ka\sin\theta)}{ka\sin\theta}\right|^2,\ \theta\lesssim\frac{1}{ka}. \tag{12.66}$$

For the total shadow-side cross-section, we may integrate using the change of variable $\zeta\equiv ka\sin\theta$:

$$\sigma_{\rm sh} \approx 2\pi\int_0^\pi \sin\theta d\theta\, k^2a^4\left|\frac{J_1(ka\sin\theta)}{ka\sin\theta}\right|^2$$
$$\approx 2\pi a^2\int_0^\infty d\zeta\,\frac{|J_1(\zeta)|^2}{\zeta}$$
$$= \pi a^2, \tag{12.67}$$

where the Bessel function integral was evaluated using an integral table.[2]

For the illuminated-side contribution, we have

$$\hat{\epsilon}^* \cdot \vec{f}(\vec{k},\vec{k}_0)\Big|_{\text{ill}} = \frac{k}{4\pi i}\int_{\text{ill}} da'\,\hat{\epsilon}^* \cdot \left[(\hat{k}-\hat{k}_0)\times(\hat{n}'\times\hat{\epsilon}_0) - (\hat{n}'\cdot\hat{\epsilon}_0)\hat{k}_0\right] e^{-i(\vec{k}-\vec{k}_0)\cdot\vec{x}\,'}. \tag{12.68}$$

In this case, when $\hat{k} \approx \hat{k}_0$ the integrand is approximately zero. For other values of $\hat{k}$, because of the rapidly-oscillating integrand the dominant contribution to the integral will occur where the phase of the complex exponential is *stationary* for $\vec{x}\,'$.

The stationary phase locations may be found as follows. The exponent $-i(\vec{k}-\vec{k}_0)\cdot\vec{x}\,'$ can be written as $-i|\vec{k}-\vec{k}_0|a\cos\alpha$, where α is the angle between $\vec{x}\,'$ and $\vec{k}-\vec{k}_0$. Stationarity requires that the partial derivative with respect to α vanishes, so that $\alpha = 0$ or π. However, when $\alpha = \pi$ then $\vec{x}\,'$ is on the shadow side, so it is not in the area of integration. On the other hand, $\alpha = 0$ corresponds to a normal vector $\hat{n}'$ that is parallel to $\vec{k}-\vec{k}_0$. This leads to an angle of incidence equal to the angle of reflection, as shown in Figure 12.3, and is exactly what we would expect from geometrical optics. So we must be on the right track!

Let us define the normal $\hat{n}_r$ at the point of reflection as

$$\hat{n}_r \equiv \frac{\hat{k}-\hat{k}_0}{|\hat{k}-\hat{k}_0|}, \tag{12.69}$$

which gives us

$$\hat{k}-\hat{k}_0 = 2\sin(\theta/2)\hat{n}_r. \tag{12.70}$$

Since the principal contribution to the integral occurs when $\vec{x}\,' \approx a\hat{n}_r$, we may set $\hat{n}' \approx \hat{n}_r$ in the integrand. Furthermore, since $\hat{\epsilon}^*$ is an outgoing polarization we have $\hat{\epsilon}^* \cdot \hat{k} = 0$, and we may rewrite the non-exponential

[2]See for instance I. S. Gradshteyn and I. M. Rhyzik, *Table of Integrals, Series, and Products*, 6.538 (2). See also J. Schwinger *et al.*, *Classical Electrodynamics*, Perseus Books (Reading, MA) 1998, Section 47.2.

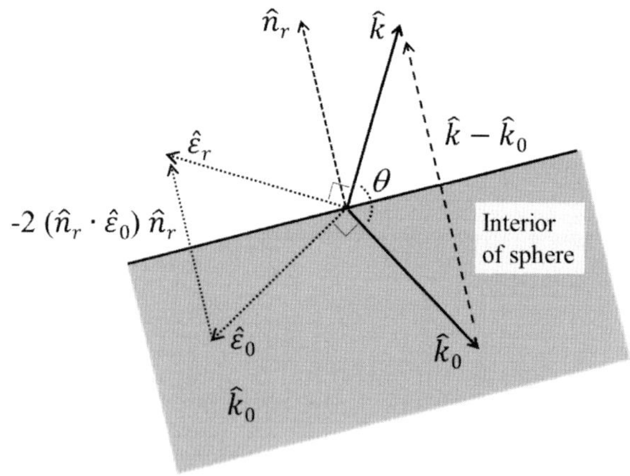

Fig. 12.3 Geometry of scattered wave (illuminated side), showing one possible polarization.

factor in the integrand of (12.68) as

$$\begin{aligned}
\hat{\epsilon}^* &\cdot \left[(\hat{k}-\hat{k}_0)\times(\hat{n}'\times\hat{\epsilon}_0)-(\hat{n}'\cdot\hat{\epsilon}_0)\hat{k}_0\right] \\
&= \hat{\epsilon}^*\cdot\left[(\hat{k}-\hat{k}_0)\times(\hat{n}'\times\hat{\epsilon}_0)+(\hat{n}'\cdot\hat{\epsilon}_0)(\hat{k}-\hat{k}_0)\right] \\
&\approx 2\sin\left(\frac{\theta}{2}\right)\hat{\epsilon}^*\cdot[\hat{n}_r\times(\hat{n}_r\times\hat{\epsilon}_0)+(\hat{n}_r\cdot\hat{\epsilon}_0)\hat{n}_r] \\
&= 2\sin\left(\frac{\theta}{2}\right)\hat{\epsilon}^*\cdot[(\hat{n}_r\cdot\hat{\epsilon}_0)\hat{n}_r-(\hat{n}_r\cdot\hat{n}_r)\hat{\epsilon}_0+(\hat{n}_r\cdot\hat{\epsilon}_0)\hat{n}_r] \\
&= 2\sin\left(\frac{\theta}{2}\right)\hat{\epsilon}^*\cdot\hat{\epsilon}_r, \qquad (12.71)
\end{aligned}$$

where the reflection vector $\hat{\epsilon}_r$ is

$$\hat{\epsilon}_r \equiv -\hat{\epsilon}_0 + 2(\hat{n}_r\cdot\hat{\epsilon}_0)\hat{n}_r. \qquad (12.72)$$

One possible linear polarization, which we shall denote $\hat{\epsilon}_{01}$, is shown in Figure 12.3; in this case, the corresponding reflection vector is denoted $\hat{\epsilon}_{r2}$. Another possibility has $\hat{\epsilon}_{02}$ perpendicular to the $\hat{k}$, $\hat{k}_0$ plane, in which case the reflection vector $\hat{\epsilon}_{r1}$ is its negative. This is consistent with $\hat{\epsilon}_{r2}\equiv\hat{k}\times\hat{\epsilon}_{r1}$, as long as $\hat{\epsilon}_{01}$ and $\hat{\epsilon}_{02}$ are chosen so that $\hat{\epsilon}_{02}\equiv\hat{k}_0\times\hat{\epsilon}_{01}$.

We may complete the evaluation of (12.68) by integrating the exponential term. Letting β denote the azimuthal angle corresponding to polar angle α (so that $d\Omega = d\alpha d\beta \, \sin\alpha$), we find

$$\begin{aligned}\int_{\text{ill}} da' \, e^{-i(\vec{k}-\vec{k}_0)\cdot\vec{x}\,'} &= a^2 \int_{\text{ill}} d\alpha d\beta \, \sin\alpha \, e^{-2ika\sin(\theta/2)\cos\alpha} \\ &= a^2 e^{-2ika\sin(\theta/2)} \int_{\text{ill}} d\alpha d\beta \, \sin\alpha \, e^{2ika\sin(\theta/2)(1-\cos\alpha)}. \end{aligned} \tag{12.73}$$

Due to the rapidly-oscillating exponential, we expect that the significant contribution to the integral occurs for small values of α. To deal with the rapid oscillations at larger α, we introduce a damping factor $(1+i\epsilon)$ in the exponent, and then let $\epsilon \to 0$ in the final evaluation:

$$\begin{aligned}&\int_{\text{ill}} da' \, e^{-i(\vec{k}-\vec{k}_0)\cdot\vec{x}\,'} \\ &\quad \approx a^2 e^{-2ika\sin(\theta/2)} \int_{\text{ill}} d\alpha d\beta \, \sin\alpha \, e^{2ika\sin(\theta/2)(1+i\epsilon)(1-\cos\alpha)}. \end{aligned} \tag{12.74}$$

The integrand is non-negligible only for small values of α, in which case the $d\beta$ integral evaluates to 2π because the full range of azimuthal angles is present. We may also make the change of variable:

$$u \equiv 2ka\sin(\theta/2)(1-\cos\alpha), \tag{12.75}$$

and extend the upper integral limit to infinity because $\theta \gg 1/ka$, hence $2ka\sin(\theta/2) \gg 1$:

$$\begin{aligned}&\int_{\text{ill}} da' \, e^{-i(\vec{k}-\vec{k}_0)\cdot\vec{x}\,'} \\ &\quad \approx 2\pi a^2 e^{-2ika\sin(\theta/2)} \int_0^\infty \frac{du}{2ka\sin(\theta/2)} \, e^{(i-\epsilon)u} \\ &\quad = -\frac{2\pi a^2 e^{-2ika\sin(\theta/2)}}{2ka\sin(\theta/2)(i-\epsilon)} \\ &\quad \underset{\epsilon\to 0}{\longrightarrow} \frac{\pi i a e^{-2ika\sin(\theta/2)}}{k\sin(\theta/2)}. \end{aligned} \tag{12.76}$$

When we include the factors $k/(4\pi i)$ from (12.68) and $2\sin(\theta/2)$ from (12.71), we get finally

$$\hat{\epsilon}^* \cdot \vec{f}(\vec{k},\vec{k}_0)\Big|_{\text{ill}} \approx \frac{a}{2} e^{-2ika\sin(\theta/2)} \hat{\epsilon}^* \cdot \hat{\epsilon}_{r1,2}. \tag{12.77}$$

An equivalent approach is to expand the argument of the oscillating exponential in small angular quantities and integrate in Cartesian coordinates. One obtains instead the product of two Gaussian integrals, which give the same result as in (12.77) above.

For the unpolarized cross-section we may choose our basis of final polarization vectors as

$$\hat{\epsilon}_1^* \equiv \hat{\epsilon}_{r1}, \quad \hat{\epsilon}_2^* \equiv \hat{\epsilon}_{r2} \equiv \hat{k} \times \hat{\epsilon}_{r1}, \tag{12.78}$$

with the result that

$$\frac{1}{2}\sum_{f,i} |\hat{\epsilon}_f^* \cdot \hat{\epsilon}_{ri}|^2 = \frac{1}{2}(1+1) = 1. \tag{12.79}$$

We obtain finally the differential cross-section for the illuminated side as

$$\left.\frac{d\sigma}{d\Omega}\right|_{\text{ill}} \approx a^2/4, \ \theta \gg \frac{1}{ka}, \tag{12.80}$$

and the integrated cross-section is

$$\sigma_{\text{ill}} \approx \pi a^2. \tag{12.81}$$

We will examine the effect of coherent addition (using scattering amplitudes (12.64) and (12.77)) and incoherent addition (using cross-sections (12.66) and (12.80)) of the short wavelength illuminated and shadow results in Exercise 12.3.4, where we will find an interesting polarization effect. Note that the total cross-section of this angular distribution, given approximately by adding Equations (12.67) and (12.81), is $2\pi a^2$. The reader might worry over this situation because this is twice the geometrical cross-section. Since the cross-section is just a power distribution divided by the incident flux, does this mean we extract energy from a scattering event? This nonintuitive result occurs because the incoming fields are both scattered from the illuminated surface and forward scattered to form the shadow by destructive interference with the incident wave. Thus half of the associated scattered energy is bound up in the shadow and is unavailable. There are no free lunches in physics!

The astute reader may note we have not yet considered the possibility of a stationary-phase contribution from the shadow side at scattering angles other than $\theta = 0$. It can be shown that this contribution works out to zero: we leave this as an exercise for the reader.

12.4 The optical theorem

Using our results of the previous section, we may compute the forward scattering amplitude for a conducting sphere as:

$$\hat{\epsilon}_0^* \cdot \vec{f} = \hat{\epsilon}_0^* \cdot (\vec{f}_{\rm sh} + \vec{f}_{\rm ill}) = -\frac{k}{2\pi i}\int_{\rm sh} da'\,(\hat{n}' \cdot \hat{k}_0)(\hat{\epsilon}_0^* \cdot \hat{\epsilon}_0) - \frac{a}{2}$$
$$= \frac{ika^2}{2} - \frac{a}{2}, \tag{12.82}$$

where the integral evaluates to πa^2 since it is the perpendicular projection of the area of the shadow-side hemisphere. Notice there is a simple relation between the imaginary part of (12.82) and the total cross-section:

$$\frac{4\pi}{k}{\rm Im}\,\hat{\epsilon}_0^* \cdot \vec{f} = \frac{ka^2}{2}\frac{4\pi}{k} = 2\pi a^2 = \sigma_{\rm total}. \tag{12.83}$$

In fact this is a general result, as we shall now show.

We compute the power flow in the scattering process, using the notation

$$\vec{E} = \vec{E}_{\rm inc} + \vec{E}_{\rm sc}, \tag{12.84}$$
$$\vec{B} = \vec{B}_{\rm inc} + \vec{B}_{\rm sc}, \tag{12.85}$$

where

$$\vec{E}_{\rm inc} = \hat{\epsilon}_0 E_0 e^{i\vec{k}_0\cdot\vec{x}}, \tag{12.86}$$
$$\vec{B}_{\rm inc} = \hat{k}_0 \times \vec{E}_{\rm inc}. \tag{12.87}$$

First, the power absorbed by the scatterer is

$$P_{\rm abs} \equiv -\frac{c}{8\pi}\oint_S da'\,{\rm Re}(\vec{E} \times \vec{B}^*) \cdot \hat{n}', \tag{12.88}$$

where S is a surface that encloses the scatterer and $\hat{n}'$ is the outward-pointing normal corresponding to area element da'. Next, the scattered power is

$$P_{\rm sc} \equiv \frac{c}{8\pi}\oint_S da'\,{\rm Re}(\vec{E}_{\rm sc} \times \vec{B}_{\rm sc}^*) \cdot \hat{n}'. \tag{12.89}$$

Also, by symmetry we may compute

$$P_{\rm inc} \equiv \frac{c}{8\pi}\oint_S da'\,{\rm Re}(\vec{E}_{\rm inc} \times \vec{B}_{\rm inc}^*) \cdot \hat{n}' = 0. \tag{12.90}$$

The total power removed from the beam is $P_{\rm abs} + P_{\rm sc}$, which we denote as $P_{\rm ext}$ (for "extinguished"):

$$P_{\rm ext} \equiv P_{\rm abs} + P_{\rm sc} = -\frac{c}{8\pi}\oint_S da'\,{\rm Re}[\vec{E}_{\rm sc}\times\vec{B}^*_{\rm inc} + \vec{E}^*_{\rm inc}\times\vec{B}_{\rm sc}]\cdot\hat{n}'$$
$$= -\frac{c}{8\pi}{\rm Re}\left[E_0^*\oint da'\,e^{-i\vec{k}_0\cdot\vec{x}'}[\hat{n}'\cdot(\vec{E}_{\rm sc}\times(\hat{k}_0\times\hat{\epsilon}_0^*)) + \hat{n}'\cdot(\hat{\epsilon}_0^*\times\vec{B}_{\rm sc})]\right]. \tag{12.91}$$

We may rewrite (12.91) using

$$\hat{n}'\cdot[\vec{E}_{\rm sc}\times(\hat{k}_0\times\hat{\epsilon}_0^*)] = (\hat{k}_0\times\hat{\epsilon}_0^*)\cdot[\hat{n}'\times\vec{E}_{\rm sc}]$$
$$= -\hat{\epsilon}_0^*\cdot[\vec{k}_0\times(\hat{n}'\times\vec{E}_{\rm sc})], \tag{12.92}$$

$$\hat{n}'\cdot(\hat{\epsilon}_0^*\times\vec{B}^*_{\rm sc}) = -\hat{\epsilon}_0^*\cdot(\hat{n}'\times\vec{B}^*_{\rm sc}), \tag{12.93}$$

which leads to

$$P_{\rm ext} = \frac{c}{8\pi}{\rm Re}\left\{E_0^*\oint da'\,e^{-i\vec{k}_0\cdot\vec{x}'}\hat{\epsilon}_0^*\cdot[\hat{n}'\times\vec{B}_{\rm sc} + \hat{k}_0\times(\hat{n}'\times\vec{E}_{\rm sc})]\right\}. \tag{12.94}$$

We also have from (12.49)

$$E_0\hat{\epsilon}_0^*\cdot\vec{f}(\vec{k}_0,\vec{k}_0) = \frac{ik}{4\pi}\oint da'\,e^{-i\vec{k}_0\cdot\vec{x}'}\hat{\epsilon}_0^*\cdot[\hat{n}'\times\vec{B}_{\rm sc} + \hat{k}_0\times(\hat{n}'\times\vec{E}_{\rm sc})]. \tag{12.95}$$

We may thus equate

$$P_{\rm ext} = \frac{c}{2k}{\rm Im}\,[E_0^*E_0\hat{\epsilon}_0^*\cdot\vec{f}(\vec{k}_0,\vec{k}_0)], \tag{12.96}$$

and rewriting $\sigma_{\rm tot}$ in terms of $P_{\rm ext}$ we have finally

$$\sigma_{\rm tot} \equiv \frac{P_{\rm ext}}{|\vec{S}_0|} = \frac{P_{\rm ext}}{\frac{c}{8\pi}|E_0|^2} = \frac{4\pi}{k}{\rm Im}[\hat{\epsilon}_0^*\cdot\vec{f}(\vec{k}_0,\vec{k}_0)]. \tag{12.97}$$

This result is known as the *optical theorem*.

12.5 Partial wave expansion for a conducting sphere

The partial wave expansion enables us to find series solutions to scattering problems. The method is similar to the multipole expansions that we used in electrostatics and magnetostatics: the difference is that now we use vector spherical harmonics, which were introduced in Section 10.11. Here we illustrate the method by applying it to our familiar example of a conducting spherical scatterer. We do not need to assume $ka \gg 1$ as in our previous solution: the expansion converges for all values of k, but convergence is more rapid when $ka \ll 1$.

We begin with some of the results derived in Section 10.11. With the definitions

$$g_{lm}(kr) \equiv A_{lm} j_l(kr) + B_{lm} n_l(kr), \tag{12.98}$$

$$f_{lm}(kr) \equiv A'_{lm} j_l(kr) + B'_{lm} n_l(kr), \tag{12.99}$$

$$\vec{L} \equiv -i\vec{r} \times \vec{\nabla}, \tag{12.100}$$

and using (e) and (m) to denote radial TE and TM modes respectively, we found for TE modes (see (10.312) and (10.313))

$$\vec{E}^{(e)}_{lm}(\vec{x}, \omega) = g_{lm}(kr) \vec{L} Y_{lm}(\theta, \phi), \tag{12.101}$$

$$\vec{B}^{(e)}_{lm}(\vec{x}, \omega) = -\frac{i}{k} \vec{\nabla} \times \vec{E}^{(e)}_{lm}(\vec{x}, \omega), \tag{12.102}$$

and for TM modes (see (10.317) and (10.318))

$$\vec{B}^{(m)}_{lm}(\vec{x}, \omega) = f_{lm}(kr) \vec{L} Y_{lm}(\theta, \phi), \tag{12.103}$$

$$\vec{E}^{(m)}_{lm}(\vec{x}, \omega) = \frac{i}{k} \vec{\nabla} \times \vec{B}^{(m)}_{lm}(\vec{x}, \omega). \tag{12.104}$$

We may express general fields as linear combinations of these modes:

$$\vec{E} = \sum_{l,m} \left[\frac{i}{k} \vec{\nabla} \times (f_{lm} \vec{L} Y_{lm}) + g_{lm} \vec{L} Y_{lm} \right], \tag{12.105}$$

$$\vec{B} = \sum_{l,m} \left[f_{lm} \vec{L} Y_{lm} - \frac{i}{k} \vec{\nabla} \times (g_{lm} \vec{L} Y_{lm}) \right]. \tag{12.106}$$

As in (10.321) we may rewrite the curl term in (12.105) using

$$\vec{\nabla} \times (f_{lm} \vec{L} Y_{lm}) = \frac{df_{lm}}{dr} \hat{r} \times \vec{L} Y_{lm} + f_{lm} \vec{\nabla} \times \vec{L} Y_{lm}, \tag{12.107}$$

and similarly for (12.106). To evaluate $\vec{\nabla} \times \vec{L} Y_{lm}$, we have from (10.322)

$$\vec{\nabla} \times \vec{L} Y_{lm} = \frac{1}{i} [-2\vec{\nabla} + \vec{r} \nabla^2 - (\vec{r} \cdot \vec{\nabla}) \vec{\nabla}] Y_{lm},$$

which we may further simplify using

$$\vec{\nabla}^2 Y_{lm} = -\frac{l(l+1)}{r^2} Y_{lm}, \tag{12.108}$$

$$(\vec{r} \cdot \vec{\nabla}) \vec{\nabla} Y_{lm} = \vec{\nabla}(\vec{r} \cdot \vec{\nabla}) Y_{lm} - \vec{\nabla} Y_{lm} = -\vec{\nabla} Y_{lm}. \tag{12.109}$$

This gives us

$$\vec{\nabla} \times \vec{L} Y_{lm} = \frac{1}{i} [-\vec{\nabla} + \vec{r}\, \nabla^2] Y_{lm}. \tag{12.110}$$

Recalling the definition in (10.296)

$$\vec{X}_{lm} \equiv \frac{1}{\sqrt{l(l+1)}}\vec{L}Y_{lm}, \tag{12.111}$$

we obtain finally

$$\vec{E} = \sum_{l,m}\left[g_{lm}\vec{X}_{lm} + \frac{i}{kr}\frac{d}{dr}(rf_{lm})\hat{r}\times\vec{X}_{lm} - \frac{f_{lm}}{kr}\sqrt{l(l+1)}\hat{r}\,Y_{lm}\right], \tag{12.112}$$

$$\vec{B} = \sum_{l,m}\left[f_{lm}\vec{X}_{lm} - \frac{i}{kr}\frac{d}{dr}(rg_{lm})\hat{r}\times\vec{X}_{lm} + \frac{g_{lm}}{kr}\sqrt{l(l+1)}\hat{r}\,Y_{lm}\right], \tag{12.113}$$

where we have redefined the constants in f_{lm} and g_{lm} to accommodate (12.111).

It is convenient to work with circularly-polarized plane waves. We may do both polarizations at the same time by defining

$$\vec{E}^{\pm}_{\text{inc}} = \frac{\hat{e}_1 \pm i\hat{e}_2}{\sqrt{2}}e^{ikz}, \tag{12.114}$$

$$\vec{B}^{\pm}_{\text{inc}} = \hat{e}_3 \times \vec{E}^{\pm}_{\text{inc}} = \mp i\vec{E}^{\pm}_{\text{inc}}. \tag{12.115}$$

(This corresponds to using $E_0 = 1$ with a circular polarization vector in (12.6).) In order for the fields to be regular in all space, we must have $B_{lm} = B'_{lm} = 0$ in (12.98) and (12.99). Anticipating the final result, we will also use the leading notation $a_\pm$, $b_\pm$ to replace A_{lm}, A'_{lm} when representing the $\vec{E}^{\pm}_{\text{inc}}$, $\vec{B}^{\pm}_{\text{inc}}$ fields, with the understanding that $a_\pm$ and $b_\pm$ depend on l, m. This gives us the general regular form

$$\vec{E}^{\pm}_{\text{inc}} = \sum_{l,m}\left[a_\pm j_l\vec{X}_{lm} + i\frac{b_\pm}{kr}\frac{d}{dr}(rj_l)\hat{r}\times\vec{X}_{lm} - b_\pm\frac{j_l}{kr}\sqrt{l(l+1)}\hat{r}Y_{lm}\right], \tag{12.116}$$

$$\vec{B}^{\pm}_{\text{inc}} = \sum_{l,m}\left[b_\pm j_l\vec{X}_{lm} - i\frac{a_\pm}{kr}\frac{d}{dr}(rj_l)\hat{r}\times\vec{X}_{lm} + a_\pm\frac{j_l}{kr}\sqrt{l(l+1)}\hat{r}Y_{lm}\right]. \tag{12.117}$$

The orthonormality of $\vec{X}_{lm}$, $\hat{r}\times\vec{X}_{lm}$ (see (10.305)) leads to

$$\int d\Omega\vec{E}^{\pm}_{\text{inc}}\cdot\vec{X}^*_{lm} = a_\pm j_l(kr), \tag{12.118}$$

and using the operators (see Exercise 10.11.1 reference)

$$L_\pm = L_x \pm iL_y, \tag{12.119}$$

we have

$$a_\pm j_l(kr) = \int d\Omega\, e^{ikz} \frac{L_\pm Y_{lm}^*}{\sqrt{2l(l+1)}} = \frac{\sqrt{(l\pm m)(l\mp m+1)}}{\sqrt{2l(l+1)}} \int d\Omega\, e^{ikz} Y_{l,m\mp 1}^*. \tag{12.120}$$

To evaluate the remaining integral, we revisit our generating function expression (10.274)

$$e^{ikz} = \sum_l i^l (2l+1) j_l(kr) P_l(\cos\theta), \tag{12.121}$$

which we rewrite using (4.190) as

$$e^{ikz} = \sum_l i^l \sqrt{4\pi(2l+1)} j_l(kr) Y_{l0}(\cos\theta). \tag{12.122}$$

It follows from the orthogonality of the Y_{lm}'s that

$$a_\pm j_l(kr) = \frac{\sqrt{(l\pm m)(l\mp m+1)}}{\sqrt{2l(l+1)}} \times i^l \sqrt{4\pi(2l+1)} j_l(kr) \delta_{m,\pm 1}, \tag{12.123}$$

which leads to

$$a_\pm = i^l \sqrt{2\pi(2l+1)} \delta_{m,\pm 1}. \tag{12.124}$$

Now with Equation (12.115) we have $b_\pm = \mp i a_\pm$, so that (12.116) and (12.117) give

$$\vec{E}_{\rm inc}^\pm = \sum_l i^l \sqrt{2\pi(2l+1)} \left[j_l \vec{X}_{l,\pm 1} \pm \frac{1}{kr}\frac{d}{dr}(r j_l)\hat{r} \times \vec{X}_{l,\pm 1} \right.$$
$$\left. \pm i \frac{j_l}{kr} \sqrt{l(l+1)}\hat{r}\, Y_{l,\pm 1} \right], \tag{12.125}$$

$$\vec{B}_{\rm inc}^\pm = \sum_l i^l \sqrt{2\pi(2l+1)} \left[\mp i j_l \vec{X}_{l,\pm 1} - \frac{i}{kr}\frac{d}{dr}(r j_l)\hat{r} \times \vec{X}_{l,\pm 1} \right.$$
$$\left. + \frac{j_l}{kr} \sqrt{l(l+1)}\hat{r}\, Y_{l,\pm 1} \right]. \tag{12.126}$$

So we have succeeded in expanding our simple vector plane waves in terms of vector spherical harmonics.

Now we may also expand the scattered fields as a sum of modes:

$$\vec{E}_{\rm sc}^\pm = \frac{1}{2}\sum_l i^l \sqrt{2\pi(2l+1)} \left[\alpha_\pm h_l^{(1)} \vec{X}_{l,\pm 1} \right.$$
$$\left. \pm \frac{\beta_\pm}{kr}\frac{d}{dr}(r h_l^{(1)})\hat{r} \times \vec{X}_{l,\pm 1} \pm \frac{i\beta_\pm}{kr} h_l^{(1)} \sqrt{l(l+1)}\hat{r} Y_{l,\pm 1} \right], \tag{12.127}$$

$$\vec{B}_{\rm sc}^\pm = \frac{1}{2}\sum_l i^l \sqrt{2\pi(2l+1)} \left[\mp i\beta_\pm h_l^{(1)} \vec{X}_{l,\pm 1} \right.$$
$$\left. - \frac{i\alpha_\pm}{kr}\frac{d}{dr}(r h_l^{(1)})\hat{r} \times \vec{X}_{l,\pm 1} + \frac{\alpha_\pm}{kr} h_l^{(1)} \sqrt{l(l+1)}\hat{r} Y_{l,\pm 1} \right]. \tag{12.128}$$

The $\alpha_\pm$ and $\beta_\pm$ coefficients again depend on l and m and are of the general form found in Equations (12.116) and (12.117), but the signs are chosen on a pattern based on (12.125), (12.126). Note that we have excluded terms with $h_l^{(2)}(kr)$ since these correspond to incoming spherical waves according to (10.270): the included terms all depend on $h_l^{(1)}(kr)$, which by (10.269) correspond to outgoing spherical waves.

We define the normalized scattering amplitudes $\vec{f}_\pm$ via

$$\vec{E}_{\text{sc}}^{\pm} \equiv \frac{e^{ikr}}{r}\vec{f}_\pm, \tag{12.129}$$

and you will show in Exercise 12.5.4 that

$$\vec{f}_\pm = \frac{1}{ik}\sqrt{\frac{\pi}{2}}\sum_l \sqrt{2l+1}\left[\alpha_\pm \vec{X}_{l,\pm 1} \pm i\beta_\pm \hat{r}\times\vec{X}_{l,\pm 1}\right]. \tag{12.130}$$

We may obtain the differential cross-sections corresponding to the two circular polarizations using Equation (12.36):

$$\left.\frac{d\sigma}{d\Omega}\right|_{f,\pm} = |\hat{\epsilon}_f^* \cdot \vec{f}_\pm|^2 \tag{12.131}$$

$$= \frac{\pi}{2k^2}\left|\sum_l \sqrt{2l+1}\;\hat{\epsilon}_f^*\cdot\left[\alpha_\pm \vec{X}_{l,\pm 1} \pm i\beta_\pm \hat{r}\times\vec{X}_{l,\pm 1}\right]\right|^2. \tag{12.132}$$

The coefficients $\alpha_\pm$, $\beta_\pm$ are determined (as usual) by boundary conditions. In the current case of a conducting sphere, these are

$$\hat{r}\times\vec{E}\Big|_{r=a} = 0, \tag{12.133}$$

$$\hat{r}\cdot\vec{B}\Big|_{r=a} = 0. \tag{12.134}$$

Let $x \equiv ka$. Then (12.133) with $\vec{E} = \vec{E}_{\text{inc}} + \vec{E}_{\text{sc}}$ gives

$$\vec{E} \sim \vec{X}_{l,\pm 1} \text{ terms} \implies j_l(x) + \frac{1}{2}\alpha_\pm h_l^{(1)}(x) = 0, \tag{12.135}$$

$$\vec{E} \sim \hat{r}\times\vec{X}_{l,\pm 1} \text{ terms} \implies \frac{1}{x}\frac{d}{dx}(xj_l) + \frac{\beta_\pm}{2x}\frac{d}{dx}(xh_l^{(1)}) = 0. \tag{12.136}$$

The reader may verify that (12.134) with $\vec{B} = \vec{B}_{\text{inc}} + \vec{B}_{\text{sc}}$ also gives (12.135). We find

$$\alpha_\pm = -2\frac{j_l(x)}{h_l^{(1)}(x)} = -1 - \frac{h_l^{(2)}(x)}{h_l^{(1)}(x)}, \tag{12.137}$$

$$\beta_\pm = -2\frac{\frac{d}{dx}(x\,j_l(x))}{\frac{d}{dx}(x\,h_l^{(1)}(x))} = -\frac{\frac{d}{dx}(x\,h_l^{(2)}(x))}{\frac{d}{dx}(x\,h_l^{(1)}(x))} - 1. \tag{12.138}$$

Note that

$$|\alpha_\pm + 1| = 1, \ |\beta_\pm + 1| = 1, \tag{12.139}$$

and in Exercise 12.5.1 you will show that this implies that the absorption cross-section $\sigma_{\text{abs}} = 0$.

The scattering phase shifts δ_l, δ'_l are defined via

$$\alpha_\pm \equiv e^{2i\delta_l} - 1 \implies \delta_l \equiv \frac{1}{2}\text{Arg}[\alpha_\pm + 1], \tag{12.140}$$

$$\beta_\pm \equiv e^{2i\delta'_l} - 1 \implies \delta'_l \equiv \frac{1}{2}\text{Arg}[\beta_\pm + 1], \tag{12.141}$$

and we may compute

$$\tan\delta_l = \frac{j_l(x)}{n_l(x)}; \quad \tan\delta'_l = \frac{\frac{d}{dx}(xj_l(x))}{\frac{d}{dx}(xn_l(x))}. \tag{12.142}$$

As it turns out, these same phase shifts occur in quantum scattering (using the Schrödinger equation) from a finite range potential with Dirichlet and Neumann boundary conditions.[3]

In the $ka \ll 1$ limit, we may use the small-argument expansions for $j_l(x)$ and $n_l(x)$ in (10.259) and (10.260) to obtain

$$\alpha_\pm \approx \frac{-2i(ka)^{2l+1}}{(2l+1)[(2l-1)!!]^2}; \quad \beta_\pm = -\alpha_\pm\left(\frac{l+1}{l}\right). \tag{12.143}$$

In this case only the $l = 1$ term is significant, and we have

$$\alpha_\pm = -\frac{1}{2}\beta_\pm \approx -\frac{2i}{3}(ka)^3 \quad (l = 1), \tag{12.144}$$

which gives

$$\left(\frac{d\sigma}{d\Omega}\right)_\pm = \frac{2\pi}{3}a^2(ka)^4\left|\vec{X}_{1,\pm 1} \mp 2i\hat{r} \times \vec{X}_{1,\pm 1}\right|^2, \tag{12.145}$$

when final polarization is summed. We may simplify the absolute value term using (see Exercise 10.11.1)

$$\vec{X}_{1,\pm 1} = \frac{\vec{L}Y_{1,\pm 1}}{\sqrt{2}} = \frac{1}{\sqrt{2}}\left(\hat{\epsilon}_\pm Y_{10} \pm \hat{e}_z Y_{1,\pm 1}\right), \tag{12.146}$$

which gives

$$|\hat{r} \times \vec{X}_{1,\pm 1}|^2 = |\vec{X}_{1,\pm 1}|^2 = \frac{3}{8\pi}\cos^2\theta + \frac{3}{16\pi}\sin^2\theta = \frac{3}{16\pi}(1 + \cos^2\theta). \tag{12.147}$$

[3] W. Wilcox, *Quantum Principles and Particles*, CRC Press (Boca Raton) 2012, Sec. 10.11.

The cross term is proportional to the imaginary part of $\vec{X}_{1,\pm1} \cdot (\hat{r} \times \vec{X}^*_{1,\pm1})$, which is equal to $(\vec{X}^*_{1,\pm1} \times \vec{X}_{1,\pm1}) \cdot \hat{r}$. We have

$$\vec{X}^*_{11} \times \vec{X}_{11} = \frac{1}{2}(\hat{\epsilon}_- Y^*_{10} + \hat{e}_z Y^*_{11}) \times (\hat{\epsilon}_+ Y_{10} + \hat{e}_z Y_{11}), \tag{12.148}$$

and using

$$\hat{\epsilon}_- \times \hat{e}_z = -i\hat{\epsilon}_-, \quad \hat{e}_z \times \hat{\epsilon}_+ = -i\hat{\epsilon}_+, \quad \hat{\epsilon}_- \times \hat{\epsilon}_+ = i\hat{e}_z, \tag{12.149}$$

we find

$$\begin{aligned}(\vec{X}^*_{11} \times \vec{X}_{11}) \cdot \hat{r} &= i\frac{Y_{10}}{2}\left[Y_{10}\hat{e}_z - Y_{11}\hat{\epsilon}_- - Y^*_{11}\hat{\epsilon}_+\right] \cdot \hat{r} \\ &= i\frac{Y_{10}}{2}\left[\frac{1}{2}\sqrt{\frac{3}{\pi}}\,\hat{r}\right] \cdot \hat{r} \\ &= \frac{3i}{8\pi}\cos\theta. \end{aligned} \tag{12.150}$$

Likewise we find

$$(\vec{X}^*_{1-1} \times \vec{X}_{1-1}) \cdot \hat{r} = -\frac{3i}{8\pi}\cos\theta. \tag{12.151}$$

Thus, for either initial polarization:

$$\frac{d\sigma}{d\Omega} \approx a^2(ka)^4\left[\frac{5}{8}(1 + \cos^2\theta) - \cos\theta\right]. \tag{12.152}$$

The k^4 dependence is characteristic of low frequency dipole scattering.

12.6 Scalar diffraction theory

We now come to the important and interesting subject of electromagnetic diffraction. Diffraction is the tendency for light or other electromagnetic waves to spread out as they propagate. This follows from *Huygens' Principle*, which is the idea that every point of a wave front is a source of further waves, interacting and interfering with one another. Mathematically, this is still a scattering problem using the wave equation in the form of the Helmholtz equation at its core, except one considers apertures or holes in screens rather than individual scatterers like spheres. Many of the approximations and techniques in the two cases are the same. In this section we will consider a mathematically simplified situation where a scalar field substitutes for the full electromagnetic one. This often suffices in cases where one is simply interested in predicted intensities rather than polarization phenomenon. This approach will illustrate some of the major steps and approximations in the formalism. From there we will build up to the more realistic vector field case.

The idealized geometry used for the diffraction discussion is indicated in Figure 12.4, where it is understood that the surface S_∞ is located a large distance R_∞ away from the origin. It will be convenient to refer to the entire $z' = 0$ plane as the surface S. We will also refer in these sections to the barrier (or obstruction) B and apertures (or gaps) A on S. The field equations for the sourceless scalar field and the Green function are:

$$(\vec{\nabla}^2 + k^2)\Phi(\vec{x}) = 0, \tag{12.153}$$

$$(\vec{\nabla}^2 + k^2)G^{\text{out}}(\vec{x}, \vec{x}\,'; k) = -4\pi\delta(\vec{x} - \vec{x}\,'). \tag{12.154}$$

We should be familiar with the usual "song and dance" to construct the field everywhere from the boundary values; see Section 8.6 for time dependent scalar fields and also Section 12.2 for the scattered electric field in a spherical region. This leads to the construction:

$$\Phi(\vec{x}) = \frac{1}{4\pi}\oint da' \left(\partial_{n'}G^{\text{out}}(\vec{x}, \vec{x}\,'; k)\Phi(\vec{x}\,') - \partial_{n'}\Phi(\vec{x}\,')G^{\text{out}}(\vec{x}, \vec{x}\,'; k)\right), \tag{12.155}$$

where $\partial_{n'} \equiv \hat{n}' \cdot \vec{\nabla}$, and we are using the outward normal $-\hat{n}'$ ($\hat{n}' = \hat{z}$ on the barrier). We may delete the surface at infinity S_∞ from the complete surface in (12.155) using outgoing radiation boundary conditions. In this context it would be geometrically inconsistent to choose the infinite space

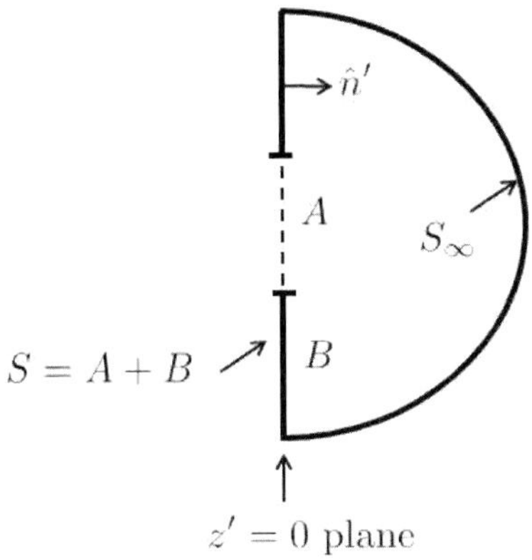

Fig. 12.4 Geometry for diffraction.

Green function,

$$G^{\text{out}}(\vec{x}, \vec{x}\,'; k) = \frac{e^{ik|\vec{x}-\vec{x}\,'|}}{|\vec{x}-\vec{x}\,'|}, \tag{12.156}$$

for this problem. Instead we need the Green function for Dirichlet or Neumann boundary conditions:

$$G^{\text{out}}_D\Big|_S = 0, \quad \partial_n G^{\text{out}}_N\Big|_S = 0. \tag{12.157}$$

We have from the image solution,

$$G^{\text{out}}_{\substack{D\\N}}(\vec{x}, \vec{x}\,'; k) = \frac{e^{ik|\vec{x}-\vec{x}\,'|}}{|\vec{x}-\vec{x}\,'|} \mp \frac{e^{ik|\vec{x}-\vec{x}\,''|}}{|\vec{x}-\vec{x}\,''|}, \tag{12.158}$$

where $\vec{x}\,'' = (x', y', -z')$. This gives

$$\Phi_D(\vec{x}) = \frac{z}{2\pi}\int_S da'\,\Phi(\vec{x}\,')\frac{G^{\text{out}}(\vec{x}, \vec{x}\,'; k)}{|\vec{x}-\vec{x}\,'|}\left(ik - \frac{1}{|\vec{x}-\vec{x}\,'|}\right), \tag{12.159}$$

$$\Phi_N(\vec{x}) = -\frac{1}{2\pi}\int_S da'\,\partial_{n'}\Phi(\vec{x}\,')\,G^{\text{out}}(\vec{x}, \vec{x}\,'; k). \tag{12.160}$$

Because we will be working in the $kd \gg 1$ approximation, the second term in (12.159) makes little difference. Note that the integrals in (12.159) and (12.160) are over the entire surface S. Notice also the regions $z < 0$ and $z > 0$ are completely separate mathematically. The question arises as to what to assume for the values or normal derivative of Φ on the barrier in an actual diffraction setup where the two sides communicate. If there were no apertures in the barrier, we know that the correct boundary condition would be applied to the incident plus reflected waves for $z < 0$. If we assume the same holds when there are apertures, we need only consider the condition on the diffracted field. This separate consideration implies we should apply the same Dirichlet or Neumann boundary condition we impose on the whole field to the diffracted field as well. Therefore in the following we will take $\Phi|_B = 0$ or $\partial_{n'}\Phi|_B = 0$ on the barrier in the appropriate case, and the integration will be limited to the apertures.

In applications we will need approximate evaluations of the power of the exponential in $G^{\text{out}}(\vec{x}, \vec{x}\,'; k)$:

$$\begin{aligned} |\vec{x}-\vec{x}\,'| &= r|\hat{n} - \frac{\vec{x}\,'}{r}| \\ &\approx r\left(1 - \frac{r'}{r}\hat{n}\cdot\hat{x}' + \frac{1}{2}\frac{r'^2}{r^2}(1-(\hat{n}\cdot\hat{x}')^2)\right). \end{aligned} \tag{12.161}$$

Taking the first two terms only is called *Fraunhofer diffraction*. Taking the third and higher terms into account gives the near-field or *Fresnel diffraction* result. In the Fraunhofer case we have

$$e^{ik|\vec{x}-\vec{x}\,'|} \longrightarrow e^{ikr}e^{-i\vec{k}\cdot\vec{x}\,'}, \tag{12.162}$$

where $\vec{k} \equiv k\,\hat{n}$. The third term in (12.161) becomes important in the exponential when

$$\frac{kd^2}{r} \approx 1, \tag{12.163}$$

where d is a typical spatial dimension of the aperture. A significant change in the diffraction pattern occurs for this approximate screen distance: one transitions from a geometrically determined angular diffraction width $\sim d/r$ to a wave form one $\sim 1/(kd)$.

With no more to-do, and specializing to a single aperture, we assume that the field in the aperture is approximately given by the incident field, so that

$$\Phi|_A \approx \Phi_0\, e^{i\vec{k}_0\cdot\vec{x}\,'}, \tag{12.164}$$

giving in the Fraunhofer approximation

$$\Phi(\vec{x}) = \frac{ikFe^{ikr}}{2\pi r}\Phi_0 \int_A da'\, e^{-i(\vec{k}-\vec{k}_0)\cdot\vec{x}\,'}. \tag{12.165}$$

This is essentially a shadow amplitude integral (cf. Equation (12.62)), with a multiplicative factor

$$F \equiv \begin{cases} \cos\theta, & \text{Dirichlet} \\ -\cos\alpha, & \text{Neumann} \end{cases} \tag{12.166}$$

where $\cos\alpha \equiv \hat{n}' \cdot \vec{k}_0$. Notice how the Dirichlet F factor enforces the appropriate boundary condition for our diffracted field at $\theta = \pi/2$. Our use of the incident field as the exact diffracted field in the aperture is a crude guess. Intuitively, we expect this kind of assumption to best hold at high energies, $kd \gg 1$. Effectively, we are making a short-wavelength diffraction assumption analogous to the one in Section 12.3 for the conducting sphere. Actually finding the correct field to insert in (12.159) or (12.160) in a given geometry is a difficult problem, indeed. Note that in special geometries, such as oblate spheroidal for circular diffraction, there are partial wave formalisms like the last section which allow one to sum over the solutions of the Helmholtz equation which satisfy the boundary conditions of the problem which do not require the $kd \gg 1$ assumption.

The differential transition rate is defined by normalizing the differential power by the incoming power:

$$\frac{dT}{d\Omega} \equiv \frac{1}{P^{\rm in}}\frac{dP^{\rm out}}{d\Omega}. \tag{12.167}$$

We have incoming total and outgoing differential power fluxes defined from using the appropriate harmonic component of the scalar Poynting vector (see upcoming Exercise 13.10.4) for an opening area A:

$$P_\Phi^{\rm in} = \frac{ck^2}{2}|\Phi_0|^2 A\cos\alpha, \qquad \frac{dP_\Phi^{\rm out}}{d\Omega} = r^2\frac{ck^2}{2}|\Phi(\vec{x})|^2. \tag{12.168}$$

This gives

$$\frac{dT}{d\Omega} = \frac{1}{A\cos\alpha}\left|\frac{r}{\Phi_0}\right|^2 |\Phi(\vec{x})|^2. \tag{12.169}$$

For high energies $kd \gg 1$ where our considerations will hold, we expect for the integrated transition that

$$T = \int \frac{dT}{d\Omega} \approx 1, \tag{12.170}$$

simply from geometrical considerations. Notice that for $\alpha \to \pi/2$ we are prevented from normalizing the Dirichlet amplitude by $P^{\rm in}$, although the differential power is finite. This means some power leaks through the barrier even though the beam is parallel to the interface. We will see a similar phenomenon in the electromagnetic case.

12.7 Scalar diffraction examples

We will illustrate the scalar formalism with two examples: non-normally incident plane waves on a circular aperture and normal incidence on an infinitely long straight slit.

We begin with an evaluation of the exponential factor needing to be included in the integral in (12.165) in the circular case. We have

$$(\vec{k}-\vec{k}_0)\cdot\vec{x}\,'|_{z'=0} = k\rho'[\sin\theta\cos(\phi-\phi') - \sin\alpha\cos\phi'], \tag{12.171}$$

where $\vec{k}_0 = (\sin\alpha, 0, \cos\alpha)$. In order to do the ϕ' integral in (12.165), we need to consolidate the ϕ' dependence. This can be done via the assumption:

$$\Delta\cos(\phi'+\delta) \equiv \sin\theta\cos(\phi-\phi') - \sin\alpha\cos\phi'. \tag{12.172}$$

We may solve for Δ and δ. For the former we have

$$\Delta = (\sin^2\theta + \sin^2\alpha - 2\sin\alpha\sin\theta\cos\phi)^{1/2}. \tag{12.173}$$

It turns out the latter angle is immaterial to the actual integral. We may now write

$$\int_A da'\, e^{-i(\vec{k}-\vec{k}_0)\cdot\vec{x}\,'} = \int_0^a d\rho'\rho' \int_0^{2\pi} d\phi'\, e^{-ik\rho'\Delta\cos(\phi'+\delta)}. \tag{12.174}$$

This gives

$$\begin{aligned}\int_0^{2\pi} d\phi' e^{-ik\rho'\Delta\cos(\phi'+\delta)} &= \int_0^{2\pi} d\phi''\, e^{-ik\rho'\Delta\cos\phi''}\\ &= 2\pi J_0(k\rho'\Delta),\end{aligned} \tag{12.175}$$

where $\phi'' \equiv \phi' + \delta$. Using the Bessel function identity

$$\frac{d}{dx}(x\,J_1(x)) = x\,J_0(x), \tag{12.176}$$

allows us to integrate in ρ'. We obtain

$$\int_A da'\, e^{-i(\vec{k}-\vec{k}_0)\cdot\vec{x}\,'} = 2\pi\frac{a}{k\Delta}J_1(ka\Delta). \tag{12.177}$$

This gives our final result for the field

$$\Phi(\vec{x}) = \frac{iFa\,e^{ikr}}{r}\Phi_0\frac{J_1(ka\Delta)}{\Delta}, \tag{12.178}$$

as well as the angular transition rate

$$\frac{dT}{d\Omega} = \frac{F^2}{\pi\cos\alpha}\left(\frac{J_1(ka\Delta)}{\Delta}\right)^2. \tag{12.179}$$

Note that the condition $\Delta = 0$, given that the initial plane wave direction was along $\phi = 0$ (π), says that $\sin\theta = \sin\alpha$ $(-\sin\alpha)$, picking out the correct geometrical optics direction for the maximum amplitude.

Our second example is diffraction from a long slit of size d at normal incidence. In this case we will include the next term in the expansion of $k|\vec{x} - \vec{x}\,'|$ to examine mathematical issues encountered in a more accurate calculation. Here we back up to Equations (12.159) (without the extra $1/R$ term) and (12.160) and consider

$$\Phi(\vec{x}) = \frac{ikFe^{ikr}}{2\pi\rho}\Phi_0 \int_{\text{slit}} da' e^{-i\vec{k}\cdot\vec{x}\,' + ik\rho'^2/2\rho}. \tag{12.180}$$

We have switched to cylindrical coordinates and use $\vec{x} = \rho(\sin\phi, 0, \cos\phi)$ implying $\hat{k} = (\sin\phi, 0, \cos\phi)$. This will characterize forward scattering as

$\phi = 0$. The only difference between the Dirichlet and Neumann cases is the factor F, which in this case is:

$$F = \begin{cases} \cos\phi, & \text{Dirichlet} \\ -1, & \text{Neumann} \end{cases} \tag{12.181}$$

We will find the transition rate as a function of ϕ. We need the integral

$$\int_{\text{slit}} da' e^{-i\vec{k}\cdot\vec{x}'+ik\rho'^2/2\rho} = I_y I_x, \tag{12.182}$$

where

$$I_y \equiv \int_{-\infty}^{\infty} dy'\, e^{iky'^2/2\rho}, \qquad I_x \equiv \int_{-d/2}^{d/2} dx' e^{ikx'^2/2\rho - ikx'\sin\phi}. \tag{12.183}$$

An important theoretical tool for these evaluations are the *Fresnel integrals*:

$$C(z) \equiv \int_0^z dt\, \cos(\pi t^2/2), \tag{12.184}$$

$$S(z) \equiv \int_0^z dt\, \sin(\pi t^2/2). \tag{12.185}$$

Some important properties are:

$$C(\infty) = \frac{1}{2},\ S(\infty) = \frac{1}{2}, \tag{12.186}$$

$$C(-z) = -C(z),\ S(-z) = -S(z). \tag{12.187}$$

The I_y integral is straightforward and is just the same as a complete Gaussian one:

$$I_y = \sqrt{2}\, e^{i\pi/4}\sqrt{\frac{\pi\rho}{k}}. \tag{12.188}$$

For the other integral, our first step is to complete the square in the argument of the exponential:

$$\frac{1}{2\rho}x'^2 - x'\sin\phi = \frac{1}{2\rho}(x' - \rho\sin\phi)^2 - \frac{\rho}{2}\sin^2\phi. \tag{12.189}$$

Defining

$$t' = \sqrt{\frac{k}{\pi\rho}}\,(x' - \rho\sin\phi)\,, \tag{12.190}$$

allows us to put the I_x integral in the form

$$I_x = \sqrt{\frac{\pi\rho}{k}}\, e^{-i\frac{k\rho}{2}\sin^2\phi} \int_{-\frac{d}{2}\sqrt{\frac{k}{\pi\rho}}-\sqrt{\frac{k\rho}{\pi}}\sin\phi}^{\frac{d}{2}\sqrt{\frac{k}{\pi\rho}}-\sqrt{\frac{k\rho}{\pi}}\sin\phi} dt'\, e^{i\pi t'^2/2}$$
$$= \sqrt{\frac{\pi\rho}{k}}\, e^{-i\frac{k\rho}{2}\sin^2\phi} \times$$
$$\left[-C\left(-\frac{d}{2}\sqrt{\frac{k}{\pi\rho}} - \sqrt{\frac{k\rho}{\pi}}\sin\phi\right) + C\left(\frac{d}{2}\sqrt{\frac{k}{\pi\rho}} - \sqrt{\frac{k\rho}{\pi}}\sin\phi\right)\right.$$
$$\left. -iS\left(-\frac{d}{2}\sqrt{\frac{k}{\pi\rho}} - \sqrt{\frac{k\rho}{\pi}}\sin\phi\right) + iS\left(\frac{d}{2}\sqrt{\frac{k}{\pi\rho}} - \sqrt{\frac{k\rho}{\pi}}\sin\phi\right)\right]. \tag{12.191}$$

Introducing the *Fresnel number* (see (12.163))

$$N_F \equiv \frac{kd^2}{8\pi\rho}, \tag{12.192}$$

allows us to simplify the result for the field to:

$$\Phi(\vec{x}) = \frac{iFe^{ikr}}{\sqrt{2}}\Phi_0 e^{i\pi/4} e^{-i\frac{k\rho}{2}\sin^2\phi}$$
$$\times\left[C\left(\sqrt{2N_F}\left(\frac{2x}{d}+1\right)\right) - C\left(\sqrt{2N_F}\left(\frac{2x}{d}-1\right)\right)\right.$$
$$\left. + iS\left(\sqrt{2N_F}\left(\frac{2x}{d}+1\right)\right) - iS\left(\sqrt{2N_F}\left(\frac{2x}{d}-1\right)\right)\right], \tag{12.193}$$

where $x = \rho\sin\phi$. The angular differential transition rate is defined by:

$$\frac{dT}{d\phi} \equiv \frac{\rho}{d\,|\Phi_0|^2}|\Phi(\vec{x})|^2. \tag{12.194}$$

Although we have not composed a formal exercise on this transition rate, one may use numerical means to examine the angular transition patterns arising from (12.194) for various values of λ, d and r. However, in Exercise 12.7.2 you will consider near-field Fresnel diffraction patterns from circular diffraction, which have similar features.

Finally, for $\rho \to \infty$ one may use the asymptotic expansions (equivalent to Gradshteyn and Ryzhik, *op. cit.*, 8.255 (1) and (2))

$$C(z) \approx \frac{z}{2|z|} + \frac{1}{\pi z}\sin(\pi z^2/2), \tag{12.195}$$

$$S(z) \approx \frac{z}{2|z|} - \frac{1}{\pi z}\cos(\pi z^2/2), \tag{12.196}$$

for $z \to \pm\infty$, to find the Fraunhofer diffraction limit. The sign of each Fresnel integral in (12.193) is determined by the sign of x in this limit. In Exercise 12.7.3 you will show that for $\rho \to \infty$ one obtains the beautiful and simple Fraunhofer slit diffraction result:

$$\frac{dT}{d\phi} = \frac{2F^2}{\pi k d} \left(\frac{\sin\left(\frac{kd}{2}\sin\phi\right)}{\sin\phi} \right)^2 . \tag{12.197}$$

Exact forward scattering is defined as a limit, similar to the forward scattering circular case.

12.8 Vector diffraction theory

We have covered many of the issues in the theory of diffraction in our scalar considerations. The new aspect in the electromagnetic case is of course the two polarization states and their associated angular patterns.

Using the field equations for $\vec{A}(\vec{x})$ (analogous to (12.153)) and $G^{\text{out}}(\vec{x}, \vec{x}\,'; k)$, we may again do the "song and dance" to write down the expression for the vector potential everywhere outside the source for the above geometry. In this process we may consistently assume $A_z = 0$ since any currents on the barrier are transverse. Adopting Neumann boundary conditions and assuming the fields fall off fast enough at spatial infinity as in the Kirchhoff integral discussion, we have

$$\begin{aligned} \vec{A}(\vec{x}) &= -\frac{1}{2\pi} \int_S da'\, \partial_{n'} \vec{A}(\vec{x}\,') G^{\text{out}}(\vec{x}, \vec{x}\,'; k) \\ &= \frac{1}{2\pi} \int_S da'\, \hat{n}' \times \vec{B}(\vec{x}\,') G^{\text{out}}(\vec{x}, \vec{x}\,'; k), \end{aligned} \tag{12.198}$$

giving

$$\vec{B}(\vec{x}) = \frac{1}{2\pi} \vec{\nabla} \times \int_S da'\, \hat{n}' \times \vec{B}(\vec{x}\,') G^{\text{out}}(\vec{x}, \vec{x}\,'; k). \tag{12.199}$$

In source free regions the Maxwell equations are invariant under the duality transformation $\vec{E} \to \vec{B}$, $\vec{B} \to -\vec{E}$. This immediately implies that one expression for the electric field is

$$\vec{E}(\vec{x}) = \frac{1}{2\pi} \vec{\nabla} \times \int_S da'\, \hat{n}' \times \vec{E}(\vec{x}\,') G^{\text{out}}(\vec{x}, \vec{x}\,'; k). \tag{12.200}$$

This derivation follows the one in Jackson, *op. cit.*, Section 10.7. We will use this expression, which assures a transverse electric field in the Fraunhofer limit, in an upcoming example.

It is interesting to try to recover Equation (12.200) from other points of view. Duality was employed in writing down (12.200). This implies one way it may be derived is by writing down Maxwell's equations for fictitious magnetic currents and charges and evaluating the electric field as the curl of a dual vector potential, as was done by Bethe.[4] Another method, due to Smythe,[5] sums over infinitesimal solenoids in the apertures whose magnetic moments are proportional to $\hat{n}' \times \vec{E}$. We seek a more direct way of establishing this result. We find that we may do so by introducing a field averaging model with interior fields, and no internal currents or charges, which obey Maxwell's equations as an intermediate step. This allows us to use infinite space boundary conditions so that integration by parts is defined. We will also need two electric field properties for the diffracted field:

(1) $\vec{\nabla} \cdot \vec{E}\big|_S = 0$

(2) E_z is continuous across the interface and odd in z.

The source of the diffracted field is a divergenceless incoming plane wave for $z < 0$. Condition (1) does not say there are no surface charge densities induced on the interface, but simply that the diffracted field is not the source. The second condition simply recognizes the symmetry of the situation: the $z > 0$ diffraction fields are paired with a symmetric $z < 0$ diffraction shadow field: there is no reason to suppose the actual field favors one side or the other!

Let us proceed. From the Maxwell equations

$$(\vec{\nabla}'^2 + k^2)\vec{E}(\vec{x}\,') = -\frac{4\pi i k}{c}\vec{J}(\vec{x}\,'), \tag{12.201}$$

where $\vec{J}(\vec{x}\,')$ is the spatial current density. We know from (12.199) that

$$\vec{K}(\vec{x}\,')\big|_S = \frac{c}{2\pi}\hat{n}' \times \vec{B}(\vec{x}\,'), \tag{12.202}$$

gives the surface current on the $z' = 0$ plane:

$$\int dz'\, \vec{J}_{\text{mod}}(\vec{x}\,') = \vec{K}(\vec{x}\,')\big|_S. \tag{12.203}$$

We will take

$$\vec{J}_{\text{mod}}(\vec{x}\,') \equiv \frac{c}{2\pi\boldsymbol{\Delta}}\,\hat{n}' \times \vec{\mathcal{B}}(\vec{x}\,'), \tag{12.204}$$

---

[4] H. A. Bethe, "Theory of Diffraction by Small Holes", Phys. Rev. **66**, 163 (1944).
[5] W. R. Smythe, "The Double Current Sheet in Diffraction", Phys. Rev. **72**, 1066 (1947).

everywhere inside the barrier and aperture ($\vec{\mathcal{B}}(\vec{x}\,') = \vec{B}(\vec{x}\,')$ in the aperture). We introduce fields $\vec{\mathcal{E}}(\vec{x}\,')$ and $\vec{\mathcal{B}}(\vec{x}\,')$ which satisfy Maxwell's equations in this region, which will be assigned a thickness $\boldsymbol{\Delta}$. We assume the average over the internal fields acts as the "value" on the surface S:

$$\frac{1}{\boldsymbol{\Delta}}\int dz'\vec{\mathcal{E}}(\vec{x}\,') = \vec{E}(\vec{x}\,')\big|_S,\ \frac{1}{\boldsymbol{\Delta}}\int dz'\vec{\mathcal{B}}(\vec{x}\,') = \vec{B}(\vec{x}\,')\big|_S. \tag{12.205}$$

First, in the interior of the barrier we use

$$\vec{\mathcal{B}}(\vec{x}\,') = -\frac{i}{k}\vec{\nabla}\times\vec{\mathcal{E}}(\vec{x}\,'), \tag{12.206}$$

in (12.204). Next, we employ a Jacobi identity:

$$\hat{n}'\times(\vec{\nabla}'\times\vec{\mathcal{E}}(\vec{x}\,')) = \vec{\nabla}'\times(\hat{n}'\times\vec{\mathcal{E}}(\vec{x}\,')) + (\hat{n}'\times\vec{\nabla}')\times\vec{\mathcal{E}}(\vec{x}\,'). \tag{12.207}$$

The last term is:

$$(\hat{n}'\times\vec{\nabla}')\times\vec{\mathcal{E}}(\vec{x}\,') = \vec{\nabla}'(\hat{n}'\cdot\vec{\mathcal{E}}(\vec{x}\,')) - \hat{n}'(\vec{\nabla}'\cdot\vec{\mathcal{E}}(\vec{x}\,')). \tag{12.208}$$

The second term on the right vanishes for no internal charges, consistent with condition (1) above. Our volume current is:

$$\vec{J}_{\text{mod}}(\vec{x}\,') = -\frac{ic}{2\pi k\boldsymbol{\Delta}}\left(\vec{\nabla}'\times(\hat{n}'\times\vec{\mathcal{E}}(\vec{x}\,')) + \vec{\nabla}'(\hat{n}'\cdot\vec{\mathcal{E}}(\vec{x}\,'))\right). \tag{12.209}$$

We can now finish up by multiplying left and right hand sides by the infinite space Green function and integrating over all space. On the left-hand side we have

$$\begin{aligned}
&\int d^3x'\,(\vec{\nabla}'^2 + k^2)\vec{E}(\vec{x}\,')G^{\text{out}}(\vec{x},\vec{x}\,';k)\\
&\quad = \int d^3x'\,\vec{E}(\vec{x}\,')\underbrace{(\vec{\nabla}'^2 + k^2)G^{\text{out}}(\vec{x},\vec{x}\,';k)}_{-4\pi\delta(\vec{x}-\vec{x}\,')}\\
&\quad = -4\pi\vec{E}(\vec{x}).
\end{aligned} \tag{12.210}$$

On the other hand

$$\begin{aligned}
&-\frac{4\pi ik}{c}\int d^3x'\,\vec{J}_{\text{mod}}(\vec{x}\,')G^{\text{out}}(\vec{x},\vec{x}\,';k)\\
&\quad = -\frac{2}{\boldsymbol{\Delta}}\int d^3x'\,\left(\vec{\nabla}'\times(\hat{n}'\times\vec{\mathcal{E}}(\vec{x}\,')) + \vec{\nabla}'(\hat{n}'\cdot\vec{\mathcal{E}}(\vec{x}\,'))\right)G^{\text{out}}(\vec{x},\vec{x}\,';k)\\
&\quad = -\frac{2}{\boldsymbol{\Delta}}\vec{\nabla}\times\int d^3x'\,\hat{n}'\times\vec{\mathcal{E}}(\vec{x}\,')G^{\text{out}}(\vec{x},\vec{x}\,';k)\\
&\qquad -\frac{2}{\boldsymbol{\Delta}}\vec{\nabla}\int d^3x'\,\hat{n}'\cdot\vec{\mathcal{E}}(\vec{x}\,')G^{\text{out}}(\vec{x},\vec{x}\,';k)\\
&\quad = -2\vec{\nabla}\times\int_S da'\,\hat{n}'\times\vec{E}(\vec{x}\,')G^{\text{out}}(\vec{x},\vec{x}\,';k).
\end{aligned} \tag{12.211}$$

The second term on the right after the second equality in (12.211) involves $\mathcal{E}_z$ which is an odd function in z'. Although the $G^{\rm out}(\vec{x}, \vec{x}\,'; k)$ function multiplying it in the barrier integral is strictly neither even nor odd, in the limit of an infinitesimally thin barrier the remaining factors become constant in z' and the integral can be considered zero from symmetry. Finally, putting (12.210) and (12.211) together yields the result (12.200), as desired. We can also use this averaging model to reconcile different expressions for the diffracted electric field, which gives us more confidence in the approach; this is left as Exercise 12.8.1.

Equation (12.200) determines the value of the diffracted field in terms of itself, evaluated over the entire surface. This is essentially the same situation as for the scalar results for Dirichlet and Neumann boundary conditions. There, we argued that we may apply the desired boundary condition to the diffracted field separately. We will do the same thing here; namely, we assume that for a perfect conductor $\hat{n}' \times \vec{E}$ on the barrier is zero and integrate only over the diffracted field in the aperture. We will check that the deduced field is consistent with this boundary condition in our example application.

Let us finish with the classic example of high energy circular aperture electromagnetic diffraction at non-normal incidence. We should be fairly familiar with the mathematical steps to be taken from our scalar experience. We start with (12.200) and take the Fraunhofer $r \gg a$ limit, where a is the circle's radius. We recognize that we may decompose the incoming wave into two plane wave polarizations for the non-normal incidence case, as shown in Figure 12.5. One, which is called $B_\perp$ polarization in analogy to the interface situation in Section 10.2, has the electric field inclined at an angle α with respect to the interface, as shown in (a). The other polarization, called $E_\perp$, inclines only the magnetic field and is shown in (b). Note both inclinations are in the $x - z$ plane. Turning the crank on the formalism, we have

$$\vec{E}(\vec{x}) = \frac{i\vec{F} a\, e^{ikr}}{r} E_0 \frac{J_1(ka\Delta)}{\Delta}, \tag{12.212}$$

where

$$\vec{F} \equiv \begin{cases} \cos\alpha\,(\hat{k} \times \hat{y}), & B_\perp \text{ polarization} \\ \hat{k} \times \hat{x}, & E_\perp \text{ polarization.} \end{cases} \tag{12.213}$$

These polarization factors are reminiscent of those for the scalar Dirichlet and Neumann cases in Equation (12.166). Note the similar factor of $\cos\alpha$

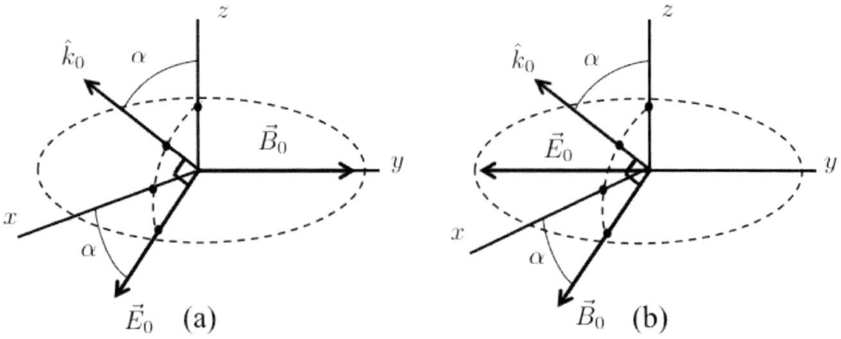

Fig. 12.5 Two incoming polarization states for electromagnetic diffraction from a circular gap: (a) $B_\perp$ polarization; (b) $E_\perp$ polarization.

present in the Neumann scalar case and the $B_\perp$ vector case. Notice also, however, new azimuthal angular dependence in the vector case. One can check that both polarizations obey the correct boundary conditions on the tangential $\vec{E}$ and normal $\vec{B}$ field; one can also work out the ratio of the square of the amplitudes at a given angular point (Exercise 12.8.2).

The expressions for the incoming and outgoing powers in this case are

$$P^{\text{in}} = \frac{c}{8\pi}|E_0|^2 A \cos\alpha, \qquad \frac{dP^{\text{out}}}{d\Omega} = r^2 \frac{c}{8\pi}|\vec{E}|^2, \tag{12.214}$$

giving

$$\frac{dT}{d\Omega} \equiv \frac{1}{A\cos\alpha}\left|\frac{r}{E_0}\right|^2 |\vec{E}|^2. \tag{12.215}$$

In this particular case one has

$$\frac{dT}{d\Omega} = \frac{\vec{F}^2}{\pi\cos\alpha}\left(\frac{J_1(ka\Delta)}{\Delta}\right)^2. \tag{12.216}$$

This expression should be compared to Equation (12.179) above for the scalar case, where there are similarities and differences in the F and $\vec{F}$ factors as noted above. Notice that for $\alpha \to \pi/2$ we are prevented from normalizing the $E_\perp$ amplitude by P^{in}, similar to the scalar Dirichlet case above. Some energy "leaks" into the $z > 0$ region even though the initial plane wave propagation direction is completely parallel to the interface. This is considered in more detail in Exercise 12.8.3.

12.9 Going Deeper

12.9.1 *Scattering and diffraction theory*

(1) C. F. Bohren and D. R. Huffman, *Absorption and Scattering of Light by Small Particles*, John Wiley & Sons (New York) 1983.

(2) M. Born and E. Wolf, *Principles of Optics*, 7th corrected ed., Cambridge University Press (Cambridge) 2002; Section 14.5.

(3) J. J. Bowman, T. B. A. Senior and P. L. E. Uslenghi, *Electromagnetic and Acoustic Scattering by Simple Shapes*, revised reprint, North Holland (Amsterdam) 1987.

(4) C. A. Brau, *Modern Problems in Classical Electrodynamics*, Oxford University Press (Oxford) 2004; Section 9.3.

(5) C. J. Bouwkamp, *Diffraction Theory*, Rep. Prog. Phys. **17**, 35 (1954).

(6) R. W. P. King and T. T. Wu, *The Scattering and Diffraction of Waves*, Harvard University Press (Cambridge) 1959.

(7) J. Van Bladel, *Electromagnetic Fields*, 2nd ed., John Wiley-IEEE Press (Piscataway, NJ) 2007; Chapters 9 and 11–14.

12.10 Exercises

Exercise 12.1.1. Use the model of Section 9.5 and the harmonic results in (11.112) to discuss the scattering from free or bound electrons in a medium in electric dipole approximation. Using a linearly polarized incoming wave and integrating over angles, show that one obtains the cross-section

$$\sigma(\omega) \equiv \int d\Omega \left(\frac{dP}{d\Omega}\right)_{\text{avg}} / |\vec{S}_0| = \frac{8\pi}{3} r_e^2 f(\omega),$$

where $r_e \equiv e^2/(mc^2)$ is called the *classical electron radius* (encountered again in Chapter 14), and where

$$f(\omega) \equiv \frac{\omega^4}{[(\omega_0^2 - \omega^2)^2 + \gamma^2\omega^2]},$$

which gives a resonance in the cross-section at $\omega \approx \omega_0$. The low and high energy forms of this cross-section are

$$\omega \ll \omega_0, \frac{\omega_0^2}{\gamma}: \quad \sigma(\omega) \approx \frac{8\pi}{3} r_e^2 \left(\frac{\omega}{\omega_0}\right)^4,$$

$$\omega \gg \omega_0, \gamma: \quad \sigma(\omega) \approx \frac{8\pi}{3} r_e^2.$$

The first result gives rise to so-called *Rayleigh scattering* and the second is called the *Thomson cross-section*. In particular, Rayleigh scattering implies that sunlight is preferentially scattered at higher frequencies, resulting in our blue sky and red sunsets!

Exercise 12.1.2. Let us investigate the polarization content of the synchrotron power results from Section 11.9. Considering that the radiation pattern is azimuthally symmetric and the synchrotron is in the $x-y$ plane, let us observe in the $\hat{n} = (\sin\theta, 0, \cos\theta)$ direction. Starting from the explicit form (11.263) and using the rule (12.29), use the final linear polarization vectors,

$$\hat{\epsilon}_1 = (0, 1, 0), \quad \hat{\epsilon}_2 = (-\cos\theta, 0, \sin\theta),$$

as well as the final circular polarization vectors,

$$\hat{\epsilon}_\pm = \frac{1}{\sqrt{2}}(\hat{\epsilon}_1 \pm i\hat{\epsilon}_2),$$

to form the fractions of the angular power emitted from synchrotron radiation in the linear and circular polarization cases:

$$F_{1,m}(\theta) \equiv \left(\frac{dP_m}{d\Omega}\right)_1 \Big/ \left(\frac{dP_m}{d\Omega}\right)_{\text{tot}}, \quad F_{2,m}(\theta) \equiv \left(\frac{dP_m}{d\Omega}\right)_2 \Big/ \left(\frac{dP_m}{d\Omega}\right)_{\text{tot}},$$

for linear polarization and

$$F_{+,m}(\theta) \equiv \left(\frac{dP_m}{d\Omega}\right)_+ \Big/ \left(\frac{dP_m}{d\Omega}\right)_{\text{tot}}, \quad F_{-,m}(\theta) \equiv \left(\frac{dP_m}{d\Omega}\right)_- \Big/ \left(\frac{dP_m}{d\Omega}\right)_{\text{tot}},$$

for circular polarization; the denominator is the unpolarized or total power. Using a mathematical graphics package, plot as a function of θ for: $m = 1$, $\beta = 0.1$; $m = 10$, $\beta = 0.1$; $m = 1$, $\beta = 0.99$; $m = 10$, $\beta = 0.99$. At which values of θ do the linear polarization fractions have equal magnitude? Which circular polarizations dominate near the endpoints $\theta \approx 0$ and $\theta \approx \pi$?

Exercise 12.1.3. Establish that

$$\sum_f \left|\hat{\epsilon}_f^* \cdot \vec{f}\right|^2 = \left|\hat{k} \times \vec{f}\right|^2,$$

where $\hat{\epsilon}_f^*$ is a final polarization vector and $\vec{f}$ is any scattering amplitude (note: $\hat{k} = \hat{n} = \hat{r}$ all denote the outward unit normal).

Exercise 12.3.1. The induced charges and currents on the surface of the spherical conductor are given by (12.58) and (12.59), using standard harmonic incident plane wave fields $\vec{E}_{\text{inc}}$ and $\vec{B}_{\text{inc}}$ with initial wave number $\vec{k}_0$ and polarization vector $\hat{\epsilon}_0$.

(a) Make sure that (12.58) and (12.59) form a conserved combination by showing the continuity equation is satisfied for the given incident fields.
(b) Use an adapted Equation (12.34) for the scattering amplitude to calculate the polarized scattering amplitude induced by the surface current $\vec{K}(\vec{r}\,')$. Show that it gives the shadow (forward) scattering amplitude, Equation (12.64). Thus, we can regain the entire scattering amplitude by integrations over just the illuminated portion of the sphere!

Exercise 12.3.2. Consider high-energy scattering of electromagnetic plane waves (initial direction $\hat{k}_0$, initial polarization vector $\hat{\epsilon}_0$) off of a perfect conductor in the shape of a flat disk of radius a. Consider only the special case of $\hat{k}_0$ perpendicular to the plane of the disk, as shown in Figure 12.6:

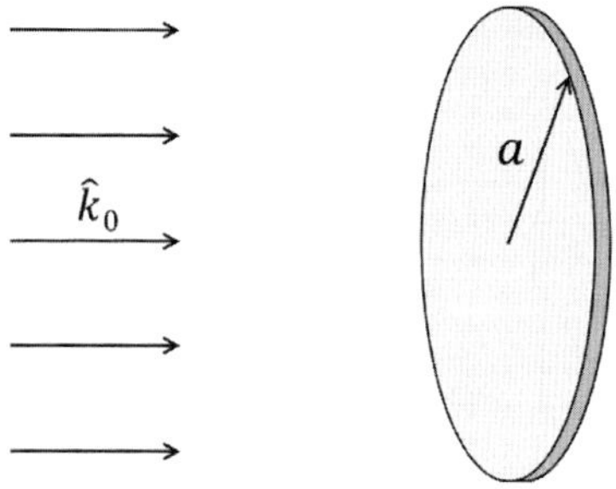

Fig. 12.6 Reference figure for Exercise 12.3.2.

(a) Show that the *illuminated* side unpolarized differential cross-section is $(\hat{k}_0 \cdot \hat{k} = \cos\theta)$

$$\left.\frac{d\sigma}{d\Omega}\right|_{\text{ill}} = \frac{a^2 J_1^2(ka\sin\theta)}{\sin^2\theta}\sin^4(\theta/2),$$

where $J_1(x)$ is an integer Bessel function.

(b) Plot this for increasingly large values of $ka \gg 1$. Notice the cross-section peaks in the backward direction, as one would expect.

Exercise 12.3.3. Consider short wavelength electromagnetic scattering off of a perfectly conducting cylinder of radius a and length $2L$ oriented with its long axis perpendicular to the incoming plane wave. See Figure 12.7. ($\hat{n}'$ is the surface outward normal.) Choose your z-axis along the axis of the rod and your x-axis pointing in the direction of $\vec{k}_0$. Evaluate the illuminated

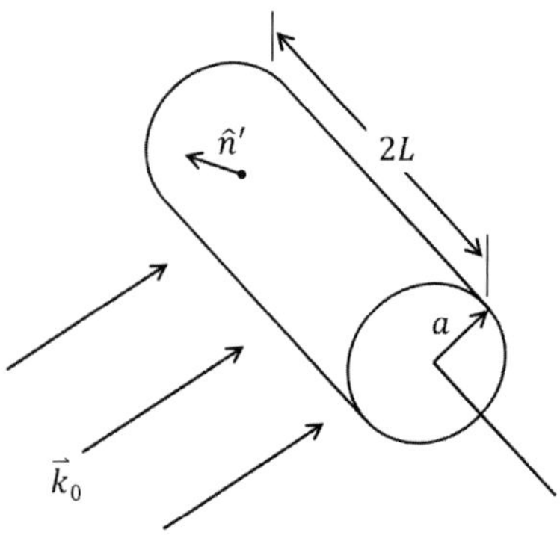

Fig. 12.7 Reference figure for Exercise 12.3.3.

side scattering amplitude for arbitrary $\vec{k}$ and show that the unpolarized cross-section is

$$\left.\frac{d\sigma}{d\Omega}\right|_{\text{ill}} = \frac{ka}{\pi}\sin(\phi/2)\left(\frac{\sin(kL\cos\theta)}{k\cos\theta}\right)^2,$$

where ϕ and θ are the usual spherical coordinate angles.

Exercise 12.3.4. There is interesting linear polarization dependence in the coherent short wavelength cross-sections for the conducting sphere which is not captured in the incoherent sums of the same quantities. Using unsummed initial polarization vectors $\hat{\epsilon}_1 = \hat{x}$ and $\hat{\epsilon}_2 = \hat{y}$, and by summing over final polarizations, numerically examine the cross-sections for scattering into the $\phi = 0$ half-plane, where ϕ is the azimuthal angle, with $\hat{n} = (\sin\theta, 0, \cos\theta)$ and $ka = 100$. Show that when the coherent sum of shadow and illuminated cross-sections are considered, the two polarization cross-sections are alternately enhanced or decreased over the other. Demonstrate that this does not occur with the incoherent sum with either initial polarization. In addition, you should find that the average of the two coherent cross-sections gives a result that is indistinguishable from either incoherent piece. Using a graphics software package, do $\log_{10}$ plots of $4\left(\frac{d\sigma}{d\Omega}\right)\Big|_{\text{sh+ill}}/a^2$ for these 4 cross-sections (two initial polarizations, coherent or incoherent) in the polar angle θ for $\theta \le .12$.

Helps: The two coherent polarized cross-sections for the conducting sphere of radius a (where the final polarization is summed on and $\hat{k}_0 = (0,0,1)$) are

$$\left(\frac{d\sigma}{d\Omega}\right)_{x-pol} = \left|ia\frac{J_1(ka\sin\theta)}{\sin\theta}\cos\theta + \frac{a}{2}e^{-2ika\sin\left(\frac{\theta}{2}\right)}\right|^2,$$

for $\hat{\epsilon}_0 = (1,0,0)$ initial polarization, and

$$\left(\frac{d\sigma}{d\Omega}\right)_{y-pol} = \left|ia\frac{J_1(ka\sin\theta)}{\sin\theta} - \frac{a}{2}e^{-2ika\sin\left(\frac{\theta}{2}\right)}\right|^2,$$

for $\hat{\epsilon}_0 = (0,1,0)$ initial polarization.

Exercise 12.3.5. Consider scattering from a tilted circular conducting plane of radius R. A cut through the middle of the plane is shown in

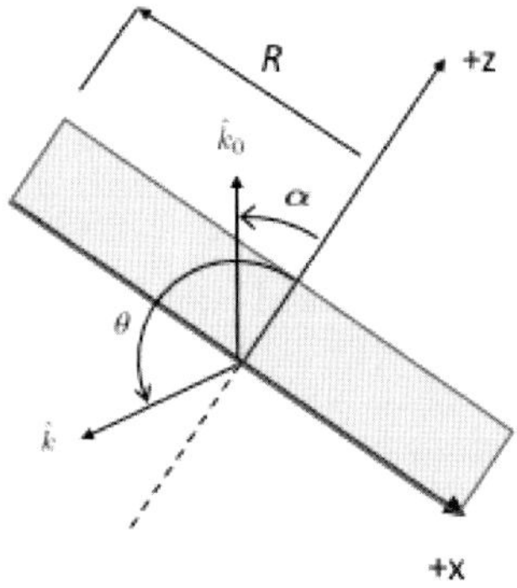

Fig. 12.8 Scattering situation in Exercise 12.3.5.

Figure 12.8. The incoming plane wave momentum direction is $\hat{k}_0$ and the outgoing momentum is given by $\hat{k}$. The angle the incoming momentum direction $\hat{k}_0$ makes with the z-axis is α. Limit your scattering evaluations to the $\phi = \pi$ half-plane ($\hat{k} = (-\sin\theta, 0, \cos\theta)$). Show that the *unpolarized* illuminated cross-section is

$$\left(\frac{d\sigma_{ill}}{d\Omega}\right)_{unpol} = \frac{R^2 J_1(kR(\sin\theta - \sin\alpha))^2}{4(\sin\theta - \sin\alpha)^2}(\cos\theta - \cos\alpha)^2.$$

Exercise 12.5.1.

(a) Derive the expression for the total cross-section,

$$\sigma_{\text{tot}} = -\frac{\pi}{k^2}\sum_{l=1}(2l+1)\text{Re}\{\alpha_\pm + \beta_\pm\},$$

from the optical theorem, Equation (12.97).

(b) Obtain a general formula for the scattering cross-section σ_{sc}, directly by integrating the unpolarized version of Equation (12.132).

(c) Given that $\sigma_{\text{tot}} = \sigma_{\text{abs}} + \sigma_{\text{sc}}$ (or other means) show that

$$\sigma_{\text{abs}} = \frac{\pi}{2k^2}\sum_{l=1}(2l+1)[2 - |\alpha_\pm + 1|^2 - |\beta_\pm + 1|^2],$$

and verify the statement under Equation (12.139).

Exercise 12.5.2. In the long wavelength limit of scattering from a perfect conducting sphere, we reached the conclusion that the phase shifts which determined $\alpha_\pm$ and $\beta_\pm$,

$$\alpha_\pm = e^{2i\delta_l} - 1, \quad \beta_\pm = e^{2i\delta_{l'}} - 1,$$

were given by $(x = ka)$

$$\tan\delta_l = \frac{j_l(x)}{n_l(x)}, \quad \tan\delta_{l'} = \frac{\frac{d}{dx}(x\, j_l(x))}{\frac{d}{dx}(x\, n_l(x))}.$$

(a) Using the Exercise 12.5.1 expression for σ_{tot} and the above phase shifts, find the total cross-section to lowest nonvanishing order.

(b) Confirm your result by integrating the angular cross-section

$$\frac{d\sigma}{d\Omega} = a^2(ka)^4\left[\frac{5}{8}(1+\cos^2\theta) - \cos\theta\right].$$

Exercise 12.5.3. Referring again to scattering off the perfect conducting sphere of radius a in the long wavelength limit, show that if the incident radiation is linearly polarized with incident field

$$\vec{E}^1_{\text{inc}} = \hat{x}\, e^{ikz},$$

the cross-section is given by

$$\left(\frac{d\sigma}{d\Omega}\right)_1 = k^4a^6\left[\frac{5}{8}(1+\cos^2\theta) - \cos\theta - \frac{3}{8}\sin^2\theta\cos 2\phi\right],$$

where the azimuthal angle ϕ is measured as usual starting from the x-axis. Knowing this, what cross-section do you get by linearly polarizing along $\hat{y}$?

Exercise 12.5.4. Show Equation (12.130) using (12.127) and asymptotic expansions of the spherical Hankel functions $h_l^{(1)}(x)$ (see Section 10.10).

Exercise 12.5.5. Use Equation (12.132) (or make the reverse replacement (12.29)) to obtain an "exact" result for the unpolarized scattering cross-section from a conducting sphere with $ka = 10$. Note that either incoming circular polarization will give you the same result, so averaging over initial polarization is not necessary. A representation of the vector spherical harmonics may be found in Exercise 10.11.1. Plot your result. Compare with the approximate unpolarized differential cross-section which

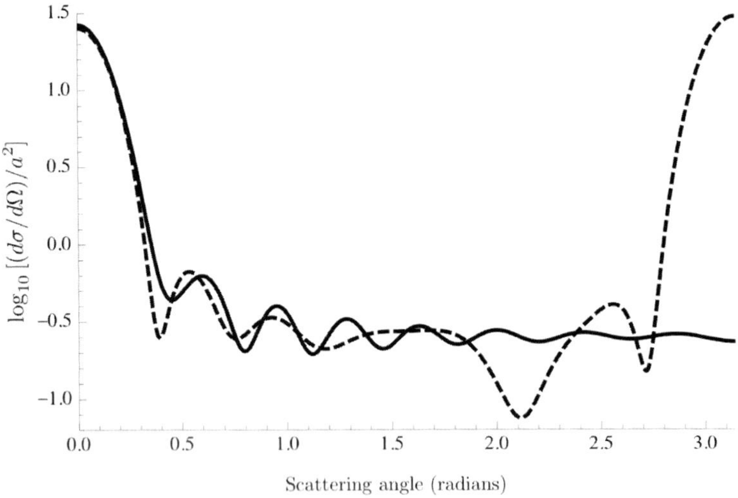

Fig. 12.9 Unpolarized differential cross-sections for Exercise 12.5.5 plotted on a $\log_{10}$ scale as a function of polar angle. The solid curve is the "exact" result, the dotted curve the short wavelength approximation.

results from combining the short wavelength scattering amplitudes from Equations (12.64) and (12.77) in Section 12.3 coherently. The final results should look like Figure 12.9. Comment on the results.

Exercise 12.5.6. Use Equation (12.130) to derive and plot "exact" results for the circularly polarized scattering cross-sections from a conducting sphere with $ka = 10$. For simplicity, azimuthal symmetry allows one to choose to construct the cross-section for the $\phi = 0$ half-plane. One may also use the $\hat{n} = \hat{k}$ outgoing unit vector as well as the final polarization vectors from Exercise 12.1.2. The cross-sections for equal initial and final circular polarizations (++ and −−) should be the same; likewise for different polarizations (+− and −+). The interesting part occurs for forward and backward scattering in these two cross-sections. Describe your findings!

Exercise 12.5.7. One may recover the short wavelength ($ka \gg 1$) shadow side scattering amplitude from Section 12.3 for the conducting sphere from the Section 12.5 formalism. (We connect with Sakurai and Napolitano, *Modern Quantum Mechanics*, 2nd ed., Addison-Wesley (Boston) 2011, Section 6.5.)

(a) Let's start by defining

$$\left(\vec{f}_{\pm}\right)_{\text{sh}} \equiv \frac{1}{ik}\sqrt{\frac{\pi}{2}}\sum_{l}^{l_{\max}}\sqrt{2l+1}\left[-\vec{X}_{l\pm1} \mp i\,\hat{r}\times\vec{X}_{l\pm1}\right].$$

Notice we cap the sum with $l_{\max} = ka$, where one would expect a classical contribution cutoff. Show for the unpolarized amplitude that

$$\int d\Omega \left|\left(\vec{f}_{\pm}\right)_{\text{sh}}\right|^2 \approx \pi a^2.$$

(b) Now given that (Gradshteyn and Ryzhik, *op. cit.*, 8.722 (2))

$$\lim_{l\to\infty} P_l\left(\cos\frac{x}{l}\right) = J_0(x),\ \ x \geq 0,$$

show that, by changing a sum into an approximate integral, we obtain (note the useful vector spherical harmonic results in Section 10.11 and Exercise 10.11.1)

$$\hat{\epsilon}_{\pm}^{*}\cdot\left(\vec{f}_{\pm}\right)_{\text{sh}} \approx ia\frac{J_1(ka\theta)}{\theta},$$

analogous to (12.64).

Exercise 12.5.8. Consider *linearly* polarized electromagnetic scattering from a perfectly conducting sphere of radius a for scattering into the $\phi = 0$ azimuthal plane (the $x - z$ half-plane) using the partial wave formalism of Section 12.5. Show the formulas for scattering cross-section divided by a^2,

$$\left(\frac{d\sigma}{d\Omega}\right)_{(f,i)} / \, a^2,$$

for the (f, i) cross-sections (2,1) and (1,2), using (final pol., initial pol.) notation. Demonstrate *analytically* that the (1,1) and (2,2) cross-sections are exactly zero. Use the final direction vector $\hat{k} = (\sin\theta, 0, \cos\theta)$, the final polarization vectors $\hat{\epsilon}_{f1} = (0, 1, 0)$ and $\hat{\epsilon}_{f2} = (-\cos\theta, 0, \sin\theta)$, and the initial polarization vectors $\hat{\epsilon}_{i1} = (1, 0, 0)$ and $\hat{\epsilon}_{i2} = (0, 1, 0)$. Using a graphics package plot the cross-sections for the (1,2) and (2,1) cases as a function of θ for $ka = 10$ and $0 < \theta < \pi$ using a $\log_{10}$ plot. The results should look like Figure 12.10.

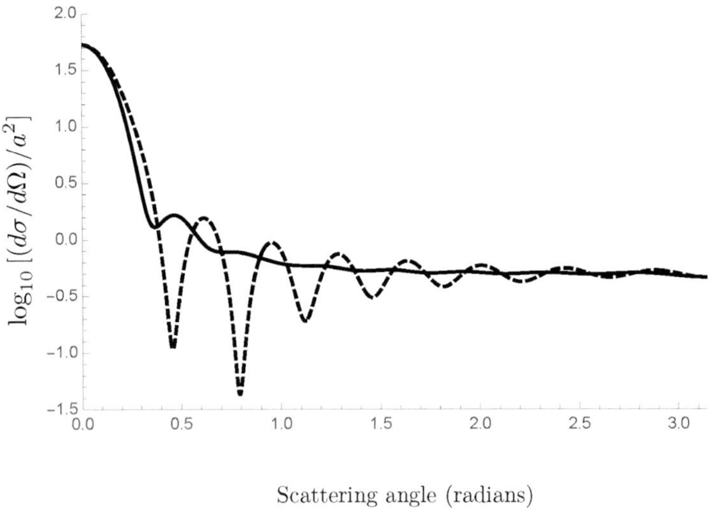

Fig. 12.10 Log_{10} plot of conducting sphere scattering cross-sections from Exercise 12.5.8 into the $\phi = 0$ half-plane as a function of polar angle. The solid curve is the (2,1) polarization case, the dotted is the (1,2) polarization case.

Exercise 12.7.1. Use (12.179) to evaluate and numerically plot the transmission coefficient T as a function of ka in the range $0 < ka < 10$ for the two cases of normal incidence Dirichlet and Neumann scalar circular diffraction in the Fraunhofer limit.

Exercise 12.7.2.

(a) By including the third term in the expansion of $k|\vec{x} - \vec{x}\,'|$ from (12.161), derive the Fresnel result

$$\frac{dT}{d\Omega} = \frac{F^2}{\pi}\frac{|I|^2}{(ka)^2},$$

for the angular diffraction transmission coefficient for normal incidence on a circular slit of radius a where

$$I \equiv k^2 \int_0^a d\rho' \rho' \, e^{ik\rho'^2/2r} \, J_0(k\rho' \sin\theta).$$

(b) For the Dirichlet case, using aperture $a = 0.5$ cm and wavelength $\lambda = 5.5 \times 10^{-5}$ cm (green light), plot the angular transmission for spherical screen distance:

(1) $r = 10^5$ cm
(2) $r = 10^4$ cm
(3) $r = 10^3$ cm.

Monitor the Fresnel number, $a^2k/(8\pi r)$ (just using the radius a instead of slit width d in (12.192)). Judge the angular spread by the first node and decide if each is better described by geometrical $\sim a/r$ or wave $\sim 1/(ka)$ width. Try to make sure your results are normalized correctly.

Exercise 12.7.3. Derive the Fraunhofer slit diffraction result Equation (12.197) from (12.193) and (12.194) using the asymptotic forms of the Fresnel integrals, (12.195) and (12.196).

Exercise 12.7.4. Consider a double-slit diffraction screen with infinitely long slits of opening size d separated by a central blocking strip of width a, as seen in Figure 12.11. Given normal incidence of a *scalar* plane wave with wave number k on the screen, find the differential transition rate $\frac{dT}{d\phi}$ as a function of ϕ in the $\rho \to \infty$ limit. Simplify your expression as much as possible.

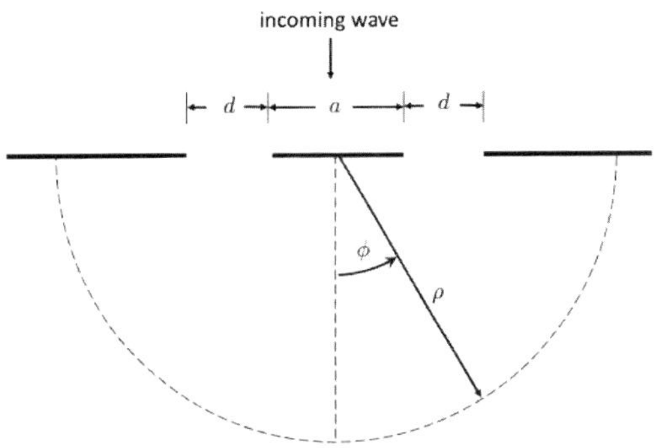

Fig. 12.11 Diffraction screen setup for Exercise 12.7.4.

Exercise 12.7.5. Using a plotting package simulate the differential transition rate for diffraction from a slit of width d using Equation (12.194) with Neumann boundary conditions. Numerically evaluate $\frac{dT}{d\phi}$ and plot as a function of ϕ for three cases (all for $\rho = 0.5$ cm, $\lambda = 5.5 \times 10^{-5}$ cm): $d = 0.1$ cm, .01 cm and .001 cm. Also numerically integrate and find the total transition probability T. Fill out a table with the headings (N_F is the Fresnel number):

d(cm)	N_F	$\frac{d}{\rho}$	$\frac{2\pi}{kd}$	*Approx. 1st node(radians)*	T

Comment on your findings relative to the wave/geometrical aspects of the diffraction patterns found. Are the values of T *exactly* correct? If not, why not?

Exercise 12.8.1. The immediate implication from Equation (12.199) is the alternate expression

$$\vec{E}(\vec{x}) = \frac{i}{k}\vec{\nabla} \times \vec{B}(\vec{x}) = \frac{i}{2\pi k}\vec{\nabla} \times \left(\vec{\nabla} \times \int_S da'\, \hat{n}' \times \vec{B}(\vec{x}\,')G^{\rm out}(\vec{x}, \vec{x}\,'; k)\right)$$

for the electric field everywhere outside the source region. It is interesting to try to reconcile this with (12.200) using our averaging approach. In the above equation we modify the interior integration, giving a model form for the electric field:

$$\vec{E}_{\rm mod}(\vec{x}) \equiv \frac{i}{k}\vec{\nabla} \times \vec{B}_{\rm mod},$$
$$\vec{B}_{\rm mod}(\vec{x}) \equiv \frac{1}{2\pi\boldsymbol{\Delta}}\vec{\nabla} \times \int d^3x'\, \hat{n}' \times \vec{\boldsymbol{B}}(\vec{x}\,')G^{\rm out}(\vec{x}, \vec{x}\,'; k).$$

This also gives a model form for the magnetic field in (12.199). Use a Jacobi identity and consider the interior to have no internal currents or charges. With these assumptions, show that

$$\vec{B}_{\rm mod}(\vec{x}) = -\frac{ik}{2\pi}\int_S da'\, \hat{n}' \times \vec{E}(\vec{x}\,')G^{\rm out}(\vec{x}, \vec{x}\,'; k)$$
$$-\frac{1}{2\pi}\vec{\nabla}\int_S da'\, \hat{n}' \cdot \vec{B}(\vec{x}\,')G^{\rm out}(\vec{x}, \vec{x}\,'; k)$$

when the z' integral is done. Finally, we see that the gradient term above does not contribute in the context of the alternate $\vec{E}(\vec{x})$ equation, and we recover (12.200). Notice the dual of this equation gives the electric field correctly as the time derivative of the originally introduced vector potential Equation (12.198) plus a gradient of the scalar potential (which actually would vanish by symmetry as in (12.211)).

Exercise 12.8.2.

(a) Show that the amplitudes for the electric and magnetic fields from circular diffraction respect the boundary conditions

$$\hat{n}' \times \vec{E}\big|_S = 0, \quad \hat{n}' \cdot \vec{B}\big|_S = 0.$$

at the surface of a conductor.

(b) Show that the ratio of intensities of the two polarizations at each point in angular space is

$$\left|\frac{\vec{E}_{B_\perp}}{\vec{E}_{E_\perp}}\right|^2 = \cos^2\alpha \left(\frac{\cos^2\theta + \sin^2\theta\cos^2\phi}{\cos^2\theta + \sin^2\theta\sin^2\phi}\right).$$

Exercise 12.8.3. Using Equation (12.214), numerically plot the normalized $E_\perp$ polarization angular power

$$\frac{8\pi}{ca^2|E_0|^2}\frac{dP^{\text{out}}_{E_\perp}}{d\Omega}\bigg|_{\alpha=\theta=\pi/2}$$

for circular vector diffraction as a function of azimuthal angle ϕ for various values of ka. Show that there is no diffraction in the exact forward direction and that interesting symmetric backscattering patterns result. Characterize the polarization state of this diffracted field.

Chapter 13

Relativistic Formulations of Electrodynamics

Electromagnetic theory is deeply bound up with the theory of relativity. Relativity first arose following the (unsuccessful) attempt to reconcile Maxwell's equations with Galilean invariance. When recast in relativistic language, the Maxwell equations take on an elegant form that brings deep insights into the nature of the electromagnetic field. Just as space and time or mass and energy are interrelated, so the electric and magnetic fields are shown to be different components derived from a single vector potential. Indeed, the form of the electromagnetic field equations are uniquely determined by some straightforward assumptions, as we shall see in Section 13.9. The effects of the electromagnetic field on matter may also be summarized as a single equation of momentum-energy conservation, as shown in Section 13.10. In addition, the practical importance of the theory is highlighted in Section 13.7, where we talk about radiation effects in particle accelerators.

13.1 The exact relativistic transformation between comoving frames

In Section 7.1 we were unable to find a transformation that preserved Maxwell's equations under Galilean transformations between reference frames in uniform relative motion:

$$\vec{x}\,' \to \vec{x} - \vec{\beta}ct; \qquad t' \to t.$$

Physicists' failure to find such a transformation led to the breakdown of classical physics in the early 20th century.

The root problem of course is that the Galilean transformation itself is faulty. The key to the solution was Einstein's insight that the correct transformation must preserve the speed of light. Based on this assumption, we may derive the correct transformation by taking full advantage of the symmetries inherent in the situation.

Consider first the restricted case of two frames in one spatial dimension in uniform relative motion. The "primed" frame (with spacetime coordinates $\{t', x'\}$) moves with speed $v > 0$ in the $+x$ direction relative to the "unprimed" frame (with spacetime coordinates $\{t, x\}$). A particular point in spacetime is chosen as the origin of both frames, so $\{0, 0\}$ in both frames refer to the same point. We may then derive a change of coordinates mapping between frames as follows.

An object moving at constant velocity u in the unprimed frame corresponds to a straight line in the $\{t, x\}$ plane of slope u. It follows that a beam of light passing through $\{0, 0\}$ at speed c moving in the $+x$ direction corresponds to the line of slope c in *both* frames. Similarly, a beam of light through the origin moving in the $-x$ direction corresponds to the line of slope $-c$ in *both* frames. It follows that lines of slope $\pm c$ must be left invariant under the change of coordinates. If we suppose the transformation $\{t, x\} \to \{t', x'\}$ is linear (which is required by translation invariance), this implies

$$\{1, c\} \to f(v) \cdot \{1, c\}; \quad \{1, -c\} \to g(v) \cdot \{1, -c\}, \tag{13.1}$$

where $f(v)$, $g(v)$ are functions to be determined.

Next, we characterize the inverse transformation using the symmetry between the two frames. The primed observer sees the unprimed frame as moving in the $-x'$ direction at speed v. Therefore under the inverse transformation $\{t', x'\} \to \{t, x\}$ we obtain

$$\{1, c\} \to f(-v) \cdot \{1, c\}; \quad \{1, -c\} \to g(-v) \cdot \{1, -c\}. \tag{13.2}$$

Combining (13.1) and (13.2) gives

$$f(v)f(-v) = 1; \qquad g(v)g(-v) = 1. \tag{13.3}$$

We now exploit the symmetry between $+x$ and $-x$. Since the $\pm x$ directions are physically indistinguishable, replacing $\{1, c\} \leftrightarrow \{1, -c\}$ and $v \leftrightarrow -v$ should leave the equations unchanged. If we make these replacements in (13.1) and compare with the original, we find

$$f(v) = g(-v); \qquad g(v) = f(-v). \tag{13.4}$$

Combining (13.3) and (13.4) gives

$$f(v) = \frac{1}{g(v)}. \tag{13.5}$$

Since a stationary observer in the unprimed frame is moving at speed v in the $-x'$ direction in the primed frame, it follows that the transformation $\{t, x\} \to \{t', x'\}$ gives

$$\{1, 0\} \to \gamma(v) \cdot \{1, -v\}, \tag{13.6}$$

where $\gamma(v)$ is an unknown function of v. However, from Galilean transformations we know that $\gamma(v) \to 1$ for $v \to 0$.

This puts us in a position to solve for $f(v)$. Since

$$\{1, 0\} = \frac{1}{2}\left[\{1, c\} + \{1, -c\}\right], \tag{13.7}$$

by linearity {1,0} transforms under $\{t, x\} \to \{t', x'\}$ as

$$\begin{aligned}\{1, 0\} &\to \frac{1}{2}\left[f(v) \cdot \{1, c\} + g(v)\{1, -c\}\right] \\ &= \left\{\frac{f(v) + 1/f(v)}{2}, c \cdot \frac{f(v) - 1/f(v)}{2}\right\}.\end{aligned} \tag{13.8}$$

Comparing (13.6) and (13.8), we find

$$f(v) = \sqrt{\frac{1-\beta}{1+\beta}}; \qquad \gamma = \frac{1}{\sqrt{1-\beta^2}}, \tag{13.9}$$

where

$$\beta \equiv \frac{v}{c}. \tag{13.10}$$

Finally we may obtain the general transformation $\{t, x\} \to \{t', x'\}$ via linear combination:

$$\{t, x\} = \frac{t + x/c}{2}\{1, c\} + \frac{t - x/c}{2}\{1, -c\}, \tag{13.11}$$

which leads via (13.1) to the familiar one dimensional *Lorentz boost* (details of the calculation are left as an exercise)

$$ct' = \gamma(ct - \beta x); \qquad x' = \gamma(x - \beta ct). \tag{13.12}$$

Transformation (13.12) not only leaves the speed of light invariant: it also preserves the quantity $(ct)^2 - x^2 = (ct')^2 - x'^2$. If we extend (13.12) to three dimensions by leaving the other two space dimensions invariant,

$$x' = \gamma(x - \beta ct); \qquad y' = y; \qquad z' = z; \qquad ct' = \gamma(ct - \beta x), \tag{13.13}$$

then the resulting transformation preserves the *4-interval*:

$$(ct')^2 - x'^2 - y'^2 - z'^2 = (ct)^2 - x^2 - y^2 - z^2. \tag{13.14}$$

We see that invariance of the speed of light in different reference frames is equivalent to the invariance of 4-intervals of length zero.

We may immediately rewrite (13.13) in rotationally-invariant form as (see Exercise 13.1.2):

$$\begin{cases} \vec{x}\,' = \vec{x} + \dfrac{(\gamma-1)}{\beta^2}\vec{\beta}(\vec{\beta}\cdot\vec{x}) - \gamma\vec{\beta}(ct), \\ ct' = \gamma(ct - \vec{\beta}\cdot\vec{x}). \end{cases} \tag{13.15}$$

At this point we may notice a similarity between these equations and the approximate invariances that were established in Exercise 7.1.1, which may be rewritten as:

$$\begin{cases} \vec{J}\,' = \vec{J} - \vec{\beta}\rho c, \\ \rho' c = \rho c - \vec{\beta}\cdot\vec{J}; \end{cases} \qquad \begin{cases} \vec{\nabla}\,' = \vec{\nabla} + \vec{\beta}\left(\dfrac{1}{c}\dfrac{\partial}{\partial t}\right), \\ \dfrac{1}{c}\dfrac{\partial}{\partial t'} = \dfrac{1}{c}\dfrac{\partial}{\partial t} + \vec{\beta}\cdot\vec{\nabla}. \end{cases} \tag{13.16}$$

In fact, if we replace γ with 1 in (13.15), we exactly reproduce the form of the equations in (13.16). It is therefore not unreasonable to infer that the following are *exact* transformation laws:

$$\begin{cases} \vec{J}\,' = \vec{J} + \dfrac{(\gamma-1)}{\beta^2}\vec{\beta}(\vec{\beta}\cdot\vec{J}) - \gamma\vec{\beta}(c\rho), \\ c\rho' = \gamma(c\rho - \vec{\beta}\cdot\vec{J}); \end{cases} \tag{13.17}$$

$$\begin{cases} \vec{\nabla}' = \vec{\nabla} + \dfrac{(\gamma-1)}{\beta^2}\vec{\beta}(\vec{\beta}\cdot\vec{\nabla}) + \dfrac{\gamma\vec{\beta}}{c}\dfrac{\partial}{\partial t}, \\ \dfrac{1}{c}\dfrac{\partial}{\partial t'} = \gamma\left(\dfrac{1}{c}\dfrac{\partial}{\partial t} + \vec{\beta}\cdot\vec{\nabla}\right). \end{cases} \tag{13.18}$$

Equations (13.15), (13.17), and (13.18) summarize the effect of Lorentz boosts on coordinates, charges and coordinate derivatives.

13.2 Lorentz transformations: matrix and index notation

In Section 8.8 we showed how the mathematical language of vectors, scalars, and tensors could be used to describe the effect of (passive or active) rotations on electromagnetic sources and fields. We will now do the same thing for a larger set of transformations which includes Lorentz boosts as well as rotations. We will discover that in this case there are two types of vectors which obey different transformation laws. This will motivate the introduction of index (or tensor) notation, which is very well-suited for dealing with these complications.

We begin by noting that since the Lorentz boost (13.13) is a linear transformation, it can be described by a matrix. We will use the notation

$$\boldsymbol{x} = \left(x^0\ x^1\ x^2\ x^3\right)^{\mathrm{T}} \equiv (ct\ x\ y\ z)^{\mathrm{T}}, \tag{13.19}$$

to denote a column vector. Then we may write

$$\boldsymbol{x}' = A_{\text{boost}}\boldsymbol{x}, \tag{13.20}$$

where

$$A_{\text{boost}} \equiv \begin{pmatrix} \gamma & -\gamma\beta & 0 & 0 \\ -\gamma\beta & \gamma & 0 & 0 \\ 0 & 0 & 1 & 0 \\ 0 & 0 & 0 & 1 \end{pmatrix}. \tag{13.21}$$

We have shown that transformation (13.20) satisfies the 4-interval invariance condition (13.14), which can be written in vector form as:

$$\boldsymbol{x}'^{\mathrm{T}}\eta\boldsymbol{x}' = \boldsymbol{x}^{\mathrm{T}}\eta\boldsymbol{x}, \tag{13.22}$$

where

$$\eta \equiv \begin{pmatrix} 1 & 0 & 0 & 0 \\ 0 & -1 & 0 & 0 \\ 0 & 0 & -1 & 0 \\ 0 & 0 & 0 & -1 \end{pmatrix}. \tag{13.23}$$

The matrix η is called the *flat-space metric* or *Minkowski metric*. Here we have chosen to put the minus signs on the space components; the alternative choice, which makes the time component negative is also possible. "Check your metric convention!" is a good piece of advice for initial relativistic considerations. The term "metric" reflects the fact that η is used to define the invariant 4-interval, which can be thought of as a generalization of distance. The "flat-space" designation indicates that η is a special case of the more general metric tensor used in general relativity which represents the curvature of spacetime due to gravity.

In general, the linear transformation

$$\boldsymbol{x}' = A\boldsymbol{x}, \tag{13.24}$$

satisfies (13.22), if and only if the matrix A satisfies

$$\boldsymbol{x}'^{\mathrm{T}}\eta\boldsymbol{x}' = (A\boldsymbol{x})^{\mathrm{T}}\eta(A\boldsymbol{x}) = \boldsymbol{x}^{\mathrm{T}}\eta(\eta A^{\mathrm{T}}\eta A)\boldsymbol{x}. \tag{13.25}$$

Since (13.22) holds for any $\boldsymbol{x}$, we may conclude

$$\eta A^{\mathrm{T}}\eta A = I \quad \text{or} \quad (A^{-1})^{\mathrm{T}} = \eta A \eta. \tag{13.26}$$

We will use (13.26) as the defining condition for *Lorentz transformations* (LT). It can be shown directly that the product of any two LT's is also a LT (see Exercise 13.2.1).

Next we derive the transformation law for the gradient operator defined by

$$\frac{\partial}{\partial \boldsymbol{x}} \equiv \left(\frac{\partial}{\partial x^0} \;\; \frac{\partial}{\partial x^1} \;\; \frac{\partial}{\partial x^2} \;\; \frac{\partial}{\partial x^3} \right), \tag{13.27}$$

under the LT with matrix A. Since $\boldsymbol{x}' = A\boldsymbol{x}$, it follows that $\boldsymbol{x} = A^{-1}\boldsymbol{x}'$ and hence

$$\frac{\partial x^{\mu}}{\partial x'^{\nu}} = (A^{-1})_{\mu\nu} \quad \text{where } \mu, \nu = 0, 1, 2, 3, \tag{13.28}$$

from taking primed partial derivatives. From the chain rule we have

$$\begin{aligned} \frac{\partial}{\partial x'^{\nu}} &= \sum_{\mu=0}^{3} \frac{\partial}{\partial x^{\mu}} \left(\frac{\partial x^{\mu}}{\partial x'^{\nu}} \right) \\ \Longrightarrow \frac{\partial}{\partial \boldsymbol{x}'} &= \frac{\partial}{\partial \boldsymbol{x}} A^{-1} = \frac{\partial}{\partial \boldsymbol{x}} (\eta A^{\mathrm{T}} \eta), \end{aligned} \tag{13.29}$$

where the last equality comes from (13.26). The final expression in (13.29) should be thought of as a linear combination of derivatives; the derivatives do not act on the entries of $(\eta A^{\mathrm{T}}\eta)$, which are constants.

The 4-interval invariance condition (13.22) may be viewed from an alternate point of view as the invariance of the scalar product of $\boldsymbol{x}^{\mathrm{T}}\eta$ with $\boldsymbol{x}$. The transformation law for $\boldsymbol{x}^{\mathrm{T}}\eta$ is exactly the same as that for $\partial/\partial\boldsymbol{x}$:

$$\boldsymbol{x}'^{\mathrm{T}}\eta = (\boldsymbol{x}^{\mathrm{T}}\eta)(\eta A^{\mathrm{T}}\eta). \tag{13.30}$$

This motivates the definition of two different types of vectors, with two different transformation laws.

- *Contravariant* vectors are column vectors that transform via the matrix A, as in (13.20). The vectors $\boldsymbol{x}$ and $\boldsymbol{J} \equiv (c\rho \; J^1 \; J^2 \; J^3)^{\mathrm{T}}$ as defined above are contravariant vectors.

- *Covariant* vectors are row vectors that transform via multiplication on the right by the matrix $\eta A^{\mathrm{T}}\eta = A^{-1}$, as in (13.29), (13.30). Examples of covariant vectors include $\boldsymbol{x}^{\mathrm{T}}\eta$ and $\partial/\partial\boldsymbol{x}$.

We have already seen how the quantity $\boldsymbol{x}^{\mathrm{T}}\eta\boldsymbol{x}$ is invariant under LT's. Notice that this is the matrix product of a covariant vector with a contravariant vector. In fact, we may show that the product of a covariant vector with *any* contravariant vector is also invariant. Take for instance the product of the covariant vector $\boldsymbol{y}$ with the contravariant vector $\boldsymbol{x}$. Under LT, we have:

$$\boldsymbol{y}'\boldsymbol{x}' = \boldsymbol{y}(\eta A^{\mathrm{T}}\eta)A\boldsymbol{x} = \boldsymbol{y}\boldsymbol{x}. \tag{13.31}$$

In the same way, we may show that $\boldsymbol{x}^{\mathrm{T}}\eta\boldsymbol{J}$ and $(\partial/\partial\boldsymbol{x})\boldsymbol{J}$ are Lorentz invariants. (Lorentz invariants are also called (Lorentz) *scalars.*)

For computations involving covariant and contravariant vectors, it is convenient to represent them using index notation. Contravariant and covariant vectors are represented using superscripts and subscripts, respectively: for example, the contravariant 4-vector $\boldsymbol{x}$ is written as x^μ ($\mu = 0,1,2,3$), while the covariant 4-vector $\boldsymbol{x}^{\mathrm{T}}\eta$ translates as x_μ. To maintain consistency with this convention, the covariant 4-gradient $\partial/\partial\boldsymbol{x}$ is written as ∂_μ, while the contravariant gradient $(\partial/\partial\boldsymbol{x})^{\mathrm{T}}\eta$ is written as

$$\partial^\mu \equiv \left(\frac{\partial}{\partial x_0}, \frac{\partial}{\partial x_1}, \frac{\partial}{\partial x_2}, \frac{\partial}{\partial x_3}\right) = \left(\frac{1}{c}\frac{\partial}{\partial t}, -\vec{\nabla}\right). \tag{13.32}$$

The LT with matrix A (13.24) is represented in index notation by the symbol $\Lambda^\mu{}_\nu$, where

$$\Lambda^\mu{}_\nu = (A)_{\mu\nu} \quad \text{for } \mu,\nu = 0,1,2,3. \tag{13.33}$$

This translates as

$$x'^\mu = \Lambda^\mu{}_\nu x^\nu. \tag{13.34}$$

Here the summation over the repeated index ν is understood: this is called the *Einstein summation convention.* Notice how the subscript ν "contracts" with the superscript ν, leaving only a vector with superscript μ. Note that for a contraction to be valid, one index must be a subscript (covariant) and the other a superscript (contravariant); so for instance the expression $x_\mu y_\mu$ does not produce a Lorentz invariant object.

The flat-space metric η is represented in index notation as either $\eta_{\mu\nu}$ or $\eta^{\mu\nu}$. ($\eta_{\mu\nu} = \eta^{\mu\nu}$ holds in the special case of flat space; in general relativity the lower indexed metric tensor, usually denoted by $g_{\mu\nu}$, has $g_{\mu\nu} \neq g^{\mu\nu}$ in general.) Using the metric tensor, it is possible to raise or lower vector indices at will; so for example

$$x_\mu = \eta_{\mu\nu} x^\nu, \qquad x^\mu = \eta^{\mu\nu} x_\nu. \tag{13.35}$$

Similarly, we define

$$\Lambda_\mu{}^\nu \equiv \eta_{\mu\gamma} \Lambda^\gamma{}_\rho \eta^{\rho\nu}, \tag{13.36}$$

where the repeated indices γ and ρ are both summed. The transformation equation (13.30) is written in index notation as

$$x'_\mu = \Lambda_\mu{}^\nu x_\nu. \tag{13.37}$$

Notice that the summation involving $\Lambda_\mu{}^\nu$ in (13.37) is over the second index, and corresponds to matrix multiplication of $\boldsymbol{x}^{\mathrm{T}}\eta$ on the right by $\eta A^{\mathrm{T}}\eta = A^{-1}$. Therefore from either (13.36) or (13.37) we may identify

$$\Lambda_\mu{}^\nu = ((A^{-1})^{\mathrm{T}})_{\mu\nu} \quad \text{for } \mu, \nu = 0, 1, 2, 3. \tag{13.38}$$

The Lorentz transformation condition $A^{\mathrm{T}}\eta A = \eta$ can be transcribed directly in index language as

$$\Lambda^\lambda{}_\mu \eta_{\lambda\kappa} \Lambda^\kappa{}_\nu = \eta_{\mu\nu}. \tag{13.39}$$

The identity $A^{-1}A = I$ appears as

$$\Lambda_\gamma{}^\mu \Lambda^\gamma{}_\nu = \delta^\mu_\nu, \tag{13.40}$$

where δ^μ_ν is the index representation of the identity matrix. As another example, $(A^{-1})^{\mathrm{T}} A^{\mathrm{T}} = I$ appears as

$$\Lambda_\mu{}^\gamma \Lambda^\nu{}_\gamma = \delta^\nu_\mu. \tag{13.41}$$

Summarizing, the statements

$$\frac{\partial x'^\mu}{\partial x^\nu} = (A)_{\mu\nu} = \Lambda^\mu{}_\nu, \tag{13.42}$$

$$\frac{\partial x^\mu}{\partial x'^\nu} = ((A^{-1})^{\mathrm{T}})_{\nu\mu} = \Lambda_\nu{}^\mu, \tag{13.43}$$

give the relations between the transformational matrix and index quantities for the LT in (13.24).

In Section 8.8 we introduced the notion of *tensor* as the direct product of vectors. So we may form more general tensor objects by taking direct products (also called *tensor products*) of covariant and/or contravariant vectors. For instance, the tensor product of contravariant vectors x^μ and y^ν is a *second-rank tensor* which can be identified with a 4×4 matrix whose

μ, ν entry is $x^\mu y^\nu$. Any sum of such products is also a second-rank tensor. In general, the transformation law for the second-rank tensor $F^{\mu\nu}$ is

$$F'^{\mu\nu} = \Lambda^\mu{}_\rho \Lambda^\nu{}_\sigma F^{\rho\sigma} \quad \text{(matrix notation: } F' = AFA^{\mathrm{T}}\text{)}. \tag{13.44}$$

We may also raise and lower indices as before, for example:

$$G_\mu{}^\nu \equiv \eta_{\mu\rho} G^{\rho\nu}; \quad H_{\mu\nu} \equiv \eta_{\mu\rho}\eta_{\nu\sigma} H^{\rho\sigma}.$$

The transformation laws for these tensors are:

$$G'_\mu{}^\nu = \Lambda_\mu{}^\rho \Lambda^\nu{}_\sigma G_\rho{}^\sigma \quad \text{(matrix notation: } G' = (A^{-1})^{\mathrm{T}} G A^{\mathrm{T}}\text{)}; \tag{13.45}$$

$$H'_{\mu\nu} = \Lambda_\mu{}^\rho \Lambda_\nu{}^\sigma H_{\rho\sigma} \quad \text{(matrix notation: } H' = (A^{-1})^{\mathrm{T}} H A^{-1}\text{)}. \tag{13.46}$$

But why stop with two indices? We can form tensors of arbitrary order by multiplying an arbitrary number of vectors, such as:

$$T^{\mu\nu}{}_{\rho\sigma}\cdots \equiv x^\mu y^\nu z_\rho w_\sigma \cdots,$$

and we may raise and lower indices at will, for example:

$$T^\mu{}_{\nu\rho}{}^\sigma \cdots = \eta_{\nu\kappa}\eta^{\sigma\tau} T^{\mu\kappa}{}_{\rho\tau} \cdots .$$

The transformation laws for such tensors follow the same format as (13.45)–(13.46), for example

$$T'^\mu{}_{\nu\rho}{}^\sigma \cdots = \Lambda^\mu{}_\gamma \Lambda_\nu{}^\lambda \Lambda_\rho{}^\kappa \Lambda^\sigma{}_\tau T^\gamma{}_{\lambda\kappa}{}^\tau \cdots .$$

For equations involving tensors of rank 3 or more, matrix notation breaks down. This is one reason why index notation is preferred for relativistic calculations. It follows from the above that contraction of two tensor indices gives another tensor with rank reduced by two from the sum of the two ranks (if a single contraction). For a tensor equation with contracted indices to be valid, the uncontracted indices must agree in type on both sides, so for instance

$$F_\mu{}^\lambda \equiv G_{\mu\nu} H^{\lambda\nu},$$

although the ordering of the indices on $F_\mu{}^\lambda$ is arbitrary. If all indices are contracted in pairs, the result is a scalar. For example, the invariant scalar product of covariant vector $\boldsymbol{x}$ with contravariant vector $\boldsymbol{y}$ is written in index notation as $y_\mu x^\mu$ $(= y^\mu x_\mu)$.

13.3 Relativistic form of Maxwell's equations

We have successfully "relativized" the approximate transformations of the contravariant

$$J^\mu = (c\rho, \vec{J}),$$

and covariant

$$\partial_\mu = \left(\frac{1}{c}\frac{\partial}{\partial t}, \vec{\nabla}\right),$$

vectors from Exercise 7.1.1. In this section, we will do the same thing for electric and magnetic fields.

The approximate transforms in (7.27) suggest that $\vec{E}$ and $\vec{B}$ are not 4-vectors. We may obtain some clues about the nature of these fields by rewriting Maxwell's equations in terms of electromagnetic potentials. Recall that

$$\vec{B} = \vec{\nabla} \times \vec{A}, \tag{13.47}$$

$$\vec{E} = -\vec{\nabla}\Phi - \frac{1}{c}\frac{\partial \vec{A}}{\partial t}. \tag{13.48}$$

The two sourceless Maxwell's equations follow immediately from (13.47) and (13.48). The other two Maxwell equations may be rewritten as:

$$\nabla^2\Phi + \frac{1}{c}\frac{\partial}{\partial t}(\vec{\nabla} \cdot \vec{A}) = -4\pi\rho, \tag{13.49}$$

$$\nabla^2\vec{A} - \frac{1}{c^2}\frac{\partial^2 \vec{A}}{\partial t^2} - \vec{\nabla}\left(\vec{\nabla} \cdot \vec{A} + \frac{1}{c}\frac{\partial \Phi}{\partial t}\right) = -\frac{4\pi}{c}\vec{J}. \tag{13.50}$$

The right-hand sides of (13.49) and (13.50) form a contravariant vector, as we saw in the previous section. It stands to reason that we should attempt to identify invariant and contravariant combinations on the left-hand side as well. Note that the gauge transformation (Λ arbitrary)

$$(\Phi, \vec{A}) \to (\Phi,\ \vec{A}) + \left(\frac{1}{c}\frac{\partial \Lambda}{\partial t},\ -\vec{\nabla}\Lambda\right) \tag{13.51}$$

leaves (13.49) and (13.50) unchanged. Here we may recognize the contravariant vector $\partial^\mu \Lambda$, where ∂^μ is defined in (13.32). This suggests that $(\Phi,\ \vec{A})$ is also a contravariant vector: so we write

$$A^\mu \equiv (\Phi,\ \vec{A}). \tag{13.52}$$

By adding and subtracting $(1/c)^2\partial^2\Phi/\partial t^2$ in (13.49) we may recognize that both equations contain the invariant scalar operator "$\Box$" that was defined in (8.86):

$$\Box \equiv \partial_\alpha \partial^\alpha = \frac{1}{c^2}\frac{\partial^2}{\partial t^2} - \vec{\nabla}^2.$$

This enables us to write (13.49) and (13.50) as a single 4-vector equation that is form-invariant under Lorentz transformation:

$$-\Box A^\mu + \partial^\mu(\partial_\nu A^\nu) = -\frac{4\pi}{c} J^\mu. \tag{13.53}$$

(A form-invariant equation is also called *covariant*; note however this sense of covariance differs from that used above.) The form-invariance of (13.53) substantiates our "leap of faith" that A^μ is a contravariant 4-vector.

We still have the gauge freedom expressed in (13.51), and with a judicious choice of gauge we may simplify (13.53) even further. If we adopt the form-invariant Lorenz gauge given in (7.109),

$$\partial_\mu A^\mu = 0,$$

then (13.53) becomes (compare (8.84) and (8.85)):

$$\Box A^\mu = \frac{4\pi}{c} J^\mu. \tag{13.54}$$

We are now ready to find the LT equations for electric and magnetic fields. Since A^μ is a contravariant vector, it has the same transformation law as x^μ, which was given in (13.15):

$$\begin{cases} \vec{A}' = \vec{A} + \dfrac{(\gamma - 1)}{\beta^2}\vec{\beta}(\vec{\beta}\cdot\vec{A}) - \gamma\vec{\beta}\Phi, \\ \Phi' = \gamma(\Phi - \vec{\beta}\cdot\vec{A}). \end{cases} \tag{13.55}$$

At this point we may write the components of $\vec{B}$ and $\vec{E}$ in terms of Φ and $\vec{A}$. The reader should note that the equations we are about to write for $\vec{B}$ and $\vec{E}$ are not tensor equations, and the subscripts that we use to index the components of $\vec{B}$ and $\vec{E}$ should not be thought of as covariant indices. With this caveat in mind, we write

$$B_i = (\vec{\nabla}\times\vec{A})_i = \partial_j A^k - \partial_k A^j = -(\partial^j A^k - \partial^k A^j) \ (i, j, k \text{ cyclic})$$

$$\Longrightarrow B_i = -\frac{\epsilon_{ijk}}{2}(\partial^j A^k - \partial^k A^j), \tag{13.56}$$

$$E_i = \left(-\frac{1}{c}\frac{\partial}{\partial t}\vec{A} - \vec{\nabla}A^0\right)_i = -\frac{1}{c}\frac{\partial}{\partial t}A^i - \partial_i A^0$$

$$\Longrightarrow E_i = -(\partial^0 A^i - \partial^i A^0). \tag{13.57}$$

We may define an antisymmetric second-rank tensor $F^{\mu\nu}$ via

$$F^{\mu\nu} \equiv \partial^\mu A^\nu - \partial^\nu A^\mu = -F^{\nu\mu}, \tag{13.58}$$

which transforms form invariantly as in Equation (13.44). Expression (13.53), which gives the two sourced Maxwell equations, can now be written in terms of $F^{\mu\nu}$ as

$$\partial_\nu F^{\mu\nu} = -\frac{4\pi}{c} J^\mu. \tag{13.59}$$

$F^{\mu\nu}$ is called the *electromagnetic field strength tensor.*

From (13.56) and (13.57) we find the $\vec{E}$ and $\vec{B}$ fields are related to $F^{\mu\nu}$ as follows:

$$F^{0i} = -E_i; \qquad F^{jk} = -B_i \qquad (i, j, k \text{ cyclic}). \tag{13.60}$$

These identifications, together with the transformation laws for fields, Equation (13.55), and operators, Equation (13.18), are sufficient to derive the transformation laws for $\vec{E}$ and $\vec{B}$. Alternatively, the tensor transformation law (13.44) and the explicit transformation (13.15) may be used to derive identical results. The reader is asked to complete the argument in Exercise 13.3.2; the result is

$$\vec{E}\,' = \gamma(\vec{E} + \vec{\beta} \times \vec{B}) - \frac{\gamma^2}{\gamma+1}\vec{\beta}(\vec{\beta} \cdot \vec{E}), \tag{13.61}$$

$$\vec{B}\,' = \gamma(\vec{B} - \vec{\beta} \times \vec{E}) - \frac{\gamma^2}{\gamma+1}\vec{\beta}(\vec{\beta} \cdot \vec{B}). \tag{13.62}$$

In the above discussion, the two sourceless Maxwell equations were obtained as identities resulting from the expression of $\vec{E}$ and $\vec{B}$ in terms of vector and scalar potentials. It is also possible to find a form-invariant expression of these equations in terms of $F^{\mu\nu}$, in analogy to (13.59). To do this, we recall the completely antisymmetric symbol defined in (8.162), which for 4-indices is

$$\epsilon^{\mu\nu\lambda\kappa} = \begin{cases} 0 & \text{if any two indices are equal,} \\ 1 & \text{if } (\mu, \nu, \lambda, \kappa) \text{ is an even permutation of } (0, 1, 2, 3), \\ -1 & \text{if } (\mu, \nu, \lambda, \kappa) \text{ is an odd permutation of } (0, 1, 2, 3). \end{cases} \tag{13.63}$$

The observant reader will note that we are using superscripts on the symbol instead of subscripts as in the original definition. This is because we are going to treat these as contravariant indices. (Of course, in the original definition we didn't have to worry about the covariant/contravariant distinction.)

Let us investigate the transformation law for the completely antisymmetric symbol. For the 4×4 LT matrix A, the mathematical definition of determinant gives

$$\det A \equiv \epsilon^{\mu\nu\lambda\kappa} A_{0\mu} A_{1\nu} A_{2\lambda} A_{3\kappa}. \tag{13.64}$$

The condition that $A^{\mathrm{T}} \eta A \eta = I$ together with well-known properties of determinants imply that

$$\det A = \pm 1 \quad \text{for any LT } A. \tag{13.65}$$

We write (13.64) in index notation as (defining $\det \Lambda \equiv \det A$)

$$\det \Lambda = \epsilon^{\mu\nu\lambda\kappa} \Lambda^0{}_\mu \Lambda^1{}_\nu \Lambda^2{}_\lambda \Lambda^3{}_\kappa. \tag{13.66}$$

By permuting the $(0, 1, 2, 3)$ superscripts in (13.66) and using the fact that $\det \Lambda = \pm 1$, we may show that

$$\epsilon^{\alpha\beta\gamma\delta} = (\det \Lambda) \Lambda^\alpha{}_\mu \Lambda^\beta{}_\nu \Lambda^\gamma{}_\lambda \Lambda^\delta{}_\kappa \epsilon^{\mu\nu\lambda\kappa}. \tag{13.67}$$

This transformation law characterizes $\epsilon^{\alpha\beta\gamma\delta}$ as a 4th-order contravariant *pseudotensor*. We may define the corresponding covariant pseudotensor in the usual way:

$$\begin{aligned} \epsilon_{\alpha\beta\gamma\delta} &= \eta_{\alpha\mu} \eta_{\beta\nu} \eta_{\gamma\lambda} \eta_{\delta\kappa} \epsilon^{\mu\nu\lambda\kappa} = (\det \eta) \epsilon^{\alpha\beta\gamma\delta} \\ &= -\epsilon^{\alpha\beta\gamma\delta}. \end{aligned} \tag{13.68}$$

(This is another example of a flat-space result that is not true in general.) Naturally, the covariant pseudotensor also has a factor of $\det \Lambda$ in its transformation law.

Using $\epsilon^{\alpha\beta\gamma\delta}$ we may now define the *dual pseudotensor* to the electromagnetic field tensor $F^{\mu\nu}$ as

$$\tilde{F}^{\alpha\beta} \equiv \frac{1}{2} \epsilon^{\alpha\beta\gamma\delta} F_{\gamma\delta} = \epsilon^{\alpha\beta\gamma\delta} \partial_\gamma A_\delta. \tag{13.69}$$

From the transformation laws (13.46) and (13.67) together with the identity (13.40), it follows that $\tilde{F}$ does indeed transform as a pseudotensor:

$$\tilde{F}'^{\mu\nu} = (\det \Lambda) \Lambda^\mu{}_\alpha \Lambda^\nu{}_\beta \tilde{F}^{\alpha\beta}. \tag{13.70}$$

Applying the contravariant 4-gradient to $\tilde{F}$ gives

$$\partial_\alpha \tilde{F}^{\alpha\beta} = -\partial_\alpha \tilde{F}^{\beta\alpha} = -\underbrace{\epsilon^{\beta\alpha\gamma\delta}}_{\text{antisym.}} \underbrace{\partial_\alpha \partial_\gamma}_{\text{sym.}} A_\delta, \tag{13.71}$$

which implies

$$\partial_\nu \tilde{F}^{\mu\nu} = 0. \tag{13.72}$$

Let us express these equations in terms of $\vec{E}$ and $\vec{B}$. Using our results from (13.69) as well as (13.60) we find

$$\tilde{F}^{0i} = \frac{1}{2}\epsilon^{0i\gamma\delta}F_{\gamma\delta} = \frac{1}{2}\epsilon_{ijk}F_{jk} = \frac{1}{2}\epsilon_{ijk}F^{jk} = \frac{1}{2}\epsilon_{ijk}(-\epsilon_{jk\ell}B_\ell) = -B_i. \tag{13.73}$$

Here we have used that $\epsilon^{0i\gamma\delta} \neq 0$ only when γ and δ are spatial indices (which we rename as j and k), and also that $\epsilon^{0ijk} = \epsilon_{ijk}$ where ϵ_{ijk} is the 3-dimensional antisymmetric symbol. Similarly, we find

$$\begin{aligned}\tilde{F}^{jk} &= \frac{1}{2}\epsilon^{jk\gamma\delta}F_{\gamma\delta} = \frac{1}{2}\epsilon_{ijk}(F_{0i} - F_{i0}) \\ &= \epsilon_{ijk}F_{0i} = \epsilon_{ijk}(-F^{0i}) = \epsilon_{ijk}E_i \\ &= E_i \quad (i, j, k \text{ cyclic}).\end{aligned} \tag{13.74}$$

By comparison with (13.60) we see that the dual operation which takes $F \to \tilde{F}$ also takes $\vec{E} \to \vec{B}$ and $\vec{B} \to -\vec{E}$. (Note this corresponds to the duality transformation (7.199) and (7.200) with $\xi = \pi/2$.) Since we know already that (13.59) gives the sourced Maxwell equations

$$\vec{\nabla} \cdot \vec{E} = 4\pi\rho; \qquad \vec{\nabla} \times \vec{B} - \frac{1}{c}\frac{\partial \vec{E}}{\partial t} = -\frac{4\pi}{c}\vec{J},$$

it follows immediately that (13.72) gives the dual equations

$$\vec{\nabla} \cdot \vec{B} = 0; \qquad -\vec{\nabla} \times \vec{E} - \frac{1}{c}\frac{\partial \vec{B}}{\partial t} = 0,$$

which are indeed the sourceless Maxwell equations as anticipated.

The reader might question whether it is possible to go a step farther and derive further physical equations by taking the dual of the dual of $F^{\alpha\beta}$. We leave as an exercise the proof that $\tilde{\tilde{F}}^{\alpha\beta} = -F^{\alpha\beta}$, so we have exhausted the possibilities afforded by the dual operation.

13.4 Interval invariance, restricted Lorentz transformations and causality

In Section 13.3 we stated that the defining property for LT's was that they left the 4-interval invariant. In this section we explore some physical consequences of this statement.

The invariance of the 4-interval implies that any surface of the form

$$(x^0)^2 - (x^1)^2 - (x^2)^2 - (x^3)^2 = s^2 \tag{13.75}$$

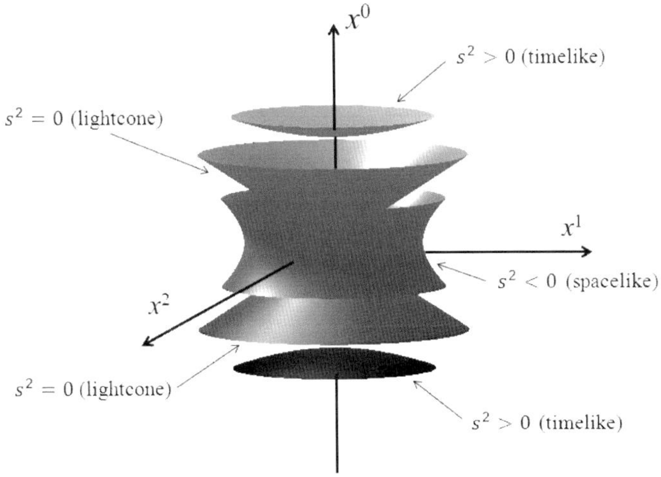

Fig. 13.1 Invariant surfaces under Lorentz transformation (constant 4-interval).

is invariant under Lorentz transformation, where s^2 is a fixed constant. (Although it is common practice to write the constant as s^2, note that s^2 can be negative. Note also that some authors use different sign conventions for s^2; others use τ^2 instead of s^2.) These surfaces are hyperboloids, as is shown in Figure 13.1 for two space dimensions plus time.

Based on Figure 13.1 we may distinguish four types of 4-vectors:

- 4-vectors that lie on the *light cone* defined by $s^2 = 0$ are called *lightlike*.
- 4-vectors that lie on hyperboloids with $s^2 < 0$ are called *spacelike*. Any two spacelike 4-vectors having the same 4-interval s^2 lie on the same hyperboloid.
- 4-vectors that lie on hyperboloids with $s^2 > 0$ are called *timelike*. From the figure, we may distinguish two types of timelike 4-vectors: those with $x^0 > 0$, and those with $x^0 < 0$. These two types correspond to two disconnected sheets associated with each 4-interval s^2.

This categorization has several remarkable physical consequences. Consider for instance a unit time vector for a particular observer at rest, which has invariant 4-interval $s^2 = c^2 > 0$ and thus is timelike. We have found in Section 13.1 that the same interval viewed in a moving frame with velocity

v along the $+x$ direction is proportional to $\{1, -v\}$ in (t', x') coordinates, which corresponds to $(c, -\vec{v})$ in 4-vector notation. Since the transformed vector is also timelike, we have that $c^2 - \vec{v}^2 > 0$. It follows that *the relative velocity between two observers can never exceed the speed of light.*

We may gain more insights by applying invariance to intervals between spacetime events. If x^μ and y^μ are spacetime events then $\Delta x^\mu \equiv y^\mu - x^\mu$ is also a 4-vector. If Δx^μ is lightlike, then the two events may be connected by a light signal wavefront. If Δx^μ is timelike then it is possible to travel from the earlier event to the other at sub-light speed, but if Δx^μ is spacelike this is impossible. Furthermore, if Δx^μ is spacelike then consistent with (13.75) it is always possible to find a reference frame in which $\Delta x'^0 = 0$. In such a reference frame, the two events are perceived as simultaneous. It follows that there is no meaningful sense in which one of these events can be said to occur "before" the other. We may conclude that in order for relativity to be a *causal* theory (i.e., one in which the present is determined by the past) it must be impossible for spacelike-separated events to influence each other. In other words, *no physical information can travel faster than the speed of light.*

In order to characterize the possible causal relationships between timelike-separated events, we first define *restricted Lorentz transformations* (RLT) as the set of all LT's that are continuously connected to the identity. Physically, RLT's represent the set of all reference frames that can achieved by continuously changing an observer's orientation and state of motion. So if a particular timelike interval is represented as Δx^μ and $\Delta x'^\mu$ respectively by two different observers, then $\Delta x'^\mu$ is related to Δx^μ via a RLT. Since RLT's are continuously connected to the identity, it follows that $\Delta x'^\mu$ must be connected to Δx^μ via a curve in 4-space, such that every point on the curve is also the transformation of Δx^μ under some RLT. Since s^2 is invariant under RLT's, this implies $\Delta x'^\mu$ and Δx^μ are joined by a continuous curve that lies in the same invariant hyperboloid. If Δx^μ is timelike, we saw from Figure 13.1 that its invariant hyperboloid has two disconnected sheets: but $\Delta x'^\mu$ must lie in the *same* sheet as Δx^μ. Since the two sheets are characterized by $x^0 > 0$ and $x^0 < 0$ respectively, we may conclude that the *time-ordering between timelike events is the same for all observers.*

The set of RLT's that we have used in this proof have a precise mathematical characterization. Clearly Lorentz boosts are RLT's, because we can let $\vec{\beta}$ vary continuously from zero to any fixed value. Rotations are also RLT's, as are any series of products involving Lorentz boosts and rotations.

In fact, there is a decomposition theorem[1] which states that every RLT can be written in a unique fashion as a rotation followed by a boost: in matrix notation, this means that $A_{\mathrm{RLT}} = A_{\mathrm{boost}}(\vec{\beta})R$, where we use R to represent a 4×4 rotation matrix,

$$R = \begin{pmatrix} 1 & 0 & 0 & 0 \\ 0 & & & \\ 0 & & a & \\ 0 & & & \end{pmatrix}, \tag{13.76}$$

and where a is the 3×3 rotation matrix from Section 8.8. This representation implies that an RLT can be uniquely specified by 6 real parameters (3 for the boost, 3 for the rotation). On the other hand, we know that the matrix A_{RLT} has 16 real entries. However, these entries are related by $(A_{\mathrm{RLT}})^{\mathrm{T}}\eta A_{\mathrm{RLT}} = \eta$, which amounts to 10 independent conditions since $(A_{\mathrm{RLT}})^{\mathrm{T}}\eta A_{\mathrm{RLT}}$ is symmetric. We are left with $16 - 10 = 6$ independent real parameters, as we would expect.

The above discussion suggests the question of whether or not there are other LT's in addition to RLT's. The answer is yes, as we will see in the following section.

13.5 Discrete transformations and the full Lorentz group

In our discussion of three-dimensional transformations in Section 8.8, we found spatial inversion to be an orthogonal transformation that is disconnected from the group of rotations. For inversion we found that the determinant of the transformation matrix is -1. A similar situation exists for LT's: in fact, the reader may verify that spatial inversion is also a LT. So we may distinguish between LT's with $\det \Lambda = 1$ and those with $\det \Lambda = -1$. (According to (13.65), these are the only two possibilities.) We denote these two types as *proper* and *improper* LT's respectively. It is clear that the product of any two proper LT's is also proper. Since Lorentz boosts and rotations are both proper, in view of the decomposition theorem for RLT's we have the result that *all RLT's are proper LT's.*

[1]See F. Scheck, *Mechanics*, 2nd ed., Springer-Verlag (Berlin) 1994, Section 4.5.

We may find another classification of LT's by considering the (0,0) component of the defining equation (13.39) for LT's:

$$\begin{aligned}\eta_{00}(\Lambda^0{}_0)^2 + \eta_{11}(\Lambda^1{}_0)^2 + \eta_{22}(\Lambda^2{}_0)^2 + \eta_{33}(\Lambda^3{}_0)^2 = 1\\ \Longrightarrow (\Lambda^0{}_0)^2 = 1 + \sum_i (\Lambda^i{}_0)^2 \geq 1.\end{aligned} \tag{13.77}$$

Similarly, we find

$$(\Lambda^0{}_0)^2 = 1 + \sum_i (\Lambda^0{}_i)^2 \geq 1. \tag{13.78}$$

So we may distinguish between *orthochronous* LT's, for which $\Lambda^0{}_0 \geq 1$, and *nonorthochronous* LT's, for which $\Lambda^0{}_0 \leq -1$. One example of a nonorthochronous LT is time inversion, which corresponds to a diagonal matrix with diagonal entries $(-1, 1, 1, 1)$. On the other hand, both Lorentz boosts and rotations are orthochronous. In view of the decomposition theorem for RLT's we may show that all RLT's are orthochronous if we can prove that the product of two orthochronous LT's is orthochronous. To show this, consider two orthochronous LT's Λ and Λ', and suppose that Λ'' is the resulting transformation when Λ is followed by Λ'. We may write this in index notation as

$$\Lambda''^{\mu}{}_{\nu} = \Lambda'^{\mu}{}_{\lambda}\Lambda^{\lambda}{}_{\nu}. \tag{13.79}$$

Taking the (0,0) component, we have

$$\Lambda''^0{}_0 = \Lambda'^0{}_0\Lambda^0{}_0 + \sum_i \Lambda'^0{}_i\Lambda^i{}_0. \tag{13.80}$$

From the Cauchy-Schwartz inequality we have

$$\left(\sum_i \Lambda'^0{}_i\Lambda^i{}_0\right)^2 \leq \left(\sum_i (\Lambda'^i{}_0)^2\right)\left(\sum_i (\Lambda^0{}_i)^2\right), \tag{13.81}$$

so using (13.77) and (13.78) it follows that

$$\begin{aligned}\left|\sum_i \Lambda'^0{}_i\Lambda^i{}_0\right| &\leq \sqrt{((\Lambda'^0{}_0)^2 - 1)\,((\Lambda^0{}_0)^2 - 1)}\\ &< \left|\Lambda'^0{}_0\Lambda^0{}_0\right|.\end{aligned} \tag{13.82}$$

Comparing (13.82) with (13.80), we find that $\Lambda''^0{}_0$ has the *same sign* as $\Lambda'^0{}_0\Lambda^0{}_0$. In particular, the product of two orthochronous LT's is an orthochronous LT. This completes the proof that *all RLT's are orthochronous.*

We have shown that RLT's are proper and orthochronous. It turns out that these characteristics provide a comprehensive specification of RLT's:

The set of RLT's is equal to the set of proper orthochronous Lorentz transformations. The previous discussion makes this plausible, but we will not provide a formal proof.

From our discussion we may conclude that there are four disconnected subsets within the group of LT's. Within each connected subset the sign of the determinant is constant, as is the sign of ${\Lambda^0}_0$. Furthermore, it can be shown that any two LT's within the same connected subset are related by an RLT. In fact, if Λ_1, Λ_2 are two LT's within the same subset, then $\Lambda_1 {\Lambda_2}^{-1}$ has determinant 1 and is orthochronous, so we may write

$$\Lambda_1 {\Lambda_2}^{-1} = \Lambda_{RLT} \implies \Lambda_1 = \Lambda_{RLT}\Lambda_2.$$

It follows that we can obtain all LT's within a given connected subset by choosing one particular LT within that subset, then multiplying it by all possible RLT's. In other words, each subset is characterized by a single representative transformation. These representative transformations are identified as *discrete transformations*. The conventional choices for these discrete transformations are given below.

For the case $\det \Lambda = -1$, ${\Lambda^0}_0 \geq 1$ we may choose the following representative transformation:

$$\Lambda_{PR} \equiv \begin{pmatrix} 1 & & & \\ & -1 & & \\ & & -1 & \\ & & & -1 \end{pmatrix}.$$

In component form, this gives

$$t' = t, \ \vec{x}\,' = -\vec{x}.$$

This is *complete inversion* (also known as *parity reversal*), which has already been discussed in Section 8.8. Of course observers cannot spatially invert themselves, so this transformation doesn't correspond to an active change of reference frame as RLT's do. Nonetheless, parity reversal is an important consideration in particle theories, as we shall see shortly. Space inversion symmetry is simply called *parity* and is determined by whether wave functions are even or odd under $\vec{x} \to -\vec{x}$ in quantum theories. As we saw previously, spatial inversion distinguishes between vectors and pseudovectors.

For the case $\det \Lambda = -1$, ${\Lambda^0}_0 \leq -1$ we may choose:

$$\Lambda_{TR} \equiv \begin{pmatrix} -1 & & & \\ & 1 & & \\ & & 1 & \\ & & & 1 \end{pmatrix}.$$

In component form, this gives

$$t' = -t, \ \vec{x}\,' = \vec{x}.$$

This discrete transformation corresponds to *time reversal*, which was also discussed in Section 8.8. Unlike space inversion, time reversal exchanges the disconnected timelike regions shown in Figure 13.1.

For the case $\det \Lambda = 1, \Lambda^0{}_0 \leq -1$ we may use the total space-time inversion:

$$\Lambda_{PT} \equiv \begin{pmatrix} -1 & & & \\ & -1 & & \\ & & -1 & \\ & & & -1 \end{pmatrix},$$

which in component form gives

$$t' = -t, \ \vec{x}\,' = -\vec{x}.$$

This transformation may be considered as combining both parity and time reversal.

The equations of electrodynamics are all invariant under both space and time inversions. This implies in the quantum version that the states and/or interactions also respect these symmetries. This has far-reaching consequences for allowed transitions in such theories. For example, an even-parity atomic state cannot transition to an odd-parity state via electromagnetic interactions. As another example, meson states in quantum chromodynamics or QCD (which is also invariant under space and time inversions except for the famous, but small, "θ" term, which breaks time reversal) are classified by their parity. An interesting distinction between these two discrete transformations which arises in quantum theory is that space inversion invariance gives rise to a conserved quantum number (parity), while time reversal invariance does not, although it does constrain the form of the interactions. An example here is that invariance under time reversal strictly rules out a neutron electric dipole moment. Actually, current theories predict a violation of this at a small level, and this is being sought experimentally. (Both time reversal and parity are violated by the weak interaction.) These are just a few examples of the enormously important role that discrete symmetries play in quantum theories.

13.6 Energy/momentum aspects of Lorentz transformations

So far we've been talking about the space/time implications of relativity. In this section, we consider what relativity has to say about momentum and energy.

In Section 13.1 we introduced the notion of invariant 4-intervals. The invariant interval between two infinitesimally-separated events is

$$ds^2 = c^2 dt^2 - d\vec{x}^{\,2} = dx_\mu dx^\mu. \tag{13.83}$$

By rescaling we obtain an interval with dimensions of time:

$$d\tau \equiv \frac{1}{c} ds = \frac{1}{c}\sqrt{c^2 dt^2 - d\vec{x}^{\,2}}, \tag{13.84}$$

which is known as *proper time*. The proper time interval between two space-time points separated by the 4-interval dx^μ has a physical interpretation. If the "world line" of a particle (or observer) passes through both points, then the proper time is the elapsed time experienced by the particle (or observer) in its own reference frame. Since $d\tau$ is an invariant scalar under Lorentz transformation, it follows that the quantity

$$u^\mu \equiv \frac{dx^\mu}{d\tau}, \tag{13.85}$$

called the *4-velocity* transforms as a contravariant 4-vector. u^μ must be a timelike vector if it is to describe the motion of a material particle. (Note that the proper time corresponding to a spacelike interval is imaginary.) Indeed, we may compute

$$u_\mu u^\mu = \frac{dx_\mu dx^\mu}{(d\tau)^2} = \left(\frac{ds}{d\tau}\right)^2 = c^2 > 0. \tag{13.86}$$

We may relate the spatial components of the 4-velocity to our conventional (nonrelativistic) notion of velocity as follows:

$$\vec{u} = \frac{d\vec{x}}{d\tau} = \frac{dt}{d\tau}\frac{d\vec{x}}{dt} \equiv \frac{dt}{d\tau}\vec{v}, \tag{13.87}$$

and using

$$\frac{d\tau}{dt} = \frac{1}{c}\sqrt{c^2 - \left(\frac{d\vec{x}}{dt}\right)^2} = \sqrt{1-\beta^2}, \tag{13.88}$$

we find that (using the definition of γ from (13.9))

$$\frac{d\vec{x}}{d\tau} = \frac{\vec{v}}{\sqrt{1-\beta^2}} = \gamma\vec{v} \underset{\beta \ll 1}{\rightarrow} \vec{v}. \tag{13.89}$$

Similarly, the time component of the 4-velocity in the nonrelativistic limit is

$$u^0 = \frac{dx^0}{d\tau} = c\frac{dt}{d\tau} = c\gamma \underset{\beta\ll 1}{\rightarrow} c. \tag{13.90}$$

Since the space components of u^μ reduce in the $\beta \ll 1$ limit to the ordinary velocity $\vec{v}$, a natural definition for *relativistic momentum* is

$$p^\mu \equiv mu^\mu, \tag{13.91}$$

which has components

$$\vec{p} = \gamma m\vec{v}; \qquad p^0 = \gamma mc. \tag{13.92}$$

We have an interpretation of $\vec{p}$, but what about p^0? Since p^μ is a timelike vector, $p^0 > 0$ in any (restricted) Lorentz frame. Thus it seems plausible that p^0 should be related to the particle's kinetic energy. However, notice that

$$p^0 = mc\left(1 + \frac{1}{2}\beta^2 + \ldots\right) = \frac{1}{c}\left(mc^2 + \frac{1}{2}mv^2 + \ldots\right). \tag{13.93}$$

This forces upon us the realization that there is an energy associated *mass* in addition to the (kinetic) energy associated with *motion*. We are thus motivated to define

$$\begin{aligned} \text{Total energy: } & E \equiv p^0 c; \\ \text{Rest energy: } & E_0 \equiv mc^2; \\ \text{Kinetic energy: } & E_{\text{kin}} \equiv E - E_0 = p^0 c - mc^2. \end{aligned}$$

(Note that the '0' subscript on E_0 is not a 4-index: rather, E_0 is the smallest possible total energy for a particle of mass m.) We have thus arrived at Einstein's crowning insight: "Mass and energy are therefore essentially alike; they are only different expressions for the same thing."[2]

Since p^μ in component form is $(E/c, \vec{p})$ it follows that conservation of momentum and energy can be combined into a single conservation law for the 4-vector p^μ. Naturally, p^μ transforms as

$$p^\mu = \Lambda^\mu{}_\nu p^\nu. \tag{13.94}$$

We may find an alternative expression for energy using the scalar Lorentz invariant:

$$p_\mu p^\mu = E^2/c^2 - \vec{p}^{\,2}. \tag{13.95}$$

[2] A. Einstein, *The Meaning of Relativity: Four lectures delivered at Princeton University* (trans. Edwin Plimpton Adams), Project Gutenberg EBook: `http://www.gutenberg.org/files/36276/36276-pdf.pdf`.

On the other hand, from (13.91) and (13.86) we have

$$p_\mu p^\mu = m^2 u_\mu u^\mu = m^2 c^2 \tag{13.96}$$

$$\Longrightarrow E = +\sqrt{\vec{p}^{\,2} c^2 + m^2 c^4}, \tag{13.97}$$

where we have taken the positive square root in (13.97) since $E = p^0 c > 0$. From (13.97) we see that one can have energy without mass ($E = |\vec{p}|c$ when $m = 0$), but one cannot have mass without energy.

In the following, we will also make use of the *relativistic acceleration* given by

$$\alpha^\mu \equiv \frac{d^2 x^\mu}{d\tau^2} = \frac{du^\mu}{d\tau}, \tag{13.98}$$

which by virtue of its definition is a 4-vector.

13.7 Relativistic kinematics in the context of linear and circular particle accelerators

Now that we know how to transform between inertial reference frames, we may gain a deeper understanding of some of our earlier results about radiation. Consider for instance the power radiated by an accelerating particle, which we discussed in Section 11.6. In the (instantaneous) rest frame of the radiating particle, the radiated power is given by the Larmor formula (11.99):

$$P(t) = \frac{2}{3}\frac{e^2}{c^3}[\ddot{\vec{x}}(t_0)]^2. \tag{13.99}$$

The generalization of the Larmor formula to arbitrary reference frames is the Liénard result for $P(t_r)$, Equation (11.160), which we originally took several pages to derive. But if we show that $P(t_r)$ is a Lorentz invariant, the Liénard result follows almost immediately from the Larmor formula. Exercise 13.7.1 shows that this relativistic generalization is given by

$$P_L(t_r) = -\frac{2}{3}\frac{e^2}{m^2 c^3}\frac{dp^\mu}{d\tau}\frac{dp_\mu}{d\tau}. \tag{13.100}$$

We may also revisit our earlier analysis of (also in Section 11.6) energy loss via radiation due to acceleration parallel and perpendicular to the direction of the particle's motion. Previously we found that energy losses were greater for parallel acceleration if $|\dot{\vec{v}}|$ and β are the same in both cases. However in practical applications such as in particle accelerators, the design is limited by the forces which can be applied. So let us consider the radiative losses associated with a given force applied parallel or perpendicular to the direction of motion.

In linear particle accelerators $\vec{v}$ and $\dot{\vec{v}}$ are parallel, and we obtain from (11.160)

$$P_{\rm lin} = \frac{2}{3}\frac{e^2}{c}\gamma^6\dot{\beta}^2. \tag{13.101}$$

We may re-express this in terms of $\dot{\vec{p}}$ by using

$$\frac{\vec{v}}{c} = \frac{\vec{p}c}{E}, \quad |\vec{p}c| = \sqrt{E^2 - m^2c^4}$$
$$\Longrightarrow \beta^2 = 1 - \left(\frac{mc^2}{E}\right)^2 \Longrightarrow \gamma^6\dot{\beta}^2 = \frac{\dot{E}^2}{\beta^2(mc^2)^2}.$$

Since acceleration and velocity are collinear, we may consider this as a one dimensional problem. Defining $p \equiv |\vec{p}|$, we have from taking a time derivative of $E^2 = p^2c^2 + m^2c^4$,

$$\dot{E} = \dot{p}\frac{pc^2}{E} = \dot{p}v \Longrightarrow \gamma^6\dot{\beta}^2 = \left(\frac{\dot{\vec{p}}}{mc}\right)^2,$$

which can be substituted into (13.101) to give

$$P_{\rm lin} = \frac{2}{3}\frac{e^2}{m^2c^3}\left(\frac{d\vec{p}}{dt}\right)^2. \tag{13.102}$$

On the other hand, for circular accelerators we have $\vec{v}$ and $\dot{\vec{v}}$ are perpendicular, and (11.160) gives

$$P_{\rm cir} = \frac{2}{3}\frac{e^2}{c}\gamma^4\dot{\beta}^2 = \frac{2}{3}\frac{e^2}{c^3}\gamma^4(\dot{\vec{v}})^2. \tag{13.103}$$

In this case γ is constant (since $|\vec{v}|$ is constant), and thus

$$\vec{p} = \gamma m\vec{v} \Longrightarrow \dot{\vec{p}} = \gamma m\dot{\vec{v}},$$

which gives us

$$P_{\rm cir} = \frac{2}{3}\frac{e^2}{c^3}\gamma^4(\dot{\vec{v}})^2 = \frac{2}{3}\frac{e^2}{m^2c^3}\gamma^2\left(\frac{d\vec{p}}{dt}\right)^2. \tag{13.104}$$

Note that under the same force, the radiative losses from circular acceleration are a factor of γ^2 times larger than those from linear acceleration. This ratio is reversed from what we found in Section 11.6, where we considered radiative losses associated with the same β and $\dot{\beta}$.

Let's get some realistic estimates for these radiation losses for linear and circular particle accelerators.

First we consider linear accelerators. We suppose that the accelerator exerts a constant force on the particle, so that $|\dot{\vec{p}}|$ is a constant. (It is possible to show that this models the case of a constant electric field acting on charged particle; see Exercise 13.7.3.) If the particle's time of flight is T, then the total radiated energy is

$$E_{\rm rad} = \frac{2}{3}\frac{e^2}{m^2c^3}\left(\frac{dp}{dt}\right)^2 T. \tag{13.105}$$

But $E^2 = p^2c^2 + m^2c^4$, so

$$\frac{dE}{dx} = \frac{pc^2}{Ev}\frac{dp}{dt} = \frac{dp}{dt} \implies E_{\rm rad} = \frac{2}{3}\frac{e^2}{m^2c^3}\left(\frac{dE}{dx}\right)^2 T. \tag{13.106}$$

If we suppose the input energy $E_{\rm lin}$ is sufficient to accelerate the particle to near-light speeds within a short initial distance, then the accelerator's length is $L \approx Tc$ and

$$E_{\rm in} = \frac{dE}{dx}L \approx \frac{dE}{dx}Tc \implies \frac{E_{\rm rad}}{E_{\rm in}} \simeq \frac{2}{3}\frac{e^2}{m^2c^4}\left(\frac{dE}{dx}\right). \tag{13.107}$$

For a typical accelerator, $dE/dx \sim 10\,\text{MeV/m}$: at such energies, electrons are relativistic within a few centimeters, and protons are relativistic within about 100 meters. In this case, our rough calculation for energy losses gives:

$$\text{electron:}\ \frac{E_{\rm rad}}{E_{\rm in}} \sim 10^{-14}; \qquad \text{proton:}\ \frac{E_{\rm rad}}{E_{\rm in}} \sim 10^{-20},$$

which shows that radiation effects are far too small to represent a significant source of energy loss.

For circular accelerators, we may obtain the energy supplied per cycle from (11.171):

$$\Delta E_{cy} = \frac{4\pi e^2}{3R}\frac{\beta^3}{(1-\beta^2)^2} = \frac{4\pi e^2}{3R}\left(\frac{E}{mc^2}\right)^4\beta^3. \tag{13.108}$$

Let's consider some practical situations. The world's highest energy particle collider, the circular accelerator/storage ring at the European Organization for Nuclear Research (CERN) called the Large Hadron Collider (LHC) spans the border of France and Switzerland near Geneva in an underground tunnel, and has a radius of $R \approx 4.3 \times 10^5$ cm (4.3 kilometers). It accelerates and maintains beams of counter circulating protons using strong magnets. For its ultimate design energy of 7 TeV protons per beam, we obtain $\Delta E_{cy} \approx 4.5 \times 10^3$ eV for the energy radiated per cycle of revolution, which is relatively small. To put this in perspective, it would take about 1.5×10^9 revolutions for a proton to radiate its original energy E, if the system is constantly resupplied. Each revolution takes $2\pi R/c \approx 10^{-4}$ seconds, so this process would take $1.5 \times 10^9 \times 10^{-4}$ sec or about 1.7 days to occur. Thus, radiation losses are not a limiting factor on achieving higher energies at the LHC. However, other factors impose practical limitations. The nonrelativistic expression for the orbital radius,

$$R = \frac{|\vec{p}|c}{|e\vec{B}|}, \tag{13.109}$$

for a general charge e turns out to be relativistically correct as well (see Exercise 13.7.4). In the extreme relativistic limit $|\vec{p}|c \approx E$, so for a fixed radius R we have

$$|\vec{B}| \approx \frac{E}{|e|R}. \tag{13.110}$$

For the LHC with 7 TeV protons, this gives $|\vec{B}| \approx 5.4 \times 10^4$ Gauss $=$ 5.4 Tesla. (Since the entire circumference of the LHC tunnel is not filled with bending magnets, a higher field in each magnet is actually needed.) So in practice, proton energies in storage rings are limited by available magnets. The situation with electrons is somewhat different, as you will investigate in Exercise 13.7.5.

13.8 Relativistic Lagrangian formalism

Before presenting the Lagrangian for electrodynamics, we first provide a quick review of the variational formulation of classical mechanics, also known as *Hamilton's principle*. The equations of motion for a system of point particles acted on by conservative, time-independent forces can be formulated as the solution to an extremum problem as follows. Suppose the state of the system is specified by coordinates q_i which are functions of time t. (The index i characterizes both the different spatial coordinates and the different particles in the system.) Given a Lagrangian function $L(q_i, \dot{q}_i)$, we may define the associated *action* as

$$S(q_i(t)) \equiv \int_{t_1}^{t_2} L(q_i(t), \dot{q}_i(t))dt. \tag{13.111}$$

The action is a *functional*, that is, it is a real-valued function defined on all possible functions $q_i(t)$. The problem of minimizing S among all possible

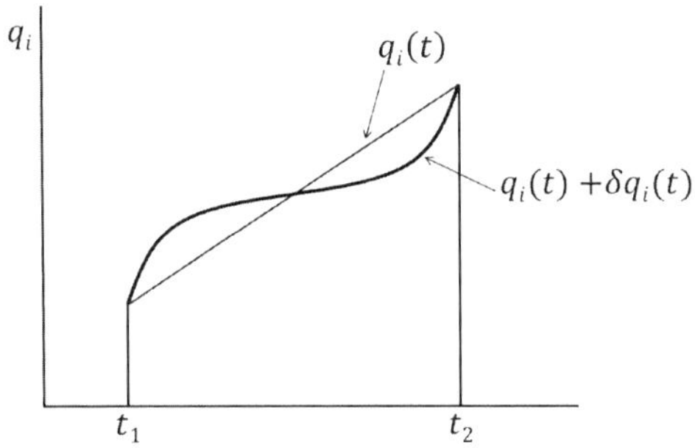

Fig. 13.2 Path variation for Hamilton's principle.

paths $\{q_i(t)\}_{i=1,\ldots,N}$ with fixed endpoints (so that $q_i(t_1)$ and $q_i(t_2)$ are fixed for all i) is expressed mathematically as:

$$\delta S = 0. \tag{13.112}$$

That is, when S is minimized it is *stationary* with respect to variations in the path $\{q_i(t)\}_{i=1,\ldots,N}$ such as shown in Figure 13.2. To solve this variational problem, we introduce the following notation for path variations:

$$\delta q_i = \eta_i(t) d\alpha, \qquad \text{where} \quad \eta_i(t_1) = \eta_i(t_2) = 0.$$

The condition (13.112) then implies

$$0 = \left.\frac{dS(\alpha)}{d\alpha}\right|_{\alpha=0} = \int_{t_1}^{t_2} dt \left[\sum_i \frac{\partial L}{\partial q_i}\frac{\partial q_i}{\partial \alpha} + \sum_i \frac{\partial L}{\partial \dot{q}_i}\frac{\partial \dot{q}_i}{\partial \alpha}\right]_{\alpha=0}. \tag{13.113}$$

Using the fact that

$$\frac{\partial \dot{q}}{\partial \alpha} = \frac{d}{dt}\frac{\partial q}{\partial \alpha}, \tag{13.114}$$

we may integrate the last term in (13.113) by parts to get

$$0 = \left.\frac{dS(\alpha)}{d\alpha}\right|_{\alpha=0} = \int_{t_1}^{t_2} dt \sum_i \left(\frac{\partial L}{\partial q_i} - \frac{d}{dt}\left(\frac{\partial L}{\partial \dot{q}_i}\right)\right)\eta_i(t). \tag{13.115}$$

Since this must be true for arbitrary $\eta_i(t)$, we obtain the *Euler-Lagrange equations*:

$$\frac{\partial L}{\partial q_i} - \frac{d}{dt}\left(\frac{\partial L}{\partial \dot{q}_i}\right) = 0. \tag{13.116}$$

Notice that different Lagrangians can give rise to the same equations of motion. In particular, if we modify any Lagrangian by adding a total time derivative term of the form

$$L' \equiv L + \frac{d}{dt}F(q_i) \tag{13.117}$$

where F is an arbitrary function of the coordinates q_i, then the action corresponding to L' is

$$S' \equiv \int_{t_1}^{t_2} \left(L + \frac{d}{dt} F(q_i) \right) dt = S + F(q_i(t_2)) - F(q_i(t_1)). \qquad (13.118)$$

It follows that $\delta S' = \delta S$, so the Euler-Lagrange equations are unchanged.

Recall that in mechanics the q_i can be used to represent an ensemble of particles with locations $\{\vec{x}_\alpha\}$ and masses m_α, where α indexes the particles. In this case, $L = T - U$ where $T = \frac{1}{2} \sum_\alpha m_\alpha \dot{\vec{x}}_\alpha^2$ and $U = \sum_{\beta<\gamma} U_{\beta\gamma}(|\vec{x}_\beta - \vec{x}_\gamma|)$, and the Euler-Lagrange equations become Newton's equations for the particles indexed by α:

$$m_\alpha \frac{d^2 \vec{x}_\alpha}{dt^2} = -\vec{\nabla}_\alpha \sum_{\beta<\gamma} U_{\beta\gamma}. \qquad (13.119)$$

The above derivation of the Euler-Lagrange equations can be re-expressed in the following simplified notation:

$$\delta S = \delta \int_{t_1}^{t_2} dt L = \int_{t_1}^{t_2} dt \delta L, \qquad (13.120)$$

$$\delta L = \sum_i \frac{\partial L}{\partial q_i} \delta q_i + \sum_i \frac{\partial L}{\partial \dot{q}_i} \delta \dot{q}_i, \qquad (13.121)$$

$$\delta \dot{q}_i = \delta \frac{dq_i}{dt} = \frac{d(q_i + \delta q_i)}{dt} - \frac{dq_i}{dt} = \frac{d}{dt} \delta q_i. \qquad (13.122)$$

Using this new notation, we may write

$$\delta L = \sum_i \left(\frac{\partial L}{\partial q_i} - \frac{d}{dt} \left(\frac{\partial L}{\partial \dot{q}_i} \right) \right) \delta q_i + \sum_i \frac{d}{dt} \left(\frac{\partial L}{\partial \dot{q}_i} \delta q_i \right). \qquad (13.123)$$

When δL is integrated to form δS, the last term in (13.123) vanishes due to the condition that $\delta q_i(t_1) = \delta q_i(t_2) = 0$. Then since δq_i is arbitrary, we obtain the Euler-Lagrange equations as before.

We now consider whether it is possible to obtain a similar variational formulation for the electromagnetic field equations. In this case, we are trying to get an equation for fields which are functions of space and time. This indicates that fields should be the dependent variables, and spacetime positions x^μ should be the independent (vector) variable. So we should look for an action of the general form

$$S(\phi_k) = \int_\Omega d^4x \, \mathcal{L}(\phi_k, \partial_\mu \phi_k), \qquad (13.124)$$

where Ω is an invariant region in spacetime, $d^4x = cdt\, d^3x$, ϕ_k are fields (functions of x^μ) and k is an index that may include spacetime indices as well as possible internal degree of freedom. (It turns out this formalism covers scalar, vector and tensor fields, which correspond to spins 0, 1, and 2 in field theory.) $\mathcal{L}$ is the Lagrangian density, but we will revert to simply calling it the Lagrangian in the following.

Let's go through the argument again for the Euler-Lagrange equations with this new action. We obtain

$$\delta S = \delta \int_\Omega d^4x\, \mathcal{L} = \int_\Omega d^4x\, \delta\mathcal{L}, \tag{13.125}$$

where $\delta\mathcal{L}$ is that change in the Lagrangian due to a variation in the fields $\delta\phi_k$ such that

$$\delta\phi_k|_{\partial\Omega} = 0, \tag{13.126}$$

and $\partial\Omega$ denotes the boundary of Ω. In analogy to (13.121) we have

$$\delta\mathcal{L} = \frac{\delta\mathcal{L}}{\delta\phi_k}\delta\phi_k + \frac{\delta\mathcal{L}}{\delta(\partial_\mu\phi_k)}\delta(\partial_\mu\phi_k), \tag{13.127}$$

and in analogy to (13.122) we have

$$\delta(\partial_\mu\phi_k) = \partial_\mu(\delta\phi_k). \tag{13.128}$$

Then

$$\delta\mathcal{L} = \left(\frac{\delta\mathcal{L}}{\delta\phi_k} - \partial_\mu\left(\frac{\delta\mathcal{L}}{\delta(\partial_\mu\phi_k)}\right)\right)\delta\phi_k + \partial_\mu\left(\frac{\delta\mathcal{L}}{\delta\partial_\mu\phi_k}\delta\phi_k\right). \tag{13.129}$$

At this point, we may recall the divergence theorem (a.k.a. Gauss' theorem) in 3 dimensions:

$$\int_V d^3x\, \vec{\nabla}\cdot\vec{F}(\vec{x}) = \int_S d^2\sigma\, \hat{n}\cdot\vec{F}(x), \tag{13.130}$$

where $\hat{n}$ is the outward unit normal to the surface S that bounds volume V. The 4-dimensional version (which is proved similarly) is

$$\int_\Omega d^4x\, \partial_\mu F^\mu = \int_{\partial\Omega} d^3\sigma\, \hat{n}_\mu F^\mu, \tag{13.131}$$

where $\hat{n}_\mu$ is the outward unit normal to the boundary $\partial\Omega$. Applying this theorem with $F^\mu \equiv (\delta\mathcal{L}/\delta\partial_\mu\phi_k)\delta\phi_k$ and using (13.126), we conclude that the final term in (13.129) makes no contribution to the action. The variational condition $\delta S = 0$ thus gives

$$\frac{\delta\mathcal{L}}{\delta\phi_k} - \partial_\mu\left(\frac{\delta\mathcal{L}}{\delta(\partial_\mu\phi_k)}\right) = 0. \tag{13.132}$$

As an example of this formalism we consider Lagrangian densities that depend on a single scalar field, $\phi_k \to \Phi$. We assume that $\mathcal{L}$ is real and can be expressed as the sum of Lorentz-invariant terms that are at most second order in the field and its first derivative. We further assume a linear coupling between the field Φ and a source S. Any overall multiplicative constant will have no effect on the equations of motion and can be chosen by convention. Another constant will determine the relative field/source normalization. Given these choices, the most general $\mathcal{L}$ can be written in the form

$$\mathcal{L} = \frac{1}{2}(\partial_\lambda \Phi \, \partial^\lambda \Phi - \mu^2 \Phi^2) + \Phi S = \frac{1}{2}(\eta^{\lambda\kappa} \partial_\lambda \Phi \, \partial_\kappa \Phi - \mu^2 \Phi^2) + \Phi S. \quad (13.133)$$

In this case, we have

$$\frac{\delta \mathcal{L}}{\delta \Phi} = -\mu^2 \Phi + S, \quad (13.134)$$

$$\partial_\mu \left(\frac{\delta \mathcal{L}}{\delta(\partial_\mu \Phi)} \right) = \partial_\mu \left[\frac{1}{2} \eta^{\lambda\kappa} \delta^\mu_\lambda \, \partial_\kappa \Phi + \frac{1}{2} \eta^{\lambda\kappa} \partial_\lambda \Phi \, \delta^\mu_\kappa \right] = \partial_\mu [\eta^{\lambda\mu} \partial_\lambda \Phi]$$
$$= \partial^\lambda \partial_\lambda \Phi = \Box \Phi, \quad (13.135)$$

so the Euler-Lagrange equations are

$$(\Box + \mu^2)\Phi = S. \quad (13.136)$$

This is the *Klein-Gordon equation* (seen before in Exercise 9.8.1) with a source given by $S(\vec{x}, t)$. One must choose $\mu > 0$ for a particle interpretation ($\mu = mc/\hbar$ where m is the mass) with the correct relativistic dispersion relation (cf. Exercise 13.9.1(b)).

13.9 Construction of Lagrangian for the electromagnetic field

After the scalar field, the next simplest example of a classical field Lagrangian is for vector fields. In this section we will demonstrate the striking result that the Maxwell equations are *uniquely* specified by the Lagrangian formalism for a vector field, given a few simplifying conditions that parallel the conditions we imposed on the scalar field case. As in the scalar field case, we assume that the Lagrangian is real and can be expressed as the sum of Lorentz-invariant terms that are at most second order in the field A^μ and its first derivatives $\partial^\nu A^\mu$. We also assume a linear coupling between the field A^μ and a vector source, which we write suggestively as J^μ/c. These conditions imply the following form for the Lagrangian:

$$\mathcal{L} = c_1\partial_\mu A^\nu\,\partial^\mu A_\nu + c_2\partial_\mu A^\nu\,\partial_\nu A^\mu + c_3(\partial_\mu A^\mu)^2 + c_4 A_\mu A^\mu - \frac{1}{c}A_\mu J^\mu + c_5\partial_\mu A^\mu. \tag{13.137}$$

Note that the c_3 term differs from the c_2 term by a total divergence. Using an argument similar to that used in (13.117)–(13.118) above, we conclude that this total divergence term makes no contribution to the equation of motion. Thus we may set $c_3 = 0$ without loss of generality.

At this point we impose the additional condition that the action be gauge-invariant; that is, the transformation

$$A_\mu \to A_\mu + \partial_\mu\Lambda, \tag{13.138}$$

should leave the action unchanged (up to boundary terms that do not affect the field equations). Applying this transformation to lowest order (which is equivalent to considering infinitesimal $\Lambda \to \delta\Lambda$) to the Lagrangian gives

$$\mathcal{L} \to \mathcal{L} + 2(c_1+c_2)[\partial_\mu\partial_\nu\Lambda\,\partial^\mu A^\nu] + 2c_4\partial_\mu\Lambda A^\mu + \frac{1}{c}\partial_\mu\Lambda\,J^\mu + c_5\Box\Lambda. \tag{13.139}$$

Gauge invariance requires $c_1 = -c_2$ from the first set of additional terms. At first sight it might seem that gauge invariance also implies that the source term must be set to zero. Fortunately we can avoid this by imposing appropriate conditions on the source J^μ. The gauge transformation (13.138) introduces the following source-dependent terms in to the action:

$$\frac{1}{c}\int_\Omega d^4x\,[\partial_\mu(\Lambda J^\mu) - \Lambda\partial_\mu J^\mu].$$

The second term in the integrand vanishes if we require that $\partial_\mu J^\mu = 0$. The variation of the first term also vanishes given this requirement, as can be shown from the divergence theorem (13.131) and the fact that Λ is constant on the boundary (which follows from Hamilton's principle). We thus obtain a beautiful result: current conservation is a consequence of gauge invariance!

The remaining terms in (13.139) must vanish to ensure gauge invariance, which gives $c_4 = c_5 = 0$. Given the already fixed value of the current coupling term, the conventional value of the last unfixed constant is $c_1 = -1/(8\pi)$. This yields

$$\mathcal{L} = -\frac{1}{8\pi}(\partial_\mu A_\nu\partial^\mu A^\nu - \partial_\mu A_\nu\partial^\nu A^\mu) - \frac{1}{c}A_\mu J^\mu, \tag{13.140}$$

which in view of (13.58) may be written very compactly as

$$\mathcal{L} = -\frac{1}{16\pi} F_{\mu\nu} F^{\mu\nu} - \frac{1}{c} A_\mu J^\mu. \tag{13.141}$$

From (13.60) we may obtain

$$F_{\mu\nu} F^{\mu\nu} = -2\vec{E}^2 + 2\vec{B}^2, \tag{13.142}$$

and the above can be written in more explicit terms as

$$\mathcal{L} = \frac{1}{8\pi}(\vec{E}^2 - \vec{B}^2) - \Phi\rho + \frac{1}{c}\vec{A}\cdot\vec{J}, \tag{13.143}$$

where we have used the familiar expressions for A_μ and J^μ.

This Lagrangian is not unique; in analogy to (13.117), we may define an alternative Lagrangian $\mathcal{L}'$ as

$$\mathcal{L}' = \mathcal{L} + \partial_\mu M^\mu(A^\nu, \partial_\lambda A^\nu), \tag{13.144}$$

and as before the Euler-Lagrange equations are unchanged. These equations may be computed explicitly from (13.140), which may be rewritten as

$$\mathcal{L} = -\frac{1}{8\pi}[\partial_\mu A_\beta \partial_\alpha A_\nu \eta^{\beta\nu}\eta^{\mu\alpha} - \partial_\mu A_\beta \partial_\alpha A_\nu \eta^{\mu\nu}\eta^{\alpha\beta}] - \frac{1}{c} A_\mu J^\mu. \tag{13.145}$$

Now

$$\frac{\delta\mathcal{L}}{\delta A_\lambda} = -\frac{1}{c} J^\lambda, \tag{13.146}$$

and

$$\begin{aligned} -\partial_\kappa\left(\frac{\delta\mathcal{L}}{\delta(\partial_\kappa A_\lambda)}\right) &= \frac{1}{8\pi}\partial_\kappa[\delta^\kappa_\mu\delta^\lambda_\beta\,\partial_\alpha A_\nu \eta^{\beta\nu}\eta^{\mu\alpha} + \delta^\kappa_\alpha\delta^\lambda_\nu\,\partial_\mu A_\beta \eta^\beta \eta^{\beta\nu}\eta^{\mu\alpha} \\ &\quad - \delta^\kappa_\mu\delta^\lambda_\beta\,\partial_\alpha A_\nu \eta^{\mu\nu}\eta^{\alpha\beta} - \delta^\kappa_\alpha\delta^\lambda_\nu\,\partial_\mu A_\beta \eta^{\mu\nu}\eta^{\alpha\beta}] \\ &= \frac{1}{4\pi}\partial_\kappa[\partial^\kappa A^\lambda - \partial^\lambda A^\kappa]. \end{aligned} \tag{13.147}$$

This gives us

$$\frac{\delta\mathcal{L}}{\delta A_\lambda} - \partial_\kappa\left(\frac{\delta\mathcal{L}}{\delta(\partial_\kappa A_\lambda)}\right) = -\frac{1}{c}J^\lambda + \frac{1}{4\pi}[\Box A^\lambda - \partial^\lambda\partial_\kappa A^\kappa] = 0, \tag{13.148}$$

or

$$\Box A^\lambda - \partial^\lambda(\partial_\kappa A^\kappa) = \frac{4\pi}{c} J^\lambda. \tag{13.149}$$

This equation is only consistent if $\partial_\lambda J^\lambda = 0$, which we saw was required by invariance of the action under gauge transformations.

We encourage the reader to take a few moments to fully appreciate what we have shown. Starting with the most general possible scalar Lagrangian that is second-order in fields and first-order in sources, and assuming gauge invariance, we have shown that the usual form of the 4-potential is the *only* possible solution, up to the numerical value of c! In some sense, Maxwell's equations are the "only possible" equations they could be. Actually, if we allow pseudoscalar terms we could add terms to $\mathcal{L}$ of the form $\tilde{F}^{\mu\nu}F_{\mu\nu}$. (Note $F^{\mu\nu}\tilde{F}_{\mu\nu} = \tilde{F}^{\mu\nu}F_{\mu\nu}$, and $\tilde{F}^{\mu\nu}\tilde{F}_{\mu\nu} = -F^{\mu\nu}F_{\mu\nu}$ so these terms are already included.) However, in Exercise 13.9.3 you will show that

$$\tilde{F}^{\mu\nu}F_{\mu\nu} = -4\vec{E}\cdot\vec{B} = \partial_\lambda S^\lambda. \tag{13.150}$$

Since this is a total differential, as we have seen before it has no effect on the variation of the action and hence doesn't effect the equations of motion.

13.10 Covariant form of the energy-momentum tensor

In Section 8.2 we showed how the Lorentz force law could be interpreted as a statement of momentum conservation. In order to do this we introduced the Maxwell stress tensor, which contains the momentum information associated with an electromagnetic field. In this section, we derive a relativistically form-invariant generalization of the Maxwell stress tensor, and use it to give a conservation of momentum equation that is also form-invariant. Continuing in the spirit of the previous section, we will show that this relativistic Maxwell stress tensor (denoted by $T^{\mu\nu}$) is uniquely specified by a small number of conditions. These conditions, which are similar to those we imposed on the Lagrangian in the previous section, are as follows.

First, we assume that $T^{\mu\nu}$ is a true Lorentz form invariant tensor with no pseudotensor terms.

Second, we assume that $T^{\mu\nu}$ depends only on terms that are at most second order in the field derivatives $\partial_\mu A_\nu$.

Third, we assume that $T^{\mu\nu}$ is symmetric in μ, ν. This assumption is justified by two reasons. First, it provides us with a conserved angular momentum, as you will show in Exercise 13.10.1. Also, it has the right number of independent components:

	stress		*energy*		*momentum*		*Total*
	T_{ij}		T_{00}		T_{0i}		$T_{\alpha\beta}$
$\Longrightarrow$	(6)	+	(1)	+	(3)	=	(10)

Fourth, we assume that $T^{\mu\nu}$ is gauge invariant under the transformation $A_\lambda \to A_\lambda + \partial_\lambda \Lambda$ for arbitrary infinitesimal Λ.

Fifth, we assume that $T^{\mu\nu}$ is conserved, so that $\partial_\mu T^{\mu\nu} = 0$ when $J^\nu = 0$. For $\mu = 0, 1, 2, 3$ this gives us four equations, which correspond to energy conservation ($\mu = 0$) and momentum conservation ($\mu = 1, 2, 3$).

Finally, we assume that $T^{\mu\nu}$ reduces correctly to our previous expression (8.37) for T_{ij} when $\mu, \nu = 1, 2, 3$. (This condition gives us the overall normalization.)

Now let's roll up our sleeves and begin with the derivation. Symmetry and either conservation or gauge invariance prohibit any terms that are first order in the derivatives of A_μ. The remaining possible gauge-invariant terms that are second-order in derivatives are

$$\begin{aligned} T^{\mu\nu} &= c_1 \eta^{\mu\nu} F^{\lambda\kappa} F_{\lambda\kappa} + c_2 \partial^\mu A_\alpha \partial^\nu A^\alpha + c_3 \partial^\mu A_\alpha \partial^\alpha A^\nu \\ &\quad + c_4 \partial^\nu A_\alpha \partial^\alpha A^\mu + c_5 \partial^\mu A^\nu (\partial_\lambda A^\lambda) + c_6 \partial^\nu A^\mu (\partial_\lambda A^\lambda) + c_7 \partial_\lambda A^\mu \partial^\lambda A^\nu. \end{aligned} \tag{13.151}$$

To ensure symmetry of $T^{\mu\nu}$, we compare

$$\begin{aligned} T^{\nu\mu} &= c_1 \eta^{\mu\nu} F^{\lambda\kappa} F_{\lambda\kappa} + c_2 \partial^\nu A_\alpha \partial^\mu A^\alpha + c_3 \partial^\nu A_\alpha \partial^\alpha A^\mu \\ &\quad + c_4 \partial^\mu A_\alpha \partial^\alpha A^\nu + c_5 \partial^\nu A^\mu (\partial_\lambda A^\lambda) + c_6 \partial^\mu A^\nu (\partial_\lambda A^\lambda) + c_7 \partial_\lambda A^\nu \partial^\lambda A^\mu, \end{aligned} \tag{13.152}$$

and obtain $c_3 = c_4,\ c_5 = c_6$. We then have

$$\begin{aligned} T^{\mu\nu} &= c_1 \eta^{\mu\nu} F^{\lambda\kappa} F_{\lambda\kappa} + c_2 \partial^\nu A_\alpha \partial^\mu A^\alpha + c_3 [\partial^\nu A_\alpha \partial^\alpha A^\mu + \partial^\mu A_\alpha \partial^\alpha A^\nu] \\ &\quad + c_5 [\partial^\nu A^\mu + \partial^\mu A^\nu] \partial_\lambda A^\lambda + c_7 \partial_\lambda A^\nu \partial^\lambda A^\mu. \end{aligned} \tag{13.153}$$

Now applying the gauge transformation $A_\lambda \to A_\lambda + \partial_\lambda \Lambda$ gives us

$$\begin{aligned} T^{\mu\nu} \to T^{\mu\nu} &+ c_2 [\overbrace{\partial^\mu \partial_\alpha \Lambda \partial^\nu A^\alpha}^{(1)} + \overbrace{\partial^\mu A_\alpha \partial^\nu \partial^\alpha \Lambda}^{(2)}] \\ &+ c_3 [\overbrace{\partial^\mu \partial_\alpha \Lambda \partial^\alpha A^\nu}^{(3)} + \overbrace{\partial^\mu A_\alpha \partial^\alpha \partial^\nu \Lambda}^{(2)} + \overbrace{\partial^\nu \partial_\alpha \Lambda \partial^\alpha A^\mu}^{(4)} + \overbrace{\partial^\nu A_\alpha \partial^\alpha \partial^\mu \Lambda}^{(1)}] \\ &+ c_5 [\partial^\mu \partial^\nu \Lambda + \partial^\nu \partial^\mu \Lambda] \partial_\lambda A^\lambda + c_5 [\partial^\mu A^\nu + \partial^\nu A^\mu] \Box \Lambda \\ &+ c_7 [\overbrace{\partial_\lambda \partial^\mu \Lambda \partial^\lambda A^\nu}^{(3)} + \overbrace{\partial_\lambda A^\mu \partial^\lambda \partial^\nu \Lambda}^{(4)}]. \end{aligned} \tag{13.154}$$

Cancellation of like terms leads us to $c_2 + c_3 = 0$, $c_3 + c_7 = 0$, and $c_5 = 0$. What's left is

$$T^{\mu\nu} = c_1\eta^{\mu\nu}F^{\lambda\kappa}F_{\lambda\kappa} + c_2[\overbrace{\partial^\mu A_\alpha\partial^\nu A^\alpha}^{(1)} - \overbrace{\partial^\mu A_\alpha\partial^\alpha A^\nu}^{(2)}$$
$$- \overbrace{\partial^\nu A_\alpha\partial^\alpha A^\mu}^{(3)} + \overbrace{\partial_\alpha A^\mu\partial^\alpha A^\nu}^{(4)}]. \tag{13.155}$$

Notice that:

$$F^{\mu\alpha}F_\alpha{}^\nu = (\partial^\mu A^\alpha - \partial^\alpha A^\mu)(\partial_\alpha A^\nu - \partial^\nu A_\alpha)$$
$$= \overbrace{\partial^\mu A^\alpha\partial_\alpha A^\nu}^{(2)} - \overbrace{\partial^\mu A^\alpha\partial^\nu A_\alpha}^{(1)} + \overbrace{\partial^\alpha A^\mu\partial^\nu A_\alpha}^{(3)} - \overbrace{\partial^\alpha A^\mu\partial_\alpha A^\nu}^{(4)}, \tag{13.156}$$

so we may rewrite (13.155) as

$$T^{\mu\nu} = c_1\eta^{\mu\nu}F^{\lambda\kappa}F_{\lambda\kappa} - c_2F^{\mu\alpha}F_\alpha{}^\nu. \tag{13.157}$$

Next, we form the conservation equation $\partial_\mu T^{\mu\nu} = 0$ (assuming $J^\mu = 0$):

$$\partial_\mu T^{\mu\nu} = c_1\partial^\nu(F^{\lambda\kappa}F_{\lambda\kappa}) - c_2\underbrace{\partial_\mu F^{\mu\alpha}}_{\frac{4\pi}{c}J^\alpha}F_\alpha{}^\nu - c_2\underbrace{F^{\mu\alpha}\partial_\mu F_\alpha{}^\nu}_{(\partial^\lambda F^{\kappa\nu})F_{\lambda\kappa}}. \tag{13.158}$$

The first and last terms in $\partial_\mu T^{\mu\nu}$ above are related, since

$$\partial^\nu(F^{\lambda\kappa}F_{\lambda\kappa}) = 2(\partial^\nu F^{\lambda\kappa})F_{\lambda\kappa}. \tag{13.159}$$

Furthermore, it is possible to show directly that

$$\partial^\alpha F^{\beta\gamma} + \partial^\beta F^{\gamma\alpha} + \partial^\gamma F^{\alpha\beta} = 0, \tag{13.160}$$

which leads to

$$(\partial^\nu F^{\lambda\kappa})F_{\lambda\kappa} = -(\partial^\lambda F^{\kappa\nu} + \partial^\kappa F^{\nu\lambda})F_{\lambda\kappa} = -2(\partial^\lambda F^{\kappa\nu})F_{\lambda\kappa}. \tag{13.161}$$

With these changes, (13.158) becomes

$$\partial_\mu T^{\mu\nu} = -4c_1(\partial^\lambda F^{\kappa\nu})F_{\lambda\kappa} - c_2(\partial^\lambda F^{\kappa\nu})F_{\lambda\kappa} - c_2\frac{4\pi}{c}J^\alpha F_\alpha{}^\nu. \tag{13.162}$$

Since we are requiring that this vanish for $J^\alpha = 0$, we have $4c_1 + c_2 = 0$ and thus

$$\begin{cases} T^{\mu\nu} = -c_2[F^{\mu\alpha}F_\alpha{}^\nu + \frac{1}{4}\eta^{\mu\nu}F^{\lambda\kappa}F_{\lambda\kappa}], \\ \partial_\mu T^{\mu\nu} = -c_2\dfrac{4\pi}{c}J^\alpha F_\alpha{}^\nu. \end{cases} \tag{13.163}$$

We may determine the value of c_2 by comparing (13.163) with our previous expression for the Maxwell stress tensor in (8.37). Using (13.142) and

(13.60) we may rewrite the space-space components of $T^{\mu\nu}$ in terms of $\vec{E}$ and $\vec{B}$ as

$$\begin{aligned} T^{ij} &= -c_2\left[-\frac{1}{4}\delta^{ij}(-2\vec{E}^2+2\vec{B}^2)+F^{i0}F_0{}^j+F^{ik}F_k{}^j\right], \\ &= -c_2\left[\frac{1}{2}\delta^{ij}(\vec{E}^2+\vec{B}^2)-E_iE_j-B_iB_j\right]. \end{aligned} \tag{13.164}$$

By comparison with (8.37), we have $-c_2 = 1/4\pi$, so

$$T^{\mu\nu} = \frac{1}{4\pi}\left[F^{\mu\alpha}F_\alpha{}^\nu + \frac{1}{4}\eta^{\mu\nu}F^{\lambda\kappa}F_{\lambda\kappa}\right], \tag{13.165}$$

$$\partial_\mu T^{\mu\nu} = \frac{1}{c}J^\alpha F_\alpha{}^\nu, \tag{13.166}$$

where (13.166) comprises both energy and momentum statements. The explicit connection with the previously defined quantities in Sections 8.1 and 8.2 are

$$T^{00} = u(\vec{x},t) = \frac{1}{8\pi}(\vec{E}^2+\vec{B}^2), \tag{13.167}$$

$$T^{0i} = cg_i(\vec{x},t) = \frac{1}{4\pi}(\vec{E}\times\vec{B})_i. \tag{13.168}$$

It is possible to derive (13.165) also using a Hamiltonian-type variational construction for $T^{\mu\nu}$ (see the example in Exercise 13.10.4). In the electromagnetic case this derivation results in an energy momentum tensor which differs from (13.165) by a total divergence. When the tensor is symmetrized, the divergence is eliminated and the result is identical with (13.165).

Using $T^{\mu\nu}$, we may define a total field energy-momentum 4-vector:

$$cP^\mu_{\text{field}} \equiv \left(\int d^3x\, T^{00}, \int d^3x\, T^{0i}\right). \tag{13.169}$$

It is possible to show (Exercise 13.10.5) that (13.169) does indeed define a contravariant 4-vector, given $\partial_\mu T^{\mu\nu} = 0$. However, when sources are present this construction fails. We will see a consequence of this failure in the next chapter!

13.11 Going Deeper

13.11.1 *Relativistic formulations*

(1) W. E. Baylis, *Electrodynamics, A Modern Geometric Approach*, Birkhaïser (Boston) 1999.

(2) J. D. Jackson, *Classical Electrodynamics*, 3rd ed., John Wiley & Sons (New York) 1999; Chapters 11 and 12.

(3) L. D. Landau and E. M. Lifshitz, *Classical Theory of Fields*, 4th ed. with corrections, Butterworth-Heinemann (Oxford) 2000; Chapters 1–4.

(4) S. Parrott, *Relativistic Electrodynamics and Differential Geometry*, Springer (New York) 1986.

(5) W. Rindler, *Introduction to Special Relativity*, 2nd ed., Oxford University Press (Oxford) 1991.

(6) M. Schwartz, *Principles of Electrodynamics*, Dover Publications (New York) 1987.

13.11.2 *Particle accelerators*

(1) A. Chao and M. Tigner, Eds., *Handbook of Accelerator Physics and Engineering*, World Scientific (Singapore) 1999.

(2) M. Conte and W. W. MacKay, *Introduction to the Physics of Particle Accelerators*, 2nd ed., World Scientific (Singapore) 2008.

(3) L. R. Evans, Ed., *The Large Hadron Collider: A Marvel of Technology*, CRC Press (Boca Raton) 2009.

(4) K. Olive *et al.* (Particle Data Group), Chin. Phys. C **38**, 090001 (2014): `http://pdg.lbl.gov`; Review, "Accelerator Physics of Colliders".

(5) H. Wiedemann, *Particle Accelerator Physics*, 3rd ed., Springer (Heidelberg) 2007.

(6) K. Wille, *The Physics of Particle Accelerators: An Introduction*, Oxford University Press (Oxford) 2001.

(7) E. J. N. Wilson, *An Introduction to Particle Accelerators*, Oxford University Press (Oxford) 2001.

13.12 Exercises

Exercise 13.1.1. Complete the derivation of the Lorentz transformation (13.12).

Exercise 13.1.2. The full form of the Lorentz transformation for boosts of a 4-vector Equation (13.15) is actually determined by our simple forms, Equation (13.13), linearity, and the assumption of the isotropy of space. One may show this as follows. Given that the only vectors available are $\vec{x}$ and $\vec{\beta}$, and given that the transformation is linear, the most general expression for ct' and $\vec{x}\,'$ one can write down is,

$$ct' = A(ct) + B(\vec{\beta} \cdot \vec{x}),$$

$$\vec{x}\,' = C\vec{x} + D\vec{\beta}(\vec{\beta} \cdot \vec{x}) + E\vec{\beta}(ct) + F(\vec{\beta} \times \vec{x}),$$

where $A, \ldots, F$ are scalars (functions of $\vec{\beta}^2$ only). Comparing these with (13.13), and taking specifically $\vec{\beta} = \beta\hat{i}$, show that the transformation is completely determined, consistent with (13.15).

Exercise 13.2.1. The defining relation for Lorentz transformations can be written as

$$A^{\mathrm{T}} \eta A = \eta.$$

(a) If A_1 and A_2 are Lorentz transformations, then so is $A_{12} \equiv A_1 A_2$.
(b) Show that A^{-1} is a Lorentz transformation: $(A^{-1})^{\mathrm{T}} \eta A^{-1} = \eta$. Write this directly in index notation.
(c) Show that A^{T} is a Lorentz transformation: $A \eta A^{\mathrm{T}} = \eta$. Write this in index notation also.

Exercise 13.2.2.

(a) Show that rotations satisfy the Lorentz transformation condition (13.26).
(b) Show that any transformation A of the form $A = R' A_{\text{boost}} R$ satisfies the 4-interval invariance condition, where R and R' are rotation matrices.

(c) Explain why the 4×4 matrix for the transformation defined in (13.15) (which we will denote as $A_{\text{boost}}(\vec{\beta})$) can be expressed as

$$A_{\text{boost}}(\vec{\beta}) = R^{\mathrm{T}} A_{\text{boost}}(\beta_1) R$$

for some rotation matrix R, where $A_{\text{boost}}(\beta_1)$ is defined in (13.21).

(d) Use the representation in part (c) to show that

$$\left(A_{\text{boost}}(\vec{\beta})\right)^{-1} = A_{\text{boost}}(-\vec{\beta}).$$

Exercise 13.2.3. For tensors $B^{\mu\nu}$ and $C_{\mu\nu}$, show explicitly that $B^{\mu\nu}C_{\mu\nu}$ transforms as a Lorentz scalar. Use both matrix and index notation.

Exercise 13.3.1. [Adapted from D. Griffiths, "Introduction to Electrodynamics" 3rd ed., Prentice Hall (1999), Problem 12.64.]

(a) In a certain inertial reference frame, S, the electric ($\vec{E}$) and magnetic fields ($\vec{B}$) are neither parallel nor perpendicular at a particular space-time point. Show that in a different inertial system, S', moving relative to S with velocity $\vec{v} = \vec{\beta}c$ given by

$$\vec{\beta} = \hat{\beta}(y - \sqrt{y^2 - 1}),$$

$$\hat{\beta} = \frac{\vec{E} \times \vec{B}}{|\vec{E} \times \vec{B}|}, \quad y = \frac{\vec{E}^2 + \vec{B}^2}{2|\vec{E} \times \vec{B}|},$$

the fields $\vec{E}\,'$ and $\vec{B}\,'$ are parallel at that point. [*Hint*: Take $\vec{E} \times \vec{B}$ along x, say, and break $\vec{E}\,'$, $\vec{B}\,'$ up into components along and perpendicular to x.]

(b) Show that the maximum and minimum values of y correspond to the minimum and maximum values of β, respectively. Is there a frame in which $\vec{E}\,'$ and $\vec{B}\,'$ are rendered perpendicular? [*Hint*: $\tilde{F}^{\mu\nu}F_{\mu\nu} = -4\vec{E} \cdot \vec{B}$; $\tilde{F}^{\mu\nu}$ is the dual of $F^{\mu\nu}$.]

Exercise 13.3.2. Follow the guidance in the text to derive the transformation laws for $\vec{E}$ and $\vec{B}$, Equations (13.61) and (13.62).

Exercise 13.3.3. Knowing the relativistic transformation between frames can simplify a derivation. We obtained the nonrelativistic relation (7.63) for an object in motion,

$$4\pi\vec{M}' = 4\pi\vec{M} + (\vec{D} - \vec{E}) \times \frac{\vec{v}}{c},$$

by a macroscopic averaging process. Now obtain the exact relativistic relation between $\vec{M}'$ and $\vec{M}$ simply as a consequence of relativistic transformations and deduce the above in the nonrelativistic limit. Explain the meaning of primed and unprimed field in this case. [*Hint*: Assume the ($\vec{D}$, $\vec{H}$) fields behave the same under Lorentz transformations as ($\vec{E}$, $\vec{B}$); see Equations (13.61), (13.62).]

Exercise 13.3.4. Show that:

(a) $\epsilon^{\alpha\beta\lambda\kappa}\epsilon_{\alpha\beta\sigma\tau} = 2\left(\delta^{\lambda}_{\tau}\delta^{\kappa}_{\sigma} - \delta^{\lambda}_{\sigma}\delta^{\kappa}_{\tau}\right)$

(b) $F^{\tau\sigma} = -\frac{1}{2}\epsilon^{\alpha\beta\tau\sigma}\tilde{F}_{\alpha\beta}$

(c) $\tilde{F}^{\alpha\beta}\tilde{F}_{\alpha\beta} = -F^{\alpha\beta}F_{\alpha\beta}$.

Exercise 13.3.5.

(a) Show that the quantity $F^{\mu\nu}F_{\nu\lambda}F^{\lambda\kappa}F_{\kappa\mu}$ ($F^{\mu\nu}$ is the field strength tensor) transforms as a scalar using both matrix and index formalisms.
(b) Evaluate

$$F^{\mu\nu}F_{\nu\lambda}F^{\lambda\kappa}F_{\kappa\mu} = a\left(\vec{E}^2\right)^2 + b\left(\vec{B}^2\right)^2 + c\,\vec{E}^2\vec{B}^2 + d\left(\vec{E}\cdot\vec{B}\right)^2.$$

Determine the coefficients a, b, c, and d. (Note that the form $(\vec{E}\times\vec{B})^2$ can be reduced to the last two possibilities above.)

Exercise 13.5.1.

(a) Show Equation (13.78): $(\Lambda^0{}_0)^2 = 1 + \sum_i(\Lambda^0{}_i)^2$. ($\Lambda^0{}_i \neq \Lambda^i{}_0$ in general, but they are equal for pure velocity boosts)
(b) Assuming that A^μ is a timelike 4-vector, show that the sign of its 4th component ($\mu = 0$) is preserved under a restricted Lorentz transformation. [*Hint*: Use the Cauchy-Schwartz inequality and part (a).]

Exercise 13.6.1. [Adapted from M. Schwartz, *Principles of Electrodynamics*, Dover Publications (New York) 1987, Problem 3-9.] Consider an electron of mass m and charge $-e$ in a classical circular orbit of radius R about a proton. Ignore the possibility of radiation for purposes of this problem.

(a) Using the correct relativistic formulation for force, solve for the velocity of the electron as a function e, m and R.
(b) Consider an inertial observer whose path and velocity coincides instantaneously with the electron at a point. Using the transformation equations (13.61) and (13.62) and the geometry in Figure 13.3, find the magnetic field seen at this point. Examine the ultrarelativistic and nonrelativistic limits of your field, and show that $|\vec{B}| \sim 1/R^3$ in one case, $|\vec{B}| \sim 1/R^{5/2}$ in the other.

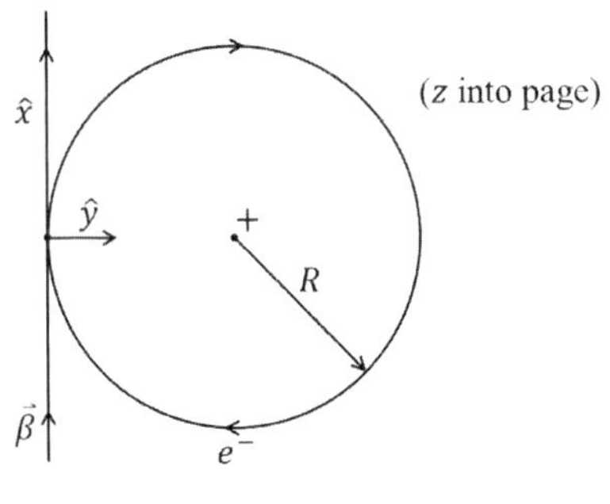

Fig. 13.3 For Exercise 13.6.1.

Exercise 13.6.2. The relativistic acceleration, α^μ, is

$$\alpha^\mu \equiv \frac{du^\mu}{d\tau} = \frac{d^2x^\mu}{d\tau^2},$$

where τ is proper time and $x^\mu = (ct, \vec{x})$. Show that:

(a) $\alpha^\mu u_\mu = 0$.
(b) $\alpha^\mu \alpha_\mu < 0$ (assuming $\alpha^\mu \neq 0$). [*Hint*: Use (a) and construct a proof by contradiction.]

Exercise 13.6.3. The following is laboratory data on pion-proton collisions, yielding a resonance known as a delta: $\pi^+ + p \to \Delta^{++}$. The pion kinetic energies (KE) are in MeV, the proton is initially at rest in the lab, and the cross-section $\sigma\,(\pi^+ + p \to \Delta^{++})$ of the resonance is in mb ("millibarns"; $1\ mb = 10^{-31}\ m^2$). (See J. R. Haskins, *Am. J. Phys.* **53**, 988 (1985).)

$\pi^+ KE$ (MeV)	$\sigma\,(\pi^+ + p \to \Delta^{++})$ (mb)
71.6	22.18
97.4	50.04
118.9	91.07
120.4	95.79
136.0	135.51
138.7	142.85
155.8	183.50
161.2	192.09
168.0	197.81
182.5	196.12
205.3	166.19
228.5	127.02
254.1	92.44
282.8	65.17

(a) For this interaction, show from relativistic dynamics that

$$p_{cm} = \frac{p_\pi}{\sqrt{1 + \dfrac{m_\pi^2}{m_p^2} + \dfrac{2}{m_p}\sqrt{m_\pi^2 + \dfrac{p_\pi^2}{c^2}}}},$$

$$E_{cm} = \sqrt{m_\pi^2 c^4 + p_{cm}^2 c^2} + \sqrt{m_p^2 c^4 + p_{cm}^2 c^2},$$

where p_π is the laboratory momentum of the pion, p_{cm} is the center of mass (cm) momentum of the pion or proton and E_{cm} is the cm energy of the system. (The cm of the system is the one in which the *total* spatial momentum of the particles is zero.)

(b) Using part (a), convert the 14 data points above to the following form. The first two value pairs are filled in below. The masses of the charged pion and proton are $m_\pi = 139.57\ MeV/c^2$ and $m_p = 938.27\ MeV/c^2$.

E_{cm} (MeV)	$\dfrac{\sigma\,(\pi^+ + p \to \Delta^{++})\,k^2}{8\pi}$
1138.5	0.03866
1159.5	0.12279

Note: multiplying the cross-section by $k^2/8\pi$ ($k = \dfrac{p_{cm}}{\hbar}$) gives a dimensionless quantity which should peak at 1 if the Δ^{++} decays only into these same two particles (an "elastic" resonance), which it does to an excellent approximation.

(c) This data may be fit by a non-symmetric *Breit-Wigner* resonance formula given by[3]

$$\frac{\sigma\,(\pi^+ + p \to \Delta^{++})\,k^2}{8\pi} = \frac{\left(\dfrac{\Gamma}{2}\cos\delta + (M_p c^2 - E_{cm})\sin\delta\right)^2}{(M_p c^2 - E_{cm})^2 + \Gamma^2/4},$$

where Γ is the decay width, M_p is called the pole mass (not the proton mass!), E_{cm} is the cm energy, and the angle δ measures the extent of mixing with a resonant background contribution. Try fitting your (b) data set with the given functional form using a nonlinear fitting package. Note that the pole mass, M_p, is related to the Breit-Wigner mass, M_{BW}, listed in particle tables by

$$M_{BW}c^2 = M_p c^2 - \frac{\Gamma}{2}\tan\delta.$$

What values for Γ, M_p, δ, and M_{BW} do you get from the fit? Compare relevant parameters with values from the Particle Data Group review pages for the $\Delta(1232)$ resonance. Our graph looks like:

Exercise 13.6.4. The so-called *GZK* energy limit on cosmic ray proton energies is caused by the production of neutral pions in collisions with cosmic ray background photons of approximate energy $E_\gamma = kT$, where k is Boltzmann's constant and T is cosmic background temperature $T \sim 2.7°$ Kelvin. In the center of momentum frame the pion and proton are both produced at rest at production threshold. Show by doing a Lorentz transformation boost from the center of momentum frame to the frame in

[3]S. Ceci, M. Korolija, and B. Zauner, *Phys. Rev. Lett.* **111**, 112004 (2013).

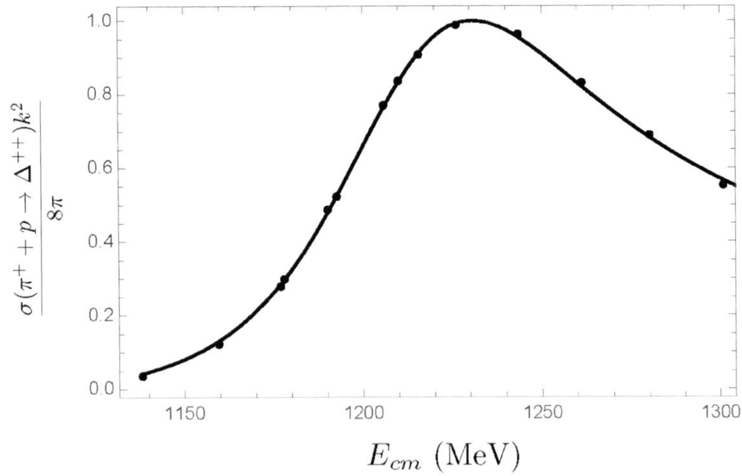

Fig. 13.4 Dimensionless cross-section for the interaction $\pi^+ + p \to \Delta^{++}$ as a function of center of mass energy, E_{cm}.

which the photon has energy, E_γ, that the proton threshold energy for pion production, E_p^{th}, is given by ($c = 1$ units)

$$E_p^{th} = \frac{m_\pi(m_\pi + 2m_p)}{4E_\gamma} + E_\gamma \frac{m_p^2}{m_\pi(m_\pi + 2m_p)},$$

where m_π is the pion mass and m_p is the proton mass. (The second term above is much smaller than the first.) When values are put in this formula, the threshold energy comes out as $E_p^{th} \approx 3 \times 10^{20}$ electron volts. A better model for the photon energies gives this cutoff as $\sim 5 \times 10^{19}$ electron volts.

Exercise 13.7.1.

(a) Carefully argue that power, $P(t) = \frac{dE}{dt}$, is an invariant by considering that the numerator and denominator transform similarly. Note in particular that t could stand for retarded time t_r. *Extra*: Generalize this result to time derivatives of the form $\frac{dT^{\mu\nu\cdots 0}}{dt}$.

(b) By working out the explicit components of the relativistic acceleration, α^μ, show that

$$\alpha^0 = \gamma^4 \vec{\beta}\cdot\dot{\vec{v}}, \quad \vec{\alpha} = \gamma^4\left(\dot{\vec{v}} + \vec{\beta}\times(\vec{\beta}\times\dot{\vec{v}})\right)$$
$$\implies \alpha^\mu\alpha_\mu = -c^2\gamma^6\big[\dot{\vec{\beta}}^2 - (\vec{\beta}\times\dot{\vec{\beta}})^2\big].$$

(c) On the basis of parts (a) and (b) above, argue that the relativistically correct version of the nonrelativistic Larmor formula (13.99) is given by (13.100), in agreement with the Liénard result (11.160).

Exercise 13.7.2.

(a) A particle of mass m and charge e is accelerated uniformly in a constant electric field, $E_0\hat{x}$, starting from rest at $x = 0$, $t = 0$. Show that the particle's position as a function of time $x(t)$ is given by

$$\left(x(t) + \frac{mc^2}{eE_0}\right)^2 - (ct)^2 = \left(\frac{mc^2}{eE_0}\right)^2.$$

This is a hyperbola in the $x - t$ plane. [*Hint*: One must use the Lorentz force equation and the relativistic expression for momentum.]

(b) Show that for a person located at the origin the relationship between t and exact retarded time t_R is

$$t_R = \frac{t}{2}\left(\frac{t + 2\zeta}{t + \zeta}\right).$$

where $\zeta \equiv mc/(|eE_0|)$. Examine the limits $t \ll \zeta$ and $t \gg \zeta$. Describe the physical meaning of these results.

Exercise 13.7.3.

(a) Assume that the relativistically form-invariant equation

$$\frac{dp^\mu}{d\tau} = \frac{e}{c}F^\mu{}_\nu u^\nu,$$

correctly describes the motion of a particle of charge e in an electromagnetic field, where $u^\nu = dx^\nu/d\tau$, τ is proper time, and $p^\nu \equiv mu^\nu$. (See also Exercise 13.10.2.) Show that the Lorentz force law

$$\vec{F} = e\left(\vec{E} + \frac{\vec{v}}{c} \times \vec{B}\right),$$

holds even for relativistic velocities, where $\vec{v} = d\vec{x}/dt$ and $\vec{F} = d\vec{p}/dt$. Derive also the equation,

$$\frac{dE}{dt} = e(\vec{v} \cdot \vec{E}),$$

where $E = p^0c$.

(b) Using (a) and given the radiation loss formula

$$P = -\frac{2}{3}\frac{e^2}{m^2c^3}\frac{dp^\mu}{d\tau}\frac{dp_\mu}{d\tau},$$

show that one obtains

$$P_E = \frac{2}{3}\frac{e^4}{m^2c^3}\vec{E}^2, \ \vec{B} = 0, \ \vec{E} \,||\, \vec{v};$$

$$P_B = \frac{2}{3}\frac{e^4}{m^2c^3}\gamma^2\beta^2\vec{B}^2, \ \vec{E} = 0 \text{ and } \vec{B} \perp \vec{v}.$$

Exercise 13.7.4.

(a) Using the Lorentz force law, show that the relativistically correct expression for the orbital radius R of a particle with charge e in a magnetic field $\vec{B}$ is

$$R = \frac{|\vec{p}|c}{|e\vec{B}|}.$$

(b) Using the result for P_B from Exercise 13.7.3, recover the result that in the case of uniform circular motion ($\vec{B}$ is constant in space; R is radius of the circle) the energy loss per cycle is given by

$$E_{\text{cycle}} = \frac{4\pi e^2}{3R}\left(\frac{E}{mc^2}\right)^4 \beta^3.$$

Exercise 13.7.5. The accelerating tunnel currently used by the Large Hadron Collider (LHC) was previously used for the Large Electron-Positron Collider (LEP). In this case, the maximum electron energies obtained were about 100 GeV. Compute the radiation losses dE/dx ($R \approx 4.3 \times 10^5$ cm). Given that klystron RF gains are limited to about 50 MeV/m, calculate the maximum possible energy for electrons in the LEP/LHC tunnel.

Exercise 13.7.6. Consider the motion of a particle with charge $q > 0$ and mass m in constant $\vec{E}$ and $\vec{B}$ fields Assume both initial position $\vec{x}(0)$ and velocity $\dot{\vec{x}}(0)$ of the particle are zero. Show that the Lorentz force equation,

$$\frac{d\vec{p}}{dt} = \frac{d}{dt}(\gamma m\vec{v}) = q\left(\vec{E} + \frac{\vec{v}}{c} \times \vec{B}\right),$$

where $\gamma = \frac{1}{\sqrt{1-\beta^2}}$, $\beta = v/c$, may be integrated to yield the proper time equations $\left(\frac{dt}{d\tau} = \gamma\right)$,

$$m\frac{d\vec{x}(\tau)}{d\tau} = q\left(\vec{E}\,t + \frac{\vec{x}(\tau)}{c} \times \vec{B}\right),$$

where

$$\gamma(\tau') = \left(1 + \frac{q}{mc^2}\vec{E}\cdot\vec{x}(\tau')\right),$$

and

$$t = \int_0^\tau d\tau'\,\gamma(\tau').$$

Exercise 13.7.7.

(a) Continue the solution of the proper time equations from Exercise 13.7.6. Specialize to $\vec{E} = E_1\,\hat{x} + E_2\,\hat{y}$ and $\vec{B} = B\,\hat{x}$, with E_1, E_2, $B > 0$. Show that the second order proper time equations of motion are

$$\frac{d^2x(\tau)}{d\tau^2} = q\frac{E_1}{m} + q^2\frac{E_1^2}{m^2c^2}x(\tau) + q^2\frac{E_1E_2}{m^2c^2}y(\tau),$$

$$\frac{d^2y(\tau)}{d\tau^2} = q\frac{E_2}{m} + q^2\frac{E_1E_2}{m^2c^2}x(\tau) + \frac{q^2}{m^2c^2}(E_2^2 - B^2)y(\tau),$$

and that

$$z(\tau) = -\frac{qB}{mc}\int_0^\tau d\tau\,'\,y(\tau\,').$$

(b) Disentangle and solve these equations to show the complete motion is given by

$$x'\,(\tau) = -\frac{mc^2E_1'}{q\,Q_1^2}\left(1 - \cosh\left(\frac{qQ_1}{mc}\,\tau\right)\right),$$

$$y'\,(\tau) = \frac{mc^2E_2'}{q\,Q_2^2}\left(1 - \cos\left(\frac{qQ_2}{mc}\,\tau\right)\right),$$

where

$$x\,(\tau) = \cos\theta\,x\,'\,(\tau) + \sin\theta\,y\,'\,(\tau),$$
$$y\,(\tau) = -\sin\theta\,x\,'\,(\tau) + \cos\theta\,y\,'\,(\tau),$$

$$E_1' = \cos\theta\,E_1 - \sin\theta\,E_2\,,$$
$$E_2' = \sin\theta\,E_1 + \cos\theta\,E_2\,,$$

$$\tan 2\theta = -\frac{2E_1E_2}{E_1^2 - E_2^2 + B^2},$$

$$Q_1^2 \equiv \frac{1}{2}\left[-(E_1^2 + E_2^2 - B^2) + \sqrt{(E_1^2 + E_2^2 - B^2)^2 + (2E_1B)^2}\right],$$

$$Q_2^2 \equiv \frac{1}{2}\left[(E_1^2 + E_2^2 - B^2) + \sqrt{(E_1^2 + E_2^2 - B^2)^2 + (2E_1B)^2}\right].$$

Exercise 13.8.1. As an example of a classical tensor theory, consider a linearized gravitational Lagrangian density, $\mathcal{L}$, using the symmetric field potential $h_{\mu\nu} = h_{\nu\mu}$ (the analog of the A_μ 4-potential in electrodynamics). It is given by[4]

$$\mathcal{L} = \frac{1}{2}[\partial^\lambda h^{\mu\nu}\partial_\lambda h_{\mu\nu} - \partial^\lambda h^\mu{}_\mu \partial_\lambda h^\nu{}_\nu] + \partial_\mu h^{\mu\nu}\partial_\nu h^\lambda{}_\lambda - \partial_\mu h^{\mu\nu}\partial^\lambda h_{\lambda\nu}.$$

The action is

$$W = \frac{1}{c}\int d^4x \left[T^{\mu\nu}h_{\mu\nu} + \frac{1}{\kappa}\mathcal{L}\right].$$

(κ is the gravitational coupling constant, $8\pi G$, where G is Newton's gravitational constant, and $T_{\mu\nu}$ is the symmetric conserved energy momentum tensor.) Derive the gravitational field equations for the fields $h^{\alpha\beta}$ from variation of the action, δW. Show that:

$$\begin{aligned}\Box(h^{\alpha\beta} - \eta^{\alpha\beta}h^\mu{}_\mu) - \partial^\alpha\partial^\mu h_\mu{}^\beta - \partial^\beta\partial^\mu h_\mu{}^\alpha \\ + \partial^\alpha\partial^\beta h^\mu{}_\mu + \eta^{\alpha\beta}\partial^\mu\partial^\nu h_{\mu\nu} = \kappa T^{\alpha\beta}.\end{aligned}$$

Note: The Euler-Lagrange equation must be symmetrized in its two free indices in this case since $h_{\mu\nu}$ is a symmetric tensor.

Exercise 13.9.1. The "Proca" Lagrangian density is given by

$$\mathcal{L} = -\frac{1}{16\pi}F^{\alpha\beta}F_{\alpha\beta} + \frac{\mu^2}{8\pi}A^\alpha A_\alpha - \frac{1}{c}J^\alpha A_\alpha,$$

where

$$F^{\alpha\beta} = \partial^\alpha A^\beta - \partial^\beta A^\alpha.$$

[4]See J. Schwinger, *Particles, Sources, and Fields*, Vol. I, Addison-Wesley (Reading. MA) 1970, Sections 3-5, 3-17.

(a) Defining as usual,

$$\vec{E} \equiv -\vec{\nabla}\Phi - \frac{1}{c}\frac{\partial \vec{A}}{\partial t},$$

$$\vec{B} \equiv \vec{\nabla} \times \vec{A},$$

$\left(A^\mu = (\Phi, \vec{A}); \text{ these satisfy } \vec{\nabla} \cdot \vec{B} = 0,\ \vec{\nabla} \times \vec{E} = -\frac{1}{c}\frac{\partial \vec{B}}{\partial t}\right)$ show that the field equations which follow from this Lagrangian are

$$\vec{\nabla} \cdot \vec{E} + \mu^2 \Phi = 4\pi\rho,$$

$$\vec{\nabla} \times \vec{B} - \frac{1}{c}\frac{\partial \vec{E}}{\partial t} + \mu^2 \vec{A} = \frac{4\pi}{c}\vec{J}.$$

$(J^\mu = (c\rho, \vec{J}).)$

(b) Find the analog to the wave equation for this system for the $\vec{E}$ and $\vec{B}$ fields. (You may set $J^\mu = 0$.) Are plane waves still solutions? (That is, can we write $\vec{E} = \mathrm{Re}\,(\vec{E}_0 e^{i(\vec{k}\cdot x - \omega t)})$?) From this analog wave equation deduce the dispersion relation between $\vec{k}^2$, ω and μ.

Exercise 13.9.2. Referring to the "Proca" Lagrangian in Exercise 13.9.1, derive the energy conservation equation,

$$\frac{\partial u}{\partial t} + \vec{\nabla} \cdot \vec{S} = -\vec{J} \cdot \vec{E},$$

assuming conservation of charge, $\partial^\alpha J_\alpha = 0$. Find the general forms for the energy density, u, and Poynting vector, $\vec{S}$, in terms of the fields $\vec{E}$, $\vec{B}$, Φ, and $\vec{A}$.

Exercise 13.9.3.

(a) Show that $\tilde{F}^{\mu\nu} F_{\mu\nu}$ is a pure divergence. That is, demonstrate that

$$\tilde{F}^{\mu\nu} F_{\mu\nu} = \partial_\lambda S^\lambda.$$

Exhibit S^λ.

(b) Derive *explicitly* the Euler-Lagrange equations for the Lagrange density ($\theta =$ constant, *general* gauge A^μ)

$$\mathcal{L} = -\frac{1}{16\pi} F^{\mu\nu} F_{\mu\nu} + \theta \tilde{F}^{\mu\nu} F_{\mu\nu} - \frac{1}{c} A_\mu J^\mu.$$

Are the Maxwell equations modified by the presence of the θ-term?

Exercise 13.9.4.

(a) Construct the relativistically correct Hamiltonian

$$H_R = \sum_k p^k \dot{x}^k - L_R,$$

where

$$p^k \equiv \frac{\partial L_R}{\partial \dot{x}^k},$$

for the Lagrangian

$$L_R = -\frac{1}{\gamma}\left(mc^2 + \frac{q}{c}u^\mu A_\mu\right).$$

This describes particle motion for a particle of charge q in a given background field: $\Phi \equiv \Phi(\vec{x},t)$, $\vec{A} \equiv \vec{A}(\vec{x},t)$. Make sure you express H_R in the correct variables. Note that $\int dt\, L_R = \int d\tau \gamma L_R \implies \gamma L_R$ forms a Lorentz invariant. [*Ans.:*

$$H_R = c\sqrt{(mc)^2 + \left(\vec{p} - \frac{q}{c}\vec{A}\right)^2} + q\Phi.]$$

(b) Verify that Hamilton's equations,

$$\dot{x}^k = \frac{\partial H_R}{\partial p^k}, \; \dot{p}^k = -\frac{\partial H_R}{\partial x^k},$$

give

$$m\gamma\dot{x}^k = p^k - \frac{q}{c}A^k,$$
$$\frac{d}{dt}\left(m\gamma\dot{x}^k\right) = q\left(\left(\frac{\dot{\vec{x}}}{c} \times \vec{B}\right)^k + E^k\right),$$

with the usual definitions for $\vec{E}$ and $\vec{B}$.

Exercise 13.10.1. Recall that in Section 8.3 we obtained angular momentum conservation by multiplying the terms in the momentum conservation Equation (8.42) (with free index i changed to k) by $\epsilon_{ijk}x_j$ and summing over j, k; see Equation (8.55). This is not a form-invariant operation on vectors under Lorentz transformations; however, if we multiply a vector v_k by $\epsilon_{mni}\epsilon_{ijk}x_j$ and sum over i, j, k we obtain:

$$\epsilon_{mni}\epsilon_{ijk}x_j v_k = (\delta_{mj}\delta_{nk} - \delta_{mk}\delta_{nj})x_j v_k = x_m v_n - x_n v_m.$$

This suggests that we may obtain a covariant version of angular momentum conservation by applying the operation

$$x^\mu[\ldots]^\nu - x^\nu[\ldots]^\mu,$$

where $[\ldots]$ represents the terms of (13.166). Show that in this way we may obtain a Lorentz form-invariant conservation equation

$$\partial_\alpha M^{\alpha\mu\nu} \equiv \frac{1}{c} J^\alpha \mathcal{T}_\alpha{}^{\mu\nu},$$

as long as $T^{\mu\nu}$ is symmetric. In this case, the four-dimensional form-invariant version of the shear tensor is:

$$M^{\alpha\mu\nu} = T^{\alpha\mu} x^\nu - T^{\alpha\nu} x^\mu.$$

Give an explicit expression for $\mathcal{T}_\alpha{}^{\mu\nu}$.

Exercise 13.10.2. From the energy momentum divergence equation (13.166) for a point particle of charge e,

$$\rho(\vec{x}, t) = e\,\delta(\vec{x} - \vec{x}(t)); \quad \vec{J}(\vec{x}, t) = e\vec{v}\,\delta(\vec{x} - \vec{x}(t)),$$

derive the covariant form of the Lorentz force law seen in Exercise 13.7.3:

$$\frac{dp^\mu}{d\tau} = \frac{e}{c} F^\mu{}_\nu u^\nu,$$

($u^\nu = dx^\nu/d\tau$). [*Hint*: This equation applies to particle momentum, not field momentum. Be sure to conserve *total* momentum!]

Exercise 13.10.3.

(a) Using the angular momentum conservation equation that was derived in Exercise 13.10.1, show that in source free regions the field angular momentum is conserved:

$$\frac{d\vec{L}}{dt} = 0,$$

where

$$\vec{L} \equiv \int d^3x\, \vec{x} \times \vec{g}(\vec{x}, t).$$

(b) Show that

$$\frac{d\vec{N}}{dt} = 0,$$

which expresses conservation of center of energy position, where

$$N^i \equiv \frac{1}{\mathcal{E}} \int d^3x [T^{00}x^i - T^{0i}x^0], \ \mathcal{E} \equiv \int d^3x\, T^{00}.$$

[*Note*: We may write the $\vec{N}$ equation as

$$< \vec{x} > \equiv \frac{1}{\mathcal{E}} \int d^3x [T^{00}(\vec{x}, t)\, \vec{x}] = \vec{N} + \vec{v}t \implies \frac{d < \vec{x} >}{dt} = \vec{v},$$

where $\vec{v} \equiv c^2 \vec{P}_{\text{field}}/\mathcal{E}$.]

Exercise 13.10.4. It was mentioned at the end of Section 13.10 that there is another way to find the energy momentum tensor from a variational point of view. Here it is:

$$T^{\mu\nu} = \sum_k \frac{\delta \mathcal{L}}{\delta(\partial_\mu \phi_k)} \partial^\nu \phi_k - \eta^{\mu\nu} \mathcal{L}.$$

Use this on the simple scalar Lagrangian,

$$\mathcal{L} = \frac{1}{2}(\partial_\lambda \Phi\, \partial^\lambda \Phi - \mu^2 \Phi^2),$$

for the single field $\phi_1 \to \Phi$. Write out expressions for the energy density T^{00} and the Poynting vector cT^{0i}. Then using the field equation (13.136) evaluate $\partial_\mu T^{\mu\nu}$ for $S \neq 0$.

Exercise 13.10.5. Prove that Equation (13.169) transforms covariantly, given $\partial_\mu T^{\mu\nu} = 0$. The four dimensional Gauss theorem for $T^{\mu\nu}$ formally reads

$$\int_\Omega d^4x\, \partial_\mu T^{\mu\nu} = \int_{\partial\Omega} d^3\sigma\, \hat{n}_\mu T^{\mu\nu},$$

similar to (13.131), where $\hat{n}_\mu$ is the outward 4-dimensional unit vector and $d^3\sigma$ is the invariant 3-area. For a consistent explanation of this equation, see for example Misner, Thorne and Wheeler, "Gravitation", W. H. Freeman (1973), Box 5.3. [*Hint*: Take $\hat{n}_\mu = (1, 0, 0, 0)$ in the rest frame as the "top" of the 4-volume.]

Exercise 13.10.6.

(a) Construct the scalar energy-momentum tensor $T^{\mu\nu}$. Starting with

$$T^{\mu\nu} = c_1\left(\partial^\mu\Phi\,\partial^\nu\Phi\right) + c_2(\eta^{\mu\nu}\partial_\lambda\Phi\,\partial^\lambda\Phi) + c_3(\eta^{\mu\nu}\Phi^2),$$

where $c_{1,2,3}$ are constants, and the field equation

$$\left(\Box + \mu^2\right)\Phi = S,$$

require that $T^{\mu\nu}$ be (it is already symmetric):

1. Conserved: $\partial_\mu T^{\mu\nu} = 0$ for $S = 0$.
2. Normalized: Energy density reduces to the form

$$T^{00} = \frac{1}{2}\left(\left(\frac{1}{c}\frac{\partial\Phi}{\partial t}\right)^2 + (\vec{\nabla}\Phi)^2 + \mu^2\,\Phi^2\right).$$

Show that a unique form results. (*Comment*: We are not requiring gauge invariance here because the above action actually describes a neutral particle.)

(b) Construct the angular momentum conservation statement for scalar fields,

$$\partial_\alpha M^{\alpha\mu\nu} = ?,$$

where $M^{\alpha\mu\nu}$ is the shear tensor.

Chapter 14

Special Topics

Having made it this far, you should be rewarded! This chapter is a smorgasbord of three interesting topics. We present points of view on: the electromagnetic radiation reaction force, the ideas of electromagnetic field theory and some developments in quasinormal radiation theory. All are fascinating, urgent and deep. Our treatments here are introductory: more complete presentations are indicated in the "Going Deeper" section. Let's hurry and get started....

14.1 Radiation reaction force

According to the Lorentz force equation,

$$\vec{F} = e\left(\vec{E} + \frac{\vec{v}}{c} \times \vec{B}\right), \tag{14.1}$$

a charged particle e moving in a magnetic field will continue in perpetual circular motion. However, we saw in the previous chapter that in reality the particle loses energy through radiation. So an additional force term is needed, which is commonly called the *radiation reaction* force. In this section, we derive an expression for this force. In order to proceed we will have to make an assumption concerning the charge density of this particle, which we will term a *classical electron*; we will attempt to take it to be a spherically symmetric charge distribution.

We begin with the expression for the force on a point charge e used in Exercise (13.7.3) part (a):

$$\frac{dp^\mu}{d\tau} = \frac{e}{c}F^{\mu\nu}u_\nu, \tag{14.2}$$

and we obtain the force density on a continuous charge by replacing eu_ν with j_ν. It follows that the total force on a continuous charge is

$$F^k_{\text{self}}(t) = \frac{1}{c}\int d^3x\, F^{k\nu}(\vec{x},t) j_\nu(\vec{x},t), \tag{14.3}$$

and writing out $F^{k\nu}$ in terms of j^μ gives

$$F^k_{\text{self}}(t) = \frac{1}{c^2}\iint d^3x\, d^3\xi\, \frac{1}{|\vec{\xi}|}\left[\frac{\partial j^\nu(\vec{x}-\vec{\xi},t')}{\partial x_k} - \frac{\partial j^k(\vec{x}-\vec{\xi},t')}{\partial x_\nu}\right]\Bigg|_{t'=t-|\vec{\xi}|/c} j_\nu(\vec{x},t). \tag{14.4}$$

Let us calculate the force in the instantaneous rest frame at time t of the distributed charge, so that

$$j_\nu(\vec{x},t) = j^\nu(\vec{x},t) = (c\rho(\vec{x},t),\vec{0}). \tag{14.5}$$

We suppose that the charge distribution is centered at the origin and is spherically symmetric, so that

$$\rho(\vec{x},t) = \rho(|\vec{x}|,t). \tag{14.6}$$

We may then decompose the integral in (14.4) into two parts:[1]

$$F^k_{\text{self}}(t) = \text{(I)}^k + \text{(II)}^k,$$

where

$$\text{(I)}^k = -\int d^3x\, \rho(\vec{x},t)\int d^3\xi\, \frac{1}{|\vec{\xi}|}\left(\frac{\partial\rho(\vec{x}-\vec{\xi},t')}{\partial x^k}\right)\Bigg|_{t'=t-|\vec{\xi}|/c}, \tag{14.7}$$

$$\text{(II)}^k = -\frac{1}{c^2}\int d^3x\, \rho(\vec{x},t)\int d^3\xi\, \frac{1}{|\vec{\xi}|}\left(\frac{\partial j^k(\vec{x}-\vec{\xi},t')}{\partial t}\right)\Bigg|_{t'=t-|\vec{\xi}|/c}. \tag{14.8}$$

A very long calculation follows; here we will just try to give a flavor of the manipulations necessary. We start with a multipole expansion (which is justified because we are in the instantaneous rest frame of the charge, so velocities are nonrelativistic):

$$\rho(\vec{x}-\vec{\xi},t') \approx \rho(\vec{x}-\vec{\xi},t) - \frac{|\vec{\xi}|}{c}\frac{\partial\rho(\vec{x}-\vec{\xi},t)}{\partial t} + \frac{1}{2}\frac{|\vec{\xi}|^2}{c^2}\frac{\partial^2\rho(\vec{x}-\vec{\xi},t)}{\partial t^2} - \frac{1}{6}\frac{|\vec{\xi}|^3}{c^3}\frac{\partial^3\rho(\vec{x}-\vec{\xi},t)}{\partial t^3} + \dots, \tag{14.9}$$

[1]Here we are using the same notation as in B. DiBartolo, *Classical Theory of Electromagnetism*, 2nd ed., World Scientific (Singapore) 2004.

which enables us to write

$$\textcircled{\scriptsize I}^k = \textcircled{\scriptsize 1}^k + \textcircled{\scriptsize 2}^k + \textcircled{\scriptsize 3}^k + \textcircled{\scriptsize 4}^k. \tag{14.10}$$

(We will examine the effect of higher order terms in the expansion in Exercise 14.1.2.) Taking these one at a time, we have first

$$\begin{aligned}
\textcircled{\scriptsize 1}^k &= -\int\int d^3x\, d^3\xi\, \rho(\vec{x},t)\frac{1}{|\vec{\xi}|}\frac{\partial\rho(\vec{x}-\vec{\xi},t)}{\partial x^k} \\
&= \int\int d^3x\, d^3\xi\, \rho(\vec{x},t)\frac{1}{|\vec{\xi}|}\frac{\partial}{\partial\xi^k}\rho(\vec{x}-\vec{\xi},t) \\
&= -\int\int d^3x\, d^3\xi\, \rho(\vec{x},t)\rho(\vec{x}-\vec{\xi},t)\frac{\partial}{\partial\xi^k}\frac{1}{|\vec{\xi}|} \\
&= \int d^3x\, \rho(\vec{x},t)\underbrace{\int d^3\xi\, \frac{\rho(\vec{x}-\vec{\xi},t)\xi^k}{|\vec{\xi}|^3}}_{=0} = 0,
\end{aligned} \tag{14.11}$$

where the integral over $d^3\xi$ vanishes because the integrand is an antisymmetric function of ξ.

Next, we have

$$\textcircled{\scriptsize 2}^k = \int\int d^3x\, d^3\xi\, \rho(\vec{x},t)\frac{1}{|\vec{\xi}|}\frac{\partial}{\partial x^k}\Big[\frac{|\vec{\xi}|}{c}\underbrace{\frac{\partial\rho}{\partial t}(\vec{x}-\vec{\xi},t)}_{-\vec{v}\cdot\vec{\nabla}\rho(\vec{x}-\vec{\xi},t)}\Big], \tag{14.12}$$

which vanishes since $\vec{v}=0$ at time t.

The third term is

$$\textcircled{\scriptsize 3}^k = -\frac{1}{2}\int\int d^3x\, d^3\xi\, \rho(\vec{x},t)\frac{1}{|\vec{\xi}|}\frac{\partial}{\partial x^k}\left[\frac{|\vec{\xi}|^2}{c^2}\frac{\partial^2\rho(\vec{x}-\vec{\xi},t)}{\partial t^2}\right], \tag{14.13}$$

which may be simplified through the following sequence of mathematical operations:

(1) Change $\partial/\partial x^k$ into $-\partial/\partial\xi^k$, and integrate by parts to move the $-\partial/\partial\xi^k$ over to act on $1/|\vec{\xi}|$;

(2) Make the substitution

$$\frac{\partial^2\rho}{\partial t^2} = \left(\frac{\partial\vec{v}}{\partial t}\cdot\vec{\nabla}_\xi\right)\rho(\vec{x}-\vec{\xi},t) \tag{14.14}$$

(which follows from the equation of continuity), and integrate by parts so that $(\partial\vec{v}/\partial t)\cdot\vec{\nabla}_\xi$ acts on $\xi^k/|\vec{\xi}|^3$ from step (2);

(3) Use the fact that

$$\int d^3\xi\, \frac{\xi^i \xi^k}{|\vec{\xi}|^3} \times f_{\text{even}}(\vec{\xi}) = \frac{1}{3}\delta_{ik} \int \frac{d^3\xi}{|\vec{\xi}|} \times\ f_{\text{even}}(\vec{\xi}), \tag{14.15}$$

which holds for any even function $f_{\text{even}}(\vec{\xi})$.

The result of these operations is

$$\textcircled{3}^k = \frac{2}{3c^2}\dot{v}^k U_{\text{el}}, \tag{14.16}$$

where $\dot{v}^k \equiv \partial v^k/\partial t$ and the *electrostatic self-energy* U_{el} is defined as

$$U_{\text{el}} \equiv \frac{1}{2}\iint d^3x\, d^3\xi\, \frac{\rho(\vec{x},t)\rho(\vec{x}-\vec{\xi},t)}{|\vec{\xi}|}. \tag{14.17}$$

Similar steps applied to $\textcircled{4}^k$ give:

$$\textcircled{4}^k = -\frac{1}{3c^3}\ddot{v}^k \iint d^3x\, d^3\xi\, \rho(\vec{x},t)\rho(\vec{x}-\vec{\xi},t) = -\frac{e^2}{3c^3}\ddot{v}^k, \tag{14.18}$$

with the final result that

$$\textcircled{\text{I}}^k = \frac{2}{3c^2}\dot{v}^k U_{\text{el}} - \frac{e^2}{3c^3}\ddot{v}^k. \tag{14.19}$$

Similar computations lead to

$$\textcircled{\text{II}}^k = -\frac{2}{c^2}\dot{v}^k U_{\text{el}} + \frac{e^2}{c^3}\ddot{v}^k. \tag{14.20}$$

Putting everything together, we have

$$F^k_{\text{self}}(t) = -\frac{4}{3}m_{\text{el}}\dot{v}^k + \frac{2}{3}\frac{e^2}{c^3}\ddot{v}^k, \tag{14.21}$$

where the "electrostatic mass" m_{el} of the particle is given by

$$m_{\text{el}} \equiv \frac{U_{\text{el}}}{c^2}. \tag{14.22}$$

The first term in (14.21) gives a positive electrostatic contribution $\frac{4}{3}m_{\text{el}}$ to the electron's inertial mass when taken on the other side of the equation, although the factor of 4/3 is puzzling. In order to understand how to deal with this term properly, we must confront the fact that additional forces must be present in the electron to keep it from flying apart due to electrostatic repulsion. These forces are called *Poincaré stresses*,[2] and should

[2]H. Poincaré, "Sur la dynamique de l'électron", Comptes Rendus de l'Académie des Sciences **140**, 1504 (1905).

be included in our description of the energy-momentum of the complete system (particle plus field). This brings us back to the problem, mentioned at the end of Section 13.10, concerning the non-covariance of the integrated stress tensor when electromagnetic sources J^μ are present. The problem is resolved if we can consistently define a total energy momentum tensor

$$S^{\mu\nu} \equiv T^{\mu\nu} + P^{\mu\nu}, \tag{14.23}$$

such that

$$\partial_\mu S^{\mu\nu} = 0, \tag{14.24}$$

compensating for the nonzero divergence from $T^{\mu\nu}$ seen in Equation (13.166) and allowing the new integrated form to transform covariantly. The additional term, $P^{\mu\nu}$, is called the *Poincaré stress tensor*. For our simple model, one can show that the contribution of the Poincaré stress to the inertia of a particle is (see Exercise 14.1.8)

$$m_{\rm p} \equiv \frac{1}{c^2}\int d^3x\, P^{00} = \frac{1}{3}m_{\rm el}, \tag{14.25}$$

giving the total

$$m_{\rm tot} = m_{\rm el} + m_{\rm p} = \frac{4}{3}m_{\rm el}. \tag{14.26}$$

This solves both the covariance and stability problems and explains the mysterious 4/3 factor in (14.21). In passing we note that the model we are using can be generalized and that the actual value of the inertial mass term coming from the total $S^{\mu\nu}$ is not unique.[3]

With this in mind for our particular model, we will rewrite (14.21) as

$$\vec{F}_{\rm self}(t) = -m_{\rm tot}\dot{\vec{v}} + \frac{2}{3}\frac{e^2}{c^3}\ddot{\vec{v}}. \tag{14.27}$$

The second factor in (14.27) is called the *Abraham-Lorentz force*, and is related to the Larmor formula (11.99) as we will discuss below.

Notice that the calculation for self-force depends on the choice of the retarded Green's function in (14.4). Although this choice seems reasonable because of our practical experience with radiation, the mathematics allows for other choices. If we used the advanced Green's function (defined in (8.113)), then we would have to replace $t' = t - |\vec{\xi}|/c$ in (14.4) with $t' = t + |\vec{\xi}|/c$. To get the result we need only change the sign of the odd terms

[3]See for example, J. Schwinger, "Electromagnetic Mass Revisited", Found. Phys. **13**, 373 (1983).

in $|\vec{\xi}|$ in the multipole expansion (14.9). In this case, the expression for self-force becomes

$$F^k_{\text{self}} = -m_{\text{tot}}\dot{v}^k - \frac{2}{3}\frac{e^2}{c^3}\ddot{v}^k. \qquad \text{[from advanced Green function]} \quad (14.28)$$

It is also possible to take linear combinations of the advanced and retarded Green functions, as long as the coefficients sum to one. (For such linear combinations, the Green function defining equation is still satisfied.) From (14.21) and (14.28) it is evident that the average of advanced and retarded Green functions leads to

$$F^k_{\text{self}} = -m_{\text{tot}}\dot{v}^k. \qquad \text{[from averaged Green function]} \quad (14.29)$$

Notice that the radiation reaction force is absent!

14.1.1 *Classical renormalization and relativistic generalization*

We may formally absorb the electromagnetic+stress mass into an observed mass m via

$$m = m_0 + m_{\text{tot}}, \quad (14.30)$$

where m_0 is the assumed pre-existing *bare mass*. This is a type of classical *renormalization* in which we attempt to synchronize the values of calculated parameters in the theory such as mass with observed values. Of course we already know from our point particle electrostatic calculations in Section 2.10 that anything containing m_{el} is infinite, so there is an additional motivation for introducing electron structure. With this interpretation, the Abraham-Lorentz force equation for a classical charged particle in electrodynamics reads

$$\dot{\vec{v}}(t) = \vec{F}/m + \tau\ddot{\vec{v}}(t), \quad (14.31)$$

where

$$\tau \equiv \frac{2e^2}{3mc^3}, \quad (14.32)$$

is a characteristic time. For an electron, $\tau \approx 6.27 \times 10^{-24}$ s. The distance that a light pulse travels in this time is given by $c\tau = \frac{2}{3}r_e$, where $r_e = e^2/mc^2$ is termed the classical electron radius (introduced previously in Exercise 12.1.1); this is the approximate radius of a hypothetical spherically shaped classical electron whose rest mass is entirely given by its electrostatic self-energy (see Exercise 2.10.4). Numerically, $c\tau \approx 1.88 \times 10^{-13}$ cm. In comparison, the electromagnetic charge radius of the proton is smaller: $\sim 0.8 \times 10^{-13}$ cm. Also note that $c\tau$ is much smaller than the corresponding quantum mechanical Compton wavelength $\lambda_C = h/mc$ where h is Planck's constant: $c\tau = \frac{2}{3}\alpha\lambda_C/(2\pi)$, where $\alpha \approx 1/137$ is the fine structure constant.

Dotting Equation (14.31) into the velocity $\vec{v}$ and rearranging gives us the energy equation,

$$\frac{d}{dt}\left(\frac{1}{2}mv^2\right) = \vec{F}\cdot\vec{v} - \frac{2e^2}{3c^3}\dot{\vec{v}}^{\,2} + \tau\frac{d^2}{dt^2}\left(\frac{1}{2}mv^2\right). \tag{14.33}$$

Here we encounter the nonrelativistic Larmor energy loss as the second term on the right.

There is no problem putting Equations (14.31) and (14.33) into relativistic form, which will emerge as the four components of a power/force four-vector equation.[4] One way to approach this generalization is to consider the inertial frame ($\vec{v} = 0$) limit of terms which might be encountered in the process of constructing this equation. The reader may check that:

$$p^\mu|_{\vec{v}=0} = (mc, 0), \tag{14.34}$$

$$\left.\frac{dp^\mu}{d\tau}\right|_{\vec{v}=0} = mc\,(0, \dot{\vec{\beta}}\,), \tag{14.35}$$

$$\left.\frac{d^2p^\mu}{d\tau^2}\right|_{\vec{v}=0} = mc\,(\dot{\vec{\beta}}^{\,2}, \ddot{\vec{\beta}}\,). \tag{14.36}$$

Note the $\dot{\vec{\beta}}^2$ term in the energy component in (14.36) which must be subtracted off in the $\vec{v} = 0$ limit since all energy contributions vanish in this frame of reference. Now, using only the four-vectors p^μ, $\dfrac{dp^\mu}{d\tau}$ and $\dfrac{d^2p^\mu}{d\tau^2}$ one may construct the relativistic generalization of (14.31) and (14.33) as

$$\frac{dp^\mu}{d\tau} = K^\mu + \tau\left[\frac{d^2p^\mu}{d\tau^2} + \frac{p^\mu}{(mc)^2}\frac{dp^\nu}{d\tau}\frac{dp_\nu}{d\tau}\right], \tag{14.37}$$

where $K^\mu = \gamma F^\mu$ is the relativistic power/force. If we apply (14.37) to a purely electromagnetic situation, we would use

$$F^\mu_{\text{Lor}} \equiv \left(e\vec{\beta}\cdot\vec{E},\ e(\vec{E} + \vec{\beta}\times\vec{B})\right). \tag{14.38}$$

F^μ_{Lor} gathers together the familiar (noncovariant) Lorentz power/force terms. Since

$$\frac{d^2}{d\tau^2}\left(p^\mu p_\mu\right) = 0, \tag{14.39}$$

[4]See the interesting history of this development in Ref. 5 of the *Going Deeper* Section 14.4.1. It involves H. Lorentz, M. Abraham and P. A. M. Dirac.

we may put (14.37) into an alternate form:

$$\frac{dp^\mu}{d\tau} = K^\mu + \tau \frac{d^2 p_\nu}{d\tau^2}\left(\eta^{\mu\nu} - \frac{p^\mu p^\nu}{(mc)^2}\right). \tag{14.40}$$

Note that this may also be written as

$$\frac{dp^\mu}{d\tau} = K^\mu - \frac{1}{c}P^\mu + \tau \frac{d^2 p^\mu}{d\tau^2}, \tag{14.41}$$

where

$$P^\mu \equiv P_{\rm L}\left(\frac{p^\mu}{mc}\right). \tag{14.42}$$

The relativistic Larmor power $P_{\rm L}$ is given by (13.100). We may term P^μ the relativistic power flow. The zero$^{\rm th}$ component of (14.41) has a one-to-one relationship with the terms seen in (14.33). The space component comparison reveals we are missing a small momentum flow term in Equation (14.31). However, given the radiation term on the power side, such a term is necessary to maintain momentum conservation. It is understood that the Abraham-Lorentz-Dirac power/force equation (14.37), (14.40) or (14.41), pertains to a charged point particle.

We are briefly pleased with the result. Unfortunately, we have been blithely developing the mathematics and ignoring the physics in this discussion! The extra term in (14.31) includes a third order time derivative of the particle's position. This classical equation has unexpected types of solutions. In particular it allows so-called *runaway* and *pre-acceleration* solutions. The runaway solutions involve terms proportional to τ, which acquire a time dependence $\sim e^{t/\tau}$ for motions confined to $t > 0$. As you will see in Exercise 14.1.6, we may banish the runaway solutions in (14.31) via boundary conditions, only to encounter the pre-acceleration problem. The pre-acceleration involves the motion of the particle before the classical force has acted. This effect is also proportional to τ, and has the behavior $\sim e^{t/\tau}$ before the force acts. Both of these problems are also embedded in the relativistic approach (although we might expect the runaway speeds to be limited to the speed of light). Given the banishment of runaways but the existence of pre-acceleration in either context, how do we proceed from here?

One approach is decidedly drastic: we amputate the offending terms! Nonrelativistically, this is accomplished with an approximate evaluation

$$m\ddot{\vec{v}} \approx \dot{\vec{F}}, \tag{14.43}$$

resulting in the modified force equation:

$$\dot{\vec{v}} = \frac{1}{m}\left(\vec{F} + \tau\dot{\vec{F}}\right). \tag{14.44}$$

(14.44) is called the Landau-Lifshitz formula.[5] This should be a good description as long as

$$|\tau\dot{\vec{F}}| \ll |\vec{F}|. \tag{14.45}$$

The advantage of this approach is that (14.44) is a second-order differential equation in position, restoring a type of Newtonian mechanics and banishing the pre-acceleration solutions. Following the same path relativistically, we use

$$\frac{d^2p^\mu}{d\tau^2} \approx \frac{dK^\mu}{d\tau}, \tag{14.46}$$

in (14.40), giving the generalization of (14.44) as

$$\frac{dp^\mu}{d\tau} = K^\mu + \tau\frac{dK_\nu}{d\tau}\left(\eta^{\mu\nu} - \frac{p^\mu p^\nu}{(mc)^2}\right). \tag{14.47}$$

The approximation here is

$$\left|\tau\frac{dK_\nu}{d\tau}\left(\eta^{\mu\nu} - \frac{p^\mu p^\nu}{(mc)^2}\right)\right| \ll |K^\mu|, \tag{14.48}$$

interpreted in a component by component manner. We may again apply (14.47) to the case of pure electromagnetic forces using (14.38). After a great churning of derivatives, one may finally show that

$$\frac{d\vec{p}}{dt} = \vec{F}_{\text{Lor}} + \tau\gamma\left[\frac{d\vec{F}_{\text{Lor}}}{dt} + \gamma^2\dot{\vec{\beta}}\times\left(\vec{F}_{\text{Lor}}\times\vec{\beta}\right)\right], \tag{14.49}$$

$$\frac{dp^0}{dt} = \vec{\beta}\cdot\frac{d\vec{p}}{dt}. \tag{14.50}$$

Equation (14.50) is not too hard to establish given that $p_\mu\dfrac{dp^\mu}{d\tau} = 0$ is maintained in (14.47) in the electromagnetic case. The extra terms in (14.49) amend the original Lorentz force and power equations in a way which makes eminent good sense; for example, see Exercise 14.1.7. However, it is clear that this approach is an approximation; we have just kept the first term in an infinite series in τ.

A deeper insight into the issues faced in evaluating self-forces in classical electrodynamics has emerged from a study of classical charged particle

[5]L. D. Landau and E. M. Lifshitz, *Classical Theory of Fields*, 4th ed. with corrections, Butterworth-Heinemann (Oxford) 2000, Secs. 75, 76.

models. This theoretical development involves the connection found between the "electron" size and pre-acceleration in the context of a spherical shell distribution. If the classical electron we are studying has a radius a greater than $c\tau = \frac{2}{3}r_e$, it can be shown[6] that the motion of our classical particle is both causal and without runaway solutions. The proof uses the idea from Section 9.7 regarding causality in Fourier transformed quantities being associated with the absence of poles in the upper half complex plane. The critical radius $c\tau$ is picked out by the requirement that the bare mass m_0 in (14.30) be positive, or equivalently, that the electromagnetic contribution to the classical electron's mass not exceed its physical mass. This idea presumably holds for other spherically symmetric charge distributions as well. This is certainly very reasonable and reassuring.

What are we to make of the general situation? We should be pleased that we are now in possession of a practical and effective differential equation that accurately accounts for radiation reaction, even if it may not be exact. Should we stop here in our development of classical electrodynamics? Are we at the end of the classical electrodynamic road? One can not escape the feeling that there is something more to the situation. Of course, the real electron has a radius, if it exists, much smaller than $c\tau$. However, its structure clearly can not be treated classically (as attested by its Compton radius) and we should not expect our considerations to constrain it in any way. But note that if we take a classical electron radius with $a \lesssim c\tau$, we have a finite, covariant theory, but with pre-acceleration. One may view the classical electron model in this case simply as a *regularization* procedure. Regularization procedures are common in field theories which have formal infinities, which one wants to control without violating any important properties or symmetries in the theory. Is this classical theory unreasonable? There is another development which may be relevant in interpreting such a theory which we will now consider.

14.1.2 *Wheeler-Feynman electrodynamics*

Let us consider going "all the way" and setting $a = 0$. In this case we have that m_{el} is infinite, which implies that the self-force is infinite as well. This was the electrodynamic situation that confronted John Wheeler and

[6]E. J. Moniz and D. H. Sharp, "Absence of Runaways and Divergent Self-Mass in Nonrelativistic Quantum Electrodynamics", Phys. Rev. D **10**, 1133 (1974); "Radiation Reaction in Nonrelativistic Quantum Electrodynamics", Phys. Rev. D **15**, 2850 (1977).

Richard Feynman before the development of quantum renormalization.[7] One possible way around this dilemma is to postulate that elementary point particles do not interact with the field that they produce. But this solution amounts to throwing the baby out with the bathwater: besides getting rid of the inertial force term in (14.27), it also gets rid of the Larmor radiation term, which has a physically demonstrable effect. Notice however that if we use the average of retarded and advanced Green functions, then the self force expression (14.29) does not include the Larmor term. So if we use this averaged Green function and postulate that elementary particles do not interact with their own fields, then we remove only the infinite term.

The averaged Green function has the additional advantage that it is symmetric in time, which corresponds to the fact that the differential equations for the 4-potential A^μ are symmetric in time. This leads to an alternative picture of radiation as a symmetric interaction at a distance between charged particles, which is analogous to the Newtonian picture of gravity. In Newton's theory, the gravitational field is not an independent dynamical entity. Rather, it is essentially a bookkeeping device that describes potential interactions between masses. The gravitational field at a point in space-time only tells what the force would be on a mass located at that point, *if* such a mass were present. If no mass is there, the field has no physical consequence. For this same idea to work in electromagnetism, then all radiation that is emitted must also be absorbed. This idea was first advanced by Tetrode in 1922.[8]

These circumstances led Wheeler and Feynman to suggest that the average of advanced and retarded Green functions may indeed be the physical one. However, this choice is not without its own difficulties. It implies that the influence of accelerating charged particles extends to the past as well as the future! This is certainly a violation of common sense, but that doesn't mean that it is wrong. Time after time, new physical insights have destroyed common-sense notions. The real test is whether the theory makes correct predictions. To investigate this we will work out a practical (but somewhat idealized) example, using first the conventional retarded Green function, and then alternatively using the averaged Green function.

---

[7]J. Wheeler and R. P. Feynman, "Interaction with the Absorber as the Mechanism of Radiation", Rev. Mod. Phys. **17**, 157 (1945); "Classical Electrodynamics in Terms of Direct Interparticle Action", Rev. Mod. Phys. **21**, 3 (1949).

[8]H. Tetrode, "Über den Wirkungszusammenhang der Welt. Eine Erweiterung der Klassischen Dynamik" Zeits. f. Physik **10**, 317 (1922).

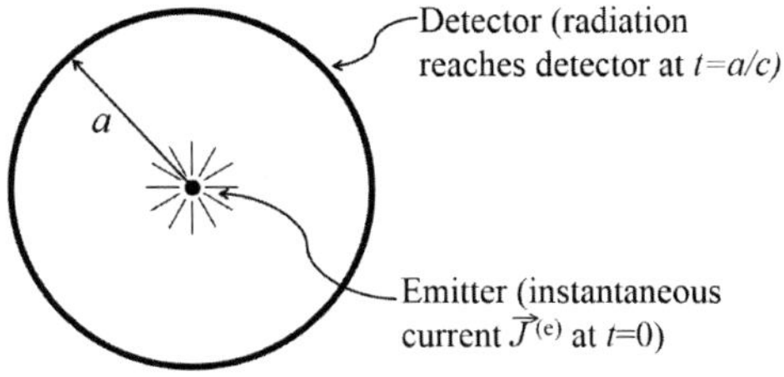

Fig. 14.1 Source and detector for scenario illustrating Wheeler-Feynman electrodynamics.

Consider the situation shown in Figure 14.1 of a positive point charge located at $\vec{x} = 0$, which at time $t = 0$ experiences a sudden, infinitesimal displacement. Since the displacement is infinitesimal, the charge remains virtually constant, so the scalar potential Φ is constant over time. The vector potential produced by this instantaneous current (which we designate as the "emitter" current, or $\vec{J}^{(e)}$) may be computed using the free-space retarded Green function with source at $\vec{x}' = 0$, $t' = 0$ as from (8.115):

$$\vec{A}^{(e)}_{\text{ret}}(\vec{x}, t) = \frac{1}{c} \int d^3x' \, \frac{\vec{J}^{(e)}(\vec{x}', t - \frac{|\vec{x}|}{c})}{|\vec{x}|}, \tag{14.51}$$

where the integral is over an infinitesimal volume around the origin. Now suppose that the charge is located at the center of a spherical detector of radius a, as shown in the figure. At time $t = a/c$, the pulse of radiation from the emitter reaches the detector; and we suppose that charges on the detector move so as to completely absorb the radiation. By linearity, the retarded vector potential due to the motion of these charges completely cancels out the retarded vector potential from the original instantaneous current. It follows that

$$\vec{A}^{(a)}_{\text{ret}}(\vec{x}, t) = -\vec{A}^{(e)}_{\text{ret}}(\vec{x}, t) \qquad (t \geq a/c), \tag{14.52}$$

where the superscript (a) refers to the charges in the detector that absorb the radiation.

Let us consider now the advanced vector potentials. At this point, we may use the time-reversibility of the equations for the vector potential. If we took a movie of this situation and ran the movie backwards in time (starting at $t = a/c$), we would still see a valid electromagnetic process. The retarded potential for the time-reversed process is exactly equal to the advanced potential for the original process. It follows that

$$\vec{A}^{(a)}_{\text{adv}}(\vec{x}, t) = \vec{A}^{(e)}_{\text{ret}}(\vec{x}, t), \qquad (0 \leq t \leq a/c). \tag{14.53}$$

Now, how about continuing the advanced vector potential to $t < 0$? In analogy to (14.52), this advanced potential must completely cancel out the advanced field of the point emitter. Therefore we have

$$\vec{A}^{(\mathrm{e})}_{\mathrm{adv}}(\vec{x},t) = -\vec{A}^{(\mathrm{a})}_{\mathrm{adv}}(\vec{x},t), \qquad (t \le 0). \tag{14.54}$$

Let us now describe the same situation using the vector potentials obtained using the averaged Green function:

$$\vec{A}^{(\mathrm{e})}_{\mathrm{avg}}(\vec{x},t) \equiv \frac{1}{2}[\vec{A}^{(\mathrm{e})}_{\mathrm{ret}}(\vec{x},t) + \vec{A}^{(\mathrm{e})}_{\mathrm{adv}}(\vec{x},t)], \tag{14.55}$$

$$\vec{A}^{(\mathrm{a})}_{\mathrm{avg}}(\vec{x},t) \equiv \frac{1}{2}[\vec{A}^{(\mathrm{a})}_{\mathrm{ret}}(\vec{x},t) + \vec{A}^{(\mathrm{a})}_{\mathrm{adv}}(\vec{x},t)]. \tag{14.56}$$

Using relations (14.52)–(14.54), we have for the total vector potential due to absorber + emitter:

$$\begin{aligned}
&\vec{A}^{(\mathrm{e})}_{\mathrm{avg}}(\vec{x},t) + \vec{A}^{(\mathrm{a})}_{\mathrm{avg}}(\vec{x},t) \\
&\quad = \frac{1}{2}(\vec{A}^{(\mathrm{e})}_{\mathrm{adv}}(\vec{x},t) + \vec{A}^{(\mathrm{a})}_{\mathrm{adv}}(\vec{x},t)) = 0 && (t < 0), && (14.57)\\
&\quad = \frac{1}{2}(\vec{A}^{(\mathrm{e})}_{\mathrm{ret}}(\vec{x},t) + \vec{A}^{(\mathrm{a})}_{\mathrm{adv}}(\vec{x},t)) = \vec{A}^{(\mathrm{e})}_{\mathrm{ret}}(\vec{x},t) && (0 \le t \le a/c), && (14.58)\\
&\quad = \frac{1}{2}(\vec{A}^{(\mathrm{e})}_{\mathrm{ret}}(\vec{x},t) + \vec{A}^{(\mathrm{a})}_{\mathrm{ret}}(\vec{x},t)) = 0 && (t > a/c). && (14.59)
\end{aligned}$$

In other words, the vector potentials derived from the averaged Green function agree exactly with those derived from the retarded Green function!

Can this picture explain the force of radiation reaction on the emitting particle? Yes, since $\vec{A}^{(\mathrm{e})}_{\mathrm{ret}} = \vec{A}^{(\mathrm{e})}_{\mathrm{avg}} + \vec{A}^{(\mathrm{a})}_{\mathrm{avg}}$ and the $\vec{A}^{(\mathrm{e})}_{\mathrm{avg}}$ term produces the inertial force term in (14.21), it must be the case that $\vec{A}^{(\mathrm{a})}_{\mathrm{avg}}$ produces the radiation reaction force.

So it seems that Wheeler and Feynman have presented an elegant solution to the infinite self-force problem. Given that a particle's field represents its interactions with other particles (and does not affect the particle itself), and supposing that the time-symmetrized Green function is the correct one, we arrive at a finite expression for forces on elementary particles. The pre-acceleration experienced for point particles in their view emerged as the "witness" to the noncausal propagation. There is just one problem: particles *do* interact with their own fields! The Lamb shift in hydrogen (measured by Lamb and Retherford in 1947[9]) results from the interaction

[9]W. Lamb and R. Retherford, "Fine Structure of the Hydrogen Atom by a Microwave Method", Phys. Rev. **72**, 241 (1947). Lamb shared the 1955 Nobel Prize in physics for this research.

of the electron with its own field in the vicinity of the atom. This does not mean that the time-symmetrized Green function is invalid: it still provides a mathematically consistent picture of electromagnetic interactions between particles. However, the Wheeler-Feynman theory is undercut by the fact that the original motivating assumption (that particles do not interact with their own field) has been proven to be incorrect. Since the theory provides no advantage in practical computations, it has fallen by the wayside. Although Wheeler-Feynman electrodynamics is not currently an active field of physics research, there are subtle hints that the noncausal picture it paints may be more than just a mathematical fiction. John G. Cramer has developed an interpretation of quantum mechanics based on similar assumptions.[10] His proposal, called "Transactional Quantum Mechanics", provides a natural interpretation of the Born probability rule, and provides a plausible explanation for quantum paradoxes such as nonlocal correlations (that is, "quantum entanglement") and the so-called collapse of the wave packet. But Cramer's theory, like Wheeler-Feynman's, is far from universally accepted by the physics community. Despite this lukewarm response, there is one sense in which Feynman has had the last laugh. The "Feynman propagator", one of the foundational tools of quantum field theory, is the space-time generalized analog of the retarded and advanced averaged Green function described above. In this case, the additional parts of the propagator are intimately associated with the presence of independent (e.g., electron) or non-independent (e.g., photon) particles with the opposite set of quantum numbers, called antiparticles. We will study this propagator and other matters in the next section.

14.2 Connections between classical electrodynamics and quantum field theory

The classical theory of electromagnetism is grounded in a more fundamental field theory known as *Quantum Electrodynamics*, or QED. The development of QED is one of the greatest achievements of intellectual thought. QED is just one example of a quantum field theory; such theories postulate the existence of certain point-like particles or quanta. Although our focus in

[10] J. G. Cramer, "The Transactional Interpretation of Quantum Mechanics", Reviews of Modern Physics **58**, 647 (1986); "An Overview of the Transactional Interpretation", International Journal of Theoretical Physics **27**, 227 (1988). See also Cramer's web site at `http://www.npl.washington.edu/npl/int_rep/tiqm/TI_toc.html`

this text is on classical electrodynamics, in this section we will point out a few connections with quantum field theories. In particular, we will show some of the relations between the Feynman propagator used in field theory and the Green functions used in classical electromagnetism.

Quantum field theory arose as a fusion and extension of the ideas of relativistic fields and quantum mechanics. In effect, a quantum field is a field of quantum-mechanical operators. As in quantum mechanics, physical observables are expressed in terms of matrix elements of operators (constructed from the field operators) which act on an underlying Hilbert space. Unlike the quantum-mechanical wavefunction, a quantum field does not merely represent a single particle but rather multiple particles, and allows for the possibility that these particles are created or destroyed.

14.2.1 *Massless scalar field and the Feynman propagator*

We begin our mathematical description with the simplest of quantum fields: the massless scalar field, which describes a type of massless particle or field which carries no spin quantum number. We will use the notation $\phi(\vec{x},t)$ to denote the field operator associated with space-time location $(\vec{x},t)$. These operators act on a Hilbert space of quantum states; in particular, the *vacuum state* is represented as a "ket" vector $|0\rangle$, while its dual is the "bra" vector $\langle 0|$ (so that $\langle 0\,|\,0\rangle = 1$). These operators transform as ordinary scalars, so that under the Lorentz transformation $(\vec{x},t) \to (\vec{x}\,',t')$ we have $\phi'(\vec{x}\,',t') = \phi(\vec{x},t)$. Since the $\phi(\vec{x},t)$ are operators rather than numbers or vectors, they do not commute in general, so that $\phi(\vec{x}_1,t_1)\phi(\vec{x}_2,t_2) \neq \phi(\vec{x}_2,t_2)\phi(\vec{x}_1,t_1)$.

In electromagnetism we used Green functions; in quantum field theory, the analogous mathematical objects are called *propagators*. Let us define the *Feynman propagator*[11] for a scalar field theory as

$$G_+(\vec{x},t;\vec{x}\,',t') \equiv \langle 0 \mid T[\phi(\vec{x},t)\phi^\dagger(\vec{x}\,',t')] \mid 0\rangle, \tag{14.60}$$

where T is the *time-ordering operator*, which acts on fields as follows:

$$T[\phi(\vec{x},t)\phi^\dagger(\vec{x}\,',t')] = \begin{cases} \phi(\vec{x},t)\phi^\dagger(\vec{x}\,',t') & t > t' \\ \phi^\dagger(\vec{x}\,',t')\phi(\vec{x},t) & t' > t \\ \frac{1}{2}\left(\phi(\vec{x},t)\phi^\dagger(\vec{x}\,',t) + \phi^\dagger(\vec{x}\,',t)\phi(\vec{x},t)\right) & t = t'. \end{cases} \tag{14.61}$$

The $\dagger$ symbol in (14.60) and (14.61) represents Hermitian conjugation.

[11]The Feynman propagator was actually first developed by E.C.G. Stueckelberg.

Recall that in Section 3.2 we represented Green functions as Fourier transforms in order to obtain solutions. For a scalar field, $\phi(\vec{x}, t)$, the analogous Fourier representation is (a factor of $\sqrt{\hbar}$ is suppressed for convenience)

$$\phi(\vec{x}, t) = c \sum_j \sqrt{\frac{2\pi}{\omega_j L^3}} \left\{ a_j e^{i(\vec{k}_j \cdot \vec{x} - \omega_j t)} + a_j^\dagger e^{-i(\vec{k}_j \cdot \vec{x} - \omega_j t)} \right\}. \tag{14.62}$$

In (14.62), j indexes momentum eigenstates of the field (only certain momenta are allowed, as described below); $\vec{k}_j$ and ω_j are the wave number and angular frequency respectively of the indicated eigenstate, corresponding to momentum $\hbar\vec{k}_j$ and energy $\hbar\omega_j$ ($\omega_j \equiv c|\vec{k}_j|$ for a massless field); and $a_j^\dagger$ and a_j are *creation* and *destruction* operators which are described in more detail below (note $a_j^\dagger$ is the Hermitian conjugate of a_j).

From (14.62) it may be seen that ϕ is a Hermitian field:

$$\phi^\dagger(\vec{x}, t) = \phi(\vec{x}, t). \tag{14.63}$$

The so-called periodic box normalization conditions for a box of size L^3 yield momenta with discretized x, y, and z components:

$$\vec{k} = (k_x, k_y, k_z) = \frac{2\pi}{L} \cdot (\ell_1, \ell_2, \ell_3), \qquad \ell_i = 0, \pm 1, \pm 2, \ldots. \tag{14.64}$$

When spatial integrals are done with this normalization, one has

$$\int d^3x \, e^{i\vec{k}_j \cdot \vec{x}} e^{-i\vec{k}_{j'} \cdot \vec{x}} = L^3 \delta_{j,j'}, \tag{14.65}$$

where j and j' denote (ℓ_1, ℓ_2, ℓ_3) and $(\ell_1', \ell_2', \ell_3')$ respectively.

The eigenstates of the system are specified as $|n_1 \; n_2 \; n_3 \; \ldots\rangle$ ($n_j = 0, 1, 2, \ldots$), where n_j signifies the number of particles of momentum $\vec{k}_j$. The $a_j^\dagger$ and a_j operators in (14.62) respectively create and destroy particles of momentum $\vec{k}_j$ as follows:

$$a_j^\dagger \mid n_1, n_2, \ldots, n_j \ldots\rangle = \sqrt{n_j + 1} \mid n_1 \; n_2 \; \ldots n_j + 1 \ldots\rangle, \tag{14.66}$$

$$a_j \mid n_1, n_2, \ldots, n_j \ldots\rangle = \sqrt{n_j} \mid n_1 \; n_2 \; \ldots n_j - 1 \ldots\rangle, \tag{14.67}$$

$$a_j \mid n_1, n_2, \ldots, n_j \ldots\rangle = 0, \; n_j = 0. \tag{14.68}$$

The operator, $a_j^\dagger a_j$ is known as the *number operator* because

$$a_j^\dagger a_j \mid \ldots, n_j, \ldots\rangle = a_j^\dagger \sqrt{n_j} \mid \ldots n_j - 1 \ldots\rangle = n_j \mid \ldots, n_j, \ldots\rangle. \tag{14.69}$$

Each possible momentum in the system obeys the same mathematical relations as for a harmonic oscillator in quantum mechanics, where $a_j^\dagger$ and a_j

play the role of raising and lowering operators corresponding to the particular momentum $\vec{k}_j$. This identification is reflected in the commutators:

$$\left[a_j, a_{j'}^\dagger\right] = \delta_{j,j'}; \qquad [a_j, a_{j'}] = 0; \qquad [a_j^\dagger, a_{j'}^\dagger] = 0, \tag{14.70}$$

which are the same as the commutation relations for raising and lowering operators in quantum mechanics.

Let us now evaluate the scalar Feynman propagator in this formalism. When $t > t'$, we have

$$G_+(\vec{x}, t; \vec{x}\,', t') = \langle 0 \mid \phi(\vec{x}, t)\phi^\dagger(\vec{x}\,', t') \mid 0\rangle. \tag{14.71}$$

Using (14.62) this yields

$$G_+(\vec{x}, t; \vec{x}\,', t') =$$

$$c^2\langle 0 \mid \sum_j \sum_{j'} \sqrt{\frac{2\pi}{\omega_j L^3}} \sqrt{\frac{2\pi}{\omega_{j'} L^3}} \{a_j a_{j'}^\dagger e^{i(\vec{k}_j\cdot\vec{x} - \omega_{\vec{k}_j} t)} e^{-i(\vec{k}_{j'}\cdot\vec{x}\,' - \omega_{j'} t')} + \ldots\}|0\rangle. \tag{14.72}$$

Since the expectation is evaluated between vacuum states, the only term which contributes is the $a_j a_{j'}^\dagger$ term (shown explicitly above) for $j = j'$. This leads to the discrete form,

$$G_+(\vec{x}, t; \vec{x}\,', t') = c^2 \sum_j \left(\frac{2\pi}{\omega_j L^3}\right) e^{i(\vec{k}_j\cdot(\vec{x} - \vec{x}\,') - \omega_j(t - t'))}. \tag{14.73}$$

As the box size $L \to \infty$ we can use the continuum correspondence,

$$\sum_j \;\to\; L^3 \int \frac{d^3k}{(2\pi)^3}, \tag{14.74}$$

to do the momentum sums. This results in (recall that $\omega = ck$ where $k \equiv |\vec{k}|$)

$$G_+(\vec{x}, t; \vec{x}\,', t') = 4\pi c \int \frac{d^3k}{(2\pi)^3} \frac{1}{2k} e^{i(\vec{k}\cdot(\vec{x} - \vec{x}\,') - ck(t - t'))}, \; (t > t'). \tag{14.75}$$

Considering the other time ordering $t' > t$ gives us the general expression

$$G_+(\vec{x}, t; \vec{x}\,', t') = 4\pi c \int \frac{d^3k}{(2\pi)^3} \frac{1}{2k} e^{i(\vec{k}\cdot(\vec{x} - \vec{x}\,') - ck|t - t'|)}. \tag{14.76}$$

A convergence factor of $(1 - i\epsilon)$ is now inserted to multiply the $ck|t - t'|$ factor in the exponential, producing exponential damping for large k. In Exercise 14.2.1 the reader will prove that (14.76) is identical (up to an overall factor of $-i$) to the classical electromagnetic Green function (8.105) but with a different pole prescription. This exercise also leads us to the final form, which is valid for arbitrary t and t':

$$G_+(\vec{x},t;\vec{x}\,',t') = \frac{c}{\pi}\frac{1}{R^2 - c^2T^2 + i\epsilon}, \tag{14.77}$$

where $T \equiv t - t'$ and $R \equiv |\vec{x} - \vec{x}\,'|$. We may use the familiar delta function limit expression

$$\lim_{\epsilon\to 0^+} \frac{\epsilon}{x^2+\epsilon^2} \to \pi\delta(x), \tag{14.78}$$

to obtain an expression which is easier to interpret:

$$G_+(\vec{x},t;\vec{x}\,',t') = \frac{c}{\pi}\left\{P\frac{1}{R^2 - c^2T^2} - i\pi\delta\left(R^2 - c^2T^2\right)\right\}. \tag{14.79}$$

The second term in (14.79) only contributes when $R^2/c^2 = T^2$, which has two solutions: $T = R/c$ and $T = -R/c$. It equally weights these two possibilities, as in Wheeler and Feynman's classical theory. In addition, the first term contributes for all time relations between T and R/c including those with $T < R/c$. Clearly, the Feynman propagator is not causal!

14.2.2 *Composition rule for Feynman propagator*

Let us give another example of this quantum field-theoretic formalism. This time we will derive a composition rule for the scalar Feynman propagator which is similar to the non-relativistic composition property in quantum mechanics.[12] Recall that the composition (or completeness) rule (for position-space eigenvectors) can be expressed as (assuming $t'' > t' > t$)

$$\langle\vec{x}\,'',t''\mid\vec{x},t\rangle = \int d^3x'\,\langle\vec{x}\,'',t''\mid\vec{x}\,',t'\rangle\,\langle\vec{x}\,',t'\mid\vec{x},t\rangle, \tag{14.80}$$

where $t'' > t' > t$. In the scalar quantum field case, we consider the similar expression (also assuming $t'' > t' > t$)

$$\int d^3x'\,\left\{\langle 0|\phi(\vec{x}\,'',t'')\phi^\dagger(\vec{x}\,',t')|0\rangle\frac{\partial}{\partial t'}\langle 0|\phi(\vec{x}\,',t')\phi^\dagger(\vec{x},t)|0\rangle\right.$$
$$\left.-\langle 0|\phi(\vec{x}\,',t')\phi^\dagger(\vec{x},t)|0\rangle\frac{\partial}{\partial t'}\langle 0|\phi(\vec{x}\,'',t'')\phi^\dagger(\vec{x}\,',t')|0\rangle\right\}. \tag{14.81}$$

[12]The composition rule or property can be found in many texts on quantum mechanics, including J. J. Sakurai and J. Napolitano, *Modern Quantum Mechanics*, 2nd ed., Addison-Wesley (San Francisco) 2011, Section 2.6.

Using (14.71) and (14.73), we may rewrite (14.81) as

$$c^4 \int d^3x' \sum_j \sum_{j'} \left(\frac{2\pi}{\omega_j L^3}\right)\left(\frac{2\pi}{\omega_{j'} L^3}\right)$$
$$\left\{e^{i(\vec{k}_{j'}\cdot(\vec{x}\,''-\vec{x}\,')-\omega_{j'}(t''-t'))}(-i\omega_j)e^{i(\vec{k}_j\cdot(\vec{x}\,'-\vec{x})-\omega_j(t'-t))}\right.$$
$$\left.-e^{i(\vec{k}_j\cdot(\vec{x}\,'-\vec{x})-\omega_j(t'-t))}(i\omega_{j'})e^{i(\vec{k}_{j'}\cdot(\vec{x}\,''-\vec{x}\,')-\omega_{j'}(t''-t'))}\right\}. \qquad (14.82)$$

The spatial integral may be done using (14.65) leaving a single sum which we may recognize:

$$-4\pi i c^4 \sum_j \left(\frac{2\pi}{\omega_j L^3}\right) e^{i(k_j\cdot(x''-x)-\omega_j(t''-t))} = -4\pi i c^2 G_+(\vec{x}\,'', t''; \vec{x}, t). \qquad (14.83)$$

Equating (14.83) with (14.81) and using (14.71) yields

$$G_+(\vec{x}\,'', t''; \vec{x}, t) = \frac{i}{4\pi c^2}\int d^3x' \left\{G_+(\vec{x}\,'', t''; \vec{x}\,', t')\frac{\partial}{\partial t'}G_+(\vec{x}\,', t'; \vec{x}, t)\right.$$
$$\left. - G_+(\vec{x}\,', t'; \vec{x}, t)\frac{\partial}{\partial t'}G_+(\vec{x}\,'', t''; \vec{x}\,', t')\right\}, \qquad (14.84)$$

for $t'' > t' > t$.

Expression (14.84) may be interpreted physically as representing the propagation of a scalar particle or wave from $(\vec{x}, t)$ to an intermediate position and time $(\vec{x}\,', t')$, after which another Green function continues the propagation from $(\vec{x}\,', t')$ to $(\vec{x}\,'', t'')$. In fact, expression (14.82) shows that the two terms on the right-hand side of (14.84) give the same overall contribution; this makes (14.84) similar to the completeness rule (14.80), except for the presence of the time derivative. Exercise 14.2.3 outlines a similar derivation for retarded Green functions; in this case, the two terms on the right-hand side of the analog of (14.82) make different contributions.

14.2.3 *The photon field*

In the quantum version of electrodynamics, the scalar field used above is replaced by the *photon field*, which is a vector field of the form (this time the factor of $\sqrt{\hbar}$ is included)

$$\vec{A}(\vec{x}, t) = c\sum_{j,\alpha}\sqrt{\frac{2\pi\hbar}{\omega_j L^3}}\left\{a_{j,\alpha}\hat{\epsilon}_{j,\alpha}e^{i(\vec{k}_j\cdot x-\omega_j t)} + a^\dagger_{j,\alpha}\hat{\epsilon}_{j,\alpha}e^{-i(\vec{k}_j\cdot x-\omega_j t)}\right\}. \qquad (14.85)$$

Here we see a new aspect of the formalism in the insertion of polarization vectors, $\hat{\epsilon}_{j,\alpha}$, and the addition of a polarization sum over α. This corresponds to the fact that classical electromagnetic waves are also polarized, as we discussed in Section 9.1. Here we will regard α as having two values, specifying the possible linear polarization states, and that these real vectors satisfy the relations,

$$(\hat{\epsilon}_{\{1,2\}})^2 = 1, \quad \hat{\epsilon}_{j,1} \times \hat{\epsilon}_{j,2} = \hat{k}_j, \quad \hat{\epsilon}_{\{1,2\}} \cdot \hat{k}_j = 0, \tag{14.86}$$

where $\hat{k}_j \equiv \vec{k}_j/|\vec{k}_j|$. As before, the vector $\vec{A}(\vec{x}, t)$ is interpreted as a Hermitian field operator in number space. Using such constructions, one may build a theory of the quantized electromagnetic field and study its measurement properties (see Exercise 14.2.4).

In a generalized field theory formalism, the fundamental variables in the theory are integrated over to produce what is called the Feynman path integral, analogous to the partition function in statistical mechanics. This is done in the arena known as Euclidean space-time, where the space and time coordinates enter on the same footing. One may also introduce numerical analogs of the independent field theory variables using a weighted Monte Carlo approach in a computer simulation of such theories. Such a numerical approach is called a "lattice field theory" because an Euclidean space-time lattice is used to define the fields, which have degrees of freedom associated with the lattice points. The Euclidean approach has the advantage of forming observables out of decaying exponentials, rather than phases of oscillatory quantities. Lattice field theories are helping to unlock the mysteries of the interactions in the strong interaction dynamics of quarks and gluons in the theory called *Quantum Chromodynamics* (QCD); the numerical application is called *Lattice QCD*. Many more concepts and types of technology are necessary to accomplish the goals of this research than can or should be described here, even in a superficial manner. A few helpful resources are listed in the *Going Deeper* section.

14.3 Radiating cavities, scattering resonances, and quasinormal modes[13]

The topic introduced in this section builds on the material in Chapters 9–12. In Chapter 9, we discussed plane wave propagation in dielectric materials.

[13]This section is written by Lamis Al Sheikh.

In Chapter 10, we examined normal modes in dielectric filled cavities with both perfectly conducting and imperfectly conducting boundaries, treating the latter case as a perturbation of the former. However, we did not consider dielectrics with spatially varying properties, nor the situation where a localized dielectric was surrounded by a homogeneous medium (as would be the case for instance for a suspended particle of dielectric material). Physically, this means that fields within the dielectric will interact with the field in the surrounding medium. This leads to radiation and scattering, which were discussed in Chapters 11 and 12. Such systems therefore include both localized mode and asymptotic wave-like behavior, and they require new mathematical techniques for analysis, as well as special techniques for numerical solution.[14] One particularly useful notion is *quasinormal modes*, which are an extension of the previously introduced idea of normal modes.

14.3.1 *Practical applications*

Our topic in this section is of considerable practical importance, thanks to recent advances in technology and theoretical physics. Recently developed technology has made it possible to fabricate nanoparticles of various shapes from various materials with dimensions between 10–100 nm. Such nanoparticles are increasingly used in a variety of applications. In medicine they enable targeted drug delivery to specific locations within the body via the bloodstream. In manufacturing, they can be dispersed in industrial coatings to protect wood, plastic, and textiles from exposure to ultraviolet rays. Sunlight concentrated on nanoparticles suspended in water can produce steam to power steam turbines with high energy efficiency. For these applications a thorough understanding of the optical response of nanoparticles is required, and a number of recent studies have explored the electromagnetic properties of nanopartiles from a theoretical point of view.[15]

[14] Lamis Al Sheikh, Scattering resonances and pseudospectrum: stability and completeness aspects in optical and gravitational systems, PhD Thesis, Burgundy University - France, Jan. 2022. *Supervisor: José Luis Jaramillo*

[15] K. Ando and H. Kang, "Analysis of plasmon resonance on smooth domains using spectral properties of the Neumann–Poincaré operator", Journal of Mathematical Analysis and Applications **435(1)**, 162–178 (2016); H. Ammari, M. Ruiz, S. Yu, and H. Zhang, "Mathematical analysis of plasmonic resonances for nanoparticles: the full Maxwell equations", Journal of Differential Equations **261(6)**, 3615–3669 (2016); H. Ammari, P. Millien, M. Ruiz, and H. Zhang, "Mathematical analysis of plasmonic nanoparticles: the scalar case", arXiv preprint arXiv:1506.00866 (2015).

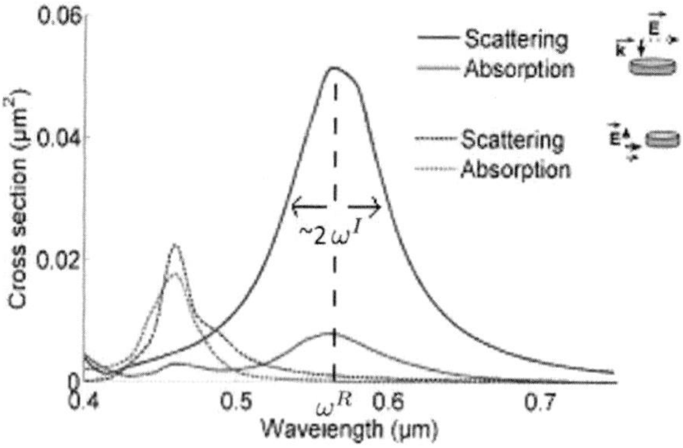

Fig. 14.2 Silver nanodisk optical response.

Physically, a nanoparticle can be modeled as an open cavity which has a discrete set of resonant frequencies. These frequencies show up as peaks in the nanoparticles' scattering and absorption cross-sections. For example, Figure 14.2 shows the optical response of a silver nanodisk with radius $= 0.35$ μm, and thickness $= 0.30$ μm. The peak frequencies and peak width at half maximum correspond to the real and approximately twice the imaginary part of a complex resonance frequency (compare Figure 10.11).

On the theoretical side, we have the recent discovery of gravitational waves emitted when two black holes merge. These waves can be modeled as the "ringing down" of waves formed in an open cavity. The equations are identical to the electromagnetic case, where the gravitational potential can be absorbed into an effective "dielectric constant" which is space-dependent (compare Section 9.5, in which an effective dielectric constant is defined based on the damped motion of constitutive particles within a macroscopic medium).

Because this section is only an introduction to the topic, we will focus on the case of one dimension which illustrates the main ideas. In order to make the discussion self-contained, we will first review the mathematics of normal modes previously discussed in Chapter 9. Then we will show different mathematical and numerical techniques for finding fields associated with dielectric regions in free space.

Conventions: In contrast to the rest of the book, in this section we use the complex representation $e^{i(\omega t-kx)}$ instead of $e^{i(kx-\omega t)}$ (the real parts are the same).

14.3.2 *Definition of scattering field and quasinormal modes*

In one dimension, a nanoparticle can be modeled mathematically as an interval $[0, L]$ on which a variable permittivity $\epsilon(x, t)$ is defined, while $\epsilon(x, t) = 1$ for $x \notin [0, L]$. In this case the time-dependent equation for any electric field component within the nanoparticle (denoted by $\phi_p(x, t)$) is:

$$\left[\epsilon(x, t) * \frac{1}{c^2}\frac{\partial^2}{\partial t^2} - \frac{\partial^2}{\partial x^2}\right] \phi_p(x, t) = 0, \tag{14.87}$$

where $*$ represents convolution. This equation is the inverse Fourier transform of usual wave equation in (x, ω) space, with a multiplicative frequency-dependent dielectric constant: the multiplication becomes convolution under such a transform.

In order to isolate the modal behavior we decompose the total field into *background field* and *scattering field*, denoted by ϕ_b and ϕ_s respectively. The background field is due to external sources that stimulate response within the nanoparticle, and satisfies the free wave equation in the region under consideration. Hence the equation for the background field ϕ_b is

$$\left[\frac{1}{c^2}\frac{\partial^2}{\partial t^2} - \frac{\partial^2}{\partial x^2}\right] \phi_b(x, t) = 0. \tag{14.88}$$

Subtracting (14.88) from (14.87) and equating $\phi_s \equiv \phi_p - \phi_b$, we obtain

$$\left[\epsilon(x, t) * \frac{1}{c^2}\frac{\partial^2}{\partial t^2} - \frac{\partial^2}{\partial x^2}\right] \phi_s(x, t) = \left[(1 - \epsilon(x, t)) * \frac{1}{c^2}\frac{\partial^2}{\partial t^2}\right] \phi_b(x, t). \tag{14.89}$$

After Fourier transformation we get

$$\left[\epsilon(x, \omega)\frac{\omega^2}{c^2} + \frac{\partial^2}{\partial x^2}\right] \phi_s(x, \omega) = \left[(\epsilon(x, \omega) - 1)\frac{\omega^2}{c^2}\right] \phi_b(x, \omega). \tag{14.90}$$

Identifying $[(\epsilon(x, \omega) - 1)\frac{\omega^2}{c^2}]\phi_b(x, \omega)$ as a source $I(x, \omega)$, the equation of the scattered field becomes

$$\left[\epsilon(x, \omega)\frac{\omega^2}{c^2} + \frac{\partial^2}{\partial x^2}\right] \phi_s(x, \omega) = I(x, \omega). \tag{14.91}$$

According to the optics literature, *quasinormal modes* (QNMs) are solutions $\phi(x, \omega)$ to the homogeneous equation (14.91), with $I(x, \omega) = 0$ that satisfy so-called *outgoing boundary conditions*

$$\phi \sim e^{-i\omega|x|/c}, x \to \pm\infty. \tag{14.92}$$

(Note these boundary conditions are consistent with our convention that $e^{i\omega(t-x/c)}$ and $e^{i\omega(t+x/c)}$ represent waves moving in the $+x$ and $-x$ directions, respectively.) These boundary conditions are quite different from the Dirichlet and Neumann boundary conditions that we have seen before, and lead to eigensolutions with quite different properties. We will explore these differences further in Section 14.3.4. But first to motivate the study of QNMs, we discuss three important physical applications.

14.3.3 *Physical applications of quasinormal modes*

Drude model of a conductor

In this section, we derive the specific form of the model presented in Section 14.3.2 for conducting nanoparticles. In Section 9.6 we gave the following phenomenological model for the dielectric constant ((9.148), repeated here for convenience):

$$\epsilon(\omega) = 1 + \frac{4\pi e^2}{m}\sum_i \frac{n_{\mathrm{b},i}}{\omega_{0,i}^2 - \omega^2 - i\omega\gamma_i} + i\frac{4\pi}{\omega}\left(\frac{e^2 n_{\mathrm{c}}}{m}\frac{1}{\gamma - i\omega}\right). \tag{14.93}$$

This is known as the *Lorentz model* for permittivity. In the case where the material has no bound electrons, the sum in (14.93) vanishes and we obtain the *Drude model*:

$$\epsilon_r = \epsilon_\infty - \frac{\omega_p^2(x)}{\omega^2 + i\gamma\omega}, \tag{14.94}$$

where $\omega_p \equiv \dfrac{4\pi e^2 n_c}{m}$ is the *plasmon frequency* and γ represents the absorption of the conducting electrons. We have also replaced 1 with ϵ_∞ to describe the case where the surrounding medium is not a vacuum.

For frequencies ω such that $\gamma \ll \omega$, we may linearize the fractional term in (14.94):

$$\epsilon_r(\omega) \approx \epsilon_\infty - \frac{\omega_p^2}{\omega^2}\left(1 - i\frac{\gamma}{\omega}\right). \tag{14.95}$$

Substituting (14.95) into (14.89) and setting $I = 0$ gives the equation for resonance modes:

$$\left[\frac{\partial^2}{\partial x^2} - V(x) + (\epsilon_\infty - \beta(x,\omega))\,\omega^2\right]\phi_s(x,\omega) = 0, \tag{14.96}$$

where

$$V(x) \equiv \omega_p^2 > 0, \qquad \beta \equiv \frac{i\gamma\omega_p^2}{\omega^3} \ll \epsilon_\infty. \tag{14.97}$$

In the case of a very good conductor, we may take $\beta \approx 0$ in (14.96). The equation for finding the quasinormal modes becomes

$$\left[\frac{\partial^2}{\partial x^2} - V(x)\right] \phi_q(x,\omega) = \epsilon_\infty \omega_q^2 \phi_q(x,\omega). \tag{14.98}$$

This may be recognized as the wave equation with a potential, which has numerous applications across multiple fields of physics. In particular, the well-developed theory of Lax and Phillips can be applied.[16]

Spontaneous emission and the Purcell factor

Spontaneous emission is a process in which a quantum mechanical system (such as a molecule, an atom or a subatomic particle) transitions from an excited energy state to a lower energy state (e.g., its ground state) and emits a quantum of energy in the form of a photon. The rate of spontaneous emission may be estimated by *Fermi's golden rule*.[17] This rate changes due to the environment: for example, the spontaneous emission rate for a dipole can be increased by placing it in a resonating optical cavity. This effect may be quantified using the so-called *Purcell factor* F, which gives the amplification in decay rate for an ideal coupling between the emitter and the cavity mode. Purcell's expression for F is

$$F = \frac{3}{4\pi^2}\left(\frac{\lambda_0}{n}\right)^3 \frac{Q}{V}, \tag{14.99}$$

where λ_0/n is the resonance wavelength in the material surrounding the emitter; Q is the quality factor, explained in Section 10.12 and V is the *mode volume*. Purcell originally identified V as the geometrical volume of the cavity. However, following extensive research on optical micro-cavities in the 1990's, the mode volume definition evolved into the following expression:[18]

$$V \equiv \frac{1}{n^2}\int \epsilon(r)|\vec{E}(r)|^2 d^3r, \tag{14.100}$$

[16] P. D. Lax and R. S. Phillips, "Numerical relativity: a review", Archive for Rational Mechanics and Analysis **40**, 268–280 (1971).

[17] Fermi's golden rule is discussed in most textbooks on advanced quantum mechanics: see for example J. J. Sakurai and J. Napolitano, *Modern Quantum Mechanics*, Cambridge, Cambridge University Press, 2020.

[18] Alternative expressions for mode volume are still being developed: see for example C. Sauvan, J.-P. Hugonin, I. Maksymov, and P. Lalanne, "Theory of the spontaneous optical emission of nanosize photonic and plasmon resonators", Physical Review Letters **110(23)**, 237401 (2013).

where $\vec{E}(r)$ is the field for one of the quasinormal modes. Since QNMs are not normalizable (as shown below in Section 14.3.4), the volume integral in (14.100) cannot be computed directly. This shows a need for techniques to normalize QNMs, such as those described below in Section 14.3.5.

Gravitational physics of black holes

In gravitational physics, quasinormal modes represent the resonant modes in black holes and neutron stars. As such, they have received a great deal of attention in the astrophysics literature (see "Going Deeper" for several references). In particular, the ringing down behavior of gravitational waves emitted from the merger of two black holes can be described in terms of these modes. If the gravitational system is spherically symmetric, then the problem becomes one-dimensional.

In the following we will consider a one dimensional model in which the propagating waves are gravitational waves and the resonant structure is described by a gravitational potential, which in general is a smooth function throughout space. The system can be described using a scalar field $\phi(t,x)$ of one variable that satisfies a wave equation with potential:

$$\left[-c^{-2}\partial_t^2 + \partial_x^2 - V(x)\right]\phi(t,x) = 0. \tag{14.101}$$

The appropriate initial conditions are

$$\phi(0,x) = \phi_0(x) \quad \text{and} \quad \frac{\partial}{\partial t}\phi(t,x)|_{t=0} = \phi_1(x). \tag{14.102}$$

Under Laplace transform in the variable ct, (14.101) becomes

$$\left[-s^2 + \partial_x^2 - V(x)\right]\phi(s,x) = -s\phi_0(x) - \phi_1(x), \tag{14.103}$$

which has a similar mathematical form as (14.91), where the terms $-s\phi_0 - \phi_1$ function as a source. As in the optical case, outgoing boundary conditions are physically appropriate.

14.3.4 *Mathematical properties of quasinormal modes*

In Section 14.3.3 we established the physical applicability of QNM's. In this section, we will examine their properties. In order to provide a context and highlight the contrasts between normal and quasinormal modes, we will begin by reviewing normal modes for a simple geometry,

1-d normal modes in a perfectly reflecting cavity

The simplest example of a one-dimensional optical cavity consists of two perfectly reflecting planar mirrors at $x = 0$ and $x = L$ respectively, with a dielectric medium in between. Physically this corresponds to an infinitely-extended dielectric slab between two parallel mirrors. We assume that fields are independent of y and z. Since the mirrors are perfectly reflecting, no fields will escape to the outside, so all fields at the two boundaries of the cavity are zero (i.e. homogeneous Dirichlet boundary conditions — compare Section 1.4). In this case all electromagnetic field components decouple, and can be treated as independent scalar fields governed by the wave equation (compare Equations (9.5), (9.6)) As in Chapter 9, we express $\phi(\vec{x}, t)$ in terms of frequency components $\Phi(\vec{x}, \omega)$, where (compare (9.48)):

$$\phi(\vec{x}, t) = \frac{1}{2\pi} \int_{-\infty}^{\infty} e^{-i\omega t} \Phi(\vec{x}, \omega) d\omega, \tag{14.104}$$

and $\Phi(\vec{x}, \omega)$ satisfies the equation

$$\left(\nabla^2 + \frac{\mu\epsilon\omega^2}{c^2}\right) \Phi(\vec{x}, \omega) = 0. \tag{14.105}$$

Taking the boundary conditions into account (along with the assumption that ϕ is independent of y and z), we arrive at the following eigenvalue problem:

$$-\frac{\partial^2}{\partial x^2} \Phi(x, \omega) = \lambda \Phi(x, \omega); \quad \Phi(0, \omega) = \Phi(L, \omega) = 0, \tag{14.106}$$

with

$$\lambda = \mu\epsilon \frac{\omega^2}{c}. \tag{14.107}$$

The operator $-\frac{\partial^2}{\partial x^2}$ is a *self-adjoint* operator on the Hilbert space of square-integrable functions defined on the interval $[0, L]$ satisfying homogeneous boundary conditions (see Exercise 14.3.2). In this case we can rely on the spectral theory of self-adjoint operators, which leads to the following:

(a) The eigenvalues $\{\lambda_n, n = 1, 2, 3, \ldots\}$ are real and discrete.
(b) The corresponding modes $\{\Phi_n\}$ provide an orthonormal basis for the solutions of (14.105) in the appropriate functional space. In particular the expansion coefficients are determined through taking scalar products with these orthonormal modes.

The above conclusions still hold if μ and/or ϵ are real functions of the frequency ω. In this case, the resonant frequencies are obtained as implicit solutions of (14.107).

Quasinormal modes with reflecting boundaries removed

Let us now remove the conducting plates, so that the dielectric interfaces directly with the surrounding space (for which we assume $\mu = \epsilon = 1$). We impose outgoing boundary conditions as discussed above, and look for traveling wave solutions propagating normal to the boundary, as shown in Figure 14.3 (compare Figure 9.4). We saw in Chapter 9 that the tangential electric field and normal magnetic field are continuous and have continuous derivatives at a dielectric interface. These conditions together with the requirement of no incoming waves leads to the following condition on the free space wave number k_0 when $n \neq 1$ (see Exercise 14.3.3):

$$\left(\frac{n-1}{n+1}\right)^2 e^{2ink_0L} = 1, \tag{14.108}$$

where $n \equiv \sqrt{\mu\epsilon}$ as in (9.11).

If $Re(n) \neq 0$, then all solutions to (14.108) must have non-real values for nk_0. Also, since k_0 is determined implicitly via (14.108) as a function of the complex variable n, in general k_0 will also be complex. We therefore write $k_0 = k^R + ik^I$ and $\omega = \omega^R + i\omega^I$, where in general k^I and ω^I are nonzero. In the case of a dielectric where n is real and positive (which is the normal physical case), we can show that $k^I > 0$, and k^R has both positive and negative solutions.

The plane wave solutions in the free space region $x \notin [0, L]$ have the form

$$e^{i(\omega t - k_0|x|)} = e^{i(\omega^R t - k^R|x|)} e^{-\omega^I t + k^I|x|}. \tag{14.109}$$

When $k^I > 0$, the solutions blow up at $x \to \pm\infty$, while $\omega^I > 0$ causes exponential time decay. Hence QNM's are not square-integrable, so they cannot be considered as elements of a Hilbert space. This non-normalizability leads to mathematical difficulties which both complicate the theory and cause

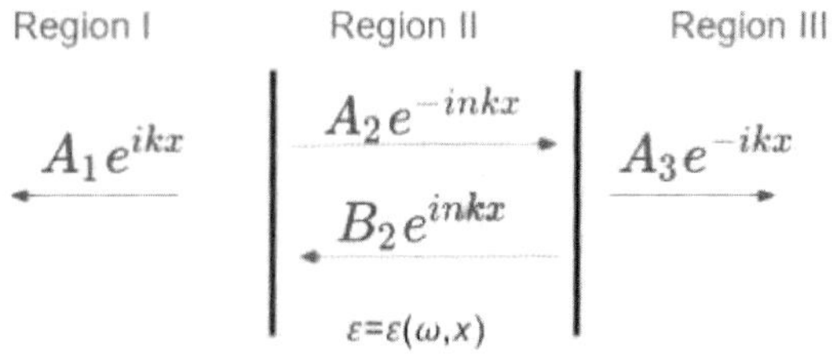

Fig. 14.3 Quasinormal mode fields in three regions of a cavity.

problems in calculations of physical quantities (such as the calculation of the Purcell factor described in Section 14.3.3).

The nice properties of normal modes, including real eigenvalues and eigenfunctions that are orthogonal and complete, are due to the fact that the differential operator with Dirichlet or Neumann boundary conditions is *self-adjoint*. In contrast, the operator with outgoing boundary conditions is not self adjoint, and the corresponding eigenfunctions and eigenvalues lack all these nice properties. Nonetheless, QNMs play an important role in optics, and methods have been developed to deal with the mathematical difficulties. In the next section we will introduce two of these methods.

14.3.5 *Solution methods for quasinormal modes*

The divergence of QNMs at $\pm\infty$ causes difficulties in both theoretical and numerical analysis. In this subsection we briefly describe three approaches to circumventing the mathematical difficulties posed by QNMs. Two of the methods (complex scaling and hyberboloidal coordinates) employ a change of variable to regularize the problem. The other method (perfectly matched layer) modifies the effective potential outside the physical region of interest to mimic absorption of outgoing waves.

Complex scaling approach to QNMs

The complex scaling method was suggested first by Aguilar, Balslev and Combes.[19] Later, Zworski further developed the method in the case of compactly supported potentials, enabling the estimation of the multiplicities of resonances as well as the normalization of modes.[20]

Complex scaling translates the 1-dimensional problem into another differential problem on an infinite domain where the fields do not explode at infinity. This method can be used for compactly supported potentials as well as exponentially decaying potentials. In the compactly-supported case, complex scaling defines a new non self-adjoint operator whose discrete

[19] J. Aguilar and J. M. Combes, "A class of analytic perturbations for one-body Schrödinger Hamiltonians", Communications in Mathematical Physics (1971); E. Balslev and J. M. Combes, "Spectral properties of many-body Schrödinger operators with dilatation-analytic interactions", Communications in Mathematical Physics (1971).
[20] M. Zworski, "Mathematical study of scattering resonances", Bulletin of Mathematical Sciences **7(1)**, 1–85 (2017).

spectrum is the same as the physical operator $P_V \equiv -\partial_x^2 + V(x)$, but whose eigenfunctions are different outside the support of the potential. This is accomplished by changing the domain from the real line to a contour in the complex plane.[21]

As a specific example, consider the piecewise linear contour Γ shown in Figure 14.4, where the three linear parts are denoted as Γ^-, Γ', and Γ^+ respectively. The support of V is the real interval $[-R, R] \subset \Gamma'$. We may parametrize $\Gamma^\pm$ as $z = \pm R + (x \mp R)e^{i\theta}$, where θ is sufficiently small that there are no resonances between $\Gamma^\pm$ and the real axis. We now construct the modified operator

$$P_{V,z} \equiv -\partial_z^2 + V = \begin{cases} -\partial_z^2 + V & z \in \Gamma', \\ -\partial_z^2 & z \in \Gamma^\pm. \end{cases} \tag{14.110}$$

On $[-R, R]$ we have $P_{V,z} = P_V$. For $z \in \Gamma^\pm$, we have the free space operator, and take outgoing wave solutions $\phi^\pm = \alpha^\pm e^{\mp i\lambda z}$, where $\alpha^\pm$ are determined by continuity conditions. We may then express the mode solution as a function of x as

$$\phi(x) = \begin{cases} \phi^+(x), x > R \\ \phi(x), x \in [-R, R] \\ \phi^-(x), x < -R, \end{cases} \tag{14.111}$$

and $\phi(x)$ is an eigenfunction of the operator $P_{V,z}$.

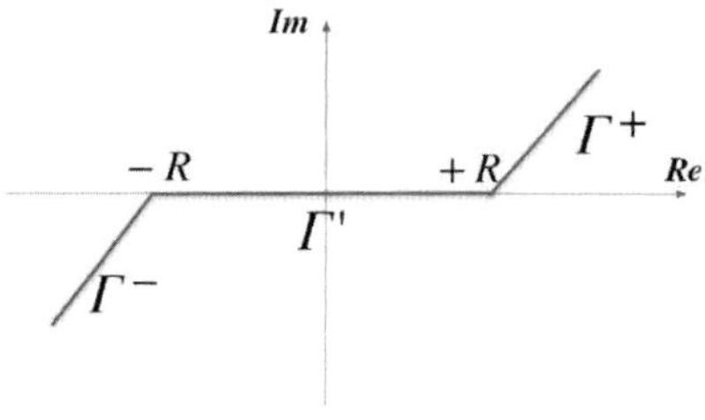

Fig. 14.4 Contour Γ used for complex scaling.

Unfortunately, the complex scaling method as presented above has serious limitations. First of all, the method is limited to one dimension. It is also difficult to apply to the permittivity case, in which the effective potential depends on both x and ω, and a different θ must be used for different values of ω. Finally, the complex scaled problem is still defined

[21]Readers familiar with *Wick rotation* in quantum field theory may note the similarity.

on an infinite domain, so it is not very suitable for numerical calculations. In the next subsection we consider a widely-used alternative method that addresses these problems.

Perfectly matched layer (PML)

The perfectly matched layer (PML) method was proposed by Berenger as a way to avoid infinite domain calculations.[22] The method can be seen from a physical point of view as adding a totally absorbent layer around the resonator in the far field region. In his original paper, Berenger calculated electric and magnetic conductivities within the layer that produce zero reflection for the outgoing waves from the resonator. Electric and magnetic field components are further split, allowing control of the reflection.

Berenger's method was derived in 2-dimensional Cartesian coordinates for a rectangular-shaped artificial boundary. Later many studies were made to improve PMLs and make them more flexible with respect to coordinates and the shape of the scatterer. For additional details on Berenger's method, readers may consult the "Going Deeper" section at the end of this chapter. PML is to date the go-to method for practical calculations. Still one must be careful in its application, because it sometimes introduces spurious resonances.

There is yet a third method for computing QNMs, which as yet has been applied mostly to gravitational physics and in particular the study of gravity waves associated with black holes. We will present this third method in greater depth in the next section.

14.3.6 *Hyperboloidal slices approach to quasinormal modes*

Like complex scaling, the hyperboloidal slices approach also depends on a coordinate transformation. However the transformation involves both space and time, not just space. Writing the problem in the new coordinates system converts it to a problem in a Hilbert space, where the fields can be normalized. However this method is not yet developed for higher dimensions. In this section we will first show how the transformation is used in gravitational physics. Next we apply it to the Lorentz-Drude model for

[22] J.-P. Berenger, "Perfectly matched layer for the FDTD solution of wave-structure interaction problems." IEEE Transactions on antennas and propagation **44**, no. 1 (1996): 110–117.

a single region in space, then for three regions. After that we introduce numerical methods for solution, followed by a brief presentation of results.

Hyperboloidal coordinate transformation

The transformed coordinates are called *Bizoń-Mach coordinates*, which we shall represent as (y, τ). The coordinate transformation from (x, t) to (y, τ) maps the x axis to the finite interval $-1 < y < 1$, so that the QNM problem in x is mapped to a finite domain.

The transformation defines τ such that the fields can be normalized. The coordinate transformation that accomplishes this is

$$\begin{aligned} \tau &= t - \ln(2\cosh(x)), \\ y &= \tanh(x). \end{aligned} \tag{14.112}$$

To see how this works, consider an electromagnetic plane wave $\phi(t, x)$ in free space moving in the $+x$ direction: $\phi(t, x) = Ae^{i\omega t}e^{-i\omega x}$. With this sign convention, quasinormal mode frequencies always have a positive imaginary part, so we may write $\omega = \omega^R + i\omega^I$ with $\omega^I > 0$. Plugging this into the plane wave equation gives

$$\phi(t, x) = Ae^{i\omega^R(t-x)}e^{\omega^I(x-t)}. \tag{14.113}$$

When $x \to +\infty$ and $t \to +\infty$ such that $x - t \to$ constant, then (14.113) shows that $\phi(t, x)$ approaches a constant. Figure 14.5 shows the hyperbolic shape of a $\tau =$ constant curve, which approaches $t - x =$ constant asymptotically. Thus the field ϕ has regular behavior along curves of constant τ.

A field $\phi(t, x)$ can be expressed in terms of frequency components, where each component is proportional to $e^{i\omega t}\phi(\omega, x)$ in Cartesian coordinates. In Bizoń-Mach coordinates, it can be shown (see Exercise 14.3.4) that the same field component is written as $e^{i\omega\tau}\tilde{\phi}(\omega, y)$ in the new coordinates, where

$$\tilde{\phi}(\omega, y) = (e^x + e^{-x})^{i\omega}\phi(\omega, x), \tag{14.114}$$

where x and y are related according to (14.112). Taking ω equal to a mode frequency $\omega_n = \omega_n^R + i\omega_n^I$ with $\omega_n^I > 0$, we obtain

$$\tilde{\phi}(\omega_n, y) = (e^x + e^{-x})^{-\omega_n^I}(e^x + e^{-x})^{i\omega_n^R}\phi(\omega, x). \tag{14.115}$$

The factor $(e^x + e^{-x})^{-\omega_n^I}$ cancels out the exponential growth in $\phi(\omega, x)$, so that the product does not explode as $x \to \pm\infty$.

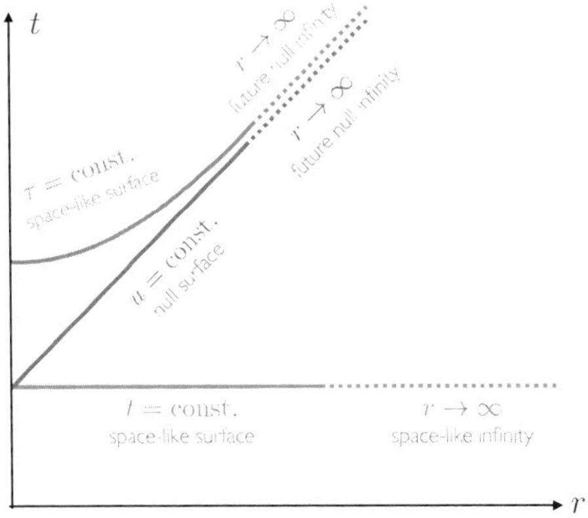

Fig. 14.5 τ=constant curve, where τ is the second Bizoń-Mach coordinate. Here $u \equiv t - x$.

Electromagnetic problem in hyperboloidal coordinates

According to the Lorentz model, the electron-nucleus system with external field $\vec{E}$ is modeled as a harmonic oscillator. The equation for the electron-nucleus separation vector $\vec{r}$ is:

$$\frac{\partial^2 \vec{r}}{\partial t^2} + \gamma \frac{\partial \vec{r}}{\partial t} + \omega_0^2 \vec{r} = -\frac{e\vec{E}}{m_e}, \tag{14.116}$$

where γ is the damping loss rate, ω_0 is the characteristic frequency, m_e is the mass of the electron, and $\vec{E}$ is the local electric field acting on the electron. This gives the following equation for the polarization, $\vec{P} \equiv -Ne\vec{r}$ (Cf. (9.112)):

$$\frac{\partial^2 \vec{P}}{\partial t^2} + \gamma \frac{\partial \vec{P}}{\partial t} + \omega_0^2 \vec{P} = \omega_p^2 \vec{E}, \tag{14.117}$$

where $\omega_p^2 \equiv Ne^2/m_e$, and N is the electron density. Note that γ, ω_0, ω_p are the same parameters in Lorentz-Drude model.

From Maxwell's equations, we obtain:

$$\begin{aligned} &\frac{\partial \vec{B}}{\partial t} = -\nabla \times \vec{E}; \quad \frac{\partial \vec{E}}{\partial t} = \nabla \times \vec{H} - \frac{\partial \vec{P}}{\partial t} \\ &\Longrightarrow \frac{\partial^2 \vec{E}}{\partial t^2} = -\nabla^2 \vec{E} - \frac{\partial^2 \vec{P}}{\partial t^2}. \end{aligned} \tag{14.118}$$

Since $\vec{P}$ is in the same direction as $\vec{E}$, we may take the x axis along the direction of $\vec{E}$ and write $\vec{E} = E\hat{x}, \vec{P} = P\hat{x}$. Then we can rewrite (14.117) and (14.118) as equations for the scalar fields E and P. Writing (14.117) and (14.118) in hyperboloidal coordinates, using (14.112) leads to:

$$\begin{aligned}
&\partial_\tau^2 P + \gamma\partial_\tau P + \omega_0^2 P = \omega_p^2 E,\\
&(1-y^2)\partial_\tau^2 E + 2y(1-y^2)\partial_y \partial\tau E\\
&\quad + (1-y^2)\partial_\tau E + 2y(1-y^2)\partial_y E - (1-y^2)^2\partial_y^2 E\\
&\quad + \partial_\tau^2 P = 0.
\end{aligned} \tag{14.119}$$

Now we introduce two auxiliary scalar fields U and V, where:

$$U \equiv \partial_\tau P \text{ and } V := \partial_\tau E. \tag{14.120}$$

Rewriting (14.119) using (14.120), we get:

$$\begin{aligned}
&\partial_\tau U + \gamma U + \omega_0^2 P = \omega_p^2 E,\\
&(1-y^2)\partial_\tau V + 2y(1-y^2)\partial_y V\\
&\quad + (1-y^2)V + 2y(1-y^2)\partial_y E - (1-y^2)^2\partial_y^2 E\\
&\quad + \partial_\tau U = 0.
\end{aligned} \tag{14.121}$$

Taking the Fourier transformation with respect to τ (noting that $\partial_t = \partial_\tau$) produces the following matrix equation for the fields E, V, P, U (see Exercise 14.3.4):

$$\begin{bmatrix} L1 & L2 & 0 & 0\\ 0 & 1 & 0 & 0\\ 0 & 0 & 0 & 1\\ \omega_p^2 & 0 & -\omega_0^2 & -\gamma \end{bmatrix} \begin{bmatrix} E\\ V\\ P\\ U \end{bmatrix} = i\omega \begin{bmatrix} 0 & 1-y^2 & 0 & 1\\ 1 & 0 & 0 & 0\\ 0 & 0 & 1 & 0\\ 0 & 0 & 0 & 1 \end{bmatrix} \begin{bmatrix} E\\ V\\ P\\ U \end{bmatrix}, \tag{14.122}$$

where

$$\begin{aligned}
L1 &= -2y(1-y^2)\partial_y + (1-y^2)^2\partial_y^2,\\
L2 &= -2y(1-y^2)\partial_y - (1-y^2).
\end{aligned} \tag{14.123}$$

Now (14.122) has the form of a generalized eigenvalue problem, but still lacks boundary conditions. (Note that although the problem is defined on a finite interval, the boundary $y = \pm 1$ physically corresponds to $x \to \pm\infty$.) Instead of imposing boundary conditions, we divide the space into three regions, where the resonating cavity is located within the middle region (see Figure 14.6). For each region we write an equation of the form (14.122), taking into account the different permittivity parameters. We then impose continuity conditions on E and its derivatives. This is sufficient to give a well-posed eigenvalue problem. We will show how boundary conditions are implemented numerically in Section 14.3.7.

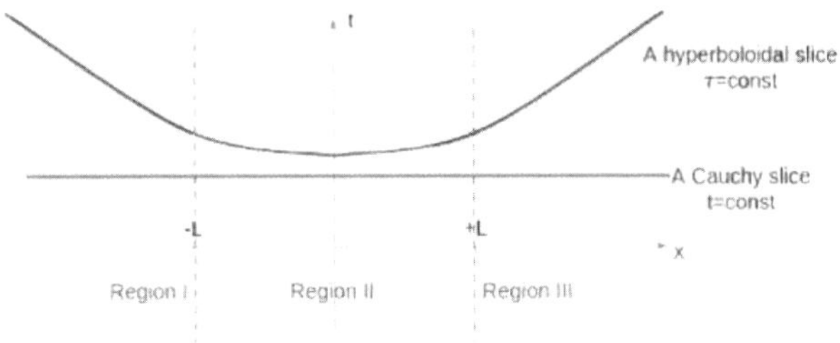

Fig. 14.6 Three regions, with one hyperboloidal slice.

14.3.7 *Numerical calculation of QNMs via the spectral method*

In this section we introduce some numerical tools for calculating QNMs and the corresponding frequencies. We then show how the problem is converted to a matrix eigenvalue problem by means of these tools.

The choice of numerical methods

The most elementary (and most familiar) method for numerical solution of partial differential equations is the finite difference method, which imposes a uniform rectangular grid on the region of interest and solves for function values at grid points using a matrix equation, where derivatives are expressed as sparse matrices. Another common method is the finite element method, which decomposes the region into small polyhedral "elements", and approximates the function piecewise using different polynomials on the different elements. A less well-known and more recently developed approach utilizes spectral methods. Like the finite element method, spectral methods approximate the function using a basis of differentiable functions (which can be polynomials or sinusoids). But with spectral methods, the basis functions are defined on the entire domain instead of individual elements, and the solution is specified globally rather than piecewise. For problems on domains with simple shapes, which have smooth solutions, spectral methods tend to be both more efficient and more accurate than the other methods.

In any numerical method, *discretization errors* must be considered. These errors perturb the differential operator, and if the spectrum is not stable under perturbations the frequencies may shift significantly. With spectral methods these errors are reduced compared to the other methods. Furthermore the errors may be controlled using machine precision. The importance of accuracy in estimating QNMs will be underscored in Section 14.3.8.

The concept of differential matrix

Discretization of differential equations involves changing a calculus problem into a linear algebra problem. Functions become vectors, and operators become matrices. In the case of finite difference, a function $\phi(x)$ on the interval $[a, b]$ is represented as a vector $(\phi(x_0), \ldots \phi(x_N)$, where $a = x_0 < x_1 < \ldots < x_n = b$ and $x_{k+1} - x_k = h$. Finite difference uses the centered difference approximation to the derivative: denoting the discretized derivative vector as $(\psi_0, \ldots, \psi_N)$, we have:

$$\psi_k = \frac{\phi_{k+1} - \phi_{k-1}}{2h} + \mathcal{O}(h^2), k = 1, \ldots n-1. \tag{14.124}$$

If the problem is periodic then

$$\psi_0 = \frac{\phi_1 - \phi_N}{2h} \quad \text{and} \quad \psi_N = \frac{\phi_0 - \phi_{N-1}}{2h}. \tag{14.125}$$

And the differentiation matrix will be as in the following equation:

$$\begin{bmatrix} \psi_0 \\ \\ \vdots \\ \\ \psi_N \end{bmatrix} = h^{-1} \begin{bmatrix} 0 & \frac{1}{2} & & & -\frac{1}{2} \\ -\frac{1}{2} & 0 & \ddots & & \\ & & \ddots & & \\ & & \ddots & 0 & \frac{1}{2} \\ \frac{1}{2} & & & -\frac{1}{2} & 0 \end{bmatrix} \begin{bmatrix} \phi_0 \\ \\ \vdots \\ \\ \phi_N \end{bmatrix}. \tag{14.126}$$

Note that the boundary conditions must be taken into consideration when formulating the matrix (see Exercise 14.3.5).

This simple approximation formula for the derivative is relatively inaccurate. In the following we show that by using a nonuniform grid and making use of polynomial expansions, we obtain better approximations to the derivatives which leads to higher accuracy and faster convergence.

Spectral basis: Chebyshev polynomials

The spectral method requires a set of basis functions $\{\phi_k(x)\}_{k=1,2,\ldots}$ that ideally should satisfy the following conditions:

- The partial sum $\sum_{k=0}^{N} c_k \phi_k(x)$ should converges rapidly to the solution as $N \to \infty$;
- Given a function, the calculation of the coefficients c_k should be computationally inexpensive. Reconstructing a function given the coefficients should also be inexpensive.
- Given a function $\phi(x)$ approximated by $\sum_{k=0}^{N} c_k \phi_k(x)$, it should be easy to find another set of coefficients $\{c'_k\}$ such that the derivative $\phi'(x)$ is approximated by $\sum_{k=0}^{N} c'_k \phi_i(x)$.

The *Chebyshev polynomials* meet all three requirements. Chebyshev polynomials of the first kind are defined for $|x| < 1$ as:

$$T_k(x) = \cos(k \arccos(x)). \tag{14.127}$$

The first few Chebyshev polynomials are:

$$T_0(x) = 1; \quad T_1(x) = x; \quad T_2(x) = 2x^2 - 1; \quad T_3(x) = 4x^3 - x. \tag{14.128}$$

Chebyshev polynomials satisfy the recursion relation:

$$T_{k+1} = 2xT_k(x) - T_{k-1}(x), \tag{14.129}$$

and the orthogonality relation (with weighting function $(1 - x^2)^{-1/2}$):

$$\int_{-1}^{+1} \frac{T_n(x)\, T_m(x)}{\sqrt{1 - x^2}} = \int_0^{\pi} \cos(n\phi)\, \cos(m\phi) d\phi = \frac{\pi}{2}\delta_{nm}(1 + \delta_{n0}). \tag{14.130}$$

Any square integrable function $\psi(x)$ can be written as:

$$\psi(x) = \frac{c_0}{2} + \sum_{k=1}^{\infty} c_k T_k(x), \tag{14.131}$$

where the coefficients $\{c_k\}$ may be conveniently determined using (14.130).

Chebyshev polynomials can be used to determine optimal sets of grid points $x_0, \ldots x_N$. Results from numerical analysis indicate that the following are good choices for grid points:

- *Chebyshev-Lobatto:* $x_j = \cos\left(\frac{j\pi}{N}\right), j = 0, \ldots N$;
- *Left Chebyshev-Radau:* $x_j = -\cos\left(\frac{j\pi}{2N+1}\right), j = 0, \ldots N$;
- *Right Chebyshev-Radau:* $x_j = \cos\left(\frac{j\pi}{2N+1}\right), j = 0, \ldots N$.

Chebyshev-Lobatto corresponds to the extrema of T_N, and includes the boundary points ± 1. The left and right Chebyshev-Radau grids both have one point in each interval between successive roots of T_N, and include the left and right boundary points $+1$ and -1, respectively. All of these grids have a greater density of points near the edges of the interval $[-1, 1]$ as compared to the center. This reduces the magnitude of oscillatory errors, which tend to be larger near the boundary (the so-called *Runge phenomenon*). The different grids are used to accommodate different types of boundary conditions.

Chebyshev differential matrices may be computed by the following procedure:[23]

(a) Construct the invertible matrix C that obtains the vector of coefficients $(c_0, \dots c_N)$ in (14.131) from the vector of values $(\phi(x_0), \dots \phi(x_N))$, where $x_0, \dots x_N$ are the grid points.
(b) Construct the matrix P that gives Chebyshev coefficients of the polynomial derivatives $T'_0, \dots T'_N$. Explicitly,

$$T'_k = \sum_{j=0}^{N} P_{kj} T_j. \qquad (14.132)$$

(c) The Chebyshev differential matrix D is then given by:

$$D = C^{-1} P C. \qquad (14.133)$$

Higher derivative matrices may be constructed similarly.

Implementation

With a grid of $N+1$ points, the vector $\left[E\ V\ P\ U\right]^T$ has length $4N+4$. The two matrices in (14.122) have dimension $(4N+4) \times (4N+4)$. (Note that the scalar entries in (14.122) become matrices: for example, entries denoted as '1' become $N+1 \times N+1$ identity matrices.) Denoting these matrices as L and M respectively, we have:

$$L\left[E\ V\ P\ U\right]^T = i\omega M \left[E\ V\ P\ U\right]^T. \qquad (14.134)$$

[23]Although this procedure is mathematically correct, in practice, it is difficult to compute the matrix P in step (b). There is a much more efficient method for computing differential matrices that uses Fourier series (Al Sheikh, op. cit.).

Since we are solving in three domains, we need three matrix equations with the same ω:

$$L_k \left[E_k\ V_k\ P_k\ U_k\right]^T = i\omega M_k \left[E_k\ V_k\ P_k\ U_k\right]^T, k = 1, 2, 3. \tag{14.135}$$

The entries in L_k and M_k in (14.135) differ due to the different permittivities, and also due to boundary conditions.

In order to accommodate boundary conditions, we choose different grids in the different regions so that the boundaries between regions are included as grid points. Assuming that the cavity occupies the interval $[-a, b]$, in the left, central, and right regions we use right Ridau, Lobatto, and left Ridau grids respectively. Altogether, (14.135) can be written as a single matrix equation:

$$\boldsymbol{L\Phi} = i\omega \boldsymbol{M\Phi}, \tag{14.136}$$

where

$$\boldsymbol{L} \equiv \begin{bmatrix} L_1 & 0 & 0 \\ 0 & L_2 & 0 \\ 0 & 0 & L_3 \end{bmatrix}; \quad \boldsymbol{M} \equiv \begin{bmatrix} M_1 & 0 & 0 \\ 0 & M_2 & 0 \\ 0 & 0 & M_3 \end{bmatrix}, \tag{14.137}$$

and

$$\boldsymbol{\Phi} \equiv [\Phi_1, \Phi_2, \Phi_3]^T \text{ where } \Phi_k \equiv [E_k, V_k, P_k, U_k]. \tag{14.138}$$

The matrices $\boldsymbol{L}$ and $\boldsymbol{M}$ both have size $12(N+1) \times 12(N+1)$.

The final step before implementing (14.136) in software is to impose continuity conditions:

$$\begin{aligned} E_1\big|_{y_1=\tanh(-a)} &= E_2\big|_{y_2=\tanh(-a)}; \quad E_1'\big|_{y_1=\tanh(-a)} = E_2'\big|_{y_2=\tanh(-a)} \\ E_2\big|_{y_1=\tanh(b)} &= E_3\big|_{y_3=\tanh(b)}; \qquad E_2'\big|_{y_1=\tanh(b)} = E_3'\big|_{y_3=\tanh(b)}. \end{aligned} \tag{14.139}$$

This involves replacing rows $N+1$, $4N+5$, $5N+5$, and $8N+9$ (taking the first row as row 1) in both matrices in (14.136) by rows that are equivalent to (14.139) (see Exercise 14.3.8).

Finally we have arrived at a matrix eigenvalue problem that can be solved by conventional software packages. By way of illustration, we have implemented numerically (14.136) with the continuity conditions in (14.139) for a one-dimensional optical cavity with a permittivity that follows the Drude model, with $\epsilon_\infty = 1$, $\omega_P = 1$, $\gamma = 0.1$ and $\omega_0 = 0$. The cavity occupies the interval $x \in [-1, 1]$, which in Bizoń-Mach coordinates

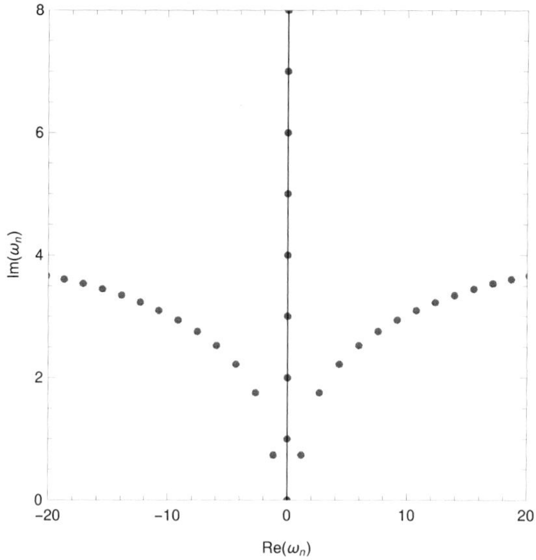

Fig. 14.7 Eigenvalues obtained by implementing hyperboloidal slices approach in three space regions.

becomes $y \in [-\tanh(1), \tanh(1)]$. Figure 14.7 shows the QNM frequencies that are obtained by the implementation. Only the frequencies in the upper complex half plane are shown, because those in the lower half plane give unphysical modes that explode with time. Evidently there are two kind of eigenvalues, namely those situated on a two symmetric logarithmic curves, and those that are purely imaginary. The imaginary frequencies represent "overdamped" modes that exhibit pure exponential damping in time, while the others with nonzero real part have both damping and oscillatory behavior.

Of course it is necessary to verify the accuracy of these results. An appropriate measure of accuracy is the *logarithmic relative error* denoted by ϵ_N, where

$$\epsilon_N \equiv \log_{10} \left| \frac{\omega^{(N)} - \omega^{(N_{Max})}}{\omega^{(N_{Max})}} \right|, \tag{14.140}$$

where $\omega^(N_{max}$ is the value computed from the grid with the maximum number of points. Figure 14.8 shows ϵ_N for the fundamental mode as a function of N, where N denotes the number of grid points in each of the three grids (all three grids have the same number of points). The slope is

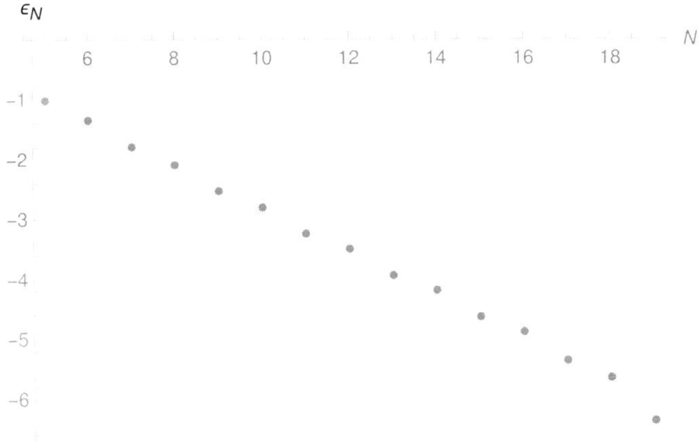

Fig. 14.8 The logarithmic relative error $\epsilon_n^{(N)}$ for the fundamental mode, as a function of number of grid points.

linear, indicating exponential convergence. Increasing the number of points in each grid from N to $N+3$ decreases the error by roughly 90%, regardless of the value of N.

14.3.8 *Pseudospectrum and the stability of eigenvalues*

Spectral stability is an important issue to consider in numerical estimation of eigenvalues (such as finding QNM frequencies). If an operator is not spectrally stable, then perturbations introduced by the discretization can cause the numerical solutions to deviate from their actual values. Physically, this means that resonances are not estimated accurately. In this section we will introduce the *ϵ-pseudospectrum*, which is a powerful tool for characterizing and visualizing the stability of eigenvalues with respect to operator perturbations of order ϵ (such as those introduced by discretization).

Let A be a linear operator on a Hilbert space with scalar product $\langle \cdot, \cdot \rangle$, and denote its adjoint by $A^\dagger$: $\langle A^\dagger u, v \rangle = \langle u, Av \rangle$. The operator A is called *normal* if and only if. In particular, a self-adjoint operator $A^\dagger = A$ is normal. In this setting, the 'spectral theorem' (under the appropriate functional space assumptions) states that a normal operator A is diagonalizable via unitary transformation, and the eigenfunctions of A form an orthonormal basis. Furthermore (crucially to the present discussion), the eigenvalues are stable under perturbations of A. The lack of such a 'spectral theorem' for non-normal operators means that eigenfunction completeness may not be guaranteed, and the spectrum may not be stable with respect to operator perturbations.

One way to characterize the spectral (in)stability of the operator A is to consider the set of complex numbers $\lambda \in \mathbb{C}$ that are eigenvalues of "some" small perturbation $A + \delta A$, with $||\delta A|| < \epsilon$, where $|| \cdot ||$ can represent any matrix norm such as the operator norm. If there are values of λ that stray far from the original spectrum of A, then this indicates that A is spectrally unstable. These considerations lead to the following definition:

Definition 1 (*Pseudospectrum: perturbative approach*). Let $M_n(\mathbb{C})$ denote the $n \times n$ complex matrices, equipped with a matrix norm $|| \cdot ||$. Given $A \in M_n(\mathbb{C})$ with spectrum $\sigma(A)$, and given $\epsilon > 0$, then the *ϵ-pseudospectrum* $\sigma^\epsilon(A)$ of A is the set of complex numbers defined as follows:

$$\sigma^\epsilon(A) \equiv \{\lambda \in \mathbb{C}, \exists\, \delta A \in M_n(\mathbb{C}), ||\delta A|| < \epsilon : \lambda \in \sigma(A + \delta A)\}. \qquad (14.141)$$

A second characterization of the ϵ-pseudospectrum is based on the notion of the *resolvent* $R_A(\lambda) = (\lambda I - A)^{-1}$ of the operator A. The complex number λ is an eigenvalue of A if and only if $R_A(\lambda)$ is singular.

Definition 2 (*Pseudospectrum: resolvent norm approach*). Given $A \in M_n(\mathbb{C})$, its resolvent $R_A(\lambda) \equiv (\lambda I - A)^{-1}$ and $\epsilon > 0$, the ϵ-pseudospectrum $\sigma^\epsilon(A)$ of A is characterised as

$$\sigma^\epsilon(A) = \{\lambda \in \mathbb{C} : ||R_A(\lambda)|| = ||(\lambda I - A)^{-1}|| > 1/\epsilon\}. \qquad (14.142)$$

The ϵ-pseudospectrum for the full range of ϵ's can be visually represented by means of a "pseudospectrum map". An example is shown in Figure 14.9 in the case of a self-adjoint operator A. In the figure, the eigenvalues of A are represented as points, while white lines are boundaries of the sets $\sigma^\epsilon(A)$ for different values of ϵ. The figure shading indicates the size ϵ of the perturbation, and each white line traces along a constant shading level. The sets $\{\sigma^\epsilon(A)\}$ form a nested family, in the sense that $\sigma^{\epsilon_1}(A) \subset \sigma^{\epsilon_2}(A)$, for $\epsilon_1 < \epsilon_2$, and the actual spectrum is the intersection, $\sigma(A) = \cap_{\epsilon>0}\sigma^\epsilon(A)$. For small ϵ, $\sigma^\epsilon(A)$ consists of disjoint near-circles around each eigenvalue. As described previously, a change δA with a norm of order ϵ in the operator A entails a maximum change in the eigenvalues of the same order ϵ. Specifically, ϵ-pseudospectra sets show concentric circles around the spectra that quickly reach large-epsilon values, i.e. $\epsilon \approx O(1)$, when moving away from eigenvalues. As a consequence, one would need perturbations in the

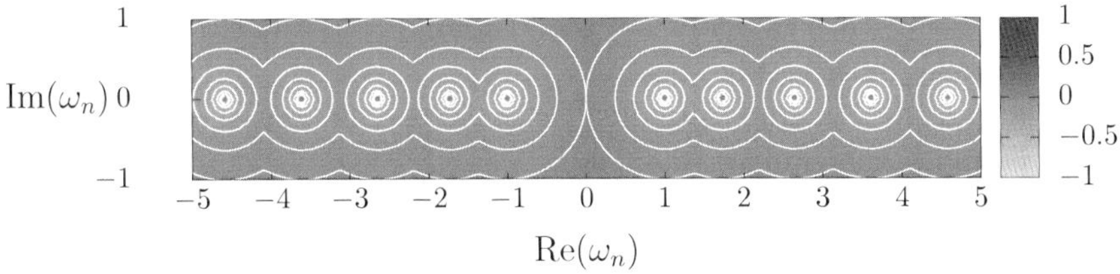

Fig. 14.9 Pseudospectrum for a self-adjoint operator. The shading scale corresponds to $\log_{10}\epsilon$, and white lines indicate the boundaries of ϵ-pseudospectra sets σ^ϵ.

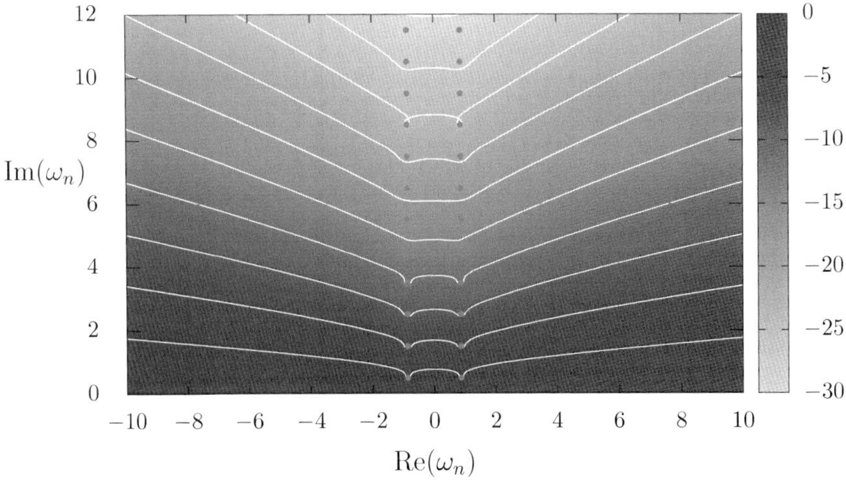

Fig. 14.10 Pseudospectrum for the Pöschl-Teller potential. The points located at $\text{Re}(\omega) \approx \pm 1$ are the QNM frequencies. The shading scale corresponds to $\log_{10}\epsilon$, and white lines indicate the lower boundaries of ϵ-pseudospectra sets σ^ϵ, whose interiors extend upwards in the ω-complex plane.

operator of the same order to dislodge the eigenvalues slightly away from their original values; we say then that A is spectrally stable.

Figure 14.10 shows the pseudospectrum for a non-self-adjoint operator. One immediately observes a striking qualitative difference from the self-adjoint case depicted in Figure 14.9. The specific operator depicted in Figure 14.10 is the potential operator $P_{V,z} \equiv -\partial_z^2 + V$ with the *Pöschl-Teller potential*:

$$V(\bar{x}) \equiv \frac{V_0}{\cosh^2(\bar{x})} = V_0, \text{sech}^2(\bar{x}), \quad \bar{x} \in (-\infty, \infty). \tag{14.143}$$

The Pöschl-Teller potential has been widely used as a benchmark for the study of QNMs in the context of black hole perturbations. In this case, all eigenvalues have nonzero real and imaginary parts, so all eigenfunctions experience both attenuation and oscillation. Eigenvalues with smaller imaginary parts are much more stable.

As in Figure 14.9, the white lines in Figure 14.10 denote curves of constant ϵ, and the shading scale indicates the value of $\log_{10} \epsilon$. Each white line corresponding to a particular value ϵ_0 serves as a lower boundary for the set σ^ϵ: all eigenvalues above the line will remain above the line for perturbations with magnitude less than or equal to ϵ_0. On the other hand, eigenvalues below the line are essentially unchanged by perturbations of this size. (If the scale of the figure were expanded, one would see white near-circles surrounding the lower eigenvalues, but these circles are invisible on this scale.)

The figure underscores the difficulty of computing higher QNM's for non-self-adjoint operators, for they can deviate wildly, even when miniscule errors are introduced into the operator. Approaches towards addressing this difficulty are described in Al Sheikh (op. cit.).

This closes out our examination of QNM's, which have provided yet another example of the amazing richness of mathematical behavior to be found in a deceptively simple set of equations, and the astounding range of physical phenomena which they describe.

14.4 Going Deeper

14.4.1 *Radiation reaction*

(1) C. A. Brau, *Modern Problems in Classical Electrodynamics*, Oxford University Press (Oxford) 2004; Section 11.1.

(2) D. J. Griffths, T. C. Proctor and D. F. Schroeter, "Abraham-Lorentz Versus Landau-Lifshitz", Am. J. Phys. **78**, 391 (2010).

(3) J. D. Jackson, *Classical Electrodynamics*, 3rd ed., John Wiley & Sons (New York) 1999; Chapter 16.

(4) L. D. Landau and E. M. Lifshitz, *Classical Theory of Fields*, 4th ed. with corrections, Butterworth-Heinemann (Oxford) 2000; Sections 75 and 76.

(5) F. Rohrlich, "The Self-Force and Radiation Reaction", Am. J. Phys. **68**, 1109 (2000).

(6) A. D. Yaghjian, *Relativistic Dynamics of a Charged Sphere: Updating the Lorentz-Abraham Model*, 2nd ed., Springer (New York) 2006.

14.4.2 *Quantum field theory*

(1) I. J. R. Aitchison and A. J. G. Hey, *Gauge Theories in Particle Physics, Vols. I and II*, 3rd ed., I (From Relativistic Quantum Mechanics to QED), II (Non-Abelian Gauge Theories: QCD and the Electroweak Theory), Taylor and Francis (New York) 2003, IOP Publishing (Bristol) 2004.

(2) V. B. Berestetskii, L. P. Pitaevskii and E. M. Lifshitz, *Quantum Electrodynamics*, 2nd ed., Butterworth-Heinemann (Oxford) 1982.

(3) M. Creutz, *Quarks, Gluons and Lattices*, Cambridge University Press (Cambridge) 1983.

(4) C. Itzykson and J.-B. Zuber, *Quantum Field Theory*, McGraw-Hill (New York) 1980.

(5) M. E. Peskin and D. V. Schroeder, *An Introduction to Quantum Field Theory*, Addison-Wesley (Reading, MA) 1995.

(6) H. J. Rothe, *Lattice Gauge Theories: An Introduction*, 4th ed., World Scientific Publishing (Singapore) 2012.

(7) M. D. Schwartz, *Quantum Field Theory and the Standard Model*, Cambridge University Press (Cambridge) 2013.

(8) J. Schwinger, *Particles, Sources and Fields, Vols. I, II, and III*, I (Particles, Sources, Fields), II (Electrodynamics I), III (Electrodynamics II), Addison-Wesley (Reading, MA) 1970, 1973, 1973.

(9) S. Weinberg, *The Quantum Theory of Fields, Vols. I, II, and III*, I (Foundations), II (Modern Applications), III (Supersymmetry), Cambridge University Press (Cambridge) 1995, 1995, 2000.

14.4.3 *Radiating cavities and quasinormal modes*

The main reference for this section is the PhD thesis of Lamis Al Sheikh. This work is interdisciplinary, examining both mathematical tools and concepts on the one hand and physical problems involving resonances in gravitational physics and optics on the other. For readers who are curious about numerical methods or about the analogy between electromagnetic and gravitational cases, this thesis is helpful.

- Lamis Al Sheikh, Scattering resonances and pseudospectrum: Stability and completeness aspects in optical and gravitational systems, PhD Thesis, Burgundy University - France, Jan. 2022. *Supervisor: José Luis Jaramillo.*

To go deeper into the optical aspects, see:

- P. Lalanne, W. Yan, K. Vynck, C. Sauvan, and J. Hugonin, Light Interaction with Photonic and Plasmonic Resonances, Laser and Photonics Reviews **12**(5), 1700113 (2018).
- B. Stout, R. Colom, N. Bonod, and R. C. McPhedran, Spectral expansions of open and dispersive optical systems: Gaussian regularization and convergence, **23**(8), 083004 (August 2021).
- Boucher, Y. G., Al Sheikh, L. Universal coupled-mode theory formulation of quasi-normal modes in a 1D photonic crystal. Opt Quant Electron **52**, 453 (2020). https://doi.org/10.1007/s11082-020-02577-1.

A review article that summarizes improvements in PML is

- F. Pled and C. Desceliers, "Review and Recent Developments on the Perfectly Matched Layer (PML) Method for the Numerical Modeling and Simulation of Elastic Wave Propagation in Unbounded Domains", Archives of Computational Methods in Engineering, pp. 1–48 (2021).

An interesting article on PML with a relation to complex scaling method

- F. Collino and P. B. Monk, "Optimizing the perfectly matched layer", Computer methods in applied mechanics and engineering **164**(1-2), 157–171 (1998).

For applications of QNMs to gravitational physics, see

- H. Beyer, "The spectrum of radial adiabatic stellar oscillations", Journal of Mathematical Physics **36**(9), 4815–4825 (1995).
- H.-P. Nollert, "Topical Review: Quasinormal modes: the characteristic 'sound' of black holes and neutron stars", Class. Quant. Grav. **16**, R159–R216 (1999).
- C. F. Sopuerta, "Time Domain Simulations of EMRIs using Finite Element Methods", AIP Conf. Proc. **873**, 274–278 (2006).
- E. Berti, V. Cardoso, and C. M. Will, "On gravitational-wave spectroscopy of massive black holes with the space interferometer LISA", Phys. Rev. D **73**, 064030 (2006).

For additional references on the hyperboloidal slices approach, pseudospectrum, and applications to gravitational physics see

- R. Panosso Macedo, Hyperboloidal framework for the Kerr spacetime, Class. Quant. Grav. **37**(6), 065019 (2020).
- R. Panosso Macedo, J. L. Jaramillo, and M. Ansorg, Hyperboloidal slicing approach to quasi-normal mode expansions: the Reissner-Nordström case, Phys. Rev. D **98**(12), 124005 (2018).
- J. L. Jaramillo, R. Panosso Macedo, and L. Al Sheikh, Pseudospectrum and Black Hole Quasinormal Mode Instability, Phys. Rev. X **11**(3), 031003 (2021).
- J. L. Jaramillo, R. P. Macedo, and L. Al Sheikh, Gravitational wave signatures of black hole quasi-normal mode instability, Phys. Rev. Lett. **128**, 211102 (2022).
- J. L. Jaramillo, R. Panosso Macedo, Oscar Meneses-Rojas, Bernard Raffaelli and L. Al Sheikh: A Weyl's law for black holes. https://doi.org/10.48550/arXiv.2212.05570 (2022).

14.5 Exercises

Exercise 14.1.1. Finish the reduction of the term $\textcircled{4}^k$ in the multipole expansion calculation of the self force, $F^{\rm self}$. Show that

$$\textcircled{4}^k = \iint d^3x\, d^3\xi\, \rho(\vec{x},t)\frac{1}{|\vec{\xi}|}\frac{\partial}{\partial x^k}\left[\frac{1}{6c^3}|\vec{\xi}|^3\frac{\partial^3\rho(\vec{x}-\vec{\xi},t)}{\partial t^3}\right],$$

can be written as

$$\textcircled{4}^k = -\frac{e^2}{3c^3}\ddot{v}^k.$$

Exercise 14.1.2. Consider the next term $\textcircled{5}^k$ in the multipole expansion calculation of the "I" self force. Let

$$\textcircled{5}^k \equiv -\iint d^3x\, d^3\xi\, \rho(\vec{x},t)\frac{1}{|\vec{\xi}|}\frac{\partial}{\partial x^k}\left[\frac{1}{24c^4}|\vec{\xi}|^4\frac{\partial^4\rho(\vec{x}-\vec{\xi},t)}{\partial t^4}\right].$$

Evaluate for: (a) a point charge; (b) a uniform surface charge density on a spherical shell of radius a; (c) a uniform volume charge density inside a sphere of radius a.

Exercise 14.1.3. An oscillatory electric field $\sim e^{i\omega t}$ will induce a term proportional to $\omega\tau$ in the Abraham-Lorentz equation. This force term becomes comparable with the applied force when $\omega\tau \approx 1$. What is the energy $E = \hbar\omega$ (in MeV) of a photon whose angular frequency obeys this approximate condition?

Exercise 14.1.4.

(a) Show Equations (14.34)–(14.36).
(b) Show that

$$p_\mu K^\mu_{\rm react} = 0,$$

where

$$K^\mu_{\rm react} \equiv \left[\frac{d^2p^\mu}{d\tau^2} + \frac{p^\mu}{(mc)^2}\frac{dp^\nu}{d\tau}\frac{dp_\nu}{d\tau}\right].$$

Exercise 14.1.5. Starting from (14.37), (14.40) or (14.41), show that the relativistic Abraham-Lorentz form for momentum change is unfortunately very complicated:

$$\frac{d\vec{p}}{dt} = \vec{F} + \tau mc\gamma^2 \left[\ddot{\vec{\beta}} + 3\gamma^2(\vec{\beta}\cdot\dot{\vec{\beta}})\left(\dot{\vec{\beta}} + \gamma^2(\vec{\beta}\cdot\dot{\vec{\beta}})\vec{\beta}\right) + \gamma^2\vec{\beta}(\vec{\beta}\cdot\ddot{\vec{\beta}})\right],$$

where $\vec{F}$ is the ordinary force. It clearly reduces to (14.31) in the nonrelativistic limit but has many more unexpected terms. ($\frac{dp^0}{dt}$ follows from this by the identity in Exercise 14.1.4(b).)

Exercise 14.1.6.

(a) Show that the general solution of (14.31) for a time dependent force $\vec{F}(t)$, given $\vec{F}(t) = 0$ for $t < 0$, is

$$\dot{\vec{v}}(t) = e^{t/\tau}\left[\dot{\vec{v}}(0) - \frac{1}{\tau m}\int_0^t dt'\, e^{-t'/\tau}\vec{F}(t')\right].$$

(b) Assuming

$$\vec{v}(0) = 0,\ \dot{\vec{v}}(0) = 0,$$

solve for the complete velocity profile $\vec{v}(t), t \in [-\infty, \infty]$ produced for $\vec{F}(t) = \vec{F}_0\, H(t)$, where $H(t)$ is the Heaviside step function, $H(t) = 0$ for $t < 0$ and $H(t) = 1$ for $t > 0$.

(c) Assuming

$$\dot{\vec{v}}(0) = \frac{1}{\tau m}\int_0^\infty dt'\, e^{-t'/\tau}\vec{F}(t'),$$

now show that

$$\dot{\vec{v}}(t) = \frac{1}{m}\int_0^\infty ds\, e^{-s}\vec{F}(s\tau + t).$$

(d) Again, solve explicitly for the complete velocity profile $\vec{v}(t)$ produced by $\vec{F}(t) = \vec{F}_0\, H(t)$.

Exercise 14.1.7.

(a) Use Equations (14.49) and (14.50) to calculate the relativistic power loss for a particle of charge e in a spatially uniform electric field $\vec{E}$. Also assume the field turns on at $t = 0$ and shuts off at $t = T$ in an instantaneous manner,

$$\vec{E}(t) = \vec{E}[H(t) - H(t-T)],$$

where $H(t)$ is the Heaviside step function as defined in the last exercise. Show the radiative power loss is

$$\Delta p^0\Big|_{\text{rad}} = -\frac{2e^4}{3m^2c^4}\vec{E}^2 T = P_E T/c,$$

in agreement with the total from the constant power result from Exercise 13.7.3(b). Comment on the time dependence of the signal.

(b) Now consider the case of energy loss for a charged particle in a space and time uniform magnetic field $\vec{B}$. Assuming motion in a plane perpendicular to the magnetic field and approximate instantaneous circular motion, show from (14.49) and (14.50) that one recovers the radiation rate

$$\left.\frac{dp^0}{dt}\right|_{\text{rad}} = -\frac{2e^4}{3m^2c^4}\gamma^2\beta^2\vec{B}^2 = P_B/c,$$

also from Exercise 13.7.3(b).

Exercise 14.1.8.

(a) The simplest relativistically covariant form for the Poincaré stresses is

$$P^{\mu\nu} = p(r)\,\eta^{\mu\nu},$$

where $\eta^{\mu\nu}$ is the metric tensor and $p(r)$ is the energy density. Using Equations (13.166), (14.23) and (14.24), and assuming a continuous spherically symmetric charge distribution of radius a at rest, show that the force density (or differential pressure) dp/dr is given by

$$\frac{dp(r)}{dr} = -\frac{\rho(r)Q(r)}{r^2},$$

where

$$Q(r) \equiv \int_r d^3x'\rho(r'),$$

is the amount of charge contained in a sphere of radius r. We have made use of Gauss's theorem to determine the radial electric field.

(b) Using

$$m_{\rm p} = \frac{1}{c^2}\int d^3x\, p(r),$$

$$m_{\rm el} = \frac{1}{c^2}\int d^3x\, T^{00} = \frac{1}{2c^2}\int\int d^3x\, d^3x'\, \frac{\rho(r)\rho(r')}{|\vec{x}-\vec{x}'|},$$

where $\rho(r)$ is the charge density, and assuming that $p(a) = 0$, show that any spherically symmetric charge distribution for which $Q(r)$ is finite has

$$m_{\rm p} = \frac{1}{3}m_{\rm el}.$$

(A shell of charge gives the same result and can be considered separately.) [*Hint*: Note there is a tricky integration by parts necessary: $m_{\rm p} = -\frac{4\pi}{3c^3}\int dr\, r^3 \frac{dp(r)}{dr}$.]

Exercise 14.1.9.

(a) Look up the two original Wheeler-Feynman papers from footnote 7. (Note the appropriate violation of causality: Part III (1945) was published before Part II (1949)!) Describe one of the derivations done to obtain the force of radiative reaction.

(b) Describe Wheeler and Feynman's point of view toward the pre-acceleration manifest in their theory. For example, did they simply think light "sped up" in the vicinity of the particle? Did they find the effect could be magnified by other particles?

(c) Discuss whether or not you think pre-acceleration of charged particles is a testable hypothesis at the level "predicted" in the theory. Consider quantum uncertainty principles, for example.

Exercise 14.1.10. We calculated the leading terms in the self-force, $F^k_{\rm self}(t)$, using a Taylor series (or multipole) expansion of the sources $\rho(\vec{x}-\vec{\xi},t')$ and $j^k(\vec{x}-\vec{\xi},t')$, which appear in Equations (14.7) and (14.8). These were assumed to be spherically symmetric. The more complete calculation is described for example in Jackson's book, Section 16.3. The proposed expression for the self-force from Jackson's Equation (16.28) in our notation reads

$$F^k_{\rm self}(t) = -\frac{2}{3c^2}\sum_{n=0}^{\infty}\frac{(-1)^n}{c^n n!}\frac{\partial^{n+1}v^k(t)}{\partial t^{n+1}}\int d^3x\int d^3\xi\, \rho(\vec{x})|\vec{\xi}|^{n-1}\rho(\vec{x}-\vec{\xi}).$$

(a) Show that the first two terms ($n = 0, 1$) of this expression give back the two terms in Equation (14.21).

(b) Do a Fourier transform of these time-dependent functions. Define

$$v^k(\omega) \equiv \frac{1}{\sqrt{2\pi}} \int_{-\infty}^{\infty} dt\, v^k(t)\, e^{i\omega t}.$$

Show that a Fourier transform of the equation,

$$m_0 \dot{v}^k - F^k_{\text{self}}(t) = F^k(t),$$

which is a generalization of Equation (14.31), gives

$$-i\omega M(\omega) v^k(\omega) = F^k(\omega),$$

where the "effective mass" $M(\omega)$ is given by

$$M(\omega) = m_0 + \frac{2e^2}{3c^2} \int d^3x \int d^3\xi\, \rho(\vec{x})\, G^{\text{out}}(\vec{\xi}; k) \rho(\vec{x} - \vec{\xi}),$$

with

$$G^{\text{out}}(\vec{\xi}; k) = \frac{e^{ik|\vec{\xi}|}}{|\vec{\xi}|},$$

where $k = \omega/c$. This infinite space outgoing Green function was encountered previously in Equation (12.11). Note that these results follow from the use of the retarded Green function in Equation (14.4).

Exercise 14.1.11. The plane wave Helmhotz expansion,

$$\frac{e^{ikR}}{R} = 4\pi ik \sum_{\ell,m} j_\ell(kr_<) h^{(1)}_\ell(kr_>) Y_{\ell m}(\theta, \phi) Y^*_{\ell m}(\theta', \phi'),$$

where $R = |\vec{x} - \vec{x}'|$ and $h^{(1)}_\ell(x)$ is the spherical Hankel function of Equation (10.267), is a generalization of the expansion found in Equation (10.275). Using this form in the effective mass expression from Exercise 14.1.10, and assuming a thin spherical shell of charge of radius a,

$$\rho(r) = \frac{e}{4\pi a^2} \delta(r - a),$$

show that

$$M(\omega) = m + \frac{2e^2}{3c^2} \left(\frac{e^{2ika} - 1 - 2ika}{2ika} \right),$$

where $m = m_0 + m_{\text{tot}}$ and $m_{\text{tot}} = \dfrac{2e^2}{3ac^2}$.

Exercise 14.1.12.

(a) Using the results from Exercise 14.1.10, show that the solution for the acceleration $a^k(t)$ from a given force $F^k(t)$ can be written as ($\tau = \dfrac{2e^2}{3mc^2}$)

$$a^k(t) = \int_{-\infty}^{\infty} dt' \, \frac{F^k(t')}{\tau M(t-t')},$$

where

$$M(t-t') = \frac{2\pi}{\tau \displaystyle\int_{-\infty}^{\infty} d\omega \, \frac{e^{-i\omega(t-t')}}{M(\omega)}}.$$

This identifies $1/M(\tau)$ as the response function for acceleration. As we learned in Chapter 9 for the response function $G(\tau)$ for the electric field, this function will only have causal behavior if $1/M(\tau)$ ($M(\tau)$) does not have poles (zeros) in the upper half complex plane.

(b) For the model introduced in Exercise 14.1.11, show that the real and imaginary parts of $ika\, M(\omega)$ are

$$\text{Real: } \frac{2e^2}{3ac^2}\left[e^{-y}\cos x - 1 + y\left(1 - \frac{a}{c\tau}\right)\right],$$
$$\text{Imag: } \frac{2e^2}{3ac^2}\left[e^{-y}\sin x - x\left(1 - \frac{a}{c\tau}\right)\right],$$

using $x = \text{Re}(2ka)$ and $y = \text{Im}(2ka)$.

(c) For $a > \tau c$ show there are no zeros in the upper half plane. Since $m = m_0 + m\dfrac{c\tau}{a}$, this condition is equivalent to $m_0 > 0$.

Exercise 14.1.13. Building on Exercise 14.1.12, now evaluate the self-force on the same shell of charge using the Feynman half-advanced, half-retarded propagator,

$$G^{avg} \equiv \frac{1}{2}(G^{ret} + G^{adv}).$$

Show that the real and imaginary parts of

$$ika\, M^{avg}(\omega) \equiv \frac{ika}{2}\left(M^{ret}(\omega) + M^{adv}(\omega)\right),$$

using $x = \text{Re}(2ka)$ and $y = \text{Im}(2ka)$ give

$$\text{Real: } \frac{2e^2}{3ac^2}\left[-\cos x\ \sinh y + y\left(1-\frac{a}{c\tau}\right)\right],$$

$$\text{Imag: } \frac{2e^2}{3ac^2}\left[\sin x\ \cosh y - x\left(1-\frac{a}{c\tau}\right)\right].$$

Using the above, show that there are now no zeros of $ikaM^{avg}(\omega)$ on the imaginary y-axis (i.e., along $x=0$) except the trivial one at $x=0$, $y=0$, even when $a<\tau c$. (Are there any poles in general? We do not know!)

Exercise 14.1.14. The most practical result of the radiation reaction considerations in Chapter 14 is the approximate modified force equation. This is given by Equation (14.49) of the text.

(a) Starting from Equation (14.47), and applying it to a pure electromagnetic 4-force K^μ acting on a particle of charge e, derive Equation (14.49).
(b) Now show that

$$\frac{d\vec{p}}{dt} = e\left(\vec{E}+\vec{\beta}\times\vec{B}\right) +$$
$$e\,\tau\left[\gamma^3\left(\vec{\beta}\cdot\dot{\vec{\beta}}\right)\left(\vec{E}+\vec{\beta}\times\vec{B}\right)+\gamma\dot{\vec{\beta}}\times\vec{B}-\gamma^3\vec{\beta}\left(\dot{\vec{\beta}}\cdot\left(\vec{E}+\vec{\beta}\times\vec{B}\right)\right)\right],$$

is an explicit form of Equation (14.49) (14.49) for $\vec{E}$ and $\vec{B}$ fields which are uniform and time constant. (Dots denote time derivatives and $\vec{\beta}=\vec{v}/c$, $\gamma=1/\sqrt{1-\beta^2}$.)
(c) In addition, starting with Equation (14.47), show that

$$\frac{dp^0}{dt} = e\vec{E}\cdot\vec{\beta}+e\,\tau\gamma^3\vec{\beta}\cdot\left[\dot{\vec{\beta}}\times\vec{B}+\left(\vec{\beta}\times\dot{\vec{\beta}}\right)\times\vec{E}\right],$$

is an explicit form of Equation (14.50) for $\vec{E}$ and $\vec{B}$ fields which are again uniform and time constant.

Exercise 14.2.1. There are other prescriptions for Green function solutions to the wave equation that are not used in classical electrodynamics, as discussed in Section 14.2. Try evaluating the integral,

$$G_+(\vec{x},t;\vec{x}\,',t') = -4i\pi\int\frac{d^3k}{(2\pi)^3}\int_{-\infty}^{\infty}\frac{d\omega}{2\pi}\frac{e^{i\vec{k}\cdot(\vec{x}-\vec{x}\,')-i\omega(t-t')}}{k^2-\frac{\omega^2}{c^2}(1+i\epsilon)},$$

consistently holding the infinitesimal quantity $\epsilon > 0$ (we suggest doing the ω-integral first). Show first that (ϵ must be rescaled)

$$G_+ = \frac{c}{\pi}\frac{1}{R^2 - c^2T^2 + i\epsilon},$$

and then that (see Exercise 8.6.6)

$$\mathrm{Im}(G_+) = -\frac{1}{2}(G^{\mathrm{ret}} + G^{\mathrm{adv}}),$$

where $T = t - t'$ and $R = |\vec{x}\,' - \vec{x}|$.

Exercise 14.2.2. Solve for the one-dimensional Feynman (massless) propagator,

$$G_+(x,t;x',t') = -4i\pi \int_{-\infty}^{\infty}\frac{dk}{2\pi}\int_{-\infty}^{\infty}\frac{d\omega}{2\pi}\frac{e^{ik(x-x')-i\omega(t-t')}}{k^2 - \frac{\omega^2}{c^2}(1+i\epsilon)}.$$

[You may need an integral table, or computer algebra.]

Exercise 14.2.3. Derive the analog of the "composition rule" in Section 14.2 for retarded Dirichlet or Neumann Green functions (with or without surface boundaries), expressing $G(\vec{x}\,'',t'';\vec{x},t)$ in terms of $G(\vec{x}\,'',t'';\vec{x}\,',t')$ and $G(\vec{x}\,',t';\vec{x},t)$ by using the fundamental wave equations these Green functions obey. [*Hint*: One possible procedure is as follows: Multiply $G(\vec{x}\,'',t'';\vec{x}\,',\bar{t}')$ times the differential equation satisfied by $G(\vec{x}\,',\bar{t}';\vec{x},t)$, and $G(\vec{x}\,',\bar{t}';\vec{x},t)$ by the differential equation satisfied by $G(\vec{x}\,'',t'';\vec{x}\,',\bar{t}')$. Subtract and integrate over all $\vec{x}\,'$, and $\bar{t}'$ over a time range from t' to ∞, holding $t'' > t' > t$. It is also possible to prove this using a momentum representation of the retarded Green function.]

Exercise 14.2.4. Using the expression for $\vec{A}$ in (14.85), the polarization vector properties, (14.86), and the formalism for creation and destruction operators, prove that the Hamiltonian energy operator,

$$H = \frac{1}{8\pi}\int d^3x'\{\vec{E}^2 + \vec{B}^2\},$$

where all fields have arguments $(\vec{x}\,',t)$ and (note the scalar potential Φ gives no contribution in radiation gauge)

$$\vec{E} = -\frac{1}{c}\frac{\partial\vec{A}}{\partial t},\quad \vec{B} = \vec{\nabla}\times\vec{A},$$

gives

$$H = \sum_{n,\alpha} \hbar\omega_n \left(N_{n,\alpha} + \frac{1}{2} \right) \equiv \sum_{n,\alpha} \left(H_{n,\alpha} + \frac{1}{2}\hbar\omega_n \right),$$

where $N_{n,\alpha} \equiv a^\dagger_{n,\alpha} a_{n,\alpha}$ is the number operator. The factor of $\frac{1}{2}\hbar\omega_n$ is a real energy termed the "zero-point" energy of the electromagnetic fields, ultimately due to Heisenberg's uncertainty principle.

Extra: Show that the momentum field operator,

$$\vec{P} \equiv \frac{1}{4\pi c} \int d^3x'\, \vec{g} = \frac{1}{4\pi c} \int d^3x'\, \vec{E} \times \vec{B},$$

leads to

$$\vec{P} = \sum_{n,\alpha} \hbar\vec{k}_n \left(N_{n,\alpha} + \frac{1}{2} \right) \equiv \sum_{n,\alpha} \left(\vec{P}_{n,\alpha} + \frac{1}{2}\hbar\vec{k}_n \right).$$

The $\frac{1}{2}\hbar\vec{k}_n$ term above can be dropped when summed over $\vec{k}_n$ and $-\vec{k}_n$. Thus for each mode,

$$H_{n,\alpha} = |\vec{P}_{n,\alpha}|c,$$

for $|\vec{k}_n| = \omega_n/c$ consistent with the interpretation after (8.40).

Exercise 14.2.5.

(a) Evaluate the photon propagator,

$$G^+_{ab}(\vec{x}, t; \vec{x}\,', t\,') \equiv \langle 0|\mathrm{T}\left[A_a(\vec{x}, t)\, A^+_b(\vec{x}\,', t\,')\right]|0\rangle,$$

where a and b are spatial indices ($a,b = 1,2,3$), T is the time ordering operator and $A_a(\vec{x}, t)$ is the photon field given by Equation (14.85). Show that it may be written as

$$G^+_{ab}(\vec{x}, t; \vec{x}\,', t\,') = 4\pi c \int \frac{d^3k}{(2\pi)^3} \frac{1}{2k} e^{i(\vec{k}\cdot(\vec{x}-\vec{x}')-ck|t-t'|(1-i\epsilon))} \left(\delta_{ab} - \frac{k_a k_b}{k^2} \right),$$

for arbitrary time ordering. The convergence factor $(1 - i\epsilon)$ has been inserted by hand.

(b) Now show that the expression for $G^+_{ab}(\vec{x}, t; \vec{x}\,', t\,')$ in position space may be written as

$$G^+_{ab}(\vec{x}, t; \vec{x}\,', t\,') = \left(\delta_{ab} - c^2 \int_T^\infty dT' \int_{T'}^\infty dT'' \frac{\partial}{\partial x_a} \frac{\partial}{\partial x_b} \right) G_+(R, T''),$$

for $R = |\vec{x} - \vec{x}'|$, $T = |t - t'|$, $T' > 0$, and where $G_+(R, T'')$ is the massless scalar propagator given in Equations (14.76) and (14.77):

$$G_+(R, T'') = 4\pi c \int \frac{d^3k}{(2\pi)^3} \frac{1}{2k} e^{i(\vec{k}\cdot(\vec{x}-\vec{x}') - ckT''(1-i\epsilon))},$$
$$= \frac{c}{\pi} \frac{1}{R^2 - c^2T''^2 + i\epsilon},$$

where $T'' > 0$.

Exercise 14.3.1. Given the 1-dimensional perfectly reflecting cavity described in Section 14.3.4.

- What are the directions of the electrical field and the magnetic field do these directions change inside the cavity when reflecting, and why?
- Write the equation of E and H inside this cavity. Can these be considered as scalar fields?

Exercise 14.3.2. Given an interval $[a, b]$ and a Hilbert space $\mathcal{H}$ of complex square-integrable functions defined on the interval. An operator $P : \mathcal{H} \to \mathcal{H}$ is *self-adjoint* if

$$\int_a^b f(Pg)dx = \int_a^b g(Pf)dx, \tag{14.144}$$

for all $f, g \in \mathcal{H}$.

(a) Prove the operator $P_V \equiv -\frac{d^2}{dx^2} + V(x)$ is self-adjoint on any interval, where the Hilbert space consists of square-integrable functions satisfying any fixed set of Dirichlet or Neumann boundary conditions.
(b) Show that P_V is *not* self-adjoint if the functions in H satisfy outgoing boundary conditions.

Exercise 14.3.3.

(a) Prove the condition (14.108) for the free space wave number for the dielectric cavity with outgoing boundary conditions depicted in Figure 14.3.
(b) Show that the solution of (14.108) for k_0 must be non-real if $n \neq 1$.

Exercise 14.3.4.

(a) Prove (14.114).
(b) Prove (14.122).

Exercise 14.3.5. Modify the differential matrix in (14.126) to accommodate homogeneous boundary conditions instead of periodic boundary conditions.

Exercise 14.3.6.

(a) Prove the recursion relation (14.129).
(b) Prove the orthogonality relation (14.130).

Exercise 14.3.7. Give Chebyshev derivative matrices for the Lobatto and left and right Radau grids on $[-1, 1]$ (see Section 14.3.7) for the cases:

(a) $N = 3$
(b) $N = 4$
(c) $N = 5$.

Exercise 14.3.8. Give the replacements for rows $N+1$, $4N+5$, $5N+5$, and $8N+9$ for the matrices $\boldsymbol{L}, \boldsymbol{M}$ in (14.136) so as to implement the continuity conditions (14.139). For example, row $N+1$ of $\boldsymbol{L}$ is all zeros except entries $N+1$ and $4N+5$ are $+1$ and -1 respectively, while row $N+1$ of $\boldsymbol{M}$ is all zeros.

Exercise 14.3.9. Given a simple cavity with Dirichlet boundary conditions, where the Fourier-transformed wave equation is:

$$\left[\partial_x^2 - 1\right] \phi(\omega, x) = -\omega^2 \phi(\omega, x). \tag{14.145}$$

(a) Show how to formulate this problem numerically using a second order Chebyshev differentiation matrix with the appropriate grid, taking boundary conditions into account.
(b) *(Advanced)* Using the Chebyshev differentiation matrices in (Al Sheikh, op. cit.), implement the problem and calculate the eigenvalues, what do you notice and why?

Appendix A

Appendix on Electromagnetic Units

One can consider unit systems as a set of tools to apply to different applications. One should not expect one tool set to efficiently apply to all circumstances. There will always be a need for alternate approaches to calculate and communicate physical results. Both Gaussian and SI approaches have their appropriate places in physics.

As mentioned in Chapter 1, SI units are more practical and popular. Electrical units are especially familiar and convenient in SI: volts, amperes, ohms, farads, etc. However, the simplifications inherent in Gaussian units are of great advantage in discussions of fundamental relationships involving relativistic electrodynamics formulations as well as connections to quantum mechanics and other advanced topics. There are no conversion factors introduced for charges (ϵ_0) and currents (μ_0) and all electromagnetic units are simply combinations of the gram (mass), the centimeter (length) and the second (time). Because of this, Gaussian units have a distinct pedagogical advantage.

Most students use SI units in undergraduate mechanics and electrodynamic courses, and so may be unfamiliar with Gaussian units. Because of this likely transition, we feel it is incumbent upon us give more units "helps" in this text than in most. One of these helps is the accompanying extended conversion table in four parts. It is much more detailed than similar tables, and gives symbol conversions, unit conversions and name designations for 30 physical quantities. The other help we are providing is the computer programs/applications we have developed for symbol conversion, discussed below. Using these programs allows the student to convert most of the equations of this text immediately to SI without consulting the tables. We have not found similar conversion utilities available, which is surprising.

This seems to be a neglected topic given the fundamental importance of measurement systems. We believe the issue of units usage in textbooks will eventually become less significant as more computer applications are developed for convenient conversion.

In the last column of Tables A.1 and A.2 the quantity "c" is simply a numerical factor without units and represents the *exact* value

$$c = 299,792,458 \approx 3 \times 10^8.$$

Of course, this originates from the numerical value of the speed of light in SI units. The speed of light is associated with the exact SI statements (here the c *has* physical units):

$$\begin{aligned} c &= \frac{1}{\sqrt{\epsilon_0 \mu_0}} = 299,792,458 \text{ m s}^{-1}, \\ \epsilon_0 &= \frac{1}{4\pi}\frac{10^7}{c^2} \text{ kg}^{-1} \text{ m}^{-3} \text{ s}^4 \text{ A}^2, \\ \mu_0 &= 4\pi\, 10^{-7} \text{ kg m s}^{-2} \text{ A}^{-2}. \end{aligned}$$

Following Ref. (5) of Section 1.6.2, the quantities α and β represent the conversion factors for distance and energy, respectively:

$$\alpha = 10^2 \text{ cm/m}, \quad \beta = 10^7 \text{ erg/J}.$$

The tables are set up to provide the connections from Gaussian to SI units by replacing the Gaussian quantities listed with SI quantities on both sides of an equation. Of course, the tables can be used in the opposite conversion sense as well.

A few words concerning SI units for orientation purposes are necessary. The SI units which come into play in electrodynamics are the kilogram, the meter, the second and the ampere (current). These are called *base units*. There are seven base units in the SI, the additional ones being the kelvin (thermodynamic temperature), the mole (amount of substance) and the candela (luminous intensity). The ampere is defined to be the current maintained in each of two straight parallel conductors of infinite length and negligible circular cross-section placed one meter apart which produces a force of 2×10^{-7} newtons/m. This uses the SI equation

$$\frac{F}{L} = \frac{\mu_0 I^2}{2\pi r},$$

for the magnitude of the force, F, the SI analog of Equation (6.15). In practice coils of wire are used instead of straight ones, and the separation is on the order of centimeters. This sensitive measurement device is called a *current balance.*

The long history of the evolution of unit systems in physics is fascinating. The second of time has evolved from a certain fraction of the solar day to a more precise atomic standard, which was adopted in 1967. The second is now defined by the duration of exactly 9,192,631,770 periods of a certain electromagnetic transition between two hyperfine levels of the ground state of the cesium-133 atom. Likewise, the original definition of the meter was in terms of the distance along the meridian through Paris from the North Pole to the equator of the Earth, defined to be 10^7 meters. In 1960 an atomic standard was also adopted for the meter. It was defined as a certain number of wavelengths of the light emitted by atoms of krypton-86 in a specific atomic transition. Using this definition, the speed of light in vacuum was determined to be 299,792,458 m/s. In 1983 the meter was re-defined to be the distance traveled by light in vacuum in a time of 1/299,792,458 of a second. Tying it to the second in this manner provides a much more precise standard than the one based on a wavelength. Mass, however, is still not related to an atomic standard. It is defined to be the mass of a particular cylinder of platinum-iridium alloy kept at the International Bureau of Weights and Measures near Paris. Related to this, see Ref. (1) of Section 1.6.2.

Note that all 30 quantities listed in Tables A.3–A.4 are designated with different unit names, although many of them have the same dimensions in Gaussian units. There are 4 exceptions to the different designation rule: electric field ($\vec{E}$) and displacement or D-field ($\vec{D}$); magnetic field ($\vec{B}$) and H-field ($\vec{H}$). In Gaussian units there is no dimensional difference in these quantities and no need to distinguish the units. Note also that the names chosen for these quantities (column 1) would be different in texts where the emphasis is on SI units. These pairs have different engineering dimensions in SI units, and it is natural that different names should be adopted. In SI leaning texts, $\vec{B}$ is called magnetic induction, and $\vec{H}$ is called the magnetic field; $\vec{E}$ is called the electric field in both traditions, while $\vec{D}$ is called displacement. Also note on Parts 3 and 4: column 4 gives the present authors' version of Gaussian unit names, which certainly are not unique. Finally, note that surface charge density (σ, Σ) is designated as Σ only in Chapter 10. Otherwise, it is distinguished by context from conductivity, which is designated only as σ.

Heaviside-Lorentz units is another system which avoids conversion factors for charges and currents and is favored in particle physics and other research areas. This unit system is defined by dropping the factors of 4π in the Gaussian Maxwell equations, Table 1.1, using the connections,

$$\vec{D} = \vec{E} + \vec{P},$$
$$\vec{H} = \vec{B} - \vec{M},$$

and keeping the Gaussian expression for the Lorentz force, Equation (1.1). Both Heaviside-Lorentz and SI are so-called *rationalized* systems of units, that is, factors of 4π do not appear in their Maxwell equations. Assuming CGS units, the numerical Gaussian to Heaviside-Lorentz conversion is simply given by the factors of 4π in Tables A.1 and A.2. This system is extended to include *natural units*, which in addition sets $c = 1$ and $\hbar = 1$.

As mention above, we have developed Gaussian/SI conversion applications, available on the *Open Text Project*™ website at Baylor University, as additional conversions resources. These applications use computer languages known as *Sage* and *Mathematica*™. The Sage cell conversion utility attaches automatically to the free Sage symbolic mathematics program (Gnu Public License). It can easily be modified by the user if desired. The Mathematica notebook file of course requires the Mathematica computer program which is a commercial product. The conversions in the programs are based upon the "Symbol Conversion" columns in Tables A.1 and A.2. Please consult the programs for instructions on use and limitations. We hope that others will see the need and improve upon our units conversion contributions.

A.1 Exercises

Exercise A.1.1. Explain how the amount conversion factor in column 4 of Tables A.1 and A.2 is determined.

Exercise A.1.2. The Gaussian system is not necessarily tied to CGS units. Discuss the effects of adopting MKS units within the Gaussian system. Are there advantages? Are there any new disadvantages?

Table A.1 Electromagnetic variables and conversion factors.

Part 1	Symbol Conversion		Amount Conversion to SI	
Quantity	**Gaussian**	**SI**	**Factor**	**Value**
Mass	m	m	$\alpha^2\beta^{-1}$	10^{-3}
Length	l	l	α^{-1}	10^{-2}
Time	t	t	1	1
Force	F	F	$\alpha\beta^{-1}$	10^{-5}
Energy	W	W	β^{-1}	10^{-7}
Energy density	w	w	$\alpha^3\beta^{-1}$	10^{-1}
Power	P	P	β^{-1}	10^{-7}
Power flow density	$\vec{S}$	$\vec{S}$	$\alpha^2\beta^{-1}$	10^{-3}
Charge	q	$\frac{q}{\sqrt{4\pi\epsilon_0}}$	$\left(\frac{4\pi\epsilon_0}{\alpha\beta}\right)^{1/2}$	$10^{-1}\,c^{-1}$
Surface charge density	$\sigma,\ \Sigma$	$\frac{\sigma,\ \Sigma}{\sqrt{4\pi\epsilon_0}}$	$\left(\frac{4\pi\epsilon_0\alpha^3}{\beta}\right)^{1/2}$	$10^3\,c^{-1}$
Charge density	ρ	$\frac{\rho}{\sqrt{4\pi\epsilon_0}}$	$\left(\frac{4\pi\epsilon_0\alpha^5}{\beta}\right)^{1/2}$	$10^5\,c^{-1}$
Current	I	$\frac{I}{\sqrt{4\pi\epsilon_0}}$	$\left(\frac{4\pi\epsilon_0}{\alpha\beta}\right)^{1/2}$	$10^{-1}\,c^{-1}$
Current density	$\vec{J}$	$\frac{\vec{J}}{\sqrt{4\pi\epsilon_0}}$	$\left(\frac{4\pi\epsilon_0\alpha^3}{\beta}\right)^{1/2}$	$10^3\,c^{-1}$
Polarization	$\vec{P}$	$\frac{\vec{P}}{\sqrt{4\pi\epsilon_0}}$	$\left(\frac{4\pi\epsilon_0\alpha^3}{\beta}\right)^{1/2}$	$10^3\,c^{-1}$
Electric dipole moment	$\vec{p}$	$\frac{\vec{p}}{\sqrt{4\pi\epsilon_0}}$	$\left(\frac{4\pi\epsilon_0}{\alpha^3\beta}\right)^{1/2}$	$10^{-3}\,c^{-1}$

Table A.2 Electromagnetic variables and conversion factors.

Part 2	**Symbol Conversion**		**Amount Conversion to SI**	
Quantity	**Gaussian**	**SI**	**Factor**	**Value**
Electric field	$\vec{E}$	$\sqrt{4\pi\epsilon_0}\,\vec{E}$	$\left(\frac{\alpha^3}{4\pi\epsilon_0\beta}\right)^{1/2}$	$10^{-4}\,c$
Potential (Emf, Voltage)	$\Phi,\ V$	$\sqrt{4\pi\epsilon_0}\,\Phi,\ V$	$\left(\frac{\alpha}{4\pi\epsilon\beta}\right)^{1/2}$	$10^{-6}\,c$
D-field	$\vec{D}$	$\sqrt{\frac{4\pi}{\epsilon_0}}\,\vec{D}$	$\left(\frac{\epsilon_0\alpha^3}{4\pi\beta}\right)^{1/2}$	$\frac{1}{4\pi}\times 10^3\,c^{-1}$
Magnetic field	$\vec{B}$	$\sqrt{\frac{4\pi}{\mu_0}}\,\vec{B}$	$\left(\frac{\mu_0\alpha^3}{4\pi\beta}\right)^{1/2}$	10^{-4}
Vector potential	$\vec{A}$	$\sqrt{\frac{4\pi}{\mu_0}}\,\vec{A}$	$\left(\frac{\mu_0\alpha^5}{4\pi\beta}\right)^{1/2}$	10^{-2}
H-field	$\vec{H}$	$\sqrt{4\pi\mu_0}\,\vec{H}$	$\left(\frac{\alpha^3}{4\pi\mu_0\beta}\right)^{1/2}$	$\frac{1}{4\pi}\times 10^3$
Magnetization	$\vec{M}$	$\sqrt{\frac{\mu_0}{4\pi}}\,\vec{M}$	$\left(\frac{4\pi\alpha^3}{\mu_0\beta}\right)^{1/2}$	10^3
Magnetic moment	$\vec{m}$	$\sqrt{\frac{\mu_0}{4\pi}}\,\vec{m}$	$\left(\frac{4\pi}{\mu_0\alpha^3\beta}\right)^{1/2}$	10^{-3}
Magnetic flux	F_{ij}	$\sqrt{\frac{4\pi}{\mu_0}}\,F_{ij}$	$\left(\frac{\mu_0}{4\pi\alpha\beta}\right)^{1/2}$	10^{-8}
Conductivity	σ	$\frac{\sigma}{4\pi\epsilon_0}$	$4\pi\epsilon_0$	$10^7\,c^{-2}$
Dielectric constant	ϵ	$\frac{\epsilon}{\epsilon_0}$	ϵ_0	$\frac{1}{4\pi}\times 10^7\,c^{-2}$
Magnetic permeability	μ	$\frac{\mu}{\mu_0}$	μ_0	$4\pi\times 10^{-7}$
Resistance	R	$4\pi\epsilon_0 R$	$\frac{\alpha}{4\pi\epsilon_0}$	$10^{-5}\,c^2$
Inductance	L	$4\pi\epsilon_0 L$	$\frac{\alpha}{4\pi\epsilon_0}$	$10^{-5}\,c^2$
Capacitance	C	$\frac{C}{4\pi\epsilon_0}$	$\frac{4\pi\epsilon_0}{\alpha}$	$10^5\,c^{-2}$

Table A.3 Electromagnetic variables, unit dimensions and names.

Part 3	**Dimensions**		**Unit Name/ Designation**	
Quantity	**Gaussian**	**SI**	**Gaussian**	**SI**
Mass	gm	kg	gram (gm)	kilogram (kg)
Length	cm	m	centimeter (cm)	meter (m)
Time	s	s	second (s)	second (s)
Force	gm cm s^{-2}	kg m s^{-2}	dyne	newton (N)
Energy	gm cm^2 s^{-2}	kg m^2 s^{-2}	erg	joule (J)
Energy density	gm cm^{-1} s^{-2}	kg m^{-1} s^{-2}	erg cm^{-3}	J m^{-3}
Power	gm cm^2 s^{-3}	kg m^2 s^{-3}	erg s^{-1}	watt (W)
Power flow density	gm s^{-3}	kg s^{-3}	erg cm^{-2} s^{-1}	W m^{-2}
Charge	$gm^{1/2}$ $cm^{3/2}$ s^{-1}	s A	statcoulomb (statcoul)	coulomb (C)
Surface charge density	$gm^{1/2}$ $cm^{-1/2}$ s^{-1}	m^{-2} s A	statcoul cm^{-2}	C m^{-2}
Charge density	$gm^{1/2}$ $cm^{-3/2}$ s^{-1}	m^{-3} s A	statcoul cm^{-3}	C m^{-3}
Current	$gm^{1/2}$ $cm^{3/2}$ s^{-2}	A	statampere (statamp)	ampere (A)
Current density	$gm^{1/2}$ $cm^{-1/2}$ s^{-2}	m^{-2} A	statamp cm^{-2}	A m^{-2}
Polarization	$gm^{1/2}$ $cm^{-1/2}$ s^{-1}	m^{-2} s A	dipole moment cm^{-3}	C m^{-2}
Electric dipole moment	$gm^{1/2}$ $cm^{5/2}$ s^{-1}	m s A	statcoul cm	C m

Table A.4 Electromagnetic variables, unit dimensions and names.

Part 4	**Dimensions**		**Unit Name/ Designation**	
Quantity	**Gaussian**	**SI**	**Gaussian**	**SI**
Electric field	$gm^{1/2}\ cm^{-1/2}\ s^{-1}$	$kg\ m\ s^{-3}\ A^{-1}$	statvolt cm^{-1}	$V\ m^{-1}$
Potential (Emf, Voltage)	$gm^{1/2}\ cm^{1/2} s^{-1}$	$kg\ m^2\ s^{-3}\ A^{-1}$	statvolt	volt (V)
D-field	$gm^{1/2}\ cm^{-1/2}\ s^{-1}$	$m^{-2}\ s\ A$	statvolt cm^{-1}	$C\ m^{-2}$
Magnetic field	$gm^{1/2}\ cm^{-1/2}\ s^{-1}$	$kg\ s^{-2}\ A^{-1}$	gauss (G)	tesla (T)
Vector potential	$gm^{1/2}\ cm^{-3/2}\ s^{-1}$	$kg\ m^{-1}\ s^{-2}\ A^{-1}$	$G\ cm^{-1}$	$T\ m^{-1}$
H-field	$gm^{1/2}\ cm^{-1/2}\ s^{-1}$	$m^{-1}\ A$	gauss *or* oersted (Oe)	$A\ m^{-1}$
Magnetization	$gm^{1/2}\ cm^{-1/2}\ s^{-1}$	$m^{-1}\ A$	magnetic moment cm^{-3}	$A\ m^{-1}$
Magnetic moment	$gm^{1/2}\ cm^{5/2}\ s^{-1}$	$m^2\ A$	gauss cm^3	$A\ m^2$
Magnetic flux	$gm^{1/2}\ cm^{3/2}\ s^{-1}$	$kg\ m^2\ s^{-2}\ A^{-1}$	gauss cm^2 *or* maxwells	weber (Wb)
Conductivity	s^{-1}	$kg^{-1}\ m^{-3}\ s^3\ A^2$	s^{-1}	siemens m^{-1} ($S\ m^{-1}$) *or* mho m^{-1}
Dielectric constant	dimensionless	$kg^{-1}\ m^{-3}\ s^4\ A^2$	–	$F\ m^{-1}$
Magnetic permeability	dimensionless	$kg\ m\ s^{-2}\ A^{-2}$	–	$H\ m^{-1}$
Resistance	$cm^{-1}\ s$	$kg\ m^2\ s^{-3}\ A^{-2}$	$cm^{-1}\ s$	ohm (Ω)
Inductance	$cm^{-1}\ s^2$	$kg\ m^2\ s^{-2}\ A^{-2}$	$cm^{-1}\ s^2$	henry (H)
Capacitance	cm	$kg^{-1}\ m^{-2}\ s^4\ A^2$	cm	farad (F)

Appendix B

Fundamental Concepts in Electrodynamics

B.1 Preface

Perhaps the main purpose for Appendix A: *Appendix on Electromagnetic Units* was to give deep "helps" for beginning graduate students in making the transition from the *International System of Units* (SI), used in most undergraduate treatments, to *Gaussian* units, used in advanced treatments and scientific articles, for electromagnetic quantities and equations. We feel that an additional help or pedagogical tool that we can provide for such students is a concise summary of fundamental ideas, equations and solutions from undergraduate level treatments of electrodynamics. Such a summary will also serve to further acquaint beginning graduate students with equations and quantities in Gaussian units as well as to help familiarize students with our mathematical notations. Thus, this appendix should help prepare students for the advanced insights in *Macroscopic Electrodynamics* as well as serve as a fundamental topics reference. We will follow a summary approach rather than an explicitly pedagogical one here; we will not attempt to justify in detail the steps that are taken or the concepts that are introduced. Our goal is not to *teach* undergraduate electrodynamics here, but simply to *reconnect* students to previous course materials.

B.2 Electrostatics

Electrostatics concerns the situation where only an electric field, denoted $\vec{E}(\vec{x})$, exists in a given geometry. It results from a static charge distribution in a given reference frame. We present a gentle introduction to the fundamental concepts and equations in this section, including: field equations, forces, energy and boundary condition considerations, electric dipoles, capacitance, boundary value problems, as well as changed media characteristics due to a macroscopic polarization field.

Field equations and point charges

The equations of electrostatics can be stated in differential form using standard differential operators. Given a continuous and static *charge density*, $\rho(\vec{x})$, the differential form for the electric field, known as *Gauss' law*, is given by

$$\vec{\nabla} \cdot \vec{E}(\vec{x}) = 4\pi\rho(\vec{x}). \tag{B.1}$$

The electric charge, q, of the system is

$$q = \int d^3x' \, \rho(\vec{x}). \tag{B.2}$$

The dot product between the gradient operator $\vec{\nabla}$ and the electric field in Equation (B.1) is called the divergence operation. (See Section 1.8 of the book for some useful differential operator identities.) The gradient operator can be represented in Cartesian coordinates as

$$\vec{\nabla} = \hat{e}_1 \frac{\partial}{\partial x_1} + \hat{e}_2 \frac{\partial}{\partial x_2} + \hat{e}_3 \frac{\partial}{\partial x_3}, \tag{B.3}$$

where the $\hat{e}_i$ are unit vectors in the 3 spatial coordinate directions. Written out in full, Equation (B.1) actually reads

$$\frac{\partial E_1(\vec{x})}{\partial x_1} + \frac{\partial E_2(\vec{x})}{\partial x_2} + \frac{\partial E_3(\vec{x})}{\partial x_3} = 4\pi\rho(\vec{x}). \tag{B.4}$$

Integrating Equation (B.1) over a volume V bounded by surface S produces the integral form of Gauss' law,

$$\oint_S da \; \hat{n} \cdot \vec{E}(\vec{x}) = 4\pi \int d^3x \, \rho(\vec{x}) = 4\pi q_{enc}. \tag{B.5}$$

$\hat{n}$ is an unit vector normal to the closed surface S directed out of volume V, and q_{enc} is the amount of electric charge which is enclosed in the volume. (Deriving Equation (B.5) from Equation (B.1) involves the use of the *divergence theorem*, a statement of which can be found in Chapter 1). The left hand side of Equation (B.5) is called the electric flux. This is a measure of the electric field entering or leaving the volume through its surface. In a situation with a known symmetry, Equation (B.5) is the fast and easy way to solve for the electric field.

The electric field due to a general continuous charge density, $\rho(\vec{x}\,')$, is given by the general expression,

$$\vec{E}(\vec{x}) = \int d^3x' \rho(\vec{x}\,') \frac{\vec{x} - \vec{x}\,'}{|\vec{x} - \vec{x}\,'|^3}, \tag{B.6}$$

where the three-dimensional integral indicated by the volume element d^3x' is over all space, or at least wherever the charge density is nonzero. Using Equation (B.6), one may show that it is always possible to write the electric field as the gradient of a *scalar potential* (or just potential) $\Phi(\vec{x})$,

$$\vec{E}(\vec{x}) = -\vec{\nabla}\Phi(\vec{x}), \tag{B.7}$$

which ensures that

$$\vec{\nabla} \times \vec{E}(\vec{x}) = 0, \tag{B.8}$$

is automatic in static conditions. The cross product between the gradient $\vec{\nabla}$ and the magnetic field $\vec{E}(\vec{x})$ is the differential operation called the curl. On the other hand, the scalar potential between a reference point $\mathcal{O}$ and final position, $\vec{x}$, may be expressed as

$$\Phi(\vec{x}) = -\int_{\mathcal{O}}^{\vec{x}} \vec{E} \cdot d\ell. \tag{B.9}$$

The integral expression for the scalar potential in infinite space is given by

$$\Phi(\vec{x}) = \int d^3x' \frac{\rho(\vec{x}\,')}{|\vec{x} - \vec{x}\,'|^2}. \tag{B.10}$$

When stated in terms of the scalar potential, Gauss' law becomes

$$\nabla^2\Phi(\vec{x}) = -4\pi\rho(\vec{x}), \tag{B.11}$$

($\nabla^2 = \vec{\nabla} \cdot \vec{\nabla}$) a result which is known as *Poisson's equation*. When there is no charge present, $\rho(\vec{x}) = 0$, this equation is termed *Laplace's equation*.

As pointed out above, one can use the integral form of Gauss' law, Equation (B.5), to derive the electric field for simple symmetrical situations. For a point charge at the origin of coordinates, one may assume the solution is radial and spherically symmetric. Establishing a sphere of radius r about the charge q, one has for this radial component, E_r

$$\oint_S da\, \hat{n} \cdot \vec{E}(\vec{x}) = 4\pi r^2 E_r = 4\pi q, \tag{B.12}$$

$$\Longrightarrow \vec{E}(\vec{x}) = q\,\frac{\hat{e}_r}{r^2}, \tag{B.13}$$

where $\hat{e}_r$ is a unit vector pointing in the outward radial direction. The associated potential may be deduced from (B.7) or (B.9) (with the origin $\mathcal{O}$ at spatial infinity) to be

$$\Phi(\vec{x}) = \frac{q}{r}. \tag{B.14}$$

We prefer to use standard notation position vectors $\vec{x}$ and $\vec{x}\,'$ to indicate field and source positions. However, when the scalar distance $|\vec{x}| = \sqrt{x^2 + y^2 + z^2}$ is considered, we simplify this in the main text as "r" rather than "x", which could be misidentified as an x-component.

Another, more mathematical, way of looking at the point charge starts with Equation (B.6). The source of a point charge, q, is a three-dimensional *Dirac delta function*. A one-dimensional Dirac delta function satisfies

$$\int dx\, \delta(x - x') = 1, \tag{B.15}$$

for any interval containing the point x'. A three-dimensional Dirac delta function is represented in Cartesian coordinates as

$$\delta(\vec{x} - \vec{x}\,') = \delta(x - x')\delta(y - y')\delta(z - z'), \tag{B.16}$$

and satisfies

$$\int d^3x\, \delta(\vec{x} - \vec{x}\,') = 1, \tag{B.17}$$

where the integration volume includes the point $\vec{x}\,'$. Setting the charge density to

$$\rho(\vec{x}) = q\, \delta(\vec{x} - \vec{x}\,'), \tag{B.18}$$

represents a point charge at $\vec{x}\,'$ with total charge

$$\int d^3x\, \rho(\vec{x}) = q. \tag{B.19}$$

The associated scalar potential is

$$\Phi(\vec{x}) = \frac{q}{|\vec{x} - \vec{x}\,'|}, \tag{B.20}$$

and the electric field is

$$\vec{E}(\vec{x}) = q\frac{\vec{x} - \vec{x}\,'}{|\vec{x} - \vec{x}\,'|^3}. \tag{B.21}$$

This is called *Coulomb's law.*

Finally, the force, $\vec{F}$, on a point charge q at position $\vec{x}$ is given by

$$\vec{F} = q\vec{E}(\vec{x}). \tag{B.22}$$

Equations (B.21) and (B.22) actually constitute the first step in our development of the electric field formalism in the main text (Section 2).

Boundary conditions, surface pressure and work

The boundary conditions

$$E_{2n} - E_{1n} = 4\pi\sigma, \tag{B.23}$$

$$E_{2t} - E_{1t} = 0, \tag{B.24}$$

hold on an idealized interface or surface between two different media. In this expression σ is the local surface charge density, $E_{1,2n}$ and $E_{1,2t}$ are *normal* (*i.e.* perpendicular to the surface) and *tangential* (*i.e.* parallel to the surface) components of the electric field in the two regions, respectively, and the unit surface normal $\hat{n}$ points from region 1 to 2.

The local normal surface force per area (pressure), denoted $\mathcal{F}_n$, reads

$$\mathcal{F}_n = \sigma E_n^{avg} = \frac{\sigma}{2}\left(E_{1n} + E_{2n}\right). \tag{B.25}$$

The average electric field E_n^{avg} is necessary because it eliminates the self-field contribution of the surface element. When Equation (B.23) is used, this becomes

$$\mathcal{F}_n = \frac{1}{8\pi}\left(E_{2n}^2 - E_{1n}^2\right). \tag{B.26}$$

The quasi-static work, W, done in *slowly* moving a point charge, q, from position $\vec{x}_1$ to $\vec{x}_2$ is

$$W = -q\int_{\vec{x}_1}^{\vec{x}_2} \vec{E}\cdot d\ell = q\left(\Phi(\vec{x}_2) - \Phi(\vec{x}_1)\right). \tag{B.27}$$

In free space one may choose $\lim_{\vec{x}_1\to\infty} \Phi(\vec{x}_1) = 0$ for the point at infinity. A discrete set of n point charges has the interaction energy ($n \geq 2$),

$$W = \sum_{i=1}^{n} q_i\Phi(\vec{x}_i), \tag{B.28}$$

or

$$W = \frac{1}{2}\sum_{i=1}^{n}\sum_{j\neq i}^{n} \frac{q_i q_j}{|\vec{x}_i - \vec{x}_j|}. \tag{B.29}$$

The factor of 1/2 in (B.29) compensates for double counting. The energy of assembly of a *continuous* charge density, $\rho(\vec{x})$, is given by the expressions

$$W = \frac{1}{2}\int d^3x\,\rho(\vec{x})\Phi(\vec{x}), \tag{B.30}$$

$$= \frac{1}{2}\int\int d^3x\,d^3x'\frac{\rho(\vec{x})\rho(\vec{x}\,')}{|\vec{x} - \vec{x}\,'|}, \tag{B.31}$$

$$= \frac{1}{8\pi}\int d^3x\,\vec{E}^2, \tag{B.32}$$

in an infinite volume. The passage from Equation (B.30) to (B.31) utilizes Equation (B.11). Equation (B.31), which is the total energy of the continuous system, contains electromagnetic self-energies, whereas (B.28), which is the interaction energy of point charges, does not. Equation (B.31) is always positive, whereas Equation (B.28) can be positive or negative.

Electric dipole field

The field of given charge density $\rho(\vec{x})$ possesses a so-called *multipole expansion* for the scalar potential involving charge or monopole ($\sim 1/r$), dipole ($\sim 1/r^2$), quadrupole ($\sim 1/r^3$) and higher order terms. The first two terms in the expansion of the electric potential are

$$\Phi(\vec{x}) = \frac{q}{r} + \frac{\vec{p}\cdot\vec{x}}{r^3}, \tag{B.33}$$

where the charge is given by Equation (B.2) and electric dipole vector, $\vec{p}$, is defined as

$$\vec{p} \equiv \int d^3x'\, \vec{x}\,'\rho(\vec{x}\,'). \tag{B.34}$$

The dipole field far away from the charges has the form,

$$\vec{E}^d(\vec{x}) = -\vec{\nabla}\left(\frac{\vec{x}\cdot\vec{p}}{r^3}\right) = \frac{3(\vec{x}\cdot\vec{p})\vec{x} - r^2\vec{p}}{r^5}. \tag{B.35}$$

If we place a charge $-q$ at position $-d/2\,\hat{z}$ and a charge q at position $d/2\,\hat{z}$, the electric dipole vector $\vec{p}$ is given by $\vec{p} = qd\,\hat{z}$, and Equation (B.35) represents the field for $r \gg d$. If we define the charges as $q \equiv p_0/d$ and consider the limit $d \to 0$, then Equation (B.35) represents the field of a *point* dipole $\vec{p} = p_0\hat{z}$ located at the origin of coordinates.

The force $\vec{F}$ and torque $\vec{N}$ on an elemental electric dipole, $\vec{p}$, in an electric field are

$$\vec{F} = (\vec{p}\cdot\vec{\nabla})\vec{E} = \vec{\nabla}(\vec{p}\cdot\vec{E}), \tag{B.36}$$

$$\vec{N} = \vec{p}\times\vec{E}. \tag{B.37}$$

Simple electric solutions

Planar geometry

The electric field generated by a plane with a uniformly distributed surface charge density σ located at $z = 0$ is given by

$$\vec{E}(z) = \begin{cases} 2\pi\sigma\hat{z}, & z > 0 \\ -2\pi\sigma\hat{z}, & z < 0. \end{cases} \tag{B.38}$$

A flat planar conductor of nonvanishing thickness with total charge per area σ will have surface charges $\sigma/2$ on the two surfaces. This can be argued either from symmetry or from Newton's third law applied to the surface pressures. For parallel planes at $z = -d/2$ and $z = d/2$ with opposite surface charge densities σ and $-\sigma$, respectively, one has

$$\vec{E}(z) = \begin{cases} 0, & z > d/2 \\ 4\pi\sigma\hat{z}, & -d/2 < z < d/2 \\ 0, & z < -d/2. \end{cases} \tag{B.39}$$

For parallel conductors of nonvanishing thickness one may show from the divergence theorem that the interior surface charges cancel and the exterior ones are equal. So, if the total surface charges on the two conductors are equal and opposite, the exterior charges are zero and the field outside vanishes.

Spherical geometry

The electric field generated by a sphere of radius R with its center at the origin of coordinates and a charge q uniformly distributed over its surface, is given by

$$\vec{E}(r) = \begin{cases} q\dfrac{\hat{e}_r}{r^2}, & r > R \\ 0, & r < R. \end{cases} \tag{B.40}$$

The $r > R$ branch may be found in exactly the same manner as we solved for a point charge field above. The lack of an electric field on the interior of this closed spherical surface is not a special case, but occurs whenever an interior space is enclosed by an equipotential surface as provided, for example, by a conductor (unless there are additional interior conducting surfaces).

Cylindrical geometry

The electric field generated by a circular cylinder of radius R with its axis along the z-coordinate and a uniformly distributed surface charge density σ is given by

$$\vec{E}(\rho) = \begin{cases} \dfrac{4\pi R\sigma}{\rho}\hat{e}_\rho, & \rho > R \\ 0, & \rho < R. \end{cases} \tag{B.41}$$

Let λ be a uniform linear line charge density positioned along the z-axis. The scalar potential is

$$\Phi(\rho) = -2\lambda \ln(\rho/\rho_0), \tag{B.42}$$

where ρ_0 can be considered the finite distance to a surface where the zero of potential is established. The associated electric field is

$$\vec{E}(\rho) = \frac{2\lambda\hat{e}_\rho}{\rho}. \tag{B.43}$$

Conductors and capacitors

Conductors are materials in which one or more electrons per atom can be considered free particles. Under static conditions, the surface of the conductor will be equipotential and the interior will have no electric field. Given this, the electric field lines emerging from the conductor will be normal to the surface, which is where the free charge exists. A conductor can be isolated or grounded. An isolated conductor has a given fixed charge whereas a grounded conductor has a zero of potential and can exchange charge with the ground under non-static conditions. Since $E_{1n} = 0$ in the interior of the conductor, Equation (B.23) gives

$$E_{2n} = -\frac{\partial\Phi}{\partial n} = 4\pi\sigma, \tag{B.44}$$

just outside the conductor, where n is now the local outward normal coordinate to the surface. Using this in Equation (B.26) then gives the outward normal force per area on a conducting surface with a surface charge σ as

$$\mathcal{F}_n = 2\pi\sigma^2. \tag{B.45}$$

Capacitance is a measure of how much charge can be stored on isolated conductors or systems of conductors for fixed voltages. Given conductors

a and b with charges $-Q$ and Q, respectively, the capacitance C of the system is defined as

$$C \equiv \frac{Q}{V}, \tag{B.46}$$

where V is the potential difference

$$V \equiv \Phi_b - \Phi_a. \tag{B.47}$$

For an isolated conductor, the surface at infinity serves as the second conductor with $\Phi_a = 0$. Thus, for a sphere of radius R one has

$$C_{sph} = \frac{Q}{V} = R. \tag{B.48}$$

For a large parallel plate capacitor of area A and plate separation d, the capacitance C is

$$C_{pp} = \frac{A}{4\pi d}. \tag{B.49}$$

Likewise, the capacitance of two concentric spheres with radii a and b ($b > a$) is

$$C_{2sph} = \frac{ab}{(b-a)}, \tag{B.50}$$

and the capacitance of long concentric cylinders also with radii a and b ($b > a$) as well as length L is

$$C_{2cyl} = \frac{L}{2\ln(b/a)}. \tag{B.51}$$

The summed energy of assembly from Equations (B.27) and (B.46) in every case is

$$W = \int_0^Q dq\,\frac{q}{C} = \frac{1}{2}\frac{Q^2}{C} = \frac{1}{2}CV^2. \tag{B.52}$$

Boundary value problems

An important class of problems in electrostatics concerns the solution for the electric potential when the values on the boundaries of certain geometries are given. These techniques begin with the mathematical technique called *separation of variables*.

Cartesian coordinates

The Laplace equation ((B.11) with $\rho(\vec{x}) = 0$) in Cartesian coordinates (x, y, z) reads

$$\frac{\partial^2 \Phi(x,y,z)}{\partial x^2} + \frac{\partial^2 \Phi(x,y,z)}{\partial y^2} + \frac{\partial^2 \Phi(x,y,z)}{\partial z^2} = 0. \tag{B.53}$$

Assuming $\Phi(x,y,z) = X(x)Y(y)Z(z)$ and dividing Equation (B.53) by Φ gives

$$\frac{1}{X(x)}\frac{d^2 X(x)}{dx^2} + \frac{1}{Y(y)}\frac{d^2 Y(y)}{dy^2} + \frac{1}{Z(z)}\frac{d^2 Z(z)}{dz^2} = 0. \tag{B.54}$$

The equation has been separated into three functions which depend on independent variables x, y, z. The most general solution of this is

$$\frac{1}{X(x)}\frac{d^2 X(x)}{dx^2} = C_1, \quad \frac{1}{Y(y)}\frac{d^2 Y(y)}{dy^2} = C_2, \quad \frac{1}{Z(z)}\frac{d^2 Z(z)}{dz^2} = C_3, \tag{B.55}$$

where the $C_{1,2,3}$ are termed separation constants. There are only two independent values since these satisfy $C_1 + C_2 + C_3 = 0$. The general solution of the second order differential X-equation for example is

$$X(x) = \begin{cases} A \sinh\left(\sqrt{C_1}\, x\right) + B \cosh\left(\sqrt{C_1}\, x\right), \; C_1 > 0, \\ A \sin\left(\sqrt{-C_1}\, x\right) + B \cos\left(\sqrt{-C_1}\, x\right), \; C_1 < 0, \\ Ax + B, \quad C_1 = 0, \end{cases} \tag{B.56}$$

where A and B are constants. Similarly for the $Y(y)$ and $Z(z)$ functions. For example, the conducting box problem in the text in Section 3.5 has $C_1 < 0$, $C_2 < 0 \implies C_3 > 0$. Application of boundary conditions on the surfaces then yields an infinite series solution for this problem. The methods of Fourier series are employed in such cases.

Spherical coordinates

The Laplace equation in spherical coordinates for the potential $\Phi(r, \theta, \phi)$ reads

$$\frac{1}{r^2}\frac{\partial}{\partial r}\left(r^2 \frac{\partial \Phi}{\partial r}\right) + \frac{1}{r^2 \sin\theta}\frac{\partial}{\partial \theta}\left(\sin\theta \frac{\partial \Phi}{\partial \theta}\right) + \frac{1}{r^2 \sin^2\theta}\frac{\partial^2 \Phi}{\partial \phi^2} = 0. \tag{B.57}$$

Problems which have no ϕ dependence, $\Phi(r,\theta,\phi) \to \Phi(r,\theta)$, are said to have azimuthal symmetry. Assuming

$$\Phi(r,\theta) = R(r)\Theta(\theta), \tag{B.58}$$

and multiplying by r^2/Φ gives the equation

$$\frac{1}{R}\frac{\partial}{\partial r}\left(r^2\frac{\partial R}{\partial r}\right)+\frac{1}{\Theta\sin\theta}\frac{\partial}{\partial\theta}\left(\sin\theta\frac{\partial\Theta}{\partial\theta}\right)=0, \tag{B.59}$$

which has separated functions for r and θ. The terms are separated with

$$\frac{1}{R}\frac{\partial}{\partial r}\left(r^2\frac{\partial R}{\partial r}\right)=-\frac{1}{\Theta\sin\theta}\frac{\partial}{\partial\theta}\left(\sin\theta\frac{\partial\Theta}{\partial\theta}\right)=\ell(\ell+1), \tag{B.60}$$

where $\ell=0,1,2,3,\ldots$ are the only bounded solutions. The solutions are

$$R(r)=Ar^{\ell}+\frac{B}{r^{\ell+1}}, \tag{B.61}$$

$$\Theta(\theta)=C\,P_\ell(\cos\theta). \tag{B.62}$$

$P_\ell(\cos\theta)$ is called a *Legendre polynomial*. It is a polynomial of order ℓ in variable $\cos\theta$. (A second solution called a Legendre function of the second kind is infinite at the endpoints $x=\pm1$.) It is given by the *Rodrigues formula* (using $x=\cos\theta$)

$$P_\ell(x)=\frac{1}{2^\ell\ell!}\left(\frac{d}{dx}\right)^\ell\left(x^2-1\right)^\ell. \tag{B.63}$$

The first 5 of these polynomials are given in Table 4.1 of the main text. The general solution of such azimuthally symmetric problems can be expressed as

$$\Phi(r,\theta)=\sum_{\ell=0}^{\infty}\left(A_\ell+\frac{B_\ell}{r^{\ell+1}}\right)P_\ell(\cos\theta). \tag{B.64}$$

The expansion coefficients can be solved for using the boundary conditions on the spherical surfaces and the inversion formula

$$\int_0^\pi d\theta\sin\theta P_\ell(\cos\theta)P_{\ell'}(\cos\theta)=\frac{\delta_{\ell\ell'}}{2\ell+1}. \tag{B.65}$$

$\delta_{\ell\ell'}$ is called a *Kronecker delta*:

$$\delta_{\ell\ell'}=\begin{cases}0, & \ell\neq\ell',\\ 1, & \ell=\ell'.\end{cases} \tag{B.66}$$

Cylindrical coordinates

The two-dimensional Laplace equation in cylindrical coordinates for the potential $\Phi(\rho,\phi)$ is

$$\frac{1}{\rho}\frac{\partial}{\partial\rho}\left(\rho\frac{\partial\Phi}{\partial\rho}\right)+\frac{1}{\rho^2}\frac{\partial^2\Phi}{\partial\phi^2}=0. \tag{B.67}$$

When the full azimuthal range for ϕ is allowed ($0 \leq \phi < 2\pi$), the separation equation for

$$\Phi(\rho, \phi) = R(\rho)F(\phi), \tag{B.68}$$

reads

$$\frac{\rho}{R}\frac{\partial}{\partial\rho}\left(\rho\frac{dR}{d\rho}\right) = -\frac{1}{F}\frac{d^2F}{d\phi^2} = m^2. \tag{B.69}$$

One requires the F functions to be periodic and linearly independent, giving $m = 0, 1, 2, 3, \ldots$ as a sufficient set. The solutions are

$$m \neq 0: \qquad R(\rho) = a_m\rho^m + b_m\rho^{-m}, F(\phi) = A_m \sin(m\phi + \alpha_m), \tag{B.70}$$

$$m = 0: \qquad R(\rho) = a_0 + b_0 \ln\frac{\rho}{\rho_0}, F(\phi) = A_0 + B_0\phi. \tag{B.71}$$

The general solution is

$$\Phi(\rho, \phi) = a_0 + b_0 \ln\frac{\rho}{\rho_0} + \sum_{m=1}^{\infty} a_m\rho^m \sin(m\phi + \alpha_m)$$
$$+ \sum_{m=1}^{\infty} b_m\rho^{-m} \sin(m\phi + \beta_m). \tag{B.72}$$

The solution is again reduced to a Fourier series problem using boundary values.

Dielectrics

A *dielectric* is a piece of material which contains only "bound electrons". There are localized in the vicinity of atomic nuclear centers. Additional "free" charges can be positioned at will in or on this material. We assume the electric field $\vec{E}$ still satisfies Equation (B.1) where $\rho(\vec{x})$ is the total charge density. We postulate that

$$\vec{\nabla} \cdot \vec{E} = 4\pi\left(\rho_f + \rho_b\right), \tag{B.73}$$

where the f and b subscripts refer to bound and free charges, respectively, and assign

$$\rho_b = -\vec{\nabla} \cdot \vec{P}, \tag{B.74}$$

where $\vec{P}$ is called the polarization field. This motivates the definition of the *D-field*,

$$\vec{D} \equiv \vec{E} + 4\pi\vec{P}, \tag{B.75}$$

which satisfies

$$\vec{\nabla} \cdot \vec{D} = 4\pi \rho_f. \tag{B.76}$$

One needs to postulate a relationship between $\vec{D}$ and $\vec{E}$. The simplest possible relationship is the linear one

$$\vec{D} = \epsilon \vec{E}, \tag{B.77}$$

where the (dimensionless) dielectric constant or permittivity ϵ is 1 in the vacuum and a constant greater than 1 in the material. Equations (B.77) and (B.74) imply

$$\vec{P} = \frac{\epsilon - 1}{4\pi} \vec{E}. \tag{B.78}$$

There is a Gauss' law for the D-field,

$$\oint_S da\ \hat{n} \cdot \vec{D} = 4\pi q_f^{enc}, \tag{B.79}$$

($\hat{n}$ points out of the surface) where q_f^{enc} is the free charge enclosed in the volume. The boundary conditions for $\vec{P}$ and $\vec{D}$ give the local bound and free surface charge,

$$P_{2n} - P_{1n} = -\sigma_b, \tag{B.80}$$

$$D_{2n} - D_{1n} = 4\pi \sigma_f. \tag{B.81}$$

We remind the reader that the "normal" surface unit vector $\hat{n}$ here points from region 1 to 2. For tangential components, Equation (B.24) continues to hold. If the vacuum bounds the polarized 1-material ($P_{2n} = 0$), then the bound surface charge is

$$\sigma_b = \frac{\epsilon - 1}{4\pi} E_{1n}, \tag{B.82}$$

from Equations (B.78) and (B.80).

The assembly energy of a dielectric system is given as

$$W = \frac{1}{8\pi} \int d^3x\, \vec{D} \cdot \vec{E}. \tag{B.83}$$

If a dielectric with constant permittivity ϵ is inserted in a parallel plate capacitor and fills up the entire space between the plates, the new capacitance C_{pp}^{ϵ} is

$$C_{pp}^{\epsilon} = \epsilon C_{pp}. \tag{B.84}$$

Simple dielectric solutions

Planar geometry

Consider a flat dielectric slab of infinite extent and thickness d with constant permittivity ϵ. It is oriented in the xy-plane with its center at z-axis origin. When it is placed in a normal external uniform electric field $\vec{E} = E_0\hat{z}$, the electric field profile is given by

$$\vec{E}(z) = \begin{cases} E_0\hat{z}, & z > d/2 \\ \dfrac{E_0}{\epsilon}\hat{z}, & -d/2 < z < d/2 \\ E_0\hat{z}, & z < -d/2. \end{cases} \tag{B.85}$$

The surface charge densities are

$$\sigma_b = \begin{cases} \left(1 - \dfrac{1}{\epsilon}\right)\dfrac{E_0}{4\pi}, & z = d/2 \\ \left(\dfrac{1}{\epsilon} - 1\right)\dfrac{E_0}{4\pi}, & z = -d/2. \end{cases} \tag{B.86}$$

Note that the $\vec{D} = \epsilon\vec{E}$ field is continuous across the interface since there is no free charge present.

Spherical geometry

Consider a dielectric sphere with constant permittivity ϵ and radius R in an initially uniform electric field $\vec{E} = E_0\hat{z}$. The resultant potential $\Phi(r,\theta)$ is given in spherical coordinates by

$$\Phi(r,\theta) = \begin{cases} -E_0 r\cos\theta + \left(\dfrac{\epsilon-1}{\epsilon+2}\right)E_0\dfrac{R^3}{r^2}\cos\theta, & r > R \\ -\left(\dfrac{3}{\epsilon+2}\right)E_0 r\cos\theta, & r < R. \end{cases} \tag{B.87}$$

Note the implied electric field inside is uniform and the induced field outside has an electric dipole form as seen from Section B.2. The induced dipole moment is

$$\vec{p} = \frac{\epsilon-1}{\epsilon+2}E_0 R^3\hat{z}. \tag{B.88}$$

The spherical coordinate techniques of Section B.2 may be used to establish this result. Only terms $\sim rP_1(\cos\theta)$, $r^{-2}P_1(\cos\theta)$ will contribute after the external field is supplied for $r > R$ and the tangential $\vec{E}$ and normal $\vec{D}$ boundary conditions are formulated on the sphere.

Cylindrical geometry

Consider a dielectric circular cylinder with constant permittivity ϵ and radius R, with its axis along the z-coordinate in an initially uniform electric field $\vec{E} = E_0\hat{y}$. The resultant potential $\Phi(\rho, \phi)$ is given in cylindrical coordinates by

$$\Phi(\rho, \phi) = \begin{cases} -E_0\rho\sin\phi + \left(\dfrac{\epsilon - 1}{\epsilon + 1}\right) E_0 \dfrac{R^2}{\rho} \sin\phi, & \rho > R \\ -\dfrac{2}{(1+\epsilon)} E_0\rho\sin\phi, & \rho < R. \end{cases} \tag{B.89}$$

The cylindrical coordinate techniques of Section B.2 may be used to establish this result. Only terms $\sim \rho\sin\phi$, $\rho^{-1}\sin\phi$ will contribute when the external field is supplied for $\rho > R$ and the tangential $\vec{E}$ and normal $\vec{D}$ boundary conditions are formulated on the cylinder.

B.3 Magnetostatics

Magnetostatics concerns the situation where only a magnetic field, denoted $\vec{B}(\vec{x})$, exists in a given geometry. It results from a steady current flow in a given inertial reference frame. We again present a gentle introduction to the fundamental concepts here, including: field equations and forces, energy and boundary condition considerations, magnetic dipoles, inductance, as well as changed media characteristics due to a macroscopic magnetization field.

Field equations and currents

The differential field equation for the magnetic field $B(\vec{x})$ is given by *Ampére's law*:

$$\vec{\nabla} \times \vec{B}(\vec{x}) = \frac{4\pi}{c}\vec{J}(\vec{x}). \tag{B.90}$$

c is the speed of light. We saw the cross product earlier in the electric case in Equation (B.8). The source of the field, $\vec{J}(\vec{x})$, is called the *current density*. In magnetostatics, $\vec{J}(\vec{x})$ is not arbitrary, but satisfies the local requirement

$$\vec{\nabla} \cdot \vec{J}(\vec{x}) = 0, \tag{B.91}$$

or more explicitly

$$\frac{\partial J_x}{\partial x} + \frac{\partial J_y}{\partial y} + \frac{\partial J_z}{\partial z} = 0. \tag{B.92}$$

Written explicitly in Cartesian coordinates, Equation (B.90) is a vector equation consisting of three components,

$$\left(\frac{\partial B_z}{\partial y} - \frac{\partial B_y}{\partial z} \right) = \frac{4\pi}{c} J_x, \tag{B.93}$$

$$\left(\frac{\partial B_x}{\partial z} - \frac{\partial B_z}{\partial x} \right) = \frac{4\pi}{c} J_y, \tag{B.94}$$

$$\left(\frac{\partial B_y}{\partial x} - \frac{\partial B_x}{\partial y} \right) = \frac{4\pi}{c} J_z. \tag{B.95}$$

Integrating Equation (B.90) over oriented surface elements $d\vec{a}$ of area A, bounded by a closed contour C, gives

$$\oint_C \vec{B} \cdot d\vec{\ell} = \frac{4\pi}{c} \int_A \vec{J} \cdot d\vec{a} = \frac{4\pi}{c} I_{enc}. \tag{B.96}$$

This is the integral form of Ampére's law. (Deriving Equation (B.96) from Equation (B.90) involves the use of *Stokes' theorem*, a statement of which can be found in Chapter 1). Note that I_{enc} indicates the current flux going through A, or enclosed by the contour C. In a situation with a known symmetry, Equation (B.90) is the fast and easy way to solve for the magnetic field.

The magnetic field due to a general current density, $\vec{J}(\vec{x}\,')$, is given by

$$\vec{B}(\vec{x}) = \frac{1}{c} \int d^3x' \frac{\vec{J}(\vec{x}\,') \times (\vec{x} - \vec{x}\,')}{|\vec{x} - \vec{x}\,'|^3}, \tag{B.97}$$

where the integration is over all space. Often, the current distribution is due to a line current for which $d^3x'\, \vec{J}(\vec{x}\,') \to I\, d\vec{x}\,'$, where I is the current magnitude and $d\vec{x}\,'$ gives the local direction. For a closed current contour (*i.e.* a circuit), C', this results in

$$\vec{B}(\vec{x}) = \frac{I}{c} \oint_{C'} d\vec{x}\,' \times \frac{(\vec{x} - \vec{x}\,')}{|\vec{x} - \vec{x}\,'|^3}. \tag{B.98}$$

These are forms of the *Biot-Savart law*. Using either Equation (B.97) or (B.98), one may show that

$$\vec{B}(\vec{x}) = \vec{\nabla} \times \vec{A}(\vec{x}), \tag{B.99}$$

where

$$\vec{A}(\vec{x}) = \frac{1}{c} \int d^3x' \frac{\vec{J}(\vec{x}\,')}{|\vec{x} - \vec{x}\,'|}, \tag{B.100}$$

for a general current density, or for a circuit

$$\vec{A}(\vec{x}) = \frac{I}{c} \oint_C \frac{d\vec{x}\,'}{|\vec{x} - \vec{x}\,'|}. \tag{B.101}$$

$\vec{A}(\vec{x})$ is called the *vector potential*. The representation in Equation (B.99) results in

$$\vec{\nabla} \cdot \vec{B}(\vec{x}) = 0. \tag{B.102}$$

One may modify $\vec{A}(\vec{x})$ via a *gauge transformation*,

$$\vec{A}(\vec{x}) \to \vec{A}(\vec{x}) + \vec{\nabla}\Lambda(\vec{x}), \tag{B.103}$$

where $\Lambda(\vec{x})$ is an arbitrary function of position, without changing $\vec{B}(\vec{x})$ from Equation (B.99). Either Equation (B.100) (via Equation (B.91)) or (B.101) is consistent with

$$\vec{\nabla} \cdot \vec{A}(\vec{x}) = 0, \tag{B.104}$$

which is called the *Coulomb gauge*. (It is also called the radiation gauge or the transverse gauge in the full theory of electrodynamics.) The resulting field equation for $\vec{A}(\vec{x})$ from Equation (B.90) is then

$$\nabla^2 \vec{A} = -\frac{4\pi}{c} \vec{J}(\vec{x}\,'). \tag{B.105}$$

One can use Equation (B.96) to determine the magnetic field from a line current I directed along the z-axis. One assumes the magnetic field is circular and cylindrically symmetric. Establishing a circular path of radius $\rho = \sqrt{x^2 + y^2}$ surrounding the wire then gives

$$\oint_C \vec{B} \cdot d\vec{\ell} = 2\pi\rho\, B_\phi = \frac{4\pi}{c} I, \tag{B.106}$$

$$\implies B_\phi = \frac{2I}{c\rho}. \tag{B.107}$$

B_ϕ is the azimuthal component of the magnetic field, $\vec{B} = B_\phi \hat{e}_\phi$, where $\hat{e}_\phi$ is the unit azimuthal vector in cylindrical coordinates. Note the right hand rule for the field: *pointing your right thumb in the direction of the current, your fingers curl in the direction of the associated magnetic field.*

One may verify the result in Equation (B.107) by direct integration from the form of the Biot-Savart law in Equation (B.98). One need only take the current in the $\hat{z}$ direction and the contour C' along the complete z' axis. This results in

$$\vec{B} = \frac{I}{c}\rho\,\hat{e}_\phi \int_{-\infty}^{\infty} \frac{dz'}{(x^2+y^2+(z-z')^2)^{3/2}} = \frac{2I}{c\rho}\hat{e}_\phi. \tag{B.108}$$

Finally, the force, $\vec{F}$, on a point charge q at position $\vec{x}$ in motion with velocity $\vec{v}$ in a static magnetic field is given by

$$\vec{F} = q\frac{\vec{v}}{c} \times \vec{B}(\vec{x}). \tag{B.109}$$

Boundary conditions, surface pressure and energy

The magnetic field boundary conditions,

$$B_{2n} - B_{1n} = 0, \tag{B.110}$$

$$\hat{n} \times \left(\vec{B}_2 - \vec{B}_1\right) = \frac{4\pi}{c}\vec{K}, \tag{B.111}$$

hold for the normal, Equation (B.110), and tangential, Equation (B.111), components on an interface or surface between two different media. The $\vec{K}$ are idealized surface currents. The tangential components in Equation (B.111) are explicitly

$$(B_{2y} - B_{1y}) = -\frac{4\pi}{c}K_x, \tag{B.112}$$

$$(B_{2x} - B_{1x}) = \frac{4\pi}{c}K_y \tag{B.113}$$

for current components K_x and K_y on a z = constant surface. In this expression $B_{1x,y}$ and $B_{2x,y}$ are components of the magnetic field tangential to the surface in the two regions and the surface normal $\hat{n} = \hat{z}$ points from region 1 to 2.

Similar to the electric case, the normal force per area (pressure) in magnetostatics is given by

$$\mathcal{F}_n = \frac{1}{c}\hat{n} \cdot \left[\vec{K} \times \left(\frac{\vec{B}_1 + \vec{B}_2}{2}\right)\right], \tag{B.114}$$

where we use the average magnetic field at the surface to remove the tangential self-field. Using the boundary condition Equation (B.111) one has

$$\mathcal{F}_n = \frac{1}{8\pi}\left(B_{1||}^2 - B_{2||}^2\right), \tag{B.115}$$

where $B_{1,2||}$ denotes the local tangential field components in the two regions.

Unlike the electrostatic case, magnetic systems do no work. That is, there is no analog of Equation (B.27) for magnetic systems given the elemental force in Equation (B.109). However, the energy of assembly of a magnetic system gives expressions similar to electrostatics ones. One has

$$W = \frac{1}{2c}\int d^3x\, \vec{J}(\vec{x})\cdot\vec{A}(\vec{x}), \tag{B.116}$$

$$= \frac{1}{2c}\int\int d^3x\, d^3x'\, \frac{\vec{J}(\vec{x})\cdot\vec{J}(\vec{x}\,')}{|\vec{x}-\vec{x}\,'|}, \tag{B.117}$$

$$= \frac{1}{8\pi}\int d^3x\, \vec{B}^2. \tag{B.118}$$

One often works with closed circuits in such situations. Then, the spatial integrals in Equation (B.117) can be specialized to sums over closed contour loops:

$$W = \sum_{i,j}\frac{I_i I_j}{2c}\oint_i\oint_j \frac{d\vec{x}_i\cdot d\vec{x}_j}{|\vec{x}_i-\vec{x}_j|}. \tag{B.119}$$

Note there are both interaction energy ($i \neq j$) and self-energy ($i = j$) terms in this expression. (In the self-energy case the expression needs to go back to having primes on one of the $\vec{x}_i$ variables.)

Magnetic dipole field

Similar to the electric field, the field of a current density $\vec{J}(\vec{x})$ possesses a magnetic multipole expansion, the first term of which has the form

$$\vec{A}^d(\vec{x}) = \frac{\vec{m}\times\vec{x}}{r^3}, \tag{B.120}$$

where the magnetic dipole vector, $\vec{m}$, is defined as

$$\vec{m} \equiv \frac{1}{2c}\int d^3x\, \vec{x}\times\vec{J}(\vec{x}). \tag{B.121}$$

For a planar current loop of area S carrying current $\int \vec{J}\cdot d\vec{a} = I$ (we may choose $I > 0$ by directing $d\vec{a}$ locally along the current), the magnetic moment is

$$\vec{m} = \frac{I}{c}S\,\hat{n}. \tag{B.122}$$

The unit vector $\hat{n}$ is perpendicular to the surface. The overall direction is determined by a right hand rule: *curling the fingers of your right hand along the current, your thumb will point in direction* $\vec{m}$. The magnetic field far from the source is then given as

$$\vec{B}^d(\vec{x}) = \frac{3(\vec{x}\cdot\vec{m}\,)\vec{x} - r^2\vec{m}}{r^5}. \tag{B.123}$$

This may also be interpreted as the field of a point magnetic dipole source located at the origin (excluding the origin point).

The force $\vec{F}$ and torque $\vec{N}$ on an elemental magnetic dipole, $\vec{m}$, in a magnetic field are

$$\vec{F} = \vec{\nabla}\left(\vec{m}\cdot\vec{B}\right), \tag{B.124}$$

$$\vec{N} = \vec{m}\times\vec{B}. \tag{B.125}$$

An alternate form for Equation (B.124) in the case where the source of the field does not overlap with the dipole is $\vec{F} = (\vec{m}\cdot\vec{\nabla})\vec{B}$.

Simple magnetic solutions

Current loop

Consider a circular current loop of radius R carrying current I in the xy-plane centered on the origin. The magnetic field along the z-axis is directed along $\hat{z}$, and is given by

$$B_z(z) = \frac{2\pi I}{c}\frac{R^2}{(R^2+z^2)^{3/2}}. \tag{B.126}$$

A right hand rule is again at work: *curling the fingers of your right hand along the current, your thumb will point in direction of the on-axis field.*

Current sheet

The magnetic field generated by a planar surface current $\vec{K} = K_x\,\hat{x}$ flowing in the $z = 0$ plane is given by

$$\vec{B}(z) = \begin{cases} -\dfrac{2\pi}{c}K_x\,\hat{y}, & z > 0, \\ \dfrac{2\pi}{c}K_x\,\hat{y}, & z < 0. \end{cases} \tag{B.127}$$

If this sheet is simulated by closely spaced parallel current-carrying wires, one may make the approximate replacement $K_x \to In$ where I is the current in each wire and n the number of loops per length.

Long circular solenoid

Consider a long circular cylinder (solenoid) of radius R with its symmetry axis along the z direction. Using cylindrical coordinates, establish an ideal azimuthal circular surface current $\vec{K} = K_\phi \, \hat{e}_\phi$. Then $\vec{B}$ is along z and is given by

$$B_z(\rho) = \begin{cases} \dfrac{4\pi}{c} K_\phi, & \rho < R, \\ 0, & \rho > R. \end{cases} \tag{B.128}$$

Given that the idealized surface current may actually be approximated by tightly wound insulated loops of wire, the approximate connection $K_\phi \to In$ holds where I is the current in each wire, and n is the number of loops per length. The result for a long coil of arbitrary cross-section then has exactly the same form.

Torodial coil

A torodial coil of arbitrary cross-section has a global circular shape combined with an arbitrary, but invariant, cross-section. Given the z-axis is through the symmetry axis, the magnetic field is in the azimuthal direction everywhere and given by

$$B_\phi(\rho) = \begin{cases} \dfrac{2IN}{c\rho}, & \text{inside the toroid}, \\ 0, & \text{outside the toroid}, \end{cases} \tag{B.129}$$

where N is the total number of turns of the current I and ρ is the cylindrical distance coordinate to the z-axis. Note that for a very large torus with a fixed density of loops, n, we have $N \approx 2\pi\rho\, n$, giving an internal field that connects with Equation (B.128) when $B_\phi \to B_z$ using $K_\phi \to In$ in Equation (B.128) in some given long, approximately straight section.

Inductance

The self-inductance and mutual inductance of current carrying systems are related to the field energies displayed in Section B.3. Let us say one has two circuits interacting with one another. In this case one may write the assembly energy as

$$W = \frac{1}{2}\left(L_1^2 I_1^2 + L_2^2 I_2^2\right) + M_{12} I_1 I_2, \tag{B.130}$$

where the *self-inductance* L of a given circuit has the form

$$L = \frac{1}{c^2}\oint\oint' \frac{d\vec{x}\cdot d\vec{x}\,'}{|\vec{x}-\vec{x}\,'|}, \tag{B.131}$$

and *mutual-inductance* M_{12} of a combination is

$$M_{12} = \frac{1}{c^2}\oint_1\oint_2 \frac{d\vec{x}_1\cdot d\vec{x}_2}{|\vec{x}_1-\vec{x}_2|}. \tag{B.132}$$

Note these are purely geometrical quantities. The mutual inductance is manifestly symmetric in its indices: $M_{12} = M_{21}$. In the above forms, both of these quantities look very difficult to evaluate, but a connection to magnetic flux simplifies things. Using the symmetry of M_{12}, we may write Equation (B.130) as

$$W = \frac{I_1}{2}(L_1 I_1 + M_{12} I_2) + \frac{I_2}{2}(L_2 I_2 + M_{21} I_1). \tag{B.133}$$

Here comes the magic. Using Stokes' theorem, one may show that the magnetic flux through circuit i due to a flux from circuit j,

$$\mathcal{F}_{ij} \equiv \int da_i\, \hat{n}_i \cdot \vec{B}_j \tag{B.134}$$

($\hat{n}_i$ is a unit vector associated with current-loop surface i) satisfies

$$\mathcal{F}_{11} = cL_1 I_1, \quad \mathcal{F}_{22} = cL_2 I_2, \tag{B.135}$$

$$\mathcal{F}_{12} = cM_{12} I_2, \quad \mathcal{F}_{21} = cM_{21} I_1. \tag{B.136}$$

Thus Equation (B.133) may be written

$$W = \frac{1}{2c}\left(I_1\mathcal{F}_1 + I_2\mathcal{F}_2\right), \tag{B.137}$$

where $\mathcal{F}_1$ and $\mathcal{F}_2$ are the total fluxes through each circuit,

$$\mathcal{F}_1 = cL_1 I_1 + cM_{12} I_2, \quad \mathcal{F}_2 = cL_2 I_2 + cM_{21} I_2. \tag{B.138}$$

It is the connection to flux that provides the easy way to calculate mutual and self-inductances. Another calculational approach is to use the connection between magnetic energy in Equation (B.118) and the self-inductance part of the energy in Equation (B.130). Self-inductance is the magnetic analog of capacitance; that is, the magnetic analog to Equation (B.28) is the self-inductance in Equation (B.135). These devices store magnetic energy instead of electrical energy.

The self-inductance of a long circular solenoid of length L and radius R is

$$L_{sol} = \frac{4\pi^2 N^2 R^2}{c^2 L}. \tag{B.139}$$

The self-inductance of two long parallel conducting wires of length L and radius R separated by a distance d is

$$L_{wires} = \frac{4L}{c^2} \cosh^{-1}\left(\frac{d}{2R}\right). \tag{B.140}$$

The self-inductance of a toroid with a rectangular cross-section of height h and inner and outer radii R_1 and R_2 respectively, is

$$L_{tor} = \frac{2N^2 h}{c^2} \ln\left(\frac{R_2}{R_1}\right). \tag{B.141}$$

Magnetic materials

We again make a separation of bound and free charges in the material,

$$\vec{\nabla} \times \vec{B} = \frac{4\pi}{c}\left(\vec{J}_f + \vec{J}_b\right), \tag{B.142}$$

but continue to maintain

$$\vec{\nabla} \cdot \vec{B} = 0. \tag{B.143}$$

Now set

$$\vec{J}_b = c\vec{\nabla} \times \vec{M}. \tag{B.144}$$

Defining the *H-field*

$$\vec{H} \equiv \vec{B} - 4\pi\vec{M}, \tag{B.145}$$

we obtain the equations

$$\vec{\nabla} \times \vec{H} = \frac{4\pi}{c}\vec{J}_f. \tag{B.146}$$

It is necessary to have a constitutive relation giving $\vec{H}$ as a function of $\vec{B}$ for the material. For linear, isotropic materials we have

$$\vec{B} = \mu\vec{H}, \tag{B.147}$$

and therefore

$$\vec{M} = \frac{\mu - 1}{4\pi}\vec{H}, \tag{B.148}$$

where μ is the *magnetic permeability* of the material. If μ is constant in the material, it follows that materials which have $\mu > 1$ enhance the free current contribution, while materials with $\mu < 1$, shielding the free current. The case $\mu > 1$ is called *paramagnetism*, and the case $\mu < 1$ is called *diamagnetism*. We will refer to both as magnetic materials.

We summarize our magnetostatic boundary conditions as

$$\left(\vec{B}_2 - \vec{B}_1\right) \cdot \hat{n} = 0, \tag{B.149}$$

$$\hat{n} \times \left(\vec{H}_2 - \vec{H}_1\right) = \frac{4\pi}{c}\vec{K}_f, \tag{B.150}$$

$$\hat{n} \times \left(\vec{M}_2 - \vec{M}_1\right) = \frac{1}{c}\vec{K}_b. \tag{B.151}$$

Consider now a material/vacuum interface where region 1 corresponds to the material ($\mu_1 \neq 1$) and region 2 corresponds to the vacuum ($\mu_2 = 1$). We are assuming $\vec{B} = \mu\vec{H}$ and that there are no free currents in the volume or surface of the material. Then the bound current on the surface is given by

$$\vec{K}_b = c\vec{M}_1 \times \hat{n} = \frac{\mu - 1}{4\pi} c\vec{H}_1 \times \hat{n}. \tag{B.152}$$

The assembly energy of a linear and isotropic magnetic system is given as

$$W = \frac{1}{8\pi} \int d^3x\, \vec{B} \cdot \vec{H}. \tag{B.153}$$

Simple magnetic material solutions

Planar geometry

Consider a flat magnetic slab of infinite extent and thickness d with constant permeability μ. It is oriented in the xy-plane. When it is placed in a uniform magnetic field $\vec{B} = B_0\hat{z}$, the magnetic field profile is unchanged everywhere as seen from boundary condition (B.149), but of course $\vec{H} = \vec{B}/\mu$ inside the material. In addition, there are no bound surface currents induced.

Spherical geometry

Consider a sphere of radius R with a permanent uniform magnetization $\vec{M}$. In cases where there are no volume or surface free currents, the equation $\vec{\nabla} \times \vec{H} = 0$ implies we may write $\vec{H} = -\vec{\nabla}\Phi_m$ everywhere. Φ_m is called the *magnetic scalar potential*. Using this technique, we may establish inside the sphere that

$$\vec{H} = -\frac{4\pi}{3}\vec{M}, \quad \vec{B} = \frac{8\pi}{3}\vec{M}. \tag{B.154}$$

Outside the sphere we have a perfect dipole field,

$$\Phi_m(\vec{x}) = \frac{\vec{m} \cdot \vec{x}}{r^3}, \tag{B.155}$$

$$\vec{m} = \frac{4\pi}{3} R^3 \vec{M}. \tag{B.156}$$

In addition, the separation of variables techniques of Section B.2 may be used to obtain the analogs of the dielectric sphere and cylinder placed in uniform fields, Section B.2. With no free currents or charges, the boundary sets ((B.149), (B.150)) and ((B.81), (B.24)) are interchangeable, and we only need to make the substitutions $\vec{E} \to \vec{H}$ and $\vec{D} \to \vec{B}$. In this case the actual replacements are $\Phi \to \Phi_m$, $E_0 \to B_0$ and $\epsilon \to \mu$.

Cylindrical geometry

Consider a solenoid filled with a magnetic material with constant permeability μ. A free surface current $\vec{K}_f = K_\phi \hat{e}_\phi$ is established, with $K_\phi \to In$ in the case of wound current loops. The field inside is

$$\vec{H} = \frac{1}{\mu}\vec{B} = \frac{4\pi}{c} K_\phi \hat{z}, \tag{B.157}$$

and the bound current at the surface is

$$\vec{K}_b = (\mu - 1)\vec{K}_f. \tag{B.158}$$

B.4 Electromagnetic synthesis

We now put the electric and magnetic pieces together. Electric and magnetic fields are not separate phenomena, but their non-static time behaviors and inertial frame transformation laws show they are interrelated aspects of a single entity called the electromagnetic field. We will begin by completing both the Maxwell equations and the macroscopic Maxwell equations, and then discussing the changes made to make them fully dynamical. Based on this foundation it will then be possible to discuss conservation laws for electric charges, momentum and energy.

Field equations, energy and force

The differential electrostatic field equation is Gauss' law,

$$\vec{\nabla} \cdot \vec{E}(\vec{x}) = 4\pi\rho(\vec{x}, t), \tag{B.159}$$

which gives the Coulomb solution subject to the understood time source condition

$$\frac{\partial \rho(\vec{x},t)}{\partial t} = 0. \tag{B.160}$$

This solution form implies

$$\vec{\nabla} \times \vec{E}(\vec{x}) = 0. \tag{B.161}$$

The differential field equation of magnetostatics is Ampére's law

$$\vec{\nabla} \times \vec{B}(\vec{x}) = \frac{4\pi}{c}\vec{J}(\vec{x}), \tag{B.162}$$

which gives the Biot-Savart solution subject to the spatial source condition

$$\vec{\nabla} \cdot \vec{J}(\vec{x}) = 0. \tag{B.163}$$

This solution form implies

$$\vec{\nabla} \cdot \vec{B}(\vec{x}) = 0. \tag{B.164}$$

The fundamental equations of electrodynamics, the *Maxwell equations*, are

$$\vec{\nabla} \cdot \vec{E} = 4\pi\rho, \tag{B.165}$$

$$\vec{\nabla} \times \vec{E} + \frac{1}{c}\frac{\partial \vec{B}}{\partial t} = 0, \tag{B.166}$$

$$\vec{\nabla} \times \vec{B} - \frac{1}{c}\frac{\partial \vec{E}}{\partial t} = \frac{4\pi}{c}\vec{J}, \tag{B.167}$$

$$\vec{\nabla} \cdot \vec{B} = 0. \tag{B.168}$$

The full field equations amazingly involve only two modifications from the static forms! The first equation (B.165) is the unmodified Gauss law, and the last equation (B.168) is also unmodified. The

$$\frac{1}{c}\frac{\partial \vec{B}}{\partial t}$$

term in (B.166) was introduced by Faraday and the new equation is called *Faraday's law*. This results in a changing magnetic field creating an electric field. The

$$\vec{J}_d \equiv \frac{1}{4\pi}\frac{\partial \vec{E}}{\partial t}$$

term in (B.167) is called the *displacement current*, and was introduced by Maxwell. This new form is still called the Ampére equation. This results in a changing electric field creating a magnetic field.

In addition, the two source conditions, (B.160) and (B.163), are now replaced by

$$\frac{\partial \rho}{\partial t} + \vec{\nabla} \cdot \vec{J} = 0, \tag{B.169}$$

which expresses conservation of charge. As shown in Section 2.1 of the text, Equation (B.169) may be derived from the Maxwell equations.

The *macroscopic Maxwell equations* of electrodynamics are (the sources are the *free* charges and currents)

$$\vec{\nabla} \cdot \vec{D} = 4\pi \rho_f, \tag{B.170}$$

$$\vec{\nabla} \times \vec{E} + \frac{1}{c}\frac{\partial \vec{B}}{\partial t} = 0, \tag{B.171}$$

$$\vec{\nabla} \times \vec{H} - \frac{1}{c}\frac{\partial \vec{D}}{\partial t} = \frac{4\pi}{c}\vec{J}_f, \tag{B.172}$$

$$\vec{\nabla} \cdot \vec{B} = 0. \tag{B.173}$$

Note only the first and third equations are changed from the fundamental forms. The macroscopic form of the displacement current is now,

$$\vec{J}_d^M \equiv \frac{1}{4\pi}\frac{\partial \vec{D}}{\partial t}. \tag{B.174}$$

The connections,

$$\vec{D} - \vec{E} = 4\pi \vec{P}, \tag{B.175}$$

$$\vec{B} - \vec{H} = 4\pi \vec{M}, \tag{B.176}$$

allow the macroscopic equations with sources, Equations (B.170) and (B.172), to be cast in the forms,

$$\vec{\nabla} \cdot \vec{E} = 4\pi(\rho_{\mathrm{f}} - \vec{\nabla} \cdot \vec{P}), \tag{B.177}$$

$$\vec{\nabla} \times \vec{B} - \frac{1}{c}\frac{\partial \vec{E}}{\partial t} = \frac{4\pi}{c}\left(\vec{J}_{\mathrm{f}} + \frac{\partial \vec{P}}{\partial t} + c\vec{\nabla} \times \vec{M}\right). \tag{B.178}$$

This allows the identification of the bound charge density and current.

$$\rho_{\mathrm{b}} = -\vec{\nabla} \cdot \vec{P}, \tag{B.179}$$

$$\vec{J}_{\mathrm{b}} = c\vec{\nabla} \times \vec{M} + \frac{\partial \vec{P}}{\partial t}. \tag{B.180}$$

These are conserved, as is the free charge, as expressed by

$$\frac{\partial \rho_{\mathrm{b}}}{\partial t} + \vec{\nabla} \cdot \vec{J}_{\mathrm{b}} = 0. \tag{B.181}$$

In order to be mathematically complete, the macroscopic equations must be supplemented with *constitutive relations*, characterizing the material. Linear isotropic materials have

$$\vec{D} = \epsilon \vec{E}, \tag{B.182}$$
$$\vec{B} = \mu \vec{H}, \tag{B.183}$$

where μ and ϵ are dimensionless constants not equal to 1 in the material.

The boundary conditions on fields at interfaces are just the combined ones from the earlier sections, collected together here for convenience:

$$(\vec{D}_2 - \vec{D}_1) \cdot \hat{n} = 4\pi\sigma_f, \tag{B.184}$$
$$(\vec{B}_2 - \vec{B}_1) \cdot \hat{n} = 0. \tag{B.185}$$
$$\hat{n} \times \left(\vec{H}_2 - \vec{H}_1\right) = \frac{4\pi}{c}\vec{K}_f, \tag{B.186}$$
$$\hat{n} \times \left(\vec{E}_2 - \vec{E}_1\right) = 0. \tag{B.187}$$

Note that by crossing both sides of Equation (B.186) into $\hat{n}$ and using the double cross product rule, we get a direct form for the discontinuity in the tangential component, $\vec{H}_t$:

$$\vec{H}_{2t} - \vec{H}_{1t} = \frac{4\pi}{c}\vec{K}_f \times \hat{n}. \tag{B.188}$$

The field energy is just the sum of the two terms in Equations (B.32) and (B.118).

$$W = \frac{1}{8\pi}\int d^3x \left(\vec{D} \cdot \vec{E} + \vec{B} \cdot \vec{H}\right). \tag{B.189}$$

In addition, the *Lorentz force law* in electrodynamics is just a sum of the forms in Equations (B.22) and (B.109).

$$\vec{F} = q\left(\vec{E} + \frac{\vec{v}}{c} \times \vec{B}\right). \tag{B.190}$$

Faraday's law and displacement current

Let us discuss the two equation changes introduced above. First, we define the *magnetic flux*magnetic flux through surface S as follows:

$$\mathcal{F} \equiv \int_S da\, \hat{n} \cdot \vec{B}. \tag{B.191}$$

We also define the *electromotive force* or *emf* associated with closed loop C as follows:

$$\mathcal{E} \equiv \oint_C \vec{E} \cdot d\vec{\ell}. \tag{B.192}$$

By integrating Equation (B.166) over the oriented area of a closed loop and using Stokes' theorem, we get an integral statement of Faraday's Law,

$$\mathcal{E} = -\frac{1}{c}\frac{d\mathcal{F}}{dt}. \tag{B.193}$$

Applied to the changing flux in a current loop leads to *Lenz' Law*: The induced current in a conducting loop flows in a direction so as to produce a magnetic field that *opposes* the change in flux in the circuit.

Second, let us discuss the significance of the displacement current term in Equation (B.172). Using Stokes' theorem, this equation may be integrated over a surface bounded by the contour C to obtain the analog of Equation (B.96):

$$\oint_C \vec{H} \cdot d\vec{\ell} = \frac{4\pi}{c}\int_A (\vec{J}_f + \vec{J}_d) \cdot d\vec{a}, \tag{B.194}$$

where the free current and the displacement currents are both contributing. Now, imagine a circuit containing a parallel plate capacitor with an inserted dielectric with permittivity ϵ, across which a time-varying potential $V(t)$ is applied. The contour C is taken to surround the wire leading to the capacitor. Associated with the contour is the open area surface A bounded by C; A is not fixed but can be shaped as desired. (See for example Figure 1.1(b) of the text.) Since the contour surrounds a wire element, one can define A_{out} as a surface which cuts directly through the wire. Neglecting fringing fields, $\vec{D} = 0 \implies \vec{J}_d = 0$ on this surface. Thus,

$$\int_{A_{out}} (\vec{J}_f + \overset{0}{\cancel{\vec{J}_d}}) \cdot d\vec{a} = I_f. \tag{B.195}$$

Another surface A_{in} is one which cuts through the capacitor where $\vec{J}_f = 0$:

$$\int_{A_{in}} (\overset{0}{\cancel{\vec{J}_f}} + \vec{J}_d) \cdot d\vec{a} = I_d. \tag{B.196}$$

With a parallel plate capacitor as a model, one has

$$I_f = \frac{dq(t)}{dt} = C\frac{dV(t)}{dt} = \frac{\epsilon A}{4\pi d}\frac{dV(t)}{dt}, \tag{B.197}$$

where $q(t)$ is the charge transported to the capacitor and $V(t)$ is the capacitor voltage difference. On the other hand,

$$I_d = \frac{1}{4\pi}\int_{A_{in}} d\vec{a}\cdot\frac{\partial\vec{D}}{\partial t} = \frac{1}{4\pi}\int\frac{\partial}{\partial t}\left(\frac{\epsilon V(t)}{d}\right)da = \frac{\epsilon A}{4\pi d}\frac{dV}{dt}. \qquad \text{(B.198)}$$

Thus, either surface will yield the same current in time-varying circumstances. Maxwell recognized that this surface independence was crucial for the mathematical consistency of the theory.

Conservation laws

Besides the conservation of charge, conservation of electromagnetic energy and momentum can also be formulated using the Maxwell fields. For energy conservation one has

$$\frac{\partial u}{\partial t} + \vec{\nabla}\cdot\vec{S} + \vec{J}\cdot\vec{E} = 0, \qquad \text{(B.199)}$$

where

$$u \equiv \frac{\vec{E}^2 + \vec{B}^2}{8\pi}, \qquad \text{(B.200)}$$

$$\vec{S} \equiv \frac{c}{4\pi}\vec{E}\times\vec{B}. \qquad \text{(B.201)}$$

$\vec{J}\cdot\vec{E}$ is the power density, u is the *energy density* and the *Poynting vector* $\vec{S}$ expresses the flow of energy. In source free regions, $\vec{J} = 0$, this has the same form as the continuity equation, Equation (B.169). When macroscopic media are present a similar construction gives

$$\frac{\partial u^M}{\partial t} + \vec{\nabla}\cdot\vec{S}^M + \vec{J}_f\cdot\vec{E} = 0, \qquad \text{(B.202)}$$

where

$$u^M \equiv \frac{\vec{E}\cdot\vec{D} + \vec{B}\cdot\vec{H}}{8\pi}, \qquad \text{(B.203)}$$

$$\vec{S}^M \equiv \frac{c}{4\pi}\vec{E}\times\vec{H}. \qquad \text{(B.204)}$$

Momentum conservation is expressed as

$$\frac{\partial g_i}{\partial t} + \sum_j \nabla_j T_{ij} + f_i = 0, \qquad \text{(B.205)}$$

where the momentum density is

$$\vec{g} \equiv \frac{1}{4\pi c}\vec{E}\times\vec{B}, \qquad \text{(B.206)}$$

the *electromagnetic force density* f_i is defined by

$$f_i \equiv \left[\rho\vec{E} + \frac{1}{c}\vec{J}\times\vec{B}\right]_i, \tag{B.207}$$

and the *Maxwell stress tensor* is given as

$$T_{ij} \equiv -\frac{1}{4\pi}\left[E_iE_j + B_iB_j - \frac{1}{2}(E^2+B^2)\delta_{ij}\right]. \tag{B.208}$$

δ_{ij} is the Kronecker delta. In a static situation, the force from Equation (B.205) is

$$F_i = -\sum_j \oint_S da\,\hat{n}_j T_{ij}, \tag{B.209}$$

where we have used the divergence theorem to transform the volume integral into a closed surface integral.

As an example of the use of the stress tensor let us consider the simple situation of a point charge q in a uniform electric field taken along the $\hat{z}$ direction, $\vec{E} = E_0\hat{z}$. Let us situate the charge at position $z = a$ along the positive z-axis. For ease of evaluation and integration, we will take S to be the closed surface of a plane disc of thickness $2a$ in the xy-plane enclosing the point change in a symmetrical fashion. Then the outer normals are $\hat{n} = -\hat{z}$ on $z = 0$ and $\hat{n} = \hat{z}$ on the $z = 2a$ plane. If we take the radial extent of the plane to be infinite, the fields fall off fast enough that the vertical integration along the sides of the disc vanish. Then we only need to integrate on the $z = 0, 2a$ surfaces to get the force in the third direction. (Forces in the 1 and 2 directions will vanish by symmetry.)

$$F_3 = \oint da\,T_{33}|_{z=0} - \oint da\,T_{33}|_{z=2a}. \tag{B.210}$$

From Equation (B.208)

$$T_{33} = -\frac{1}{8\pi}\left(E_3^2 - E_1^2 - E_2^2\right). \tag{B.211}$$

We have

$$E_3|_{z=2a} = E_0 + \frac{q}{\rho^2+a^2}\frac{a}{\sqrt{\rho^2+a^2}}, \tag{B.212}$$

$$E_3|_{z=0} = E_0 - \frac{q}{\rho^2+a^2}\frac{a}{\sqrt{\rho^2+a^2}}. \tag{B.213}$$

The first term in these expressions are the external field, the second term are the electric fields of the charge, q. We need not consider the $E_{1,2}$ fields on these surfaces. The reason is that these terms only involve the field of the charge q, and these terms cancel in the integrations in Equation (B.210). (A charge can not move itself!) For the same reason the external field contributions in (B.212) and (B.213) cancel across the top and bottom of the disc. What remains in E_3^2 is the cross terms between E_0 and q. The two surface contributions just double the answer, and the final result is given as

$$F_3 = 2 \times \frac{1}{8\pi} \int_0^{2\pi} d\phi \int_0^{\infty} d\rho\rho \frac{2E_0 qa}{(\rho^2 + a^2)^{3/2}} = qE_0. \tag{B.214}$$

We had to work quite a bit to get this simple result, but it gives us some assurance that everything is working the way it should! Notice that we only used fields and surface integrations, not sources and volume integrations in this calculation.

B.5 Electromagnetic waves

We start our investigation into the properties and solutions of the wave equation by defining plane waves in one and two spatial dimensions. We study the reflection and transmission properties of waves which encounter a flat interface between materials with different propagation characteristics. Next, the electromagnetic wave equation is introduced and linear polarization states are defined for such plane waves. Reflection and transmission properties are investigated for models of both nonconductors and conductors. In addition, we discuss the practical subject of propagation properties of electromagnetic waves in wave guides.

One-dimensional wave propagation

In one spatial dimension, the wave equation is ($f = f(z, t)$)

$$\frac{\partial^2 f}{\partial z^2} = \frac{1}{v^2} \frac{\partial^2 f}{\partial t^2}. \tag{B.215}$$

v is the velocity, which can be positive or negative. z is the space coordinate and t the time. Because these are partial derivatives, this equation admits any solution of the form

$$f(z, t) = g(z - vt) + h(z + vt), \tag{B.216}$$

where g and h are arbitrary functions. If $v > 0$, the g function represents a wave with a fixed wave form traveling in the positive z direction, and the h function represents such a wave traveling in the negative z direction.

We will start with the usual connections for waves. We have

$$\lambda\nu = |v|, \tag{B.217}$$

where λ is the wavelength and ν is frequency. The angular frequency is

$$\omega \equiv 2\pi\nu. \tag{B.218}$$

Wave number is defined as

$$k \equiv \text{sign}(v)\frac{2\pi}{\lambda}. \tag{B.219}$$

Angular frequency in terms of wave number is

$$\omega = kv. \tag{B.220}$$

We will use a convenient complex representation of a sinusoidal wave as

$$\left.\begin{aligned} \cos(kz - \omega t) &= \text{Re}\left[e^{i(kz-\omega t)}\right], \\ \sin(kz - \omega t) &= \text{Re}\left[-ie^{i(kz-\omega t)}\right]. \end{aligned}\right\} \tag{B.221}$$

One may absorb a general phase factor, δ, into the wave by using

$$\begin{aligned} f(z,t) &= A\cos(kz - \omega t + \delta), \\ &= \text{Re}\left[\tilde{f}(z,t)\right], \end{aligned} \tag{B.222}$$

where

$$\tilde{f}(z,t) = \tilde{A}e^{i(kz-\omega t)}, \quad \tilde{A} = Ae^{i\delta}. \tag{B.223}$$

Consider a one-dimensional interface located at $z = 0$. To start with, we will consider the simplest possible interface which produces no discontinuities in the value of f and its first derivative:

$$\tilde{f}(0^-,t) = \tilde{f}(0^+,t), \tag{B.224}$$

$$\frac{\partial\tilde{f}(0^-,t)}{\partial z} = \frac{\partial\tilde{f}(0^+,t)}{\partial z}. \tag{B.225}$$

Equation (B.224) requires continuity of values or amplitudes across the $z = 0$ boundary and Equation (B.225) requires continuity of the derivatives. The notation 0^- specifies a mathematical approach to the boundary from slightly less than zero, whereas 0^+ means approaching the boundary from small positive values.

Let us refer to $z < 0$ values as region 2, and $z > 0$ values as region 1, as in Section 9.2 of our text. In general the wave has different velocities in the two regions, which implies different wave numbers. We will call these two wave numbers k_2 and k_1. For simplicity, we will assume k_2 and k_1 are positive. The wave has a fixed angular frequency ω in both regions.

We have three waves amplitudes. The initial (I) wave originates in region 2 and is given as

$$\tilde{f}_I(z,t) = \tilde{A}_I e^{i(k_2 z - \omega t)}, \quad (z < 0) \tag{B.226}$$

For $k_1 > 0$ it is traveling in the positive z direction and approaches region 1. There are also reflected (R) and transmitted (T) waves:

$$\tilde{f}_R(z,t) = \tilde{A}_R e^{i(-k_2 z - \omega t)}, \quad (z < 0), \tag{B.227}$$

$$\tilde{f}_T(z,t) = \tilde{A}_T e^{i(k_1 z - \omega t)}, \quad (z > 0). \tag{B.228}$$

Notice the minus sign in front of the reflected wave k_2 factor in the exponent. Also notice the different wave number in region 1, k_1. Of course one has

$$\tilde{f}(z,t) = \begin{cases} \tilde{f}_I(z,t) + \tilde{f}_R(z,t), & z < 0, \\ \tilde{f}_T(z,t), & z > 0, \end{cases} \tag{B.229}$$

in the two regions. Equations (B.224) and (B.225) lead to

$$\tilde{A}_I + \tilde{A}_R = \tilde{A}_T, \tag{B.230}$$

$$k_2 \left(\tilde{A}_I - \tilde{A}_R\right) = k_1 \tilde{A}_T, \tag{B.231}$$

and therefore

$$\tilde{A}_T = \left(\frac{2k_2}{k_1 + k_2}\right) \tilde{A}_I = \left(\frac{2v_1}{v_1 + v_2}\right) \tilde{A}_I, \tag{B.232}$$

$$\tilde{A}_R = \left(\frac{k_2 - k_1}{k_1 + k_2}\right) \tilde{A}_I = \left(\frac{v_1 - v_2}{v_1 + v_2}\right) \tilde{A}_I. \tag{B.233}$$

Notice that if $k_1 > k_2$ ($\Longrightarrow v_2 > v_1$) $\tilde{A}_R$ is negatively related to $\tilde{A}_I$. They are said to be out of phase (in the complex plane) by 180°. On the other hand, if $k_1 < k_2$, then all three amplitudes have the same complex phase.

Two-dimensional wave propagation

The one-dimensional formalism is helpful in introducing the general ideas of plane waves and transmission characteristics at interfaces, but can not model angular characteristics or the phenomenon known as wave polarization. We will need to go to a two-dimensional context to model this. The $z = 0$ line will continue to serve as our interface, and the additional spatial dimension will be called y. We will consider plane waves with a positive z

velocity component. The incident waves will start in region 2 and transmit into region 1, as before. Our vector components are:

$$\vec{x} = (y, z), \tag{B.234}$$

$$\vec{k}^{I,R,T} = (k_y^{I,R,T}, k_z^{I,R,T}). \tag{B.235}$$

The amplitudes are

$$\tilde{f}_I(\vec{x}, t) = \tilde{A}_I e^{i(\vec{k}^I \cdot \vec{x} - \omega t)}, \quad (z < 0), \tag{B.236}$$

$$\tilde{f}_R(\vec{x}, t) = \tilde{A}_R e^{i(\vec{k}^R \cdot \vec{x} - \omega t)}, \quad (z < 0), \tag{B.237}$$

$$\tilde{f}_T(\vec{x}, t) = \tilde{A}_T e^{i(\vec{k}^T \cdot \vec{x} - \omega t)}, \quad (z > 0). \tag{B.238}$$

The wave will travel from region 2 to 1 if $k_z^I > 0$. We also choose $k_z^R = -k_z^I$. The effective wave equation for each type of wave is

$$\frac{\partial^2 f_{I,R,T}}{\partial y^2} + \frac{\partial^2 f_{I,R,T}}{\partial z^2} = \frac{1}{(\vec{v}^{I,R,T})^2} \frac{\partial^2 f_{I,R,T}}{\partial t^2}, \tag{B.239}$$

where

$$(\vec{k}^{I,R,T})^2 = \frac{\omega^2}{(\vec{v}^{I,R,T})^2}, \tag{B.240}$$

respectively. Since the incident and reflected waves are in the same medium, we have $(\vec{v}^I)^2 = (\vec{v}^R)^2 \implies (\vec{k}^I)^2 = (\vec{k}^R)^2$. We simplify notation as in the main text by renaming the z-components $k_z^I = k_2$, $k_z^R = -k_2$ and $k_z^T = k_1$. We will model discontinuities in both the value of the functions $\tilde{f}_{I,R,T}$, as well as their first derivative in order to accommodate the electromagnetic situation we will encounter. We assume in general that

$$\alpha_2 \tilde{f}(y, 0^-, t) = \alpha_1 \tilde{f}(y, 0^+, t), \tag{B.241}$$

$$\beta_2 \frac{\partial \tilde{f}(y, 0^-, t)}{\partial z} = \beta_1 \frac{\partial \tilde{f}(y, 0^+, t)}{\partial z}. \tag{B.242}$$

The parameters $\alpha_{1,2}$ and $\beta_{1,2}$ will eventually model electromagnetic properties. From Equation (B.241) we obtain

$$\alpha_2(\tilde{A}_I + \tilde{A}_R) = \alpha_1 \tilde{A}_T, \tag{B.243}$$

from the amplitudes, and

$$k_y^I = k_y^R = k_y^T, \tag{B.244}$$

from the y dependence of the exponents. In addition, from Equation (B.242),

$$\beta_2 k_2(\tilde{A}_I - \tilde{A}_R) = \beta_1 k_1 \tilde{A}_T. \tag{B.245}$$

Equation (B.244) then gives ($k^{I,R,T}$ are the magnitudes of the $\vec{k}^{I,R,T}$ and $v^{I,R} = v_2$, $v^T = v_1$ are the velocity magnitudes)

$$k^I \sin i = k^R \sin r = k^T \sin t. \tag{B.246}$$

Since $k^I = k^R$, this gives

$$i = r, \tag{B.247}$$

as well as

$$\frac{\sin t}{\sin i} = \frac{k^I}{k^T} = \frac{v_1}{v_2}, \tag{B.248}$$

which determines the angle of transmission into region 1. Our final results are

$$\tilde{A}_T = \left(\frac{2}{\alpha_1/\alpha_2 + \gamma}\right) \tilde{A}_I, \tag{B.249}$$

$$\tilde{A}_R = \left(\frac{\alpha_1/\alpha_2 - \gamma}{\alpha_1/\alpha_2 + \gamma}\right) \tilde{A}_I, \tag{B.250}$$

where

$$\gamma \equiv \frac{\beta_1 k_1}{\beta_2 k_2} = \frac{\beta_1 v^I \cos t}{\beta_2 v^T \cos i}. \tag{B.251}$$

Electromagnetic plane waves

(The introductory material in this section is inevitably fairly close to Section 9.1 of the text.) The macroscopic Maxwell equations in the absence of free sources (ρ_f, $\vec{J}_f = 0$) from Equations (B.170)–(B.173) are ($\vec{B} = \mu\vec{H}$ and $\vec{D} = \epsilon\vec{E}$)

$$\vec{\nabla} \cdot \vec{D} = 0, \tag{B.252}$$

$$\vec{\nabla} \times \vec{E} + \frac{1}{c}\frac{\partial \vec{B}}{\partial t} = 0, \tag{B.253}$$

$$\vec{\nabla} \cdot \vec{B} = 0, \tag{B.254}$$

$$\vec{\nabla} \times \vec{B} - \frac{\mu\epsilon}{c}\frac{\partial \vec{E}}{\partial t} = 0, \quad \left(\vec{\nabla} \times \vec{H} - \frac{1}{c}\frac{\partial \vec{D}}{\partial t} = 0\right). \tag{B.255}$$

We decouple the $\vec{E}$ and $\vec{B}$ equations by taking the curl of Equation (B.253) and then substituting Equation (B.255):

$$\left(\nabla^2 - \frac{\mu\epsilon}{c^2}\frac{\partial^2}{\partial t^2}\right)\vec{E} = 0. \tag{B.256}$$

In the same way, we may derive

$$\left(\nabla^2 - \frac{\mu\epsilon}{c^2}\frac{\partial^2}{\partial t^2}\right)\vec{B} = 0. \tag{B.257}$$

By comparison to the form of the wave equations we have encountered, Equations (B.256) and (B.257) show that the wave velocity is

$$v = \frac{c}{n}, \tag{B.258}$$

where the *index of refraction*, n, is given as

$$n \equiv \sqrt{\mu\epsilon}. \tag{B.259}$$

In this section we will use $k \equiv |\vec{k}|$. Then we also have

$$n = \frac{ck}{\omega}. \tag{B.260}$$

We now consider plane waves,

$$\vec{E}(\vec{x}, t) = \mathrm{Re}\left[\vec{\tilde{E}}_0\, e^{i(\vec{k}\cdot\vec{x}-\omega t)}\right], \tag{B.261}$$

$$\vec{B}(\vec{x}, t) = \mathrm{Re}\left[\vec{\tilde{B}}_0\, e^{i(\vec{k}\cdot\vec{x}-\omega t)}\right]. \tag{B.262}$$

If we ride on the crest of a wave,

$$\vec{k}\cdot\vec{x} - \omega t = \text{const.}, \implies \vec{k}\cdot\frac{d\vec{x}}{dt} = \omega. \tag{B.263}$$

We define $\hat{k} \equiv \vec{k}/k$ (does not mean the z-direction!). Then the position of a phase of the wave form is given by

$$\vec{x} = v\,\hat{k}\,t, \tag{B.264}$$

with

$$v = \frac{\omega}{k} = \frac{c}{n}. \tag{B.265}$$

This is called the *phase velocity*. It is the same as the wave velocity identified from Equation (B.256) or (B.257). Note that wave equations do not always take these forms and the phase velocity is not always the speed of energy transmission.

The no source condition $\vec{\nabla}\cdot\vec{E} = 0$ demands that

$$\begin{aligned} \vec{\nabla}\cdot\vec{E} &= \mathrm{Re}\left[i\vec{\tilde{E}}_0\cdot\vec{k}\, e^{i(\vec{k}\cdot\vec{x}-\omega t)}\right] = 0, \\ \implies \mathrm{Re}\left[\vec{\tilde{E}}_0\cdot\vec{k}\right] &= \mathrm{Im}\left[\vec{\tilde{E}}_0\cdot\vec{k}\right] = 0, \\ &\implies \vec{\tilde{E}}_0\cdot\hat{k} = 0. \end{aligned} \tag{B.266}$$

From $\vec{\nabla} \times \vec{E} = -\dfrac{\partial \vec{B}}{\partial t}$, we then get

$$\tilde{\vec{B}}_0 = n\,\hat{k} \times \tilde{\vec{E}}_0 \implies \tilde{\vec{B}}_0 \cdot \hat{k} = 0. \tag{B.267}$$

Thus the $\vec{E}$ and $\vec{B}$ field directions are transverse to the direction of propagation, $\hat{k}$.

There are two independent *linear polarization* states of $\vec{E}$ or $\vec{B}$. One has

$$\vec{E}_i = \mathrm{Re}\left[\hat{\epsilon}_i \tilde{E}_{0i} e^{i(\vec{k}\cdot\vec{x}-\omega t)}\right], \tag{B.268}$$

where ($i = 1, 2$ only) with

$$\hat{\epsilon}_1 \times \hat{\epsilon}_2 = \hat{k}, \tag{B.269}$$

$$(\hat{\epsilon}_{1.2})^2 = 1. \tag{B.270}$$

Then

$$\vec{B}_i = n\,\hat{k} \times \vec{E}_i. \tag{B.271}$$

The most general plane waves can be written

$$\vec{E}(\vec{x}, t) = \sum_{i=1,2} \vec{E}_i(\vec{x}, t), \tag{B.272}$$

$$\vec{B}(\vec{x}, t) = n\,\hat{k} \times \vec{E}(\vec{x}, t). \tag{B.273}$$

Since the fields are transverse to $\hat{k}$, we also have

$$\vec{E}(\vec{x}, t) = -\frac{1}{n}\,\hat{k} \times \vec{B}(\vec{x}, t). \tag{B.274}$$

An alternate complex polarization set, called *circular polarization*, is

$$\hat{\epsilon}_{\pm} = \frac{1}{\sqrt{2}}\left(\hat{\epsilon}_1 \pm i\,\hat{\epsilon}_2\right), \tag{B.275}$$

(i is the complex number) where

$$\hat{\epsilon}^*_{\pm} \cdot \hat{\epsilon}_{\mp} = 0, \tag{B.276}$$

$$\hat{\epsilon}^*_{\pm} \cdot \hat{\epsilon}_{\pm} = 1. \tag{B.277}$$

(* denotes complex conjugation.) The most general $\vec{E}$ can also be written

$$\vec{E}_{\pm} = \mathrm{Re}\left[\hat{\epsilon}_{\pm} \tilde{E}_{0\pm} e^{i(\vec{k}\cdot\vec{x}-\omega t)}\right], \tag{B.278}$$

$$\vec{E}(\vec{x}, t) = \vec{E}_{+} + \vec{E}_{-}. \tag{B.279}$$

In the case where $\tilde{E}_{0\pm}$ are real, one has

$$\vec{E}_{+} = \frac{\tilde{E}_{0+}}{\sqrt{2}}\left(\hat{\epsilon}_1 \cos(\vec{k}\cdot\vec{x} - \omega t) - \hat{\epsilon}_2 \sin(\vec{k}\cdot\vec{x} - \omega t)\right), \tag{B.280}$$

$$\vec{E}_{-} = \frac{\tilde{E}_{0-}}{\sqrt{2}}\left(\hat{\epsilon}_1 \cos(\vec{k}\cdot\vec{x} - \omega t) + \hat{\epsilon}_2 \sin(\vec{k}\cdot\vec{x} - \omega t)\right), \tag{B.281}$$

$\vec{E}_{+}$ moves in a counterclockwise circular path in the $\hat{\epsilon}_1, \hat{\epsilon}_2$ plane and is called *left circular polarization.* $\vec{E}_{-}$ moves in a clockwise circular path in the $\hat{\epsilon}_1, \hat{\epsilon}_2$ plane and is called *right circular polarization.*

Electromagnetic interface - nonconducting case

The behavior of electromagnetic waves at interfaces are determined by the laws of geometrical optics. We have seen these laws emerge from the one and two-dimensional scattering considerations in the previous sections. We are now in a full three-dimensional context. Let us consider electromagnetic waves impinging on a flat interface between materials with different indices of refraction and permeabilities. We have that:

- The direction vectors of the incident, reflected and transmitted waves form the *plane of incidence*. This plane is perpendicular to the *interface plane* between the materials.

- Incident and reflected angles are equal: $i = r$.

- $\dfrac{\sin t}{\sin i} = \dfrac{v_1}{v_2} = \dfrac{n_2}{n_1}$.

The last statement in terms of indices of refraction is called *Snell's law*. We will study two polarization cases for linearly polarized plane electromagnetic waves interacting with an interface between two materials. We will continue to require that the waves move from region 2 into region 1.

Case I: The electric field is perpendicular to the plane of incidence and parallel to the interface plane. The magnetic field is completely contained in the plane of incidence. (This is the $E_\perp$ polarization case in Section 9.2.)

Given the plane waves

$$\vec{\tilde{E}}(\vec{x}, t) = \vec{\tilde{E}}_0\, e^{i(\vec{k}\cdot\vec{x}-\omega t)}, \tag{B.282}$$

$$\vec{\tilde{B}}(\vec{x}, t) = \vec{\tilde{B}}_0\, e^{i(\vec{k}\cdot\vec{x}-\omega t)}, \tag{B.283}$$

one has $\vec{k} = \vec{k}_I, \vec{k}_R, \vec{k}_T$ for the various traveling waves. Now require $\vec{\tilde{E}}_0 = \tilde{E}_0\hat{x}$ and set up the magnetic field direction vectors:

- $\vec{\tilde{B}}{}_0^I = -\tilde{B}_0^I\, \hat{k}_I \times \hat{x}$,
- $\vec{\tilde{B}}{}_0^R = \tilde{B}_0^R\, \hat{k}_R \times \hat{x}$,
- $\vec{\tilde{B}}{}_0^T = -\tilde{B}_0^T\, \hat{k}_T \times \hat{x}$.

The $\tilde{E}_0^{I,R,T}$ are now related to the $\tilde{B}_0^{I,R,T}$ field amplitudes via Equation (B.274):

$$\tilde{E}_0^I = -\frac{1}{n_2}\tilde{B}_0^I,\ \tilde{E}_0^R = \frac{1}{n_2}\tilde{B}_0^R,\ \tilde{E}_0^T = -\frac{1}{n_1}\tilde{B}_0^T. \tag{B.284}$$

The four specific boundary conditions on the $z = 0$ surface are:

(1) $\epsilon_2 \tilde{E}_{2z} = \epsilon_1 \tilde{E}_{1z}$,
(2) $\tilde{B}_{2z} = \tilde{B}_{1z}$,
(3) $\tilde{E}_{2x} = \tilde{E}_{1x}$,
(4) $\frac{1}{\mu_2}\tilde{B}_{2y} = \frac{1}{\mu_1}\tilde{B}_{1y}$.

It turns out that any two of conditions (2), (3) and (4) above can be used to solve for the reflected and transmitted amplitudes. (Condition (1) gives no information in this case.) We will find it convenient to consider (3) and (4) in the context of the two-dimensional scattering considerations of Section B.5. The Maxwell Equation (B.253) implies in this case that

$$\tilde{B}_y = -i\frac{c}{\omega}\frac{\partial \tilde{E}_x}{\partial z}. \tag{B.285}$$

We can now solve for $\tilde{E}_0^{R,T}$ in terms of $\tilde{E}_0^I$ using conditions (3) and (4), where the latter condition uses Equation (B.285). Employing Equations (B.241) and (B.242) for the $\tilde{f} \to \tilde{E}_x$ field, one identifies:

$$\alpha_1 = 1,\ \alpha_2 = 1,\ \beta_1 = \frac{1}{\mu_1},\ \beta_2 = \frac{1}{\mu_2}, \tag{B.286}$$

$$\gamma = \frac{\mu_2 k_1}{\mu_1 k_2} = \frac{\mu_2}{\mu_1}\frac{n_1}{n_2}\frac{\cos t}{\cos i}. \tag{B.287}$$

Equations (B.249) and (B.250) then give immediately

$$\frac{\tilde{E}_0^T}{\tilde{E}_0^I} = \frac{2}{\left(1 + \dfrac{\mu_2 k_1}{\mu_1 k_2}\right)} = \frac{2}{\left(1 + \dfrac{\mu_2}{\mu_1}\dfrac{n_1}{n_2}\dfrac{\cos t}{\cos i}\right)}, \tag{B.288}$$

$$\frac{\tilde{E}_0^R}{\tilde{E}_0^I} = \left(\frac{1 - \dfrac{\mu_2 k_1}{\mu_1 k_2}}{1 + \dfrac{\mu_2 k_1}{\mu_1 k_2}}\right) = \left(\frac{1 - \dfrac{\mu_2}{\mu_1}\dfrac{n_1}{n_2}\dfrac{\cos t}{\cos i}}{1 + \dfrac{\mu_2}{\mu_1}\dfrac{n_1}{n_2}\dfrac{\cos t}{\cos i}}\right). \tag{B.289}$$

The first forms are given for comparison with the results in Section 9.2 when $\mu_1 = \mu_2 = 1$. The $\tilde{B}_0^{R,T}/\tilde{B}_0^I$ ratios are given by the relations in (B.284).

Case II: The magnetic field is perpendicular to the plane of incidence and parallel to the interface plane. The electric field is completely contained in the plane of incidence. (This is the $B_\perp$ case in Section 9.3.)

Choosing $\vec{\tilde{B}}_0 = \tilde{B}_0 \hat{x}$, we also set up:

- $\vec{\tilde{E}}_0^I = \tilde{E}_0^I \, \hat{k}_I \times \hat{x}$,
- $\vec{\tilde{E}}_0^R = -\tilde{E}_0^R \, \hat{k}_R \times \hat{x}$,
- $\vec{\tilde{E}}_0^T = \tilde{E}_0^T \, \hat{k}_T \times \hat{x}$.

The $\tilde{B}_0^{I,R,T}$ are related to the $\tilde{E}_0^{I,R,T}$ field amplitudes via Equation (B.284) again. The boundary conditions (1) and (2) are the same as before. We simply have different Cartesian components for the other two:

(3)′ $\tilde{E}_{2y} = \tilde{E}_{1y}$,
(4)′ $\dfrac{1}{\mu_2}\tilde{B}_{2x} = \dfrac{1}{\mu_1}\tilde{B}_{1x}$.

This time it turns out that any two of conditions (1), (3)′ and (4)′ above can be used to solve for the reflected and transmitted amplitudes. (Condition (2) gives no information in this case.) The Maxwell Equation (B.255) implies the connection

$$\tilde{E}_y = i\frac{c}{\omega n^2}\frac{\partial \tilde{B}_x}{\partial z}. \tag{B.290}$$

We can now solve for $\tilde{B}_0^{R,T}$ in terms of $\tilde{B}_0^I$ using conditions (3)′ and (4)′, where the former uses Equation (B.290). Once again employing Equations (B.241) and (B.242), but for the $\tilde{f} \to \tilde{B}_x$ field, one identifies:

$$\alpha_1 = \frac{1}{\mu_1},\ \alpha_2 = \frac{1}{\mu_2},\ \beta_1 = \frac{1}{(n_1)^2},\ \beta_2 = \frac{1}{(n_2)^2}, \tag{B.291}$$

$$\gamma = \frac{(n_2)^2\, k_1}{(n_1)^2\, k_2} = \frac{n_2}{n_1}\frac{\cos t}{\cos i}. \tag{B.292}$$

Using Equations (B.249) and (B.250) now gives

$$\frac{\tilde{B}_0^T}{\tilde{B}_0^I} = \frac{2}{\left(\dfrac{\mu_2}{\mu_1} + \dfrac{(n_2)^2\, k_1}{(n_1)^2\, k_2}\right)} = \frac{2}{\left(\dfrac{\mu_2}{\mu_1} + \dfrac{n_2}{n_1}\dfrac{\cos t}{\cos i}\right)}, \tag{B.293}$$

$$\frac{\tilde{B}_0^R}{\tilde{B}_0^I} = \left(\frac{\dfrac{\mu_2}{\mu_1} - \dfrac{(n_2)^2\, k_1}{(n_1)^2\, k_2}}{\dfrac{\mu_2}{\mu_1} + \dfrac{(n_2)^2\, k_1}{(n_1)^2\, k_2}}\right) = \left(\frac{\dfrac{\mu_2}{\mu_1} - \dfrac{n_2}{n_1}\dfrac{\cos t}{\cos i}}{\dfrac{\mu_2}{\mu_1} + \dfrac{n_2}{n_1}\dfrac{\cos t}{\cos i}}\right). \tag{B.294}$$

The first forms are given for comparison with the results in Section 9.3 when $\mu_1 = \mu_2 = 1$. (The relation between index of refraction n and permittivity

ϵ given in Equation (B.259) is $n^2 = \mu\epsilon$.) The $\tilde{E}_0^{R,T}/\tilde{E}_0^I$ ratios are given by the relations in (B.284).

We point out some common features of both cases. First, one has

$$\frac{\tilde{E}_0^T}{\tilde{B}_0^T} = -\frac{1}{n_1}, \tag{B.295}$$

which has nothing to do with scattering, but simply reflects the in-medium transmission properties in Equations (B.284). However, this implies that the interior fields are mostly magnetic in the $n_1 \gg 1$ limit. In the same limit we also have (to leading order)

$$\frac{\tilde{E}_0^T}{\tilde{E}_0^I} \xrightarrow{n_1 \gg 1} 2\frac{\mu_1}{\mu_2}\frac{n_2}{n_1}, \quad \frac{\tilde{E}_0^R}{\tilde{E}_0^I} \xrightarrow{n_1 \gg 1} -1, \tag{B.296}$$

$$\frac{\tilde{B}_0^T}{\tilde{B}_0^I} \xrightarrow{n_1 \gg 1} 2\frac{\mu_1}{\mu_2}, \quad \frac{\tilde{B}_0^R}{\tilde{B}_0^I} \xrightarrow{n_1 \gg 1} 1, \tag{B.297}$$

for both cases. In addition, for normal incidence $(i, r, t = 0)$ both cases reduce to

$$\frac{\tilde{E}_0^T}{\tilde{E}_0^I} = \frac{2n_2}{n_2 + \frac{\mu_2}{\mu_1}n_1}, \quad \frac{\tilde{E}_0^R}{\tilde{E}_0^I} = \frac{n_2 - \frac{\mu_2}{\mu_1}n_1}{n_2 + \frac{\mu_2}{\mu_1}n_1}, \tag{B.298}$$

$$\frac{\tilde{B}_0^T}{\tilde{B}_0^I} = \frac{2n_1}{n_2 + \frac{\mu_2}{\mu_1}n_1}, \quad \frac{\tilde{B}_0^R}{\tilde{B}_0^I} = -\frac{n_2 - \frac{\mu_2}{\mu_1}n_1}{n_2 + \frac{\mu_2}{\mu_1}n_1}. \tag{B.299}$$

For $n_1 > \dfrac{\mu_1}{\mu_2}n_2$ there is a phase reversal in the reflected $\tilde{\vec{E}}_0^R$ field but not the $\tilde{\vec{B}}_0^R$ field.

Electromagnetic interface - conducting case

We start with the two curl equations from (B.171) and (B.172):

$$\vec{\nabla} \times \vec{H} = \frac{1}{c}\frac{\partial \vec{D}}{\partial t} + \frac{4\pi}{c}\vec{J}_f,$$

$$\vec{\nabla} \times \vec{E} = -\frac{1}{c}\frac{\partial \vec{B}}{\partial t}.$$

We model the free currents with *Ohm's law*:

$$\vec{J}_f = \sigma\vec{E} = \frac{\sigma}{\epsilon}\vec{D}. \tag{B.300}$$

In this case we introduce a single frequency,

$$\vec{E}(\vec{x},t) = \mathrm{Re}\left[\tilde{\vec{E}}(\vec{x},\omega)e^{-i\omega t}\right], \tag{B.301}$$

$$\vec{B}(\vec{x},t) = \mathrm{Re}\left[\tilde{\vec{B}}(\vec{x},\omega)e^{-i\omega t}\right]. \tag{B.302}$$

With the introduction of the relations $\vec{D} = \epsilon\vec{E}$ and $\vec{H} = \mu\vec{B}$ for constant ϵ and μ, we obtain

$$\vec{\nabla} \times \tilde{\vec{B}} = -\mu\omega\epsilon_{\mathrm{eff}}(\omega)\tilde{\vec{E}}, \tag{B.303}$$

$$\vec{\nabla} \times \tilde{\vec{E}} = i\omega\tilde{\vec{B}}, \tag{B.304}$$

where

$$\epsilon_{\mathrm{eff}}(\omega) \equiv \epsilon + i\frac{\sigma}{\omega}, \tag{B.305}$$

and we obtain the wave equations

$$\left(\nabla^2 + \mu\epsilon_{\mathrm{eff}}(\omega)\right)\tilde{\vec{E}} = 0, \tag{B.306}$$

$$\left(\nabla^2 + \mu\epsilon_{\mathrm{eff}}(\omega)\right)\tilde{\vec{B}} = 0. \tag{B.307}$$

Let us consider the case of plane waves traveling in the z-direction. Then $\nabla^2 \to \dfrac{\partial^2}{\partial z^2}$ and the above wave equations have solutions ($\tilde{\vec{E}}(\vec{x},\omega) \to \tilde{\vec{E}}(z,\omega)$, $\tilde{\vec{B}}(\vec{x},\omega) \to \tilde{\vec{B}}(z,\omega)$)

$$\tilde{\vec{E}}(z,\omega) = \tilde{\vec{E}}_0\, e^{i\tilde{k}_z z}, \tag{B.308}$$

$$\tilde{\vec{B}}(z,\omega) = \tilde{\vec{B}}_0\, e^{i\tilde{k}_z z}, \tag{B.309}$$

where

$$\tilde{k}_z^2 = \mu\frac{\omega^2}{c^2}\epsilon_{\mathrm{eff}}(\omega) = \mu\frac{\omega^2}{c^2}\left(\epsilon + i\frac{\sigma}{\omega}\right), \tag{B.310}$$

and

$$\tilde{\vec{B}}_0 = c\frac{\tilde{k}_z}{\omega}\hat{z} \times \tilde{\vec{E}}_0. \tag{B.311}$$

$\tilde{k}_z$ has both real and imaginary parts. One may show that

$$\tilde{k}_z = k + i\kappa, \tag{B.312}$$

$$k^2 = \tfrac{1}{2}\mu\epsilon\omega^2\left(\sqrt{1 + \frac{\sigma}{\omega\epsilon}} + 1\right), \tag{B.313}$$

$$\kappa^2 = \tfrac{1}{2}\mu\epsilon\omega^2\left(\sqrt{1 + \frac{\sigma}{\omega\epsilon}} - 1\right). \tag{B.314}$$

Note that the imaginary part of $\tilde{k}_z$ produces an exponential distance suppression factor $\sim e^{-\kappa z}$ for transmission in the fields in Equations (B.308) and (B.309). The "skin depth" d of a material is defined as $d \equiv 1/\kappa$ for positive κ.

For normal incidence on conductor surface 1 one may simply replace n_1 with $\tilde{n}_1$,

$$\tilde{n}_1 = \frac{c}{\omega}\tilde{k}_z, \tag{B.315}$$

in Equations (B.298) and (B.299). For conductors, the magnitude of $\tilde{n}_1$ becomes large at low frequencies. Its phase is also purely imaginary. Thus the fields are mostly magnetic in the interior of the conductor, but are confined to within the skin depth of the surface. The limits (B.296) and (B.297) hold for the transmitted and reflected waves with the Equation (B.315) replacement.

Waveguides

One may transmit electromagnetic signals through waveguides. Assuming no sources ($\vec{J}_{\text{eff}} = 0$) and that the material is nonohmic ($\sigma = 0$), the harmonic fields inside the waveguide will again originate in Equations (B.172), (B.171):

$$\vec{\nabla} \times \vec{\tilde{H}} = -i\frac{\omega\epsilon}{c}\vec{\tilde{E}}, \tag{B.316}$$

$$\vec{\nabla} \times \vec{\tilde{E}} = i\frac{\omega\mu}{c}\vec{\tilde{H}}. \tag{B.317}$$

We assume a straight axis for transmission, which we will choose to be the z direction, and a closed invariant cross-section. We also assume the inner surface to be perfectly conducting. This leads us to the forms,

$$\vec{\tilde{E}}(\vec{x},\omega) = \vec{E}(\vec{x}_\perp,k,\omega)\,e^{ikz}, \tag{B.318}$$

$$\vec{\tilde{B}}(\vec{x},\omega) = \vec{B}(\vec{x}_\perp,k,\omega)\,e^{ikz}. \tag{B.319}$$

Note that although the fields on the right of (B.318) and (B.319) lack the *tilde* designation (for convenience), they remain complex. In this context we may make the replacements

$$\vec{\nabla} \rightarrow \vec{\nabla}_\perp + ik\,\hat{z}, \tag{B.320}$$

$$\vec{H} \rightarrow \vec{H}_\perp + \hat{z}H_z, \tag{B.321}$$

$$\vec{E} \rightarrow \vec{E}_\perp + \hat{z}E_z. \tag{B.322}$$

This results in

$$\text{Equation (B.316)} \rightarrow \begin{cases} \dfrac{\partial H_z}{\partial y} - ikH_y = -i\dfrac{\omega\epsilon}{c}E_x, \\ -\dfrac{\partial H_z}{\partial x} + ikH_x = -i\dfrac{\omega\epsilon}{c}E_y, \\ \dfrac{\partial H_y}{\partial x} - \dfrac{\partial H_x}{\partial y} = -i\dfrac{\omega\epsilon}{c}E_z, \end{cases} \tag{B.323}$$

as well as

$$\text{Equation (B.317)} \rightarrow \begin{cases} \dfrac{\partial E_z}{\partial y} - ikE_y = i\dfrac{\omega\mu}{c}H_x, \\ -\dfrac{\partial E_z}{\partial x} + ikE_x = i\dfrac{\omega\mu}{c}H_y, \\ \dfrac{\partial E_y}{\partial x} - \dfrac{\partial E_x}{\partial y} = i\dfrac{\omega\mu}{c}H_z. \end{cases} \tag{B.324}$$

Using

$$\gamma^2 \equiv \mu\epsilon\frac{\omega^2}{c^2} - k^2, \tag{B.325}$$

we may solve for

$$E_x = \frac{i}{\gamma^2}\left[k\frac{\partial E_z}{\partial x} + \frac{\omega\mu}{c}\frac{\partial H_z}{\partial y}\right], \tag{B.326}$$

$$E_y = \frac{i}{\gamma^2}\left[k\frac{\partial E_z}{\partial y} - \frac{\omega\mu}{c}\frac{\partial H_z}{\partial x}\right], \tag{B.327}$$

$$H_x = \frac{i}{\gamma^2}\left[k\frac{\partial H_z}{\partial x} - \frac{\omega\epsilon}{c}\frac{\partial E_z}{\partial y}\right], \tag{B.328}$$

$$H_y = \frac{i}{\gamma^2}\left[k\frac{\partial H_z}{\partial y} + \frac{\omega\epsilon}{c}\frac{\partial E_z}{\partial x}\right]. \tag{B.329}$$

Equation (B.326) for example may be derived from substituting the middle equation of (B.324) into the top equation of (B.323). The others may be derived similarly.

From Equations (B.256) and (B.257)

$$\left(\nabla^2 + \frac{\mu\epsilon\omega^2}{c^2}\right)\{E_z, H_z\} = 0. \tag{B.330}$$

However, using

$$\nabla_z^2 \rightarrow -k^2 = \gamma^2 - \frac{\mu\epsilon\omega^2}{c^2}$$

in this context yields

$$\left(\vec{\nabla}_{\perp}^{2}+\gamma^{2}\right)\{E_z, H_z\}=0, \tag{B.331}$$

where

$$\vec{\nabla}_{\perp} \equiv \hat{x}\frac{\partial}{\partial x}+\hat{y}\frac{\partial}{\partial y}. \tag{B.332}$$

In the case where $\gamma = 0$, one may show that the $\vec{E}_{\perp}$ electric fields inside the conductor satisfy

$$\vec{\nabla}_{\perp} \times \vec{E}_{\perp}=0, \quad \vec{\nabla}_{\perp} \cdot \vec{E}_{\perp}=0, \tag{B.333}$$

$$\Longrightarrow \vec{\nabla}_{\perp}^{2}\{E_x, E_y\}=0. \tag{B.334}$$

This is just an electrostatics problem in the interior area with perfect conductor boundary conditions on the waveguide boundary: $\hat{n}\times\vec{E}_{\perp}|_s = 0$. The potential on this surface is a constant. However, the electric field inside is trivially zero unless there are additional interior surfaces at a different potential. For TEM modes $\gamma_{\text{TEM}} = 0$, which means that $k = \pm\sqrt{\mu\epsilon}\,\omega$, the same as for a wave in free space traveling in the $\pm z$-direction. The associated $H_{x,y}$ fields are then given by the top two equations in (B.324).

There are therefore three sets of modes in general: Transverse electric (TE), transverse magnetic (ME) and transverse electric and magnetic (TEM).

Mode type	**Field condition**	**Boundary condition**
TE	$E_z = 0, H_z \neq 0$	$\hat{n}\cdot\vec{\nabla}H_z\|_s = 0$
TM	$H_z = 0, E_z \neq 0$	$E_z\|_s = 0.$
TEM	$E_z = H_z = 0, \{\vec{E}_{\perp}, \vec{H}_{\perp}\} \neq 0$	$\hat{n}\times\vec{E}_{\perp}\|_s = 0$

One may define the "cutoff" angular frequency as

$$\omega_{\{n\}} \equiv \frac{\gamma_{\{n\}}}{\sqrt{\mu\epsilon}}, \tag{B.335}$$

where $\{n\}$ symbolizes the model labels. Note that Equation (B.325) may be re-written as

$$k^2 = \frac{\mu\epsilon}{c^2}\left(\omega^2-\omega_{\{n\}}^2\right). \tag{B.336}$$

Thus, when

$$\omega < \omega_{\{n\}},$$

the forward traveling modes acquire a positive imaginary value for the wavenumber k. These "cutoff" modes are space-attenuated and have an associated penetration length of $\sim 1/|k|$. There is no cutoff condition for TEM waves because $\gamma_{\text{TEM}} = 0$.

B.6 Potentials and fields

We continue our survey of fundamental concepts in electrodynamics with the re-formulation of the Maxwell equations in terms of scalar and vector potentials. We learn that these solutions are not unique, but allow a fundamental change called *gauge transformations*. The physical solutions of these equations take the finite speed of light into account and are called *retarded potentials*. We will look in particular at the retarded fields from a traveling point particle.

Gauge transformations

The Maxwell equations represent a first order formulation (in derivatives). We can obtain a second order formulation involving the scalar potential Φ and the vector potential $\vec{A}$ as follows. Using Equation (B.99)

$$\vec{B} = \vec{\nabla} \times \vec{A}.$$

Then we have from Faraday's differential law, Equation (B.166),

$$\vec{\nabla} \times \vec{E} + \frac{1}{c}\frac{\partial \vec{B}}{\partial t} = 0,$$

that

$$\vec{\nabla} \times \left(\vec{E} + \frac{1}{c}\frac{\partial \vec{A}}{\partial t}\right) = 0.$$

This implies

$$\Longrightarrow \vec{E} + \frac{\partial \vec{A}}{\partial t} = -\vec{\nabla}\Phi, \tag{B.337}$$

$$\text{or} \quad \vec{E} = -\vec{\nabla}\Phi - \frac{\partial \vec{A}}{\partial t}. \tag{B.338}$$

Now substitute these potentials into the other Maxwell equations. Then the electric field equation, Equation (B.165),

$$\vec{\nabla} \cdot \vec{E} = 4\pi\rho,$$

becomes

$$\nabla^2\Phi + \frac{\partial}{\partial t}(\vec{\nabla} \cdot \vec{A}) = -4\pi\rho. \tag{B.339}$$

Likewise the other Maxwell equation, Equation (B.167),

$$\vec{\nabla} \times \vec{B} - \frac{1}{c}\frac{\partial \vec{E}}{\partial t} = \frac{4\pi}{c}\vec{J},$$

becomes

$$\nabla^2 \vec{A} - \frac{\partial^2 \vec{A}}{\partial t^2} - \vec{\nabla}\left(\vec{\nabla} \cdot \vec{A} + \frac{1}{c}\frac{\partial \Phi}{\partial t}\right) = -\frac{4\pi}{c}\vec{J}. \tag{B.340}$$

Notice that

$$\vec{A} \to \vec{A} + \vec{\nabla}\Lambda, \tag{B.341}$$

$$\Phi \to \Phi - \frac{\partial \Lambda}{\partial t}, \tag{B.342}$$

where Λ is an arbitrary continuous scalar function of position and time, $\Lambda = \Lambda(\vec{x}, t)$, leaves $\vec{E}$ and $\vec{B}$ unchanged. We can use this to simplify Equations (B.339) and (B.340).

If one chooses

$$\vec{\nabla} \cdot \vec{A} + \frac{1}{c}\frac{\partial \Phi}{\partial t} = 0 \tag{B.343}$$

called the *Lorenz gauge* then one obtains

$$\nabla^2 \Phi - \frac{1}{c^2}\frac{\partial^2 \Phi}{\partial t^2} = -4\pi\rho, \tag{B.344}$$

$$\nabla^2 \vec{A} - \frac{1}{c^2}\frac{\partial^2 \vec{A}}{\partial t^2} = -\frac{4\pi}{c}\vec{J}. \tag{B.345}$$

If one chooses instead

$$\vec{\nabla} \cdot \vec{A} = 0 \tag{B.346}$$

called the *Coulomb gauge* then one obtains

$$\nabla^2 \Phi = -4\pi\rho, \tag{B.347}$$

$$\nabla^2 \vec{A} - \frac{1}{c^2}\frac{\partial^2 \vec{A}}{\partial t^2} = -\frac{4\pi}{c}\vec{J} + \frac{1}{c}\frac{\partial \vec{\nabla}\Phi}{\partial t}. \tag{B.348}$$

Equation (B.347) has the same form as in electrostatics, Equation (B.11). It is coupled to $\vec{A}$ through Equation (B.348) and produces an additional current-like term. The combination

$$\vec{J}_T \equiv \vec{J} - \frac{1}{4\pi}\frac{\partial \vec{\nabla}\Phi}{\partial t}, \tag{B.349}$$

is called the *transverse current*.

Retarded potentials

Let us work in the Lorenz gauge. We assume solutions to Equations (B.344) and (B.345) are

$$\Phi(\vec{x},t) = e\int d^3x\,dt'\,G^{\rm ret}(\vec{x},t;\vec{x}\,',t')\rho(\vec{x}\,',t'), \tag{B.350}$$

and

$$\vec{A}(\vec{x},t) = e\int d^3x\,dt'\,G^{\rm ret}(\vec{x},t;\vec{x}\,',t')\vec{J}(\vec{x}\,',t'). \tag{B.351}$$

These will satisfy (B.344) and (B.345) if $G^{\rm ret}(\vec{x},t;\vec{x}\,',t')$ satisfies

$$\left(\nabla^2 - \frac{1}{c^2}\frac{\partial^2}{\partial t^2}\right)G^{\rm ret}(\vec{x},t;\vec{x}\,',t') = -4\pi\delta(\vec{x}\,'-\vec{x})\delta(t-t'). \tag{B.352}$$

One solution to this is

$$G^{\rm ret}(\vec{x},t;\vec{x}\,',t') = \frac{\delta((t-t')-R/c)}{|\vec{x}-\vec{x}\,'|}, \tag{B.353}$$

where $R \equiv |\vec{x}-\vec{x}\,'|$. This is called the *retarded Green function*. Notice the nonzero contribution in $\delta((t-t')-R/c)$ occurs when $(t-t') = R/c$, i.e., when the time interval between field and source times, $t-t'$, is equal to the time the signal takes to transverse from the source. This is *causal* behavior where effect follows cause.

One way to show that Equation (B.353) is a solution is by considering the time Fourier transform of Equation (B.352),

$$\left(\nabla^2 + \frac{\omega^2}{c^2}\right)G^{\rm ret}(\omega;\vec{x}) = -4\pi\delta(\vec{x}), \tag{B.354}$$

where

$$G^{\rm ret}(\omega;\vec{x}) \equiv \int_{-\infty}^{\infty} dt\,e^{i\omega t}G^{\rm ret}(\vec{x},t;0,0). \tag{B.355}$$

We are temporarily setting $t' = \vec{x}\,' = 0$. One solution to Equation (B.354) is

$$G^{\rm ret}(\omega;\vec{x}) = \frac{e^{-i\omega|\vec{x}|/c}}{|\vec{x}|}. \tag{B.356}$$

This can be confirmed using the vector identity (see *Second-order differential identities* in Section 1.8)

$$\nabla^2(\phi\psi) = \phi\nabla^2\psi + \psi\nabla^2\phi + 2\vec{\nabla}\phi\cdot\vec{\nabla}\psi,$$

where $\phi = \dfrac{1}{|\vec{x}|}$ and $\psi = e^{i\omega|\vec{x}|/c}$. The inverse transform gives back the time dependence,

$$\begin{aligned} G^{\rm ret}(\vec{x},t;0,0) &= \frac{1}{2\pi}\int_{-\infty}^{\infty} d\omega\, e^{i\omega t} G^{\rm ret}(\vec{x},\omega) \\ &= \frac{1}{2\pi|\vec{x}|}\int_{-\infty}^{\infty} d\omega\, e^{i\omega(t-|\vec{x}|/c)} \\ &= \frac{\delta(t-|\vec{x}|/c)}{|\vec{x}|}. \end{aligned} \tag{B.357}$$

Shifting $\vec{x} \to \vec{x} - \vec{x}\,'$ and $t \to t - t'$ then gives Equation (B.353).

We are actually making a choice of solution in Equation (B.356). The other choice

$$G^{\rm adv}(\omega;\vec{x}) = \frac{e^{i\omega|\vec{x}|/c}}{|\vec{x}|}, \tag{B.358}$$

gives the *advanced Green function*. This other linearly independent solution is *noncausal* because the effect actually precedes the cause.

Our retarded or causal solutions are

$$\begin{aligned} \Phi(\vec{x},t) &= \int d^3x' dt' \frac{\delta((t-t')-R/c)}{|\vec{x}-\vec{x}\,'|}\rho(\vec{x}\,',t'), \\ &= \int d^3x' \frac{\rho(\vec{x}\,',t-R/c)}{|\vec{x}-\vec{x}\,'|}, \end{aligned} \tag{B.359}$$

and

$$\begin{aligned} \vec{A}(\vec{x},t) &= \int d^3x' dt' \frac{\delta((t-t')-R/c)}{|\vec{x}-\vec{x}\,'|}\vec{J}(\vec{x}\,',t'), \\ &= \int d^3x' \frac{\vec{J}(\vec{x}\,',t-R/c)}{|\vec{x}|}. \end{aligned} \tag{B.360}$$

Due to the finite speed of light, the sources that contribute to the potentials at field time t are those arrayed throughout space and time which acted at the earlier time $t - R/c$. This is called the *retarded time*.

Point particle fields

If we simply substitute the charge density and current for a moving point particle (e is the charge, $\vec{x}(t)$ is the given path and $\vec{v}(t) = \dot{\vec{x}}(t)$; in this notation $\vec{x}$ is a spatial coordinate but $\vec{x}(t)$ is the path)

$$\rho(\vec{x},t) = e\,\delta(\vec{x}-\vec{x}(t)), \tag{B.361}$$

$$\vec{J}(\vec{x},t) = e\,\vec{v}(t)\,\delta(\vec{x}-\vec{x}(t)), \tag{B.362}$$

into Equation (B.359) we obtain

$$\Phi(\vec{x},t) = \int d^3x' \frac{\delta(\vec{x}\,' - \vec{x}(t - R/c))}{|\vec{x} - \vec{x}\,'|}. \tag{B.363}$$

It is essentially impossible to integrate Equation (B.363) since there is now $\vec{x}\,'$ dependence in the given path function: $\vec{x}(t - |\vec{x} - \vec{x}\,'|/c)$. We can cure this problem by going back to the first part of Equation (B.359), before the t' integral has been done. This then frees the $\vec{x}'$ integral from the above problem. The first step is

$$\Phi(\vec{x},t) = e \int d^3x'\, dt' \frac{\delta^{(3)}(\vec{x}' - \vec{x}(t'))\, \delta(t' - t + |\vec{x} - \vec{x}'|/c)}{|\vec{x} - \vec{x}'|}. \tag{B.364}$$

It is now possible to do the $\vec{x}'$ integral first, yielding

$$\Phi(\vec{x},t) = e \int dt' \frac{\delta(t' - t + |\vec{x} - \vec{x}(t')|/c)}{|\vec{x} - \vec{x}(t')|}. \tag{B.365}$$

The way to proceed now is to use a property of Dirac delta functions when one is considering a function rather than a variable as its argument. Consider the linear function $f(x) = ax + b$ and the delta function,

$$\delta(f(x)) = \delta(ax + b) = \delta\left(a\left(x + \frac{b}{a}\right)\right).$$

A scaling property of delta functions which can be readily proven by a change in variables and using the fact that the delta function is an even function of its argument, is that

$$\delta(ax') = \frac{1}{|a|}\delta(x').$$

Applying this to the above gives

$$\delta(ax + b) = \frac{1}{|a|}\delta\left(x - \frac{b}{a}\right). \tag{B.366}$$

Note that a is the derivative of the function $f(x)$. The generalization of this result is

$$\delta(f(x)) = \frac{1}{\left|\frac{df}{dx}\big|_{x=x'}\right|}\delta(x - x'), \tag{B.367}$$

where x' is the single zero of $f(x)$, assuming it can be linearized in its vicinity. In this case

$$f(t') = t' - t + |\vec{x} - \vec{x}(t')|/c. \tag{B.368}$$

We calculate that

$$\frac{df(t')}{dt'} = 1 + \frac{1}{c}\frac{d|\vec{x} - \vec{x}(t')|}{dt'} = 1 - \frac{1}{c}\frac{(\vec{x} - \vec{x}(t')) \cdot \vec{v}(t')}{|\vec{x} - \vec{x}(t')|}. \tag{B.369}$$

This yields

$$\Phi(\vec{x}, t) = e \int dt' \frac{\delta(t - t_r(t'))}{|\vec{x} - \vec{x}(t')| \left(1 - \frac{1}{c}\frac{(\vec{x} - \vec{x}(t')) \cdot \vec{v}(t')}{|\vec{x} - \vec{x}(t')|}\right)}, \tag{B.370}$$

where

$$t_r(t') \equiv t' - |\vec{x} - \vec{x}(t')|/c. \tag{B.371}$$

This gives

$$\Phi(\vec{x}, t) = \frac{e}{R - \vec{R} \cdot \vec{v}/c}, \tag{B.372}$$

with all quantities being evaluated at the approximate retarded time, $t_r(t)$. Very similar reasoning gives the form of the vector potential,

$$\vec{A}(\vec{x}, t) = \frac{e\vec{v}}{R - \vec{R} \cdot \vec{v}/c}. \tag{B.373}$$

Equations (B.372) and (B.373) are called the *Liénard-Wiechert potentials.*

As an example, consider a charge, e, moving on a straight line with constant velocity $\vec{v}$. We take $\vec{x} = \vec{v}t$. The retarded time from Equation (B.371) for this example is given by solving the quadratic equation for $t_r(t)$ implied by Equation (B.371):

$$t_r(t) = \frac{t - \vec{x} \cdot \vec{v}/c^2 - \sqrt{(\vec{x} - \vec{v}\,t)^2 - (\vec{x} \times \vec{v})^2/c^2}}{1 - v^2/c^2}. \tag{B.374}$$

The relativistic scalar potential following from Equation (B.372) is

$$\Phi(\vec{x}, t) = \frac{e}{\sqrt{(\vec{x} - \vec{v}\,t)^2 - (\vec{x} \times \vec{v})^2/c^2}}. \tag{B.375}$$

The vector potential can be found similarly.

B.7 Radiation

We begin the subject of electromagnetic radiation with an understanding of the various approximations necessary to define radiation fields at large distances from arbitrary sources. There are two formalisms presented in the text for radiation problems. One is called the *harmonic source formalism,* where one works with a single frequency and complex sources. This allows one to get to the heart of the matter quicker for such circumstances and leads to an immediate understanding of the various zones of approximation for radiation type problems. A more general approach can also be developed that avoids complex sources and deals with arbitrary motion of sources and particles which we call the *real source formalism*. We will present a streamlined version of this formalism here. Both formalisms also include a necessary approximation called a multipole expansion. In this discussion we will be limited to the lowest order multipole result for radiation called *electric dipole radiation*.

The instantaneous differential radiated power is given by calculating the energy delivered in the outward direction $\hat{n} = \vec{r}/r$ using the Poynting vector from Equation (B.201):

$$\frac{dP(t)}{d\Omega} = \lim_{r\to\infty}\left(r^2\frac{c}{4\pi}\hat{n}\cdot(\vec{E}\times\vec{B})\right). \tag{B.376}$$

From this, we see that it is only necessary to keep the long-distance fields that fall off like $\vec{E},\vec{B}\sim\frac{1}{r}$. These are called the *radiation fields*.

Real source radiation formalism

We begin with the retarded potentials

$$\Phi(\vec{x},t) = \int d^3x'\,\frac{\rho(\vec{x}\,',t-R/c)}{R}, \tag{B.377}$$

$$\vec{A}(\vec{x},t) = \frac{1}{c}\int d^3x'\,\frac{\vec{J}(\vec{x}\,',t-R/c)}{R}, \tag{B.378}$$

where $R = |\vec{x}-\vec{x}\,'|$. We specialize to fields far away from the source. Formally, this means $r \gg d$, where d symbolizes the source size.

$$|\vec{x}-\vec{x}\,'| = r - \hat{n}\cdot\vec{x}\,' + \mathcal{O}\left(\frac{d^2}{r}\right).$$

To lowest order we obtain

$$\Phi(\vec{x},t) \approx \frac{1}{r}\int d^3x'\,\rho(\vec{x}\,',t_r), \tag{B.379}$$

$$\vec{A}(\vec{x},t) \approx \frac{1}{cr}\int d^3x'\,\vec{J}(\vec{x}\,',t_r). \tag{B.380}$$

t_r is the first-order approximation to the retarded time in the far zone following from Equation (B.7):

$$t_r \equiv t - \frac{r}{c} + \frac{1}{c}\hat{n}\cdot\vec{x}\,'. \tag{B.381}$$

We call t_r the "approximate retarded time". In order to calculate the angular distribution of radiated power, we need not just Φ and $\vec{A}$, but the $\vec{E}$ and $\vec{B}$ fields. They are given by (all fields functions of $\vec{x}$ and t)

$$\vec{B} = \vec{\nabla} \times \vec{A}; \qquad \vec{E} = -\vec{\nabla}\Phi - \frac{1}{c}\frac{\partial \vec{A}}{\partial t}. \tag{B.382}$$

Both Φ and $\vec{A}$ are functions of t_r. By the chain rule, for an arbitrary function $f(t_r)$ we have

$$\vec{\nabla} f(t_r) = \frac{\partial f(t_r)}{\partial t_r}\vec{\nabla} t_r = \frac{\partial f(t_r)}{\partial t}\vec{\nabla} t_r, \tag{B.383}$$

where for $|\vec{x}\,'| \leq d$ and $r \gg d$ one has

$$\vec{\nabla} t_r = \vec{\nabla}\left(t - \frac{r}{c} + \frac{1}{c}\hat{n}\cdot\vec{x}\,'\right) \approx -\frac{\hat{n}}{c}. \tag{B.384}$$

On the other hand for a primed gradient

$$\vec{\nabla}' t_r \approx \frac{\hat{n}}{c}. \tag{B.385}$$

Then to first order

$$\vec{B} = \vec{\nabla} \times \vec{A} \approx -\frac{\hat{n}}{c} \times \frac{\partial \vec{A}}{\partial t} \tag{B.386}$$

$$\approx -\frac{1}{rc^2}\,\hat{n} \times \int d^3x'\, \frac{\partial \vec{J}(\vec{x}\,', t_r)}{\partial t}. \tag{B.387}$$

Likewise

$$\begin{aligned}\vec{E} &= -\vec{\nabla}\Phi - \frac{1}{c}\frac{\partial \vec{A}}{\partial t} \approx \frac{\hat{n}}{c}\frac{\partial \Phi}{\partial t} - \frac{1}{rc^2}\int d^3x'\, \frac{\partial \vec{J}(\vec{x}\,', t_r)}{\partial t}\\ &\approx \frac{1}{rc}\left[\hat{n}\int d^3x' \frac{\partial \rho(\vec{x}\,', t_r)}{\partial t} - \frac{1}{c}\int d^3x' \frac{\partial \vec{J}(\vec{x}\,', t_r)}{\partial t}\right].\end{aligned} \tag{B.388}$$

We may re-express the $\partial\rho/\partial t$ term using the continuity equation as follows:

$$\begin{aligned}\frac{\partial \rho(\vec{x}\,', t_r)}{\partial t} &= -\left.\vec{\nabla}' \cdot \vec{J}(\vec{x}\,', t_r)\right|_{t_r=\text{const.}}\\ &= -\left(\vec{\nabla}' \cdot \vec{J}(\vec{x}\,', t_r(\vec{x}\,')) - \frac{\partial \vec{J}(\vec{x}\,', t_r)}{\partial t}\cdot\vec{\nabla}' t_r\right)\\ &\approx -\vec{\nabla}' \cdot \vec{J}(\vec{x}\,', t_r(\vec{x}\,')) + \frac{\hat{n}}{c}\cdot\frac{\partial \vec{J}(\vec{x}\,', t_r)}{\partial t},\end{aligned} \tag{B.389}$$

where we have re-asserted the non-constant $\vec{x}\,'$ dependence of t_r in the middle step. Since the current distribution is bounded, we have by Gauss' theorem on a large spherical surface $R \gg d$:

$$\int d^3x'\,\vec{\nabla}'\cdot\vec{J}(\vec{x}\,',t_r(\vec{x}\,')) = \oint_R da'\,\hat{n}\cdot\vec{J}(\vec{x}\,',t_r(\vec{x}\,')) = 0. \tag{B.390}$$

Therefore

$$\vec{E} \approx \frac{1}{rc^2}\left[\hat{n}\int d^3x'\,\hat{n}\cdot\frac{\partial\vec{J}(\vec{x}\,',t_r)}{\partial t} - \int d^3x'\,\frac{\partial\vec{J}(\vec{x}\,',t_r)}{\partial t}\right]$$

$$= \hat{n}\times\left[\frac{\hat{n}}{c}\times\frac{1}{cr}\int d^3x'\,\frac{\partial\vec{J}(\vec{x}\,',t_r)}{\partial t}\right], \tag{B.391}$$

$$\approx \hat{n}\times\left[\frac{\hat{n}}{c}\times\frac{\partial\vec{A}}{\partial t}\right] \approx \vec{B}\times\hat{n}. \tag{B.392}$$

This establishes that the $\vec{E}, \vec{B}$ radiation fields are transverse to the outward direction of motion, $\hat{n}$.

The instantaneous angular distribution of power is now seen to be (our equalities are in the large r radiation limit)

$$\frac{dP(t)}{d\Omega} = \frac{c\,r^2}{4\pi}\hat{n}\cdot(\vec{E}\times\vec{B}) = \frac{c\,r^2}{4\pi}\vec{B}^2$$

$$= \frac{r^2}{4\pi c}\left[\hat{n}\times\frac{\partial\vec{A}(\vec{x},t)}{\partial t}\right]^2, \tag{B.393}$$

where we have used (B.386) to re-express $\vec{B}^2$.

This is the beginning point of the multipole formalism discussed in Section 11.4, which is really just a Taylor series expansion in the quantity $\hat{n}\cdot\vec{x}\,'/c$ contained in the approximate retarded time, t_r:

$$t_r = t_0 + \frac{\hat{n}\cdot\vec{x}\,'}{c}, \tag{B.394}$$

where

$$t_0 \equiv t - \frac{r}{c}, \tag{B.395}$$

is called the "origin retarded time". This is the time associated with a source located at the chosen coordinate origin. So, the approximation being made is that whatever origin we choose to describe the radiation source is actually viewed as the source for all radiation. Section 11.4 shows that this will be satisfactory for sufficiently small sources or sufficiently slow particles. Here we will forgo this development and just concentrate on the lowest order term in the expansion.

Electric dipole radiation

From (B.380) with $t_r \to t_0$, we have

$$\vec{A}(\vec{x}, t) \approx \frac{1}{rc} \int d^3x' \, \vec{J}(\vec{x}\,', t_0). \tag{B.396}$$

We may simplify the right-hand side using

$$\vec{J}(\vec{x}\,', t_0) = \sum_i \nabla_i' \left(\vec{x}\,' J_i(\vec{x}\,', t_0)\right) - \vec{x}\,' \vec{\nabla}' \cdot \vec{J}(\vec{x}\,', t_0)$$

$$\Longrightarrow \int d^3x' \vec{J}(\vec{x}\,', t_0) = \int d^3x' \sum_i \nabla_i' \left(\vec{x}\,' J_i(\vec{x}\,', t_0)\right) - \int d^3x' \, \vec{x}\,' \vec{\nabla}' \cdot \vec{J}(\vec{x}\,', t_0). \tag{B.397}$$

The first term on the right in Equation (B.397) may be evaluated using Gauss' theorem and put on a large spherical surface R. The assumption of a bounded charge distribution shows this term is zero. Then one has

$$\int d^3x' \vec{J}(\vec{x}\,', t_0) = \int d^3x' \, \vec{x}\,' \frac{\partial \rho(\vec{x}\,', t_0)}{\partial t}, \tag{B.398}$$

where we used the continuity equation, Equation (B.169). Now recall that the electric dipole vector, Equation (B.34), is given by

$$\vec{p}(t_0) = \int d^3x' \, \vec{x}\,' \rho(\vec{x}\,', t_0). \tag{B.399}$$

We obtain finally that to lowest order

$$\vec{A}(\vec{x}, t) \approx \frac{\dot{\vec{p}}(t_0)}{rc}, \tag{B.400}$$

where the dot symbolizes the time derivative. From (B.393) we obtain the angular distribution of power emitted as

$$\frac{dP(t)}{d\Omega} = \frac{1}{4\pi c^3} \left[\hat{n} \times \ddot{\vec{p}}(t_0)\right]^2 = \frac{\left[\ddot{\vec{p}}(t_0)\right]^2}{4\pi c^3} \sin^2 \theta(t_0). \tag{B.401}$$

This displays the so-called dipole form. The total instantaneous power is

$$P(t) = \frac{2[\ddot{\vec{p}}(t_0)]^2}{3c^3}. \tag{B.402}$$

Larmor formula for point particles

In the case of a single nonrelativistic particle of charge e, from Equation (B.362) we have the source

$$\vec{J}(\vec{x},t) = e\,\vec{v}(t)\,\delta(\vec{x}-\vec{x}(t)),$$

for use in (B.380) with $t_r \to t_0$. We then have ($\vec{v}(t_0) = \dot{\vec{x}}(t_0)$)

$$\vec{A}(\vec{x},t) \approx \frac{e\vec{v}(t_0)}{rc}. \tag{B.403}$$

This is the same as using the dipole source $\vec{p}(t_0) = e\,\vec{x}(t_0)$ in Equation (B.400). We obtain then

$$\frac{dP(t)}{d\Omega} = \frac{e^2}{4\pi c^3}\left(\hat{n}\times\dot{\vec{v}}(t_0)\right)^2. \tag{B.404}$$

Doing the angular integral then gives the total instantaneous power,

$$P(t) = \frac{2e^2[\dot{\vec{v}}(t_0)]^2}{3c^3}. \tag{B.405}$$

This expression for $P(t)$ is called the *Larmor formula*. The form for $P(t)$ is actually independent of the particle velocity in the nonrelativistic approximation. It may be generalized to arbitrary velocities where this is no longer the case. But first we must study the relativity of motion!

B.8 Special relativity

Physics at the end of the 19th century was in a smug place. Newton's equations for mechanical phenomena, including gravity, waves and fluids provided an effective framework for understanding physical behaviors of systems. In the words of physicist Albert Michelson in 1894, "most of the grand underlying principles have been firmly established."

It was understood that light was a manifestation of Maxwell's equations. Like all wave phenomena it displayed wave characteristics like diffraction and interference. It was assumed that there was a medium for its propagation, the ether, and that the form of the electromagnetic equations given by Maxwell were valid only for a frame of reference at rest with respect to this medium. This was believed to be true because Maxwell's equations in fact were not invariant in form under coordinate transformations that were presumed to hold between inertial reference systems; these are called *Galilean transformations*. It was thought that reference frames in relative motion were completely equivalent as far as mechanical properties were concerned, but not with respect to electromagnetic phenomena. There was only one frame of reference for which, for example, it was thought that the velocity of light had the value predicted by the "rest form" of Maxwell's equations. There was therefore a search for a change in the speed of light when there was motion with respect to the ether frame. It is rather ironic that the person referenced above concerning firm establishment of grand underlying principles, along with Edward Morley, was also the person who helped to show that this point of view was quite mistaken. The *Michelson-Morley experiment* of 1887 was designed to test for motion of the Earth relative to the ether. It failed to detect any. It has been called the most famous failed experiment in history.[1]

We will begin with two postulates:

(*i*) The speed of light is the same for all inertial observers.

(*ii*) The form of the laws of electromagnetism are the same for all inertial observers.

The change from one inertial reference frame to another is accomplished by coordinate transformations called *velocity boosts*. Using this concept, (*i*) is stating that the speed of light is unchanged, or *invariant*, under velocity boosts, while (*ii*) states that the form of the laws of electromagnetism are unchanged, or *covariant*, under velocity boosts. Indeed, all the laws of nature are thought to be invariant under inertial reference frame changes.

Lorentz transformations

Let us do some setups to make our transformations easier to understand. Consider inertial coordinate systems K and K', whose axes are parallel but displaced along the 1-coordinate. The K' coordinate system is moving with steady velocity v with respect to K along the 1-coordinate. The origins and coordinates of the two coordinate systems coincide at a single instant of time. At this moment, a pulse of light is emitted from the common origin. Using a bar above the coordinates to denote the position of the light pulse, according to postulate (i) the wavefront of the light pulse is described by

[1] See for example: https://ssec.si.edu/stemvisions-blog/most-famous-failed-experiment. A fairer characterization would substitute "null" for "failed".

the two observers as

$$\left.\begin{aligned} K : c^2\bar{t}^2 - \sum_{i=1}^{3}(\bar{x}^i)^2 = 0, \\ K' : c^2\bar{t}'^2 - \sum_{i=1}^{3}(\bar{x}'^i)^2 = 0. \end{aligned}\right\} \tag{B.406}$$

Introduce

$$\bar{x}^0 \equiv c\,\bar{t},\ \bar{x}'^0 \equiv \bar{t}'. \tag{B.407}$$

Notice that spatial (i) and time (0) indices are all superscripts at this point. We will need a way of building in the minus sign between space and time coordinates in Equations (B.406). Therefore, also define

$$\bar{x}_\mu \equiv \sum_{\nu=0}^{3} \eta_{\mu\nu}\bar{x}^\nu, \tag{B.408}$$

using

$$[\eta]_{\mu\nu} \equiv \begin{pmatrix} 1 & 0 & 0 & 0 \\ 0 & -1 & 0 & 0 \\ 0 & 0 & -1 & 0 \\ 0 & 0 & 0 & -1 \end{pmatrix}. \tag{B.409}$$

The matrix η is called the *flat-space metric* or *Minkowski metric*. Our light front characterization now reads

$$\left.\begin{aligned} K : \sum_{\mu=0}^{3}\bar{x}^\mu\bar{x}_\mu = 0, \\ K' : \sum_{\mu=0}^{3}\bar{x}'^\mu\bar{x}'_\mu = 0. \end{aligned}\right\} \tag{B.410}$$

We usually use Latin letters to sum over spatial indices, as in $\{i, j, k = 1, 2, 3\}$, but Greek letters to sum over both space and time indices, such as $\{\alpha, \beta, \gamma = 0, 1, 2, 3\}$. $\bar{x}^\mu$ is called a *contravariant 4-vector* and $\bar{x}_\mu$ is termed a *covariant 4-vector*. Notice that when one sums on an index such as the ν index in Equation (B.408) or the μ index in (B.410), one index is a superscript and one is a subscript.

The form of the equation for the light pulse front is the same form in K or K'. We want to find the linear transformation describing velocity boosts which is consistent with this relation, but which reduces to the known Galilean transformation for low velocities. The statements in Equation (B.410) are not covariant under Galilean transformation.

Explicitly, our linear transformation looks like:

$$\left.\begin{aligned} x'^{\,0} &= \Lambda^0{}_0\, x^0 + \Lambda^0{}_1\, x^1 + \Lambda^0{}_2\, x^2 + \Lambda^0{}_3\, x^3, \\ x'^{\,1} &= \Lambda^1{}_0\, x^0 + \Lambda^1{}_1\, x^1 + \Lambda^1{}_2\, x^2 + \Lambda^1{}_3\, x^3, \\ x'^{\,2} &= \Lambda^2{}_0\, x^0 + \Lambda^2{}_1\, x^1 + \Lambda^2{}_2\, x^2 + \Lambda^2{}_3\, x^3, \\ x'^{\,3} &= \Lambda^3{}_0\, x^0 + \Lambda^3{}_1\, x^1 + \Lambda^3{}_2\, x^2 + \Lambda^3{}_3\, x^3, \end{aligned}\right\} \tag{B.411}$$

where the $\Lambda^\mu{}_\nu$ coefficients are not yet known. Notice we have dropped the bars on the coordinates. We are not necessarily talking about light pulse coordinates at this point, but general "events" which have $\sum_{\mu=0}^{3} x^\mu x_\mu \neq 0$. This will be called a *Lorentz transformation*. One can think of the coordinates x^μ and $x'^{\,\mu}$ as representing the space and time coordinates of an arbitrary space-time "event" from the point of view of two different inertial coordinate systems, the common origin of which is the place/time where/when the K, K' spacial origins coincide.

Our Lorentz transformation is described as

$$x'^{\,\mu} = \sum_\nu \Lambda^\mu{}_\nu x^\nu. \tag{B.412}$$

In matrix language,

$$x' = Ax, \tag{B.413}$$

where the matrix A elements are given as

$$[A]_{\mu\nu} \equiv \Lambda^\mu{}_\nu. \tag{B.414}$$

Given the form of the transformation on the coordinates in Equation (B.412), one has

$$\sum_\mu x'^{\,\mu} x'_\mu = \sum_{\gamma,\kappa} \eta_{\gamma\kappa} x'^{\,\gamma} x'^{\kappa} = \sum_{\mu,\nu,\gamma,\kappa} \Lambda^\mu{}_\gamma \eta_{\mu\nu} \Lambda^\nu{}_\kappa x^\gamma x^\kappa. \tag{B.415}$$

We want to transform the statement $\sum_\mu \bar{x}'^{\,\mu} \bar{x}'_\mu = 0$ into $\sum_\mu \bar{x}^\mu \bar{x}_\mu = 0$ for light pulses. The only way this can be done when general event intervals are also considered is if the two sums, $\sum_\mu x^\mu x_\mu$ and $\sum_\mu x'^{\,\mu} x'_\mu$ are directly proportional to one another. Invertibility between primed and unprimed coordinates implies this proportionality factor is either 1 or -1. If we also require that the transformation be continuously connected to the identity transformation, $x^\mu = x'^{\,\mu}$, then these intervals must be equal:

$$\sum_\mu x^\mu x_\mu = \sum_\mu x'^{\,\mu} x'_\mu. \tag{B.416}$$

Comparison of the right part of (B.415) and the left part of (B.416) yields

$$\sum_{\gamma,\kappa} \eta_{\gamma\kappa} x^{\gamma} x^{\kappa} = \sum_{\mu,\nu,\gamma,\kappa} \Lambda^{\mu}{}_{\gamma} \eta_{\mu\nu} \Lambda^{\nu}{}_{\kappa} x^{\gamma} x^{\kappa}. \tag{B.417}$$

The coefficients of the individual $x^{\gamma}x^{\kappa}$ terms must be equal:

$$\Rightarrow \sum_{\mu,\nu} \Lambda^{\mu}{}_{\gamma} \eta_{\mu\nu} \Lambda^{\mu}{}_{\kappa} = \eta_{\gamma\kappa}. \tag{B.418}$$

Using matrix language the above identity reads:

$$(A)^{T} \eta A = \eta, \tag{B.419}$$

where ^{T} denotes the matrix transpose. The following A is the simplest possible one-parameter solution to Equation (B.419) involving only the time "0" and spatial "1" components:

$$[A]_{\mu\nu} \equiv \begin{pmatrix} \cosh\alpha & -\sinh\alpha & 0 & 0 \\ -\sinh\alpha & \cosh\alpha & 0 & 0 \\ 0 & 0 & 1 & 0 \\ 0 & 0 & 0 & 1 \end{pmatrix}. \tag{B.420}$$

It is a candidate for a velocity boost. Then in matrix language this transformation used in Equation (B.413) gives

$$\begin{pmatrix} x'^{0} \\ x'^{1} \end{pmatrix} = \begin{pmatrix} \cosh\alpha & -\sinh\alpha \\ -\sinh\alpha & \cosh\alpha \end{pmatrix} \begin{pmatrix} x^{0} \\ x^{1} \end{pmatrix}. \tag{B.421}$$

Or equivalently,

$$\left.\begin{aligned} x'^{0} &= x^{0}\cosh\alpha - x^{1}\sinh\alpha, \\ x'^{1} &= -x^{0}\sinh\alpha + x^{1}\cosh\alpha. \end{aligned}\right\} \tag{B.422}$$

Let us use this transformation and tie it in with our K, K' setup now. Since the origin point of K', $x'^{1} = 0$, is moving with velocity v along the x^{1} axis, this point is characterized by $x^{1} = vt$ by K. Then the lower equation in (B.422) reads

$$0 = -ct\sinh\alpha + vt\cosh\alpha \implies \frac{v}{c} = \tanh\alpha. \tag{B.423}$$

We then have

$$\sinh^{2}\alpha = \frac{v/c}{1-\left(\frac{v}{c}\right)^{2}}, \ \cosh^{2}\alpha = \frac{1}{1-\left(\frac{v}{c}\right)^{2}}. \tag{B.424}$$

We take the positive square root for the sinh in Equation (B.424) to agree with our physical setup. This gives the Lorentz transformation,

$$\left.\begin{aligned} t' &= \gamma\left(t - x^{1}\frac{\beta}{c}\right), \\ x'^{1} &= \gamma\left(x^{1} - vt\right), \end{aligned}\right\} \tag{B.425}$$

where

$$\gamma \equiv \frac{1}{\sqrt{1-\beta^2}}, \; \beta \equiv v/c. \tag{B.426}$$

In the formal limit $c \to \infty$, Lorentz boosts reduce to

$$\left.\begin{aligned} t' &= t, \\ x'^1 &= x^1 - vt, \end{aligned}\right\} \tag{B.427}$$

which is just the nonrelativistic Galilean boost formula.

The *form* of this transformation was known before Einstein by Hendrik Lorentz, who was attempting to understand and explain experimental results which did not detect Earth's motion relative to the ether. However, it was Einstein who gave the transformation the correct physical interpretation. This is in fact the transformation law that is consistent with the form invariance of Maxwell's equations, fulfilling postulate (*ii*)! The contravariant 4-vector potential in electrodynamics is the combination,

$$A^\mu \equiv \left(\Phi, \vec{A}\right), \tag{B.428}$$

($A^0 = \Phi$) and the 4-vector of sources is the combination of charge density and current,

$$J^\mu \equiv \left(c\rho, \vec{J}\right). \tag{B.429}$$

($J^0 = c\rho$). These 4-vectors change under Lorentz transformations as in Equation (B.412).

Two important concepts in relativistic evaluations are *invariant distance* intervals

$$ds \equiv \sqrt{c^2 dt^2 - d\vec{x}^{\,2}} = c\,|dt|\sqrt{1-\beta^2} = c\frac{|dt|}{\gamma}, \tag{B.430}$$

and *proper time* intervals,

$$d\tau \equiv \frac{1}{c} ds = \frac{|dt|}{\gamma}. \tag{B.431}$$

These quantities are invariant under Lorentz boosts (relativistic scalars) and help to form relativistic 4-vector quantities like momentum, velocity and acceleration.

Just as Lorentz transformations unite time and space transformations, they also unit energy and momentum in a single 4-vector, p^μ:

$$p^\mu = \left(\frac{E}{c}, \vec{p}\right) \Longrightarrow \sum_\mu p^\mu p_\mu = \frac{E^2}{c^2} - \vec{p}^{\,2}. \tag{B.432}$$

Let us call E_0 the energy in the rest frame of the particle. If we do a Lorentz transformation on a particle at rest in K, then in K' we have

$$\left.\begin{aligned} \frac{E}{c} &= \gamma\frac{E_0}{c}, \\ p^1 &= -\gamma\beta\frac{E_0}{c}. \end{aligned}\right\} \tag{B.433}$$

If we require that $p^1 = -mv$ for $|\beta| \ll 1$, then one identifies $E_0 = mc^2$, a rather famous result of Einstein's. This is the particle's *rest energy*. Thus, the four-vector of a particle in motion with velocity v has a four-momentum given by

$$p^\mu = m\gamma\,(c, \vec{v})\,, \tag{B.434}$$

and a relativistic energy-momentum relationship given by

$$\sum_\mu p^\mu p_\mu = (mc)^2 \implies E^2 = (mc^2)^2 + (c\vec{p})^2. \tag{B.435}$$

Of course for nonrelativistic motion ($|\vec{p}| \ll mc$), this gives

$$E \approx mc^2 + \frac{\vec{p}^2}{2m}, \tag{B.436}$$

which represents the rest energy plus kinetic energy of the particle.

Relativistic generalization of the Larmor radiation formula

In attempting to generalize the Larmor radiation result for a point particle, the most important point to make is that although the previous result, Equation (B.405), was deduced in the multipole approximation, which we learned implied $|v| \ll c$, it is in fact an exact result in an appropriate frame of reference. This frame of reference has $\vec{x}(t_r) = 0$ (which implies $t_r = t_0$), as well as $\vec{v}(t_0) = 0$ (one condition does not imply the other), which renders the multipole expansion superfluous. Since this is an exact result, we should be able to generalize it to any other frame of reference using a Lorentz transformation.

Another crucial point to make in generalizing the Larmor result is that the value of the energy lost by the particle in a given amount of time is a quantity that all inertial observers will agree upon, once the correction for the particle's motion is made, just as all observers agree on a particle's mass. This means that there is a Lorentz invariant instantaneous power, $P(t)$, for which Equation (B.405) is the value in the $\vec{x}(t_r) = 0$, $\vec{v}(t_0) = 0$

frame. We can now generalize to an arbitrary Lorentz frame, $\vec{v}(t_0) \neq 0$, by simply letting

$$\dot{\vec{v}}^2 \to -\sum_\mu \alpha^\mu \alpha_\mu, \tag{B.437}$$

(note that $\sum_\mu \alpha^\mu \alpha_\mu \leq 0$) where

$$\alpha^\mu \equiv \frac{du^\mu}{d\tau}, \tag{B.438}$$

is the relativistic acceleration. The proper time interval $d\tau$ was defined in B.8. It transforms as a scalar and makes α^μ transform as a 4-vector. In addition

$$u^\mu \equiv \frac{p^\mu}{m} = \gamma\,(c, \vec{v})\,, \tag{B.439}$$

is the relativistic velocity. Then

$$P(t) = -\frac{2e^2}{3c^2}\sum_\mu \alpha^\mu \alpha_\mu = \frac{2e^2}{3c^2}\left(\vec{\alpha}^{\,2} - (\alpha^0)^2\right). \tag{B.440}$$

One has that

$$\alpha^0 = \frac{du^0}{d\tau} = \gamma^4 \frac{\vec{v}\cdot\dot{\vec{v}}}{c}, \tag{B.441}$$

$$\vec{\alpha} = \frac{d\vec{u}}{d\tau} = \gamma^4 \left(\vec{v}\left(\frac{\vec{v}\cdot\dot{\vec{v}}}{c^2}\right) + \dot{\vec{v}}\,(1-\beta^2)\right), \tag{B.442}$$

so that the result may be written

$$P(t) = \frac{2e^2\gamma^6}{3c^2}\left(\dot{\vec{v}}^2 - \left(\frac{\vec{v}\times\dot{\vec{v}}}{c^2}\right)^2\right). \tag{B.443}$$

This relativistic generalization is termed the *Liénard formula*, first derived in 1898 by the French physicist Alfred-Marie Liénard.

Fundamental Concepts Index

4-vector, 925

Ampére's law, 881

Biot-Savart law, 882

charge density, 868
circular polarization, 904
constitutive relations, 894
Coulomb gauge, 883, 914
Coulomb's law, 870
current density, 882

D-field, 878
diamagnetism, 889
dielectric, 878
Dirac delta function, 870
divergence theorem, 868

electric dipole radiation, 919
energy density, 896

Faraday's law, 892

gauge transformation, 883
Gauss' law, 868
Green function
  advanced, 916
  retarded, 915

H-field, 889

index of refraction, 903
invariant distance, 928

Kronecker delta, 877

Laplace's equation, 869
Larmor formula, 923
Legendre polynomial, 877
Lenz' Law, 895
Liénard formula, 930
Liénard-Wiechert potentials, 918
linear polarization, 904
Lorentz force law, 894
Lorentz transformation, 926
Lorenz gauge, 914

magnetic scalar potential, 890
Maxwell equations, 892
  macroscopic, 893
Minkowski metric, 925
multipole expansion, 872
mutual-inductance, 888

Ohm's law, 908

paramagnetism, 889
phase velocity, 903
Poisson's equation, 869
Poynting vector, 896
proper time, 928

radiation fields, 919
rest energy, 929
retarded time, 916
Rodrigues formula, 877

scalar potential, 869
self-inductance, 888
separation of variables, 875
skin depth, 910
Snell's law, 905
Stokes' theorem, 882
System of Units
  Gaussian, 867
  International, 867

transverse current, 914

vector potential, 883
velocity boosts, 924

Index

4-gradient, contravariant and covariant, 753
4-interval, 750
4-vector
  contravariant, 752
  covariant, 753
  current density, 752
  electromagnetic potential, 358, 756
  lightlike, 761
  spacelike, 761
4-velocity, 767

Acceleration
  field, of moving point particle, 684
  invariance under time reversal, 446
  of particle
    and radiated power, 648, 770
  relativistic, 769
Accelerator
  particle, 769–772
Action
  classical, 772
Action at a distance
  and Wheeler-Feynman electrodynamics, 811
Addition theorem, 572
  of spherical harmonics, 169
Admittance, 416, 548
  relative shunt, 549
Ampère's law, 274–275
Angles of incidence, reflection, and refraction, 471–473, 475
Angular distribution of radiation from oscillating dipole, 632–633
Angular momentum
  local conservation of, 422
  operator, 567–568
Angular momentum density
  of electromagnetic field, 422
Antenna
  circular loop, 619
  linear, 621
Aperture, 721
Attenuation
  in waveguides, 541
  constant, 559

Bare mass, 806
Bessel functions

integral representation, 138
of imaginary argument, 150
orthogonality of, 146
completeness, 142
generating function, 137
Hankel transform, 142
non-integer order, 140
recurrence relations, 139
spherical, 569
Bessel's equation, 139
Biot-Savart law, 270
Bizoń-Mach coordinates, 832
Black hole
perturbations, 844
Born probability rule, 814
Boundary
of open surface, 9
Boundary conditions
at surface of, good conductor, 534
on fields, 8
perturbation of, 560–565
Dirichlet, 49
for TE and TM waves in waveguide, 538–539
mixed, 50
Neumann, 50
Bra-ket notation, 815
Breathing mode, 635
Bremsstrahlung, 648–655
differential energy density, 652
Brewster's angle, 476

Capacitance, 65
Capacitive induction
coefficients, 65
Capacitor
cylindrical, fluid level in, 239
parallel plate, 67
two hemisphere with gap, 194
Capacitors
system of, 82
Cauchy-Riemann equations, 113
Causality, 428
speed of light and, 762
Charge
conservation, 230
density, 30
effective, of atoms, 215
free, 355
Charged particle
dynamics of, 497–501
Chebyshev polynomials, 837
Classical electron
definition, 801
mass, 810
self force, 811
Clausius-Mossotti relation, 249
Collisions
free electrons, 478
Commutation relations
of quantum operators, 817
Completely antisymmetric
symbol, 441
Completeness
of spherical harmonics, 169
relation, 142
of TE, TM, TEM waves in waveguides, 538
complex scaling, 829
Composition
property, in quantum mechanics, 818
rule, for Feynman propagator, 818
Compton wavelength, 807
Conduction current
macroscopic, 481

Conductivity
effective, in media, 481
Conductor, 6
"good", 530
attenuation in, 541, 559
boundary conditions, 50–51
perfect electric, 529
perfect magnetic, 530
skin depth, 483, 531
Confinement
for quarks, 233
Connective derivative, 339
Constitutive relations, 7
Contravariant
4-vector, 752
vector, 752
Conversion table of SI and Gaussian units, 862
Cooper pair, 304
Coulomb expansion, 168
alternate form, 169
Coulomb gauge, 358
Coulomb's law, 29
Covariance
of electrodynamics, 757
under Galilean transformation, 342
Covariant
4-vector, 753
expression, 757
Cramer, J.G., 814
Cross-section
for bremsstrahlung, 652
polarized, 697
Thomson, 734
unpolarized, 702
Current
effective, 300
free, 355
displacement, 340, 892
Current density
as 4-vector, 756
of point particle, 348
surface, 13
volume, 273
Cutoff
modes, in waveguides, 543
frequency
and nonideal waveguides, 563
in general waveguides, 559
in rectangular waveguides, 541
modes, in waveguides, 553
wavenumber, 536
Cylindrical coordinates
vector operators in, 25–29
boundary value problems, 148–149

D'Alembertian, 427
Damping
constant, 399, 497
factor, 711
response, 487
Delta function, 31
integral form, 36
membrane, 38, 279
and line integrals, 38
periodic, 658
Density
torque, 422
Determinant
properties, 441
Diamagnetism, 301
Dielectric constant, 7, 216
effective, 482
frequency-dependent, 480

Dielectric tensor, 7
Dimensions
  of SI and Gaussian units, 861
Dipole
  electrostatic, 293
  field, 206
  induced moment, 225
  magnetic current, 295
  moment, 632
Dipole layer, 47
Direction cosine, 440
Dirichlet boundary conditions
  homogeneous, 14
Dispersion relations
  Kramers-Kronig, 489
  for plane waves, 462
Dispersive media, 466
Displacement current, 340, 892
  macroscopic, 481
Displacement field (D-field), 6, 215
Distribution
  Gaussian, 32
Divergence
  linear, 62
Divergence theorem, 9
Drude model, 824
Duality transform
  of $\vec{E}$ and $\vec{B}$ fields, 375
Dyadic, 699
dyon, 384

Earth-ionosphere system as resonant cavity, 579, 609
Effective current, 300
Eigenfunctions
  of ∇^2, 102, 136
Eigenstates
  for scalar quantum field, 816
Einstein summation convention, 753
Electric dipole vector, 205
Electric field, 1
  derivable from potentials, 42
  of point charge, 29
Electromagnetic field
  strength tensor, 758
Electromagnetic momentum
  density, 418
Electromotic force, 343
Electromotive force, 895
Electron
  classical (definition), 801
  radius (classical), 734, 806
Electrostatic mass
  of classical electron, 804
Electrostatic potential, 74
Electrostatic self-energy
  of charge distribution, 804
Energy
  density, 61, 413
  energy-momentum 4-vector, 782
  flux, 413
  conservation, 795
  electrostatic, 60–63
  electrostatic, of atoms in a lattice, 106
  magnetic, 359
  of magnetostatic field, 359–368
  relativistic (of a particle), 768
Energy loss
  and bremsstrahlung, 652–655
  in materials, 482, 497–501
  in particle accelerators, 771
  in resonant cavities, 579–584
Euler's constant, 151
Euler-Lagrange equations, 773

Exterior problem, 54

Far zone, 699
  radiation, 618
  scattering, 699
Faraday's Law, 344
Faraday's law, 343
  induction, 344
Fermi's golden rule, 825
Ferromagnetism, 301
Feynman
  path integral, 820
  propagator, 814, 815
Field
  energy, 61
  electromagnetic, 411
  photon, 819
  scalar quantum, 815
Fine structure constant, 655, 807
Flat-space metric, 751
Force
  Abraham-Lorentz, 805
  Coulombs law of, 29
  Lorentz, 3
  between dielectric sphere and a point charge, 235
  between point charges, 29
  between two wires, 270–272
  radiation reaction, 801
Force density
  electromagnetic, 419, 897
Fourier
  sine series, 102
  transform, 36
  series, 102
Fourier transform
  alternative normalization, 627
Fraunhofer diffraction, 723
Free
  charge, 355
  current, 355
Frequency shift in a resonant cavity, 584
Fresnel
  diffraction, 723
  integrals, 726
  number, 727

Galilean Transformations, 924
Gamma function, 140
Gauge
  Coulomb, 358
  Lorenz, 358, 615, 757
Gauss' theorem, 9
Gauss's law, 234
Glueball, 608
Gluons, 608
Gradient
  in rectangular, cylindrical, and spherical coordinates, 25
Green function
  advanced, 433
  definition, 51
  in magnetostatics, 276
  reduced, 91
  retarded, 432
  source variable and field variable, 51
  symmetry property, 54
  two dimensional, 89
Green's first identity, 49
Green's reciprocation theorem, 77
Green's second identity, 52
Group velocity, 496
Gyromagnetic ratio
  of electron and muon, 5

H-field, 6, 301
Hadrons
  bag model, 579
Hamilton's principle, 772
Hankel functions, 150
  spherical, 570
Harmonic formalism, 615
Heaviside function, 456
Heaviside-Lorentz units, 862
Helicity states, of photon, 465
Helmholtz equation, 135, 537
Hermitian conjugation
  of field operators, 815
hyperboloidal slices, 831
Hysteresis, 302
  limiting curve, 302

ICRU, 501
Image method, 94
  in electrostatics, 89
  for a conducting sphere, 94
  in magnetostatics, 304
Impact parameter, 649
Impedance, 416
  wave, 539, 546
Index notation
  vectors and tensors, 753
Index of refraction, 463, 472
Inductance
  mutual, 362, 364
  self, 362
Induction
  capacitive, 65
  magnetic, 861
Integral rule
  for bound electrons, 481
  for conduction electrons, 480
Inversion
  complete, 765
  complete (or point), 442
  spatial, 442, 763
Ionization energy
  in materials, 500
Ionosphere, 609

Jacobian matrix, 35

Kirchhoff identity, 705
Klein-Gordon equation, 776
Klystron, 579
Kramers-Kronig relations, 488

Lagrangian density, 775
Laplace's equation, 100
  eigenvalue equation, 175
  cylindrical coordinates, 140, 148
Larmor formula, 633
Lattice QCD, 820
Legendre functions
  associated, 163
  second kind, 174
Legendre polynomials, 162
  generating function, 191
Lenz' Law, 344
Levi-Civita symbol, 441
Liénard-Wiechert potentials, 456, 639
Light cone, 761
Lorentz force law, 3
  generalized, 375
Lorentz model, 824
Lorentz transformation
  boost, 749–750
  definition, 752
  discrete, 765
  of scalar quantum fields, 815
  proper and improper, 763

restricted (RLT), 762

Madelung constant, 106
Magnetic
domains, 302
moment, 291, 634
Magnetic field, 1
solid angle formula for, 278–281
Magnetic monopole, 5
Magnetization, 300
microscopic model, 354
Mass
effective, of electron, 480
Maxwell equations, 1
macroscopic, 6
Maxwell stress tensor, 418, 897
Media
homogeneous, 7
isotropic, 7
linear, 7
local, 7
nonlocal, 7
constitutive relations, 7
Medium
isotropic, 215
Jackson's model, 352–355
linear, 215
Schwinger's model, 345–352
Metric
flat space (Minkowski), 751
Minkowski metric, 751
MIT bag model, 608
mode volume, 825
Moment
electric dipole, 211
electric quadrupole, 205
magnetic dipole, 291, 634
Momentum
relativistic, 768
Momentum density
of electromagnetic field, 418
Momentum flux
of electromagnetic field, 418
Multipole expansion
in electrostatics, 203–214
of energy, 207–211
of force, on charge distribution, 211–214
of radiation field, 630

Near zone (radiation), 617
Neumann
conductor, 51, 304
function, 140
Neumann boundary conditions, 50
homogeneous, 14
numerical methods
discretization error, 835

Oblate spheroidal coordinates, 70, 74, 723
Ohm's law, 8
generalized, 479
Ohmic medium, 528
Operator
creation, 816
destruction, 816
number, in quantum field theory, 816
raising and lowering, in quantum field theory, 817
time-ordering, 815
operator
adjoint, 841
normal, 841

Optical theorem, 714
  proof of, 713–714
Orthogonal, 440
Orthogonality
  of Bessel functions, 146
  of waveguide modes, 541

Pöschl-Teller potential, 843
Paramagnetism, 301
Parity
  reversal, 442, 765
Parseval's theorem, 627
Particle
  creation and destruction, in quantum field theory, 816
Particle accelerator
  circular, 770
  linear, 770–771
Permeability, 7
  magnetic, 301, 889
  tensor, 7
Permittivity
  tensor, 7
Perturbation
  of boundary conditions, 560–565
Phase velocity, 463
Plasma frequency, 483
plasmon frequency, 824
Poincaré stress tensor, 805
Poisson
  equation, 42
    Green function as a solution, 51
  sum rule, 660
Polarization
  EM waves, 463, 465
  electric, 214
  induced, 216
  microscopic model, 354
  orientation, 216
Polarized cross-section
  scattering, 702
Potential
  vector, 275
Power
  instantaneous radiated, 769
  Larmor formula, 633
  loss in non-ideal waveguides
    perturbative method, 560–565
  loss, in non-ideal waveguides, 556–559
Poynting vector, 413
  harmonic, 503
Pre-acceleration
  for motion of point particle, 808
Pseudoscalar, 444
pseudospectrum, 841, 842
  map, 842
Pseudotensor
  dual, 759
  transformation law, 759
Pseudovector, 443
Purcell factor, 825

Quality factor (Q)
  of an oscillator, 579
Quantization
  of Maxwell's equations, 5
  of charge, 4
Quantum chromodynamics, 820
Quantum electrodynamics
  (QED), 5, 814
Quantum field
  definition, 814

Hermitian, 816
lattice theory of, 820
massless scalar, 815
Quantum mechanics
transactional, 814
Quark
confinement, 235
leading logarithm model, 232, 530
quasinormal modes, 823

Radiated energy
angular distribution, 626–627
from impulsive scattering, 628
Radiation
coherency, of sources, 631
transition, 629
Radiation gauge, 358
Radiation power
angular distribution
electric dipole, 632
from current loop, 621
from linear antenna, 622
harmonic source, 619
magnetic dipole, 635
real source, 625–626
instantaneous
from harmonic source, 632
Rationalized units, 862
Rayleigh
scattering, 734
Reactance, 416
Reduced Green function
for conducting cylinder, 146–148
for conducting wedge, 155–157
for flat conductor, 90–94
Reflection coefficient, 472
Regularization, 810
relative error
logarithmic, 840
Renormalization, 806
Resistance, 416
resolvent, 842
Resonant cavity
cylindrical, 581
rectangular, 565
spherical, 567
TE modes, 574
TM modes, 577
Response function, 485
Retarded time, 456
approximate, 640
exact, 639
far zone approximation, 623
origin, 630
Right-hand rule, 9
Rotation
active or passive, 439
matrix, 440
4×4, 763
Rotationally invariant
length of vectors, 440
polynomials in matrix entries, 637
subspaces, and spherical harmonics , 169
Runaway solutions
for motion of point particle, 808
Runge phenomenon, 838
Rutherford scattering, 498

Scalar, 443
field, 444
quantum, 815
Lorentz, 753
Lorentz invariant, 767

triple product, 444
Scalar potential, 21
electrostatic, 42
wave equation for, 358
Maxwell equations in terms of, 359
of magnetized material, 312
Scale factors, 36
Scattering
impulsive, of particle, 628
Rayleigh, 734
Scattering amplitude, 701
for conducting sphere, 707, 709
from Kirchhoff's identity, 705
Schrödinger equation, 455, 495
Schumann resonances, 579, 609
Self force
from averaged Green's function, 806
from advanced Green's function, 433, 805
self force
of classical electron, 811
self-adjoint, 827
Self-energy, 61
magnetostatic, 360
Separation of variables
polar coordinates, 107
rectangular coordinates, 100
Shear tensor, 422
Skin depth, 483, 531
Snell's law
derivation, 471
Source
of electromagnetic field, 1
of electric field, 30
Spectral stability, 841
spectral theory
of self-adjoint operators, 827
Spherical harmonics, 160
addition theorem, 169
vector, 575
Spin systems, 298
Spontaneous emission, 825
Standing waves, 535
Stationary phase approximation, 496
Stern-Gerlach experiment, 298
Stokes' theorem, 9
Stopping power, 498
Strength tensor
electromagnetic field, 758
String
Dirac, 376–380
and charge quantization, 381–382
Schwinger, 382
Sum rule
for capacitance, 66
Poisson, 660
for ϵ, 489
Superconductors, 304
Surface
open and closed, 9
Susceptibility, 216
frequency-dependent, 480
Synchrotron radiation, 662

Tensor
3×3, 7, 205
electric quadrupole, 634
non-traceless, 348
electromagnetic field strength, 758
in four dimensions, 754
Maxwell stress, 418, 897
orthogonal transformation of, 445

permeability, 7
quadrupole
traceless, 205
second-rank, 754
shear, 422
Tensor product, 754
Test charge, 30
Thompson's theorem
for a single closed conductor, 62
Time
reversal, 428, 445, 766
proper, 767
Torque
microscopic, on an atom, 349
density, 422
on an arbitrary charge distribution, 245
on magnetic moment, 298
Transformation properties
under Lorentz transformation
4-interval, 750, 751
covariant and contravariant vectors, 752
scalars, 753
tensors, 755
under rotation
scalar, 443
scalar and vector fields, 444
vector, 440
Transition radiation, 629
Transmission coefficient, 472
Transverse electric (TE)
attenuation in waveguides, 541
modes, 584
in ideal rectangular cavity, 565–566
in ideal spherical cavity, 574–576
waves
definition, 538
field equations, 539
in ideal cylindrical waveguides, 586
in ideal rectangular waveguides, 540–541, 584–585
orthogonality properties, 541–546
Transverse electromagnetic (TEM)
waves
definition, 538
field equations, 538
not always possible, 539
Transverse gauge, 358
Transverse magnetic (TM)
attenuation in waveguides, 541
modes
in ideal rectangular cavity, 565–566
in ideal spherical cavity, 577–578
waves
definition, 538
in ideal cylindrical waveguides, 586
in ideal rectangular waveguides, 540–541, 585
orthogonality properties, 541–546
Triple product
orthogonal transformation of, 444
scalar, 444

Triple product identity, 13
Two-dimensional
  electrodynamics, 14–19

Ultraviolet transparency, 483
Units
  Gaussian (CGS), 2
  Heaviside-Lorentz, 862
  MKSA (SI), 2
  natural, 381, 862

Vacuum state
  in quantum field theory, 815
Vandermonde matrix, 193
Variational methods
  for waveguides, 552–555
  in electrostatics, 111–113
Vector potential, 21
  for two-dimensional Maxwell equations, 16
  magnetostatic, 275
    Coulomb (or radiation) gauge, 276
    far-field approximation, 290–296
    for circular current loop, 308
    for magnetized material, 311
    gauge transformation, 275
    integral expression for, 276
    wave equation for, 358
  Maxwell equations in terms of, 359

Wave
  equation, 358, 427
  number, 462
  packet, 495
  standing, 535
Waveguides, 535
  variational methods in, 552–555
  cutoff frequency, 541
  ideal rectangular, 540
  non-ideal, power loss in
    BC method, 560–565
    power method, 556–559
  with thin obstructions, 547–555
Wheeler-Feynman
  electrodynamics, 811–814
Wick rotation, 830